HEINEMANN CHEMISTRY 2

6TH EDITION

COORDINATING AUTHOR
Lanna Derry

AUTHORS
Drew Chan
Chris Commons
Penny Commons
Elizabeth Freer
Elissa Huddart
Louise Lennard
Melissa MacEoin
Mick Moylan
Pat O'Shea
Bob Ross
Kellie Vanderkruk
Paul Waldron

VCE UNITS 3 AND 4 • 2024–2027

Pearson Australia
(a division of Pearson Australia Group Pty Ltd)
459–471 Church Street
Level 1, Building B
Richmond, Victoria 3121
www.pearson.com.au

First published 2023 by Pearson Australia
2026 2025
10 9 8 7 6 5 4 3

VCE Heinemann Project Leads: Fiona Cooke, Bryonie Scott, Misal Belvedere, Malcolm Parsons
Development Editor: Leanne Peters
Schools Programme Manager: Michelle Thomas
Senior Production Editor: Casey McGrath
Production Editor: Chris Woods
Series Designer: Anne Donald
Rights & Permissions Editors: Amirah Fatin Binte Mohamed Sapi'ee
Publishing Services Analyst: Jit-Pin Chong
Design Analyst: Jennifer Johnston
Illustrators: Bruce Rankin, DiacriTech, QBS Learning
Editors: Maryanne Park, Sam Trafford
Proofreader: Margaret Trudgeon
Indexer: Max McMaster

Printed in China by Golden Cup/03 (2025)

National Library of Australia Cataloguing-in-Publication entry

A catalogue record for this book is available from the National Library of Australia

ISBN 978 0 6557 0009 8

Pearson Australia Group Pty Ltd ABN 40 004 245 943

Disclaimer
The selection of internet addresses (URLs) provided for this book was valid at the time of publication and was chosen as being appropriate for use as a secondary education research tool. However, due to the dynamic nature of the internet, some addresses may have changed, may have ceased to exist since publication, or may inadvertently link to sites with content that could be considered offensive or inappropriate. While the authors and publisher regret any inconvenience this may cause readers, no responsibility for any such changes or unforeseeable errors can be accepted by either the authors or the publisher.

Indigenous Australians
We respectfully acknowledge the traditional custodians of the lands upon which the many schools throughout Australia are located. We acknowledge traditional Indigenous Knowledge systems are founded upon a logic that developed from Indigenous experience with the natural environment over thousands of years. This is in contrast to Western science, where the production of knowledge often takes place within specific disciplines. As a result, there are many cases where such disciplinary knowledge does not reflect Indigenous worldviews, and can be considered offensive to Indigenous peoples.

Some of the images used in *Heinemann Chemistry 2* 6th edition might have associations with deceased Indigenous Australians. Please be aware that these images might cause sadness or distress in Aboriginal or Torres Strait Islander communities.

Practical activities
All practical activities, including the illustrations, are provided as a guide only and the accuracy of such information cannot be guaranteed. Teachers must assess the appropriateness of an activity and take into account the experience of their students and facilities available. Additionally, all practical activities should be trialled before they are attempted with students and a risk assessment must be completed. All care should be taken and appropriate personal protective clothing and equipment should be worn when carrying out any practical activity. Although all practical activities have been written with safety in mind, Pearson Australia and the authors do not accept any responsibility for the information contained in or relating to the practical activities, and are not liable for loss and/or injury arising from or sustained as a result of conducting any of the practical activities described in this book.

HEINEMANN CHEMISTRY 2

6TH EDITION

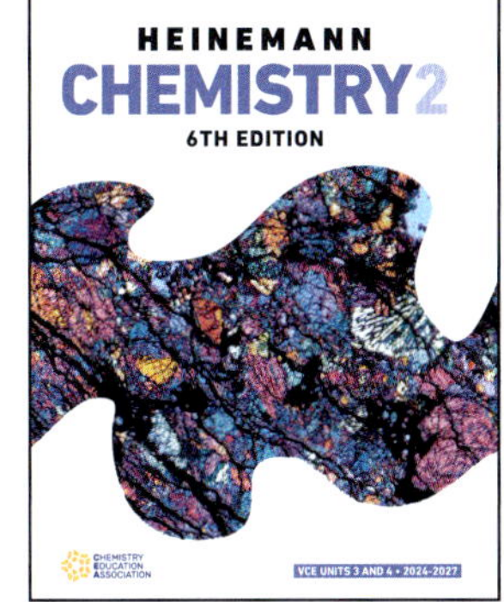

Writing and development team

We are grateful to the following people for their time and expertise in contributing to the Heinemann Chemistry 2 project.

Lanna Derry
Chemistry teacher, Head of Science
Coordinating author and reviewer

Doug Bail
Education consultant
SPARKlab developer

Reuben Bolt
Deputy Vice-Chancellor First Nations Leadership
Reviewer

Erin Bruns
Chemistry teacher, Head of Science
Digital question author

Drew Chan
Chemistry teacher, Head of Chemistry
Author

Chris Commons
Retired chemistry teacher and honorary chemistry researcher
Author and reviewer

Penny Commons
Honorary chemistry lecturer and tutor
Author

Elizabeth Freer
Chemistry teacher
Author

Renee Hill
Chemistry teacher, Head of Chemistry
Author

Elissa Huddart
Chemistry teacher, Head of Science
Author

Carissa Kelly
Chemistry teacher
Answer checker

Rowan Kidd
Chemistry teacher
Author

Louise Lennard
Chemistry teacher
Author

Peter Lewis
Chemistry teacher and lab technician
Reviewer

Melissa MacEoin
Chemistry teacher
Coordinating author (Year 11) and reviewer

Mick Moylan
Chemistry lecturer
Author

Pat O'Shea
Chemistry teacher
Author

Bob Ross
Retired chemistry teacher
Author

Kellie Vanderkruk
Environmental scientist and former chemistry lecturer
Author

Nic Volkmann
Chemistry teacher and edupreneur
VOLK video developer

Paul Waldron
Chemistry teacher
Author

Gregory White
Scientist
Answer checker

Laurence Wooding
Laboratory technician
Reviewer

The Publisher wishes to thank and acknowledge Warwick Clarke and Geoff Quinton for their contribution in creating the original content for the 5th edition and their longstanding dedicated work with Pearson and Heinemann.

The Chemistry Education Association was formed in 1977 by a group of chemistry teachers from secondary and tertiary institutions. It aims to promote the teaching and learning of chemistry in schools. The CEA has established a tradition of providing professional development opportunities for teachers and up-to-date text and electronic materials, together with other resources supporting both students and teachers. The CEA offers scholarships and bursaries to students and teachers to further their interest in chemistry, assists the Science Chemistry Teachers Association of Victoria (STAV) to present the VCE Chemistry Conference and sponsors prizes for the STAV Science Talent Search.

Contents

AREA OF STUDY 2

How can the rate and yield of chemical reactions be optimised?

AREA OF STUDY 3

Heinemann Chemistry 2 6th edition includes a comprehensive set of resources to support Area of Study 3 via your Pearson Places bookshelf.

Unit 4 How are carbon-based compounds designed for purpose?

AREA OF STUDY 1

How are organic compounds categorised and synthesised?

AREA OF STUDY 2

How are organic compounds analysed and used?

AREA OF STUDY 3

How is scientific inquiry used to investigate the sustainable production of energy and/or materials?

Heinemann Chemistry 2 6th edition includes a comprehensive set of resources to support Area of Study 3 via your Pearson Places bookshelf.

How to use this book

Heinemann Chemistry 2 6th edition

Heinemann Chemistry 2 6th edition has been written to the new VCE Chemistry Study Design 2024–2027. The book covers Units 3 and 4. Explore how to use this book below.

Case study

Case studies place chemistry in an applied situation or relevant context. Text and artwork refer to the nature and practice of chemistry, applications of chemistry and associated issues, and the historical development of chemical concepts and ideas.

CASE STUDY

Boab trees: a food source for Indigenous Australians

The *larrkardiy*, or Australian Boab tree (*Adansonia gregorii*) (Figure 13.2.6a), is found only in the Kimberley region of Western Australia and east into the Northern Territory. The tree comes into bloom in the wet season, producing fruit and flowers which fall from the tree during the dry season. The fruits are hard-shelled and usually referred to as nuts (Figure 13.2.6b). Some boab trees are more than 1500 years old, making them among the oldest living organisms in Australia. As boabs age they become hollow; they use this hollow to store water within the trunk to help survive the harsh drought conditions of the dry season.

For thousands of years Indigenous Australians have used boab trees in many ways. They obtained water from hollows in the tree and used the white powder that fills the seed pod as a food; the leaves were used medicinally. Decorative paintings or carvings were sometimes made on the outer surface of the fruit and the trunk of the tree.

The dry powdery pulp inside the boab nut is rich in vitamin C (ascorbic acid), with 40 g of the powder providing 84–100% of the recommended daily intake.

FIGURE 13.2.6 (a) The boab tree is a defining feature in the Kimberley region of Western Australia. (b) Boab nuts, traditionally eaten by Indigenous Australians, are rich in vitamin C and other essential nutrients.

It is also rich in other vitamins (A, B1, B2 and B6), minerals (calcium, iron, magnesium and potassium), dietary fibre and amino acids. The nuts are lightweight and keep easily for over a year, making them a convenient food source for Indigenous Australians when moving around an area. The roots can be chewed to extract moisture and the iron-rich leaves used as a leafy vegetable.

Volumetric analysis can be used to determine the vitamin C content of the powder in the boab nut. An acid–base titration using a standardised sodium hydroxide solution can be used or a redox titration as described in this section.

Case study: Analysis

These case studies include real-world data that can be analysed and evaluated.

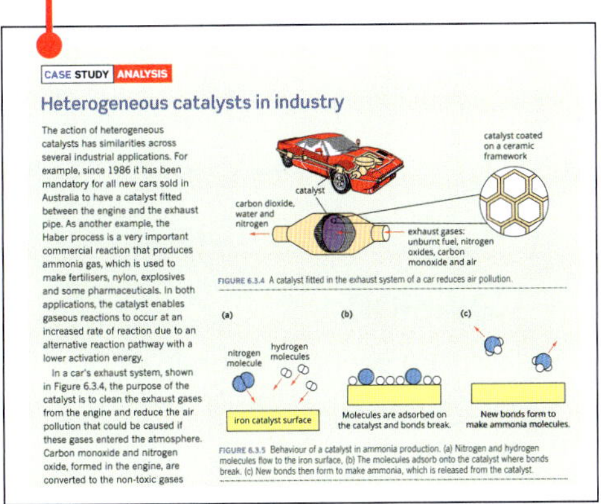

CASE STUDY ANALYSIS

Heterogeneous catalysts in industry

The action of heterogeneous catalysts has similarities across several industrial applications. For example, since 1986 it has been mandatory for all new cars sold in Australia to have a catalyst fitted between the engine and the exhaust pipe. As another example, the Haber process is a very important commercial reaction that produces ammonia gas, which is used to make fertilisers, nylon, explosives and some pharmaceuticals. In both applications, the catalyst enables gaseous reactions to occur at an increased rate of reaction due to an alternative reaction pathway with a lower activation energy.

In a car's exhaust system, shown in Figure 6.3.4, the purpose of the catalyst is to clean the exhaust gases from the engine and reduce the air pollution that could be caused if these gases entered the atmosphere. Carbon monoxide and nitrogen oxide, formed in the engine, are converted to the non-toxic gases

FIGURE 6.3.4 A catalyst fitted in the exhaust system of a car reduces air pollution.

FIGURE 6.3.5 Behaviour of a catalyst in ammonia production. (a) Nitrogen and hydrogen molecules flow to the iron surface, (b) The molecules adsorb onto the catalyst where bonds break. (c) New bonds then form to make ammonia, which is released from the catalyst.

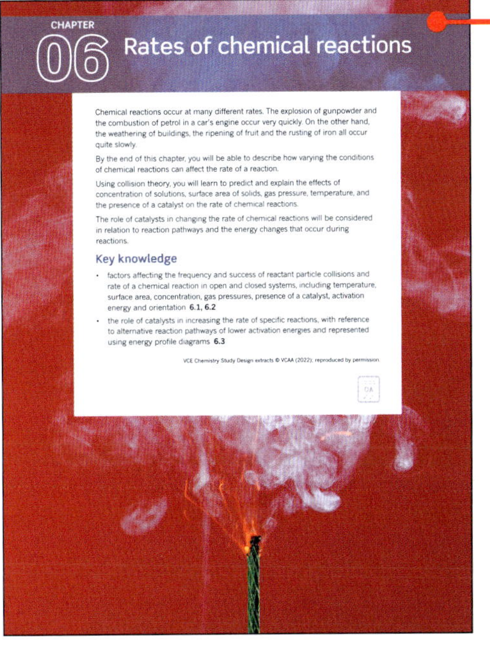

CHAPTER 06 Rates of chemical reactions

Chemical reactions occur at many different rates. The explosion of gunpowder and the combustion of petrol in a car's engine occur very quickly. On the other hand, the weathering of buildings, the ripening of fruit and the rusting of iron all occur quite slowly.

By the end of this chapter, you will be able to describe how varying the conditions of chemical reactions can affect the rate of a reaction.

Using collision theory, you will learn to predict and explain the effects of concentration of solutions, surface area of solids, gas pressure, temperature, and the presence of a catalyst on the rate of chemical reactions.

The role of catalysts in changing the rate of chemical reactions will be considered in relation to reaction pathways and the energy changes that occur during reactions.

Key knowledge

- factors affecting the frequency and success of reactant particle collisions and rate of a chemical reaction in open and closed systems, including temperature, surface area, concentration, gas pressures, presence of a catalyst, activation energy and orientation **6.1, 6.2**
- the role of catalysts in increasing the rate of specific reactions, with reference to alternative reaction pathways of lower activation energies and represented using energy profile diagrams **6.3**

VCE Chemistry Study Design extracts © VCAA (2022); reproduced by permission.

Chapter opener

Chapter opening pages link the study design to the chapter content. Key knowledge addressed in the chapter is clearly listed. To help you find where each outcome is covered in the chapter, the relevant section numbers are written in bold.

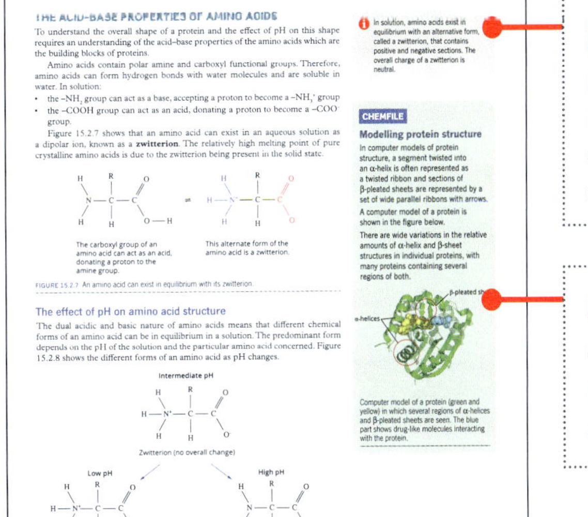

THE ACID–BASE PROPERTIES OF AMINO ACIDS

To understand the overall shape of a protein and the effect of pH on this shape requires an understanding of the acid–base properties of the amino acids which are the building blocks of proteins.

Amino acids contain polar amine and carboxyl functional groups. Therefore, amino acids can form hydrogen bonds with water molecules and are soluble in water. In solution:

- the $-NH_2$ group can act as a base, accepting a proton to become a $-NH_3^+$ group
- the $-COOH$ group can act as an acid, donating a proton to become a $-COO^-$ group.

Figure 15.2.7 shows that an amino acid can exist in an aqueous solution as a dipolar ion, known as a **zwitterion**. The relatively high melting point of pure crystalline amino acids is due to the zwitterion being present in the solid state.

FIGURE 15.2.7 An amino acid can exist in equilibrium with its zwitterion.

The effect of pH on amino acid structure

The dual acidic and basic nature of amino acids means that different chemical forms of an amino acid can be in equilibrium in a solution. The predominant form depends on the pH of the solution and the particular amino acid concerned. Figure 15.2.8 shows the different forms of an amino acid as pH changes.

In solution, amino acids exist in equilibrium with an alternative form, called a zwitterion, that contains positive and negative sections. The overall charge of a zwitterion is neutral.

CHEMFILE

Modelling protein structure

In computer models of protein structure, a segment twisted into an α-helix is often represented as a twisted ribbon and sections of β-pleated sheets are represented by a set of wide parallel ribbons with arrows.

A computer model of a protein is shown in the figure below.

There are wide variations in the relative amounts of α-helix and β-sheet structures in individual proteins, with many proteins containing several regions of both.

Computer model of a protein (green and yellow) in which several regions of α-helices and β-pleated sheets are seen. The blue part shows drug-like molecules interacting with the protein.

Highlight

Highlight boxes focus on important information such as key definitions and summary points.

ChemFile

ChemFiles include interesting information and real-world examples.

Icons

This icon is used to alert you to engage with auto-corrected questions through Pearson Places.

These icons indicate when it is the best time to engage with a worksheet (WS), a practical activity (PA) or exam questions (EQ) in the *Heinemann Chemistry 2 Skills and Assessment* book.

Worked example

Worked examples are set out in steps that show thinking and working. Each Worked example is followed by a Worked example: Try yourself that allows you to test your understanding.

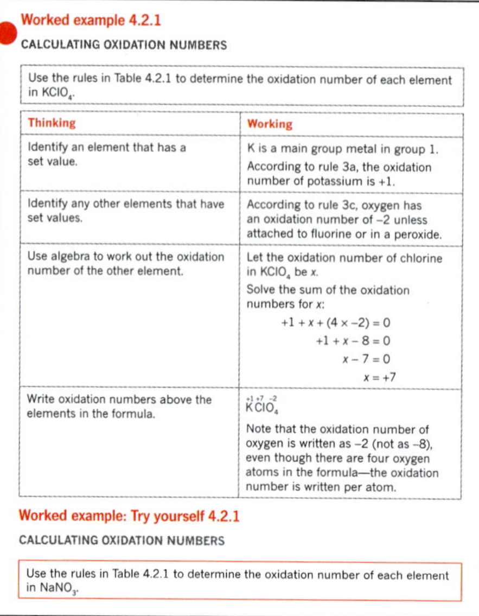

Worked example 4.2.1

CALCULATING OXIDATION NUMBERS

Use the rules in Table 4.2.1 to determine the oxidation number of each element in $KClO_4$.

Thinking	Working
Identify an element that has a set value.	K is a main group metal in group 1. According to rule 3a, the oxidation number of potassium is +1.
Identify any other elements that have set values.	According to rule 3c, oxygen has an oxidation number of −2 unless attached to fluorine or in a peroxide.
Use algebra to work out the oxidation number of the other element.	Let the oxidation number of chlorine in $KClO_4$ be x. Solve the sum of the oxidation numbers for x: $+1 + x + (4 \times -2) = 0$; $+1 + x - 8 = 0$; $x - 7 = 0$; $x = +7$
Write oxidation numbers above the elements in the formula.	$\overset{+1}{K}\overset{+7}{Cl}\overset{-2}{O}_4$ Note that the oxidation number of oxygen is written as −2 (not as −8), even though there are four oxygen atoms in the formula—the oxidation number is written per atom.

Worked example: Try yourself 4.2.1

CALCULATING OXIDATION NUMBERS

Use the rules in Table 4.2.1 to determine the oxidation number of each element in $NaNO_3$.

Section summary

Each section includes a summary to help you consolidate key points and concepts.

Section review

Each section concludes with questions that test your ability to recall, explain and apply key concepts.

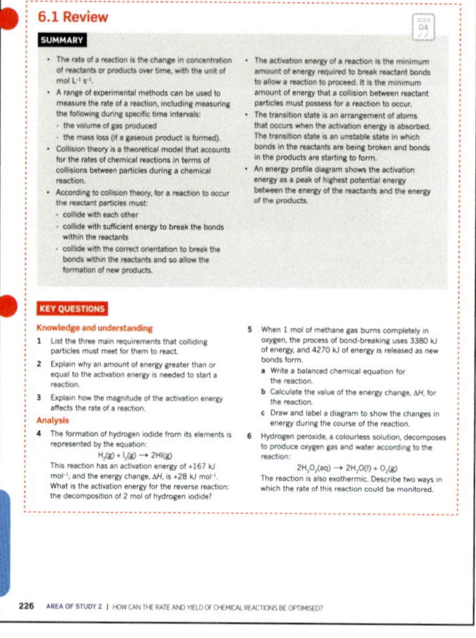

6.1 Review

SUMMARY

KEY QUESTIONS

Area of Study review

Each Area of Study concludes with a comprehensive set of exam-style questions, including multiple choice and short answer, to support you in your exam preparation.

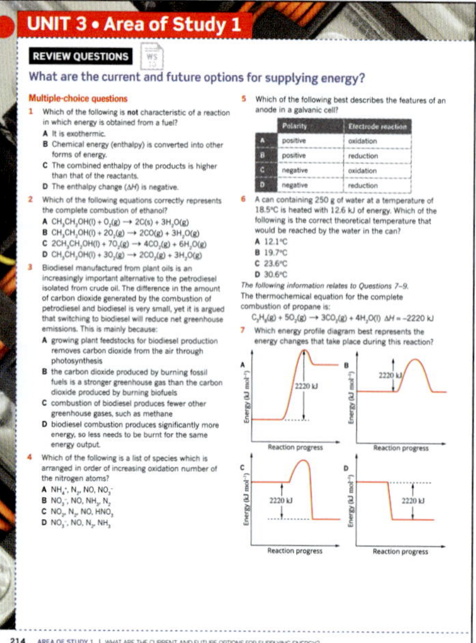

UNIT 3 • Area of Study 1

REVIEW QUESTIONS

What are the current and future options for supplying energy?

Glossary

Key terms are shown in **bold** throughout and listed at the end of each chapter. A comprehensive glossary at the end of the book defines all key terms.

Answers

Comprehensive answers for all section review, chapter review and Area of Study review questions are provided via the *Heinemann Chemistry 2 6th edition eBook + Assessment.*

Chapter review

Each chapter concludes with a list of key terms and questions that test your understanding of the key knowledge covered in the chapter.

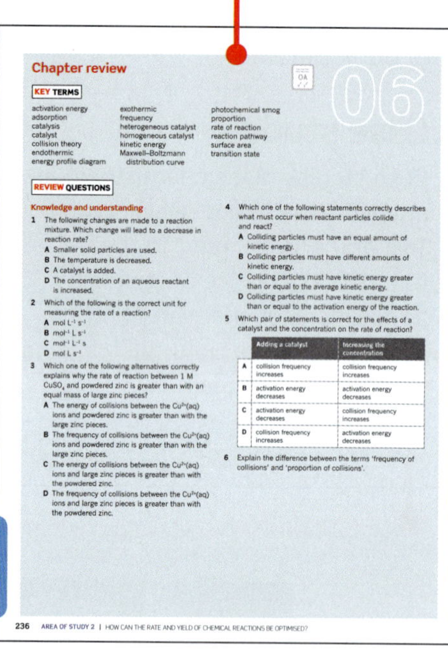

Chapter review

KEY TERMS

REVIEW QUESTIONS

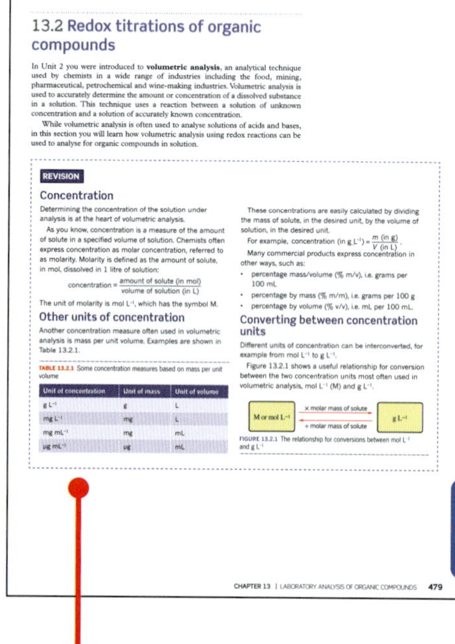

13.2 Redox titrations of organic compounds

REVISION

Concentration

Revision

Revision sections contain vital information from Year 11.

CHAPTER

01 Scientific investigation

The development of a set of key science skills is a core component of the study of VCE Chemistry and applies across Units 1 to 4. Chapter 1 scaffolds the development of these skills. The opportunity to develop, use and demonstrate these skills in a variety of contexts is important before you undertake investigations and when you are evaluating the research of others.

Although this chapter can be read as a whole, it is best to refer to it and use it when the need arises as you work through other chapters. For example, you may need a refresher on what is required to be included in a conclusion. This chapter also contains useful checklists to help you plan investigations, graph results and complete aspects of a scientific report, such as that required for Unit 4 Area of Study 3, which is undertaken in Units 3 and/or 4 in Unit 4. Similarly, when performing a practical investigation, as part of your chemistry studies, refer to this chapter to make sure that you collect data properly and that your data is of high quality.

Key science skills

Develop aims and questions, formulate hypotheses and make predictions

- identify, research and construct aims and questions for investigation **1.1, 1.2**
- identify independent, dependent and controlled variables in experiments **1.2**
- formulate hypotheses to focus investigations **1.2**
- predict possible outcomes of investigations **1.2**

Plan and conduct investigations

- determine appropriate investigation methodology: case study; classification and identification; controlled experiment; fieldwork; literature review; modelling; product, process or system development; simulation **1.1**
- design and conduct investigations; select and use methods appropriate to the selected investigation methodology, including consideration of sampling technique and size, equipment and procedures, taking into account potential sources of error and causes of uncertainty; determine the type and amount of qualitative and/or quantitative data to be generated or collated **1.1, 1.2, 1.3**
- work independently and collaboratively as appropriate and within identified research constraints, adapting or extending processes as required and recording such modifications in a logbook **1.3**

Comply with safety and ethical guidelines

- demonstrate safe laboratory practices when planning and conducting investigations by using risk assessments that are informed by safety data sheets (SDS), and accounting for risks **1.2**
- apply relevant occupational health and safety guidelines while undertaking practical investigations **1.2**
- demonstrate ethical conduct when undertaking and reporting investigations **1.2**

Generate, collate and record data

- systematically generate and record primary data, and collate secondary data, appropriate to the investigation, including use of databases and reputable online data sources **1.3**
- record and summarise both qualitative and quantitative data, including use of a logbook as an authentication of generated or collated data **1.3**
- organise and present data in useful and meaningful ways, including schematic diagrams, flow charts, tables, bar charts, line graphs, and calibration curves **1.3, 1.4**

Analyse and evaluate data and investigation methods

- process quantitative data using appropriate mathematical relationships and units, including calculations of ratios, percentages, percentage change and mean **1.4**
- use appropriate numbers of significant figures in calculations **1.3, 1.4**
- plot graphs involving two variables that show linear and non-linear relationships **1.4**
- identify and analyse experimental data qualitatively, handling, where appropriate, concepts of: accuracy, precision, repeatability, reproducibility, resolution, and validity of measurements; and errors (random and systematic) **1.3**
- identify outliers, and contradictory, provisional or incomplete data **1.4**
- repeat experiments to evaluate the precision of data **1.3**
- evaluate investigation methods and suggest ways to improve precision, and to reduce the likelihood of errors **1.3, 1.5**

Construct evidence-based arguments and draw conclusions

- distinguish between opinion, anecdote and evidence, and between scientific and non-scientific ideas **1.1, 1.2**
- evaluate data to determine the degree to which the evidence supports the aim of the investigation, and make recommendations, as appropriate, for modifying or extending the investigation **1.5**
- evaluate data to determine the degree to which the evidence supports or refutes the initial prediction or hypothesis **1.5**
- use reasoning to construct scientific arguments, and to draw and justify conclusions consistent with evidence and relevant to the question under investigation **1.5**
- identify, describe and explain the limitations of conclusions, including identification of further evidence required **1.5**
- discuss the implications of research findings and proposals **1.5**

Analyse, evaluate and communicate scientific ideas

- use appropriate chemical terminology, representations and conventions, including standard abbreviations, graphing conventions, algebraic equations, units of measurement and significant figures **1.3, 1.4, 1.5, 1.6**
- discuss relevant chemical information, ideas, concepts, theories and models and the connections between them **1.1, 1.2, 1.3, 1.4, 1.5**
- analyse and explain how models and theories are used to organise and understand observed phenomena and concepts related to chemistry, identifying limitations of selected models/theories **1.1, 1.2, 1.3, 1.4**
- critically evaluate and interpret a range of scientific and media texts (including journal articles, mass media communications and opinions in the public domain), processes, claims and conclusions related to chemistry by considering the quality of available evidence **1.2, 1.5, 1.6**
- apply sustainability concepts (green chemistry principles, development goals and the transition from a linear towards a circular economy) to analyse and evaluate responses to chemistry-based scenarios, case studies, issues and challenges **1.2**
- identify and explain when judgments or decisions associated with chemistry-related issues may be based on sociocultural, economic, political, legal and/or ethical factors and not solely on scientific evidence **1.1**
- use clear, coherent and concise expression to communicate to specific audiences and for specific purposes in appropriate scientific genres, including scientific reports and posters **1.6**
- acknowledge sources of information and assistance, and use standard scientific referencing conventions **1.6**

OA
✓✓

1.1 The nature of scientific investigations

Chemistry is the study of matter, how it behaves and interacts with other matter. As scientists, chemists extend their understanding using the scientific method, which involves investigations that are carefully designed, conducted and reported. Well-designed research is based on a sound knowledge of what is already understood about a subject, as well as careful preparation, measurements and observations.

OBSERVATIONS

FIGURE 1.1.1 During the electrolysis of tin(II) chloride, distinct metallic crystals are formed at the negative electrode. These can be shown to be tin crystals.

When you make **observations** during a scientific investigation, you use your senses and a wide variety of instruments and laboratory techniques. **Qualitative** observations provide information about what is present. For example, the formation of crystals on the negative electrode in an electrolysis investigation, as shown in Figure 1.1.1, can indicate that the metal ions are being reduced. In contrast, **quantitative** observations provide numerical information to answer questions such as how many, how much and how often; for example, by taking a reading on an electronic balance, or measuring the initial and final volumes on a burette. Quantitative observations are accompanied by relevant units such as grams (g) or millilitres (mL) (Figure 1.1.2).

FIGURE 1.1.2 The measurement of masses is an example of quantitative observations. Masses are measured using an electronic balance.

● *You will now be able to answer key question 1.*

Interpreting observations

A student might observe that when a solution of tin(II) chloride is electrolysed, metallic crystals are formed at the negative electrode. Similarly, the electrolysis of copper(II) chloride produces a red-brown coating on the negative electrode and metallic crystals are again formed on the negative electrode during the electrolysis of lead(II) nitrate.

How observations are interpreted depends on past experience and knowledge, but to an enquiring mind the above observations could provoke further questions, such as:

- How can the student know the crystals/coating were of a metal?
- Does the electrolysis of all metal compounds result in a solid metal being formed?
- Why do some metals make a coating, while others form crystals at the negative electrode?
- Does the concentration of the aqueous solution affect the size of the metal crystals being formed?

Many of these questions cannot be answered through observation alone, but they can be answered through scientific investigations. Good scientists have acute powers of observation and enquiring minds; they rely on evidence and trends in evidence and they make the most of chance opportunities. They gather and record their observations carefully, so that they can be referred to in the future.

THE SCIENTIFIC METHOD

Scientists observe, consider what is already known by consulting the work of other scientists, and then they ask questions.

Scientific inquiry is not a linear process. Figure 1.1.3 illustrates the **scientific method**, in which a research question is investigated by forming a **hypothesis** (a prediction based on scientific reasoning that can be tested experimentally) and then testing it. Scientists will not necessarily complete these steps in the order shown below, and some steps may need to be repeated or altered to more accurately address the **research question** that they are trying to answer.

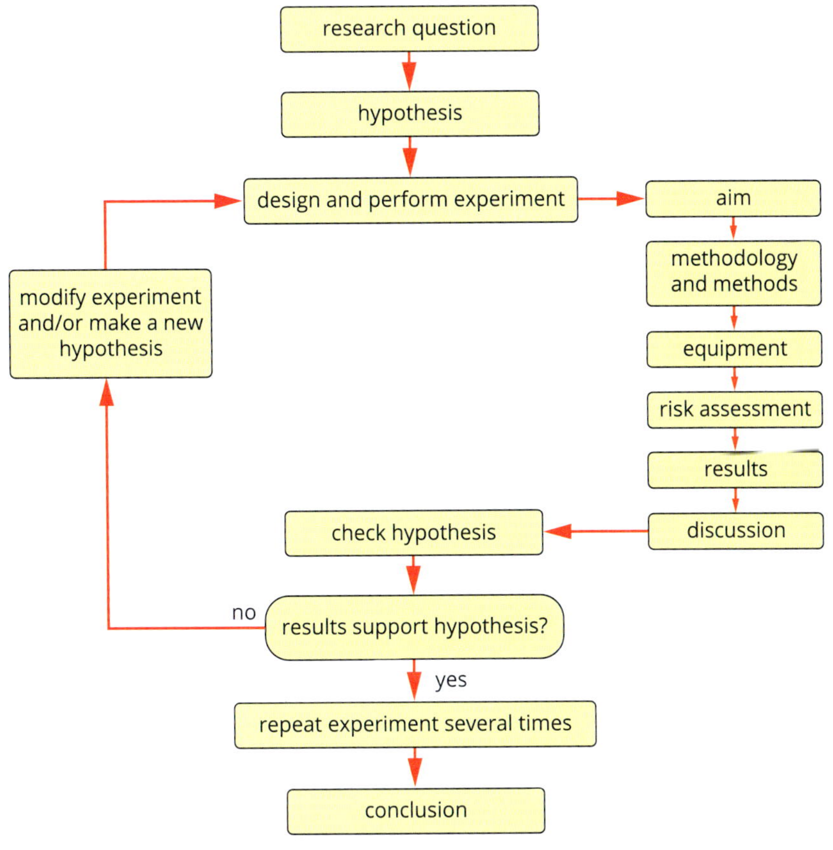

FIGURE 1.1.3 The scientific method

Scientific investigation methodologies

Scientists test their ideas about their research question using various investigation **methodologies**. The methodology is a brief description of the general approach taken to investigate the research question and the reasons why this approach is taken. Using the evidence gained from those investigations, scientists suggest possible explanations for the things they have observed.

Practical investigations involve direct experiences or hands-on activities. Suitable methodologies for a practical investigation would be controlled experiments, simulations, fieldwork or modelling. In comparison, a research investigation could follow the methodology of a literature review or could start with a case study.

The different approaches that you could use for investigations are outlined in Table 1.1.1.

The methodology is the general approach used to investigate the research question. The method (procedure) is the set of specific steps that are taken to collect data during an investigation.

TABLE 1.1.1 Some scientific investigation methodologies

Type of methodology	Explanation
case study	investigation of a real or hypothetical situation, such as an activity, event, or problem, often involving analysis of data within a real-world context
classification and identification	using features or properties to classify or identify a substance
controlled experiment	experimental investigation that involves formulating a hypothesis and testing the effect of an independent variable on the dependent variable while controlling all other variables in the experiment
fieldwork	collecting data outside the laboratory
literature review	critical analysis of what has already been investigated and published, using secondary data from other people's investigations to explain events or propose new ideas or relationships
modelling	using models as representations of objects, systems or processes to aid understanding or make predictions
product, process or system development	using scientific understanding and advances in technology to design a new tool, method or process to meet the demands or needs of society
simulation	using mathematical models or computer simulations to test hypotheses or conduct virtual experiments

LIMITATIONS OF THE SCIENTIFIC METHOD

The scientific method can be applied only to hypotheses that can be tested. A hypothesis that is not testable can be neither supported nor disproved by the scientific method.

It is important to understand that although a hypothesis may be supported by experimental data, the same hypothesis may not be supported in all circumstances—it has only been found to be true under the conditions that have been tested.

The scientific method cannot be used to test morality or ethics. These judgements belong to the fields of philosophy, history, politics and law. Science can, however, provide valuable information that people can consider when making these judgements. For example, science can be used to predict the environmental consequences of pollution and the medical consequences of chemical weapons.

- *You will now be able to answer key questions 2, 3 and 6.*

NON-SCIENTIFIC FACTORS THAT INFLUENCE SCIENTIFIC INVESTIGATIONS

When you investigate questions or issues related to applications of the principles of chemistry in society, it is important to be able to distinguish between factors that are scientific and those that are non-scientific. Often an investigation into a chemical issue must consider non-scientific information, including sociocultural, economic, political, legal or ethical factors.

For example, a discussion of the issue of climate change must involve scientific data, such as the concentration of carbon dioxide in the atmosphere, but it must also consider political factors, such as governmental support of the United Nations Paris Agreement (2015) on climate change.

Non-scientific factors may be classified under more than one category.

Sociocultural factors

Sociocultural factors are those related to individuals, communities, cultures and society. Questions these may raise include:

- Who is directly involved?
- Who will benefit?
- Who might be negatively affected?

Economic factors

Economic factors are those related to costs and resources. Examples might include:

- Who will pay for the research?
- Who will pay for the development, production and application?
- Will this cause a loss of profit for another stakeholder?
- Will there be any costs involved for the government?

Political factors

Political factors are those related to government or public affairs. Examples might include:

- What is the relevant government policy?
- Is there a difference in opinion between political parties?

Legal factors

Legal factors are those connected to law (legislation) or rules. Examples might include:

- What state legislation covers this area?
- What federal legislation covers this area?

Ethical factors

Ethical factors are those related to moral principles, and the need to determine what is right and what is wrong. Examples might include:

- Is one group of people advantaged over another by the desired outcome?
- Does the desired outcome prevent anyone from meeting their basic needs?
- Has the investigation been reported honestly, or have some results been omitted because they didn't support the desired outcome?

Table 1.1.2 gives examples of non-scientific factors that may have been considered during the rapid development of COVID-19 vaccines in 2020–21.

TABLE 1.1.2 Examples of non-scientific factors affecting the development of COVID-19 vaccines

Sociocultural	Economic	Political	Legal	Ethical
Why is vaccination necessary?	Who will pay for the research, development and production?	What agreements have been made for the supply of various vaccines?	Have the vaccines been approved for use in Australia by the Therapeutic Goods Administration?	Are there vulnerable groups who should be prioritised for vaccination?

Non-scientific factors, such as sociocultural, economic, political, legal or ethical factors, may influence a chemical issue and should be considered as part of some investigations.

You will now be able to answer key questions 4 and 5.

1.1 Review

OA ✓✓

SUMMARY

- Examples of useful methodologies for an investigation include case studies, classification and identification, a literature review, modelling, simulations and controlled experiments.
- Well-designed experiments are based on a sound knowledge of what is already understood or known, and on careful observation.
- The scientific method is an accepted procedure for conducting investigations.
- A hypothesis is a possible explanation for a set of observations that can be used to make predictions, which can then be tested experimentally.
- Controlled experiments allow us to examine one factor at a time: they are a commonly used methodology for testing hypotheses.
- Science helps us to understand a situation or phenomenon. It is used in conjunction with other considerations, such as sociocultural, economic, political. legal and ethical factors.

KEY QUESTIONS

Knowledge and understanding

1 Rewrite the following table, identifying which of the following pieces of information about data collected in an experiment are qualitative and which are quantitative. Place a tick in the appropriate column.

Information	Quantitative	Qualitative
colour of electrode		
temperature of electrolyte		
presence of bubbles		
change in mass of electrode		
concentration of electrolyte		

2 The scientific method has certain limitations. Which of the following cannot be tested using the scientific method?

A a difficult hypothesis

B the moral issues involved in an investigation

C a complicated, multistep methodology

D the environmental aspects of an investigation

3 A student wishes to investigate how the dissolved oxygen concentration in water changes along the length of a creek from its source in the hills to where it joins the Yarra River. Identify the methodologies that would be involved in this investigation.

4 Copy this table into your workbook. Classify each specific factor or resource related to climate change as scientific or non-scientific, and for non-scientific factors, identify the category.

Factor	Scientific (yes or no)	Non-scientific (include category)
concentration of carbon dioxide in the atmosphere (ppm)		
federal governmental support of the United Nations Paris Agreement on climate change		
UNESCO 2017 *Declaration of Ethical Principles in relation to Climate Change*		
university modelling for the impact of costs of climate change		
use of hydrogen fuel cells to limit carbon dioxide emissions		
Commonwealth *Renewable Energy (Electricity) Act 2000*		

Analysis

5 Economic factors should be considered when establishing industrial chemical processes. Consider the commercial production of hydrogen for use as a transport fuel. Describe two economic factors that are considered when trying to maximise the production of hydrogen.

6 A student wanted to investigate how to produce a low-alcohol wine from grapes. By referring to Table 1.1.1, suggest three different methodologies the student could use in this investigation.

1.2 Planning investigations

Taking the time to carefully plan and design an investigation before you begin will help you maintain a clear and concise focus throughout. In this section you will learn about some of the key steps to take when planning an investigation:

- choosing a topic
- developing and refining your investigation (i.e. determining your research question, hypothesis, aim, methodology and methods, and variables)
- modifying an existing investigation
- complying with safety guidelines
- applying ethical principles
- sourcing and evaluating information.

CHOOSING A TOPIC

When you choose a topic, consider the following:

- Do you find the research topic interesting?
- Is there background information on your topic that relates to your course?
- Does your school laboratory have the resources for you to perform the investigation?
- Can you collect clear, measurable data?

Several practical research areas are suggested in Table 1.2.1. You will learn more about useful research techniques for topics such as these later in this section.

TABLE 1.2.1 Potential research areas that could be investigated

Production of energy	Production of chemicals
efficiency of galvanic cells	changes that can be made to electrolytic cells
factors affecting the operation of fuel cells	factors affecting the oxidation of ethanol to make ethanoic acid

DEVELOPING AND REFINING YOUR INVESTIGATION

When you begin a scientific investigation, you should develop and evaluate a research question, decide on an appropriate investigation methodology and method, determine the associated variables, define the aim and formulate a hypothesis. Each of these can be refined as the planning of your investigation continues.

Determining your research question

Once you have decided upon a topic or idea of interest, the first thing you need to do is conduct a search of the relevant literature; that is, you must read scientific reports and other articles on the topic to find out what is already known, and what is not known or not yet agreed upon. The literature also gives you important information for the introduction to your report and ideas for experimental methods. Use this information to generate potential research questions. Figure 1.2.1 on the following page outlines steps for developing ideas for a proposed research question.

Your research question must be focused on and limited to ideas that are within your abilities to investigate, or the resources and equipment you have available.

For example, for an experimental investigation you might form the research question: How does the distance between the electrodes affect the mass of copper deposited at the cathode during the electrolysis of copper(II) sulfate? You will be able to investigate this research question using a compound, copper(II) sulfate and copper or graphite electrodes, which are commonly found in schools.

In a **controlled experiment**, the question should refer to the relationship between the independent variable and dependent variable. For example: In the research question: 'How does the distance between the electrodes affect the mass of copper deposited at the cathode during the electrolysis of copper(II) sulfate?', the independent variable is the distance between the electrodes and the dependent variable is the mass of copper deposited at the cathode.

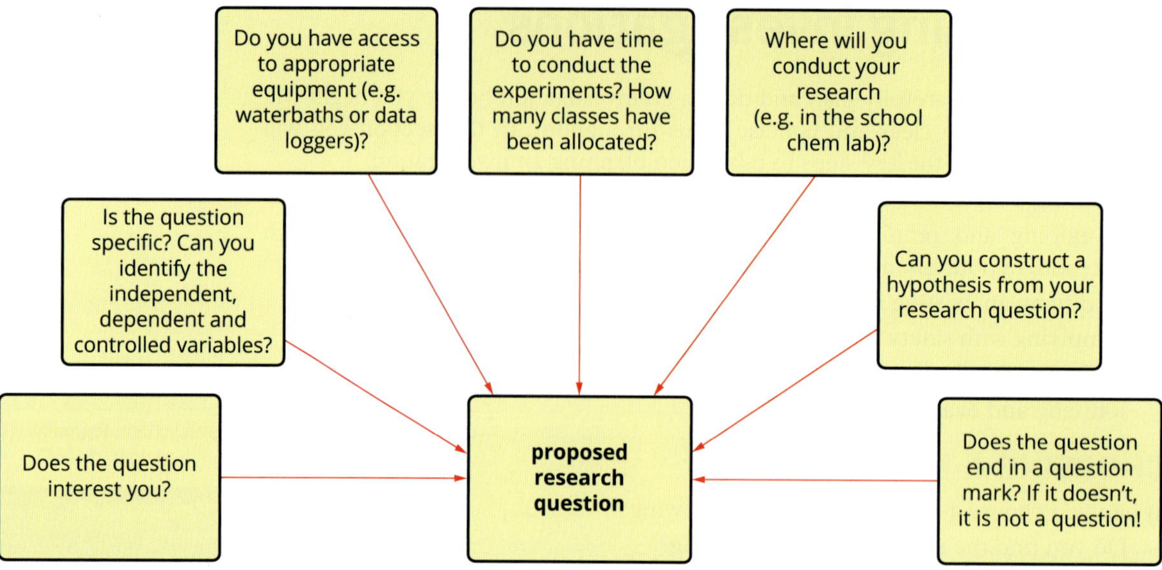

FIGURE 1.2.1 Developing ideas for a proposed research question

Consider using the following checklist for the research question:

- ❑ Relevance—make sure your question is related to your chosen topic.
- ❑ Clarity and measurability—make sure your prediction can be framed as a clear hypothesis that can be measured.
- ❑ Knowledge and skills—make sure you have a level of knowledge and laboratory skills that will allow you to explore the question. Keep the question simple and achievable.
- ❑ Advice—seek advice from your teacher on your question. Their experience may lead them to consider aspects of the question that you have not thought about.

Determining the aim

The **aim** is a statement describing in detail what will be investigated. For example: To determine how the distance between the electrodes affects the mass of copper deposited at the cathode during the electrolysis of copper(II) sulfate.

Making predictions and constructing a hypothesis

A hypothesis is a prediction based on scientific reasoning that can be tested experimentally. Carefully designed experiments are conducted to determine whether the predictions made in a hypothesis are accurate or not. If the results of an experiment do not fall within an acceptable range, the hypothesis is rejected. If the predictions are found to be accurate, the hypothesis is supported. If, after many different experiments, one hypothesis is supported by all the results obtained so far and is considered to have been verified using the scientific method, then this explanation can be given the status of a **theory** or **principle**.

In a practical investigation, a hypothesis defines a proposed relationship between two variables and takes the form of cause and effect.

For example, an investigation is made into how the distance between the electrodes affects the mass of copper deposited at the cathode during the electrolysis of copper(II) sulfate. A potential hypothesis could be: If the distance between the electrodes decreases, then the mass of copper deposited on the cathode will increase because a smaller distance will result in a lower resistance within the solution and a greater current will flow through the circuit. Faraday's first law of electrolysis relates current to charge according to the equation, $Q = It$, so if current increases, then the charge will increase, the amount of electrons moving through the circuit will increase, and the mass of copper deposited will increase.

● *You will now be able to answer key questions 3 and 4.*

Selecting an appropriate methodology and method

When you plan a scientific investigation, you will need to think about the best way to address the research question. The scientific investigation methodology (Table 1.1.1) and the method/s (also known as procedure/s) selected will depend on the aim of the investigation and the research question.

Factors to consider when selecting an appropriate methodology include:

- Do you have access to a laboratory, materials and chemicals (for a controlled experiment or product/process/system development)?
- Do you have access to a school or public library (for a literature review or case study)?
- Do you have computer access (e.g. international databases for classification and identification or for access to simulations)?

The **method** is the set of specific steps that are taken to collect data during the investigation. You will learn more about determining your method/s in Section 1.3.

Defining your variables

The **variables** are the factors that change during your experiment. An experiment or investigation determines the relationship between variables.

There are three categories of variable:

- The **independent variable** is the variable that is controlled by the researcher.
- The **dependent variable** is the variable that may change in response to a change in the independent variable. This is the variable that you will measure or observe.
- **Controlled variables** are all the variables that must be kept constant during the investigation.

A **valid** experiment should have only one independent variable. If it had more than one, you could not be sure which independent variable was responsible for the changes observed in the dependent variable. Table 1.2.2 gives examples of research questions and potential independent and dependent variables.

When writing a research question for a controlled experiment, include the independent and dependent variables. For example, what is the effect of *[the independent variable]* on *[the dependent variable]?*

TABLE 1.2.2 Examples of research questions and corresponding independent and dependent variables

Research question	Independent variable	Dependent variable
How does the surface area of the electrodes affect the current transferred through a galvanic cell made up of a $Cu(s)/Cu^{2+}(aq)$ and a $Zn(s)/Zn^{2+}(aq)$ half-cell?	surface area of the electrodes —this could be varied by keeping a constant length and changing the width of the electrodes or keeping the size of the electrode constant and dipping a larger depth of the electrode into the half-cells for each variation	current going through a simple circuit made up of the galvanic cell and an ammeter
How does the number of carbon atoms in alcohol affect the enthalpy of combustion of the alcohol?	number of carbon atoms in the alcohol that is combusted	enthalpy of combustion as determined using simple calorimetry
How does the vegetable oil from which biodiesel is synthesised affect the enthalpy of combustion of the biodiesel?	the type of vegetable oil (such as peanut oil, sunflower oil, sesame oil, etc.) used in the synthesis of biodiesel samples	enthalpy of combustion as determined using simple calorimetry

CHEMFILE

Swapping independent and controlled variables

When you have found an experiment that you want to change to an investigation, consider the controlled variables that were used for the initial experiment. Often a controlled variable for one investigation can become the independent variable for a new investigation. For example, in many rates investigations, the possible independent variables include concentration, temperature and sometimes the use of a catalyst. You may have already carried out an experiment, in which you changed one independent variable (e.g. the concentration). When designing a new investigation, you could make that independent variable one of the controlled variables, and choose a new independent variable (e.g. temperature) for your investigation. You will now have a new investigation, ready to go!

Linking the planning parts of the investigation together

Your research question, hypothesis, aim and variables should all link together. Table 1.2.3 provides an example of a research question, variables, potential hypothesis and aim that link together.

TABLE 1.2.3 Example of research question, variables, potential hypothesis and aim

Research question	How does the number of carbon atoms in an alcohol affect the enthalpy of combustion of the alcohol?
Independent variable	the number of carbon atoms in the alcohol that is combusted
Dependent variable/s	the enthalpy of combustion as determined using simple calorimetry
Controlled variables	the spirit burner used to burn the hydrocarbon, the distance between the flame and the calorimetry can, the calorimetry can used to hold the water that is heated, the initial mass of alcohol in the spirit burner (affects how well the wick is soaked), the initial temperature of the water in the calorimetry can
Potential hypothesis	As the number of carbon atoms increases, the enthalpy of combustion will increase because there are more carbon–carbon bonds to break and more products (CO_2 and H_2O) are formed, so *H*(products) − *H*(reactants) will increase.
Aim	To determine how the number of carbon atoms in an alcohol affects the enthalpy of combustion of the alcohol.

- *You will now be able to answer key questions 1, 5, 6 and 8.*

MODIFYING AN EXISTING INVESTIGATION

When you are designing a research question for a controlled experiment, it is often easiest to modify an investigation that you have already conducted in class. For example, you could use an existing method and choose a different independent variable (which might change the dependent variable). Table 1.2.4 shows possible changes that could be made to a practical investigation that you have already undertaken in class.

TABLE 1.2.4 Examples of how to change an existing practical investigation into a student-designed investigation

Existing practical investigation	Potential student experiment
synthesis of biodiesel	How does the vegetable oil from which biodiesel is synthesised affect the enthalpy of combustion of the biodiesel?
half-cells and the electrochemical series	How does the concentration of the solutions in the half-cells affect the voltage of the electrochemical cell?
preparing artificial fragrances and flavours	How does the number of carbon atoms in an ester molecule affect the boiling point of the ester?

The topics listed in Table 1.2.4 are only suggestions. Figure 1.2.2 shows how you can modify an existing research question.

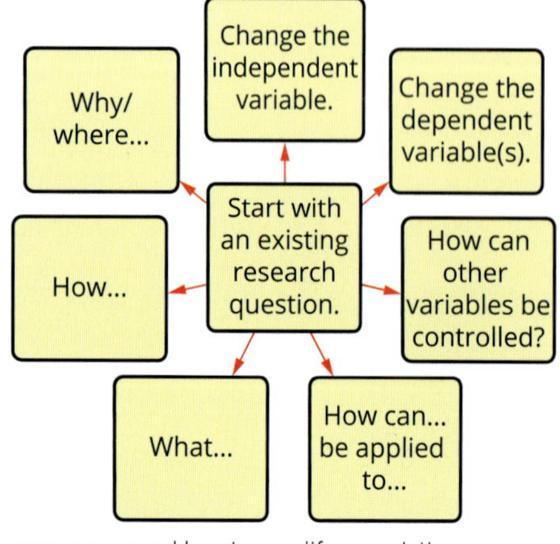

FIGURE 1.2.2 How to modify an existing investigation

COMPLYING WITH SAFETY GUIDELINES

Everything we do involves some risk. A **risk assessment** is performed for a controlled experiment to identify, assess and control hazards. Always identify the risks and control them to keep everyone safe.

To identify risks, think about:

- the activity you will be carrying out
- the equipment or chemicals you will be using or producing. For example, when hydrochloric acid reacts with sodium thiosulfate, the toxic gas sulfur dioxide, SO_2, is produced, so this reaction must be conducted in a fume cupboard.

Figure 1.2.3 shows a flow chart of how to consider and assess the risks involved in a practical investigation.

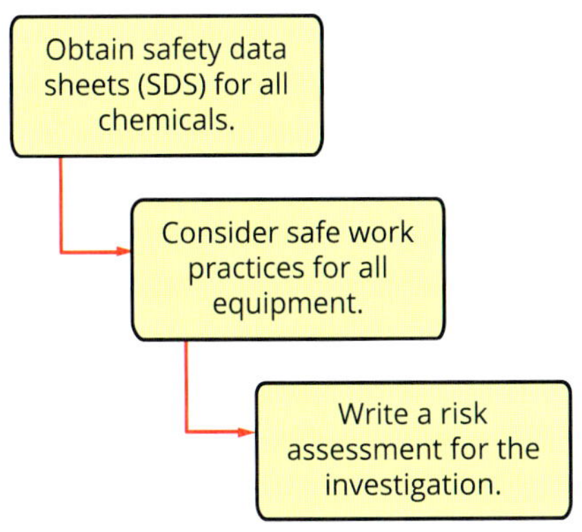

FIGURE 1.2.3 Steps involved in identifying risks

Occupational health and safety

Occupational health and safety refers to all the measures that employers need to provide to ensure their employees are safe at work. Schools must also ensure that equipment and processes used in school laboratories are safe for all students, teachers and technical staff.

Ethical and safety considerations must be the highest priority at all times during a practical investigation.

Chemical codes

The chemicals at school or at the hardware store have warning symbols on their labels. These symbols are a chemical code indicating the nature of the contents (Table 1.2.5). From 1 January 2017, the Globally Harmonized System of Classification and Labelling of Chemicals (GHS) pictograms were introduced into Australia. Some of the pictograms that you might see denote chemicals that are corrosive, pose a health hazard or are flammable.

TABLE 1.2.5 GHS pictograms used as warning symbols on chemical labels

GHS pictogram	Use	GHS pictogram	Use	GHS pictogram	Use
	flammable liquids, solids and gases; including self-heating and self-igniting substances		oxidising liquids, solids and gases, may cause or intensify fire		explosion, blast or projection hazard
	corrosive chemicals; may cause severe skin and eye damage and may be corrosive to metals		gases under pressure		fatal or toxic if swallowed, inhaled or in contact with skin
	low level toxicity; this includes respiratory, skin and eye irritation, skin sensitisers and chemicals harmful if swallowed, inhaled or in contact with skin		hazardous to aquatic life and the environment		chronic health hazards; this includes aspiratory and respiratory hazards, carcinogenicity, mutagenicity and reproductive toxicity

Safety data sheets

Each chemical substance has an accompanying document called a **safety data sheet (SDS)** (Figure 1.2.4). An SDS contains important safety and first aid information about each chemical you commonly use in the laboratory. If the products of a reaction are toxic to the environment, you must pour your waste into a special container (not down the sink). This is something to discuss with your teacher or the laboratory technician.

1. IDENTIFICATION OF THE MATERIAL AND SUPPLIER

Product identifier

Product Name	HYDROCHLORIC ACID - 20% OR GREATER
Product Code(s)	000031061101

Other means of identification

UN number	1789
Pure substance/mixture	Mixture

Recommended use of the chemical and restrictions on use

Recommended use	Precursor for generation of chlorine dioxide gas used in water treatment.
Uses advised against	No information available.

Supplier

Ixom Operations Pty Ltd
ABN: 51 600 546 512
Level 8, 1 Nicholson Street
Melbourne 3000
Australia

Telephone Number: +61 3 9906 3000

Emergency telephone number

Emergency telephone number **1 800 033 111 (ALL HOURS)**

2. HAZARDS IDENTIFICATION

GHS Classification

Classified as dangerous goods in accordance with the Australian Code for the Transport of Dangerous Goods by Road and Rail (ADG).

Classified as a hazardous chemical in accordance with the criteria of Safe Work Australia - Globally Harmonized System (GHS).

Corrosive to metals	Category 1
Skin corrosion/irritation	Category 1 Sub-category B
Serious eye damage/eye irritation	Category 1
Specific target organ toxicity (single exposure)	Category 3

SIGNAL WORD

Danger

Label elements

Corrosion
Exclamation mark

Hazard statements

H290 - May be corrosive to metals
H314 - Causes severe skin burns and eye damage
H335 - May cause respiratory irritation

Precautionary Statements - Prevention

Keep only in original container
Do not breathe fume, gas, mist, vapours, spray
Wash face, hands and any exposed skin thoroughly after handling Use only outdoors or in a well-ventilated area
Wear protective gloves / protective clothing / eye protection / face protection

Precautionary Statements - Response

Specific treatment (see First aid on this SDS)
IF IN EYES: Rinse cautiously with water for several minutes. Remove contact lenses, if present and easy to do. Continue rinsing
Immediately call a POISON CENTER or doctor/physician
IF ON SKIN (or hair): Remove/Take off immediately all contaminated clothing. Rinse skin with water/shower Wash contaminated clothing before reuse
IF INHALED: Remove victim to fresh air and keep at rest in a position comfortable for breathing
Call a POISON CENTER or doctor/physician if you feel unwell
IF SWALLOWED: Rinse mouth. DO NOT induce vomiting
IF SWALLOWED: Call a POISON CENTER or doctor/physician if you feel unwell
Absorb spillage to prevent material damage

Precautionary Statements - Storage

Store in a well-ventilated place. Keep container tightly closed Store locked up
Store in corrosive resistant container with a resistant inner liner

Precautionary Statements - Disposal

Dispose of contents/container in accordance with local, regional, national, and international regulations as applicable

Other hazards which do not result in classification

Poisons Schedule (SUSMP) 6

FIGURE 1.2.4 Extract of a safety data sheet (SDS) for concentrated hydrochloric acid

Protective equipment

Everyone who works in a laboratory should wear items that help keep them safe. This is called personal protective equipment (PPE) (Figure 1.2.5) and includes:

- safety glasses
- closed-toed shoes
- disposable gloves for handling chemicals
- an apron or a lab coat if there is risk of damage to clothes
- a fume cupboard that should be used when toxic or corrosive gases are being handled or produced.

FIGURE 1.2.5 It is important to wear appropriate personal protective equipment as identified in a risk assessment.

APPLYING ETHICAL PRINCIPLES

Ethics are a set of moral principles by which your actions can be judged as right or wrong. Every society or group of people has its own principles or rules of conduct.

Applying ethical principles in chemistry means:

- using integrity and honesty when recording and reporting the outcomes of your investigation and when using other people's data (such as in a literature review)
- recognising the importance of social, economic and political values when forming conclusions using scientific understanding (Section 1.1)
- acknowledging the work of others by including in-text citations and details in a list of references (Section 1.6).

This means you should always record your data in a bound logbook, rather than scraps of paper, using the correct number of significant figures/decimal places for the reported values (Section 1.3).

You should NEVER make up raw or processed data and you should record all data that you gathered in the investigation, even if it is not used in later calculations.

Green chemistry

Green chemistry is an approach to chemistry that aims to design products and processes that efficiently use renewable raw materials, and minimise hazardous effects on human health and the environment. The goals of green chemistry are described in 12 principles (summarised in Table 1.2.6).

Many of the 12 principles of green chemistry can be employed or considered by VCE Chemistry students.

TABLE 1.2.6 Some of the 12 principles of green chemistry that are most relevant to VCE Chemistry

Waste prevention	It is better to prevent waste than to treat it or clean it up after it has been produced.
Atom economy	Chemical processes should be designed to maximise incorporation of all reactant materials used in the process into the final product.
Less hazardous chemical syntheses	Wherever practicable, synthetic methods should be designed to use and generate substances that possess little or no toxicity to human health and the environment.
Designing safer chemicals	Chemical products should be designed to achieve their intended function while minimising toxicity.
Design for energy efficiency	Chemical processes should be designed for maximum energy efficiency and with minimal negative environmental and economic impacts.
Use of renewable feedstocks	Raw materials or feedstocks should be made from renewable (mainly plant-based) materials, rather than from fossil fuels whenever practicable.
Catalysis	Catalysts should be selected to generate the same desired product(s) with less waste and using less energy and reagents in chemical reactions.
Design for degradation	Chemical products should be designed so that at the end of their use they break down into harmless degradation products and do not persist in the environment.

Since they were developed in 1991, interest in applying the 12 principles of green chemistry to chemical processes across the world has continued to increase. Of these, several principles are particularly relevant to VCE Chemistry. The replacement of a complex multi-step process for making an important cholesterol-reducing medication with a much less complex process which produces less hazardous waste (see the case study on the next page) is just one recent example of how green chemistry is being applied to solving real-world problems. In the school laboratory, we can apply the principles of green chemistry by using minimum quantities of chemicals in experiments, thus reducing the quantity of product that must be disposed of after the experiment is finished.

CASE STUDY

Green chemistry: A greener way to make simvastatin

Heart attacks and stroke are among the major causes of ill-health and premature death in Australia. One of the leading risk factors for heart attacks and stroke is high blood cholesterol which leads to fatty deposits in arteries blocking blood flow to the heart, brain and other parts of the body. The drug simvastatin is a prescription medication consumed by millions of people worldwide to help control their blood cholesterol levels.

Originally, simvastatin was produced in multiple steps from a compound called lovastatin. Lovastatin is found naturally in the mould, *Aspergillus terreus*, and also in oyster mushrooms (*Pleurotus ostreatus*), shown in Figure 1.2.6. The original synthetic pathway to produce simvastatin from lovastatin involved between four and six individual steps. As you will see in Chapter 11, the more steps that are involved in a synthetic process, the greater the chance of wastage and production of unwanted biproducts. This original synthetic pathway for making simvastatin was indeed wasteful and used large amounts of hazardous reagents.

FIGURE 1.2.6 Oyster mushrooms contain the compound lovastatin, which is the precursor to simvastatin.

In 2007 Professor Yi Tang and his colleagues from the University of California, Los Angeles, designed a two-step green chemistry pathway to synthesise simvastatin. This process was found to be much more efficient, leading to a greater percentage yield, while also greatly reducing hazards during manufacturing and resulting in less toxic waste. As a result, the improved process is more cost-effective and has a higher atom economy compared to the original pathways.

SOURCING INFORMATION

When you are sourcing information during your search of the literature, such as when researching experimental methods or investigating a broader issue, consider whether that information is from primary or secondary sources. You should also consider the advantages and disadvantages of using resources such as books or the internet.

Primary and secondary sources

Primary and secondary sources provide valuable information for research.

Primary sources of information are created by a person directly involved in an investigation. Examples of primary sources are results from research and **peer-reviewed** scientific articles. Peer-reviewed journal articles (articles that have been reviewed by other experts in the field) are likely to be up to date and reliable. Examples of **secondary sources** of information include textbooks, newspaper articles and websites that present a synthesis, review or interpretation of primary sources. All sources of information may have a **bias** (a focus on only one part or one direction of the data or evidence) so you need to determine if they are reliable sources of information. You will learn about assessing the quality of data in Section 1.3. Table 1.2.7 compares characteristics and examples of primary sources and secondary sources.

TABLE 1.2.7 Summary of primary and secondary sources

	Characteristics	Examples
Primary sources	• first-hand records of events or experiences • written at the time the event happened • original documents	• results of experiments • scientific journal or magazine articles • reports of scientific discoveries • photographs, specimens, maps and artefacts • interviews with experts • websites (if they meet the criteria above)
Secondary sources	• interpretations of primary sources • written by people who did not see or experience the event • use information from original documents but rework it	• textbooks • biographies • newspaper articles • magazine articles • radio and television documentaries • websites that interpret the scientific work of others • podcasts

Primary sources of information are created by a person directly involved in an investigation. Secondary sources of information are a synthesis, review or interpretation of primary sources.

You will now be able to answer key question 7.

Using books and the internet

Peer-reviewed scientific journals are the best sources of information, but you are unlikely to have access to many of these, and much of the information is difficult to interpret if you are not an expert in the field.

As books and internet searches will most likely be your most commonly used resources for information, you should be aware of their advantages and disadvantages (Table 1.2.8).

TABLE 1.2.8 Advantages and disadvantages of book and internet resources

	Book resources	Internet resources
Advantages	• written by experts • authoritative information • reviewed to ensure information is accurate • logical, organised layout • content is relevant to the topic • contain a table of contents and index to help you find relevant information	• quick and easy to access • allow access to hard-to-find information • allow access to a vast amount of information from around the world • up-to-date information • may be interactive and use animations to enhance understanding
Disadvantages	• may not have been published recently • usable by only a finite number of people at a time	• time-consuming looking for relevant information • search engines may not display the most useful sites • cannot always tell how up-to-date information is • difficult to tell if information is accurate • hard to tell who has responsibility for authorship and if they are biased • information may not be well ordered • only a small proportion of sites are educational

CRITICALLY EVALUATING INFORMATION

Not all sources of information are credible. **Critical thinking** involves asking questions when evaluating the content and its origin, including:

- Who created this message? What are the qualifications, expertise, reputation, and **affiliation** (who they work for or are associated with) of the author/s?
- Why was the information written?
- Where was the information published?
- When was the information published?
- How often has the information been referred to by other researchers?
- Are the conclusions supported by data or evidence?
- What is implied?
- What has been omitted?

Peer review

Peer review is a process in which other researchers who work in the same field review your work and provide feedback about your methodology and whether your conclusions are justified. Scientists are expected to publish their findings in peer-reviewed journals. Some examples of peer-reviewed chemistry journals are:

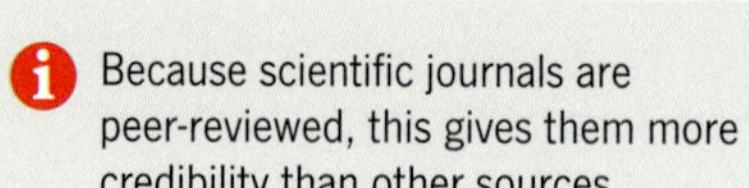

Because scientific journals are peer-reviewed, this gives them more credibility than other sources.

- *Australian Journal of Chemistry*
- *Journal of the American Chemical Society*
- *Nature Chemistry*
- *Chemical Reviews*
- *Green Chemistry*

Evaluating books and journals

Your textbook is an excellent starting point for reliable information. Information that you find elsewhere should be consistent with the information in your textbook. Articles published in newspapers and magazines often present findings of new research, which may or may not be confirmed later, so be careful not to treat such sources of information as established fact. Peer-reviewed journal articles are likely to be up to date and reliable.

Evaluating websites

Remember that anyone can publish anything on the internet, so it is important to evaluate the credibility, currency and content of online information. To evaluate online information, follow this checklist.

❏ Credibility—consider who the author is, their qualifications and expertise. Check for their contact information and for a trusted abbreviation in the web address, such as .gov or .edu. Websites using .com may have a bias towards selling a product (although this product could be a reputable science magazine or journal), and .org sites might have a bias towards one point of view (although these sites can be a good starting point for general information).

❏ Currency—check the date the information you are using was last revised.

❏ Content—consider whether the information presented is fact or opinion. Check for properly referenced sources, and compare information to other reputable sources, including books and science journals.

Recognising evidence compared to opinion and anecdote

An important aspect of the scientific method is to make conclusions based on data and evidence.

Data that is authoritative has been published by a credible source—for example, VCAA's *VCE Chemistry Data Book*.

Data should be **objective** (free of personal bias) and may be used as evidence, whereas opinion and anecdote are **subjective** (influenced by personal views). An anecdote is a story that might not be typical or representative of a pattern. Table 1.2.9 compares evidence versus opinion and anecdote.

TABLE 1.2.9 Comparison of evidence versus opinion and anecdote

Data and evidence	Opinion and anecdote
objective	subjective
measured, unbiased, replicable, systematic, representative	personal, individual story, non-systematic, not necessarily representative or part of a pattern
example: Scientific consensus among the majority of experts is that greenhouse gases are contributing to climate change.	example: Today is a cold day, therefore global warming is not true.

● *You will now be able to answer key question 2.*

Identifying information that is not scientifically valid

Beware of publications or sources of information that are presented as science but that are not scientifically valid. Non-scientific ideas and writing can be identified by:

- a lack of data or evidence
- bias—only part of the data or evidence is considered (usually the data supporting the claim)
- poorly collected data or evidence; for example, basing data or evidence on a group that is too small or not representative of the whole
- invalid conclusions (that is, not supported by evidence)
- lack of objectivity—appealing to emotion rather than presenting facts and evidence impartially.

1.2 Review

OA ✓✓

SUMMARY

- The first step in planning an investigation is to identify the topic.
- A research question is a statement that broadly defines what is being investigated.
- A hypothesis:
  - is a statement that can be tested and is based on previous knowledge and evidence or observations, and addresses the research question
  - often takes the form of a proposed relationship between two or more variables in a cause-and-effect relationship
  - must be testable; that is, able to be supported (verified) or not supported (falsified) by investigation.
- The aim of an investigation is a statement describing in detail what will be investigated.
- The scientific investigation methodology and method/s depend on the aim of the investigation and the research question.
- The three types of variables are:
  - independent—the variable that is selected and manipulated by the researcher
  - dependent—a variable that may change in response to a change in the independent variable and is measured or observed
  - controlled variables—variables that are kept constant during the investigation because they may affect the dependent variable.
- Ethical and safety considerations must be the highest priority at all times during a practical investigation.
- Many of the 12 principles of green chemistry can be employed or considered by VCE Chemistry students
- Primary sources of information are created by a person directly involved in an investigation. Secondary sources of information are a synthesis, review or interpretation of primary sources.
- Use critical thinking to evaluate sources of information by asking questions such as:
  - Who created this message?
  - Why was the information written?
  - When was the information published?
  - Are the conclusions supported by data or evidence?
  - What is implied?
  - What has been omitted?

KEY QUESTIONS

Knowledge and understanding

1 Distinguish between the terms ‘independent variable’, ‘dependent variable’ and ‘controlled variables’.

2 Explain what is meant by the terms ‘objective’ and ‘subjective’.

3 Identify each of the following statements as either a theory, a hypothesis or an observation.
   a Adding more hydrochloric acid caused a colour change from blue to yellow.
   b Le Châtelier’s principle states that when a change is made to a system at equilibrium, the system will adjust itself to partially oppose the change.
   c If the concentration of the reactant, hydrochloric acid, is increased, the rate of the reaction with Mg ribbon will increase due to an increased frequency of collisions.

4 Write a hypothesis for each of the following scenarios:
   a A student wondered whether the surface area of liver would affect the rate of reaction when liver is used as a catalyst for the decomposition of hydrogen peroxide.
   b A student wanted to know if the temperature of half-cell solutions affected the current moving through a $Zn(s)/Zn^{2+}(aq)//Cu^{2+}(aq)/Cu(s)$ galvanic cell.
   c Enzymes in washing powder are denatured in water at 50°C.

Analysis

5 A student conducted a controlled experiment to investigate whether increasing the temperature of the acid would increase the initial rate of the reaction between hydrochloric acid and marble chips.

For each of the following aspects of the investigation, indicate the type of variable described and explain your choice.

a mass lost by the system (flask, including contents) after the first 30 seconds of reaction.

b concentration of hydrochloric acid, size of marble chips

c temperature of hydrochloric acid (20°C, 30°C, 40°C)

6 A pair of students developed the hypothesis: As the concentration of a copper(II) sulfate solution increases, there will be an increase in the mass of copper metal deposited at the negative electrode during electrolysis using graphite electrodes.

a Write a research question for this investigation.

b Identify the independent variable.

c Identify the dependent variable.

d Suggest possible controlled variables.

7 State whether each of the following is a primary or a secondary source of information.

a a newspaper article about using green chemistry to synthesise simvastatin

b experiments with calculations of atom economy of two different pathways to synthesise simvastatin

c an interview with a research pharmacist regarding the use of simvastatin

d a journal article written by the researchers of an alternative pathway to synthesise simvastatin published in a peer-reviewed journal

8 A student has the following research question: How does the temperature of the electrolyte affect the charge flowing in a copper-plating electrolytic cell?

a State the independent variable.

b State the dependent variable.

c Describe the equipment that could be used in a laboratory to monitor the independent variable and to measure the dependent variable.

1.3 Data collection and quality

In this section you will learn about data collection in a controlled experiment. You will learn how to record both quantitative and qualitative data. You will also learn about the various factors that contribute to data quality, and the importance of controlled experiments in producing valid results.

KEEPING A LOGBOOK

Throughout Units 3 and 4, and particularly during your practical investigation for Unit 4 Area of Study 3, you must keep a **logbook**. This is generally thought of as a bound book that includes every detail of your research (Figure 1.3.1 on the following page). The reason for keeping a logbook, rather than recording information on random pieces of paper, is primarily to stay organised and to prevent loss of details. The best place for your logbook to be kept is in the laboratory, so that it is always accessible when you are doing your experimental research. Keeping track of all observations, qualitative and quantitative raw data, modifications to the method and your plans to analyse the data will ensure that you don't forget what you did or spend extra time repeating work because you have lost your results.

The following checklist will help you remember what to include in your logbook:

- ❑ your ideas when planning the research
- ❑ tables ready for raw qualitative and quantitative data entry
- ❑ notes about what data you will collect and how you intend to analyse it (refer to Section 1.4 page 34 for further details)
- ❑ records of all materials, methods, details of the experiment and raw data
- ❑ all notes, sketches, photographs and results
- ❑ records of any incidents or errors that may influence the results.

Make sure to date all entries in your logbook. Figure 1.3.1 shows a sample page of a student logbook.

Determining the method

Record any difficulties you encounter in the investigation and any changes that you make to the method as a result of the difficulties.

The method for a controlled scientific experiment must consider how the independent variable will be changed, how the dependent variable will be measured and determine the controlled variables that will not change. Any decisions to repeat measurements and change to the next variation (e.g. a different concentration) of the independent variable should be included in the method.

Ensure that you have outlined a detailed step-by-step method that other people could follow if they were to replicate your experiment. Sometimes a diagram of your equipment set-up is a valuable addition to your method.

Modifying the method

The method may need modifying as the investigation is carried out. If this is the case:

- record everything
- note any difficulties you encountered and the ways they were overcome.

If the expected data is not obtained, don't worry. Every test carried out can contribute to the understanding of the investigation as a whole, no matter how much of a disaster it may appear at first. Do your best to critically evaluate the data and identify the limitations of the investigation. Based on this information you should be able to propose further investigations.

Research question - How does the concentration of the electrolyte in the Daniell cell affect the voltage achieved by that cell?

Construct a Daniell cell

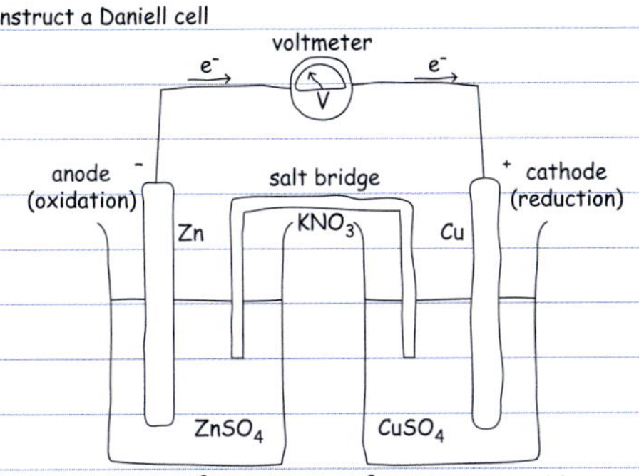

voltmeter

e^- e^-

V

anode (oxidation) -

salt bridge

+ cathode (reduction)

Zn

KNO_3

Cu

$ZnSO_4$

$CuSO_4$

$Zn(s) \rightarrow Zn^{2+}(aq) + 2e^-$ $Cu^{2+}(aq) + 2e^- \rightarrow Cu(s)$

Independent variable - concentration of $ZnSO_4$ and concentration of $CuSO_4$

Dependent variable - voltage

Controlled variables - anode and cathode, wires, volume of solutions, temperature of solutions, concentration of KNO_3 for salt bridge, thickness and size of salt bridge

Concentrations to use:

$ZnSO_4$ - solubility in water at 20°C = 57.7 g/100 g (Wikipedia) = 3.57 M

$CuSO_4 \cdot 5H_2O$ - solubility in water at 20°C = 1.26 M (Wikipedia) *** Not very soluble - use low concentrations

Materials: - start with 0.5 M solutions of $ZnSO_4$ and $CuSO_4$ and dilute (80 mL per beaker)

Dilutions - use $c_1V_1 = c_2V_2$ to calculate e.g. $0.5 \times V_1 = 0.4 \times 80$, $V_1 = 64$ mL (round to 65 mL and recalculate concentration)

Concentration (M)	Volume of 0.5 M $ZnSO_4$ or $CuSO_4$ (mL)	Volume of water (mL)
0.50	80	0
0.41	65	15
0.31	50	30
0.22	35	45
0.13	20	60

Results

Concentration of $ZnSO_4$ or $CuSO_4$ (M)	Trial	Voltage (V)	Average voltage (V)
	1		
	2		
	3		

FIGURE 1.3.1 Sample page from a student logbook showing development of ideas in planning a practical investigation

FIGURE 1.3.1 Sample page from a student logbook showing development of ideas in planning a practical investigation

CASE STUDY

Antoine and Marie-Anne Lavoisier

The French scientist Antoine Lavoisier (1743–94) is remembered for his experimental work on the chemical reactivity of oxygen and he is credited with the law of conservation of mass: in a chemical reaction, matter can neither be created nor destroyed. In his experiments, Lavoisier sought evidence that his theory about the conservation of mass was correct, but he had difficulty proving that it was universally valid. With such a broad hypothesis, it is, perhaps, no surprise that it was difficult to collect enough evidence to unequivocally prove it.

Lavoisier's methods were based on experimental work. He collected and weighed the substances involved in his reactions, and used expensive apparatus that was built to his own designs. It is fortunate that he was independently wealthy so that he could support his work financially.

Lavoisier married Marie-Anne Paulze (1758–1826) when she was only 13. Marie-Anne was well educated and she played a major role in Lavoisier's work. She translated the work of English scientists such as Richard Kirwan, Joseph Priestley and Henry Cavendish into French for her husband, who did not speak English. Access to this work on the theory of phlogiston, a fire-like material which was thought to be gained or lost during combustion, enabled Antoine to investigate the ideas using his experimental techniques. He not only showed that phlogiston theory was incorrect, but most importantly, he discovered that it was oxygen gas that enabled combustion.

FIGURE 1.3.2 Antoine and Marie-Anne Lavoisier working together in the laboratory, with Marie-Anne keeping notes in their logbook

Marie-Anne was also Antoine's laboratory assistant. She wrote details of his work in his laboratory logbook, keeping meticulous records of the procedures he used, thus lending validity to the findings published by Lavoisier. With her superior skills in drawing, Marie-Anne was able to sketch diagrams of his experimental designs, drawing the equipment accurately and precisely. Thirteen of her drawings featured in his 1789 publication of the *Elementary Treatise on Chemistry*, which is considered to be pivotal in the progression of chemistry as a field of study.

Many of Lavoisier's notebooks, which had been saved by Marie-Anne, can still be seen in a collection at Cornell University, New York, USA.

DATA COLLECTION

Before you begin your experiment, you should construct data collection tables in your logbook so that you can fill in the data as you conduct the experiment.

The measurements or observations that you collect during your investigation are your **primary data**. This primary data is also known as **raw data**. When you start to do calculations with that raw data, such as calculating an average, then the data is known as **processed data**. Processed data is raw data that has been organised, altered or analysed to produce meaningful information (Section 1.4, page 34).

> Record all raw and processed data in tables in your logbook. Photographs of observations are also useful.

> Raw data is the data you measure or observe during the course of the investigation. Processed data is data obtained by applying a calculation or formula to raw data.

Raw and processed data

The data you record in your logbook during the investigation is raw data. This data often needs to be processed or analysed before it can be presented.

Raw data that should be recorded includes:

- all quantitative data (measurements)
- all qualitative observations and other notes.

The raw data may be presented in the form of:

- tables of results
- diagrams and/or photographs of results
- graphs from datalogging equipment.

Processed data may include:

- graphs generated from raw data
- calculations in which raw data is used to determine other quantities
- graphs of processed data showing trends in that data.

Qualitative and quantitative data

Qualitative data can be observed and relates to a type or category, such as colour, or state (such as solid, liquid, gas or aqueous).

Measured numeric values are **quantitative data**. Quantitative data should be accompanied by a relevant unit such as mass (g) or time (s). Numeric data can be discrete or continuous:

- **Discrete data** are values that can be counted or measured, but which can only have certain values. An example is the number of chemistry students in a class or the number of hydrogen atoms in a molecule of water.
- **Continuous data** may be any number value within a given range that can be measured. Examples are time, temperature, mass, pH and concentration.

Determining whether a solution is acidic or basic can be qualitative or quantitative. Figure 1.3.3 shows qualitative and quantitative tests for acidity. Litmus paper gives a qualitative observation indicating the nature of the solution (Figure 1.3.3a), whereas a calibrated pH probe or pH meter gives a quantitative measurement indicating the nature of the solution (Figure 1.3.3b).

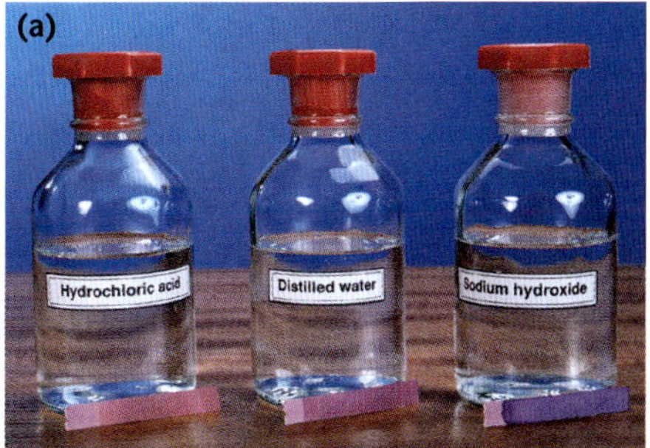

FIGURE 1.3.3 (a) Litmus paper can be used to indicate the acidity of a solution qualitatively. (b) A pH probe measures pH quantitatively.

Recording qualitative raw data

Examples of qualitative data that might be observed in an investigation, such as the one being performed in Figure 1.3.4 are:

- disappearance of a substance
- appearance of a solid (precipitate)
- production of a gas
- colour change (e.g. from colourless to purple)
- temperature change (e.g. increase or decrease without using a thermometer).

When recording qualitative data, focus on the observation. For example, a gas may be produced at the cathode during electrolysis of an aqueous solution. However, you cannot assume that the gas is hydrogen unless you have confirmed this; for example, by collecting the gas and performing the 'pop' test.

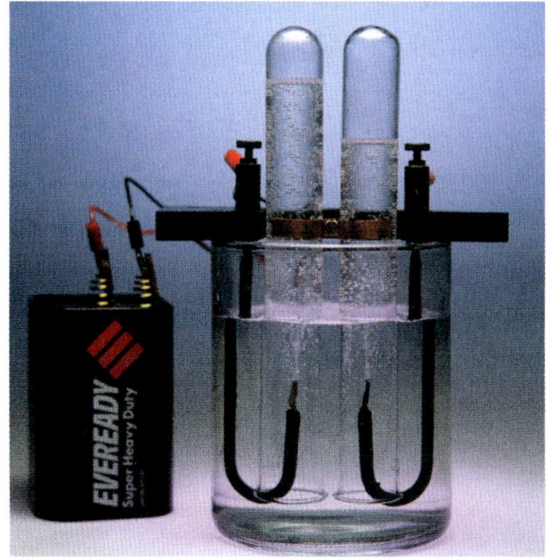

FIGURE 1.3.4 Qualitative observations include the production of a gas at an anode or cathode during electrolysis.

Recording quantitative raw data

Have your logbook ready, with prepared data collection tables, before you begin your experiment. Accurately record the correct number of significant figures and decimal places for measurements. For example, burettes should be read to two decimal places. Mass measurements should be recorded to the number of decimal places shown on the electronic balance; for example, 2.034 g.

Presenting raw data in tables

Tables organise data into rows and columns, and they vary in complexity according to the nature of your data. Tables can be used to organise raw data and processed data, and to summarise results. Results from all trials should be listed in tables of raw data.

Figure 1.3.5 shows an example of how to present quantitative raw data in a table. Note that the table has the following features:

- a descriptive title
- column headings (including the unit)
- the independent variables (distance between the electrodes) placed in the left column
- the dependent variable(s) placed in the right column(s).

Table 1 The time taken for black cross to disappear in reaction between 0.25 M sodium thiosulfate solution and 2 M hydrochloric acid when the temperature is varied

Initial temperature of $Na_2S_2O_3$ solution (°C)	Trial	Time taken for the cross to disappear (s)
15.0	1	125.0
	2	130.0
	3	127.0
30.0	1	60.0
	2	55.0
	3	59.0
45.0	1	40.0
	2	48.0
	3	38.0
60.0	1	25.0
	2	22.0
	3	24.0

FIGURE 1.3.5 Features of a good raw data table

● *You will now be able to answer key question 7.*

Units

Each column in a table of quantitative data should include relevant units. Commonly, these will be SI (International System of Units) units. Examples of commonly used units are shown in Table 1.3.1.

TABLE 1.3.1 Examples of units for various quantities

Quantity	SI unit	SI unit symbol
mass	kilogram	kg (but chemists usually work in g)
time	second	s
temperature	kelvin	K
electric current	ampere	A
amount of substance	mole	mol
volume	litre	L
charge	coulomb	C
energy	joule	J

DATA QUALITY

The results of your data analysis will only be as good as the quality of the data. You should consider this when collecting primary data in your investigations, and also when you evaluate the quality of secondary data from other sources.

Accuracy and precision

If repeated measurements of the same quantity give values that are in close agreement, the measurement is said to be **precise**. To obtain precise results you must minimise random errors. A set of precise measurements will have values very close to the mean value of the measurements.

If the average of a set of measurements of a quantity is very close to the **true value**, the value that would be obtained under perfect conditions, then the measurement is said to be **accurate**. To obtain accurate results you must minimise systematic errors. **Precision**, how close two or more measurements are to each other, gives no indication of how close the measurements are to the true value and is therefore a separate consideration to **accuracy**. Figure 1.3.6 shows the difference between accuracy and precision.

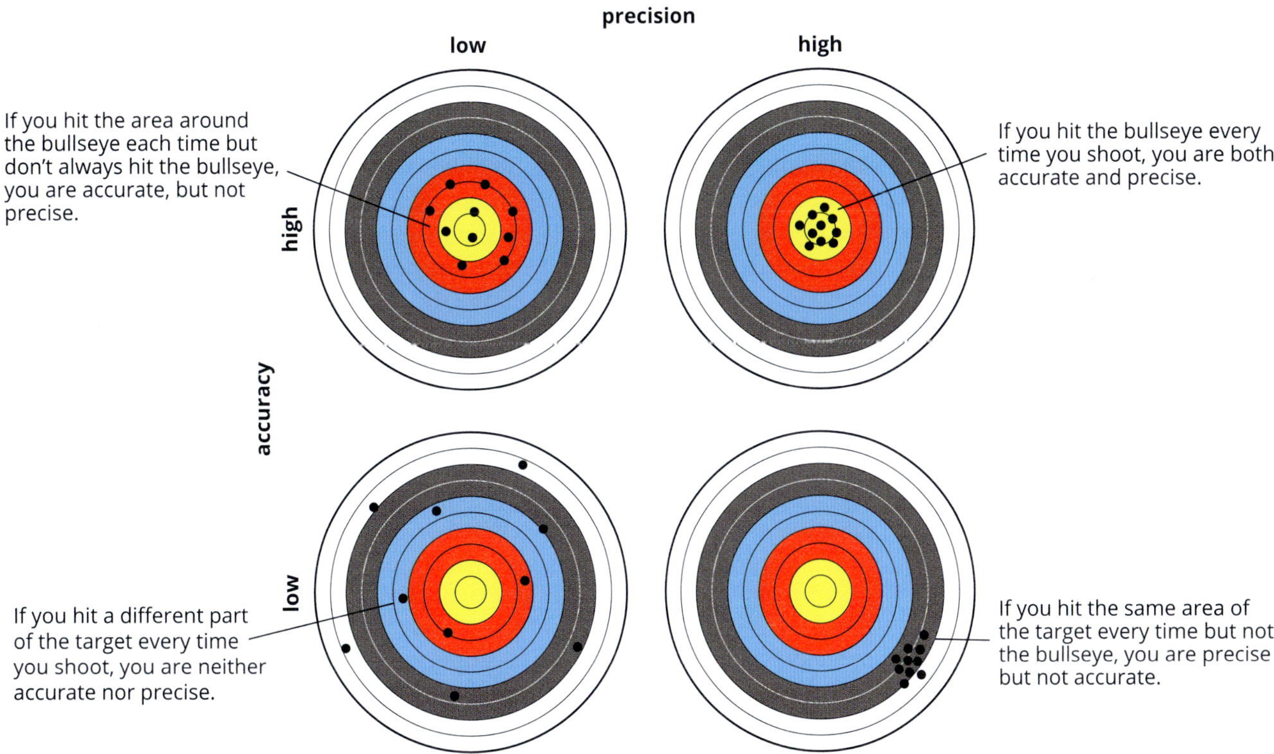

FIGURE 1.3.6 An example of a target illustrates the difference between accuracy and precision.

If the true value is known, then the difference between the true value and a measured value is called the **measurement error**. In many cases, because the true value is unknown, the measurement error will also be unknown.

- *You will now be able to answer key question 5.*

Significant figures

When reporting measurements in raw data tables, you should report a measurement that is as accurate as possible within the limitations of the equipment being used. In doing so, you will be correctly stating the number of **significant figures** for the measurement. The significant figures indicate the precision of the instrument used and indicate a range within which the value can be considered to exist. For example, a measurement of 56.0 g has 3 significant figures and indicates that the value is somewhere between 55.95 g and 56.04 g, whereas a measurement of 56 g, which has 2 significant figures, indicates that the value is somewhere between 55.5 g and 56.4 g.

When giving an answer to a calculation, it is important to take care with the number of significant figures that you use. An answer cannot be more precise than the data or the measuring device used to calculate it. For example, if an electronic balance that measures to the nearest 0.01 g shows that an object has a mass of 1.17 g, then the mass should be recorded exactly as 1.17 g (not as 1.170 g). The number 56 has 2 significant figures, because it is assumed that the decimal point occurs at the end of the number. Recording to 3 significant figures (e.g. 56.0 g or 55.8 g) would not be scientifically ethical (Section 1.2, page 15). If this mass of 56 g is used to calculate another value, it would also not be ethical to give an answer that has more than 2 significant figures. Similarly, it is not correct to cut off decimal places if the equipment is more precise than you require.

Working out the number of significant figures

When counting significant figures, the following rules apply.

- All non-zero digits count as significant figures.
- Zeros may or may not be significant, depending on whether they are part of the measured value (zeros after the last non-zero digit) or are simply used to position a decimal point (zeros before the first non-zero digit).
- Whole numbers written without a decimal point will have the same number of significant figures as the number of digits, with the assumption that the decimal point occurs at the end of the number, for example 500 has 3 significant figures. Therefore, a stated volume of '500 mL' will be considered as having 3 significant figures.
- Significant figures are only applied to measurements and calculations involving measurements (Section 1.4). They do not apply to quantities that are inherently integers or fractions (e.g. a stoichiometric ratio such as 2 or $^{1}/_{2}$), defined quantities (e.g. 1 metre equals 100 centimetres), or conversion factors (multiplying by 100 to get a percentage or adding 273 to convert °C to K). For example, the accuracy of a temperature reading of 12°C cannot be increased from 2 significant figures to 3 significant figures by converting to kelvin (285 K).

● *You will now be able to answer key question 1.*

Types of errors

Most practical investigations have errors associated with them. The main errors that you will encounter are described below.

Systematic errors

A **systematic error** produces a constant bias in a measurement that cannot be eliminated by repeating the measurement. Systematic errors that affect a practical investigation could include:

- a pH meter that has not been calibrated, so it is consistently recording pH values that are greater than they should be
- an unsuitable indicator being used in a **titration**
- heat being lost to the surroundings due to lack of insulation during an experiment in which the temperature is being measured

- a person reading the scale on a measuring cylinder or burette with a constant **parallax error**. In this case, the parallax error occurs in reading a volume where the **meniscus**, the curved surface of the liquid, is not at eye level. This is shown in Figure 1.3.7.

Parallax error can also occur with **analogue** (non-digital) meters, such as voltmeters and ammeters, and any instrument in which a line or needle has to be compared to a fixed scale.

Whatever the cause, the resulting error is in the same direction for every measurement and the average will either be too high or too low as a result. Repeating an experiment will not remove systematic error.

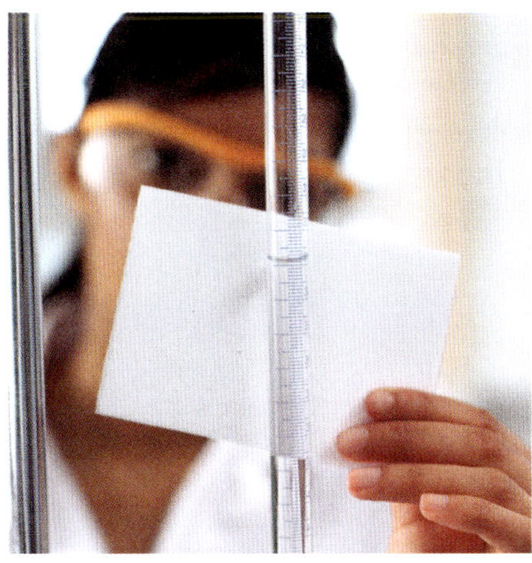

FIGURE 1.3.7 It is important to read the bottom of the meniscus of a solution in a burette at eye level to avoid parallax error.

Random errors

Random errors follow no regular pattern. The measurement is sometimes too large and sometimes too small. Random errors in a practical investigation could include:

- error in estimating the second decimal place in the volume read from a burette
- temperature variation within a solution that has not been mixed consistently
- error in estimating the colour of an indicator.

The effects of random errors can be reduced by taking multiple measurements of the same quantity, then calculating a **mean**, or **average**. For example, when you repeat an experiment to find out how much solid is produced in three separate trials, the three masses of the solid should be added together and divided by three. In this case, the mean of the three masses could be known as the **measurement result**, just as a single (not repeated) result could be known as a measurement result.

Another way to reduce random error is to refine a measurement technique by repeating it until you are more skilled in the technique.

Mistakes

Mistakes or personal errors are avoidable errors. Mistakes made during a practical investigation could include:

- misreading the numbers on a scale
- not labelling a sample adequately
- spilling a portion of a sample.

A measurement that involves a mistake must be repeated if possible, or rejected and not included in any calculations. Care in carrying out the experiment will prevent the majority of mistakes occurring.

● *You will now be able to answer key question 6.*

Repeatability

Repeatability refers to the consistency of the results when the experiment is repeated many times. Maintain your investigation's repeatability by ensuring there is sufficient **replication** of the experiment. In general, you should repeat each experiment at least three times (i.e. perform three trials). Three is not a magic number, and if you can do more trials, your results should become more precise, in part because you are becoming more familiar with the method and therefore better at carrying it out.

Reproducibility

Reproducibility is the closeness of the agreement between measurements of the same quantity, carried out under changed conditions of measurements. Reproducibility links closely to validity (described below). Data is reproducible if similar results are obtained by different operators in different laboratories, repeating an experimental procedure and obtaining similar results. A benefit of analysing the reproducibility of data is to help identify possible systematic errors that would affect the accuracy of the experiment.

CHEMFILE

The glass used to make laboratory glassware

The glass that we use in school laboratories is not the same type of glass used for bottles and jars. It is borosilicate glass. Like all types of glass, it is a mixture of several different compounds; in this case, silica, SiO_2, boron oxide, B_2O_3, sodium oxide, Na_2O, and aluminium oxide, Al_2O_3. It was developed for use in laboratories by Otto Schott, a glass-maker working in Germany in the late 1800s.

In comparison to the sodium-based glass used for windows, borosilicate glass is notable for its high tolerance to heat and substantial resistance to sudden changes in temperature. It is also tolerant to a wide range of chemicals, even strong acids and alkalis. The notable exception to this is hydrofluoric acid, HF, which reacts with glass and must be stored in plastic containers.

Glassware that is used in the laboratory is made of borosilicate glass.

Resolution

Various types of equipment can be used to measure quantitative data. For example, time could be measured using a stopwatch, phone or using a clock. **Resolution** is the smallest change in the measured quantity that causes a perceptible change in the value shown by the measuring instrument and this has implications for the number of decimal places that should be quoted.

Resolution is determined differently for analogue and digital instruments. For example, if the resolution of a burette is 0.05 mL, then measurement readings of 10.50 mL or 10.55 mL are possible, but not a measurement reading of 10.53 mL. The meniscus of the liquid will either be on the burette line marking, in which case the reading would be 10.50 mL, or it will lie between 10.50 and 10.60, in which case it is measured as 10.55 mL.

Digital measuring equipment has a resolution to the last decimal place. For example, an electronic balance that measures to 0.001 g would have a resolution of 0.001 g.

If we compare analogue and digital devices, the resolution of an analogue clock might be 1 s, the resolution of a stopwatch might be 0.01 s, and the resolution of a phone app might be 0.001 s. Unfortunately, the reaction time of a human is commonly considered to be 0.2 s, so the apparent accuracy of the phone app is limited by the human who is recording the time.

Validity

Validity refers to whether an experiment or investigation is testing the set hypothesis and aims.

To ensure an investigation is valid, it should be designed so that only one variable is being changed at a time. The other variables must remain constant, so that meaningful conclusions can be drawn about the effect of each variable.

● *You will now be able to answer key questions 2, 3, 4 and 8.*

TECHNIQUES FOR IMPROVING DATA QUALITY

Designing the method carefully, including the selection and use of equipment, will help reduce random and systematic errors. Once you have chosen the appropriate equipment, you may need to calibrate the equipment to increase the accuracy of your measurements. To increase precision with the same instrument, use a larger sample size.

Calibration

Calibration is the process of adjusting an instrument using a reference or standard. Some equipment, such as pH probes, need calibrating before use; for example, to check that a pH probe is giving a true reading of pH 4.0. By using calibrated equipment, you can be more certain that your measured values are accurate. In colorimetry a **calibration curve** is constructed showing the relationship between the concentration of known standards and absorbance.

Equipment

To minimise errors, check the precision of the equipment that you intend to use. Pipettes, burettes and volumetric flasks have greater precision than measuring cylinders for measuring volumes of liquids. However, you must still use all equipment correctly to reduce error.

Sample size

In general, the larger the sample (e.g. the mass of starting material that you are using), the more precise the measured values will be. While wasteful experiments are contrary to the principles of green chemistry, your results will be more precise if you use 5 g of starting material rather than 0.5 g, given that you are likely to be using the same electronic balance to determine mass in both cases. Even though any error in the balance will be the same, the percentage error will be smaller for the larger mass.

TECHNIQUES TO SUPPORT YOUR PRACTICAL INVESTIGATION

A variety of techniques can be used for conducting practical investigations. The choice of equipment is important to minimise error, to ensure your measurements are accurate and your results are reproducible and reliable, and to minimise **uncertainty** (the range of values between which a measurement may fall). Techniques that you might use when conducting practical investigations are outlined in Table 1.3.2.

TABLE 1.3.2 Techniques that might be appropriate for a practical investigation

Technique	Purpose	Dependent variable	Example application
titration	to calculate the unknown concentration of a solution	volume of solution (titre)	determining concentration of reagents in redox reactions
data logging	to collect data over a defined period of time	change in pH, temperature, conductivity, colour, carbon dioxide concentration, pressure, time	determination of temperature or pH to see if a change is occurring
calorimetry	to determine the energy being released or absorbed by a chemical reaction	temperature change of a known volume of water in a calibrated calorimeter, or a simple calorimeter	determining the enthalpy of combustion of fuels or food, or the enthalpy of reaction for other reactions
electrochemistry	to measure the voltage or current of a particular galvanic or fuel cell	voltage or current	compare the voltage of a galvanic cell when the concentration of the electrolyte is varied

Data logging

In recent times, the ability to connect independent sensors to devices such as a computer, tablet or phone (Figure 1.3.8) has enabled simple and highly effective electronic data acquisition (data logging) in schools. The software will generally graph and allow advanced calculations directly without the need for an additional program. Some probes can enable data collection to occur over an extended period of time. Examples include pH probes, temperature probes, carbon dioxide or oxygen concentration probes, conductivity probes, colorimeters and automatic mass balance equipment. Data logging plays an important role in improving data collection in a range of investigations.

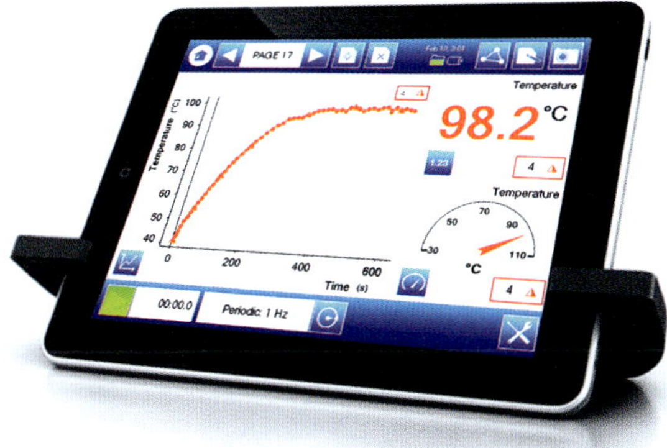

FIGURE 1.3.8 Data acquisition software produces real-time graphs that can be downloaded or printed.

1.3 Review

OA ✓✓

SUMMARY

- Record all information objectively in your logbook, including your data and method of investigation.
- The method is a step-by-step record of what you did in the experiment, written in such a way that another person can reproduce the experiment.
- Data may be in one of a number of forms:
  - Raw data is the data you collect in your logbook.
  - Processed data is raw data that has been mathematically analysed.
  - Qualitative data is observed and relates to a type or category, such as colour.
  - Quantitative data consists of measured numeric values and includes units.
  - Discrete data consists of values that can be counted or measured, but which can only have certain values.
  - Continuous data may be any number value within a given range that can be measured.
- The quality of data relies on a number of factors:
  - Validity refers to whether your results measure what the investigation set out to measure.
  - Repeatability is the consistency of your results when they are repeated many times as trials under the exact same set of experimental conditions.
  - Reproducibility is the ability to obtain the same results if an experiment is replicated.
  - Accuracy is how close a measurement is to the true value.
  - Precision is how closely a set of measurements agree with each other.
  - Resolution is the smallest change in the measured quantity that causes a perceptible change in the value shown by the measuring instrument.
- Errors in data may be classified as systematic or random
  - Systematic errors result in errors in the same direction for every measurement—either too high or too low.
  - Random errors follow no regular pattern.
- Mistakes or personal errors are avoidable and should not be included in further analysis of results.

KEY QUESTIONS

Knowledge and understanding

1 For each of the following masses, state the number of significant figures.

a 200 g

b 0.0410 g

c 20.450 g

d 0.000 35 g

2 Fill in the gaps in the following sentences using the terms:

repeatability; reproducibility; resolution; accuracy; precision; true value; validity

a ________ is how close a measurement is to the ________ , whereas ____________ is how closely a set of measurements agree with each other.

b ____________ refers to whether your results measure what the investigation set out to measure.

c ______________ is the consistency of your results when they are repeated many times as trials under the exact same set of experimental conditions, whereas __________ is the ability for another experimenter to obtain the same results if they replicate your experiment.

d ____________ is the smallest change in the measured quantity that causes a perceptible change in the value shown by the measuring instrument.

3 A pair of students conducted an experiment as part of their Units 3 and 4 Chemistry studies. The experiment was repeated three times, and all values were reported in the results section of their report. Which of the following aspects of the data quality will not be improved by repeating an experiment?

A random error

B reproducibility

C precision

D resolution

4 State the correct term or terms (accurate, reproducible, precise or valid) that describes an experiment with each of the following conditions.

a The experiment addresses the hypothesis and aim.

b The experiment is repeated and consistent results are obtained.

c Appropriate equipment is chosen for the measurements required.

Analysis

5 Two students submitted their tables of raw data for the same titration experiment of the same sample of diluted ethanoic acid with the same solution of 0.105 M sodium hydroxide, using the same indicator. From calculations based on her own careful analysis, their teacher expected that the titrations would have an average volume of 19.50 mL. The students' tables of data are shown below.

Student A:

	Trial 1	Trial 2	Trial 3	Average
Final volume (mL)	19.25	38.80	19.75	
Initial volume (mL)	0.00	19.25	0.05	
Volume of titre (mL)	19.25	19.55	19.70	19.50

Student B:

	Trial 1	Trial 2	Trial 3	Average
Final volume (mL)	19.55	38.95	19.55	
Initial volume (mL)	0.05	19.55	0.10	
Volume of titre (mL)	19.50	19.40	19.45	19.45

Discuss the two sets of results in terms of their accuracy and their precision.

6 Classify the following as systematic or random errors and propose how to minimise the limitation.

Error	Type of error	Suggested improvement
finding it difficult to finish at the same colour in successive titres during a titration		
using a stopwatch while chatting with your partner and not quite concentrating on the reaction being timed		
reading the initial volume of a burette from below the level of the meniscus		
loss of heat to environment during a combustion calorimetry experiment		

7 Consider the following table of raw data for an experiment in which the heat of combustion of ethanol is determined. List three mistakes that you can identify in the table and describe how they could be corrected.

Trial	Volume of water in can (mL)	Initial temper-ature of water	Final temper-ature of water	Initial mass of ethanol and spirit burner	Final mass of ethanol and spirit burner (g)
1	200 mL	15.00	45.0	250.50	250.50
2	200 mL	14.0	30	243.35	241.20
3	200 mL	14.551	95.0	241.50	239.00

8 Which of the following methods is repeatable and reproducible? Justify your response, critiquing the less correct method.

Method A

1 Measure 25 mL of 0.10 M nitric acid into a solution calorimeter.
2 Measure the temperature of the nitric acid and record the result.
3 Measure the temperature of 25 mL of 0.10 M potassium hydroxide solution and record the result.
4 When the temperature is stable, add the potassium hydroxide solution to the nitric acid, place the lid on the calorimeter and stir.
5 Record the temperature of the mixture every 10 seconds for 3 minutes.

Method B

1 Pour some nitric acid into a beaker.
2 Place a thermometer into the beaker.
3 When the temperature is stable, add some potassium hydroxide solution and stir.
4 Record the final temperature of the mixture.

1.4 Data analysis and presentation

In Section 1.3 you learnt about different types of data and the factors that affect data quality. In this section, you will learn how to process and analyse your data and present analysed data in graphs.

PROCESSING DATA

Whenever you process data, it is important to show the formula you used and a sample calculation. Even if you don't include all of the information in your report or scientific poster, everything should be recorded in your logbook.

In an experiment to investigate Faraday's first law you might have measured the initial and final masses of the cathode which was used in the electrolysis of copper(II) sulfate using different currents. You can then process this data further.

For example,

$$\text{change in mass} = \text{final mass} - \text{initial mass}$$

The mean change in mass is found by adding all the changes in mass for a particular current and dividing the sum by the number of trials.

$$\text{mean change in mass} = \frac{(\text{change in mass}_1 + \text{change in mass}_2 + \text{change in mass}_3)}{3}$$

When summarising processed data in a table, record the independent variable in the left-hand column and the dependent variable(s) in the column(s) that follow. The processed or calculated values should be recorded in the final column.

In Table 1.4.1, the data in the final two columns is processed data (the change in mass and the mean change in mass).

TABLE 1.4.1 Summary results table indicating the increase in mass of the cathode when varying the current during the electrolysis of copper(II) sulfate

Current (A)	Trial number	Change in mass (g)	Mean change in mass (g)
0.20	1	0.125	0.128
	2	0.122	
	3	0.138	
0.40	1	0.220	0.224
	2	0.230	
	3	0.223	
0.60	1	0.384	0.380
	2	0.380	
	3	0.375	
0.80	1	0.512	0.520
	2	0.520	
	3	0.528	

Transforming decimal notation into scientific notation

Scientists use **scientific notation** to express very large and very small numbers.

For example, instead of writing 0.000 003 5, scientists would write 3.5×10^{-6}.

A number in scientific notation (also called **standard form** or power of ten notation) is written as $a \times 10^n$, where:

- a is a number equal to or greater than 1 and less than 10: that is, $1 \leq a < 10$
- n, the **index** (plural indices), is an integer (a positive or negative whole number) and is the power to which 10 is raised.

Two examples of how to transform decimal notation into scientific notation are shown in Worked example 1.4.1.

When large numbers are written in scientific notation, the 10 has a positive index. When very small numbers are written in scientific notation, the 10 has a negative index.

Worked example 1.4.1

TRANSFORMING DECIMAL NOTATION INTO SCIENTIFIC NOTATION

Transform each of the following numbers into scientific notation:

a 6243

b 0.000325

Thinking	Working
a Place a decimal point after the first non-zero digit, so that the original number is written as a decimal number greater than 1 but less than 10.	6243 becomes 6.243.
Determine the index number by counting the number of places the decimal point needs to be moved and noting the direction.	6243. The decimal is moved 3 places to the left.
Multiply the decimal number by the appropriate power of 10, as determined by counting in the step above. If the decimal point was moved to the left, the index will be a positive number. If the decimal point was moved to the right, the index will be a negative number.	The decimal was moved 3 places to the left, so the index number will be 3. 6.243×10^3
b Place a decimal point after the first non-zero digit, so that the original number is written as a decimal number greater than 1 but less than 10.	0.000325 becomes 3.25.
Determine the index number by counting the number of places the decimal point needs to be moved and noting the direction.	0.000325 The decimal is moved 4 places to the right.
Multiply the decimal number by the appropriate power of 10, as determined by counting in the step above. If the decimal point was moved to the left, the index will be a positive number. If the decimal point was moved to the right, the index will be a negative number.	The decimal was moved 4 places to the right, so the index number will be –4. 3.25×10^{-4}

Worked example: Try yourself 1.4.1

TRANSFORMING DECIMAL NOTATION INTO SCIENTIFIC NOTATION

Transform each of the following numbers into scientific notation:

a 96 500

b 0.000 0625

You will now be able to answer key question 1.

Managing significant figures in calculations

- If you are multiplying or dividing, use the smallest number of significant figures provided in the initial values.
- If you are adding or subtracting, use the smallest number of decimal places provided in the initial values.

The number of significant figures to which an answer should be quoted depends on what kind of calculation you are doing.

- If you are multiplying or dividing, use the smallest number of significant figures provided in the initial values.
- If you are adding or subtracting, use the smallest number of decimal places provided in the initial values, e.g. 12.78 + 10.0 = 22.78 = 22.8.

When you are processing data using your scientific calculator, you do not have to report every decimal place stated on the calculator just because it is there. For example, if the least accurate piece of data in your calculation involving multiplication or division is to 3 significant figures, then the calculated result should likewise be reported to 3 significant figures, even if the calculator gives you 6 or more significant figures. However, if you are undertaking a calculation with several steps, keep the more accurate value with extra decimal places in your calculator and don't round your answer off before the calculation is complete.

The use of scientific notation will help you express a final answer to the correct number of significant figures.

The number 2500, in scientific notation, would be written as:

- 2.5×10^3 (2 significant figures)
- 2.50×10^3 (3 significant figures)
- 2.500×10^3 (4 significant figures)

depending on the actual data used to calculate this value.

Note that the significant figures in a molar mass should not limit the significant figures in an answer. The molar mass is not experimental data and, in the VCE Chemistry course, all relative atomic masses are quoted to one decimal place. So, unless you are working with very small elements, such as hydrogen and helium, relative atomic masses will have at least 3 significant figures.

Only the final calculated result should be reported to the appropriate number of significant figures, as shown in Worked example 1.4.2.

Worked example 1.4.2

PRESENTING CALCULATIONS TO THE CORRECT NUMBER OF SIGNIFICANT FIGURES

Calculate the enthalpy of combustion for ethanol, given that the temperature of 150.42 g of water increased from 12.0°C to 43.52°C when the mass of the spirit burner in which the ethanol was burning decreased from 200.500 g to 199.788 g. (Remember that the specific heat capacity of water = 4.18 J g^{-1} °C^{-1}.)

Thinking	Working
Record all data, taking note of the significant figures. Remember when subtracting numbers, the answer should have no more decimal places than the value with the smallest number of decimal places.	mass (water) = 150.42 g (5 sig figs) temperature change (water) = 43.52 − 12.0 = 31.5 (to 1 decimal place and 3 sig figs) specific heat capacity (water) = 4.18 J g^{-1} °C^{-1} (3 sig figs) mass change (ethanol) = initial mass of spirit burner − final mass of spirit burner = 200.500 − 199.788 = 0.712 g (3 sig figs)
Calculate the energy transferred, q, using the equation $q = mc\Delta T$. If the energy value is large, convert to kJ by dividing by 1000. Do not round off this answer and keep all figures in the calculator for the next step.	$q = mc\Delta T$ $= 150.42 \times 4.18 \times 31.5$ $= 19805.8$ J $= 19.8058$ kJ
Calculate the amount, in mol, of ethanol that was burned using $n = \frac{m}{M}$. Note that the significant figures in the molar mass should not limit your significant figures in the answer. Do not round off this answer and keep all figures in the calculator for the next step.	$n(C_2H_5OH) = \frac{m}{M}$ $= \frac{0.712}{46.0}$ $= 0.015478$ mol
Calculate the enthalpy of combustion for ethanol using the equation $\Delta H = \frac{q}{n}$.	$\Delta H = \frac{q}{n}$ $= \frac{19.805}{0.015478}$ $= 1280$ kJ mol^{-1}
Round the calculated answer to the lowest number of significant figures (3 significant figures in the temperature), using scientific notation if necessary and use correct sign to indicate enthalpy change.	ΔH_c(ethanol) $= -1.28 \times 10^3$ kJ mol^{-1}

Worked example: Try yourself 1.4.2

PRESENTING CALCULATIONS TO THE CORRECT NUMBER OF SIGNIFICANT FIGURES

Calculate the enthalpy of combustion for methanol, given that the temperature of 200.445 g of water increased from 15.0°C to 50°C when the mass of the spirit burner in which the methanol was burning, decreased from 183.450 g to 182.157 g. (Remember that the specific heat capacity of water = 4.18 J g^{-1} °C^{-1}.)

You will now be able to answer key questions 2 and 3.

- If you are multiplying or dividing, use the smallest number of significant figures provided in the initial values.
- If you are adding or subtracting, use the smallest number of decimal places provided in the initial values.

DATA ANALYSIS

In the discussion section of your scientific investigation report, the findings of your investigation need to be analysed and interpreted. The following factors should be considered and discussed.

- Whether a pattern, trend or relationship was observed between the independent and dependent variables.
- What kind of pattern it was and under what conditions it was observed.
- Whether there were data points that did not match the trend (**outliers**).
- How did the values you were monitoring change over the course of the experiment?
- How did your results compare to the theoretical results, or those recorded in literature sources?

Graphs

It is easier to observe trends and patterns in data in graphical form than in tabular form. An example of a **line graph** is shown in Figure 1.4.1.

The independent variable should always be plotted along the horizontal axis and the dependent variable on the vertical axis of a graph.

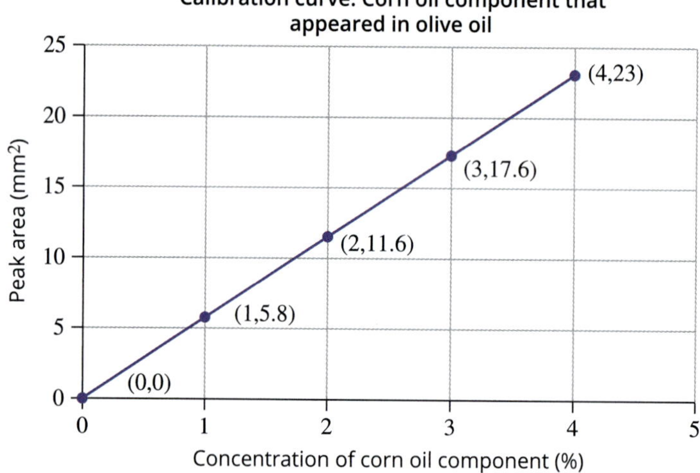

FIGURE 1.4.1 A graph is a good way to observe trends and patterns in data.

General rules to follow when presenting a line graph include the following.

- Use a descriptive title.
- Represent the independent variable on the horizontal axis and the dependent variable on the vertical axis.
- Clearly label the axes with both the variable and the unit in which they are measured.
- Choose a scale on the axes that will be appropriate for the data and that is consistent.
- If values between plotted points need to be read, then the values between marked points on the scale should be as easy to read as possible. For example, mark the scale off in multiples of 2 or 5, rather than 3 or 4.
- If there is a data point at (0,0), then the line of best fit can go through the origin. Otherwise draw the line of best fit only through the data points on the graph.

Types of graphs

A **scatter graph**, such as Figure 1.4.1, is used when two variables are being considered and one variable is dependent on the other. The independent variable is plotted along the horizontal axis and the dependent variable is plotted along the vertical axis. When an appropriate **line of best fit** is fitted to the data points, the graph should show the relationship between the two variables. When the line of best fit is drawn on the scatter graph, it becomes a line graph.

Column graphs, also called **bar graphs** (Figure 1.4.2), also clearly illustrate relationships between variables, but are most appropriate for discrete rather than continuous data (Section 1.3, page 25), whereas **pie charts** illustrate percentages effectively (Figure 1.4.3).

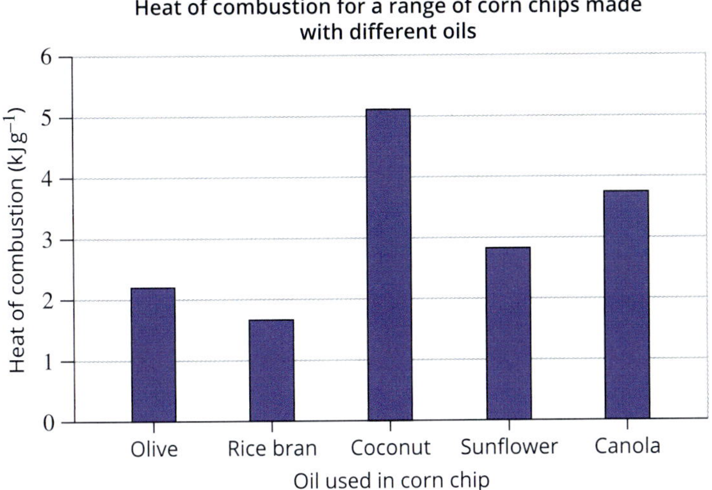

FIGURE 1.4.2 A column graph can be used to compare discrete data. This graph shows how the heat of combustion varied when corn chips made with different oils were burnt.

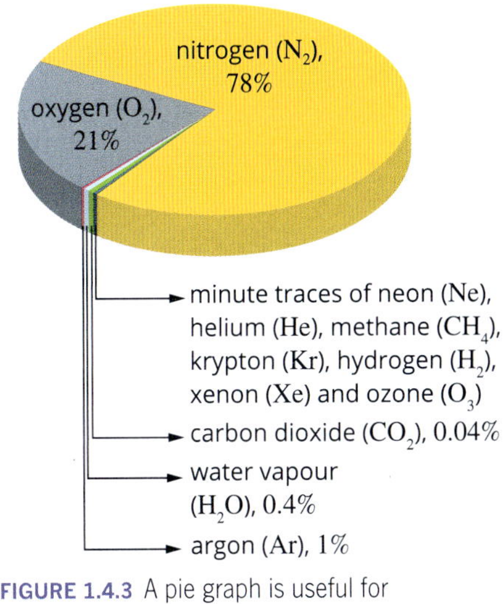

FIGURE 1.4.3 A pie graph is useful for representing proportional data; e.g. the chemical composition of air.

Trend lines in graphs

Line graphs are drawn to show the relationship, or **trend**, between two variables.

- Variables that change in direct proportion to each other produce a straight trend line, as shown in Figure 1.4.4. This is described as a **linear trend** or relationship.

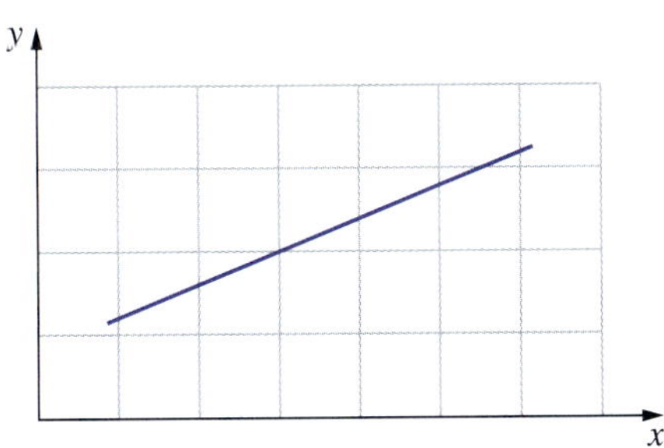

FIGURE 1.4.4 Trend line for variables that change in direct response to each other

- Variables that change non-linearly in proportion to each other may produce a curved trend line, as shown in Figure 1.4.5. This graph reflects an **exponential relationship**; that is, as x increases, y increases non-linearly. This would correspond to the equation $y = a^x$, where a is a positive number other than 1.

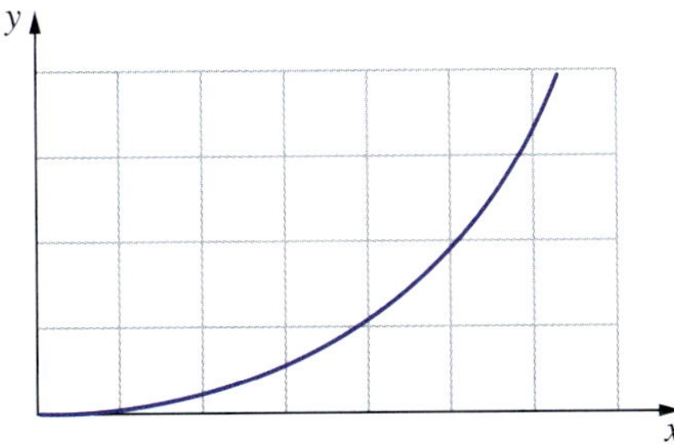

FIGURE 1.4.5 Variables that change in response to each other in a non-linear way

- When there is an **inverse relationship**, one variable increases as the other variable decreases (Figure 1.4.6).

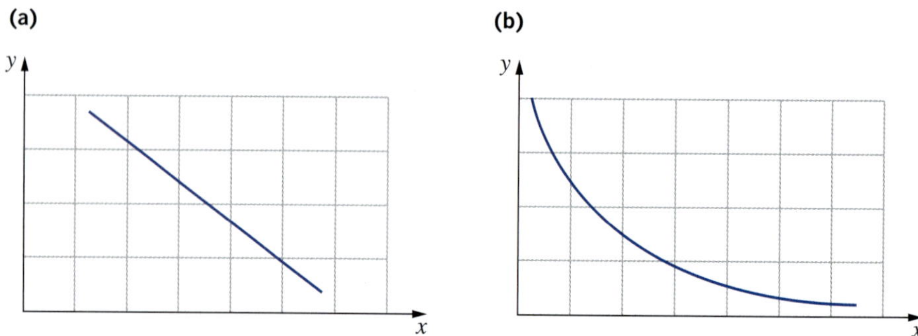

FIGURE 1.4.6 An inverse relationship in which one variable decreases as the other variable increases. The relationship may be (a) direct or (b) exponential.

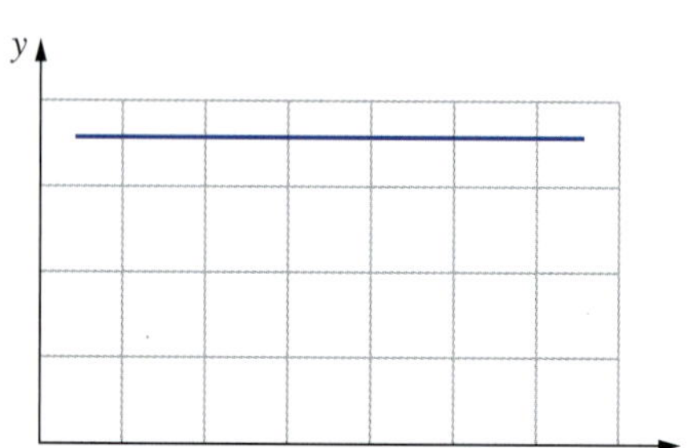

FIGURE 1.4.7 When two variables show no relationship, there is no trend in the graph.

- When there is no relationship between two variables, one variable will not change even if the other changes. A graph in which there is no relationship can be seen in Figure 1.4.7.

Scatter graphs and lines of best fit

Scatter graphs are commonly used to display data and can be used to plot raw or processed data. They are used to show the relationship between two variables when one variable is dependent on the other.

The independent variable, which is set by the experimenter, is usually shown on the horizontal axis. The dependent variable, which is the variable measured in the experiment, is shown on the vertical axis. The data is plotted on the graph as a series of points.

In a scatter graph, a trend line is then fitted to the data (Figure 1.4.8). The trend line might be a straight line of best fit or a curved line. It is used to show the overall trend in the data and can be used to predict values between the data points. A line of best fit usually does not pass through every data point. Its position can be estimated by eye (ensuring an equal number of data points are above and below the line when drawn). Alternatively, mathematical graphing programs can be used to accurately determine the line of best fit.

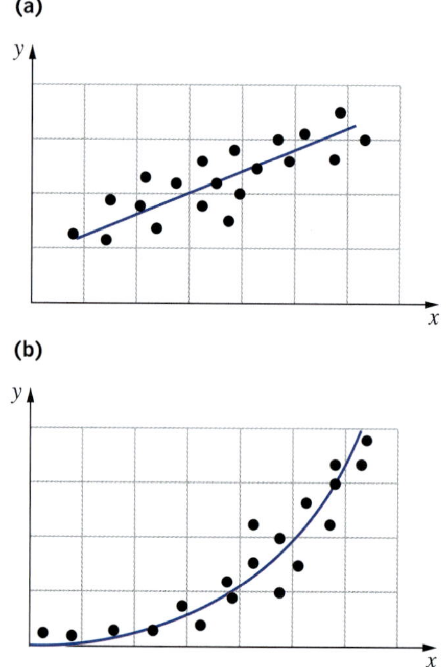

FIGURE 1.4.8 Scatter graphs showing (a) straight and (b) curved trend lines

Outliers

Sometimes when you collect data there may be points or observations that differ significantly from other data points and do not fit the trend observed in the data. These are referred to as outliers. An outlier is often caused by a mistake made when data was measured or recorded. To work with your results ethically, any outliers in your data must be further analysed and explained, rather than being automatically dismissed. Repeating trials may be useful in further examining an outlier—for example, to determine whether the outlier is the result of a mistake.

A common experiment in which outliers are often observed is a titration. When using a burette with a resolution of 0.05 mL, students are taught to aim for three **concordant titres** in their results. These are most commonly defined as titres which have a range of no more than 0.10 mL, that is, the difference between the largest and smallest of the titres is no more than 0.10 mL. Any titre with a value beyond that range is regarded as an outlier and is not used in the calculation of the mean titre. You will learn about titrations in Chapter 13.

The graph shown in Figure 1.4.9 is a calibration curve for determining the concentration of phosphate in a water sample. The data point that does not fit the trend of the other data points is an outlier. A mistake may have occurred when preparing that particular phosphate standard solution. In this case, you should show the data point, but do not use this point when drawing a line of best fit.

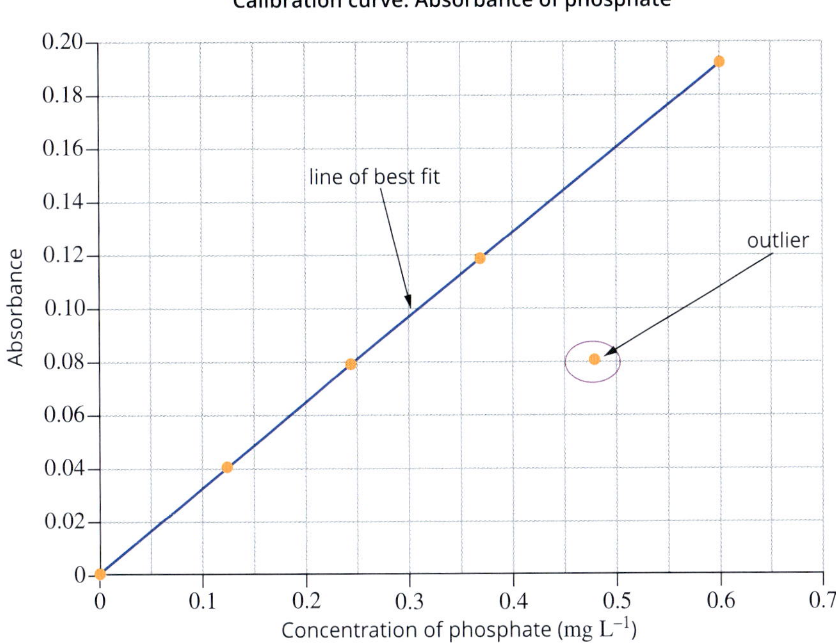

FIGURE 1.4.9 A calibration curve with an outlier that has not been included in the line of best fit

You will now be able to answer key questions 4, 5, 6 and 8.

Comparing your data to other values

While you may have taken care in your observations and recording your data, and you have every confidence in your efforts to be precise, you cannot claim your results are accurate unless you compare them to **literature** or **accepted values**. A literature value is one which has been found in published scientific reports, such as journals, or reliable websites, such as those of universities. Accepted values include values that are obtained from published sources and they also include those that you may calculate using a correctly balanced equation for the reaction and stoichiometry.

If you wish to discuss how your results changed over the duration of the experiment, then a calculation of the percentage change might be appropriate. To make a comparison between your results and the accepted or theoretical value, a percentage difference is calculated and can form part of your discussion.

Percentage difference

Often you might have a theoretical value to compare to your experimental value. **Percentage difference** is very useful for showing how accurate your experimental result is.

$$\text{percentage difference} = \frac{(\text{experimental value} - \text{theoretical value})}{\text{theoretical value}} \times 100$$

An example of how to calculate and express percentage difference is shown in Worked example 1.4.3.

Worked example 1.4.3

CALCULATING PERCENTAGE DIFFERENCE

A student electrolysed a solution of copper(II) sulfate using a current of 0.95 A for 10.0 minutes. The initial mass of the cathode was 0.712 g, and the final mass of the cathode was 0.879 g. Using Faraday's laws, the theoretical mass gain of the cathode was calculated to be 0.188 g.

Calculate the percentage difference between the mass gain of the cathode predicted by Faraday's laws and the mass gain obtained in the experiment.

Thinking	Working
Calculate the mass gain from the experimental results.	Mass gain = final mass – initial mass = 0.879 – 0.712 = 0.167 g
Identify the two masses which are to be compared, noting which one is larger.	The theoretical mass gain of the cathode, 0.188 g is larger than the experimental mass gain of the cathode, 0.167 g.
Substitute the masses into the formula: $\text{percentage difference} = \frac{(\text{experimental value} - \text{theoretical value})}{\text{theoretical value}} \times 100$	$\text{percentage difference} = \frac{(0.167 - 0.188)}{0.188} \times 100$ $= \frac{-0.021}{0.188} \times 100$
Ensure that significant figures are correct.	The answer should be expressed to 2 significant figures, due to the subtraction in the third step. $= -11\%$
State percentage difference, using the sign (+ or –) to help you compare the experimental and theoretical values clearly.	The experimental value is 11% less than the theoretical mass gain of the cathode.

Worked example: Try yourself 1.4.3

CALCULATING PERCENTAGE DIFFERENCE

A student electrolysed a solution of copper(II) sulfate using a current of 1.05 A for 10.0 minutes. The initial mass of the anode was 0.820 g, and the final mass of the anode was 0.626 g. Using Faraday's laws, the theoretical mass loss of the anode was calculated to be 0.207 g.

Calculate the percentage difference between the mass loss of the anode predicted by Faraday's laws and the mass loss obtained in the experiment.

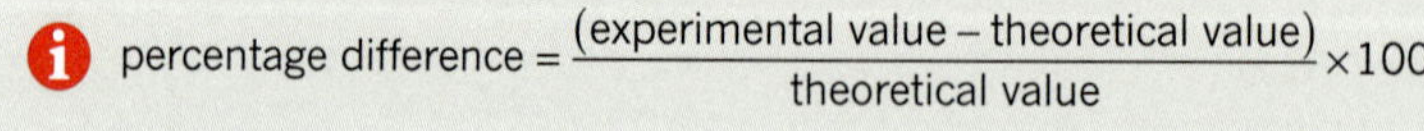

$$\text{percentage difference} = \frac{(\text{experimental value} - \text{theoretical value})}{\text{theoretical value}} \times 100$$

● *You will now be able to answer key question 7.*

Percentage change

Percentage change can be a useful measure of the difference between the initial and final values. The percentage change is calculated using the following mathematical formula:

$$\text{percentage change} = \frac{(\text{final value} - \text{initial value})}{\text{initial value}} \times 100$$

An example of how to calculate and express percentage change is shown in Worked example 1.4.4.

Worked example 1.4.4

CALCULATING PERCENTAGE CHANGE

A student was monitoring the mass loss due to production of CO_2 during the reaction:

$$CaCO_3(s) + 2HCl(aq) \longrightarrow CaCl_2(aq) + H_2O(l) + CO_2(g)$$

The initial mass of the hydrochloric acid and marble chips ($CaCO_3$) was 43.45 g and the mass after 120 seconds was 36.90 g. Calculate the percentage change in mass to the correct number of significant figures.

Thinking	Working
Identify the two masses, noting whether there has been an increase or decrease in mass.	The initial mass, 43.45 g, is larger than the final mass, 36.90 g, so the mass has decreased.
Substitute the masses into the formula: $\text{percentage change} = \frac{(\text{final value} - \text{initial value})}{\text{initial value}} \times 100$	$\text{percentage change} = \frac{(36.90 - 43.45)}{43.45} \times 100$ $= \frac{-6.55}{43.45} \times 100$
Ensure that significant figures are correct.	The answer should be expressed to 3 significant figures, due to the subtraction in the first step. $= -15.1\%$
State percentage change, using the sign (+ or –) and your initial observation of mass change, to guide you as you indicate in words whether there has been an increase or a decrease.	There has been a 15.1% *decrease* in the mass.

Worked example: Try yourself 1.4.4

CALCULATING PERCENTAGE CHANGE

A student was monitoring the mass loss due to production of O_2 during the reaction:

$$2H_2O_2(aq) \longrightarrow 2H_2O(l) + O_2(g)$$

The initial mass of the hydrogen peroxide was 25.10 g and the mass after 180 seconds was 23.05 g. Calculate the percentage change in mass to the correct number of significant figures.

$$\text{percentage change} = \frac{(\text{final value} - \text{initial value})}{\text{initial value}} \times 100$$

1.4 Review

OA ✓✓

SUMMARY

- The mean is the average of a set of quantitative data.
- When performing calculations using raw data, it is important to correctly state the number of significant figures.
  - When adding or subtracting, the final result should be reported to the least number of decimal places.
  - When multiplying or dividing, the final result should be reported to the least number of significant figures.
- Tables allow the presentation of summarised calculations, whereas graphs allow trends to be shown more clearly.
- Graphs can be described according to their trend—for example, proportional, linear or inverse.
- Some important rules to remember for graphs include:
  - The independent variable is usually plotted along the *x*-axis and the dependent variable on the *y*-axis.
  - Line graphs are useful for presenting continuous quantitative data.
  - Scatter graphs are useful when showing quantitative data where one variable is dependent on another variable. A line of best fit should be included.
  - Outliers should be reported and considered, but are not usually included in drawing a line of best fit.
  - Bar graphs are useful for comparing discrete data.
  - Pie diagrams are useful for showing proportional data.
- Percentage difference can be used as a measure of the difference between an experimental value and the published value.
- Percentage change can be used to describe the change in the value of a measurement during an experiment compared to the initial value.

KEY QUESTIONS

Knowledge and understanding

1 Write each of the following values in scientific notation using the appropriate number of significant figures.

a 0.002 05

b 96 500

c 100.4

d 0.000 43

2 A student carried out the calculations to find the expected mass deposited at the cathode during the electrolysis of copper(II) sulfate. The equation for these calculations was

$$m(Cu) = \frac{1}{2} \times \frac{0.950 \times (15 \times 60)}{96500} \times 63.5.$$

Which of the options below shows the result of this calculations to the correct number of significant figures?

A 0.2813

B 0.281

C 0.28

D 0.3

3 During an electrolysis experiment, a student accidentally measured the mass of a copper cathode before and after the electrolysis on different electronic balances. The initial mass of the copper cathode was 2.550 g. The final mass of the copper cathode was 3.15 g.

Which of the following is the mass of the copper that was deposited on the cathode in this experiment, given to the correct number of significant figures?

A 0.6000 g

B 0.600 g

C 0.60 g

D 0.6 g

4 Explain the meaning of the term 'trend' when considering your results for a scientific investigation and list the types of trends that might be illustrated by a graph of results.

Analysis

5 A titration was carried out by a student during a research investigation. These titres were recorded as follows: 15.40 mL, 15.18 mL, 15.24 mL, and 15.26 mL.

   a Identify all concordant titres.

   b Identify all outliers.

   c Use the concordant titres from part **b** to calculate the average (mean) concordant titre in mL.

6 Which graph from the following list would be best to use with each set of data listed here?

   pie graph; scatter graph (with line of best fit); column graph; line graph

   a The concentration of iron in a range of green leafy vegetables such as silver beet, kale and broccoli

   b The volume of gas produced at 10 second intervals when hydrochloric acid reacts with magnesium metal

   c The concentration of vitamin C in a set of samples of orange juice that have been heated to a range of temperatures.

7 A student used electrolysis to calculate an experimental value for Faraday's constant and hence for Avogadro's constant. The student's experimental value for Avogadro's constant was 5.13×10^{23}. The published value for Avogadro's constant is 6.02×10^{23}. Use this to calculate the percentage difference in the experimental value compared to the published value.

8 Describe four ways in which the following graph, showing the change in mass of a rates experiment over a period of time, measured in minutes, should be improved.

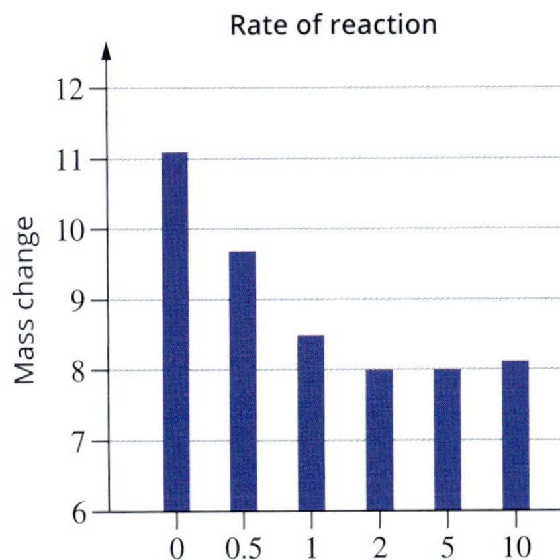

1.5 Evaluation and conclusion

In this section, you will learn how to evaluate the method that you used in your investigation and draw evidence-based conclusions in relation to your hypothesis and research question. When you consider the accuracy of your results, ideally by comparing them to the literature or accepted value, you should find a focus for your evaluation of the method.

EVALUATING THE METHOD

It is important to evaluate the method used in the controlled experiment to determine whether it was completely appropriate for your experiment. You may have designed your method with all good intentions, but now your results are not as you expected.

Consider the following:

- What were the sources of systematic errors?
- What were the sources of random errors?
- Was the method valid?
- Was the experiment repeated?
- Was the data precise?
- Would a larger sample or further variations in the independent variable lead to a stronger **conclusion**?
- What could be done next time to improve the method?

The method should be evaluated to identify its strengths and weaknesses and to consider validity, precision, accuracy and reproducibility.

For example, when measuring the volume of gas produced when calcium carbonate (marble chips) reacted with hydrochloric acid in a rates experiment, the following gas collection system was used, which involved the displacement of water (Figure 1.5.1a).

The method was valid—it aimed to measure the volume of carbon dioxide gas produced. However, the method was not accurate. Some of the carbon dioxide gas would dissolve in the water, so the volume of carbon dioxide measured would be less than the true value. This would be a systematic error. This method could be improved by measuring the mass loss (due to production and escape of carbon dioxide) on an electronic balance (Figure 1.5.1b).

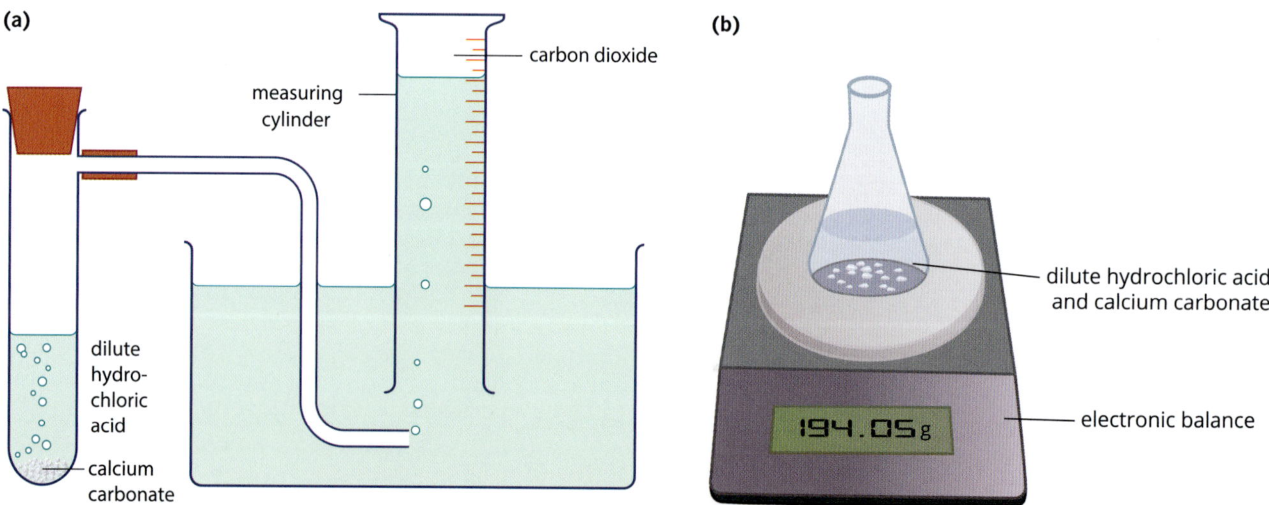

FIGURE 1.5.1 Different methods of carbon dioxide gas collection: (a) using displacement of water and (b) using an electronic balance. Using an electronic balance to directly measure mass of carbon dioxide produced is usually considered to be more accurate.

IDENTIFYING ERRORS

Most practical investigations have errors associated with them. Errors can be classified as either systematic errors or random errors, with mistakes being considered quite separately. An important part of evaluating your method is to consider errors which may have occurred because of the method you have chosen and have impacted on your results. Systematic and random errors were discussed in Section 1.3 (pages 28–29), as were mistakes and outliers (page 40). Systematic errors will cause your results to be consistently greater than or less than the accepted value, so these should be identified in your method when your results consistently do not match the accepted value. Random errors may not be as easy to identify, but they will contribute to less precision in your results and may be evident when not enough trials have been conducted, or you are not experienced in using the equipment.

Systematic errors are often caused by a problem with part of the method. Random errors affect the precision of the results and may be due to a lack of experience.

What to do about errors

As you consider your results and how they differ from what you predicted in your hypothesis, one of the most important steps is to suggest what could be done to overcome errors. Random errors can be minimised by repeating trials and averaging the results. Systematic errors can often be avoided by taking care with the use of equipment, but not always. Mistakes must not be used as valid results.

Your qualitative observations often hold the key to a systematic error. For example, when a hydrocarbon is burning, a yellow, sooty flame suggests that incomplete combustion is occurring. You can use the information gained from such an observation to identify issues with the method.

When you are analysing your data, be sure to acknowledge contradictions in the data. Rather than ignoring results that don't fit your hypothesis, look for the reasons which explain why these results occurred by working carefully through the method.

Sometimes your experimental findings will lead you to formulate new research questions and develop new hypotheses. These can be listed as ways in which the investigation can be extended in the future.

Improving the method

For each methodological issue that you identify, you should suggest a way in which the method could be improved to prevent that issue. While these issues will be particular to any investigation, Table 1.5.1 shows a selection of issues and improvements possible for an investigation with the research question, 'How does the carbon chain length of an alcohol affect the heat of combustion of the alcohol?'.

When evaluating the method, make sure that for every weakness you identify, you have suggested a way to improve the method.

TABLE 1.5.1 Methodological weaknesses and suggestions for improvement of a controlled experiment to determine the heat of combustion of alcohols

Methodological weakness	How this may have influenced the results	Suggestion for improvement
The simple calorimeter was not insulated and did not have a lid.	Once the water in the can was raised above room temperature, heat may have been lost to the surroundings through the metal can and from the surface of the water. As a result, the calculated heat of combustion would be smaller than it should be. This would be a systematic uncertainty.	Making a lid for the can using one or more layers of aluminium foil and loosely covering the sides of the can with the foil will provide non-flammable insulation.
The alcohol continued to evaporate from the wick after the flame had been blown out.	The mass loss of alcohol during the experiment would be greater than was correct, which would decrease the calculated heat of combustion.	Snuff the flame using a metallic snuffer/cap rather than blowing out the flame, and remember to weigh the spirit burner with the snuffer before and after each trial.
The three spirit burners had wicks that were different lengths.	The longer wicks would result in larger flames, which could transfer heat with different efficiencies to the can. With more efficient heat transfer, the water would heat up more quickly and less heat loss would occur because the experiment was finished more quickly. The different wick lengths would introduce a random uncertainty to the experiment.	Measure and record the length of the wick before starting to burn the alcohol in each trial, and try to match that length for subsequent trials.

You will now be able to answer key question 4.

DRAWING EVIDENCE-BASED CONCLUSIONS

A conclusion is a summary of your investigation. It should reflect your aim, results, the validity of the results and whether your hypothesis was supported.

A conclusion is a summary of your investigation. It should be possible for a reader to read your conclusion and understand what you did in the investigation, what results you found and how valid the results were. A good conclusion links the collected evidence to your aim and your hypothesis. It uses your results as evidence to provide a justified and relevant response to your research question.

An example of how a conclusion would compare to the aim of the experiment is shown in Table 1.5.2.

TABLE 1.5.2 Writing a conclusion for the energy from different fuels experiment

Part of the report		Example of what might be written
Aim:		To measure and compare the heat of combustion of three different alcohols, methanol, ethanol and propan-1-ol, using a spirit burner and a simple calorimeter.
Conclusion:	**Restatement of the aim:** Instead of starting with 'to' the aim is now rearranged to be in the past tense. It serves as an introduction to the conclusion.	Three different alcohols, methanol, ethanol and propan-1-ol, were burnt in a spirit burner and the thermal energy released was calculated using the change in temperature of the water in the calorimeter. These values were then used to calculate the heat of combustion.
	Statement of your results: These ideally should be the mean of multiple trials' values and should include results of all parts of the experiment. Sometimes this will be quite lengthy and you will have to try hard to keep it concise.	The heat of combustion of the three alcohols was calculated from the average of three trials for each alcohol and was found to be 550 kJ mol^{-1} for methanol, 1020 kJ mol^{-1} for ethanol and 1650 kJ mol^{-1} for propan-1-ol.
	A comparison of your results to the accepted value or your hypothesis: Some quantitative evaluation of how well your results compared to the accepted value is useful, and this is where a percentage difference may be useful.	These three values were considerably lower than the literature values of 726 kJ mol^{-1} for methanol, 1360 kJ mol^{-1} for ethanol (*VCE Chemistry Data Book*) and 2021 kJ mol^{-1} for propan-1-ol (NIST Chemistry webbook). The percentage differences for these values were 24.2%, 25.0% and 18.4% respectively.
	A brief suggestion of what may have caused the results to be different to what was expected	It is likely that these experimental values are low because heat energy was lost to the surroundings during the experiment due to lack of insulation on the simple calorimeter, and the lack of a lid.

Language required for conclusions

You will find a full discussion of the language used in a scientific report in Section 1.6, but students often find that the conclusion poses a challenge for them in terms of the style of language required.

Traditionally, the conclusion is written in the past tense and in the third person. This means that you should write the conclusion as though someone else did the work. For example, the sentence 'I found that the heat of combustion of ethanol was less than it should be' should be written as 'The heat of combustion of ethanol was less than was expected'. The conclusion can be easily be introduced by changing the aim into the past tense and writing it in the third person.

- *You will now be able to answer key questions 1, 2, 3 and 5.*

1.5 Review

OA ✓✓

SUMMARY

- The method should be evaluated to identify its strengths and weaknesses and to consider validity, precision, accuracy and reproducibility.
- An evaluation of the method should:
  - consider the effects of random and systematic errors
  - identify issues that may have affected validity, accuracy and precision, repeatability and reproducibility
  - make recommendations for improving the investigation method.
- A conclusion should:
  - be introduced by referring to the research question or aim
  - quote the results of the investigation
  - indicate whether the hypothesis was supported or refuted.
  - suggest what may have caused the investigation to differ from the expected value.

KEY QUESTIONS

Knowledge and understanding

1 Which of the following options is a suitable conclusion for the research question: How does the type of catalyst affect the rate of decomposition of hydrogen peroxide solution?

A I completed the experiment and the results were as I expected. The different catalysts did affect the rate of decomposition.

B Some catalysts made the hydrogen peroxide decompose faster than other catalysts.

C A 6% m/v hydrogen peroxide solution was decomposed at 22°C by three different catalysts, X, Y and Z. Catalyst X resulted in a greater rate of decomposition, with catalyst Y being the second most effective and Z the least effective. This data supports the hypothesis.

D Catalysts increase the rate of reaction. Catalysts X, Y and Z all increased the rate of reaction.

2 For the research question in Question **1**, which of the following would be a suitable link to chemical theory to explain the results?

A Catalyst X was most effective in lowering the activation energy of decomposition of hydrogen peroxide to oxygen gas and water vapour.

B According the collision theory, catalyst X was a reactant, so when there was more catalyst X, the concentration of reactants increased.

C According to Le Châtelier's principle, catalysts increase the rate of the forward reaction, so this is why the hydrogen peroxide decomposed more quickly.

D Catalyst X was most effective because it increased the energy of the collisions between hydrogen peroxide molecules.

3 For each of the following aims, construct an introductory sentence for a conclusion which is in the third person and the past tense.

a To investigate the effect of concentration changes on an aqueous equilibrium between Fe^{3+} ions, SCN^- ions and $FeSCN^{2+}$ ions.

b To measure the rate of reaction between calcium carbonate and hydrochloric acid and investigate the effect of particle size and concentration on the rate.

c To calibrate a calorimeter by measuring the increase in temperature that results from a measured input of electrical energy.

Analysis

4 Consider the diagram below, which shows the method that is used to determine the rate of the reaction between hydrochloric acid and pieces of marble (calcium carbonate).

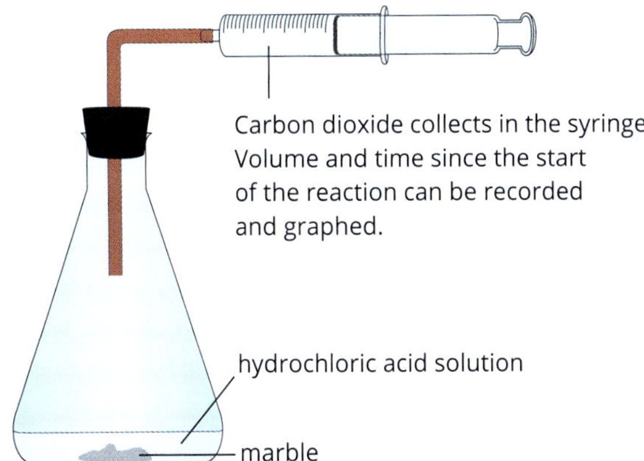

a List two systematic or random errors which could affect the results for this experiment. Identify whether the error is systematic or random as part of your answer.

continued over page

1.5 Review *continued*

b Use the table below to evaluate the method using your errors listed in part **a** as the listed methodological weaknesses.

Methodological weakness	How this may have influenced the results	Suggestion for improvement

5 A scientist designed and completed an experiment to test the following hypothesis:

'Decreasing the distance between copper electrodes will increase the current that flows through a copper-plating electrolytic cell and will result in an increase in the mass of copper metal deposited at the cathode.'

a Write a possible aim for this scientist's experiment.

b State the independent, dependent and controlled variables in this investigation.

i independent

ii dependent

iii controlled

c What type of quantitative data would be collected in this investigation?

d Assuming that the scientist's hypothesis was correct, write a short conclusion for this investigation. Without actual results, this cannot be a full conclusion.

e Considering this research question and other electrolysis experiments that you may have encountered, suggest two weaknesses in the method for this investigation and how you might improve the method to overcome these weaknesses.

6 A pair of students had developed the following research question for an investigation: 'How does the concentration of the copper(II) sulfate electrolyte affect the mass of copper deposited on the cathode when electrolysis is carried out using a power pack set on 6 volts for 10 minutes per trial?'

The students listed their variables as follows:

- Independent variable—concentration of copper(II) sulfate (0.1 M, 0.2 M, 0.4 M, 0.5 M, 0.8 M)
- Dependent variable—increase in mass of cathode
- Controlled variables—voltage set on power pack, volume of copper(II) sulfate solution, time for which each trial was carried out (10 minutes), size of the copper electrodes (2 cm × 8 cm) used for each trial.

To speed up the process the students decided to use two power packs, two beakers (student A used a 250 mL beaker and student B used a 100 mL beaker) and four copper electrodes. Each student carried out one electrolysis experiment with each of the five variations of the independent variable.

During the electrolysis, the electrodes were clipped to the inside of the beaker using an alligator clip and the two wires lead directly to the power pack. The electrodes were submerged in the electrolyte to a depth of 4 cm.

After the electrolysis was finished, both students air-dried their cathodes before weighing them.

a The students wanted to determine their percentage yield to see how accurate their experiment was. Explain why this was not possible.

b Evaluate the method used by the students, finding three problems with their method.

c Using your answer to part **b** as a guide, suggest three improvements that these students could make to their method to reduce the random and systematic errors in their experiment.

1.6 Reporting investigations

Now that you have thoroughly researched your topic, formulated a research question and hypothesis, conducted experiments and collected data, it is time to bring it all together. The final part of an investigation involves summarising the findings in an objective, clear and concise manner for your audience.

Scientists report their findings in a number of ways, including in written peer-reviewed journal articles, on web pages, and with short oral presentations or scientific posters (Figures 1.6.1 and 1.6.2).

FIGURE 1.6.1 Posters at a scientific conference

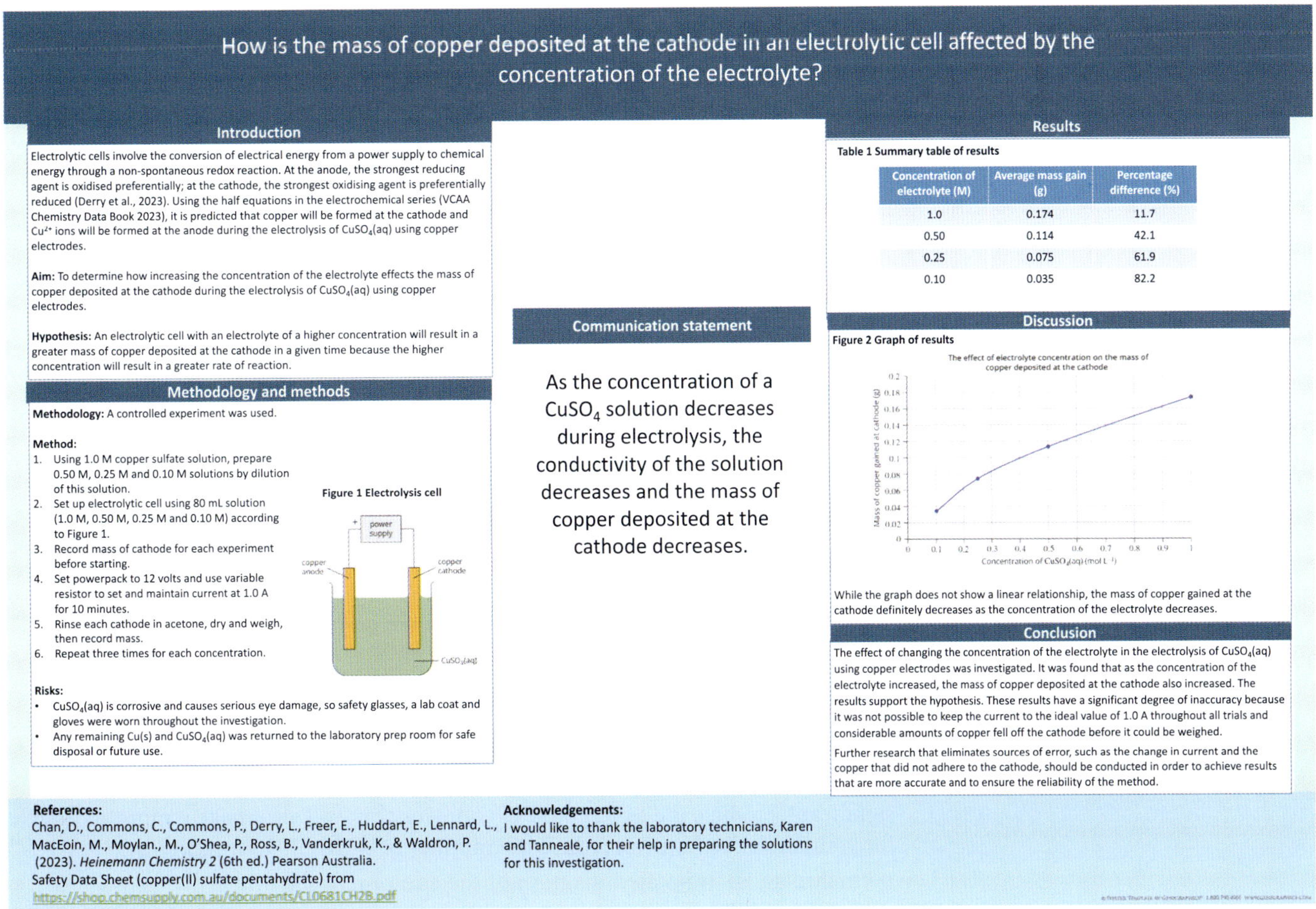

How is the mass of copper deposited at the cathode in an electrolytic cell affected by the concentration of the electrolyte?

Introduction

Electrolytic cells involve the conversion of electrical energy from a power supply to chemical energy through a non-spontaneous redox reaction. At the anode, the strongest reducing agent is oxidised preferentially; at the cathode, the strongest oxidising agent is preferentially reduced (Derry et al., 2023). Using the half equations in the electrochemical series (VCAA Chemistry Data Book 2023), it is predicted that copper will be formed at the cathode and Cu^{2+} ions will be formed at the anode during the electrolysis of $CuSO_4(aq)$ using copper electrodes.

Aim: To determine how increasing the concentration of the electrolyte effects the mass of copper deposited at the cathode during the electrolysis of $CuSO_4(aq)$ using copper electrodes.

Hypothesis: An electrolytic cell with an electrolyte of a higher concentration will result in a greater mass of copper deposited at the cathode in a given time because the higher concentration will result in a greater rate of reaction.

Methodology and methods

Methodology: A controlled experiment was used.

Method:

1. Using 1.0 M copper sulfate solution, prepare 0.50 M, 0.25 M and 0.10 M solutions by dilution of this solution.
2. Set up electrolytic cell using 80 mL solution (1.0 M, 0.50 M, 0.25 M and 0.10 M) according to Figure 1.
3. Record mass of cathode for each experiment before starting.
4. Set powerpack to 12 volts and use variable resistor to set and maintain current at 1.0 A for 10 minutes.
5. Rinse each cathode in acetone, dry and weigh, then record mass.
6. Repeat three times for each concentration.

Figure 1 Electrolysis cell

Risks:

- $CuSO_4(aq)$ is corrosive and causes serious eye damage, so safety glasses, a lab coat and gloves were worn throughout the investigation.
- Any remaining Cu(s) and $CuSO_4(aq)$ was returned to the laboratory prep room for safe disposal or future use.

Communication statement

As the concentration of a $CuSO_4$ solution decreases during electrolysis, the conductivity of the solution decreases and the mass of copper deposited at the cathode decreases.

Results

Table 1 Summary table of results

Concentration of electrolyte (M)	Average mass gain (g)	Percentage difference (%)
1.0	0.174	11.7
0.50	0.114	42.1
0.25	0.075	61.9
0.10	0.035	82.2

Discussion

Figure 2 Graph of results

While the graph does not show a linear relationship, the mass of copper gained at the cathode definitely decreases as the concentration of the electrolyte decreases.

Conclusion

The effect of changing the concentration of the electrolyte in the electrolysis of $CuSO_4(aq)$ using copper electrodes was investigated. It was found that as the concentration of the electrolyte increased, the mass of copper deposited at the cathode also increased. The results support the hypothesis. These results have a significant degree of inaccuracy because it was not possible to keep the current to the ideal value of 1.0 A throughout all trials and considerable amounts of copper fell off the cathode before it could be weighed.

Further research that eliminates sources of error, such as the change in current and the copper that did not adhere to the cathode, should be conducted in order to achieve results that are more accurate and to ensure the reliability of the method.

References:

Chan, D., Commons, C., Commons, P., Derry, L., Freer, E., Huddart, E., Lennard, L., MacEoin, M., Moylan., M., O'Shea, P., Ross, B., Vanderkruk, K., & Waldron, P. (2023). *Heinemann Chemistry 2* (6th ed.) Pearson Australia.

Safety Data Sheet (copper(II) sulfate pentahydrate) from https://shop.chemsupply.com.au/documents/CL0681CH28.pdf

Acknowledgements:

I would like to thank the laboratory technicians, Karen and Tanneale, for their help in preparing the solutions for this investigation.

FIGURE 1.6.2 An example of a student's scientific poster

WRITING A SCIENTIFIC REPORT

Whether an investigation is presented as a poster, written report or oral presentation, an article for a scientific publication or a multimedia presentation, the same key elements are included in the same sequence: Title, Introduction, Methodology and methods, Results, Discussion, Conclusion, and References and acknowledgements, as summarised in Figure 1.6.3. All of these key elements are described in detail in Sections 1.2, 1.3, 1.4 and 1.5.

A scientific poster format will include the same elements as a scientific report, as outlined in this section. However, a poster is meant to be more succinct and direct in its approach to presenting your findings and should be appropriate for both technical and non-technical audiences. You should pick the information that you want to present carefully so that the impact of your investigation is best communicated. For example, in analysing your data, you may have produced large tables of raw and processed data, as well as a graph. In the poster, it may be best to simply show a summary results table and a graph which will clearly represent any trends in your data.

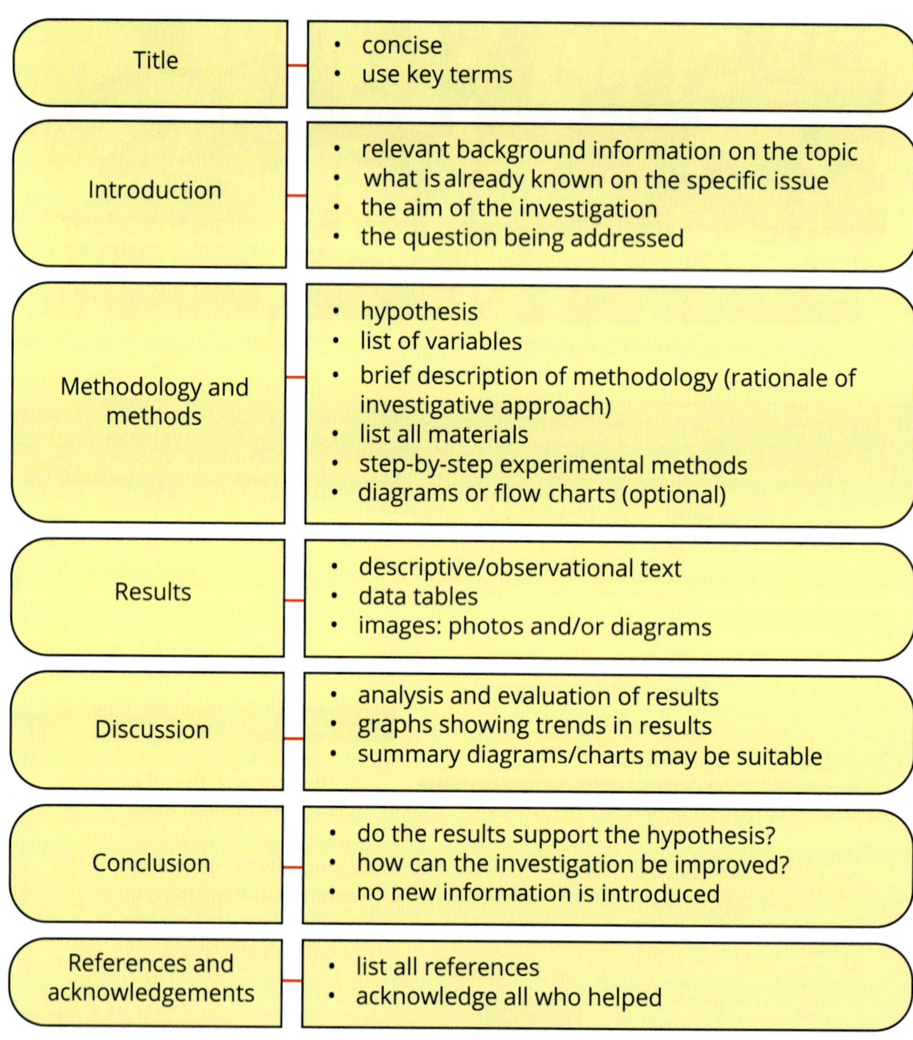

FIGURE 1.6.3 Elements of a scientific report or presentation

Visual support

You may be able to identify concepts that can be explained using visual models and information that can be presented in graphs or diagrams. This will not only significantly reduce the word count of your work but will also make it more accessible for your audience. You can reduce the number of words in your report by:

- including an annotated diagram of the equipment set-up
- using a flow chart to summarise the key steps of the method (a detailed method can appear in the logbook)

Visual support, such as flow charts and graphs, can help to convey scientific concepts and processes efficiently.

- using tables, graphs and schematic diagrams (diagrams that show something complex such as a process in a simple way, often using standardised symbols). Ensure you include:
- a descriptive title
- labels, captions or descriptions
- numbering; for example, Figure 1, Figure 2 or Table 1, Table 2
- a citation (source) if the work is not your own or is adapted from work that is not your own.

You will now be able to answer key question 1.

WRITING FOR SCIENCE

Scientific reports are usually written in an objective or unbiased style. This contrasts with writing in your English class, which may use subjective techniques, such as rhetoric or persuasion. In addition, be careful of words that are absolute, such as always, never, shall, will and proven. Sometimes it may be more accurate and appropriate to use qualifying words, such as may, might, possible, probably, likely and suggests.

Scientific reports should be written concisely, particularly if you want to engage and maintain the interest of your audience. Use shorter sentences that are less verbose (less wordy).

Consistent reporting narrative

Scientific writing can be written either in first-person or third-person narrative, although traditionally only the third person was used. You should check whether your teacher prefers first or third person before writing an investigation report that is going to be assessed. In either case, ensure that you keep the narrative point of view consistent. Read the examples of first-person and third-person narrative in Table 1.6.1.

Scientific writing uses unbiased, objective, accurate, formal language. It should be concise and qualified.

TABLE 1.6.1 Examples of first-person and third-person narrative

First person	Third person
I added 5 mL of HCl to the beaker and started timing.	First, 5 mL of HCl was added and then timing was commenced.
After I observed the reaction, I found that ...	After the reaction was completed, the results showed ...

You will now be able to answer key questions 2 and 3.

AVOIDING PLAGIARISM

Plagiarism is using other people's work without acknowledging them as the author or creator. It is very important to always write using your own words, even if you think that the words used by the original author are the best possible. To avoid plagiarism, include a reference every time you report the work of others; for example, at the end of a sentence or following a diagram. If you use a direct quotation from a source, enclose it in quotation marks. This will ensure you give credit to the original author and it will enable the reader to find the original source.

References and bibliography

A bibliography is a list of resources you have consulted during your research, whereas a reference list is a detailed list of references that you have cited in your work. Both lists includes details about each resource. To avoid plagiarising the work of others, it is very important that you acknowledge where you have found information. All sources must be listed at the end of the report in alphabetical order (by last name of author or organisation name). The referencing should include enough information for the reader to easily locate the source.

As you gather resources, you should also begin compiling your references in your logbook. This will prevent you from wasting time later trying to find your sources and will be the basis of your reference list.

APA (American Psychological Association) style is a commonly used referencing style. In the bibliography at the end of your work, you should represent the reference in the following order:

Author (year of publication). Title (edition). Publisher.

This is illustrated in Figure 1.6.4.

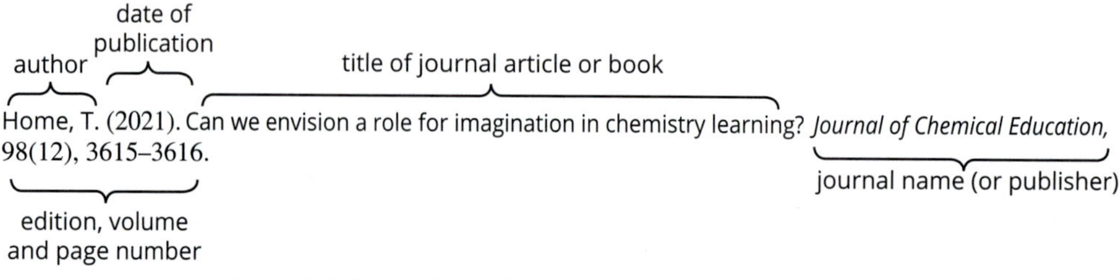

FIGURE 1.6.4 An annotated example of APA referencing of a journal article for a bibliography

In APA referencing, no distinction is made between books, journal articles or internet documents, except where electronic documents do not provide page numbers. Table 1.6.2 gives examples of how to write citations and references of different types of sources using APA academic referencing style.

TABLE 1.6.2 APA academic referencing style

Resource type	Information required	Example
print book	• author's surname and initials • date of publication • title • edition number • page number • publisher's name	Chan, D., Commons, C., Commons, P., Derry, L., Freer, E., Huddart, E., Lennard, L., MacEoin, M., Moylan., M., O'Shea, P., Ross, B., Vanderkruk, K., & Waldron, P. (2023). *Heinemann Chemistry 2* (6th ed., p. 426). Pearson Australia.
journal article	• author's surname and initials • date of publication • title of article • journal/magazine title • volume/issue • page numbers	Kizil, N., Soylak, M. & Tüzen, M. (2017). Spectrophotometric detection of rhodamine B in tap water, lipstick, rouge, and nail polish samples after supramolecular solvent microextraction. *Turkish Journal of Chemistry, 41*, 987–994.
internet	• author's surname and initials, or name of organisation or title • year website was written or last revised (if this cannot be found, write 'n.d.' for 'no date') • website title, italicised, or description • website address	University of Canterbury. (n.d.). *Determination of vitamin C concentration by titration.* https://www.canterbury.ac.nz/media/documents/science-outreach/vitaminc_iodine.pdf

In-text citations

Each time you write about the findings of other people or organisations, you need to provide an in-text citation and provide full details of the source in a reference list. In APA style, in-text citations include the first author's surname, the year of publication, and a page, chapter or section number if you need to be specific, in brackets (author, year, pages). List the full details in your bibliography as shown in Table 1.6.3.

The following examples show the use of in-text citation.

- An earlier study (Kizil et al., 2017) reported a method for determining the concentration of rhodamine B by UV–visible spectroscopy.
- The procedure for determining the concentration of vitamin C in lemon juice was adapted from the University of Canterbury method (University of Canterbury, n.d.)

The intention is that in-text citations can be linked to the information in your bibliography, but are more concise, for inclusion in your work at the point where they are most relevant. In Table 1.6.3, three bibliography entries and their corresponding in-text citations are compared.

References and acknowledgements should give credit to the original authors of the information you are using and should follow a standard format.

TABLE 1.6.3 Bibliography entries compared to in-text citations for the same resources

Resource type	Bibliography entry	In-text citation
print book	Chan, D., Commons, C., Commons, P., Derry, L., Freer, E., Huddart, E., Lennard, L., MacEoin, M., Moylan., M., O'Shea, P., Ross, B., Vanderkruk, K., & Waldron, P. (2023). *Heinemann Chemistry 2* (6th ed., p. 426). Pearson Australia.	(Derry et al., 2023, p. 426)
journal article	Kizil, N., Soylak, M. & Tüzen, M. (2017). Spectrophotometric detection of rhodamine B in tap water, lipstick, rouge, and nail polish samples after supramolecular solvent microextraction. *Turkish Journal of Chemistry, 41*, 987–994.	(Kizil et al., 2017)
internet	University of Canterbury. (n.d.). *Determination of vitamin C concentration by titration*. https://www.canterbury.ac.nz/media/documents/science-outreach/vitaminc_iodine.pdf	(University of Canterbury, n.d.)

You will now be able to answer key questions 4, 5 and 6.

Acknowledgements

It is important to acknowledge the contribution of others in your investigation. This might include acknowledging support from your teacher and preparation of chemicals and materials by the laboratory technician.

For example, statements similar to the following could be used.

- This research was supported by the staff of the Science Faculty at Northern Secondary College, Melbourne.
- Special thanks to Mrs Smith for preparing stock solutions.

1.6 Review

OA ✓✓

SUMMARY

- A practical report should include the following sections:
  - title
  - introduction
  - materials and method
  - results
  - discussion
  - conclusion
  - references
  - acknowledgements.
- The title should give a clear idea of what the report is about, without being too long.
- The introduction sets the context of your report. It should outline relevant chemical ideas, concepts, theories and models, and how they relate to your specific question and hypothesis.
- The methodology and method section should:
  - outline the methodology used and the rationale for using this approach
  - clearly state the materials required and the methods used to collect data during your investigation
  - be presented in a clear, logical order that accurately reflects how you conducted your study.
- The results section should state your results and present them using graphs, figures and tables, but it should not interpret the results.
- The discussion should:
  - interpret data
  - evaluate the investigative method and make recommendations for improving the method
  - explain the link between investigative findings and relevant chemical concepts.
- The conclusion should succinctly link the evidence collected to the hypothesis and research question, indicating whether the hypothesis was supported or refuted.
- References and acknowledgements should be present in an appropriate format.
- Scientific writing uses unbiased, objective, accurate, formal language. It should be concise and qualified.
- Visual support such as flow charts and graphs can help to convey scientific concepts and processes efficiently.

KEY QUESTIONS

Knowledge and understanding

1 List three techniques you might use to reduce the number of words in your experimental report to present it in a poster.

2 Which of the following statements taken from a conclusion is written in scientific style?
   A The experiment was fun…
   B The results were unbelievable…
   C We believe that…
   D The results demonstrated that…

3 Which of the following statements is written in third-person narrative suitable for scientific writing?
   A Three trials were conducted…
   B We found that…
   C George's results demonstrated that…
   D I showed that…

Analysis

4 Which of the following would be the correct way to cite in text the following source in APA style?
Simurdiak, M., Olukoga, O. & Hedberg, K. (2016). Obtaining the Iodine Value of Various Oils via Bromination with Pyridinium Tribromide. *Journal of Chemical Education*, 93(2), 322–325.
   A A method of obtaining the iodine value of various oils (Simurdiak et al., 2016) was modified to create the method in this report.
   B The method used by Simurdiak et al. in which they obtained iodine value of various oils was modified for this report.
   C The method of obtaining the iodine value of various oils reported by Simurdiak, M., Olukoga, O. & Hedberg, K. (2016). Obtaining the Iodine Value of Various Oils via Bromination with Pyridinium Tribromide. *Journal of Chemical Education*, 93(2), 322–325 was modified.
   D Simurdiak, M., Olukoga, O. & Hedberg, K. (2016) used a method of obtaining the iodine value of various oils and this was modified to use in this report.

5 In the table below, all the details of three references are described. Rewrite this information as it should appear in a bibliography using the APA referencing format.

Resource type	Information about the reference	Correct format in a bibliography, using APA referencing format
print book	Title of book: Molecules with silly or unusual names Edition: 1st edition Author: Paul W. May Date published: 2008 Page: 64 Publisher: Imperial College Press	
journal article	Title of article: The nature hackers Author: Stenmark, I. E. Journal name: Education in Chemistry Volume: Vol. 56 Pages: 12–14 Date published: January 2019	
internet	Website owner: Royal Society of Chemistry Website title/description: The rate of reaction of magnesium with hydrochloric acid Date posted: 2016 Website address: www.edu.rsc.org/experiments/the-rate-of-reaction-of-magnesium-with-hydrochloric-acid/1916.article	

6 The Australia New Zealand Food Standards code requires that vegetable juices for infants under 12 months of age must contain no less than 25 mg vitamin C per 100 g. Shelf-life is the length of time for which a food material remains useable or fit for consumption (legal requirement), e.g. 7 days.

Consider the following research question: 'How does the age of commercially available vegetable juices affect the concentration of vitamin C?'

a Suggest a hypothesis for this experiment.

b Identify each of the following as either independent, dependent or controlled variables.

  i Temperature of juice, analytical technique and equipment used for measuring volumes and masses.

  ii Concentration of vitamin C per 100 g

  iii Proximity of the juice to its use-by date

c When the students researched the Australian food law, they found the following reference:

Food Standards Australia New Zealand (FSANZ) (2016). *Australia New Zealand Food Standards Code: Standard 2.9.2 Food for Infants.* https://www.foodstandards.gov.au/

  i Write an in-text citation for this reference.

  ii How should the full citation be included in the reference list using APA format?

  iii Is this source reliable? If so, why?

Chapter review

01

KEY TERMS

accepted value
accuracy
accurate
affiliation
aim
analogue
average
bar graph
bias
calibration
calibration curve
column graph
concordant titres
continuous data
controlled experiment
controlled variable
critical thinking
dependent variable
discrete data
ethics
exponential relationship
green chemistry
hypothesis
independent variable
index
inverse relationship
line graph
line of best fit
linear trend
literature value
logbook
mean
measurement error
measurement result
meniscus
method
methodology
mistake
objective
observation
outlier
parallax error
peer-reviewed
percentage change
percentage difference
pie chart
precise
precision
primary data
primary source
principle
processed data
qualitative
qualitative data
quantitative
quantitative data
random error
raw data
repeatability
replication
reproducibility
research question
resolution
risk assessment
safety data sheet (SDS)
scatter graph
scientific method
scientific notation
secondary source
significant figures
standard form
subjective
systematic error
theory
titration
trend
true value
uncertainty
valid
validity
variable

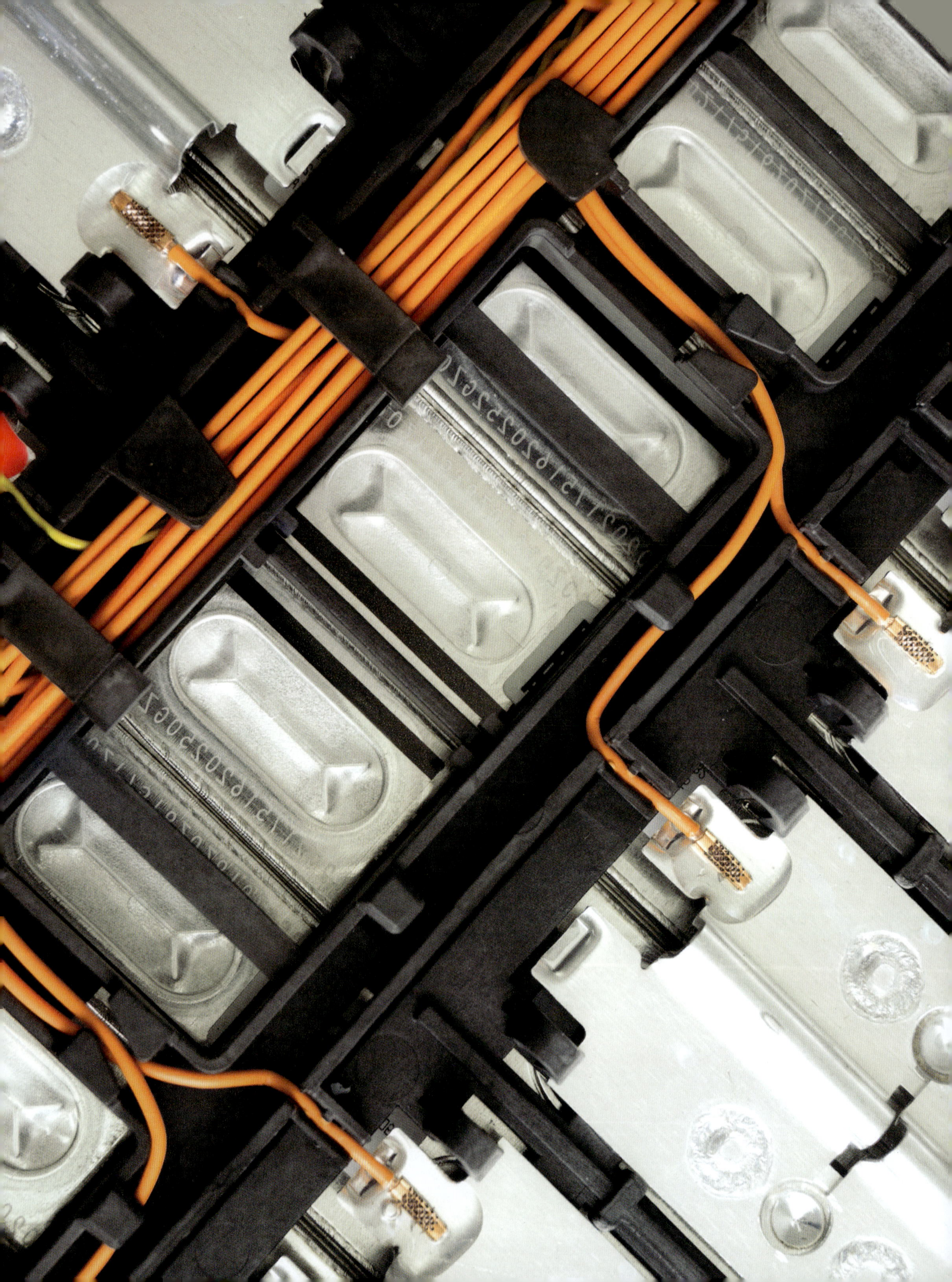

UNIT 3 How can design and innovation help to optimise chemical processes?

To achieve the outcomes in Unit 3, you will draw on key knowledge outlined in each area of study and the related key science skills on pages 11 and 12 of the study design. The key science skills are discussed in Chapter 1 of this book.

AREA OF STUDY 1

What are the current and future options for supplying energy?

Outcome 1: On completion of this unit, the student should be able to compare fuels quantitatively with reference to combustion products and energy outputs, apply knowledge of the electrochemical series to design, construct and test primary cells and fuel cells, and evaluate the sustainability of electrochemical cells in producing energy for society.

AREA OF STUDY 2

How can the rate and yield of chemical reactions be optimised?

Outcome 2: On completion of this unit, the student should be able to experimentally analyse chemical systems to predict how the rate and extent of chemical reactions can be optimised, explain how electrolysis is involved in the production of chemicals, and evaluate the sustainability of electrolytic processes in producing useful materials for society.

CHAPTER

02 Carbon-based fuels

In this chapter, you will learn how fuels are used to meet global energy needs and you will gain an appreciation of the chemistry that underpins decisions about the use of fuels. Combustion reactions are used to release useful heat energy from the chemical energy stored in fuels. You will consider different ways in which the energy content of a fuel is represented and be introduced to calculations involving energy changes. You will also learn about the energy changes occurring when the human body processes the food it consumes.

You will consider the environmental impact of using different types of fuels, as the phasing-out of fossil fuels is arguably the most significant factor in addressing climate change fears. Current research being conducted into the production of biofuels will also be discussed.

Key knowledge

- the definition of a fuel, including the distinction between fossil fuels (coal, natural gas, petrol) and biofuels (biogas, bioethanol, biodiesel), with reference to their renewability (ability of a resource to be replaced by natural processes within a relatively short period of time) **2.2**
- fuel sources for the body measured in kJ g^{-1}: carbohydrates, proteins and lipids (fats and oils) **2.3**
- photosynthesis as the process that converts light energy into chemical energy and as a source of glucose and oxygen for respiration in living things: $6CO_2(g) + 6H_2O(l) \longrightarrow C_6H_{12}O_6(aq) + 6O_2(g)$ **2.3**
- oxidation of glucose as the primary carbohydrate energy source, including the balanced equation for cellular respiration: $C_6H_{12}O_6(aq) + 6O_2(g) \longrightarrow 6CO_2(g) + 6H_2O(l)$ **2.3**
- production of bioethanol by the fermentation of glucose and subsequent distillation to produce a more sustainable transport fuel: $C_6H_{12}O_6(aq) \longrightarrow 2C_2H_5OH(aq) + 2CO_2(g)$ **2.4**
- comparison of exothermic and endothermic reactions, with reference to bond making and bond breaking, including enthalpy changes (ΔH) measured in kJ, molar enthalpy changes measured in kJ mol^{-1} and enthalpy changes for mixtures measured in kJ g^{-1}, and their representations in energy profile diagrams **2.1**
- combustion (complete and incomplete) reactions of fuels as exothermic reactions: the writing of balanced thermochemical equations, including states, for the complete and incomplete combustion of organic molecules using experimental data and data tables. **2.5**

2.1 Exothermic and endothermic reactions

> A fuel is a substance that can release stored energy relatively easily.
> A combustion reaction is a reaction in which a substance reacts with oxygen gas, releasing energy.

Few issues are more relevant to modern society than energy. Our affluent lifestyle generates a huge demand for energy for electricity, heating and transport. The production of this energy, in turn, is placing serious stresses on our environment and climate. An understanding of energy involves an understanding that all chemicals contain stored energy and that any chemical reaction will produce an energy change. In this section, you will learn that fuels are substances that can release stored chemical energy relatively easily.

To release stored energy, fuels must react, often with oxygen. The reactions with oxygen are examples of **combustion** reactions. Reactions that release energy to the environment are called **exothermic** reactions, while reactions that absorb energy are **endothermic** reactions.

When **fuels** undergo combustion, energy is released in the form of heat. It is useful for chemists to be able to show the energy changes that occur as a reaction proceeds. All reactions absorb some energy before they can proceed, even if energy is released overall. An energy profile diagram represents the energy changes that occur during the course of a reaction.

ENTHALPY CHANGE

The **chemical energy** of a substance is sometimes called its **heat content** or **enthalpy**. It is given the symbol H. The enthalpy of the reactants in a chemical reaction is given the symbol H_r and the enthalpy of the products is given the symbol H_p.

Most chemical processes take place in open systems under a constant pressure (usually atmospheric). The exchange of heat energy between the system and its surroundings under constant pressure is referred to as the **enthalpy change**, or heat of reaction, and is given the symbol ΔH. The capital delta symbol (Δ) is commonly used in chemistry to represent 'change in'. For example, ΔT is the symbol for change in temperature.

> Enthalpy change is a measure of the quantity of energy absorbed or released during chemical reactions. It is given the symbol ΔH and is determined by subtracting the enthalpy of the reactants (H_r) from the enthalpy of the products (H_p).
> Enthalpy change, $\Delta H = H_p - H_r$

Fuels provide you with energy by undergoing exothermic combustion reactions. Knowing the precise enthalpy change per mole or gram that occurs during the combustion of different fuels helps you to decide which fuel might be most suitable for a particular purpose.

For the general reaction:

$$\text{reactants} \longrightarrow \text{products}$$

the enthalpy change (ΔH) is calculated by:

$$\Delta H = H_p - H_r$$

For reactions to occur, the bonds in the reactants must break and new bonds must be formed when the products are formed. It requires energy to break the bonds in the reactants, and energy is released when bonds form in the products. The enthalpy of the products will differ from the enthalpy of the reactants. In some reactions, energy is released to the surroundings, while in others, energy is absorbed from the surroundings.

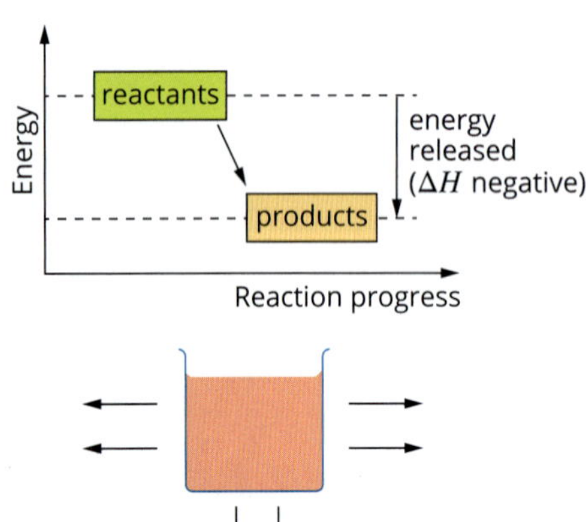

FIGURE 2.1.1 In an exothermic reaction, the energy of the reactants is greater than the energy of the products, so energy is released to the surroundings during the reaction and ΔH is negative.

Enthalpy change in exothermic reactions

When H_p is less than H_r, energy is released from the system into the surroundings, so the reaction is exothermic. The system has lost energy, so ΔH has a negative value.

Therefore, for combustion reactions (which are exothermic reactions), $\Delta H < 0$ (Figure 2.1.1).

In an exothermic reaction:

- the total chemical energy of the products is less than the total chemical energy of the reactants
- the energy released when the bonds in the products form is greater than the energy required to break the bonds in the reactants

- excess energy is released to the surroundings
- ΔH is negative.

Enthalpy change in endothermic reactions

When H_p is greater than H_r, energy must be absorbed from the surroundings, so the reaction is endothermic. The system has gained energy, so ΔH has a positive value, i.e. $\Delta H > 0$ (Figure 2.1.2).

In an endothermic reaction:

- the total chemical energy of the products is greater than the total chemical energy of the reactants
- the energy released when the bonds in the products form is less than the energy required to break the bonds in the reactants
- the energy required is absorbed from the surroundings
- ΔH is positive.

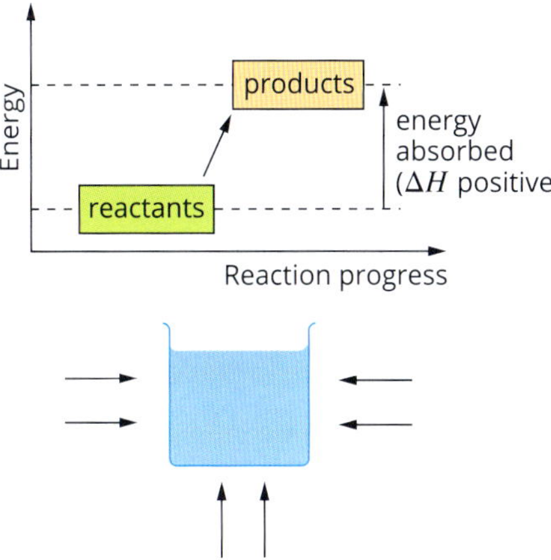

FIGURE 2.1.2 In an endothermic reaction, the energy of the reactants is less than the energy of the products, so energy is absorbed from the surroundings during the reaction and ΔH is positive.

Activation energy

The energy required to break the bonds of reactants so that a reaction can proceed is called the **activation energy**. The activation energy is an energy barrier that must be overcome before a reaction can commence.

An activation energy barrier exists for both exothermic and endothermic reactions. If the activation energy for a reaction is very low, the chemical reaction can be initiated as soon as the reactants come into contact because the reactants already have sufficient energy for a reaction to take place. Special conditions are not always required for reactions to occur. An example of this can be seen in the reaction between zinc and hydrochloric acid in Figure 2.1.3.

The reaction between zinc and hydrochloric acid produces hydrogen gas:

$$Zn(s) + 2HCl(aq) \longrightarrow ZnCl_2(aq) + H_2(g)$$

As you can see in the figure, bubbles of hydrogen gas are vigorously produced as soon as zinc is added to the acid.

FIGURE 2.1.3 When zinc comes into contact with hydrochloric acid, it reacts almost immediately. The reactants have sufficient energy to 'overcome' the activation energy barrier.

Units of energy

It is useful to know the magnitude of ΔH values so that fuels can be compared. The International System of Units (SI units) is a widely used system of measurement that specifies units for a range of quantities. The SI unit for energy is the **joule**, with the symbol J. As 1 J of energy is a relatively small quantity, it is common to see the following units in use.

- kilojoules, 1 kJ = 10^3 J
- megajoules, 1 MJ = 10^6 J
- gigajoules, 1 GJ = 10^9 J
- terajoules, 1 TJ = 10^{12} J.

Figure 2.1.4 provides a conversion guide for different energy units.

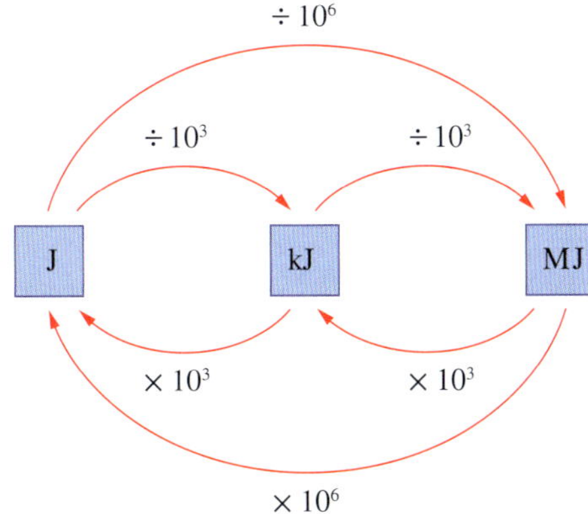

FIGURE 2.1.4 Converting between different energy units

Writing a thermochemical equation

The combustion reaction for the fuel methane is:

$$CH_4(g) + 2O_2(g) \longrightarrow CO_2(g) + 2H_2O(l)$$

Activation energy is required to break the covalent bonds in both methane and oxygen molecules for this reaction to proceed. Methane is a widely used fuel as the energy released when bonds form in CO_2 and H_2O is significantly higher than the activation energy required.

The quantity of energy released in the combustion of 1 mole of methane is 890 kJ. If the value of ΔH is added to a balanced equation, it is referred to as a **thermochemical equation**. The thermochemical equation for the combustion of methane is:

$$CH_4(g) + 2O_2(g) \longrightarrow CO_2(g) + 2H_2O(l) \qquad \Delta H = -890 \text{ kJ}$$

A thermochemical equation is a balanced chemical equation that includes the enthalpy change, ΔH.

This equation tells you that the combustion of 1 mole of methane and 2 moles of oxygen gas to produce 1 mole of carbon dioxide and 2 moles of liquid water will release 890 kJ of energy.

ENERGY PROFILE DIAGRAMS

The energy changes that occur during the course of a chemical reaction can be shown on an **energy profile diagram** or energy profile.

The energy profile diagram for an exothermic reaction like the one shown in Figure 2.1.5 indicates that the enthalpy of the products is less than the enthalpy of the reactants. Overall, energy is released and so the ΔH value is negative. The energy profile also shows that, even in exothermic reactions, activation energy must first be absorbed to start the reaction.

The energy profile diagram for an endothermic reaction (Figure 2.1.6) shows that the enthalpy of the products is greater than the enthalpy of the reactants. Overall, energy is absorbed and so the ΔH value is positive. The energy profile also shows the absorption of the activation energy before the release of energy as bonds form in the products.

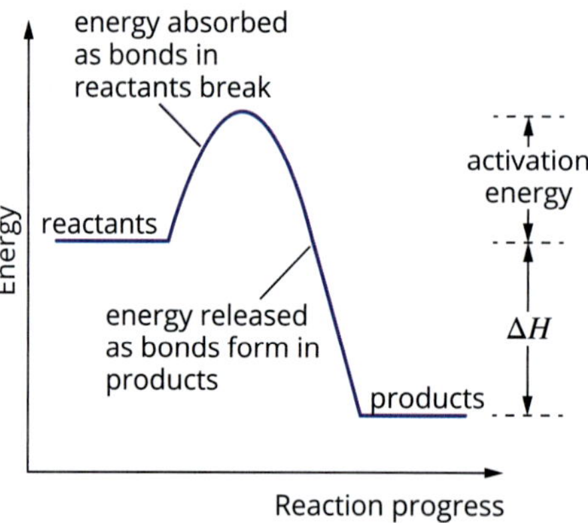

FIGURE 2.1.5 The characteristic shape of an energy profile diagram for an exothermic reaction

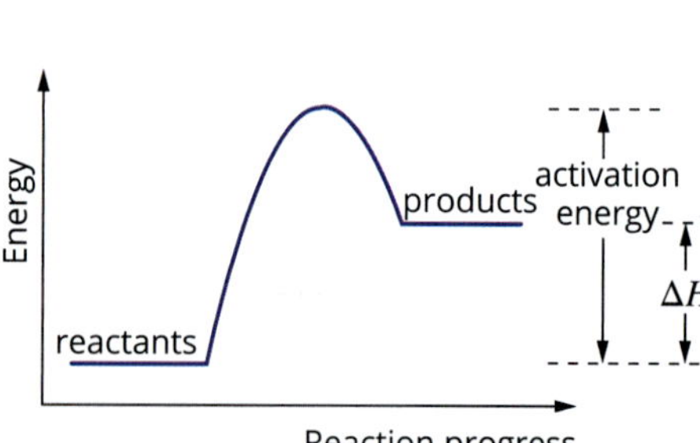

FIGURE 2.1.6 The characteristic shape of an energy profile diagram for an endothermic reaction

2.1 Review

WS 2 OA ✓✓

SUMMARY

- A fuel is a substance with stored energy that can be released relatively easily for use as heat or power.
- All chemical reactions involve a change in energy due to the breaking of reactant bonds and the formation of new bonds.
- The chemical energy of a substance is sometimes called its heat content or enthalpy.
- The net enthalpy change in a reaction is represented as ΔH.
- Exothermic reactions have a net release of energy to the surroundings. ΔH is negative.
- Endothermic reactions have a net absorption of energy from the surroundings. ΔH is positive.
- The combustion of fuels is an exothermic process.
- Energy profile diagrams are used to represent the energy changes in a reaction.

KEY QUESTIONS

Knowledge and understanding

1 Which one of the following is a correct thermochemical equation for the combustion of methane?

A $CH_4(g) + 2O_2(g) \rightarrow CO_2(g) + 2H_2O(l)$

B $CH_4(g) + O_2(g) \rightarrow CO_2(g) + H_2O(l)$ $\Delta H = -890$ kJ

C $CH_4(g) + 2O_2(g) \rightarrow CO_2(g) + 2H_2O(l)$ $\Delta H = -890$ kJ

D $CH_4(g) + 2O_2(g) \rightarrow CO_2(g) + 2H_2O(l)$ $\Delta H = +890$ kJ

2 Complete the following sentences by selecting the correct term for each of the alternatives in bold:

In an exothermic reaction, the chemical potential of the **products/reactants** is lower than that of the **products/reactants.** Energy is **absorbed from/released to** the surroundings. The sign of ΔH will be **positive/negative.**

3 Explain what a negative ΔH value indicates about a chemical reaction, in terms of the relative enthalpies of the reactants and products.

4 Explain why both exothermic and endothermic reactions require activation energy to proceed.

5 Use Figure 2.1.4 to convert each of the following to kJ.

a 25.8 J

b 26.3 MJ

c 6.6 GJ

Analysis

6 Ammonium chloride, NH_4Cl, is readily soluble in water. The reaction with water is endothermic and can be represented by the equation:

$$NH_4Cl(s) \rightarrow NH_4^+(aq) + Cl^-(aq)$$

5.0 g of NH_4Cl is added to water and stirred until it dissolves.

a What bonds need to be broken for this reaction to occur?

b Will this reaction absorb or release energy to the environment?

c Will the temperature of the surroundings increase or decrease when this reaction occurs?

7 The energy profile diagram below shows the energy changes during the reaction of hydrogen and iodine to form hydrogen iodide:

$$H_2(g) + I_2(g) \rightarrow 2HI(g)$$

a Is the reaction exothermic or endothermic?

b Describe how the enthalpies of the reactants and products compare to one another.

c Comment on the size of the activation energy compared with ΔH.

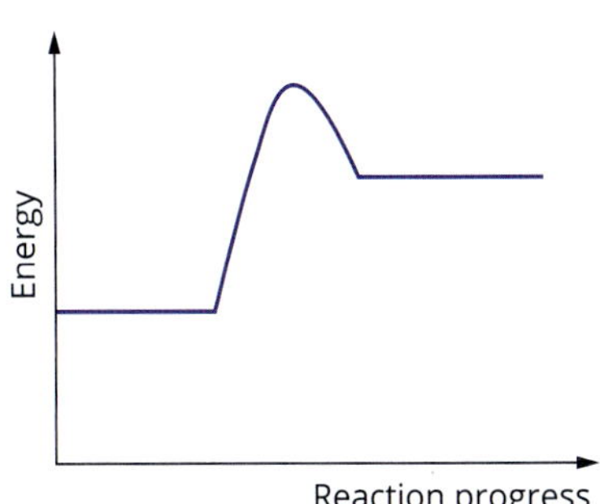

2.2 Types of fuels

Fuels provide you with energy. They are substances that have chemical energy stored within them. All chemicals contain stored energy. What makes a fuel special is that this stored chemical energy can be released relatively easily.

Sugar is an example of a common fuel (Figure 2.2.1). A cube of table sugar (sucrose) can provide your body with 82 kilojoules of energy. This is about 1% of your daily energy needs. If sucrose is burnt, this energy is released as heat. The combustion of 1 kilogram of sucrose releases sufficient energy to melt more than 5 kilograms of ice and then boil all the liquid water produced.

FIGURE 2.2.1 Sugars, such as sucrose, are fuels for your body.

Although sugars provide energy for your body, you do not heat your home, power cars or produce electricity by burning sugar. A range of other fuels, such as wood, coal, oil, natural gas, LPG, ethanol and petrol (Figure 2.2.2), are used for these energy needs.

In this section, you will explore the range of fuels available and how they are sourced.

FIGURE 2.2.2 Petrol is just one type of fuel that is used each day to meet our energy needs.

THE NEED FOR FUELS

A fuel is a substance with stored energy that can be released relatively easily for use as heat or power. Although this chapter will focus on fuels with stored chemical energy, the term 'fuel' is also applied to sources of nuclear energy, such as uranium.

The use of fuels by society can be considered from a number of points of view, including at a:

- local level (e.g. the type of petrol used in your car)
- national level (e.g. whether Australia's use of energy resources is **sustainable**)
- global level (e.g. whether the use of **fossil fuels**—coal, oil and natural gas—is contributing to the enhanced greenhouse effect).

These are not separate issues. Choices made locally have regional and global effects. The decisions of global and national governments affect how and which fuels are used.

Use of energy in Australia and the world

World energy consumption is around 4×10^{20} joules per year. The United States consumes a quarter of the world's energy. Australia consumes about 1% of the world's energy. However, energy consumption per person in Australia is only just below that of the United States. Figure 2.2.3 shows the ways in which Australians use energy. As you can see, heating and transportation account for 87% of Australia's total energy consumption.

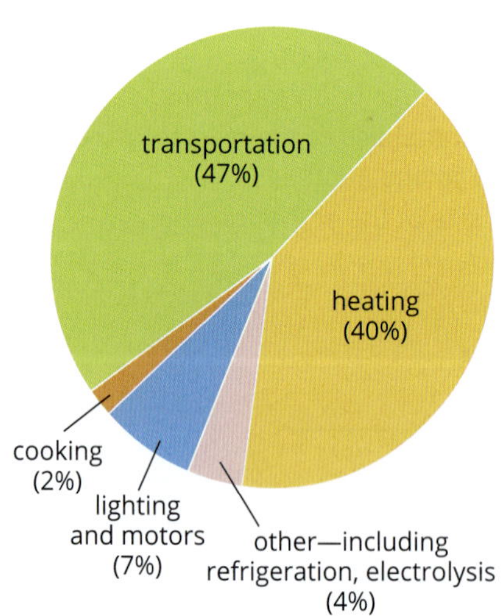

FIGURE 2.2.3 This pie chart shows where energy is used in Australia.

In Australia and around the world, most of the energy used for heating, electricity generation and powering vehicles comes from fossil fuels. In 2021, approximately 70% of Australian electricity was generated from these fuels, with around 50% from coal and the remainder from natural gas and oil. The other approximately 30% of Australia's electricity came from renewable energy sources. Hydroelectricity contributed 6% of total electricity, and wind, biofuels and solar energy made up the remaining 24%.

Future energy needs

World energy supply is in a state of flux and change. For most of the 20th century, world energy consumption grew exponentially, and the coal and oil industry expanded to meet this demand. Over the last decades, however, it has become apparent that reserves of fossil fuels are finite and that the emissions from fossil fuels are threatening climate stability.

Given the limited reserves and concerns about the link between fossil fuels and climate change, there is considerable interest in identifying and developing new energy sources. Most countries are managing a transition to alternative sources of energy, but the use of **renewable fuels** (also known as **renewables**), fuels that can be replaced at a sustainable rate for large-scale energy production, is not a simple task. Replacement energy sources need to meet a range of requirements, such as being reliable, sustainable and cost-effective. Figure 2.2.4 provides a breakdown of how the world's energy is currently sourced.

Fuels are considered to be renewable if they can be replaced at a sustainable rate.

Non-renewable fuels are those that are consumed at a faster rate than they can be replaced.

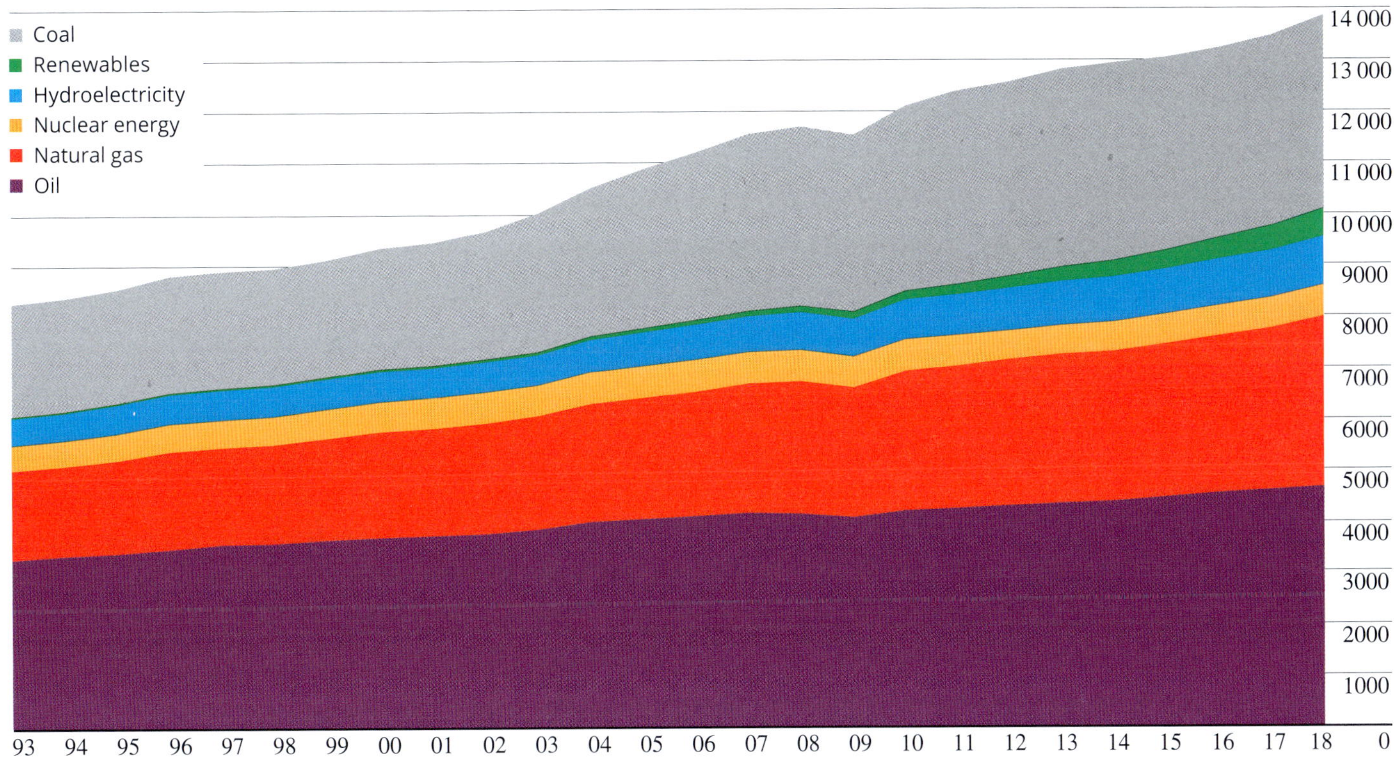

FIGURE 2.2.4 As the world's energy requirements have grown, renewable energy sources have provided an increasing, but still small, proportion of our overall energy usage.

FOSSIL FUELS

Non-renewable resources are those that are used faster than they can be replaced. Coal, oil and natural gas are non-renewable fuels. Reserves of fossil fuels are limited and they could eventually be exhausted.

Fossil fuels are naturally occurring fuels such as coal, oil or gas, that were formed in the geological past from the remains of living organisms.

Formation of fossil fuels

Coal, oil and natural gas were formed from ancient plants, animals and microorganisms. Buried under tonnes of mud, sand and rock, this once biological material has undergone complex changes to become the fossil fuels used by societies today. The organic matter still retains some of the chemical energy the plants originally accumulated by carrying out **photosynthesis**. Chemical energy in fossil fuels can be considered to be trapped solar energy.

Fossil fuel formation occurs over millions of years. This is why these fuels are considered non-renewable. Once reserves of the fossil fuels have been used, they will not be replaced in the foreseeable future.

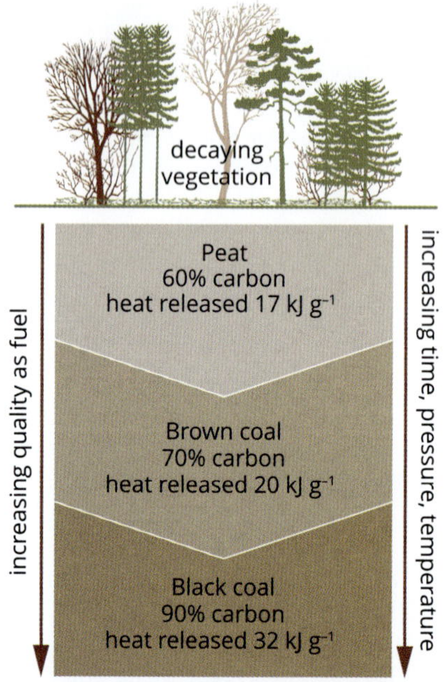

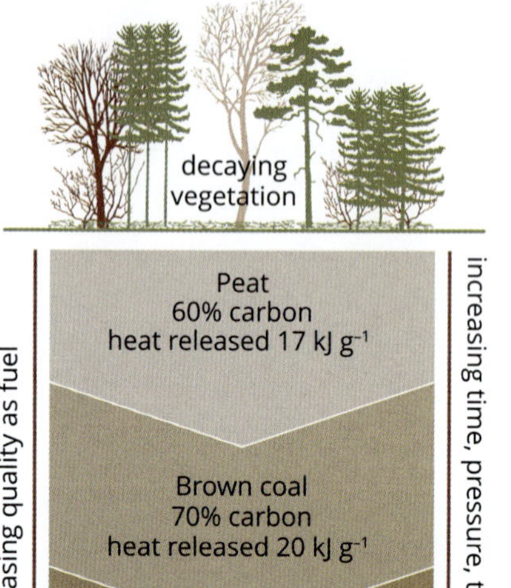

FIGURE 2.2.5 Steps in the formation of coal. Values of the carbon content and heat released upon combustion are for dried coal.

Coal

As wood and other plant material is converted into coal, the carbon content increases and the proportion of hydrogen and oxygen decreases. The wood progressively becomes peat, brown coal and then black coal (Figure 2.2.5). Coal is a mixture of large molecules made from carbon, hydrogen, nitrogen, sulfur and other elements.

Electricity from coal

Chemical energy from most fuels is harnessed through combustion of the fuel. Thermal energy released from the combustion of fuels can be converted into electrical energy. In Australia, electrical energy is produced from several different fuels.

The combustion of coal generates over 50% of Australia's electricity. Rather than transport coal to every factory, office and household, the chemical energy is converted to electrical energy at a power station. Electricity is transmitted easily from the power station by metal cables and wires to other regions. The reaction occurring when coal burns can be written as:

$$C(s) + O_2(g) \rightarrow CO_2(g)$$

The energy released from the combustion of coal is about 32 kJ g^{-1}.

Figure 2.2.6 illustrates how thermal energy is released when the coal is converted to electrical energy.

flue gas

10% of the coal's chemical energy is lost as heat in the chimney gases.

chimney

steam

35% of the coal's chemical energy is converted to electricity.

50% of the coal's chemical energy is lost as heat in steam.

electricity

generator

turbine

5% of the coal's chemical energy is lost in various ways in the power plant.

boiler

condenser

pump

pump

cooling tower

coal

chemical energy in coal → thermal energy of burning coal → thermal energy of steam → mechanical energy of turbine → electrical energy from generator

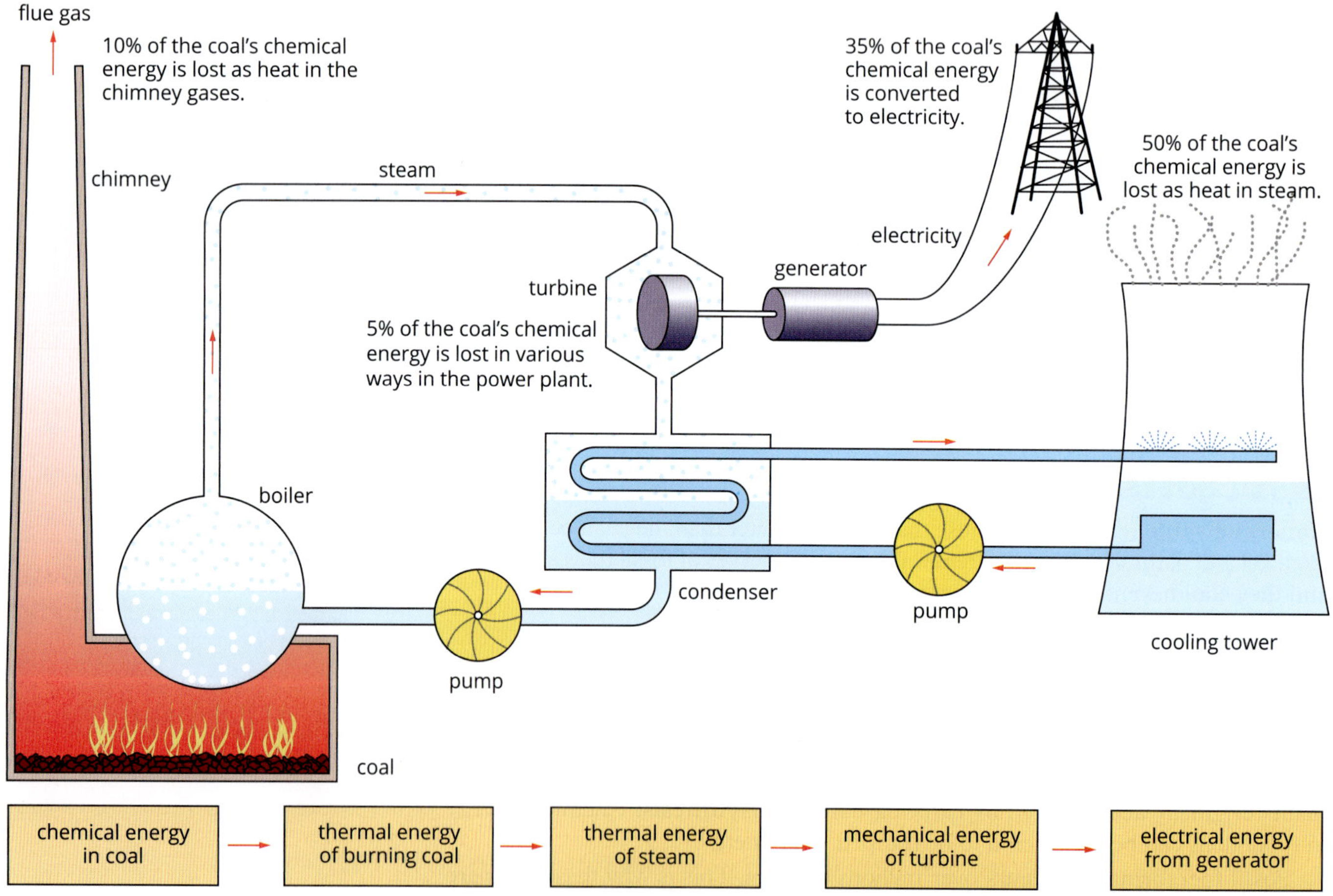

FIGURE 2.2.6 A representation of the process of burning coal to produce electricity. The chemical energy in the coal undergoes several transformations before electricity is produced.

Most Australian states are trying to retire their large-scale coal-fired electricity plants ahead of schedule in response to concerns about their greenhouse emissions. For example, Hazelwood Power Station, which had been producing around 20% of Victoria's electricity since it was commissioned in 1964, was closed in 2017.

Petrol

Crude oil (petroleum) is a mixture of **hydrocarbon** molecules that are mostly members of the **homologous series** of **alkanes**. Crude oil itself is of no use as a fuel, but it contains many useful compounds. These useful compounds are separated into a range of fractions by **fractional distillation**. Fractional distillation does not produce pure substances. Each fraction is still a mixture of hydrocarbon compounds. These fractions can be used as fuels or treated further to produce more specific products through chemical processes.

Petrol is one of the fractions obtained from crude oil. It includes the compound octane and other alkanes with a similar boiling point. The equation for the combustion of octane is:

$$2C_8H_{18}(l) + 25O_2(g) \longrightarrow 16CO_2(g) + 18H_2O(l)$$

Combustion occurs in the cylinder of a car engine. The hot gases formed push the pistons in the engine up and down, enabling the car to move.

Diesel, or more correctly, **petrodiesel**, is also a fraction of crude oil. The alkanes in petrodiesel are slightly longer molecules than those in petrol.

Natural gas

Natural gas is another fossil fuel found in deposits in the Earth's crust. It is mainly composed of methane (CH_4) together with small amounts of other hydrocarbons, such as ethane (C_2H_6) and propane (C_3H_8). Water, sulfur, carbon dioxide and nitrogen may also be present in natural gas.

Natural gas can be found:

- in gas reservoirs trapped between layers of rocks
- as a component of petroleum deposits
- in coal deposits, where it is bonded to the surface of the coal.

Coal seams usually contain water and the pressure of the water can keep the gas adsorbed to the coal surface. Natural gas found in this way is known as coal seam gas or CSG. It is a major component of Queensland's energy supplies.

Natural gas is accessed by drilling, as with crude oil; drilling allows the natural gas to flow to the surface (Figure 2.2.7).

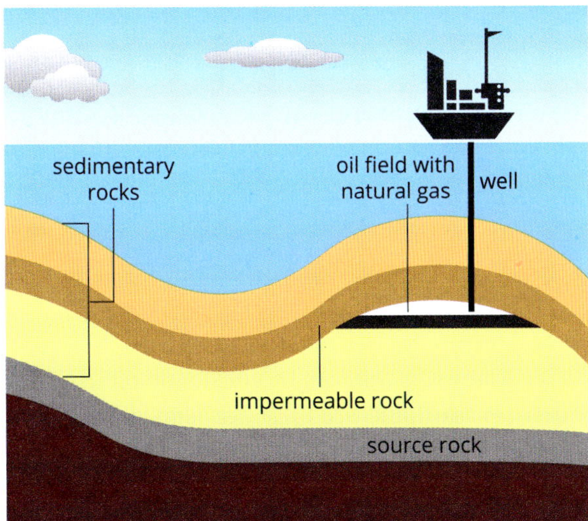

FIGURE 2.2.7 Natural gas deposits are often found trapped above crude oil. Once a well is sunk into the deposit, the natural gas flows to the surface.

In 2018 Australia replaced Qatar as the world's leading exporter of natural gas. In November 2018 alone, Australia exported 6.5 million tonnes of natural gas. Natural gas is exported as a liquid, referred to as LNG. The conversion to a liquid increases the energy density, but the liquification consumes a significant quantity of energy.

Electricity from natural gas

Natural gas is used in Victoria to generate electricity for the power grid. In a gas-fired power plant, methane and other small alkanes are burnt to release energy. As shown in Figure 2.2.8, the hot gases produced by combustion cause air to expand in a combustion turbine to generate electrical energy. This is a simpler process than in a coal-fired plant where the thermal energy is used to produce steam.

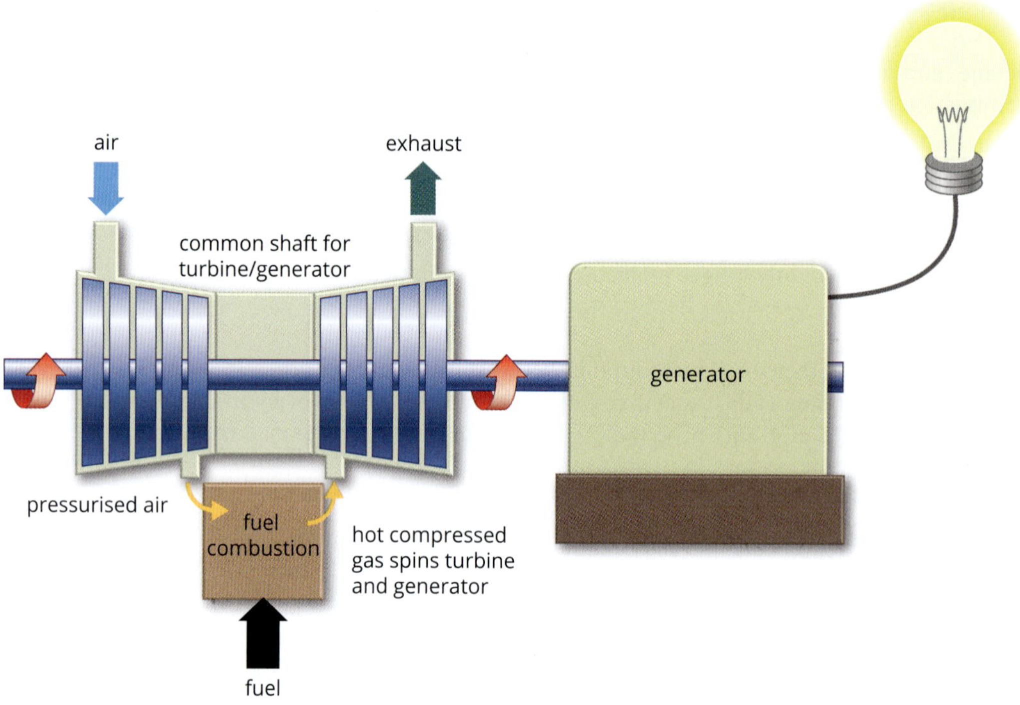

FIGURE 2.2.8 In a gas-fired power plant, the hot gases produced expand air in a combustion turbine to generate electricity.

FIGURE 2.2.9 Many homes use natural gas for cooking and/or heating.

> Biofuels are fuels derived from plants or animals. They can usually be replenished at a sustainable rate, hence are considered forms of renewable energy.

The composition of natural gas varies, but the main combustion reaction involves methane. The equation is:

$$CH_4(g) + 2O_2(g) \longrightarrow CO_2(g) + 2H_2O(l)$$

Natural gas has many domestic applications. A network of pipes is used to deliver natural gas to most cities in Victoria where it is used for cooking (Figure 2.2.9), hot water and home heating.

BIOFUELS

Governments and industry are exploring alternatives to fossil fuels to meet our future energy needs and limit the impact of fossil fuels on the environment. Ideally, new sources of energy will be renewable. Renewable energy is energy that can be obtained from natural resources that can be constantly replenished.

Biochemical fuels (or **biofuels**) are fuels derived from plant materials, such as grains (maize, wheat, barley or sorghum), sugar cane (Figure 2.2.10 on the following page) and vegetable waste, and vegetable oils. The three main biofuels are **bioethanol**, **biogas** and **biodiesel**. They can be used alone or blended with fossil fuels such as petrol and diesel. Bioethanol will be discussed in detail in Section 2.4.

FIGURE 2.2.10 Harvesting sugar cane in Queensland. Sugar cane can be a source of the raw materials for the production of bioethanol.

As well as being renewable, biofuels are predicted to have less impact on the environment than fossil fuels. The plant materials used in the generation of biofuels are produced by photosynthesis, which removes carbon dioxide from the atmosphere and produces glucose ($C_6H_{12}O_6$) in the following reaction:

$$6CO_2(g) + 6H_2O(l) \longrightarrow C_6H_{12}O_6(aq) + 6O_2(g)$$

The plants convert the glucose into **cellulose** and starch. Although carbon dioxide is released back into the atmosphere when the biofuel is burnt, the net impact should be less than for fossil fuels. In theory, a biofuel could be **carbon neutral**, absorbing the same amount of CO_2 in its formation as is released in its combustion. In practice, energy is required to farm, fertilise and transport biofuels, so they are not likely to be fully carbon neutral.

Biogas

Biogas is gas that is released in the breakdown of organic waste by **anaerobic** bacteria. These bacteria decompose the complex molecules contained in substances such as carbohydrates and proteins into the simple molecular compounds, carbon dioxide and methane. A digester (Figure 2.2.11) is a large tank filled with the anaerobic bacteria that digest (consume) the complex molecules to form biogas.

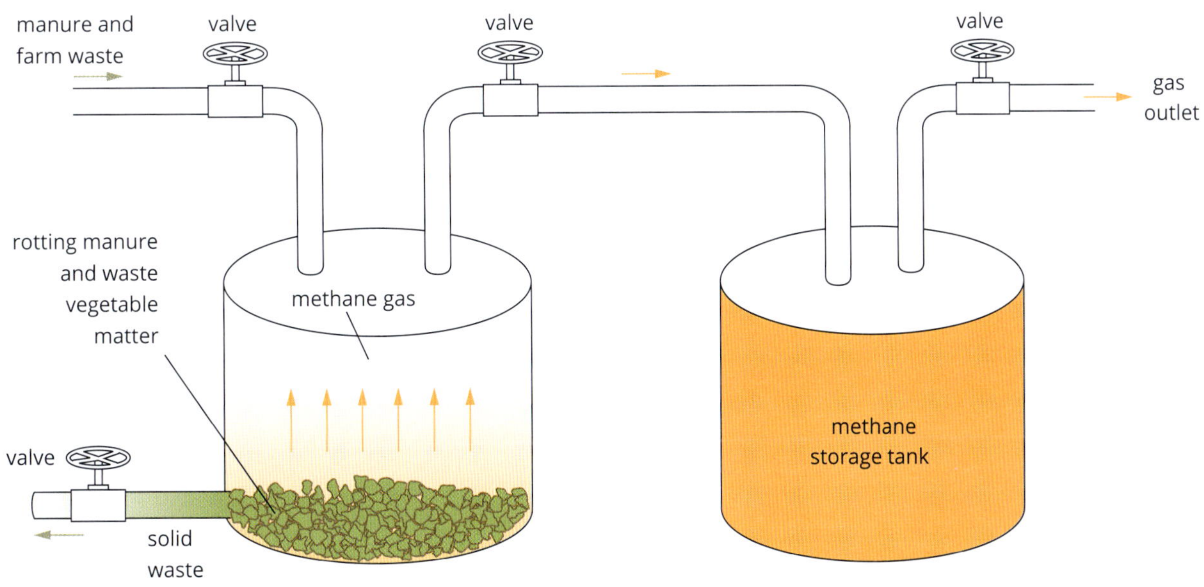

FIGURE 2.2.11 A digester is used in the production of biogas.

CHEMFILE

Berrybank piggery biogas

Berrybank Farm near Ballarat is an example of the innovative use of biogas. Over \$2 million has been spent on building infrastructure to collect the manure from 20 000 pigs. The manure is fed into a digester, shown in the figure below, which produces two useful products: biogas and fertiliser. The biogas is used to fire generators that produce an estimated \$250 000 of electricity annually. Some of the energy produced is used to maintain the temperature of the biodigester above 30°C—not always easy in a Ballarat winter.

Large digestor tanks similar to those used at Berrybank Farm for biogas production

There are now several Victorian farms that have installed biodigesters, but the Berrybank operation was the first; it was established over 30 years ago. The cost at that time was around \$2 million, but the savings in electricity and water consumption have ensured the venture is profitable.

> Biogas is formed in anaerobic conditions by the action of bacteria. The main component of biogas that acts as a fuel is methane.

A range of materials, including rotting rubbish and decomposing plant material, can be used to produce biogas. The composition of biogas depends on the original material from which it is obtained and the method of decomposition. The typical composition of a sample of biogas is shown in Table 2.2.1.

TABLE 2.2.1 Typical percentage composition of different gases found in biogas

Gas	Formula	Percentage composition (by volume)
methane	CH_4	60
carbon dioxide	CO_2	32
nitrogen	N_2	4.5
hydrogen sulfide	H_2S	2
oxygen	O_2	1
hydrogen	H_2	0.5

As you can see from Table 2.2.1, biogas consists mainly of methane and carbon dioxide. Biogas can be used for heating and to power homes and farms. There are more than 30 million biogas generators in China. Biogas generators are particularly suited to farms, as the waste from a biogas generator makes a rich fertiliser.

In the future, it is likely more energy will be obtained from biogas generated at sewage works, chicken farms, piggeries and food-processing plants. Your local rubbish tip also has the potential to supply biogas (Figure 2.2.12). The gas can be used directly for small-scale heating or to generate electricity.

FIGURE 2.2.12 Pipes buried in this rubbish tip collect biogas.

Electricity from biogas

Biogas can be used to generate electricity, usually in small-scale electricity generators rather than large power plants. These smaller generators are often located at the site where the biogas is produced. For example, sewage works commonly burn biogas produced in a generator to supply some of their power needs.

The main reaction that occurs in the combustion of biogas is the same reaction as that of a gas-fired power station, the combustion of methane. The energy released per gram of biogas is less than that of natural gas because the methane content in biogas is significantly lower.

Biodiesel

Biodiesel is a mixture of organic compounds called esters. These esters are produced by a chemical reaction between vegetable oils or animal fats and an alcohol (most commonly methanol, CH_3OH).

The usual raw material for the production of biodiesel is vegetable oil from sources such as soybean, canola or palm oil. Recycled vegetable oil or animal fats can also be used. Fats and oils are **triglycerides** with a molecular structure consisting of three hydrocarbon chains, each attached by an ester functional group to a backbone of three carbon atoms, as shown in Figure 2.2.13. The triglyceride is converted into biodiesel by warming it with an alcohol, usually methanol or ethanol, in a process known as **transesterification**. Concentrated KOH is used as a **catalyst** in the reaction.

In the transesterification reaction, the triglyceride is converted into a small molecule called glycerol and three ester molecules with long carbon chains. The ester molecules are the biodiesel product. This biodiesel can be used as the fuel for some diesel engines.

$$\begin{array}{l} \text{H}_2\text{C--O--C(=O)--(CH}_2)_{14}\text{CH}_3 \\ \text{HC--O--C(=O)--(CH}_2)_{14}\text{CH}_3 \\ \text{H}_2\text{C--O--C(=O)--(CH}_2)_{14}\text{CH}_3 \end{array} + 3CH_3OH \xrightarrow{KOH} \begin{array}{l} \text{H}_2\text{C--OH} \\ \text{HC--OH} \\ \text{H}_2\text{C--OH} \end{array} + 3CH_3(CH_2)_{14}COOCH_3$$

triglyceride (palm oil) — methanol — glycerol — fatty acid methyl esters (biodiesel)

FIGURE 2.2.13 The reaction of a triglyceride with methanol and a KOH catalyst to form fatty acid methyl esters (biodiesel) and glycerol

> Biodiesel molecules are esters of fatty acids, formed from the reaction between triglycerides and small alcohol molecules.

The structure of a typical biodiesel molecule is shown in Figure 2.2.14.

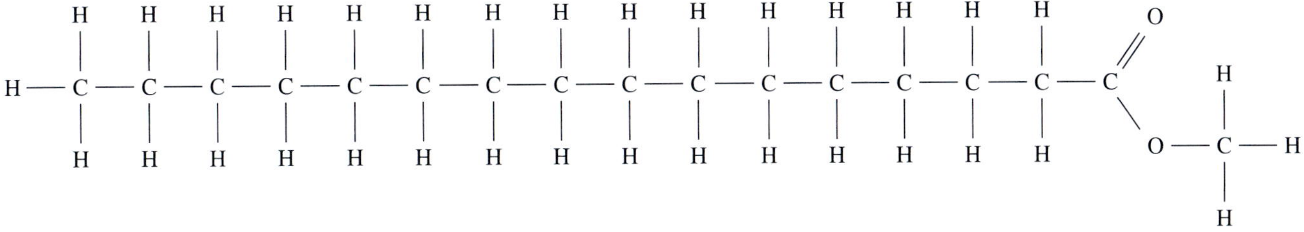

FIGURE 2.2.14 Structural formula of a typical biodiesel molecule

RENEWABILITY

Fossil fuels are a non-renewable source of energy. It took millions of years for their formation in the Earth's crust, so the likelihood of new deposits forming cannot be considered.

Biofuels are renewable and current production levels are sustainable. However, currently biofuels produce only a small percentage of Australia's fuel needs. For biofuel production to increase significantly, we would need to grow crops specifically for this purpose. This would present a number of issues, including land degradation, clearing of forest and bushland, and ensuring food supplies are maintained.

It is mainly wind and solar investments that are steadily replacing fossil fuels. Figure 2.2.15 from the Australian Clean Energy Council shows that in 2021, renewables accounted for over 32% of our electricity generation and this figure is increasing each year.

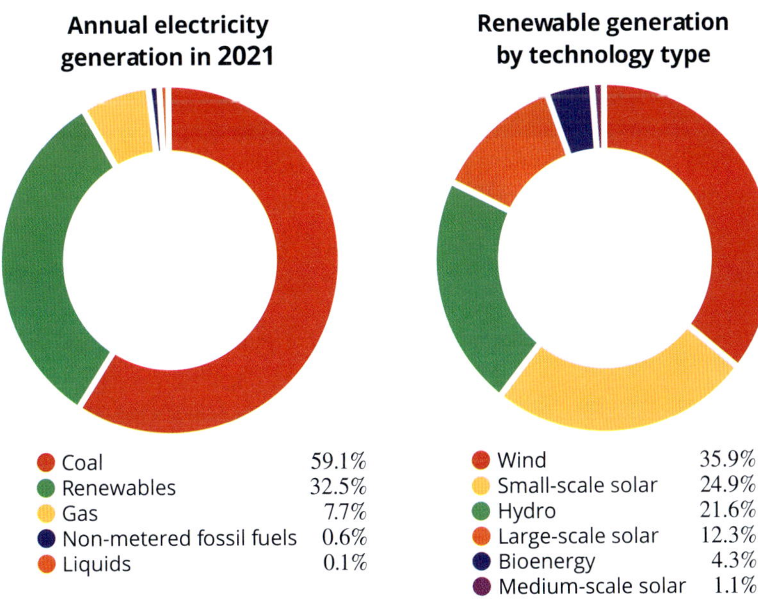

FIGURE 2.2.15 A 2021 breakdown of how Australia's electrical energy is sourced

It is also interesting to see in Figure 2.2.16, the divergent paths of various Australian states in their energy production. Tasmania produced nearly all of its energy in 2021 from renewable sources, while Queensland and New South Wales were below 30%.

FIGURE 2.2.16 The level of renewable energy use varies significantly between the Australian states.

Table 2.2.2 compares the advantages and disadvantages of some fuels.

TABLE 2.2.2 Advantages and disadvantages of fuels described in this section

Fuel	Advantages	Disadvantages
Coal	• Large reserves • More easily transported than liquid or gaseous fuels • Relatively high energy content	• Non-renewable • High level of emissions
Natural gas	• More efficient than coal for electricity production • Easy to transport through pipes • Relatively high energy content	• Non-renewable • Limited reserves • Polluting, but less than coal and petrol
Biogas	• Renewable • Made from waste • Reduces waste disposal • Low running costs • CO_2 absorbed during photosynthesis	• Low energy content • Supply of waste raw materials limited
Petrol	• High energy content • Ease of transport	• Non-renewable • Polluting, but less than coal • Limited reserves
Bioethanol	• Renewable • Can be made from waste • CO_2 absorbed during photosynthesis • Burns smoothly • Fewer particulates produced than petrol	• Limited supply of raw materials from which to produce it • Lower energy content than petrol • May require use of farmland otherwise used for food production

2.2 Review

OA ✓✓

SUMMARY

- A fuel is considered to be non-renewable if it cannot be replenished at the rate at which it is consumed. Fossil fuels, such as coal, oil and natural gas, are non-renewable.
- Fuels such as petrol, natural gas and petrodiesel undergo combustion reactions in excess oxygen to form carbon dioxide and water.
- The combustion reactions of fuels are used to produce electricity and to power vehicles.
- Fuels are considered to be renewable if they can be replenished at a sustainable rate.
- Biogas is formed by the anaerobic breakdown of organic waste. It is usually burnt to generate electrical energy.
- Biodiesel is produced from triglycerides in animal fats or plant oils. The triglycerides react with methanol in a transesterification reaction to form biodiesel.
- Some of the non-renewable and renewable fuels in use in Australia:

Non-renewable fuels	Renewable fuels
coal oil, petrol, petrodiesel natural gas	bioethanol biogas biodiesel

- Biofuels offer several environmental advantages: CO_2 is absorbed during the growth of crops used in their production, they can be replenished, and they can be produced from material that would otherwise have been waste.
- A shift to large-scale production of biofuels could place a strain on resources and available farmland.

KEY QUESTIONS

Knowledge and understanding

1 Use natural gas as an example to explain what a fuel is.

2 What is the difference between a renewable and non-renewable fuel?

3 A large dairy farm in country Victoria has built a digestor to process the cow manure.

a Which biofuel would be produced from the animal waste?

b How is this fuel most likely to be used?

4 There are several small plants in Australia that collect the waste oil from businesses such as fish-and-chip shops.

a Which biofuel is produced from the waste oil?

b How is this fuel most likely to be used?

Analysis

5 Biodiesel and petrodiesel can both be used as a fuel for trucks. The two main products of combustion of these fuels are carbon dioxide and water. Explain why biodiesel is considered more sustainable than petrodiesel, even though the products of combustion are very similar.

6 The production of biogas from animal waste provides a fuel from a renewable source and disposes of a nuisance material. Despite the obvious benefits of biogas production, the volume used in Australia is still relatively low. Suggest reasons for this limited use.

7 Despite the fact that biodiesel can be produced from a renewable resource such as canola crops, there are still concerns with the production and use of biodiesel. Discuss what these concerns might be.

2.3 Fuel sources for the body

In Section 2.2 you learnt about the various fuels developed by humans for their external energy needs. In this section, you will focus on the energy needs of the body's internal processes. Humans require energy for warmth, movement and for the synthesis of necessary biomolecules, such as hormones, enzymes, carbohydrates and triglycerides.

PHOTOSYNTHESIS

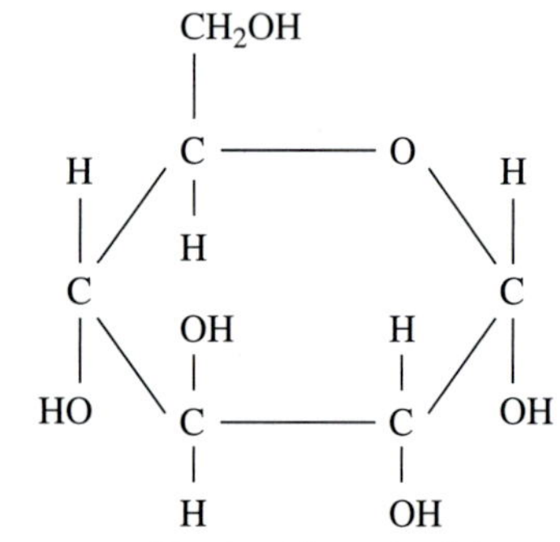

FIGURE 2.3.1 Structural formula of glucose

One of the most important biomolecules is glucose, shown in Figure 2.3.1.

Glucose is found in all living things, especially in the sap of plants and the blood and tissue of animals. Plants can use glucose as a monomer to form important natural polymers such as starch and cellulose. Both glucose and its polymer, starch, are more rapidly digested than other forms of food. They are the main sources of energy in most diets and the human body uses them for energy in preference to fats and proteins.

Green plants, such as the one seen in Figure 2.3.2 carry out photosynthesis, one of the key chemical reactions that supports life. Plants use photosynthesis to make their own food in the form of glucose. Photosynthesis is an endothermic chemical reaction carried out in chloroplasts in the cells of green leaves. Energy from sunlight is used to produce glucose from carbon dioxide and water. Oxygen gas is the other product.

FIGURE 2.3.2 Green plants carry out photosynthesis during the day. Photosynthesis occurs in the leaves of the plant.

Glucose contains chemical energy, which feeds the plant and is also stored and becomes food for animals that eat it.

The thermochemical equation for photosynthesis is:

$$6CO_2(g) + 6H_2O(l) \rightarrow C_6H_{12}O_6(aq) + 6O_2(g) \qquad \Delta H = +2803 \text{ kJ}$$

CELLULAR RESPIRATION

Glucose is the primary energy source for the cells of plants and animals and is used to obtain energy by a process known as **cellular respiration**. The main form of respiration is also referred to as aerobic respiration, as oxygen gas is required.

In aerobic respiration, glucose is oxidised to carbon dioxide and water through a sequence of reactions. The overall equation is:

$$C_6H_{12}O_6(aq) + 6O_2(g) \rightarrow 6CO_2(g) + 6H_2O(l) \qquad \Delta H = -2803 \text{ kJ}$$

Note that the products of photosynthesis are the reactants for aerobic respiration. Figure 2.3.3 shows that the energy from the Sun stored in glucose molecules is released to the plant or animal through respiration. It should also be apparent that the respiration reaction is the same as the reaction that would occur if glucose were to undergo complete combustion in a flame. The glucose stores energy from the Sun as chemical energy.

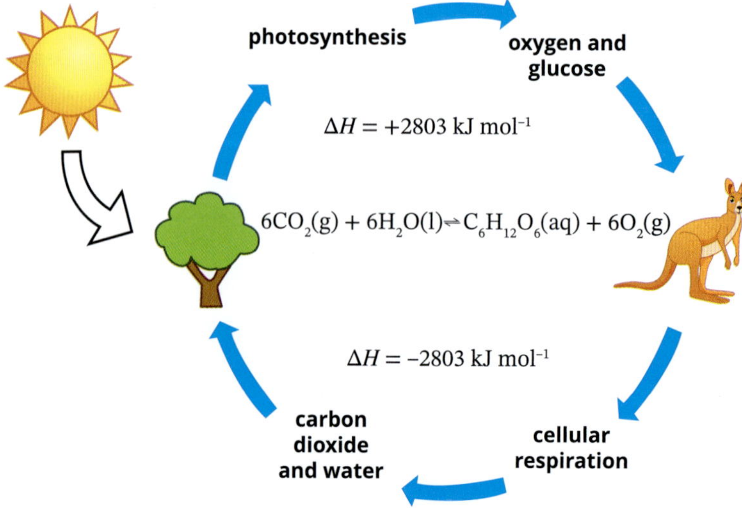

FIGURE 2.3.3 Photosynthesis in plants can store energy from the Sun in plants. Respiration of glucose releases that energy for the plant or animal to use.

Although the equations for photosynthesis and respiration are the reverse of each other, the actual processes are very different. Both reactions occur in stages, requiring the presence of a range of other biomolecules. Photosynthesis occurs in plant cells, while respiration occurs in both animal and plant cells.

The thermochemical equations for photosynthesis and cellular respiration are:

photosynthesis $6CO_2(g) + 6H_2O(l) \longrightarrow C_6H_{12}O_6(aq) + 6O_2(g)$ $\Delta H = +2803$ kJ

respiration $C_6H_{12}O_6(aq) + 6O_2(g) \longrightarrow 6CO_2(g) + 6H_2O(l)$ $\Delta H = -2803$ kJ

CASE STUDY ANALYSIS

Lavoisier and combustion

Antoine Lavoisier was a famous French chemist. He published a systematic way of naming chemicals based on their composition, he discovered the composition of water, and he clarified the role of oxygen in combustion and oxidation, among his many achievements. From the early 1780s Lavoisier proposed that combustion and respiration were the same thing; that all respiration is a form of combustion. Given that oxygen had only recently been discovered, this was quite an amazing insight.

To test his theory, he designed an apparatus that included both a living guinea pig and ice. The vigorousness of the movement of the guinea pig gave an indication of the oxygen content of the air and the rate of melting of the ice provided a measure of the energy released. His experiments confirmed that respiration was a reaction used by the body to release energy.

Lavoisier next measured the volume of oxygen inhaled by a human volunteer, and he demonstrated that the volume increased with the rate of exercise. The volunteer's heart rate and pulse rate also increased. Figure 2.3.4 shows a sketch made by Lavoisier's wife Marie-Anne Lavoisier of this experiment. The volunteer can be seen sitting near the table and the apparatus to measure the air volume is on the table. A translation of Lavoisier's insightful conclusion reads:

Respiration is nothing but a slow combustion of carbon and hydrogen, similar in all respects to that of a lamp or a lighted candle, and from this point of view, animals which breathe are really combustible substances burning and consuming themselves.

One of Lavoisier's conclusions, referring to the location of respiration, did need to be modified about 100 years later. Lavoisier assumed incorrectly that it was in the lungs themselves rather than in body cells.

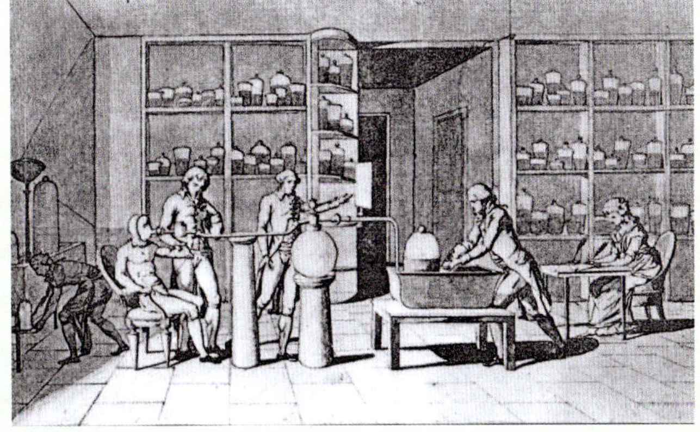

FIGURE 2.3.4 Lavoisier's experiment to measure the oxygen intake and energy output of a volunteer

Analysis

1 Lavoisier wanted to show that respiration and combustion of glucose are the same thing. Use an equation to explain what this means.

2 Lavoisier used ice in his apparatus as an indicator of how much heat his volunteer generated. Explain how ice could be used to do this.

3 Lavoisier used a live guinea pig as an indication of air quality. Explain how the behaviour of the guinea pig could indicate air quality.

4 Discuss some of the variables Lavoisier might have considered monitoring in his investigation of the respiration process of a volunteer.

> One of the main reasons humans consume nutrients is to obtain energy. The quantity of energy obtained depends upon the bonding in the nutrient.

ENERGY FROM CARBOHYDRATES, FATS AND PROTEINS

Glucose is not the only molecule to provide energy to living things. A balanced diet is made up of a variety of foods containing **nutrients** such as **carbohydrates** (glucose is an example), **proteins** and **fats**. Nutrients are substances used by an organism to survive, grow and reproduce. Food supplies the energy required for the millions of chemical reactions that occur in your body. Energy is needed for physical activity, for warmth and for functions such as breathing. Energy is also required to build the large molecules in our systems and this energy can be released as the molecules are digested, or **metabolised**, back to smaller units.

Carbohydrates

Carbohydrates are made from the elements carbon, hydrogen and oxygen, and usually have the general formula, $C_x(H_2O)_y$. Many important carbohydrates are polymers of glucose, such as starch, which is used as an energy storage molecule in plants. Figure 2.3.5 shows the repeating glucose units in a polymer of starch. During digestion, **enzymes** in our saliva and small intestine break the starch molecules back into glucose. Enzymes are organic catalysts that alter the rate of biochemical reactions. The glucose is transported in the blood to body cells where respiration can occur. Energy is required to form the many bonds in these large carbohydrates and that energy is released during digestion.

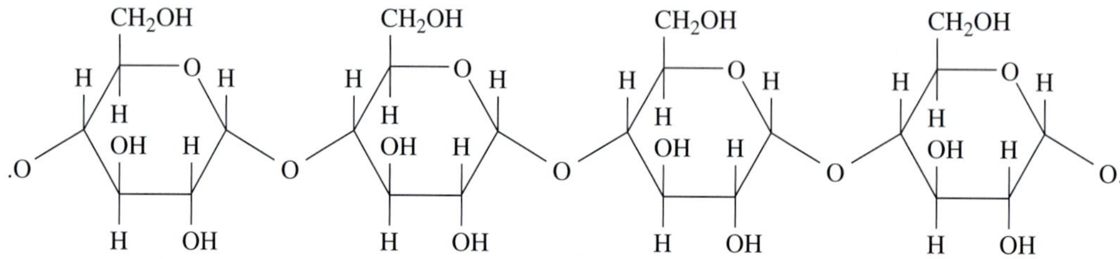

FIGURE 2.3.5 Starch is a natural polymer formed from the polymerisation of glucose monomers. Note: Due to the complexity of the structure the carbon atoms in the rings have been omitted for clarity.

Fats and oils

Fats and oils are examples of triglycerides, large non-polar molecules with three long hydrocarbon chains attached to a glycerol molecule. The general structure of a triglyceride is shown in Figure 2.3.6. Fats play an important role in providing and storing energy in the body. Digestion breaks down fats and components of this breakdown can be oxidised in body cells to carbon dioxide and water, releasing large quantities of energy.

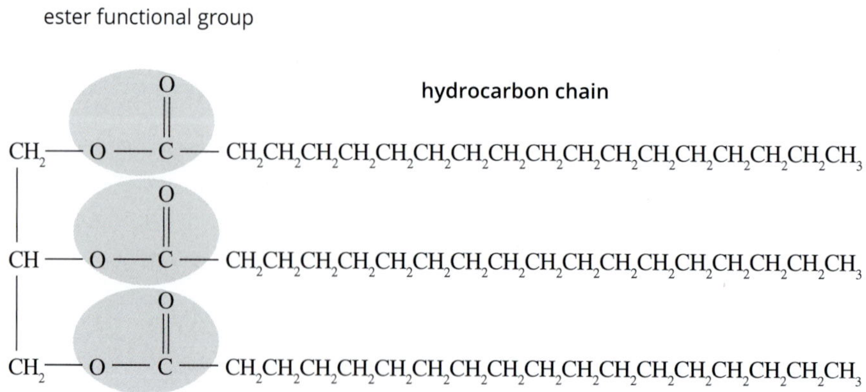

FIGURE 2.3.6 General structure of a triglyceride. Note the three long hydrocarbon chains and the three ester functional groups.

Proteins

A segment of a typical protein molecule is shown in Figure 2.3.7. Proteins are rarely used by the body as an energy source as they have so many other important roles in the body. If, however, intensive exercise depletes the body's stores of glycogen and fat, protein can be used as an alternative. When this occurs, it is important to replace the protein quickly to ensure the functioning of other processes.

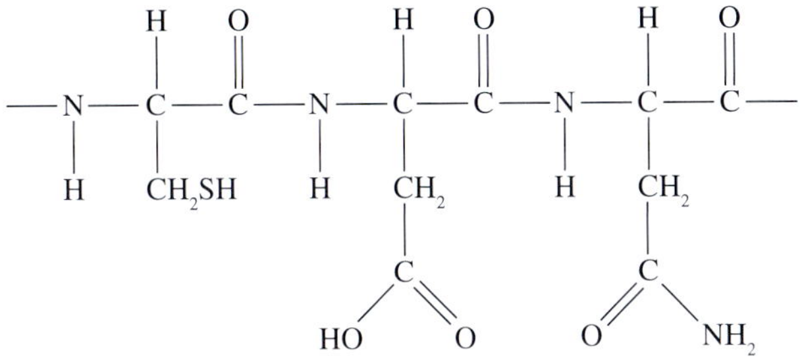

FIGURE 2.3.7 A section of a protein chain

Energy content of carbohydrates, fats and proteins

Each of these three major nutrients provides a different quantity of energy per gram.

The **energy content** of foods—the amount of energy a food or fuel can supply—is measured in kJ g^{-1}, kJ/100 g or even kJ mol^{-1} if the food is a pure substance such as glucose. For most foods, the energy released on combustion is similar to the energy released when the food is oxidised during respiration.

For convenience, each of the major food nutrients—carbohydrates, fats and proteins—are considered to have a particular heat of combustion, although there is a range of values for different members of these food groups. For example, carbohydrates are considered to have a heat of combustion of 16 kJ g^{-1}, whereas the heat of combustion of glucose is 15.7 kJ g^{-1} and that of polysaccharides is 17.6 kJ g^{-1}.

Table 2.3.1 compares the heats of combustion (energy content) and energy available to the body for each of the major food groups. The energy available to the body from a nutrient or food is called its **energy value**. The energy value of different foods will be discussed further in Chapter 3.

TABLE 2.3.1 Comparison of the heat of combustion and energy available to the human body of the three main nutrients

Nutrient	Energy content (heat of combustion kJ g^{-1})	Energy value (energy available for the body kJ g^{-1})
carbohydrates	16	16
fats and oils	39	37
proteins	24	17

Note that fats and oils have a significantly higher energy value than carbohydrates and proteins. This is essentially due to the degree to which these molecules can be oxidised. Carbohydrates tend to contain more oxygen atoms than fats and oils do. At a simple level, the carbon atoms in carbohydrate molecules have a higher 'degree of oxidation'. Therefore, fats and oils have greater potential for oxidation and release more energy on combustion.

The energy released when food is burned is often greater than the energy that is available for the human body to use after the food has been digested. For example, dietary fibre is mainly the carbohydrate cellulose. Humans cannot digest most fibre, so the energy it contains is not available to humans.

2.3 Review

OA ✓✓

SUMMARY

- Green plants use the energy from the Sun for photosynthesis, the conversion of carbon dioxide and water to glucose and oxygen. The equation for photosynthesis is:

$$6CO_2(g) + 6H_2O(l) \rightarrow C_6H_{12}O_6(aq) + 6O_2(g) \quad \Delta H = +2803 \text{ kJ}$$

- Plants and animals use cellular respiration to oxidise glucose to obtain energy. The equation is the reverse of photosynthesis:

$$C_6H_{12}O_6(aq) + 6O_2(g) \rightarrow 6CO_2(g) + 6H_2O(l) \quad \Delta H = -2803 \text{ kJ}$$

- The metabolism in the body of carbohydrates, fats and proteins is a source of energy for animals.
- Fats produce the most energy per gram of the major food groups followed by proteins, then carbohydrates.
- The human body will often obtain less energy from a food item than the theoretical quantity based on direct combustion of the food.

KEY QUESTIONS

Knowledge and understanding

1 Name three conditions (including chemicals) that are essential for photosynthesis to occur.

2 It is stated in this section that glucose is a key molecule for the energy processes in living things.

 a Explain why glucose is considered a starting point for energy systems.

 b Explain the role of glucose in energy storage for plants.

 c Explain the role of glucose in energy processes for humans.

3 Complete the following sentences by selecting the correct term for each of the alternatives in bold:

In **cellular/anaerobic** respiration, glucose is used by cells to obtain energy. Aerobic respiration is an **endothermic/exothermic** process in which the glucose is **oxidised/reduced** by **carbon dioxide/oxygen**. A relatively large quantity of energy is **absorbed/released** during aerobic respiration and can be used by the cells of the body.

Analysis

4 State whether each of the following is endothermic or exothermic.

 a respiration

 b digestion of fish oil

 c formation of cellulose from glucose in plants

 d photosynthesis

 e digestion of starch.

5 Explain how the molecular formulas of fats and carbohydrates indicate which will release the most energy.

6 Cellular respiration is an example of an oxidation reaction. What definition of oxidation best describes this reaction?

2.4 Bioethanol

Ethanol is used as an energy source in combustion engines, fuel cells and electricity generators. Ethanol can be produced from ethene obtained from crude oil, but this adds to the problems already mentioned relating to fossil fuels. If, however, the ethanol is produced from **biomass** the environmental concerns are not as significant. Given Australia's strong farming and forestry industries, there is considerable interest in producing bioethanol from the biomass wastes. In this section, you will look at how bioethanol is produced and learn about the potential sources of biomass.

FERMENTATION

Bioethanol is produced from glucose and other sugars in a **fermentation** process. Various enzymes and microorganisms catalyse or facilitate the reactions involved. The process is carried out at 35°C because at higher temperatures, the microorganisms and enzymes would be destroyed. The main reaction is:

$$C_6H_{12}O_6(aq) \rightarrow 2CH_3CH_2OH(aq) + 2CO_2(g)$$

Fermentation is a natural process in which an organism converts a carbohydrate such as starch or sugar into an alcohol.

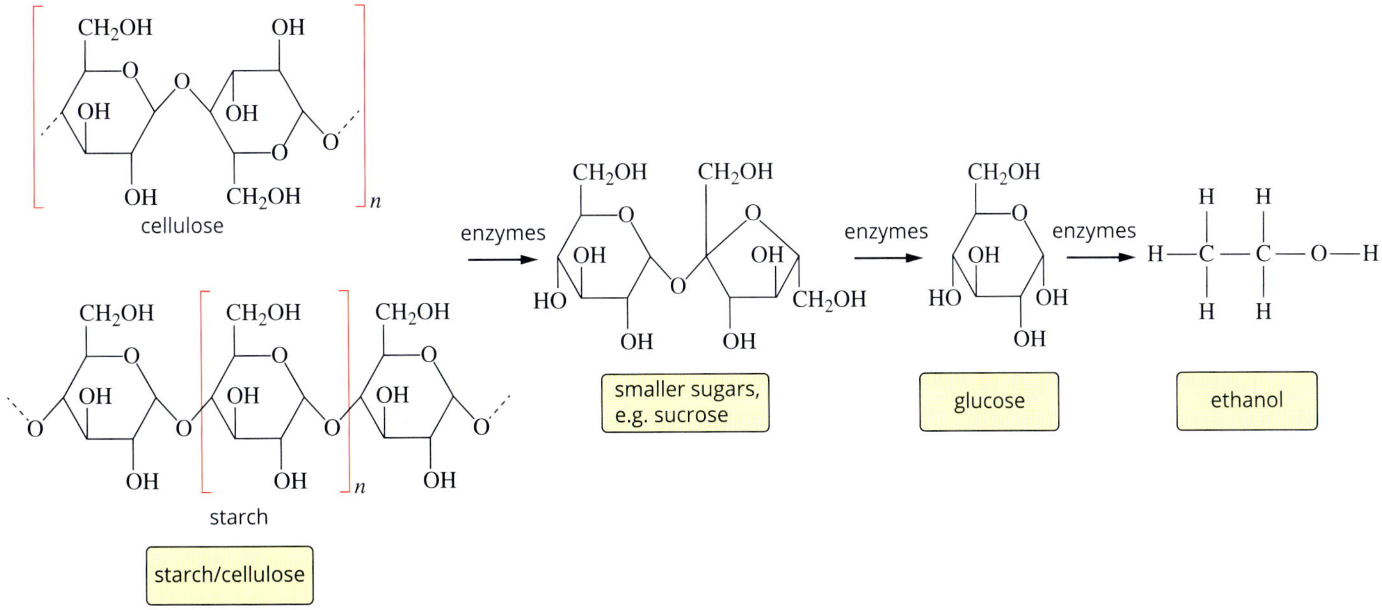

Glucose is present in plants as glucose itself and also as a component of larger carbohydrates. The bioethanol industry therefore needs abundant carbohydrate sources that can be broken down to glucose. In the future, the three main bioethanol feedstocks in Australia are likely to be:

- sugar cane (high in sucrose)
- wheat (high in starch)
- forest waste (high in cellulose).

Figure 2.4.1 is an outline of the chemical changes occurring in the bioethanol manufacturing process. The carbohydrates are pulped with water, a process that involves blending the biomass in water to break up the cell or plant structures. Various enzymes are then added to the mixture to break the carbohydrate molecules down to form glucose. Enzymes catalyse the fermentation of glucose to produce an ethanol solution.

FIGURE 2.4.1 Enzymes are used to break large carbohydrates down to glucose. The complexity of this process depends upon the raw material used.

Figure 2.4.2 provides a more detailed flow chart of the fermentation process. The stronger the bonding in the carbohydrate feedstock, the greater the pre-treatment required to ensure the enzymes are effective. Cellulose, in particular, is abundant as forest waste but difficult to process as the hydrogen bonds holding the cellulose molecules together are relatively strong. In contrast, sugar cane contains smaller and highly soluble sucrose molecules that require no pre-treatment.

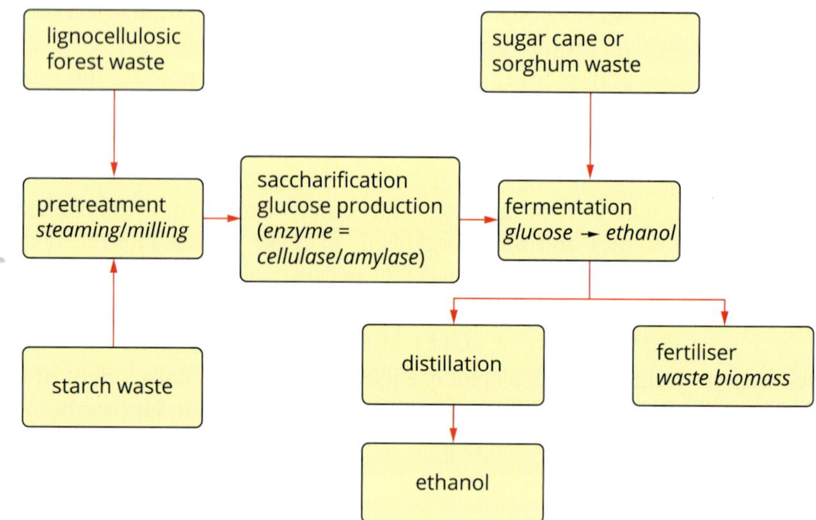

FIGURE 2.4.2 Stages of bioethanol production. The feedstock used varies with available carbohydrate resources.

DISTILLATION

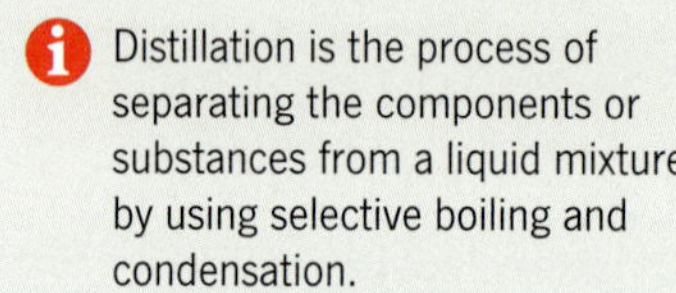
Distillation is the process of separating the components or substances from a liquid mixture by using selective boiling and condensation.

Fermentation produces a solution that is about 10 (%v/v) ethanol. The ethanol needs to be separated from the water for ethanol to be useful as a fuel. **Distillation**, a process that utilises the different boiling points of the liquids, is used to separate the two liquids. The solution is heated to boiling, then fed into tall distillation columns, such as the ones shown in Figure 2.4.3. The temperature of the column is carefully controlled so that water (boiling point is 100°C) falls to the bottom of the column while the ethanol gas (boiling point is 79°C) is collected from the top of the column (Figure 2.4.4 on the following page). The ethanol, at this point, still contains traces of water, which are then removed using micro-filtration and dehydrating agents. You will learn more about distillation in Section 13.1.

A small percentage of methanol is usually added to ethanol sold in Australia to make it unfit for human consumption, hence its common name 'methylated spirits'.

The distillation process is relatively straightforward, but it requires a significant quantity of energy to boil the ethanol solution. The energy required means bioethanol production is not a carbon-neutral process.

FIGURE 2.4.3 Tall distillation columns such as these ones at Manildra in NSW are common to all bioethanol plants.

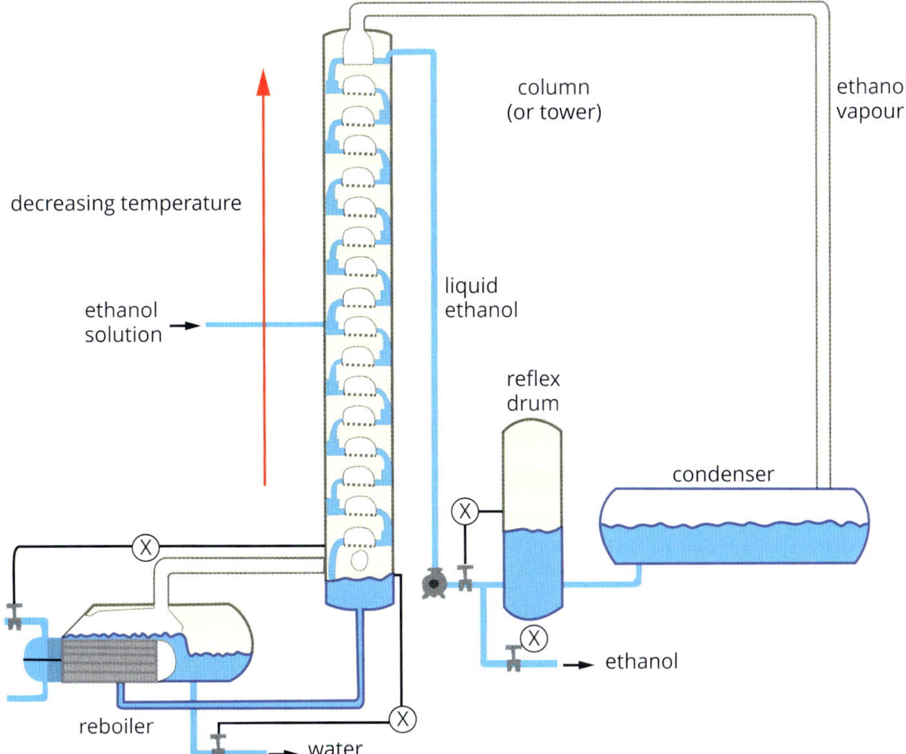

FIGURE 2.4.4 Ethanol is collected from the top of the distillation column and water from the bottom.

THE AUSTRALIAN BIOETHANOL INDUSTRY

Australia has three commercial bioethanol manufacturing plants. Table 2.4.1 summarises the feedstock and capacity of each operation. If more biomass was readily available, this volume could easily be increased as the resultant bioethanol could be blended with petrol.

TABLE 2.4.1 Australia's commercial bioethanol plants

Plant	Location	Feedstock	Annual production (ML)
Manildra	Nowra NSW	wheat waste	300
Wilmar	Sarina QLD	molasses from sugar cane	60
United Petroleum	Dalby QLD	sorghum syrup	46

All three plants are located in a particular region where the feedstock crop is grown, so the waste materials can be easily collected for fermentation. Figure 2.4.5a shows the Wilmar bioethanol distillary in Sarina, Queensland, and Figure 2.4.5b shows its raw material, a sugarcane crop ready for harvest. Sorghum is grown to provide feed for cattle, pigs and poultry. Its high starch content makes it ideal for bioethanol production.

FIGURE 2.4.5 Waste materials from sugarcane crops are used in Wilmar's bioethanol plant at Sarina, QLD.

CHEMFILE

Solving the ethanol crisis during COVID-19

As COVID-19 reached Australia in 2020, many new sources of ethanol emerged.

The arrival of COVID-19 in 2020 was accompanied by universal advice for people to wash and sanitise their hands frequently. The major component of hand sanitiser is alcohol, either ethanol or propan-2-ol. This advice quickly led to a world shortage of ethanol. The solution to this problem came from an unlikely source, Australia's gin-making companies. The ethanol in gin is the same as that used in hand sanitiser, but hand sanitiser does not require ageing or oak storage! Australian distilleries such as Archie Rose Distilling Co. quickly adapted production to launch a whole new range of sanitising products. In a concerning twist to the story, the Apollo Bay Distillery in Victoria had to issue a recall of nine bottles of gin in 2020 when the bottles were found to incorrectly contain hand sanitiser.

FIGURE 2.4.6 Bioethanol can be produced from corn, but is the corn more valuable as a food?

BIOETHANOL POTENTIAL

While sugar cane is an obvious source of sugar to use for fermentation, it is also needed for table sugar production, so there are limits to the amounts of bioethanol that can be produced in this way. The same dilemma extends to other foods such as corn. Brazil is one of the world's leading producers of corn, but corn is also an important food staple. Figure 2.4.6 suggests that a balance needs to be maintained between ensuring that basic nutrition needs are met and that vehicles are powered in an environmentally responsible manner.

Chemists are trialling less valuable sources of sugar and starch for bioethanol production. Forest waste is an obvious target. Australia has a large timber industry and there is little use for the bark and trimmings produced. The cellulose in forest waste is a polymer of glucose, but it takes harsh pre-treatment with steam and specialised enzymes to break the cellulose into glucose.

Comparison of E10 and petrol

Bioethanol is used extensively in Australia, often in a blend with petrol. Australian government regulations limit the proportion of ethanol in petrol to 10%. This petrol blend is labelled E10 and sold at most Australian service stations. Motoring bodies provide advice on the suitability of E10 fuel for each model of car. Regulations in Queensland are different to other states as they have attempted to support local sugarcane growers by mandating that the volume of ethanol sold in fuel be at least 4% of the total petrol volume sold. The presence of ethanol reduces the emissions of particulates and gases such as oxides of nitrogen, but high levels of ethanol can damage engines, especially in older vehicles. An overall increase in the volume of bioethanol blended into petrol should produce less overall emissions and less demand on scarce fossil fuel reserves.

The equation for the combustion of ethanol is:

$$C_2H_5OH(l) + 3O_2(g) \rightarrow 2CO_2(g) + 3H_2O(l) \qquad \Delta H = -1360 \text{ kJ}$$

The combustion of 1 mole of ethanol releases 1360 kJ of energy, equivalent to 29.6 kJ g^{-1}. The energy content of ethanol is about 62% that of petrol, so a larger mass (and volume) of ethanol is required to provide the same quantity of energy. At a simple level, the lower energy content of ethanol can be regarded as the result of the carbon atoms in an ethanol molecule being partly oxidised ('partly burnt'). Some oxygen is already present in the ethanol molecules. E10 fuel and petrol are compared in Table 2.4.2.

TABLE 2.4.2 Comparison of E10 and petrol

E10	Petrol
lower energy density, 46 kJ g^{-1} cheaper cleaner combustion can be a renewable fuel absorbs CO_2 during formation to negate the impact of CO_2 emissions during combustion—lower net CO_2 emissions generally safer production	higher energy density, 48 kJ g^{-1} more widely distributed less CO_2 produced per km when driving

CHEMFILE

Ethtec: Ethanol plants of the future

Ethtec (Ethanol Technologies) has established a bioethanol pilot plant near Mussellbrook in the Hunter Valley of New South Wales. This pilot is a 'next generation' plant because it is not focused on taking one source of waste and converting it into one product. Instead, it is looking at taking any available biomass in the district and producing a range of useful products.

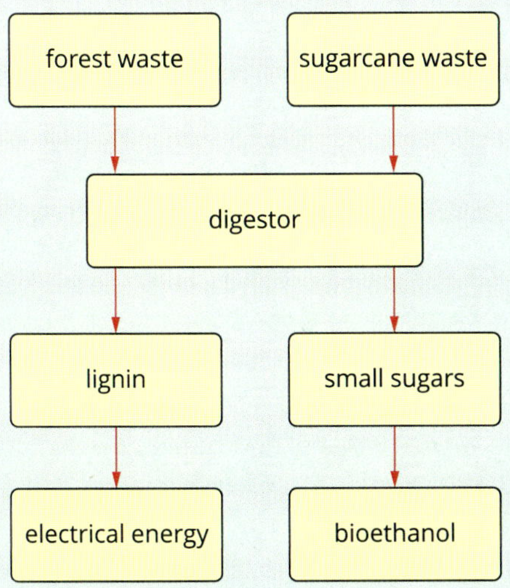

FIGURE 2.4.7 A flow chart showing how multiple sources of biomass can be used to produce multiple useful products

Mussellbrook was chosen as both sugar cane and timber are grown in the region. The waste from both industries is collected for the pilot plant. The forest waste is referred to as lignocellulosic waste as it contains high proportions of both cellulose and lignin. Lignin is a natural polymer present in the cells of all plants. The forest waste is pulped and heated with acid to break down (hydrolyse) cellulose into smaller sugars and to remove the lignin. The process being trialled is outlined in Figure 2.4.7 as a flow chart, and Figure 2.4.8 shows the equipment in the plant itself.

FIGURE 2.4.8 Ethtec trial plant, showing biomass fermentation tanks on the left

2.4 Review

OA ✓✓

SUMMARY

- Bioethanol can be produced by fermentation of carbohydrates and sugars.
  The fermentation equation is:
  $$C_6H_{12}O_6(aq) \rightarrow 2CH_3CH_2OH(aq) + 2CO_2(g)$$
- Bioethanol solutions are distilled to separate the ethanol from the water. Ethanol has a lower boiling point than water.
- Australia has commercial plants producing bioethanol. The carbohydrate sources are biomass from sugar cane, sorghum and wheat.
- Bioethanol is sold as E10, which is composed of petrol blended with up to 10% ethanol.
- The energy density of ethanol is lower than that of petrol.

KEY QUESTIONS

Knowledge and understanding

1. a Draw a molecule of ethanol.
   b Explain why ethanol is soluble in water.
   c Write a balanced equation for the complete combustion of ethanol.
2. Classify each of the following as an advantage or a disadvantage of the use of bioethanol compared to petrol as an energy source.
   a Can absorb small amounts of water
   b Lower energy content (kJ g^{-1})
   c Can be produced from potato peel waste at a commercial chip manufacturing plant.
   d Renewable resource
   e Greater amount of CO_2 emitted to travel a set distance
   f Lubricates engine moving parts
3. State two reasons why the production of bioethanol from the molasses in sugar cane is a simpler process than from forest waste.
4. Fermentation is typically conducted at temperatures around 35°C. It is a relatively slow reaction. Explain why high temperatures are not used to increase the reaction rate.

Analysis

5. Bioenergy supporters would like to see the volume of E10, in which the ethanol is bioethanol, increase in use significantly in Australia. Describe the impact on CO_2 levels of an increase in E10 use.
6. Explain why bioethanol and biodiesel are more likely to be transported from where they are made than biogas.
7. Explain why distillation is required in the manufacture of bioethanol, but not biodiesel or biogas.

2.5 Energy from the combustion of fuels

Bushfires are an example of uncontrolled combustion. They can destroy homes and lives, devastating huge areas of bush and damaging the habitat of many animals. As the impact of climate change leads to increased weather extremes in Australia, the threat posed by bushfires has also increased, and fire-control experts are turning to Indigenous Australian experts for assistance with fire prevention.

Figure 2.5.1a shows a ranger from the Kija Indigenous Ranger team, managing a controlled burn in the Kimberley region in northern Australia. Indigenous rangers use traditional knowledge and techniques, together with modern science and technology, to fight fire with fire to reduce the likelihood of wildfires. The Kimberley Land Council's Indigenous Fire Management Program is extremely important to the biodiversity of this remote region.

Similarly, in 2021, Indigenous Australian leaders and University of Melbourne researchers collaborated on a submission to the Inquiry into the 2019–20 Victorian Fire Season recommending a return to cultural burning practices to limit the likelihood of further bushfires (Figure 2.5.1b).

Fire results from the combustion of substances. Combustion reactions need three things:

- fuel to burn
- oxygen for the fuel to burn in
- energy to get the process started.

Fire can be understood and controlled by applying your knowledge of chemistry. In this section, you will learn about combustion reactions and their various representations in more detail.

FIGURE 2.5.1 (a) A ranger from the Kija Indigenous Ranger team, overseeing a controlled burn. (b) In the summer of 2019–20, large areas of Eastern Victoria were devastated by bushfires.

Combustion reactions are exothermic reactions in which the reactant combines with oxygen to produce oxides. This type of reaction is often referred to as an **oxidation** reaction. The combustion of a hydrocarbon produces carbon dioxide and water, provided there is enough oxygen present.

In redox reactions that involve oxygen as a reactant, oxidation can be defined as the addition of oxygen to form oxides, such as in a combustion reaction.

COMPLETE AND INCOMPLETE COMBUSTION

Combustion reactions can be described as complete or incomplete. The difference between the two is due to the amount of oxygen available to react with the fuel.

Complete combustion occurs when oxygen is plentiful. The only products are carbon dioxide and water.

An example is the complete combustion of methane:

$$CH_4(g) + 2O_2(g) \longrightarrow CO_2(g) + 2H_2O(l)$$

When oxygen is not plentiful, incomplete combustion of fuels occurs. The products of incomplete combustion are carbon monoxide and/or carbon and water.

CHEMFILE

Carbon monoxide poisoning

Carbon monoxide is a highly poisonous gas. It combines readily with haemoglobin, the oxygen carrier in blood. When attached to carbon monoxide, haemoglobin cannot transport oxygen around the body, which leads to oxygen starvation of tissues.

Even at concentrations as low as 10 parts per million (ppm), carbon monoxide can cause drowsiness, dizziness and headaches. At about 200 ppm, carbon monoxide can lead to death. The average carbon monoxide concentration in large cities, mostly due to incomplete combustion of fuels in cars, is now 7 ppm, but it can be as high as 120 ppm at busy intersections in heavy traffic.

Car exhaust gases can contain high levels of carbon monoxide as a result of incomplete combustion of fuels.

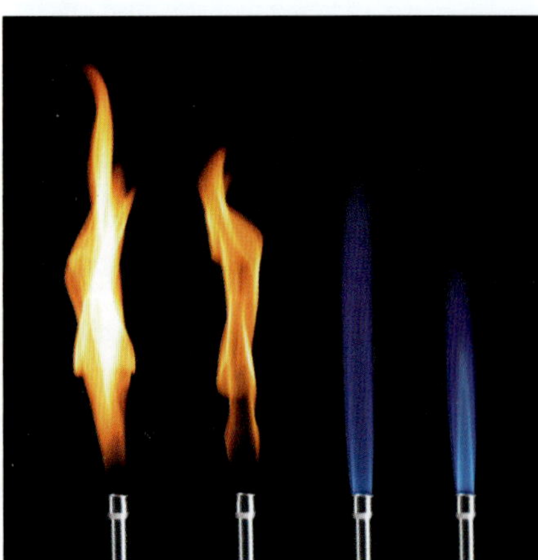

FIGURE 2.5.2 The yellow flame of a Bunsen burner is due to incomplete combustion and produces carbon as a product. The blue flame is a hotter flame that occurs when the air hole is open and more oxygen is allowed to react, resulting in compete combustion.

When the oxygen supply is limited, **incomplete combustion** occurs. As less oxygen is available, not all of the carbon can be converted into carbon dioxide. Carbon monoxide and/or carbon are produced instead. The hydrocarbon burns with a yellow, smoky or sooty flame due to the presence of glowing carbon particles. Figure 2.5.2 shows the appearance of the different flames of a Bunsen burner due to incomplete and complete combustion.

The equation for the incomplete combustion of methane to form carbon monoxide is:

$$2CH_4(g) + 3O_2(g) \longrightarrow 2CO(g) + 4H_2O(l)$$

Writing equations for complete combustion of fuels

It is important to write chemical equations correctly because they tell you a lot about chemical reactions. Writing equations for the complete combustion reactions of fuels containing carbon and hydrogen is relatively straightforward, because the products are always carbon dioxide and water.

Perhaps the most important of all combustion reactions involving fuels are those that occur when petrol is burnt. Petrol is a mixture of hydrocarbons, including octane.

The combustion reactions of octane (C_8H_{18}) and the other hydrocarbons in petrol power the internal combustion engines in most of Australia's motor vehicles.

Worked example 2.5.1

WRITING EQUATIONS FOR COMPLETE COMBUSTION OF HYDROCARBON FUELS

Write the equation, including state symbols, for the complete combustion of butane (C_4H_{10}).

Thinking	Working
Add oxygen as a reactant and carbon dioxide and water as the products.	$C_4H_{10} + O_2 \longrightarrow CO_2 + H_2O$
Balance carbon and hydrogen atoms, based on the formula of the hydrocarbon.	$C_4H_{10} + O_2 \longrightarrow 4CO_2 + 5H_2O$
Find the total number of oxygen atoms on the product side.	Total O $= (4 \times 2) + 5$ $= 13$
If this is an odd number, multiply all of the coefficients in the equation by 2, except for the coefficient of oxygen.	$2C_4H_{10} + O_2 \longrightarrow 8CO_2 + 10H_2O$
Balance oxygen by adding the appropriate coefficient to the reactant side of the equation.	$2C_4H_{10} + 13O_2 \longrightarrow 8CO_2 + 10H_2O$
Add state symbols.	$2C_4H_{10}(g) + 13O_2(g) \longrightarrow 8CO_2(g) + 10H_2O(l)$

Worked example: Try yourself 2.5.1

WRITING EQUATIONS FOR COMPLETE COMBUSTION OF HYDROCARBON FUELS

Write the equation, including state symbols, for the complete combustion of hexane (C_6H_{14}).

A similar series of steps can also be used to write the combustion equations for other carbon-based fuels that contain oxygen, for example, alcohols.

Worked example 2.5.2

WRITING EQUATIONS FOR COMBUSTION REACTIONS OF ALCOHOLS

Write the equation, including state symbols, for the complete combustion of liquid ethanol (C_2H_5OH).

Thinking	Working
Add oxygen as a reactant and carbon dioxide and water as the products.	$C_2H_5OH + O_2 \rightarrow CO_2 + H_2O$
Balance carbon and hydrogen atoms, based on the formula of the alcohol.	$C_2H_5OH + O_2 \rightarrow 2CO_2 + 3H_2O$
Find the total number of oxygen atoms on the product side. Then subtract the one oxygen atom in the alcohol molecule from the total number of oxygen atoms on the product side.	Total O on product side $= (2 \times 2) + 3$ $= 7$ Total O on product side – 1 in alcohol $= 7 - 1 = 6$
If this is an odd number, multiply all the coefficients in the equation by 2, except for the coefficient of oxygen.	$C_2H_5OH + O_2 \rightarrow 2CO_2 + 3H_2O$
Balance oxygen by adding the appropriate coefficient to the reactant side of the equation.	$C_2H_5OH + 3O_2 \rightarrow 2CO_2 + 3H_2O$
Add state symbols.	$C_2H_5OH(l) + 3O_2(g) \rightarrow 2CO_2(g) + 3H_2O(l)$

Worked example: Try yourself 2.5.2

WRITING EQUATIONS FOR COMBUSTION REACTIONS OF ALCOHOLS

Write the equation, including state symbols, for the complete combustion of liquid methanol (CH_3OH).

Writing equations for incomplete combustion of fuels

When the supply of oxygen is insufficient, incomplete combustion of fuels occurs. Equations can also be written to represent this. In general, for the incomplete combustion of hydrocarbons, as well as carbon-based fuels that contain oxygen, the products are carbon monoxide and/or carbon and water.

Worked example 2.5.3

WRITING EQUATIONS FOR INCOMPLETE COMBUSTION OF FUELS

Write an equation, including state symbols, for the incomplete combustion of ethane gas (C_2H_6). The only carbon product is carbon monoxide.

Thinking	Working
Add oxygen as a reactant and carbon monoxide and water as the products.	$C_2H_6 + O_2 \rightarrow CO + H_2O$
Balance the carbon and hydrogen atoms, based on the formula of the hydrocarbon.	$C_2H_6 + O_2 \rightarrow 2CO + 3H_2O$
Balance oxygen by adding the appropriate coefficient to the reactant side of the equation.	$C_2H_6 + \frac{5}{2}O_2 \rightarrow 2CO + 3H_2O$
If oxygen gas has a coefficient that is half of a whole number, multiply all of the coefficients in the equation by 2.	$2C_2H_6 + 5O_2 \rightarrow 4CO + 6H_2O$
Add state symbols.	$2C_2H_6(g) + 5O_2(g) \rightarrow 4CO(g) + 6H_2O(l)$

Worked example: Try yourself 2.5.3

WRITING EQUATIONS FOR INCOMPLETE COMBUSTION OF FUELS

Write an equation, including state symbols, for the incomplete combustion of liquid methanol (CH_3OH). The only carbon product is carbon monoxide.

The heat of combustion is usually measured at conditions of 298 K and 100 kPa. This means the water produced should be shown in the liquid state.

Only fuels that exist as pure substances can have their heat of combustion measured in kJ mol^{-1}.

HEAT OF COMBUSTION

The **heat of combustion** of a fuel can be defined as the heat energy released when a specified amount (e.g. 1 g, 1 L, 1 mol) of a substance burns completely in oxygen and is therefore a positive value. It is usually measured under the standard conditions of 298 K and 100 kPa, which means that the water produced should be shown in the liquid state. In comparison, the **enthalpy of combustion**, ΔH_c, which is found in a thermochemical equation, would have a negative value, indicating the exothermic nature of the combustion reaction.

Many fuels, including wood, coal and kerosene, are mixtures of chemicals and do not have a specific chemical formula or molar mass. This means their heat of combustion cannot be expressed in kJ mol^{-1}. Therefore, the heat of combustion of these fuels are measured only as kJ g^{-1} or kJ L^{-1}.

The heats of combustion for some common elements and compounds present in fuels are listed in Table 2.5.1. Notice that the heats of combustion are written as positive values, showing the quantity of energy released.

TABLE 2.5.1 Heats of combustion for some common elements and compounds

Substance	Heat of combustion (kJ mol^{-1})
methane	890
ethane	1560
propane	2220
butane	2880
octane	5460
methanol	725
ethanol	1360
hydrogen	282
carbon (graphite)	394

Another term associated with the energy content of fuels is molar enthalpy. **Molar enthalpy** is the enthalpy of a substance given per mole and is essentially the same as enthalpy of combustion for a fuel.

As Worked example 2.5.4 shows, you can use heat of combustion data to calculate the energy released on combustion of a specified mass of one of the fuels. The energy released when n mol of a fuel burns is given by the equation:

$$\text{energy} = n \times \Delta H_c$$

Worked example 2.5.4

CALCULATING ENERGY RELEASED BY A SPECIFIED MASS OF A PURE FUEL

Calculate the quantity of energy released when 3.60 kg of pentane (C_5H_{12}) is burnt in an unlimited supply of oxygen. The molar heat of combustion of pentane is 3510 kJ mol^{-1}.

Thinking	Working
Calculate the number of moles of the compound using: $n = \frac{m(\text{in grams})}{M}$	$n(C_5H_{12}) = \frac{m}{M} = \frac{3.60 \times 10^3}{72} = 50.0$ mol
Multiply the number of moles by the heat of combustion per mole.	Energy $= n \times \Delta H_c = 50.0 \times 3510 = 1.76 \times 10^5$ kJ

Worked example: Try yourself 2.5.4

CALCULATING ENERGY RELEASED BY A SPECIFIED MASS OF A PURE FUEL

Calculate the quantity of energy released when 5.40 kg of propanol (C_3H_8O) is burnt in an unlimited supply of oxygen. The molar heat of combustion of propanol is 2020 kJ mol^{-1}.

Energy content

The energy content of a fuel is often expressed in units of kilojoules per gram.

For a pure substance, the heat of combustion per gram can be calculated by simply dividing the heat of combustion per mole (kJ mol^{-1}) by the molar mass of the substance.

For example, for ethanol:

Heat of combustion per mole = 1360 kJ mol^{-1}

Molar mass = 46.0 g mol^{-1}

$$\text{Heat of combustion per gram} = \frac{1360}{46.0} = 29.6 \text{ kJ g}^{-1}$$

For fuels that are mixtures, approximate values for the heat of combustion per gram are shown in Table 2.5.2.

TABLE 2.5.2 Approximate heat of combustion values for some fuel mixtures

Substance	State	Heat of combustion (kJ g^{-1}) (approx.)
kerosene	liquid	46.2
wood	solid	18.0
diesel	liquid	45.0
black coal	solid	35.0
natural gas	gas	54.0

Energy content per gram can be a useful comparison when determining the suitability of transport fuels.

Petrol has a higher energy per gram than bioethanol. This means more E10 fuel is required to produce the same amount of energy as petrol.

THERMOCHEMICAL EQUATIONS

As you learnt in Section 2.1, thermochemical equations can be used to compare energy changes. Thermochemical equations include a sign and numerical value for the energy change that occurs in the reaction. The thermochemical equation for the complete combustion of propane is:

$$C_3H_8(g) + 5O_2(g) \rightarrow 3CO_2(g) + 4H_2O(l) \qquad \Delta H = -2220 \text{ kJ}$$

The above equation tells us:

- the complete combustion of 1 mole of propane gas and 5 moles of oxygen gas to carbon dioxide and water releases 2220 kJ of energy
- the molar enthalpy of combustion for propane is –2220 kJ mol^{-1}
- the reaction is exothermic
- propane is likely to be a useful fuel as the magnitude of the energy change is relatively high
- the use of propane as a fuel will be associated with greenhouse gas emission issues, since CO_2 is a product.

Thermochemical equations and mole ratios

The ΔH value in a thermochemical equation corresponds to the mole amounts specified by the equation. If the coefficients in the equation are changed, the ΔH value will also change.

For example, the thermochemical equation for the combustion of methanol can be written as:

$$CH_3OH(l) + \frac{3}{2}O_2(g) \rightarrow CO_2(g) + 2H_2O(l) \qquad \Delta H = -726 \text{ kJ}$$

This means that 726 kJ of energy is released when 1 mole of methanol reacts with 1.5 moles of oxygen gas, to produce 1 mole of carbon dioxide and 2 moles of water.

If twice as much methanol were to react, then twice as much energy would be released. So, if the coefficients of the equation are doubled, the ΔH value is also doubled:

$$2CH_3OH(l) + 3O_2(g) \rightarrow 2CO_2(g) + 4H_2O(l) \qquad \Delta H = -1452 \text{ kJ}$$

If the mole amounts are tripled, the ΔH value is also tripled:

$$3CH_3OH(l) + 4\frac{1}{2}O_2(g) \rightarrow 3CO_2(g) + 6H_2O(l) \qquad \Delta H = -2178 \text{ kJ}$$

Effect on ΔH of reversing a chemical reaction

Reversing a chemical equation changes the sign but not the magnitude of ΔH.

For example, methane reacts with oxygen gas to produce carbon dioxide gas and water in an exothermic reaction:

$$CH_4(g) + 2O_2(g) \rightarrow CO_2(g) + 2H_2O(l) \qquad \Delta H = -890 \text{ kJ}$$

If this reaction is reversed, the magnitude of ΔH remains the same because the enthalpies of the individual chemicals have not changed, but the sign changes to indicate that energy must be absorbed for this reaction to proceed.

$$CO_2(g) + 2H_2O(l) \rightarrow CH_4(g) + 2O_2(g) \qquad \Delta H = +890 \text{ kJ}$$

Worked example 2.5.5

CALCULATING ΔH FOR ONE EQUATION USING ANOTHER EQUATION

Iron reacts with oxygen according to the equation:

$$3Fe(s) + 2O_2(g) \rightarrow Fe_3O_4(s) \qquad \Delta H = -1121 \text{ kJ}$$

Calculate ΔH for the reaction represented by the equation:

$$2Fe_3O_4(s) \rightarrow 6Fe(s) + 4O_2(g)$$

Thinking	Working
The reaction has been reversed in the second equation, so the sign for ΔH is changed to the opposite sign.	ΔH for the second equation is positive.
Identify how the coefficients in the equation have changed.	The coefficient for Fe_3O_4 has changed from 1 to 2, O_2 has changed from 2 to 4, and Fe has changed from 3 to 6. They have all doubled.
Identify how the magnitude of ΔH will have changed for the second equation.	The coefficients in the equations have all doubled, so ΔH will also have doubled.
Calculate the new magnitude of ΔH. (You will write the sign of ΔH in the next step.)	2×1121 $= 2242$
Write ΔH for the second equation, including the sign.	$\Delta H = +2242$ kJ

Worked example: Try yourself 2.5.5

CALCULATING ΔH FOR ONE EQUATION USING ANOTHER EQUATION

Carbon reacts with hydrogen according to the equation:

$$6C(s) + 3H_2(g) \rightarrow C_6H_6(g) \qquad \Delta H = +49 \text{ kJ}$$

Calculate ΔH for the reaction represented by the equation:

$$3C_6H_6(g) \rightarrow 18C(s) + 9H_2(g)$$

THE IMPORTANCE OF STATES

Enthalpy changes also occur during physical changes, so thermochemical equations can be written for physical changes. Boiling water is an example of a physical change. It is an endothermic process, because heat must be applied to liquid water to convert it into steam. The thermochemical equation for this reaction is:

$$H_2O(l) \rightarrow H_2O(g) \qquad \Delta H = +40.7 \text{ kJ}$$

The ΔH value is positive because this is an endothermic reaction.

The thermochemical equation for the combustion of propane was shown above. You can see in the two equations below that the enthalpy value quoted depends upon whether the water produced is shown as a gas or a liquid.

$$C_3H_8(g) + 5O_2(g) \rightarrow 3CO_2(g) + 4H_2O(g) \qquad \Delta H = -2057 \text{ kJ}$$
$$C_3H_8(g) + 5O_2(g) \rightarrow 3CO_2(g) + 4H_2O(l) \qquad \Delta H = -2220 \text{ kJ}$$

By convention, heats of combustion are calculated at standard laboratory conditions (298 K and 100 kPa) with combustion products being $CO_2(g)$ and $H_2O(l)$.

CASE STUDY ANALYSIS

Explosives—a blast of chemical energy

Humans have been using chemicals to make explosions since 919 BCE. In China, people first mixed saltpetre (potassium nitrate), sulfur and charcoal with explosive results. They quickly realised that there were many uses for this mixture, which later became known as gunpowder. It was put to military use and eventually led to the development of bombs, cannons and guns.

Today, explosives are an essential tool for mining and other engineering works, such as road construction, tunnelling, building and demolition (Figure 2.5.3).

Explosives transform chemical energy into large quantities of thermal energy very quickly. Although thermal energy is also released when fuels such as petrol and natural gas burn, the rate of combustion in these reactions is limited by the availability of oxygen gas to the fuel. In contrast, the compounds making up an explosive contain sufficient oxygen for a complete (or almost complete) reaction to occur very quickly.

When chemical explosives, such as ammonium nitrate, trinitrotoluene (TNT) and nitroglycerine decompose, they release large quantities of energy and gaseous products very quickly.

This is the thermochemical equation for the decomposition of nitroglycerine:

$$4C_3H_5N_3O_9(l) \rightarrow 12CO_2(g) + 10H_2O(g) + 6N_2(g) + O_2(g)$$
$$\Delta H = -1456 \text{ kJ}$$

The equation above highlights a common feature of most explosives—a solid or liquid reactant being converted to gaseous products. The gaseous products form very quickly, requiring a much greater space between the particles and subsequently 'exploding' the fuel and surroundings.

FIGURE 2.5.3 An old bridge is demolished with the help of explosives.

At atmospheric pressure, the products of this reaction would expand to fill a volume more than 10000 times larger than the volume of the reactants! During a blast, this gas is usually produced within a small hole into which the liquid explosive has been placed, creating huge pressures that shatter the surrounding rock or structure.

Analysis

1. Is this reaction exothermic or endothermic? Justify your answer.
2. What is the ratio of moles of products to moles of reactants? How does this make the reaction suitable for a chemical explosion?
3. How much energy would be produced if 8 moles of nitroglycerine were reacted?
4. Explain why this highly explosive reaction can only be started using some type of trigger.

2.5 Review

WS 3 OA ✓✓

SUMMARY

- The products of the complete combustion of hydrocarbons and carbon-based fuels containing oxygen are carbon dioxide and water.
- Incomplete combustion occurs when hydrocarbons and carbon-based fuels containing oxygen undergo combustion in a limited supply of oxygen. The products of incomplete combustion include carbon monoxide and/or carbon and water.
- A number of related terms are used when comparing fuels.

  General terms

  Enthalpy: the chemical energy of a substance

  Enthalpy change: difference in chemical energy of the products compared to the reactants

  Molar enthalpy: the enthalpy of a substance given per mole

  Specific term for combustion reactions

  Heat of combustion (of a fuel): the heat energy released when a specified amount (e.g. 1 g, 1 L, 1 mol) of a substance burns completely in oxygen

  Enthalpy of combustion, ΔH_c: a negative value which has the same numerical value as the heat of combustion
- Heats of combustion indicate the maximum quantity of energy that can be released when a specified amount of fuel undergoes complete combustion. Common units are kJ mol^{-1} and kJ g^{-1}. Because these values are a measure of the quantity of energy released, they have a positive value when tabulated.
- The quantity of energy released by different fuels can be compared using their heats of combustion.
- For a pure fuel with a heat of combustion measured in kJ mol^{-1}, the energy released when n mol of the fuel burns can be calculated by the equation:

  $$\text{Energy} = n \times \text{heat of combustion}$$
- Combustion reactions are exothermic reactions that can be represented by balanced thermochemical equations. Combustion reactions have a negative ΔH value.
- Reversing an equation causes the sign of ΔH to change, as the reaction changes from exothermic to endothermic, or vice versa.
- Doubling the coefficients in a chemical reaction causes the ΔH value to also double, as twice as many reactants react to produce or absorb twice as much energy.
- States of matter must be included in thermochemical equations because changes of state involve enthalpy changes.
- Heat of combustion data tables can be used to determine the enthalpy change, ΔH, in a thermochemical equation.

KEY QUESTIONS

Knowledge and understanding

1 Write a balanced chemical equation for the complete combustion of liquid benzene (C_6H_6).

2 Write a balanced equation for the incomplete combustion of ethanol (C_2H_5OH) when carbon monoxide is formed.

3 When pentane, C_5H_{12}, combusts, it reacts with oxygen to form carbon dioxide and water. When 1 mole of pentane combusts, 3509 kJ of energy is released. Write a balanced thermochemical equation for this reaction.

4 A data book lists the heat of combustion of propan-1-ol as 2021 kJ mol^{-1}. Determine the heat of combustion of propan-1-ol in kJ g^{-1}.

5 Use the information in Table 2.5.1 to calculate the quantity of energy released when the following quantities of fuel undergo complete combustion.

a 250 g of methane

b 9.64 kg of propane

c 403 kg of ethanol

6 Define the term 'heat of combustion'.

continued over page

2.5 Review *continued*

Analysis

7 **a** Which will release the most energy — complete or incomplete combustion of propane?

b Explain your answer to part **a**.

8 The thermochemical equation for photosynthesis can be represented as:

$$6CO_2(g) + 6H_2O(l) \rightarrow C_6H_{12}O_6(aq) + 6O_2(g)$$

$$\Delta H = +2803 \text{ kJ}$$

a Determine whether this reaction is endothermic or exothermic. Give a reason for your answer.

b Consider a new reaction that reversed and halved the above reaction. Calculate the energy change of the new reaction that is written below:

$$\frac{1}{2}C_6H_{12}O_6(aq) + 3O_2(g) \rightarrow 3CO_2(g) + 3H_2O(l)$$

9 The combustion of octane to form carbon dioxide and liquid water can be written as:

$$C_8H_{18}(g) + 12\frac{1}{2}O_2(g) \rightarrow 8CO_2(g) + 9H_2O(l)$$

$$\Delta H = -5450 \text{ kJ}$$

The combustion of octane to form carbon dioxide and steam can be written as:

$$C_8H_{18}(g) + 12\frac{1}{2}O_2(g) \rightarrow 8CO_2(g) + 9H_2O(g)$$

The thermochemical equation for the reaction in which liquid water turns into steam is:

$$H_2O(l) \rightarrow H_2O(g) \ \Delta H = +40.7 \text{ kJ}$$

How would the energy released by the combustion of 1 mole of octane to form steam compare with the energy released by 1 mole of octane to form liquid water?

Chapter review

KEY TERMS

activation energy
alkane
anaerobic
biodiesel
bioethanol
biofuel
biogas
biomass
carbohydrate
carbon neutral
catalyst
cellular respiration
cellulose
chemical energy
combustion
complete combustion
distillation
endothermic
energy content
energy profile diagram
energy value
enthalpy
enthalpy change
enthalpy of combustion, ΔH_c
enzyme
exothermic
fats
fermentation
fossil fuel
fractional distillation
fuel
heat content
heat of combustion
homologous series
hydrocarbon
incomplete combustion
joule
metabolise
molar enthalpy
natural gas
non-renewable
nutrient
oxidation
petrodiesel
photosynthesis
protein
renewable fuel
renewables
respiration
sustainable
thermochemical equation
transesterification
triglyceride

REVIEW QUESTIONS

Knowledge and understanding

1 Which of the following fuels is used more commonly for transport than electricity generation?

A coal
B biodiesel
C natural gas
D biogas

2 Which one of the following is a renewable form of energy?

A natural gas
B petrol
C gas generated from animal manure
D electrical energy produced in a coal-fired power station where the emissions are stored underground

3 Select the correct statement about E10 fuel.

A E10 does not produce carbon-containing emissions.
B E10 will be less soluble in water than petrol.
C Reserves of E10 are limitless.
D E10 has a lower energy density than petrol.

4 Which one of the following is correct about the energy profile diagrams of both endothermic and exothermic reactions?

A There is always less energy absorbed than released.
B The enthalpy of the products is always less than the energy of the reactants.
C Some energy is always absorbed to break bonds in the reactants.
D The ΔH value is the difference between the enthalpy of the reactants and the highest energy point reached on the energy profile.

5 Label each of the following as exothermic or endothermic.

a ice melting to water
b candle burning
c reaction that causes the surroundings to drop in temperature
d reaction with a negative value of ΔH

6 Explain why fossil fuels are considered to be non-renewable.

7 Calculate the quantity of energy released from complete combustion of:

a 0.740 mol of ethanol
b 12.0 g of ethanol.

8 The combustion of hydrogen is an exothermic reaction:
- **a** Identify the bonds which need to be broken for a reaction to occur.
- **b** Identify the bonds which are formed in the reaction.
- **c** State which quantity is greater for this reaction—the energy required to break bonds or the energy produced from bonds that are formed?
- **d** Explain whether this reaction will increase or decrease the temperature of the surroundings.

9
- **a** Write a balanced overall equation for the photosynthesis reaction.
- **b** Name the important molecule produced in plants by photosynthesis.
- **c** Write a balanced overall equation for the process of aerobic respiration.

10
- **a** Name two common feedstocks that can be used to produce bioethanol.
- **b** Write a balanced equation for the fermentation process.
- **c** What is the role of a distillation column in the production of bioethanol?

11 Write a thermochemical equation for the complete combustion of ethanol.

12 Refer to Table 2.2.1 on page 74, which lists the typical composition of biogas.
- **a** Which two gases form the largest percentage of a typical biogas sample?
- **b** Why might there be variations in the percentage of gases making up different samples of biogas?

13 Identify whether each of the following statements related to activation energy is true or false.
- **a** Activation energy is the energy required to break bonds in the reactants.
- **b** The activation energy of the forward reaction equals that of the reverse reaction.
- **c** The magnitude of the activation energy is always greater than the enthalpy change.
- **d** All fuels have very low activation energies.

14 Explain why reversing a chemical reaction reverses the sign of ΔH.

15 The combustion of butane gas in portable stoves can be represented by the thermochemical equation:

$$2C_4H_{10}(g) + 13O_2(g) \rightarrow 8CO_2(g) + 10H_2O(l)$$
$$\Delta H = -5772 \text{ kJ}$$

- **a** How does the overall energy of the bonds in the reactants compare with the overall energy of the bonds in the products?
- **b** Draw an energy profile diagram for the reaction, labelling ΔH and activation energy.

16
- **a** Write a balanced chemical equation for the complete combustion of propanol (C_3H_7OH).
- **b** Write a balanced chemical equation for the incomplete combustion of pentane (C_5H_{12}) where carbon monoxide is formed.

Application and analysis

17 A sample of biodiesel has been prepared using palmitic acid ($C_{15}H_{31}COOH$) and methanol. Write a balanced equation for the complete combustion of this biodiesel.

18 The combustion reaction of ethyne gas that occurs during welding can be represented by the thermochemical equation:

$$2C_2H_2(g) + 5O_2(g) \rightarrow 4CO_2(g) + 2H_2O(l) \quad \Delta H = -2619 \text{ kJ}$$

- **a** Is this reaction endothermic or exothermic?
- **b** What would the new value of ΔH be if the equation was now written as follows?

$$4C_2H_2(g) + 10O_2(g) \rightarrow 8CO_2(g) + 4H_2O(l)$$

19 According to a data table, the heat of combustion of hydrogen is 286 kJ mol^{-1}. Write a balanced thermochemical equation for the complete combustion of hydrogen.

20 Explain why biodiesel made from canola oil is sometimes described as a 'carbon-neutral' fuel. Use a chemical equation to support your answer.

21 Use the terms 'methane', 'oxygen', 'bacteria' and 'carbon dioxide' to explain the formation and composition of biogas.

22 Trials are being conducted to source biogas from cheese. A prototype plant in England uses the whey left over from cheese production. Bacteria act on this cheese waste in anaerobic conditions to produce biogas. Classify the following as advantages or disadvantages of large-scale production of biogas from cheese waste.
- **a** fewer particulate emissions than natural gas
- **b** renewable
- **c** less reliance on fossil fuels
- **d** can only be used on-site
- **e** fewer net CO_2 emissions than natural gas
- **f** raw material needs to be heated in winter
- **g** lower sulfur content than natural gas
- **h** the cheese waste was previously used as a fertiliser.

23 The energy profile diagram below shows the relative enthalpies, on an arbitrary scale, of the reactants and products of a chemical reaction.

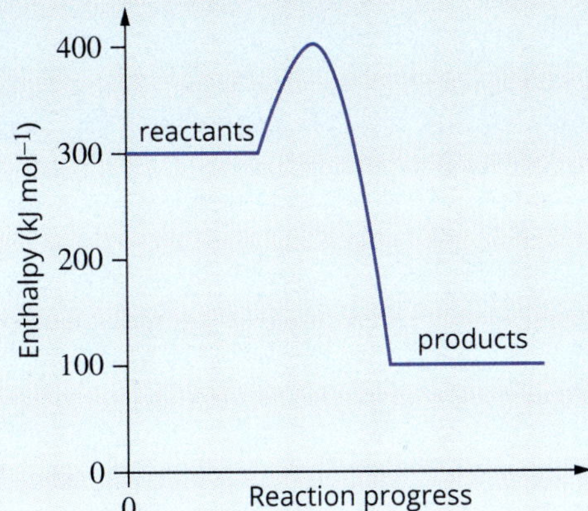

a Is the reaction exothermic or endothermic?

b What is the value of the activation energy of the forward and reverse reactions?

c What is the value of ΔH of the reverse reaction?

24 a Calculate the energy available from 1 kg of each of the following fuels, using the information provided in Table 2.5.1 on page 92.

i octane, C_8H_{18}

ii butane, C_4H_{10}

iii ethane, C_2H_6

b Use the results of your calculations in part **a** to describe the trend in heats of combustion (in kJ/kg) in the alkane homologous series.

25 In a steelworks, carbon monoxide present in the exhaust gases of the blast furnace can be used as a fuel elsewhere in the plant. It reacts according to the equation:

$$CO(g) + \frac{1}{2}O_2(g) \rightarrow CO_2(g) \qquad \Delta H = -283 \text{ kJ}$$

a Which has the greater total enthalpy: 1 mol of CO(g) and 0.5 mol of O_2(g), or 1 mol of CO_2(g)?

b Write the value of ΔH for the following equations:

i $2CO(g) + O_2(g) \rightarrow 2CO_2(g)$

ii $2CO_2(g) \rightarrow 2CO(g) + O_2(g)$

OA ✓✓

CHAPTER 03

Obtaining energy from fuels

Energy released by the combustion of fuels is used for everyday activities like heating, cooking and the generation of electricity. Humans also consume food each day to obtain fuel for movement and all the functions of the body.

While we depend on the energy obtained from fuels, their combustion also releases gases such as carbon dioxide and water vapour, which are both greenhouse gases. An increase of these in the atmosphere has serious environmental implications.

In this chapter, you will learn about the energy content in fuels and food. You will learn how to calculate the amount of useful energy produced when a certain mass of a fuel or food undergoes combustion, and you will also learn how to calculate the mass and volume of products such as the major greenhouse gases.

Key knowledge

- determination of limiting reactants or reagents in chemical reactions **3.2**
- calculations related to the application of stoichiometry to reactions involving the combustion of fuels, including mass–mass, mass–volume and volume–volume stoichiometry, to determine heat energy released, reactant and product amounts and net volume or mass of major greenhouse gases (CO_2, CH_4 and H_2O), limited to standard laboratory conditions (SLC) at 25°C and 100 kPa **3.1**
- the use of specific heat capacity of water to approximate the quantity of heat energy released during the combustion of a known mass of fuel and food **3.3**
- the principles of solution calorimetry, including determination of calibration factor and consideration of the effects of heat loss; analysis of temperature–time graphs obtained from solution calorimetry **3.4**
- energy from fuels and food:
  - calculation of energy transformation efficiency during combustion as a percentage of chemical energy converted to useful energy **3.5**
  - comparison and calculations of energy values of foods containing carbohydrates, proteins and fats and oils. **3.5**

3.1 Stoichiometry involving combustion of fuels

The ability to measure and predict quantities involved in chemical reactions is a very important part of chemistry. One way in which scientists evaluate fuels is through comparison of quantities of chemicals involved in combustion reactions, as well as quantities of energy released. For example, the ethanol produced from corn in the processing plant shown in Figure 3.1.1 can be compared to petroleum through calculations of how much oxygen is required for **complete combustion**, the amount of atmospheric pollution the fuels will produce and the heat energy released per gram of fuel. In this section, you will learn about these types of calculations, all of which rely on an understanding of the mole concept.

FIGURE 3.1.1 A corn ethanol processing plant. Ethanol is a biofuel that is used as an alternative to petroleum. The concepts taught in this section allow you to calculate the mass of oxygen required to combust a given mass of ethanol, the mass of carbon dioxide (a greenhouse gas) produced, as well as the energy released.

STOICHIOMETRY AND QUANTITIES OF CHEMICALS

The calculations involved in determining the quantity of a reactant or product from another reactant or product use **stoichiometry**. Stoichiometry is the study of mole ratios of substances and is based on the law of conservation of mass.

A chemical reaction involves the rearrangement of pre-existing atoms; atoms are neither created nor destroyed. This means the total mass of all products is equal to the total mass of all reactants.

Consider the equation for the reaction that occurs when methane (CH_4) burns in oxygen:

$$CH_4(g) + 2O_2(g) \rightarrow CO_2(g) + 2H_2O(g)$$

The **coefficients** used to balance the equation show the **mole ratio** between the reactants and products involved in the reaction. The equation indicates that 1 mole of $CH_4(g)$ reacts with 2 moles of $O_2(g)$ to form 1 mole of $CO_2(g)$ and 2 moles of $H_2O(g)$. Examples of what this means in more general terms are:

- the amount of oxygen required to react with the methane will be double the amount of methane used
- the amount of carbon dioxide produced will be the same as the amount of methane that reacted.

These mole ratios can be expressed in formulas:

$$\frac{n(O_2)}{n(CH_4)} = \frac{2}{1} \text{ and } \frac{n(CO_2)}{n(CH_4)} = \frac{1}{1}$$

MOLE THEORY

The mole

The mole is a unit used by chemists for counting particles. One mole contains exactly $6.022\,140\,772 \times 10^{23}$ particles. We usually round this off to 6.02×10^{23} particles. This number is called Avogadro's constant and is given the symbol N_A.

The symbol for the amount of a substance measured in moles is n. The unit is mol.

A useful relationship links the amount of a substance (n), in mol, Avogadro's constant (N_A) and the number of particles in a substance (N):

$$n = \frac{N}{N_A}$$

Where needed, this relationship can be rearranged to:

$$N = n \times N_A$$

Mass

Chemists often use mass, measured in grams, to measure an exact amount, in mol, of an element or compound.

The molar mass of an element or compound is the mass of 1 mole of the element or compound. The amount of a substance (n), in mol, is related to the mass of the substance (m), in g, and its molar mass (M) in g mol^{-1}, by the formula.

$$n = \frac{m}{M}$$

This relationship can be rearranged to:

$$m = n \times M \text{ or } M = \frac{m}{n}$$

Volume of gases

Quantities of gases are often measured using volume, measured in litres. The same amount, in mol, of different types of gases will occupy the same volume if pressure and temperature are the same. The **molar volume** of a gas is the volume of 1 mole of the gas. At standard laboratory conditions (SLC) of 100 kPa and 25°C, the molar volume has been determined to be 24.8 L mol^{-1}. A useful relationship links the amount of a gas (n), in mol, its molar volume (V_m) in L mol^{-1}, and the volume occupied by the gas (V) in L:

$$n = \frac{V}{V_m}$$

Where needed, this relationship can be rearranged to:

$$V = n \times V_m$$

24.8 L mol^{-1} can be substituted into either formula for V_m if the gas is at SLC.

In general, for stoichiometric calculations you will be given, or will be able to work out, the number of moles of one chemical in the reaction (called the 'known chemical'). This can then be used with the mole ratio to determine the number of moles of one of the other reactants or products involved in the reaction (called the 'unknown chemical').

The coefficients used to balance chemical equations show the mole ratio between the reactants and products involved in the reaction.

The mole ratio can be written:

$$\frac{n(\text{unknown chemical})}{n(\text{known chemical})} = \frac{\text{coefficient of unknown chemical}}{\text{coefficient of known chemical}}$$

MASS–MASS STOICHIOMETRY

When a reaction is carried out in a laboratory, quantities of chemicals are often measured in grams, not moles. As stoichiometry uses a mole ratio, the amount, in mol, of the known chemical must first be calculated from its mass using the formula:

$$n = \frac{m}{M}$$

After the mole ratio has been used to determine the amount, in mol, of the unknown chemical, the mass of the unknown chemical is calculated by rearranging the formula to:

$$m = n \times M$$

When the unknown and the known in a stoichiometric calculation are both masses, the calculation is known as **mass–mass stoichiometry**.

Calculating the mass of carbon dioxide produced in a combustion reaction

Stoichiometry can be combined with your knowledge of combustion reactions to find the mass of carbon dioxide produced.

There are several steps involved in calculating the mass of carbon dioxide produced based on the mass of fuel that undergoes combustion. The mass of fuel is the known chemical and carbon dioxide is the unknown.

1 Write a balanced chemical equation for the reaction.
2 Calculate the number of moles of the fuel from its mass.
3 Find the mole ratio from the coefficients in the chemical equation to calculate the number of moles of carbon dioxide produced.
4 Calculate the mass of carbon dioxide produced.

Figure 3.1.2 provides a flow chart that summarises this process. Worked example 3.1.1 will help you understand these steps.

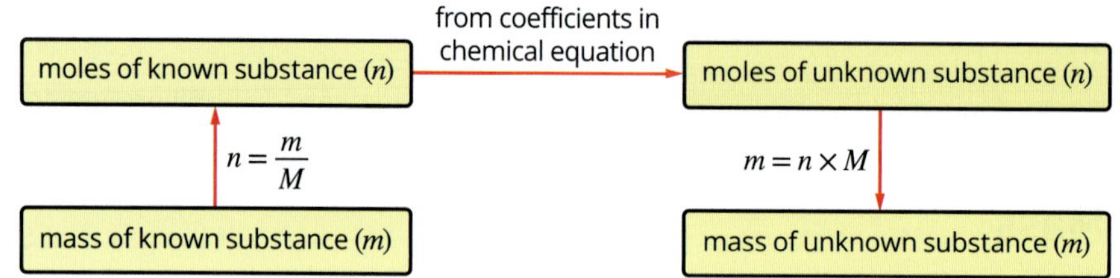

FIGURE 3.1.2 A flow chart for mass–mass stoichiometric calculations is helpful when solving these types of problems.

Worked example 3.1.1

MASS–MASS STOICHIOMETRIC PROBLEMS

Calculate the mass of carbon dioxide, in kg, produced when 540 g of propane (C_3H_8) burns completely in oxygen.

Thinking	Working
Write a balanced equation for the reaction.	$C_3H_8(g) + 5O_2(g) \rightarrow 3CO_2(g) + 4H_2O(g)$
As the quantity of the known substance is given in g, calculate the number of moles of the known substance using: $n = \frac{m}{M}$	$n(C_3H_8) = \frac{540}{44.0}$ $= 12.3$ mol
Find the mole ratio: $\frac{\text{coefficient of unknown}}{\text{coefficient of known}}$	$\frac{n(CO_2)}{n(C_3H_8)} = \frac{3}{1}$
Calculate the number of moles of the unknown substance using: n(unknown) = mole ratio × n(known)	$n(CO_2) = \frac{3}{1} \times 12.3$ $= 36.8$ mol
Calculate the mass of the unknown substance using: $m = n \times M$	$m(CO_2) = 36.8 \times 44.0$ $= 1620$ g $= 1.62$ kg

Worked example: Try yourself 3.1.1

SOLVING MASS–MASS STOICHIOMETRIC PROBLEMS

Calculate the mass of carbon dioxide, in kg, produced when 3.60 kg of butane (C_4H_{10}) burns completely in oxygen.

MASS–VOLUME STOICHIOMETRY

Stoichiometric calculations that follow the same general pattern can also be used to calculate the volume of oxygen required to combust a fuel and the volumes of gases produced by the reactions. When the known in a stoichiometric calculation is a mass and the unknown is a volume of a gas, the calculation is known as **mass–volume stoichiometry**. The number of moles of the known substance (often mass of fuel) is calculated from data that is given to you, the mole ratio from the coefficients in the equation is used to find the number of moles of the unknown chemical, and the desired quantity of the unknown substance is then calculated. This is summarised in Figure 3.1.3.

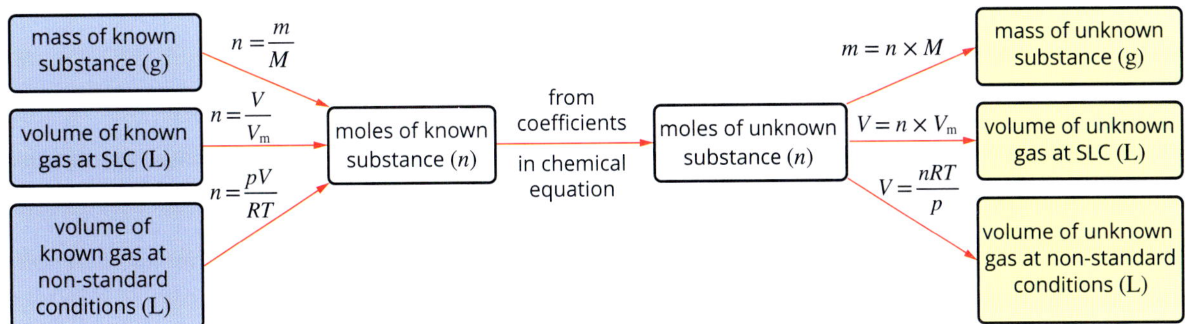

FIGURE 3.1.3 Stoichiometric calculations generally follow the steps shown in the flow chart. Calculating the number of moles and using a mole ratio from a balanced chemical equation are central to all stoichiometric calculations.

Worked example 3.1.2 shows how to calculate the volume of carbon dioxide under **standard laboratory conditions, SLC**, produced from a known mass of fuel. Standard laboratory conditions for a gas indicate that the gas has a pressure of 100 kPa and that the temperature is 298 K (25°C).

Worked example 3.1.2

MASS–VOLUME STOICHIOMETRIC CALCULATIONS AT SLC

Calculate the volume of carbon dioxide, in L, produced when 2.00 kg of propane (C_3H_8) is burned completely in oxygen. The gas volume is measured at SLC.

Thinking	Working
Write a balanced chemical equation for the reaction.	$C_3H_8(g) + 5O_2(g) \longrightarrow 3CO_2(g) + 4H_2O(l)$
As the quantity of the known substance is a mass, calculate the number of moles of the known substance using: $n = \frac{m}{M}$	$n(C_3H_8) = \frac{2000}{44.0}$ $= 45.5$ mol
Find the mole ratio: $\frac{\text{coefficient of unknown}}{\text{coefficient of known}}$	$\frac{n(CO_2)}{n(C_3H_8)} = \frac{3}{1}$
Calculate the number of moles of the unknown substance using: n(unknown) = mole ratio × n(known)	$n(CO_2) = \frac{3}{1} \times 45.5$ $= 136$ mol
Calculate the volume of the unknown substance using: $V = n \times V_m$	$V(CO_2) = 136 \times 24.8$ $= 3382$ L $= 3.38 \times 10^3$ L

Worked example: Try yourself 3.1.2

MASS–VOLUME STOICHIOMETRIC CALCULATIONS AT SLC

Calculate the volume of carbon dioxide, in L, produced when 300 g of butane (C_4H_{10}) is burned completely in oxygen. The gas volume is measured at SLC.

GAS VOLUME–VOLUME CALCULATIONS

For chemical reactions where all reactants and products are gases, it is often more convenient to measure reactant volumes rather than masses. When the unknown and the known in a stoichiometric calculation are both gas volumes, the calculation is known as **volume–volume stoichiometry**.

All gases occupy equal volumes measured at the same temperature and pressure. Therefore, the mole ratios of gases will also be volume ratios if temperature and pressure are kept constant.

For example, the reaction between propane gas and oxygen can be represented by the equation:

$$C_3H_8(g) + 5O_2(g) \rightarrow 3CO_2(g) + 4H_2O(g)$$

This equation tells us that 1 mole of propane reacts with 5 moles of oxygen gas to produce 3 moles of carbon dioxide and 4 moles of water vapour. As every reactant and product is in the gaseous state, the ratio also tells us that 1 litre of propane gas reacts with 5 litres of oxygen gas to produce 3 litres of carbon dioxide and 4 litres of water vapour at constant temperature and pressure.

When all species in a chemical equation are in the gaseous state, and the temperature and pressure are constant, the mole ratio will also be a volume ratio.

Worked example 3.1.3

GAS VOLUME–VOLUME CALCULATIONS

Methane gas (CH_4) is burned in a gas stove according to the following equation:

$$CH_4(g) + 2O_2(g) \rightarrow CO_2(g) + 2H_2O(g)$$

If 50 mL of methane is burned, calculate the volume of O_2 gas required for complete combustion of the methane under constant temperature and pressure conditions.

Thinking	Working
Find the mole ratio: $\frac{\text{coefficient of unknown}}{\text{coefficient of known}}$	$\frac{n(O_2)}{n(CH_4)} = \frac{2}{1}$
The temperature and pressure are constant, so volume ratios are the same as mole ratios.	$\frac{V(O_2)}{V(CH_4)} = \frac{2}{1}$
Calculate the volume of the unknown substance using: $V(\text{unknown}) = \text{mole ratio} \times V(\text{known})$	$V(O_2) = \frac{2}{1} \times 50$ $= 100$ mL

Worked example: Try yourself 3.1.3

GAS VOLUME–VOLUME CALCULATIONS

Methane gas (CH_4) is burned in a gas stove according to the following equation:

$$CH_4(g) + 2O_2(g) \rightarrow CO_2(g) + 2H_2O(g)$$

If 50 mL of methane is burned in air, calculate the volume of CO_2 gas produced under constant temperature and pressure conditions.

STOICHIOMETRY AND QUANTITY OF ENERGY RELEASED

The quantity of energy released when a fuel undergoes combustion is directly proportional to the amount of fuel used. In Figure 3.1.4, the bonfire contains more fuel than the match, so the bonfire releases far more energy. We also know that doubling the quantity of petrol in a car allows a car to travel twice as far.

FIGURE 3.1.4 The amount of energy released by a bonfire is far greater than the amount of energy released by the burning of a single match.

We can use a thermochemical equation and the principles of stoichiometry to calculate the energy released by the combustion of specified quantities of fuel. The use of a thermochemical equation means energy released can be calculated and compared under different conditions.

Combustion and energy

The quantity of energy obtained from the combustion of a fuel depends on:

- the type of fuel
- the amount of fuel
- whether complete combustion or incomplete combustion is involved.

The thermochemical equations for the complete combustion of methane and pentane are shown below. Complete combustion occurs when there is a plentiful oxygen supply.

methane: $CH_4(g) + 2O_2(g) \longrightarrow CO_2(g) + 2H_2O(l) \quad \Delta H = -890 \text{ kJ}$

pentane: $C_5H_{12}(g) + 8O_2(g) \longrightarrow 5CO_2(g) + 6H_2O(l) \quad \Delta H = -3509 \text{ kJ}$

The equations show that the equal amounts of different fuels release different quantities of energy. The combustion of 1 mole of pentane releases more energy than the combustion of 1 mole of methane.

> Complete combustion produces carbon dioxide and water as products. Incomplete combustion forms carbon monoxide and sometimes carbon.

Incomplete combustion occurs when the oxygen supply is limited, and carbon monoxide is formed rather than carbon dioxide. The complete combustion of a fuel releases more energy than incomplete combustion of the same fuel. For example, for ethane:

complete combustion: $2C_2H_6(g) + 7O_2(g) \longrightarrow 4CO_2(g) + 6H_2O(l)$
$\Delta H = -3120 \text{ kJ}$

incomplete combustion: $2C_2H_6(g) + 5O_2(g) \longrightarrow 4CO(g) + 6H_2O(l)$
$\Delta H = -1989 \text{ kJ}$

Calculating energy change from thermochemical equations

The coefficients of the reactants in a thermochemical equation indicate the amount, in moles, of each substance that reacts or is produced to give the specified enthalpy change, ΔH.

For example, the thermochemical equation for the complete combustion of ethane indicates that 3120 kJ of energy is released by the reaction of 2 moles of ethane with 7 moles of oxygen.

Worked examples 3.1.4 and 3.1.5, on the following pages, show you how to use a thermochemical equation to calculate the energy released by the combustion of different amounts of fuels or the amount of fuel required to produce a specified quantity of energy.

> The coefficients of the reactants in a thermochemical equation indicate the amount, in moles, of each substance that react to give the enthalpy change, ΔH, specified in the equation.

Worked example 3.1.4

USING A THERMOCHEMICAL EQUATION TO CALCULATE ENERGY RELEASED BY THE COMBUSTION OF A FUEL

Calculate the heat energy released, in MJ, when 10.0 kg of octane undergoes complete combustion.

$$2C_8H_{18}(l) + 25O_2(g) \rightarrow 16CO_2(g) + 18H_2O(l) \qquad \Delta H = -10\,900 \text{ kJ}$$

Thinking	Working
As the quantity of the known substance is a mass, calculate the number of moles of the known substance using: $n = \frac{m}{M}$	$n(C_8H_{18}) = \frac{10.0 \times 10^3}{114.0}$ $= 87.7$ mol
Using the thermochemical equation, find the relationship between the number of moles of fuel burned and energy released.	x kJ is released by 87.7 mol 10900 kJ is released by 2 mol
Calculate the energy released by the fuel in kilojoules.	By proportion: $\frac{x}{10900} = \frac{87.7}{2}$ $x = \frac{87.7}{2} \times 10900$ $= 478070$ kJ
Convert the energy released to MJ using: $1 \text{ MJ} = 10^3 \text{ kJ}$ Express the answer to the correct number of significant figures. 10.0 kg of octane in question has 3 significant figures. Therefore, 3 significant figures are given in the answer.	478070 kJ = 478 MJ (to 3 significant figures)

Worked example: Try yourself 3.1.4

USING A THERMOCHEMICAL EQUATION TO CALCULATE ENERGY RELEASED BY THE COMBUSTION OF A FUEL

Calculate the heat energy released, in MJ, when 10.0 kg of ethane undergoes complete combustion.

$$2C_2H_6(g) + 7O_2(g) \rightarrow 4CO_2(g) + 6H_2O(l) \qquad \Delta H = -3120 \text{ kJ}$$

Worked example 3.1.5

USING A THERMOCHEMICAL EQUATION TO CALCULATE THE AMOUNT OF FUEL THAT MUST BE BURNED TO PRODUCE A PARTICULAR AMOUNT OF ENERGY

What volume of methane, measured at SLC, burns completely to provide 4.00×10^4 kJ?

$$CH_4(g) + 2O_2(g) \rightarrow CO_2(g) + 2H_2O(l) \qquad \Delta H = -890 \text{ kJ}$$

Thinking	Working
Using the thermochemical equation, write a relationship between the number of moles of fuel burned and energy released.	x mol releases 4.00×10^4 kJ 1 mol releases 890 kJ
Calculate the amount of the fuel that was burned to produce the energy.	By proportion: $\frac{x}{1} = \frac{4.00 \times 10^4}{890}$ $= \frac{4.00 \times 10^4}{890} \times 1$ $= 44.9$ mol
Calculate the volume of the fuel, at SLC.	$V(CH_4) = n \times V_m$ $= 44.9 \times 24.8$ $= 1115$ L $= 1.11 \times 10^3$ L

Worked example: Try yourself 3.1.5

USING A THERMOCHEMICAL EQUATION TO CALCULATE THE AMOUNT OF FUEL THAT MUST BE BURNED TO PRODUCE A PARTICULAR AMOUNT OF ENERGY

What volume of methane, measured at SLC, burns completely to provide 5.00×10^3 kJ?

$CH_4(g) + 2O_2(g) \rightarrow CO_2(g) + 2H_2O(l) \qquad \Delta H = -890$ kJ

COMBUSTION AND GREENHOUSE GASES

All carbon-based fuels undergo combustion to produce the greenhouse gases carbon dioxide and water. A **greenhouse gas** is a gas that can absorb infrared radiation. Human activities such as combustion of fuels are increasing the presence of greenhouse gases in the atmosphere, which is contributing to an enhanced greenhouse effect that causes global warming. As the amount of greenhouse gases in the atmosphere increases, more heat is trapped closer to Earth, as seen in Figure 3.1.5.

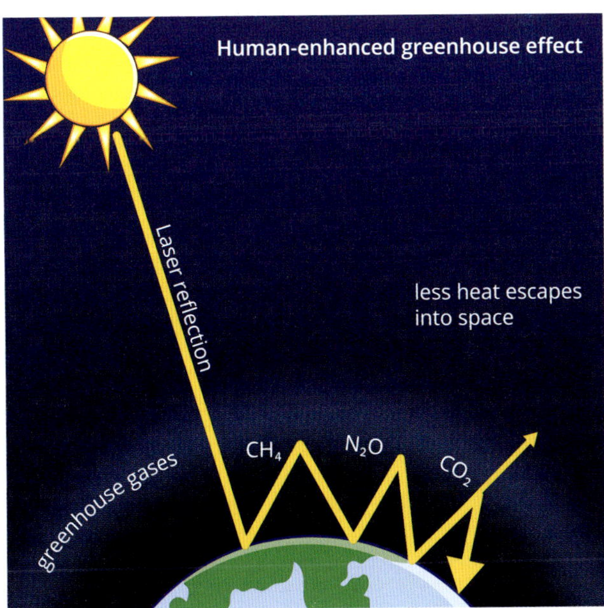

FIGURE 3.1.5 Human activities are increasing the amount of greenhouse gases in the atmosphere, leading to an enhanced greenhouse effect.

CHEMFILE

Energy production in Australia

Many Australian power stations, like the one in the figure below, produce energy from the combustion of brown and black coal. These combustion reactions produce tonnes of carbon dioxide on a daily basis, which is released into the atmosphere. Recent global climate change negotiations among developed nations, including Australia, have focused on pledges to reduce the amount of carbon dioxide emitted during the production of energy due to carbon dioxide being a significant greenhouse gas. Stoichiometry allows chemists to usefully predict and compare the mass and volume of carbon dioxide emissions produced by similar quantities of different fuels.

In recent years, a number of Australian states have accelerated plans to close coal-fired power stations due to pressure from cleaner and lower-cost energy generation, including renewable options like solar, wind and batteries. Although these are economic pressures, they have the added benefit of reducing gases released to the atmosphere.

It is useful to be able to compare the amount of carbon dioxide emissions produced during the production of energy from the combustion of different fuels.

For the majority of fuels, the quantities of carbon dioxide and water produced represent the net mass or volume of greenhouse gases added to the atmosphere when those fuels undergo combustion.

Methane, CH_4, a fuel which is a constituent of both natural gas and biogas, is also a greenhouse gas. In fact, methane is far more potent than both carbon dioxide and water, having a significantly greater warming power than carbon dioxide over the first 20 years after it reaches the atmosphere. Recent observations of methane levels from satellites have shown substantial quantities of methane escaping from landfills and coal mines around the world, as well as from agriculture and even home kitchens.

CO_2, CH_4 and H_2O are major greenhouse gases.

Capturing the methane produced by agricultural animals or decomposition of organic matter in landfill and using it as a biogas in combustion reactions does increase the net volume of carbon dioxide released to the atmosphere, but it also reduces the volume of the more potent methane. Worked example 3.1.6 shows you how to calculate the net volume of greenhouse gases released when methane undergoes combustion.

Worked example 3.1.6

CALCULATING THE VOLUME OF GREENHOUSE GASES RELEASED

What is the volume of greenhouse gases released when 1.5 L of methane (CH_4) undergoes complete combustion in oxygen? Assume all reactants and products are in the gaseous state, and the temperature and pressure remain constant.

Thinking	Working
Write a balanced chemical equation for the reaction.	$CH_4(g) + 2O_2(g) \rightarrow CO_2(g) + 2H_2O(g)$
Identify the volume ratio between methane and greenhouse gases produced (CO_2 and H_2O).	$\frac{V(\text{greenhouse gases})}{V(CH_4)} = \frac{3}{1}$
Calculate the volume of greenhouse gases produced.	$V(\text{greenhouse gases}) = \frac{3}{1} \times V(CH_4)$ $= \frac{3}{1} \times 1.5$ $= 4.5$ L
Establish the final volumes of greenhouse gases released.	$V(CH_4) = 0$ L $V(CO_2 + H_2O) = 4.5$ L

Worked example: Try yourself 3.1.6

CALCULATING THE VOLUME OF GREENHOUSE GASES RELEASED

What is the volume of greenhouse gases released when 10.8 L of methane (CH_4) undergoes complete combustion in oxygen? Assume all reactants and products are in the gaseous state, and the temperature and pressure remain constant.

CASE STUDY **ANALYSIS**

Bioethanol as a fuel for cars to reduce greenhouse gases

Ethanol, C_2H_5OH, can be used as a fuel in cars as an alternative to octane (petrol) or as part of a blend with petrol. As with octane, C_8H_{18}, ethanol produces the greenhouse gases carbon dioxide and water when it undergoes complete combustion. However, when the ethanol is obtained from plant sources it is known as bioethanol and its use has greatly reduced net greenhouse gas emissions. As you learnt in Chapter 2, bioethanol is produced from the fermentation of glucose derived from cropping plants such as sugarcane or corn. As the crops grow, they remove carbon dioxide from the atmosphere for the process of photosynthesis, the chemical reaction that produces glucose in plants according to the equation:

$$6CO_2(g) + 6H_2O(l) \rightarrow C_6H_{12}O_6(aq) + 6O_2(g)$$

This removal of carbon dioxide from the atmosphere offsets the release of carbon dioxide when the ethanol is burnt in cars. The net greenhouse emissions are lower than when petrol obtained from fossil fuels is burnt.

The largest bioethanol producer in Australia is in Nowra on the NSW South Coast (Figure 3.1.6). Australia has the capacity to make approximately 440 million litres of ethanol annually, with nearly 70% of this manufactured at the Nowra facility. The manufacturing process uses a distillery that produces ethanol through the fermentation of glucose sourced from starch. At the Nowra facility, starch is a by-product of the production of flour from wheat, which is the main manufacturing process at the Nowra site.

The enthalpy of combustion of ethanol is –1360 kJ mol^{-1} at SLC.

Analysis

1 Write a balanced thermochemical equation that represents the complete combustion of ethanol.

2 a What mass of carbon dioxide, in kg, would be formed if 1.00 kg of ethanol reacts?

b What volume of carbon dioxide would be formed from 1.00 kg of ethanol at SLC?

3 a If the density of ethanol is 0.785 g mL^{-1}, calculate the mass of ethanol, in kg, in a 50.0 L tank of the fuel using the formula:

$$\text{density} = \frac{\text{mass (g)}}{\text{volume (mL)}}$$

b Calculate the energy, in MJ, that can be obtained from the complete combustion of 50.0 L of ethanol.

4 Calculate the mass, in g, of carbon dioxide formed when 1.00 kJ of energy is released.

FIGURE 3.1.6 An ethanol plant at Nowra in New South Wales

3.1 Review

OA ✓✓

SUMMARY

- The coefficients in a balanced equation show the mole ratio of the amount, in moles, of reactants and products involved in the reaction.
- Stoichiometric calculations follow the general steps:
  1 Calculate the amount, in moles, of a known substance from the data given.

  Use $n = \frac{m}{M}$ or $n = \frac{V}{V_m}$

  2 Use the mole ratio from a balanced chemical equation to determine the amount, in moles, of the unknown substance.

$$\frac{n(\text{unknown chemical})}{n(\text{known chemical})} = \frac{\text{coefficient of unknown chemical}}{\text{coefficient of known chemical}}$$

  3 Find the desired quantity of the unknown substance from its amount, in moles, using $m = n \times M$ or $V = n \times V_m$.
- Stoichiometric calculations can be used to calculate the mass of carbon dioxide and water released during the combustion of a carbon-based fuel and the mass of oxygen required for complete combustion.
- The mole ratio in a balanced equation is also a volume ratio if all reactants and products are in the gaseous state and the temperature and pressure are kept constant.
- Stoichiometric calculations based on thermochemical equations can be used to calculate the amount of heat energy released during the combustion of a carbon-based fuel.
- The coefficients of the reactants in a thermochemical equation indicate the amounts, in moles, of each substance that react to give the enthalpy change specified in the equation.
- Fuels can be compared in terms of energy released and total greenhouse gases produced.

KEY QUESTIONS

Knowledge and understanding

1 Consider the following balanced equation that shows the formation of carbon dioxide from carbon monoxide.

$$2CO(g) + O_2(g) \longrightarrow 2CO_2(g)$$

a Calculate the amount, in mol, of oxygen required to completely react with 2 mol of carbon monoxide.

b Calculate the amount, in mol, of oxygen required to completely react with 6 mol of carbon monoxide.

c Calculate the volume of oxygen needed to completely react with 1.5 L of carbon monoxide. Assume all volumes are measured at the same temperature and pressure.

2 Consider the following thermochemical equation that shows the combustion of methane.

$$CH_4(g) + 2O_2(g) \longrightarrow CO_2(g) + 2H_2O(g) \quad \Delta H = -890 \text{ kJ}$$

a Determine the amount, in mol, of methane that must undergo combustion in order to release 890 kJ of energy.

b Calculate the amount, in mol, of carbon dioxide and water that are produced when 890 kJ of energy is released.

c Calculate the net mass of greenhouse gases released when 890 kJ of energy is released.

3 Calculate the energy released, in MJ, when the following quantities of ethane gas burn according to the equation:

$$2C_2H_6(g) + 7O_2(g) \longrightarrow 4CO_2(g) + 6H_2O(l)$$
$$\Delta H = -3120 \text{ kJ}$$

a 3.00 mol

b 100 g

c 10.0 L at SLC

Analysis

4 Octane (C_8H_{18}) is a component of petrol. 200 g of octane burns in oxygen to produce carbon dioxide and water. The equation for this reaction is:

$$2C_8H_{18}(g) + 25O_2(g) \longrightarrow 16CO_2(g) + 18H_2O(g)$$

a Calculate the mass of oxygen required to completely react with 200 g of octane.

b Calculate the mass of carbon dioxide produced when 200 g of octane is burnt.

5 8.00 g of propane was burned in excess oxygen.

a Write a balanced chemical equation for the complete combustion of propane.

b What mass of carbon dioxide would be produced?

c What volume, at SLC, of oxygen would be consumed in the reaction?

6 Calculate the energy released, in MJ, when 250.0 g of petrol burns completely in a car engine. Assume petrol is all octane (C_8H_{18}) and burns according to the equation:

$$2C_8H_{18}(l) + 25O_2(g) \longrightarrow 16CO_2(g) + 18H_2O(l) \qquad \Delta H = -10\,900 \text{ kJ}$$

7 Butane is used as the fuel in some portable camping stoves. It is a liquid when stored under pressure in a butane cylinder but vaporises when the valve is opened. Combustion of butane is represented by the equation:

$$2C_4H_{10}(g) + 13O_2(g) \longrightarrow 8CO_2(g) + 10H_2O(l) \qquad \Delta H = -5772 \text{ kJ}$$

a How much energy is produced when 10.0 g of butane burns completely?

b How much energy is produced when 0.100 L of butane, measured at SLC, burns completely?

8 Methane and methanol both burn in air according to the thermochemical equations:

$$CH_4(g) + 2O_2(g) \longrightarrow CO_2(g) + 2H_2O(l) \qquad \Delta H = -890 \text{ kJ}$$

$$2CH_3OH(l) + 3O_2(g) \longrightarrow 2CO_2(g) + 4H_2O(l) \qquad \Delta H = -1450 \text{ kJ}$$

a If 1 mole of methane and 1 mole of methanol are completely burned in separate experiments, which experiment will release the most energy?

b If each of the above reactions is used to produce 2000 kJ of energy, which fuel will release the most carbon dioxide?

3.2 Determination of limiting reactants or reagents

In the previous section, you learnt how to calculate the amount of product that could be formed, or energy that could be released, given a specified amount of a fuel. It was assumed for each combustion reaction that there was sufficient oxygen present for the fuel to be completely consumed.

In this section, you will learn an approach that you can take when it is not immediately obvious if a reactant or **reagent** is completely consumed in a reaction. A reactant is a starting material that undergoes change during a chemical reaction whereas a reagent is a substance added to a system to cause a chemical reaction. Both reactants and reagents can be a limiting substance in a reaction.

In these scenarios, you are given quantities of both reactants present. Before you can calculate the amount of product formed, you will need to work out which reactant is completely consumed.

LIMITING AND EXCESS REACTANTS

When two reactants are mixed to create a chemical reaction, it is possible they will be combined in just the right mole ratio, as indicated in the equation, for each to be completely consumed. However, it is more likely that they are not present in exactly the right mole ratio, meaning that one of the reactants will be used up before the other. Some of the other reactant will remain unreacted and be left over, or **in excess**, once the reaction has ceased.

To illustrate this situation simply, consider a problem in which you have been given some skateboard decks and wheels and you want to make as many complete skateboards as you can. As shown in Figure 3.2.1, a complete skateboard is made up of one deck and four wheels.

FIGURE 3.2.1 Construction of a complete skateboard requires one deck and four wheels.

If you were given two decks and ten wheels, as shown in Figure 3.2.2, how many complete skateboards can you make from these materials?

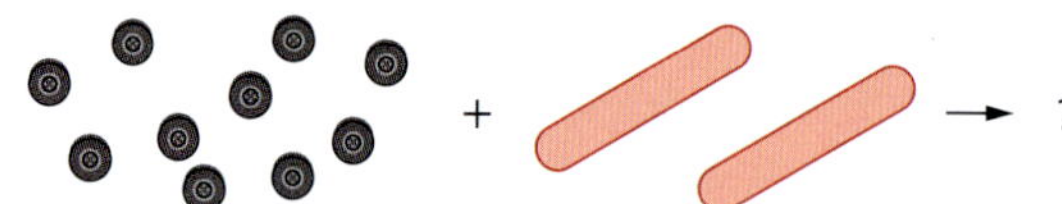

FIGURE 3.2.2 When provided with ten wheels and two skateboard decks, how many complete skateboards can be made?

The answer is that you could make two complete skateboards and there would be two wheels left over (Figure 3.2.3).

FIGURE 3.2.3 When supplied with two decks and ten wheels the maximum number of skateboards that can be made is two. There will be two wheels that are not used.

We can see that the number of skateboards that could be made was limited by the number of decks available. The decks were the limiting factor. The wheels were not completely used up and can be said to be in excess.

A similar situation arises in chemical reactions if the quantities of reactants present do not match the exact same ratio as that shown in the equation for the reaction.

When this happens:

- the reactant that is completely consumed is called the **limiting reactant**
- the reactant that is not completely consumed is called the **excess reactant**.

Figure 3.2.4 shows three different scenarios for the reaction in which hydrogen gas and oxygen gas combine to form water, according to the equation:

$$2H_2(g) + O_2(g) \rightarrow 2H_2O(l)$$

Each of the diagrams provides examples to illustrate the concepts of limiting and excess reactants.

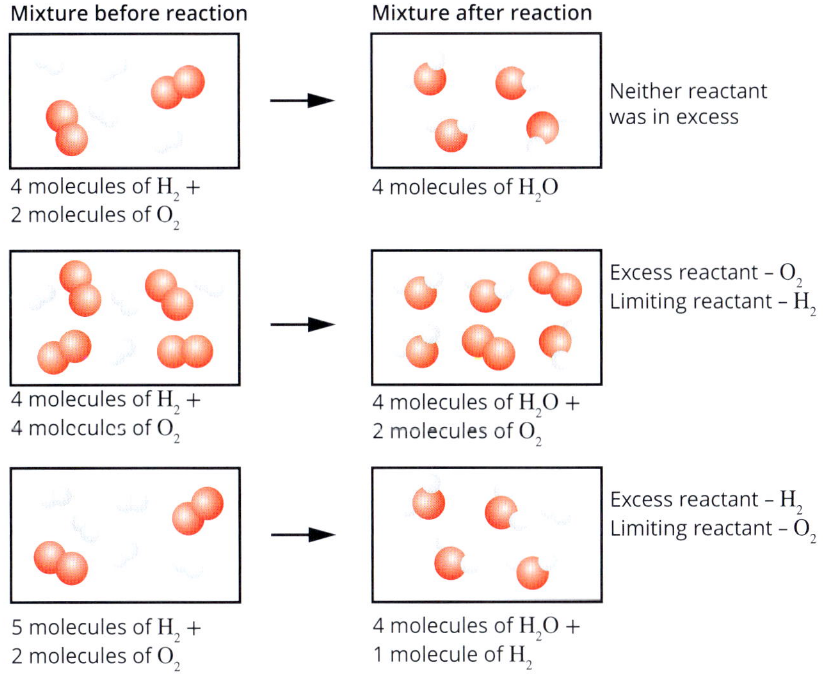

FIGURE 3.2.4 Different scenarios showing the concept of limiting reactant for the reaction $2H_2(g) + O_2(g) \rightarrow 2H_2O(l)$

Note that in each of the examples shown in Figure 3.2.4, the amount of product formed is determined by the amount of the limiting reactant present in the reaction mixture. The amount of product formed cannot be determined from the amount of excess reactant. This means it is essential to always identify the limiting reactant for use in stoichiometry.

Steps in solving stoichiometry problems involving excess reactants

When attempting to solve a problem in which you are required to work out the limiting reactant and use stoichiometry to calculate the amount of product, there are three main steps:

1 Calculate the number of moles of each reactant.

2 Identify which reactant is the limiting reactant.

3 Use the amount of limiting reactant to determine the amount of product formed.

Each step is illustrated in Worked examples 3.2.1 and 3.2.2.

In a chemical reaction, the limiting reactant is the reactant that is completely consumed in the reaction.

CHEMFILE

Excess reactants in the chemical industry

Many chemical reactions are used in the chemical industry to make useful products from soap to pesticides and pharmaceuticals. The design of these processes will consider cost, product yield and minimising waste. If, in a particular chemical reaction, one reactant is more expensive than another, it will be important to ensure the expensive reactant is fully utilised. One way to assist with this is making sure the other reactant is in excess. Stoichiometric calculations help to determine how much reactant is required to ensure it is in excess. However, having an excess reactant may cause the disadvantage of requiring that reactant to be recovered and recycled.

A chemist designing an industrial manufacturing process will consider which reactants or reagents to have in excess to maximise yield, while minimising costs and waste.

The amount of the limiting reactant must always be used to determine the amount of product that will be formed.

Worked example 3.2.1

IDENTIFYING LIMITING AND EXCESS REACTANTS

A gaseous mixture of 25.0 g of hydrogen gas (H_2) and 150 L of oxygen gas at SLC are mixed and ignited. The equation for the reaction is:

$$2H_2(g) + O_2(g) \rightarrow 2H_2O(l)$$

a Identify which reactant is the limiting reactant.

b Calculate the amount, in mol, of the excess reactant that remains unreacted.

Thinking	Working
a Calculate the number of moles of each reactant using $n = \frac{m}{M}$ or $n = \frac{V}{V_m}$ or $n = \frac{pV}{RT}$. The formula chosen will depend on the quantity and conditions provided for each reactant.	$n(H_2) = \frac{m}{M}$ $= \frac{25.0}{2.0}$ $= 12.5$ mol $n(O_2) = \frac{V}{V_m}$ $= \frac{150}{24.8}$ $= 6.05$ mol
Choose one reactant and use the coefficients in the equation to find the amount of the other reactant needed for it to completely react.	$n(H_2)$ needed to react $= 2 \times n(O_2)$ $= 2 \times 6.05$ $= 12.1$ mol
Compare the values for the amount required of the second reactant and the amount actually present to determine which is the limiting factor.	12.1 mol of H_2 is needed for all of the O_2 to react. There is 12.5 mol of H_2 present. The H_2 is in excess. O_2 is the limiting reactant.
b Use the mole ratio and amount of limiting reactant to determine the amount of excess reactant involved in the reaction.	$n(O_2)$ that reacted = 6.05 mol $\frac{n(H_2)}{n(O_2)} = \frac{2}{1}$ $n(H_2O) = 2 \times 6.05$ $= 12.1$ mol
Determine the amount of excess reactant left over after the limiting reactant has been consumed.	$n(H_2)$ in excess $= n(H_2)$ initially $- n(H_2)$ reacted $= 12.5 - 12.1$ $= 0.4$ mol

Worked example: Try yourself 3.2.1

IDENTIFYING LIMITING AND EXCESS REACTANTS

8.00 g of butane (C_4H_{10}) is burned completely in 20.0 L of oxygen at SLC. The equation for the reaction is:

$$2C_4H_{10}(g) + 13O_2(g) \rightarrow 8CO_2(g) + 10H_2O(l)$$

a Which reactant is the limiting reactant?

b Calculate the amount, in mol, of the excess reactant that remains unreacted.

Worked example 3.2.2

STOICHIOMETRY CALCULATIONS INVOLVING LIMITING AND EXCESS REACTANTS

100.0 g of ethanol (C_2H_5OH) burns in 250.0 g of oxygen gas. The equation for the reaction is:

$$C_2H_5OH(l) + 3O_2(g) \rightarrow 2CO_2(g) + 3H_2O(l)$$

Calculate the volume, in L, of carbon dioxide formed at 100 kPa and 15°C.

Thinking	Working
Calculate the number of moles of each reactant using $n = \frac{m}{M}$ or $n = \frac{V}{V_m}$ or $n = \frac{pV}{RT}$. The formula chosen will depend on the quantity and conditions provided for each reactant.	$n(C_2H_5OH) = \frac{m}{M}$ $= \frac{100}{46.0}$ $= 2.17$ mol $n(O_2) = \frac{m}{M}$ $= \frac{250.0}{32.0}$ $= 7.81$ mol
Identify the limiting reactant.	$n(O_2)$ needed to react $= 3 \times n(C_2H_5OH)$ $= 3 \times 2.17$ $= 6.52$ mol As there is 7.81 mol of O_2 present, the O_2 is in excess. Ethanol is the limiting reactant.
Find the mole ratio: $\frac{\text{coefficient of unknown}}{\text{coefficient of known (limiting reactant)}}$	$\frac{n(CO_2)}{n(C_2H_5OH)} = \frac{2}{1}$
Calculate the number of moles of the unknown substance using: n(unknown) = mole ratio × n(known)	$n(CO_2) = \frac{2}{1} \times 2.17$ $= 4.35$ mol
Calculate the required quantity of the unknown substance using $m = n \times M$, $V = n \times V_m$ or $V = \frac{nRT}{p}$ as appropriate. The choice will depend on the required quantity and provided conditions.	The volume of a gas at non-standard conditions needs to be calculated, so the formula chosen is $V = \frac{nRT}{p}$ $V(CO_2) = \frac{4.35 \times 8.31 \times 288}{100}$ $= 104$ L

Worked example: Try yourself 3.2.2

STOICHIOMETRY CALCULATIONS INVOLVING LIMITING AND EXCESS REACTANTS

150 g of propanol (C_3H_7OH) burns in 200.0 g of oxygen gas. The equation for the reaction is:

$$2C_3H_7OH(l) + 9O_2(g) \rightarrow 6CO_2(g) + 8H_2O(l)$$

Calculate the volume, in L, of carbon dioxide formed at 120 kPa and 20°C.

3.2 Review

WS 4 OA ✓✓

SUMMARY

- If quantities of more than one reactant or reagent are provided for a chemical reaction, the limiting reactant or reagent must be identified and used in stoichiometry calculations.
- A reactant is a starting material that undergoes change during a chemical reaction, whereas a reagent is a substance added to a system to cause a chemical reaction.
- The limiting reactant or reagent is the reactant or reagent that is completely consumed in the reaction.
- The reactant or reagent that is not completely used up and has some remaining when the reaction has stopped, is said to be in excess.

KEY QUESTIONS

Knowledge and understanding

1 Iron reacts with oxygen gas to form iron oxide. The balanced equation for the reaction is:

$$4Fe(s) + 3O_2(g) \longrightarrow 2Fe_2O_3(s)$$

In a particular reaction, 25 g of iron reacts with 60 g of oxygen gas. List the following steps in order to determine the mass of iron oxide that will form.

A Calculate the mass of iron oxide that forms.
B Refer to the balanced equation.
C Calculate the number of moles of iron oxide that form.
D Calculate the number of moles of iron and oxygen.
E Use mole ratios to determine which reactant is limiting.

2 In three different experiments, different amounts of propane (C_3H_8) and oxygen undergo combustion according to the equation:

$$C_3H_8(g) + 5O_2(g) \longrightarrow 3CO_2(g) + 4H_2O(g)$$

The following table shows the amount of reactants and products in each experiment. Complete the table to indicate the amount of each product remaining at the end of the reaction.

Propane molecules available	Oxygen molecules available	Carbon dioxide molecules produced	Propane molecules in excess	Oxygen molecules in excess
2	15			
300	1200			
2.5 mol	10 mol			

Analysis

3 Fluorine gas reacts with solid phosphorus to form phosphorus trifluoride (PF_3) according to the following reaction:

$$3F_2(g) + 2P(s) \longrightarrow 2PF_3(g)$$

42.0 g of fluorine gas is reacted with 29.0 g of phosphorus. Calculate the mass of phosphorus trifluoride that will form in the reaction.

4 7.0 g of ethanol (C_2H_5OH) and 30.0 g of O_2 are mixed in a closed vessel and allowed to react according to this equation:

$$C_2H_5OH(l) + 3O_2(g) \longrightarrow 2CO_2(g) + 3H_2O(g)$$

a Which reactant is in excess and by what mass, in g?
b What mass of CO_2 forms?
c What mass of H_2O forms?
d What is the total mass of the mixture after the reaction is complete?

5 23.0 g of ethanol (C_2H_5OH) is reacted with 10.0 L of oxygen at SLC. Calculate the volume of carbon dioxide produced in this reaction.
The equation for the reaction is:

$$C_2H_5OH(g) + 3O_2(g) \longrightarrow 2CO_2(g) + 3H_2O(l)$$

6 8.50 g of propanol (C_3H_7OH) is reacted with 20.0 L of oxygen at 120 kPa and 30°C. Calculate the volume of carbon dioxide produced in this reaction.
The equation for the reaction is:

$$2C_3H_7OH(g) + 9O_2(g) \longrightarrow 6CO_2(g) + 8H_2O(l)$$

3.3 Calculating heat energy released

Knowing the energy released by similar quantities of different fuels helps you to compare fuels and determine their suitability for specific purposes. For example, the fuel used to power an aeroplane (Figure 3.3.1) is different from the fuel used to power a car or bus.

In Chapter 2 you learnt how thermochemical equations include an ΔH value, which shows the amount of heat released when a fuel undergoes combustion. The value of ΔH is based on the stoichiometric ratios in the equation.

In this section, you will learn how knowledge of the specific heat capacity of water can be used to obtain an experimental estimate of **heat of combustion**, as well as how the heat of combustion can be used to write thermochemical equations.

FIGURE 3.3.1 F15 jets being refuelled by a Boeing 707. The type of fuel suitable for use in aircraft is different from the type of fuel suitable for use in cars or buses.

SPECIFIC HEAT CAPACITY OF WATER

The **specific heat capacity** of a substance is a measure of the amount of energy (usually in joules) needed to increase the temperature of a specific quantity of that substance (usually 1 gram) by 1°C.

Specific heat capacity is given the symbol *c* and is usually expressed in joules per gram per degrees Celsius, i.e. $J\ g^{-1}\ °C^{-1}$. It can also be expressed in joules per grams per kelvin, i.e. $J\ g^{-1}\ K^{-1}$ (an increase of 1°C is the same as an increase of 1 K).

The specific heat capacities of some common substances are listed in Table 3.3.1. You can see that the value for water is relatively high.

TABLE 3.3.1 Specific heat capacities of some common liquids

Substance	Specific heat capacity ($J\ g^{-1}\ °C^{-1}$)
water	4.18
glycerine	2.43
ethanol	2.46
hexane	2.26
olive oil	1.97
paraffin oil	2.13

The specific heat capacity of a substance reflects the types of bonds holding the molecules, ions or atoms together in the substance. Water has a specific heat capacity of $4.18\ J\ g^{-1}\ °C^{-1}$. This means that 4.18 joules of heat energy are needed to increase the temperature of 1 gram of water by 1°C. This relatively high value is due to the hydrogen bonds between the water molecules. The higher the specific heat capacity, the more effectively a material stores heat energy.

When a substance is being heated, its temperature rises. The temperature of 1 g of water increases by 1°C when it is supplied with 4.18 J of heat energy. In comparison, 2.43 J of heat energy is required to increase the temperature of 1 g of glycerine by 1°C. The effect of the different specific heat capacities of water and glycerine on their temperatures when heated can be seen in Figure 3.3.2.

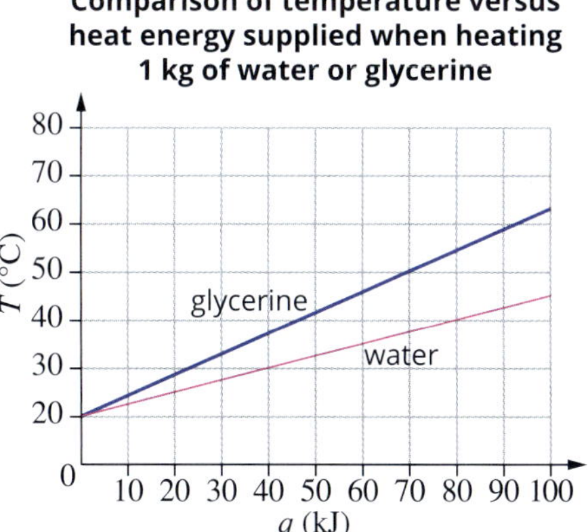

FIGURE 3.3.2 A comparison of the effect of the different specific heat capacities of water and glycerine on the increase in temperature (*T*). Water has a very high heat capacity, so it requires more heat energy (*q*) to increase its temperature by 1°C.

TRANSFERRING HEAT ENERGY TO WATER

The specific heat capacity of water can be used to calculate the heat energy in joules needed to increase the temperature of a given mass of water by a particular amount. Heat energy is given the symbol q, as seen in Figure 3.3.2 on the previous page.

The heat energy transferred to the mass of water can be calculated by using measurements of the:

- initial temperature of the water
- highest temperature of the water
- mass of water.

The relationship used to calculate the energy that has been transferred to the water is:

Heat energy = mass of water × specific heat capacity × temperature change

Using symbols, the equation can be written as:

$$q = m \times c \times \Delta T \quad \text{or} \quad q = mc\Delta T$$

where q is the amount of heat energy (in J), m is the mass (in g), c is the specific heat capacity of the water (4.18 J $g^{-1}\,°C^{-1}$) and ΔT is the temperature change (in °C or K):

$$\Delta T = \Delta T_{final} - \Delta T_{initial}$$

Recall that the density of water is 1.0 g mL^{-1}, so 1 mL of water has a mass of 1 g. Sometimes the density of water is quoted as 0.997 g mL^{-1}; however, this value, to 3 significant figures, is only accurate at 25°C.

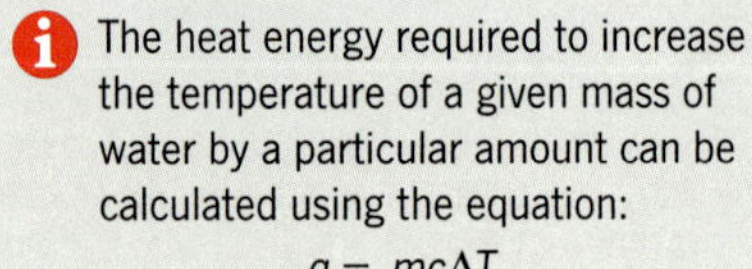

The heat energy required to increase the temperature of a given mass of water by a particular amount can be calculated using the equation:

$q = mc\Delta T$

Worked example 3.3.1

CALCULATING THE ENERGY REQUIRED TO HEAT A MASS OF WATER USING SPECIFIC HEAT CAPACITY

Calculate the heat energy, in kJ, needed to increase the temperature of 500 mL of water by 15.0°C.

Thinking	Working
Change the volume of water, in mL, to mass of water, in g. Remember that the density of water is 1.0 g mL^{-1}, so 1 mL of water has a mass of 1 g.	500 mL of water has a mass of 500 g.
Find the specific heat capacity of water from the data in Table 3.3.1.	The specific heat capacity of water is 4.18 J $g^{-1}\,°C^{-1}$.
To calculate the quantity of heat energy in joules, use the formula: $q = mc\Delta T$	$q = 500 \times 4.18 \times 15.0$ $= 3.14 \times 10^4$ J
Express the quantity of energy in kJ. Remember that to convert from J to kJ, you multiply by 10^{-3}.	$q = 3.14 \times 10^4 \times 10^{-3}$ $= 31.4$ kJ

Worked example: Try yourself 3.3.1

CALCULATING THE ENERGY REQUIRED TO HEAT A MASS OF WATER USING SPECIFIC HEAT CAPACITY

Calculate the heat energy, in kJ, needed to increase the temperature of 375 mL of water by 45.0°C.

CHEMFILE

Hot beach, cool bay

The different heat capacities of water and sand are evident at the beach on a hot summer's day. As a cloudless day dawns, both sand and water sit under the hot sun, as seen in the figure below. Sand that is not in contact with water from waves or tides can heat very quickly to the point where it can cause burns to the bottom of your feet. In contrast, the water, even in rock pools where constant mixing is not occurring, remains cool and refreshing. The specific heat capacity of water is 4.18 J $g^{-1}\,°C^{-1}$, while that of sand is 0.48 J $g^{-1}\,°C^{-1}$, so water can absorb nearly 10 times as much energy as sand for the same temperature increase. Each gram of water will absorb 4.18 J before it increases by 1°C. Each gram of sand will absorb only 0.48 J before it increases by 1°C.

At this Mornington Peninsula beach, the sand on a sunny day can be extremely hot while the water remains cool.

Experimental determination of heat of combustion

When an **exothermic** chemical reaction, such as the combustion of a pure substance, is carried out underneath a container of water, such as a metal can or a test tube, some of the heat released by the combustion reaction is transferred to the water. If you measure the temperature change of the water, it can be used to determine the approximate amount of energy released by the substance.

An experimental arrangement for estimating the heat of combustion of an organic liquid, such as ethanol, is shown in Figure 3.3.3. This experimental determination is suitable for many organic liquids, such as alcohols and some alkanes and alkenes. Because these liquids release a lot of energy when they are burnt, some of them could be regarded as fuels.

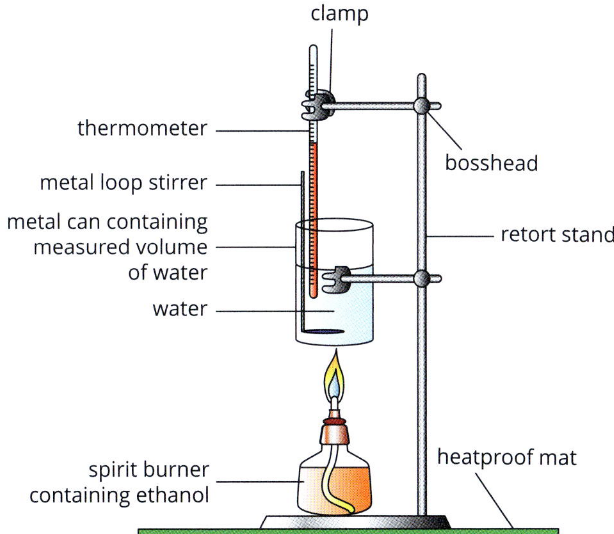

FIGURE 3.3.3 Apparatus for measuring heat of combustion of an organic liquid fuel (e.g. ethanol). A metal can containing a measured volume of water is held above the wick of a spirit burner.

Figure 3.3.4 summarises the steps followed in this experiment.

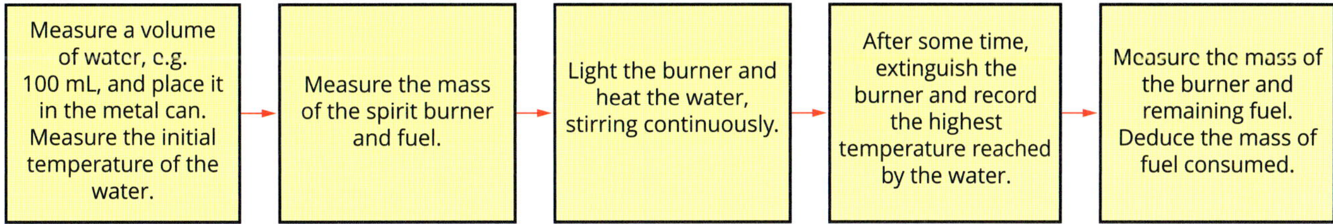

FIGURE 3.3.4 Flow chart of the steps followed when using the specific heat capacity of water to determine the heat of combustion of a pure organic liquid fuel

Three key pieces of information collected from this procedure are the:

- mass of water
- change in temperature of the water, ΔT
- mass of organic liquid fuel consumed, from which the amount, in mol, of the substance can be calculated.

In Chapter 2 you learnt that the heat of combustion of a fuel can be determined when the energy released by that fuel is divided by the amount, in mol, of the fuel that is burnt. You also learnt that the heat of combustion of a fuel is the heat energy released when a specified amount of the substance burns completely and is a positive value, while the enthalpy of combustion of a fuel reflects its exothermic nature with a negative sign and has the symbol, ΔH_c.

The heat of combustion can be calculated using the equation:

$$\text{heat of combustion} = \frac{q}{n}$$

where q, the energy absorbed by the water in the can, is calculated using the equation:

$$q = mc\Delta T$$

and n is the amount of fuel, in this case the pure organic liquid, that has been burnt.

Recall that the heat of combustion of a fuel is reported as a positive value due to its definition, while the enthalpy of combustion, ΔH_c is written in a thermochemical equation and is negative.

This calculation of heat of combustion assumes that all the energy is transferred from the burning pure organic liquid fuel to the mass of water. Substantial heat losses occur when the experiment described above is performed, so the values calculated for the heat of combustion will be less than the actual values. At best, this experiment could only be used to obtain an estimate of the heat of combustion or perhaps to compare the energy released by two or more pure substances, fuels or foods.

This is shown in Worked example 3.3.2.

Worked example 3.3.2

CALCULATING THE HEAT OF COMBUSTION OF A PURE ORGANIC LIQUID FUEL FROM EXPERIMENTAL DATA

0.355 g of methanol (CH_3OH) undergoes complete combustion in a spirit burner. The heat energy released is used to heat 100 g of water. The temperature of the water rose from 20.24°C to 37.65°C. Calculate the heat of combustion of methanol in kJ mol^{-1} and write the thermochemical equation for the reaction.

Thinking	Working
Calculate the temperature change of the water.	$\Delta T = 37.65 - 20.24$ $= 17.41°C$
Use the specific heat capacity of water to determine the energy used to heat the water. Use the formula: $q = mc\Delta T$	$q = 100 \times 4.18 \times 17.41$ $= 7277$ J
Express the quantity of energy in kJ. Remember that to convert from J to kJ, you multiply by 10^{-3}.	$q = 7277 \times 10^{-3}$ $= 7.277$ kJ
Calculate the amount, in moles, of methanol using the formula: $n = \frac{m}{M}$	$n = \frac{0.355}{32.0}$ $= 0.0111$ mol
Determine the heat of combustion of methanol, in kJ mol^{-1}. Heat of combustion = $\frac{\text{heat energy released by sample}}{\text{amount of sample (in mol)}}$ Or heat of combustion = $\frac{q}{n}$	Heat of combustion $= \frac{7.277}{0.0111}$ $= 656$ kJ mol^{-1}
The thermochemical equation for this reaction is a balanced equation, including the enthalpy of combustion, ΔH, which has a negative sign.	$CH_3OH(l) + \frac{3}{2}O_2(g) \rightarrow$ $CO_2(g) + 2H_2O$; $\Delta H = -656$ kJ or $2CH_3OH(l) + 3O_2(g) \rightarrow$ $2CO_2(g) + 4H_2O$; $\Delta H = -1312$ kJ

Worked example: Try yourself 3.3.2

CALCULATING THE HEAT OF COMBUSTION OF A PURE ORGANIC LIQUID FUEL FROM EXPERIMENTAL DATA

0.295 g of ethanol (C_2H_5OH) undergoes complete combustion in a spirit burner. The heat energy released is used to heat 100 g of water. The temperature of the water rose from 19.56°C to 38.85°C. Calculate the heat of combustion of ethanol in kJ mol^{-1} and write the thermochemical equation for the reaction.

To calculate the heat of combustion of a fuel, the following two equations are used.

$q = mc\Delta T$

heat of combustion = $\frac{q}{n}$

ESTIMATING THE ENERGY CONTENT OF A FUEL OR FOOD

Because foods and some fuels, such as diesel, are mixtures, rather than a pure substance, the heat energy released during the combustion of a known mass of these fuels or food is best described as an **energy content**, in kJ g^{-1}.

If the mass of fuel or food burned to produce this energy is measured, the energy content of the fuel or food can be calculated as follows:

$$\text{energy content} = \frac{\text{energy transferred to the water}}{\text{change in mass of the fuel/food during combustion}} = \frac{q}{\Delta m}$$

The mass of fuel or food burned is determined by measuring the initial mass of fuel or food, and then subtracting the final mass. It is not necessary to burn all the fuel or food sample, as long as the mass that is burned is determined.

In this type of experiment, heat loss is a consistent problem, so the values calculated for the energy content are an estimate only and will be less than the actual values.

Worked example 3.3.3 describes how to calculate energy content from simple experimental data.

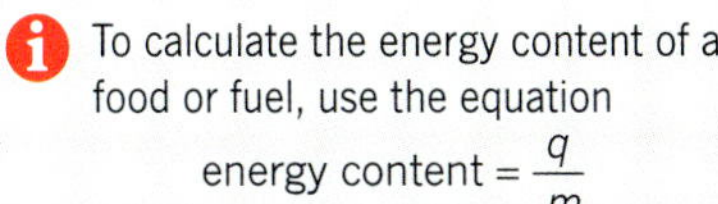

To calculate the energy content of a food or fuel, use the equation

$$\text{energy content} = \frac{q}{m}$$

where q is the energy transferred to the water and Δm is the change in the mass of the food or fuel.

Worked example 3.3.3

ESTIMATING THE ENERGY CONTENT OF A SAMPLE OF FOOD

A 1.670 g sample of cheese biscuit was burned under a steel can containing 200 g of water. After the flame went out, the mass of the cheese biscuit was 0.300 g and the temperature of the water had risen by 29.8°C. Calculate the energy content of the biscuit in kJ g^{-1}.

Thinking	Working
Calculate the heat energy absorbed by the water in joules, using the formula: $q = mc\Delta T$	$q = mc\Delta T$ $= 200 \times 4.18 \times 29.8$ $= 24913$ J
Express the quantity of energy in kJ. Remember that to convert from joules to kilojoules, you divide by 10^3 or multiply by 10^{-3}.	$q = \frac{24913}{1000}$ $= 24.913$ kJ
Calculate the mass of the food that was burned by subtracting the final mass from the initial mass: $\Delta m = \Delta m_{initial} - \Delta m_{final}$	$\Delta m = \Delta m_{initial} - \Delta m_{final}$ $= 1.670 - 0.300$ $= 1.370$ g
Calculate the energy content of the food by dividing the energy transferred to the water by the change in mass during combustion: Energy content $= \frac{q}{\Delta m}$	Energy content $= \frac{q}{\Delta m}$ $= \frac{24.913}{1.370}$ $= 18.2$ kJ g^{-1} (3 significant figures)

Worked example: Try yourself 3.3.3

ESTIMATING THE ENERGY CONTENT OF A SAMPLE OF FOOD

A 2.500 g sample of a corn chip was burned under a steel can containing 200 g of water. After the flame went out, the mass of the corn chip sample was 1.160 g and the temperature of the water had risen by 35.0°C. Calculate the energy content of the corn chip in kJ g^{-1}.

Heat loss

When energy is transferred from burning a sample of fuel or food, across an open space, heat is lost to the surroundings such as the air around the burning fuel. Similarly, if there is no lid on a container of water, heat will be lost from the surface of the water. This heat loss is illustrated in Figure 3.3.5.

When some heat energy from the burning fuel or food is transferred to the surrounding air, the temperature of the water does not increase as much as it would if all the energy was used to heat the water. A lower change in temperature, ΔT, of the water results in a lower energy value, q.

There are several ways to reduce heat loss during the experiment shown in Figure 3.3.5, including:

- putting a lid on the container holding the water
- insulating the beaker of water (with flameproof material)
- placing insulation around the burning fuel, although sufficient oxygen must reach the fuel for combustion to be complete.

As mentioned earlier, this loss of heat energy represents a significant systematic error in the calculation of heat of combustion by this experimental method. More accurate determinations of energy content require a more sophisticated piece of equipment called a bomb calorimeter.

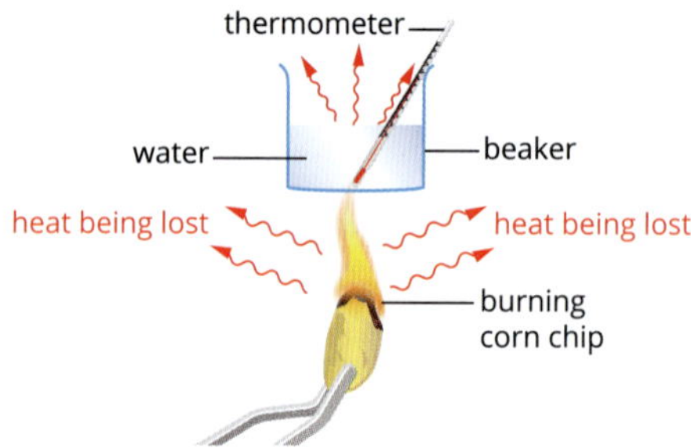

FIGURE 3.3.5 Heat is lost to the surroundings when a sample of food is burned to heat some water.

Bomb calorimetry

Figure 3.3.6 shows the components of a **bomb calorimeter**, which is a piece of equipment used for measuring the energy released by combustion reactions that involve gaseous reactants or products. The reaction vessel in a bomb calorimeter is designed to withstand the high pressures that may build up during reactions. An actual bomb calorimeter can be seen in Figure 3.3.7.

The energy content of fuels is measured by burning them in a bomb calorimeter. Sufficient oxygen is required to completely combust the fuel so that all the available energy is released. Insulation around the calorimeter prevents heat escaping and the change in temperature is measured with a thermometer. The stirrer ensures that the temperature of the water is uniform. You will learn more about calorimetry in Section 3.4.

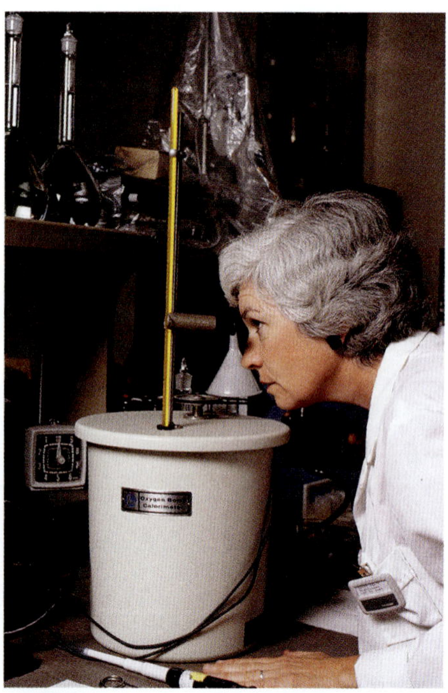

FIGURE 3.3.7 Bomb calorimeter temperature changes being observed by a laboratory technician

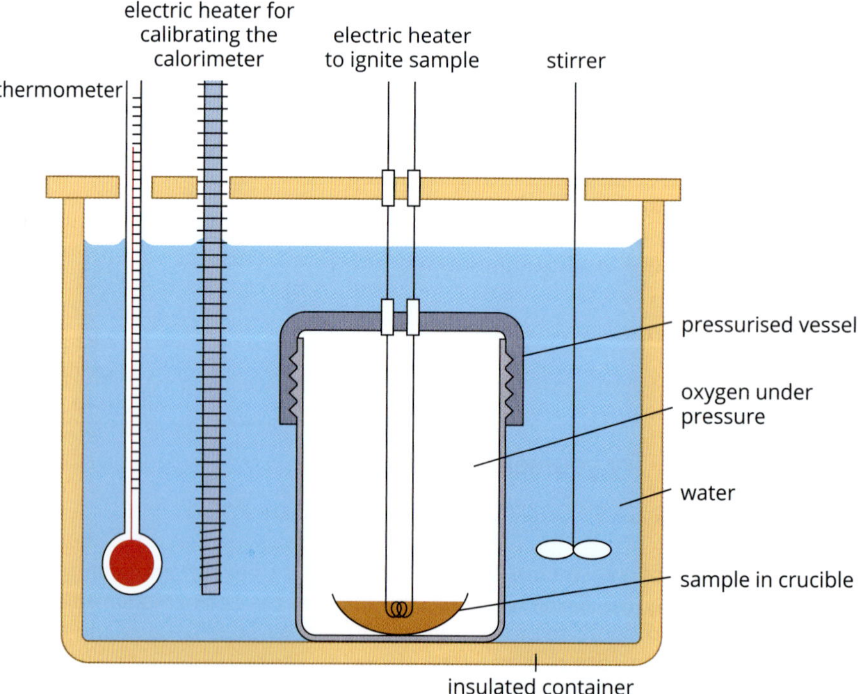

FIGURE 3.3.6 A bomb calorimeter used for reactions that involve gases, including the combustion of fuels

CASE STUDY **ANALYSIS**

The energy of candlelight

Little is known about the origins of candles, but it is thought that they have been around for over 5000 years of human civilisation. Candle wax is made of large molecules composed of carbon and hydrogen atoms, such as eicosane, $C_{20}H_{42}$, and it has an approximate energy content of 47.1 kJ g^{-1}. Candles were used by the ancient Romans as early as 500 BCE. However, Romans used oil lamps in which olive oil was burnt as their main light source. The burning of whale oil was common practice in the use of lamps from the 16th to the 19th centuries. Whale oil is a mixture containing large quantities of oleic acid which has an energy content of 39 kJ g^{-1}.

One such lamp was developed in 1780 (see Figure 3.3.8) by François-Pierre-Amédée (also known as Ami) Argand, a student of Antoine Lavoisier. These lamps were considered to have about the same light output as 6 to 8 candles. In fact, a 'candela' or candle power is a measure of the intensity of light, now known as lumens. A modern 10 Watt LED globe has an output of over 300 lumens. Argand further developed his lamp into a burner that was used until the development of the Bunsen burner.

Cheaper alternatives to whale oil were available at that time, but they did not burn as cleanly, producing large quantities of carbon soot. Demand for whale oil vastly increased in the 17th century and whale hunting reached its peak in the 1820s. After that time there was a rapid decline in use of whale oil, as cheaper and cleaner-smelling combustible materials became more common. Camphine, also known as 'oil of turpentine', a mixture of turpentine and alcohol, was used until kerosene rapidly dominated the market from the 1860s, making the use of camphine insignificant by 1866.

Now, in the 21st century, whale hunting is banned across most of the globe, and we use electricity to power our lights. Just as humans have moved away from the use of whale oil for lighting, we have the potential to explore more renewable fuel sources for our energy needs, so that we can preserve our environment.

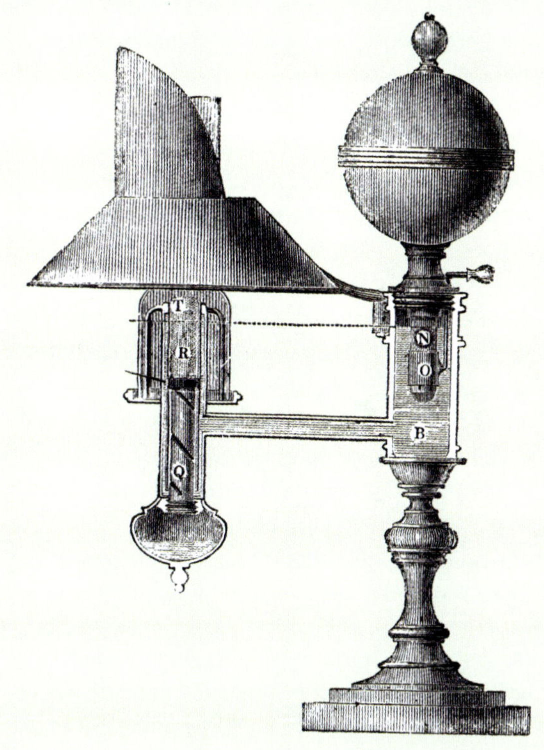

FIGURE 3.3.8 An engraving depicting a sectional view of an Argand lamp, a kind of oil lamp, that was invented and patented in 1780 by Ami Argand

Analysis

1 Write a balanced chemical reaction for the complete combustion of eicosane, $C_{20}H_{42}$.

2 What is occurring when a fuel, such as oil, is burning? Refer to states of matter, energy absorbed and released, and bonding in your answer.

3 A student decided to use a candle made only of eicosane to heat a beaker containing 100 mL of water. Assuming that there is no heat loss, calculate the mass of candle that would be used to heat the water from 18.0°C to 100.0°C.

3.3 Review

OA ✓✓

SUMMARY

- The specific heat capacity of a substance measures the quantity of energy (in joules) needed to increase the temperature of a specified quantity of that substance (usually 1 gram) by 1°C.
- The specific heat capacity of water is 4.18 J g^{-1} °C^{-1}.
- The heat energy required to increase the temperature of a given mass of water by a particular amount can be calculated using the equation:

$$q = mc\Delta T$$

where q is heat energy (in J), m is mass of water to be heated (in g), c is the specific heat capacity (in J g^{-1} °C^{-1}) and ΔT is the temperature change (in °C).
- In the experimental determination of the heat energy released in the combustion of a fuel, the specific heat capacity of water is used to determine the heat energy absorbed by a measured mass of water placed above the burning fuel.
- The heat of combustion of a pure substance (in kJ mol^{-1}) can be calculated using the experimental results of the energy absorbed by the measured mass of water (using $q = mc\Delta T$) and the amount of the substance (in mol) whose combustion released that energy using the equation:

$$\text{heat of combustion} = \frac{q}{n}$$

where q is heat energy (in kJ) and n is the amount (in mol) of fuel that was burnt.
- Because foods and some fuels are mixtures, their energy content is measured in kJ g^{-1}. This is determined using the equation energy content = $\frac{q}{m}$, where energy is measured in kJ.
- Calculations of heat of combustion of a pure substance or energy content of a food or fuel from experimental data can be very inaccurate due to extensive heat loss to the surroundings during the experiment.
- A bomb calorimeter is an insulated container in which a sealed, oxygen-filled reaction vessel is surrounded by a known volume of water.
- Combustion reactions are carried out in the reaction vessel and the heat from the reaction is transferred to the surrounding water.

KEY QUESTIONS

Knowledge and understanding

1 Calculate the energy, in kJ, required to increase the temperature of:

a 100 g of water by 25.1°C

b 500 g of water from 17.0°C to 80.0°C

c 1.50 kg of water from 20.0°C to 30.0°C.

2 A beaker containing 200 g of water at a temperature of 21.0°C is heated with 10.0 kJ of energy. Calculate the temperature reached by the water.

Analysis

3 A temperature rise of 1.78°C was observed when 1.00×10^{-3} mol of propane gas (C_3H_8) was burned and used to heat 300 g of water. Calculate the heat of combustion of propane in kJ mol^{-1}, using this experimental data.

4 A temperature rise of 11.5°C was observed when 0.500 g of butane gas (C_4H_{10}) was burned and used to heat 500 g of water.
Calculate the heat of combustion, in kJ g^{-1}, for butane, assuming all the heat released is used to heat the water.

5 The heat of combustion of methane (CH_4) is 890 kJ mol^{-1}. Determine the mass of methane, in g, which needs to be burned to heat 500 g of water from 20.0°C to boiling. Assume all the heat released is used to heat the water.

6 If the heat of combustion of a sample of sub-bituminous coal is 20.7 kJ g^{-1}, what is the mass of coal that is burned when 62.7 kJ of energy is transferred to a container of water, assuming all the energy from the burning coal is absorbed by the water?

3.4 Solution calorimetry

Calorimetry is the experimental method of measuring the heat energy released or absorbed by a chemical reaction or physical process, such as by the combustion of a fuel.

Energy changes that occur during chemical and physical changes are measured with a device called a calorimeter. **Calorimeters** are constructed in such a way that the energy losses that occur in the simple experimental apparatus described earlier are minimised, enabling more accurate results. In a calorimeter, almost all of the heat energy released or absorbed is transferred directly to or from a measured volume of water.

Two types of calorimeters are designed for measuring the energy changes in different types of reactions.

- In solution calorimeters, the reaction takes place in a solution.
- In bomb calorimeters, the reaction takes place in a sealed bomb vessel. This was described briefly in Section 3.3.

USING A SOLUTION CALORIMETER

Energy changes for reactions that occur in solution can be measured with a **solution calorimeter**. A solution calorimeter may be as simple as a polystyrene foam coffee cup with a lid, as shown in Figure 3.4.1.

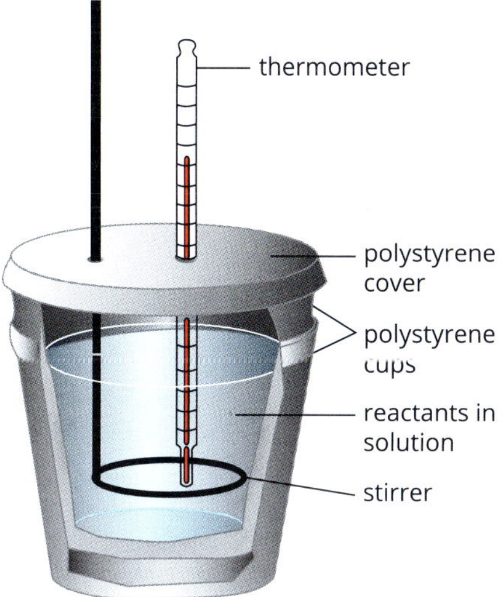

FIGURE 3.4.1 A simple 'coffee-cup' solution calorimeter

The insulation provided by the polystyrene foam prevents the transfer of heat to or from the surroundings of the calorimeter. The reaction is carried out in the calorimeter with an accurately known volume of water. The initial and final temperatures are measured and recorded, as are the amounts of reactants used.

If the temperature of the water in the calorimeter increases, the reaction occurring in the calorimeter is an exothermic reaction. The reaction has released heat energy, the water in the calorimeter has absorbed that energy and the temperature of the water has increased.

Similarly, if the temperature of the water in the calorimeter decreases, the reaction occurring in the calorimeter has absorbed energy from the water. In this case, the reaction is an **endothermic** reaction.

A coffee cup solution calorimeter has some limitations. The polystyrene container absorbs some heat, so the temperature change is lower than it would otherwise be and the calculated value for the heat released or absorbed by the reaction is lower than it should be.

A solution calorimeter cannot be used to measure the energy content of fuels or foods because the reaction in this case is a combustion reaction in which the fuel burns in oxygen. However, a solution calorimeter can be used to determine the energy change that occurs when a substance such as glucose dissolves in water. Solution calorimetry is used in the laboratory to determine the enthalpy changes that occur when acids react with bases, metals react with acids, and solids dissolve in water.

The construction of a laboratory solution calorimeter is shown in Figure 3.4.2 on the following page. The stirrer is used to ensure the temperature of the water is uniform. If the temperature of the solution is not uniform, then the change in temperature measured as a result of the dissolution will not be accurate. The use of the electrical heater for calibrating the calorimeter is described in the next section.

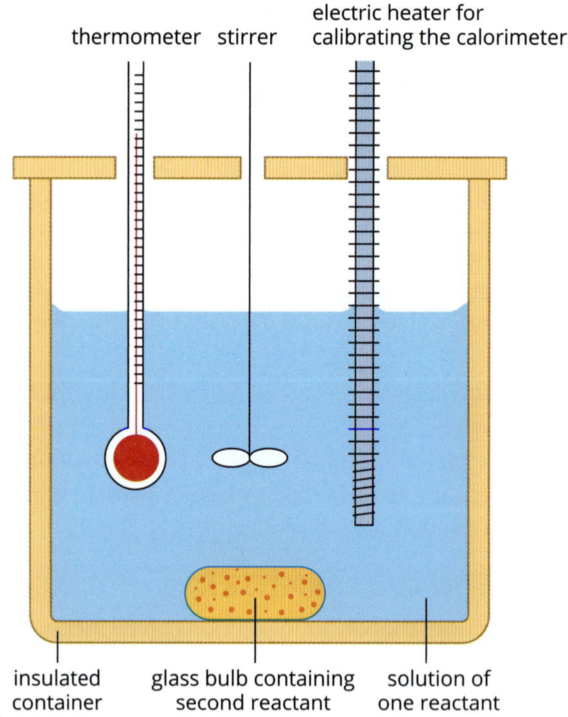

FIGURE 3.4.2 A solution calorimeter: breaking the glass bulb starts the reaction.

CALIBRATION OF CALORIMETERS

When a reaction takes place in a calorimeter, the heat change causes a rise or fall in the temperature of the contents of the calorimeter. Although the formula $q = mc\Delta T$ can be used to determine the relationship between heat energy and temperature change, the heat losses that occur can lead to inaccurate results.

For more accurate measurements of the heat produced in a reaction, you have to first determine how much energy is required to change the temperature of the water by 1°C for the particular calorimeter in use. This is known as the **calibration factor** (CF) of the calorimeter. For energy measured in joules, the calibration factor has the unit J °C^{-1}. Once the calibration factor of a calorimeter is known, it is said to be **calibrated**.

Both solution and bomb calorimeters are usually calibrated before use.

Electrical calibration of calorimeters

A calorimeter can be calibrated using the electric heater to release a known quantity of thermal energy and measuring the resultant rise in temperature of the water in the calorimeter.

The thermal energy released when an electric current passes through the heater can be calculated from the formula:

energy (in joules) = voltage (volts) × current (amps) × time (seconds),

or

$$E = V \times I \times t$$

If the temperature change, ΔT, caused by the addition of the known amount of energy is measured, the calibration factor for the calorimeter can be calculated from the expression:

$$CF = \frac{E}{\Delta T} = \frac{VIt}{\Delta T}$$

Temperature–time graphs in calorimetry

Measuring the temperature change, ΔT, that occurs during a calorimetry experiment is not always as simple as calculating the difference between the final temperature and the initial temperature of the water in the calorimeter. If a calorimeter is not perfectly insulated, it may slowly lose heat during and after the heater is operating. A more accurate determination of the temperature change (ΔT) can be achieved by plotting a graph of temperature against time before, during and after the calibration.

Two temperature–time graphs for the calibration of different calorimeters are shown in Figure 3.4.3. Figure 3.4.3a shows the results for a calorimeter with 'perfect' insulation and no heat loss. Figure 3.4.3b is typical of the results obtained using school laboratory calorimeters.

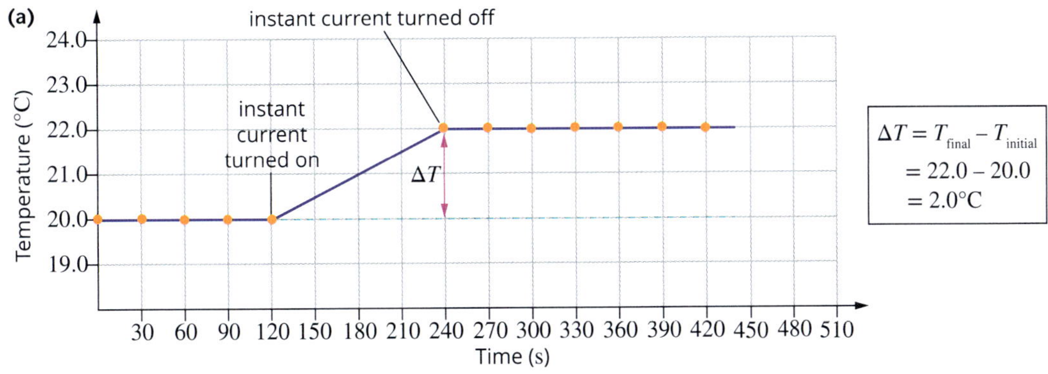

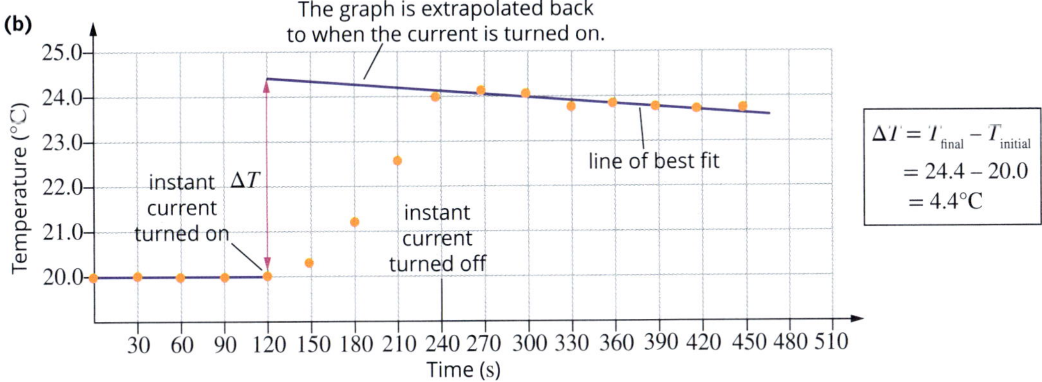

FIGURE 3.4.3 Graphs of temperature in a solution calorimeter before, during and after calibration. (a) These results would be obtained when a perfectly insulated calorimeter is calibrated. (b) These results might be obtained for a different calorimeter when the temperature readings are affected by heat loss and non-instantaneous heat transfer.

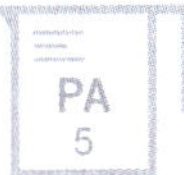

In Figure 3.4.3b, heat loss causes a negative slope to the line of the graph after the heater has been turned off. A more accurate estimate of the value of ΔT can be found by **extrapolating** the line back to the time when heating commenced. The temperature change between this value and the initial temperature can be measured. A delay in the transfer of heat through the water can be observed by the continuing increase in temperature after the heater has been turned off. This can also be accounted for by the extrapolation method, although it is important to understand that the accuracy of this approach is limited.

Worked example 3.4.1

CALCULATING THE CALIBRATION FACTOR OF A CALORIMETER BY ELECTRICAL CALIBRATION

A solution calorimeter was calibrated by passing 1.50 A through the electric heater for 50.5 s at a potential difference of 6.05 V. The temperature of the water in the calorimeter was initially 18.05°C and rose to 19.38°C during the calibration. Determine the calibration factor of the calorimeter.

Thinking	Working
Calculate the thermal energy released by the heater in the calorimeter when the electric current was passed through it. Use the equation: $E = VIt$	$E = VIt$ $= 6.05 \times 1.50 \times 50.5$ $= 458.29$ J
Calculate the temperature change during the calibration. Use the equation: $\Delta T = \Delta T_{final} - \Delta T_{initial}$	$\Delta T = \Delta T_{final} - \Delta T_{initial}$ $= 19.38 - 18.05$ $= 1.33$°C
Calculate the calibration factor by dividing the energy by the change in temperature. Use the equation: $CF = \frac{E}{\Delta T} = \frac{VIt}{\Delta T}$	$CF = \frac{E}{\Delta T}$ $= \frac{458}{1.33}$ $= 345$ J °C^{-1}

Worked example: Try yourself 3.4.1

CALCULATING THE CALIBRATION FACTOR OF A CALORIMETER BY ELECTRICAL CALIBRATION

A solution calorimeter was calibrated by passing 1.05 A through the electric heater for 120 s at a potential difference of 5.90 V. The temperature of the water in the calorimeter was initially 15.20°C and rose to 17.50°C during the calibration. Determine the calibration factor of the calorimeter.

If you know the calibration factor for a calorimeter, you can use it to determine the energy change that is responsible for a temperature change during a reaction in the calorimeter. These calculations are discussed later in this section.

Chemical calibration of solution calorimeters

A solution calorimeter may also be calibrated by performing a chemical reaction in the calorimeter that releases a known quantity of thermal energy and then measuring the resultant rise in temperature.

A highly soluble salt such as potassium nitrate (KNO_3) could be used to calibrate a solution calorimeter. The **enthalpy of solution**, ΔH, of potassium nitrate is known to be +34.9 kJ mol^{-1}.

Calibration is achieved by dissolving a known amount (in moles) of potassium nitrate in the calorimeter. The change in temperature (ΔT) is measured and used to calculate the calibration factor, as shown in Worked example 3.4.2.

The energy released or absorbed during calibration is calculated using ΔH for the reaction and the amount of reactant (n):

$$E = n \times \Delta H$$

The calibration factor is then calculated using:

$$CF = \frac{E}{\Delta T}$$

Worked example 3.4.2

CALCULATING THE CALIBRATION FACTOR OF A CALORIMETER BY CHEMICAL CALIBRATION

A solution calorimeter was calibrated by completely dissolving 15.0 g of potassium nitrate in 200 mL of water in a calorimeter. ($M(KNO_3)$ = 101.3 g mol^{-1}; ΔH = +34.9 kJ mol^{-1}). The temperature of the water in the calorimeter decreased from 20.5°C to 10.0°C during the calibration. Determine the calibration factor of the calorimeter.

Thinking	Working
Determine the amount, in mol, of potassium nitrate that dissolves. Use the equation: $n = \frac{m}{M}$	$n = \frac{m}{M}$ $= \frac{15.0}{101.3}$ = 0.148 mol
Calculate the thermal energy absorbed from the water when the potassium nitrate dissolves in the calorimeter. Use the equation $E = n \times \Delta H$, noting that the energy will be in kJ.	$E = n \times \Delta H$ $= 0.148 \times 34.9$ = 5.17 kJ
Calculate the temperature change during the calibration. Use the equation: $\Delta T = \Delta T_{final} - \Delta T_{initial}$	$\Delta T = \Delta T_{final} - \Delta T_{initial}$ = 20.5 − 10.0 = 10.5°C
Calculate the calibration factor, CF, by dividing the energy by the change in temperature. Use the equation: $CF = \frac{E}{\Delta T}$	$CF = \frac{E}{\Delta T}$ $= \frac{5.17}{10.5}$ = 0.492 kJ °C^{-1} = 492 J °C^{-1}

Worked example: Try yourself 3.4.2

CALCULATING THE CALIBRATION FACTOR OF A CALORIMETER BY CHEMICAL CALIBRATION

A solution calorimeter was calibrated by completely dissolving 8.20 g of potassium nitrate in 200 mL of water in a calorimeter. ($M(KNO_3)$ = 101.3 g mol^{-1}; ΔH = +34.9 kJ mol^{-1}.) The temperature of the water in the calorimeter decreased from 20.0°C to 13.5°C during the calibration. Determine the calibration factor of the calorimeter.

USING A CALIBRATED CALORIMETER TO DETERMINE THE ENTHALPY OF REACTION

A solution calorimeter may be calibrated by either the electrical or the chemical calibration methods described earlier in this section.

While it is inappropriate for a solution calorimeter to be used to determine the enthalpy of combustion for either a food or fuel, the enthalpy of reaction for many other reactions may be determined. These include acid–base reactions, reactions of a metal with an acid and dissolving a solid in water.

The calibration factor is used to determine the energy (E) that is responsible for the temperature change that occurs during the reaction:

$$E = CF \times \Delta T$$

Note that the value of ΔT is the change in temperature that occurs when the reaction is carried out in the calorimeter, not when the calorimeter is calibrated.

The enthalpy change, ΔH, in kJ mol^{-1}, is calculated by dividing the energy change, in kJ, by the amount of the limiting reactant.

$$\Delta H = \frac{E}{n}$$

Worked example 3.4.3

CALCULATING THE ENTHALPY OF SOLUTION USING A SOLUTION CALORIMETER

A solution calorimeter has a calibration factor of 400 J °C^{-1}. If the temperature decreases by 1.375°C when 9.00 g of glucose (M = 180.0 g mol^{-1}) is dissolved in water in the calibrated calorimeter, calculate the enthalpy of solution of glucose.

Thinking	Working
Calculate the amount of reactant in moles, using the equation: $n = \frac{m}{M}$	$n = \frac{m}{M}$ $= \frac{9.00}{180.0}$ $= 0.0500$ mol
Calculate the heat energy released or absorbed, in kJ, using the equation: $E = CF \times \Delta T$	$E = CF \times \Delta T$ $= 400 \times 1.375$ $= 550$ J $= 0.550$ kJ
Calculate the energy released or absorbed per mole. $\Delta H = \frac{E}{n}$	$\Delta H = \frac{E}{n}$ $= \frac{0.550}{0.0500}$ $= 11.0$ kJ mol^{-1}
State the enthalpy of solution with the correct sign, remembering that a temperature increase indicates an exothermic process and a temperature decrease indicates an endothermic process.	$\Delta H = +11.0$ kJ mol^{-1} (3 significant figures)

Worked example: Try yourself 3.4.3

CALCULATING THE ENTHALPY OF SOLUTION

A solution calorimeter has a calibration factor of 396.4 J °C^{-1}. If the temperature decreases by 1.65°C when 41.587 g of sucrose, $C_{12}H_{22}O_{11}$ (M = 342.0 g mol^{-1}), is dissolved in water in the calibrated calorimeter, calculate the enthalpy of solution of sucrose.

3.4 Review

SUMMARY

- Calorimetry is the experimental method by which the heat energy released or absorbed in a chemical reaction or physical process is measured.
- A calorimeter is an instrument that measures energy changes in a reaction. It is made up of an insulated container of water in which the reaction occurs, with a stirrer and thermometer to measure the temperature change during the reaction. A lid is an important part of the insulation.
- A solution calorimeter is an insulated container that holds a known volume of water and in which a reaction in solution, such as dissolution of a solid or a neutralisation reaction, can be carried out.
- Calorimeters can be calibrated to establish the relationship between the energy transferred to the water and the temperature change in the calorimeter. Calibration involves adding a known quantity of heat energy from an electrical source or from a chemical source.
- For electrical calibration of a calorimeter, the calibration factor is calculated using the equation $CF = \frac{E}{\Delta T}$ where $E = VIt$. (V is the voltage, in volts, I is the current, in amps, and t is the time, in seconds.)
- During electrical calibration, the value of the temperature change, ΔT, may be affected by heat loss from the calorimeter, so a temperature–time graph may be used to determine a more accurate value of ΔT.
- When a reaction is carried out in a calibrated calorimeter, the energy released by the reaction is determined using the equation $E = CF \times \Delta T$.

KEY QUESTIONS

Knowledge and understanding

1 Use the following terms to complete the statements about calorimeters:

absorbed; accuracy; precision; decreases; endothermic; exothermic; increases; lid; lost; released; required; stays the same; thermometer

Insulating a calorimeter improves the __________ of measurement of the quantity of energy __________ or absorbed by a chemical reaction.

Heat energy can be __________ from a calorimeter, so a __________ is a useful form of insulation.

If the reaction occurring in a calorimeter is exothermic, the temperature of the water __________. If the reaction occurring is __________ the temperature of the water in a calorimeter decreases.

2 A solution calorimeter was calibrated by completely dissolving 6.52 g of ammonium chloride in 200 mL of water in a calorimeter. ($M(NH_4Cl) = 53.5$ g mol^{-1}; $\Delta H = +14.8$ kJ mol^{-1}.) The temperature of the water in the calorimeter decreased from 19.5°C to 14.3°C during the calibration. Determine the calibration factor of the calorimeter.

3 A solution calorimeter has a calibration factor of 389 J °C^{-1}. If the temperature decreases by 2.34°C when 9.60 g of citric acid, $C_6H_8O_7$ ($M = 192.0$ g mol^{-1}), is dissolved in water in the calibrated calorimeter, calculate the enthalpy of solution of citric acid.

4 The enthalpy of solution of sodium hydrogen carbonate, $NaHCO_3$ ($M = 84.0$ g mol^{-1}), can be determined using solution calorimetry. When 15.00 g of sodium hydrogen carbonate was dissolved in water in a calibrated calorimeter, the temperature decreased by 4.20°C. The calibration factor of the calorimeter was 674.3 J °C^{-1}. Calculate the enthalpy of solution of sodium hydrogen carbonate, including the sign to indicate whether it is an exothermic or endothermic reaction.

continued over page

3.4 Review *continued*

Analysis

5 A simple polystyrene calorimeter was set up for an experiment. A 100 g mass of deionised water was used during the electrical calibration. The temperature of the calorimeter increased by 9.5°C when 3.8 A of current was passed through it at a potential of 6.0 V for 5 minutes.

a Determine the calibration factor for the calorimeter.

b What would be the effect on the calibration factor obtained from this experiment if 50 g of water is used instead of 100 g?

6 A solution calorimeter was calibrated by the electrical method, using a current of 4.10 A for 150 s with a voltage of 5.75 V. The temperature–time graph for the calibration is shown in the figure below. The heater was turned on at $t = 120$ s and turned off at $t = 270$ s. Calculate the calibration factor for this calorimeter.

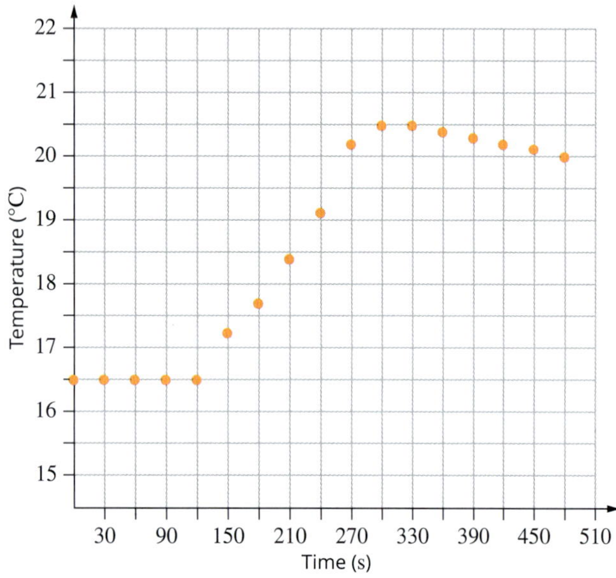

Graph of temperature against time during electrical calibration of a solution calorimeter

7 Several mistakes or errors that could occur when calibrating a solution calorimeter are listed in the following table. For each mistake or error, indicate the likely effect on the calculated calibration factor by placing a tick in the appropriate column, and write an explanation for your answer in the final column.

Mistake or error	Calibration factor is too small	Calibration factor would be correct	Calibration factor is too large	Explanation for answer
The lid is left off during calibration.				
During electrical calibration, the voltmeter had a systematic error, making it read too low.				
During chemical calibration, not all of the solid KNO_3 dissolved in the water of the calorimeter.				

3.5 Energy from fuels and food

Just as fossil fuels and biofuels supply you with energy for purposes such as heating, lighting and transport, food supplies the energy required for the millions of chemical reactions that occur in your body. You learnt in Chapter 2 that some fuels provide more energy per gram than others, and similarly, different foods provide different quantities of energy. When the energy contained in these fuels and food is transformed into forms of energy that we can use, much energy is lost in the transformation process.

In this section, you will learn how to calculate the efficiency of energy transformations and you will learn about the different energy values of food.

ENERGY TRANSFORMATIONS

When fuels are used or food is digested, the chemical energy is converted into a different form of energy. For example, a competitive cyclist may eat energy bars. Much of the chemical energy of the food is converted in the cyclist's body to mechanical and kinetic energy.

The conversion of the chemical energy of the food to the kinetic energy of the cyclist is an example of an energy transformation—energy is converted from one form to a different form. The use of fuels, including food, involves energy transformations. The chemical energy in a log on a campfire can be converted into thermal energy to heat a kettle of water over the fire.

When energy transformations occur, the total amount of energy is unchanged because energy cannot be created or destroyed. However, not all of the energy is converted into one specific form. In the case of the burning logs shown in Figure 3.5.1, very little of the logs' chemical energy is converted into heat energy in the water.

FIGURE 3.5.1 The chemical energy in the logs of the campfire is converted to thermal energy to boil the water in the kettle.

Energy transformation efficiency

The first law of thermodynamics tells us that energy is always conserved. However, when energy is being converted from one form into others, only some of the converted energy is likely to be useful. Scientists are interested in the efficiency of energy transformations. Many devices that we use every day can be described as **energy converters**. They transform energy from non-renewable or renewable sources into the form of energy we require, such as electrical or mechanical energy.

The term **energy transformation efficiency** is used to describe the percentage of energy from a source that is converted to useful energy. For example, the engine in a petrol-fuelled car converts the chemical energy in petrol into thermal energy by combustion. You learnt in Chapter 2 that the combustion of octane, the main component of petrol, occurs according to the equation:

$$C_8H_{18}(l) + 25O_2(g) \rightarrow 16CO_2(g) + 18H_2O(l)$$

Some of the heat released by the reaction expands the gases in the engine's combustion chamber. As you can see in Figure 3.5.2a on the following page, these gases push the pistons along the cylinders, causing the crankshaft to rotate and drive the wheels of the car.

In Figure 3.5.2b the distribution of energy from the combustion of the petrol is illustrated. Only 20 J out of every 100 J is used to propel the car. The majority of the energy is transferred through the exhaust and the cooling system to the outside air as waste heat.

(a)

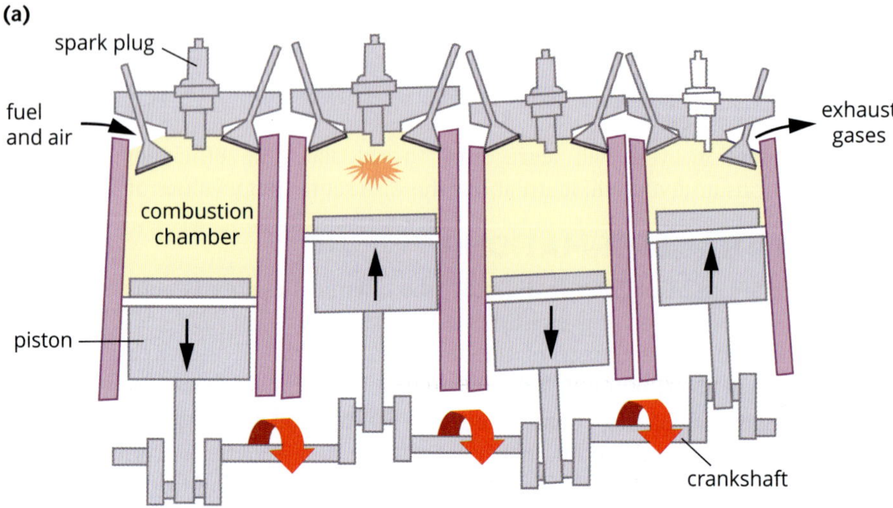

(b)

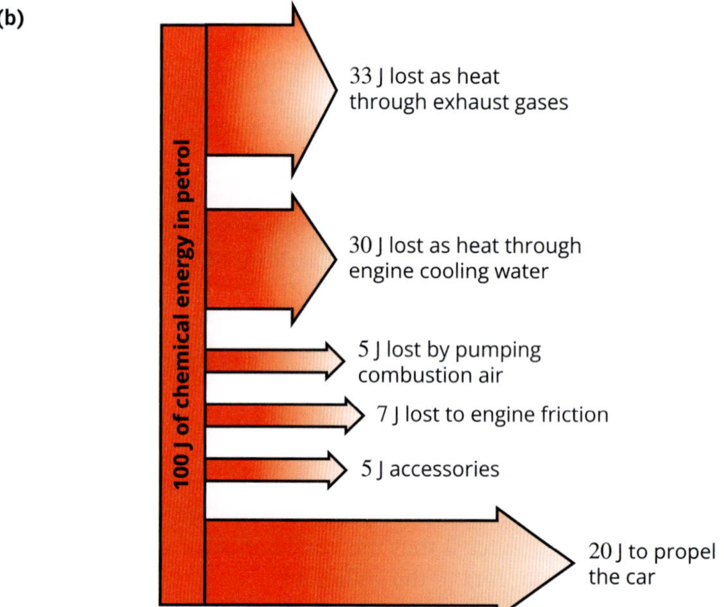

FIGURE 3.5.2 Energy transformations in a petrol-fuelled car. (a) Combustion of the petrol generates heat and moves the pistons. (b) Only 20% of the energy released by the petrol is used to move the car.

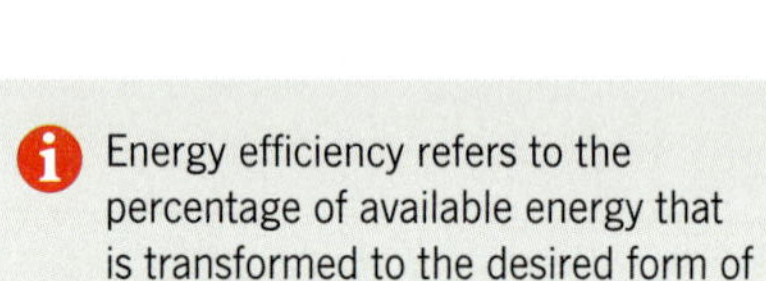

TABLE 3.5.1 Energy efficiency of some energy converters

Device	Energy transformation	Typical efficiency
coal-fired power station	chemical to electrical	30%
lead–acid battery	chemical to electrical	70%
electric motor (as found in an electric car)	electrical to mechanical	77%
petrol engine	chemical to mechanical	20%
diesel engine	chemical to mechanical	30%
hydrogen fuel cell	chemical to electrical	60%

The efficiency of an energy conversion can be viewed as an expression of the balance between energy input and useful energy produced.

$$\% \text{ energy efficiency} = \frac{\text{useful energy}}{\text{energy input}} \times \frac{100}{1}$$

The maximum value of **energy efficiency** is 100%, but this value is not considered attainable in real-world situations.

In the case of the petrol-fuelled car described above, the energy efficiency is only 20%.

In comparison to the various devices described in Table 3.5.1, the energy efficiency of the human body is around 25%.

Throughout this book you will encounter further discussion of energy efficiencies as researchers are always striving to improve the sustainability of a process by increasing its efficiency.

In Section 3.3 you learnt how to determine the enthalpy of combustion of a fuel, such as an alcohol (see Figure 3.3.4, page 123). When this experiment is carried out in a school laboratory, the results are rarely equal to the tabulated values due to heat loss to the surroundings. The energy efficiency of this experiment can be calculated using the theoretical value for the enthalpy of combustion and the calculated value from your results. This is illustrated in Worked example 3.5.1.

Worked example 3.5.1

CALCULATING THE ENERGY TRANSFORMATION EFFICIENCY OF AN EXPERIMENT

A pair of students carried out an experiment to measure the enthalpy of combustion of methanol. They burned methanol in a spirit burner under a can of water and measured the temperature change of the water and the mass change of the spirit burner. From this experiment they determined the enthalpy of combustion of methanol to be –375 kJ mol^{-1}.

The theoretical enthalpy of combustion of methanol was recorded in their data book as –726 kJ mol^{-1}. Calculate the percentage efficiency of the energy transformation from chemical energy (of the methanol) to thermal energy (in the water).

Thinking	Working
Identify which energy best matches the description of useful energy and which is the energy input.	The 'useful energy' quantity is the energy that heated the water and enabled the calculation of the experimental value, –375 kJ mol^{-1}. The energy input quantity is the theoretical enthalpy of combustion: –726 kJ mol^{-1}
Use the formula $\%\text{ energy efficiency} = \frac{\text{useful energy}}{\text{energy input}} \times \frac{100}{1}$ to calculate the percentage energy transformation efficiency.	$\%\text{ energy efficiency} = \frac{\text{useful energy}}{\text{energy input}} \times \frac{100}{1}$ $= \frac{-375}{-726} \times \frac{100}{1}$ $= 51.7\%$

Worked example: Try yourself 3.5.1

CALCULATING THE ENERGY TRANSFORMATION EFFICIENCY OF AN EXPERIMENT

A pair of students carried out an experiment to measure the enthalpy of combustion of ethanol. They burned ethanol in a spirit burner under a can of water and measured the temperature change of the water and the mass change of the spirit burner. From this experiment they determined the enthalpy of combustion of ethanol to be –820 kJ mol^{-1}.

The theoretical enthalpy of combustion of ethanol was recorded in their data book as –1360 kJ mol^{-1}. Calculate the percentage efficiency of the energy transformation from chemical energy (of the ethanol) to thermal energy (in the water).

In some cases, you will know the percentage efficiency of a process. Worked example 3.5.2 shows you how to include this in your calculations.

Worked example 3.5.2

ALLOWING FOR % EFFICIENCY IN ENERGY TRANSFORMATIONS

A propane burner is used to heat a saucepan containing water. The % efficiency of this process is 35.0%. Calculate the energy transferred to the water from the complete combustion of 0.320 mol of propane. The heat of combustion is 2220 kJ mol^{-1}.

Thinking	Working
Calculate the amount, in mol, of the fuel.	0.320 mol (given in the question)
Calculate the quantity of energy released (the energy input) by the combustion of the fuel.	Energy released $= n \times \Delta H$ $= 0.320 \times 2220$ $= 710.4$ kJ
Rearrange the formula $\% \text{ energy efficiency} = \frac{\text{useful energy}}{\text{energy input}} \times \frac{100}{1}$ to Useful energy = energy input × % energy efficiency so that the energy available to heat the water can be determined.	Useful energy = energy input × % energy efficiency $= 710 \times \frac{35}{100}$ $= 249$ kJ

Worked example: Try yourself 3.5.2

ALLOWING FOR % EFFICIENCY IN ENERGY TRANSFORMATIONS

The % efficiency of octane (petrol) in a particular car model is 42.0%. Calculate the energy released from the complete combustion of 22.0 g of octane. The heat of combustion of octane is 5460 kJ mol^{-1}.

In other cases, you will know the percentage efficiency of a process and the energy that was transferred to the water, but you will be asked to determine the theoretical heat of combustion of the fuel. Worked Example 3.5.3 shows you how to include this in your calculations.

Worked example 3.5.3

CALCULATING HEAT OF COMBUSTION USING % EFFICIENCY IN ENERGY TRANSFORMATIONS

A student burnt 0.0250 mol of propan-2-ol in a spirit burner to heat 200 g of water in a can. The % efficiency of this process was 65.0%. During the experiment, the temperature of the water in the can increased by 39.0°C. Calculate the heat of combustion of propan-2-ol. (Remember that the specific heat capacity of water is 4.18 J °C^{-1} g^{-1}.)

Thinking	Working
Calculate the quantity of energy that heated the water—this is the useful energy. Use the formula $q = mc\Delta T$ to calculate the useful energy, because the can does not have a calibration factor.	$q = mc\Delta T$ $= 200 \times 4.18 \times 39.0$ $= 32\,604$ J $= 32.604$ kJ
Convert the % energy efficiency to a decimal to assist with calculations. Divide % energy efficiency by 100.	$\frac{\%\text{ energy efficiency}}{100}$ $= \frac{65.0}{100}$ $= 0.650$
Rearrange the formula $\%\text{ energy efficiency} = \frac{\text{useful energy}}{\text{energy input}} \times \frac{100}{1}$ to $\text{energy input} = \frac{\text{useful energy}}{\%\text{ energy efficiency (as a decimal)}}$ so that the energy released by the fuel can be determined.	$\text{energy input} = \frac{32.604}{0.650}$ $= 50.16$ kJ
Remembering that the energy released is represented by the energy input, rearrange the formula: $\text{energy released} = n \times \Delta H$ to: $\Delta H = \frac{\text{energy released}}{n}$	$\Delta H = \frac{50.16}{0.0250}$ $= 2006$ $= 2.01 \times 10^3$ kJ mol^{-1} The heat of combustion of propan-2-ol is 2.01×10^3 kJ mol^{-1}

Worked example: Try yourself 3.5.3

CALCULATING HEAT OF COMBUSTION USING % EFFICIENCY IN ENERGY TRANSFORMATIONS

A student burnt butan-1-ol in a spirit burner to heat 200 g of water in a can. The % efficiency of this process was 57.0%. During the combustion of 0.0105 mol of butan-1-ol, the temperature of the water in the can increased by 19.2°C. Calculate the heat of combustion of butan-1-ol. (Remember that the specific heat capacity of water is 4.18 J °C^{-1} g^{-1}.)

CHEMFILE

The importance of cellulose

Even though humans are unable to digest and obtain the energy available from cellulose, a diet high in cellulose provides 'bulk' to aid the passage of food through the digestive system. Such 'bulk' helps prevent constipation and reduces the risk of bowel cancer, also known as colorectal cancer. Bowel cancer is the second most common cancer in men and women in Australia and is more common in people over the age of 50. High red meat consumption (in excess of 500 g per week), especially processed meats, increases the risk of developing bowel cancer. This risk can be reduced by eating a healthy diet with plenty of fresh fruit and vegetables, as shown in the figure below.

A diet high in fruit and vegetables reduces the risk of bowel cancer.

The energy content of the three major nutrients is:

carbohydrates	16 kJ g^{-1}
fats	37 kJ g^{-1}
proteins	17 kJ g^{-1}

ENERGY AVAILABLE TO THE BODY

Food supplies your body with **nutrients** that are needed to keep you alive. These nutrients are large **biomolecules** that are used by the body to provide energy, regulate growth, and maintain and repair body tissue. The three main nutrient classes are carbohydrates, proteins and fats. The **energy value** or energy content of these nutrients was discussed in Section 2.5. The energy released when food is burned is often greater than the energy that is available for the human body to use after the food has been digested. This can be due to:

- incomplete absorption of nutrients by the body after digestion of the food
- incomplete oxidation of nutrients, such as proteins and insoluble fibre
- heat loss; not all of the energy released by the digestion of food is available for use in cells as some is lost as waste heat.

Even within a particular nutrient class, such as the carbohydrates, the energy available to the body can vary. Starch is readily digested by humans and is the primary source of glucose, so most of the energy from starch is obtained during digestion. On the other hand, humans lack the enzyme cellulase, which is required to break down cellulose so that energy can be obtained. Cellulose in foods such as vegetables is often referred to as **dietary fibre** and provides no energy.

A balanced diet is made up of a variety of foods containing carbohydrates, proteins and fats. Each of these three major nutrients provides a different quantity of energy per gram. In Section 2.3 you learnt that the energy available to the body from carbohydrates is 16 kJ g^{-1}, from fats and oils is 37 kJ g^{-1}, and from proteins is 17 kJ g^{-1}.

You can use these values to calculate the energy value of different foods.

CALCULATING THE ENERGY VALUE OF FOODS

While you may distinguish between different foods based on how delicious they are to you, a more scientific difference is the variation in percentages of carbohydrates, fat and protein in each food. In Figure 3.5.3 you can see that foods with a high percentage of carbohydrates and fat also have a high energy content.

Since it is important to know how much energy is available from particular foods, processed foods are labelled with their energy content, as well as percentages of carbohydrates, fats and protein.

Food	Carbohydrate (g/100g)	Fat (g/100g)	Protein (g/100g)	Energy content (kJ/100g)
milk chocolate	52	30	8	2130
doughnut	51	23	5	1803
apple	14	0.2	0.3	247

FIGURE 3.5.3 Foods that have a high percentage of carbohydrate and fat have a high energy content.

When the percentage composition of a food is known, its energy value can be calculated using the percentages and the energy available to the body for each nutrient. The energy value of foods may be expressed in kJ/100 g or kJ g^{-1}. The composition and energy values of a range of foods are given in Table 3.5.2.

TABLE 3.5.2 The composition and energy value of a range of foods

Food	Carbohydrate (%)	Protein (%)	Fats and oils (%)	Energy value (kJ g^{-1})
white rice	79	7	negligible	15.2
wholemeal bread	39	11	4	9.7
avocados	6	2	17	7.2
roasted peanuts	18	26	50	24.3
pizza (with cheese)	31	11	8	10.0
apples	53	4	8	11.7
almonds	19	19	54	25.4

Worked example 3.5.4

CALCULATING THE ENERGY VALUE OF FOODS

The labelling on a sample of unsalted cashews indicates they contain 29.0% carbohydrates, 18.0% protein and 46.0% fat. The remaining 7.0% is water, which does not supply energy.

Calculate the energy value of the cashews, in kJ g^{-1}.

Thinking	Working
Use the information on page 142 to determine the available energy for each nutrient type.	carbohydrate: 16 kJ g^{-1} protein: 17 kJ g^{-1} fat: 37 kJ g^{-1}
Assuming that there is 100 g of the sample, multiply each percentage of the nutrient by the available energy per gram for the nutrient type.	carbohydrate: 29.0 g × 16 kJ g^{-1} = 464 kJ protein: 18.0 g × 17 kJ g^{-1} = 306 kJ fat: 46.0 g × 37 kJ g^{-1} = 1702 kJ
Find the sum of the energies for the three nutrient types and divide by 100 to find the energy value in kJ g^{-1}.	$\text{energy value} = \frac{464 + 306 + 1702}{100}$ $= 24.7 \text{ kJ g}^{-1}$

Worked example: Try yourself 3.5.4

CALCULATING THE ENERGY VALUE OF FOODS

The labelling on a sample of white bread indicates it contains 53.0% carbohydrates, 8.0% protein and 4.0% fat. The remaining 35.0% is water, which does not supply energy.

Calculate the energy value of the bread, in kJ g^{-1}.

3.5 Review

OA ✓✓

SUMMARY

- Energy efficiency refers to the percentage of available energy that is transformed to the desired form of energy, as calculated by:

$$\% \text{ energy efficiency} = \frac{\text{useful energy}}{\text{energy input}} \times \frac{100}{1}$$

- The energy value of fuels and foods is measured in kJ g^{-1}.
- The energy value of a food can be calculated using data for the available energy of the various components of the food and the percentages of each component of the food.

KEY QUESTIONS

Knowledge and understanding

1 Calculate the mass of each of the following nutrients that supplies the same energy as that released by 10.0 g of carbohydrates.
  a protein
  b fat

2 The use of coal to produce electrical energy requires several energy transformations.
  a What is an energy transformation?
  b How do energy transformations impact on the usefulness of a fuel?

3 An 18.0 g chocolate frog contains 5.3 g fat, 11.0 g of carbohydrates and 1.4 g protein.
  a Calculate the total energy in the chocolate frog.
  b Calculate the percentage of total energy that is a consequence of the fat content.

Analysis

4 a Butanol has a heat of combustion of 2670 kJ mol^{-1}. Calculate the mass, in kg, of butanol that is required to produce 200 MJ of energy.
  b A 375 g sample of butanol is used to heat a drum of water. The efficiency of this process is 38.0%. Calculate the quantity of energy transferred to the water.

5 Labelling on a jar of pesto indicates it contains 3.3% carbohydrates, 3.9% protein and 21.4% fat. The remaining mass is water.
Calculate the energy value of the pesto, in kJ g^{-1}.

6 A 15.0 g sample of butter is reported as having 425 kJ of energy. If all of this energy came from fat sources in the butter, calculate the percentage by mass of fat in the butter sample.

7 A student conducted a practical investigation to experimentally determine the heat of combustion of propan-1-ol using the apparatus shown in Figure 3.3.3 (page 123). The following data is obtained:

mass of spirit burner before burning	3.565 g
mass of spirit burner after burning	3.284 g
mass of water	100 g
initial temperature of water	18.6°C
final temperature of water	30.8°C

  a Determine the quantity of energy, in kJ, that was used to heat the water.
  b Calculate the heat of combustion, in kJ mol^{-1}, of propan-1-ol according to this data.
  c The heat of combustion of propan-1-ol obtained from a secondary source is 2021 kJ mol^{-1}. What was the percentage energy efficiency of this calculation of the heat of combustion of propan-1-ol?
  d Describe three ways in which the student could improve this experiment to increase the accuracy of the result.

8 A student burnt 0.0325 mol of methanol in a spirit burner under 250 g of water in a can. The temperature of the water increased by 16.9°C during the experiment. If the % efficiency of this process was 75.0%, calculate the heat of combustion of methanol.

Chapter review

KEY TERMS

amount
biomolecule
bomb calorimeter
calibrated
calibration factor
calorimeter
calorimetry
coefficient
complete combustion
dietary fibre
endothermic
energy content
energy converter
energy efficiency
energy transformation
energy transformation efficiency
energy value
enthalpy of solution
excess reactant
exothermic
extrapolating
greenhouse gas
heat of combustion
in excess
incomplete combustion
limiting reactant
mass–mass stoichiometry
mass–volume stoichiometry
molar volume
mole ratio
nutrient
reagent
solution calorimeter
specific heat capacity
standard laboratory conditions, SLC
stoichiometry
volume-volume stoichiometry

REVIEW QUESTIONS

Knowledge and understanding

1 Which one of the following nutrients will produce the most energy when consumed?

- **A** 15 g protein
- **B** 8 g fat
- **C** 20 g carbohydrate
- **D** 5 g fat and 10 g protein

2 The enthalpy of combustion of propan-1-ol is determined experimentally to be -1800 kJ mol^{-1}, while the theoretical value is -2018 kJ mol^{-1}. Which of the following is the correct value of the percentage energy transformation efficiency in this experiment?

- **A** 10.8%
- **B** 12.1%
- **C** 50.0%
- **D** 89.2%

3 Propane (C_3H_8) burns in oxygen according to the equation:

$$C_3H_8(g) + 5O_2(g) \longrightarrow 3CO_2(g) + 4H_2O(g)$$

6.70 g of propane was burned in excess oxygen.

- **a** What mass of carbon dioxide would be produced?
- **b** What mass of oxygen would be consumed in the reaction?
- **c** What mass of water would be produced?

4 5.0 g of magnesium burns completely in 20 g of oxygen gas according to the equation:

$$2Mg(s) + O_2(g) \longrightarrow 2MgO(s).$$

What mass of oxygen gas is in excess?

5 Large quantities of coal are burned in Australia to generate electricity, in the process generating significant amounts of the greenhouse gas carbon dioxide. The equation for this combustion reaction can be written as:

$$C(s) + O_2(g) \longrightarrow CO_2(g)$$

Determine the mass of carbon dioxide produced, in tonnes, by the combustion of 1.0 tonne (10^6 g) of coal, assuming that the coal is pure carbon.

6 Octane is one of the main constituents of petrol. It burns according to the equation:

$$2C_8H_{18}(g) + 25O_2(g) \rightarrow 16CO_2(g) + 18H_2O(g)$$

Calculate the volume of octane used if 50.0 L of carbon dioxide was produced, and all gas volumes were measured at SLC.

7 Calculate the quantity of energy that has been transferred to 50.0 g of water in a test tube when the temperature of the water increases from 18.5°C to 29.8°C.

8 Calculate the quantity of energy released, in kJ, when 806 g of butan-1-ol reacts with excess air according to the equation:

$$C_4H_9OH(g) + 6O_2(g) \rightarrow 4CO_2(g) + 5H_2O(l)$$
$$\Delta H = -2677 \text{ kJ}$$

9 Propane (C_3H_8) burns in oxygen according to the equation:

$$C_3H_8(g) + 5O_2(g) \rightarrow 3CO_2(g) + 4H_2O(g)$$

300 g of propane reacts with 2.00 kg of oxygen.

- **a** Determine which reactant is in excess and the mass by which it is in excess.
- **b** Calculate the volume of carbon dioxide produced, at SLC.
- **c** Calculate the volume of carbon dioxide produced if the conditions were 15°C and 150 kPa.

10 The equation $q = mc\Delta T$ describes the relationship between heat energy and the temperature change of a substance being heated. It is used to determine how much energy has been absorbed by a material, such as a mass of water, when its temperature increases by a measured amount.
Which one of the following statements about this equation is correct?
- **A** The specific heat capacity, *c*, depends on the material that is generating the heat energy.
- **B** The mass, *m*, represents the mass of the material that is being heated.
- **C** *q* is measured in kJ $°C^{-1}$.
- **D** ΔT refers to the change in temperature of the fuel generating the heat.

11 Which one of the following is most likely to be the temperature change in 100.0 mL of water in a beaker with no insulation and no lid, when 2.10 kJ of energy is produced by a burning biscuit under the beaker?
- **A** 3.20°C
- **B** 5.02°C
- **C** 5.50°C
- **D** 6.30°C

12 Pressed peat has a lower energy content than coal, and produces more carbon dioxide, and less smoke. It has a lower sulfur content, and is considered renewable.
A sample of pressed peat has an energy content of 12.8 kJ g^{-1}. Calculate the mass of water, in kg, that could be heated from 20.0°C to 60.0°C by the transfer of all of the energy from the combustion of a sample of peat with a mass of 17.0 g.

13 When 1.01 g of solid potassium nitrate was dissolved in a calorimeter, the temperature dropped by 0.672°C. When the calorimeter was calibrated, a current of 1.50 A applied at a potential difference of 5.90 V for 60.0 s caused a temperature rise of 0.456°C.
- **a** Write an equation for the dissolution of potassium nitrate.
- **b** Determine the calibration factor of the calorimeter.
- **c** Calculate the energy change during the reaction.
- **d** Calculate the enthalpy change of reaction for the equation written in part **a**.

14 Labelling for a stuffed pepper indicates it contains 16% carbohydrates, 13% protein and 6.0% fat. The rest of the pepper is water.
Calculate the energy value of the stuffed pepper, in kJ g^{-1}.

Application and analysis

15 A calorimeter is calibrated by the electrical calibration method. The temperature rise was measured as the difference between the highest temperature reached and the initial temperature.
Describe the most likely effect on the calibration factor if the quantity of current passing through the calorimeter was higher than measured.

16 Energy bars are often popular with runners and cyclists due to their portability and high energy derived largely from carbohydrates and proteins. One particular energy bar contains 4.0 g fat, 44.0 g carbohydrates and 10.0 g of protein.
- **a** Calculate the total energy in one energy bar.
- **b** An athlete running with moderate intensity will consume approximately 2400 kJ per hour of running. Calculate how many energy bars they would need consume to sustain their energy needs if they were going for a 1.5 hour-long run.

17 Methane burns in excess oxygen according to the equation:

$$CH_4(g) + 2O_2(g) \longrightarrow CO_2(g) + 2H_2O(g)$$

This reaction produces 5.0 L of carbon dioxide at SLC. Assuming all volumes are measured at the same temperature and pressure, calculate the:
- **a** volume of methane used
- **b** volume of oxygen used
- **c** mass of greenhouse gases produced.

18 Octane (C_8H_{18}), the principal component of petrol, undergoes combustion with a plentiful supply of oxygen according to the equation:

$$2C_8H_{18}(l) + 25O_2(g) \longrightarrow 16CO_2(g) + 18H_2O(l) \quad \Delta H = -10\ 900 \text{ kJ}$$

- **a** What mass of octane must be burnt in order to yield 100 kJ of energy?
- **b** Given that density is calculated as $\frac{\text{mass}}{\text{volume}}$ and the density of octane is 0.698 g mL^{-1}, determine the energy released when 50.0 L of liquid octane undergoes complete combustion.

19 An indoor gas heater burns propane (C_3H_8) at a rate of 12.7 g per minute. Calculate the minimum mass of oxygen per minute, in g, that needs to be available for the complete combustion of propane.

20 Propane (C_3H_8) burns completely in oxygen according to the equation:

$$C_3H_8(g) + 5O_2(g) \rightarrow 3CO_2(g) + 4H_2O(g)$$

a When 80 mL propane and 500 mL of oxygen are reacted, both at SLC, which is in excess and by what volume?

b What is the total volume of greenhouse gases produced in the reaction?

c What change in the total volume of all of the gases has occurred as a result of this reaction?

d The amount of energy released per tonne of carbon dioxide produced is determined to be 1.68×10^4 MJ t^{-1}. Determine the ΔH value, in kJ mol^{-1}, for the thermochemical equation. (1 tonne = 10^6 g). Show your working.

21 In a sealed reaction to produce energy from the combustion of ethanol, 80.0 L of carbon dioxide was collected at SLC. The reaction is

$$C_2H_5OH(g) + 3O_2(g) \rightarrow 2CO_2(g) + 3H_2O(g)$$

If 10.0 g of ethanol was in the sealed container when it was ignited, was the ethanol limiting or in excess? Explain your answer.

22 A peanut with a mass of 1.340 g was burned under a steel can containing 200 g of water. After the flame went out, 0.640 g of the peanut remained and the temperature of the water had risen by 17.3°C. Calculate the energy content of the peanut in kJ g^{-1}, assuming all the energy from the burning nut is transferred to the water.

23 0.254 g of black coal was burnt and used to heat 300 g of water. The temperature of the water rose from 18.25°C to 24.92°C. Calculate the heat of combustion, in kJ g^{-1}, of the coal. Assume all of the heat released during combustion was used to heat the water.

24 A temperature rise of 5.52°C was observed when 0.0450 g of ethane (C_2H_6) was burnt and used to heat 100.0 g of water initially at 25°C. Use this information to write a balanced thermochemical equation for the complete combustion of ethane.

25 A 500 g mass of water in a beaker was heated using the energy released by the combustion of a 3.00 g piece of wood. The observed temperature rise was 22.9°C.

a Calculate the heat of combustion of wood in kJ g^{-1}.

b Would the heat of combustion you calculated in part **a** be higher or lower than the actual heat of combustion of the wood? Give reasons for your answer.

26 A solution calorimeter was calibrated electrically. A potential of 9.80 V was applied for 270 s with a current of 3.60 A and the temperature increased by 3.49°C. 3.00 g of sucrose (molar mass 342.0 g mol^{-1}) was dissolved completely in the same calibrated solution calorimeter. During the dissolving of the sucrose, the temperature of the water decreased by 18.14°C. What is the heat of solution of sucrose in kJ mol^{-1}?

27 The following figure shows the temperature–time graph plotted by the data-logging equipment used by a student to determine the change in temperature during the calibration of a solution calorimeter. The heater was turned on at t = 120 s and turned off at t = 300 s.

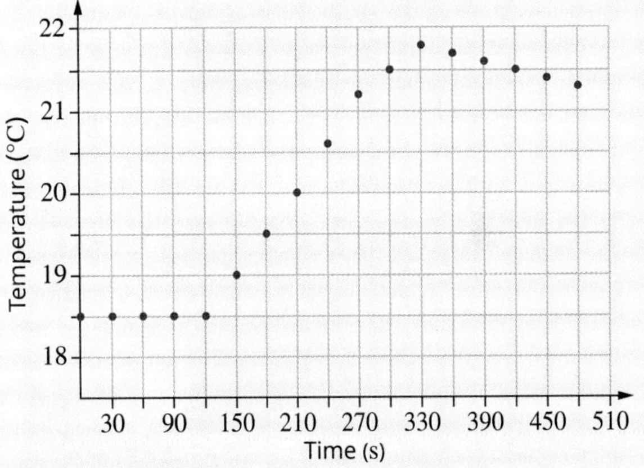

Temperature–time graph used for calibration of a calorimeter

The student ignored the graph and instead used the formula $\Delta T = T_{final} - T_{initial}$ to determine ΔT and hence the calibration factor.

The student then correctly made the additional measurements needed to determine the enthalpy change of a reaction.

Describe the effect of *not* compensating for the heat loss on the following.

a the value of ΔT

b the calibration factor calculated for the calorimeter

c the value of the enthalpy change measured by the calorimeter

28 A student is having difficulty with the calculations for the calibration of a solution calorimeter and the calculations that follow. The student wants to determine the heat energy per gram required to dissolve a packet serving of jelly crystals in water. The student's data for the calibration are shown in Tables 3.6.1 and 3.6.2.

TABLE 3.6.1 Calibration data

Volume of water in calorimeter (mL)	250
Voltage (V)	6.30
Current (amps)	2.40
Time (s)	210
Initial temperature (°C)	17.5
Final temperature (°C)	20.3

TABLE 3.6.2 Data from experiment in which jelly crystals are dissolved in water

Volume of water in calorimeter (mL)	250
Mass of jelly crystals (g)	85.0
Initial temperature (°C)	20.3
Final temperature (°C)	19.8

The student's calculations are as follows:

$$CF = 6.30 \times 2.40 \times 210 \times 2.8$$
$$= 8890 \text{ J}$$

Heat energy per gram to dissolve the jelly crystals:

$$E = CF \times \Delta T$$
$$= 8890 \times 0.5$$
$$= 4445 \text{ J}$$

$$\text{energy per gram} = \frac{85.0}{4445}$$
$$= 0.019 \text{ J g}^{-1}$$

0.019 J of energy is given out when 1 gram of jelly crystals dissolves. This is an endothermic reaction.

Describe any errors that the student has made in the calculations, and then complete the calculation correctly.

OA ✓✓

CHAPTER

04 Redox reactions

Some of the most colourful and exothermic reactions are classified as redox reactions. This group of reactions, such as combustion and respiration, includes reactions that provide the energy for our modern lifestyle and are vitally important to our existence.

In this chapter, you will learn how redox reactions involve simultaneous oxidation and reduction reactions, how they can be defined in terms of electron transfer, and how this definition can be extended by using the concept of oxidation number. You will find out how to write balanced half-equations that describe the transfer of electrons, and then how to combine these half-equations to create an overall equation for the reaction.

Key knowledge

- redox reactions as simultaneous oxidation and reduction processes, and the use of oxidation numbers to identify the reducing agent, oxidising agent and conjugate redox pairs **4.1, 4.2**
- the writing of balanced half-equations (including states) for oxidation and reduction reactions, and the overall redox cell reaction in both acidic and basic conditions **4.3**

4.1 Oxidation and reduction

Your everyday life depends on a large number of chemical reactions. Many of these are redox reactions (from *red*uction and *ox*idation). The bleaching of hair, corrosion of metals, extraction of metals from their ores, combustion of fuels, and reactions in batteries that produce electrical energy, as well as respiration and photosynthesis, are all redox reactions. The highly exothermic reaction between glycerol and potassium permanganate shown in Figure 4.1.1 is just one example of a redox reaction.

FIGURE 4.1.1 When glycerol is added to potassium permanganate, a vigorous reaction occurs in which the glycerol is oxidised by the potassium permanganate.

In this section, you will revise and extend the definitions of oxidation and reduction that were covered in Unit 2 Chemistry, and you will learn how they may be applied.

Oxidation and reduction occur simultaneously as electrons are transferred between the reactants.

TRANSFER OF ELECTRONS IN REDOX REACTIONS

In your study of Unit 2 Chemistry, you learnt that **redox reactions** involve the transfer of electrons from one chemical species to another. Redox reactions can be considered as occurring in two parts.

In these reactions:

- one of the reactants loses electrons in a process called **oxidation**
- one of the reactants gains electrons in a process called **reduction**.

Oxidation and reduction occur simultaneously as electrons are transferred between the reactants.

The mnemonic OIL RIG is a way to remember the difference between oxidation and reduction processes in terms of electron movement (Figure 4.1.2).

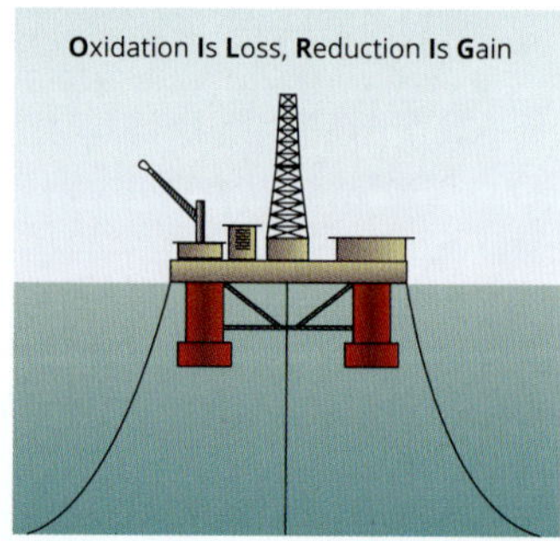

FIGURE 4.1.2 The mnemonic OIL RIG is a useful way to remember that oxidation is the loss of electrons and reduction is the gain of electrons.

Reduction and oxidation reactions

You will recall from Unit 1 Chemistry that metals usually have low electronegativities and their atoms contain 1, 2 or 3 electrons in their valence shells. When a metal atom loses the electrons in its valence shell, the atom becomes a stable positive ion with the electronic configuration of the nearest noble gas. This process is called oxidation. In comparison, non-metals have relatively high electronegativities and their atoms generally need to gain 1, 2 or 3 electrons to achieve the stability of a noble gas electronic configuration. When non-metal atoms gain electrons, the atoms become negative ions. This process is called reduction.

The mnemonic 'OIL RIG' reminds you that:

- oxidation is defined as the loss of electrons
- reduction is defined as the gain of electrons.

Consider the reaction shown in Figure 4.1.3 between aluminium and iodine. Aluminium is in group 13 and each atom has 3 electrons in its valence shell, whereas iodine is in group 17 and each atom has 7 electrons in its valence shell. When these two elements react together, each aluminium atom loses 3 electrons and becomes an Al^{3+} ion. The aluminium is **oxidised**. At the same time, each iodine atom gains 1 electron and becomes an I^- ion. The iodine is **reduced**. These two ions form the ionic compound aluminium iodide.

FIGURE 4.1.3 In a vigorously exothermic reaction, aluminium is oxidised by iodine upon the addition of a few drops of water, to form aluminium iodide. The heat that is released during the reaction causes iodine to sublime from a solid to a purple gas.

This reaction can be represented by the equation:

$$2Al(s) + 3I_2(s) \longrightarrow 2AlI_3(s)$$

The oxidation and reduction processes can be clearly seen when you break the equation into two **half-equations**. One half-equation describes the oxidation (loss of electrons) of the aluminium atoms:

$$Al(s) \longrightarrow Al^{3+}(s) + 3e^-$$

The other half-equation describes the reduction (gain of electrons) of the iodine atoms:

$$I_2(s) + 2e^- \longrightarrow 2I^-(s)$$

The electrons that are gained by the iodine atoms have been lost by the aluminium atoms.

When you write the overall redox equation, the number of electrons produced during the oxidation process must be the same as the number of electrons consumed in the reduction process. You will learn how to write overall redox equations later in this section.

WRITING REDOX EQUATIONS

Equations for redox reactions are often written as half-equations, in which the movement of electrons is shown, before they are written as overall equations. The half-equations enable you to rapidly identify the oxidation and reduction processes, based on the position of the electrons in the equations. Furthermore, the half-equations can be combined to find an overall equation by ensuring that all the electrons lost in the oxidation process are gained in the reduction process.

CHEMFILE

The thermite reaction

The thermite reaction, between a powdered mixture of aluminium and iron(III) oxide, is used for welding and iron foundry work. When the thermite mixture is ignited, the reaction shown below is rapid and very exothermic and the iron formed is molten (see figure below).

$$Fe_2O_3(s) + 2Al(s) \longrightarrow 2Fe(s) + Al_2O_3(s) + heat$$

In apparatus used for welding railway tracks in remote locations, the thermite reaction occurs in the upper chamber of the reaction vessel and the molten iron flows down through the bottom of the reaction vessel into the gap between the ends of the two rails, welding the tracks together upon cooling.

The thermite reaction between iron(III) oxide and aluminium produces a huge quantity of heat and molten iron which flows down from the reaction vessel.

In an oxidation half-equation, the electrons are 'lost' or produced, so they are always written on the right-hand side of the equation.

Checking which side of the equation the electrons appear on allows you to determine whether a process is oxidation or reduction.

States must be included in both redox half-equations and overall equations for each species. Electrons do not have a state and should always cancel in the overall equation.

Writing simple half-equations

Half-equations show what is happening as electrons are transferred in a redox reaction. Like other chemical equations, half-equations must be balanced so that there is the same number of atoms of each element on each side of the arrow. Similarly, charge must also be balanced. Half-equations and overall equations should also indicate states.

Consider the reaction between magnesium and bromine in aqueous solution as an example. The half-equation for the oxidation of Mg(s) to Mg^{2+}(aq) is written as:

$$Mg(s) \longrightarrow Mg^{2+}(aq) + 2e^-$$

Charge is balanced by the addition of two electrons to the right-hand side of the equation.

The half-equation for the reduction of Br_2(aq) to Br^-(aq) is written as:

$$Br_2(aq) + 2e^- \longrightarrow 2Br^-(aq)$$

Charge is balanced by the addition of two electrons to the left-hand side of the equation.

Writing an overall redox equation

After half-equations have been written for a redox reaction and the electrons have been balanced, they are added together to obtain an overall, or full, equation.

The overall equation does not show any electrons transferred; that is, no electrons appear in a properly balanced full equation. All the electrons lost in the oxidation reaction are gained in the reduction reaction. You may need to multiply one, or perhaps both, of the half-equations by a factor to ensure that the electrons balance and can be cancelled out in the overall equation.

The reaction of chlorine and iron is shown in Figure 4.1.4. In this reaction, each Fe atom is oxidised and loses 3 electrons. Each Cl_2 molecule is reduced and gains 2 electrons.

$$Fe(s) \longrightarrow Fe^{3+}(s) + 3e^-$$

$$Cl_2(g) + 2e^- \longrightarrow 2Cl^-(s)$$

FIGURE 4.1.4 The reaction between chlorine and iron is a redox reaction. It is so exothermic that the iron burns as it reacts.

To write an overall equation for this reaction, the number of electrons lost during oxidation must equal the number of electrons gained during reduction. Three electrons are produced by the oxidation of an iron atom, but only 2 electrons are required to reduce a chlorine molecule. The lowest common multiple of 3 and 2 is 6, so this is the number of electrons that are transferred from the iron to the chlorine.

The half-equation involving the oxidation of Fe is multiplied by a factor of 2 to bring the total number of electrons to 6, and the half-equation involving the reduction of Cl_2 is multiplied by a factor of 3 to also bring the total number of electrons to 6.

$$2Fe(s) \longrightarrow 2Fe^{3+}(s) + 6e^-$$
$$3Cl_2(g) + 6e^- \longrightarrow 6Cl^-(s)$$

The two half-equations are then added to find the overall equation:

$$2Fe(s) + 3Cl_2(g) \longrightarrow 2FeCl_3(s)$$

You can see that the electrons have been cancelled from each side of the equation to give the overall equation with no electrons.

Oxidising agents and reducing agents

Substances that cause oxidation to occur are called **oxidising agents** or **oxidants**. In the reaction between aluminium and iodine, the iodine is the oxidising agent:

$$2Al(s) + 3I_2(s) \longrightarrow 2AlI_3(s)$$

An oxidising agent *causes* oxidation, but is always reduced itself.

Substances that cause reduction are called **reducing agents** or **reductants**. Aluminium is the reducing agent in our example. A reducing agent causes reduction, but is always oxidised itself.

Figure 4.1.6 shows you how to identify the oxidising and reducing agents for the overall reaction between aluminium and iodine.

> When multiplying half-equations, it is essential to multiply the coefficients of all of the species in the equation by the same factor.

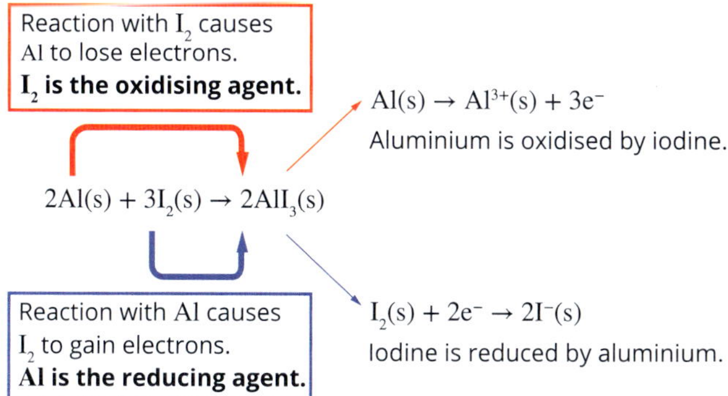

FIGURE 4.1.6 In the reaction between aluminium and iodine, aluminium is the reducing agent and iodine is the oxidising agent.

Metals, such as zinc and magnesium, and negatively charged non-metal ions, such as Br^- and I^-, can lose electrons, so they tend to act as reducing agents. For example:

$$Zn(s) \longrightarrow Zn^{2+}(aq) + 2e^-$$
$$2Br^-(aq) \longrightarrow Br_2(aq) + 2e^-$$

Non-metals, such as fluorine and chlorine, and positively charged metal ions, such as Cu^{2+} and Ag^+, can gain electrons, so they tend to act as oxidising agents. For example:

$$Cl_2(g) + 2e^- \longrightarrow 2Cl^-(aq)$$
$$Ag^+(aq) + e^- \longrightarrow Ag(s)$$

> Species that can lose electrons will act as reducing agents and will cause another reactant to gain electrons.

> Species that can gain electrons will act as oxidising agents and will cause another reactant to lose electrons.

CASE STUDY

Using redox chemistry to keep tree roots out of pipes

In gardens where trees grow close to stormwater or sewerage pipes, the tree roots can create a major problem. They grow towards the water coming from slightly permeable clay water pipes. Fine roots are able to penetrate and eventually form a tight root mass which blocks the pipes. This blockage causes all sorts of problems, including flooding and even an unpleasant overflow of sewage.

One company in the United Kingdom has created a root barrier system: a fabric made up of a layer of copper pressed between two layers of woven polypropene material, as seen in Figure 4.1.5. This geocomposite fabric, known as CuTex, is permeable (lets water through) and flexible, so it can be wrapped around the pipes that need protecting. The 18 mm thickness of copper foil reacts with water and oxygen from the soil, forming Cu^{2+} ions:

$$2Cu(s) + O_2(aq) + 2H_2O(l) \rightarrow 2Cu^{2+}(aq) + 4OH^-(aq)$$

You will see from this equation that oxygen is the oxidising agent and copper is the reducing agent.

Root cells need oxygen to respire and only grow in soil that has both water and oxygen present, so where there are roots, there will be water and oxygen. The tip of the root is where the growth occurs. As the root tips grow towards the water and the pipes, they meet a region containing a low concentration of copper(II) ions. Copper(II) ions are toxic to plants, so the dividing cells of the root tips die off.

The roots grow away from that area and the plant itself is not adversely affected. The concentration of copper(II) ions has been shown in testing at the University of Leeds, England, to be so small that the ecosystem in general and nearby groundwater are not affected.

Unfortunately the solution of wrapping the geocomposite material around a pipe requires the pipe to be dug up and re-laid, but for many households that have a nasty ongoing problem of roots in their sewerage pipes, this could be a sustainable solution.

FIGURE 4.1.5 An 18 mm thickness of copper is pressed between two layers of polypropene material in this geocomposite fabric.

4.1 Review

SUMMARY

- Redox (reduction–oxidation) reactions involve the transfer of electrons from one species to another.
- Oxidation and reduction always occur at the same time.
- Half-equations are used to represent oxidation and reduction. They show what is happening as electrons are transferred in a redox reaction.
- Oxidation is defined as the loss of electrons, e.g. $Mg(s) \rightarrow Mg^{2+}(aq) + 2e^-$.
- Reduction is defined as the gain of electrons, e.g. $Cl_2(g) + 2e^- \rightarrow 2Cl^-(aq)$.
- The reducing agent loses electrons to another substance, causing that substance to be reduced.
- The oxidising agent gains electrons from another substance, causing that substance to be oxidised.
- Half-equations are added together to obtain the overall equation. Before adding the half-equations, it may be necessary to multiply them by a factor to balance the electrons.
- State symbols should always be included in half-equations and balanced equations for overall redox reactions.

KEY QUESTIONS

Knowledge and understanding

1 Copy and complete the summary about the following redox reaction by using the following terms:

gains; loses; Cl_2; Cl^-; reduced; oxidised

$$Zn(s) + Cl_2(g) \rightarrow ZnCl_2(s)$$

When a reducing agent, such as Zn, reacts with an oxidising agent, such as ________, an ionic compound is formed. The reducing agent, Zn, ________ electrons (is ________) and at the same time the oxidising agent, Cl_2, ________ electrons (is ________). In this case, the products are Zn^{2+} and ________, which form $ZnCl_2$.

2 Classify each of the following half-equations as oxidation or reduction.

a $2Cl^-(aq) \rightarrow Cl_2(aq) + 2e^-$

b $Fe^{3+}(aq) + e^- \rightarrow Fe^{2+}(aq)$

c $Br_2(g) + 2e^- \rightarrow 2Br^-(aq)$

d $Li(s) \rightarrow Li^+(aq) + e^-$

e $Sn^{4+}(aq) + 2e^- \rightarrow Sn^{2+}(aq)$

f $Zn^{2+}(aq) + 2e^- \rightarrow Zn(s)$

3 When nickel metal reacts with copper(II) ions in an aqueous solution, nickel(II) ions are formed. Write the oxidation half-equation for the reaction.

Analysis

4 The overall equation for the reaction between aluminium metal and liquid sulfur is:

$$2Al(s) + 3S(l) \rightarrow Al_2S_3(s)$$

From the overall equation, write the half-equations for this reaction and label each one as either reduction or oxidation.

5 For each of the following reactions, write the oxidation and reduction half-equations, and identify the oxidising and the reducing agents.

a $2Zn(s) + O_2(g) \rightarrow 2ZnO(s)$

b $Ca(s) + Cl_2(g) \rightarrow CaCl_2(s)$

c $2Al(s) + 3Br_2(l) \rightarrow 2AlBr_3(s)$

6 Metal *M* is found to react with oxygen to form the compound *M*O. In a separate reaction, metal *M* reacted with an aqueous solution of silver nitrate ($AgNO_3$). The unbalanced and incomplete equations for the reaction between metal *M* and $AgNO_3$ are given below.

$Ag^+(aq) +$ _____ $\rightarrow Ag(s)$

$M(s) \rightarrow$ _____$(aq) +$ ___e^-

$M(s) +$ _____$(aq) \rightarrow$ _____$(aq) +$ ___$Ag(s)$

a Complete the half-equations and the overall equation for the reaction between metal *M* and silver ions, $Ag^+(aq)$.

b Identify each half-equation as either reduction or oxidation.

c Identify the oxidising and reducing agents for the reaction.

4.2 Oxidation numbers

FIGURE 4.2.1 These compounds of manganese have different oxidation states and are different colours.

In this section, you will learn the set of rules that chemists have created to allow a wider range of reactions to be classified as redox reactions. This involves assigning **oxidation numbers**, also known as oxidation states, to the atoms in a reaction.

You will see how oxidation numbers can be used to determine whether a reaction that does not involve the formation of ions could be classified as a redox reaction and which substances in a redox reaction have been oxidised or reduced.

You will also discover that many transition metals can have multiple oxidation states and that compounds of these metals with different oxidation states can have different colours, such as in the solutions shown in Figure 4.2.1.

DETERMINING OXIDATION NUMBERS

In Section 4.1 you learnt about redox reactions that involved the production of ionic compounds from their elements. In these cases, it was relatively easy to deduce which element gained or lost electrons by considering the charge on the ions produced in the reaction. The species that was reduced gained electrons, becoming less positive, whereas the species that was oxidised lost electrons, becoming more positive.

For some redox reactions, it is harder to identify the species that are oxidised and reduced. For example, the reaction that occurs during the wet corrosion of iron metal can be represented by the following half-equations:

$$Fe(s) \rightarrow Fe^{2+}(aq) + 2e^-$$

$$O_2(g) + 2H_2O(l) + 4e^- \rightarrow 4OH^-(aq)$$

The first half-equation shows that the iron loses electrons and therefore is oxidised. The second half-equation shows that electrons have been gained, and therefore represents a reduction reaction. However, it is not obvious whether the oxygen or the water has gained these electrons.

It is possible to determine which element has been oxidised and which has been reduced in this reaction by determining their oxidation numbers.

Oxidation numbers have no physical meaning—they do not necessarily indicate a formal charge, or the physical or chemical properties of the substance. However, they are a useful tool for identifying which atoms have been oxidised and which atoms have been reduced.

> Oxidation numbers are useful for identifying the reducing agent and the oxidising agent in a redox reaction.

Oxidation number rules

Oxidation numbers are assigned to elements involved in a reaction by following a specific set of rules. In applying these rules, we regard all compounds and polyatomic ions as though they are composed of individual ions.

Table 4.2.1 on the following page describes the rules for determining oxidation numbers. In the examples, the oxidation number of an element is placed above its symbol. A plus or minus sign precedes the number and so distinguishes the oxidation number from the charge on an ion, in which the sign is generally placed after the number. For example, the oxide ion (O^{2-}) has a charge of 2– and an oxidation number of –2. Although the values are the same in this instance, it is important to remember that oxidation states do not always indicate an ionic species (a species that carries a charge).

> Oxidation numbers are written with their sign (plus or minus) *before* the number, whereas ionic charges have the sign *after* the number; e.g. for Mg^{2+}, the oxidation number is +2 and charge is 2+.

TABLE 4.2.1 Rules for determining oxidation numbers of elements in compounds

Rule	Examples
1 The oxidation number of a free element is zero.	$\overset{0}{Na}$, $\overset{0}{C}$, $\overset{0}{Cl}$, $\overset{0}{Cl_2}$, $\overset{0}{P_4}$
2 The oxidation number of a simple ion is equal to the charge on the ion.	$\overset{+1}{Na^+}$, $\overset{-1}{Cl^-}$, $\overset{+2}{Mg^{2+}}$, $\overset{-2}{O^{2-}}$, $\overset{+3}{Al^{3+}}$, $\overset{-3}{N^{3-}}$
3 In compounds, some elements have oxidation numbers that are regarded as fixed, except in a few exceptional circumstances.	
(a) Main group metals have an oxidation number equal to the charge on their ions.	Ionic compounds: $\overset{+1}{K}Cl$, $\overset{+2}{Mg}SO_4$
(b) Hydrogen has an oxidation number of +1 when it forms compounds with non-metals. Exception: In metal hydrides the oxidation number of hydrogen is –1.	Compounds of H: $\overset{+1}{H_2}O$ Metal hydrides: $Na\overset{-1}{H}$, $Ca\overset{-1}{H_2}$
(c) Oxygen usually has an oxidation number of –2. Exceptions: In peroxides, oxygen has an oxidation number of –1. In compounds with fluorine, oxygen has a positive oxidation number.	Compounds of O: $H_2\overset{-2}{O}$ Peroxides: $H_2\overset{-1}{O_2}$, $Ba\overset{-1}{O_2}$ Fluorine compound: $\overset{+2}{O}F_2$
4 The sum of the oxidation numbers in a neutral compound is zero.	$\overset{+4}{C}\overset{-2}{O_2}$ Note that in CO_2, the oxidation number of each oxygen atom is written as –2. It is not written as –4 for two O atoms.
5 The sum of the oxidation numbers in a polyatomic ion is equal to the charge on the ion.	$\overset{+6}{S}\overset{-2}{O_4}{}^{2-}$, $\overset{-3}{N}\overset{+1}{H_4}{}^{+}$
6 The most electronegative element is assigned the negative oxidation number.	$\overset{+2}{O}\overset{-1}{F_2}$

Common oxidation states of the first 36 elements in their compounds are shown in the section of the periodic table in Figure 4.2.2. The asterisk (*) indicates that transition metals and some non-metals can have a range of oxidation states. These are usually calculated after applying the rules for all other elements in the compound.

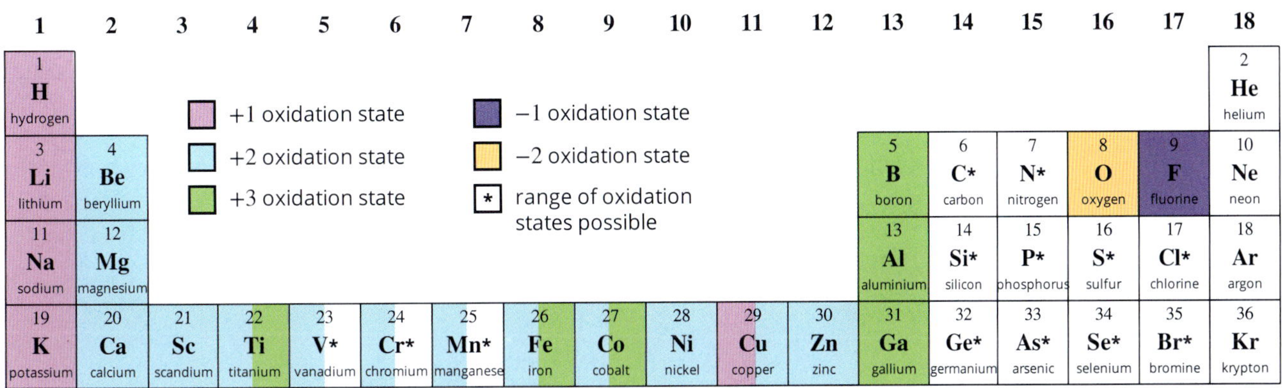

FIGURE 4.2.2 Part of the periodic table showing the most common oxidation states of some elements

CHEMFILE

Colourful oxidation states

A characteristic property of the transition elements is that they form brightly coloured compounds. Different oxidation states can result in different colours for the same transition metal.

Vanadium is a very colourful transition element that has a wide variety of colours depending on its oxidation state. You can see this in the figure below. When its oxidation state is +5, vanadium is yellow; when its oxidation state is +4, it is light blue; when its oxidation state is +3, it is green; and when its oxidation state is +2, it is magenta.

These colourful solutions have been made by adding a reducing agent (mercury–zinc amalgam) to a yellow solution containing vanadium(V) ions.

Calculating oxidation numbers

For a compound containing several elements, you can use algebra and the rules given in Table 4.2.1 on page 157 to calculate the oxidation number of each element.

For example, to find the oxidation number of sulfur in H_2SO_4, there is a rule for hydrogen and for oxygen, which leaves sulfur as the only unknown. If the oxidation number of sulfur equals x, the following expression can be written to solve for x:

$$(2 \times +1) + x + (4 \times -2) = 0$$
$$2 + x - 8 = 0$$
$$x - 6 = 0$$
$$x = +6$$

You will learn later on in this section how to use oxidation numbers to determine if a substance has been oxidised or reduced.

Worked example 4.2.1

CALCULATING OXIDATION NUMBERS

Use the rules in Table 4.2.1 to determine the oxidation number of each element in $KClO_4$.

Thinking	Working
Identify an element that has a set value.	K is a main group metal in group 1. According to rule 3a, the oxidation number of potassium is +1.
Identify any other elements that have set values.	According to rule 3c, oxygen has an oxidation number of −2, unless attached to fluorine or in a peroxide.
Use algebra to work out the oxidation number of the other element.	Let the oxidation number of chlorine in $KClO_4$ be x. Solve the sum of the oxidation numbers for x: $+1 + x + (4 \times -2) = 0$ $+1 + x - 8 = 0$ $x - 7 = 0$ $x = +7$
Write oxidation numbers above the elements in the formula.	$\overset{+1}{K}\overset{+7}{Cl}\overset{-2}{O_4}$ Note that the oxidation number of oxygen is written as −2 (not as −8), even though there are four oxygen atoms in the formula—the oxidation number is written per atom.

Worked example: Try yourself 4.2.1

CALCULATING OXIDATION NUMBERS

Use the rules in Table 4.2.1 to determine the oxidation number of each element in $NaNO_3$.

USING OXIDATION NUMBERS TO NAME CHEMICALS

Since **transition elements** can form ions with several different charges, many transition elements have variable oxidation numbers. For example, there are two compounds that can be called iron chloride: $FeCl_2$ and $FeCl_3$.

Using the rules in Table 4.2.1 (on page 157), you can see that the chloride ion has an oxidation number of −1. In $FeCl_2$ this means the oxidation number of iron is +2, whereas in $FeCl_3$ the oxidation number of iron is +3.

To distinguish between the two iron chlorides, insert Roman numerals representing the appropriate oxidation number in the name.

- $FeCl_2$ is named iron(II) chloride.
- $FeCl_3$ is named iron(III) chloride.

When naming non-metal compounds, you can also use Roman numerals to show the oxidation number of an element such as nitrogen, which has several possible oxidation states. Nitrogen dioxide (NO_2) is called nitrogen(IV) oxide, whereas nitric oxide (NO) is nitrogen(II) oxide. This method of naming makes it much easier to determine the formula from the name of the oxide.

For elements that can have variable oxidation states, the use of a Roman numeral in the name indicates the specific oxidation state of the element.

USING OXIDATION NUMBERS TO IDENTIFY OXIDATION AND REDUCTION

You can use the concept of oxidation numbers to extend the definition of oxidation and reduction.

In this new definition, a change in oxidation numbers indicates that a redox reaction has taken place. This can be used as an alternative definition of oxidation and reduction instead of our earlier definition involving loss and gain of electrons. It is particularly useful for non-ionic compounds when it is difficult to determine whether electrons have been transferred.

It can now be stated that:

- oxidation involves an increase in oxidation number
- reduction involves a decrease in oxidation number.

Remember that oxidation and reduction always occur together in a redox reaction. One process cannot happen without the other. If there is no change in oxidation number for *all* elements in a reaction, then the reaction is not a redox reaction. Oxidation numbers can be used to analyse the equation for a reaction and determine whether it represents a redox process.

An increase in oxidation number indicates the element was oxidised. A decrease in oxidation number indicates the element was reduced.

At the beginning of this chapter you were alerted to a number of everyday processes that are redox reactions. Combustion reactions, where fuel is burned to produce heat while releasing carbon dioxide and water, are examples of redox reactions that are very important to our society. The equation for the burning of carbon in excess oxygen is:

$$C(s) + O_2(g) \rightarrow CO_2(g)$$

At first glance, the reaction may not seem like a redox reaction because none of the species are ionic compounds, so it is not clear which reactant is losing or gaining electrons. However, using oxidation numbers you can identify both an oxidation process and a reduction process for the reaction.

$$\overset{0}{C}(s) + \overset{0}{O}_2(g) \rightarrow \overset{+4}{C}\overset{-2}{O}_2(g)$$

The carbon is oxidised because its oxidation number increases from 0 to +4, and the oxygen is reduced because its oxidation number decreases from 0 to −2.

The use of oxidation numbers allows you to look at a chemical reaction and determine whether it is a redox process.

Worked example 4.2.2

USING OXIDATION NUMBERS TO IDENTIFY OXIDATION AND REDUCTION IN AN EQUATION

Use oxidation numbers to determine which element has been oxidised and which has been reduced in the following equation:

$$CH_4(g) + 2O_2(g) \rightarrow CO_2(g) + 2H_2O(l)$$

Thinking	Working
Determine the oxidation numbers of one of the elements on each side of the equation.	Choose carbon as the first element. $\overset{-4}{C}H_4(g) + 2O_2(g) \rightarrow \overset{+4}{C}O_2(g) + 2H_2O(l)$
Assess if the oxidation number has changed. If so, identify if it has increased (oxidation) or decreased (reduction).	The oxidation number of C has increased from –4 to +4, so the carbon in the CH_4 has been oxidised.
Determine the oxidation numbers of a second element on the each side of the equation.	Choose oxygen as the second element. $CH_4(g) + 2\overset{0}{O}_2(g) \rightarrow C\overset{-2}{O}_2(g) + 2H_2\overset{-2}{O}(l)$
Assess if the oxidation number has changed. If so, identify if it has increased (oxidation) or decreased (reduction).	The oxidation number of O has decreased from 0 to –2, so the O_2 has been reduced.
Continue this process until the oxidation numbers of all elements have been determined.	Determine the oxidation numbers of hydrogen. $C\overset{+1}{H}_4(g) + 2O_2(g) \rightarrow CO_2(g) + 2\overset{+1}{H}_2O(l)$ The oxidation number of H has not changed.

Worked example: Try yourself 4.2.2

USING OXIDATION NUMBERS TO IDENTIFY OXIDATION AND REDUCTION IN AN EQUATION

Use oxidation numbers to determine which element has been oxidised and which has been reduced in the following equation:

$$CuO(s) + H_2(g) \rightarrow Cu(s) + H_2O(l)$$

USING OXIDATION NUMBERS TO IDENTIFY CONJUGATE REDOX PAIRS

When a half-equation is written for an oxidation reaction, the reactant, a reducing agent, loses electrons. The product is an oxidising agent. We refer to the reactant and the product that it forms as a **conjugate redox pair**.

For example, in the half-equation for the oxidation of zinc:

$$Zn(s) \longrightarrow Zn^{2+}(aq) + 2e^-$$

zinc metal (Zn) is a reducing agent and it forms $Zn^{2+}(aq)$, an oxidising agent. Zn(s) and $Zn^{2+}(aq)$ form a conjugate redox pair, $Zn(s)/Zn^{2+}(aq)$.

In the $Zn(s)/Zn^{2+}(aq)$ conjugate redox pair, the oxidation number of zinc increases from 0 to +2. The increase in the oxidation number of zinc indicates that it is an oxidation half-reaction.

For a reduction half-equation, the reactant is an oxidising agent and will gain electrons. The product formed is a reducing agent. Therefore, another conjugate redox pair is present in the reduction half-equation.

For example, consider the half-equation for the reduction of $Ag^+(aq)$:

$$Ag^+(aq) + e^- \longrightarrow Ag(s)$$

$Ag^+(aq)$ is an oxidising agent and forms Ag(s), which is a reducing agent. $Ag^+(aq)$ and Ag(s) are a conjugate redox pair. In this case, the oxidation number of silver decreases from +1 to 0, indicating that this is a reduction half-equation.

The relationship between changes in oxidation numbers and conjugate redox pairs can be seen by following the colour-coding in the equation in Figure 4.2.3. One conjugate redox pair is red and the other one is blue.

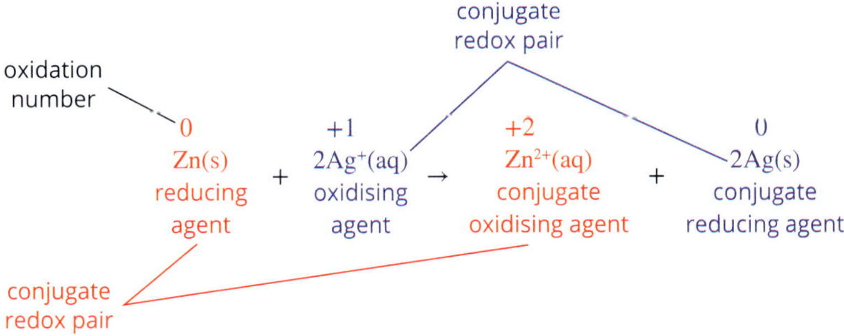

FIGURE 4.2.3 The change in oxidation numbers seen in conjugate redox pairs. For each redox reaction, there are two conjugate redox pairs.

When listing conjugate redox pairs, remember to include the states of both the oxidising and the reducing agent.

In a conjugate redox pair, the oxidation number of the oxidising agent is always more positive than that of the reducing agent.

4.2 Review

OA ✓✓

SUMMARY

- Oxidation numbers are calculated according to a set of rules.
  - Free elements have an oxidation number of 0.
  - In ionic compounds composed of simple ions, the oxidation number is equal to the charge on the ion.
  - Oxygen in a compound usually has an oxidation number of −2.
  - Hydrogen in a compound usually has an oxidation number of +1.
  - The sum of the oxidation numbers in a neutral compound is 0.
  - The sum of the oxidation numbers in a polyatomic ion is equal to the charge on the ion.
- Transition metals and some non-metals have variable oxidation numbers that can be calculated using the rules above.
- An increase in the oxidation number of an element in a reaction indicates oxidation has occurred.
- A decrease in the oxidation number of an element in a reaction indicates reduction has occurred.
- For oxidation to occur, there must be a corresponding reduction.
- If there is no change in the oxidation number of all elements in the equation for a reaction, then the reaction is not a redox reaction.
- A conjugate redox pair consists of an oxidising agent (a reactant) and the reducing agent (a product) that is formed when the oxidising agent gains electrons. In this case, the oxidation number of the oxidising agent decreases.
- The other conjugate redox pair in a redox reaction is made up of a reducing agent (a reactant) and the oxidising agent (a product) that is formed when the reducing agent loses electrons. In this case, the oxidation number of the reducing agent increases.
- The diagram below summarises the redox terms that you need to understand from this section.

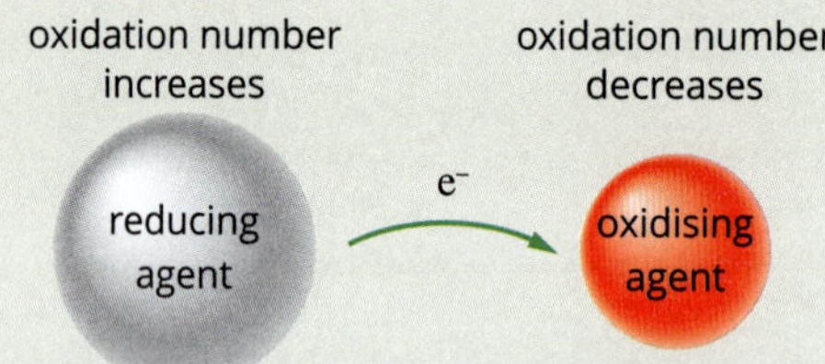

Reducing agent
- loses electrons
- reduces the oxidising agent
- undergoes oxidation
- is the reactant in the oxidation half-reaction:

reducing agent → conjugate oxidised form + ne^-

Oxidising agent
- gains electrons
- oxidises the reducing agent
- undergoes reduction
- is the reactant in the reduction half-reaction:

oxidising agent + ne^- → conjugate reduced form

KEY QUESTIONS

Knowledge and understanding

1 State the oxidation number of carbon in:

a CO

b CO_2

c CH_4

d C (graphite)

e HCO_3^-

2 Which one of the following substances contains manganese in the +6 oxidation state?

A $MnCl_2$

B $KMnO_4$

C MnO_2

D K_2MnO_4

3 Find the oxidation number of each element in the following compounds or ions. Hint: For ionic compounds, use the overall charge on each ion to help you.

a CaO
b $CaCl_2$
c HSO_4^-
d MnO_4^-
e F_2
f SO_3^{2-}
g $NaNO_3$
h $K_2Cr_2O_7$

Analysis

4 For each of the following equations:

i assign an oxidation number to each element
ii identify the oxidising agent and reducing agent.

a $Mg(s) + Cl_2(g) \longrightarrow MgCl_2(s)$
b $2SO_2(g) + O_2(g) \longrightarrow 2SO_3(g)$
c $Fe_2O_3(s) + 3CO(g) \longrightarrow 2Fe(s) + 3CO_2(g)$
d $2Fe^{2+}(aq) + H_2O_2(aq) + 2H^+(aq) \longrightarrow 2Fe^{3+}(aq) + 2H_2O(l)$

5 For each of the following redox reactions, assign oxidation numbers to the elements in the equations and complete the table to show the conjugate redox pairs.

Redox reaction	Conjugate redox pair (oxidation process)	Conjugate redox pair (reduction process)
$Na(s) + Ag^+(aq) \longrightarrow Na^+(aq) + Ag(s)$		
$Zn(s) + Cu^{2+}(aq) \longrightarrow Zn^{2+}(aq) + Cu(s)$		
$2K(s) + Cl_2(g) \longrightarrow 2KCl(s)$		

6 Consider each of the following statements and, using oxidation numbers to justify your answer, decide whether they are true or false. If false, correct the statement.

a In the following reaction, sodium is reduced and hydrogen has an oxidation number of +1 throughout the equation.
$2Na(s) + 2H_2O(l) \longrightarrow 2NaOH(aq) + H_2(g)$

b When the dichromate ion, $Cr_2O_7^{2-}$, reacts to form Cr^{3+}, the chromium is oxidised.

c In the following reaction, the oxidising agent is HCl(aq), and $KMnO_4$ is oxidised.
$2KMnO_4(aq) + 5H_2S(aq) + 6HCl(aq) \longrightarrow 2MnCl_2(aq) + 5S(s) + 2KCl(aq) + 8H_2O(l)$

CHEMFILE

Vinegary wine or winey vinegar?

The oxidation of ethanol forms ethanoic acid. Ethanoic acid is the main ingredient of vinegar and is responsible for its sourness. If a bottle of wine is left open to the atmosphere for a few days, it becomes 'vinegary' and undrinkable.

This reaction is put to good use when specialist vinegars such as apple cider vinegar, red wine vinegar and even malt (beer) vinegar (see figure below) are made by deliberately oxidising the appropriate alcoholic beverage under the right conditions.

The oxidation half-equation for the production of ethanoic acid from ethanol is given by:

$$C_2H_5OH(aq) + H_2O(l) \longrightarrow CH_3COOH(aq) + 4H^+(aq) + 4e^-$$

A selection of vinegars. From left to right: distilled malt vinegar, cider vinegar, white wine vinegar and malt vinegar.

When balancing redox half-equations in acidic solutions, first balance oxygens with H_2O, then hydrogens with H^+, then charge with electrons.

4.3 Writing complex redox equations

Not all oxidation and reduction half-equations involve simple ions and their elements. Many interesting redox reactions, such as the iodine clock reaction shown in Figure 4.3.1, involve reactants and products that have oxygen and hydrogen in their formulas. In this section, you will learn how to balance more-complex half-equations in a few simple steps.

FIGURE 4.3.1 This redox reaction involving colour changes is known as a clock reaction. The black compound is a starch–iodine complex.

BALANCING HALF-EQUATIONS IN ACIDIC SOLUTIONS

Half-equations that involve atoms or simple ions can be written quite easily. For example, knowing that magnesium metal is oxidised to form Mg^{2+} ions in solution, you can readily write the half-equation as:

$$Mg(s) \longrightarrow Mg^{2+}(aq) + 2e^-$$

However, half-equations involving polyatomic ions are usually less obvious. The anaesthetic nitrous oxide or laughing gas (N_2O) can be prepared by the reduction of nitrate ions in an acidic solution, also known as **acidic conditions**, in the presence of $H^+(aq)$ ions:

$$2NO_3^-(aq) + 10H^+(aq) + 8e^- \longrightarrow N_2O(g) + 5H_2O(l)$$

Such equations can be developed using the following steps. The reduction of nitrate ions will be used to illustrate this process.

1 Balance all elements except hydrogen and oxygen in the half-equation.

$$2NO_3^- \longrightarrow N_2O$$

2 Balance the oxygen atoms by adding water molecules.

$$2NO_3^- \longrightarrow N_2O + 5H_2O$$

3 Balance the hydrogen atoms by adding H^+ ions (which are present in an acidic solution).

$$2NO_3^- + 10H^+ \longrightarrow N_2O + 5H_2O$$

4 Balance the charge in the equation by adding electrons. In this case, the total charge on the left-hand side is $(2 \times -1) + (10 \times +1) = +8$. The total charge on the right-hand side is 0. Make the charges equal by adding 8 electrons to the left-hand side.

$$2NO_3^- + 10H^+ + 8e^- \longrightarrow N_2O + 5H_2O$$

5 Add states to complete the half-equation.

$$2NO_3^-(aq) + 10H^+(aq) + 8e^- \longrightarrow N_2O(g) + 5H_2O(l)$$

When writing half-equations and overall equations, it is important that they are fully balanced (Figure 4.3.2 on the following page). The number of each element must be equal on each side, just as with any other chemical equation. The charge for each side of the equation must also be equal. It is important to remember that the charges being equal does not mean they must be zero.

FIGURE 4.3.2 In balanced half-equations and overall equations, the number of atoms of each element is equal on both sides and the total charge on each side is equal.

The steps for balancing complex redox half-equations must be carried out in sequence for the process to work.

When balancing half-equations for charge, remember that a coefficient in front of a charged species will multiply the charge as well as the atoms.

Worked example 4.3.1 shows how to balance a complex redox half-equation in acidic solutions. The final equation in this example has an overall charge of +6 for each side. Note that an acidic solution, with $H^+(aq)$ added, is often referred to as an **acidified** solution.

Worked example 4.3.1

BALANCING A HALF-EQUATION IN AN ACIDIC SOLUTION

Write the half-equation for the reduction of an acidified solution of $Cr_2O_7^{2-}$ to aqueous Cr^{3+}.

Thinking	Working
Balance all elements except hydrogen and oxygen in the half-equation.	There are 2 Cr atoms in $Cr_2O_7^{2-}$, so 2 Cr atoms are needed on the right-hand side. $Cr_2O_7^{2-} \rightarrow 2Cr^{3+}$
Balance the oxygen atoms by adding water.	There are 7 O atoms in $Cr_2O_7^{2-}$, so 7 H_2O molecules are added to the right-hand side. $Cr_2O_7^{2-} \rightarrow 2Cr^{3+} + 7H_2O$
Balance the hydrogen atoms by adding H^+ ions. Acids provide a source of H^+ ions.	There are now 14 H atoms on the right-hand side and none on the left-hand side, so 14 H^+ ions are added to the left-hand side. $Cr_2O_7^{2-} + 14H^+ \rightarrow 2Cr^{3+} + 7H_2O$
Balance the charge in the equation by adding electrons.	The charge on the left-hand side is (–2) + (+14) = +12 and on the right-hand side is 2 × +3 = +6, so 6 electrons are added to the left-hand side to make the charges equal. $Cr_2O_7^{2-} + 14H^+ + 6e^- \rightarrow 2Cr^{3+} + 7H_2O$
Add states to complete the half-equation.	All states are (aq), except for water, which is (l). $Cr_2O_7^{2-}(aq) + 14H^+(aq) + 6e^- \rightarrow 2Cr^{3+}(aq) + 7H_2O(l)$

Worked example: Try yourself 4.3.1

BALANCING A HALF-EQUATION IN AN ACIDIC SOLUTION

Write the half-equation for the reduction of an acidified solution of MnO_4^- to solid MnO_2.

OVERALL REDOX EQUATIONS UNDER ACIDIC CONDITIONS

To write an overall redox equation, you add the oxidation half-equation to the reduction half-equation, making sure that the number of electrons used in reduction equals the number of electrons released during oxidation. For redox reactions under acidic conditions, H^+ ions, which are provided by the acid, and H_2O molecules may also be present as reactants or products, so these will also need to be cancelled out.

> When combining two half-equations to write an overall redox equation, remember to cancel out any species, such as H_2O or H^+ so that they only appear on one side of the equation.

Worked example 4.3.2

COMBINING HALF-EQUATIONS TO WRITE OVERALL REDOX EQUATIONS UNDER ACIDIC CONDITIONS

Write balanced oxidation and reduction half-equations for the reaction in which permanganate ions, $MnO_4^-(aq)$ and aqueous iodine, $I_2(aq)$ react to form $Mn^{2+}(s)$ and iodate ions, $IO_3^-(aq)$.Then write the overall redox equation for the reaction.

Thinking	Working
Identify one reactant and the product it forms, and write the balanced half-equation.	$MnO_4^-(aq) + 8H^+(aq) + 5e^- \rightarrow Mn^{2+}(s) + 4H_2O(l)$
Identify the second reactant and the product it forms, and write the balanced half-equation.	$I_2(aq) + 6H_2O(l) \rightarrow 2IO_3^-(aq) + 12H^+(aq) + 10e^-$
Multiply one or both equations by a suitable factor to ensure that the number of electrons on both sides of the arrow is equal.	The lowest common multiple is 10, so multiply the reduction half-equation by 2 and keep the oxidation equation as it is. $2 \times [MnO_4^-(aq) + 8H^+(aq) + 5e^- \rightarrow Mn^{2+}(s) + 4H_2O(l)]$ $2MnO_4^-(aq) + 16H^+(aq) + 10e^- \rightarrow 2Mn^{2+}(s) + 8H_2O(l)$ $I_2(aq) + 6H_2O(l) \rightarrow 2IO_3^-(aq) + 12H^+(aq) + 10e^-$
Add the oxidation and the reduction half-equations together, cancelling electrons so that none appear in the final equation. Also cancel H_2O and H^+ if these occur on both sides of the arrow.	$2MnO_4^-(aq) + \overset{4H^+}{\cancel{16H^+}}(aq) + \cancel{10e^-} \rightarrow 2Mn^{2+}(s) + \overset{2H_2O}{\cancel{8H_2O}}(l)$ $I_2(aq) + \cancel{6H_2O}(l) \rightarrow 2IO_3^-(aq) + \cancel{12H^+}(aq) + \cancel{10e^-}$
Write the overall redox equation for the reaction.	$2MnO_4^-(aq) + 4H^+(aq) + I_2(aq) \rightarrow 2Mn^{2+}(s) + 2H_2O(l) + 2IO_3^-(aq)$

CHEMFILE

Elephant's toothpaste

Hydrogen peroxide is such a strong oxidising agent that it can even oxidise itself, decomposing to form water and oxygen gas:

$$H_2O_2(aq) + H_2O_2(aq) \rightarrow 2H_2O(l) + O_2(g)$$

In this reaction the two half-equations are:

$$H_2O_2(aq) + 2H^+(aq) + 2e^- \rightarrow 2H_2O(l)$$
$$H_2O_2(aq) \rightarrow O_2(g) + 2H^+(aq) + 2e^-$$

The decomposition reaction of hydrogen peroxide is the basis for the impressive 'foam column' or 'elephant's toothpaste' demonstration that is commonly shown by chemistry teachers to their classes.

When detergent and a catalyst are added to concentrated hydrogen peroxide solution, a column of foam is formed, as in the figure below.

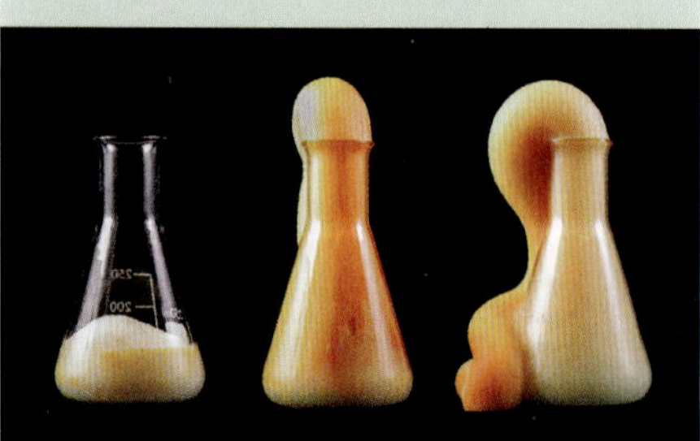

In this reaction, the hydrogen peroxide acts as both the reducing agent and the oxidising agent. In this image, a solid potassium iodide catalyst is added to a hydrogen peroxide solution. Red colouring has also been added for effect.

Worked example: Try yourself 4.3.2

COMBINING HALF-EQUATIONS TO WRITE OVERALL REDOX EQUATIONS UNDER ACIDIC CONDITIONS

Write balanced oxidation and reduction half-equations for the reaction in which $SO_3^{2-}(aq)$ and $ClO^-(aq)$ react to form $H_2S(g)$ and $ClO_3^-(aq)$. Then write the overall equation for the reaction.

Some strong oxidising agents, such as potassium dichromate and iodine, are highly coloured. The change in colour that occurs as the redox process proceeds can be used for a number of chemical analyses—for example, the change from the yellow-orange colour of dichromate ($Cr_2O_7^{2-}$) ions to the green colour of chromium(III) (Cr^{3+}) ions when it is used to oxidise a colourless reactant, such as ethanol, as shown in Figure 4.3.3.

FIGURE 4.3.3 The orange colour of dichromate ($Cr_2O_7^{2-}$) ions changes to the green colour of chromium(III) (Cr^{3+}) ions when it is used to oxidise a colourless reactant, such as ethanol.

BALANCING HALF-EQUATIONS IN BASIC CONDITIONS

If a redox reaction occurs in **basic**, or **alkaline conditions** (in an alkaline solution), OH^- rather than H^+ can be used to balance the equation. The steps are very similar to balancing redox processes under acidic conditions, although there are two different methods that suit different situations. Always remember to balance the half-equations for elements other than oxygen and hydrogen before you begin.

Balancing basic half-equations in which OH^- is present in a reactant or product

Consider the oxidation of cadmium, Cd(s), to cadmium hydroxide, $Cd(OH)_2(s)$ in a basic solution. Note that the product, $Cd(OH)_2$, has hydroxide ions in the formula.

1 Balance all elements except hydrogen and oxygen in the half-equation.

$$Cd \rightarrow Cd(OH)_2$$

2 Balance the hydroxide ions by adding $OH^-(aq)$ on the left-hand side of the equation.

$$Cd + 2OH^- \rightarrow Cd(OH)_2$$

3 Balance the charge by adding electrons on the right-hand side of the equation.

$$Cd + 2OH^- \rightarrow Cd(OH)_2 + 2e^-$$

4 Add states to complete the half-equation.

$$Cd(s) + 2OH^-(aq) \rightarrow Cd(OH)_2(s) + 2e^-$$

Aqueous hydroxide ions can be used directly to balance a half-equation if there are hydroxide ions in a reactant or product, or both.

Worked example 4.3.3

BALANCING A HALF-EQUATION IN A BASIC SOLUTION BY DIRECT ADDITION OF HYDROXIDE IONS

Write the half-equation for the oxidation of iron(II) hydroxide, $Fe(OH)_2$, to iron(III) hydroxide, $Fe(OH)_3$, in an alkaline (basic) solution.

Thinking	Working
Note that there are hydroxide ions, OH^-, in the formulas, so the equation can be balanced by the direct addition of hydroxide ions.	$Fe(OH)_2 \rightarrow Fe(OH)_3$
Balance all elements except hydrogen and oxygen in the half-equation.	$Fe(OH)_2 \rightarrow Fe(OH)_3$
Balance the hydroxide ions by adding $OH^-(aq)$.	$Fe(OH)_2 + OH^- \rightarrow Fe(OH)_3$
Balance the charge by adding electrons.	$Fe(OH)_2 + OH^- \rightarrow Fe(OH)_3 + e^-$
Add states to complete the half-equation.	$Fe(OH)_2(s) + OH^-(aq) \rightarrow Fe(OH)_3(s) + e^-$

Worked example: Try yourself 4.3.3

BALANCING A HALF-EQUATION IN A BASIC SOLUTION BY DIRECT ADDITION OF HYDROXIDE IONS

Write the half-equation for the oxidation of copper, Cu(s), to copper hydroxide, $Cu(OH)_2(s)$, in an alkaline (basic) solution.

Balancing half-equations under basic conditions when OH^- is not present in a reactant or product

Sometimes it is not obvious where the hydroxide ions should be placed; for example, in the reduction of oxygen to form water. In this case, you can balance the half-equation as though it occurs in an acidic solution, then add hydroxide ions later. Follow the steps below to see how this is done.

1 Balance the oxygen atoms by adding water.

$$O_2 \rightarrow 2H_2O$$

2 Balance the hydrogen atoms by adding H^+ ions.

$$O_2 + 4H^+ \rightarrow 2H_2O$$

3 Balance the charge by adding electrons.

$$O_2 + 4H^+ + 4e^- \rightarrow 2H_2O$$

After step 3, the hydroxide ions are introduced.

4 Add enough hydroxide ions to both sides of the equation to neutralise the H^+ ions on the left-hand side of the equation.

$$O_2 + 4H^+ + 4OH^- + 4e^- \rightarrow 2H_2O + 4OH^-$$

5 The neutralisation reaction between $H^+(aq)$ and $OH^-(aq)$ produces water, so cancel out the water molecules on the side where there are fewer water molecules.

$$O_2 + \overset{2}{\cancel{4}}H_2O + 4e^- \rightarrow \cancel{2H_2O} + 4OH^-$$

$$O_2 + 2H_2O + 4e^- \rightarrow 4OH^-$$

6 Add states to complete the half-equation.

$$O_2(g) + 2H_2O(l) + 4e^- \rightarrow 4OH^-(aq)$$

To balance a half-equation for a reaction performed in basic conditions, start by balancing as you would for an acidic solution, then add hydroxide ions to neutralise the acid and cancel out extra water molecules.

Worked example 4.3.4

BALANCING A HALF-EQUATION IN A BASIC SOLUTION WHEN OH^- IS PRESENT ONLY AS A REACTANT OR PRODUCT

Write the half-equation for the reduction of hydrogen peroxide, H_2O_2, to hydroxide ions, OH^-, in a basic solution.

Thinking	Working
Balance all elements except hydrogen and oxygen in the half-equation.	No other elements are present in this case. $H_2O_2 \rightarrow OH^-$
Balance the oxygen atoms by adding water.	$H_2O_2 \rightarrow OH^- + H_2O$
Balance the hydrogen atoms by adding H^+.	$H_2O_2 + H^+ \rightarrow OH^- + H_2O$
Balance the charge by adding electrons.	$H_2O_2 + H^+ + 2e^- \rightarrow OH^- + H_2O$
Add enough hydroxide ions to both sides of the equation to neutralise the H^+.	$H_2O_2 + H^+ + OH^- + 2e^- \rightarrow OH^- + H_2O + OH^-$
The neutralisation reaction between $H^+(aq)$ and $OH^-(aq)$ produces water, so cancel out the water molecules on the side where there are fewer water molecules and reduce the water molecules on the other side by the same number.	$H_2O_2 + \cancel{H_2O} + 2e^- \rightarrow 2OH^- + \cancel{H_2O}$ $H_2O_2 + 2e^- \rightarrow 2OH^-$
Add state symbols.	$H_2O_2(aq) + 2e^- \rightarrow 2OH^-(aq)$

Worked example: Try yourself 4.3.4

BALANCING A HALF-EQUATION IN A BASIC SOLUTION WHEN OH^- IS PRESENT ONLY AS A REACTANT OR PRODUCT

Write the half-equation for the oxidation of hydroxide ions, $OH^-(aq)$, to form oxygen gas, $O_2(g)$, in a-basic solution.

OVERALL REDOX EQUATIONS UNDER BASIC CONDITIONS

The process for combining half-equations to write an overall redox equation under alkaline conditions is similar to that for acidic conditions.

Worked example 4.3.5

COMBINING HALF-EQUATIONS TO WRITE OVERALL REDOX EQUATIONS UNDER BASIC CONDITIONS

Combine the following half-equations for the reaction between iron(II) hydroxide and oxygen gas in a basic solution to give the overall redox equation:

$$Fe(OH)_2(s) + OH^-(aq) \rightarrow Fe(OH)_3(s) + e^-$$
$$O_2(g) + 2H_2O(l) + 4e^- \rightarrow 4OH^-(aq)$$

Thinking	Working
Multiply one or both equations by a suitable factor to ensure that the number of electrons on both sides of the arrow is equal.	The half-equation for the reduction of oxygen is multiplied by 4. $4Fe(OH)_2(s) + 4OH^-(aq) \rightarrow 4Fe(OH)_3(s) + 4e^-$ $O_2(g) + 2H_2O(l) + 4e^- \rightarrow 4OH^-(aq)$
Add the oxidation and the reduction half-equations together, cancelling electrons so that none appear in the final equation. Also cancel H_2O and OH^- if these occur on both sides of the arrow.	$4Fe(OH)_2(s) + \cancel{4OH^-}(aq) \rightarrow 4Fe(OH)_3(s) + \cancel{4e^-}$ $O_2(g) + 2H_2O(l) + \cancel{4e^-} \rightarrow \cancel{4OH^-}(aq)$
Write the overall redox equation for the reaction.	$4Fe(OH)_2(s) + O_2(g) + 2H_2O(l) \rightarrow 4Fe(OH)_3(s)$

When combining two half-equations to write an overall redox equation, remember to cancel out any species, such as H_2O or OH^-, so that they only appear on one side of the equation.

Worked example: Try yourself 4.3.5

COMBINING HALF-EQUATIONS TO WRITE OVERALL REDOX EQUATIONS UNDER BASIC CONDITIONS

Combine the following half-equations for the reaction between sulfite ions, $SO_3^{2-}(aq)$ and permanganate, $MnO_4^-(aq)$, ions in a basic solution to give the overall redox equation:

$$SO_3^{2-}(aq) + 2OH^-(aq) \rightarrow SO_4^{2-}(aq) + H_2O(l) + 2e^-$$
$$MnO_4^-(aq) + 2H_2O(l) + 3e^- \rightarrow MnO_2(s) + 4OH^-(aq)$$

CASE STUDY ANALYSIS

Alcohol and lives lost on the road

Drinks such as wine and beer contain ethanol (CH_3CH_2OH). Ethanol acts as a depressant, slowing the functioning of the brain. When alcoholic drinks are consumed in excess, their intoxicating qualities can lead to antisocial behaviour and can damage a person's health.

The ethanol content of alcoholic drinks varies, as shown in Table 4.3.1.

TABLE 4.3.1 Typical ethanol content of selected alcoholic drinks

Drink	Typical ethanol content (%v/v)
spirits (brandy, bourbon, gin, rum, vodka, whiskey etc.)	37
port, sherry	18
wines (including sparkling wines)	12
pre-mixed vodka-based drinks	9
alcoholic cider	5–10
beer	5
'light' beer	2.5

Regulations require that the ethanol content of alcoholic drinks be specified on their labels, because the ethanol content determines how much drink can be consumed without adverse effects.

Because ethanol slows down reaction time, it seriously affects a person's driving skills. It is estimated that alcohol has contributed to nearly 40% of all road accidents. Governments have responded by introducing penalties, such as fines and licence disqualification, for drivers whose blood alcohol concentration exceeds a certain level—typically 0.05%(m/v). Probationary (or 'P-plate') drivers are required to have a blood alcohol level of zero.

The introduction of penalties, a policy of randomly testing motorists' blood alcohol levels, and various advertising campaigns have all helped to increase public awareness of the link between alcohol consumption and road accidents. Figure 4.3.4 shows how the road toll in Victoria changed when these various safety measures, including those designed to discourage drink-driving, were introduced.

FIGURE 4.3.4 Continual improvements and action on road safety, including crackdowns on drink-driving, have led to a decrease in the fatality rate on Victorian roads over time.

Chemists were involved in the invention of the breathalyser, an instrument designed for police to use to estimate blood alcohol content (Figure 4.3.5). Rather than analysing samples of blood, this instrument measures the concentration of alcohol in a person's breath, which is closely related to the concentration of alcohol in their blood.

FIGURE 4.3.5 Breathalysers are used to analyse the blood alcohol content of motorists.

If this screening test indicates that a driver's blood alcohol content is over 0.05%(m/v), more accurate measurements are taken either in a 'booze bus' or at a police station. The blood alcohol content may then be confirmed by techniques such as infrared spectroscopy or gas–liquid chromatography. Some police departments also use alcohol fuel-cell sensors.

The first breathalysers were very simple devices that operated by detecting the colour change that occurs when ethanol reacts with an acidified solution of potassium dichromate ($K_2Cr_2O_7$), forming Cr^{3+} ions and ethanoic acid. Modern breathalysers are more sophisticated, using fuel-cell technology, which you will learn about in Chapter 5.

Analysis

1 **a** Write the half-equation for the oxidation of ethanol, $C_2H_5OH(aq)$, to ethanoic acid, $CH_3COOH(aq)$, in an acidic solution.

b Write the half-equation for the reduction of the dichromate ion $Cr_2O_7^{2-}(aq)$ to $Cr^{3+}(aq)$ in an acidic solution.

c Use your answers to parts **a** and **b** to write a balanced equation for the overall reaction that occurs in a breathalyser that uses this reaction.

2 The first breathalysers depended on the police officer observing a colour change that showed alcohol in the driver's breath. Describe that colour change, stating which chemicals were coloured and what colour they were.

3 If a fully licenced driver registered a blood alcohol reading of 0.075% (m/v), they would be charged with drink-driving, lose their licence for 6 months and suffer other penalties. Calculate the mass of ethanol that is estimated to be present in the blood of an adult with 5.5 L of blood in their body if their blood alcohol reading is 0.075%.

4.3 Review

WS 6

SUMMARY

- To balance redox half-equations under acidic conditions:
  1 Balance all elements except hydrogen and oxygen.
  2 Balance the oxygen atoms by adding water molecules.
  3 Balance the hydrogen atoms by adding H^+ ions.
  4 Balance the charge by adding electrons.
  5 Add the states.

  To balance redox half-equations under basic conditions by direct addition of hydroxide ions:
  1 Balance all elements except hydrogen and oxygen.
  2 Balance the hydroxide ions by adding OH^- (aq).
  3 Balance the charge by adding electrons.
  4 Add the states.
- To balance redox half-equations under basic conditions where OH^- is not present in a reactant or product:
  1 Balance the half-equation as though it was under acidic conditions (see summary list for acidic conditions at left).
  2 Add enough hydroxide ions to both sides of the equation to neutralise the H^+.
  3 The neutralisation reaction between H^+(aq) and OH^-(aq) produces water, so cancel out the water molecules on the side where there are fewer water molecules and reduce the water molecules on the other side by the same number.
  4 Add the states.
- To write an overall equation, add the oxidation half-equation and the reduction half-equation, making sure that the number of electrons used in reduction equals the number of electrons released during oxidation.
- When combining oxidation and reduction half-equations, any H^+(aq), OH^-(aq) or H_2O(l) that appear on both sides of the arrow should be cancelled.

KEY QUESTIONS

Knowledge and understanding

1 Write half-equations under acidic conditions for the following reactions:

a reduction of VO_2^+(aq) to VO^{2+}(aq)

b reduction of MnO_4^-(aq) to MnO_2(s)

c oxidation of SO_3^{2-}(aq) to SO_4^{2-}(aq)

d oxidation of $S_2O_3^{2-}$(aq) to SO_2(g)

e oxidation of NH_4^+(aq) to NO_3^-(aq)

f reduction of SO_4^{2-}(aq) to H_2S(g)

2 Use your answers to parts **a**, **b** and **c** of question **1** to write the half-equations under *basic* (alkaline) conditions for the reactions.

3 When a piece of copper is placed into an *acidified* solution of potassium nitrate, a reaction occurs that produces copper(II) ions and nitrogen(II) oxide, NO(g).

a Write the oxidation half-equation for the reaction.

b Write the reduction half-equation for the reaction.

c Use your answers to parts **a** and **b** to write a balanced equation for the overall reaction.

Analysis

4 Write the half-equations and the balanced overall equation for the reaction in which:

a a solution containing iron(II) ions is oxidised by an *acidified* solution containing dichromate ions ($Cr_2O_7^{2-}$). The products include iron(III) and chromium(III) ions

b a solution containing sulfite ions (SO_3^{2-}) reacts with an *acidified* solution of permanganate ions (MnO_4^-) to produce a colourless solution containing sulfate ions and manganese(II) ions

c manganese dioxide (MnO_2) reacts with concentrated hydrochloric acid to form chlorine gas and a solution containing manganese(II) ions.

5 The following redox equations *under acidic conditions* are not balanced. For each equation:

  i identify the species that has been reduced and the species that has been oxidised
  ii write balanced half-equations for the oxidation and reduction reactions
  iii combine the half-equations to write a balanced overall equation.

a $Ce^{4+}(aq) + H_2S(g) \longrightarrow Ce^{3+}(aq) + S(s) + H^+(aq)$

b $NO_3^-(aq) + H^+(aq) + Fe^{2+}(aq) \longrightarrow NO(g) + H_2O(l) + Fe^{3+}(aq)$

c $H_2O_2(aq) + Br^-(aq) + H^+(aq) \longrightarrow Br_2(l) + H_2O(l)$

d $MnO_2(s) + H^+(aq) + S(s) \longrightarrow Mn^{2+}(aq) + H_2O(l) + SO_2(g)$

6 The following redox equations *under basic conditions* are not balanced. For each equation:

  i identify the species that has been reduced and the species that has been oxidised
  ii write balanced half-equations for the oxidation and reduction reactions under basic conditions
  iii combine the half-equations to write a balanced overall equation.

a $MnO_4^-(aq) + H_2O(l) + Br^-(aq) \longrightarrow MnO_2(s) + OH^-(aq) + BrO_3^-(aq)$

b $Al(s) + OH^-(aq) + NO_3^-(aq) + H_2O(l) \longrightarrow Al(OH)_4^-(aq) + NH_3(aq)$

7 Write the half-equations and the balanced overall equation for the reaction in which a basic solution containing iron(II) hydroxide, $Fe(OH)_2(aq)$, is oxidised to $Fe(OH)_3(aq)$ by a solution of hydrogen peroxide.

Chapter review

04

KEY TERMS

acidic conditions
acidified
alkaline conditions
basic conditions
conjugate redox pair
half-equation
oxidant
oxidation
oxidation number
oxidised
oxidising agent
redox reaction
reduced
reducing agent
reductant
reduction
transition element

REVIEW QUESTIONS

Knowledge and understanding

1 State whether each of the statements is true or false.

a Group 1 and 2 metal ions, such as Na^+, are reducing agents because they tend to lose electrons.

b Non-metal ions, such as Cl^-, can be reducing agents because they can lose electrons.

c Metals, such as Cu, can be oxidising agents because they can gain electrons.

2 Which of the following is the correct balanced overall equation for the oxidation of lead metal by reaction with silver ions in solution?

A $Pb(s) + Ag^{2+}(aq) \longrightarrow Pb^{2+}(aq) + Ag(s)$

B $Pb^{2+}(aq) + Ag(s) \longrightarrow Pb(s) + Ag^+(aq)$

C $Pb(s) + Ag^+(aq) \longrightarrow Pb^{2+}(aq) + Ag(s)$

D $Pb(s) + 2Ag+(aq) \longrightarrow Pb^{2+}(aq) + 2Ag(s)$

3 In which of the following compounds of sulfur is the oxidation number of sulfur equal to −2?

A H_2SO_4

B SO_2

C H_2S

D $Na_2S_2O_3$

4 Identify the reducing agent in each of these redox reactions.

a $4Ag(s) + O_2(g) \longrightarrow 2Ag_2O(s)$

b $Mg(s) + Cl_2(g) \longrightarrow MgCl_2(s)$

c $2Fe^{3+}(aq) + Zn(s) \longrightarrow Zn^{2+}(aq) + 2Fe^{2+}(aq)$

d $3Ag^+(aq) + Al(s) \longrightarrow Al^{3+}(aq) + 3Ag(s)$

5 Which of the following half-equations does not have a mistake in it?

A $Fe^{3+}(aq) + e^- \longrightarrow Fe^{2+}(aq)$

B $Ag(s) + e^- \longrightarrow Ag^+(aq)$

C $I_2(aq) + e^- \longrightarrow I^-(aq)$

D $Na^+(aq) - e^- \longrightarrow Na(s)$

6 When the following half-equation for the oxidation of Mn^{2+} to MnO_4^- is correctly balanced, which of the following is the correct coefficient for the electrons?

$$Mn^{2+}(aq) + H_2O(l) \longrightarrow MnO_4^-(aq) + H^+(aq) + e^-$$

A 8

B 6

C 5

D 2

7 Fill in the gaps to complete the summary relating to the following redox equation.

$$Cu(s) + 2NO_3^-(aq) + 4H^+(aq) \longrightarrow Cu^{2+}(aq) + 2NO_2(g) + 2H_2O(l)$$

The oxidation number of copper ________ from ___ to ___ and the oxidation number of nitrogen ______ from ___ to ___.

This means that the copper is the ______________ agent because the nitrogen is ____________ and the nitrogen (in NO_3^-) is the ______________ agent because the copper is ______________.

8 Complete the following table.

Compound	Element	Oxidation number
$CaCO_3$	Ca	
HNO_3	O	
H_2O_2		−1
HCO_3^-		+4
HNO_2	N	
$KMnO_4$	Mn	
H_2S	S	
Cr_2O_3	Cr	
N_2O_4		+4

9 Place the following substances in order of increasing oxidation states of nitrogen.

NO, K_3N, N_2O_4, N_2O, $Ca(NO_3)_2$, N_2O_3, N_2

10 Complete the summary below as you balance the half-equation for the reduction of NO_3^- to NO_2 in an acidic solution.

$$NO_3^-(aq) \rightarrow NO_2(g)$$

Step	Task	How it's done	Half-equation
1	Balance the nitrogen.		$NO_3^-(aq) \rightarrow NO_2(g)$
2	Balance the oxygen by adding ______.	Add _____ H_2O molecule(s) to the right-hand side of the equation.	
3	Balance the hydrogen by adding _____.	Add ______ H^+ ion(s) to the ________ of the equation.	
4	Balance the charge by adding ______.	Charge on left-hand side = ______ Charge on right-hand side = ______ Add e^- to the _______ of the equation.	
5	Add state symbols to give the final half-equation.	Give the appropriate states for each reactant and product in the equation.	

11 Complete the following table, identifying the conjugate redox pairs for each of the reactions.

Equation	Conjugate redox pair (oxidation)	Conjugate redox pair (reduction)
$Sn(s) + Br_2(aq) \rightarrow SnBr_2(aq)$		
$Zn(s) + CuCl_2(aq) \rightarrow ZnCl_2(aq) + Cu(s)$		
$Br^-(aq) + 3F_2(g) + 3H_2O(l) \rightarrow 6F^-(aq) + 6H^+ + BrO_3^-(aq)$		
$3Cu(s) + Cr_2O_7^{2-}(aq) + 14H^+(aq) \rightarrow 3Cu^{2+}(aq) + 2Cr^{3+}(aq) + 7H_2O(l)$		

Application and analysis

12 Which of the following reactions are redox reactions? Justify your answers using oxidation numbers.

a $BaCl_2(aq) + H_2SO_4(aq) \rightarrow BaSO_4(s) + 2HCl(aq)$

b $2Ag(s) + Cl_2(g) \rightarrow 2AgCl(s)$

c $2FeCl_3(aq) + SnCl_2(aq) \rightarrow 2FeCl_2(aq) + SnCl_4(aq)$

d $ZnCO_3(s) \rightarrow ZnO(s) + CO_2(g)$

e $HPO_3^{2-}(aq) + I_2(aq) + OH^-(aq) \rightarrow H_2PO_4^-(aq) + 2I^-(aq)$

f $2Cu^+(aq) \rightarrow Cu^{2+}(aq) + Cu(s)$

g $CaF_2(aq) \rightarrow Ca^{2+}(aq) + 2F^-(aq)$

h $P_4(s) + 6H_2(g) \rightarrow 4PH_3(s)$

13 Copper bowls and trays can be decorated by etching patterns on them using concentrated nitric acid. The overall reaction is:

$$Cu(s) + 4HNO_3(aq) \rightarrow Cu(NO_3)_2(aq) + 2NO_2(g) + 2H_2O(l)$$

a Deduce the oxidation number of copper:

i before the reaction

ii after the reaction.

b Deduce the oxidation number of nitrogen:

i before the reaction

ii after the reaction.

c Name the oxidising agent and reducing agent in this process.

14 The unbalanced half-equation for the reduction of the iodate ion (IO_3^-) is:

$$2IO_3^-(aq) + xH^+(aq) + ye^- \rightarrow I_2(aq) + zH_2O(l)$$

Complete the equation by inserting the correct coefficients for *x*, *y* and *z*.

15 When sulfur dioxide is bubbled through a solution of potassium permanganate, $KMnO_4$, the purple colour fades to a colourless solution containing $Mn^{2+}(aq)$ and sulfate, SO_4^{2-}, ions. Write the ionic half-equations for this reaction and then deduce the balanced overall equation.

16 As a result of a traffic accident, residents in a Melbourne suburb had to be evacuated when toxic fumes leaked from a container of sodium dithionite ($Na_2S_2O_4$). The dithionite ion reacts with water according to the equation:

$$2S_2O_4^{2-}(aq) + H_2O(l) \rightarrow S_2O_3^{2-}(aq) + 2HSO_3^-(aq)$$

a State the oxidation number of the sulfur in the following ions.

i $S_2O_4^{2-}$

ii $S_2O_3^{2-}$

iii HSO_3^-

b Write ionic half-equations for the oxidation and reduction reactions that occur when sodium dithionite is mixed with water.

17 During each of the following analyses, redox reactions occurred. Write ionic half-equations for the oxidation and reduction reactions. Use these half-equations to write an overall equation for each reaction.

a Zinc was added to a solution of Pb^{2+} ions. Lead metal was precipitated and Zn^{2+} ions were formed.

b The amount of Fe^{2+} ions in iron tablets was determined by oxidising them to Fe^{3+} ions, using an acidified solution of MnO_4^- ions. The MnO_4^- ions were reduced to Mn^{2+} ions during the reaction.

c Sulfur dioxide (SO_2), a preservative in dried fruit, was determined by oxidation to SO_4^{2-} using a solution of I_2. Iodide (I^-) ions were produced.

d An acidified solution of bleach, which contains OCl^- ions, was titrated against a solution of I^- ions. The reaction products included Cl^- and I_2.

18 Vitamin C is an essential vitamin to maintain good health. The amount of vitamin C present in food can be determined by titration against a solution of iodine, using starch as an indicator. The reaction is a redox process and can be represented by the following overall equation:

$C_6H_8O_6(aq) + I_2(aq) \longrightarrow C_6H_6O_6(aq) + 2H^+(aq) + 2I^-(aq)$

a Write the two half-equations for the reaction occurring in acidic conditions.

b Identify each reaction as either oxidation or reduction.

19 Consider the reaction of $H_3AsO_4(aq)$ with $Br^-(aq)$. The unbalanced equation for this reaction is:

$H_3AsO_4(aq) + Br^-(aq) \longrightarrow As_2O_3(s) + BrO_3^-(aq) + H_2O(l)$

a Write the two half-equations for the reaction.

b Write a balanced overall equation using your two half-equations from part **a**.

20 Solid ammonium dichromate decomposes to form chromium(III) oxide, nitrogen gas and steam.

a Complete and balance the following half-equations.

i $NH_4^+(s) \longrightarrow N_2(g)$

ii $Cr_2O_7^{2-}(s) \longrightarrow Cr_2O_3(s)$

b Identify each of the half-equations in part **a** as either reduction or oxidation processes and explain your answer in terms of oxidation numbers.

c Write the balanced overall equation for the reaction that occurs.

d Identify the conjugate redox pairs for the reaction.

21 The following overall equation for the reaction between $H_2SO_3(aq)$ and $Sn^{4+}(aq)$ is not balanced. Missing components may include coefficients, reactants or products.

$H_2SO_3(aq) + Sn^{4+}(aq) \longrightarrow Sn^{2+}(aq) + HSO_4^-(aq) + H^+(aq)$

Balance the equation by writing and balancing the two half-equations and then combining them to give the overall equation.

22 The thermite process can be used to weld lengths of railway track together. A mould placed over the ends of the two rails to be joined is filled with a charge of aluminium powder and iron(III) oxide. When the mixture is ignited, a redox reaction occurs to form molten iron, which joins the rails together, and aluminium oxide.

a Write a half-equation for the conversion of iron(III) oxide to metallic iron and oxide ions.

b Is the half-equation you wrote for part **a** an oxidation or a reduction process?

c Write the overall equation for the thermite process.

d What mass of iron(III) oxide must be present in the charge if each joint requires 3.70 g of iron to weld it together?

23 The following redox equations *under basic conditions* are not balanced. For each equation:

i identify the species that has been reduced and the species that has been oxidised

ii write balanced half-equations for the oxidation and reduction reactions under basic conditions

iii combine the half-equations to write a balanced overall equation.

a $Cr(OH)_3(s) + IO_3^-(aq) + OH^-(aq) \longrightarrow CrO_4^{2-}(aq) + I^-(aq) + H_2O(l)$

b $Al(OH)_4^-(aq) + Br^-(aq) \longrightarrow Al(s) + OH^-(aq) + BrO_3^-(aq) + H_2O(l)$

24 In alkaline dry cells commonly used in torches, an electric current is produced from the reaction of zinc metal with solid MnO_2. During this reaction, solid zinc hydroxide, $Zn(OH)_2$, and solid Mn_2O_3 are formed. Remembering that the solution is alkaline, write half-equations and an overall equation for the reaction.

OA ✓✓

CHAPTER

05 Galvanic cells and fuel cells as sources of energy

By the end of this chapter, you will know how galvanic cells generate electricity from chemical reactions. You will see how experimental data from galvanic cells can be used to compare the relative strengths of oxidising and reducing agents.

You will compare an exothermic redox reaction in a test tube with the same redox reaction where the reactants are in separate half-cells and electron transfer is used to produce a portable energy source.

The relative strengths of oxidising and reducing agents are conveniently summarised in a table known as the electrochemical series. You will learn how to use the electrochemical series to predict the likelihood of individual redox reactions occurring and to calculate the potential differences of different galvanic cells.

Finally, you will learn how cells may be developed to meet the future energy needs of society. One type of galvanic cell, called a fuel cell, is a component of a vision for a sustainable-energy future called the 'hydrogen economy', in which hydrogen becomes the major source of energy and replaces fossil fuels.

Key knowledge

- the common design features and general operating principles of non-rechargeable (primary) galvanic cells converting chemical energy into electrical energy, including electrode polarities and the role of the electrodes (inert and reactive) and electrolyte solutions (details of specific cells not required) **5.1**
- the use and limitations of the electrochemical series in designing galvanic cells and as a tool for predicting the products of redox reactions, for deducing overall equations from redox half-equations and for determining maximum cell voltage under standard conditions **5.2**
- the common design features and general operating principles of fuel cells, including the use of porous electrodes for gaseous reactants to increase cell efficiency (details of specific cells not required) **5.3**
- contemporary responses to challenges and the role of innovation in the design of fuel cells to meet society's energy needs, with reference to green chemistry principles: design for energy efficiency, and use of renewable feedstocks **5.4**

OA

5.1 Galvanic cells

FIGURE 5.1.1 Our way of life depends on cells and batteries.

Electronic devices such as mobile phones, notebook computers, cameras and hearing aids all depend on small portable sources of electricity: cells and batteries (Figure 5.1.1). Portable energy in the form of cells and batteries enables you to operate electrical equipment without the restriction of a power cord.

The demand for electronic devices has stimulated the production of a variety of cells, from tiny button cells used in watches and calculators, to the huge batteries used to operate lighthouses. The energy provided by cells and batteries may be more expensive than energy from other sources, such as fossil fuels, but this cost is offset by convenience.

In this section, you will find out how cells are constructed and how they provide a source of electrical energy.

INTRODUCING GALVANIC CELLS

A galvanic cell is an electrochemical cell in which chemical energy is converted into electrical energy.

An **electrochemical cell** is a device in which chemical energy is converted into electrical energy, or vice versa. A **galvanic cell** (which is also known as a **voltaic cell**) is a type of electrochemical cell in which chemical energy is converted into electrical energy. The cells in your mobile phone and laptop are galvanic cells.

CASE STUDY ANALYSIS

A technological leap beginning with a frog's leg

The history of electrochemistry began in 1791, when Italian biologist Luigi Galvani and his assistant were experimenting with dissected frogs. They were startled to see a frog's leg that was hanging on a copper hook twitch when it touched an iron rail. The muscles of the frog were stimulated by an electric shock. Galvani had just discovered how to generate a current. His 'error' was in thinking that the current was some sort of life force, a perfectly reasonable idea for the time.

Other scientists set out to investigate the possibility that metals were involved in this phenomenon. After several years' work, in 1800, Alessandro Volta developed a device that used a chemical reaction to produce an electric current. Figure 5.1.2 shows the device, now called a 'voltaic pile'. It consisted of a stack of alternating copper and zinc discs separated by cardboard soaked in salt water.

The voltaic pile was the first battery. Its invention quickly led to the discovery that electricity could be used to decompose compounds into their constituent elements (a process called electrolysis). Scientists found that water could be decomposed into the elements hydrogen and oxygen, and English chemist Humphry Davy used electrolysis to isolate the elements sodium, potassium, calcium, boron, barium, strontium and magnesium.

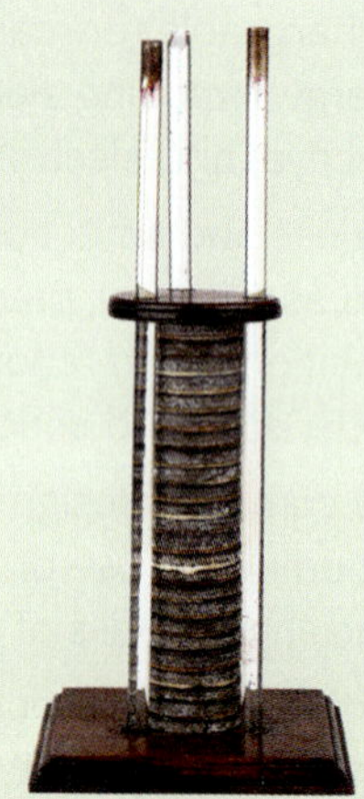

FIGURE 5.1.2 An original voltaic pile. The pile is 15 cm high and contains alternating copper and zinc discs separated by cardboard.

Analysis

In a voltaic pile, electricity is produced by a redox reaction involving zinc metal, water and oxygen gas dissolved in the water. The products of the reaction are Zn^{2+} and OH^- ions.

1 a Write a half-equation for the oxidation of zinc to form Zn^{2+} ions.
  b Write a half-equation for the reduction of O_2 to form OH^- ions.
  c Write a balanced overall equation for the reaction.

2 Identify the oxidising and reducing agents in the reaction in the voltaic pile.

If we connect several cells in series to obtain a higher potential difference or 'voltage', the combination of cells is called a **battery**. The term 'battery' strictly only applies to a combination of cells, but it is in everyday use to describe cells as well.

Figure 5.1.3 shows how you can produce a galvanic cell from simple laboratory equipment.

In Figure 5.1.4, you can see a diagram of a cell called the Daniell cell, named after the scientist who invented it in 1836, John Daniell. The cell produces an electric current that flows through the wire and light globe. This part of the cell is called the **external circuit**. The globe converts the electrical energy of the current into light and heat.

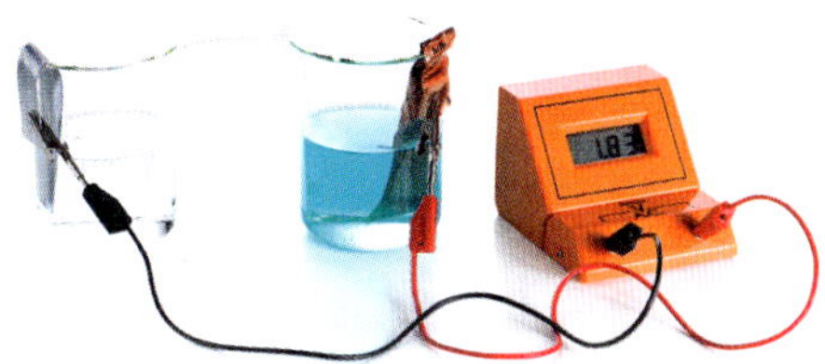

FIGURE 5.1.3 Construction of a galvanic cell from simple laboratory equipment

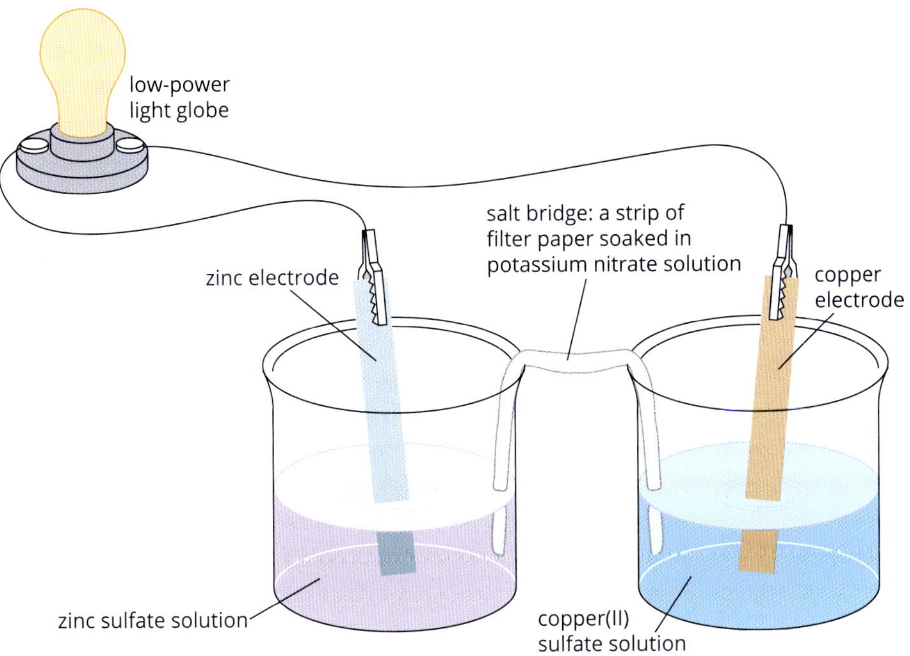

FIGURE 5.1.4 The Daniell cell is a type of galvanic cell.

CHEMFILE

The Daniell cell

In 1836, English chemist John Daniell invented a device that could supply a useful electric current. The device, which became known as the Daniell cell, was used almost exclusively to power the early English and US telegraph systems because of its reliable output.

His original design consisted of a copper cylinder with the gullet (oesophagus) from an ox hanging in its centre. A zinc rod hung inside this gullet. The cylinder was filled with sulfuric acid solution saturated with copper sulfate and the ox gullet was filled with sulfuric acid solution. The ox gullet acted as a porous membrane, allowing passage of ions.

The invention of the Daniell cell was a great advance in battery technology because it was the first reliable source of electric current.

The current flows because a chemical reaction is taking place in the cell. If you leave the cell with a light globe connected for several hours, you may see evidence of this reaction occurring: the zinc metal corrodes, the copper metal becomes covered with a furry dark-brown deposit and the blue copper(II) sulfate solution loses some of its colour.

If you replace the light globe with a **galvanometer** (an instrument for detecting electric current), the galvanometer will indicate that electrons flow from the zinc **electrode** through the wire to the copper electrode. Current flows only if the two halves of the cell are connected by a **salt bridge**. A salt bridge is often made from filter paper soaked in a relatively unreactive **electrolyte** (a chemical substance that conducts electric current), such as a solution of potassium nitrate.

These observations lead to the following explanations about what is occurring in a galvanic cell.

- The reaction in the cell is a redox reaction because electrons are being transferred.
- The zinc electrode corrodes because the zinc metal forms zinc ions in solution:
$$Zn(s) \longrightarrow Zn^{2+}(aq) + 2e^-$$
- The oxidation of the zinc metal releases electrons, which flow through the wire to the copper electrode.
- Electrons are accepted by copper(II) ions in the solution when the ions collide with the copper electrode:
$$Cu^{2+}(aq) + 2e^- \longrightarrow Cu(s)$$
- The copper metal that is formed deposits on the electrode as a dark-brown coating.

In this cell, there is a transformation of chemical energy to electrical energy. Figure 5.1.5 shows the processes that occur during the operation of the cell. The equation for the overall reaction is found by adding the two half-equations:

$$Cu^{2+}(aq) + Zn(s) \longrightarrow Cu(s) + Zn^{2+}(aq)$$

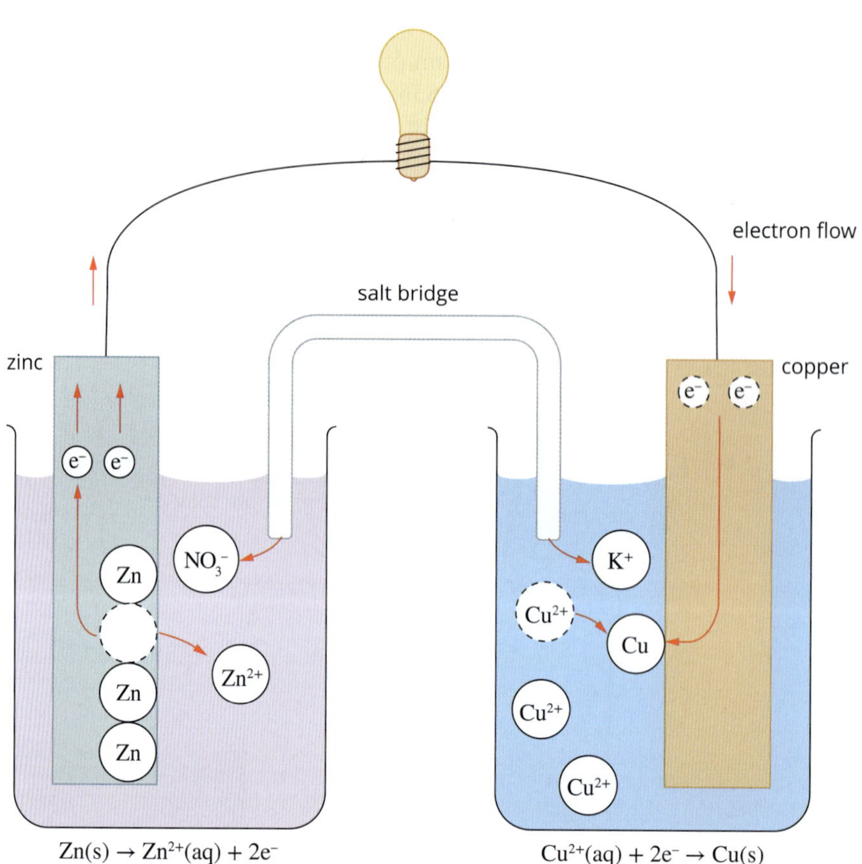

FIGURE 5.1.5 A representation of the electrode reactions in a Daniell cell. The salt bridge contains potassium nitrate solution.

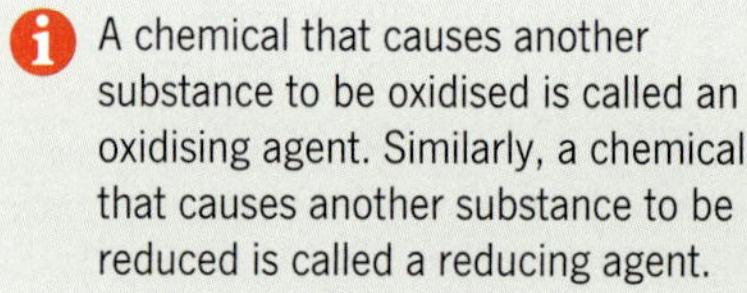

A chemical that causes another substance to be oxidised is called an oxidising agent. Similarly, a chemical that causes another substance to be reduced is called a reducing agent.

This redox reaction is described as a **spontaneous reaction** because it does not need to be driven by an external source of energy (i.e. the reaction occurs naturally).

Copper(II) ions act as the **oxidising agent**, or oxidant, and zinc metal acts as the **reducing agent**, or reductant.

FIGURE 5.1.6 Cu^{2+}(aq) ions in copper(II) sulfate solution reacting directly with a strip of zinc metal

ENERGY TRANSFORMATIONS IN DIRECT REACTIONS

You may have seen a similar reaction to the one that occurs in a Daniell cell if you have copper-plated a piece of metal, such as zinc. When zinc is immersed in an aqueous solution containing Cu^{2+}(aq) ions, the metal becomes coated in dark-brown copper (Figure 5.1.6). At the same time, thermal energy is produced, which escapes into the surrounding environment as heat.

The overall equation for this metal displacement reaction is:

$$Cu^{2+}(aq) + Zn(s) \longrightarrow Cu(s) + Zn^{2+}(aq)$$

This reaction is an example of a spontaneous exothermic reaction. If the reactants are allowed to come into direct contact with each other, their chemical energy is transformed directly to thermal energy. However, in a galvanic cell, the half-reactions occur in separate containers, and the electrons are transferred by the external circuit, so that chemical energy is transformed into electrical energy.

The energy changes that occur in galvanic cells and when reactants undergo direct reaction are summarised in Figure 5.1.7.

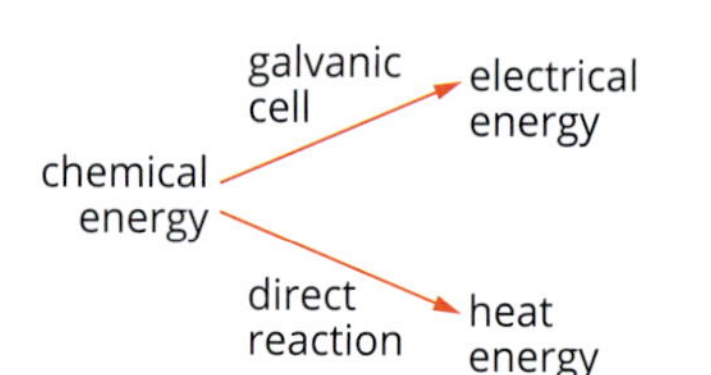

FIGURE 5.1.7 Energy transformations that occur when reactants are separated in a galvanic cell and when the reactants react directly

HOW A GALVANIC CELL OPERATES

A galvanic cell is designed so that half-reactions occur in two separate compartments of the cell. Because the oxidising agent and reducing agent do not come into direct contact with each other, electrons can only be transferred through an external circuit connecting the negative and positive electrodes.

This flow of electrons creates an electric current. Therefore, the chemical energy of the reactants is transformed into electrical energy.

Half-cells

A galvanic cell can be regarded as consisting of two **half-cells**. Each half-cell contains an electrode in contact with a solution (Figure 5.1.8). In the Daniell cell, one half-cell contains Cu(s) and Cu^{2+}(aq); the other contains Zn(s) and Zn^{2+}(aq). The species (chemicals) present in each half-cell form a conjugate redox pair (an oxidising agent and its corresponding reduced form).

If one member of the conjugate pair in a half-cell is a metal, as in Figure 5.1.8a, it is usually used as the electrode. Some redox pairs, such as Br_2(aq)/Br^-(aq) and Fe^{3+}(aq)/Fe^{2+}(aq), do not involve solid metals. If no metal is present, an inert (unreactive) electrode, such as platinum or graphite, is used, as shown in Figure 5.1.8b. Because platinum metal is expensive, graphite is usually used for inert electrodes in school experiments.

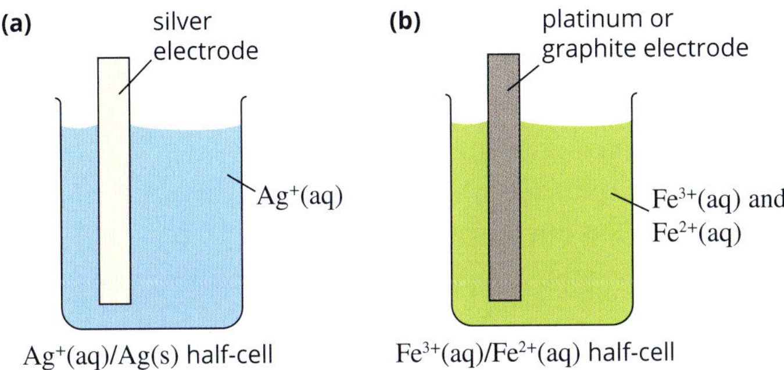

FIGURE 5.1.8 Typical half-cells. (a) The silver metal is used as the electrode. (b) There is no metal in the redox pair in the Fe^{3+}(aq)/Fe^{2+}(aq) half-cell, so an inert platinum or graphite electrode has to be added.

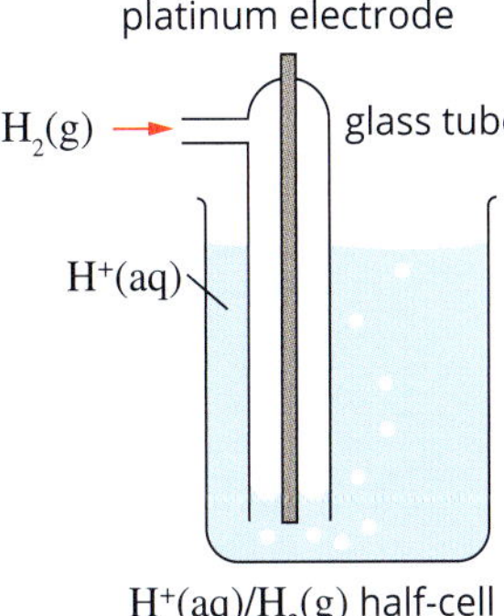

FIGURE 5.1.9 An H^+(aq)/H_2(g) half-cell consists of a platinum electrode in a solution of H^+(aq) with H_2 gas bubbling through the solution.

In some half-cells, one of the conjugate pairs may be a gas. In such cases, a special 'gas electrode', like the one shown in Figure 5.1.9 for a H^+(aq)/H_2(g) half-cell, is used. Note that half-cells usually contain other species not involved in the reaction, such as spectator ions and the solvent.

The electrode at which oxidation occurs is called the **anode**. In a galvanic cell, the anode, where electrons are released, is described as the negative terminal. The electrode at which reduction occurs is called the **cathode**. The cathode, where electrons are gained, is the positive terminal in a galvanic cell.

> The electrode in the half-cell in which oxidation occurs is called the anode; the electrode in the half-cell in which reduction occurs is called the cathode. In a galvanic cell, the anode is negative and the cathode is positive.

The purpose of the salt bridge

The salt bridge contains ions that are free to move, so that they can balance charges formed in the two compartments. Cations (positive ions) move towards the cathode and anions (negative ions) move towards the anode.

Without a salt bridge, the solution in one compartment in the galvanic cell would accumulate negative charge and the solution in the other compartment would accumulate positive charge as the reaction proceeded. Such accumulation of charge would stop the reaction very quickly and, hence, prevent further reaction.

The salt bridge is part of the **internal circuit** of a galvanic cell.

> Cations in the salt bridge move towards the cathode, and anions in the salt bridge move towards the anode.

> In balanced half-equations and overall equations, the:
> - numbers of atoms of each element are equal on both sides
> - total charge on each side is equal.

> The number of electrons lost in the oxidation reaction must equal the number of electrons gained in the reduction reaction.

WRITING HALF-CELL EQUATIONS

If a conjugate redox pair consists of an element and its corresponding ion, then the half-equation is relatively easy to write. For example, knowing that a reduction reaction involves the conjugate redox pair of Zn^{2+} ions and Zn, you can quickly write the half-cell equation as:

$$Zn^{2+}(aq) + 2e^{-} \longrightarrow Zn(s)$$

Half-cell equations involving polyatomic ions may be more complex to write. The equation for the reduction reaction in a half-cell containing the dichromate ion ($Cr_2O_7^{2-}$) and Cr^{3+} ion redox pair is:

$$Cr_2O_7^{2-}(aq) + 14H^{+}(aq) + 6e^{-} \longrightarrow 2Cr^{3+}(aq) + 7H_2O(l)$$

You learnt to write half-equations such as these in Chapter 4.

Writing an overall equation for a cell reaction

The half-equations for the oxidation and reduction reactions that occur in a cell can be added together to obtain an overall, or full, equation. Recall from Chapter 4 that an overall equation does not show any electrons; all the electrons lost in the oxidation reaction are gained in the reduction reaction. You may need to multiply one or both half-equations by a factor to ensure that the electrons balance and can be cancelled out in the overall equation.

DRAWING AND LABELLING A DIAGRAM OF A GALVANIC CELL

If you know what reaction is occurring in a galvanic cell, then you can draw a diagram of the cell, identifying key features such as the anode, cathode, electrode polarity, direction of electron flow and direction of the flow of ions.

For example, consider a cell with the cell reaction:

$$Cu(s) + Cl_2(g) \longrightarrow Cu^{2+}(aq) + 2Cl^{-}(aq)$$

From this reaction, you can see that:

- copper metal is at the anode (because copper is oxidised and oxidation occurs at the anode)
- chlorine gas is present at the cathode (because chlorine is reduced and reduction occurs at the cathode).

As in all galvanic cells:

- electrons flow through the external circuit from the anode (negative) to the cathode (positive)
- anions flow in the internal circuit to the anode and cations flow towards the cathode.

This information can be used to draw and label a diagram of the cell, as shown in Figure 5.1.10.

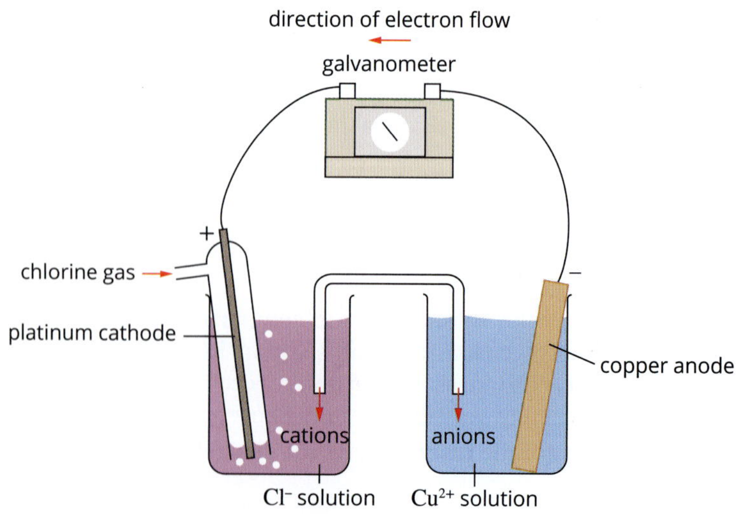

FIGURE 5.1.10 A diagram of the cell containing $Cl_2(g)/Cl^{-}(aq)$ and $Cu^{2+}(aq)/Cu(s)$ half-cells

Worked example 5.1.1

DRAWING AND LABELLING A GALVANIC CELL

A cell has the cell reaction: $Pb^{2+}(aq) + Zn(s) \longrightarrow Pb(s) + Zn^{2+}(aq)$

Write the two half-equations.

Draw a diagram of the cell, labelling the anode and cathode, electrode polarities (which electrode is positive and which is negative) and the direction of electron flow.

Thinking	Working
Identify the two relevant half-equations.	$Pb^{2+}(aq) + 2e^- \longrightarrow Pb(s)$ $Zn(s) \longrightarrow Zn^{2+}(aq) + 2e^-$
Identify the anode and the cathode in this cell. The anode is the electrode at which oxidation occurs. The cathode is the electrode at which reduction occurs.	The electrode in the $Zn^{2+}(aq)/Zn(s)$ half-cell is the anode and the electrode in the $Pb^{2+}(aq)/Pb(s)$ half-cell is the cathode.
Determine the polarities of the electrodes and thc direction of electron flow in the cell. The anode is negative and the cathode is positive.	Electrons flow from the negative electrode (anode) to the positive electrode (cathode) as shown in the diagram of the cell below. 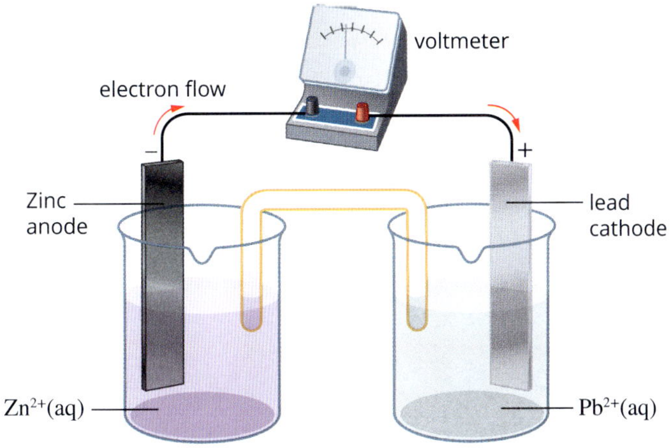

Worked example: Try yourself 5.1.1

DRAWING AND LABELLING A GALVANIC CELL

A cell has the cell reaction: $Sn^{2+}(aq) + Fe(s) \longrightarrow Sn(s) + Fe^{2+}(aq)$

Write the two half-equations.

Draw a diagram of the cell, labelling the anode and cathode, electrode polarities (which electrode is positive and which is negative) and the direction of electron flow.

Primary cells are galvanic cells that cannot be recharged, whereas secondary cells are galvanic cells that are rechargeable.

FIGURE 5.1.11 Non-rechargeable alkaline cells are used to power many devices, including torches, smoke detectors and calculators. They are discarded once they go flat.

CHEMFILE

Powering the Overland Telegraph

One of the first primary cells in widespread use in the 19th century was called the Leclanché cell. It had a similar overall cell reaction to the reaction in alkaline cells.

Leclanché cells were one of the sources of power for the Australian Overland Telegraph Line, one of 19th-century Australia's great engineering feats (see figure below). The 3200 km telegraph line was completed in 1872 and linked Port Augusta in South Australia with Darwin in the Northern Territory. In Darwin, the line was connected to an undersea telegraph cable that allowed fast communication between Australia and the rest of the world. An additional section to Western Australia was added in 1877.

The Alice Springs Telegraph Station, built in 1872

PRIMARY CELLS

More than two centuries after the battery was invented by Alessandro Volta, cells and batteries are a common power source for many household and industrial applications. Cells and batteries can be used as fixed energy storage systems, such as in solar energy systems, burglar alarms and smoke detectors. They are also used extensively for portable devices, including mobile phones, watches, digital cameras and laptop computers. The portability of these devices relies on cells and batteries as sources of electrical energy.

There are two basic types of galvanic cells:

- **primary cells**, which are disposable and designed not to be recharged
- **secondary cells**, which are **rechargeable cells** and designed to be reused many times. You will learn more about secondary cells in Chapter 9.

Common commercial **alkaline cells**, such as those you would usually use in a torch or a remote control, are primary cells (Figure 5.1.11). Primary cells are often referred to as **non-rechargeable cells**; they 'go flat' when the cell reaction reaches **equilibrium** (a point when there is no tendency for the quantities of reactants and products to change), and you have to buy a replacement. In primary cells, the products slowly migrate away from the electrodes or are consumed by side reactions occurring in the cell, preventing the cells from being recharged.

An example of a primary cell: The alkaline cell

The alkaline cell was developed in the late 1960s to meet a growing need for small, high-capacity sources of power for portable electrical appliances.

Figure 5.1.12 shows the construction of an alkaline cell and the reactions that occur in the cell. This cell is similar to the simple galvanic cells discussed earlier in this section, but it has been designed so the two half-reactions occur in separate places within one container. A potassium hydroxide electrolyte performs the same function as a salt bridge in the simple cells.

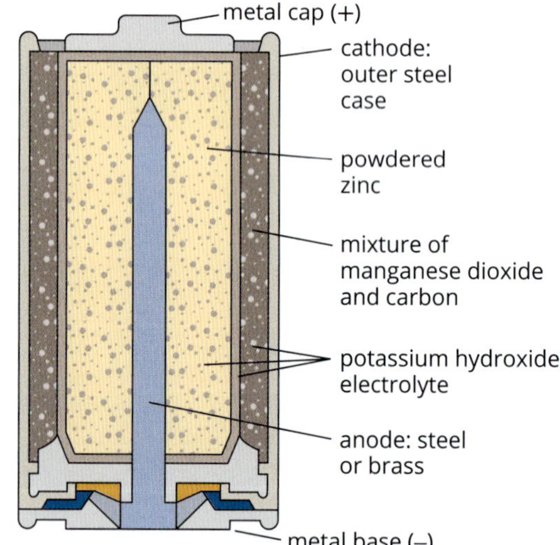

FIGURE 5.1.12 Construction of an alkaline cell

Alkaline cells are especially cost-effective in torches and motorised toys, where high currents are needed intermittently. These cells typically produce about 1.5 V. As with all primary cells, once the reactants in the cell have been consumed the cell is 'flat', and cannot be used again.

5.1 Review

OA ✓✓

SUMMARY

- The reactions that occur in galvanic cells are spontaneous and exothermic.
- In a galvanic cell, chemical energy is converted directly into electrical energy in a redox reaction.
- A cell is made from two half-cells. Each half-cell contains a conjugate redox pair.
- An oxidation reaction occurs in one half-cell and a reduction reaction occurs in the other half-cell.
- The electrode in the half-cell in which oxidation occurs is called the anode; the electrode in the half-cell in which reduction occurs is called the cathode.
- In galvanic cells, the anode is negative and the cathode is positive.
- Electrons flow through the external circuit from the anode to the cathode.
- A salt bridge allows a cell to produce electricity by allowing the movement of ions between the two half-cells. Cations in the salt bridge move towards the cathode and anions move towards the anode.
- If the reactants in a galvanic cell reaction are allowed to come into direct contact, chemical energy is converted into heat energy rather than electrical energy.
- Galvanic cells can be categorised as primary or secondary cells. Primary cells cannot be recharged; secondary cells are rechargeable.
- Key concepts about a galvanic cell are summarised in the following diagram.

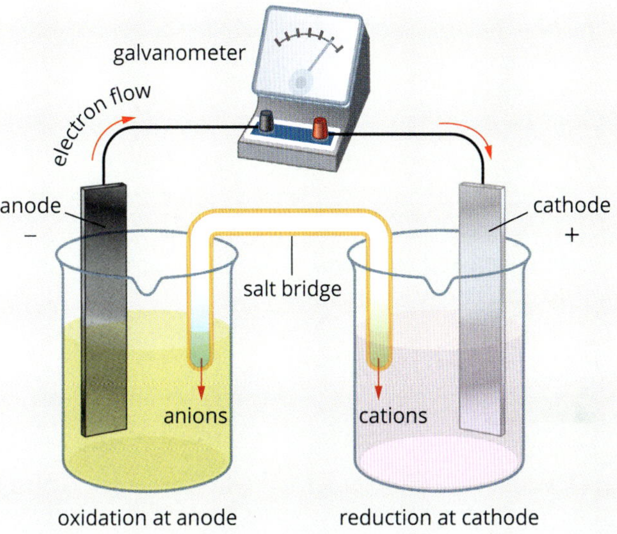

KEY QUESTIONS

Knowledge and understanding

1 State the meaning of the following terms:
- **a** galvanic cell
- **b** electrochemical cell
- **c** salt bridge
- **d** battery
- **e** primary cell.

2 Describe how to construct a galvanic cell.

3 Draw labelled diagrams of the following half-cells.
- **a** $Ni^{2+}(aq)/Ni(s)$
- **b** $Sn^{4+}(aq)/Sn^{2+}(aq)$
- **c** $H^{+}(aq)/H_2(g)$

4 Complete the following sentences by selecting the correct term for each of the alternatives in bold.
In a galvanic cell, electrons move through the external circuit from the **positive/negative** terminal to the **positive/negative** terminal. Cations in the salt bridge move towards the **anode/cathode** and anions in the salt bridge move towards the **anode/cathode**.

5 State the electrode material you would choose if you were constructing the following half-cells.
- **a** $Fe^{2+}(aq)/Fe(s)$
- **b** $Fe^{3+}(aq)/Fe^{2+}(aq)$
- **c** $Cl^{-}(aq)/Cl_2(g)$

continued over page

5.1 Review *continued*

Analysis

6 The overall equation for the reaction that occurs in a cell made up of $Al^{3+}(aq)/Al(s)$ and $Sn^{2+}(aq)/Sn(s)$ half-cells is:

$$2Al(s) + 3Sn^{2+}(aq) \rightarrow 2Al^{3+}(aq) + 3Sn(s)$$

a Identify the species that are oxidised and those that are reduced during the reaction.

b Write half-equations for the reaction occurring at the:

i cathode

ii anode.

c Draw a diagram of the cell and label the:

i half-equation for the reaction occurring in each half-cell

ii anode and cathode

iii direction of electron flow

iv electrode polarities (which electrode is positive and which is negative)

v direction of anion and cation flow from the salt bridge.

7 Copy the diagram of the galvanic cell below in your workbook and label the:

i anode

ii cathode

iii positive electrode

iv negative electrode

v reduction half-equation

vi oxidation half-equation

vii salt bridge.

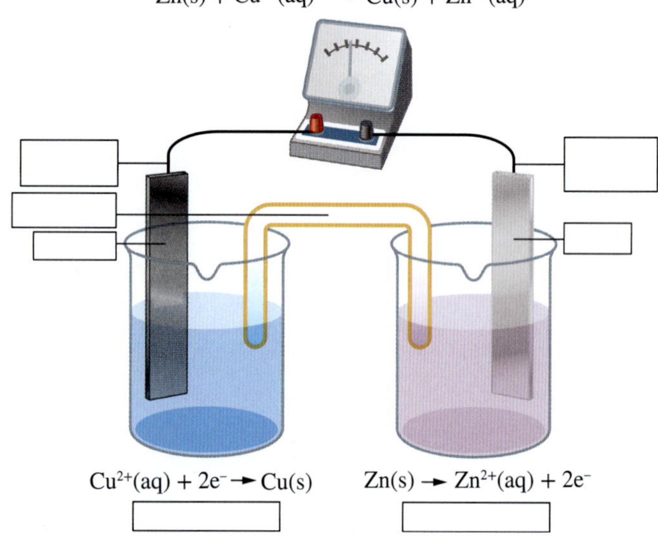

5.2 The electrochemical series

Metals vary in their reactivity. Platinum and gold are unreactive and are widely used for jewellery. Other metals are very reactive. For example, sodium reacts so readily with oxygen and water to produce sodium hydroxide and hydrogen gas (Figure 5.2.1) that it must be stored in paraffin oil.

Galvanic cells can help you compare the relative reactivity of metals. Galvanic cells can be constructed from various combinations of half-cells. The experimental data collected from these combinations allow chemists to determine the oxidising and reducing strengths of many different substances. This information is very useful because it allows scientists to predict the products of various reactions, calculate the voltages of cells and develop more powerful and longer-lasting batteries.

FIGURE 5.2.1 Water dropped onto sodium metal. Sodium is highly reactive.

RELATIVE OXIDISING AND REDUCING STRENGTHS

As you saw in Section 5.1, a half-cell contains a conjugate redox pair. The reactions that can occur in a half-cell can be written as reversible reactions, showing the relationship between the two chemicals in the redox pair. For example, the reaction in a half-cell containing the:

- $H^+(aq)/H_2(g)$ redox pair may be written as:

$$2H^+(aq) + 2e^- \rightleftharpoons H_2(g)$$

- $Zn^{2+}(aq)/Zn(s)$ redox pair may be written as:

$$Zn^{2+}(aq) + 2e^- \rightleftharpoons Zn(s)$$

Figure 5.2.2 shows a diagram of a cell constructed from $H^+(aq)/H_2(g)$ and $Zn^{2+}(aq)/Zn(s)$ half-cells.

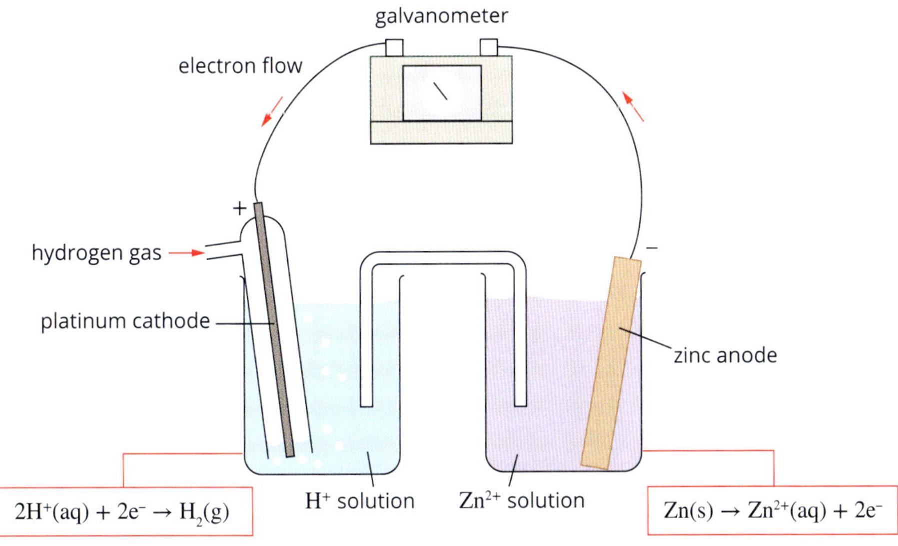

FIGURE 5.2.2 A diagram of a cell constructed from $H^+(aq)/H_2(g)$ and $Zn^{2+}(aq)/Zn(s)$ half-cells

Notice that in this cell the zinc electrode (the anode) is negative. The reactions that are occurring are:

$$2H^+(aq) + 2e^- \longrightarrow H_2(g)$$
$$Zn(s) \longrightarrow Zn^{2+}(aq) + 2e^-$$

Zn is oxidised to Zn^{2+} and H^+ is reduced to H_2. Zinc can be described as a reducing agent because it causes the reduction of H^+ to occur. The H^+ is described as an oxidising agent because it causes the Zn to be oxidised to Zn^{2+}.

Because electrons flow from the Zn^{2+}/Zn half-cell to the H^+/H_2 half-cell, we can infer that:

- zinc is a stronger reducing agent than H_2
- H^+ is a stronger oxidising agent than Zn^{2+} ions.

In a galvanic cell, the stronger reducing agent is in the half-cell with the negative electrode (anode). The stronger oxidising agent is in the half-cell with the positive electrode (cathode).

POTENTIAL DIFFERENCE

A current flows in a galvanic cell because one half-cell has a greater tendency to push electrons into the external circuit than the other half-cell. Chemists say that a **potential difference** exists between the two half-cells. The potential difference of a cell is sometimes also called the **electromotive force**, or emf, and is commonly referred to as the voltage.

The potential difference of a cell, symbol E, has the unit of the **volt** (V) and is measured with a **voltmeter**.

Potential differences of cells are usually measured under the **standard conditions** of:

- a pressure of 100 kPa
- 1 M concentration of solutions.

The potential difference of a cell under standard conditions is given the symbol $E°$. Potential differences are usually measured at 25°C.

Standard electrode potentials

It is impossible to measure the potential difference of an isolated half-cell because both oxidation and reduction must take place for a potential difference to exist. However, you can assign a **standard electrode potential** ($E°$) to each half-cell by connecting the cells to a standard reference half-cell and measuring the voltage produced.

A hydrogen half-cell, $H^+(aq)/H_2(g)$, under standard conditions, is used for this purpose and its $E°$ value is arbitrarily assigned as zero. This half-cell is known as the **standard hydrogen half-cell** or **standard hydrogen electrode (SHE)**.

The standard electrode potential of other cells may then be measured by connecting them to the standard hydrogen half-cell, as shown in Figure 5.2.3 for an $Fe^{2+}(aq)/Fe(s)$ half-cell.

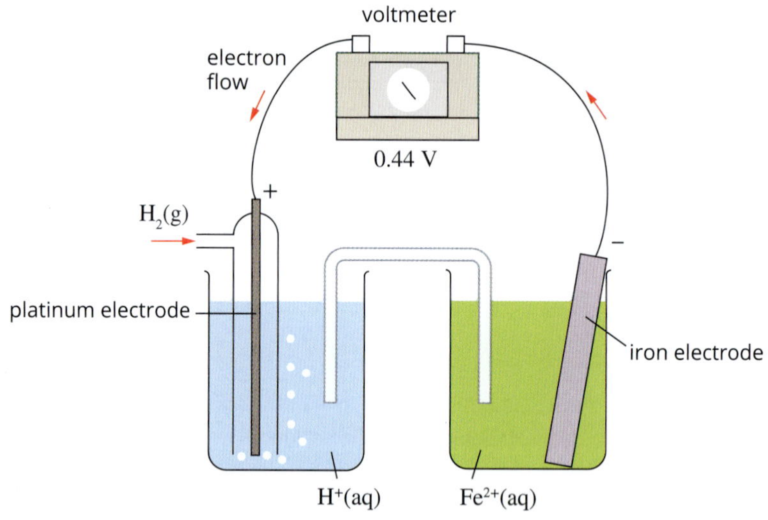

FIGURE 5.2.3 Measuring the standard electrode potential of an $Fe^{2+}(aq)/Fe(s)$ half-cell. From the voltmeter reading and negative polarity of the iron electrode, the $E°$ of this half-cell is −0.44 V.

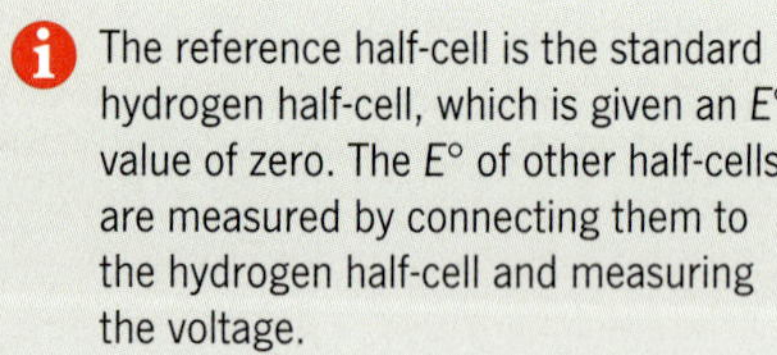
The reference half-cell is the standard hydrogen half-cell, which is given an $E°$ value of zero. The $E°$ of other half-cells are measured by connecting them to the hydrogen half-cell and measuring the voltage.

We can summarise the information obtained from the measurement shown in Figure 5.2.3 as follows:

$$Fe^{2+}(aq) + 2e^- \rightleftharpoons Fe(s) \quad E° = -0.44\,V$$

The negative sign indicates that the electrode in the half-cell was negative when connected to the hydrogen half-cell. Oxidation is occurring in the $Fe^{2+}(aq)/Fe(s)$ half-cell, and the electrons that are produced from the iron electrode move towards the hydrogen half-cell. The value of −0.44 V is known as both the standard electrode potential and the **standard reduction potential**. The standard electrode potential gives a numerical measure of the tendency of a half-cell reaction to occur as a reduction reaction.

USING THE ELECTROCHEMICAL SERIES

By connecting the standard hydrogen electrode to different half-cells and measuring their standard electrode potentials, chemists have developed a table called the **electrochemical series** (Table 5.2.1).

TABLE 5.2.1 The electrochemical series. The strongest oxidising agents are at the top left of the table and the strongest reducing agents are at the bottom right of the table.

	Oxidising agents		Reducing agents		E° (V)
Increasing oxidising strength ↑	$F_2(g) + 2e^-$	⇌	$2F^-(aq)$	Increasing reducing strength ↓	+2.87
	$H_2O_2(aq) + 2H^+(aq) + 2e^-$	⇌	$2H_2O(l)$		+1.77
	$Au^+(aq) + e^-$	⇌	$Au(s)$		+1.68
	$Cl_2(g) + 2e^-$	⇌	$2Cl^-(aq)$		+1.36
	$O_2(g) + 4H^+(aq) + 4e^-$	⇌	$2H_2O(l)$		+1.23
	$Br_2(l) + 2e^-$	⇌	$2Br^-(aq)$		+1.09
	$Ag^+(aq) + e^-$	⇌	$Ag(s)$		+0.80
	$Fe^{3+}(aq) + e^-$	⇌	$Fe^{2+}(aq)$		+0.77
	$O_2(g) + 2H^+(aq) + 2e^-$	⇌	$H_2O_2(aq)$		+0.68
	$I_2(s) + 2e^-$	⇌	$2I^-(aq)$		+0.54
	$O_2(g) + 2H_2O(l) + 4e^-$	⇌	$4OH^-(aq)$		+0.40
	$Cu^{2+}(aq) + 2e^-$	⇌	$Cu(s)$		+0.34
	$Sn^{4+}(aq) + 2e^-$	⇌	$Sn^{2+}(aq)$		+0.15
	$S(s) + 2H^+(aq) + 2e^-$	⇌	$H_2S(g)$		+0.14
	$2H^+(aq) + 2e^-$	⇌	$H_2(g)$		0.00
	$Pb^{2+}(aq) + 2e^-$	⇌	$Pb(s)$		−0.13
	$Sn^{2+}(aq) + 2e^-$	⇌	$Sn(s)$		−0.14
	$Ni^{2+}(aq) + 2e^-$	⇌	$Ni(s)$		−0.25
	$Co^{2+}(aq) + 2e^-$	⇌	$Co(s)$		−0.28
	$Cd^{2+}(aq) + 2e^-$	⇌	$Cd(s)$		−0.40
	$Fe^{2+}(aq) + 2e^-$	⇌	$Fe(s)$		−0.44
	$Zn^{2+}(aq) + 2e^-$	⇌	$Zn(s)$		−0.76
	$2H_2O(l) + 2e^-$	⇌	$H_2(g) + 2OH^-(aq)$		−0.83
	$Mn^{2+}(aq) + 2e^-$	⇌	$Mn(s)$		−1.18
	$Al^{3+}(aq) + 3e^-$	⇌	$Al(s)$		−1.66
	$Mg^{2+}(aq) + 2e^-$	⇌	$Mg(s)$		−2.37
	$Na^+(aq) + e^-$	⇌	$Na(s)$		−2.71
	$Ca^{2+}(aq) + 2e^-$	⇌	$Ca(s)$		−2.87
	$K^+(aq) + e^-$	⇌	$K(s)$		−2.93
	$Li^+(aq) + e^-$	⇌	$Li(s)$		−3.04

Strong reducing agents donate electrons more readily than weak ones.
Strong oxidising agents accept electrons more readily than weak ones.

Strong reducing agents have weak conjugate oxidising agents.
Strong oxidising agents have weak conjugate reducing agents.

Oxidising agents react with reducing agents that are lower in the electrochemical series.

Under non-standard conditions, the order of the half-cells may change.

Notice the value of 0.00 V given for the $H^+(aq)/H_2(g)$ half-equation. All other E° values are relative to this arbitrary standard. The strongest oxidising agent, F_2, is at the top left of the table and the strongest reducing agent, Li, is at the bottom right of the table.

In a galvanic cell, the stronger reducing agent is oxidised, so it is in the half-cell with the negative electrode (anode). The stronger oxidising agent is reduced, so it is in the half-cell with the positive electrode (cathode).

Predicting cell reactions

The electrochemical series can be used to predict the reactions that are likely to occur when any pair of half cells is connected to make a galvanic cell. The series also allows you to predict the maximum potential difference of the cell under standard conditions.

The strongest oxidising agent in the cell will react with the strongest reducing agent. Another way to predict the electrode reactions is to remember that the half-reaction that is higher in the electrochemical series goes forward and the lower one is reversed. Because of this:

- a reduction reaction will occur in the half-cell with the higher E° value, whereas an oxidation reaction will occur in the half-cell with the lower E° value
- the positive electrode will be in the half-cell with the higher E° value, whereas the negative electrode will be in the half-cell with the lower E° value.

The equation for the overall cell reaction is found by adding the two half-equations. Worked example 5.2.1 shows you how to use the electrochemical series to predict a cell reaction.

Worked example 5.2.1

PREDICTING THE OPERATION OF A GALVANIC CELL

A cell is made from $Ag^+(aq)/Ag(s)$ and $Fe^{2+}(aq)/Fe(s)$ half-cells under standard conditions at 25°C. Use the electrochemical series to predict the overall cell reaction, identify the anode and cathode, and determine the direction of electron flow.

Thinking	Working
Identify the two relevant half-equations in the electrochemical series.	$Ag^+(aq) + e^- \rightleftharpoons Ag(s)$ $E^\circ = 0.80$ V $Fe^{2+}(aq) + 2e^- \rightleftharpoons Fe(s)$ $E^\circ = -0.44$ V
Identify the strongest oxidising agent (the species on the left of the series with the most positive E° value) and the strongest reducing agent (the species on the bottom right).	Because Ag^+ is higher on the left side of the table than Fe^{2+}, it is the stronger oxidising agent. Fe is lower on the right side of the table than Ag, so it is the stronger reducing agent.
Write the two half-equations that will occur. The strongest oxidising agent will react with the strongest reducing agent. (Hint: The reduction equation has the most positive E° value and the oxidation equation has the most negative E° value.) Remember that the oxidation reaction should be written in reverse.	Reduction: $Ag^+(aq) + e^- \rightarrow Ag(s)$ Oxidation: $Fe(s) \rightarrow Fe^{2+}(aq) + 2e^-$
Write the overall cell equation.	Multiply the Ag^+/Ag half-cell equation by 2 so that the number of electrons in each half-equation is equal, and then add the two equations together: $[Ag^+(aq) + e^- \rightarrow Ag(s)] \times 2$ $Fe(s) \rightarrow Fe^{2+}(aq) + 2e^-$ $2Ag^+(aq) + Fe(s) \rightarrow 2Ag(s) + Fe^{2+}(aq)$

Identify the anode and the cathode in this cell. The anode is the electrode at which oxidation occurs. The cathode is the electrode at which reduction occurs.	The silver electrode is the cathode and the iron electrode is the anode.
Determine the polarities of the electrodes and the direction of electron flow in the cell. The anode is negative and the cathode is positive.	Electrons flow from the negative electrode (anode) to the positive electrode (cathode) as shown in the diagram of the cell below. voltmeter; electron flow; –; +; iron anode; silver cathode; $Fe^{2+}(aq)$; $Ag^{+}(aq)$

Worked example: Try yourself 5.2.1

PREDICTING THE OPERATION OF A GALVANIC CELL

A cell is made from $Sn^{2+}(aq)/Sn(s)$ and $Ni^{2+}(aq)/Ni(s)$ half-cells under standard conditions at 25°C. Use the electrochemical series to predict the overall cell reaction, identify the anode and cathode, and determine the direction of electron flow.

Calculating the voltage of a cell

The maximum potential difference of a cell under standard conditions is the difference between the $E°$ values of its two half-cells. It is defined as follows:

cell potential difference = $E°$ of half-cell containing the oxidising agent – $E°$ of half-cell containing the reducing agent

An easy way to remember this for galvanic cells is:

cell potential difference = higher half-cell $E°$ – lower half-cell $E°$

For example, the maximum cell voltage of a cell constructed from $Ag^{+}(aq)/Ag(s)$ and $Fe^{2+}(aq)/Fe(s)$ half-cells under standard conditions can be calculated as follows:

$$
\begin{aligned}
\text{cell potential difference} &= \text{higher half-cell } E° - \text{lower half-cell } E° \\
&= E°(Ag^{+}(aq)/Ag(s)) - E°(Fe^{2+}(aq)/Fe(s)) \\
&= 0.80 - (-0.44) \\
&= 1.24\,V
\end{aligned}
$$

Different values for the cell voltage are obtained under non-standard conditions.

As a galvanic cell **discharges**, the cell voltage eventually drops to zero and the cell is referred to as 'flat'. Equilibrium has then been reached.

The easiest way to calculate the cell potential difference is to remember: cell potential difference = higher half-cell $E°$ – lower half-cell $E°$

Limitations of predictions

As stated above, the standard electrode potentials given in the electrochemical series are measured under standard conditions. As you might expect, electrode potentials can vary under other conditions.

When conditions are very different from standard conditions, the order of half-reactions in the electrochemical series may also be different, and predictions of reactions based on the standard electrode potentials may not be reliable.

It is also important to remember that the electrochemical series gives no information about the rate at which reactions occur. Just because a reaction is predicted to occur does not mean it will occur rapidly or even noticeably.

DIRECT REDOX REACTIONS

If the contents of the half-cells of a galvanic cell were mixed, the reactants would react directly. Energy would be released as heat rather than as electrical energy. Reactions that occur in galvanic cells or when chemicals are directly mixed are described as naturally occurring reactions, or spontaneous reactions.

Earlier in this section, you saw that in a galvanic cell the strongest oxidising agent in the cell reacts with the strongest reducing agent. In other words, the higher half-reaction (the one with the most positive E° value) in the electrochemical series occurs in the forward direction (as reduction) and the lower half-reaction (the one with the most negative E° value) occurs in the reverse direction (as oxidation).

This principle applies equally to redox reactions that occur when reactants are mixed directly.

As shown in Figure 5.2.4, for a spontaneous reaction to occur, an oxidising agent (on the left of the electrochemical series) must react with a reducing agent (on the right) that is lower in the electrochemical series.

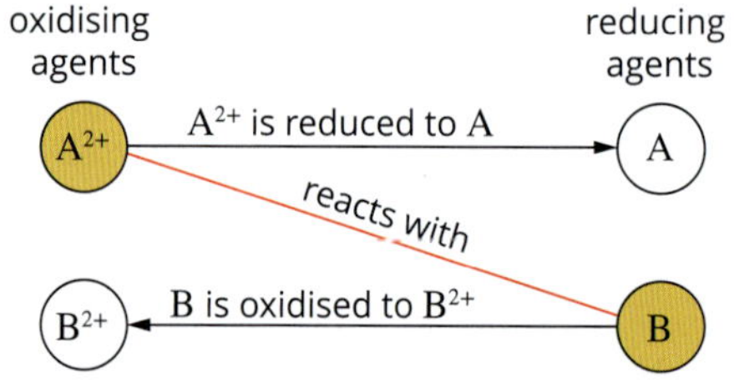

A^{2+} reacts with B.
No reaction occurs for the other combinations: A and B, A^{2+} and B^{2+}, A^{2+} and A, A and B^{2+}, B^{2+} and B.

FIGURE 5.2.4 Oxidising agents only react significantly with reducing agents that are lower in the electrochemical series.

CHEMFILE

Biodegradable batteries

Much of the current research in battery technology is focused on the development of secondary cells, rather than primary cells, because rechargeable cells are more efficient and minimise waste. However, some researchers are developing biodegradable primary cells that can be directly implanted into the body to drive low-power medical devices that are used for wound healing, bone regeneration and monitoring for diagnosis. Such devices need only operate for a short time and using a biodegradable battery would avoid the need for surgical retrieval after its use. The battery dissolves after completing its task, without causing harm to the body. Body fluids contain various ions, which the batteries can use as the electrolyte (see figure). These ions can pass through the porous polymer casing that separates the anode and cathode.

At anode: $Mg(s) + 2OH^-(aq) \rightarrow Mg(OH)_2(s) + 2e^-$

At cathode: $2H_2O(l) + 2e^- \rightarrow H_2(g) + 2OH^-(aq)$

A biodegradable cell designed to be implanted in the body for medical use

The development of biodegradable cells is still in the very early stages, but researchers have reported success with Mg–Fe, Mg–MoO_3, Mg–Au and Mg–Mo cell designs.

Worked example 5.2.2

PREDICTING DIRECT REDOX REACTIONS

Consider the following equations that appear in the order shown in the electrochemical series.

$$Br_2(aq) + 2e^- \rightleftharpoons 2Br^-(aq) \quad E^\circ = +1.09\ V$$
$$Ni^{2+}(aq) + 2e^- \rightleftharpoons Ni(s) \quad E^\circ = -0.23\ V$$
$$Mg^{2+}(aq) + 2e^- \rightleftharpoons Mg(s) \quad E^\circ = -2.34\ V$$

Use the electrochemical series to predict the effect of mixing:

a $Br_2(aq)$ and $Mg^{2+}(aq)$

b $Ni^{2+}(aq)$ and $Br^-(aq)$

c $Ni^{2+}(aq)$ and $Mg(s)$.

Thinking	Working
Identify the two relevant half-equations in the electrochemical series. Predict whether or not a reaction occurs. A chemical species on the left of the electrochemical series (an oxidising agent) reacts with a chemical species on the right (a reducing agent) that is lower in the series. If a reaction will occur, write the overall equation.	**a** $Br_2(aq) + 2e^- \rightleftharpoons 2Br^-(aq)\ \ E^\circ = +1.09\ V$ $Mg^{2+}(aq) + 2e^- \rightleftharpoons Mg(s)\ \ E^\circ = -2.34\ V$ No reaction occurs because both $Br_2(aq)$ and $Mg^{2+}(aq)$ are oxidising agents. **b** $Br_2(aq) + 2e^- \rightleftharpoons 2Br^-(aq)\ \ E^\circ = +1.09\ V$ $Ni^{2+}(aq) + 2e^- \rightleftharpoons Ni(s)\ \ E^\circ = -0.23\ V$ No reaction occurs because the oxidising agent, Ni^{2+}, is below the reducing agent, Br^-, in the electrochemical series. **c** $Ni^{2+}(aq) + 2e^- \rightleftharpoons Ni(s)\ \ E^\circ = -0.23\ V$ $Mg^{2+}(aq) + 2e^- \rightleftharpoons Mg(s)\ \ E^\circ = -2.34\ V$ A reaction occurs because the oxidising agent, Ni^{2+}, is above the reducing agent, Mg, in the electrochemical series. The higher half-equation occurs in the forward direction: $Ni^{2+}(aq) + 2e^- \rightarrow Ni(s)$ The lower half-equation occurs in the reverse direction: $Mg(s) \rightarrow Mg^{2+}(aq) + 2e^-$ The overall reaction equation is found by adding the half-equations: $Ni^{2+}(aq) + Mg(s) \rightarrow Ni(s) + Mg^{2+}(aq)$

With half-equations listed in order of increasing reducing agent strength, you can predict whether a spontaneous reaction will occur by looking for substances that are arranged in a top-left/bottom-right arrangement.

Worked example: Try yourself 5.2.2

PREDICTING DIRECT REDOX REACTIONS

Consider the following equations that appear in the order shown in the electrochemical series:

$$Cl_2(g) + 2e^- \rightleftharpoons 2Cl^-(aq) \quad E^\circ = +1.36\ V$$
$$I_2(s) + 2e^- \rightleftharpoons 2I^-(aq) \quad E^\circ = +0.54\ V$$
$$Pb^{2+}(aq) + 2e^- \rightleftharpoons Pb(s) \quad E^\circ = -0.13\ V$$

Use the electrochemical series to predict the effect of mixing:

a $I_2(s)$ and $Pb^{2+}(aq)$

b $Cl^-(aq)$ and $I_2(s)$

c $Cl_2(g)$ and $Pb(s)$.

5.2 Review

SUMMARY

- The hydrogen half-cell is used as the standard reference half-cell; its value is arbitrarily assigned as zero.
- The standard electrode potential or standard reduction potential (E°) of a half-cell is measured by connecting the half-cell to a standard hydrogen half-cell and measuring the voltage produced.
- The standard electrode potential gives a numerical measure of the tendency of a half-cell reaction to occur as a reduction reaction.
- Standard electrode potentials are used as the basis of the electrochemical series.
- In the electrochemical series, half-reactions are listed in order so that the strongest oxidising agent is at the top left of the series (with the most positive E° value) and the strongest reducing agent is at the bottom right (with the most negative E° value).
- The electrochemical series is valid for standard conditions; that is, gas pressures of 100 kPa and solution concentrations of 1 M. Standard electrode potentials are usually measured at 25°C.
- The relative strengths of oxidising and reducing agents can be determined by comparing standard electrode potentials, and these can be used to predict half-cell and overall cell reactions, as well as the tendency for a reaction to occur if the oxidising and reducing agents are mixed directly.
- The maximum potential difference of a cell under standard conditions can be calculated from standard electrode potentials:

  cell potential difference = E° of half-cell containing the oxidising agent – E° of half-cell containing the reducing agent

  A simple way of remembering this is:

  cell potential difference = higher half-cell E° – lower half-cell E°
- For a spontaneous reaction to occur, an oxidising agent (on the left of the electrochemical series) must react with a reducing agent (on the right) that is lower in the series.
- The standard electrode potentials in the electrochemical series are measured under standard conditions. Under other conditions, the order of half-reactions may be different and predictions based on the electrochemical series may not be reliable.
- When reactants react in a galvanic cell, chemical energy is transformed into electrical energy. However, when they react directly, their chemical energy is transformed into heat energy.

KEY QUESTIONS

Knowledge and understanding

1 State the meaning of the following terms:
  - **a** potential difference
  - **b** standard conditions
  - **c** standard electrode potential.

2 Complete the following sentences by selecting the correct term for each of the alternatives in bold.
  In a galvanic cell, the stronger reducing agent is in the half-cell with the **negative/positive** electrode. This electrode is the **anode/cathode**. The stronger oxidising agent is in the half-cell with the **negative/positive** electrode. This electrode is the **anode/cathode**.

3 Describe how the standard electrode potential (E°) of a half-cell is measured using a standard hydrogen electrode. In your answer, describe how the sign of the electrode potential is determined.

4 Explain why predictions of reactions based on the standard electrode potentials do not always correspond to what may be observed when the reactants are mixed.

5 Use the electrochemical series (page 189) to determine whether:
  - **a** sodium is a stronger reducing agent than lithium
  - **b** Fe^{2+}(aq) is a stronger oxidising agent than Ca^{2+}(aq)
  - **c** Sn^{2+}(aq) can act as both an oxidising agent and a reducing agent.

Analysis

6 A standard cell has the following cell reaction:

$$2H^+(aq) + 2Cr^{2+}(aq) \longrightarrow 2Cr^{3+}(aq) + H_2(g)$$

If the voltage of this cell is 0.40 V, determine the E° for the following half-reaction:

$$Cr^{3+}(aq) + e^- \rightleftharpoons Cr^{2+}(aq)$$

7 A galvanic cell was constructed from $Al^{3+}(aq)/Al(s)$ and $Pb^{2+}(aq)/Pb(s)$ half-cells. Use the electrochemical series to predict the strongest oxidising agent and reducing agent, and write the:

a oxidation and reduction half-equations

b overall cell reaction

c identities of the anode and cathode.

8 Draw a labelled diagram of a cell formed from $Cl_2(g)/Cl^-(aq)$ and $Sn^{2+}(aq)/Sn(s)$ half-cells. Use the electrochemical series to predict the strongest oxidising agent and reducing agent, and write the:

a half-cell reactions

b anode and cathode

c direction of electron flow

d electrode polarities (which electrode is positive and which is negative)

e directions of flow of the anions and cations in the salt bridge

f overall reaction.

9 Calculate the cell potential difference for each of the cells in questions **7** and **8**.

10 Use the electrochemical series to predict the strongest oxidising agent and reducing agent, and hence whether a reaction will occur in the following situations. If a reaction does occur, write the overall equation for the reaction.

a Chlorine gas is bubbled into a solution containing bromide ions.

b Chlorine gas is bubbled into a solution containing iodide ions.

c A bromine solution is added to a solution containing chloride ions.

d A bromine solution is added to a solution containing iodide ions.

5.3 Fuel cells

Fuel cells are a type of galvanic cell, converting the chemical energy of reactants directly into electrical energy.

Around the world, governments, businesses and academic institutions are spending billions of dollars to develop and produce fuel cells. A **fuel cell** is a type of galvanic cell that generates electricity from redox reactions. Fuel cells use the chemical energy of hydrogen or other fuels to cleanly and efficiently generate electricity, and with almost no pollution. For example, almost the only products from a fuel cell powered by hydrogen are electricity, heat and water. Unlike the cells you studied earlier in this chapter, fuel cells do not run down or need recharging. Electricity is available for as long as fuel is supplied to them.

The National Aeronautics and Space Administration (NASA) was the first organisation to use fuel cells, to generate power for satellites and space capsules.

Fuel cells are an important component of what is called the **hydrogen economy**, in which hydrogen could become a major source of energy and replace fossil fuels. Major technological and cost breakthroughs are needed before the hydrogen economy can become a reality, and there is dispute among scientists over its viability.

Even though the technology is still being developed, fuel cells are already being used in numerous applications. These uses include as a source of power for transport (Figure 5.3.1) such as cars, buses, bicycles and forklifts, and as a source of primary and emergency back-up for buildings and communities.

In this section, you will learn how fuel cells operate, their advantages, and some of the challenges scientists face in bringing the potential of their use to reality.

FIGURE 5.3.1 A hydrogen fuel cell for an electric bicycle. The hydrogen bicycle operates like a standard electric bicycle, but the battery lasts longer.

HYDROGEN FUEL CELLS

The major limitation of the galvanic cells described in the previous sections is that they contain relatively small amounts of reactants. Furthermore, when the reactants have been consumed, the cell must be discarded or recharged. In fuel cells the reactants are supplied continuously, allowing constant production of electrical energy. A key difference between a fuel cell and a primary cell is that the reactants are not stored in the fuel cell. They must be continuously supplied from an external source.

Although the basic principles behind the operation of fuel cells were discovered in 1838, it was not until the 1950s that fuel cells were used for small-scale power production. Fuel cells were the main onboard power supply units and source of water for the Apollo space program that put humans on the Moon, and for the subsequent Space Shuttle program (Figure 5.3.2). An explosion in a fuel cell was responsible for the failure of the *Apollo 13* mission.

FIGURE 5.3.2 Fuel cells like this were used as the primary source of electrical energy for the NASA Space Shuttle program.

Fuel cell design

Figure 5.3.3 shows a simplified diagram of the key parts of a hydrogen–oxygen fuel cell. The fuel cell has two compartments: one for the hydrogen gas and the other for the oxygen gas. The gas compartments are separated from each other by two **porous** electrodes and an electrolyte solution. The electrode at the hydrogen compartment is the anode; the electrode at the oxygen compartment is the cathode. The electrolyte in a fuel cell carries ions from one electrode to the other.

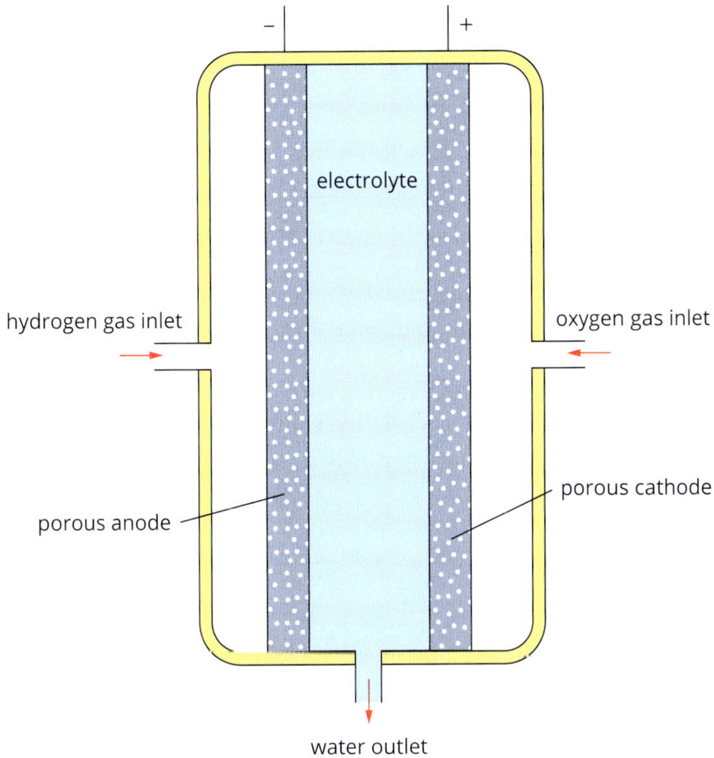

FIGURE 5.3.3 A simplified diagram showing the key parts of a hydrogen–oxygen fuel cell

OPERATION OF FUEL CELLS

Fuel cells are classified primarily by the kind of electrolyte they employ. As examples, we will look at two types of fuel cells: alkaline fuel cells and acidic fuel cells.

Oxidation of a fuel, such as hydrogen, takes place at the anode of a fuel cell, and reduction of oxygen occurs at the cathode.

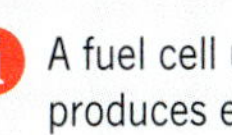

A fuel cell using hydrogen as a fuel produces electricity, water, heat and very few other emissions.

Alkaline fuel cells

In the type of fuel cell used in the NASA Apollo program, the electrolyte was potassium hydroxide solution. Because this type of fuel cell uses potassium hydroxide as the electrolyte, it is commonly referred to as an alkaline fuel cell. The reactions involved in alkaline fuel cells are as follows.

- At the anode (–), hydrogen gas (the 'fuel') is oxidised by reacting with hydroxide ions from the electrolyte:

$$H_2(g) + 2OH^-(aq) \longrightarrow 2H_2O(l) + 2e^-$$

- At the cathode (+), oxygen gas is reduced:

$$O_2(g) + 2H_2O(l) + 4e^- \longrightarrow 4OH^-(aq)$$

The overall equation for the reaction is:

$$2H_2(g) + O_2(g) \longrightarrow 2H_2O(l)$$

Note that the OH^- ions present in the electrolyte appear in the half-equations for the fuel cell, but not in the overall equation.

Each cell produces about 1 volt. Higher voltages are obtained by connecting a number of fuel cells in series to form a battery or fuel cell stack. The only by-products are water and heat.

Acidic fuel cells

The most common type of acidic fuel cell is a phosphoric acid fuel cell. The reactions involved in acidic fuel cells are as follows.

At the anode, hydrogen gas (the 'fuel') is oxidised:

$$2H_2(g) \rightarrow 4H^+(aq) + 4e^-$$

At the cathode, oxygen gas is reduced:

$$O_2(g) + 4H^+(aq) + 4e^- \rightarrow 2H_2O(l)$$

The overall equation for the reaction is:

$$2H_2(g) + O_2(g) \rightarrow 2H_2O(l)$$

Note that the H^+ ions present in the electrolyte appear in the half-equations for the fuel cell but not in the overall equation, and that this is the same overall equation as for alkaline fuel cells.

A fuel (e.g. hydrogen) is oxidised at the anode of a fuel cell, and oxygen is reduced at the cathode. Different electrolytes can be used in cells.

Efficiency of fuel cells

Like other galvanic cells, fuel cells transform chemical energy directly into electrical energy, enabling efficient use of the energy released by spontaneous redox reactions. They are a useful source of continuous electricity and could be used to provide energy for vehicles, buildings and even cities. Power generation in coal-fired power stations and combustion engines involves a series of different energy transformations, each with associated energy losses. Fuel cells do not have the same losses, however, so there is a reduction in the volume of greenhouse gases produced.

Fuel cells are generally quoted as being 40–60% efficient, compared with efficiencies of 30–40% for thermal power stations and 25–30% for car engines. In addition, modern fuel cells use the waste heat that they produce to make steam. This steam can be used for heating or to operate a turbine, thus raising the efficiency of the cells up to 85%.

Electricity generation using fuel cells is more efficient than if the electricity were generated by the combustion of the same fuel.

A range of different fuel cells has been developed using different electrolytes, electrodes and operating temperatures, as shown in Table 5.3.1. The fuel cells are often named according to the electrolyte used in the cell.

TABLE 5.3.1 A comparison of different types of fuel cells

Fuel cell	Electrolyte	Efficiency (%)	Operating temperature (°C)	Application	Typical half-equations
alkaline fuel cell	potassium hydroxide	60	90–100	space vehicles	anode: $H_2(g) + 2OH^-(aq) \rightarrow 2H_2O(l) + 2e^-$ cathode: $O_2(g) + 2H_2O(l) + 4e^- \rightarrow 4OH^-(aq)$
methanol fuel cell	polymer membrane	40	50–100	mobile phones laptop computers	anode: $CH_3OH(l) + H_2O(l) \rightarrow CO_2(g) + 6H^+(aq) + 6e^-$ cathode: $4H^+(aq) + O_2(g) + 4e^- \rightarrow 2H_2O(l)$
molten carbonate fuel cell	molten lithium, sodium or potassium carbonates	50–60	600–1000	electricity utility	anode: $H_2(g) + CO_3^{2-}(l) \rightarrow H_2O(l) + CO_2(g) + 2e^-$ cathode: $O_2(g) + 2CO_2(g) + 4e^- \rightarrow 2CO_3^{2-}(l)$
phosphoric acid fuel cell	liquid phosphoric acid	40–45	175–200	electricity utility transportation	anode: $H_2(g) \rightarrow 2H^+(aq) + 2e^-$ cathode: $O_2(g) + 4H^+(aq) + 4e^- \rightarrow 2H_2O(l)$
proton-exchange membrane fuel cell	solid polyperfluoro-sulfonic acid	60	60–100	portable power transportation	anode: $H_2(g) \rightarrow 2H^+(aq) + 2e^-$ cathode: $O_2(g) + 4H^+(aq) + 4e^- \rightarrow 2H_2O(l)$
solid oxide fuel cell	solid zirconium oxide	50–60	600–1000	electricity utility	anode: $H_2(g) + O^{2-}(s) \rightarrow H_2O(l) + 2e^-$ cathode: $O_2(g) + 4e^- \rightarrow 2O^{2-}(s)$

The nature of the electrodes is crucial to the efficiency of a fuel cell. The electrodes must be both conducting and porous to allow the hydrogen and oxygen to come into contact with the ions in the electrolyte and to allow the redox half-reactions to occur at their surface. The size of the current that can be drawn from a fuel cell depends on the surface area of the electrodes.

Catalysts are used to enhance the rate of reaction and the current that can be produced from a cell. The catalyst incorporated in the anode increases the rate of oxidation of the fuel gas. Platinum metal is commonly used as a catalyst at this electrode. The cathode catalyst, which increases the rate of the reduction half-reaction, can be made from a different material, such as nickel powder.

Fuel cells contain porous electrodes that allow reactants to diffuse through them to react with ions in the electrolyte. The electrodes often contain catalysts to increase the rate of reaction.

The electrolyte in a fuel cell carries ions from one electrode to the other. Electrolytes that have been employed in fuel cells include:

- aqueous alkaline solutions, usually KOH
- liquids such as phosphoric acid (H_3PO_4)
- molten carbonate salts, such as Na_2CO_3 or $MgCO_3$
- permeable polymer membranes, which only allow the passage of positive ions
- ceramics, made from oxides of metals such as calcium and zirconium.

APPLICATIONS OF FUEL CELLS

Most major vehicle manufacturers are developing the use of fuel cells as an alternative to the internal combustion engine, because fuel cells have better fuel efficiency and lower emissions of greenhouse gases and other pollutants than traditional fuels. As of December 2022, more than 56 000 fuel cell cars had been sold worldwide and sales are growing rapidly. Honda, Toyota and Hyundai have launched the first mass-produced hydrogen fuel cell vehicles.

On the other hand, sales of battery-powered electric vehicles have seen much more rapid growth, and it seems likely that the use of hydrogen fuel cells for transport will have a more niche role in the immediate future (Figures 5.3.4 and 5.3.5). Some other examples of fuel cell use include the following:

- A 59 MW power plant began operation in South Korea in 2014, providing power and heat for local homes.
- 430 000 fuel cells have been installed in buildings in Japan to supply heat and power.
- A fleet of hydrogen fuel cell trains began operation in Northern Germany in 2022.
- A 2 MW proton-exchange membrane fuel cell using by-product hydrogen generates 20% of the energy needed for a chlor-alkali electrolytic plant in Yingkou, China. The plant produces chlorine and sodium hydroxide by passing an electric current through concentrated salt solution.
- Technology companies in Silicon Valley, California, USA, are installing fuel cells as a stable, sustainable source of power.

Some scientists predict that fuel cells will play a key role in the transition from a dependence on fossil fuels to a sustainable supply of energy.

Challenges around cost and performance remain, and improvements are still required for hydrogen fuel cells to become truly competitive.

FIGURE 5.3.4 Type 212 class submarines are being constructed for the German and Italian navies. They use proton-exchange membrane fuel cells that operate on compressed hydrogen. The submarines can remain submerged for up to 3 weeks and are very quiet and virtually undetectable.

FIGURE 5.3.5 Proton-exchange membrane fuel cells are being investigated as a source of power for the *Mars Flyer*, a robotic aircraft designed to explore the planet Mars.

CHEMFILE

Fuel cells around us

Some types of breathalysers use fuel cell technology to measure blood alcohol levels. In this fuel cell, ethanol is oxidised and the electricity generated indicates the blood alcohol level.

Fuel cells are also found in nature. The electric eel, which grows to more than 1 metre in length, has a natural battery made of specialised fuel cells within its body (see figure below). The fuel for each cell is the food consumed by the eel, and the oxidising agent is oxygen. Each cell generates about 1.5 V, giving a total voltage from the head of the eel to its tail of more than 300 V. Electric eels use electricity for several reasons. Low pulses of electrical discharge are emitted by the eel and then bounce back from passing objects to be detected by special electroreceptors on the skin. In this way, the eel navigates and locates its prey. High-intensity electrical discharges can be used to stun or kill prey or as a form of defence.

Electric eels contain specialised fuel cells.

ADVANTAGES AND DISADVANTAGES OF FUEL CELLS

Scientists are trying to reduce the overall costs of fuel cells and to improve the electric current that can be drawn from them by increasing the rate of reaction at the electrodes. As described above, various types of fuel cells have been developed and tested, using different fuels, electrolytes and catalysts.

Table 5.3.2 lists some of the advantages and disadvantages of fuel cells.

TABLE 5.3.2 Advantages and disadvantages of fuel cells

Advantages	Disadvantages
• Fuel cells convert chemical energy directly to electrical energy. This is more efficient than the series of energy conversions that takes place in fossil fuel power stations: chemical energy → heat energy → mechanical energy → electrical energy. • Hydrogen fuel cells produce water and heat as by-products when hydrogen is used as a fuel; no greenhouse gases, such as carbon dioxide, are released. • Fuel cells generate electricity as long as the fuel is supplied; conventional batteries need to be recharged or replaced. • Fuel cells can use a variety of fuels. • Electricity can be generated on-site and users are not reliant on connection to an electricity grid; waste heat can be used to heat water for a hot-water system or provide heating for a home during winter.	• Fuel cells require a constant fuel supply. • Fuel cells are expensive. They are still a developing technology and until they are made in large numbers they will not have economies of scale. • Some types of cells use expensive electrolytes and catalysts. • The use of fuel cells in transport requires an extensive network of hydrogen filling stations before it can become widespread. • The hydrogen used in many fuel cells is mainly sourced from fossil fuels at present, with energy losses and greenhouse gas emissions. • There are significant issues with the storage and safety of hydrogen fuel. • Some types of cells use toxic electrolytes, and their electrodes contain rare, expensive or harmful materials.

5.3 Review

SUMMARY

- Fuel cells are a type of galvanic cell, converting chemical energy directly into electrical energy.
- Fuel cells are devices in which the reactants (usually gases) are supplied continuously, allowing constant production of electrical energy.
- The electrodes used in fuel cells allow direct contact between the gases and the electrolyte. They have a high surface area and are porous to ensure high cell efficiency. Catalysts are often part of an electrode to increase the rate of reaction.
- Oxidation of a fuel, such as hydrogen, takes place at the anode of a fuel cell and reduction of oxygen occurs at the cathode.
- The electrolyte in a fuel cell carries ions from one electrode to the other. Different electrolytes can be used in fuel cells.

KEY QUESTIONS

Knowledge and understanding

1 Describe a fuel cell and state how it differs from a primary cell.

2 Which one or more of the following features do primary cells and fuel cells have in common as they produce electricity?

A A catalyst is used to increase reaction rate.
B Cations in the electrolyte move towards the cathode.
C The anode is negative.
D The cathode is constructed from a porous material.
E Oxidation occurs at the cathode.
F The oxidising agent is a gas.
G Chemical energy is converted into electrical energy.

3 Why is it desirable to use porous electrodes in fuel cells?

Analysis

4 The diagram below represents a fuel cell that uses methane gas as the fuel. Complete the labels in the figure.

5 Hydrogen gas and oxygen gas are reacted in a phosphoric acid fuel cell, which is a type of fuel cell that uses phosphoric acid as an electrolyte.

a Write equations for the half-reactions that occur at the anode and cathode.
b Based on the two half-equations from part **a**, determine the overall cell equation for this reaction.

6 An experimental fuel cell that uses methanol as the fuel has the following half-equations:

$$CH_3OH(g) + 6OH^-(aq) \rightarrow CO_2(g) + 5H_2O(l) + 6e^-$$
$$O_2(g) + 2H_2O(l) + 4e^- \rightarrow 4OH^-(aq)$$

a Write the equation for the overall cell reaction.
b Which reaction occurs at the positive electrode of the cell?
c Suggest a suitable electrolyte for the cell.
d When the cell begins to produce electricity, the pH of the electrolyte near the cathode increases and eventually reaches a constant value. Explain why this occurs.
e Electricity could be obtained from thermal energy produced by combustion of methanol. What is the main advantage of using a fuel cell to produce electricity?

5.4 Supplying energy sustainably

Burning fossil fuels is the dominant source of the greenhouse gases and other pollutant emissions that are polluting the atmosphere and driving human-caused climate change (Figure 5.4.1). It is widely recognised that there is a need to optimise the efficiency of energy conversions and reduce fuel consumption and pollutant emissions. In recent decades, there has also been a search for alternative sources of energy in an effort to protect our planet from further deterioration.

FIGURE 5.4.1 Burning fossil fuels is the source of much of the emissions from human activities.

Biofuels (fuel from plants or organic waste) can be used as an alternative energy source to fossil fuels. Bioethanol produces about 12% less greenhouse gas emissions than its fossil fuel alternatives, and biodiesel produces about 10–20% less.

However, the use of biofuels has limitations. Bioethanol and biodiesel are manufactured from crops, using land that would otherwise be used for the production of food for people and animals. It is estimated that replacing just 10% of the world's liquid transportation fuels (e.g. petrol and diesel) with biofuels in the year 2050 would require nearly 30% of the world's current production of crops. In short, fossil fuels are likely to be only partially replaced by bioethanol and biodiesel.

Although the biomass used as solid biofuels (e.g. wood, dried animal waste) is inedible, and therefore does not necessarily contribute to food shortages or higher food prices, the fuel produced is typically of a low quality that does not produce much energy when burned. Biogas has a similar limitation; the energy released per gram of biogas is less than that of natural gas because the methane content in biogas is significantly lower.

For our energy supplies to be sustainable and to minimise carbon dioxide emissions, it is likely that developed nations will need to look beyond fossil fuels and biofuels and change the way energy is harnessed and used. Some of these changes are already taking place, with increasing use of renewable energy for national electricity grids and rising sales of electric vehicles. It is predicted that fuel cell technology will have an important part to play in this energy future.

The use of fuel cells avoids the energy losses described in Chapter 2 that occur in coal-fired power stations and combustion engines, so there is a reduction in the volume of greenhouse gases produced. From Section 5.3, you will recall that the energy efficiency of fuel cells (up to 85%) is considerably higher than the efficiency of thermal power stations (30–40%) and car engines (25–30%). Only a single energy transformation is involved in the operation of a fuel cell (chemical energy $\rightarrow$ electrical energy), whereas there are multiple transformations in power stations and internal combustion engines.

> ℹ Electricity generation using fuel cells is more efficient than if the electricity were generated by the combustion of the same fuel, resulting in substantially lower greenhouse gas emissions.

Although fuel cells can be designed to use a range of different fuels, including natural gas, methanol, ethanol and ammonia, most are designed to use hydrogen. The only product from a hydrogen–oxygen fuel cell is water.

HYDROGEN—A FUEL FOR THE FUTURE

Some scientists predict that in the future hydrocarbons will be replaced by hydrogen as the principal source of energy for transport and other purposes. This is the hydrogen economy mentioned in Section 5.3. In the hydrogen economy, fuel cells are seen as a replacement for the internal combustion engine. Currently, hydrogen is produced using fossil fuels. However, if the source of hydrogen were renewable, our energy supplies would become more sustainable and there would be a drastic reduction in the production of greenhouse gases and other pollutants without affecting our quality of life.

The concept of the hydrogen economy has fallen in and out of favour over past decades. However, improvements in technology and manufacturing mean that commercial products powered by fuel cells are becoming more widely available and utilised. Combined with the increased global resolve to limit the extent of climate change, there is growing interest in the development of hydrogen-based fuel cell technologies.

The widespread use of hydrogen as a source of energy presents special challenges and issues in terms of its production, distribution, storage and safety.

Hydrogen production

A fuel cell using hydrogen can be described as a 'zero-emission' device because water is almost the only product apart from electricity and heat. However, unless the hydrogen fuel is produced using renewable energy, the production of hydrogen can result in significant levels of greenhouse gases and other pollutants. To some extent, this negates the benefits of the use of fuel cells.

At present, approximately 95% of hydrogen is produced from fossil fuels, such as natural gas, oil and coal, through the process of **steam reforming**. In this process, steam reacts with the fossil fuel at high temperature in the presence of a nickel catalyst. The reaction for methane (CH_4) is represented by:

$$CH_4(g) + H_2O(g) \xrightarrow{\text{Ni}} CO(g) + 3H_2(g)$$

The carbon monoxide (CO) generated can be used to generate further hydrogen, using a copper or iron catalyst:

$$CO(g) + H_2O(g) \xrightarrow{\text{Cu or Fe}} CO_2(g) + H_2(g)$$

The hydrogen produced by steam reforming has a lower energy content than the original fuel, as some of the original chemical energy of these exothermic reactions is lost as waste heat during production. Furthermore, steam reforming leads to carbon dioxide emissions, in the same way as the use of the fossil fuel in a power station or car engine would do.

Advocates of the use of steam reforming for generating hydrogen for fuel cells suggest that, because the carbon dioxide is produced at the site of the steam-reforming process, the greenhouse gas could be captured at its source and stored, preventing its release to the atmosphere. Hydrogen from this source has been proposed as a bridging solution until the cost of producing hydrogen from renewable power decreases, to help meet climate objectives.

At present, there are only two practical methods of generating hydrogen sustainably:

- Using electrical energy to convert water to hydrogen. Electricity can be generated from renewable sources such as solar-power farms and wind farms. (You will look at the chemical principles involved in the process of converting electrical energy to chemical energy in Chapter 8 and the production of 'green' hydrogen in Chapter 9.)
- Collecting biogas from landfill sites and converting the methane in the gas to hydrogen by steam reforming.

Hydrogen sourced from renewable sources is the most desirable method of production because the energy produced from the hydrogen would be sustainable, but it costs about twice as much as hydrogen from coal and natural gas. Although there is relatively little hydrogen production from renewable sources at present, the costs of producing 'green' hydrogen are falling rapidly and the quantities produced are projected to grow quickly in the coming years.

Hydrogen supply, storage and safety

Widespread adoption of hydrogen as the primary fuel for powering vehicles and providing electricity for industrial and domestic applications would require massive expenditure on infrastructure. Changes to pipelines and filling stations, as well as improvements in hydrogen storage methods, would be required. Figure 5.4.2 shows a vision for the future use of hydrogen.

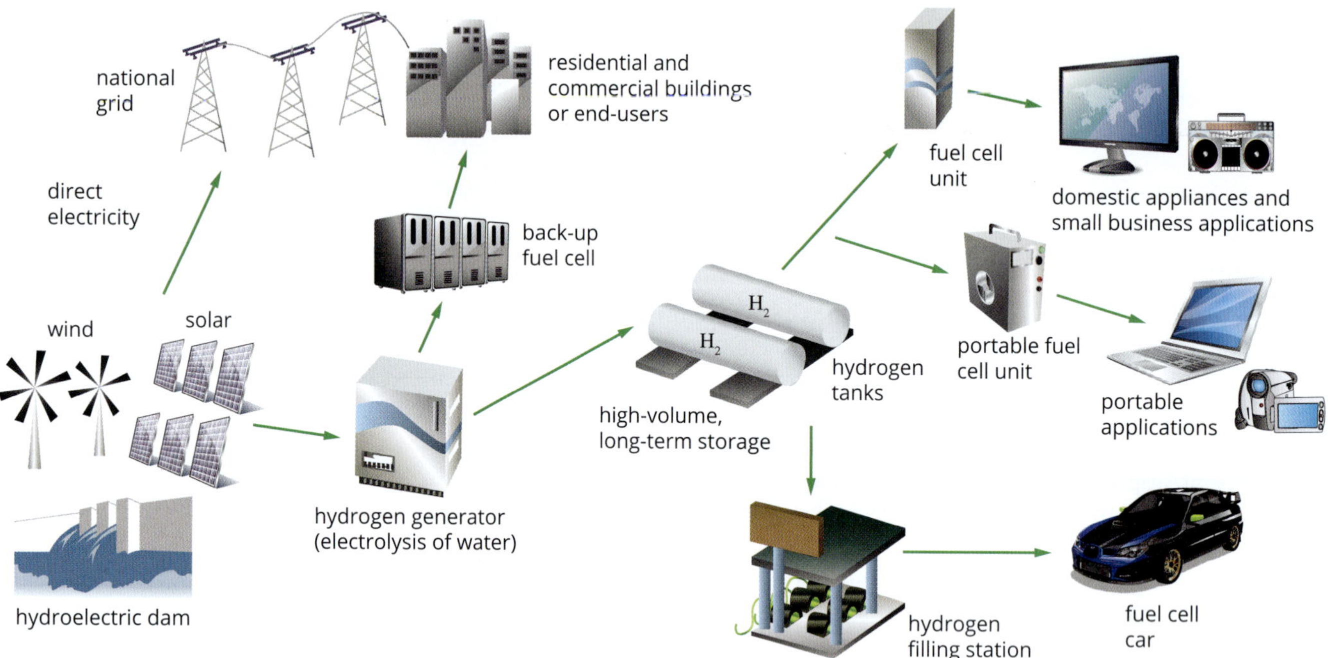

FIGURE 5.4.2 A view of future energy production based on the use of hydrogen

Although hydrogen has a very high energy content by mass (143 kJ g^{-1}) compared with fuels such as petrol (44 kJ g^{-1}), there are issues with storage, particularly within vehicles. This is because hydrogen is a gas at room temperature. The established methods of hydrogen storage in cars, trucks and buses are as liquid hydrogen or compressed hydrogen. However, the energy available per litre of liquid or high-pressure hydrogen is much less than for a liquid hydrocarbon fuel such as petrol, so a hydrogen vehicle requires a much larger fuel tank to achieve a satisfactory driving range.

Scientists are investigating alternative methods of storing hydrogen, known generally as **materials-based storage**, which offer considerable promise but need further development. This technique uses materials—solids or liquids—that can absorb or react with hydrogen and then release it when it is needed. By these means, it may be possible to store hydrogen in smaller volumes at low pressure and at temperatures close to room temperature.

The use of hydrogen poses unique safety challenges, since hydrogen is highly flammable and potentially explosive. Strict codes and standards will be required for its use, including the need for hydrogen sensors to detect leaks. However, a number of studies have concluded that, although there are different safety concerns, the use of hydrogen as a fuel is no more dangerous than petrol. Experts predict that, in the future, hydrogen will be used with at least the same level of safety and convenience as today's hydrocarbon fuels.

DESIGNING BETTER FUEL CELLS

Fuel cells are likely to have a critical role in the current and future development of technologies for renewable energy, energy storage, energy management and greenhouse gas reduction. A large number of different cells have already been developed and scientists continue to optimise electrode materials, electrolytes and general cell design to improve characteristics such as cost, energy storage, safety, lifetime and performance.

One consideration is the avoidance of the use of materials that pose environmental and humanitarian risks. These materials include heavy metals such as cobalt and nickel. Cobalt in particular, which is largely found in Central Africa, has come under scrutiny for the unsafe and exploitative practices involved in its extraction. The metal is mined underground by workers, including children, who labour in conditions that are harsh and dangerous, without appropriate protective equipment. Furthermore, the mining activity exposes local communities to levels of cobalt that may give rise to health problems such as birth defects. Fortunately, the platinum group metals (e.g. platinum, rhodium, palladium) widely used in modern fuel cells are considered to have fewer risks associated with them than some other metals.

Our future energy use is likely to involve increased use of mobile devices, including laptop computers and mobile phones, as well as increasing use of transport, which is one of the largest consumers of non-renewable energy. Renewable energy sources such as wind and solar will play an increasing role in the supply of energy. As higher levels of renewable energy are integrated into electricity grids, there is a need to balance the intermittent nature of the power generated from these sources (i.e. during windy weather, or during the day only), with electrochemical energy generation and storage.

The more established electrochemical devices in current use are being joined by emerging technologies, such as new types of batteries and fuel cells. These technologies are expected to substantially benefit the environment and the way we produce and use energy in the future.

Society is less and less prepared to use products that are discarded after use (a **linear economy**) and there is a higher expectation of optimal use and reuse of resources (a **circular economy**). The development of new battery and fuel cell technologies is part of the move from a linear economy to a circular economy. Their development also provides examples of the importance of creativity and innovation in the application of green chemistry principles to the design of products that are safer and more sustainable than the traditional ones. The green chemistry principles that are of particular importance in this context are:

- design for energy efficiency
- use of renewable feedstocks.

Two examples of emerging fuel cell technologies are the microbial fuel cell (described below) and redox flow batteries (described in the case study).

Microbial fuel cells

A microbial fuel cell (MFC) is a device that converts organic material to electrical energy by the action of microorganisms. The construction of an MFC is shown in Figure 5.4.3 on the following page. Microorganisms form a film on the surface of the anode and oxidise organic material, usually to produce CO_2, protons and electrons. These microorganisms then transfer electrons to the anode of the fuel cell. The cathode reaction can use a variety of oxidising agents, often oxygen (O_2). Most MFCs contain a membrane to separate the anode and cathode compartments and allow the movement of ions to balance charge. Energy efficiencies are in the 21–35% range.

CHEMFILE

Metal organic frameworks

Metal organic frameworks (MOFs) are a promising new class of materials for the capture, storage and delivery of gases, including hydrogen and carbon dioxide. These highly porous, sponge-like materials are made up of two components: metal ions and organic molecules known as 'linkers' (see the figure on the right). The choice of metal ion and linker allows the particular gases they capture and the gas storage capacity of the MOFs to be controlled. The materials have a large surface area per gram and can store significant quantities of CO_2, H_2, CH_4 and other gases.

When MOFs are first made, their pores are usually occupied by solvents, but in some cases these can be eliminated and refilled with compounds such as hydrogen and carbon dioxide. Other possible applications of MOFs are in gas separation, catalysis, drug delivery and as conducting solids.

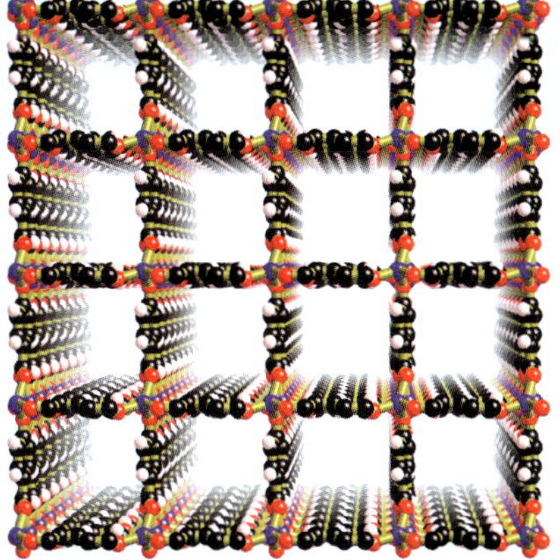

A perspective view of the atomic arrangement of an MOF made by the research team of Professors Brendan Abrahams and Richard Robson at the University of Melbourne.

Fuel cells are being developed in response to society's energy needs. Innovations in the designs of these cells can be viewed in terms of the principles of green chemistry.

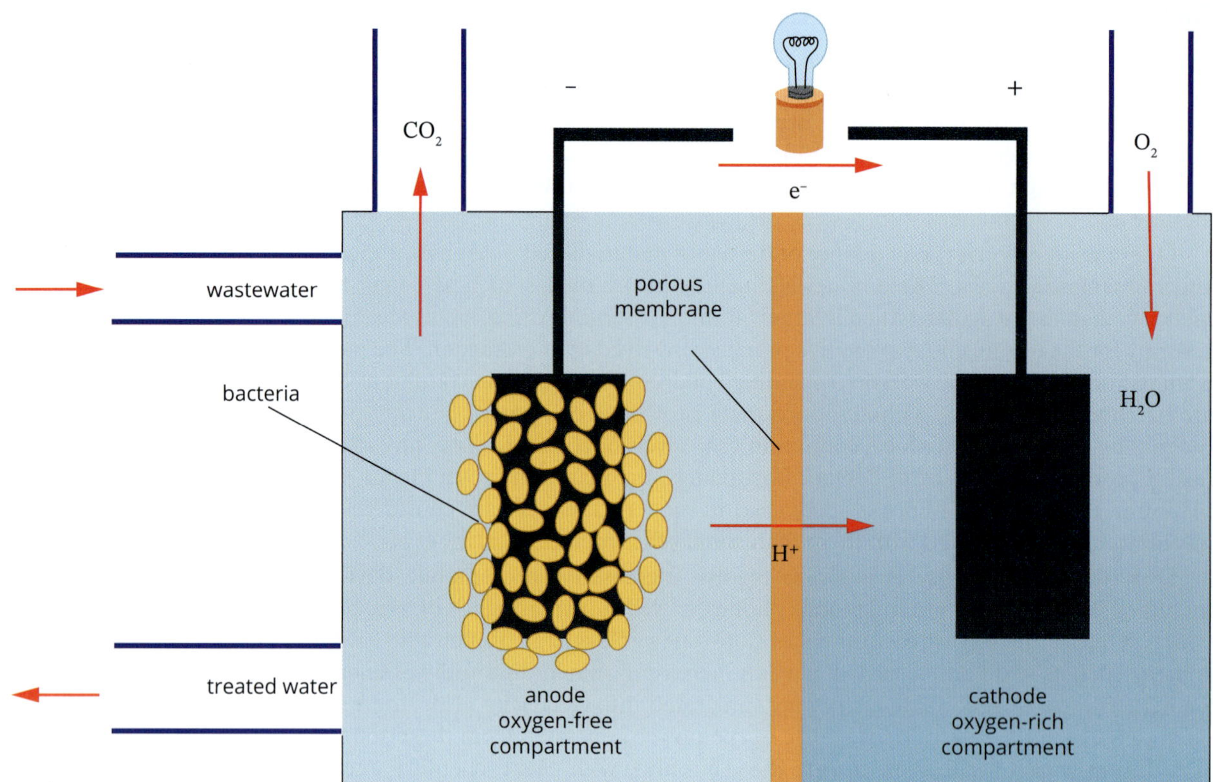

FIGURE 5.4.3 A microbial fuel cell converts chemical energy to electrical energy by the action of microorganisms on organic material.

The reactions can be represented as follows.

At the anode:

$$\text{organic material (s or aq)} \rightarrow CO_2(g) + H^+(aq) + e^-$$

At the cathode:

$$O_2(g) + 4H^+(aq) + 4e^- \rightarrow 2H_2O(l)$$

MFCs are considered promising as they operate at or near room temperature and can use low-grade waste materials, such as soils and sediments, wastewater and agricultural waste. They can be used for sustainable wastewater treatment and contaminant removal, as well as generating low-power electricity.

CASE STUDY **ANALYSIS**

Redox flow batteries

An interesting type of fuel cell under development is the redox flow battery, which has the potential of cheaper energy storage than devices such as lead–acid batteries, without containing potentially toxic metals like lead, cadmium, zinc or nickel.

As shown in Figure 5.4.4, a redox flow battery has two separate compartments of solutions containing the reactants. A wide range of different reactants has been used in experimental cells. The solutions are pumped into the cell from storage tanks. A membrane between the compartments allows exchange of ions while preventing contact between the two solutions. A redox flow battery can be recharged by applying an opposite voltage to the cell, which causes the reverse cell reaction to occur.

Although these batteries could provide cheaper energy storage, their storage tanks, pumps and flow control units add to their weight and make their operation complicated. Flow batteries have the advantages of low cost, easy transportability and high efficiency. They also have the capacity to be joined together so they can be deployed on a large scale. As a result, their main development at present is focused on remote area power systems or grid energy storage/generation that compensates for the fluctuations in supply that occur with renewable energy sources.

Analysis

1 Vanadium redox flow batteries were first developed in Australia almost 40 years ago and offer immediate energy release with long lifespans. Vanadium is much less toxic than other chemicals used in batteries, such as lead and cadmium. The batteries can achieve energy efficiencies of up to 85%.

The half-cell reactions involve vanadium ions in different oxidation states. These can be written as:

$VO_2^+(aq) + 2H^+(aq) + e^- \rightleftharpoons VO^{2+}(aq) + H_2O(l) \quad E^\circ = +1.00\ V$

$V^{3+}(aq) + e^- \rightleftharpoons V^{2+}(aq) \quad E^\circ = -0.26\ V$

- **a** Write a half-equation for the reaction that occurs at the anode when the battery discharges.
- **b** Write the overall equation for the reaction that occurs when the battery discharges.
- **c** Calculate the operating voltage of the battery under standard conditions.
- **d** Vanadium redox flow batteries have been described as being far greener than other batteries. With reference to green chemistry principles, suggest reasons for this description.

2 An Australian company is a major supplier of zinc–bromine redox flow batteries, which offer sustained energy output for a lifetime of more than 10 years. These batteries have been described as a type of electroplating machine, because zinc metal is plated onto one of the electrodes during one phase of the charging–discharging cycle.

The half-cell reactions can be written as:

$Br_2(aq) + 2e^- \rightleftharpoons 2Br^-(aq) \quad E^\circ = +1.09\ V$

$Zn^{2+}(aq) + 2e^- \rightleftharpoons Zn(s) \quad E^\circ = -0.76\ V$

- **a** Write a half-equation for the reaction that occurs at the cathode when the battery discharges.
- **b** Write the overall equation for the reaction that occurs when the battery discharges.
- **c** Zinc metal in plated onto the cathode at one stage when the battery is in use. Does this occur during discharging or recharging?
- **d** The two compartments of each cell are divided by a porous membrane. Why is the membrane required?

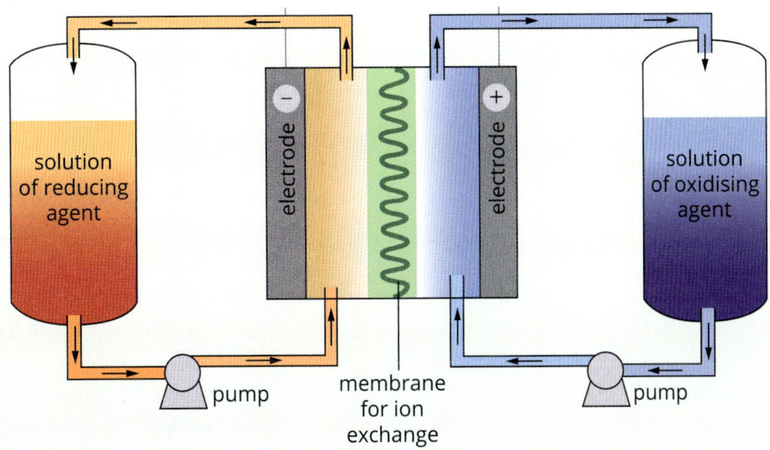

FIGURE 5.4.4 A representation of a redox flow battery

5.4 Review

WS 9 OA ✓✓

SUMMARY

- Scientists are developing galvanic cells such as fuel cells to meet society's future energy needs.
- Electricity generation using fuel cells is more efficient than if the electricity were generated by the combustion of the same fuel.
- The greenhouse gas emissions from a fuel cell are much less than if that fuel were burnt in a power station or internal combustion engine.
- Some scientists predict that fuel cells will play a key role in the transition from a dependence on fossil fuels for energy to a hydrogen economy.
- Developments in battery technology can be viewed in terms of the green chemistry principles: design for energy efficiency, and use of renewable feedstocks.

KEY QUESTIONS

Knowledge and understanding

1 Why is large-scale production of electricity using fuel cells suggested as part of a solution to limiting greenhouse gas emissions?

2 Explain why fuel cells have a greater efficiency than thermal power stations.

3 List three industrial methods for generating hydrogen gas and identify the process by which most of the world's hydrogen is produced.

Analysis

4 Two green chemistry principles that are of particular importance in the design of fuel cells to meet society's future energy are: design for energy efficiency, and use of renewable feedstocks. Read the description of microbial fuel cells on page 206 and identify the relevance of green chemistry principles to the development of this cell.

5 Early versions of a fuel cell called a solid oxide fuel cell (SOFC) were developed in Australia by CSIRO. The diagram below shows the structure of the fuel cell, which uses a solid oxide material as the electrolyte, rather than a liquid, to allow the movement of oxide ions (O^{2-}) from the cathode to the anode.

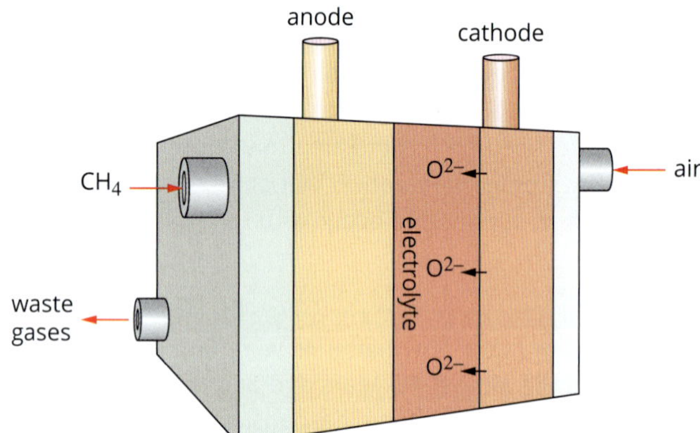

The cells operate at very high temperatures up to 1000°C, which avoids the use of expensive platinum catalysts, and they can use different fuels, including natural gas, methanol from biomass, and hydrogen. The reaction at the cathode is:

$$O_2(g) + 4e^- \rightarrow 2O^{2-}(s)$$

a If methane from natural gas is used as a fuel, write the half-reaction that occurs at the anode. (Hint: One of the reactants is $O^{2-}(s)$.)

b Explain why catalysts are often used in fuel cells and why a high operating temperature may reduce the need for one.

c Suggest three possible areas for research by scientists in their efforts to maximise the usefulness of SOFCs.

d This method of generating electricity has been described as more sustainable than traditional sources of electricity. Explain why this is the case.

Chapter review

KEY TERMS

alkaline cell
anode
battery
cathode
circular economy
discharge
electrochemical cell
electrochemical series
electrode
electrolyte
electromotive force (emf)
equilibrium
external circuit
fuel cell
galvanic cell
galvanometer
half-cell
hydrogen economy
internal circuit
linear economy
materials-based storage
non-rechargeable cell
oxidising agent
porous
potential difference
primary cell
rechargeable cell
reducing agent
salt bridge
secondary cell
spontaneous reaction
standard conditions
standard electrode potential
standard hydrogen electrode (SHE)
standard hydrogen half-cell
standard reduction potential
steam reforming
volt
voltaic cell
voltmeter

REVIEW QUESTIONS

Knowledge and understanding

1 Explain the difference between:

a an oxidising agent and a reducing agent
b an anode and a cathode
c an external circuit and an internal circuit
d a cell and a battery.

2 **a** Name the half-cell that is used as the reference for the measurement of standard electrode potential ($E°$).

b Explain the meaning of the term 'standard conditions'.

3 Use the electrochemical series to show that the Daniell cell (Figure 5.1.4, page 179) should have a cell potential difference of 1.10 V under standard conditions.

4 Explain how the electrochemical series can be used to identify the relative strengths of oxidising agents and reducing agents.

5 What are the limitations that need to be considered when using the electrochemical series to predict whether or not a certain reaction will occur?

6 Complete the following sentences by selecting the correct term for each of the alternatives in bold.

A spontaneous redox reaction will occur if an oxidising agent (on the **left/right** of the electrochemical series) reacts with a reducing agent (on the **left/right** of the series) that has a half-reaction with a **more/less** positive $E°$ value. When the reactants are in direct contact, **electrical/heat** energy is released.

7 Mains electricity typically costs about 7 cents per MJ of energy, depending on the tariff. The cost of the same amount of electrical energy produced by a cell is far more—about $1300 for a dry cell and even more for a button cell. Why are people prepared to use cells and pay such relatively high prices for electricity?

8 Fill in the blanks in the following passage about galvanic cells using terms from the following list:

negative; positive; cells; half-cells; anode; cathode; can; cannot

The redox reaction occurring in a galvanic cell takes place in two ______________. The electrode at which oxidation occurs is called the ______________, whereas the other electrode where reduction occurs is the ______________. The polarity of the anode is ______________ and the polarity of the cathode is ______________.
Galvanic cells can be classified as either primary cells, which ______________ be recharged, or secondary cells, which ______________ be recharged.

9 Which one of the following substances is most likely to be found as a reactant at the cathode of a fuel cell?

A H_2
B CO_2
C CH_3OH
D O_2

10 Complete the following information about fuel cells by choosing the correct terms from the ones in bold.
At the anode of a fuel cell the fuel (often **hydrogen/biodiesel**) undergoes **reduction/oxidation** that generates ions and electrons. The ions move through the **electrolyte/external circuit**. At the same time, electrons flow from the **positive/negative** electrode to the cathode, producing electricity. The cell converts the **electrical/chemical** energy of the fuel and an oxidising agent into **electrical/chemical** energy. The energy efficiencies of fuel cells are generally **higher/lower** than if the fuel were burnt in a power station or internal combustion engine and they have **higher/lower** levels of emissions.

11 Indicate whether the following statements about the generation, storage and safety of hydrogen are true or false.

- **a** Most of the hydrogen produced in the world is currently generated from fossil fuels.
- **b** Hydrogen is regarded by scientists as almost twice as dangerous a fuel for vehicles as petrol.
- **c** Fuel cell vehicles fitted with a tank of liquid hydrogen can travel up to four times further than the same vehicles fitted with petrol tanks of the same size.

Application and analysis

12 Which one of the following statements about a galvanic cell is correct?

- **A** The electrode where oxidation occurs is the cathode.
- **B** Electrons flow towards the electrode where oxidation occurs.
- **C** Anions flow into the half-cell containing the electrode where reduction occurs.
- **D** The electrode where oxidation occurs has a negative polarity.

13 The overall reaction for a galvanic cell constructed from $Cl_2(g)/Cl^-(aq)$ and $Pb^{2+}(aq)/Pb(s)$ half-cells is:

$$Cl_2(g) + Pb(s) \longrightarrow 2Cl^-(aq) + Pb^{2+}(aq)$$

Draw a diagram of a galvanic cell, and on your diagram show:

- **a** the direction of electron flow in the external circuit
- **b** a half-equation for the reaction at each electrode
- **c** which electrode is the anode
- **d** which electrode is positive
- **e** the direction in which cations flow in the salt bridge.

14 Use the electrochemical series to determine whether:

- **a** elemental iodine is an oxidising agent or a reducing agent
- **b** calcium metal is a strong or weak reducing agent
- **c** nickel is a better reducing agent than silver
- **d** $Cu^{2+}(aq)$ is a better oxidising agent than $Ag^+(aq)$
- **e** $Fe^{2+}(aq)$ can act as an oxidising agent and a reducing agent.

15 Two half-cells are set up. One contains a solution of magnesium nitrate with a strip of magnesium as the electrode. The other contains a solution of lead nitrate with a strip of lead as the electrode. The solutions in the two half-cells are connected by a piece of filter paper soaked in potassium nitrate solution. When the electrodes are connected by wires to a galvanometer, the magnesium electrode is shown to be negatively charged.

- **a** Sketch the galvanic cell described. Label the positive and negative electrodes. Mark the direction of the electron flow.
- **b** Write the half-equations for the reactions that occur in each half-cell and an equation for the overall reaction.
- **c** Label the anode and the cathode.
- **d** Indicate the direction in which ions in the salt bridge migrate.

16 Which one of the following reactions might be used as the basis for a fuel cell?

- **A** $SO_2(g) + H_2O(l) \longrightarrow H_2SO_3(aq)$
- **B** $N_2O_4(g) \longrightarrow 2NO_2(g)$
- **C** $NaOH(aq) + HCl(aq) \longrightarrow NaCl(aq) + H_2O(l)$
- **D** $CH_4(g) + 2O_2(g) \longrightarrow CO_2(g) + 2H_2O(g)$

17 The two galvanic cells shown in the figure below were constructed under standard conditions. On the basis of the electrode polarities, determine the order of reducing agent strength from strongest to weakest.

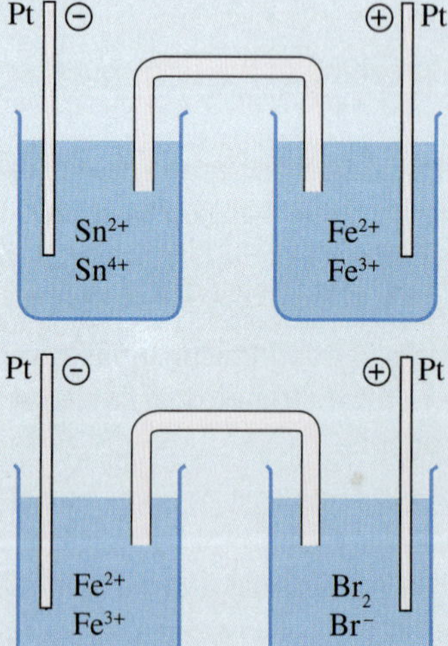

18 Each of these pairs of half-cells combines to form a galvanic cell.

- **i** $Ag^{+}(aq)/Ag(s)$ and $Zn^{2+}(aq)/Zn(s)$
- **ii** $Fe^{2+}(aq)/Fe(s)$ and $Pb^{2+}(aq)/Pb(s)$
- **iii** $Ni^{2+}(aq)/Ni(s)$ and $Cu^{2+}(aq)/Cu(s)$

Draw a diagram of each galvanic cell, and on your diagrams show:

- **a** the direction of electron flow in the external circuit
- **b** a half-equation for the reaction at each electrode
- **c** an equation for the overall reaction in the galvanic cell
- **d** which electrode is the anode
- **e** which electrode is positive
- **f** which way negative ions flow in the salt bridge
- **g** the maximum voltage they could generate under standard conditions.

19 Four half-cells $A^{2+}(aq)/A(s)$, $B^{2+}(aq)/B(s)$, $C^{2+}(aq)/C(s)$ and $D^{2+}(aq)/D(s)$ are used to make the cells shown in the diagram below. Rank the half-cells in order of their half-cell potentials, from lowest standard electrode potential (most negative) to highest electrode potential.

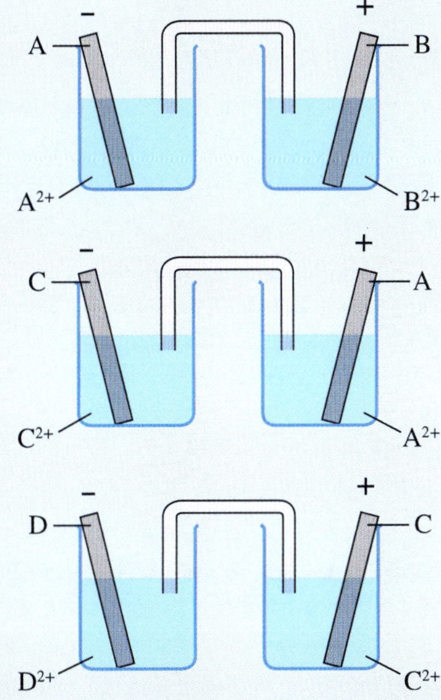

20 Many of the alkaline cells on the market contain zinc electrodes in contact with an electrolyte containing hydroxide ions. The half-cell reaction can be represented as:

$$Zn(s) + 4OH^{-}(aq) \longrightarrow Zn(OH)_4^{2-}(aq) + 2e^{-}$$

To investigate if the standard electrode potential (E°) of these half-cells is the same as that of a half-cell reaction using a zinc electrode in contact with a zinc nitrate electrolyte (half-cell reaction: $Zn(s) \longrightarrow Zn^{2+}(aq) + 2e^{-}$), a student made up the two half-cells shown in the diagram below, as well as a $Cu^{2+}(aq)/Cu(s)$ half-cell.

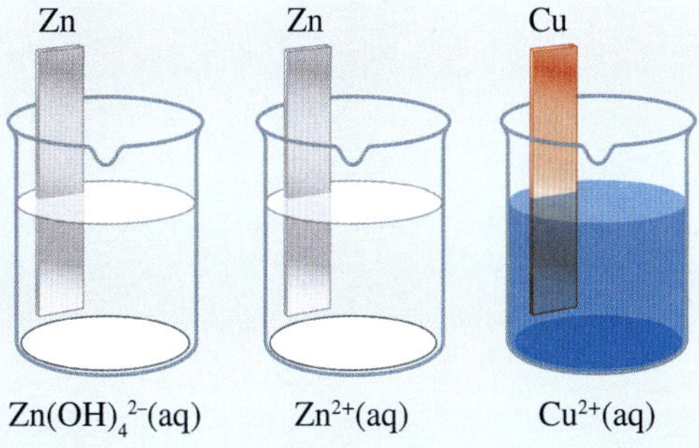

- **a** Write an overall equation for the reaction in a galvanic cell in which the alkaline zinc half-cell is connected to the Cu^{2+}/Cu half-cell.
- **b** Carefully explain how the student could use the half-cells that were provided to determine whether the two different half-cells containing zinc had the same E° value. Include fully labelled diagrams with your answer and explain how the results could be interpreted.
- **c** State the independent variable for this investigation.
- **d** List two controlled variables for this investigation.
- **e** Identify the dependent variable for this investigation.
- **f** Write an aim for this experiment, based on one of the independent variables that you listed in part **a**.

21 The following half-equations appear in the order shown in the electrochemical series:

$$Cl_2(g) + 2e^{-} \rightleftharpoons 2Cl^{-}(aq)$$
$$Ag^{+}(aq) + e^{-} \rightleftharpoons Ag(s)$$
$$Zn^{2+}(aq) + 2e^{-} \rightleftharpoons Zn(s)$$

- **i** Predict whether a redox reaction would occur in the following mixtures. Assume concentrations of 1 M, gas pressures of 100 kPa and temperatures of 25°C.

For the mixtures in which you predict a reaction will occur, write:

- **ii** separate half-equations for the oxidation and reduction reactions
- **iii** an overall equation.

- **a** $Cl_2(g)$ and $Zn(s)$
- **b** $Ag^{+}(aq)$ and $Ag(s)$
- **c** $Ag^{+}(aq)$ and $Zn(s)$
- **d** $Zn^{2+}(aq)$ and $Cl^{-}(aq)$

22 The following half-equations form part of the electrochemical series. They are ranked in the order shown.

$Ag^+(aq) + e^- \rightleftharpoons Ag(s)$

$Pb^{2+}(aq) + 2e^- \rightleftharpoons Pb(s)$

$Fe^{2+}(aq) + 2e^- \rightleftharpoons Fe(s)$

$Zn^{2+}(aq) + 2e^- \rightleftharpoons Zn(s)$

$Mg^{2+}(aq) + 2e^- \rightleftharpoons Mg(s)$

a Which species is the strongest oxidising agent and which species is the weakest oxidising agent?

b Which species is the strongest reducing agent and which species is the weakest reducing agent?

c Lead rods are placed in solutions of silver nitrate, iron(II) sulfate and magnesium chloride. In which solutions would you expect to see a coating of another metal form on the lead rod? Explain your answer.

d Which of the metals silver, zinc or magnesium might be coated with lead when immersed in a solution of lead(II) nitrate?

23 **a** Use the electrochemical series to predict what might be expected to occur if hydrogen gas were bubbled through a solution containing yellow-coloured Fe^{3+} ions.

b Write an equation for the predicted reaction.

c When the reactants were mixed in an experiment, no reaction was observed. Suggest possible reasons for this.

24 One type of breathalyser that is used to detect the presence of alcohol in a motorist's breath is based on a fuel cell.

The equation for the cell reaction is:

$$CH_3CH_2OH(g) + O_2(g) \rightarrow CH_3COOH(aq) + H_2O(l)$$

In this reaction, ethanol (CH_3CH_2OH) is oxidised to ethanoic acid (CH_3COOH) and oxygen is reduced to water in acidic conditions.

a Write an equation for the half-reaction that occurs at the anode.

b Write an equation for the half-reaction that occurs at the cathode.

c As the cell operates, will cations in the electrolyte move towards the anode or the cathode?

d Explain whether the fuel cell will generate a voltage if natural gas, which contains mostly methane, is blown into the breath inlet.

25 Coal is the most abundant fossil fuel and currently provides more than 30% of global energy production. Direct carbon fuel cells are an emerging technology that release the chemical energy in coal directly as electrical energy, with efficiencies expected to exceed 80%. One cell design uses a solid ceramic electrolyte through which oxide ions (O^{2-}) can move (see the figure below).

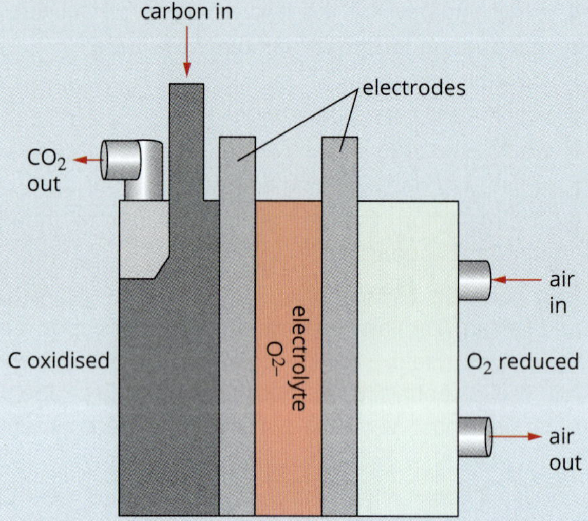

The reaction at the electrode that is in contact with carbon is:

$$C(s) + 2O^{2-}(s) \rightarrow CO_2(g) + 4e^-$$

and the overall reaction is:

$$C(s) + O_2(s) \rightarrow CO_2(g)$$

a Write a half-equation for the reaction that occurs at the other electrode of the fuel cell.

b State whether the electrode in contact with the carbon is:

i positive or negative

ii the anode or the cathode.

c One writer described the cells as 'the ultimate approach for sustainable coal energy generation'. Given the cells produce the greenhouse gas carbon dioxide, suggest how the writer might justify this claim.

26 An experimental fuel cell uses methane as the source of energy. A cobalt-based catalyst is incorporated in the anode, and it uses an acidic electrolyte operating at 70°C. The relevant half-equations in the electrochemical series are:

$CO_2(g) + 8H^+(aq) + 8e^- \rightleftharpoons CH_4(g) + 2H_2O(l)$ $E° = +0.17$ V

$O_2(g) + 4H^+(aq) + 4e^- \rightleftharpoons 2H_2O(l)$ $E° = +1.23$ V

a Write a half-equation for the reaction that occurs at the negative electrode of the fuel cell.

b If the fuel cell operates under standard conditions, what is the maximum voltage that could be generated?

c Use the internet to research and briefly summarise the social and ethical issues associated with the use of cobalt in the cell.

27 The figure below shows a version of an alkaline aluminium–air fuel cell being used to power a load, such as an electric motor.

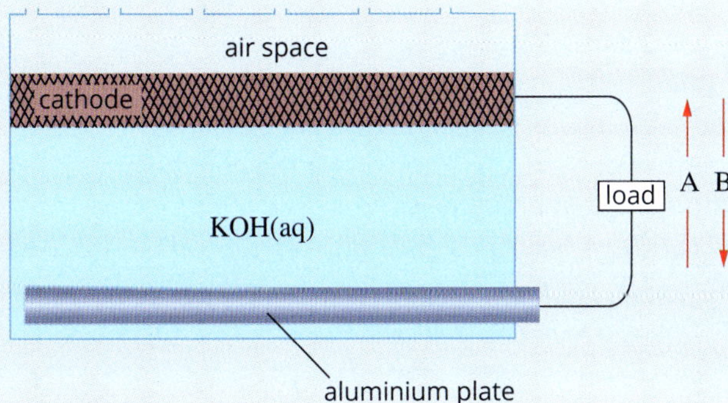

The half-reactions occurring are:

$Al(s) \longrightarrow Al^{3+}(aq) + 3e^-$

$O_2(g) + 2H_2O(l) + 4e^- \longrightarrow 4OH^-(aq)$

a State whether arrow A or arrow B shows the direction of electron flow through the load.

b Identify which species is the oxidising agent in this cell.

c Construct a balanced equation for the overall cell reaction.

d Deduce why it is important for the cathode to be porous.

e If a zinc plate is used in place of the aluminium plate, the cell would still function. However, if a silver plate is used, the cell would not function. Explain this difference by referring to the relative strengths of the oxidising agents and reducing agents involved.

f Determine whether the voltage produced by the fuel cell with the zinc plate would be larger or smaller than the fuel cell with the aluminium plate. Explain your answer.

g When the potassium hydroxide electrolyte was replaced with a sulfuric acid solution, the cell with the silver plate began to deliver a current. Deduce why changing the electrolyte had this effect.

UNIT 3 • Area of Study 1

REVIEW QUESTIONS

WS 10

What are the current and future options for supplying energy?

Multiple-choice questions

1 Which of the following is **not** characteristic of a reaction in which energy is obtained from a fuel?

- **A** It is exothermic.
- **B** Chemical energy (enthalpy) is converted into other forms of energy.
- **C** The combined enthalpy of the products is higher than that of the reactants.
- **D** The enthalpy change (ΔH) is negative.

2 Which of the following equations correctly represents the complete combustion of ethanol?

- **A** $CH_3CH_2OH(l) + O_2(g) \longrightarrow 2C(s) + 3H_2O(g)$
- **B** $CH_3CH_2OH(l) + 2O_2(g) \longrightarrow 2CO(g) + 3H_2O(g)$
- **C** $2CH_3CH_2OH(l) + 7O_2(g) \longrightarrow 4CO_2(g) + 6H_2O(g)$
- **D** $CH_3CH_2OH(l) + 3O_2(g) \longrightarrow 2CO_2(g) + 3H_2O(g)$

3 Biodiesel manufactured from plant oils is an increasingly important alternative to the petrodiesel isolated from crude oil. The difference in the amount of carbon dioxide generated by the combustion of petrodiesel and biodiesel is very small, yet it is argued that switching to biodiesel will reduce net greenhouse emissions. This is mainly because:

- **A** growing plant feedstocks for biodiesel production removes carbon dioxide from the air through photosynthesis
- **B** the carbon dioxide produced by burning fossil fuels is a stronger greenhouse gas than the carbon dioxide produced by burning biofuels
- **C** combustion of biodiesel produces fewer other greenhouse gases, such as methane
- **D** biodiesel combustion produces significantly more energy, so less needs to be burnt for the same energy output.

4 Which of the following is a list of species which is arranged in order of increasing oxidation number of the nitrogen atoms?

- **A** NH_4^+, N_2, NO, NO_3^-
- **B** NO_3^-, NO, NH_3, N_2
- **C** NO_2, N_2, NO, HNO_3
- **D** NO_3^-, NO, N_2, NH_3

5 Which of the following best describes the features of an anode in a galvanic cell?

	Polarity	Electrode reaction
A	positive	oxidation
B	positive	reduction
C	negative	oxidation
D	negative	reduction

6 A can containing 250 g of water at a temperature of 18.5°C is heated with 12.6 kJ of energy. Which of the following is the correct theoretical temperature that would be reached by the water in the can?

- **A** 12.1°C
- **B** 19.7°C
- **C** 23.6°C
- **D** 30.6°C

The following information relates to Questions 7–9.

The thermochemical equation for the complete combustion of propane is:

$$C_3H_8(g) + 5O_2(g) \longrightarrow 3CO_2(g) + 4H_2O(l) \quad \Delta H = -2220 \text{ kJ}$$

7 Which energy profile diagram best represents the energy changes that take place during this reaction?

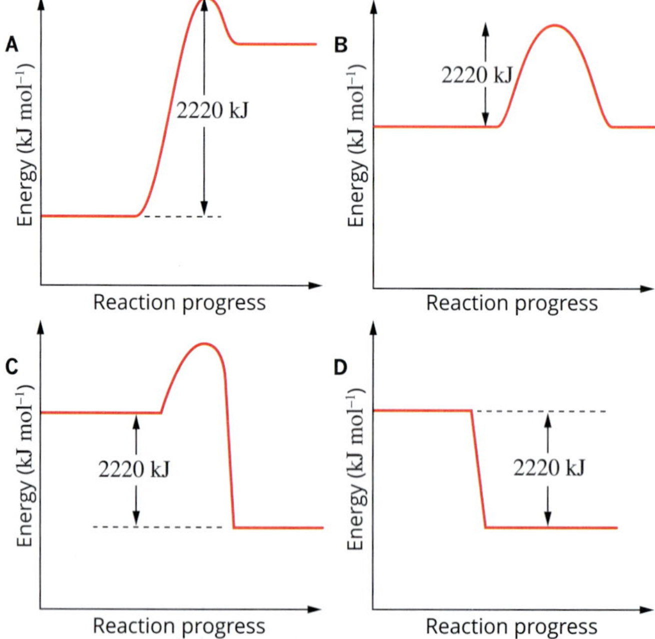

8 When 100 g of propane undergoes complete combustion:

A 5.05×10^3 kJ of energy is absorbed
B 5.05×10^3 kJ of energy is released
C 3.92×10^1 kJ of energy is absorbed
D 3.92×10^1 kJ of energy is released.

9 When 1.00 MJ of energy is released in this reaction, the volume of $CO_2(g)$ produced, at standard laboratory conditions (SLC), is:

A 3.35×10^4 L
B 33.5 L
C 1.12×10^4 L
D 11.2 L

Refer to the following equation for the reaction between hydrochloric acid and ammonia solution when answering Questions 10 and 11.

$$HCl(aq) + NH_3(aq) \rightarrow NH_4Cl(aq) \quad \Delta H = -52 \text{ kJ}$$

10 The reaction is:

A endothermic, because the energy involved in bond forming is more than that involved in bond breaking
B endothermic, because the energy involved in bond forming is less than that involved in bond breaking
C exothermic, because the energy involved in bond forming is more than that involved in bond breaking
D exothermic, because the energy involved in bond forming is less than that involved in bond breaking.

11 A solution calorimeter with a calibration factor equal to 305 J °C^{-1} was used to react 0.0250 mol of HCl(aq) and 0.0750 mol of $NH_3(aq)$. The total volume of solution was the same as that used for calibration. Which of the following would be equal to the temperature increase in the calorimeter during this reaction?

A 0.17°C
B 4.26°C
C 12.79°C
D 17.04°C

12 Hydrogen gas (H_2) is bubbled through an aqueous solution containing Sn^{4+} ions. Which of the following observations is most likely?

A No reaction will occur.
B A precipitate of $Sn(OH)_4$ will form.
C The pH of the solution will decrease.
D Oxygen gas will form.

13 The following equations involve ions of the transition metal vanadium. The half-cell potential E° is shown for each equation.

$$V^{3+}(aq) + e^- \rightleftharpoons V^{2+}(aq) \quad E^\circ = -0.25 \text{ V}$$
$$VO^{2+}(aq) + 2H^+(aq) + e^- \rightleftharpoons V^{3+}(aq) + H_2O(l) \quad E^\circ = +0.36 \text{ V}$$
$$VO_2^+(aq) + 2H^+(aq) + e^- \rightleftharpoons VO^{2+}(aq) + H_2O(l) \quad E^\circ = +1.00 \text{ V}$$

Of the ions listed, which of the following combinations shows the strongest oxidising agent and the strongest reducing agent?

	Strongest oxidising agent	Strongest reducing agent
A	$VO_2^+(aq)$	$V^{2+}(aq)$
B	$VO_2^+(aq)$	$VO^{2+}(aq)$
C	$V^{3+}(aq)$	$VO^{2+}(aq)$
D	$V^{2+}(aq)$	$VO_2^+(aq)$

14 An electrochemical cell was made by dipping a copper rod into a solution of 1 M $CuSO_4$ in one beaker and dipping a nickel rod into a solution of 1 M $NiSO_4$ in another beaker. The metals were connected with wire and the two solutions were connected by a piece of paper towel that had been soaked in a potassium nitrate solution. The cell is shown in the following diagram.

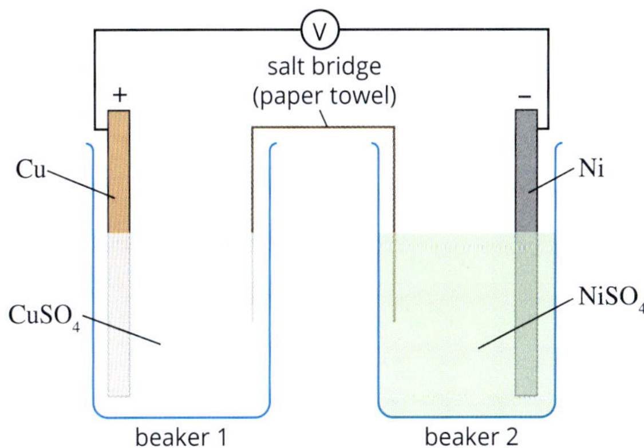

The solution in beaker 1 was initially coloured blue, owing to the presence of Cu^{2+} ions. The solution in beaker 2 was initially coloured green because of the presence of Ni^{2+} ions. Which of the following changes might it be possible to detect after the galvanic cell has been discharging for a long period of time?

A The green colour in beaker 2 has faded and the mass of the copper electrode has increased.
B The blue colour in beaker 1 has faded and the mass of the nickel electrode has increased.
C The green colour in beaker 2 has faded and the mass of the copper electrode has decreased.
D The blue colour in beaker 1 has faded and the mass of the nickel electrode has decreased.

15 Which of the following fuel cells is most likely to meet society's energy needs in terms of the green chemistry principle, use of renewable feedstocks? One or more answers may be correct.

A a microbial fuel cell which converts organic material to electrical energy and uses low-grade waste materials such as wastewater and agricultural waste

B a hydrogen fuel cell which uses hydrogen produced from a fossil fuel

C a fuel cell which uses methane collected from the decomposition of household waste at a landfill tip

D a methanol fuel cell which uses methanol produced from natural gas

Short-answer questions

16 Consider the following table of values of energy available from the combustion of a range of liquid fuels. Note that because many of these fuels are mixtures whose composition can vary, the values provided are representative only.

Liquid fuel	Typical formula	Typical energy available	
		kJ g^{-1}	kJ mL^{-1}
methanol	CH_3OH	20	16
ethanol	CH_3CH_2OH	27	21
petrol	C_8H_{18}	46	33
biodiesel	$C_{19}H_{38}O_2$	38	34

a Identify one fossil fuel present in the table.

b Identify one renewable fuel present in the table.

c When ethanol is described as bioethanol, it has been produced by a fermentation reaction.

i Write the equation for this reaction.

ii Explain why distillation of the products of this fermentation is required to make bioethanol suitable as a transportation fuel.

d The incomplete combustion of ethanol may occur when insufficient oxygen is present. Write an equation in which carbon monoxide is produced by the incomplete combustion of ethanol.

17 The combustion of ethanol occurs according to the equation:

$$2C_2H_5OH(l) + 7O_2(g) \longrightarrow 4CO_2(g) + 6H_2O(l)$$

If 4.725 g of ethanol is burnt in the presence of 10.0 L of oxygen gas at SLC:

a state which reactant is the limiting reactant

b calculate the mass of CO_2 that is produced.

18 For each of the following unbalanced redox reactions (all occurring in acidic aqueous solution):

i write separate balanced ionic equations for the oxidation and reduction half-reactions

ii write a balanced ionic equation for the overall reaction.

a $Al(s) + Br_2(aq) \longrightarrow Al^{3+}(aq) + Br^-(aq)$

b $ClO^-(aq) + S_2O_3^{2-}(aq) \longrightarrow SO_4^{2-}(aq) + Cl^-(aq)$

c $H_2O_2(aq) + MnO_4^-(aq) \longrightarrow O_2(aq) + MnO_2(s)$

19 Referring to an electrochemical series, and for each of the half-cell combinations listed below, predict:

i the maximum cell voltage expected at standard conditions

ii which of the half-cells will contain the negative electrode

iii the ionic equation for the overall reaction occurring when the cell is discharging.

a $Sn^{2+}(aq)/Sn(s)$ and $Fe^{2+}(aq)/Fe(s)$

b $Fe^{3+}(aq)/Fe^{2+}(aq)$ and $Al^{3+}(aq)/Al(s)$

c $H^+(aq)/H_2(g)$ and $I_2(aq)/I^-(aq)$

20 One of the boilers at a power station burns 75.6 kg of brown coal every second. Each kilogram of brown coal produces 9.82 MJ of energy. The electrical output of the generator connected to this boiler is 300 MJ s^{-1}.

a Calculate the amount of energy produced each second by burning 75.6 kg of coal.

b Give two reasons why the answer to part **a** is much greater than the output of the generator.

c Over three-quarters of the electricity generated in Victoria comes from brown coal, a non-renewable energy source. Explain what is meant by the term 'non-renewable energy source'.

d Many sewage treatment plants incorporate digesters that generate biogas, which can be burned to produce electricity in power plants similar to Victoria's coal-fired ones. List some of the benefits this might provide.

21 For each of the following reactions:

i identify them as redox or non-redox reactions

ii for each redox reaction, list the changes in oxidation number occurring

iii for each redox reaction, identify the oxidising agent and the reducing agent.

a $HIO_3(aq) + 3SO_2(g) \longrightarrow 3SO_3(aq) + HI(aq)$

b $SO_3(g) + Ca(OH)_2(aq) \longrightarrow CaSO_4(s) + H_2O(l)$

c $4Fe(OH)_2(s) + 2H_2O(l) + O_2(g) \longrightarrow 4Fe(OH)_3(aq)$

d $NaH(s) + H_2O(l) \longrightarrow H_2(g) + NaOH(aq)$

22 Lighter fluid is mainly butane (molar enthalpy of combustion is –2876 kJ mol^{-1}), which undergoes combustion according to the following equation:

$$2C_4H_{10}(g) + 13O_2(g) \rightarrow 8CO_2(g) + 10H_2O(g)$$

a What is the value of ΔH for the equation?

b Sketch and label an energy-level diagram (showing the relative enthalpies of reactants and products) illustrating the energy changes that occur during this reaction.

c Use your diagram from part **b** to explain why butane does not spontaneously ignite when exposed to air; that is, why a spark is necessary to begin combustion.

d Calculate the heat of combustion of butane in kJ g^{-1}.

e An inexperienced hiker wishes to use his lighter, which contains 2.00 g of butane, to heat the water in his mug.

i If his mug contains 150 mL of water initially at 20.0°C, and 70.0% of the heat energy generated by the lighter is lost to the surroundings, what will be the temperature of the water when the fuel in the lighter is completely used up?

ii In practice, the hiker found that the water was heated to 90.0°C. Calculate the energy transformation efficiency of the conversion of energy from chemical energy of the butane to heat energy of the water.

23 One manufacturer of direct methanol fuel cells (DMFCs) quotes a methanol consumption rate of 900 mL kWh (1 kWh (kilowatt hour) = 3600 kJ). The equation for the reaction in these fuel cells is identical to that for the direct combustion of methanol (with liquid water as a product):

$$2CH_3OH(l) + 3O_2(g) \rightarrow 2CO_2(g) + 4H_2O(l)$$
$$\Delta H = -1453 \text{ kJ}$$

a Given that the fuel cell contains an acidic electrolyte, write balanced half-equations for the reactions occurring at the anode and cathode respectively.

b Use the thermochemical equation given above to calculate the maximum energy (in kJ) available from the reaction of 1.00 g of methanol with oxygen.

c Use the fuel consumption figure provided to calculate the amount of electrical energy obtained from the reaction of 1.00 g of methanol in the fuel cell (density of liquid methanol is 0.79 g mL^{-1}).

d Assume the maximum energy theoretically available from the cell is the value calculated in part **a**. What is the percentage efficiency of conversion of available energy to electrical energy in the fuel cell?

e List some possible energy losses that might occur in the fuel cell to lower the efficiency of electricity production.

24 An experiment was performed to measure the enthalpy change for the reaction between zinc metal and copper sulfate solution:

$$Zn(s) + CuSO_4(aq) \rightarrow ZnSO_4(aq) + Cu(s)$$

A calorimeter containing an excess of copper sulfate in solution was allowed to reach temperature equilibrium and then 0.833 g of zinc powder was added. The temperature was monitored for several minutes. The graph of the temperature changes is shown below.

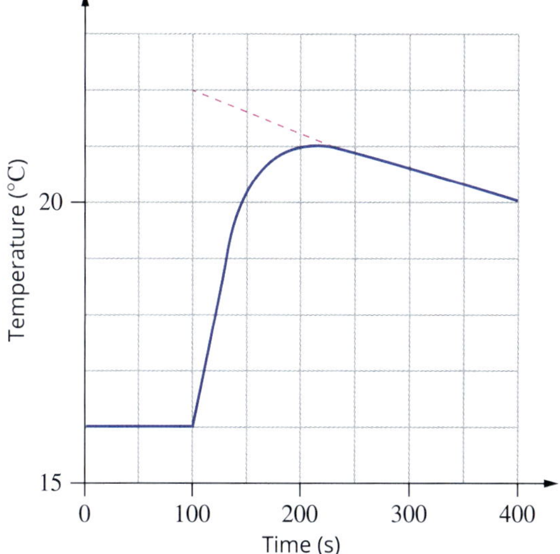

a At what time was the zinc powder added?

b Is the reaction endothermic or exothermic?

c Why does the temperature decrease after 200 s? How could the calorimeter be modified to reduce this?

d Using the dotted extrapolation line, estimate the true temperature change for the experiment.

e The calibration factor for the calorimeter used is 486 J °C^{-1}. Calculate the enthalpy change for the reaction.

25 A pair of students wanted to determine the energy content of 100 g of cheese-flavoured corn chips using an experimental set-up that was similar to that shown below.

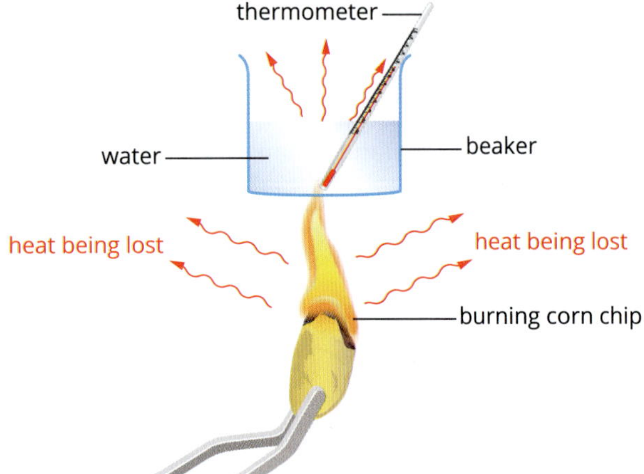

The students burnt three different corn chips from the same bag of corn chips under a beaker containing 100 g of water, measuring the mass of the chip before and after combustion, and the temperature of the water before and after it was heated by the chip. The following results were obtained:

Trial number	Initial mass of corn chip (g)	Mass after corn chip was extinguished (g)	Initial temperature of water in beaker (°C)	Final temperature of water in beaker (°C)
1	2.510	0.627	16.0	63.5
2	2.452	0.613	16.5	65.0
3	2.425	0.606	16.0	64.0

a Calculate the average mass of corn chip burnt in the experiment and the average rise in temperature during the experiment.

b Using the value for ΔT that you calculated in part **a**, calculate the energy that was transferred to the water in the beaker.

c During this experiment, it is estimated that 50% of the energy produced by the burning corn chip was transferred to the water. Calculate the actual quantity of energy that was produced when the corn chip was burnt.

d Using the average mass of corn chip burnt in the experiment, calculate the energy content, in kJ per 100 g, of the corn chips.

e According to the nutritional information on the packet of corn chips, 26% of the corn chip was fat and 60% of the corn chip was carbohydrate. Using the information in Table 2.3.1 on page 81, calculate the energy available to the body when 100 g of corn chips is consumed due to:

i fat

ii carbohydrates

f Identify the systematic error that had the greatest impact on this investigation.

g Discuss the validity, accuracy, precision and repeatability of this investigation.

h Explain how you could determine the reproducibility of this experimental design.

26 A 224 L domestic hot water system running on natural gas is filled with water at 14°C and requires combustion of 1.25 kg of fuel to heat the water to the storage temperature of 70°C (density of water is 1.00 g mL^{-1}).

a Assuming 100% conversion, how much energy is available, in MJ kg^{-1}, from combustion of the natural gas?

b A typical energy value for natural gas is around 54 MJ kg^{-1}. Explain any difference between this value and your answer in part **a**.

EQ U301

CHAPTER 06 Rates of chemical reactions

Chemical reactions occur at many different rates. The explosion of gunpowder and the combustion of petrol in a car's engine occur very quickly. On the other hand, the weathering of buildings, the ripening of fruit and the rusting of iron all occur quite slowly.

By the end of this chapter, you will be able to describe how varying the conditions of chemical reactions can affect the rate of a reaction.

Using collision theory, you will learn to predict and explain the effects of concentration of solutions, surface area of solids, gas pressure, temperature, and the presence of a catalyst on the rate of chemical reactions.

The role of catalysts in changing the rate of chemical reactions will be considered in relation to reaction pathways and the energy changes that occur during reactions.

Key knowledge

- factors affecting the frequency and success of reactant particle collisions and rate of a chemical reaction in open and closed systems, including temperature, surface area, concentration, gas pressures, presence of a catalyst, activation energy and orientation **6.1, 6.2**
- the role of catalysts in increasing the rate of specific reactions, with reference to alternative reaction pathways of lower activation energies and represented using energy profile diagrams **6.3**

6.1 Rate of reaction and collision theory

Chemical reactions are taking place all around us:

- in the soil and rocks beneath our feet
- in the air around and above us
- inside every plant and animal
- in our homes (Figure 6.1.1), schools and workplaces.

FIGURE 6.1.1 The chemical reactions involved in baking occur rapidly.

Some of these reactions are over in a flash. In a car accident, when a car's airbag needs to be inflated, the chemical reactions producing the gas that expands the airbag need to happen extremely quickly. On the other hand, if the car's painted surface is scratched to expose the metal beneath, the rusting reactions may take years.

The chemical equation for a reaction indicates the nature of the reactants and products, but it provides no information about the way in which the reaction proceeds.

The equation for the decomposition of hydrogen peroxide:

$$2H_2O_2(l) \rightarrow 2H_2O(l) + O_2(g)$$

gives no indication about whether the reaction proceeds quickly or slowly; nor can you tell how the products have been formed.

In fact, this reaction normally occurs very slowly, but when a catalyst, such as solid manganese dioxide, MnO_2, is added to the hydrogen peroxide solution, the reaction occurs rapidly, producing so much oxygen gas and heat that the reaction mixture foams and some of the liquid water vaporises (Figure 6.1.2).

FIGURE 6.1.2 The decomposition of hydrogen peroxide is usually very slow, but the addition of solid manganese dioxide, MnO_2, results in the rapid production of oxygen gas and water vapour.

MEASURING RATES OF REACTION EXPERIMENTALLY

The **rate of reaction** is defined as the change in concentration of a reactant or product per unit time. The usual unit for rate of reaction is moles per litre per second (mol L^{-1} s^{-1} or M s^{-1}).

> The rate of reaction is defined as the change in concentration of a reactant or product per unit of time:
>
> $$\text{rate} = \frac{\text{change in concentration}}{\text{time}}$$

To experimentally determine the rate of reaction, you need to measure how much of a reactant is being used up or how much of a product is being formed in a given time period. When a reaction involves gaseous products, this might involve measuring changes in gas volume or mass over time. Other reactions might involve a change in pH or of colour. It is possible to monitor these properties using, for example, data logging with a pH probe, or colorimetry.

Using volume of gas produced

The volume of hydrogen gas produced in the reaction between magnesium and hydrochloric acid can be measured using a gas syringe connected to a conical flask, as shown in Figure 6.1.3a. When the gas volume is measured and recorded at fixed time intervals, a graph similar to the one shown in Figure 6.1.3b may be obtained.

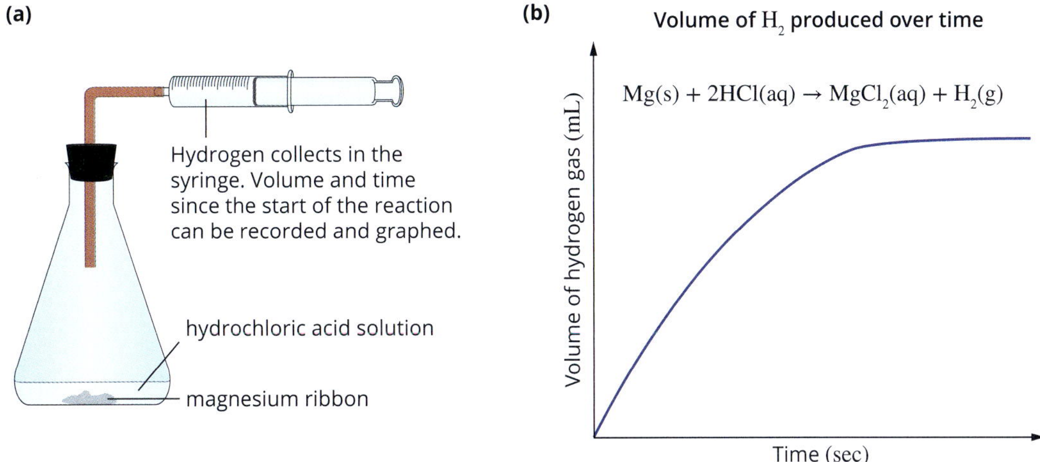

FIGURE 6.1.3 (a) Apparatus for monitoring the volume of gas produced over a period of time. (b) As hydrogen gas is produced from the reaction between the magnesium ribbon and hydrochloric acid, it is collected in a gas syringe and its volume is recorded and graphed.

The rate of reaction can be obtained by measuring the gradient of the graph. The steeper initial gradient of the graph indicates that the initial rate of production of hydrogen gas is greater than the rate of reaction later in the experiment.

This method of monitoring reaction rate could also be used for other reactions in which gases are produced, such as the decomposition of hydrogen peroxide.

Using mass loss

To monitor the production of gases during a reaction, the reaction mixture might be placed on a balance, as shown in Figure 6.1.4, for a reaction between marble chips and an acid. The cottonwool plug in the top of the flask allows carbon dioxide gas to escape, but prevents any drops of solution from splashing out.

If the mass of the apparatus is recorded at regular time intervals, the mass of gas released can be found and graphed against time, as shown in Figure 6.1.5. By drawing tangents to the curve, a measure of the rate of reaction can be calculated—the greater the gradient of the tangent, the faster the reaction rate. In this graph, the initial rate of reaction has the maximum gradient and is therefore the maximum rate. The rate of production of carbon dioxide gas is fastest initially, and progressively decreases with time as the concentration of acid in the reaction mixture decreases.

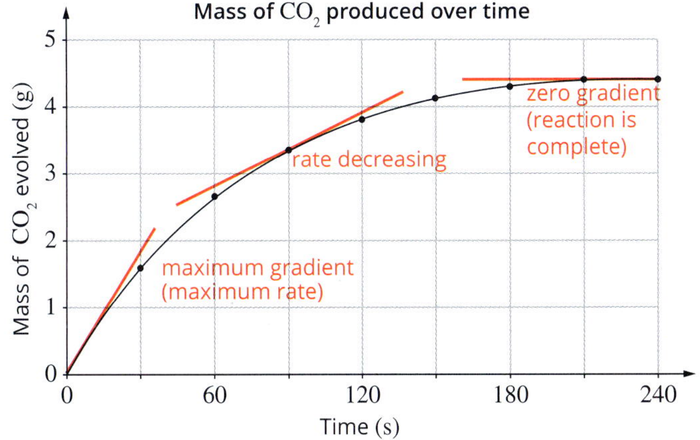

FIGURE 6.1.5 The rate of production of carbon dioxide gas is greatest initially and progressively decreases with time as the concentration of acid in the reaction mixture decreases.

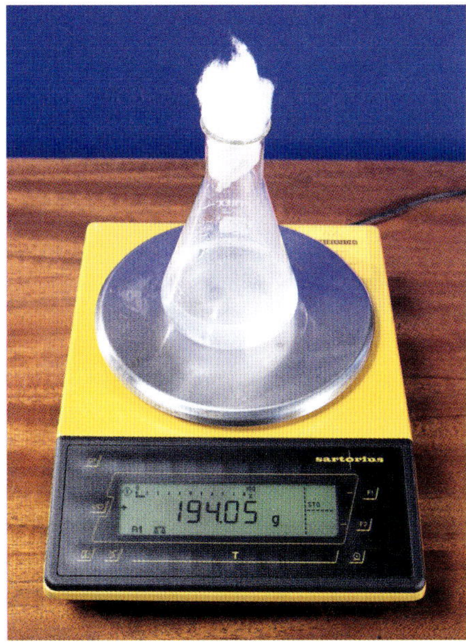

FIGURE 6.1.4 Apparatus to monitor the mass of carbon dioxide gas released during a reaction between marble chips and an acid

Changes in the rate of reaction can be found by drawing gradients of graphs of the change in mass or volume of a reacting mixture over time.

CASE STUDY ANALYSIS

Saved by a very fast chemical reaction

Imagine the scene: an 18-year-old borrows his parent's car to take his friend for a drive to celebrate gaining his driver's licence. Around one corner he encounters a wet and slippery road surface and the car begins to slide. As the driver brakes, the car starts to spin. Suddenly, the car leaves the road and collides with a large tree.

Later, the car was estimated to have been travelling at 60 km h^{-1} when it hit the tree. The collision was a 'head-on', with the front and passenger side taking most of the impact. Yet the passenger suffered just a chipped tooth, and the driver sustained only minor bruising.

This is the true story of a lucky escape, thanks to a very rapid chemical reaction. As the collision took place, the car's airbags were inflated and then deflated as the car stopped, but the travellers' momentum carried them forward towards the windscreen. The driver described it as being 'all over in a flash' and had no clear recollection of the airbags going off.

Hidden in a car's steering wheel, dashboard and windscreen pillars, special nylon bags fill with gas within 30 milliseconds of impact (Figure 6.1.6). Because of this, the car's occupants are prevented from smashing their heads against the steering wheel, dashboard, windscreen or front pillars, all within the blink of an eye. As the head and body strike the airbags, the cushion of gas is forced out of the bag through tiny vents, and within 100 milliseconds the bag has completely deflated.

Airbags contain a mixture of crystalline solids—sodium azide (NaN_3), potassium nitrate (KNO_3) and silica (SiO_2)—stored in a canister. Sensors in the front of the car detect the difference between a bump and a life-threatening impact. When a response is required, an electronic impulse 'ignites' the sodium azide. Sodium metal and hot nitrogen gas are the products of this energy-releasing redox reaction:

$$2NaN_3(s) \rightarrow 2Na(l) + 3N_2(g)$$

The pulse of hot nitrogen gas released from this reaction starts to inflate the nylon bag. The molten sodium metal immediately reduces the potassium nitrate, generating more nitrogen gas, as well as sodium oxide and potassium oxide, which are white powdery solids.

FIGURE 6.1.6 Airbags are deployed within 30 milliseconds of an impact.

The equation for this reaction is:

$$10Na(l) + 2KNO_3(s) \rightarrow K_2O(s) + 5Na_2O(s) + N_2(g)$$

A filtration system prevents any of the oxides from entering the nylon bag, and a third reaction 'captures' them to produce a harmless glassy solid. In this reaction they combine with silica:

$$K_2O(s) + 5Na_2O(s) + SiO_2(s) \rightarrow \text{alkaline silicate ('glass')}$$

Chemical reactions do save lives!

Analysis

This case study highlights a common feature of many 'explosive' reactions: they make a large volume of gas in a short space of time.

1 Identify the gas being produced in the airbags.

2 Explain why it is safe to produce this gas at the same time as a very reactive metal such as sodium.

3 The two explosive reactions which generate the gas that fills the airbags can be combined to give the equation:

$$10NaN_3(s) + 2KNO_3(s) \rightarrow K_2O(s) + 5Na_2O(s) + 16N_2(g)$$

If the volume of an airbag is 67.0 L, calculate the mass of sodium azide, NaN_3, that is required to react to fill the airbag to a pressure of 34.0 kPa at 25°C.

COLLISION THEORY AND ACTIVATION ENERGY

Chemical reactions occur as a result of collisions between the reacting particles. This idea is part of the **collision theory** of reaction rates, which will be discussed in this section.

During chemical reactions, particles (atoms, molecules or ions) collide and are rearranged to produce new particles. Consider the decomposition reaction of hydrogen peroxide:

$$2H_2O_2(l) \rightarrow 2H_2O(l) + O_2(g)$$

The collision that forms the first step of the reaction occurs between the two hydrogen peroxide molecules. If this collision is to result in the formation of molecules of water and oxygen, it must occur in such a way that the covalent bonds in the hydrogen peroxide molecules break. To break bonds, energy is required.

The collision theory of reactions explains why some collisions result in reactions and others do not. According to collision theory, for a reaction to occur, the reactant particles must:

- collide with each other
- collide with sufficient energy to break the bonds within the reactants
- collide with the correct orientation to break the bonds within the reactants and so allow the formation of new products.

If a collision does not meet all of these requirements, then no reaction occurs.

Activation energy

For a reaction to occur between reactant molecules, the molecules must collide with a certain minimum amount of energy. Unless this minimum amount of energy is met or exceeded, the colliding molecules will rebound and simply move away from each other without reacting.

The minimum energy that a collision must possess for a reaction to occur is called the **activation energy**, E_a. When the energy of a collision is greater than or equal to the activation energy, a reaction can occur.

The reactant particles must collide with energy greater than or equal to the activation energy before a reaction can occur.

Activation energy can be represented on an **energy profile diagram**. An energy profile diagram represents the potential energies of the reactants and the products over the course of the reaction.

You will recall from Chapter 2 that an **exothermic** reaction releases more heat energy during the reaction than it absorbs (Figure 6.1.7). As a result, exothermic reactions result in an increase in the temperature of the surroundings. An **endothermic** reaction absorbs more heat energy during the reaction than it releases (Figure 6.1.8), so the temperature of the surroundings decreases. The energy change during a reaction is represented on an energy profile diagram as ΔH, the difference in energy between the reactants and the products.

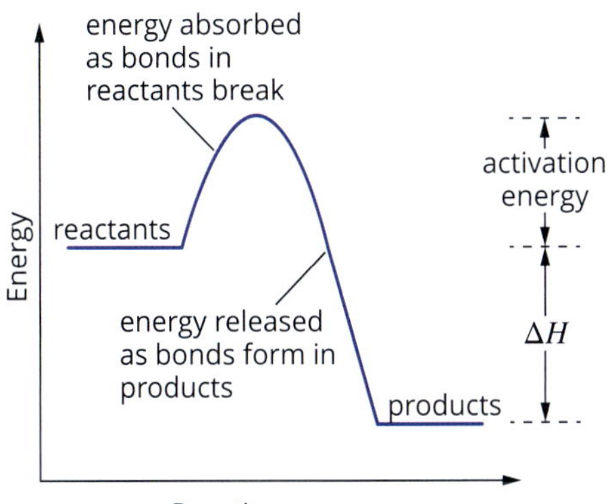

FIGURE 6.1.7 The characteristic shape of an energy profile diagram for an exothermic reaction

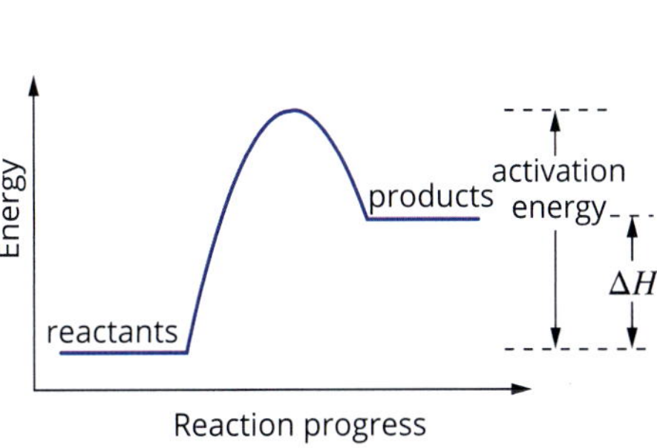

FIGURE 6.1.8 The characteristic shape of an energy profile diagram for an endothermic reaction

Energy profile diagrams for both exothermic and endothermic reactions have a peak that represents the activation energy. This represents the minimum energy that must be absorbed to break the bonds of reactants so that a chemical reaction can progress. The activation energy is measured from the energy of the reactants to the top of the peak.

Transition state

When the activation energy is absorbed, a new arrangement of the atoms, known as the **transition state**, occurs. The transition state occurs at the stage of maximum potential energy in the reaction: the activation energy (Figure 6.1.9). Bond-breaking and bond-forming are both occurring at this stage, and the arrangement of atoms is unstable. The atoms in the transition state rearrange into the products as the reaction progresses.

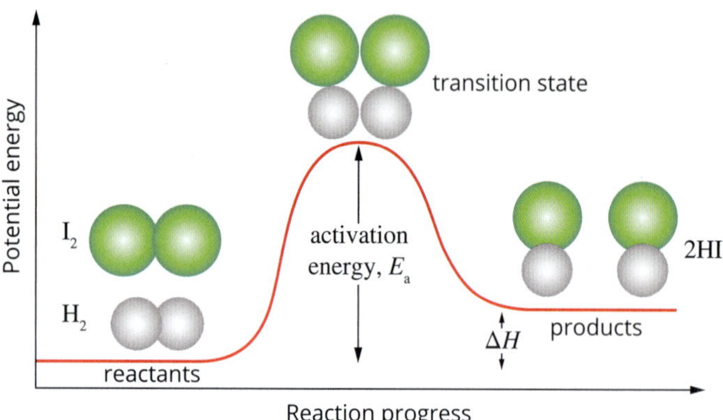

FIGURE 6.1.9 Energy profile diagram for the endothermic reaction $H_2(g) + I_2(g) \longrightarrow 2HI(g)$

Activation energy and reaction rate

The magnitude of the activation energy determines how easy it is for a reaction to occur and therefore what proportion of collisions result in a successful reaction. For this reason, the reaction rate is dependent upon the activation energy.

The existence of an activation energy for a reaction means that collisions between reactants do not always result in a chemical change. For example, nitrogen (N_2) and oxygen (O_2) molecules collide frequently in the air around us at room temperature. However, it is only when energy is provided by a spark, such as in car engines or during a lightning strike, that the energy of the collisions is increased enough to overcome the activation energy barrier. This allows nitrogen monoxide gas to be produced. The nitrogen monoxide formed in this process can then react to form brown nitrogen dioxide (NO_2), a poisonous gas that is a major contributor to the formation of the **photochemical smog** seen in Figure 6.1.10.

FIGURE 6.1.10 Photochemical smog, such as seen here over Barcelona, Spain, is made up of several gases, including NO_2, which is a brown, poisonous gas.

Orientation of colliding molecules

For a reaction to occur, reactants need to collide with enough energy to provide the activation energy. Reacting molecules must also collide with each other in the correct orientation in such a way that particular bonds in the reactants are broken and new bonds are formed in the products.

Figure 6.1.11 shows the importance of collision orientation. In the decomposition of hydrogen iodide gas into hydrogen gas and iodine gas, two hydrogen iodide molecules must collide, with hydrogen and iodine atoms oriented towards each other, for a reaction to possibly occur. If the collision orientation is incorrect, the particles simply bounce off each other, and no reaction occurs.

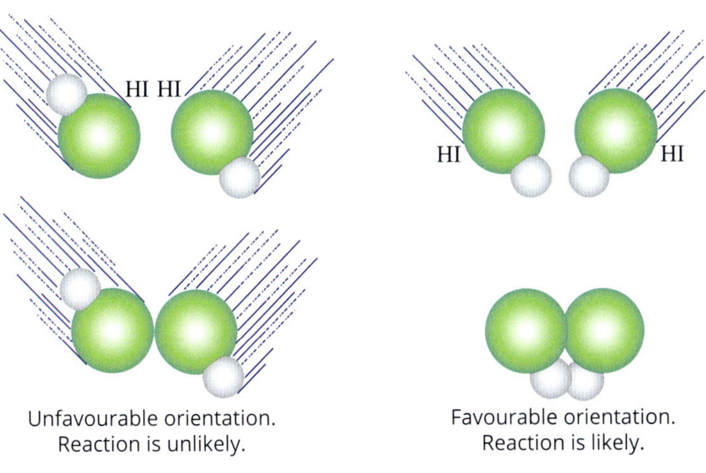

$$2HI(g) \rightarrow H_2(g) + I_2(g)$$

FIGURE 6.1.11 A reaction between colliding molecules is more likely to occur if the orientation of the collision is favourable.

CHEMFILE

A little too reactive!

In 1846, the Italian chemist Ascanio Sobrero reacted glycerol with a mixture of sulfuric and nitric acids to make the explosive liquid nitroglycerin. Nitroglycerin is so unstable that even a small bump can cause it to explode. It decomposes according to the equation:

$$4C_3H_5N_3O_9(l) \longrightarrow 12CO_2(g) + 10H_2O(g) + 6N_2(g) + O_2(g)$$

Although it was many times more powerful than conventional gunpowder, it was far too dangerous to be practical. Some years later, the Swedish scientist Alfred Nobel developed a safer way to manage nitroglycerin through his invention of dynamite. Dynamite is a mixture of nitroglycerin absorbed into an inert material, diatomaceous earth, formed into sticks wrapped in greased, waterproof paper. This modification made the explosive safer to transport.

The reason for nitroglycerin's instability is the very small activation energy for its decomposition reaction (see the figure at right).

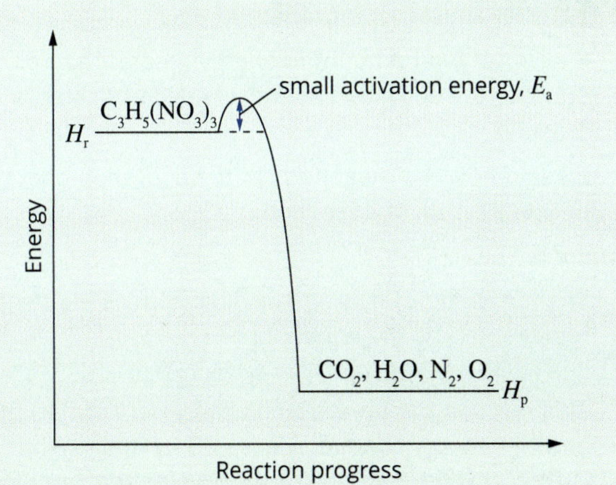

Nitroglycerin has a very low activation energy, making its rate of reaction very large.

6.1 Review

OA ✓✓

SUMMARY

- The rate of a reaction is the change in concentration of reactants or products over time, with the unit of mol L^{-1} s^{-1}.
- A range of experimental methods can be used to measure the rate of a reaction, including measuring the following during specific time intervals:
  - the volume of gas produced
  - the mass loss (if a gaseous product is formed).
- Collision theory is a theoretical model that accounts for the rates of chemical reactions in terms of collisions between particles during a chemical reaction.
- According to collision theory, for a reaction to occur the reactant particles must:
  - collide with each other
  - collide with sufficient energy to break the bonds within the reactants
  - collide with the correct orientation to break the bonds within the reactants and so allow the formation of new products.
- The activation energy of a reaction is the minimum amount of energy required to break reactant bonds to allow a reaction to proceed. It is the minimum amount of energy that a collision between reactant particles must possess for a reaction to occur.
- The transition state is an arrangement of atoms that occurs when the activation energy is absorbed. The transition state is an unstable state in which bonds in the reactants are being broken and bonds in the products are starting to form.
- An energy profile diagram shows the activation energy as a peak of highest potential energy between the energy of the reactants and the energy of the products.

KEY QUESTIONS

Knowledge and understanding

1 List the three main requirements that colliding particles must meet for them to react.

2 Explain why an amount of energy greater than or equal to the activation energy is needed to start a reaction.

3 Explain how the magnitude of the activation energy affects the rate of a reaction.

Analysis

4 The formation of hydrogen iodide from its elements is represented by the equation:

$$H_2(g) + I_2(g) \rightarrow 2HI(g)$$

This reaction has an activation energy of +167 kJ mol^{-1}, and the energy change, ΔH, is +28 kJ mol^{-1}. What is the activation energy for the reverse reaction: the decomposition of 2 mol of hydrogen iodide?

5 When 1 mol of methane gas burns completely in oxygen, the process of bond breaking uses 3380 kJ of energy, and 4270 kJ of energy is released as new bonds form.

a Write a balanced chemical equation for the reaction.

b Calculate the value of the energy change, ΔH, for the reaction.

c Draw and label a diagram to show the changes in energy during the course of the reaction.

6 Hydrogen peroxide, a colourless solution, decomposes to produce oxygen gas and water according to the reaction:

$$2H_2O_2(aq) \rightarrow 2H_2O(l) + O_2(g)$$

The reaction is also exothermic. Describe two ways in which the rate of this reaction could be monitored.

6.2 Effect of changes of conditions on rate of reaction

Experimental investigations have shown that the rate of a reaction can be altered by a change in the:

- surface area of a solid reactant
- concentration of reactants in a solution
- pressure of any gaseous reactants
- temperature of the reaction
- presence of a catalyst.

In any given reaction mixture, only a certain **proportion** of collisions between reactant particles have energy that is greater than or equal to the activation energy and the correct orientation for reaction, i.e. are successful.

To increase the reaction rate, you can increase:

- the **frequency** of successful collisions by increasing the number of collisions that can occur in a given time.
- the proportion of collisions that have energy that is greater than or equal to the activation energy by increasing the energy of all the collisions.

In this section, you will consider how various changes to conditions can affect the frequency of successful collisions that occur between reactant particles and hence change the rate of reaction.

In a successful collision, the energy must be equal to or greater than the activation energy, i.e. $E \geq E_a$, and the reactants must collide with the correct orientation to break their bonds, allowing the formation of new products.

INCREASING THE FREQUENCY OF COLLISIONS

The rate of a reaction increases as the frequency of collisions increases. The frequency of collisions between reactants can be increased by:

- increasing the concentration or pressure of the reactants, since collisions occur more frequently when particles are closer together
- increasing the surface area of a solid reactant.

Increasing concentration or pressure

The rate of a reaction can be increased by increasing the concentration of a reacting solution or the pressure of a reacting gas. This can be explained by collision theory.

The rate of a reaction increases when the frequency of collisions between reactants increases. When the concentration of a solution increases, there are more reactant particles moving randomly in a given volume of solution (Figure 6.2.1). The frequency of collisions therefore increases and so the frequency of successful collisions also increases.

When the frequency of collisions increases, there are more collisions in a given time, so the frequency of collisions with $E \geq E_a$ also increases.

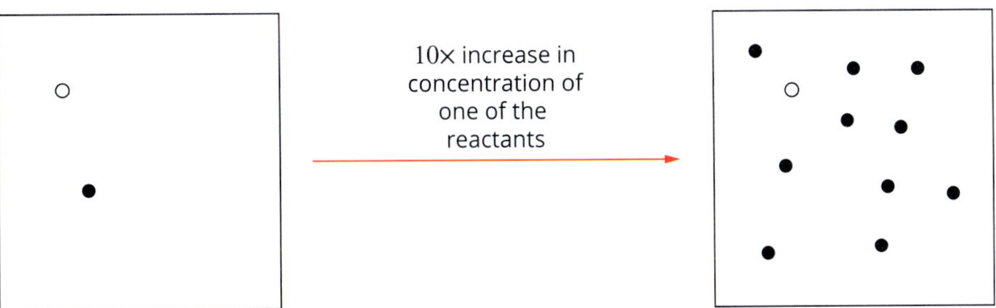

FIGURE 6.2.1 On the left, the concentration of both reactants is low. On the right, the concentration of one of the reactants has been increased 10-fold, resulting in an increase in collision frequency.

For a reaction in the gas phase, the pressure of the gases can be increased either by adding more of the reacting gas to a fixed-volume container or by decreasing the volume of a container with a variable volume, such as a gas syringe. Increasing the pressure increases the concentration of gas molecules, causing more frequent collisions and increasing the number of successful collisions in a given time.

Increasing surface area

When a solid is involved in a reaction, only the particles at the surface of the solid participate in the reaction. The number of particles at the surface depends on the **surface area** of the substance. As you can see in Figure 6.2.2, breaking a solid into smaller pieces means that more particles are present at the surface and are available to react. The surface area has increased.

Because of the greater number of exposed solid reactant particles, the frequency of collisions between reactant particles increases, and so the reaction occurs at an increased rate.

When using collision theory to explain the effect of concentration, pressure and surface area on the rate of a chemical reaction, you must discuss the effect on either collision frequency or the number of successful collisions per unit time.

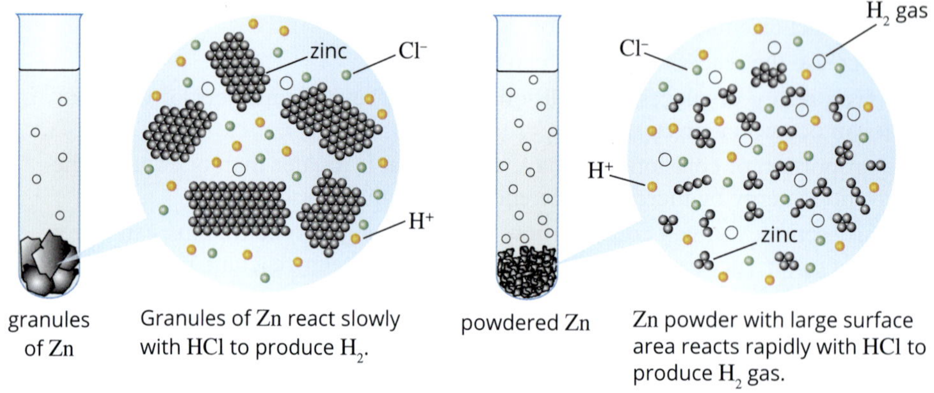

FIGURE 6.2.2 The reaction of hydrochloric acid and zinc. As the surface area of zinc increases, the rate of reaction with hydrochloric acid increases. (The H_3O^+ ion is represented as H^+ in the diagrams for simplicity.)

The effect of increasing surface area can be seen when setting up an open fire at home or in a camp. It is best to first light a pile of kindling rather than trying to directly light large logs. The kindling has a larger surface area than the logs, so it catches fire more easily and will then burn rapidly, providing enough sustained heat energy to make the logs catch fire as well.

Worked example 6.2.1

USING COLLISION THEORY TO EXPLAIN CHANGES IN RATES OF REACTIONS

There have been many explosions in underground coal mines because of the presence of coal dust. Explain this observation in terms of collision theory.

Thinking	Working
Consider the state of the reactants.	Coal is a solid. In the mine, there would be lumps of coal and also powdered coal dust.
Relate the state of the reactant to the factor that affects the reaction rate and explain it in terms of collision theory.	The surface area of powdered coal is greater than that of solid coal. When the surface area increases, the frequency of collisions increases and so the rate of reaction increases.
Return to the question to complete your answer.	An explosion is a very fast reaction. The very large surface area of the coal dust allows for an increase in the frequency of collisions with reacting particles, which increases the reaction rate so much that explosions can occur.

Worked example: Try yourself 6.2.1

USING COLLISION THEORY TO EXPLAIN CHANGES IN RATES OF REACTIONS

Iron anchors recovered from shipwrecks deep in the sea, where the concentration of oxygen is low, can show little corrosion after years in the sea, whereas anchors recovered from shallow water are badly corroded. Explain this observation in terms of collision theory.

INCREASING THE ENERGY OF COLLISIONS

As you have seen, a reaction can be made to occur more rapidly by increasing the concentration of the reactants or, for solids, increasing the surface area. A change in temperature can also have a major effect on the rate of a reaction. An increase in temperature not only increases the kinetic energy of the particles, it also increases the frequency and the energy of the collisions between reactant particles.

Effect of temperature on rate of reaction

The relationship between **kinetic energy** (the energy of motion) and velocity is given by the formula $KE = \frac{1}{2}mv^2$. As the temperature of a reaction system increases, the average kinetic energy of the particles increases and therefore the average speed of the particles in the system increases.

With increased kinetic energy, the frequency of collisions increases, so the frequency of successful collisions (with correct orientation and $E \geq E_a$) increases, and consequently the rate of reaction increases.

Collisions occurring at higher temperatures have a greater energy than those at lower temperatures, due to the increased kinetic energy of the particles. A greater proportion of particles will have energies that are greater than or equal to the activation energy ($E \geq E_a$), and so the proportion of successful collisions increases.

In fact, when the temperature increases, the increase in reaction rate due to the increased energy of the collisions significantly outweighs the increase in reaction rate due to the increased frequency of collisions.

A graph of the effect of increasing the temperature of hydrochloric acid on the rate of the reaction between hydrochloric acid and calcium carbonate can be seen in Figure 6.2.3. As the temperature increases, the rate of reaction increases. Specifically, for every temperature increase of 10°C, the rate doubles.

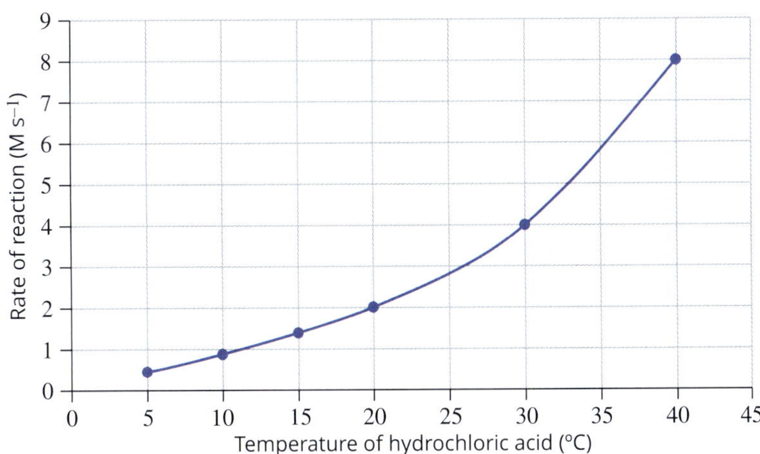

FIGURE 6.2.3 In the reaction $2HCl(aq) + CaCO_3(s) \longrightarrow CaCl_2(aq) + H_2O(l) + CO_2(g)$, the rate of the reaction doubles with every increase of 10°C in the temperature of the hydrochloric acid.

CHEMFILE

Ötzi the Iceman

In September 1991, Erika and Helmut Simon were walking in the Ötztal Alps near the border between Austria and Italy when they discovered the body of what they thought was a dead mountaineer (see the figure below). It was known that, in this region, bodies decompose very slowly because they are enclosed in ice. Closer examination of the body, and the Bronze Age items with it, eventually established that he had died approximately 5300 years ago.

Ötzi the Iceman has been so well preserved in the ice that his stomach contents could be identified. The pollen of a spring plant was also found on his clothes.

Increasing the temperature increases the rate of reaction because the average kinetic energy of the particles, the frequency of collisions and the proportion of reacting particles that have $E \geq E_a$ all increase.

While a temperature increase of just 10°C causes the rate of many reactions to double (an increase of 100%), it can be shown that this is not only because of the increased frequency of collisions. The frequency of collisions only increases by about 3% when the temperature increases by 10°C. The main reason why the reaction rate increases is that a greater proportion of particles have sufficient energy to overcome the activation energy barrier of the reaction. This can be understood by considering the range of kinetic energies that occur at each temperature.

Maxwell–Boltzmann distribution

At any particular temperature, the particles in a substance have a range of kinetic energies. Although most of the particles have similar kinetic energies, there are always some particles with a high energy or a low energy. This range of energies is shown on a graph called a **Maxwell–Boltzmann distribution curve**. Figure 6.2.4 shows how the distribution of energies of a gas at a particular temperature is represented in a Maxwell–Boltzmann distribution curve.

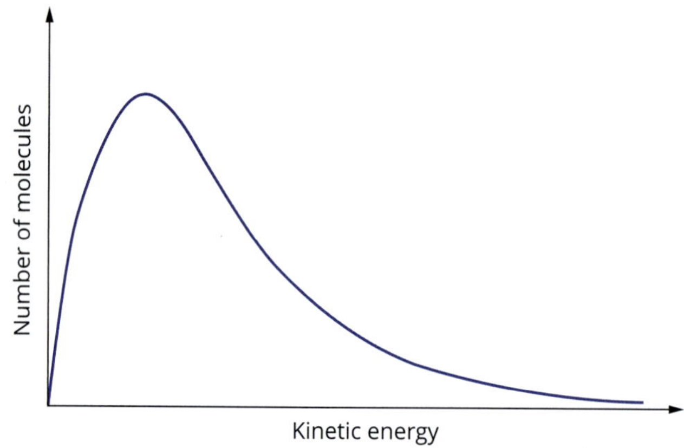

FIGURE 6.2.4 This Maxwell–Boltzmann curve shows the distribution of energies of particles in a sample at a particular temperature. The peak of the graph corresponds to the energy of the greatest number of particles.

During a reaction at a given temperature, only a small proportion of the reactant particles have kinetic energy that is greater than or equal to the activation energy and so can react. You can see this in Figure 6.2.5 as a small, shaded area to the right of the activation energy, E_a. This figure also shows the range of kinetic energies for a gas at three different temperatures. Note that the area under the curve, which is equal to the total number of particles in the sample, stays constant when the temperature is changed. The Maxwell–Boltzmann curves in Figure 6.2.5 show that, as the temperature increases, a greater proportion of the molecules have an energy that is greater than or equal to the activation energy. This explains the increase in rate as temperature increases.

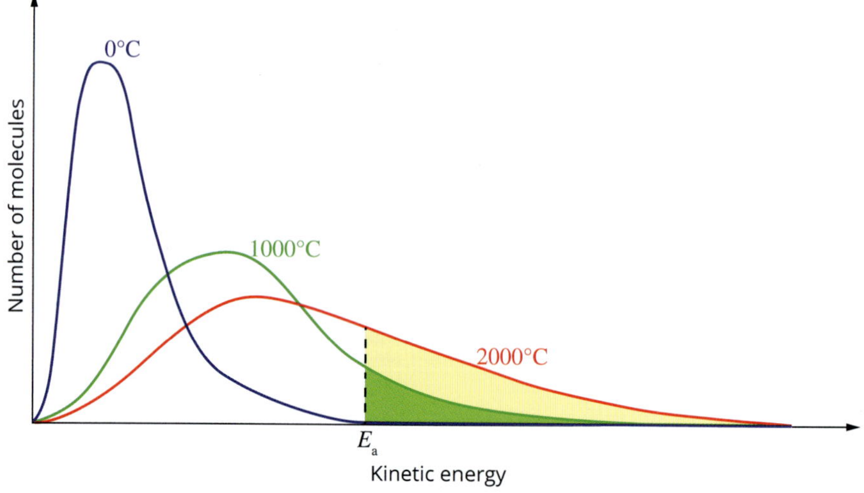

FIGURE 6.2.5 Kinetic energy distribution for a gas at a range of temperatures

6.2 Review

OA ✓✓

SUMMARY

- Collision theory can be used to explain the increase in rate of reaction by an increase in:
  - concentration of a reactant solution
  - pressure of a gaseous reactant
  - surface area of a solid reactant
  - temperature.
- An increase in concentration, pressure or surface area results in an increase in the:
  - frequency of collisions between reactants
  - frequency of collisions with correct orientation and $E \geq E_a$.
- An increase in temperature results in an increase in the:
  - average kinetic energy of the reactant particles
  - energy of collisions, so an increased proportion of collisions has $E \geq E_a$
  - frequency of collisions between reactants
  - frequency of collisions with correct orientation and $E \geq E_a$.

KEY QUESTIONS

Knowledge and understanding

1 List the changes in reaction conditions that will increase the frequency of collisions in a chemical reaction.

2 Describe the relationship between temperature and kinetic energy.

3 Describe the two ways in which an increase in temperature affects the collisions in a chemical reaction.

4 In terms of collision theory, explain why the reaction between a 1 M solution of HCl and magnesium ribbon will have a greater rate of reaction than the reaction between a 0.1 M solution of HCl and magnesium ribbon.

Analysis

5 Five experiments involving the reaction between 100 mL of hydrochloric acid and 5 g of calcium carbonate were carried out.

A Powdered $CaCO_3$ and 2 M HCl are mixed at 40°C.
B Powdered $CaCO_3$ and 1 M HCl are mixed at 25°C.
C Large pieces of $CaCO_3$ and 0.5 M HCl are mixed at 15°C.
D Powdered $CaCO_3$ and 1 M HCl are mixed at 15°C.
E Small pieces of $CaCO_3$ and 0.5 M HCl are mixed at 15°C.

Consider these five experiments and, identify the following, including an explanation based on collision theory.

a Which experiment would have the greatest reaction rate?
b Which experiment would have the smallest reaction rate?

6 Explain the following observations with reference to collision theory.

a Surfboard manufacturers find that fibreglass plastics set within hours in summer, but may take days to set in winter.
b A bottle of fine aluminium powder has a sticker warning 'Highly flammable, dust explosion possible'.

6.3 Catalysts

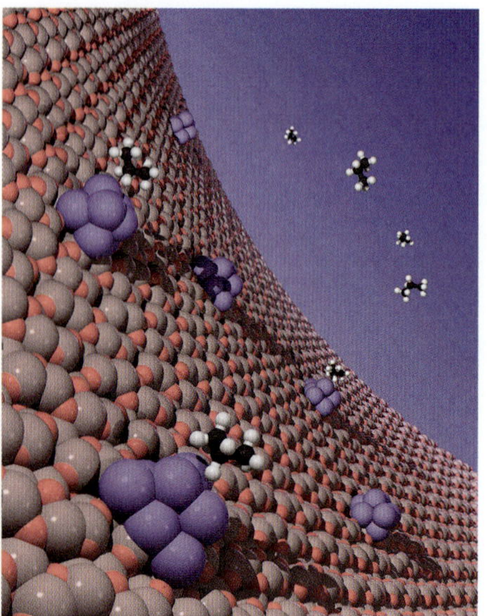

FIGURE 6.3.1 A representation of the structure of a synthetic zeolite catalyst that is widely used in petroleum refineries to break down large hydrocarbon molecules, shown in purple, into smaller, more useful molecules, shown in black and white

The change in reaction rate when a catalyst is present is often substantial. The action of a catalyst can be understood using collision theory and the changes in energy that occur during a chemical reaction. Catalysts play an important role in industrial chemistry, in limiting air pollution, as seen in their use in the catalytic converters of cars. They are also used in controlling biochemical processes—for example, in petroleum refineries to break down large hydrocarbon molecules into smaller, more useful molecules, as shown in Figure 6.3.1.

In this section, you will learn how catalysts increase the rate of reaction.

CATALYSTS AND ACTIVATION ENERGY

You have already learnt that the potential energy changes associated with a reaction can be represented as an energy profile diagram of the reaction. An energy profile diagram for an exothermic reaction is shown in Figure 6.3.2.

The activation energy, E_a, is the minimum amount of energy required for a reaction to take place. On the energy profile diagram, the activation energy is measured from the energy of the reactants to the peak of the energy profile diagram. The enthalpy change, ΔH, can also be represented on an energy profile diagram and is equal to the difference in energy between the products and the reactants.

Some reactions occur very readily at room temperature because they have very small activation energies. These reactions need only a small amount of energy to be absorbed for bonds in the reactants to be broken. Many reactions occur much more rapidly in the presence of a particular element or compound, most commonly a transition metal element or compound. These substances are known as **catalysts**. Catalysts are not consumed during the reactions they speed up, and so do not appear as either reactants or products in reaction equations.

Catalysts are able to increase the rate of a reaction because they provide an alternative **reaction pathway**, which reduces the activation energy of the overall reaction dramatically, as seen in the energy profile diagrams in Figure 6.3.3.

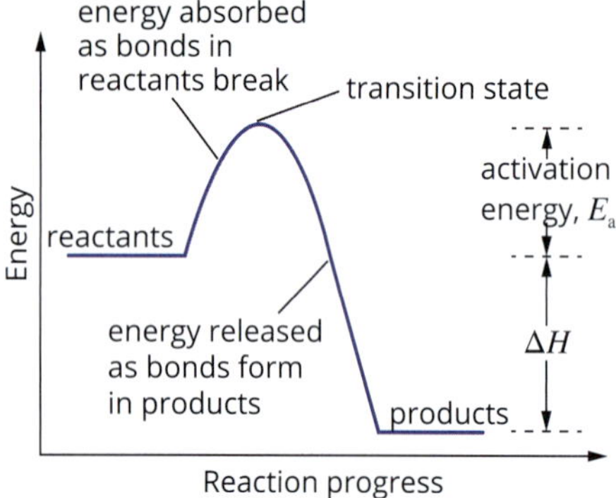

FIGURE 6.3.2 Energy profile diagram of an exothermic reaction such as the combustion of methane

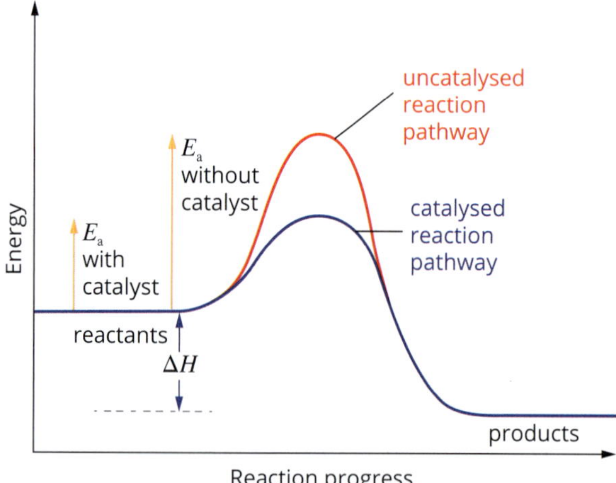

FIGURE 6.3.3 Energy profile diagrams of a catalysed and an uncatalysed reaction

Catalysts increase the rate of a reaction by providing an alternative reaction pathway with a lower activation energy.

The presence of a catalyst does not change the value of ΔH for a reaction.

With the catalyst present and therefore a lower activation energy, the colliding particles are more likely to have energies that exceed the activation energy, causing the bonds in the reactants to be broken more frequently. As a result, a greater proportion of collisions are successful; that is, they lead to the formation of products. Thus, the reaction rate is increased.

However, the enthalpies of the reactants and the products are still the same, so there is no change to ΔH for the reaction.

CHEMFILE

Routes from Melbourne to Canberra

The figure at right shows two groups of tourists travelling from Melbourne to Canberra by two different routes. The route which takes the tourists over the top of Mt Kosciuszko can be likened to the progress of an uncatalysed reaction: a lot of energy will be needed to get to Canberra. The route that goes around the foothills and does not scale tall mountains can be likened to the progress of a catalysed reaction, because less energy will be needed and the tourists taking that alternative route will get to Canberra in a shorter time.

An analogy for the role of a catalyst in a chemical reaction

TYPES OF CATALYSTS

Depending on the physical state of the chemicals involved in the reaction and the catalyst, catalysts can be divided into two groups.

- **Homogeneous catalysts** are in the same physical state as the reactants and products of the reaction.
- **Heterogeneous catalysts** are in a different physical state to the reactants and products of the reaction.

An example of homogeneous **catalysis** occurs in the upper atmosphere and has contributed to the depletion of the ozone layer. Chlorine atoms in the gaseous state act as catalysts in the decomposition of ozone gas into oxygen gas. The chlorine atoms may have come from chlorofluorocarbons (CFCs) released into the atmosphere from refrigerators or air conditioners.

Figure 6.1.2 on page 220 illustrates the catalysed decomposition of a hydrogen peroxide solution using the black powder manganese(IV) oxide (MnO_2). This is an example of the use of a heterogeneous catalyst.

CATALYSTS IN INDUSTRY

The chemical industry uses catalysts extensively. Chemists prefer to use heterogeneous catalysts for industrial processes because they are:

- more easily separated from the products of a reaction
- much easier to reuse
- able to be used at high temperatures.

Particles at the surface of some solids of high surface area tend to adsorb (form a bond with) gas molecules that strike the surface. **Adsorption** distorts bonds in the gas molecules or may even break them completely. This allows a reaction to proceed more easily than it would if the solid were absent.

These solid surfaces provide a new way for the reaction to occur (a new reaction pathway) that has a significantly lower activation energy.

A powdered or sponge-like form of a solid catalyst is often used to provide the greatest possible surface area. With a larger surface area, more reactant molecules can be adsorbed and the reaction is even faster.

CASE STUDY ANALYSIS

Heterogeneous catalysts in industry

The action of heterogeneous catalysts has similarities across several industrial applications. For example, since 1986 it has been mandatory for all new cars sold in Australia to have a catalyst fitted between the engine and the exhaust pipe. As another example, the Haber process is a very important commercial reaction that produces ammonia gas, which is used to make fertilisers, nylon, explosives and some pharmaceuticals. In both applications, the catalyst enables gaseous reactions to occur at an increased rate of reaction due to an alternative reaction pathway with a lower activation energy.

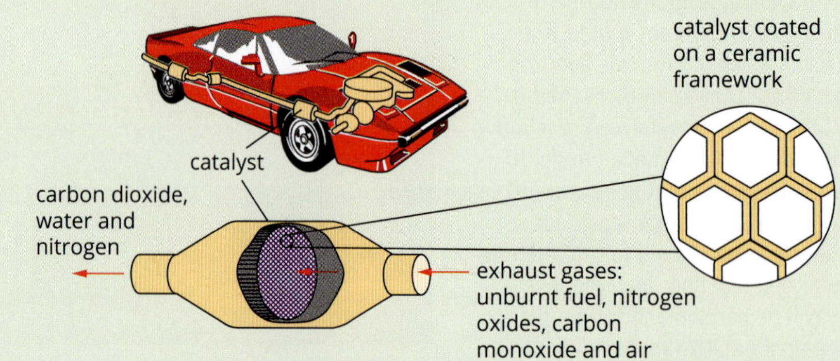

FIGURE 6.3.4 A catalyst fitted in the exhaust system of a car reduces air pollution.

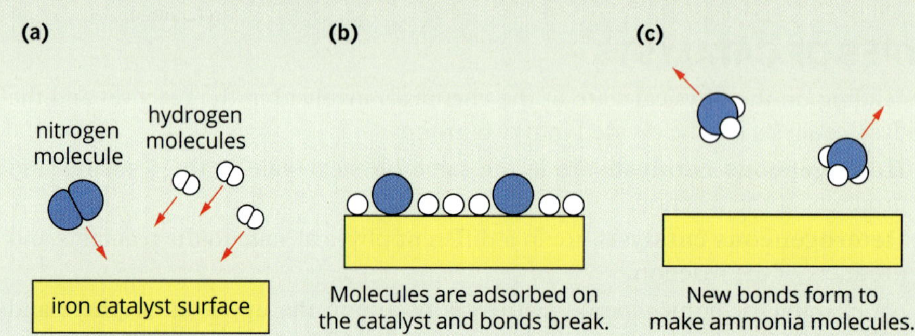

FIGURE 6.3.5 Behaviour of a catalyst in ammonia production

In a car's exhaust system, shown in Figure 6.3.4, the purpose of the catalyst is to clean the exhaust gases from the engine and reduce the air pollution that could be caused if these gases entered the atmosphere. Carbon monoxide and nitrogen oxide, formed in the engine, are converted to the non-toxic gases carbon dioxide and nitrogen via reactions, including:

$$2NO(g) \longrightarrow N_2(g) + O_2(g)$$
$$2CO(g) + O_2(g) \longrightarrow 2CO_2(g)$$

Unburnt hydrocarbons are also converted by the catalyst to carbon dioxide and water:

$$2C_8H_{18}(g) + 25O_2(g) \longrightarrow 16CO_2(g) + 18H_2O(g)$$

The catalyst is usually a mixture of platinum, palladium and rhodium metals, and it is mounted on a honeycomb-shaped ceramic support. Millions of tiny pores in the metals provide a large surface area.

Exhaust gases enter the catalyst chamber, pass quickly over the metals, and leave the exhaust pipe in a purified condition. The catalyst is unchanged by the reaction and can be used without replacement for many years.

In the Haber process, hydrogen and nitrogen gas are converted to ammonia (NH_3), using a catalyst of powdered iron.

The reaction is represented by the equation:

$$N_2(g) + 3H_2(g) \longrightarrow 2NH_3(g); \quad \Delta H = -92 \text{ kJ mol}^{-1}$$

In Figure 6.3.5 you can see that hydrogen and nitrogen molecules both adsorb onto the iron surface. As they attach themselves to the surface, the covalent bonds within their molecules break. The hydrogen and nitrogen atoms now readily combine to form ammonia molecules and move away from the iron surface, leaving the catalyst unaltered by the reaction.

Analysis

1 Without a catalyst, the reaction $N_2(g) + 3H_2(g) \longrightarrow 2NH_3(g)$ is very slow. Explain why this could be the case.

2 Explain why the catalyst in all these gaseous reactions does not change the value of ΔH for the reactions.

3 Considering sustainability concepts, explain why it is important for catalytic converters from wrecked cars to be recycled.

6.3 Review

OA ✓✓

SUMMARY

- The rate of a reaction can be increased by using a catalyst.
- A catalyst provides an alternative reaction pathway that has a lower activation energy.
- The catalyst is not used up in the reaction.
- Energy profile diagrams (see the figure at right) may be used to represent the enthalpy change and activation energy associated with both catalysed and uncatalysed reactions.
- When a catalyst is present, a greater proportion of the collisions between particles exceeds the activation energy barrier of the reaction and therefore leads to a chemical change.

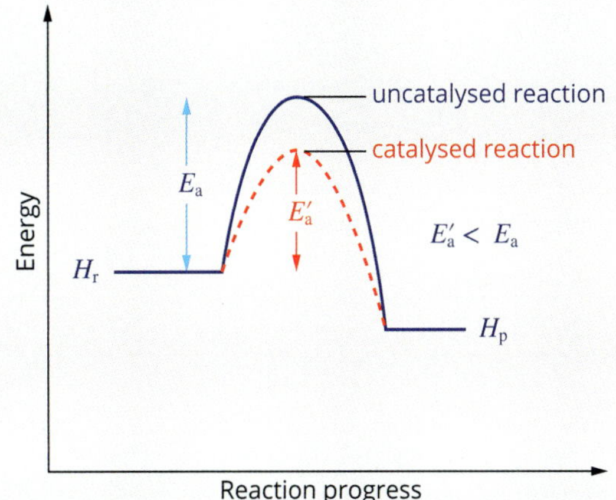

Energy profile diagram of a reaction showing the effect of a catalyst

KEY QUESTIONS

Knowledge and understanding

1 List the main features of the function of a catalyst in a chemical reaction.

2 A student wrote the following statement in their revision notes about catalysts:
'A catalyst increases the rate of a reaction because all catalysts increase the frequency of collisions.'
Explain the mistake that this student has made and rewrite the statement correctly.

3 Many industrial catalysts are made into porous pellets. What is the reason for this?

Analysis

4 If a sugar cube is held in the flame of a candle, the sugar melts and browns but does not burn. However, the cube burns if salt is first rubbed into it, even though the salt does not react. Explain the effect of the salt on the activation energy of this combustion reaction.

5 An increase in temperature and adding a catalyst both increase the rate of reaction. Compare these two ways of increasing the rate of reaction.

6 You are required to design an investigation in which the effectiveness of three catalysts (yeast, potassium iodide and manganese dioxide) on the rate of decomposition of hydrogen peroxide is compared. The equation for the reaction is:

$$2H_2O_2(aq) \rightarrow 2H_2O(l) + O_2(g)$$

List the independent, dependent and controlled variables for the investigation.

Chapter review

KEY TERMS

activation energy
adsorption
catalysis
catalyst
collision theory
endothermic
energy profile diagram
exothermic
frequency
heterogeneous catalyst
homogeneous catalyst
kinetic energy
Maxwell–Boltzmann distribution curve
photochemical smog
proportion
rate of reaction
reaction pathway
surface area
transition state

REVIEW QUESTIONS

Knowledge and understanding

1 The following changes are made to a reaction mixture. Which change will lead to a decrease in reaction rate?

A Smaller solid particles are used.
B The temperature is decreased.
C A catalyst is added.
D The concentration of an aqueous reactant is increased.

2 Which of the following is the correct unit for measuring the rate of a reaction?

A mol L^{-1} s^{-1}
B mol^{-1} L s^{-1}
C mol^{-1} L^{-1} s
D mol L s^{-1}

3 Which one of the following alternatives correctly explains why the rate of reaction between 1 M $CuSO_4$ and powdered zinc is greater than with an equal mass of large zinc pieces?

A The energy of collisions between the Cu^{2+}(aq) ions and powdered zinc is greater than with the large zinc pieces.
B The frequency of collisions between the Cu^{2+}(aq) ions and powdered zinc is greater than with the large zinc pieces.
C The energy of collisions between the Cu^{2+}(aq) ions and large zinc pieces is greater than with the powdered zinc.
D The frequency of collisions between the Cu^{2+}(aq) ions and large zinc pieces is greater than with the powdered zinc.

4 Which one of the following statements correctly describes what must occur when reactant particles collide and react?

A Colliding particles must have an equal amount of kinetic energy.
B Colliding particles must have different amounts of kinetic energy.
C Colliding particles must have kinetic energy greater than or equal to the average kinetic energy.
D Colliding particles must have kinetic energy greater than or equal to the activation energy of the reaction.

5 Which pair of statements is correct for the effects of a catalyst and the concentration on the rate of reaction?

	Adding a catalyst	Increasing the concentration
A	collision frequency increases	collision frequency increases
B	activation energy decreases	activation energy decreases
C	activation energy decreases	collision frequency increases
D	collision frequency increases	activation energy decreases

6 Explain the difference between the terms 'frequency of collisions' and 'proportion of collisions'.

7 Consider the reaction between solutions V and W that produces X and Z according to the equation:

$$V(aq) + W(aq) \longrightarrow X(aq) + Z(aq)$$

The energy profile diagram for the reaction between V(aq) and W(aq) is shown below.

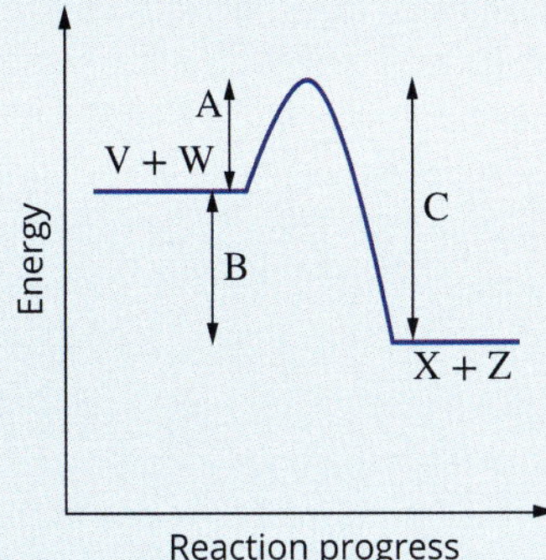

Each letter represents a quantity that can be measured from the energy profile diagram. Which quantity or quantities would change if a catalyst were used to increase the reaction rate?

8 Complete the following summary about changes in rate of reaction by using the following terms to fill in the gaps below. Note that terms may be used more than once.

activation; activation energy; collisions; frequency; gas; greater; more; proportion; rate; solid; solution; successful; time

An increase in the concentration of a __________, the surface area of a _________ or the pressure of a _______, will increase the _____________ of collisions in a reaction, meaning that an increased number of ______________ occur during a particular _____________. As a result, the frequency of ________________ collisions will also increase, and the ________of reaction will increase. Adding a catalyst to a reaction will increase the ___________ of collisions that have energy that is __________ than or equal to the ______________ ______________. This means that out of the total number of collisions, _________ collisions have enough energy to overcome the ______________ energy. Increasing the temperature of a reaction will increase both the ______________ of collisions and the _____________ of collisions that have $E \geq E_a$.

9 **a** A student defined the rate of reaction as 'the mass of products that are made during a reaction'. Explain why this answer is not accurate.

b State the correct definition for the rate of reaction.

10 **a** What are the five factors that influence the rate of a reaction?

b Construct a Venn diagram that classifies the five factors affecting the rate of reaction under the headings 'Increased frequency of collisions' and 'Increased proportion of collisions with $E \geq E_a$' and shows the factor that occurs at the intersection of the two classifications.

11 By discussing the energy and frequency of collisions, explain why the rate of reaction between 0.5 M $CuSO_4$ and zinc powder at 40°C is greater than the rate of reaction between 0.5 M $CuSO_4$ and zinc pieces at 25°C.

12 Many major car-makers have plans for hydrogen-powered cars. In the fuel cells of these cars, hydrogen reacts with oxygen from the air to produce water.

$$2H_2(g) + O_2(g) \longrightarrow 2H_2O(g)$$

Energy changes for the reaction of hydrogen and oxygen are shown in the energy profile diagram below.

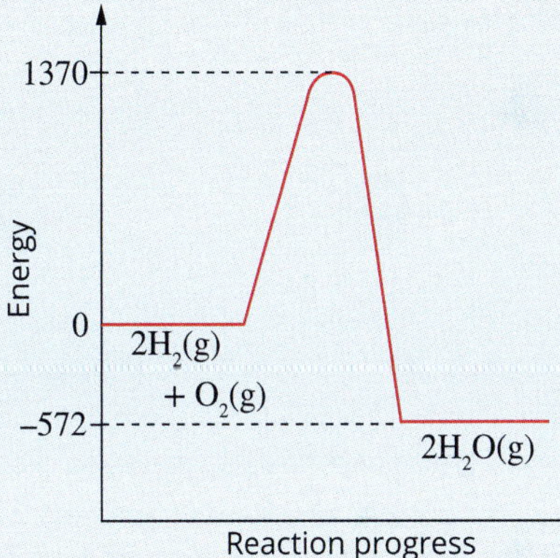

a What is the magnitude of the activation energy of this reaction?

b What is ΔH for this reaction?

Several groups of scientists claim to have split water into hydrogen and oxygen using a molybdenum catalyst:

$$2H_2O(g) \xrightarrow{Mo} 2H_2(g) + O_2(g)$$

c Sketch energy profile diagrams for this reaction with and without the presence of a catalyst.

d What is the value of ΔH for this water-splitting equation?

Application and analysis

13 The diagram below shows possible collision orientations in the substitution reaction between chloromethane and a hydroxide ion.
Using collision theory, explain why collision 1 might be successful whereas collisions 2–4 will not be successful.

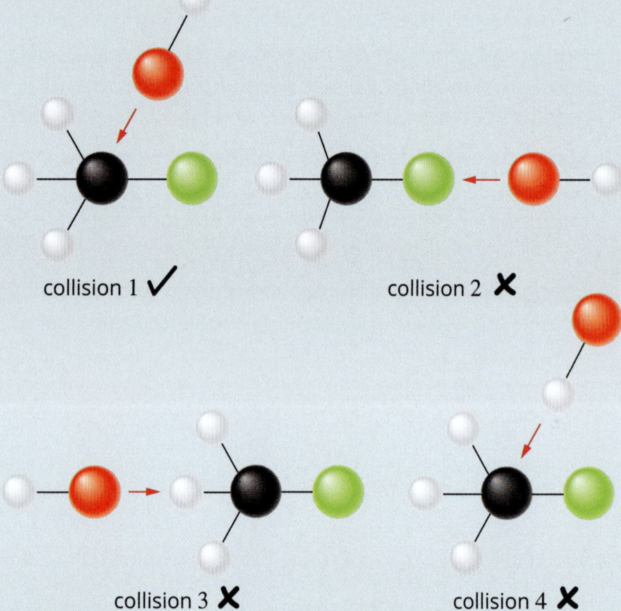

14 The first step in most toffee recipes is to dissolve about 3 cups of sugar in 1 cup of water. Although sugar is quite soluble in water, this step could be time-consuming. Use your knowledge of reaction rates to suggest at least three things you could do to increase the rate at which the sugar dissolves without ruining the toffee.

15 The graph below shows the mass of carbon dioxide gas produced during a 4-minute period from a reaction between marble chips (calcium carbonate) and 1.0 M nitric acid.

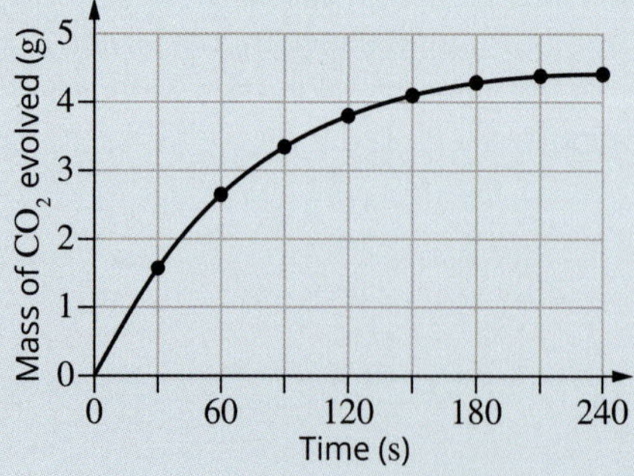

a Write a full chemical equation for this reaction.

b Explain whether the rate of this chemical reaction is increasing or decreasing over time, justifying your answer by referring to parts of the graph.

16 Consider the following points.

- Adding salt to water increases the boiling temperature of the water.
- Water boils at a lower temperature at high altitudes because of the reduced atmospheric pressure.

Use this information, together with your understanding of collision theory, to explain the following.

a Pasta cooks more quickly in salted boiling water than in plain boiling water.

b Mountaineers on Mt Everest (altitude = 8849 m above sea level) find that pasta cooked in boiling water over a camp fire cooks quite slowly.

17 The rate of the decomposition of a solution of hydrogen peroxide can be changed by changing three factors: temperature of the hydrogen peroxide, concentration of the hydrogen peroxide and using different catalysts. The equation for the reaction is:

$$2H_2O_2(aq) \rightarrow 2H_2O(l) + O_2(g)$$

a Sketch a graph which shows the expected mass of oxygen gas, O_2, produced when 20 mL of 10% hydrogen peroxide solution is decomposed over a period of 240 seconds at 25°C. There is no need to mark a scale on either axis.

b On the same set of axes as your answer to part **a**, sketch a second graph which would result when the same volume of 20% hydrogen peroxide is decomposed.

c On the same set of axes, sketch a third graph which would result when the same volume of 10% hydrogen peroxide is decomposed at 35°C.

18 Lumps of limestone, calcium carbonate, react readily with dilute hydrochloric acid. Four large lumps of limestone, mass 10.0 g, were reacted with 100 mL 0.100 M acid.

a Write a balanced equation to describe the reaction.

b Which reactant is in excess? Use a calculation to support your answer.

c Describe a technique that you could use in a school laboratory to measure the rate of the reaction.

d A mass of limestone in small lumps will react at a different rate from the same mass in large lumps. Will the rate of reaction with the smaller lumps be faster or slower? Explain your answer in terms of collision theory.

e List two other ways in which the rate of this reaction can be altered. Explain your answer in terms of collision theory.

19 A 5.00 g piece of copper was dissolved in a beaker containing 500 mL of 2.00 M nitric acid. The equation for the reaction that occurred is:

$$3Cu(s) + 8HNO_3(aq) \rightarrow 3Cu(NO_3)_2(aq) + 2NO(g) + 4H_2O(l)$$

The changing mass of the mixture was observed for a period of time, and the graph below was obtained.

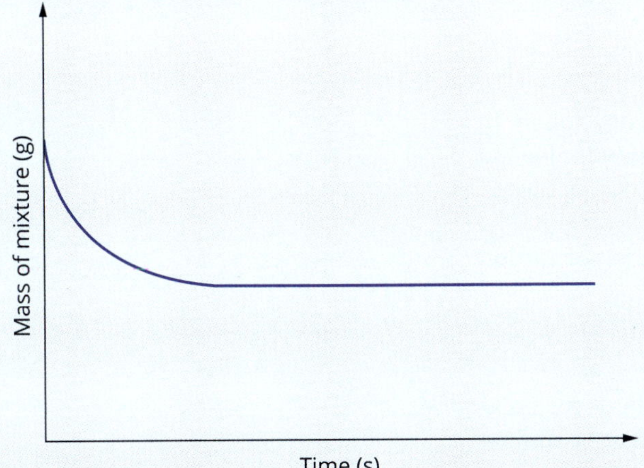

a Explain why the mass of the mixture initially decreases with increasing time.

b Based on the information provided, determine which reactant is limiting.

c Copy the graph above, then sketch in the expected curve if 500 mL of 1.00 M nitric acid had been used instead. Label your new graph line. Explain the difference in shape.

d Redraw the graph above, then sketch in the expected curve if 5.00 g of powdered copper was used instead. Label this new graph line. Explain the difference in shape.

20 The equation for the reaction between sodium thiosulfate and hydrochloric acid is:

$$Na_2S_2O_3(aq) + 2HCl(aq) \rightarrow 2NaCl(aq) + SO_2(g) + S(s) + H_2O(l)$$

The rate of this reaction can be determined by placing the reaction mixture on top of a piece of paper with a black cross on it and observing the cross through the solution until the cross can no longer be seen.

A student wanted to develop an investigation based on this reaction.

a List two independent variables that the student could consider for this investigation.

b Identify the dependent variable for this investigation.

c Describe how the dependent variable would relate to the rate of reaction.

d Write an aim for this experiment, based on one of the independent variables that you listed in part **a**.

21 A student was investigating the difference in the rate of the reaction between acidified potassium permanganate (a purple solution containing the permanganate ion) and oxalic acid ($H_2C_2O_4$) at two different temperatures. The graphs that were plotted are shown below but do not have the vertical axis labelled. The equation for the reaction involved is:

$$2MnO_4^-(aq) + 5H_2C_2O_4(aq) + 6H^+(aq) \rightarrow 2Mn^{2+}(aq) + 10CO_2(g) + 8H_2O(l)$$

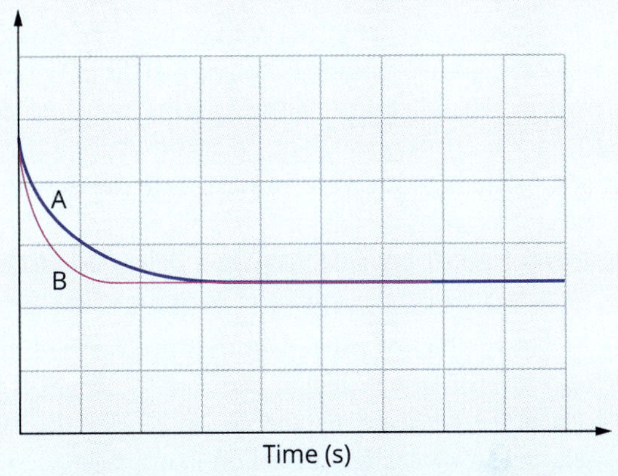

a The student wrote in their method that the mass of carbon dioxide produced was recorded during the experiment and was graphed in the figure above. Explain what is wrong with this statement and suggest a method that the student may have used to produce the graph shown.

b Identify whether graph A or B is likely to represent the higher temperature reaction.

c Referring to collision theory, explain why the initial rates of reaction for graphs A and B were so different.

22 Read the article and answer the questions that follow.

Exploding iron

In 1996, while the Turkish ship MV *B. Onal* was at anchor in Delaware Bay, near Philadelphia in the USA, a 2-tonne hatch cover suddenly blew off. As the ship was carrying a cargo of iron, the surprised crew asked themselves, 'Can iron explode?'

As you may be aware, traditionally iron oxide (Fe_2O_3) is reduced to molten iron at around 1800°C in a blast furnace.

However, a new process that uses less energy has been developed. In this process, iron oxide is converted directly to solid iron without having to heat the reactants to the melting point of iron. Iron oxide is heated to 550°C in the presence of carbon monoxide and hydrogen gas. The iron oxide is reduced to iron by both gases with the formation of carbon dioxide or water.

$$Fe_2O_3(s) + 3CO(g) \longrightarrow 2Fe(s) + 3CO_2(g) \qquad (1)$$
$$Fe_2O_3(s) + 3H_2(g) \longrightarrow 2Fe(s) + 3H_2O(g) \qquad (2)$$

The pellets of pure iron that are formed are extremely porous and full of many tiny holes, in contrast to the solid formed when the molten iron from a blast furnace cools. Under the right conditions, the iron pellets can be oxidised back to iron oxide.

In most cases, iron is oxidised slowly by oxygen back to iron oxide, and the resulting heat can readily escape. If the pellets are more than 1 metre deep, as in the hold of a ship, the heat cannot escape quickly enough and the temperature rises. This speeds up the reaction rate. If the temperature increases sufficiently and water is present, another reaction occurs and the oxidation rate speeds up 100-fold, with the release of more heat:

$$Fe(s) + H_2O(g) \longrightarrow FeO(s) + H_2(g) \qquad (3)$$

Any spark or fire will set off an explosion of hydrogen gas, and that is what happened on the MV *B. Onal*.

a What is the main reason the new reduction process uses less energy than the old process?

b Write equations showing the oxidation of iron by oxygen to form iron(II) oxide and iron(III) oxide.

c If water is present, the oxidation reaction speeds up 100-fold. Is water acting as a catalyst? Explain your answer.

d Is the reaction shown in equation 3 endothermic or exothermic?

e List the factors that increased the rate of reaction in equation 3.

f Firefighters were not able to use water to put out the fire in the cargo hold. Why not? Suggest how they could have put out the fire.

OA

CHAPTER

07 Extent of chemical reactions

The fact that many reactions do not proceed to completion has serious effects on the efficiency of chemical manufacturing industries. The presence of large amounts of unreacted starting materials in reaction mixtures is wasteful and costly. The profitability of an industry depends on the yield—the extent of conversion of reactants into products.

In this chapter, you will investigate the ideas of reversibility and irreversibility of chemical systems. This study of reversible reactions will introduce the concept of chemical equilibrium.

Various factors can change the position of a chemical equilibrium. Le Châtelier's principle enables us to understand the effects of changes in concentrations in solutions, pressures of gases and temperature on an equilibrium.

You will also learn how to write a mathematical relationship, known as an equilibrium law, for an equilibrium system. This law can be used to calculate the relative amounts of reactants and products present when a reaction is at equilibrium.

Key knowledge

- the distinction between reversible and irreversible reactions, and between rate and extent of a reaction **7.1**
- the dynamic nature of homogenous equilibria involving aqueous solutions or gases, and their representation by balanced chemical or thermochemical equations (including states) and by concentration–time graphs **7.1**
- the change in position of equilibrium that can occur when changes in temperature or species or volume (concentration or pressure) are applied to a system at equilibrium, and the representation of these changes using concentration–time graphs **7.4**
- the application of Le Châtelier's principle to identify factors that favour the yield of a chemical reaction **7.4**
- calculations involving equilibrium expressions (including units) for a closed homogeneous equilibrium system and the dependence of the equilibrium constant (K) value on the system temperature and the equation used to represent the reaction **7.3**
- the reaction quotient (Q) as a quantitative measure of the extent of a chemical reaction: that is, the relative amounts of products and reactants present during a reaction at a given point in time **7.2**

7.1 Dynamic equilibrium

In this section, you will learn that some reactions can occur in both the forward and reverse directions. These reactions are called **reversible reactions**. Reversible chemical systems are encountered in many everyday situations, including chemical manufacturing processes, the reactions of ions within individual cells in your body and the reactions which carbon dioxide undergoes in the environment.

This section also describes how some reversible reactions can reach a point where they appear to stop. At this point, the concentrations of the reactants and products remain constant, even though there are still reactants remaining. Although these reactions appear to stop, they actually continue to proceed. If you could see what was occurring at the atomic scale, you would notice that as rapidly as the reactants are forming products, the products are re-forming reactants. This situation can be likened to the queue shown in Figure 7.1.1. Although the length of the queue may seem constant, people at the front are continually leaving the queue and others are joining it at the back at the same rate.

FIGURE 7.1.1 A queue of constant length can be likened to a reaction that appears to have stopped, with people leaving the queue at one end at the same rate as others join it at the other end.

OPEN AND CLOSED SYSTEMS

Previously, you learnt that a chemical reaction can be regarded as a **system**, with everything else around it (the rest of the universe) being the **surroundings**. In an **endothermic** reaction, the system absorbs energy from the surroundings, whereas in an **exothermic** reaction, energy is released to the surroundings by the system.

Figure 7.1.2 illustrates how you can distinguish between two different types of systems:

- **open systems**—in which matter and energy can be exchanged with the surroundings
- **closed systems**—in which only energy is exchanged with the surroundings.

The most common situation in everyday life is an open system.

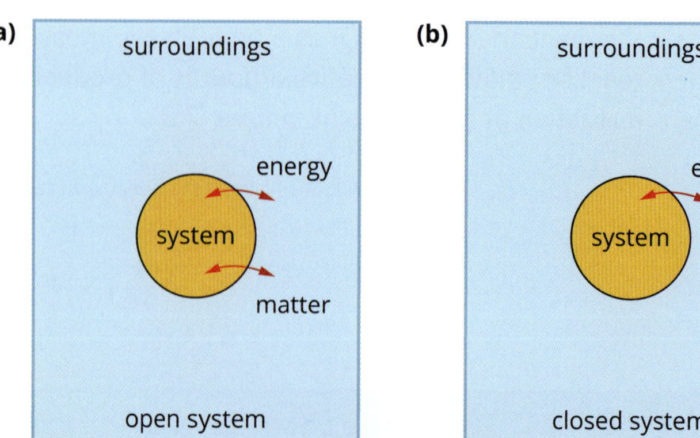

FIGURE 7.1.2 (a) Open systems exchange energy and matter with the surroundings. (b) Closed systems only exchange energy with the surroundings.

Some everyday examples of open and closed systems are illustrated in Figure 7.1.3. In the case of the bushfire, which is an open system, carbon dioxide and water vapour produced by the burning trees are released into the atmosphere. In contrast, the carefully monitored environment of the submarine in operation under water can be regarded as a closed system.

FIGURE 7.1.3 Everyday examples of open and closed systems. (a) This bushfire burning through a forest in January 2020 in Victoria is an example of an open system. (b) A nuclear submarine in operation under water is a closed system.

In a closed system, only energy can be exchanged with the surroundings.

IRREVERSIBLE AND REVERSIBLE REACTIONS

You may have thought, as a younger student, that when chemical reactions occur, the reactants form products and these products cannot be converted back to the reactants. Such reactions, which occur only in one direction, are called non-reversible or **irreversible reactions**.

Baking a cake, like the one shown in Figure 7.1.4, involves several irreversible reactions. Combustion reactions such as the burning of methane are also irreversible:

$$CH_4(g) + 2O_2(g) \rightarrow CO_2(g) + 2H_2O(g)$$

Once a fuel has burnt, the products, carbon dioxide and water, will not react with each other under normal conditions.

FIGURE 7.1.4 Baking a cake involves a series of irreversible chemical reactions.

As you will see, other reactions are reversible reactions where the products, once formed, can react again, re-forming the reactants.

You are familiar with the idea that a physical change, such as a change of state, can be reversed. The evaporation of water from lakes and rivers leading to cloud formation and eventually condensation resulting in rain are examples of such physical changes. Water can cycle between the different phases of solid, liquid and gas because each process is reversible.

In chemistry, a double arrow ($\rightleftharpoons$) is used when writing a chemical equation to show a reversible process. In this way, you can show the phase changes associated with water using the following equation:

$$H_2O(l) \rightleftharpoons H_2O(g)$$

Many reversible reactions are essential to our society, for example, the reactions that power rechargeable batteries. In reversible reactions, the formation of products as a result of collisions between reactant particles is not the end of the process. Once some products are formed, collisions between product particles can result in the reactants being re-formed.

Consider the production of ammonia from hydrogen gas and nitrogen gas known as the Haber process.

The equation for the reaction can be written as:

$$N_2(g) + 3H_2(g) \rightleftharpoons 2NH_3(g)$$

Reversible reactions can reach a state of equilibrium where the overall concentrations of reactants and products do not change over time.

Suppose you mix 1 mol of nitrogen gas and 3 mol of hydrogen gas in a sealed container. From the equation, you might expect that 2 mol of ammonia would eventually be formed. However, no matter how long you wait, the reaction seems to 'stop' when much less than 2 mol of ammonia is present, as shown in Figure 7.1.5.

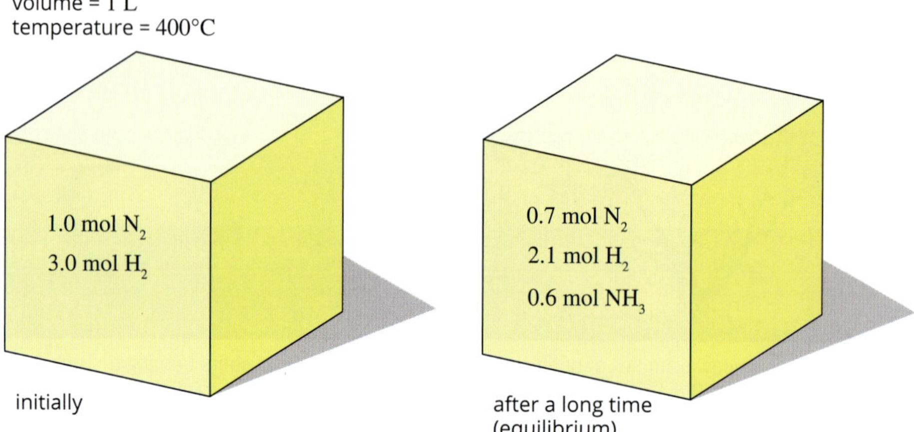

FIGURE 7.1.5 When 1 mol of nitrogen and 3 mol of hydrogen are mixed, the reaction to form ammonia appears to stop before all of the reactants are consumed. (However, notice that reaction has occurred in the same mole ratio as is present in the equation.)

The reaction vessel in which this process occurs is a closed system from which the reactants and products cannot escape. Reversible reactions in a closed system eventually reach a situation where the rate of the forward reaction and the rate of the reverse reaction are equal.

At this point, there appears to be no further change to the observer. In the case of a chemical reaction, when the reaction appears to have stopped, the system is described as having reached **equilibrium**. At equilibrium, there still may be significant amounts of reactants in the system.

EXPLAINING REVERSIBILITY

The reason why reversible reactions can occur can be understood by referring to an energy profile diagram. When particles collide, the energy associated with collisions can break bonds in the reacting particles, allowing them to rearrange to form products. As you saw in Section 6.1, the energy required to break the bonds of the reactants is known as the **activation energy** of the reaction.

You can see in the energy profile diagram, in Figure 7.1.6, that once the products form, it is possible for the reverse process to occur. If the newly formed product particles collide with enough energy to break their bonds, then it is possible to re-form the original reactants.

If the forward reaction is endothermic, the reverse reaction is exothermic, and vice versa.

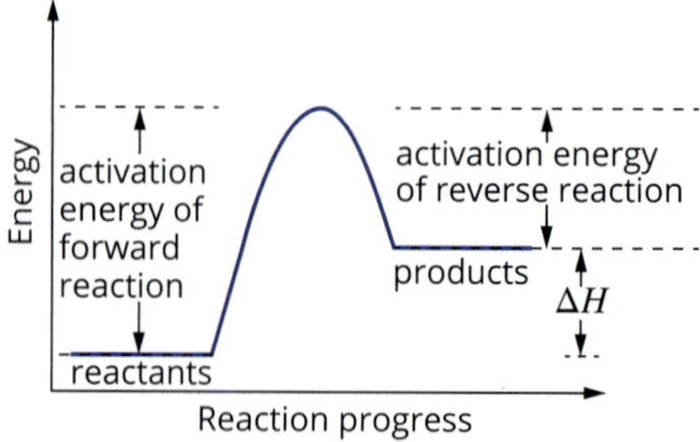

FIGURE 7.1.6 An energy profile diagram for an endothermic reaction, showing the activation energy required for both the forward (formation of products) and reverse (re-formation of reactants) reactions

CHEMFILE

Limestone caves

An example of reversibility in nature is the formation of stalactites and stalagmites in limestone caves.

The main mineral in limestone is calcite ($CaCO_3$). Water saturated with carbon dioxide from the atmosphere drips through the roof of the cave, resulting in the following reaction:

$$CO_2(g) + H_2O(l) + CaCO_3(s) \longrightarrow Ca^{2+}(aq) + 2HCO_3^-(aq)$$

As the water seeps through the rocks, it becomes saturated with Ca^{2+} ions and HCO_3^- ions. The water then evaporates, and the reverse reaction produces stalactites from the ceiling of the cave:

$$Ca^{2+}(aq) + 2HCO_3^-(aq) \longrightarrow CO_2(g) + H_2O(l) + CaCO_3(s)$$

Some of the solution drips onto the floor of the cave, where more deposits of $CaCO_3$ are produced, forming stalagmites. Stalactites and stalagmites grow in pairs and can produce beautiful columns like the ones in the Buchan cave system shown in the figure below.

Stalactites and stalagmites in the Buchan Caves in Victoria. They are made of the mineral calcite.

EXPLAINING EQUILIBRIUM

Because the reaction between nitrogen and hydrogen to form ammonia is a reversible reaction, it is best written using an equilibrium arrow:

$$N_2(g) + 3H_2(g) \rightleftharpoons 2NH_3(g)$$

Equilibrium arrows indicate that the reaction can occur in both the forward and reverse directions. These arrows should not be used where the reaction can only proceed in one direction.

The idea that processes can be reversed can be used to understand why this reaction reaches equilibrium. When nitrogen gas and hydrogen gas are added to a sealed container at a constant temperature, a sequence of events occurs that can be illustrated by a plot of reaction rate versus time, like the one shown in Figure 7.1.7.

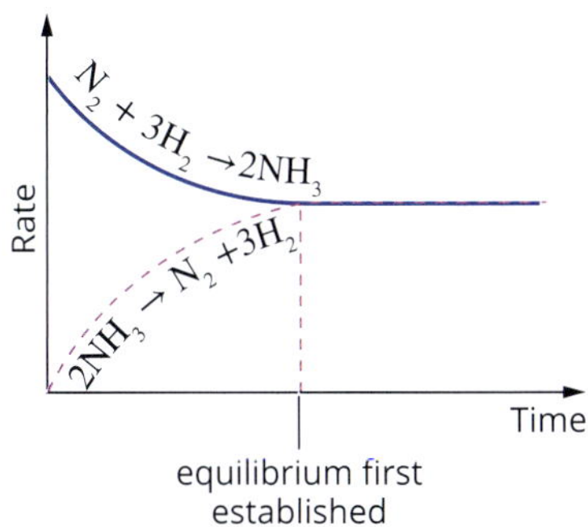

FIGURE 7.1.7 The variation of the rates of the forward and reverse reactions with time when nitrogen and hydrogen are mixed

This graph illustrates the following events that occur when the reactants are mixed.

- Nitrogen and hydrogen gas molecules collide with each other and form ammonia. As the forward reaction, $N_2(g) + 3H_2(g) \rightarrow 2NH_3(g)$, proceeds, the concentrations of nitrogen and hydrogen decrease, so the frequency of collisions between reactant molecules decreases and the rate of the production of ammonia decreases.
- At the same time as ammonia is being formed, some ammonia molecules collide and decompose to re-form nitrogen and hydrogen: $2NH_3(g) \rightarrow N_2(g) + 3H_2(g)$.
- Eventually the forward and reverse reactions proceed at the same rate, so that ammonia is formed at exactly the same rate as it is broken down. The concentrations of ammonia, nitrogen and hydrogen then remain constant. To an observer, the reaction now appears to have stopped with no observable change.

In a closed system at constant pressure and temperature, no further change will take place. The reaction has reached a point of balance—an equilibrium.

The concentration–time graph in Figure 7.1.8 shows the changes in concentrations of the chemicals with time. Equilibrium is established when there is no longer any change in any of the concentrations (i.e. the graphs are horizontal lines).

When considering graphs involving equilibrium systems, always check if the data is presented as a plot of concentration versus time or a plot of reaction rate versus time.

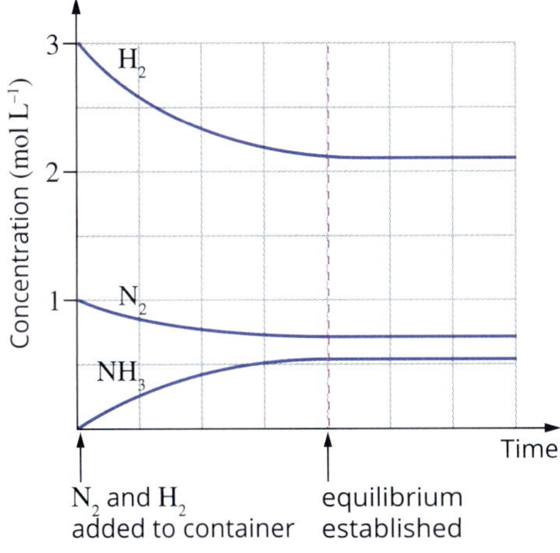

FIGURE 7.1.8 Changes in the concentrations of N_2, H_2 and NH_3 as a mixture of nitrogen and hydrogen gas reacts

Dynamic equilibrium is reached by reversible physical or chemical reactions taking place in a closed system.

DYNAMIC STATE OF EQUILIBRIUM

Chemical equilibrium can be described as being in a dynamic state because the forward and reverse reactions have not ceased. Instead, they occur simultaneously at the same rate. This is called **dynamic equilibrium**.

During dynamic equilibrium:

- the reaction is 'incomplete' and all of the substances (that is, the reactants and products) are present in the equilibrium mixture
- at the molecular level, bonds are constantly being broken and new bonds are being formed as the reactants and products continue to be converted from one to another.

The decomposition of dinitrogen tetroxide gas (N_2O_4) to nitrogen dioxide gas (NO_2) is an example of a reversible reaction that reaches a dynamic equilibrium. The progression of this reaction from pure N_2O_4 to the equilibrium mixture containing both N_2O_4 and NO_2 can be monitored through the changing colour of the gases in the reaction vessel. N_2O_4 is colourless and NO_2 is dark brown.

The reaction occurs according to the following equation:

$$\underset{\text{colourless}}{N_2O_4(g)} \rightleftharpoons \underset{\text{brown}}{2NO_2(g)}$$

Figure 7.1.9 illustrates the observations made of a reaction vessel that is injected with some pure N_2O_4. As the forward reaction proceeds, the formation of a dark brown gas is observed. The depth of colour increases until equilibrium is reached (after 8 seconds), at which point there is no further change in the colour regardless of how long the reaction is allowed to proceed.

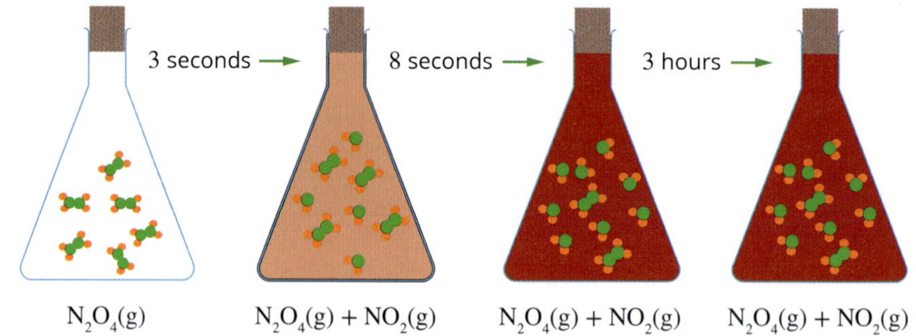

FIGURE 7.1.9 The decomposition of dinitrogen tetroxide

EXTENT OF REACTION

You have seen that some reactions are reversible, but you might wonder whether they all proceed to the same extent before they reach equilibrium. This can be answered with a simple experiment.

Both hydrogen chloride (HCl) and ethanoic acid (CH_3COOH) react with water to form ions, according to the equations:

$$HCl(aq) + H_2O(l) \rightleftharpoons H_3O^+(aq) + Cl^-(aq)$$

$$CH_3COOH(aq) + H_2O(l) \rightleftharpoons H_3O^+(aq) + CH_3COO^-(aq)$$

Solutions of both chemicals conduct electricity because they contain mobile ions. The relative conductivity of the solutions is proportional to the number of free ions in the solution. By measuring the electrical conductivity of solutions of the same concentration, you can compare how much each compound ionises in water.

Figure 7.1.10 shows the results obtained from such an experiment. You can see that the solution formed when hydrogen chloride dissolves in water (called hydrochloric acid) is a much better conductor than the ethanoic acid solution. Both solutions were formed by adding the same number of moles of acid molecules to identical volumes of water.

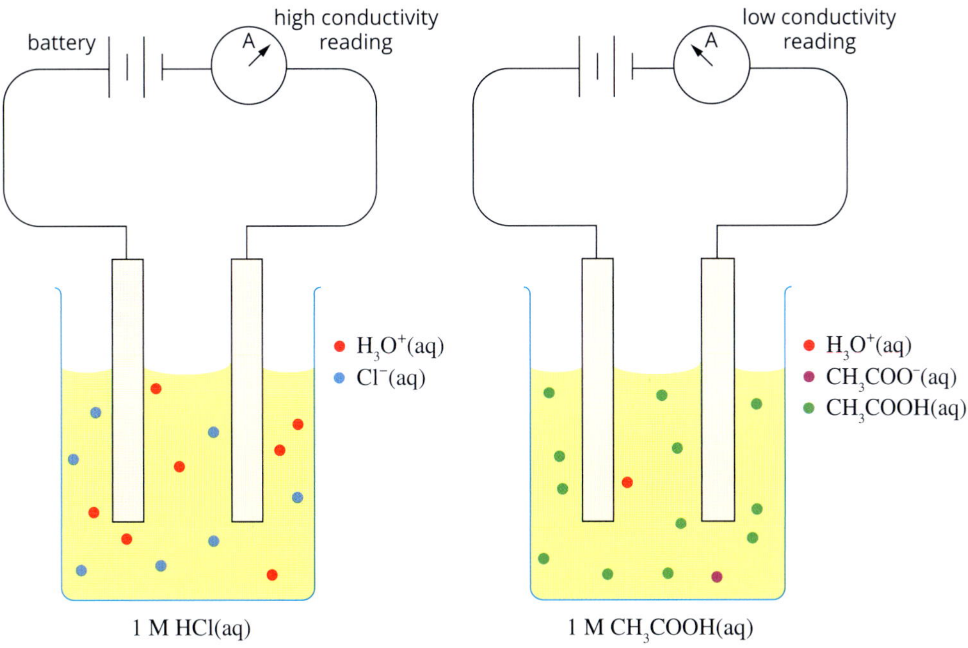

FIGURE 7.1.10 This experiment compares the electrical conductivity of 1 M solutions of hydrogen chloride and ethanoic acid.

As you will remember from Unit 2, ethanoic acid is a weak acid and will therefore only partially ionise in an aqueous solution. Hydrochloric acid is a strong acid that almost completely ionises in an aqueous solution. The concept of equilibrium allows us to better explain the idea of strong and weak acids by looking at the extent of the ionisation reaction. The difference in conductivity observed in the experiment arises because these reactions occur to remarkably different extents. At equilibrium in a 1 M solution, at 25°C, almost all the HCl molecules are ionised, whereas only approximately 1% of the CH_3COOH molecules are ionised.

We can conclude that different reactions proceed to different extents. The ratios of reactants to products are different for different equilibrium systems. It is important to note that the **extent of reaction** describes how much product is formed when the system reaches equilibrium. However, the extent of a reaction does not give any information about how fast a reaction will proceed (the rate of the reaction). The rate of reaction is a measure of the change in concentration of the reactants and products with time and is not directly related to the extent of reaction. The rates of reversible reactions range from very slow to very fast and determine how long the reaction takes to reach equilibrium.

Since HCl is almost completely ionised in water, the equation for the reaction is usually written with a single arrow:

$$HCl(aq) + H_2O(l) \longrightarrow H_3O^+(aq) + Cl^-$$

CASE STUDY **ANALYSIS**

Investigating dynamic equilibrium

Chemists can use radioactive isotopes to investigate systems in dynamic equilibrium. Radioactive isotopes behave chemically in the same way as non-radioactive atoms of the same element, but their presence and location can be easily determined by a radiation detector.

When solid sodium iodide (NaI) is added to water, it dissolves readily at first. As the concentration of dissolved sodium iodide increases, a saturated solution forms and no further solid dissolves. The concentrations of the Na^+ ions and I^- ions in solution remain constant and some solid NaI is always present.

When solid sodium iodide containing radioactive iodide ions is added to a saturated solution, the subsequent movement of these 'labelled' ions can be traced. Figure 7.1.11 shows that although solid sodium iodide is still observed in the bottom of the container, the solution quickly becomes radioactive.

The radioactivity of the solution shows that some of the radioactive sodium iodide has dissolved. The concentration of sodium iodide remains constant, so particles that were not radioactive must have crystallised from the solution at the same rate as the radioactive solid was dissolving (Figure 7.1.12).

Even though we see nothing happening, there must be continual activity at the surface of the solid. The process is a dynamic equilibrium.

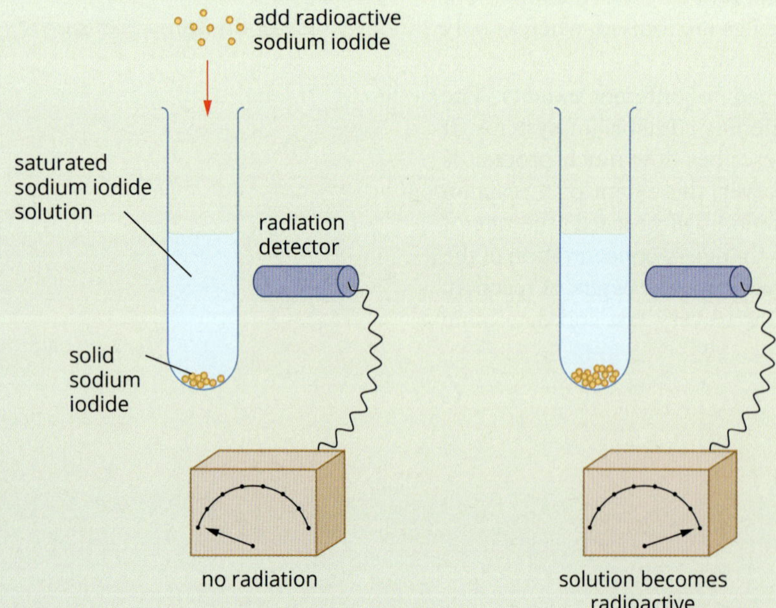

FIGURE 7.1.11 This experiment shows that a dynamic equilibrium between the solid and dissolved ions is present in a saturated solution.

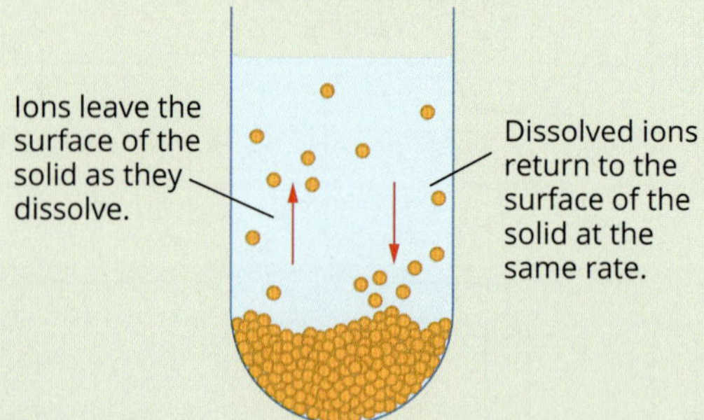

FIGURE 7.1.12 The cyclic dissolving and redepositing of ions in a saturated solution is an example of dynamic equilibrium.

Analysis

1. Explain how dynamic equilibrium is achieved from the time that solid sodium iodide is added to pure water to when no more sodium iodide will dissolve.
2. Write the equation for the solid NaI in equilibrium with its ions.
3. Explain the term 'dynamic equilibrium'.
4. Identify the parts of this mixture where radioactivity will be recorded after solid radioactive sodium iodide has been added to the existing equilibrium. Explain how this shows that the reaction is continuing, although the concentrations remain constant.

7.1 Review

OA ✓✓

SUMMARY

- In a closed system, only energy, not matter, is exchanged with the surroundings.
- In an open system, both matter and energy are exchanged between the system and the surroundings.
- A reversible reaction is a reaction in which the products can be converted back to the reactants.
- An irreversible reaction is a reaction in which the products cannot be converted back to the reactants.
- Reversible reactions can reach a point where the rate of the forward reaction and the rate of the reverse reaction are equal. At this point, a dynamic equilibrium has been achieved.
- Equilibrium can be achieved in closed systems, but not in open systems.
- Different reactions proceed to different extents.
- The relative ratios of reactants to products when equilibrium is reached are different for different reactions.
- The extent of reaction indicates how much product is formed at equilibrium, whereas the rate of reaction is a measure of the change in concentration of the reactants and products over time.

KEY QUESTIONS

Knowledge and understanding

1 Which one of the following statements about the extent of reaction is true?

A The extent of reaction indicates the rate of the reaction, i.e. the time taken to reach equilibrium.

B The extent of reaction is the point when there are equal amounts of reactants and products.

C The extent of reaction indicates how far the reaction has proceeded in the forward direction when equilibrium is achieved.

D The extent of reaction indicates the rate of reaction and is the point when the rate of the forward reaction is equal to the rate of the reverse reaction.

2 **a** Hydrogen gas is mixed with iodine gas in a sealed container. A reaction occurs according to the equation:

$$H_2(g) + I_2(g) \rightleftharpoons 2HI(g)$$

On the rate–time graph shown below for this system, label the lines for the forward and reverse reactions with the appropriate chemical equations. Also label the point when equilibrium is first established.

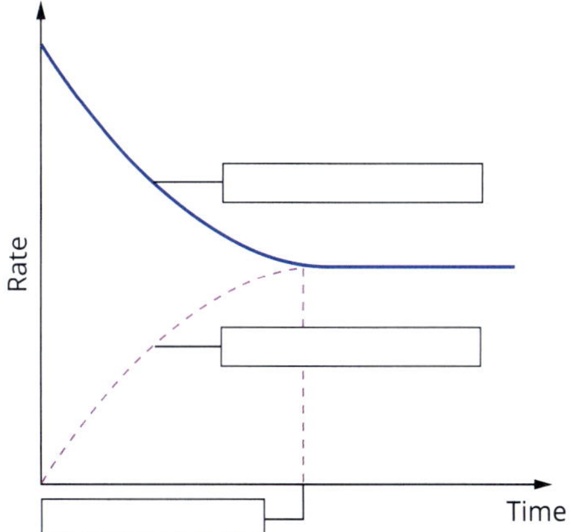

b Explain why you labelled the diagram as you did.

continued over page

7.1 Review *continued*

3 Fill in the blanks to complete the sentences about dynamic equilibrium using words from the following list:

closed; concentrations; decrease; equal; forward; increases; less; more; open; reverse

In a __________ system, as the concentrations of the reactants ____________, the rate of the forward reaction also decreases. The collisions between these reactant molecules occur __________ frequently. Once some product starts to form, the _________ reaction occurs and the frequency of collisions between product molecules __________ and between reactant molecules decreases. At equilibrium, the rates of the forward and backward reactions are __________ and the ____________ of all species do not change.

4 Describe the characteristics of a reaction system at equilibrium.

Analysis

5 The graph below shows the concentration versus time plot for the reversible decomposition of dinitrogen tetroxide: $N_2O_4(g) \rightleftharpoons 2NO_2(g)$ at 100°C in a 1.0 L reaction vessel. N_2O_4 is a colourless gas and NO_2 is brown. Use the graph to answer the following questions.

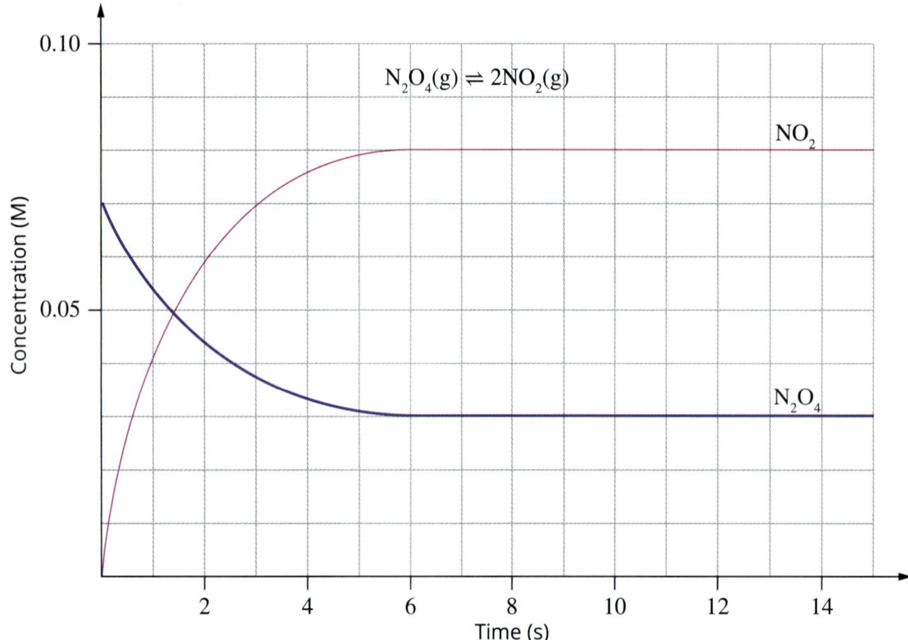

- **a** State the initial concentration of N_2O_4 in the flask.
- **b** State the initial concentration of NO_2 in the flask.
- **c** State the concentration of N_2O_4 at equilibrium.
- **d** State the concentration of NO_2 at equilibrium.
- **e** Over the course of the reaction, how many moles of N_2O_4 decompose?
- **f** What do the horizontal regions of the graph indicate?
- **g** How long does it take for equilibrium to be reached?
- **h** A student studying this reaction records her observations over time. What will she observe as the reaction proceeds?

7.2 The equilibrium law

In this section, you will investigate the relationship between the quantities of reactants and the quantities of products present when a system reaches equilibrium.

This relationship allows you to predict the relative amounts of reactants and products in individual equilibrium systems.

You will examine **homogeneous chemical systems** in which all the reactants and products are in the same state. These differ from **heterogeneous chemical systems** in which the reactants and products are in different states.

THE REACTION QUOTIENT

Consider the equilibrium system you were introduced to in Section 7.1:

$$N_2(g) + 3H_2(g) \rightleftharpoons 2NH_3(g)$$

An unlimited number of different equilibrium mixtures of the three gases—nitrogen, hydrogen and ammonia—can be prepared. Table 7.2.1 shows the concentrations of each of these gases in four different equilibrium mixtures at a constant temperature of 400°C. The values of the fraction $\frac{[NH_3]^2}{[N_2][H_2]^3}$ for each mixture are also given.

The fraction $\frac{[NH_3]^2}{[N_2][H_2]^3}$ is called the **reaction quotient** (Q) or **concentration fraction** of the mixture.

TABLE 7.2.1 Concentrations of reactants and products present in equilibrium mixtures

Equilibrium mixture	$[N_2]$ (M)	$[H_2]$ (M)	$[NH_3]$ (M)	$\frac{[NH_3]^2}{[N_2][H_2]^3}$
1	0.25	0.75	0.074	0.052
2	0.55	0.65	0.089	0.052
3	0.0025	0.0055	4.6×10^{-6}	0.051
4	0.0011	0.0011	2.7×10^{-7}	0.051

As you can see in Table 7.2.1, the reaction quotient $\frac{[NH_3]^2}{[N_2][H_2]^3}$ has an almost constant value of 0.052 for each equilibrium mixture, regardless of the concentration of each component. Note that the coefficients of the reactants and products in the chemical equation above form the indices of the respective reactant and product concentrations used in the reaction quotient.

Although the reaction quotient, Q, can be calculated for any mixture of reactants and products at any time during a reaction, it only has a constant value when the mixture is at equilibrium. At equilibrium, the value of the reaction quotient, Q, is equal to the **equilibrium constant**, K.

In general, for chemical reactions at equilibrium:

- different chemical reactions have different values of K
- the size of K indicates the proportions (relative amounts) of reactants and products in the equilibrium mixture
- for a particular reaction, K is constant for all equilibrium mixtures at a fixed temperature.

At equilibrium, $Q = K$.

K is the equilibrium constant for a reaction. The value of K is different for different reactions and at different temperatures.

THE EXPRESSION FOR THE EQUILIBRIUM LAW

By studying a large number of reversible systems, such as the one between nitrogen, hydrogen and ammonia in the previous example, chemists have been able to develop the concept of the **equilibrium law**.

The equilibrium law states that:

- the equilibrium constant, K, is the concentrations of products divided by the concentrations of reactants at equilibrium
- the index of each component concentration is the same as the coefficient for the substances in the balanced chemical equation.

For the general equation $aW + bX \rightleftharpoons cY + dZ$ at equilibrium at a particular temperature, the equilibrium expression can be written as:

Recall that [ ] should be read as 'the concentration of'.

$$K = \frac{[Y]^c[Z]^d}{[W]^a[X]^b}$$

where K is the equilibrium constant.

When you write an equilibrium expression, the concentrations of the products are always in the numerator and the concentrations of the reactants are always in the denominator.

A useful way of remembering the equilibrium law is that K can be represented as $\frac{[\text{products}]^{\text{coefficients}}}{[\text{reactants}]^{\text{coefficients}}}$. Remember that if there is more than one product or reactant, you must multiply the terms.

Units for equilibrium constants

Units for equilibrium constants depend on the expression used to represent the chemical reaction. They can be determined by substituting M or mol L^{-1} for each concentration into the concentration fraction expression. In the case of the equilibrium between nitrogen, hydrogen and ammonia above, since:

The units of K depend on the equation for the reaction, and can vary.

$$K = \frac{[NH_3]^2}{[N_2][H_2]^3}$$

the unit is given by:

$$\frac{M^2}{M \times M^3} = M^{-2} \text{ (or mol}^{-2} \text{ L}^2\text{)}$$

Worked example 7.2.1

DETERMINING THE UNITS FOR AN EQUILIBRIUM EXPRESSION

The decomposition of N_2O_4 is a reversible reaction that occurs according to the equation:

$$N_2O_4(g) \rightleftharpoons 2NO_2(g)$$

Write the expression for the equilibrium constant, K, and determine its units.

Thinking	Working
Write the expression for K. $\frac{[\text{products}]^{\text{coefficients}}}{[\text{reactants}]^{\text{coefficients}}}$	$K = \frac{[NO_2]^2}{[N_2O_4]}$
Substitute the units of concentration into the expression for K to find the units of K.	$\frac{M^2}{M} = M$ (or mol L^{-1})

Worked example: Try yourself 7.2.1

DETERMINING THE UNITS FOR AN EQUILIBRIUM EXPRESSION

Write the equilibrium expression and determine the units for K for the reversible reaction:

$$2SO_2(g) + O_2(g) \rightleftharpoons 2SO_3(g)$$

The reaction quotient and the equilibrium law

An expression for the equilibrium constant, K, can be written for any system at equilibrium. As you learnt earlier, a similar expression can be written for the reaction quotient, Q, for systems that are not necessarily at equilibrium. For:

$$aA + bB \rightleftharpoons cC + dD$$

$$Q = \frac{[C]^c[D]^d}{[A]^a[B]^b}$$

A reaction quotient can be calculated for any mixture of reactants and products at any time during a reaction. However, the reaction quotient is only a constant value when the mixture is at equilibrium. At equilibrium, the value of the reaction quotient is equal to the equilibrium constant, K.

If the reaction quotient, Q, is:

- greater than K, the system 'shifts to the left' to achieve equilibrium and more reactants are formed
- smaller than K, the system 'shifts to the right' to achieve equilibrium and more products are formed
- equal to K, the system is at equilibrium.

The relationship between Q and K is illustrated in Figure 7.2.1. By comparing the value of Q for a reaction to K at a given temperature, it is possible to predict the direction a reaction will proceed to reach equilibrium.

> If $Q < K$, system moves to right.
> If $Q > K$, system moves to left.
> If $Q = K$, system is at equilibrium.

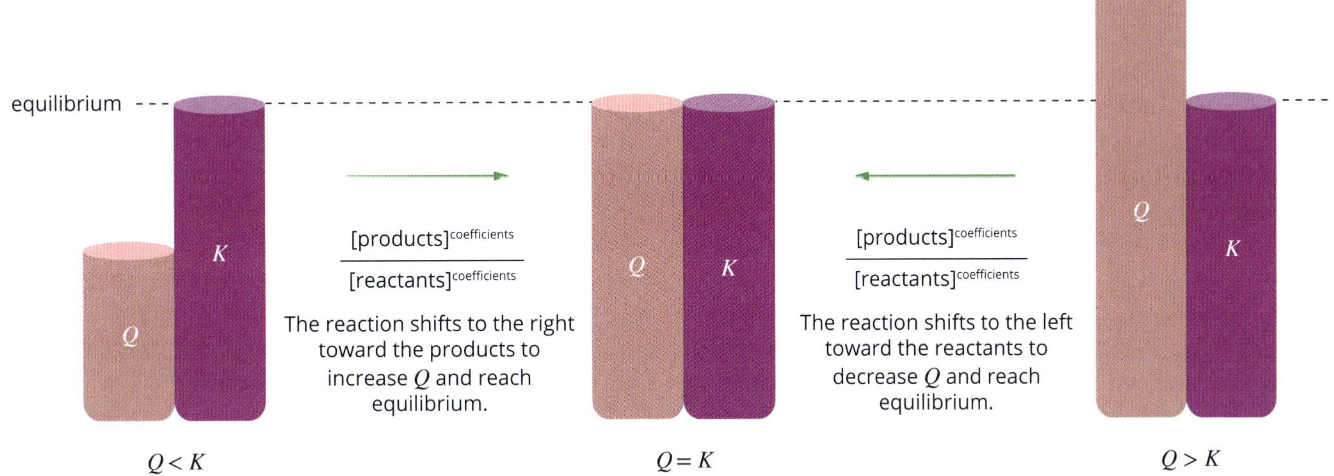

FIGURE 7.2.1 For any reversible reaction at equilibrium, $K = Q$. Comparing these two values for a reaction can indicate which way a reaction must progress to reach equilibrium.

CHEMFILE

Equilibrium in the theatre

A yellow solution containing Fe^{3+} ions reacts with a colourless solution containing SCN^- ions to form a blood-red solution containing $FeSCN^{2+}$ ions, according to the equilibrium reaction:

$$Fe^{3+}(aq) + SCN^-(aq) \rightleftharpoons FeSCN^{2+}(aq) \quad K = 9 \times 10^2\ M^{-1} \text{ at } 25°C$$

An appreciable amount of the product $FeSCN^{2+}$ is present in an equilibrium mixture, so the mixture appears blood-red (see figure at right).

This equilibrium reaction has been used in theatrical productions to make fake blood. A layer of colourless SCN^- solution can be painted onto an actor's hand before the scene. If a plastic knife that has been previously dipped in pale yellow Fe^{3+} solution is used to make a fake cut across the hand, a blood-red 'cut' appears because of the production of red $FeSCN^{2+}$.

Chemicals needed to make fake blood, from left to right: a solution containing Fe^{3+}, a solution containing SCN^-, and an equilibrium mixture of Fe^{3+}, SCN^- and $FeSCN^{2+}$

7.2 Review

WS 15 OA ✓✓

SUMMARY

- The equilibrium constant, K, is a constant for a particular chemical reaction at a particular temperature.
- The equilibrium law for the equation:

$$aW + bX \rightleftharpoons cY + dZ$$

is:

$$K = \frac{[Y]^c[Z]^d}{[W]^a[X]^b}$$

- A reaction quotient, Q, can be calculated for any stage of a chemical reaction. Q has the same mathematical expression as the equilibrium law.
- When a reaction system at a particular temperature has reached equilibrium, the value of the reaction quotient is equal to the value of the equilibrium constant.
- The unit for an equilibrium constant depends on the equation for the equilibrium. The unit can be determined by substituting M (mol L^{-1}) for each concentration into the expression for the equilibrium law.

KEY QUESTIONS

Knowledge and understanding

1 Define:

a homogenous system

b reaction quotient

c equilibrium constant.

2 Write an expression for the reaction quotient, Q, for the reaction of hydrogen and bromine according to the equation:

$$H_2(g) + Br_2(l) \rightleftharpoons 2HBr(g)$$

3 Write the expression for the equilibrium constant, K, for the following equilibrium system:

$$2Fe^{3+}(aq) + Sn^{2+}(aq) \rightleftharpoons 2Fe^{2+}(aq) + Sn^{4+}(aq)$$

Analysis

4 At a particular temperature, the equilibrium constant for the reaction represented by the following equation is 0.667 M^{-2}:

$$CO(g) + 3H_2(g) \rightleftharpoons CH_4(g) + H_2O(g)$$

At a specific point in the reaction, the reaction quotient is found to be 0.234 M^{-2}. With reference to the concentration of the products, predict which way the reaction will shift to reach equilibrium.

5 Consider the reaction represented by the equation:

$$Cu^{2+}(aq) + 4NH_3(aq) \rightleftharpoons [Cu(NH_3)_4]^{2+}(aq)$$

At 25°C the equilibrium constant is determined to be $K = 0.46$ M^{-4}. At a particular time during the reaction, the reaction quotient is 1.2 M^{-4}.

Write an expression for the reaction quotient for this reaction and predict what will happen to the system as it moves towards equilibrium.

7.3 Calculations involving equilibrium

In Section 7.2, you learnt that an equilibrium law can be written for a chemical reaction at equilibrium. The mathematical expression for the equilibrium law is a fraction (a ratio) made up of the concentrations of the products to the reactants. It has a value equal to *K*, the equilibrium constant for the reaction at equilibrium.

The value of an equilibrium constant indicates the extent of a reaction or how far a reaction will proceed towards the products. In this section, you will learn how to interpret these values in terms of the relative amounts of reactants and products present at equilibrium and discover what happens to the value of an equilibrium constant when an equation is reversed or the coefficients of the equation are changed.

You will also learn to calculate the concentration of a reactant or product using the equilibrium constant at a specified temperature.

DEPENDENCY OF AN EQUILIBRIUM CONSTANT ON THE EQUATION

The equilibrium law depends upon the chemical equation used for a particular reaction. For example, the equilibrium between the gases N_2O_4 and NO_2 can be represented by several equations, given below. For each equation the expression for the equilibrium constant, *K*, is also given.

$$N_2O_4(g) \rightleftharpoons 2NO_2(g) \quad K = \frac{[NO_2]^2}{[N_2O_4]}$$

$$2NO_2(g) \rightleftharpoons N_2O_4(g) \quad K = \frac{[N_2O_4]}{[NO_2]^2}$$

$$2N_2O_4(g) \rightleftharpoons 4NO_2(g) \quad K = \frac{[NO_2]^4}{[N_2O_4]^2}$$

$$\frac{1}{2}N_2O_4(g) \rightleftharpoons NO_2(g) \quad K = \frac{[NO_2]}{[N_2O_4]^{\frac{1}{2}}}$$

You can see from these expressions that if:

- one equation is the reverse of another, the equilibrium constants are the inverse (or reciprocal) of each other
- the coefficients of an equation are doubled, the value of *K* is squared
- the coefficients of an equation are halved, the value of *K* is the square root of the original value of *K*.

Therefore, it is important to specify the equation when quoting an equilibrium constant.

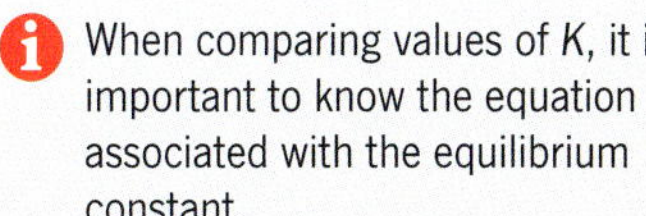
When comparing values of *K*, it is important to know the equation associated with the equilibrium constant.

THE MEANING OF THE VALUE OF AN EQUILIBRIUM CONSTANT

The value of an equilibrium constant is based on the equilibrium concentrations of the products divided by the equilibrium concentrations of the reactants. Therefore, it indicates the extent of reaction at equilibrium (how far the forward reaction proceeds to reach equilibrium and the **equilibrium yield** (the amount of products present at equilibrium).

CHEMFILE

Equilibrium in biological systems

Although the calculations in this section will be restricted to chemical equilibria, the same principles can be applied to other systems in our surroundings. For example, on the African plains of Tanzania there is a delicate balance of herbivores, such as zebras and wildebeest, and carnivores, such as lions (see the figure below). If the populations change because of drought or disease, the relative numbers change and a new balance is established. This new balance can be predicted just like the balance of a chemical equilibrium can be predicted.

The populations of zebras and lions in Africa can be understood using the principles of equilibrium.

The relationship between the value of K and the relative proportions of the reactants and products at equilibrium is shown in Table 7.3.1.

TABLE 7.3.1 The relationship between the value of K and the extent of a reaction provides information on the relative amounts of reactants and products in the reaction mixture at equilibrium.

Value of K	Extent of reaction
between about 10^{-4} and 10^{4}	The extent of reaction is significant. Appreciable concentrations of both the reactants and products are present at equilibrium. e.g. $N_2(g) + 3H_2(g) \rightleftharpoons 2NH_3(g)$ $K = 0.52\ M^{-2}$ at 400°C
very large; $>10^{4}$	Almost complete reaction occurs. The concentrations of the products are much higher than the concentrations of the reactants at equilibrium. e.g. $HCl(aq) + H_2O(l) \rightleftharpoons H_3O^+(aq) + Cl^-(aq)$ $K = 10^7\ M$ at 25°C
very small; $<10^{-4}$	Negligible reaction occurs. The concentrations of the reactants are considerably higher than the concentrations of the products at equilibrium. e.g. $CH_3COOH(aq) + H_2O(l) \rightleftharpoons H_3O^+(aq) + CH_3COO^-(aq)$ $K = 1.8 \times 10^{-5}\ M$ at 25°C

When K is very large, the numerator of the equilibrium expression must be large compared to the denominator, which means there is a large amount of products relative to the amount of reactants.

When K is very small, the numerator of the equilibrium expression must be very small compared to the denominator, which means there must be a large amount of reactants relative to the amount of products.

CASE STUDY

Equilibria in the blood: the importance of the value of the equilibrium constant

Haemoglobin (Figure 7.3.1) is a large protein molecule that is the pigment in red blood cells. It is responsible for the transport of oxygen from your lungs to the cells in your body. The haemoglobin molecule combines with oxygen to form an equilibrium system with oxyhaemoglobin:

$$\text{haemoglobin} + \text{oxygen} \rightleftharpoons \text{oxyhaemoglobin}$$

Carbon monoxide is a colourless, odourless and tasteless gas that is formed as a product in the incomplete combustion of fuels. Carbon monoxide is present in cigarette smoke and in the exhaust gases from car engines.

The high toxicity of carbon monoxide is a result of its reaction with haemoglobin:

$$\text{haemoglobin} + \text{carbon monoxide} \rightleftharpoons \text{carboxyhaemoglobin}$$

The equilibrium constant for the reaction between carbon monoxide and haemoglobin is nearly 20000 times greater than for the reaction between oxygen and haemoglobin. The larger equilibrium constant means that the forward reaction is much more likely to occur. Even small concentrations of carbon monoxide shift the position of equilibrium well to the right.

The formation of carboxyhaemoglobin reduces the concentration of haemoglobin, causing the reverse reaction of oxyhaemoglobin formation to occur. In extreme cases, almost no oxyhaemoglobin is left in the blood and carbon monoxide poisoning occurs.

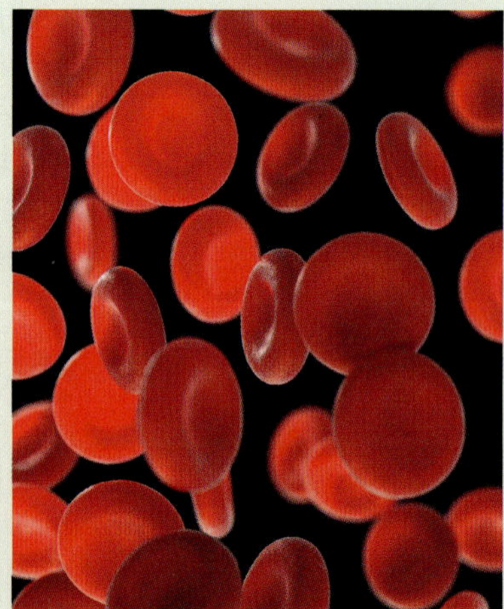

FIGURE 7.3.1 Blood cells contain a red pigment, haemoglobin, that transports oxygen from the lungs to other cells in the body.

CALCULATIONS INVOLVING THE EQUILIBRIUM CONSTANT AND CONCENTRATIONS

An equilibrium constant can be calculated from the molar concentrations of reactants and products at equilibrium, as shown in Worked example 7.3.1.

Worked example 7.3.1

CALCULATING AN EQUILIBRIUM CONSTANT

A 2.00 L vessel contains a mixture of 0.0860 mol of H_2, 0.124 mol of I_2 and 0.716 mol of HI in equilibrium at 460°C according to the equation:

$$H_2(g) + I_2(g) \rightleftharpoons 2HI(g)$$

Calculate the value of the equilibrium constant, K, at 460°C.

Thinking	Working
Find the molar concentrations for all species at equilibrium. Convert mol to mol L^{-1} using $c = \frac{n}{V}$.	The volume of the vessel = 2.00 L $[H_2] = \frac{n(H_2)}{V} = \frac{0.0860}{2.00} = 0.0430$ M $[I_2] = \frac{n(I_2)}{V} = \frac{0.124}{2.00} = 0.0620$ M $[HI] = \frac{n(HI)}{V} = \frac{0.716}{2.00} = 0.358$ M
Write the expression for K.	$K = \frac{[HI]^2}{[H_2][I_2]}$
Substitute into the expression for K to determine the value of K.	$K = \frac{0.358^2}{0.0430 \times 0.0620} = 48.1$
Determine the units of K.	$\frac{M^2}{M \times M} = \frac{M^2}{M^2}$ The units cancel, so K has no unit.

The concentrations used to determine K must be the concentration of each component of the mixture at equilibrium.

Worked example: Try yourself 7.3.1

CALCULATING AN EQUILIBRIUM CONSTANT

A 3.00 L vessel contains a mixture of 0.120 mol of N_2O_4 and 0.500 mol of NO_2 in equilibrium at 460°C according to the equation:

$$N_2O_4(g) \rightleftharpoons 2NO_2(g)$$

Calculate the value of the equilibrium constant, K, for the reaction at that temperature.

Calculating equilibrium concentrations

Another type of calculation involves 'working backwards'. In this situation, you use the equilibrium constant to determine the molar equilibrium concentration of one of the species in the reaction.

Worked example 7.3.2

CALCULATING AN EQUILIBRIUM CONCENTRATION

Consider the following equilibrium with an equilibrium constant of 0.400 M at 250°C.

$$PCl_5(g) \rightleftharpoons PCl_3(g) + Cl_2(g)$$

An equilibrium mixture contains 0.0020 M PCl_5 and 0.0010 M PCl_3 at 250°C. Calculate the equilibrium concentration of Cl_2 in this mixture.

Thinking	Working
Write the expression for *K*.	$K = \frac{[PCl_3][Cl_2]}{[PCl_5]}$
Substitute the known values into the expression for *K*.	$0.400 = \frac{0.0010 \times [Cl_2]}{0.0020}$
Rearrange the expression to make the unknown the subject and calculate the concentration of this species.	$[Cl_2] = \frac{0.400 \times 0.0020}{0.0010}$ $= 0.80$ M (or 0.80 mol L^{-1})

Worked example: Try yourself 7.3.2

CALCULATING AN EQUILIBRIUM CONCENTRATION

Consider the following equilibrium with an equilibrium constant of 0.72 M at 250°C.

$$N_2O_4(g) \rightleftharpoons 2NO_2(g)$$

An equilibrium mixture contains 0.040 M N_2O_4 at 250°C. Calculate the equilibrium concentration of NO_2 in this mixture.

Calculating an equilibrium constant using stoichiometry

For some calculations, stoichiometry (the calculation of relative amounts of reactants and products in a chemical reaction) is used to calculate the molar equilibrium concentrations of the reactants and products from the data provided. Once these are known, the equilibrium constant can then be calculated.

A popular way to set out calculations of this type is with the use of a **reaction table** (also known as an **ICE table** because the rows are labelled Initial, Change and Equilibrium, as shown in Worked example 7.3.3). The reaction table shows the initial amounts of reactants and products, the changes that occur as the system reaches equilibrium and the final values at equilibrium.

Worked example 7.3.3

USING STOICHIOMETRY TO CALCULATE AN EQUILIBRIUM CONSTANT

An equilibrium is established between A and B at a specified temperature according to the following equation:

$$A(g) \rightleftharpoons 2B(g)$$

0.540 mol of A was placed in a 2.00 L vessel. When equilibrium was achieved, 0.280 mol of B was present. Calculate the value of the equilibrium constant at this temperature.

<table>
<tr><th>Thinking</th><th>Working</th></tr>
<tr><td>Construct a reaction table using each species in the balanced equation as the headings for the columns. Insert three rows in the table labelled 'I' (Initial), 'C' (Change) and 'E' (Equilibrium):
<table><tr><th></th><th colspan="2">Reactants ⇌ Products</th></tr><tr><td>I</td><td></td><td></td></tr><tr><td>C</td><td></td><td></td></tr><tr><td>E</td><td></td><td></td></tr></table>Enter the data provided in the table.
When a species is consumed, the change is negative; when a species is produced, the change is positive.</td><td>Initially, there is:
• 0.540 mol of the reactant A(g)
• 0.00 mol of the product B(g).
If x mol of A reacts, $2x$ mol of B is produced.
At equilibrium, there is 0.280 mol of B(g).
<table><tr><th></th><th colspan="2">$A(g) \rightleftharpoons 2B(g)$</th></tr><tr><td>I</td><td>0.540 mol</td><td>0.00 mol</td></tr><tr><td>C</td><td>$-x$</td><td>$+2x$</td></tr><tr><td>E</td><td>$0.540 - x$</td><td>$2x = 0.280$ mol</td></tr></table></td></tr>
<tr><td>Using the coefficients from the equation, calculate the moles of all species at equilibrium.</td><td>Initially no B was present, so because 0.280 mol of B has been produced at equilibrium:
$2x = 0.280$ mol
$x = 0.140$ mol
Enter these values in the table:
<table><tr><th></th><th colspan="2">$A(g) \rightleftharpoons 2B(g)$</th></tr><tr><td>I</td><td>0.540</td><td>0</td></tr><tr><td>C</td><td>$-x = -0.140$</td><td>$+2x = 0.280$</td></tr><tr><td>E</td><td>$0.540 - x$
$= 0.540 - 0.140$
$= 0.400$</td><td>0.280</td></tr></table></td></tr>
<tr><td>Using the volume of the vessel, calculate the equilibrium concentrations for all species at equilibrium.
Use the formula $c = \frac{n}{V}$.</td><td>The volume of the vessel is 2.00 L.
$[A] = \frac{n}{V}$
$= \frac{0.400}{2.00}$
$= 0.200$ M
$[B] = \frac{n}{V}$
$= \frac{0.280}{2.00}$
$= 0.140$ M</td></tr>
<tr><td>Write the expression for K and substitute the equilibrium concentrations.
Calculate the equilibrium constant, K, and include the units for the equilibrium constant.</td><td>$K = \frac{[B]^2}{[A]}$
$= \frac{0.140^2}{0.200}$
$= 0.0980$ M (or 0.0980 mol L^{-1})</td></tr>
</table>

An ICE table (reaction table) is used in equilibrium calculations where the equilibrium concentration of one or more species is unknown.

Worked example: Try yourself 7.3.3

USING STOICHIOMETRY TO CALCULATE AN EQUILIBRIUM CONSTANT

At one step during the synthesis of nitric acid, nitrogen dioxide (NO_2) is in equilibrium with dinitrogen tetroxide (N_2O_4) at 60°C:

$$N_2O_4(g) \rightleftharpoons 2NO_2(g)$$

0.350 mol of N_2O_4 was placed in a 2.0 L vessel. When equilibrium was achieved, 0.120 mol of NO_2 was present. Calculate the value of the equilibrium constant at this temperature.

7.3 Review

SUMMARY

- The equilibrium constant, *K*, is specific for an equation.
- The value of *K* provides a measure of the extent of reaction and the relative concentrations of reactants and products at equilibrium.

Value of *K*	Extent of reaction
between about 10^{-4} and 10^{4}	indicates significant reaction occurs
$>10^{4}$	indicates an almost complete reaction occurs
$<10^{-4}$	indicates negligible reaction occurs

- When an equation is reversed, the new equilibrium constant is the reciprocal, or inverse, of the original *K*.
- When coefficients are doubled, *K* is squared.
- When coefficients are halved, *K* is the square root.
- An equilibrium constant at a particular temperature can be calculated from the concentrations of the reactants and products at equilibrium and the expression for the equilibrium constant.
- The concentration of a reactant or product can be calculated if the concentrations of the other reactants and products and the equilibrium constant are known.
- Stoichiometry may be used to calculate equilibrium concentrations of reactants and products and hence the value of the equilibrium constant using a reaction (ICE) table.

KEY QUESTIONS

Knowledge and understanding

1 A chemist investigated four different reactions, 1, 2, 3 and 4, and determined the value of the equilibrium constant for each. Using your understanding of the equilibrium law and the extent of reaction, in which one of the reactions at equilibrium would there be more products than reactants?

A Reaction 1: $K = 0.057$

B Reaction 2: $K = 2.5 \times 10^9$

C Reaction 3: $K = 3.1 \times 10^{-4}$

D Reaction 4: $K = 0.0068$

2 The equilibrium constant for the following reaction at 25°C is 10^{-10}.

$$2Fe^{2+}(aq) + Sn^{4+}(aq) \rightleftharpoons 2Fe^{3+}(aq) + Sn^{2+}(aq)$$

a Explain whether a significant amount of products would occur when solutions of tin(IV) chloride and iron(II) chloride are mixed.

b Determine the value of the equilibrium constant for the reaction:

$$2Fe^{3+}(aq) + Sn^{2+}(aq) \rightleftharpoons 2Fe^{2+}(aq) + Sn^{4+}(aq)$$

c Explain whether a significant amount of products would occur when solutions of tin(II) chloride and iron(III) chloride are mixed.

3 Consider the following equilibrium at 227°C:

$$2BrCl(g) \rightleftharpoons Br_2(g) + Cl_2(g)$$

a Write the expression for K for this equilibrium system.

b The value of K at 227°C for the expression in part **a** is 32. Deduce the value of the equilibrium constant for each of the following equilibria, also at 227°C.

i $BrCl(g) \rightleftharpoons \frac{1}{2}Br_2(g) + \frac{1}{2}Cl_2(g)$

ii $Cl_2(g) + Br_2(g) \rightleftharpoons 2BrCl(g)$

iii $4BrCl(g) \rightleftharpoons 2Br_2(g) + 2Cl_2(g)$

iv $\frac{1}{2}Cl_2(g) + \frac{1}{2}Br_2(g) \rightleftharpoons BrCl(g)$

4 Phosgene is a poisonous gas that was used during World War I. It can be formed by the reaction of carbon monoxide with chlorine gas in the equilibrium reaction:

$$CO(g) + Cl_2(g) \rightleftharpoons COCl_2(g)$$

In an experiment, the reaction was allowed to proceed at 74°C until equilibrium was reached. The equilibrium concentration of each species was determined and recorded as follows: $[CO] = 2.4 \times 10^{-2}$ M, $[Cl_2] = 0.108$ M and $[COCl_2] = 0.28$ M.

a Write an expression for the equilibrium constant.

b Calculate the equilibrium constant for the reaction at this temperature.

5 Calculate the equilibrium constant for the reaction represented by the equation:

$$N_2O_4(g) \rightleftharpoons 2NO_2(g)$$

if an equilibrium mixture in a 2.0 L container was found to consist of 0.80 mol of N_2O_4 and 0.40 mol of NO_2.

Analysis

6 The following reaction was allowed to reach equilibrium at a temperature of 230°C:

$$2NO(g) + O_2(g) \rightleftharpoons 2NO_2(g)$$

The value of the equilibrium constant was determined to be 6.44×10^5 M^{-1}.

If the equilibrium concentration of $[NO_2] = 15.5$ M and $[NO] = 0.0542$ M, determine the concentration of O_2 in the equilibrium mixture.

7 3.45 mol of PCl_3 and 4.50 mol of Cl_2 were mixed in a 2.00 L vessel. They reacted according to the equation:

$$PCl_3(g) + Cl_2(g) \rightleftharpoons PCl_5(g)$$

When equilibrium was reached, it was found that 0.25 mol of PCl_5 had been formed. Calculate the value of the equilibrium constant and give the units.

8 4.89 mol N_2 and 7.23 mol of H_2 were mixed in a 5.00 L vessel. They reacted according to the equation:

$$N_2(g) + 3H_2(g) \rightleftharpoons 2NH_3(g)$$

When equilibrium was reached, it was found that 0.28 mol of NH_3 had been formed. Calculate the value of the equilibrium constant and give the units.

7.4 Le Châtelier's principle

In this section, you will learn about some of the effects of changes on chemical systems at equilibrium.

Your understanding of the underlying principles of chemical equilibrium will enable you to predict the impact of changes when a reactant or product is added or removed from an equilibrium system, as seen in the cobalt system described in Figure 7.4.1. You will also learn to predict the impact of changes in gas pressure, solution concentration and temperature, and the effect of adding a catalyst on an equilibrium system.

FIGURE 7.4.1 The effect of changes on the equilibrium $Co(H_2O)_6^{2+}(aq) + 4Cl^-(aq) \rightleftharpoons CoCl_4^{2-}(aq) + 6H_2O(l)$. The addition of excess Cl^- ions causes a net forward reaction, and the solution turns blue as more $CoCl_4^{2-}$ is formed. The addition of excess water causes a net reverse reaction, and the solution returns to the original pink colour as more $[Co(H_2O)_6]^{2+}$ is formed.

The effects of changes to a chemical equilibrium are very important to the chemical industry. Conditions must be carefully selected to ensure that an optimum **yield** of products is obtained within a reasonable timeframe.

The effect of a change in conditions on an equilibrium system is summarised in a useful generalisation called **Le Châtelier's principle**. Le Châtelier's principle states that if an equilibrium system is subjected to a change so that it is no longer at equilibrium, the system will adjust itself to partially oppose the effect of the change. A net reaction will occur that partially counteracts the effect of the change, and the system will establish a new equilibrium.

As a result, the equilibrium position will change—there may be an increase in the amount of either products or reactants, depending on the nature of the change. By understanding Le Châtelier's principle, you can predict the effect of a change to an equilibrium system.

CHEMFILE

Henri Le Châtelier

Henri Le Châtelier (1850–1936) was a French chemist and engineer. He is best known for developing the principle of chemical equilibrium, which is now named after him.

Le Châtelier made an early attempt at synthesising ammonia from nitrogen and hydrogen, but an error in the design of the experiment resulted in an explosion that nearly killed one of his laboratory assistants. Later, Le Châtelier regretted not pursuing the synthesis of ammonia further, as it would have possibly prevented Fritz Haber from continuing the work. Haber's work in this area enhanced the ability of Germany to continue its participation in WW1.

Le Châtelier also developed the oxyacetylene welding torch and the thermocouple for accurate temperature measurement.

Henri Le Châtelier

CHANGES TO AN EQUILIBRIUM SYSTEM

You have seen that different reactions proceed to different extents. Because of this, the relative amounts of reactants and products differ from one reaction to another at equilibrium. The relative amounts of reactants and products at equilibrium are called the **position of equilibrium**.

The relative amounts of substances present in equilibrium mixtures depend upon reaction conditions. For any equilibrium system, the position of equilibrium may be changed by:

- adding or removing a reactant or product
- changing the pressure by changing the volume (for equilibria involving gases)
- dilution (for equilibria in solution)
- changing the temperature.

Careful control of the reaction conditions allows chemists to maximise the equilibrium yield of a desired product by moving the position of equilibrium 'to the right' (and therefore increasing the amount of products formed).

ℹ Le Châtelier's principle states that if an equilibrium system is subjected to a change, the system will adjust itself to partially oppose the effect of the change.

ℹ The position of equilibrium should not be confused with *K*. The value of *K* is only ever changed by a change in temperature.

ADDING EXTRA REACTANT OR PRODUCT

A sealed reaction vessel of hydrogen and nitrogen gases at a particular temperature will establish an equilibrium according to the equation:

$$N_2(g) + 3H_2(g) \rightleftharpoons 2NH_3(g)$$

If extra nitrogen gas were added to the container without changing the volume or temperature, the mixture would momentarily not be in equilibrium.

The events outlined in the rate–time graph in Figure 7.4.2 occur as the composition of the mixture adjusts to return to a new equilibrium. This graph uses the collision theory to explain how the rates of the forward and reverse reactions change as the system returns to equilibrium.

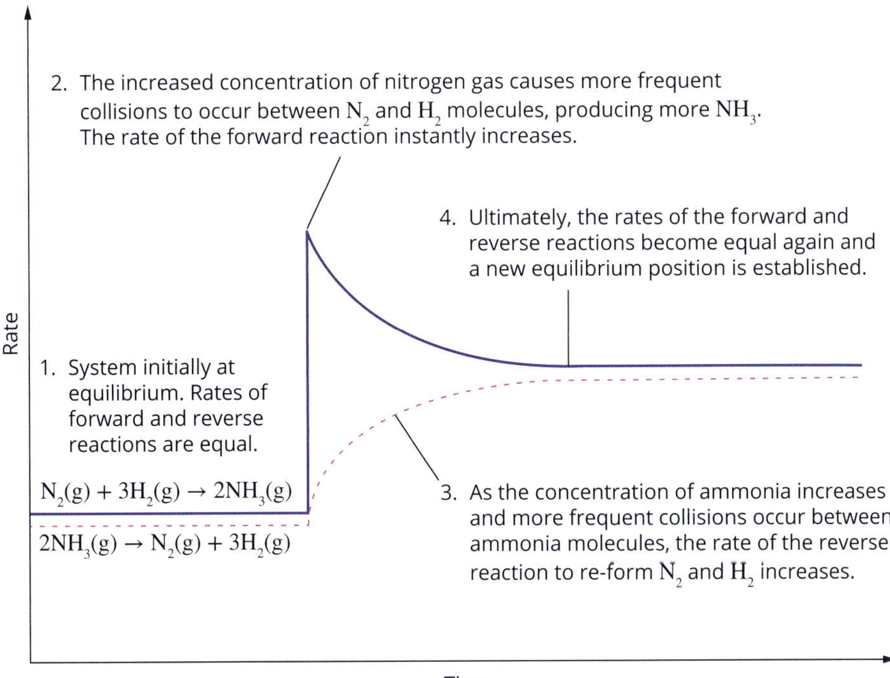

FIGURE 7.4.2 This rate–time graph shows the events that occur as a mixture of nitrogen and hydrogen gas returns to equilibrium after the addition of extra nitrogen gas.

Once the system has re-established equilibrium, the rates of the forward and reverse reactions will again be equal. Overall, though, a net forward reaction has occurred, with an increase in the concentration of ammonia at equilibrium. The equilibrium position is said to have shifted 'to the right'.

It is important to note that, even though the concentration of N_2 gas decreases as the system moves to establish the new equilibrium, its final concentration is still higher than in the original equilibrium. Le Châtelier's principle states that the change is partially opposed. The system does not return to the initial equilibrium position after the change in conditions.

The changes occurring to the system can also be shown on a **concentration–time graph**. Figure 7.4.3, on the following page, illustrates the effect on the system when N_2 gas is added as described.

The value of K for the equilibrium reaction remains unchanged as the temperature has not changed.

If you follow the same reasoning, you can see that adding extra amounts of the other reactant, H_2, to the system will also increase the concentration of ammonia produced. However, the addition of more product, NH_3, would result in a net reverse reaction and the equilibrium position shifting to the left, reducing the overall concentration of ammonia, as seen in Figure 7.4.4 on the following page.

CHEMFILE

Soft drink and equilibrium

In a sealed bottle of soft drink, $CO_2(g)$ and $CO_2(aq)$ are in equilibrium according to the equation:

$$CO_2(g) \rightleftharpoons CO_2(aq)$$

When the cap is removed from the bottle, pressure is reduced and carbon dioxide escapes to the atmosphere. According to Le Châtelier's principle, a net reverse reaction occurs and carbon dioxide comes out of solution.

This normally happens slowly, but when mints are added to the bottle of soft drink, the drink erupts violently, as you can see in the figure below. The high surface area of the mints provides many sites for bubbles to form, increasing the rate of the reaction and quickly producing an eruption of soft drink from the bottle.

Erupting soft drink when mints have been added

Remember, dynamic equilibrium is a state where the rate of formation of products is equal to the rate of formation of reactants.

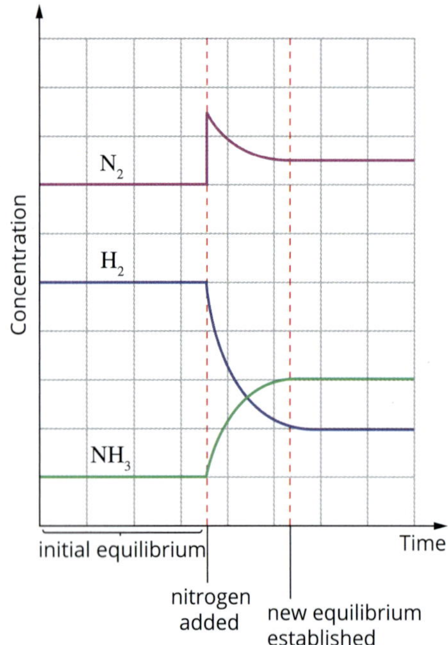

FIGURE 7.4.3 A representation of changes in concentrations that occur when additional nitrogen gas is added to the equilibrium $N_2(g) + 3H_2(g) \rightleftharpoons 2NH_3(g)$. Note that the *y*-axis shows the concentration, not rate of reaction.

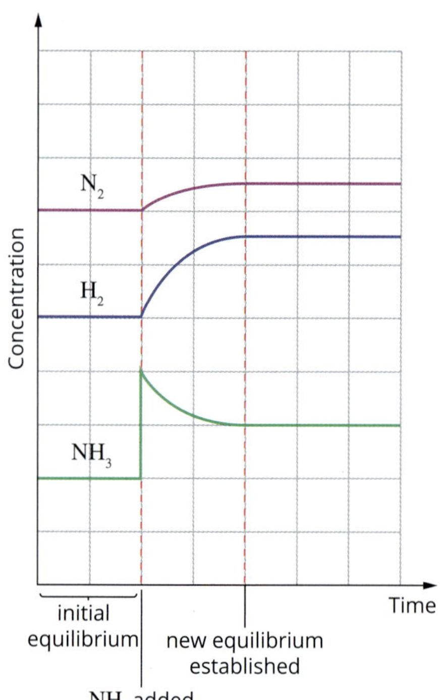

FIGURE 7.4.4 A representation of changes in concentrations that occur when additional ammonia gas is added to the equilibrium: $N_2(g) + 3H_2(g) \rightleftharpoons 2NH_3(g)$. The reaction shifts to the left, forming more N_2 and H_2.

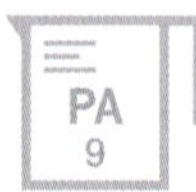

You can apply your knowledge of collision theory and reaction rates to determine the overall effect of changes on an equilibrium, as you saw above. However, Le Châtelier's principle provides a simpler way of predicting these effects. Table 7.4.1 shows how an equilibrium system acts to oppose the addition or removal of reactants and products.

TABLE 7.4.1 The general effects of a change to a system at equilibrium as predicted by Le Châtelier's principle

Change to equilibrium	Effect
adding a reactant	formation of more products—a net forward reaction: equilibrium position shifts to the right
adding a product	formation of more reactants—a net reverse reaction: equilibrium position shifts to the left
removing a product	formation of more products—a net forward reaction: equilibrium position shifts to the right

Predicting the effect of a change using the equilibrium law

The effect of adding more reactants or products can also be predicted using the equilibrium law.

If you again consider the equilibrium formed between nitrogen and hydrogen as an example:

$$N_2(g) + 3H_2(g) \rightleftharpoons 2NH_3(g)$$

The expression for the equilibrium law for this reaction can be written as:

$$K = \frac{[NH_3]^2}{[N_2][H_2]^3}$$

If extra nitrogen is added, the concentration of N_2 is increased, so the reaction quotient (Q) is momentarily less than the equilibrium constant K. The mixture is no longer at equilibrium. As you saw in Section 7.2, when $Q < K$, the reaction favours the formation of products (a net forward reaction). This increases the amount of products and decreases the amount of reactants, until the reaction quotient again becomes equal to K.

Making predictions using the mathematical expression for the equilibrium law gives the same result as using the qualitative reasoning of Le Châtelier's principle.

CASE STUDY

Ocean acidification

An example of the effects of the addition of a reactant to an equilibrium can be seen in the reactions involving carbon dioxide that take place in the ocean. (You may remember studying these reactions in Unit 2.)

Some of the carbon dioxide gas in the atmosphere dissolves in seawater to form $CO_2(aq)$:

$$CO_2(g) \rightleftharpoons CO_2(aq)$$

Although most of the carbon dissolved in seawater is present as $CO_2(aq)$, some reacts further to form carbonic acid:

$$CO_2(aq) + H_2O(l) \rightleftharpoons H_2CO_3(aq)$$

Carbonic acid is a weak diprotic acid, ionising in two steps to form hydrogen carbonate ions (HCO_3^-) and carbonate ions (CO_3^{2-}):

$$H_2CO_3(aq) + H_2O(l) \rightleftharpoons H_3O^+(aq) + HCO_3^-(aq)$$

$$HCO_3^-(aq) + H_2O(l) \rightleftharpoons H_3O^+(aq) + CO_3^{2-}(aq)$$

Using Le Châtelier's principle, scientists predict that increasing levels of carbon dioxide in the atmosphere will cause an increased concentration of carbon dioxide in the oceans. As more atmospheric carbon dioxide dissolves, there will be an increase in the concentration of H_3O^+ ions and therefore a decrease in the pH, resulting in an increase in the acidity of the ocean. This process is called **ocean acidification**. Over the past 300 years, the pH of the ocean has averaged 8.2. It is currently 8.14, and is predicted to fall to about 7.85 by 2100.

The carbonate ions generated in the last reaction above are needed for the growth of seashells, coral reefs and other marine organisms. These organisms absorb calcium ions and carbonate ions from seawater to build and maintain the calcium carbonate that forms their shells and exoskeletons in the following reaction:

$$Ca^{2+}(aq) + CO_3^{2-}(aq) \rightleftharpoons CaCO_3(s)$$

One consequence of future increases in ocean acidity is that higher concentrations of H_3O^+ ions can react with the CO_3^{2-} ions to produce more HCO_3^-, causing the equilibrium above to shift to the left and dissolve some of the solid calcium carbonate. This will put marine ecosystems and organisms at risk, including the cuttlefish in Figure 7.4.5, which has a shell made of calcium carbonate.

FIGURE 7.4.5 Cuttlefish have an internal shell made of calcium carbonate. Cuttlefish and other marine animals could be affected by the decreasing pH of oceans.

CHANGING PRESSURE BY CHANGING VOLUME

The pressure of a gas is inversely proportional to the volume of its container. This means the pressure of gases in an equilibrium mixture can be changed by increasing or decreasing the volume of the container while keeping the temperature constant.

Consider the effect of increasing the pressure on the equilibrium between sulfur dioxide gas, oxygen and sulfur trioxide gas for the following reaction:

$$2SO_2(g) + O_2(g) \rightleftharpoons 2SO_3(g)$$

3 gas particles 2 gas particles

You can see that the forward reaction involves a reduction in the number of particles of gas from 3 to 2. The formation of products would cause an overall reduction in pressure of the system. The reverse reaction involves an increase in the number of gas particles from 2 to 3. So a net reverse reaction causes an overall increase in pressure of the system.

You can predict the change in the position of this equilibrium by either applying Le Châtelier's principle or analysing the equilibrium law for the reaction.

Pressure is force per unit area and is proportional to the frequency of collisions. A change in the number of particles will change the frequency of collisions and change the pressure.

CHEMFILE

Chickens lay eggs with thinner shells in the summer

Chickens, like dogs, do not perspire. Therefore, in hot weather they must resort to panting to try to maintain a healthy temperature. This means that they exhale more carbon dioxide gas than when they are breathing normally.

This affects the following series of equilibria, which produces eggshells made from calcium carbonate:

$$CO_2(g) \rightleftharpoons CO_2(aq)$$
$$CO_2(aq) + H_2O(l) \rightleftharpoons H_2CO_3(aq)$$
$$H_2CO_3(aq) \rightleftharpoons H^+(aq) + HCO_3^-(aq)$$
$$HCO_3^-(aq) \rightleftharpoons H^+(aq) + CO_3^{2-}(aq)$$
$$CO_3^{2-}(aq) + Ca^{2+}(aq) \rightleftharpoons CaCO_3(s) \text{ (eggshell)}$$

Removing CO_2 gas shifts each equilibrium, in turn, to the left. This ultimately results in less $CaCO_3$ being made. So, in summer, chickens lay eggs with thinner shells, which are more easily broken at great cost to the famers and supermarkets.

Scientists solved the problem by giving the chickens carbonated water to drink.

Apparently the chickens like the carbonated water and they produce eggs with thicker, stronger shells. Another victory for chemistry and Le Châtelier!

The eggshell dilemma is solved with the aid of Le Châtelier's principle.

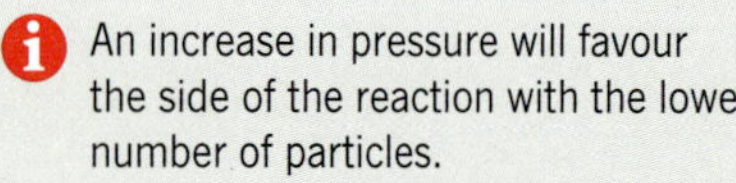

An increase in pressure will favour the side of the reaction with the lower number of particles.

Applying Le Châtelier's principle

Le Châtelier's principle tells you that an equilibrium system will respond to an increase in pressure by adjusting to reduce the pressure. The position of equilibrium will therefore move in the direction of the fewer gas particles.

In the example:

$$2SO_2(g) + O_2(g) \rightleftharpoons 2SO_3(g)$$

an increase in pressure will cause a net forward reaction to occur to reduce the overall pressure (three gaseous reactant particles become two gaseous product particles). The amount of SO_3 present at equilibrium will increase, as represented in Figure 7.4.6.

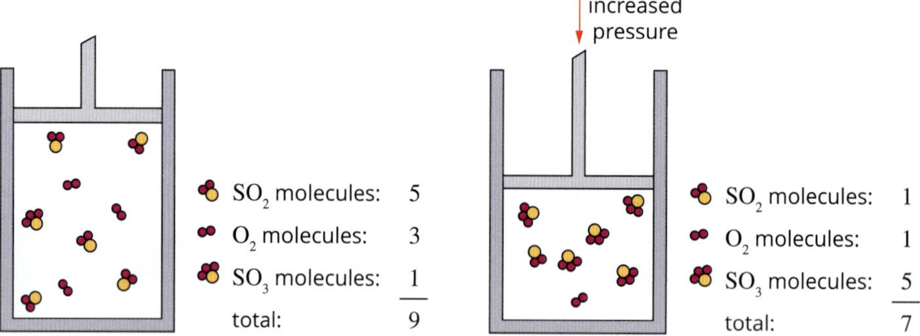

FIGURE 7.4.6 A representation of the effect of increased pressure on the equilibrium: $2SO_2(g) + O_2(g) \rightleftharpoons 2SO_3(g)$

The effect of the change can also be illustrated graphically (Figure 7.4.7). When the system is initially at equilibrium and there is an increase in pressure, the **partial pressures** of all gases increase simultaneously, as do the concentrations.

As the system adjusts, there is a gradual change in concentration of each of the species until the new equilibrium is established.

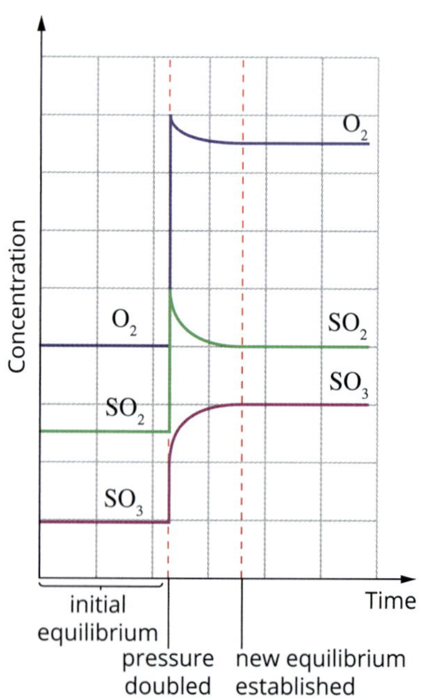

FIGURE 7.4.7 The effect of doubling the pressure on the equilibrium $2SO_2(g) + O_2(g) \rightleftharpoons 2SO_3(g)$.

At the new equilibrium position, the individual partial pressures and concentrations are different from the first equilibrium. However, the equilibrium constant, K, has not changed. The ratio of products to reactants according to the equilibrium law still equals K at the new equilibrium position.

Applying the equilibrium law

You can also predict and explain the effect of a change of pressure on the equilibrium system above in terms of the equilibrium law.

For the equilibrium system:

$$2SO_2(g) + O_2(g) \rightleftharpoons 2SO_3(g)$$

the expression for the equilibrium constant is:

$$K = \frac{[SO_3]^2}{[SO_2]^2[O_2]}$$

Suppose the volume of the closed system is halved. The partial pressures of all reactant and product gases double, as do their concentrations, as seen in Figure 7.4.7.

Momentarily, the reaction quotient, Q, becomes:

$$\frac{(2[SO_3])^2}{(2[SO_2])^2(2[O_2])} = \frac{2^2}{2^2 \times 2} \times \frac{[SO_3]^2}{[SO_2]^2[O_2]} = \frac{1}{2}K$$

The reaction quotient, Q, is now lower than K. Accordingly, there will be a net forward reaction to increase the value of the reaction quotient until it becomes equal to K, just as predicted earlier using Le Châtelier's principle.

Pressure changes do not affect the equilibrium position of systems in the liquid or solid phases. Particles in these systems are too tightly packed for an increase in pressure to have a noticeable effect on volume. This means that there is negligible change in the concentrations of the species involved and no effect on the concentration fraction.

A special case

The effect of a change of pressure or concentration, by changing the container volume, depends on the relative number of particles on both sides of the equation.

When there are equal numbers of reactant and product particles, a change in pressure will not shift the position of equilibrium. This is the case for the reaction between hydrogen and iodine in the following equilibrium:

$$H_2(g) + I_2(g) \rightleftharpoons 2HI(g)$$

2 gas particles   2 gas particles

It does not matter which way the system shifts; the number of particles in the container will remain constant. The system does not oppose the volume change applied.

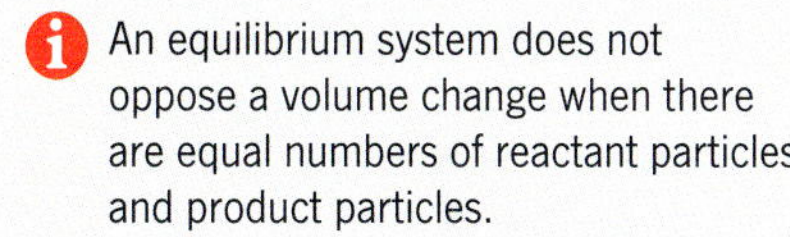
An equilibrium system does not oppose a volume change when there are equal numbers of reactant particles and product particles.

Worked example 7.4.1

USING LE CHÂTELIER'S PRINCIPLE TO DETERMINE THE SHIFT IN EQUILIBRIUM POSITION FOR A VOLUME DECREASE

Consider the equilibrium:

$$CH_4(g) + H_2O(g) \rightleftharpoons CO(g) + 3H_2(g)$$

Predict the shift in equilibrium position and the effect on the amount of CO when the volume is halved at constant temperature.

Thinking	Working
Determine the immediate effect of the change of volume on the pressure.	Halving the volume will double the pressure and therefore the concentration of all species at equilibrium.
The system will try to partially oppose the change in pressure by reducing or increasing the pressure of the system. For a volume decrease, the system will shift in the direction of the fewest particles; for a volume increase, the system will shift in the direction of the most particles. Decide how the equilibrium will respond.	There are 2 molecules of gas on the reactant side and 4 molecules of gas on the product side, so the system will shift to the left (the reactant side). This decreases the amounts of the products, including CO. (Note that the CO concentration will still be higher than it was at the initial equilibrium. The shift in equilibrium position only partially compensates for the change.)

Worked example: Try yourself 7.4.1

USING LE CHÂTELIER'S PRINCIPLE TO DETERMINE THE SHIFT IN EQUILIBRIUM POSITION FOR A VOLUME INCREASE

Consider the equilibrium:

$$PCl_3(g) + Cl_2(g) \rightleftharpoons PCl_5(g)$$

Predict the shift in equilibrium position and the effect on the amount of Cl_2 when the volume is doubled at constant temperature.

CHANGING PRESSURE BY ADDING AN INERT GAS

The total pressure of an equilibrium mixture of gases may also be changed, without changing the volume of the container, by adding a non-reacting gas such as helium, neon or argon (Figure 7.4.8).

FIGURE 7.4.8 The equilibrium of a gaseous system is unaffected by the addition of an inert gas. The total pressure of the system increases without a change in the concentrations of the reactants or products, so there is no change to the value of the reaction quotient according to the equilibrium law.

The inert gas does not appear in the reaction equation or the expression for the equilibrium constant. The presence of the additional gas does not change any of the concentrations of the reactants and products, so there is no effect on the position of equilibrium or the equilibrium constant.

DILUTION

For equilibria in solution, the situation is similar to the one you saw with pressure and gases. The focus is on the number of particles per volume of solvent.

For an equilibrium occurring in solution, dilution by adding water reduces the number of particles per volume. This results in a shift in the position of equilibrium towards the side that produces the greater number of dissolved particles.

For example, consider the equilibrium system:

$$Fe^{3+}(aq) + SCN^{-}(aq) \rightleftharpoons FeSCN^{2+}(aq)$$

2 particles in solution   1 particle in solution

If, for example, water was added so that the total volume of the solution was doubled, then the concentration of each species would be momentarily halved. In terms of Le Châtelier's principle, a net reverse reaction will occur, increasing the total concentration of particles in solution. Figure 7.4.9 shows the changes of concentrations that occur. Note that there is an instantaneous decrease in the concentration of all species at the time of dilution.

A second example, which clearly shows the effect of dilution by a change in colour, is the equilibrium below:

$$Co^{2+}(aq) + 4Cl^{-}(aq) \rightleftharpoons [CoCl_4]^{2-}(aq)$$

pink   blue

When water is added to this equilibrium system so the volume is doubled, the chloride ion concentration, as well as all other ion concentrations, is halved. This causes the equilibrium position to move in the direction of the most particles (toward the reactant side) to re-establish equilibrium, according to Le Châtelier's principle. The solution changes colour to become more pink.

Dilution of an aqueous equilibrium system has no effect on the value of K for the reaction.

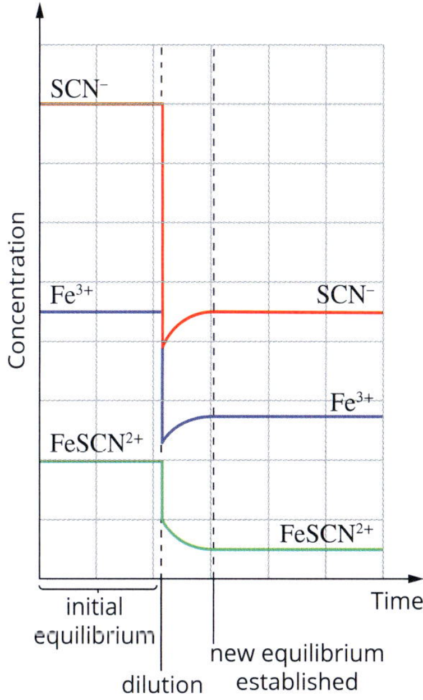

FIGURE 7.4.9 The effect of dilution on the equilibrium: $Fe^{3+}(aq) + SCN^{-}(aq) \rightleftharpoons FeSCN^{2+}(aq)$. Although the equilibrium position shifts to the left, note that the concentrations of Fe^{3+} and SCN^{-} at the new equilibrium are lower than their concentrations before dilution, as the equilibrium shift only partially opposes the change.

CHANGING TEMPERATURE AND THE EFFECT ON AN EQUILIBRIUM CONSTANT

It has been shown experimentally that the value of the equilibrium constant, K, for a particular reaction, depends only upon temperature. It is not affected by the addition of reactants or products, changes in pressure, or the use of catalysts.

The effect of a change in temperature on an equilibrium constant depends on whether the reaction is exothermic or endothermic. As temperature increases:

- for exothermic reactions, the value of K decreases and so the amount of products present at equilibrium decreases
- for endothermic reactions, the value of K increases and so the amount of products present at equilibrium increases.

Table 7.4.2 summarises the effect on K when temperature increases. The opposite is true when temperature decreases.

TABLE 7.4.2 The effect on the value of K when the temperature of the system increases

ΔH	T	K
exothermic (–)	increases	decreases
endothermic (+)	increases	increases

Because the value of K depends on temperature, it is essential to specify the temperature at which an equilibrium constant has been measured. Changing the temperature of an equilibrium in a closed system affects both the equilibrium position and the value of the equilibrium constant. The overall effect on equilibrium position because of temperature can be predicted using Le Châtelier's principle.

An example of this is the conversion of brown nitrogen dioxide gas (NO_2) to colourless dinitrogen tetroxide gas (N_2O_4). The reaction is exothermic, releasing energy to the environment. You could (but wouldn't usually) write an equation for the reaction that includes the energy released:

$$2NO_2(g) \rightleftharpoons N_2O_4(g) + \text{energy}$$

Only a change in temperature will change the value of K for a given reaction.

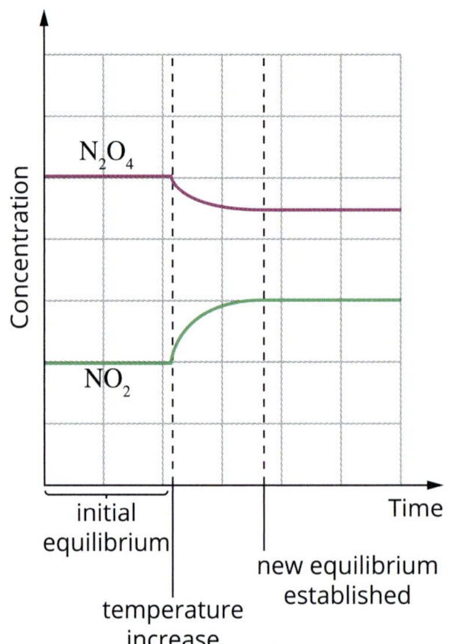

FIGURE 7.4.10 The effect of heating on the equilibrium $2NO_2(g) \rightleftharpoons N_2O_4(g)$

FIGURE 7.4.11 Equilibrium mixtures of NO_2 and N_2O_4 in hot water and ice. Heating the mixture favours the formation of brown NO_2 gas.

Increasing the temperature of the system increases the energy of the substances in the mixture. Applying Le Châtelier's principle, you can see that the reaction can 'oppose' an increase in energy by absorbing energy. As the reverse reaction is endothermic, this favours a net reverse reaction. This can be seen in Figure 7.4.10, where there is a gradual decrease in the concentration of N_2O_4 as the system moves to produce more reactants, NO_2. Note when drawing concentration–time graphs, with a change in temperature, there is no instantaneous change in concentration; the change is gradual.

Because the reactants and products of this system are different colours, you can monitor the change in this equilibrium visually. When a new equilibrium is attained, there is less dinitrogen tetroxide and more nitrogen dioxide present, so the mixture appears darker (Figure 7.4.11).

Heating an endothermic reaction causes the opposite result to occur. Applying Le Châtelier's principle, you can see that the reaction opposes an increase in energy by absorbing energy, resulting in a net forward reaction. The effect of a temperature increase on an exothermic and endothermic reaction is summarised in Table 7.4.3.

TABLE 7.4.3 Relationship between K and ΔH as temperature increases

ΔH	Effect on K	Net reaction direction
exothermic (–)	decreases	reverse reaction – less product
endothermic (+)	increases	forward reaction – more product

EFFECT OF A CATALYST ON EQUILIBRIUM

As you have seen in Chapter 6, a catalyst lowers the activation energy of both the forward and reverse reactions by providing an alternative pathway for the reaction to occur. The effect on the activation energy can be seen in the energy profile diagram shown in Figure 7.4.12.

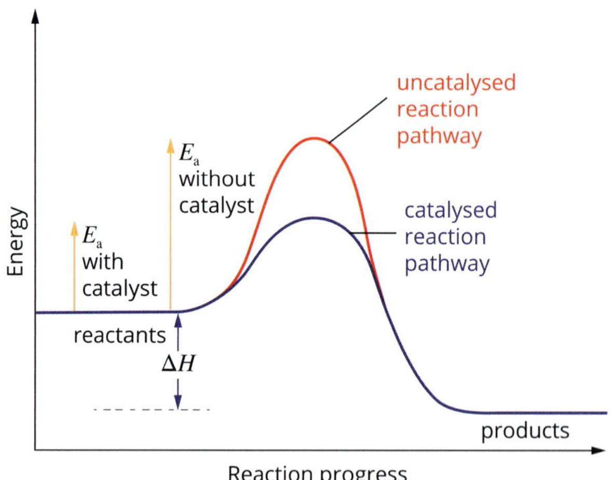

FIGURE 7.4.12 Energy changes in a catalysed and uncatalysed reaction, where E_a is the energy needed to break bonds—the activation energy.

The lower activation energy causes an increase in the number of effective collisions. As a result, there is an increase in the rate of both forward and reverse reactions. This occurs because more particles have energies greater than the activation energy barrier of the reaction.

A catalyst increases the rate of the forward and reverse reactions equally. Therefore, it will not change the relative concentrations of the reactants and products in the equilibrium law expression. As a consequence, the presence of a catalyst does not change the position of equilibrium or the value of the equilibrium constant, K. A catalyst will increase the rate at which an equilibrium is attained. It is for this reason that catalysts are used in many industrial and biological systems.

Addition of a catalyst does not change the position of equilibrium of a system or the value of the equilibrium constant, just how quickly equilibrium is attained.

CASE STUDY ANALYSIS

Equilibria in a swimming pool

The water in swimming pools is recirculated constantly. Although it is filtered, the water can become contaminated with microscopic algae and bacteria. Some interesting chemistry involving chemical equilibria is involved in keeping swimming pools safe to swim in (Figure 7.4.13).

Swimming pool water is 'chlorinated' to prevent the growth of harmful microorganisms. Chlorination produces hypochlorous acid (HOCl), a very efficient antibacterial agent and algicide, which is a chemical that kills algae.

Commercially available 'pool chlorine' powder consists of calcium hypochlorite ($Ca(OCl)_2$), which dissolves in water to release hypochlorite ions (OCl^-). The hypochlorite ions then react with hydronium ions (H_3O^+) in the water to form hypochlorous acid:

$$Ca(OCl)_2(s) \rightleftharpoons Ca^{2+}(aq) + 2OCl^-(aq)$$

$$OCl^-(aq) + H_3O^+(aq) \rightleftharpoons HOCl(aq) + H_2O(l)$$

The relative amounts of HOCl, OCl^- and H_3O^+ in a swimming pool are controlled by monitoring the pH of the swimming pool. More pool chlorine or more hydrochloric acid is added, as needed, to maintain a pH in the range 7.2–7.8. Figure 7.4.14 shows the relationship between the proportions of HOCl, OCl^- and H_3O^+ in water.

As pH increases, the concentration of $H_3O^+(aq)$ decreases. According to Le Châtelier's principle, the position of equilibrium in the second equation above will move to the left, consuming some of the HOCl. If the pH rises above 7.8, the concentration of the HOCl will be insufficient to control the growth of bacteria and algae.

As pH falls, the concentration of $H_3O^+(aq)$ increases. Le Châtelier's principle indicates the position of equilibrium will move to the right and more HOCl will be formed. The greater concentration of HOCl in the pool results in pH values below 7.2. When the pool is too acidic, the water can irritate the eyes and skin.

Thus, maintaining a pool so that it is hygienic and comfortable for swimmers involves carefully maintaining an optimum equilibrium position.

FIGURE 7.4.13 Chemical equilibria are responsible for keeping the water in swimming pools hygienic and safe for swimmers.

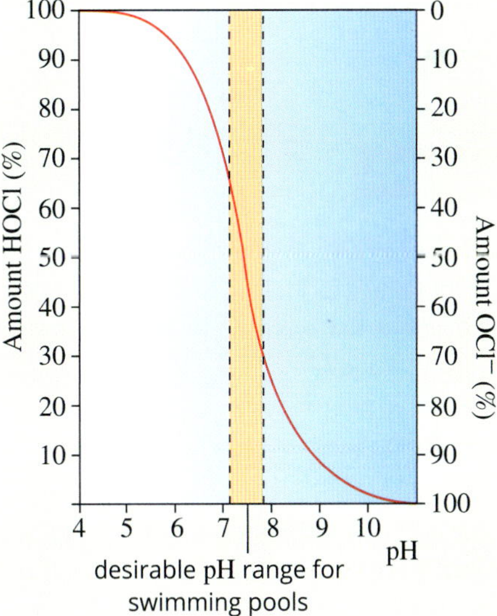

FIGURE 7.4.14 The effect of pH on the proportion of HOCl and OCl^- in water means that the position of the equilibrium can be monitored by changes in pH of the water.

Analysis

1 The H_3O^+ ions in the second equilibrium reaction are available from the self-ionisation of water. Write the equilibrium equation showing the formation of H_3O^+ ions from the self-ionisation of water.

2 An appropriate concentration of HOCl is required to control the growth of bacteria and algae. If the pH rises above 7.8, the concentration of HOCl will be inadequate. Use Le Châtelier's principle to explain how to correct this issue.

3 Care must be taken when adding hydrochloric acid to the pool because if the pH drops below 7.2, and the pool is too acidic, the water can irritate eyes and skin. Use your understanding of equilibrium to explain how to correct a pH below 7.2.

7.4 Review

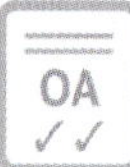

SUMMARY

- Le Châtelier's principle states that if an equilibrium system is subjected to change, the system will adjust itself to partially oppose the change.
- The effect of a change on an equilibrium can be predicted from Le Châtelier's principle. The effects of changes can also be explained by the use of collision theory and the equilibrium law.
- The effects of changes to equilibrium systems can be summarised as follows.

Change to system in equilibrium	Effect of change on equilibrium position
adding extra reactant	shifts to the right (net forward reaction)
adding extra product	shifts to the left (net reverse reaction)
decreasing pressure by increasing volume (for gases)	shifts in the direction of the most particles
adding a catalyst	no change
adding an inert gas (container volume remains constant)	no change
adding water (dilution of solutions)	shifts in the direction of the most particles
increasing the temperature for exothermic reactions	shifts to the left
increasing the temperature for endothermic reactions	shifts to the right

KEY QUESTIONS

Knowledge and understanding

1 Use Le Châtelier's principle to predict the effect of adding more of the reactant or product indicated to the following equilibria.

a Add more I_2 gas: $H_2(g) + I_2(g) \rightleftharpoons 2HI(g)$

b Add more N_2 gas: $2NH_3(g) \rightleftharpoons N_2(g) + 3H_2(g)$

c Add more CO_2 gas: $H_2(g) + CO_2(g) \rightleftharpoons H_2O(g) + CO(g)$

2 Predict the effect of the following changes on the position of each equilibrium.

a Removal of CH_3COO^- from the equilibrium:
$CH_3COOH(aq) + H_2O(l) \rightleftharpoons H_3O^+(aq) + CH_3COO^-(aq)$

b Doubling the pressure of the equilibrium system:
$$N_2(g) + 3H_2(g) \rightleftharpoons 2NH_3(g)$$

c Increasing the pressure of the equilibrium:
$$H_2(g) + I_2(g) \rightleftharpoons 2HI(g)$$

d Increasing the temperature of the endothermic equilibrium:
$$N_2(g) + O_2(g) \rightleftharpoons 2NO(g)$$

3 State whether the equilibrium constants for each of the following would be increased, decreased or unchanged by an increase in temperature:

a $2NH_3(g) \rightleftharpoons N_2(g) + 3H_2(g)$ $\Delta H = +91$ kJ mol^{-1}

b $4HCl(g) + O_2(g) \rightleftharpoons 2H_2O(g) + 2Cl_2(g)$ $\Delta H = -113$ kJ mol^{-1}

c $H_2(g) + CO_2(g) \rightleftharpoons H_2O(g) + CO(g)$ $\Delta H = +42$ kJ mol^{-1}

d $2CO(g) + O_2(g) \rightleftharpoons 2CO_2(g)$ $\Delta H = -564$ kJ mol^{-1}

Analysis

4 Consider the following equilibria.

i $H_2(g) + CO_2(g) \rightleftharpoons H_2O(g) + CO(g)$ $\Delta H = +42$ kJ mol^{-1}

ii $N_2O_4(g) \rightleftharpoons 2NO_2(g)$ $\Delta H = +58$ kJ mol^{-1}

iii $H_2(g) + F_2(g) \rightleftharpoons 2HF(g)$ $\Delta H = -536$ kJ mol^{-1}

a How could you alter the temperature of each equilibrium mixture to produce a net forward reaction?

b Explain whether it is possible to change the volume of each equilibrium mixture to produce a net forward reaction.

Explain your answer for each equilibrium.

5 An equilibrium mixture consists of the gases N_2O_4 and NO_2:
$$N_2O_4(g) \rightleftharpoons 2NO_2(g)$$
The volume of the container is doubled at constant temperature and a new equilibrium is established. Predict how each of the following quantities would change at the new equilibrium compared with the initial equilibrium.

a concentration of NO_2

b mass of NO_2

6 The International Space Station uses waste hydrogen and the carbon dioxide released by astronauts during respiration to form water according to the reaction:
$CO_2(g) + 4H_2(g) \rightleftharpoons CH_4(g) + 2H_2O(g)$ $\Delta H = -165$ kJ mol^{-1}

a How would you alter the temperature to produce a net forward reaction?

b How would you alter the volume to produce a net forward reaction?

Chapter review

07

KEY TERMS

activation energy
closed system
concentration fraction
concentration–time graph
dynamic equilibrium
endothermic
equilibrium
equilibrium constant
equilibrium law
equilibrium yield
exothermic
extent of reaction
heterogeneous chemical systems
homogeneous chemical systems
ICE table
irreversible reaction
Le Châtelier's principle
ocean acidification
open system
partial pressure
position of equilibrium
reaction quotient
reaction table
reversible reaction
surroundings
system
yield

REVIEW QUESTIONS

Knowledge and understanding

1 Consider the following equilibrium.

$$H_2O(l) \rightleftharpoons H_2O(g)$$

a Explain what is meant by the 'dynamic nature' of equilibrium and why wet clothes in a closed laundry bag do not dry.

b When the bag in part **a** is opened, the clothes begin to dry. Is this because of an equilibrium process? Explain your answer.

2 Explain the difference between the terms 'reaction quotient' (Q) and 'equilibrium constant' (K).

3 Write balanced equations for the reactions with the following expressions for the equilibrium constant.

a $K = \frac{[H_2]^2[CO]}{[CH_3OH]}$

b $K = \frac{[H_2S]^2}{[S_2][H_2]^2}$

c $K = \frac{[N_2O_4]^{\frac{1}{2}}}{[NO_2]}$

4 Complete the following statements about equilibria.

a If $K = 0.0001$ for a particular reaction, at equilibrium the concentrations of products will be _________ the concentrations of reactants.

b For the reaction with the equation:

$$2H_2(g) + 2NO(g) \rightleftharpoons 2H_2O(g) + N_2(g)$$

the expression for the equilibrium constant, K, is written as ________________.

c When the reaction quotient is smaller than K, the reaction moves to the ________ to establish equilibrium.

5 At a specified temperature, the reaction between solutions of Sn^{2+} and Fe^{3+} reaches equilibrium according to the equation:

$$2Fe^{3+}(aq) + Sn^{2+}(aq) \rightleftharpoons 2Fe^{2+}(aq) + Sn^{4+}(aq)$$

The equilibrium concentrations are 0.10 M Fe^{3+}, 0.20 M Fe^{2+}, 0.40 M Sn^{4+} and 0.30 M Sn^{2+}.
Calculate the equilibrium constant at this temperature with the correct units.

6 An equilibrium mixture at 900°C contains 0.050 mol of H_2 gas, 0.030 mol of CO_2 gas, 0.040 mol of H_2O gas and 0.020 mol of CO gas in a 2.0 L container. The gases react according to the equation:

$$H_2(g) + CO_2(g) \rightleftharpoons H_2O(g) + CO(g)$$

Calculate the equilibrium constant at 900°C.

7 Propanone (acetone) (C_3H_6O) is used to remove nail polish. It can be prepared from propan-2-ol (C_3H_8O) using a copper–zinc catalyst, according to the equation:

$$C_3H_8O(g) \rightleftharpoons C_3H_6O(g) + H_2(g)$$

If an equilibrium mixture of these gases consists of 0.21 mol of propan-2-ol, 0.21 mol of propanone and 0.21 mol of hydrogen in a 20 L vessel, calculate the value of the equilibrium constant.

8 Explain how the concentration of hydrogen gas in each of the following equilibrium mixtures would change if the mixtures are heated and kept at constant volume.

a $N_2(g) + 3H_2(g) \rightleftharpoons 2NH_3(g)$ $\Delta H = -91$ kJ mol^{-1}

b $CH_4(g) + H_2O(g) \rightleftharpoons CO(g) + 3H_2(g)$ $\Delta H = +208$ kJ mol^{-1}

9 The following equations represent reactions that are important in industrial processes. Predict the effect on the equilibrium position if the volume of each reaction mixture was increased at constant temperature.

a $C_3H_8O(g) \rightleftharpoons C_3H_6O(g) + H_2(g)$

b $CO(g) + 2H_2(g) \rightleftharpoons CH_3OH(g)$

c $N_2(g) + O_2(g) \rightleftharpoons 2NO(g)$

10 In which one of the following systems will the position of equilibrium be unaffected by a change of volume at constant temperature?

A $2NH_3(g) \rightleftharpoons N_2(g) + 3H_2(g)$

B $I_2(g) + H_2(g) \rightleftharpoons 2HI(g)$

C $2CO(g) + O_2(g) \rightleftharpoons 2CO_2(g)$

D $4CO_2(g) + 6H_2O(g) \rightleftharpoons 2C_2H_6(g) + 7O_2(g)$

11 Steam reacts with chlorine according to the equation:

$$2H_2O(g) + 2Cl_2(g) \rightleftharpoons 4HCl(g) + O_2(g)$$

At a particular temperature the value of the equilibrium constant for this reaction is determined to be 4.0×10^{-4} M. Assuming no change in temperature, calculate the value for the equilibrium constant for the following reactions.

a $4H_2O(g) + 4Cl_2(g) \rightleftharpoons 8HCl(g) + 2O_2(g)$

b $4HCl(g) + O_2(g) \rightleftharpoons 2H_2O(g) + 2Cl_2(g)$

Application and analysis

12 Methanol is manufactured for use as a fuel for racing cars. It can be made by reaction between carbon monoxide and hydrogen:

$CO(g) + 2H_2(g) \rightleftharpoons CH_3OH(g) \qquad \Delta H = -103 \text{ kJ mol}^{-1}$

As part of an investigation of this process, the concentrations of a mixture of CO, H_2 and CH_3OH were monitored continuously. The mixture was initially at equilibrium at 400°C and constant volume. After 10 minutes, additional H_2 was added to the mixture.

a Using the grid provided, sketch a graph to show how concentrations would change after H_2 was added. Make sure you indicate the mole ratios correctly in your sketch.

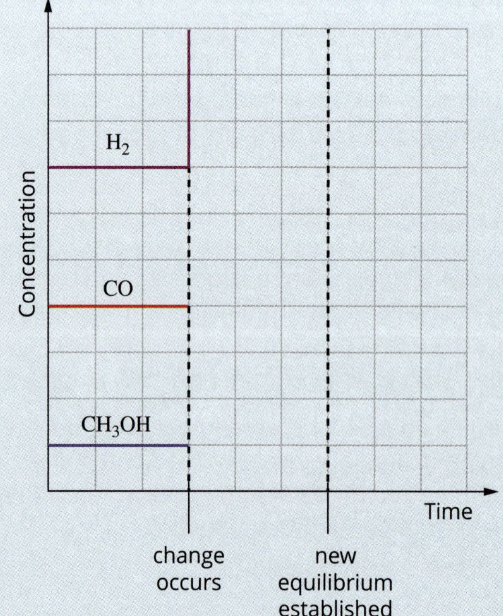

b Following the addition of the H_2, the mixture again reaches equilibrium. On the second grid provided, sketch a second graph to show the effect on the concentrations if the temperature were then decreased to 350°C. Make sure you indicate the mole ratios correctly in your sketch.

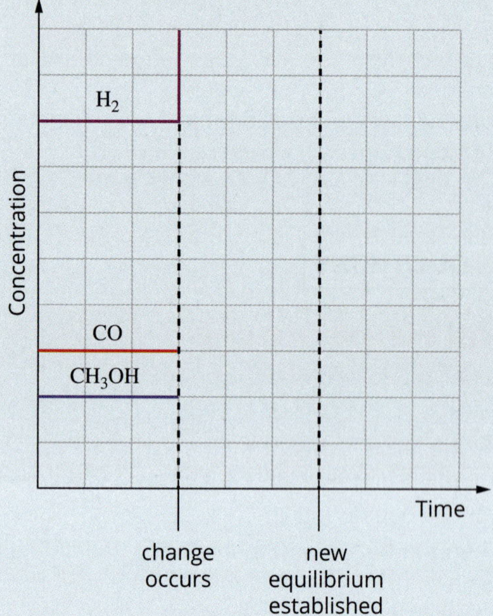

13 The value of K for the following reaction is equal to 4 at 25°C.

$C_2H_5OH(l) + CH_3COOH(l) \rightleftharpoons CH_3COOC_2H_5(l) + H_2O(l)$

At time t, the reaction quotient, Q, for a mixture of ethanol, water, ethyl ethanoate ($CH_3COOC_2H_5$) and ethanoic acid is equal to 6. Assuming that the mixture is also at 25°C, and referring to the values of K and Q, describe what will happen to the concentration of ethyl ethanoate as the system reaches equilibrium.

14 The reaction used to manufacture ammonia is represented by the equation:

$$N_2(g) + 3H_2(g) \rightleftharpoons 2NH_3(g)$$

The equilibrium constant for the reaction is 0.052 M^{-2} at 400°C.

A gas mixture contains 1.5 mol of N_2 gas, 1.5 mol of H_2 gas and 0.35 mol of NH_3 gas in a 1.0 L vessel at 400°C. Decide if the mixture is at equilibrium and, if not, predict the direction it will shift to reach equilibrium.

15 Consider the equilibrium:

$$PCl_5(g) \rightleftharpoons PCl_3(g) + Cl_2(g)$$

A 3.00 L vessel contained 6.00 mol of PCl_3, 4.50 mol of PCl_5 and 0.900 mol of Cl_2 at equilibrium at 250°C.

a Write an expression for the equilibrium constant for this reaction.

b Calculate the equilibrium constant for the reaction at 250°C.

c Another equilibrium mixture contains 0.0020 M of PCl_5 and 0.0010 M of PCl_3 at 250°C. Calculate the concentration of Cl_2 in this mixture.

d Determine the equilibrium constant at 250°C for the reaction:

$$PCl_3(g) + Cl_2(g) \rightleftharpoons PCl_5(g)$$

16 At one step during the synthesis of nitric acid, dinitrogen tetroxide is in equilibrium with nitrogen dioxide:

$$N_2O_4(g) \rightleftharpoons 2NO_2(g)$$

0.540 mol of N_2O_4 was placed in a 2.00 L vessel. When equilibrium was achieved, 0.280 mol of NO_2 was present. Calculate the value of the equilibrium constant at this temperature.

17 Consider the reaction:

$$A + 3B \rightleftharpoons 2C + D$$

Analysis of an equilibrium mixture in a 2.0 L container shows that 1.8 mol of A, 0.54 mol of B and 3.2 mol of D are present. If the equilibrium constant of the reaction is 0.024 M^{-1}, calculate the:

a concentration of A, B and D at equilibrium

b concentration of C in the equilibrium mixture

c amount of C, in mol, in the equilibrium mixture.

18 Elderly people, especially women, can become very susceptible to bone breakages. It is thought that as people age, they absorb Ca^{2+} from food inefficiently, reducing the concentration of these ions in body fluids. An equilibrium exists between calcium phosphate in bone and calcium ions in body fluids:

$$Ca_3(PO_4)_2(s) \rightleftharpoons 3Ca^{2+}(aq) + 2PO_4^{3-}(aq)$$

Use your understanding of equilibrium to explain why inefficient absorption of Ca^{2+} ions could cause weakness in bones.

19 The following reaction was studied at 330°C.

$$2HI(g) \rightleftharpoons H_2(g) + I_2(g)$$

a A mixture was prepared by placing 4.00 mol of HI in a 2.0 L vessel at 330°C. At equilibrium, 0.44 mol of H_2 and 0.44 mol of I_2 were present. Calculate the value of the equilibrium constant at this temperature.

b A second mixture consisted of 1.0 mol of HI, 0.24 mol of H_2 and 0.32 mol of I_2 in a 2.0 L container at 330°C. Decide if the mixture is at equilibrium and, if not, predict the direction the reaction will shift to reach equilibrium.

20 A mixture of 0.100 mol of NO, 0.051 mol of H_2 and 0.100 mol of H_2O was added to a reaction vessel with a volume of 1.0 L at 300°C. The reaction at equilibrium is given by the equation:

$$2NO(g) + 2H_2(g) \rightleftharpoons N_2(g) + 2H_2O(g)$$

After equilibrium was established, the concentration of NO was found to be 0.062 M. Determine the equilibrium constant, K, including units for the reaction at 300°C.

21 Carbon monoxide is used as a fuel in many industries. It reacts according to the equation:

$$2CO(g) + O_2(g) \rightleftharpoons 2CO_2(g)$$

In a study of this exothermic reaction, an equilibrium system is established in a closed vessel of constant volume at 1000°C.

a If carbon monoxide can be used as a fuel, comment on the magnitude of the equilibrium constant for the reaction.

b Predict what will happen to the equilibrium position as a result of:

i an increase in temperature

ii the addition of a catalyst

iii the removal of carbon dioxide.

c What will happen to the value of the equilibrium constant as a result of each of the changes, **i**, **ii** and **iii**, in part **b**?

22 A step during nitric acid production is the oxidation of nitrogen oxide to nitrogen dioxide:

$$2NO(g) + O_2(g) \rightleftharpoons 2NO_2(g) \qquad \Delta H = -114 \text{ kJ mol}^{-1}$$

Nitrogen dioxide is a brown gas, and nitrogen oxide and oxygen are colourless. An equilibrium mixture was prepared in a 1 L container at 350°C.

a Copy the following table, and for each of the changes listed, indicate if the reaction mixture would become darker or lighter. Give a reason for your choice.

b For change **ii** below, draw a concentration–time graph to represent the effect of this change.

Change	Colour change (lighter or darker)	Explanation
i The temperature is decreased to 250°C at constant volume.		
ii The volume of the container is halved at constant temperature.		
iii A catalyst is added at constant volume and temperature.		

23 **a** The equilibrium constant is 0.67 M^{-2} at a particular temperature for the reaction:

$$CO(g) + 3H_2(g) \rightleftharpoons CH_4(g) + H_2O(g)$$

A mixture of 0.100 M of CO, 0.200 M of H_2, 0.300 M of CH_4 and 0.400 M of H_2O is heated to this temperature. Predict whether the concentration of the following would increase, decrease or not change.

i CO

ii H_2

iii CH_4

iv H_2O

b When the temperature of the reaction mixture is increased by 10°C, the equilibrium constant for the reaction becomes 0.71 M^{-2}. What conclusion can you make about the enthalpy change of this reaction? Explain your answer.

24 Carbon disulfide gas (CS_2) is used in the manufacture of rayon. CS_2 can be made in an endothermic gas-phase reaction between sulfur trioxide gas (SO_3) and carbon dioxide. Oxygen gas is also produced in the reaction.

a Write a balanced chemical equation for the reaction.

b Write an expression for the equilibrium constant of the reaction.

c An equilibrium mixture of these gases was made by mixing sulfur trioxide and carbon dioxide. The equilibrium mixture consisted of 0.028 mol of CS_2, 0.022 mol of SO_3, 0.014 mol of CO_2 and an unknown amount of O_2 in a 20 L vessel. Calculate the:

i amount of O_2, in mol, present in the equilibrium mixture

ii value of the equilibrium constant at that temperature.

d Predict how each of the following changes to an equilibrium mixture would affect the yield of CS_2.

i removing O_2 (at constant total volume and temperature)

ii increasing the temperature

iii adding a catalyst

iv increasing the pressure by decreasing the volume of the reaction vessel (at constant temperature)

v increasing the pressure by introducing argon gas into the reaction vessel (at constant volume and temperature)

25 Sulfur dioxide gas and oxygen gas were mixed at 600°C to produce a gaseous equilibrium mixture:

$$2SO_2(g) + O_2(g) \rightleftharpoons 2SO_3(g) \quad \Delta H = -197 \text{ kJ mol}^{-1}$$

A number of changes were then made, including the addition of a catalyst, resulting in the formation of new equilibrium mixtures. The figure below shows how the concentrations of the three gases changed.

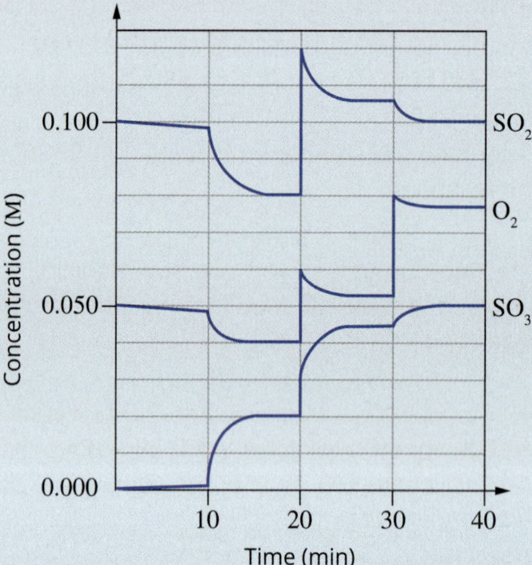

Graph of concentration versus time for sulfur dioxide and oxygen reaction

a Write an expression for the equilibrium constant, *K*, of the reaction.

b List the time intervals when the reaction is at equilibrium.

c Calculate the value of *K* at 18 minutes.

d Calculate the value of *K* at 25 minutes.

e Calculate the value of *K* at 35 minutes.

f At what time was the catalyst added? Explain your reasoning.

g What possible change(s) occurred in the system at 20 minutes? Explain your reasoning.

h What possible change(s) occurred in the system at 30 minutes? Explain your reasoning.

i If at any stage the temperature changed, indicate whether it increased or decreased, after considering the equilibrium constants.

OA ✓✓

CHAPTER 08

Production of chemicals by electrolysis

A large number of chemicals are produced through electrolysis. Some of these chemicals are often difficult to obtain by other means. This makes the process of electrolysis important. Electrolysis is used in a number of important applications in the chemical industry, such as plating a thin film of metal on the surfaces of other metals to improve their appearance or prevent corrosion, extracting reactive metals from their ores, and the production of copper and zinc.

In this chapter, you will explore the reactions that occur in electrolytic cells and compare them to the reactions that occur in galvanic cells.

Key knowledge

- the use and limitations of the electrochemical series to explain or predict the products of the electrolysis of particular chemicals, given their state (molten liquid or in aqueous solution) and the electrode materials used, including the writing of balanced equations (with states) for the reactions occurring at the anode and cathode and the overall redox reaction for the cell **8.1**
- the common design features and general operating principles of commercial electrolytic cells (including, where practicable, the removal of products as they form), and the selection of suitable electrode materials, the electrolyte (including its state) and any chemical additives that result in a desired electrolysis product (details of specific cells not required) **8.2**
- the application of Faraday's laws and stoichiometry to determine the quantity of galvanic or fuel cell reactant and product, and the current or time required to either use a particular quantity of reactant or produce a particular quantity of product **8.3**
- the application of Faraday's laws and stoichiometry to determine the quantity of electrolytic reactant and product, and the current or time required to either use a particular quantity of reactant or produce a particular quantity of product **8.3**

OA

8.1 Electrolytic cells

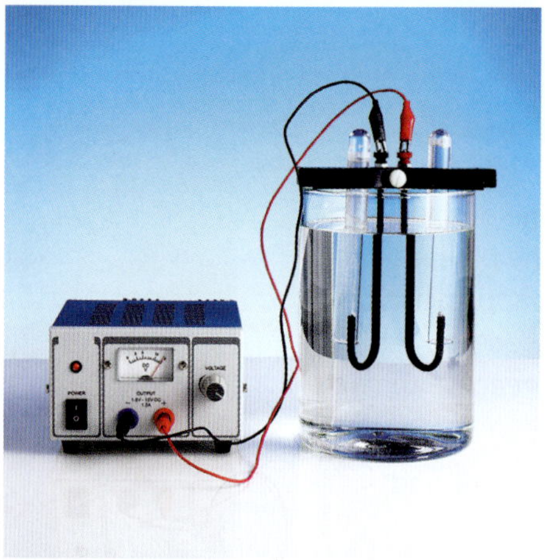

FIGURE 8.1.1 In this electrolysis set-up, electricity passing through a dilute sodium nitrate solution decomposes the water to form hydrogen gas at the left-hand electrode and oxygen gas at the right-hand electrode. The gases are collected in test tubes.

Electrolysis involves the passage of electrical energy from a direct current (DC) power supply through a conducting liquid. The electrical energy causes redox reactions that are normally non-spontaneous to occur. An example of this is shown in Figure 8.1.1, where an electric current is being passed through a dilute sodium nitrate solution, causing water to decompose into hydrogen gas and oxygen gas.

In this section, you will learn about the principles of electrolysis. You will also see that the **electrochemical series** you used in Chapter 5 can enable you to identify the reactions that occur at the electrodes in electrolytic cells.

ELECTROLYSIS OF MOLTEN IONIC COMPOUNDS

When an ionic compound is heated to above its melting point, it melts to become a liquid and is referred to as being **molten**. The electrolysis of molten sodium chloride as an example of a simple **electrolytic cell** is shown in Figure 8.1.2.

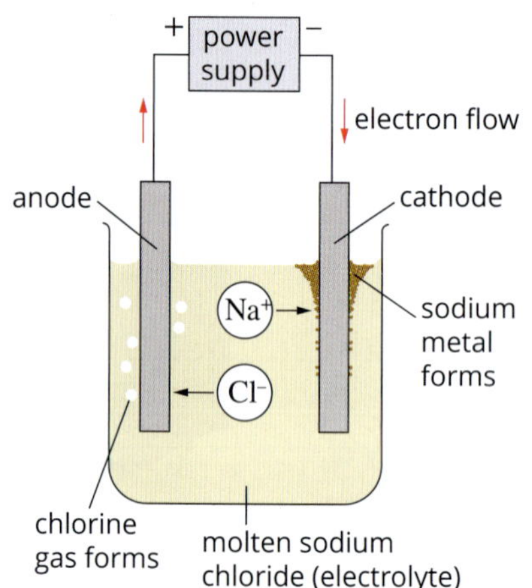

FIGURE 8.1.2 Electrolysis of molten sodium chloride

Platinum metal or graphite is used for the electrodes because they allow the passage of electrons to and from the power supply. Because platinum and graphite are **inert** (unreactive), the electrodes do not react with the contents of the cell. The molten sodium chloride is described as the **electrolyte** because it is a conducting liquid.

Sodium chloride melts at 801°C, so this electrolysis reaction must occur at a high temperature. No water is present. The species present in the electrolyte in the cell are Na^+ and Cl^- ions.

During electrolysis, reactions occur at the surface of both electrodes. The electrical energy required for these reactions to occur is provided by the power supply. The power supply can be regarded as a type of 'electron pump', pushing electrons into one electrode and withdrawing them from the other. While the negative electrode is connected to the negative terminal of the power supply, the positive electrode is connected to the positive terminal of the power supply.

At the negative electrode

The power supply pushes electrons towards the negative electrode. Na^+ ions in the electrolyte are attracted to the negative electrode, where they gain electrons and become sodium atoms:

$$Na^+(l) + e^- \rightarrow Na(l)$$

Because this is a reduction reaction, by definition this electrode is known as the **cathode**.

Sodium is solid at room temperatures, but it is liquid at the temperatures required to melt sodium chloride. It is less dense than molten sodium chloride and floats to the top of the cell.

At the positive electrode

Chloride ions in the electrolyte move towards the positive electrode, losing electrons to form molecules of Cl_2. Bubbles of chlorine gas appear at the electrode:

$$2Cl^-(l) \rightarrow Cl_2(g) + 2e^-$$

The electrons produced move through the electrode towards the power supply. Because this is an oxidation reaction, by definition this electrode is known as the **anode**.

The overall reaction that takes place in the cell is:

$$2NaCl(l) \rightarrow 2Na(l) + Cl_2(g)$$

This is a **non-spontaneous reaction**—a reaction that would not occur without the input of energy. In such reactions, the supply of electrical energy from a power supply is converted into chemical energy in the products of electrolysis. The reverse reaction, between sodium and chlorine, is a **spontaneous reaction**.

When writing equations for the electrolysis of molten ionic compounds, remember that the reactants are in the liquid state.

In galvanic cells, the chemical energy in the cells is converted to electrical energy. The opposite occurs in electrolytic cells, where electrical energy supplied from an external power source is converted to chemical energy.

In an electrolytic cell:

- anions are attracted to the anode (positive electrode)
- oxidation occurs at the anode
- cations are attracted to the cathode (negative electrode)
- reduction occurs at the cathode.

The electrolytic process usually takes place in one container. Unlike galvanic cells, there is no need to locate the electrodes in separate compartments because a non-spontaneous reaction is involved. However, the products do need to be kept apart as they are produced. Otherwise, they would spontaneously react with each other to re-form the original reactants.

Remember:

- anions are attracted to the anode (positive electrode), where oxidation occurs
- cations are attracted to the cathode (negative electrode) where reduction occurs.

COMPETITION AT ELECTRODES

Unlike the electrolysis of molten ionic compounds, in some electrolytic cells there may be several species present at each electrode that might be able to react. Water can be a potential reactant in aqueous solutions. Even the material used for the electrodes may participate in the reaction.

In these cases, you can use the electrochemical series (Table 5.2.1, page 189) to predict which of the possible reactions would be most likely to occur.

During electrolysis the:

- strongest oxidising agent present usually reacts at the negative electrode
- strongest reducing agent present usually reacts at the positive electrode.

Just as in a galvanic cell, the strongest reducing agent reacts at the anode and the strongest oxidising agent reacts at the cathode.

ELECTROLYSIS OF AQUEOUS SOLUTIONS

Consider the electrolysis of an aqueous solution such as a 1.0 M solution of sodium chloride using inert electrodes (Figure 8.1.3). With water present in the cell, as well as $Na^+(aq)$ and $Cl^-(aq)$, there are a number of species that might be able to react.

- $H_2O(l)$ could be oxidised or reduced.
- Na^+ ions could be reduced.
- Cl^- ions could be oxidised.

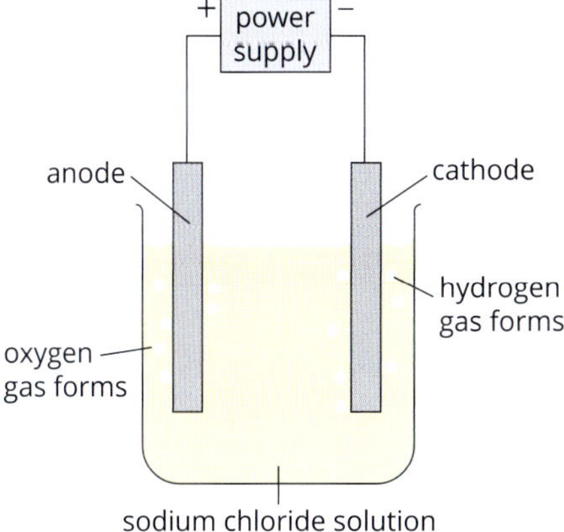

FIGURE 8.1.3 Electrolysis of aqueous sodium chloride solution

At the cathode

If you look on the left-hand side of the electrochemical series (Table 5.2.1, page 189), you will see that there are two species in the cell that can act as oxidising agents and therefore could be reduced at the cathode. These two species are H_2O and Na^+:

oxidising agents

$$2H_2O(l) + 2e^- \rightleftharpoons H_2(g) + 2OH^-(aq) \qquad E^\circ = -0.83\text{ V}$$

$$Na^+(aq) + e^- \rightleftharpoons Na(s) \qquad E^\circ = -2.71\text{ V}$$

Stronger oxidising agents appear higher in the electrochemical series and are more likely to be reduced. Therefore, water will be reduced at the cathode and hydrogen gas will form:

$$2H_2O(l) + 2e^- \rightarrow H_2(g) + 2OH^-(aq)$$

The sodium ions will not react and will remain in solution as spectator ions.

Water is a stronger oxidising agent than Mn^{2+}, Al^{3+}, Mg^{2+}, Na^+, Ca^{2+}, K^+ and Li^+. Aqueous solutions containing these metal ions cannot be electrolysed to produce the metals.

At the anode

On the right-hand side of the electrochemical series (Table 5.2.1, page 189) are two species in the cell that can act as reducing agents and could be oxidised at the anode: Cl^- and H_2O. There are three half-equations containing these species as reducing agents:

$$H_2O_2(aq) + 2H^+(aq) + 2e^- \rightleftharpoons 2H_2O(l) \qquad E^\circ = +1.77\text{ V}$$

$$Cl_2(g) + 2e^- \rightleftharpoons 2Cl^-(aq) \qquad E^\circ = +1.36\text{ V}$$

$$O_2(g) + 4H^+(aq) + 4e^- \rightleftharpoons 2H_2O(l) \qquad E^\circ = +1.23\text{ V}$$

reducing agents

CHEMFILE

Why don't nitrate and sulfate ions react during electrolysis of aqueous solutions?

Many soluble ionic compounds contain nitrate or sulfate ions. During the electrolysis of aqueous solutions containing these ions, the nitrate or sulfate ions are attracted towards the anode, but they are not oxidised. This is because the oxidation numbers of N in NO_3^- and S in SO_4^{2-} are as high as they can possibly be. For example, N in NO_3^- has an oxidation number of +5. A nitrogen atom has five valence electrons, so it is unlikely to lose further electrons (to be oxidised). A similar reason applies to S in SO_4^{2-}, which has an oxidation number of +6.

Stronger reducing agents appear lower in the electrochemical series and are more likely to be oxidised, so the predicted anode reaction is:

$$2H_2O(l) \rightarrow O_2(g) + 4H^+(aq) + 4e^-$$

The overall reaction is:

$$6H_2O(l) \rightarrow 2H_2(g) + O_2(g) + 4H^+(aq) + 4OH^-(aq)$$

When the $H^+(aq)$ ions formed at the anode come in contact with the $OH^-(aq)$ ions formed at the cathode, they react to form water. The overall equation can then be written as:

$$2H_2O(l) \rightarrow 2H_2(g) + O_2(g)$$

Therefore, water is being broken up into its elements in this process.

In practice, it is possible for chloride ions, Cl^-, to be oxidised at the anode depending on conditions such as the concentration of the solution. Recall that the electrochemical series was determined for reactions under standard conditions (solution concentration of 1.0 M, gas pressure of 100 kPa and temperature of 25°C). At non-standard conditions where the concentration of sodium chloride is high, chlorine gas is produced at the anode.

Comparison of molten and aqueous electrolytes

Table 8.1.1 summarises differences between the electrolysis of molten and aqueous sodium chloride. The products formed at each electrode depend on the state and, in the case of aqueous NaCl, the concentration of the electrolyte.

TABLE 8.1.1 A comparison of the electrolysis of molten NaCl with aqueous NaCl

	Molten NaCl	Aqueous dilute NaCl	Aqueous concentrated NaCl
Temperature	high	room temperature	room temperature
Anode reaction	$2Cl^-(l) \rightarrow Cl_2(g) + 2e^-$	$2H_2O(l) \rightarrow O_2(g) + 4H^+(aq) + 4e^-$	$2Cl^-(aq) \rightarrow Cl_2(g) + 2e^-$
Cathode reaction	$Na^+(l) + e^- \rightarrow Na(l)$	$2H_2O(l) + 2e^- \rightarrow H_2(g) + 2OH^-(aq)$	$2H_2O(l) + 2e^- \rightarrow H_2(g) + 2OH^-(aq)$
Overall reaction	$2Na^+(l) + 2Cl^-(l) \rightarrow 2Na(l) + Cl_2(g)$	$2H_2O(l) \rightarrow 2H_2(g) + O_2(g)$	$2Cl^-(aq) + 2H_2O(l) \rightarrow Cl_2(g) + H_2(g) + 2OH^-(aq)$

ELECTROLYSIS INVOLVING REACTIVE ELECTRODES

In the examples of electrolysis that you have encountered so far, the electrodes do not take part in the reaction and are described as **inert electrodes**. However, in many electrolytic cells, the electrodes are consumed in the cell reaction. Such electrodes are called **reactive electrodes**. Worked example 8.1.1 on page 281 shows how the electrochemical series can be used to predict the reactions that occur in an electrolytic cell with reactive electrodes.

The electrochemical series can be used in this way to make predictions. The series is based on standard conditions. However, most of the electrolysis reactions that occur in the laboratory and in industry are not at standard conditions. Experiments are the only sure way to find out what products form in a particular instance. In practice, reactions are affected by factors such as electrolyte concentration, gas pressures, current, voltage and the use of different electrodes.

Worked example 8.1.1

PREDICTING THE PRODUCTS OF ELECTROLYSIS

Use the electrochemical series to predict the products of the electrolysis of 1 M nickel(II) sulfate solution with copper electrodes at 25°C. A diagram of this electrolytic process is shown in Figure 8.1.4.

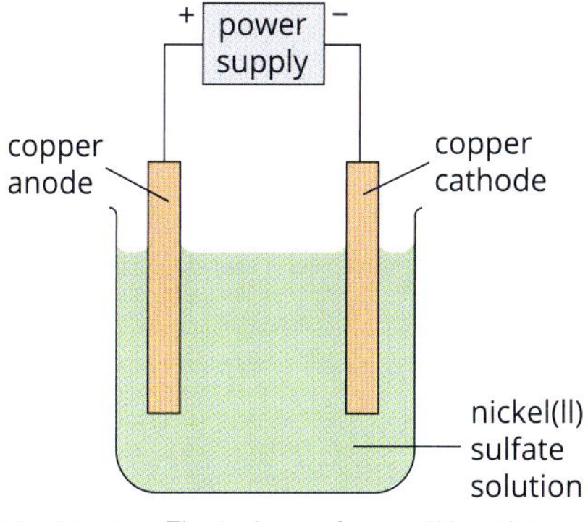

FIGURE 8.1.4 Electrolysis of nickel(II) sulfate solution using copper electrodes

Thinking	Working
Identify which species are present in the solution.	$Ni^{2+}(aq)$, $SO_4^{2-}(aq)$ and $H_2O(l)$
Identify what the electrodes are made of.	copper
Refer to the electrochemical series and identify the possible reactions. Write these half-equations in the order they appear in the series.	From the electrochemical series, you can see that a number of reactions need to be considered because the cell contains two oxidising agents, Ni^{2+} and H_2O, and two reducing agents, Cu and H_2O. $H_2O_2(aq) + 2H^+(aq) + 2e^- \rightleftharpoons 2H_2O(l)$ + 1.77 V $O_2(g) + 4H^+(aq) + 4e^- \rightleftharpoons 2H_2O(l)$ +1.23 V $Cu^{2+}(aq) + 2e^- \rightleftharpoons Cu(s)$ +0.34 V $Ni^{2+}(aq) + 2e^- \rightleftharpoons Ni(s)$ –0.24 V $2H_2O(l) + 2e^- \rightleftharpoons H_2(g) + 2OH^-(aq)$ –0.83 V The SO_4^{2-} ions are not involved in the reaction.
Determine the reactions that could occur at the anode.	At the anode (+), oxidation reactions that could occur are: $2H_2O(l) \rightarrow H_2O_2(aq) + 2H^+(aq) + 2e^-$ $2H_2O(l) \rightarrow O_2(g) + 4H^+(aq) + 4e^-$ $Cu(s) \rightarrow Cu^{2+}(aq) + 2e^-$
Determine the most likely reaction at the anode.	The strongest reducing agent, Cu(s), will be oxidised: $Cu(s) \rightarrow Cu^{2+}(aq) + 2e^-$
Determine the reactions that could occur at the cathode.	At the cathode (–), reduction reactions that could occur are: $Ni^{2+}(aq) + 2e^- \rightarrow Ni(s)$ $2H_2O(l) + 2e^- \rightarrow H_2(g) + 2OH^-(aq)$
Determine the most likely reaction at the cathode.	The strongest oxidising agent, $Ni^{2+}(aq)$, will be reduced: $Ni^{2+}(aq) + 2e^- \rightarrow Ni(s)$
Summarise your results to answer the question.	Nickel is produced at the cathode and Cu^{2+} is produced at the anode.

Worked example: Try yourself 8.1.1

PREDICTING THE PRODUCTS OF ELECTROLYSIS

Use the electrochemical series to predict the products of electrolysis of 1 M zinc sulfate solution with copper electrodes at 25°C.

CASE STUDY **ANALYSIS**

Producing hydrogen for the hydrogen economy

As discussed in Chapter 5, many scientists predict that one day society will transition to a hydrogen economy, where the main source of energy for transport and other purposes comes from hydrogen.

Currently, the major source of hydrogen is the steam reforming process, which requires fossil fuels. A greener process is needed that does not rely on fossil fuels or produce greenhouse gases.

As you have seen in this section, hydrogen is produced through the electrolysis of water with the addition of an ionic compound such as NaCl to create an electrolyte. The electrolytic process can be made greener by using renewable electricity sources. However, the efficiency of the process ranges from 50 to 70%. Efficiency can be increased by using electrodes with catalytic abilities, such as platinum, but this is very expensive. Scientists are currently working on developing cheap and effective catalytic electrodes from materials such as molybdenum sulfide, carbon nanotubes and mixtures of nickel/nickel oxide.

Electrolysis is still an energy-intensive way of producing hydrogen. Another area of scientific research is 'artificial photosynthesis', which aims to replicate the process of splitting water molecules using light, just as green plants do (Figure 8.1.5). Titanium dioxide and other titanium-based compounds form a group of materials that can split water in the presence of light to produce hydrogen.

With these scientific developments, the prospect of a future hydrogen economy is looking much brighter.

Analysis

1 Outline why the production of hydrogen through the electrolysis of water is currently not a 'green' process.

2 Write half-equations for the reactions that would occur at the anode and cathode for the electrolysis of water with the addition of NaOH using inert electrodes. Explain why there is an advantage in using NaOH rather than NaCl by referring to the electrochemical series in Table 5.2.1 (page 189).

3 Research information on the reactions that occur in the steam reforming process and evaluate its impact on the environment.

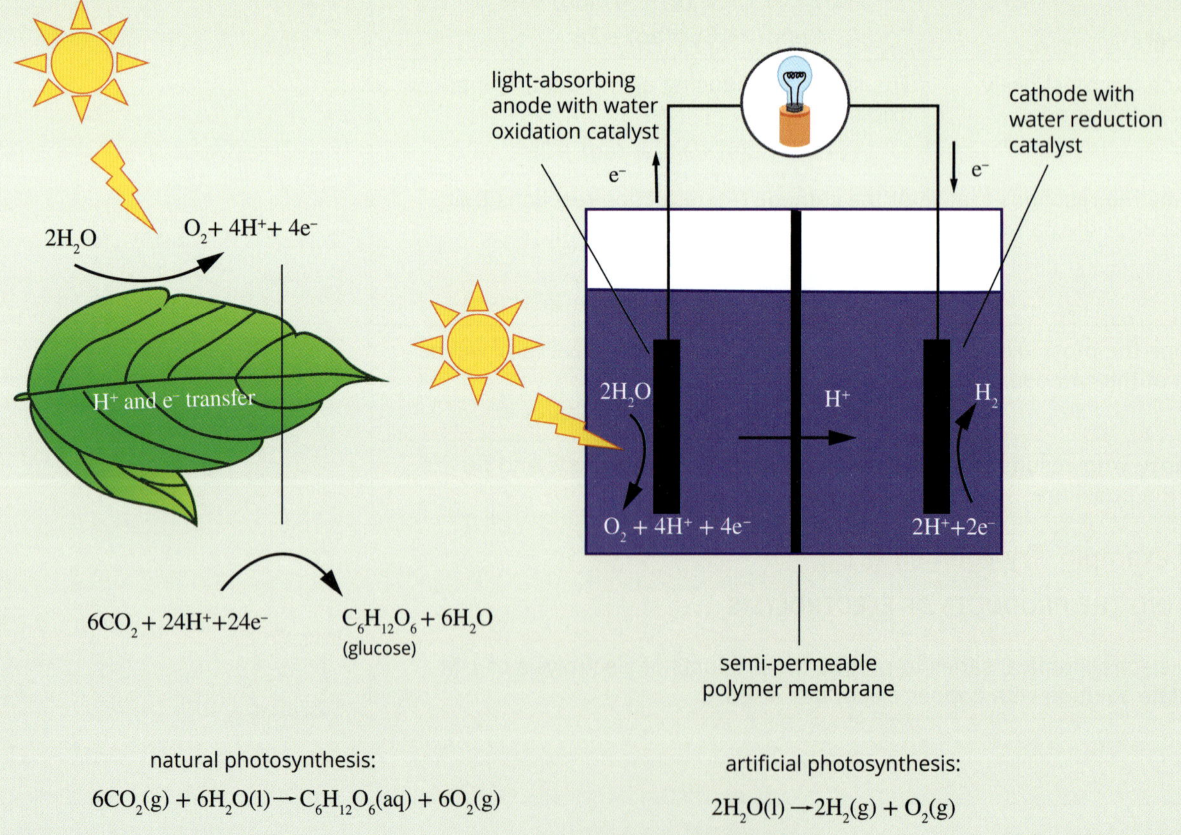

FIGURE 8.1.5 A comparison between natural photosynthesis and artificial photosynthesis in an electrochemical cell

COMPARISON OF ELECTROLYTIC AND GALVANIC CELLS

Galvanic cells and electrolytic cells are types of **electrochemical cells**. Although both cells involve the conversion between electrical energy and chemical energy, there are important differences between them.

Table 8.1.2 and Figure 8.1.6 summarise the similarities and differences between galvanic and electrolytic cells.

TABLE 8.1.2 A summary of the similarities and differences between galvanic and electrolytic cells

Galvanic cells	Electrolytic cells
produce electricity	consume electricity
have spontaneous reactions	have non-spontaneous reactions
convert chemical energy to electrical energy	convert electrical energy to chemical energy
oxidation occurs at the anode reduction occurs at the cathode	oxidation occurs at the anode reduction occurs at the cathode
the anode is negative the cathode is positive	the anode is positive the cathode is negative

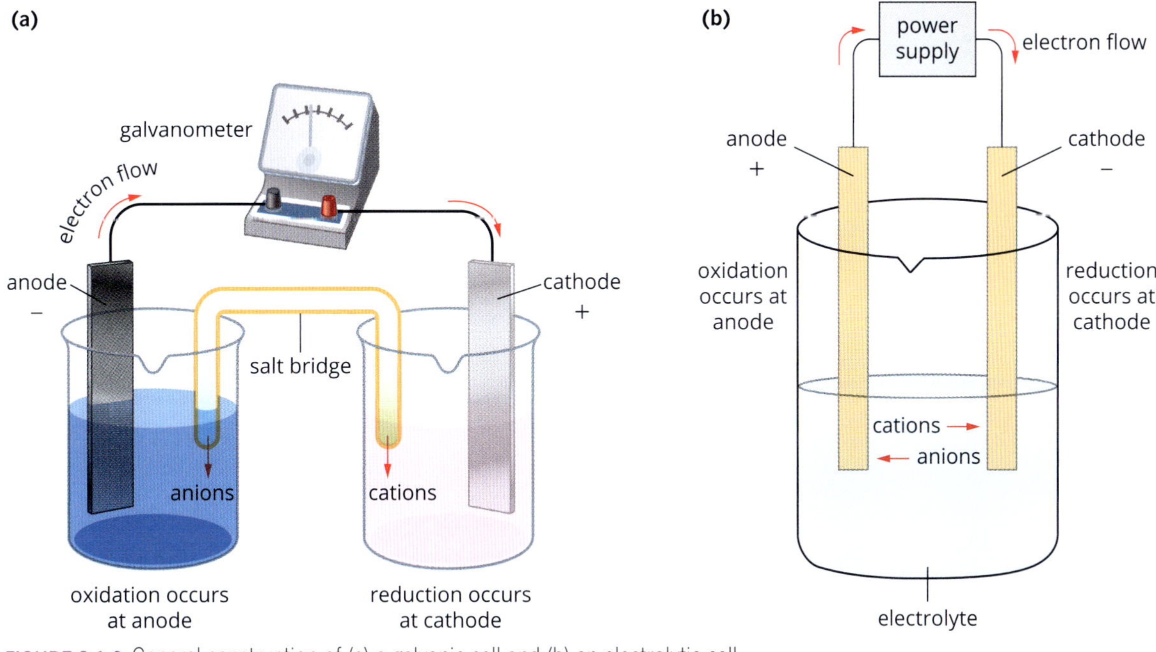

FIGURE 8.1.6 General construction of (a) a galvanic cell and (b) an electrolytic cell

8.1 Review

OA ✓✓

SUMMARY

- Electrolytic cells convert electrical energy to chemical energy, whereas galvanic cells convert chemical energy to electrical energy.
- Non-spontaneous redox reactions occur in electrolytic cells.
- The anode is defined as the electrode at which oxidation occurs; the cathode is the electrode at which reduction occurs.
- In electrolytic cells, the anode is the positive electrode and the cathode is the negative electrode.
- Although several reactions may be possible at the anode of a particular cell, generally the strongest reducing agent undergoes oxidation.
- Although several reactions may be possible at the cathode of a particular cell, generally the strongest oxidising agent undergoes reduction.
- The electrochemical series can be used to predict the products of electrolysis.
- The electrodes used in electrolysis can be either inert or reactive.
- Both molten and aqueous electrolytes can be used in electrolysis.

KEY QUESTIONS

Knowledge and understanding

1 Select the correct words from those in bold to complete the following statements.
In electrolytic cells:
  a the reactions are **spontaneous/non-spontaneous**
  b electrical energy is converted into **chemical/electrical** energy
  c the anode is **negative/positive** and the cathode is **negative/positive**
  d **oxidation/reduction** occurs at the anode and **oxidation/reduction** occurs at the cathode.

2 Identify which ions would be attracted to which electrode during the electrolysis of a solution of $CuBr_2$.

3 Identify the products of the electrolysis of molten potassium iodide using inert electrodes, and identify the electrodes at which they are formed.

Analysis

4 In an electrolysis experiment, a student is provided with an aqueous solution of silver nitrate ($AgNO_3$). The student sets up an electrolytic cell using an inert carbon rod as the positive electrode and a silver rod as the negative electrode.
  a During operation of the cell, on which rod will a silver coating appear?
  b Write the half-equations for the reactions that occur at each electrode.
  c Write an equation for the overall reaction.

5 A student sets up an electrolysis experiment using a cell made of a copper anode, a nickel cathode and a sodium nitrate solution as the electrolyte.
  a Refer to the electrochemical series and identify the possible reactions that could occur at each electrode. Write these half-equations in the order they appear in the series.
  b Identify the correct half-equations for the reactions that occur at the anode and cathode.
  c Write an equation for the overall reaction.

8.2 Commercial electrolytic cells

Although chemical industries tend to avoid using electrolysis for the manufacture of chemicals because of the relatively high cost of electrical energy, this process enables some chemicals to be produced that could not be readily produced in any other way.

In an electrolytic cell, the transformation of electrical energy into chemical energy results in non-spontaneous reactions, in which reactive chemicals are products rather than reactants.

In this section, you will learn about the use of molten and aqueous electrolytes in commercial electrolytic cells and the addition of chemical additives to achieve desired properties in an electrolyte, and you will examine the advantages and disadvantages of each type of electrolyte. You will also study the use of inert and reactive electrodes in these cells.

MOLTEN ELECTROLYTES

One of the best examples of the use of molten electrolytes in commercial electrolytic cells is in the Downs cell (Figure 8.2.1). This electrolytic cell is used to produce sodium and chlorine in commercial quantities.

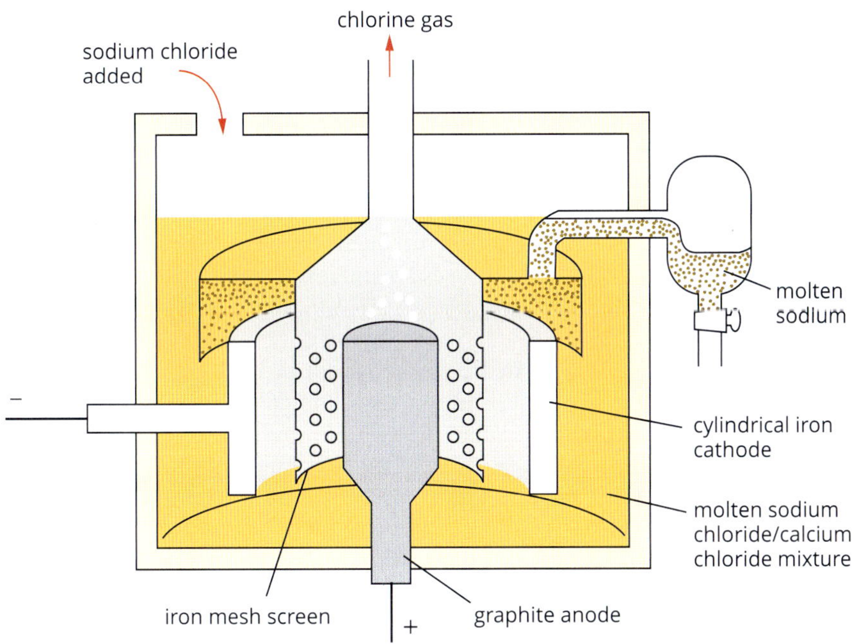

FIGURE 8.2.1 Diagram of a Downs cell, which produces chlorine gas and sodium

The electrodes used in this cell are inert (unreactive). The anode is made of a conducting material, such as graphite, and is not consumed in the reaction. The cathode is made of iron, which is effectively inert because this electrode receives a continual supply of electrons from the power supply, preventing the iron from being oxidised. The reactions occurring at each electrode were discussed in the previous section on page 278.

At the cathode (–):

$$Na^+(l) + e^- \longrightarrow Na(l)$$

At the anode (+):

$$2Cl^-(l) \longrightarrow Cl_2(g) + 2e^-$$

An overall equation may be written for the cell reaction by adding the half-equations:

$$\begin{array}{r} 2Na^+(l) + 2e^- \longrightarrow 2Na(l) \\ 2Cl^-(l) \longrightarrow Cl_2(g) + 2e^- \\ \hline 2Na^+(l) + 2Cl^- \longrightarrow 2Na(l) + Cl_2(g) \end{array}$$

In the electrolysis of a molten electrolyte:

- anions are oxidised at the anode (positive)
- cations are reduced at the cathode (negative)
- water is not present to interfere with the desired reactions.

The advantage of using a molten electrolyte in the Downs cell is that there is no water present to interfere with the desired reactions. Water is a stronger oxidising agent than Na^+ ions. If aqueous sodium chloride were used as the electrolyte instead of molten sodium chloride, water would react at the cathode and form hydrogen gas, according to the equation:

$$2H_2O(l) + 2e^- \rightarrow H_2(g) + 2OH^-(aq)$$

In general, the main disadvantage of using a molten electrolyte is that the process requires much more energy. Energy must be used to maintain the electrolyte in a molten state and at the temperature needed for the electrolysis to proceed efficiently.

Heat energy is produced when an electric current passes through a resistor. In the Downs cell, the electrolyte provides resistance and the flow of electricity keeps the electrolyte molten. A mixture of sodium chloride/calcium chloride in a 1:2 ratio is used as the electrolyte because the presence of the calcium chloride lowers the melting point of sodium chloride from 801°C to about 600°C. This saves on energy costs. The calcium ions are not reduced at the cathode because they are a weaker oxidising agent than sodium ions.

Calcium chloride is added to molten sodium chloride to lower its melting point. This does not affect the overall cell reaction because calcium ions are a weaker oxidising agent than sodium ions.

An iron mesh screen is used in the cell to keep the products at the anode and cathode apart. Because Cl_2 is a strong oxidising agent and Na is a strong reducing agent, there must be no contact between them and they must be continuously removed from the cell. Otherwise, they will react to re-form sodium chloride and the products from the electrolysis will be lost. The construction and operation of the Downs cell minimises contact between the two products.

CHEMFILE

Sodium first isolated by electrolysis

Sodium is one of the most reactive metals. Few substances are capable of reducing Na^+ ions to the metal, Na. Because of this, sodium was not isolated until 1807, when Humphry Davy (1778–1829), shown in the figure below, electrolysed molten sodium hydroxide.

The following year, Davy discovered five elements—barium, magnesium, strontium, boron and calcium—and reported news of his discoveries in two entertaining public lectures in London. Davy was renowned for the flair of his lectures and demonstrations.

Humphry Davy was the first person, using electrolysis, to isolate sodium, potassium, calcium, barium, magnesium, boron and strontium.

AQUEOUS ELECTROLYTES

Where possible, aqueous electrolytes are used in preference to molten electrolytes. It takes energy to maintain an electrolyte in a molten state, and so a cell that uses an aqueous electrolyte is more cost-effective. The membrane cell is an example of a modern industrial electrolytic cell that uses an aqueous electrolyte.

Membrane cells

Sodium hydroxide, chlorine and hydrogen are produced through the electrolysis of a concentrated sodium chloride solution, also known as a **brine**. This occurs in a membrane cell, so-named because the two compartments are separated by a **semipermeable membrane** (Figure 8.2.2). The membrane helps to prevent contact between the reactive products.

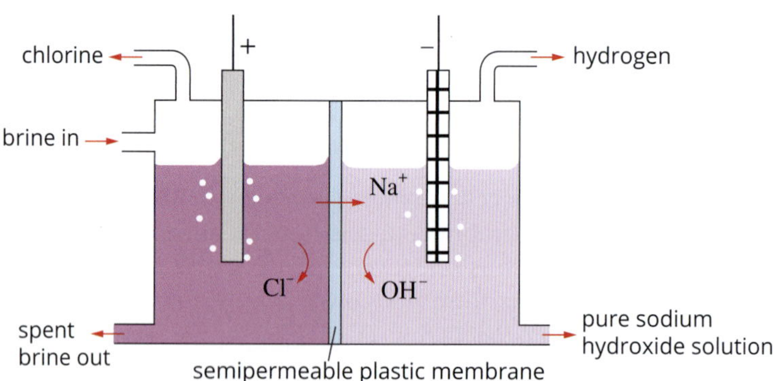

FIGURE 8.2.2 This membrane cell produces chlorine and hydrogen gas and aqueous sodium hydroxide.

The membrane is made from a polymer that only allows positive ions to pass through it. The polymer prevents the mixing of the products formed at the electrodes and allows only Na^+ ions to move from one electrode chamber to the other. Chloride and hydroxide ions are unable to move through the membrane. This results in a very pure sodium hydroxide product with very little chloride ion contamination.

At the cathode (–), water is reduced by electrons from the power supply to produce hydrogen gas and hydroxide ions:

$$2H_2O(l) + 2e^- \longrightarrow H_2(g) + 2OH^-(aq)$$

At the anode (+), even though water is a stronger reducing agent, the use of a concentrated sodium chloride solution means that chloride ions are oxidised instead:

$$2Cl^-(aq) \longrightarrow Cl_2(g) + 2e^-$$

The membrane also provides a barrier that prevents chlorine and hydrogen gases produced in the reaction from coming into contact with each other.

The advantages of the membrane in this cell are that the:

- sodium hydroxide solution produced does not become contaminated with sodium chloride
- use of an aqueous electrolyte allows the process to occur between 80°C and 90°C, so there is no need to heat the electrolyte and the cost of production is reduced.

In a membrane cell, water is not oxidised. Instead, because a concentrated sodium chloride solution is used, chloride ions are oxidised at the anode to produce chlorine gas.

CASE STUDY ANALYSIS

Electrorefining of copper

Copper metal is extracted from its ores by smelting. The process involves several stages in which the copper ore is heated strongly in air to produce impure copper metal.

As hot gases escape, the copper surface becomes blistered and it is commonly known as 'blister copper'. It contains about 2% of impurities such as sulfur, iron, antimony, silver and gold.

Blister copper is purified by electrolysis. Sheets of blister copper are placed in a large tank of sulfuric acid and thin sheets of pure copper are positioned between them. An external power source is connected so that the blister copper acts as a positive electrode (anode) and the pure copper acts as a negative electrode (cathode), as shown in Figure 8.2.3.

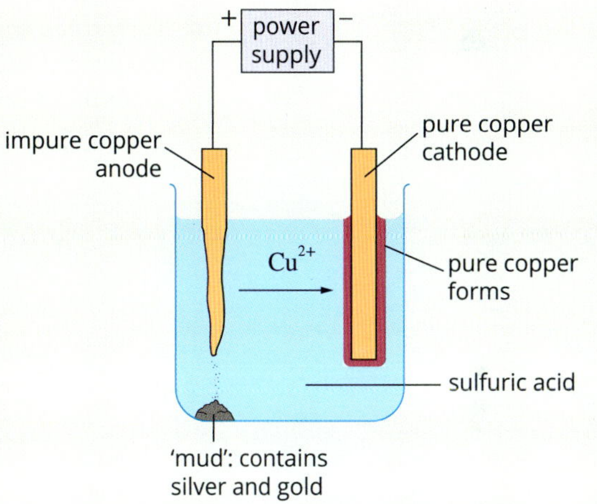

FIGURE 8.2.3 Electrorefining of copper metal by electrolysis

At the anode (+), electrons are drawn away from the blister copper anode to the positive terminal of the power source. Copper and impurities such as nickel and zinc, which are more reactive (i.e. are stronger reducing agents) than copper, are oxidised and enter the solution as ions:

$$Zn(s) \longrightarrow Zn^{2+}(aq) + 2e^-$$

$$Ni(s) \longrightarrow Ni^{2+}(aq) + 2e^-$$

Once those metal impurities have been oxidised, copper is the next strongest reducing agent, so it is oxidised to produce copper(II) ions, which enter the solution:

$$Cu(s) \longrightarrow Cu^{2+}(aq) + 2e^-$$

Impurities less reactive than copper, such as silver, gold and platinum, are not oxidised and simply fall from the anode, collecting at the bottom of the tank. This anode residue is sometimes called 'anode mud'. The precious metals are later recovered from the valuable mud.

At the pure copper cathode (–), copper metal is deposited as electrons from the power source are accepted by metal ions from solution. Because copper(II) ions are the strongest oxidising agent present in solution, copper is the only metal formed:

$$Cu^{2+}(aq) + 2e^- \longrightarrow Cu(s)$$

Analysis

1 Use online resources to identify five metal impurities found in 'blister copper'.

2 Identify the metal impurities that are oxidised during the electrorefining process. Explain why these metals are oxidised.

3 Identify the metal impurities that end up in the 'anode mud'. Explain why these metals are not oxidised.

4 Remembering that $PbSO_4$ is insoluble in aqueous solutions, determine where lead impurities will end up in this cell.

REACTIVE ELECTRODES

FIGURE 8.2.4 Boats made from aluminium are light, strong and resistant to corrosion.

The cells described so far in this section use inert electrodes. Inert electrodes are used in many commercial electrolytic cells. Note that cathodes made of metals are always inert (they cannot be oxidised) because they are connected to the negative terminal of the power supply and receive a continual supply of electrons.

Reactive electrodes have benefits in certain situations. As you saw in the case study on page 287, metals such as copper can be purified using an electrolytic cell in which the anode is made of the impure metal. The pure metal is then deposited at the cathode.

You will now examine how aluminium metal is produced by electrolysis. This process involves the use of both an inert and a reactive electrode. The low density and high strength of aluminium make it particularly useful for many applications, including cooking foil, drink cans, car engines, gutters, caravans, aeroplanes, window frames and boats (Figure 8.2.4).

Electrolytic production of aluminium

Aluminium metal is obtained from aluminium oxide (also known as alumina, Al_2O_3) by electrolysis. Pure alumina is extracted from the mineral bauxite, which is found in large deposits in the Darling Ranges in Western Australia, Weipa in northern Queensland and Gove in the Northern Territory.

Alumina melts at the very high temperature of 2050°C. However, it is soluble in molten cryolite (Na_3AlF_6). By dissolving alumina in cryolite, electrolysis can be performed at 950–1000°C, avoiding the much higher energy costs that would be involved if pure molten alumina were used as the electrolyte. Figure 8.2.5 shows a diagram of the Hall–Héroult cell, which is used in this electrolytic process.

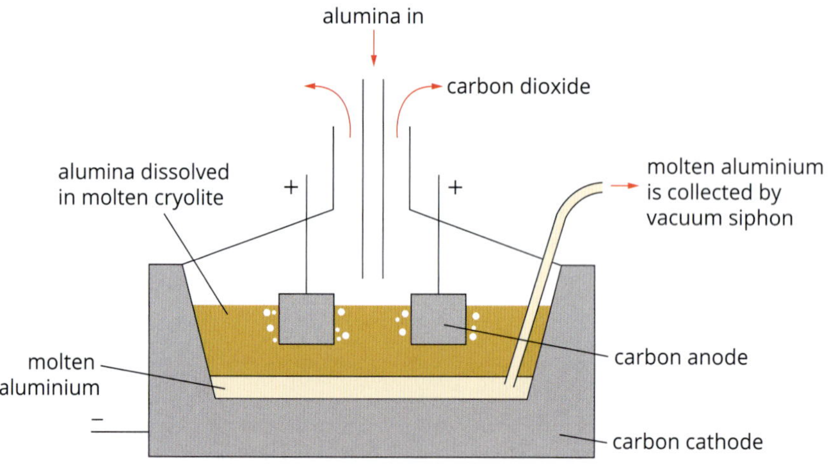

FIGURE 8.2.5 The Hall–Héroult cell used to extract aluminium from alumina

At the cathode (–), Al^{3+} ions from the alumina are reduced by electrons from the power supply to produce aluminium metal:

$$Al^{3+}(\text{in cryolite}) + 3e^- \rightarrow Al(l)$$

The aluminium, which is molten at the temperatures used, sinks to the bottom of the cell and is periodically removed by siphoning.

At the anode (+), the oxide ions of the alumina are oxidised to oxygen gas:

$$2O^{2-}(\text{in cryolite}) \rightarrow O_2(g) + 4e^-$$

The oxygen gas then immediately reacts with the carbon anodes to produce carbon dioxide gas:

$$O_2(g) + C(s) \rightarrow CO_2(g)$$

The overall electrode reaction may therefore be written as:

$$C(s) + 2O^{2-}(\text{in cryolite}) \rightarrow CO_2(g) + 4e^-$$

The overall equation for the extraction of aluminium is:

$$2Al_2O_3(\text{in cryolite}) + 3C(s) \longrightarrow 4Al(l) + 3CO_2(g)$$

Unlike in the other electrolytic cells described so far in this section, the carbon anodes take part in the reaction. Therefore, the carbon anodes need to be replaced regularly. Chemists are investigating alternative inert electrode materials to avoid the production of carbon dioxide and the cost of replacing the electrodes. Metals such as platinum and gold could be used as inert electrodes, but they are not economical on an industrial scale.

An example of a recent development in this area comes from Alcoa and Rio Tinto. From their joint venture, they have developed an electrolytic process using proprietary inert anodes that does not release carbon dioxide, but instead releases oxygen. Their process is claimed to reduce costs, increase aluminium production efficiency, and can be retrofitted to existing plants. The aluminium industry is also aware of the need to switch to renewable sources of electricity that do not produce carbon dioxide. This development is a positive example of an industry moving towards greater sustainability.

COMMON DESIGN FEATURES AND GENERAL OPERATING PRINCIPLES

Table 8.2.1 summarises the common design features and general operating principles of commercial electrolytic cells.

TABLE 8.2.1. Common design features and general operating principles of commercial electrolytic cells

Design feature or operating principle	Reason for feature or principle	Example
separation and continuous removal of products	The products formed are often relatively strong oxidising and reducing agents. If the products were to come into contact with each other, they would spontaneously react.	The semipermeable plastic membrane in the membrane cell used in sodium hydroxide production. The membrane separates chlorine and hydrogen gases, which are continuously removed.
inert or reactive electrode materials	Inert electrodes are not consumed in the reaction, whereas reactive electrodes are. The choice will depend on factors such as the cost of the electrodes (e.g. platinum electrodes are very expensive) and the ability of the electrodes to withstand cell operating conditions (e.g. electrodes must have a high melting point for use in cells with a molten electrolyte).	The carbon anode and iron cathode in a Downs cell are inert, relatively cheap and have high melting points. The reactive carbon anode in the Hall–Héroult cell used in aluminium production is also relatively cheap, so is inexpensive to replace.
molten or aqueous electrolyte	The choice of molten or aqueous electrolyte will depend mainly on whether the presence of water will interfere with the electrolytic production of the desired products. Water is a stronger oxidising agent than ions such as $Al^{3+}(aq)$, $Mg^{2+}(aq)$, $Na^{+}(aq)$ and $K^{+}(aq)$. The reduction of these ions to their respective metals cannot occur in the presence of water.	Downs cells have a molten sodium chloride electrolyte with added calcium chloride in the production of sodium metal and chlorine gas. Water is a stronger oxidising agent than $Na^{+}(aq)$. In the membrane cell, electrolysis of aqueous sodium chloride results in the production of hydrogen gas instead of sodium metal.
chemical additives to electrolyte	Chemical additives lower the melting point of a molten electrolyte or are the solvent for the compound that is electrolysed.	In the Downs cell, the addition of calcium chloride to molten sodium chloride lowers its melting point. In the Hall–Héroult cell, molten cryolite (Na_3AlF_6) is the solvent for alumina (Al_2O_3).

8.2 Review

SUMMARY

- The design features of industrial electrolytic cells must consider the:
  - chemical nature of the reactants and products
  - physical conditions under which the electrolysis takes place
  - energy requirements of operating the cell.
- To obtain the desired products from electrolysis reactions, industrial cells may use:
  - aqueous or molten electrolytes
  - a reactive or inert anode (the cathode is always inert).
- Cells that use molten electrolytes require a higher energy input than aqueous electrolytic cells because the electrolyte must be kept in a molten state.
- Cells are designed to keep the products at each electrode separate. They are then continuously removed. The products would spontaneously react if they came in contact with each other.
- The electrochemical series is useful for predicting the products and selecting the reactants for a particular industrial cell.

KEY QUESTIONS

Knowledge and understanding

1 Explain why the sodium metal and chlorine gas produced in a Downs cell must be separated, then continuously removed from the cell.

2 Explain why the iron cathode in a Downs cells does not get oxidised.

Analysis

3 Zinc can be extracted from its compounds by electrolysis at room temperature with the cell shown in the diagram below.

   a Explain whether an inert or reactive electrode would be suitable for use as an anode.

   b Explain why an aqueous solution of zinc nitrate, instead of molten zinc nitrate, is used as the electrolyte in this cell.

   c Identify the gas formed at the anode.

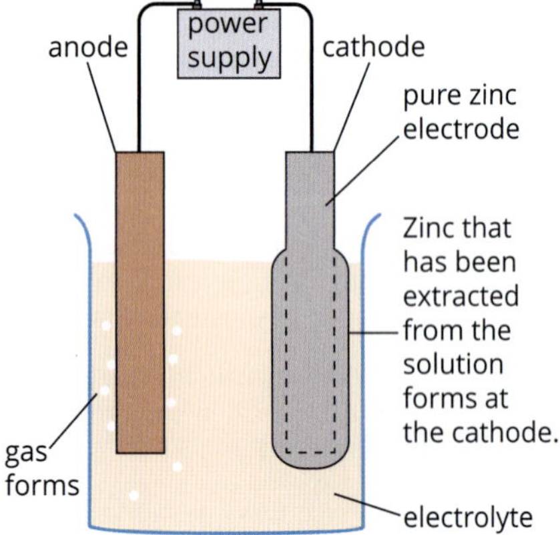

4 Explain the following.

   a Most chlorine is produced by electrolysis of a concentrated sodium chloride solution, rather than by electrolysis of the molten salt.

   b Fluorine was first isolated from fluorine compounds by electrolysis rather than using other techniques.

   c Calcium metal is not formed at the cathode in the Downs cell, even though calcium chloride is present in the electrolyte.

5 Chemists are investigating the possibility of replacing the carbon anodes in the electrolytic cell used for aluminium extraction from alumina (Al_2O_3) in a molten electrolyte. The cell is normally constructed so that the electrodes take part in the reaction and form carbon dioxide. Discuss two advantages of changing to unreactive anodes.

8.3 Faraday's laws

Industrial chemists monitoring the operation of commercial electrolytic cells need to know the relationship between the amount of product produced by the cell, the size of the electric current and the operating time of the cell.

In this section, you will learn about the laws that Michael Faraday discovered in 1834 which allow chemists to understand relationships such as these. You will also learn how these laws can be applied to an electrolytic process called **electroplating**.

ELECTROPLATING

One commercially important application of electrolysis is electroplating. During this process, a thin surface coating of metal, only a fraction of a millimetre thick, can be applied over another surface.

A common application of electroplating is the coating applied to 'tin' cans (Figure 8.3.1). Tin cans are used extensively for packaging food items as varied as soups, fruit and fish. Although they are commonly called 'tin' cans, they are mainly composed of steel, which is an alloy of iron and carbon. There is just a thin layer of tin plated over the surface of the steel can. Tin corrodes very slowly and prevents contact between the iron, moisture and air. Effectively, the tin prevents the iron from rusting.

FIGURE 8.3.1 Tin cans are used as packaging for a wide range of products. The tin coating on the can prevents it from corroding.

Electroplating cells

Electroplating is performed in electrolytic cells, such as the simplified cell shown in Figure 8.3.2 for tin plating. The object to be plated is connected by a wire to the negative terminal of a power supply. This object becomes the cathode (–) in the cell.

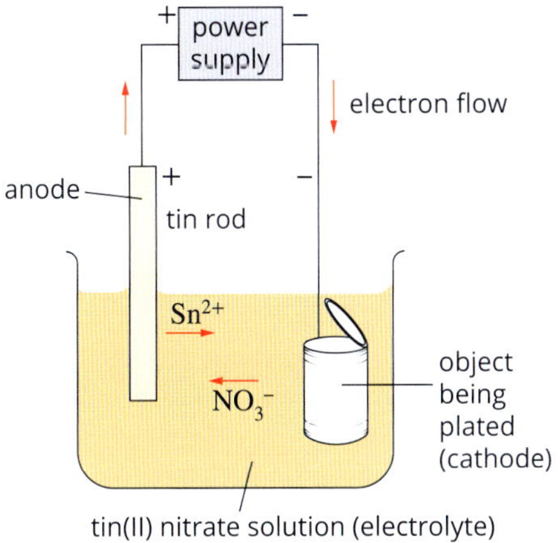

FIGURE 8.3.2 A cell used for electroplating tin

The object is immersed in an electrolyte solution, such as tin(II) nitrate solution, which contains ions of the metal to be plated. The Sn^{2+} ions in the solution move towards the negative electrode (the object to be plated) and NO_3^- ions move towards the positive electrode, allowing current to pass through the cell.

An electrode of tin metal is connected to the positive terminal of the power supply.

When the cell is in operation, the power supply acts as an 'electron pump', pushing electrons onto the negative electrode and removing electrons from the positive electrode.

At the negative electrode (cathode)

As shown in Figure 8.3.3, Sn^{2+} ions are attracted to the negative electrode, where they accept electrons and are converted to tin metal:

$$Sn^{2+}(aq) + 2e^- \longrightarrow Sn(s)$$

A coating of tin is formed on the object. Because reduction of the Sn^{2+} ions occurs here, the negative electrode is acting as a cathode.

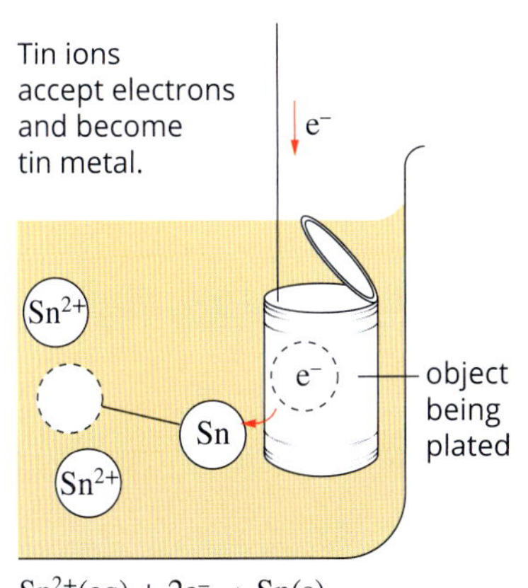

FIGURE 8.3.3 A representation of the reaction at the cathode in the process of tin plating

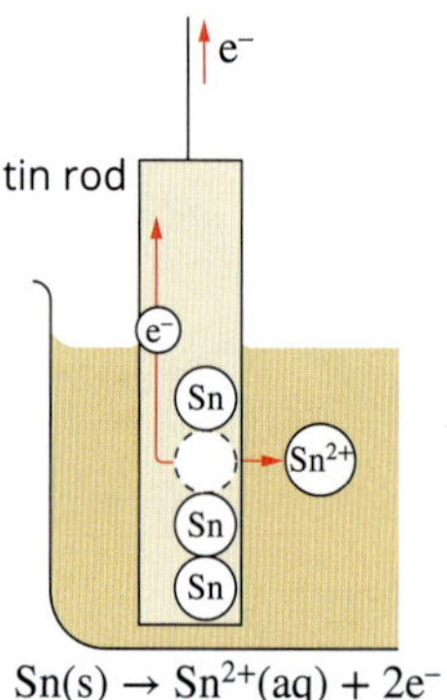

FIGURE 8.3.4 A representation of the reaction at the anode in the process of tin plating

The set-up of an electroplating cell can be summarised as follows.

- The object being plated is the cathode (negative).
- An electrode of the metal is the anode (positive).

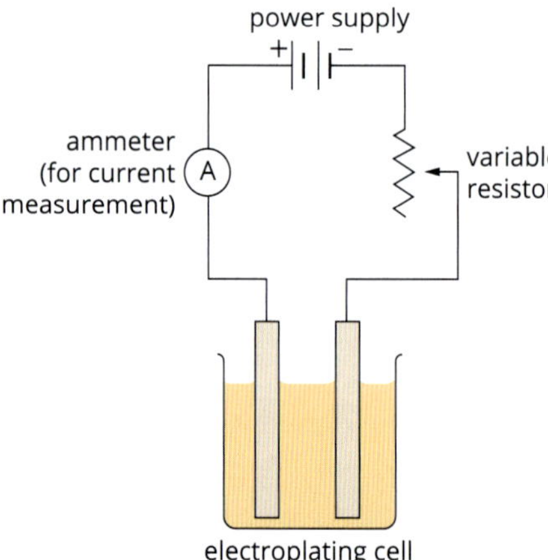

FIGURE 8.3.5 Experimental set-up used to investigate electroplating

At the positive electrode (anode)

As shown in Figure 8.3.4, the positive terminal of the power supply withdraws electrons from the tin electrode, causing an oxidation reaction to occur. Tin metal slowly dissolves as Sn^{2+} ions are formed:

$$Sn(s) \longrightarrow Sn^{2+}(aq) + 2e^-$$

This reaction replaces the Sn^{2+} ions in the solution that were consumed by the reaction at the negative electrode. The overall concentration of Sn^{2+} ions in the electrolyte remains constant.

Because an oxidation reaction is involved, the positive electrode is acting as an anode.

Several key factors determine the amounts of products that will form in an electrolytic cell, such as an electroplating cell, including the:

- charge on the ion involved in the electrode reaction
- current flowing through the cell
- length of time that the current flows.

Faraday described the relationship between these factors. You will now apply your knowledge of electroplating cells to learn about Faraday's laws.

FARADAY'S FIRST LAW OF ELECTROLYSIS

An electroplating cell can be constructed as shown in Figure 8.3.5. The variable resistor shown is used to keep the current constant.

This cell can be used to study the relationship between the mass of metal formed at the cathode and the quantity of charge that passes through the cell. Although the mass of metal formed is easily measured by weighing the cathode, the **electric charge** is determined indirectly.

Electric charge, symbol Q, is measured using the unit **coulombs**. The electric charge passing through a cell may be calculated from measurements of the current, I, through the cell and the time, t, for which the current flows. Current is measured in amperes (symbol A), commonly known as amps, and time is measure in seconds.

The relationship is:

$$\text{charge (coulombs)} = \text{current (amps)} \times \text{time (seconds)}$$

$$Q = I \times t$$

Experimental data from a silver-plating cell

An experiment was set up using the equipment shown in Figure 8.3.5 to determine the relationship between the amount of charge and the mass of silver formed in a silver-plating cell.

As Table 8.3.1 shows, the experimental data indicates that as more charge passes through the cell, more metal is formed at the cathode.

TABLE 8.3.1 The relationship between charge and the mass of silver formed in a silver-plating cell

Current (A)	Time (s)	Charge (C) $Q = I \times t$	Mass of silver formed (g)
2.0	200	400	0.45
2.0	400	800	0.91
2.0	600	1200	1.34
2.0	800	1600	1.79
2.0	1000	2000	2.24

The graph in Figure 8.3.6 shows the relationship between charge and the mass of silver formed. The mass of silver produced at the cathode is directly proportional to the electrical charge passed through the cell. This is **Faraday's first law of electrolysis**. It may be written symbolically as:

$$m \propto Q$$

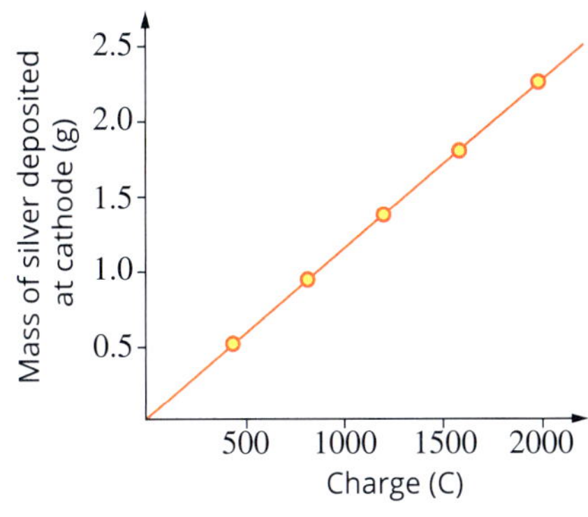

FIGURE 8.3.6 The variation with charge of mass of silver formed at the cathode in a silver-plating cell

FARADAY'S SECOND LAW OF ELECTROLYSIS

The previous experiment can be repeated using electroplating cells for metals other than silver. Results from experiments with cells using electrolytes containing Cu^{2+}, Sn^{2+}, Pb^{2+}, Cr^{3+} and Ag^{+} ions are shown in the graph in Figure 8.3.7.

Although the data for each metal obeys Faraday's first law, at first inspection there seems little relationship between the graphs for each metal. However, if the amount of each metal, measured in moles, is graphed against charge, the results are more interesting (Figure 8.3.8). All the data lies along just three lines. One mole of silver is produced by about 96 500 coulombs (C) of charge, whereas one mole of copper, tin or lead is produced by 193 000 (which is 2 × 96 500) coulombs.

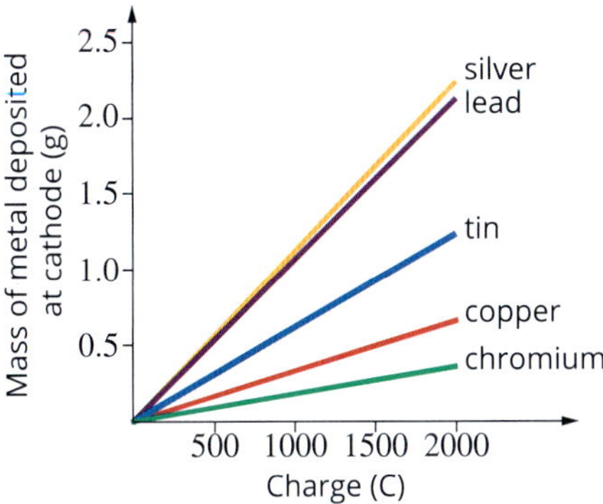

FIGURE 8.3.7 Variation with charge of the mass of metal formed at the cathode for different metal-plating cells

An underlying pattern is now evident. To deposit 1 mole of silver from a solution containing Ag^{+} ions, just 1 mole of electrons is required:

$$Ag^{+}(aq) + e^{-} \longrightarrow Ag(s)$$

The charge on 1 mole of electrons must be 96 500 C. This amount of charge is also known as a **faraday**, and has the symbol F. Therefore, 1 F = 96 500 C.

The charge on a given number of moles of electrons, $n(e^{-})$, can be readily calculated. For example, 2 moles of electrons have a charge of 193 000 C. We can express this mathematically by:

$$Q = n(e^{-}) \times F$$

or

$$Q = n(e^{-}) \times 96\,500$$

where Q is the charge (C), $n(e^{-})$ is the number of moles of electrons and F is Faraday's constant (96 500 C mol^{-1}).

The electrode reactions when copper, tin or lead are electroplated are:

$$Cu^{2+}(aq) + 2e^{-} \longrightarrow Cu(s)$$
$$Sn^{2+}(aq) + 2e^{-} \longrightarrow Sn(s)$$
$$Pb^{2+}(aq) + 2e^{-} \longrightarrow Pb(s)$$

For 1 mole of each of these metals to be deposited, 2 moles of electrons are required. Therefore, 193 000 C (2 × 96 500 C) must be passed through the cell.

To deposit 1 mole of chromium, 289 500 C (3 × 96 500 C) are required, because the equation for the electrode reaction is:

$$Cr^{3+}(aq) + 3e^{-} \longrightarrow Cr(s)$$

This is just what you can observe in the graph in Figure 8.3.8. Three different lines on the graph arise because of the different charges on the metal ions.

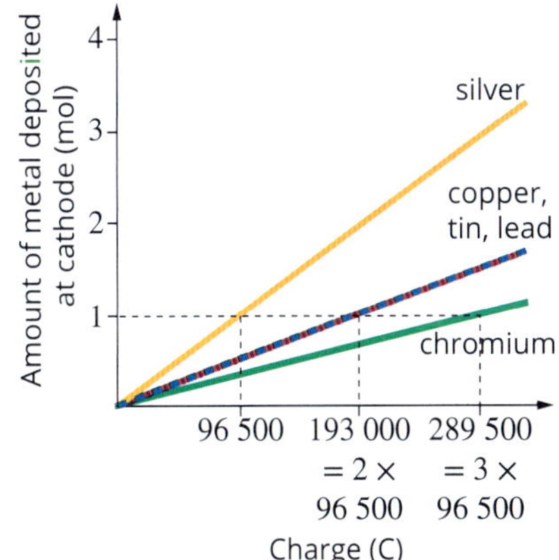

FIGURE 8.3.8 Variation with charge of amount of metal (in moles) formed at the cathode for different metal-plating cells

Faraday's second law of electrolysis states that to produce 1 mole of a metal, 1, 2, 3 or another whole number of moles of electrons must be consumed.

Although Faraday's laws were developed by studying the deposition of metal at the cathode in electroplating cells, the laws apply just as well to the production or consumption of any substance at the electrodes of either electrolytic cells or galvanic cells. Calculations are based on the two relationships:

$$Q = I \times t$$
$$Q = n(e^{-}) \times F$$

> Faraday's laws can be restated more generally as follows.
> - The mass of any substance deposited, evolved or dissolved at an electrode in an electrochemical process is directly proportional to the electrical charge passed through the cell.
> - For 1 mole of a substance to be deposited, evolved or dissolved at an electrode, the passage of 1, 2, 3 or another whole number of moles of electrons is required.

CHEMFILE

Michael Faraday's electrolysis experiments

Michael Faraday (1791–1867) was an English scientist who made great contributions to the fields of electromagnetism and electrochemistry (see figure below). His experiments on the decomposition of tin(II) chloride revealed that the amounts of tin and chlorine gas that were produced were always in proportion to the amount of electric charge that was passed through the cell. He also played a key role in popularising the use of the terms 'anode', 'cathode', 'electrode' and 'electrolyte'.

A drawing of Michael Faraday working on an electrolysis experiment

CALCULATIONS USING FARADAY'S LAWS

Worked examples 8.3.1 and 8.3.2 show you how to calculate the mass of a chemical produced at an electrode of an electrolytic cell.

Worked example 8.3.1

CALCULATING THE MASS OF A PRODUCT AT AN ELECTRODE

A chromium-plating cell operates with a steady current of 30.0 A for 20.0 minutes. What mass of chromium is plated on the object at the cathode? The equation for the reaction is:

$$Cr^{3+}(aq) + 3e^{-} \rightarrow Cr(s)$$

Thinking	Working
Calculate the quantity of charge passing through a cell using the formula: $Q = I \times t$ Remember that time must be expressed in seconds.	$Q = I \times t$ $= 30.0 \times (20.0 \times 60)$ $= 3.60 \times 10^4$ C
Calculate the amount of electrons, in mol, that passed through the cell using the formula: $n(e^-) = \frac{Q}{F}$ Recall that $F = 96\ 500$.	Rearrange to find $n(e^-)$: $n(e^-) = \frac{Q}{F}$ $= \frac{3.60 \times 10^4}{96500}$ $= 0.373$ mol
Use the mole ratio from the equation for the reaction to determine the amount, in mol, of metal plated at the cathode.	The cathode reaction is: $Cr^{3+}(aq) + 3e^{-} \rightarrow Cr(s)$ 3 moles of electrons will deposit 1 mole of chromium metal. Mole ratio: $\frac{n(Cr)}{n(e^-)} = \frac{1}{3}$ $n(Cr) = \frac{1}{3} \times n(e^-)$ $= \frac{1}{3} \times 0.373$ $= 0.124$ mol
Calculate the mass of metal plated at the cathode.	The molar mass of Cr is 52.0 g mol^{-1}. $m(Cr) = n \times M$ $= 0.124 \times 52.0$ $= 6.47$ g

Worked example: Try yourself 8.3.1

CALCULATING THE MASS OF A PRODUCT AT AN ELECTRODE

A copper-plating cell operates with a steady current of 20.0 A for 15.0 minutes. What mass of copper is plated on the object at the cathode? The equation for the reaction is:

$$Cu^{2+}(aq) + 2e^{-} \rightarrow Cu(s)$$

Faraday's laws can also be used to calculate the time it would take to produce a specific amount of product at an electrode.

Worked example 8.3.2

CALCULATING THE TIME TAKEN TO PRODUCE A PRODUCT AT AN ELECTRODE

How long would it take, in hours, to deposit 50.0 g of copper at the cathode of a copper-plating cell operating at a current of 8.00 A? The half-equation for the reaction is:

$$Cu^{2+}(aq) + 2e^- \longrightarrow Cu(s)$$

Thinking	Working
Calculate the amount of substance, in mol, that was deposited or consumed at the electrode using the formula: $n = \frac{m}{M}$	$n(Cu) = \frac{m}{M}$ $= \frac{50.0}{63.5}$ $= 0.787$ mol
Calculate the amount of electrons, in mol, that passed through the cell using the mole ratio from the equation.	The cathode reaction is: $Cu^{2+}(aq) + 2e^- \longrightarrow Cu(s)$ 2 mol of electrons will deposit 1 mol of copper metal. $\frac{n(e^-)}{n(Cu)} = \frac{2}{1}$ $n(e^-) = 2 \times n(Cu)$ $= 2 \times 0.787$ $= 1.57$ mol
Calculate the quantity of charge that passed through the cell using the formula: $Q = n(e^-) \times F$	$Q = n(e^-) \times F$ $= 1.57 \times 96\,500$ $= 1.52 \times 10^5$ C
Calculate the time required using the formula: $t = \frac{Q}{I}$	$t = \frac{Q}{I}$ $= \frac{1.52 \times 10^5}{8.00}$ $= 1.90 \times 10^4$ s
Convert seconds to hours by dividing by (60 × 60).	$t = \frac{1.90 \times 10^4}{60 \times 60}$ $= 5.28$ h

Worked example: Try yourself 8.3.2

CALCULATING THE TIME TAKEN TO PRODUCE A PRODUCT AT AN ELECTRODE

How long would it take, in hours, to deposit 20.0 g of silver at the cathode of a silver-plating cell operating at a current of 6.50 A? The half-equation for the reaction is:

$$Ag^+(aq) + e^- \longrightarrow Ag(s)$$

Although you have looked at the application of Faraday's laws in the context of electrolytic cells, the laws can also be applied to galvanic cells, such as fuel cells, which you learnt about in Chapter 5.

Worked example 8.3.3

CALCULATING THE TIME TAKEN TO USE A FIXED MASS OF FUEL IN A FUEL CELL

A fuel cell uses ethanol (C_2H_5OH) as its fuel source. How long would it take, in seconds, for 200 g of ethanol to be used in the fuel cell operating at a current of 1.25 A? The half-equation for the reaction is:

$$C_2H_5OH(l) + 3H_2O(l) \rightarrow 2CO_2(g) + 12H^+(aq) + 12e^-$$

Thinking	Working
Calculate the amount of ethanol, in mol, used in the fuel cell using the formula: $n(C_2H_5OH) = \frac{m}{M}$	$n(C_2H_5OH) = \frac{m}{M}$ $= \frac{200}{46.0}$ $= 4.35$ mol
Use the mole ratio from the equation for the reaction to determine the amount of electrons, in mol, produced in the reaction.	The reaction is: $C_2H_5OH(l) + 3H_2O(l) \rightarrow 2CO_2(g) + 12H^+(aq) + 12e^-$ 12 moles of electrons are produced for every 1 mole of ethanol that reacts. Mole ratio: $\frac{n(e^-)}{n(C_2H_5OH)} = \frac{12}{1}$ $n(e^-) = 12 \times n(C_2H_5OH)$ $= 12 \times 4.35$ $= 52.2$ mol
Calculate the quantity of charge generated by the fuel cell using the formula: $Q = n(e^-) \times F$	$Q = n(e^-) \times F$ $= 52.2 \times 96\,500$ $= 5.03 \times 10^6$ C
Calculate the time taken to use the ethanol using the formula: $Q = I \times t$	Rearrange to find t: $t = \frac{Q}{I}$ $= \frac{5.03 \times 10^6}{1.25}$ $= 4.03 \times 10^6$ s

Worked example: Try yourself 8.3.3

CALCULATING THE TIME TAKEN TO USE A FIXED MASS OF FUEL IN A FUEL CELL

A fuel cell uses propanal (CH_3CH_2CHO) as its fuel source. How long would it take, in seconds, for 375 g of propanal to be used in the fuel cell operating at a current of 5.40 A? The half-equation for the reaction is:

$$CH_3CH_2CHO(l) + 5H_2O(l) \rightarrow 3CO_2(g) + 16H^+(aq) + 16e^-$$

8.3 Review

SUMMARY

- Electroplating is an application of electrolysis. It involves the deposition of a thin surface coating of metal over another surface to improve the appearance of an object, its resistance to corrosion, or both.
- The charge passing through an electrochemical cell can be calculated from the formula:

  charge (coulombs) = current (amps) × time (seconds)

  $$Q = I \times t$$
- Faraday's first law states that the mass of any substance deposited, evolved or dissolved at an electrode in an electrochemical process is directly proportional to the electric charge passed through the cell.
- Faraday's second law states that for 1 mole of a substance to be deposited, evolved or dissolved at an electrode, 1, 2, 3 or another whole number of moles of electrons is required.
- The charge on one mole of electrons is called a faraday (*F*), where 1 *F* = 96 500 C.
- The charge on an amount of electrons can be calculated from:

  charge (coulombs)

  = amount of electrons (moles) × 96 500

  $$Q = n(e^-) \times F$$
- Faraday's laws can be used to calculate the:
  - mass of a reactant consumed or product formed in a cell, given the current and the operating time
  - current passing through a cell, given the mass of the reactant consumed or product formed and the operating time
  - time required for a cell to operate, given the mass of the reactant consumed or product needed and the current through the cell.

KEY QUESTIONS

Knowledge and understanding

1 Write half-equations for the formation of:

  a silver metal from a silver nitrate solution

  b zinc metal from a zinc nitrate solution

  c chlorine gas from molten potassium chloride

  d hydrogen gas from a sodium chloride solution

2 How much charge is required to produce the following?

  a 1 mol of silver metal from a silver nitrate solution

  b 0.2 mol of zinc metal from a zinc nitrate solution

  c 25 mol of chlorine gas from molten potassium chloride

  d 0.006 mol of hydrogen gas from sodium chloride solution

Analysis

3 A copper-plating cell with a $CuSO_4(aq)$ electrolyte operates with a current of 12.5 A for 75.0 minutes. What mass of copper, in g, is plated on the object at the cathode?

4 A car's headlights are left on and the battery discharges at a current of 9.0 A. How long would it take, in hours, for 52.9 g of lead to be lost from the anode? The reaction at the anode is represented by the half-equation:

$$Pb(s) + SO_4^{2-}(aq) \rightarrow PbSO_4(s) + 2e^-$$

5 The diagram below shows an electroplating cell that was devised by a student to place a zinc coating onto a key. An electric current is passed through a solution of zinc nitrate ($Zn(NO_3)_2$) using an inert carbon electrode as the anode.

  a Write a half-equation for the reaction that occurs at the electrode connected to the negative terminal of the power supply.

  b What current is required if 0.085 g of zinc is deposited on the surface of the key when a current is passed through the cell for 21.0 minutes?

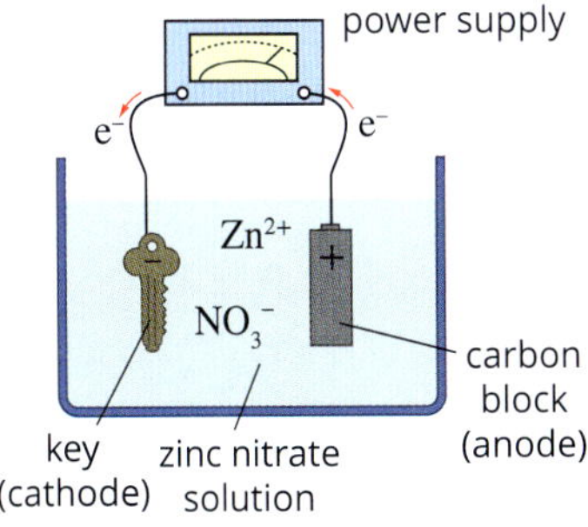

6 Electrolysis of a molten ionic compound with a current of 1.50 A for 25.0 minutes yielded 1.40 g of a metallic element at the cathode. If the element has a relative atomic mass of 118.7, calculate the charge on the metal ions.

7 Determine the mass of methanol, in kg, that would be consumed in a 24-hour period, in a methanol-oxygen fuel cell, producing a current of 3.3 kA. The oxidation half-equation is:

$$CH_3OH(l) + H_2O(l) \rightarrow CO_2(g) + 6H^+(aq) + 6e^-$$

Chapter review

KEY TERMS

anode
brine
cathode
coulomb
electric charge
electrochemical cell
electrochemical series
electrolysis
electrolyte
electrolytic cell
electroplating
faraday
Faraday's first law of electrolysis
Faraday's second law of electrolysis
inert
inert electrode
molten
non-spontaneous reaction
reactive electrode
semipermeable membrane
spontaneous reaction

REVIEW QUESTIONS

Knowledge and understanding

1 For electrolytic and galvanic cells, compare the:

a polarity of the anode and cathode

b direction of electron flow

c energy transformation occurring in the cells

d tendency of the cell reaction to occur spontaneously.

2 Write half-equations for the electrode reactions and an overall equation for the reaction that occurs when molten $SnCl_2$ is used as the electrolyte in an electrolytic cell with inert electrodes.

3 An electrochemical cell, like the one shown below, is set up to produce calcium metal. Label the parts of the cell using the following options (options may be used more than once): Cl^-, molten calcium chloride, platinum, Ca^{2+}.

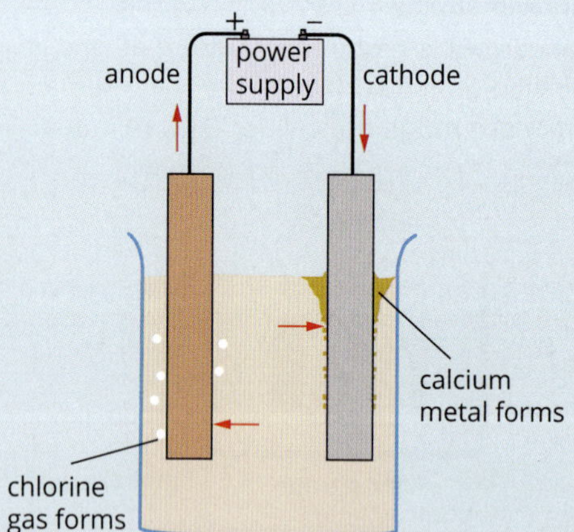

4 Hydrogen and oxygen gases are formed at the electrodes from the electrolysis of which very concentrated aqueous solution?

A KBr

B NaCl

C H_2SO_4

D $Co(NO_3)_2$

5 Which of the following are the correct products when an aqueous solution of iron(II) nitrate is electrolysed using copper electrodes?

	Cathode	Anode
A	$H_2(g)$	$Fe^{3+}(aq)$
B	Fe(s)	$Fe^{3+}(aq)$
C	Fe(s)	$Cu^{2+}(aq)$
D	$H_2(g)$	$O_2(g)$

6 Two electrolytic cells were set up for the electrolysis of aqueous lead(II) nitrate. One cell contained lead electrodes and the other contained platinum electrodes. Determine the reactions that would occur at the electrodes of both cells.

7 Sodium and chlorine are manufactured by passing direct current through molten sodium chloride. Why is it necessary for the electrolyte to be molten and not aqueous?

8 A membrane cell is used to produce sodium hydroxide through electrolysis of a concentrated sodium chloride solution (brine).

a Write half-equations for the reactions that occur at each electrode.

b State the purpose of the semipermeable membrane in the cell.

Application and analysis

9 When constructing a galvanic cell in the laboratory, why are two half-cells usually used, whereas in an electrolytic cell the reactants are often placed in a single container?

10 Construct a flow chart that includes the terms electrolysis, electrolytic cell, chemical energy, electrical energy, anode, cathode, reducing agent, oxidising agent, reduction, oxidation, electrolyte and non-spontaneous reactions.

11 A student tried to electroplate manganese onto a cobalt electrode using a cell made of a manganese anode, a cobalt cathode and a manganese(II) nitrate solution as electrolyte (see figure below) The student discovered that no manganese has plated onto the cathode, despite careful checks of all the electrical connections in the electrolytic cell. Explain why the manganese did not plate onto the cathode and what reaction occurred instead at the cobalt electrode.

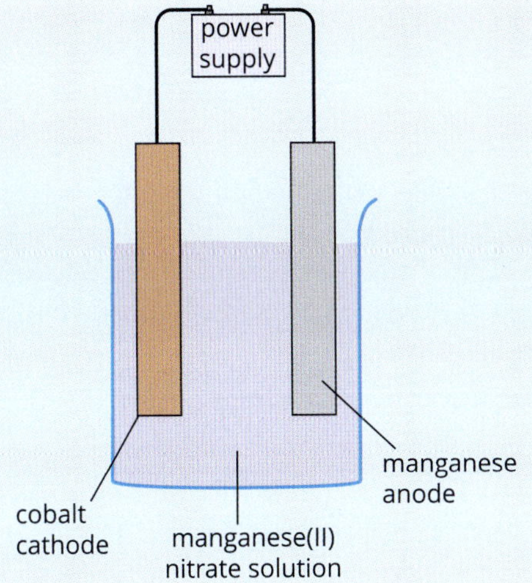

12 An aqueous zinc sulfate solution was electrolysed using copper electrodes.

a Refer to the electrochemical series and identify the possible reactions that could occur at each electrode. Write these half-equations in the order they appear in the series.

b Identify the half-equations for the reactions that occur at the anode and cathode.

c Write an equation for the overall reaction.

13 Using the electrochemical series, complete the table by predicting the initial reactions occurring at the anode and cathode in each of the following electrolysis experiments.

Experiment	Electrolyte	Electrodes	Cathode reaction	Anode reaction
a	molten potassium iodide	platinum		
b	1 M copper(II) sulfate solution	platinum		
c	1 M potassium bromide solution	copper		
d	a mixture of 1 M copper(II) nitrate and 1 M nickel(II) nitrate solutions	carbon		
e	1 M lithium fluoride solution	platinum		

14 Early attempts to produce aluminium by electrolysis of aqueous solutions of aluminium compounds were unsuccessful. Use the electrochemical series to explain why.

15 In the electrolysis of aqueous lithium sulfate using inert electrodes, 0.05 mol of a gas was formed at the anode. Which of the following is the correct description of the gas produced at the cathode?

	Gaseous product at cathode	Amount of product at cathode (mol)
A	hydrogen	0.05
B	oxygen	0.05
C	hydrogen	0.1
D	oxygen	0.1

16 0.019 *F* of electricity passes through a molten chromium compound to produce 0.50 g of chromium metal. Which of the following is most likely to be the oxidation number of chromium in the compound?

A +2
B +3
C +4
D +6

17 Determine the mass of zinc, in g, consumed in an alkaline battery used in a torch that has been turned on for 1.2 hours and draws a current of 0.45 A. The half-equation for the oxidation of zinc is:

$$Zn(s) + 2OH^-(aq) \rightarrow Zn(OH)_2(s) + 2e^-$$

A 0.011
B 0.66
C 1.3
D 2.6

18 A hydrogen-oxygen fuel cell in a car consumes 2.48 mL of hydrogen stored at 70.0 MPa and 25.0°C over a 45.0 minute period. Calculate the current, in A, generated by the fuel cell.

19 A charge of 38 600 C was passed through 1.00 L of 1.00 M copper(II) sulfate solution using carbon electrodes.

a Write half-equations for the reactions that occur at each electrode.

b Calculate the concentration of the copper(II) ions in solution after electrolysis. Assume no change in the volume of the electrolyte.

20 A number of important metals are produced by electrolysis of molten salts. In one such process a current of 25 000 A produces 272 kg of metal in 24 hours. If the cation in the salt has a 2+ charge, what is the identity of the metal?

21 Aluminium–air cells are a potential alternative to lithium-ion cells used in electric vehicles. When the cells produce current, aluminium metal is oxidised at the negative electrode and oxygen gas is reduced to water at the positive electrode. The electrodes are separated by an alkaline electrolyte.

a Write a half-equation for the reaction occurring at the negative electrode of the cell.

b An aluminium–air cell operates with a steady current of 105 mA. The cell will continue to operate until 1.15 g of aluminium at the negative electrode has been used up. Determine how many hours the cell will run for.

22 A fuel cell makes use of glucose ($C_6H_{12}O_6$) in wastewater as a fuel source. The glucose is completely oxidised to produce carbon dioxide and water. The anode and cathode containing compartments of the fuel cell are separated by a proton exchange membrane which only allows H^+ ions and water to pass through.

a Write a half-equation for the reaction occurring at the anode of the fuel cell.

b The fuel cell generates a current of 5.00 mA and operates continuously for 24 hours a day. Determine the amount of glucose, in mol, used up in the fuel cell in one hour.

c The concentration of glucose in the wastewater is 1.18×10^{-3} M. Determine the flow rate of wastewater coming into the fuel cell, in mL per hour.

23 Three electrolytic cells containing gold(I) nitrate solution, zinc sulfate solution and chromium(III) sulfate solution, respectively, were connected in series so that the same amount of electric charge passed through each cell (see the figure below). Metal was deposited at the cathode of each cell.

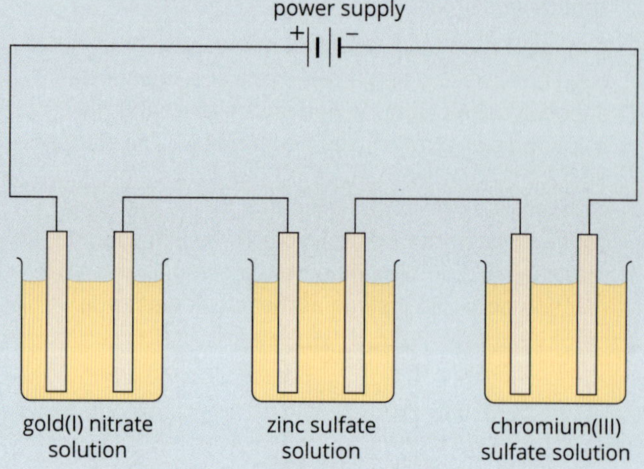

a For each cell, write the half-equation for the reduction reaction that occurs in that cell.

b If 15.0 g of gold was obtained from one cell, what mass of metal, in g, would be obtained from each of the other two cells?

24 A nickel teapot, with a surface area of 0.0900 m^2, is to be silver plated.

a Which electrode should the teapot be connected to?

b What is the polarity of this electrode?

c If the plating is to be 0.00500 cm thick and the density of silver is 10.5 g cm^{-3}, how long, in seconds, should the silver-plating cell be turned on for with a current of 0.500 A?

25 A student investigated Faraday's first law of electrolysis using an electrolytic cell where she copper-plated a nickel medallion in 250 mL of 1.00 M copper(II) sulfate solution with a large copper anode. For each run of the investigation, she turned on the cell for 20.0 minutes. Below are the results of her investigation.

Current (A)	Mass of copper deposited on the nickel medallion (g)
2.0	0.753
4.0	1.508
6.0	2.251
8.0	3.001
10.0	3.754

a Write an aim for this investigation.

b Identify the independent and dependent variables of this investigation.

c i Determine the mass of copper, in g, that was expected to be plated onto the medallion in the run with a current of 10.0 A.

ii Is the difference between the experimental and expected value caused by a random or systematic error?

iii Give a possible reason for the difference.

iv What will be the concentration of copper(II) sulfate remaining in solution? Assume there is no change in the volume of the electrolyte.

d The student then investigated the impact of replacing the copper anode with an inert graphite electrode. She ran a current of 10.0 A through the cell for 20.0 minutes.

i What gas was produced at the anode?

ii What was the concentration of copper(II) sulfate remaining in solution? Assume there was no change in the volume of the electrolyte.

26 Two electrolytic cells are connected in series, as shown below. A current of 7.50 A flows for 45.0 minutes.

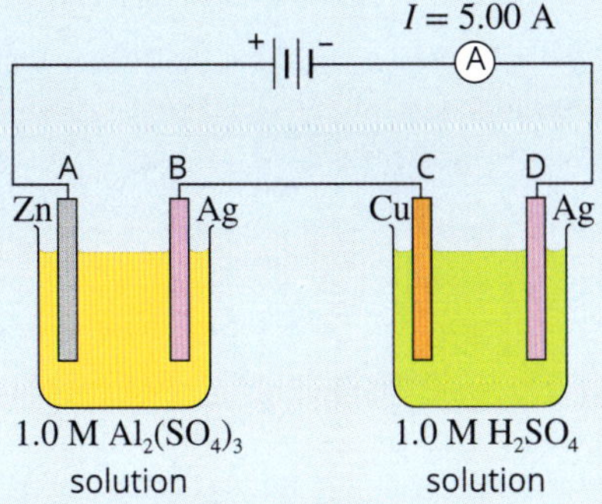

a Calculate the charge flowing through each cell.

b Write half-equations for the reactions occurring at each electrode (A–D) when the current starts to flow.

c Calculate the change in mass, in g, of electrode C after 45.0 minutes.

d Calculate the volume of gas, in L, measured under standard laboratory conditions, SLC, that is formed at electrode B after 45.0 minutes.

e As the current flows through the cells, how does the reaction at electrode D change?

27 Sodium and chlorine are produced by electrolysis of molten sodium chloride in the Downs cell. A typical Downs cell runs at 600°C, 8 V and with a current of 35 kA.

a In a 24-hour period, what mass of sodium metal, in kg, will be produced?

b In a 24-hour period, what volume of chlorine gas, measured under standard laboratory conditions, SLC, will be produced?

28 An electrolytic cell for the extraction of aluminium operates at a current of 150 000 A.

The electrode reactions are:

$$Al^{3+}\text{(in cryolite)} + 3e^- \longrightarrow Al(l)$$

$$C(s) + 2O^{2-}\text{(in cryolite)} \longrightarrow CO_2(g) + 4e^-$$

To produce 1.00 tonne of Al, calculate:

a how long the cell, in s, must operate

b the mass of carbon, in g, consumed at the anodes

c the volume of carbon dioxide gas produced in L, measured under standard laboratory conditions

d the amount of energy, in J, required if the cell operates at 5.00 V (use the formula E (in joules) = V (in volts) × Q (in coulombs))

e the cost of the electricity required if the smelter purchased electricity at the domestic rate of 28 cents per kilowatt hour (1 kilowatt hour = 3 600 000 J).

29 Rutile is a mineral composed of titanium and oxygen. In a trial electrolytic cell running at 100% efficiency, a potential difference of 4.0 V was applied and used 463.2 kJ of electricity to produce 14.37 g of titanium metal.

a Calculate the amount of titanium produced, in mol.

b Calculate the total charge of the electrons (Q) used to produce the titanium metal (use the formula E (joules) = V (volts) × Q (coulombs)).

c Calculate the amount of electrons, in mol, needed to produce the titanium metal.

d Calculate the charge of the titanium ion in rutile.

e Determine the empirical formula of rutile.

H2
HYDROGEN

CHAPTER

09 Designing for energy efficiency and sustainability

In 2015 the United Nations adopted the 2030 Agenda for Sustainable Development. At the heart of the agreement are 17 Sustainable Development Goals, which are a call to action by all countries. The goals recognise that ending poverty must go hand-in-hand with strategies that improve health and education, reduce inequality and minimise the negative impact of humankind on our natural world.

This chapter highlights the role of chemistry in achieving these goals. For a chemical industry, sustainability needs to be a priority in choosing a reaction pathway, in sourcing the raw materials needed, in minimising the energy consumption of the process and in the disposal of any waste materials. In this chapter, you will see that chemists have strategies to help meet sustainability goals.

Key knowledge

- responses to the conflict between optimal rate and temperature considerations in producing equilibrium reaction products, with reference to the green chemistry principles of catalysis and designing for energy efficiency **9.1**
- the common design features and general operating principles of rechargeable (secondary) cells, with reference to discharging as a galvanic cell and recharging as an electrolytic cell, including the conditions required for the cell reactions to be reversed and the electrode polarities in each mode (details of specific cells not required) **9.2**
- the role of innovation in designing cells to meet society's energy needs in terms of producing 'green' hydrogen (including equations in acidic conditions) using the following methods:
  - polymer electrolyte membrane electrolysis powered by either photovoltaic (solar) or wind energy **9.3**
  - artificial photosynthesis using a water oxidation and proton reduction catalyst system **9.3**

OA

9.1 Optimising the yield of industrial processes

The chemical industry is central to the world economy. The companies in this sector convert raw materials, such as oil, natural gas, air, water, metals and minerals, into tens of thousands of different products, including dyes and pigments, ammonia, chlorine, caustic soda, sulfuric and nitric acids, and organic chemicals.

Chemical manufacturing processes are not 'static'. Companies cannot establish a plant and just run the same process in the same way each year. The company must make a profit and it needs to minimise its impact on the environment. This requires research and a flexibility to adjust to changing market and social conditions.

Consider ethanol as an example. Australia manufactures large volumes of this chemical each year for use as a fuel. It can be made from crude oil or from plant waste. The ethanol industry must consider many factors, such as the following:

- the current price of crude oil—this will determine if ethanol is competitive with petrol
- whether there is more money to be made from converting sugarcane waste to fertiliser than converting it to ethanol
- if enzymes can be used effectively to allow forest waste to be utilised as a raw material rather than the more valuable sugarcane
- how to make sure waste from the plant is not leaking into local water streams
- if overall energy usage can be reduced to minimise costs.

All chemical industries face similar challenges. Research into process efficiencies and waste management is continuous.

In this section, you will learn how industrial chemists maximise the reaction rate and product **yield** of chemical reactions by manipulating the position of equilibrium, using catalysts and controlling the reaction rate (Figure 9.1.1). You will also learn about how companies apply green chemistry principles to minimise the impact of production on the environment and to ensure processes remain sustainable.

FIGURE 9.1.1 A chemical engineer at work. It is critical to select reaction conditions that carefully balance the demands of equilibrium yield, reaction rate and other factors, such as costs and safety.

CONFLICTS IN CHEMICAL MANUFACTURING

A chemical process must operate efficiently if it is to be economically viable. At the same time, companies must operate in a responsible manner that minimises hazards to employees and damage to the environment. At times, the need to be efficient can conflict with the need to be environmentally responsible.

A similar conflict can occur when an industrial chemist tries to increase the reaction rate and also increase the proportion of reactants converted into products. The production of ammonia is used below to explain this conflict in more detail.

Conflict resolution: Optimising ammonia production

Ammonia (NH_3) is one of the most commonly produced industrial chemicals. Most of the ammonia produced by industry is used in agriculture as fertiliser. Ammonia is also used in the manufacture of plastics, explosives, textiles, pesticides, dyes and other chemicals.

Ammonia is produced in large industrial plants, such as the one north of Newcastle in New South Wales (Figure 9.1.2), where Orica produces 360 kilotonnes of ammonia annually.

FIGURE 9.1.2 Orica produces ammonia at its Kooragang Island plant in New South Wales. It also produces nitric acid and ammonium nitrate.

Ensuring a high equilibrium yield gives the plant greater productivity and reduces waste and energy use. Maintaining a high reaction rate ensures the product is generated efficiently so that the plant can be economically viable.

Ammonia is manufactured from nitrogen gas (obtained from air) and hydrogen gas in the Haber process. The reaction is reversible and exothermic, and can be represented by the equation:

$$N_2(g) + 3H_2(g) \rightleftharpoons 2NH_3(g) \qquad \Delta H = -92 \text{ kJ}$$

Rate considerations

According to collision theory, the rate of reaction will be increased by increasing the frequency of collisions between reactant particles and by increasing the proportion of collisions that have energies equal to or greater than the required activation energy. This means that a higher reaction rate can be achieved when:

- the temperature is higher
- a catalyst is present
- the partial pressures of the gaseous reactants are higher (i.e there is a higher pressure overall).

CHEMFILE

Fritz Haber

Fritz Haber (1868–1934) was a German chemist who invented the process for producing ammonia that now bears his name. The reaction produces ammonia from atmospheric nitrogen and hydrogen according to the equation:

$$N_2(g) + 3H_2(g) \rightleftharpoons 2NH_3(g)$$

The Haber process made it feasible to produce fertilisers with a high nitrogen content. This allowed for better crop yields in agriculture. Easy access to reasonably priced fertilisers has reduced starvation and therefore saved lives in many parts of the world. This fact was acknowledged when Haber was awarded the Nobel Prize in Chemistry in 1918, following the end of World War I.

However, Haber's involvement in the production and use of chemical weapons by the German army in World War I meant that the award of a Nobel Prize was seen as controversial by many scientists, who considered him to be guilty of war crimes.

Fritz Haber invented a process for producing ammonia from atmospheric nitrogen and hydrogen. He was awarded the Nobel Prize in Chemistry in 1918.

> Choosing a suitable temperature for reversible, exothermic reactions can be difficult, because:
> - a high temperature can lead to a low equilibrium yield
> - a low temperature can cause the reaction rate to be too low.

Equilibrium considerations

Changing the temperature in the reactor will change the equilibrium constant of the reaction. Because the reaction in the Haber process is exothermic, lowering the temperature will increase the value of the equilibrium constant and increase the equilibrium yield of ammonia by favouring the net forward reaction.

It is evident from this ammonia example that exothermic, reversible reactions present a common problem of choosing a reaction temperature because:

- a high temperature increases the reaction rate but produces a low equilibrium yield
- a low temperature produces a high equilibrium yield but the rate is too low.

It can also be difficult to select an appropriate pressure for some processes that occur in the gas phase. High pressures that favour rapid reaction will sometimes give low equilibrium yields of product, depending on the stoichiometry of the reaction.

Table 9.1.1 summarises the conditions that favour high reaction rates and those that favour high equilibrium yields.

TABLE 9.1.1 Conditions for high rate of reaction versus high equilibrium yield

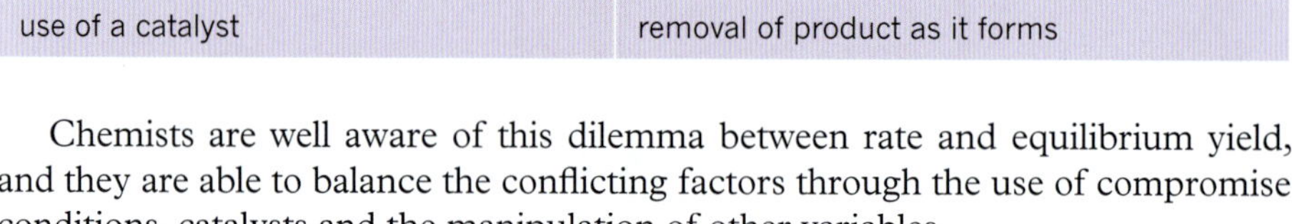

For high reaction rates	For high equilibrium yields
high concentrations/pressures	pressures depend on the relative numbers of reactant and product particles
high temperatures	low temperatures for exothermic reactions high temperatures for endothermic reactions
high surface area of solids	addition of excess reactant
use of a catalyst	removal of product as it forms

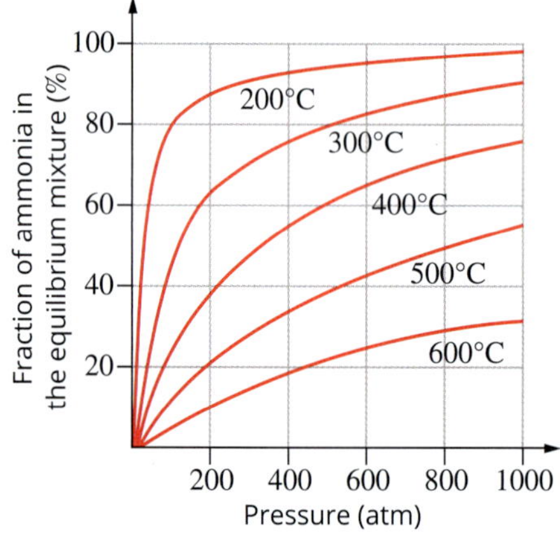

FIGURE 9.1.3 The percentage of ammonia present when a mixture of nitrogen and hydrogen has reached equilibrium

Chemists are well aware of this dilemma between rate and equilibrium yield, and they are able to balance the conflicting factors through the use of compromise conditions, catalysts and the manipulation of other variables.

In the case of ammonia, the equilibrium yield can be increased by increasing the pressure in the reactor. According to Le Châtelier's principle, increasing the pressure will cause the system to partially oppose this pressure change by shifting to the side of the equation with fewer gaseous particles to reduce the pressure.

Because the equation for the Haber process has four gaseous reactant particles and two gaseous product particles, increasing the pressure will cause a shift to the right and increase the amount of ammonia present at equilibrium.

The theoretical effects of changing temperature and pressure on the equilibrium yield of the Haber process are shown in Figure 9.1.3. As pressure is increased, the proportion of ammonia present at equilibrium increases. As the temperature is decreased, the proportion of ammonia present at equilibrium increases. Therefore, a high yield is obtained at a relatively low temperature and high pressure. (The impact of a pressure change on the position of equilibrium was discussed in Chapter 7.)

Table 9.1.2 summarises the effects of changing conditions on the reaction rate and the equilibrium yield of this particular reaction.

TABLE 9.1.2 Effect of changing reaction conditions on the equilibrium yield and rate of reaction in the Haber process

Condition	Effect on equilibrium yield	Effect on reaction rate
catalyst	no effect	increases
increasing temperature	decreases	increases
increasing pressure	increases	increases

Increasing the pressure of a system, especially on an industrial scale, is a costly and potentially hazardous process. However, in the production of ammonia, a high pressure favours both a high yield and high reaction rate. The economic benefits from the increased rate and yield outweigh the cost and safety issues of maintaining high pressures.

The actual conditions used to overcome the conflict between rate and yield in the production of ammonia are:

- high pressures of 100–250 atm (over 10^4 kPa)
- moderate temperatures of 400–450°C
- a porous iron/iron oxide (Fe_3O_4) catalyst.

Using the conditions outlined above for the production of ammonia is an example of green chemistry in practice. Optimum conditions mean that both energy consumption and wastage of raw materials are minimised. An efficient catalyst will enable the reaction temperature to be lower, further saving energy. The plant engineers will also need to ensure that pollutants are removed from gaseous emissions from the plant and that any water returned to the local water system has been treated to remove contaminants.

By careful control of reaction conditions, chemists can maximise the equilibrium yield and reaction rate for a desired product, while minimising the impact on the environment.

In industry, the best economic outcome is achieved by balancing the reaction conditions for high equilibrium yield and reaction rate.

CASE STUDY

Improving the catalyst for ammonia synthesis

The effectiveness of the catalyst in the Haber process is critical to its efficiency. Fritz Haber and his colleague Carl Bosch tested nearly 2000 different materials as catalysts when developing their process for ammonia synthesis.

By good fortune, the iron ore that they tested was from Sweden and contained traces of group 1 metal compounds that act as 'promoters', increasing the efficiency of the catalyst. A promoter creates many small pores in the catalyst, exposing iron crystals and providing a greater surface area and more sites for reaction. The catalyst widely used today is not very different—iron oxide with a potassium hydroxide promoter.

An alternative catalyst is ruthenium metal supported on high-surface-area carbon. This material began to attract considerable interest from industry in the 1990s. Although this catalyst is much more expensive, it is 45 times more active per square metre of surface, allowing the use of a lower operating temperature of about 400°C and 40 atm pressure. It has been installed in a number of ammonia plants.

The Haber process has now been used for over 100 years, but the research into suitable catalysts continues. This fact highlights the complexity of the mechanism by which catalysts work. It is now known that various combinations of metals and ionic compounds work better than pure metals, that some catalysts are efficient over a very narrow range of conditions and that the shape of the catalyst pellets is also relevant. The overall efficiency of the Haber process has not changed much in recent years, but production is at lower, more manageable, temperatures and pressures.

A dream for many chemists has been to devise a catalyst with the efficiency of nitrogenase, the enzyme found in nitrogen-fixing bacteria in legumes such as broad beans (Figure 9.1.4). Nitrogenase enzymes convert atmospheric nitrogen to ammonia without the need for the high pressures and temperatures of the industrial process. Harnessing this process would allow for the efficient production of ammonia at ambient temperatures and atmospheric pressure.

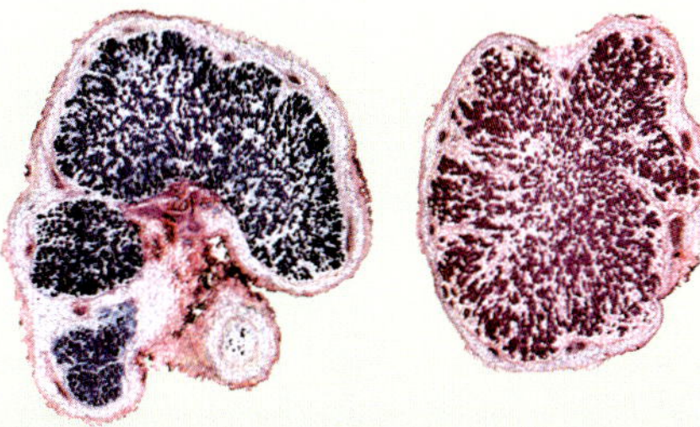

FIGURE 9.1.4 Sections through root nodules from broad beans, stained to show the presence of nitrogen-fixing bacteria

Nitrogenase contains clusters of iron, molybdenum and sulfur. The area of catalysis for the production of ammonia remains an active field; recent research includes the use of photocatalysts developed from osmium–gold nanoparticles.

9.1 Review

WS 18 OA ✓✓

SUMMARY

- Conditions for the reaction in any chemical process need to be selected carefully to ensure that both a reasonable equilibrium yield and reaction rate are achieved.
- A compromise is often needed to resolve a conflict in exothermic reactions between rate and equilibrium yield:
  - high temperature—a fast rate but a low yield
  - low temperature—a high yield but a low reaction rate.
- Chemical manufacturers have a responsibility to adhere to sustainable development goals by:
  - minimising energy use and transitioning to renewable sources where possible
  - improving the efficiency of manufacturing processes
  - using catalysts where possible to improve reaction efficiency
  - minimising gaseous emissions and wastewater contamination.

KEY QUESTIONS

Knowledge and understanding

1 In the industrial production of methanol, the conditions are adjusted to maximise the rate and the yield. The reaction is exothermic and is represented by the equation:

$$CO(g) + 2H_2(g) \rightleftharpoons CH_3OH(g)$$

Copy and complete the description of the effects of conditions on this process by selecting from the following terms:

reaction rate; temperature; pressure; high; low; catalyst.

In the production of methanol, both a higher reaction rate and a higher equilibrium yield could be achieved with a ________ pressure. In addition, a moderate ________ needs to be used to ensure that a reasonable reaction rate and equilibrium yield are achieved. A ________ is used to further increase the ________.

2 In the endothermic reaction below, specify what general conditions for temperature and pressure will maximise the equilibrium yield of NO_2.

$$N_2O_4(g) \rightleftharpoons 2NO_2(g)$$

3 **a** The yield of some reactions is improved if the pressure is increased. High pressure, however, can create other practical issues. What are these other issues?

b The yield of some reactions is improved if the temperature is increased. High temperature, however, can create other practical issues. What are these other issues?

Analysis

4 One step during sulfuric acid production is the oxidation of sulfur dioxide to sulfur trioxide:

$$2SO_2(g) + O_2(g) \rightleftharpoons 2SO_3(g) \qquad \Delta H = -197 \text{ kJ}$$

a Outline the conditions that would increase the equilibrium yield of sulfur trioxide.

b Outline the conditions that would increase the rate of production of sulfur trioxide.

c The actual conditions used for the production of sulfur trioxide are a temperature of 450°C and a pressure of 1 atm. Explain why this compromise is made for the production process.

5 During the Haber process, described on page 305 as a gas passes through the catalyst in the reactor its temperature increases.

a Why does the temperature of the gas rise?

b The gases leaving the reactor are cooled and liquid ammonia is removed before unreacted nitrogen and hydrogen are pumped back into the reactor. Why is this gas mixture cooled before re-entry into the reaction?

6 At 200°C, nearly twice as much ammonia forms when a 3 : 1 hydrogen/nitrogen mixture at 250 atm pressure reaches equilibrium than is formed at 400°C. Why are temperatures above 400°C used to manufacture ammonia?

7 The main process in the production of hydrogen from natural gas has the following equation:

$$CH_4(g) + H_2O(g) \rightleftharpoons CO(g) + 3H_2(g) \qquad \Delta H = +206 \text{ kJ}$$

a Consider the reactants in this process. Which of the reactants in the process is likely to present sustainability issues?

b CO is a product in this reaction. Describe the environmental issues associated with the production of CO.

c Is this process likely to be energy-intensive? Explain your answer.

9.2 Rechargeable cells and batteries

In Chapter 5, you learnt about primary galvanic cells. Once primary cells are discharged, they are often discarded, which is not a good outcome from a sustainability viewpoint. Secondary cells solve this problem as the discharge reactions in them can be reversed. **Rechargeable cells**, such as lithium-ion cells and nickel–metal hydride cells, are known as **secondary cells** or **accumulators**.

Rechargeable cells and batteries have become very popular and are found in mobile phones, laptop computers, cameras and portable power tools. They are also fundamental to the operation of electric vehicles and solar-power energy-storage systems.

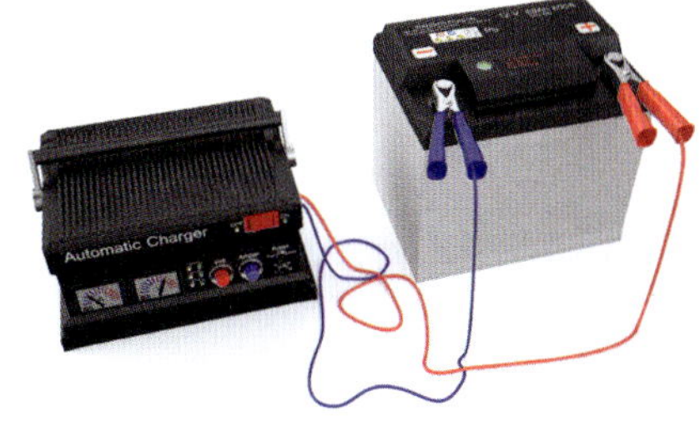

FIGURE 9.2.1 Charging a secondary battery involves connecting the positive terminal of a charger to the battery's positive electrode and the negative electrode of the charger to the battery's negative electrode.

RECHARGING PROCESS

Most types of rechargeable cells can undergo many hundreds of recharges. To recharge a cell, the cell reaction must occur in reverse: the products of the reaction must be converted back into the original reactants. This is done by connecting the cell to a charger—a source of electrical energy that has a potential difference a little greater than the potential difference of the cell. The positive terminal of the charger is connected to the cell's positive electrode and the negative electrode of the charger is connected to the cell's negative electrode (Figure 9.2.1).

A galvanic cell is rechargeable if the discharge reactions can be reversed efficiently. Reactions are more likely to be reversible when the products formed remain in contact with the electrodes.

For a cell to be rechargeable, the chemical changes occurring at both electrodes during **discharge** (when a current is being supplied by the cell) must be efficiently reversed. Reactions are more likely to be reversible if the electrodes are not damaged during discharge and the products remain in contact with the electrodes. Nickel and cadmium are popular metals in secondary cells as their hydroxides ($Cd(OH)_2$ and $Ni(OH)_2$) are not highly soluble and are easily converted back to the original metals. Zinc is less suited because $Zn(OH)_2$ will migrate from the electrode, leading to the possibility of hydrogen gas forming during recharge.

When a secondary cell discharges, it operates as a galvanic cell.
When it recharges, it operates as an electrolytic cell.

The energy transformations in a secondary cell can be summarised as follows.

- When a secondary cell discharges, it acts as a galvanic cell, converting chemical energy into electrical energy.
- When the cell is recharged, it acts as an electrolytic cell (a cell in which electrolysis can occur). Electrical energy is transformed into chemical energy in an electrolytic cell. The voltage required to recharge a cell must be slightly higher than the potential difference of the cell during discharge.

During the discharging process in a secondary cell:

- oxidation occurs at the negative terminal (the anode)
- reduction occurs at the positive terminal (the cathode).

However, when the cell is recharging, the cell reaction is reversed:

- oxidation occurs at the positive terminal (the anode)
- reduction occurs at the negative terminal (the cathode).

The energy transformations and redox processes that occur during the discharging and recharging of a secondary cell are summarised in Figure 9.2.2.

Discharging
- acts as galvanic cell
- spontaneous reaction
- negative terminal is anode

Recharging
- acts as electrolytic cell
- non-spontaneous reaction
- positive terminal is anode

FIGURE 9.2.2 Energy transformations in secondary cells

DEVELOPMENT OF RECHARGEABLE BATTERIES

Rechargeable batteries have only become popular in the last 30 years or so. Before that, batteries were made as cheaply as possible and they would be thrown out once discharged. Society had either not envisaged the benefits of many devices, such as phones, radios or calculators, being portable or did not yet have the technology to achieve it. The notable exception to this was the lead–acid car battery. Invented as far back as 1859, it can use the energy of a moving vehicle to reverse the discharge reactions and to re-form the reactants, therefore recharging itself.

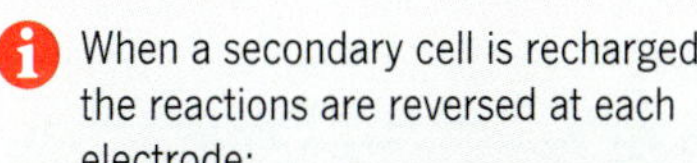

When a secondary cell is recharged, the reactions are reversed at each electrode:

- oxidation occurs at the positive terminal, which is now the anode
- reduction occurs at the negative terminal, which is now the cathode.

Towards the end of the 20th century, nickel–cadmium (Ni-Cad) cells were marketed as an alternative to the common AA and D cells favoured in many toys and small appliances.

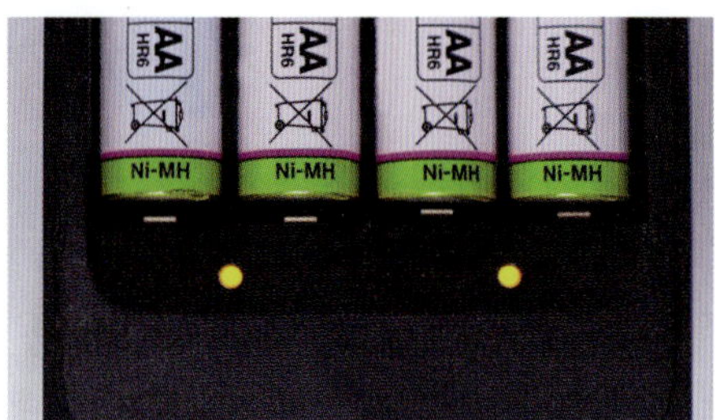

FIGURE 9.2.3 Rechargeable NiMH batteries and charger

The discharge half-equations for these cells are:

anode: $Cd(s) + 2OH^-(aq) \longrightarrow Cd(OH)_2(s) + 2e^-$

cathode: $NiO(OH)(s) + H_2O(l) + e^- \longrightarrow Ni(OH)_2(s) + OH^-(aq)$

A limitation of the Ni-Cad cells is their 'memory' problems. If the cells are not fully discharged or recharged, the voltage and capacity of the cell decrease.

Ni-Cad batteries have been mostly replaced by nickel–metal hydride batteries (NiMH), which have smaller memory effects and avoid the problems of cadmium disposal. The long-term exposure to cadmium and its compounds is associated with the formation of cancers in various organs of the body. Cadmium is replaced in the cell by a hydrogen-absorbing metal (Figure 9.2.3).

CASE STUDY ANALYSIS

Secondary batteries: Car batteries

Lead–acid batteries are the most widely used type of secondary cell. They are relatively cheap and reliable, provide high currents, and have a long lifetime.

Most people know the lead–acid battery simply as a car battery (Figure 9.2.4). It is used to start the car's engine and to operate the car's electrical accessories when the engine is not running. Once the engine starts, an alternator, which is run by the engine, provides electrical energy to operate the car's electrical system and recharge the battery.

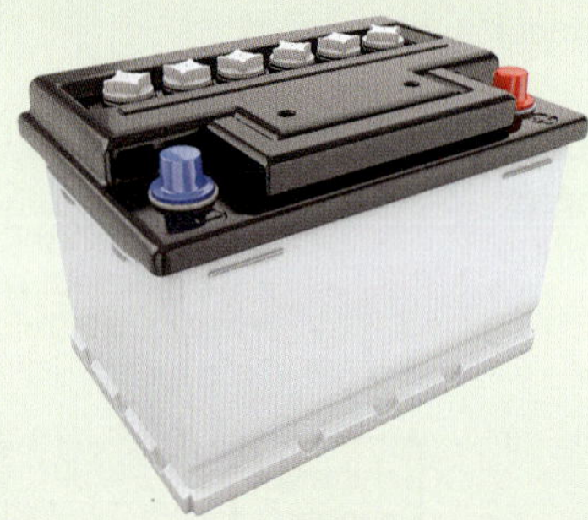

FIGURE 9.2.4 Lead–acid batteries are used to start cars, trucks and motorcycles.

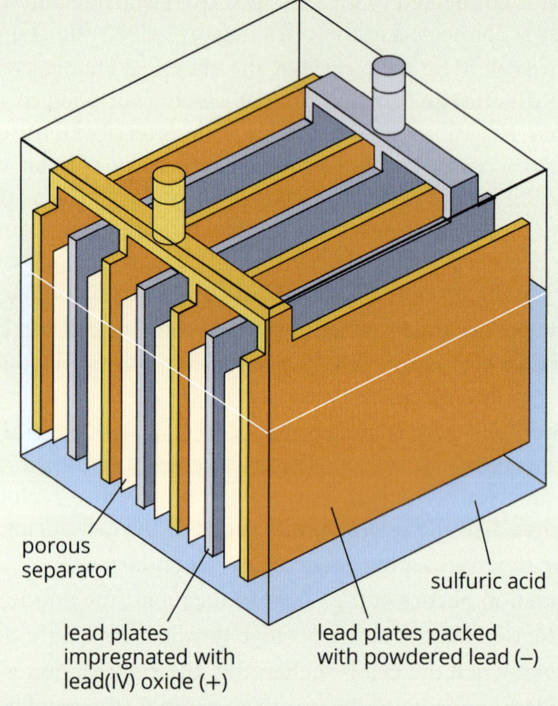

FIGURE 9.2.5 Construction of a lead–acid battery

Lead–acid batteries are also used for emergency light and power systems, for small-scale energy storage and to power some electric vehicles, such as golf buggies and small forklifts.

As shown in the diagram in Figure 9.2.5, a modern lead–acid battery is actually six separate cells connected together in series. The positive electrodes consist of lead plates packed with lead(IV) oxide (PbO_2), and the negative electrodes consist of lead plates packed with powdered lead. A solution of sulfuric acid, of about 4 M concentration, acts as the electrolyte.

Each cell has a potential difference of just over 2 V. The six cells connected in series give a total potential difference of about 12 V. When the cell is being recharged, a voltage slightly greater than 12 V needs to be used to ensure the reverse reaction proceeds. Table 9.2.1 shows more detailed chemistry of a lead–acid battery.

TABLE 9.2.1 Lead–acid cell discharge compared to recharge

Lead–acid battery discharge (galvanic cell)	Lead–acid battery recharge (electrolytic cell)
e 2.1V – anode + cathode Pb Pb^{2+} PbO_2 Pb^{2+} H^+ H^+ $PbSO_4$ H_2SO_4 $PbSO_4$	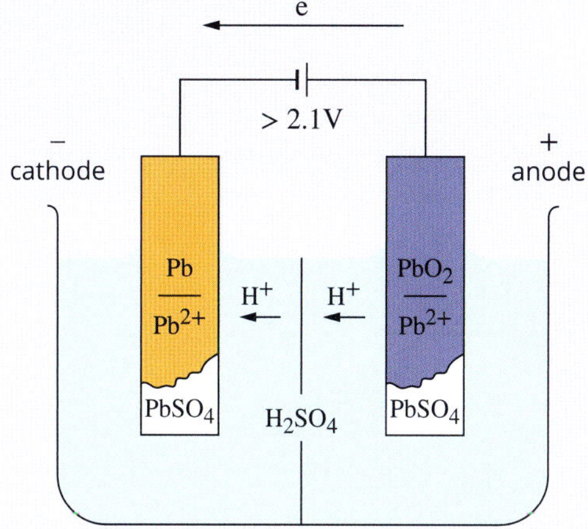
Spontaneous reaction, voltage produced 2.1 V Anode (–): Pb is oxidised to Pb^{2+} $Pb(s) + SO_4^{2-}(aq) \rightarrow PbSO_4(s) + 2e^-$	Non-spontaneous electrolytic cell with voltage > 2.1 V Anode (+): Pb^{2+} is oxidised to PbO_2 $PbSO_4(s) + 2H_2O(l) \rightarrow PbO_2(s) + SO_4^{2-}(aq) + 4H^+(aq) + 2e^-$
Cathode (+): PbO_2 is reduced to Pb^{2+} $PbO_2(s) + SO_4^{2-}(aq) + 4H^+(aq) + 2e^- \rightarrow PbSO_4(s) + 2H_2O(l)$	Cathode (–): $PbSO_4$ is reduced to Pb $PbSO_4(s) + 2e^- \rightarrow Pb(s) + SO_4^{2-}(aq)$
Overall: $Pb(s) + PbO_2(s) + 2SO_4^{2-}(aq) + 4H^+(aq) \rightarrow 2PbSO_4(s) + 2H_2O(l)$	Overall: $2PbSO_4(s) + 2H_2O(l) \rightarrow Pb(s) + PbO_2(s) + 2SO_4^{2-}(aq) + 4H^+(aq)$
The product of both electrode reactions, lead(II) sulfate, forms as a solid on the surface of the electrodes. This enables the battery to be recharged.	The original electrodes are reformed. Note that the lead electrode is the negative electrode in either discharge or recharge.

Analysis

1 Select from the following words to complete the description of a lead–acid battery. Words may be used more than once.

oxidised; reduced; cathode; anode; released; absorbed; Pb; $PbSO_4$; PbO_2; H^+

During discharge, Pb is __________ at the ______ and PbO_2 is ________ at the _____. Energy is _________ in this cycle.

During recharge, $PbSO_4$ is oxidised to ______ at the _____ and $PbSO_4$ is reduced to ______ at the _____.

2 Describe the energy conversions that occur when the battery is recharged.

3 What happens to the H^+ concentration and the pH of the battery as it produces electricity?

LITHIUM-ION BATTERIES

Low-density, rechargeable lithium-based batteries with a high energy capacity were first sold in 1991. They were not introduced to replace larger NiMH batteries, but to facilitate the miniaturisation of computers, phones and electronic devices. Their success has led to the development of portable versions of many popular appliances, such as drills and vacuum cleaners. About 37% of all batteries currently sold in the world are lithium-ion batteries. Figure 9.2.6 shows a battery from a power tool.

FIGURE 9.2.6 Rechargeable lithium-ion battery for a power drill

Although there are different types of lithium-ion batteries, a common feature is a lithium-metal oxide cathode and lithium-carbon compound anode. Lithium ions form in the reaction at the anode, and they diffuse through a membrane to the cathode (Table 9.2.2). An important property of each electrode is that lithium ions can move through them. This movement is reversed during charging. Lithium-ion batteries can produce high voltages and store relatively high quantities of charge.

TABLE 9.2.2 Movement of lithium ions through a lithium-ion battery during phone use and charging

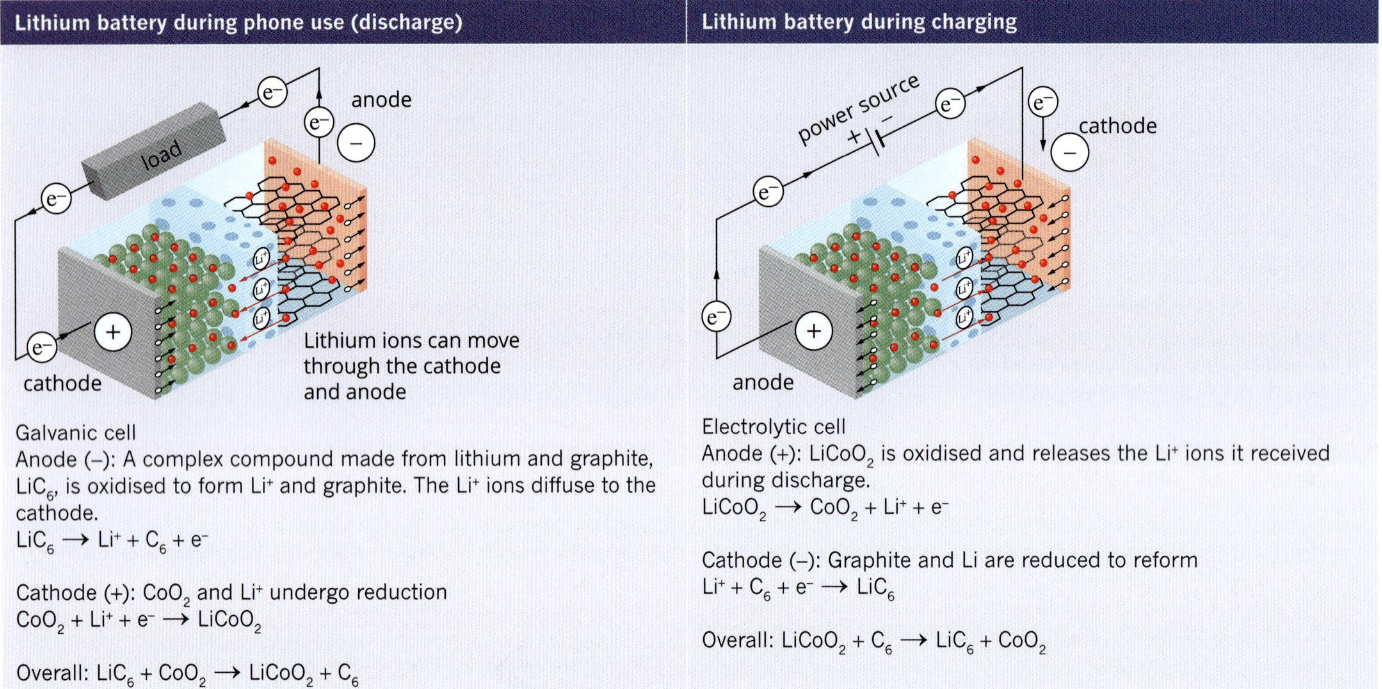

Lithium battery during phone use (discharge)	Lithium battery during charging
Galvanic cell Anode (–): A complex compound made from lithium and graphite, LiC_6, is oxidised to form Li^+ and graphite. The Li^+ ions diffuse to the cathode. $LiC_6 \rightarrow Li^+ + C_6 + e^-$ Cathode (+): CoO_2 and Li^+ undergo reduction $CoO_2 + Li^+ + e^- \rightarrow LiCoO_2$ Overall: $LiC_6 + CoO_2 \rightarrow LiCoO_2 + C_6$	Electrolytic cell Anode (+): $LiCoO_2$ is oxidised and releases the Li^+ ions it received during discharge. $LiCoO_2 \rightarrow CoO_2 + Li^+ + e^-$ Cathode (–): Graphite and Li are reduced to reform $Li^+ + C_6 + e^- \rightarrow LiC_6$ Overall: $LiCoO_2 + C_6 \rightarrow LiC_6 + CoO_2$

When you purchase a device such as a laptop that uses a lithium-ion battery, the battery is only partially charged. One of the first things you have to do is plug it into a charger and let the battery fully charge. During this process, the battery is operating as an electrolytic cell, generating the LiC_6 that will react during discharge.

The increased use of electric vehicles and large energy storage devices will ensure that battery design continues to improve. However, the latest battery technology is still expensive. For example, the batteries in a Tesla car cost up to $10 000, the high cost reflecting the need for fast recharging and to extend the range of the vehicle between charges.

The design and composition of lithium-ion batteries is continuously changing as scientists make small changes to different components to increase the **battery capacity** or efficiency. These changes may be to the electrode composition or to the conductive electrolytes used.

The use of lithium-based batteries comes with a range of issues around resources and disposal. The data in Table 9.2.3 lists some of these issues—and this data refers only to the use of lithium batteries in cars. The use of lithium batteries in domestic storage devices and in mobile appliances will add to these problems.

TABLE 9.2.3 Lithium-ion car battery statistics

Statistic	Detail
$100 billion	the value of the lithium-ion market in 2022
140 million	number of electric vehicles predicted to be on the road worldwide by 2030
11 million tonnes	mass of lithium batteries likely to be finished service by 2030
5%	percentage of lithium batteries recycled currently (of the whole battery; only the metal components are actually recycled)
100%	percentage of lead that is recycled from conventional batteries

These statistics show some obvious sustainability issues, such as the following.

- The sourcing of the lithium required to make the batteries. Australia is the world's leading producer of lithium, producing 68 000 tonnes in 2022. It will be difficult to keep expanding this level of production to meet the needs of the emerging electric vehicle market. Most of Australia's lithium is mined in Western Australia, with the Talisman mine at Greenbushes, about 250 km south of Perth, being the largest lithium mine in the world.

- The sourcing of the scarce metals contained in lithium batteries. The electrodes obviously contain lithium metal, but they also require other metals such as cobalt and nickel, and, given the number of batteries that society will require, meeting the demand for these metals will be problematic. Cobalt, in particular, is mainly found in the Democratic Republic of the Congo. There is considerable concern about child labour practices of this country, where mining is carried out by hand without appropriate safety measures in place.
- The difficulty of recycling the batteries and separating the materials. The electrodes in older-style lead–acid batteries are relatively easy to recycle as large pieces of pure lead are used as the anodes. The electrodes in lithium-ion batteries are sophisticated, composite materials, however, and it is not currently possible to isolate the individual metals from recycled batteries. The complexity of the materials in lithium-ion batteries is evident in Figure 9.2.7.
- The safety of the batteries in use. An example of safety issues is the damage, sometimes fatal, that can occur when toddlers swallow button batteries. Another example of safety concerns was highlighted by a fire in 2021 in the Geelong electricity storage complex. It took several days for firefighters to extinguish the blaze, which was emitting toxic fumes.
- The disposal and toxicity of used batteries. When a battery is added to landfill, the various metals it contains, such as cobalt, will eventually leach into the nearby soil and water system.

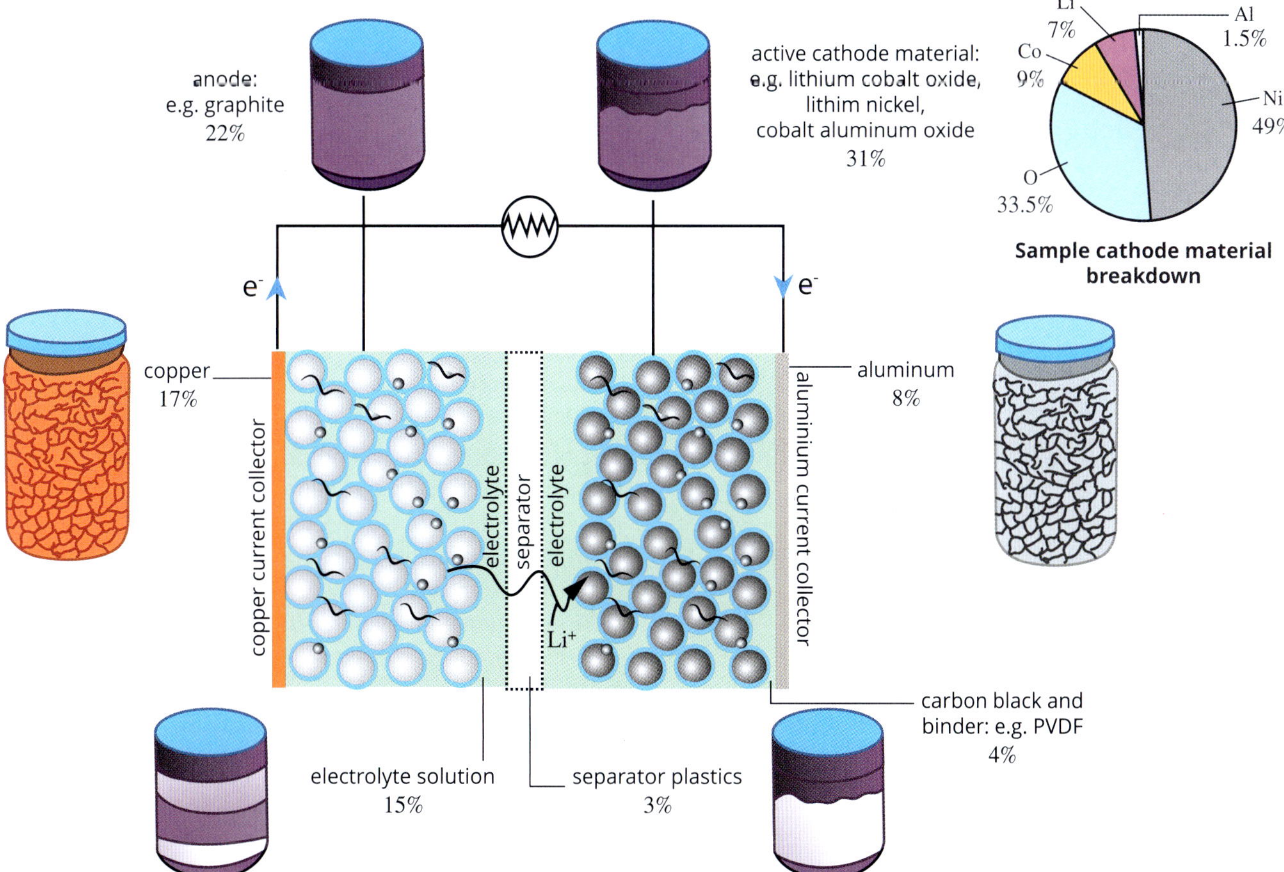

FIGURE 9.2.7 The electrodes and the electrolytes used in lithium-ion cells are complex compounds.

9.2 Review

SUMMARY

- Secondary cells can be recharged by connecting them to an external source of electricity.
- For a cell to be rechargeable, the products of the discharge reaction have to remain in contact with the electrodes.
- During the recharging of a secondary cell, the cell reaction is reversed and the products of the cell reaction are converted back into the original reactants. Therefore, the equation for the reaction that occurs during the recharging of a secondary cell is the reverse of the equation for the cell discharging.
- For a secondary cell, the anode, where oxidation occurs, is the negative terminal during discharging but the positive terminal during recharging.
- When a secondary cell discharges and produces electrical energy, it acts as a galvanic cell; when it is recharged, it acts as an electrolytic cell.
- Lead–acid batteries and lithium-ion cells are both examples of secondary cells.
- Scientists are continually trying to improve the efficiency and performance of batteries.
- The demand for lithium-ion batteries is rising sharply with the introduction of electric vehicles. This demand will cause pressure on world stocks of lithium, cobalt and nickel.

KEY QUESTIONS

Knowledge and understanding

1 State the meaning of the following terms:
  a discharge
  b recharge

2 Identify whether the cells and batteries commonly used in the following appliances are primary or secondary cells.
  a car
  b household smoke detectors
  c mobile phone

3 If a cell produces a voltage of 1.5 V during discharge, what voltage should the power supply deliver to successfully recharge the cell?

4 The demand for lithium-based batteries is growing rapidly. List three applications that are having a significant impact on this demand.

5 State whether each of the following statements about cells and batteries is true or false.
  a Oxidation will occur at the anode whether the cell is discharging or recharging.
  b Primary cells cannot be recharged effectively because their reaction products do not remain in contact with the electrodes.
  c Oxidation will be at the negative electrode for both discharge and recharge.
  d The electrodes must be unreactive in a rechargeable cell or battery.
  e The overall equation when a cell is recharged is the reverse of the discharge reaction.

Analysis

6 One size does not fit all when it comes to choosing a type of rechargeable battery. The following questions require you to consider some of the issues involved in choosing a battery.
  a Lead–acid batteries are not suitable for use in electric vehicles. Suggest two reasons they are not suitable.
  b Many conventional batteries are not suitable for use in a laptop. Suggest two reasons they are not suitable.
  c Rechargeable batteries are preferred in a laptop, but primary cells are common in basic calculators. Suggest a reason for the difference.

7 One of the discharge half-equations in a NiMH cell is shown below. M is used to represent a porous metal.

$$H_2O + M + e^- \longrightarrow OH^- + MH$$

  a What is the polarity of the electrode at which this reaction occurs?
  b Write the half-equation for the reaction at this electrode during recharge.
  c What is the polarity of this electrode during recharge?

9.3 Producing 'green' hydrogen gas

As society grapples with the issues of emissions of CO_2 from carbon-based fuels, hydrogen gas has emerged as a serious alternative. Unlike fuels such as methane, petrol and biodiesel, the sole product of the combustion of hydrogen is water. Added to that, hydrogen has a high **energy density** and it is abundant on Earth in the form of water. The production of hydrogen from water, however, requires significant amounts of energy.

The thermochemical equation for the combustion of hydrogen is:

$$2H_2(g) + O_2(g) \longrightarrow 2H_2O(l) \qquad \Delta H = -564 \text{ kJ}$$

As a comparison, the complete combustion of 1 g of hydrogen gas releases 141 kJ of energy, while the combustion of 1 g of methane releases 55.6 kJ of energy.

As you have seen in Chapter 5, a fuel cell can be used to produce electrical energy from hydrogen for vehicles, houses or industry. However, technical difficulties with its use have meant that industry has been slow to adopt this technology. Table 9.3.1 lists the advantages and disadvantages associated with the widescale use of hydrogen.

TABLE 9.3.1 Advantages and disadvantages of hydrogen gas as a fuel

Advantages of hydrogen gas as a fuel	Disadvantages of hydrogen gas as a fuel
It has a high energy density: 282 kJ mol^{-1}.	It is not found as an element, so energy is needed to produce it.
It is abundant on Earth: present in water and most carbon compounds.	It has a very low boiling point of –253°C, therefore requires significant energy to liquify it. High pressures are required if hydrogen is to be stored efficiently as a gas.
The sole product of its combustion is water: $2H_2(g) + O_2(g) \longrightarrow 2H_2O(l)$	It is explosive, so requires careful handling and storage.
	It is difficult to transport safely either as a gas or a liquid.

Hydrogen itself is a clean fuel, but its production is not necessarily clean. Historically, most hydrogen has been produced from the **steam reforming** of methane. The equation for the process is:

$$CH_4(g) + H_2O(g) \rightleftharpoons CO(g) + 3H_2(g) \qquad \Delta H = +206 \text{ kJ}$$

The carbon monoxide, CO, formed is then converted into carbon dioxide, CO_2. The hydrogen produced is referred to as **brown hydrogen**, as the release of CO_2 as a by-product is detrimental to the environment. Figure 9.3.1 shows the colour scheme that rates the relative environmental impact of methods of producing hydrogen gas. (There is no accepted definition on the colours used to classify hydrogen, but the CSIRO categories shown here are widely used.) Generally, the greener the process, the more expensive the hydrogen is to produce using current technology.

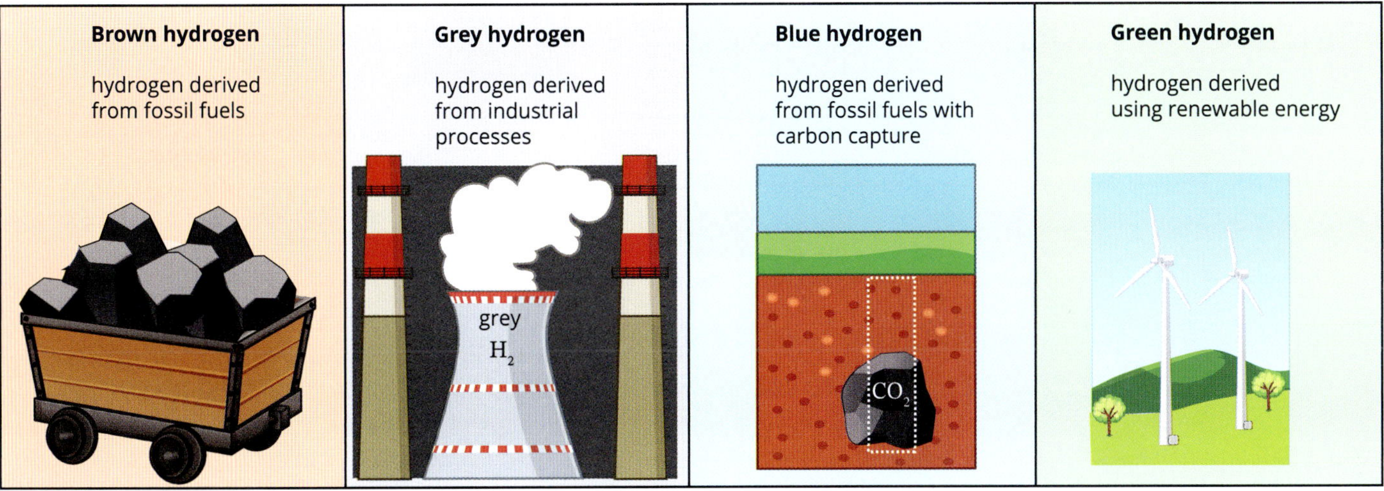

FIGURE 9.3.1 Colour scheme used to rate hydrogen sustainability

Green hydrogen is hydrogen fuel created using renewable energy instead of fossil fuels.

The focus of this section is on the production of **green hydrogen**, as it offers the greatest possibilities for emission reductions. In the production of green hydrogen, the electrical energy for electrolysis of water comes from **renewable energy sources**, such as wind or solar.

ELECTROLYSIS OF WATER

In the laboratory, the electrolysis of a dilute solution of sulfuric acid, H_2SO_4, to produce hydrogen is a simple process. Two electrodes are placed in an acid solution and a power supply is attached, as illustrated in Figure 9.3.2.

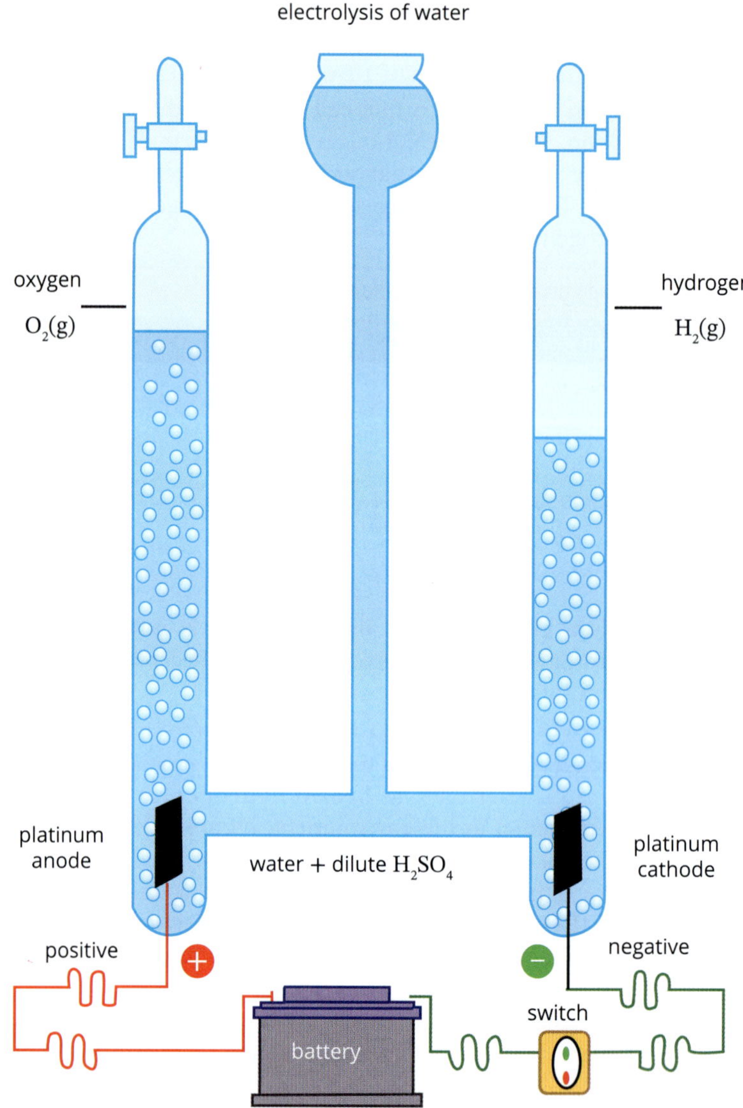

FIGURE 9.3.2 Laboratory apparatus used to produce hydrogen gas through electrolysis

The overall equation is:

$$2H_2O(l) \rightarrow 2H_2(g) + O_2(g)$$

This is an example of an acidic electrolytic cell, where the half-equations occurring are:

$$\text{anode: } 2H_2O(l) \rightarrow O_2(g) + 4H^+(aq) + 4e^-$$

$$\text{cathode: } 2H^+(aq) + 2e^- \rightarrow H_2(g)$$

Note that the process produces another useful product, oxygen gas. The overall equation shows that the volume of oxygen gas should be half the volume of hydrogen gas.

If the electrical energy for this process is obtained from renewable sources, the hydrogen fuel produced would be an example of green hydrogen.

INDUSTRIAL PRODUCTION OF GREEN HYDROGEN

Industry does not use the simple electrolysis of acidic or alkaline electrolyte solutions for hydrogen production because:

- the process is not suited to the intermittent nature of renewable energy sources, such as solar arrays and wind turbines
- the solutions are corrosive
- the hydrogen gas produced is not compressed.

Newer technology uses some form of **electrolyser**. An electrolyser is a system that uses electricity to break water down into hydrogen and oxygen via electrolysis. The knowledge gained by scientists in the development of conductive polymers and electrode design for fuel cells and secondary cells has been used in the development of efficient electrolysers.

Polymer electrolyte membrane (PEM) electrolyser

A **polymer electrolyte membrane (PEM)** electrolyser uses a solid electrolyte instead of a solution. The initials PEM are sometimes referred to as representing a proton exchange membrane, as the membrane allows the flow of protons to complete the circuit.

Figure 9.3.3 shows the design of a PEM electrolyser. Water reacts at the anode to produce oxygen gas and hydrogen ions. The hydrogen ions migrate through the conducting polymer electrolyte to the cathode, where they are reduced to hydrogen gas. The electrolyser can be designed to produce compressed hydrogen gas, removing the need for a compressor stage to be included in the production cycle.

The diagram of a PEM electrolyser, however, does not convey the high-technology nature of this cell. The electrodes are made from expensive metals, such as ruthenium and iridium, which need to be porous enough to allow for the passage of gases, while preventing the flow of liquids. These metals also act as catalysts to improve the efficiency of gas production at the electrodes. The membrane itself contains advanced polymers, known as a polysulfones, which allow the flow of protons but not electrons (Figure 9.3.4a). A large commercial PEM electrolyser set-up is shown in Figure 9.3.4b.

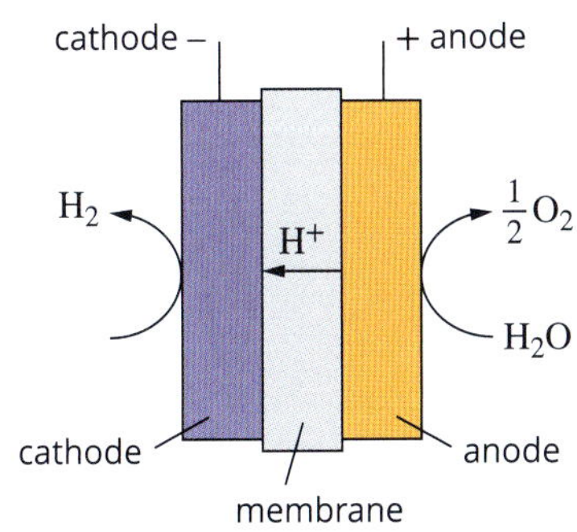

anode: $2H_2O \rightarrow 4H^+ + O_2 + 4e^-$

cathode: $2H^+ + 2e^- \rightarrow H_2$

overall equation: $2H_2O \rightarrow 2H_2 + O_2$

FIGURE 9.3.3 Design and equations for a PEM electrolyser

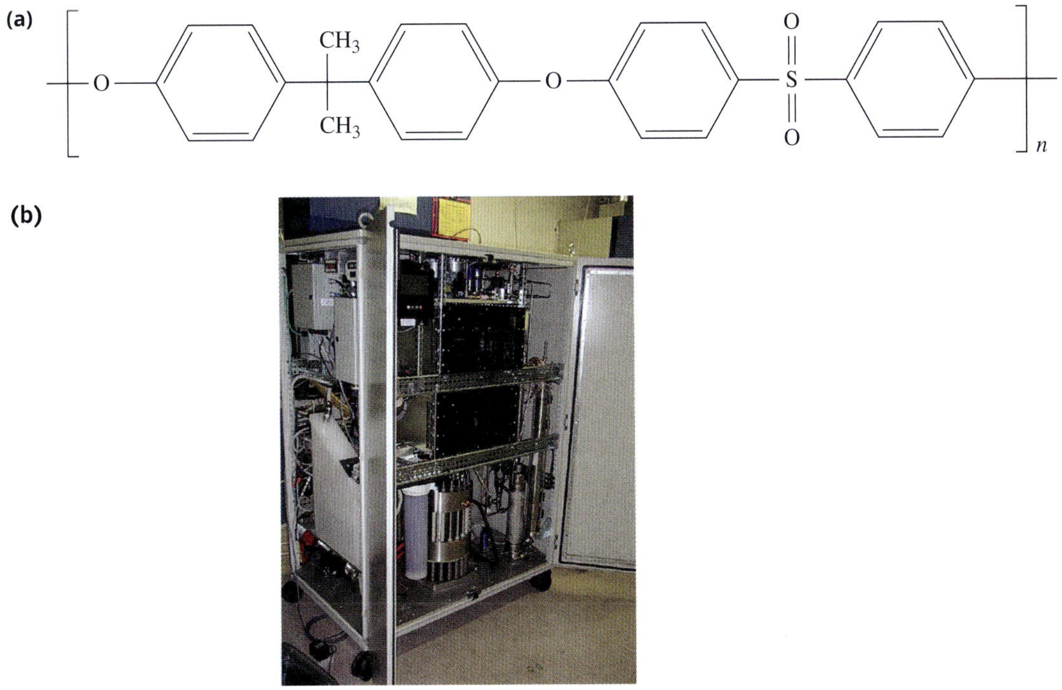

(a)

(b)

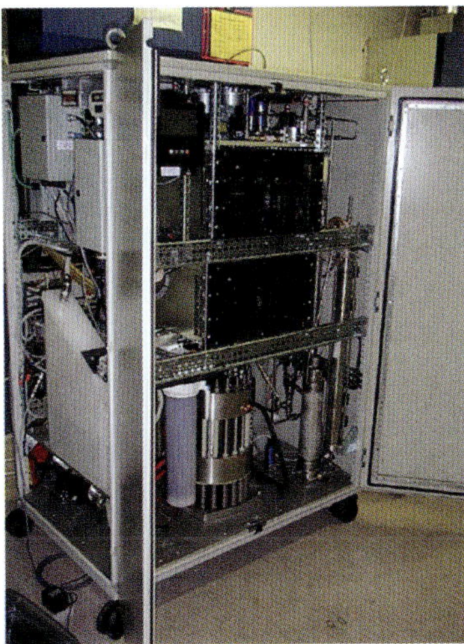

FIGURE 9.3.4 (a) The conducting polymers in PEM cells are often polymers containing sulfur: polysulfones. (b) A stack of commercial PEM electrolysers.

CHEMFILE

Microbial production of hydrogen

An interesting alternative line of research for the commercial production of hydrogen is the use of bacteria to produce hydrogen gas from biomass such as forest waste. Microorganisms can break down the cellulose in the waste into glucose at the anode of an electrolysis cell, where it is then converted to ethanoic acid. The hydrogen ions produced migrate to the cathode, where they are reduced to hydrogen gas. This process is known as dark fermentation, as no light is required for the bacteria to work. At the moment, the low reaction rate makes this process non-viable, but researchers are continuing to work on its efficiency. They expect that it will offer significant energy savings, as the voltage required for the cell to operate is much lower than the voltage required to electrolyse water.

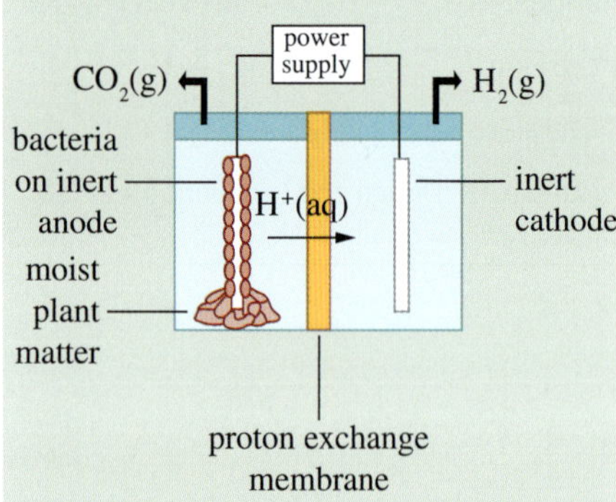

In a microbial cell, the action of bacteria on biomass leads to the production of hydrogen gas at the cathode.

A typical equation for this process is:

$$C_6H_{12}O_6(aq) + 2H_2O(l) \rightarrow 2CH_3COOH(aq) + 2CO_2(g) + 4H_2(g)$$

The CO_2 produced in the cell is not considered to be a problem as CO_2 was absorbed in the growth of the original biomass.

The rate of production of hydrogen in a PEM is high, but the inclusion of metals such as gold or indium as catalysts makes the cell expensive. To lower the costs, researchers are trialling two other forms of electrolysers, an anion exchange membrane (AEM) and a solid oxide electrolyser cell (SOEC).

The output of hydrogen gas from a single electrolyser is not high enough to justify its use. Instead, electrolysers are now being designed so that they can be assembled in a cluster, known as an **electrolyser stack**. This means the units can be assembled into a large hydrogen plant, which makes them more cost-effective. Figure 9.3.5 shows an artist's impression of a **hydrogen hub**, in which renewable energy powers an electrolyser stack to produce compressed hydrogen for storage. The idea of establishing hubs is to place as many connected industries as possible in the one precinct, so that the hydrogen can be produced and then used without the need for transport or liquification.

FIGURE 9.3.5 Hydrogen hub in which renewable electrical energy electrolyses water to produce hydrogen which is stored until required

Several Australian states are trialling hydrogen plants. Two examples are:

- Australian Gas Networks (AGN) is investing $12 million in Hydrogen Park South Australia, a demonstration project trialling a 1.25 MW electrolyser. A 50 MW plant that will be linked to South Australia's wind energy/battery storage network is being planned.
- Australian Renewable Energy Agency (ARENA) is investing over $100 million to build 10 MW electrolysers in Victoria and Western Australia.

The port of Gladstone is emerging as a potential world leader in hydrogen production. Mining billionaire Andrew Forrest, founder of Fortescue Metal Group, has commenced construction of a $3 billion dollar plant to manufacture hydrogen electrolysers in Gladstone, while a range of other companies, including the Queensland Government, is considering manufacturing and exporting hydrogen from Gladstone.

Figure 9.3.6 shows how these hubs are designed to use renewable energy to produce hydrogen gas. The gas can be used for energy storage or as a fuel in hydrogen cars, or it can be blended with domestic gas supplies. Australia also hopes to export the hydrogen to other countries. SNG, or syngas, is a mixture of hydrogen and carbon monoxide gases.

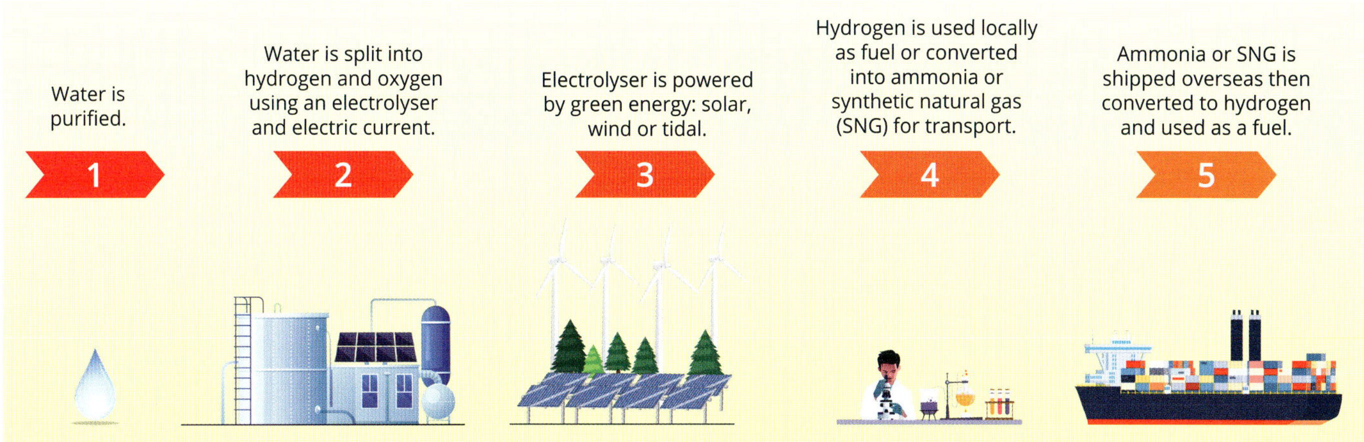

FIGURE 9.3.6 Typical hydrogen hub: renewable energy is used to electrolyse purified water. The hydrogen gas produced is blended with local gas supplies or compressed for transport.

HYDROGEN CARS

If a hydrogen fuel cell is fitted into an electric car, the hydrogen gas can generate the electrical energy needed to power the car. The chemistry of fuel cells was discussed in Chapter 5. The only emission from the car will be water. Figure 9.3.7a shows the fuel cell under the bonnet in a prototype hydrogen vehicle. For this to be viable, a hydrogen gas distribution system needs to be established and the hydrogen tank in the car needs to fitted in a way that will not lead to explosions if the vehicle crashes (Figure 9.3.7b).

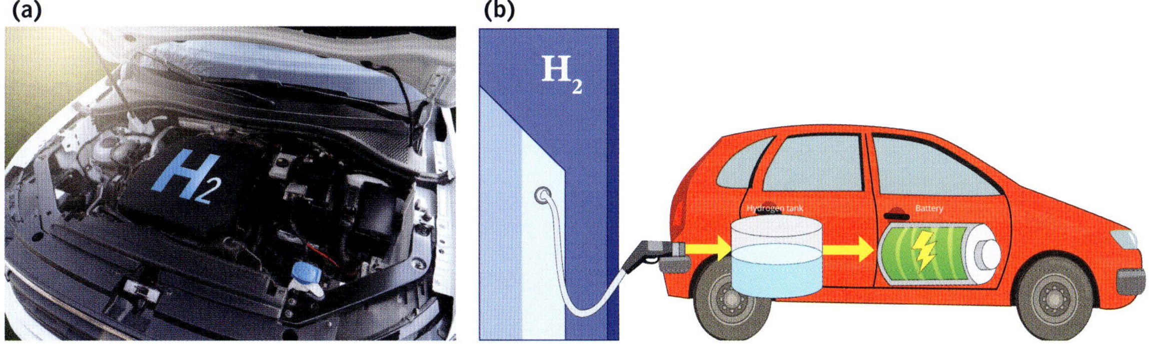

FIGURE 9.3.7 (a) Hydrogen fuel cell in a car. (b) A hydrogen distribution system needs to be set up to allow for refuelling.

The range and performance of hydrogen vehicles is comparable to that of conventional vehicles. So far, however, there has been greater investment in electric vehicles running on battery power rather than hydrogen fuel cells.

ARTIFICIAL PHOTOSYNTHESIS

The production of hydrogen gas from renewable energy represents a giant step forward in reducing emissions, but scientists are already considering a newer approach that could offer further advantages. When solar panels are used to produce electrical energy, the electrical energy needs to be stored in batteries until it is used in an electrolyser. Why not eliminate the battery and use the solar energy directly to produce the fuel?

A **photoelectrochemical cell** (shown in Figure 9.3.8) is like a solar panel operating in a solution, where:

- sunlight lands on the anode
- energy from the Sun causes excitation of the metals in the electrode
- the energy in the metals is used to oxidise water to oxygen gas and hydrogen ions:

$$2H_2O(l) \rightarrow O_2(g) + 4H^+(aq) + 4e^-$$

- hydrogen ions migrate to the cathode
- reduction of hydrogen ions to hydrogen gas occurs at the cathode

$$2H^+(aq) + 2e^- \rightarrow H_2(g)$$

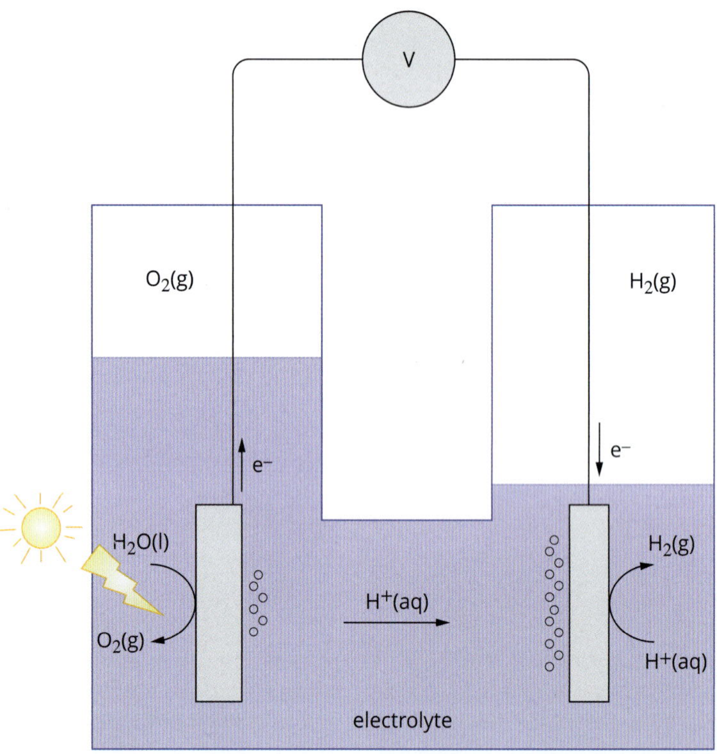

FIGURE 9.3.8 Energy from the Sun produces hydrogen gas in a photoelectrochemical cell.

The net effect is that solar radiation is directly producing hydrogen gas. This is a form of **artificial photosynthesis**. Note that the half-equations are the same as those in the PEM electrolyser; the difference, however, is that the cell is driven by direct sunlight. As light is providing the energy for this electrolytic cell, it is referred to as a photoelectrochemical cell.

In photosynthesis, plants use sunlight to convert carbon dioxide and water to glucose. Artificial photosynthesis changes the end-product from glucose into hydrogen gas.

Figure 9.3.9 shows a more detailed representation of a photoelectrochemical cell. Sunlight causes excitation of the metals in the anode, enabling oxygen gas to be produced from water. The hydrogen ions produced can then be reduced to hydrogen gas.

As with fuel cells and lithium-ion batteries, the metal combinations in the electrodes, the catalysts and the polymers in the conducting membranes are very complex. Energy efficiencies of up to 30% have been achieved for the conversion of solar energy to the chemical energy of hydrogen, but it is estimated that this technology is about 10 years away from being commercially viable.

In artificial photosynthesis, light provides the energy for the direct production of hydrogen gas.

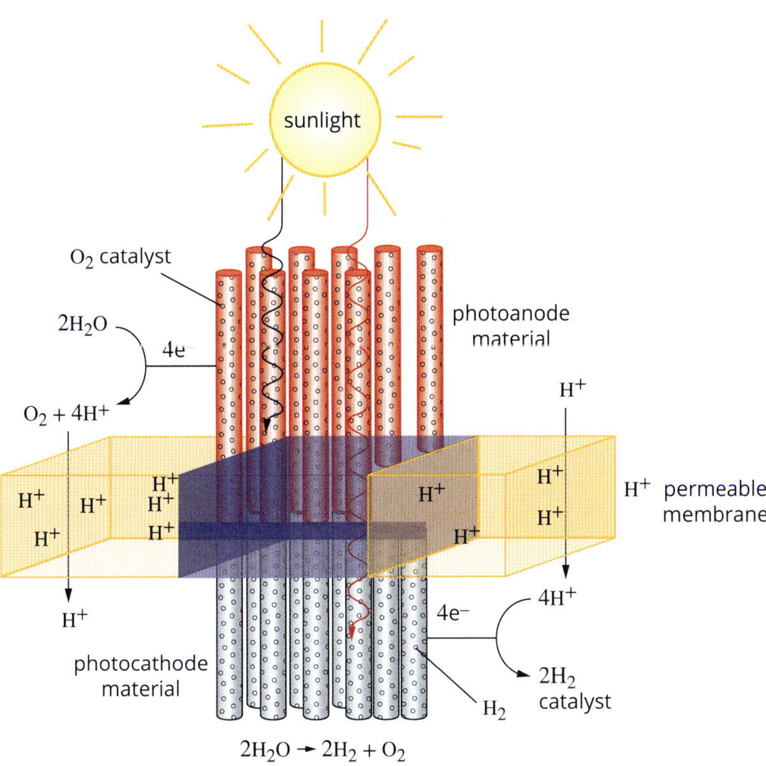

FIGURE 9.3.9 A more detailed outline of the reactions in a photoelectrochemical cell

9.3 Review

WS 20

OA ✓✓

SUMMARY

- Interest in producing hydrogen gas as a fuel is increasing, along with the capability of doing so.
- The electrolysis of water can be used to produce hydrogen and oxygen gases.
- The production of hydrogen requires a significant amount of energy.
- Hydrogen made by the steam-reforming of fossil fuels is known as brown hydrogen.
- Hydrogen made using renewable energy, such as photovoltaic (solar) or wind energy, is known as green hydrogen.
- Electrolysers are high-technology devices used to produce hydrogen gas. A PEM is a polymer electrolyte membrane electrolyser that uses a conductive polymer membrane as the electrolyte. The half-equations occurring in a PEM are:

$$2H_2O \longrightarrow 4H^+ + O_2 + 4e^-$$

$$2H^+ + 2e^- \longrightarrow H_2$$

- Hydrogen gas can be blended into domestic gas supplies or used as a fuel in electric vehicles fitted with a fuel cell.
- Scientists are investigating the commercial viability of artificial photosynthesis, a process where sunlight striking specially designed electrodes can provide the energy needed to electrolyse water.

KEY QUESTIONS

Knowledge and understanding

1 Classify each of the following as brown or green hydrogen.

a Hydrogen made from coal.

b Hydrogen produced using electrical energy from a wind turbine.

c Hydrogen produced from electrical energy from a hydroelectricity plant.

d Hydrogen produced from propane separated from LPG.

2 Hydrogen gas produced from natural gas is considered to be brown hydrogen. Explain the sustainability concerns with this process.

3 The electrolysis of water requires significant energy. Explain why this is the case.

4 **a** What is artificial photosynthesis?

b Why is this process termed 'artificial photosynthesis' when glucose is not a product?

Analysis

5 **a** Name the other product of the electrolysis of water other than hydrogen.

b Is this likely to be a useful by-product? Explain your answer.

6 The following reaction occurs at an electrode in an alkaline polymer electrolyte membrane (PEM) electrolyser:

$$4OH^- \longrightarrow 2H_2O + O_2 + 4e^-$$

a Will this reaction take place at the anode or the cathode?

b What is the polarity of this electrode?

c The OH^- ions are not shown as $OH^-(aq)$. Why is this?

7 There are several trial plants in Australia that store renewable energy in batteries, which are then used to electrolyse water to produce hydrogen.

a How is the process of artificial photosynthesis different from the process in these trial plants?

b What advantages might artificial photosynthesis offer?

Chapter review

KEY TERMS

accumulator
artificial photosynthesis
battery capacity
brown hydrogen
discharge
electrolyser
electrolyser stack
energy density
green hydrogen
hydrogen hub
photoelectrochemical cell
polymer electrolyte membrane (PEM)
rechargeable cell
renewable energy source
secondary cell
steam reforming
yield

REVIEW QUESTIONS

Knowledge and understanding

1 The conversion of SO_2 gas to SO_3 is a key step in the production of sulfuric acid.

$2SO_2(g) + O_2(g) \rightleftharpoons 2SO_3(g) \qquad \Delta H = -197\ kJ$

A manufacturer wishing to increase the rate of this reaction trials running the process at a higher temperature. As the temperature is increased in this process:

A the amount of SO_3 formed will not change, but it will form more quickly
B the amount of SO_3 formed will decrease and it will form more slowly
C the amount of SO_3 produced will decrease, but it will form more quickly
D the amounts of SO_3 formed will decrease and it will form more slowly

2 Scientists researching an endothermic equilibrium reaction for a new industrial process discover a very effective catalyst for the process. The catalyst will:

A increase the yield of product but not the rate of reaction
B decrease the amount of energy required at a particular temperature
C increase the rate of the forward reaction but not the reverse reaction
D not change the yield, but it will increase the reaction rate

3 During the recharge of a secondary cell, oxidation will occur:

A at the electrode that electrons flow to
B at the anode, which will be the positive electrode
C at the same electrode as during discharge
D at the cathode, which will be the negative electrode

4 The half-equation occurring at the anode in alkaline electrolysis of water will be:

A $4OH^-(aq) \rightarrow O_2(g) + 2H_2O(l) + 4e^-$
B $4OH^-(aq) + 4e^- \rightarrow O_2(g) + 2H_2O(l)$
C $H_2(g) + 2OH^-(aq) \rightarrow 2H_2O(l) + 2e^-$
D $2H_2O(l) + 2e^- \rightarrow H_2(g) + 2OH^-(aq)$

5 One factor that will increase the likelihood of a battery being rechargeable is that:

A Li^+ ions flow in the cell
B $Zn(OH)_2$ is one of the products of discharge
C the cell discharges at high temperatures
D the products of discharge remain in contact with the electrodes

6 Reaction conditions such as temperature and pressure can be altered to improve the yield of a reaction. However, changing the temperature or pressure does not always have the same impact on all reactions.

a Explain how you can determine the impact of a change in temperature on the equilibrium yield of a reaction.
b Explain how you can determine the impact of a change in pressure on the equilibrium yield of a reaction.

7 A catalyst does not change the yield of a reaction at a given temperature. However, it can be used to improve the efficiency of exothermic reversible reactions. Explain how a catalyst can improve reaction yield.

8 Explain the following terms:

secondary cell
electrolyser
PEM
hydrogen hub
photoelectrochemical

9 Use the following terms to complete this summary relating to secondary cells. The terms may be used more than once.

oxidation; reduction; positive; negative; galvanic; electrolytic; spontaneous; non-spontaneous

During discharge, a secondary cell acts as a __________ cell. The reaction occurring is a __________________ one. For this cycle, _________ occurs at the cathode and its polarity is _____________. During the recharge cycle, the secondary cell operates as an __________ cell and a _____-_______________ reaction occurs. For the recharge cycle, _____________ occurs at the anode, and its polarity is ____________.

10 The Haber process refers to the reaction between nitrogen and hydrogen gases to form ammonia, NH_3. The forward reaction is exothermic. The temperature of an equilibrium gas mixture is increased. Explain the effect of this change on:

a the rate of the forward reaction

b the rate of the reverse reaction

c the equilibrium yield.

11 The reactants in a lead–acid battery are PbO_2 and Pb metal. Both form $PbSO_4$ as the cell discharges. Complete the half-equations for each reactant, including the polarity of each.

$PbO_2 \qquad \rightarrow PbSO_4(s)$ polarity _____

$Pb \qquad \rightarrow PbSO_4(s)$ polarity _____

12 List three commercial uses for hydrogen gas.

13 a Write a balanced overall equation for the production of hydrogen from water.

b Does this reaction absorb or release energy? Explain your answer.

c Hydrogen produced by electrolysis might or might not be considered to be green hydrogen. How is it decided if the hydrogen is green or not?

14 Australian companies expect to be able to export the hydrogen gas they produce. Describe the challenges that hydrogen gas export presents.

Application and analysis

15 Lithium-ion rechargeable cells are safer and more widely used than cells that use pure lithium.

a Write a balanced equation for the reaction of lithium and water.

b Suggest how this reaction might affect the design of lithium-ion cells.

16 A cell currently under research is the lithium–oxygen (Li-O_2) cell. It is hoped that it might eventually replace lithium-ion cells in high energy applications. The voltage produced in this cell is 2.7 V. During discharge, the half-reactions occurring are:

$$2Li^+ + O_2 + 2e^- \rightarrow Li_2O_2$$

$$Li \rightarrow Li^+ + e^-$$

a Identify the polarity of each of the electrodes in this cell during discharge.

b Identify the oxidation state change occurring in the reduction half-equation.

c Name the ions that flow through the electrolyte in this cell.

d Write an overall equation for the discharge reaction in this cell.

e Write an overall equation for the recharge reaction in this cell.

f During recharge, which electrode needs to be connected to the positive terminal of the charger used?

17 a What is an electrolyser?

b How does the construction of a PEM electrolyser differ from the simple electrolytic cell shown in Figure 9.3.2?

c How will the half-equations differ between the simple electrolyser and a PEM electrolyser?

d Will the overall equation for both forms of electrolyser be the same?

18 Ethene gas is produced from ethane gas in an endothermic reaction represented by the equation:

$$C_2H_6(g) \rightleftharpoons C_2H_4(g) + H_2(g) \qquad \Delta H = +138 \text{ kJ}$$

a State whether the following changes will result in the percentage yield of ethene at equilibrium increasing, decreasing or not changing.

i The volume is reduced at constant temperature.

ii More hydrogen gas is added at constant temperature and volume.

iii The temperature is increased at constant volume.

iv A catalyst is added.

v Argon gas is added at constant temperature and volume.

b How will each of the changes in part **a** affect the rate at which the reaction achieves equilibrium?

19 One of the first commercial rechargeable batteries was the nickel–cadmium cell. The relevant half-equations for this cell are:

$NiOOH(aq) + H_2O(l) + e^- \rightleftharpoons Ni(OH)_2(s) + OH^-(aq)$ +0.48 V

$Cd(OH)_2(s) + 2e^- \rightleftharpoons Cd(s) + 2OH^-(aq)$ −0.82 V

- **a** Write an overall equation for the discharge cycle of this cell.
- **b** Identify the anode and the cathode for this cell.
- **c** What is the theoretical voltage produced in this cell?
- **d** Write an overall equation for the charge cycle of this cell.
- **e** Identify the anode and the cathode during recharge.

20 Lead–acid batteries used in cars are considered simple to recycle, but lithium-ion cells are not.

- **a** State reasons for this difference.
- **b** What materials could be of value if lithium batteries are recycled?
- **c** Give a reason why some materials need to be recycled even though they are not valuable.

21 A solid oxide electrolyser cell is an alternative to the PEM electrolyser. The overall equation is the same, but the half-equations are slightly different because O^{2-} ions rather than H^+ ions move from one electrode to the other. The anode half-equation is:

$$2O^{2-} \longrightarrow O_2 + 4e^-$$

What will the cathode half-equation be?

22 There is considerable research into a zinc–bromine rechargeable cell. Complete the following table to compare the operation of this cell in discharge compared to recharge. To answer this question, you may need to refer to the electrochemical series to decide what the reactants will be if a spontaneous reaction is to occur between a zinc/zinc-ion half-cell and a bromine/bromide-ion half-cell.

	Discharge	Recharge
Is the reaction spontaneous or non-spontaneous?		
reactants		
polarity of anode		
direction of electron flow		
Is the cell acting as a galvanic or electrolytic cell?		

23 In the same way that environmental considerations have led to a preference for the production of green hydrogen gas instead of brown hydrogen gas, there is a push to develop a process for the production of green ammonia gas.

The usual method used to manufacture ammonia is through the Haber process, where the reaction is:

$$N_2(g) + 3H_2(g) \rightleftharpoons 2NH_3(g)$$

The hydrogen gas is typically made from methane in an endothermic reaction known as the steam-methane reforming reaction.

$$CH_4(g) + H_2O(g) \rightleftharpoons CO(g) + 3H_2(g)$$

The nitrogen gas required can be obtained from air.

- **a** Examine both equations above to outline how the production of ammonia gas has a detrimental impact upon the environment.
- **b** One initiative being trialled is to build an ammonia plant that uses green hydrogen. Explain how this plant differs from traditional ammonia production plants and how it might lessen the impact of hydrogen production on the environment.
- **c** A breakthrough by researchers at Monash University takes the production of green ammonia further. The researchers have been able to produce ammonia in one step by electrolysing phosphorus solutions using lithium nitride, Li_3N, electrodes. In this process, the lithium ions are replaced by hydrogen atoms producing ammonia.

Discuss how this discovery might further lessen the impact on the environment of ammonia production.

REVIEW QUESTIONS

How can the rate and yield of chemical reactions be optimised?

Multiple-choice questions

1 Which of the following statements about a catalyst is **not** true?

A A catalyst is not consumed in a reaction.
B A catalyst alters the reaction pathway between reactants and products.
C A catalyst reduces the energy released or absorbed by a reaction.
D The proportion of molecules with sufficient energy to react is increased by a catalyst.

2 Which of the changes listed below increase the rate of the reaction between propane and chlorine?

$$C_3H_8(g) + 2Cl_2(g) \xrightarrow{\text{UV light}} C_3H_6Cl_2(g) + 2HCl(g)$$

I Increase in pressure
II Increase in temperature
III Addition of $Cl_2(g)$

A I and II only
B I and III only
C II and III only
D I, II and III

3 0.500 g of $CaCO_3$ reacts with 100 mL of 0.100 M hydrochloric acid according to the following equation:

$$2HCl(aq) + CaCO_3(s) \longrightarrow CaCl_2(aq) + CO_2(g) + H_2O(l)$$

In which of the following sets of conditions would the rate of reaction increase and the volume of CO_2 produced decrease?

	Volume and concentration of HCl	Mass of $CaCO_3$
A	100 mL of 0.200 M	0.500 g
B	100 mL of 0.100 M	0.250 g
C	200 mL of 0.100 M	0.500 g
D	200 mL of 0.200 M	0.250 g

4 Consider the following equation, which, at a particular temperature, has an equilibrium constant of $K = 5.0\ M^2$:

$$2W + 2X \rightleftharpoons 2Y + nZ$$

From this information, it can be concluded that the coefficient *n* in the equation is:

A 1
B 2
C 3
D 4

5 A sample of NOCl is allowed to come to equilibrium according to the following equation:

$$2NOCl(g) \rightleftharpoons 2NO(g) + Cl_2(g)$$

At a constant temperature, the volume of the mixture is halved and the mixture is allowed to come to equilibrium again. When the system comes to equilibrium again, the chlorine has:

A decreased in amount and decreased in concentration
B increased in amount and decreased in concentration
C decreased in amount and increased in concentration
D increased in amount and increased in concentration.

6 Carbon dioxide gas dissolves to a small extent in water, forming carbonic acid in an exothermic reaction.

$$CO_2(g) + H_2O(l) \rightleftharpoons H_2CO_3(aq) \qquad \Delta H < 0$$
$$H_2CO_3(aq) + H_2O(l) \rightleftharpoons HCO_3^-(aq) + H_3O^+(aq)$$

This is the reaction involved in forming fizzy drinks. Which one of the following strategies would **not** be effective in increasing the amount of dissolved carbon dioxide?

A Decreasing the pH of the solution
B Decreasing the temperature of the solution
C Increasing the concentration of carbon dioxide in the water
D Increasing the pressure of the carbon dioxide gas

7 A series of experiments is described below.

I Adding zinc granules to a solution of tin(II) chloride
II Adding small copper pieces to tin(II) chloride solution
III Electrolysis of molten tin(II) chloride
IV Electrolysis of an aqueous solution of tin(II) chloride

The experiments that would result in the production of solid tin metal are:

A all of them
B I, III and IV only
C I and II only
D I and III only

8 A nickel rod is electroplated with zinc using the nickel rod as one electrode and a zinc rod as the other electrode. Which one of the following statements is correct for this process?

A The zinc electrode is the positive electrode.
B The nickel rod is the anode.
C The electrolyte is a nickel(II) nitrate solution.
D Reduction occurs at the positive electrode.

9 The rate of a reaction between two aqueous reactants increases when the temperature is increased and a catalyst is added. Which statements are both correct for the effect of these changes on the reaction?

	Increasing the temperature	Adding a catalyst
A	activation energy increases	activation energy increases
B	collision frequency increases	activation energy does not change
C	activation energy does not change	activation energy decreases
D	collision frequency decreases	collision frequency increases

10 Which of the following changes made to a chemical reaction would result in a greater **proportion** of successful collisions between reactant particles?

I addition of a catalyst
II crushing a solid reactant into a powder
III increasing the temperature

A I and II only
B I and III only
C II and III only
D I, II and III

11 Consider the following equation:

$$2A(aq) \rightleftharpoons B(aq) + 3C(aq) \qquad \Delta H > 0$$

Which one of the following changes will decrease the concentration of the products but increase the equilibrium yield of the reaction?

A Add a catalyst
B Add water
C Decrease the temperature
D Add reactant A

12 Determine the mass of hydrogen consumed at the anode in a hydrogen-oxygen fuel cell with an acidic electrolyte. The fuel cell operates for 24 minutes with a current of 3.5 A and the half-equation for the reaction at the anode is:

$$H_2(g) \longrightarrow 2H^+(aq) + 2e^-$$

A 0.026 g
B 0.052 g
C 0.10 g
D 0.21 g

13 Three beakers contain aqueous solutions of 1.0 M chromium(III) nitrate, 1.0 M copper(II) nitrate and 1.0 M silver nitrate. Each solution has 0.60 F of electric charge passed through it, using carbon electrodes and a power pack. In each beaker a different metal is deposited onto the cathode.

In increasing order, the amount of metal, in mol, deposited onto the cathode in each beaker will be:

A Cr < Cu < Ag
B Ag < Cu < Cr
C Cr < Ag < Cu
D the same in all three beakers.

14 Molten lithium bromide and a 1.0 M aqueous solution of lithium bromide each undergo electrolysis with platinum electrodes. Which one of the following statements is correct?

A The products formed at both the anode and cathode are all the same.
B Gases form at both electrodes in the 1.0 M lithium bromide, but liquid products form in the molten lithium bromide.
C The products formed at the cathode in both cells are the same.
D The products formed at the anode are the same, but at the cathode they are different.

15 z mol of iron is deposited from $FeSO_4$ (aq) by a current, y, in time x. What is the amount of copper, in mol, deposited by electrolysis from $Cu(NO_3)_2$ (aq) by a current, $4y$, in time $\frac{x}{8}$?

A $\frac{z}{4}$
B $\frac{z}{2}$
C z
D $2z$

Short-answer questions

16 Write equations to represent the anode and cathode reactions that occur when an electric current is passed through:

a a dilute solution of sodium chloride
b a concentrated solution of sodium chloride
c molten sodium chloride.

17 Lithium metal is prepared by electrolysis of a molten mixture of lithium chloride and potassium chloride.

a Write a half-equation for the reaction occurring at the cathode in this cell.
b Write a half-equation for the reaction occurring at the anode in the cell.
c Why is it not possible to produce lithium metal by the electrolysis of an aqueous solution of lithium chloride?
d Suggest why a mixture of lithium chloride and potassium chloride is used, rather than pure lithium chloride, in this electrolytic process.
e Assuming that lithium reacts in a similar way to sodium, describe two precautions that would need to be taken in this process.

18 'Green' hydrogen can be produced from water in a PEM electrolyser powered by either solar or wind energy.

- **a** Write a half-equation for the reaction occurring at the cathode in this cell.
- **b** Write a half-equation for the reaction occurring at the anode in the cell.
- **c** Explain what is meant by the term 'green' hydrogen.
- **d** Determine the volume of hydrogen, in L, that could be produced by a PEM electrolyser in a day, running with a current of 5.25 kA. The hydrogen is stored at a pressure of 7.00×10^4 kPa and a temperature of 25°C.

19 In an electrolysis experiment, a student is provided with a solution of nickel(II) nitrate ($Ni(NO_3)_2$). The electrodes to be used are a carbon rod as the positive electrode and a metal spatula as the negative electrode.

- **a** Will the nickel coating appear on the carbon rod or on the metal spatula during the experiment described?
- **b** Write equations for the half-reactions that occur at each electrode.
- **c** A current of 2.5 A is passed through this nickel-plating cell for 15 minutes. Calculate the mass of nickel that could be plated on the cathode.

20 The decomposition of hydrogen peroxide to oxygen occurs according to the following equation:

$$2H_2O_2(aq) \rightarrow 2H_2O(l) + O_2(g)$$

The decomposition reaction normally occurs very slowly with no visible change to the hydrogen peroxide solution. The activation energy for this reaction was measured under two different conditions, I and II.

- **I** When an enzyme was added to a sample of hydrogen peroxide solution, the activation energy for the decomposition reaction was found to be 36.4 kJ mol^{-1}. Bubbles of oxygen gas were rapidly produced and the temperature of the reaction mixture increased.
- **II** When platinum was added to another sample of the hydrogen peroxide solution, the activation energy was 49.0 kJ mol^{-1}. Bubbles of oxygen gas were rapidly produced and the temperature of the reaction mixture increased.

- **a** What is the function of the enzyme and platinum in each these reactions?
- **b** Sketch, on the same set of axes, the energy profile diagrams for the decomposition of hydrogen peroxide with the enzyme and the decomposition using platinum.
- **c** Which reaction system, I or II, would have a greater rate of reaction? Explain your answer.

21 A student investigated the factors affecting the rate of reaction between a solution of sodium thiosulfate and hydrochloric acid:

$$Na_2S_2O_3(aq) + 2HCl(aq) \rightarrow 2NaCl(aq) + SO_2(g) + S(s) + H_2O(l)$$

The reaction was carried out in a conical flask placed on top of a piece of white paper with a dark cross marked on it. The rate of reaction was determined by measuring the time taken for the cross to be obscured by the suspension of sulfur formed during the reaction, as shown.

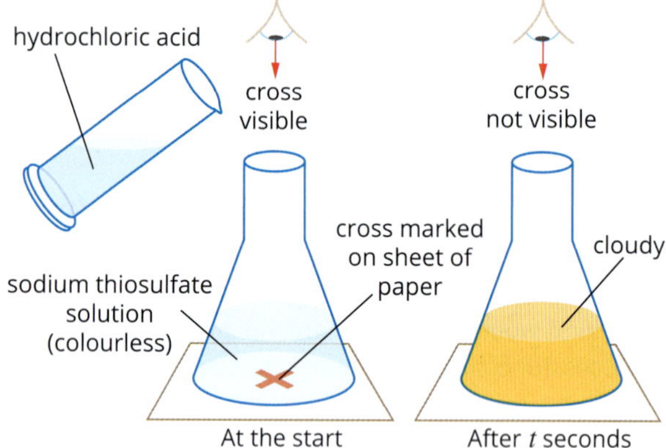

The rate was determined for different concentrations of sodium thiosulfate and for different temperatures. The volume of each solution and the concentration of hydrochloric acid was kept constant. The results are summarised in the table below.

Experiment number	$[Na_2S_2O_3(aq)]$ (M)	Temperature (°C)	Time taken for the cross to be masked (s)
1	0.1	20	36
2	0.2	20	20
3	0.1	25	28

- **a** Explain, in terms of collision theory, why the rate in experiment 2 is higher than the rate in experiment 1.
- **b** Explain, in terms of collision theory, why the rate in experiment 3 is higher than the rate in experiment 1.
- **c** Identify the independent variable when comparing experiments:
  - **i** 1 and 2
  - **ii** 1 and 3
- **d** Identify the dependent variable in this investigation.
- **e** Identify two controlled variables in this investigation.
- **f** Experiment 1 is repeated two more times for a total of three trials.
  - **i** Describe the benefits of conducting multiple trials.
  - **ii** Discuss a factor that would affect the precision of the time measurements.
- **g** For the part of the investigation comparing experiments 1 and 3 to be valid, discuss why the concentrations of the sodium thiosulfate and hydrochloric acid solutions must not be changed.

22 Methane reacts with steam to form hydrogen gas and carbon dioxide in the equilibrium reaction.

$CH_4(g) + 2H_2O(g) \rightleftharpoons 4H_2(g) + CO_2(g) \quad \Delta H = +165 \text{ kJ}$

Use your knowledge of Le Châtelier's principle to complete the table, predicting the effect (*decrease, increase* or *no change*) of the change (column 1) on the designated quantity when the system returns to equilibrium (column 2). Assume that the change listed is the only one taking place (e.g. if H_2 is added, the volume and the temperature are kept constant).

Change	Quantity	Increase/ decrease/ no change
increase temperature	K	
decrease temperature	amount of $H_2O(g)$	
add $H_2(g)$	K	
add $CO_2(g)$	amount of $H_2(g)$	
double volume	amount of CO_2	
double volume	concentration of CO_2	
remove CH_4	amount of CO_2	
add catalyst	amount of $H_2(g)$	
add argon (an inert gas)	K	

23 The reaction between hydrogen and iodine in the gaseous phase to produce hydrogen iodide is described by the following equation:

$$H_2(g) + I_2(g) \rightleftharpoons 2HI(g) \quad \Delta H < 0$$

The following diagram shows the change in concentration of gaseous hydrogen, iodine and hydrogen iodide as the equilibrium is reached and disturbed.

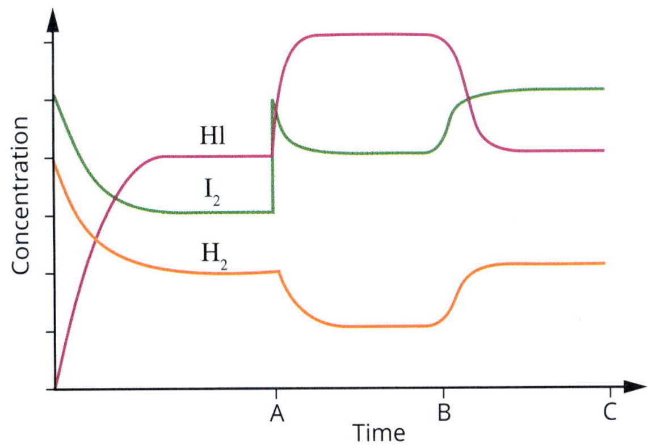

a At point A, a sudden change occurs to the system. What has happened?

b At point B, another change has occurred. What has happened?

c On the graph provided, highlight on the time axis all the time periods when the system is at equilibrium.

d At time C, the volume of the reaction vessel was doubled. Extend the graph past point C, until equilibrium is re-established, to show what would happen to the concentrations of the gases. Assume that the temperature remains constant.

e At a particular temperature, this reaction has an equilibrium constant, $K = 12.8$. This reaction was studied in a 2.00 L reaction vessel. At a certain time, the concentration of each component was determined: $[H_2] = 0.300$ M, $[I_2] = 0.100$ M, $[HI] = 0.400$ M.

- **i** Show that the system is not at equilibrium.
- **ii** Explain the direction the system will shift to reach equilibrium.
- **iii** When the system reaches equilibrium, the reaction vessel contains 0.940 mol of HI. Determine the equilibrium concentration of each gas in the reaction vessel.

24 Methanol (CH_3OH) is used as a fuel for some racing cars. The synthesis of methanol from methane involves two reactions.

- **I** A reaction of methane with steam to produce carbon monoxide gas and hydrogen
- **II** An exothermic reaction between carbon monoxide and hydrogen to produce methanol

a Write equations for the two reactions that are described.

b The reaction of carbon monoxide with hydrogen is performed at about 250°C and 100 atm pressure. Copper, zinc oxide and alumina are also present.

- **i** What is the probable function of the copper, zinc oxide and alumina in the reactor?
- **ii** How would the equilibrium yield be affected if the reaction was performed at a higher temperature?
- **iii** State two advantages of using such a high pressure.

25 A series of electrolysis experiments are carried out in a U-tube as shown in the diagram.

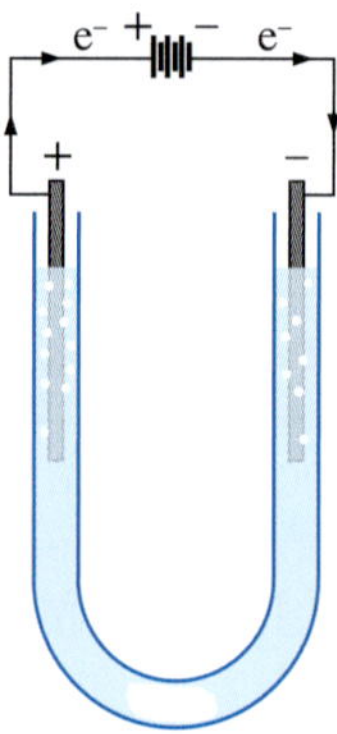

The electrolyte solution in each case, along with the electrode materials used, are listed below. Refer to the electrochemical series on page 189 and for each of the electrolysis experiments.

	Solution	Electrode materials	
		Positive electrode	Negative electrode
a	KI(aq)	carbon	carbon
b	$PbCl_2$(aq)	carbon	carbon
c	$AlCl_3$(aq)	copper	copper

i Write half-equations for the reactions predicted to occur at the positive and negative electrodes.

ii State one visible change you could look for to confirm the formation of your predicted products.

26 Some industrially important chemicals are produced by the electrolysis of a molten electrolyte, while others are produced by electrolysis of an aqueous solution. Examples of chemicals produced by the electrolysis of a molten electrolyte include sodium (and other group 1 metals), magnesium and aluminium. Examples of chemicals produced by the electrolysis of an aqueous electrolyte include hydrogen gas, chlorine gas and sodium hydroxide.

a If sodium hydroxide, chlorine and hydrogen are produced from a concentrated aqueous solution of sodium chloride:

i write an equation for the reaction occurring at the anode

ii write an equation for the cathode reaction

iii describe one feature of this electrolytic cell.

b Explain why some chemicals can be produced by the electrolysis of a molten electrolyte, but not from the electrolysis of an aqueous solution.

c Suggest three properties of graphite that account for its frequent use as an electrode material in commercial electrolytic cells.

d Steel cathodes are often used in commercial electrolytic cells but never steel anodes. Account for this.

27 A 100 mL solution containing a mixture of magnesium chloride ($MgCl_2$) and tin(II) chloride ($SnCl_2$) is prepared by dissolving 0.025 mol of each salt in water. Platinum electrodes are placed in the solution and a small current is passed through it, as shown in the diagram below.

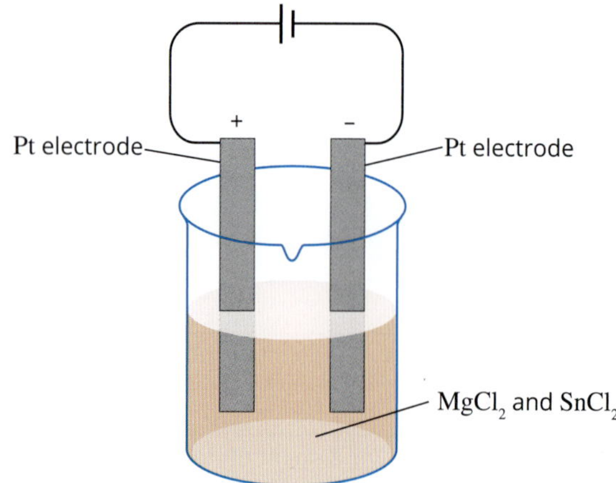

a Write half-equations for the electrode reactions that occur just after the electrolysis is started at the:

i anode

ii cathode.

b After electrolysis has been occurring for a considerable period of time, no more of the metal that was first plated on the cathode will be produced and a new electrode reaction will occur at the cathode. Write a half-equation for the next electrode reaction that occurs at the cathode.

28 The half-equations occurring in a nickel–zinc (NiZn) battery as it discharges are as follows.

cathode (+): $2H_2O(l) + 2NiOOH(s) + 2e^- \rightarrow 2Ni(OH)_2(s) + 2OH^-(aq)$

anode (−): $2OH^-(aq) + Zn(s) \rightarrow Zn(OH)_2(s) + 2e^-$

The NiZn battery is rechargeable.

a Write the half-equation for the reaction occurring at the negative and positive electrodes when the battery is being recharged.

b Why is recharging a secondary cell such as this classed as a type of electrolysis?

c One issue with this cell is that zinc hydroxide is slightly soluble, allowing small quantities to migrate away from the electrode during discharge. Explain how this will affect the performance of the cell.

d A small NiZn battery containing 2.5 g of Zn is used to power a device requiring a current of 0.83 A. How long will the battery last, in s, before it needs to be recharged? Assume the battery operates at 100% efficiency.

29 Consider the following readily reversible gas-phase reaction.

$$W(g) + 2X(g) \rightleftharpoons Y(g) + Z(g)$$

The graph below shows the changes to the rates of the forward and reverse reactions when some additional reactant, W, was added to an equilibrium mixture of these gases.

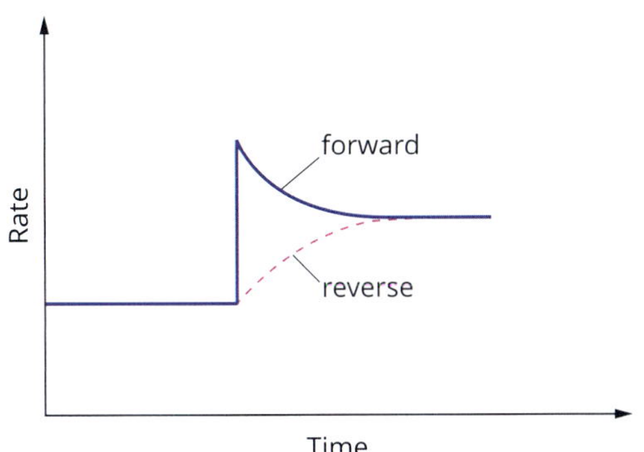

a Explain how this graph shows that the mixture was at equilibrium before the change was made.

b Use collision theory to explain the changes in the rates of the forward and reverse reactions after the change was made.

The process was repeated several times, each time with a different change occurring. Changes tested included:

- removing W from the mixture
- removing Y
- adding Y
- adding argon (an inert gas) at constant volume
- adding a catalyst
- increasing the volume of the container.

c For each of the graphs below, select a change from this list that is consistent with the rate changes depicted. In each case, briefly justify your choice.

i

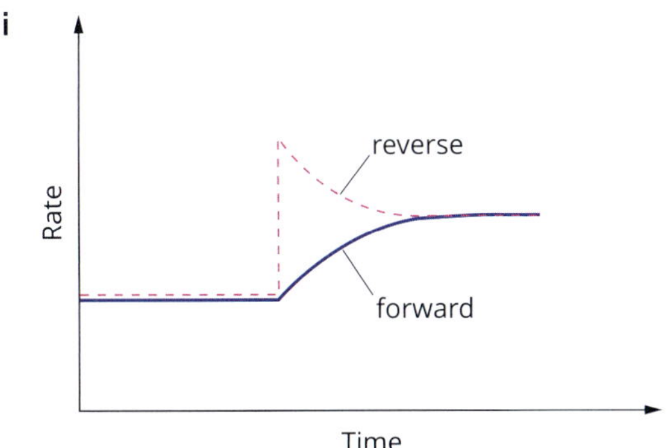

ii

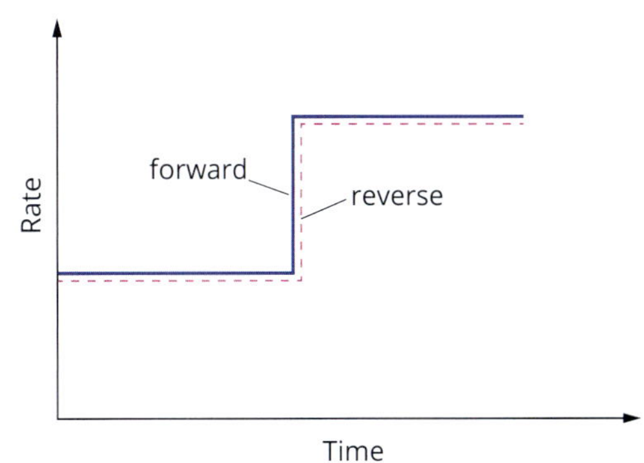

iii

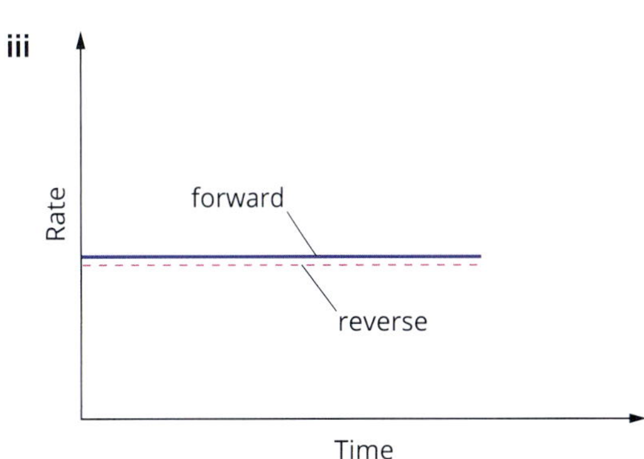

The graph below shows the rate changes after lowering the temperature.

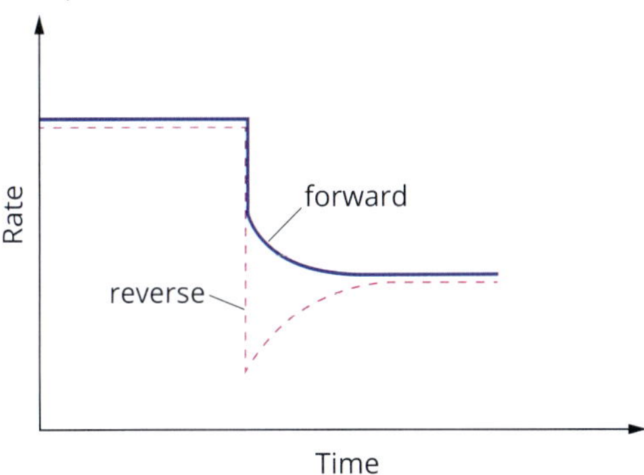

d Why does lowering the temperature cause an initial drop in both reaction rates?

e Considering the period during which the rates are unequal, in which direction does the position of equilibrium shift?

f Is the reaction exothermic or endothermic? Explain.

g At a particular temperature, this reaction has an equilibrium constant, $K = 1.50 \times 10^{-2}$ M^{-1}. A certain amount of W and 2.00 mol of X was introduced into a 0.500 L vessel. At equilibrium, there was 0.344 mol of Z present. What was the original amount of W, in mol, introduced into the vessel?

UNIT 4 How are carbon-based compounds designed for purpose?

To achieve the outcomes in Unit 4, you will draw on key knowledge outlined in each area of study and the related key science skills on pages 11 and 12 of the study design. The key science skills are discussed in Chapter 1 of this book.

AREA OF STUDY 1

How are organic compounds categorised and synthesised?

Outcome 1: On completion of this unit, the student should be able to analyse the general structures and reactions of the major organic families of compounds, design reaction pathways for organic synthesis, and evaluate the sustainability of the manufacture of organic compounds used in society.

AREA OF STUDY 2

How are organic compounds analysed and used?

Outcome 2: On completion of this unit, the student should be able to apply qualitative and quantitative tests to analyse organic compounds and their structural characteristics, deduce structures of organic compounds using instrumental analysis data, explain how some medicines function, and experimentally analyse how some natural medicines can be extracted and purified.

AREA OF STUDY 3

How is scientific inquiry used to investigate the sustainable production of energy and/or materials?

Outcome 3: On completion of this unit, the student should be able to design and conduct a scientific investigation related to the production of energy and/or chemicals and/or the analysis or synthesis of organic compounds, and present an aim, methodology and method, results, discussion and conclusion in a scientific poster.

CHAPTER

10 Structure, nomenclature and properties of organic compounds

Scientists know of more than 7 million organic compounds, and more of these compounds are being discovered or synthesised all the time. Organic compounds exhibit an enormous range of properties. This allows chemists to develop materials for applications as diverse as fuels, fibres, plastics, detergents, dyes, paints, medicines, perfumes and insecticides. Organic compounds also form the basis of every cell in all living organisms.

This chapter looks at the element carbon and why such a vast range of organic compounds is possible. It also looks at the groups, structures and names of a range of different organic compounds.

Key knowledge

- characteristics of the carbon atom that contribute to the diversity of organic compounds formed, with reference to valence electron number, relative bond strength, relative stability of carbon bonds with other elements, degree of unsaturation, and the formation of structural isomers **10.1, 10.2**
- molecular, structural and semi-structural (condensed) formulas and skeletal structures of alkanes (including cyclohexane), alkenes, benzene, haloalkanes, primary amines, primary amides, alcohols (primary, secondary and tertiary), aldehydes, ketones, carboxylic acids and non-branched esters **10.1, 10.2, 10.3, 10.4**
- the International Union of Pure and Applied Chemistry (IUPAC) systematic naming of organic compounds up to C8, with no more than two functional groups for a molecule, limited to non-cyclic hydrocarbons, haloalkanes, primary amines, alcohols (primary, secondary and tertiary), aldehydes, ketones, carboxylic acids and non-branched esters **10.2, 10.3, 10.4, 10.5**
- trends in physical properties within and between homologous series (boiling point and melting point, viscosity), with reference to structure and bonding **10.6**

WS 23

OA ✓✓

10.1 Diversity of organic compounds

FIGURE 10.1.1 Different types of organic molecules form the basis of all living systems, including the child, tomatoes, timber and plants in this photograph.

Carbon compounds make up more than 90% of all chemical compounds and form the basis of living systems. The association of carbon compounds with living systems led to the study of carbon compounds being called 'organic chemistry' and the molecules of carbon compounds being referred to as **organic molecules** (Figure 10.1.1). Organic molecules can be defined as molecules containing carbon atoms linked to each other, and to other non-metallic atoms, by covalent bonds.

In its many millions of compounds, carbon is almost always associated with hydrogen, giving organic molecules a hydrocarbon skeleton. Organic molecules also commonly contain the elements oxygen, nitrogen, sulfur and chlorine.

In this section, you will learn about the diversity of carbon compounds and the factors that contribute to carbon being the element upon which all life forms are based.

BONDING IN CARBON COMPOUNDS

Carbon atoms contain six electrons—two electrons in the first shell and four in the second. The electronic configuration of a carbon atom can be written as 2,4 or as $1s^22s^22p^2$. This means the following.

- Each carbon atom has four valence electrons, which gives it a **valence electron number** of four. The valence electron number of an element is the number of valence electrons in an atom of the element. All four of carbon's valence electrons are available for bonding with other atoms.
- A carbon atom can form four covalent bonds with up to four other carbon atoms, each of which in turn can bond with up to four other carbon atoms, potentially forming long, branched chains and rings.
- A carbon atom can form four covalent bonds with up to four other non-metal atoms.
- Single, double or triple bonds can be formed between two carbon atoms. A molecule that contains only single carbon–carbon bonds is described as a **saturated molecule**. A molecule that contains one or more double or triple carbon–carbon bonds is described as an **unsaturated molecule**.

Carbon forms a wide variety of compounds because of its ability to form four strong covalent bonds with both itself and other elements.

The wide variety of compounds formed by carbon is because of its ability to form strong covalent bonds with other carbon and hydrogen atoms and with atoms of other elements such as oxygen, nitrogen, sulfur, phosphorus and the halogens.

Carbon is also a very versatile atom because its four covalent bonds can be formed in different ways, for example:

- four single bonds with four other atoms
- one double bond and two single bonds with three other atoms
- two double bonds with two other atoms
- one triple bond and one single bond with two other atoms.

Stability of carbon bonds with other elements

The covalent bonds between carbon and other atoms each have a **bond energy**. Bond energy is the quantity of energy required to break 1 mole of covalent bonds in the gaseous state and is a measure of **bond strength**. The higher the bond energy, the stronger the bond.

The stability of a molecule is related to the strengths of the covalent bonds it contains. A molecule with strong chemical bonds (based on its bond energies) is generally more stable (less likely to react) than a molecule with weaker bonds. This relationship between strong bonds and stability helps to explain the range of chemical compounds found in nature, as substances will tend to react with one another to produce more stable substances, with stronger bonds overall.

Table 10.1.1 compares some of the bonds formed by carbon with those formed by silicon. Silicon and carbon are very similar atoms: both have a valence electron number of four and are located in group 14 of the periodic table. Firstly, you can see that silicon is not readily able to form multiple bonds with itself. Secondly, you can see that the bonds formed by carbon are in most cases stronger than the equivalent bond with silicon. One notable exception is the Si–O bond, which explains why quartz (SiO_2) is such an abundant mineral.

TABLE 10.1.1 Bond energies involving carbon and silicon atoms

Covalent bonds with carbon	Bond energy (kJ mol^{-1})	Covalent bonds with silicon	Bond energy (kJ mol^{-1})
C≡C	839	Si≡Si do not form	
C=C	614	Si=Si do not form readily	
C–C	348	Si–Si	226
C–H	413	Si–H	323
C–O	360	Si–O	466

Table 10.1.2 compares the bond energies (and thus bond strength) between atoms of the same type. As you can see, a single covalent bond between two carbon atoms, C–C, is very strong, particularly when compared to covalent bonds between other atoms of the same type. This is one of the reasons why only carbon tends to form chain-like structures at the temperatures and pressures found on Earth, and why it is carbon that forms the backbone of all living things.

TABLE 10.1.2 Bond energies involving atoms of the same type

Covalent bonds between atoms of the same type	Bond energy (kJ mol^{-1})
C–C	348
N–N	163
O–O	146
F–F	155
S–S	226

In summary, the stability and range of carbon bonds helps to explain carbon compound diversity:

- Carbon can form stable bonds with many other elements.
- Carbon can form stable single and multiple bonds with itself.
- The stability of the carbon–carbon bond means it tends to form chain and ring-like structures.

REPRESENTING ORGANIC MOLECULES

Before exploring organic molecules further, you will learn about the different ways that molecules can be represented, such as:

- molecular formulas
- structural formulas
- condensed or semi-structural formulas
- skeletal structures.

Molecular formulas

Molecular formulas, such as C_2H_6O and $C_4H_8O_2$, indicate the number and type of atoms of each element present in a molecule. However, they do not indicate how the atoms are arranged.

Structural formulas

Structural formulas show the location of atoms relative to one another in a molecule, as well as the number and location of covalent bonds.

When four single bonds are formed around a carbon atom, the pairs of electrons in each bond act as a negatively charged cloud. **Valence shell electron pair repulsion (VSEPR) theory** determines that these electron pairs repel each other, so the bonds are as far apart as possible in three dimensions, at an angle to each other of nearly 109.5°. The structure of methane, shown in Figure 10.1.2, is described as a tetrahedral shape because the four single bonds are pointing to the corners of a tetrahedron (shown in red).

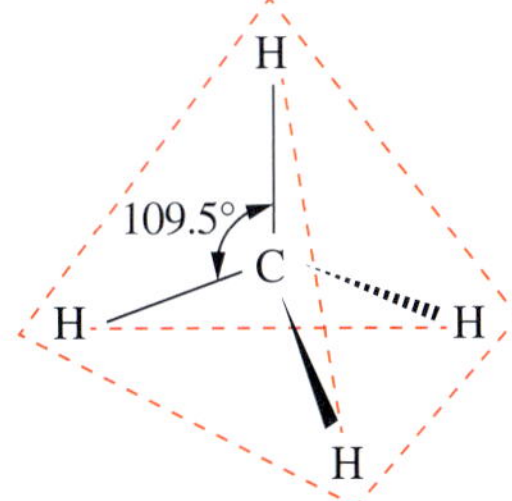

FIGURE 10.1.2 The structure of a molecule of methane (CH_4) showing the tetrahedral geometry. The solid wedge represents a bond coming out of the page, whereas the dashed wedge represents a bond going into the page.

Molecules consisting of long chains of carbon atoms joined by single bonds are often referred to as 'straight-chain' molecules. Because of the tetrahedral distribution of each carbon's bonds, the chain actually has a zigzag shape (Figure 10.1.3a). However, structural formulas showing a tetrahedral arrangement of bonds around carbon atoms can become complicated and difficult to interpret. To make the structure of these molecules clearer, the bonds are often drawn at right angles (Figure 10.1.3b).

Even though there is a tetrahedral arrangement of bonds around each carbon atom, structural formulas are often drawn with the bonds at right angles, for simplicity.

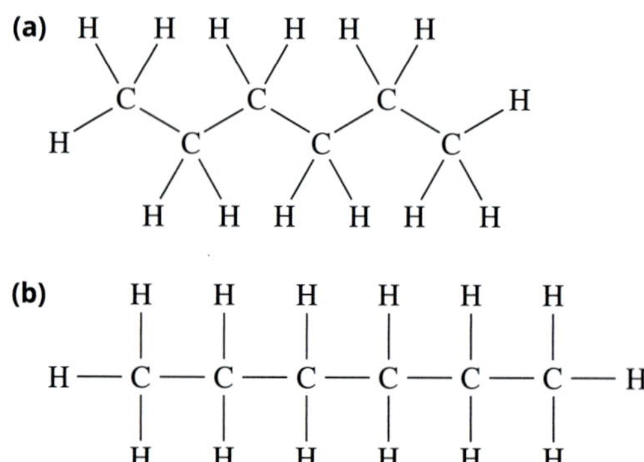

FIGURE 10.1.3 (a) The six-carbon chain of hexane (C_6H_{14}). There is a tetrahedral arrangement of bonds around each carbon atom. (b) The chain can be drawn in a straight line for convenience.

Semi-structural formulas

A **semi-structural formula** is used to indicate the connections in the structure of a compound without showing the three-dimensional arrangement of atoms. The term **'condensed formula'** (also known as 'condensed structural formula') is also used to describe this type of formula.

In a semi-structural formula, the carbon chain is represented on one line of text. The carbon atoms in the chain, and all the atoms attached to each of them, are listed in the order that they appear in the structural formula. Single bonds are not shown, but double and triple carbon–carbon bonds are often shown. Groups of atoms that form branches in a molecule are written in brackets after the carbon atom to which they are attached. Chains of $-CH_2-$ are often condensed into brackets (as can be seen for butane in Table 10.1.3 on the following page.)

Skeletal structures

A functional group is an atom or group of atoms that largely determines a molecule's properties.

A **skeletal structure** is used to represent complex organic molecules. It is a shorthand version of a **structural formula**, which shows only the arrangement of carbon–carbon bonds and **functional groups** present in a molecule. A functional group is an atom or group of atoms that largely determines a molecule's properties. In a skeletal structure, the following are left out:

- carbon atoms
- hydrogen atoms attached directly to carbon atoms
- bonds between carbon atoms and hydrogen atoms.

Thus, a skeletal structure includes only the following:

- carbon–carbon single, double and triple bonds
- bonds to functional groups
- bonds within functional groups.

Some examples of molecular, structural, semi-structural and skeletal structures are shown in Table 10.1.3 on the following page.

TABLE 10.1.3 Examples of molecular, structural and semi-structural formulas and skeletal structures

Molecular formula	Structural formula and name	Semi-structural (condensed) formula	Skeletal structure
C_4H_{10}	butane	$CH_3CH_2CH_2CH_3$ or $CH_3(CH_2)_2CH_3$	
C_4H_8	but-1-ene	$CH_2{=}CHCH_2CH_3$ or $CH_2CHCH_2CH_3$	
$C_5H_{13}N$	2-methylbutan-1-amine	$CH_3CH_2CH(CH_3)CH_2NH_2$	

ISOMERS

Isomers are molecules that contain the same number and type of atoms, arranged in different ways. The existence of isomers is a major reason why there are so many different carbon compounds. Isomers have the same **molecular formula**, but they can have different physical and chemical properties, and so behave differently.

Structural isomers form when the atoms in molecules with the same molecular formula bond together in different arrangements. To identify the structural isomers possible, it can be useful to think about two ways in which the molecule can be changed: by changing the chain length, and/or by changing the position of any functional groups.

Isomers have the same molecular formula, but the atoms are arranged differently.

Isomers formed by changing the chain length

Because branching is possible in the carbon chains that form the backbone of any large organic molecule, isomers with differing main chain lengths can be formed. These isomers of **alkanes** (the simplest hydrocarbons) can contain more than one **alkyl group** (a branch or side chain formed from an alkane); some molecules may also have more than one alkyl group attached to the same carbon atom.

Hexane is an example of an alkane with no branches. It has the molecular formula C_6H_{14} and its structure is shown in Figure 10.1.4.

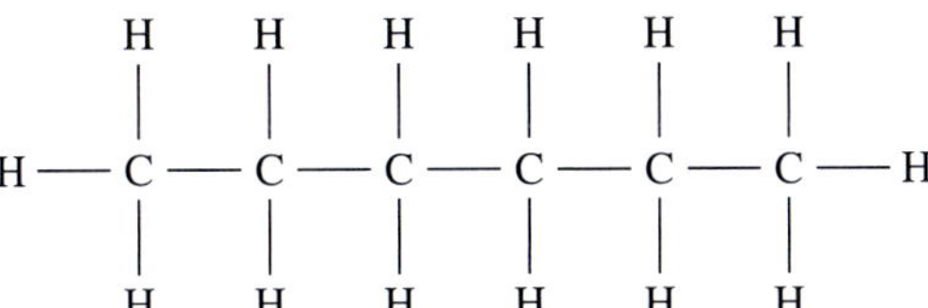

FIGURE 10.1.4 Hexane is a single-chain hydrocarbon with single bonds that contains 6 carbon atoms and 14 hydrogen atoms.

Three of the isomers of hexane that can be formed by changing the main chain length are shown in Figure 10.1.5. Each isomer has a different name that represents its exact molecular structure. The naming of these isomers is described in the next section.

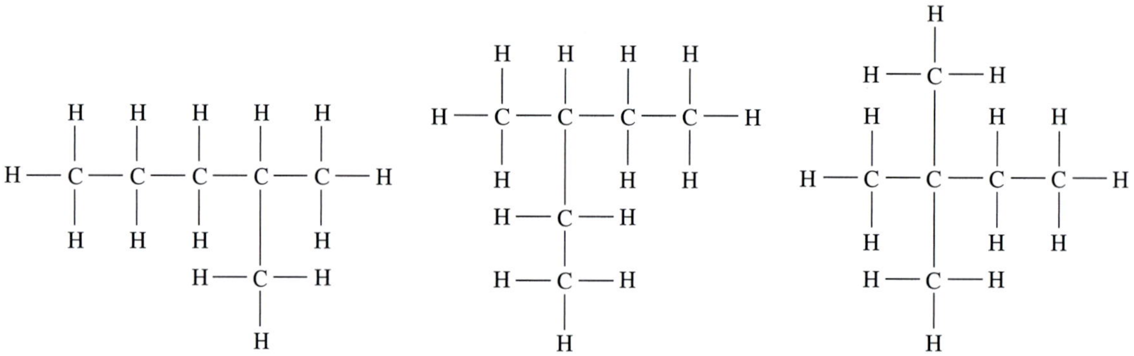

$CH_3CH_2CH_2CH(CH_3)_2$ $CH_3CH_2CH(CH_3)CH_2CH_3$ $(CH_3)_3CCH_2CH_3$

FIGURE 10.1.5 Three isomers of hexane

Isomers formed by changing the position of functional groups

Isomers can also occur for organic molecules that contain functional groups. Molecules with the same carbon chain and functional group can form isomers when the functional group is attached to a different location in the molecule.

For example, **alcohols** are organic molecules that contain a **hydroxyl functional group** (–OH). Several isomers can be drawn for an alcohol with the molecular formula $C_4H_{10}O$. Again, each isomer is distinguished by being given a different name. Figure 10.1.6 shows two isomers, butan-1-ol and butan-2-ol.

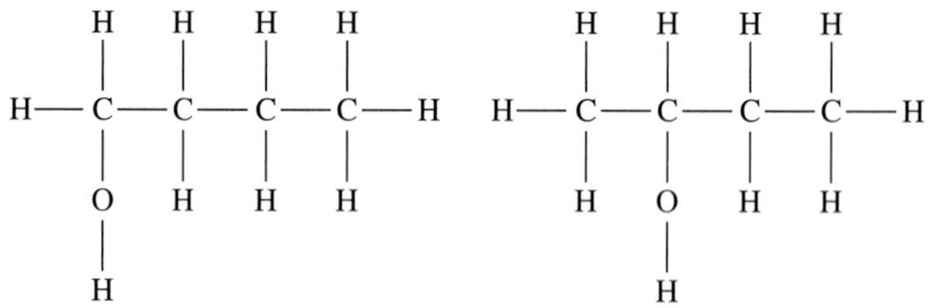

FIGURE 10.1.6 Two isomers of $C_4H_{10}O$

It is important to realise that alkenes also contain a functional group—their double carbon–carbon bond. This means that isomers can be formed when the double bond is in different locations. Figure 10.1.7 shows the two isomers of butene.

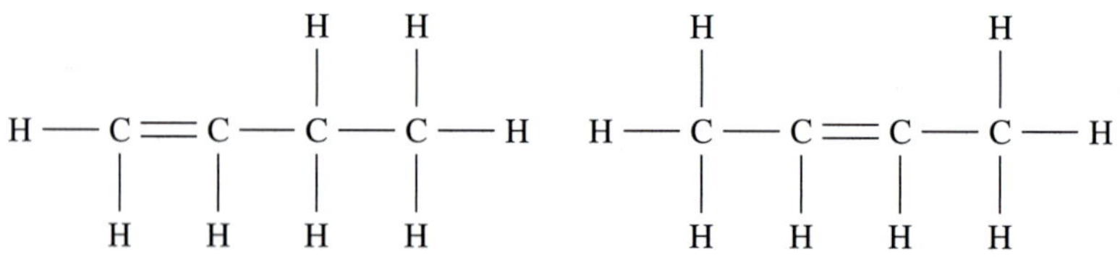

FIGURE 10.1.7 The two isomers of butene: but-1-ene and but-2-ene

Isomers only exist for molecules that contain a functional group and have a long enough carbon chain that different positions of the functional group are possible. When drawing isomers, be careful that you do not just draw the same molecule from a different perspective. You can see that the two structural formulas of ethanol in Figure 10.1.8 do not represent isomers.

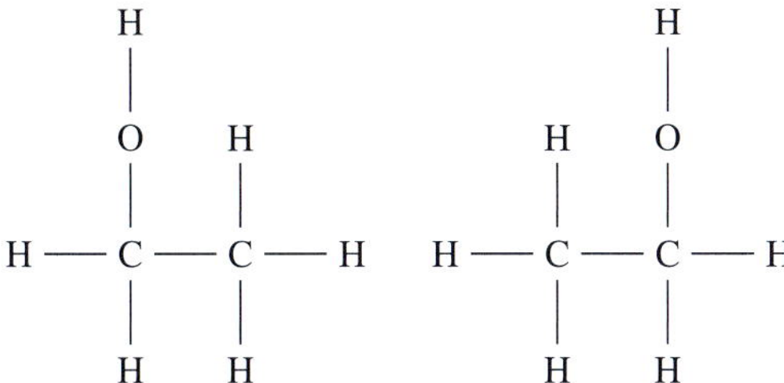

FIGURE 10.1.8 Two representations of the structure of ethanol (C_2H_6O). These are not isomers; if the first structure is flipped over, it is identical to the second.

It is possible for molecules to form isomers by changing both the length of the chain and the position of a functional group. Figure 10.1.9 shows two more isomers of an alcohol with the formula $C_4H_{10}O$. For these isomers to form, both the length of the main chain and the position of the functional group have been changed.

Structural isomers can be formed by changing the length of the main chain and/or the position of functional groups.

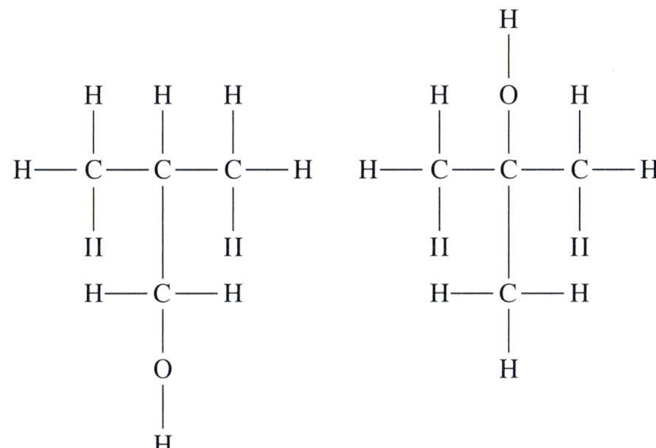

FIGURE 10.1.9 Two more molecules that are alcohols with the molecular formula $C_4H_{10}O$

10.1 Review

SUMMARY

- Carbon has a valence electron number of four.
- Carbon atoms form covalent bonds with each other and with other non-metallic atoms such as hydrogen, oxygen and nitrogen.
- Carbon atoms can form single, double or triple bonds with other carbon atoms.
- Bond energy is the amount of energy required to break a covalent bond and is a measure of bond strength.
- The bonds with the highest strength also have the highest stability. The comparatively high strength of bonds formed by carbon helps to explain carbon compound diversity.
- Molecules can be represented using molecular, structural and semi-structural formulas, and skeletal structures.
- Structural isomers can be formed by changing the length of the main chain and/or the position of functional groups.

KEY QUESTIONS

Knowledge and understanding

1 Explain why the carbon backbone in a skeletal structure might be represented as a zigzag.

2 **a** Explain the connection between bond strength and stability of a molecule.
   b Use the terms 'bond strength' and 'stability' to explain the diversity of carbon compounds.

Analysis

3 **a** Which is the shortest alkane that can have isomers?
   b Draw structural formulas for all the isomers of this alkane.

4 Complete the table by filling in the missing formulas.

Molecular formula	Structural formula	Semi-structural formula	Skeletal structure
		$CH_3(CH_2)_4CH_3$	
C_2H_6O			

10.2 Hydrocarbons

The simplest organic molecules are the hydrocarbons. Although they only contain the elements carbon and hydrogen, the different ways their atoms can be arranged to form molecules results in an enormous diversity of compounds. Crude oil, the source of many hydrocarbons used in industry, contains a mixture of different hydrocarbons and is shown in Figure 10.2.1.

In this section, you will learn about the molecules in the hydrocarbon families of alkanes and alkenes. You will learn about their molecular, structural and semi-structural formulas, as well as some of their isomers and the ways they are named.

FIGURE 10.2.1 Crude oil burning in a Petri dish

HOMOLOGOUS SERIES

The study of organic chemistry is simplified by grouping the millions of different molecules that exist into families called **homologous series**. Compounds that are members of the same homologous series have:

- similar structures
- similar chemical properties
- the same general formula
- a pattern in their physical properties.

Homologous series contain molecules that have increasingly longer chains. These chains grow by the addition of a $-CH_2-$ unit to the previous member of the series. Two examples of homologous series that are hydrocarbons are the alkanes and the alkenes.

ALKANES

Alkanes are saturated **hydrocarbons**, which means all of the carbon–carbon bonds in the molecule are single bonds. Alkanes have the general formula C_nH_{2n+2}.

Naming alkanes

In the systematic name of simple 'straight-chain' alkanes, the prefix, or first part, of the name refers to the number of carbon atoms in the main chain of the alkane (referred to as the **parent molecule**). The prefixes used are listed in Table 10.2.1.

The names of all alkanes have the suffix (ending) '-ane' to indicate that the carbon–carbon bonds are all single bonds.

Combining the prefix and suffix, you can see that methane is a hydrocarbon that contains one carbon atom, ethane is a hydrocarbon that contains two carbon atoms and propane is a hydrocarbon that contains three carbon atoms, and each of the carbon–carbon bonds are single bonds. The structures are shown in Figure 10.2.2.

TABLE 10.2.1 Prefixes used to name molecules with up to eight carbon atoms

Number of carbon atoms	Prefix (parent/ stem)
1	meth-
2	eth-
3	prop-
4	but-
5	pent-
6	hex-
7	hept-
8	oct-

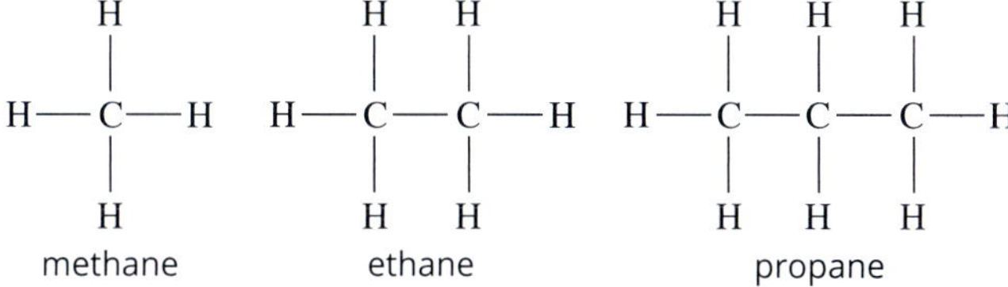

FIGURE 10.2.2 Structural formulas of methane, ethane and propane

The molecular, structural and semi-structural formulas of the first three members of the homologous series of alkanes are shown in Table 10.2.2. You can see that successive members differ by a $-CH_2-$ unit.

TABLE 10.2.2 Formulas of the first three members of the alkane homologous series

Name	Structural formula	Semi-structural formula
methane	H │ H—C—H │ H	CH_4
ethane	H H │ │ H—C—C—H │ │ H H	CH_3CH_3
propane	H H H │ │ │ H—C—C—C—H │ │ │ H H H	$CH_3CH_2CH_3$

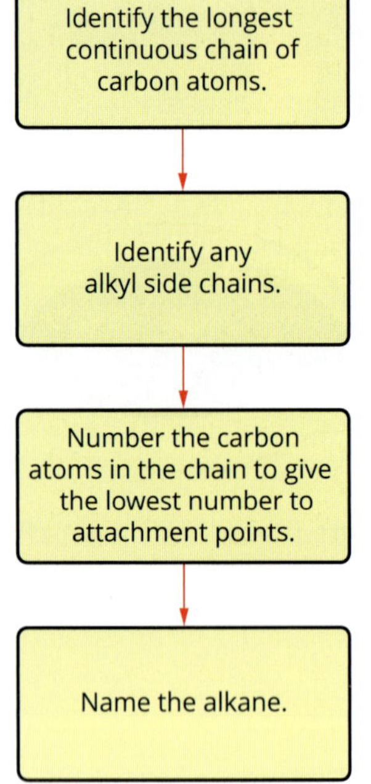

FIGURE 10.2.3 The steps used to name alkanes

> Alkyl groups are named after the alkane from which they are derived, with an -yl ending and general formula C_nH_{2n+1}.

Alkanes with more than three carbon atoms can form isomers, which are named according to the steps shown in Figure 10.2.3.

Methyl and ethyl groups are examples of alkyl groups or alkyl side chains. Alkyl groups are named after the alkane from which they are derived, with a -yl ending. They have one less hydrogen atom than the corresponding alkane of the same name, so the general formula of an alkyl group is C_nH_{2n+1}.

Table 10.2.3 lists the names of the alkyl groups with 1–3 carbon atoms.

TABLE 10.2.3 Names of the alkyl groups with 1–3 carbons

Alkyl group	Corresponding alkane	Name of alkyl group
$-CH_3$	methane	methyl
$-CH_2CH_3$	ethane	ethyl
$-CH_2CH_2CH_3$	propane	propyl

In the naming of an alkane, the following conventions are used.

- Find the longest carbon chain. This gives the parent name for the molecule. The ending of the parent name is '-ane'.
- Place the name of each alkyl group before the parent molecule name.
- If there is more than one type of alkyl group, list the groups in alphabetical order.
- If there is more than one of the same type of alkyl group, use a prefix, for example, 'di-', 'tri-' or 'tetra-'.
- Specify the carbon atom to which each alkyl group is attached by a number before the alkyl group.
- Do not have spaces in the name. Use dashes to separate numbers from words. Use commas to separate numbers from other numbers.

Worked example: 10.2.1

NAMING AN ISOMER OF AN ALKANE WITH MORE THAN ONE ALKYL BRANCH

Write the systematic name of the molecule with the following structure.

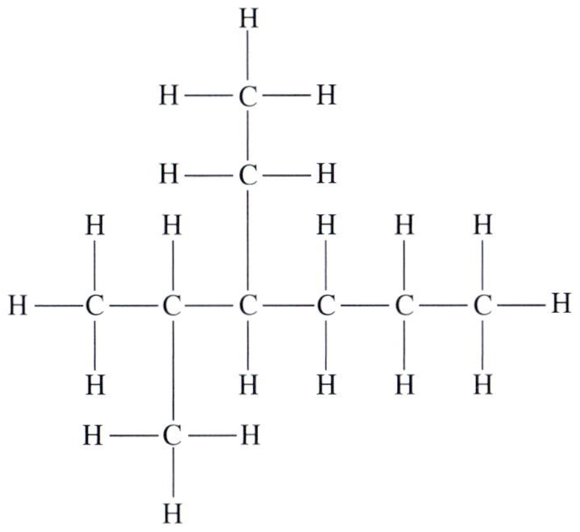

Thinking	Working
Identify the parent name by counting the longest continuous chain of carbon atoms.	longest continuous chain 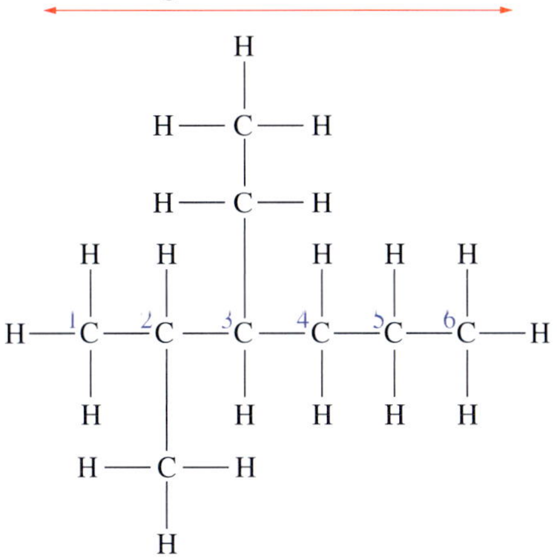 There are six carbon atoms in the longest continuous chain, so the parent name is hexane.
Identify any alkyl side chains by counting the number of carbon atoms in any branches.	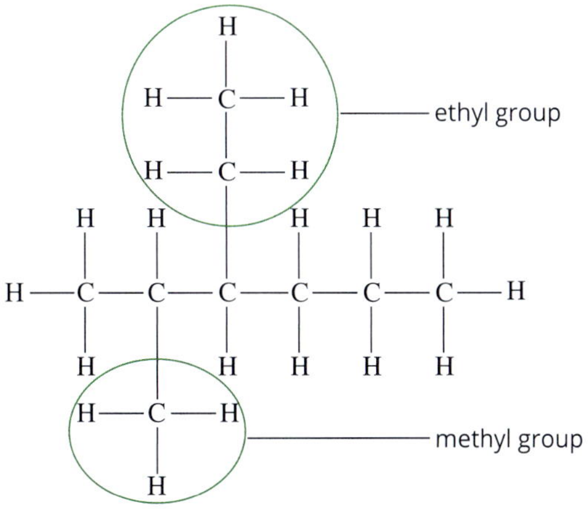 ethyl group methyl group There is one carbon atom in the first alkyl group, so it is a methyl group. The second alkyl group has two carbon atoms, so it is an ethyl group.

Identify the number(s) of the carbon atom(s) to which any alkyl group is attached. Count from the end that will give the lowest possible number to the carbons the groups are attached to.	The methyl group is attached to carbon number 2 and the ethyl group is attached to carbon number 3.
Assemble the name of the isomer, listing alkyl groups in alphabetic order with the number of the carbon atom to which they are attached before each group.	The name of the isomer is 3-ethyl-2-methylhexane.

Worked example: Try yourself 10.2.1

NAMING AN ISOMER OF AN ALKANE WITH MORE THAN ONE ALKYL BRANCH

Write the systematic name of the molecule with the following structure.

(a)

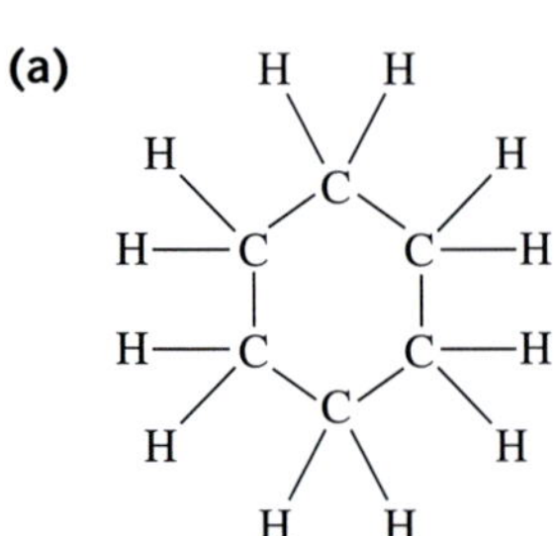

(b)

FIGURE 10.2.4 (a) The structural formula of cyclohexane (C_6H_{12}). (b) Cyclohexane is used by industry to make the starting materials for the manufacture of nylon.

CYCLOHEXANE

As you have read, carbon is able to form ring structures. The cyclic version of hexane, called cyclohexane, is a hydrocarbon that is used as a solvent. It is used in paint stripper and to make other substances such as nylon (Figure 10.2.4).

Figure 10.2.4a shows that the carbon atoms in the cyclohexane molecule form a ring. Molecules that form rings are known as **cyclic molecules**. Just like straight-chain hexane, cyclohexane is a saturated molecule that contains six carbon atoms. Because of the extra carbon–carbon bond that closes the ring structure, there are two less hydrogen atoms, making the formula of cyclohexane C_6H_{12}, whereas hexane is C_6H_{14}. Cyclohexane is a member of the cycloalkane homologous series.

ALKENES

The homologous series of hydrocarbons that contain a carbon–carbon double bond is called the **alkenes**. Alkenes are **unsaturated hydrocarbons**. Because this bond is shared by a pair of carbon atoms, all alkene molecules must have two or more carbon atoms.

Ethene, the first member of the alkene series, is used as the starting point for many polymers (Figure 10.2.5). The molecular, structural and semi-structural formulas of the first three members of the homologous series of alkenes are shown in Table 10.2.4. You can see that successive members differ by a $-CH_2-$ unit. Alkenes have the general formula C_nH_{2n}.

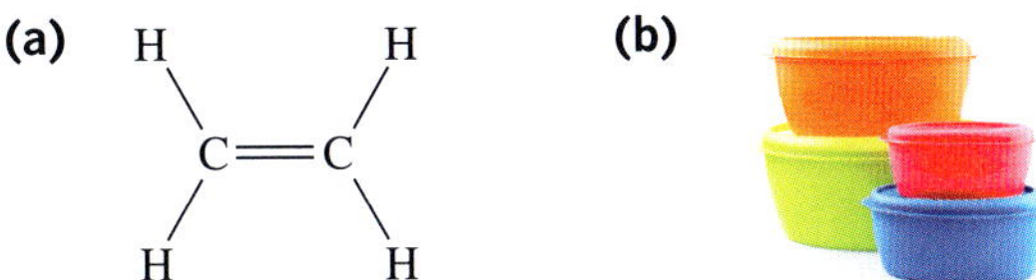

FIGURE 10.2.5 (a) Ethene (C_2H_4) is the first member of the alkene homologous series. (b) Ethene is used to make the polymers in these food-storage containers.

Alkenes with four or more carbon atoms can have structural isomers, where the location of the carbon–carbon double bond changes.

TABLE 10.2.4 Formulas of the first three members of the alkene homologous series

Name	Molecular formula	Structural formula	Semi-structural formula	Skeletal formula
ethene	C_2H_4		CH_2=CH_2 or CH_2CH_2	
propene	C_3H_6		CH_2=$CHCH_3$ or CH_2CHCH_3	
but-1-ene	C_4H_8		CH_2=$CHCH_2CH_3$ or $CH_2CHCH_2CH_3$	

Naming alkenes

The rules for naming alkenes follow the rules used for alkanes. In addition, the following conventions are applied, as shown in Figure 10.2.6.

- Use the ending '-ene' for the parent name.
- Number the carbon atoms from the end of the chain closest to the carbon–carbon double bond.
- Specify the position of the double bond by the number of the lowest-numbered carbon atom in the carbon–carbon double bond.
- Insert the number into the name immediately before '-ene'.
- The number and position of any alkyl side chain is given at the beginning of the molecule's name (as previously described for alkanes).

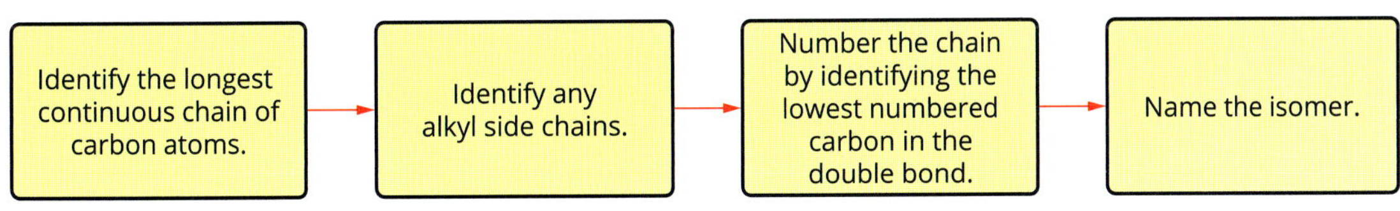

FIGURE 10.2.6 The steps used to name alkenes

DEGREE OF UNSATURATION

The degree of unsaturation identifies how many double bonds or ring structures a molecule contains.

When determining unknown chemical structures, it can be useful to know the **degree of unsaturation** of a molecule (sometimes referred to as the index of hydrogen deficiency). These terms are based on the fact that every time a molecule forms a double bond or a ring, there will be two fewer hydrogen atoms present.

For example, you may be presented with an unknown hydrocarbon molecule with the molecular formula $C_{17}H_{28}$. By comparing this to the general formula for an alkane, it is possible to work out how many double bonds or rings must be present.

The general formula for alkanes tells you that a fully saturated hydrocarbon will have the formula C_nH_{2n+2}.

So, a fully saturated hydrocarbon with 17 carbons would have the formula:

$$C_{17}H_{(34+2)} = C_{17}H_{36}$$

Since this molecule ($C_{17}H_{28}$) only contains 28 hydrogens, it has a 'hydrogen deficiency' of 36 – 28 = 8. As each degree of unsaturation decreases the number of hydrogens by two, this molecule must contain a combination of (8 ÷ 2) = 4 double bonds or rings. (There are other structures possible that are beyond the scope of this course.) The degree of unsaturation of the molecule is 4.

This concept can be summarised as:

$$\text{degree of unsaturation} = \text{number of double bond or ring equivalents}$$
$$= \frac{\text{maximum number of H possible per C} - \text{actual number of H per C}}{2}$$

Worked example 10.2.2

CALCULATING DEGREE OF UNSATURATION IN AN ORGANIC MOLECULE

Calculate the degree of unsaturation for a molecule with molecular formula C_8H_{14}.

Thinking	Working
Use the general formula for alkanes (C_nH_{2n+2}) to calculate the number of hydrogens that would be present in a fully saturated molecule.	Number of C atoms = 8 Hydrogens in fully saturated molecule = (2 × 8) + 2 = 18
Calculate the degree of unsaturation using the formula: degree of unsaturation = number of double bond or ring equivalents $= \frac{\text{maximum number of H possible per C} - \text{actual number of H per C}}{2}$	$= \frac{(18-14)}{2}$ = 2 The degree of unsaturation for C_8H_{14} is 2.

Worked example: Try yourself 10.2.2

CALCULATING DEGREE OF UNSATURATION IN AN ORGANIC MOLECULE

Calculate the degree of unsaturation for a molecule with the molecular formula C_6H_{12}.

BENZENE

Benzene is an unsaturated cyclic hydrocarbon molecule. Its structure consists of six carbon atoms arranged in a ring. Three of the four outer-shell electrons from each carbon atom form normal covalent bonds, but the fourth electron is delocalised (shared) around all six carbons (Figure 10.2.7a).

The structure of benzene is sometimes represented with alternating double and single bonds (Figure 10.2.7b), but this is not considered to be the best representation. In 1929 Kathleen Lonsdale measured the C–C bond lengths in benzene using X-ray diffraction and found that they were all equal and longer than a C=C double bond, but shorter than a C–C single bond. For this reason, each carbon is often described as having one-and-a-half bonds to each neighbour. The presence of the ring of delocalised electrons shortens the length and increases the strength of the C–C bonds.

The bonds can be represented as a ring of normal single bonds with a circle inside the ring as seen in Figure 10.2.7c.

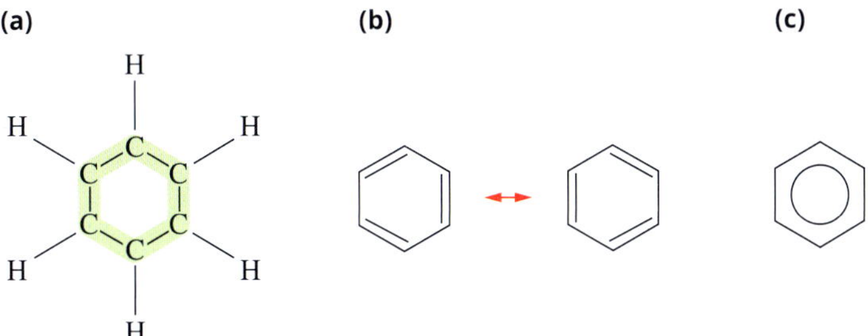

FIGURE 10.2.7 Representations of benzene's (C_6H_6) molecular structure showing (a) the delocalised electrons in green, (b) a shorthand representation with alternating double and single bonds, and (c) a representation of benzene with delocalised electrons as a circle inside the hexagon of carbon atoms.

When a benzene ring is bonded to an alkyl group or a functional group, the ring structure is known as a **phenyl functional group** and it has the formula C_6H_5–.

CHEMFILE

Kathleen Lonsdale and benzene

In 1825, Michael Faraday isolated a new hydrocarbon, named benzene (C_6H_6). The structure and properties of benzene puzzled scientists for years after that.

August Kekulé made a major contribution to the puzzle in 1865. He proposed a ring structure in which alternate carbon atoms were joined by double bonds. The idea of a ring structure came to him in a dream. However, the chemical properties of benzene more closely resemble those of an alkane than an alkene.

In 1929 Kathleen Lonsdale finally determined the three-dimensional structure of benzene using a technique called X-ray crystallography. She found that benzene had a planar, hexagonal ring structure (see figure below) in which all the bonds were the same length. We now know that there is a ring of 6 electrons (one from each carbon) which is shared (delocalised) between the 6 carbon atoms and gives the effect of a 1.5 bond between the atoms.

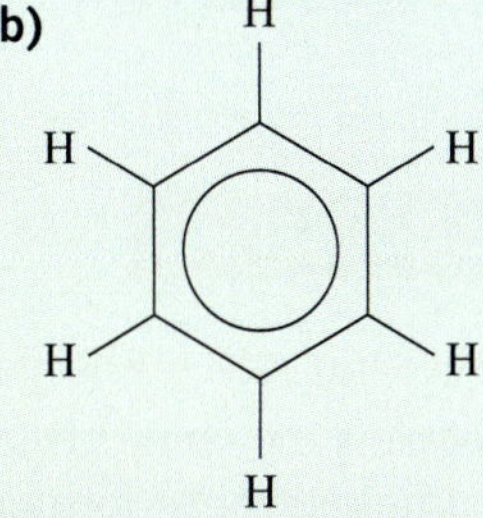

(a) Kathleen Lonsdale, the X-ray crystallographer who determined the three-dimensional structure of benzene.
(b) A representation of benzene showing a ring of delocalised electrons in the centre of a hexagon of carbon atoms, each bonded to just one hydrogen atom.

10.2 Review

OA ✓✓

SUMMARY

- Alkanes are the simplest hydrocarbons. Their molecules contain carbon and hydrogen atoms and all bonds between atoms are single covalent bonds.
- Alkanes have the general formula C_nH_{2n+2}.
- Cyclohexane is a saturated cyclic hydrocarbon with the molecular formula C_6H_{12}.
- Alkenes are a homologous series of hydrocarbons that contain one carbon–carbon double bond.
- Alkenes have the general formula C_nH_{2n}.
- Benzene is an unsaturated cyclic hydrocarbon with the formula C_6H_6.
- Hydrocarbons are named using the steps in the flow chart below.

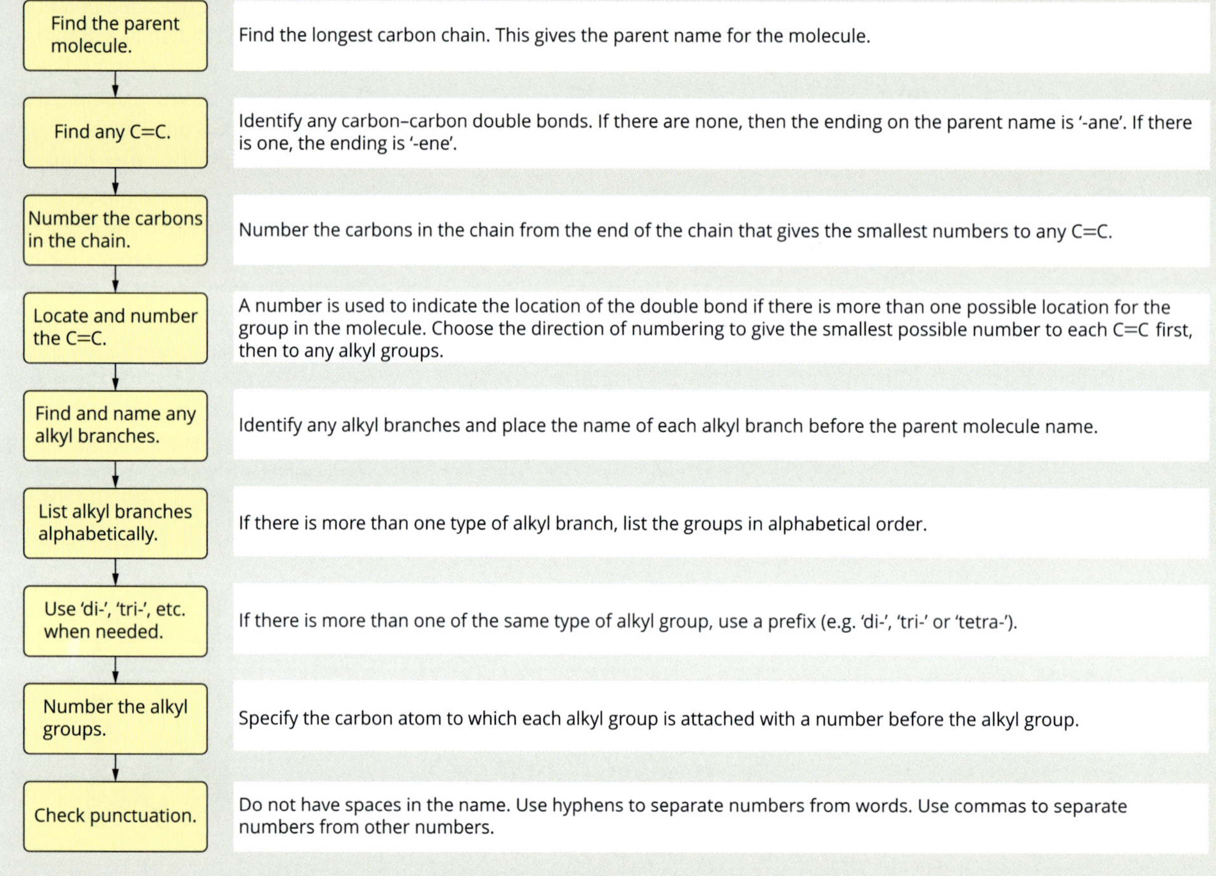

KEY QUESTIONS

Knowledge and understanding

1 What is the name given to molecules that have the same molecular formula but different structural formulas?

2 When multiple alkyl branches are present, in what order are they placed at the beginning of the systematic name?

3 Draw all the possible structural isomers with the formula C_6H_{14}.

4 Cyclohexane fits the general formula for alkenes: C_nH_{2n}. Explain why it is not a member of the alkene homologous series.

5 Name and give the molecular formula for the following two molecules:

a

b

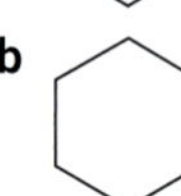

continued over page

10.2 Review *continued*

Analysis

6 Analyse the following structural formulas and determine the correct systematic name.

a

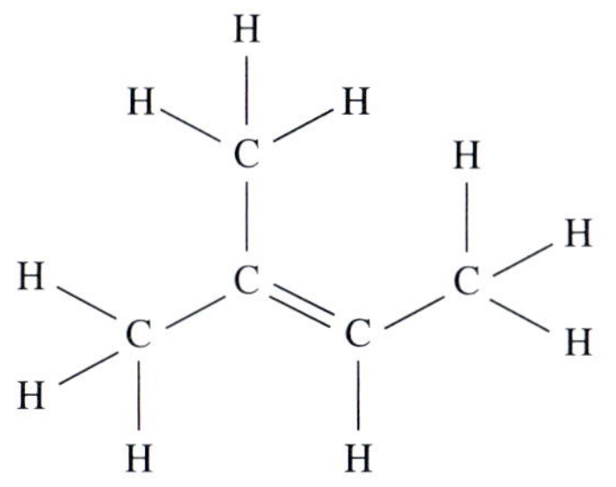

b

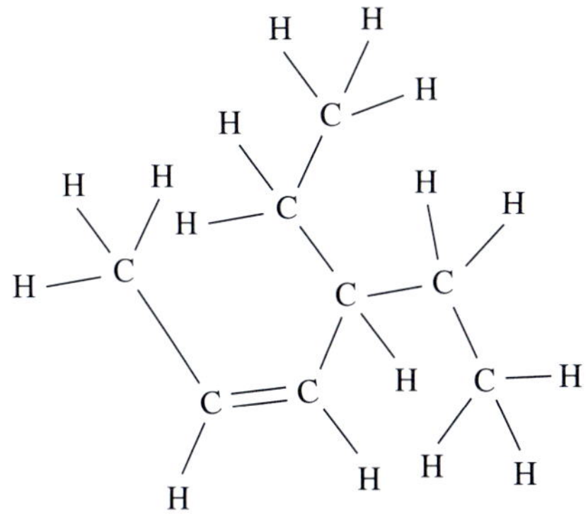

7 Draw the structure of:

a 2,3-dimethylpent-2-ene

b 4-ethylhex-2-ene.

8 a A hydrocarbon has the molecular formula $C_{10}H_{16}$. Determine the degree of unsaturation for this hydrocarbon.

b Draw two possible structures for this molecule.

10.3 Haloalkanes, alcohols and amines

Many organic compounds can be regarded as derivatives of hydrocarbons that have one or more hydrogen atoms replaced by other atoms or groups of atoms called a functional group.

The presence of a particular functional group in a molecule gives a substance certain physical and chemical properties. For example, vinegar and wine (Figure 10.3.1) both contain organic molecules based on ethane. The difference between their tastes and other properties is because of the different functional groups in each molecule.

In this section, you will explore three homologous series:

- haloalkanes
- alcohols
- primary amines.

You will learn about the functional groups in each of these families of molecules as well as how to name them.

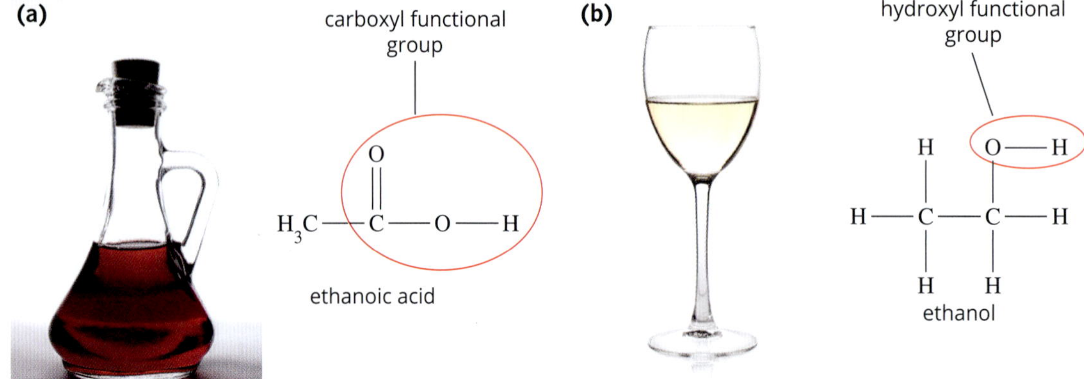

FIGURE 10.3.1 (a) Vinegar and (b) wine contain molecules with two carbon atoms. Vinegar contains ethanoic acid with a carboxyl functional group and wine contains ethanol with a hydroxyl group.

HALOALKANES

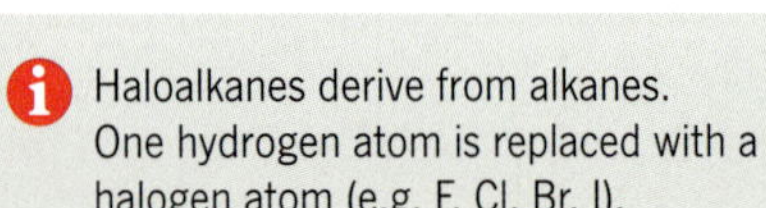

Haloalkanes derive from alkanes. One hydrogen atom is replaced with a halogen atom (e.g. F, Cl, Br, I).

Haloalkanes are a homologous series derived from alkanes, in which one hydrogen atom is replaced with a **halogen** atom.

Haloalkanes are used widely in industry as flame retardants, refrigerants, propellants, pesticides (Figure 10.3.2), solvents and pharmaceuticals. Some haloalkanes (in particular, chlorofluorocarbons or CFCs) are ozone-depleting chemicals and their use has been phased out in many applications.

FIGURE 10.3.2 Bromomethane (CH_3Br) is a colourless, odourless, non-flammable gas that was used as a pesticide in the past, but it has been phased out as it damages the ozone layer. This photo shows a field covered in plastic sheeting to minimise loss of the pesticide from the soil.

The halogen elements are in group 17 of the periodic table. All halogen atoms have seven valence electrons, which means they can form a single covalent bond with carbon atoms. The halogen elements that commonly form **halo functional groups** in organic compounds are fluorine, chlorine, bromine and iodine. Table 10.3.1 includes the names, structures and uses of some haloalkanes.

TABLE 10.3.1 Names, structures and uses of some haloalkanes

Name	Structural formula	Semi-structural formula	Use
chloromethane	H—C(Cl)(H)—H	CH_3Cl	refrigerant
bromochlorodifluoromethane	Br—C(Cl)(F)—F	$CBrClF_2$	fire extinguishers
iodoethane	H—C(I)(H)—C(H)(H)—H	CH_3CH_2I	production of chemicals
1-bromopropane	H—C(H)(H)—C(H)(H)—C(Br)(H)—H	$CH_3CH_2CH_2Br$	industrial solvent

Naming haloalkanes

The names of haloalkane functional groups are derived from the name of the halogen, as shown in Table 10.3.2.

The rules for naming haloalkanes follow the rules for naming alkanes (page 344). In addition, the following conventions are applied.

- Place the name of the specific halo functional group at the start of the parent alkane's name.
- If isomers are possible, use numbers to indicate the carbon to which the halo function group is attached.
- Number the carbons of the parent chain, beginning at the end closest to the first halo group or alkyl side chain.
- If there is more than one of the same type of halogen atom, use the prefix 'di-', 'tri-' or 'tetra-'.
- If more than one type of halo functional group or alkyl group is present, list them in alphabetical order.

TABLE 10.3.2 Names of the haloalkane functional groups

Halogen	Functional group name
fluorine	fluoro-
chlorine	chloro-
bromine	bromo-
iodine	iodo-

Table 10.3.3 gives the names and structures of some haloalkanes with the halogen atoms shown in red.

TABLE 10.3.3 Examples of haloalkanes and their names

Name	Semi-structural formula	Structural formula	Skeletal structure
bromomethane	CH_3Br	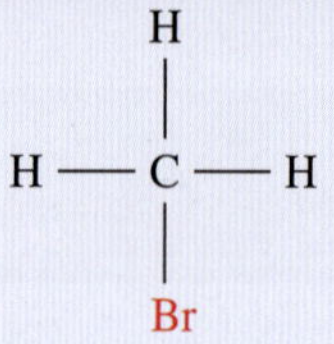	
1,1-dichloroethane	CH_3CHCl_2	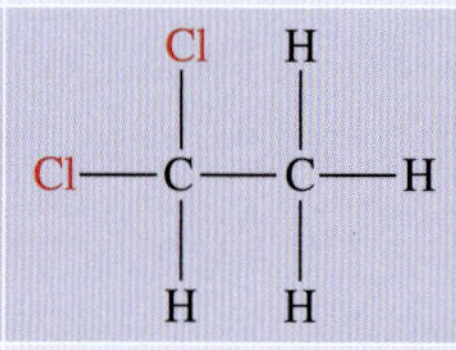	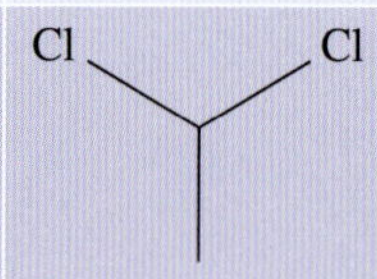
1-chlorobutane	$CH_3CH_2CH_2CH_2Cl$	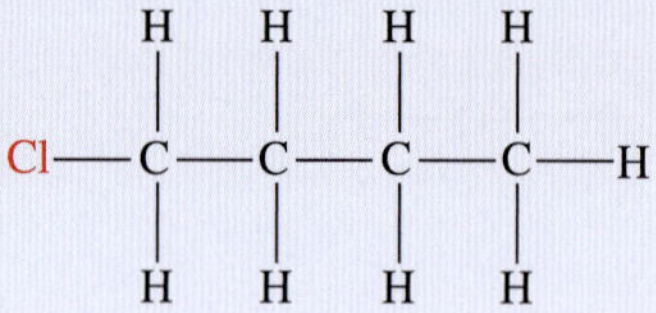	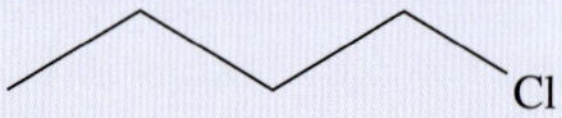
2-chloro-2-fluorobutane	$CH_3CClFCH_2CH_3$	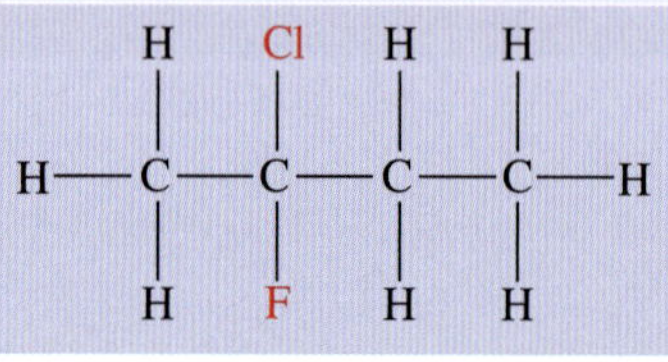	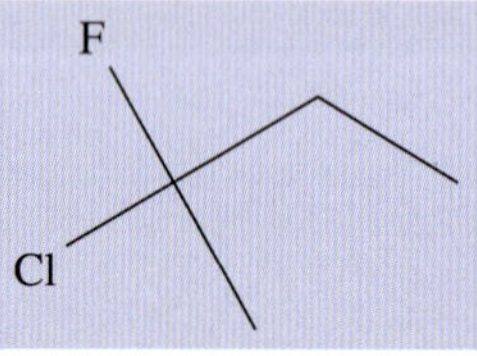

ALCOHOLS

Ethanol is a member of the homologous series of alcohols. Alcohols contain a hydroxyl (–OH) functional group in place of one of the hydrogens in a hydrocarbon chain. A representation of the hydroxyl group is shown in Figure 10.3.3.

Figure 10.3.4 shows the structural and semi-structural formulas of three alcohols.

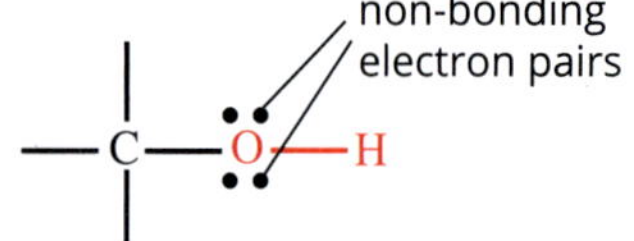

FIGURE 10.3.3 Alcohols contain the hydroxyl functional group, –OH (in red above). There are two pairs of non-bonding electrons on the oxygen atom.

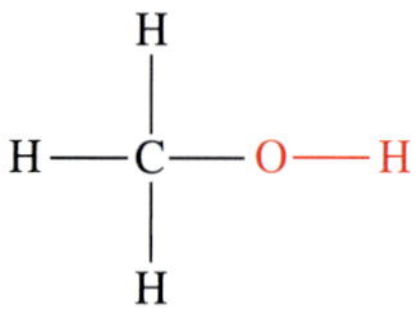

methanol

CH_3OH

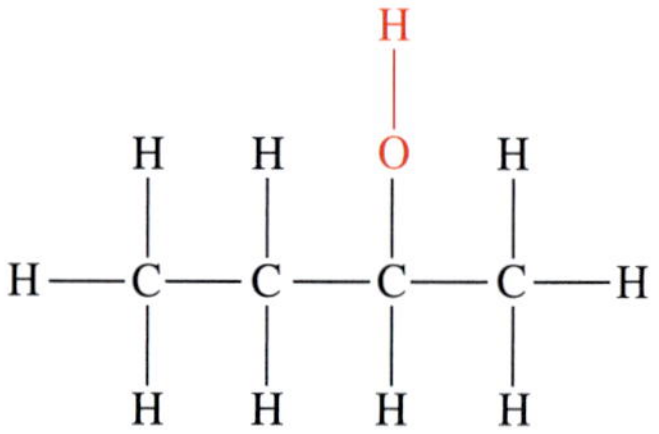

butan-2-ol

$CH_3CH_2CHOHCH_3$

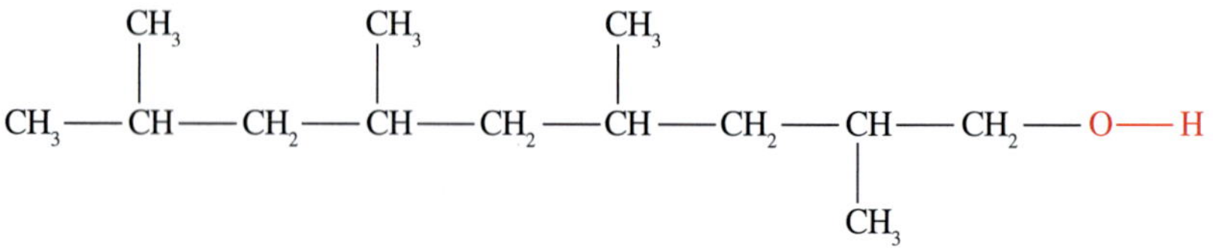

2,4,6,8-tetramethylnonan-1-ol

$(CH_3)_2CHCH_2CH(CH_3)CH_2CH(CH_3)CH_2CH(CH_3)CH_2OH$

FIGURE 10.3.4 Structural and semi-structural formulas of three alcohols. Each contains the hydroxyl functional group (shown in red). Note that the non-bonding electron pairs on the oxygen atom are usually omitted in structural formulas.

Types of alcohols

Alcohols are classified according to the number of alkyl groups attached to the carbon bonded to the hydroxyl group. The three different types of alcohols are:

- **primary alcohols**
- **secondary alcohols**
- **tertiary alcohols**.

The definition of each type and examples are shown in Table 10.3.4. Alkyl groups are represented by the general symbol R, and the hydroxyl groups are shown in red.

TABLE 10.3.4 The three different types of alcohols

Type of alcohol	Definition	General formula	Example
primary	The carbon bonded to the –OH group is only bonded to one alkyl group.	H R—C—OH H	H H H—C—C—O—H H H
secondary	The carbon bonded to the –OH group is also bonded to two alkyl groups.	R H—C—OH R	H H O H H—C—C—C—H H H H
tertiary	The carbon atom bonded to the –OH group is also bonded to three alkyl groups.	R R—C—OH R	H H—C—H H H H H—C—C—C—C—H H H O H H

Naming alcohols

Alcohol names follow the rules used for alkanes (page 344), except that the '-e' at the end of the parent alkane's name is replaced with the suffix '-ol'.

The following rules also apply.

- Identify the parent name from the longest carbon chain containing the hydroxyl group.
- If isomers are possible, insert a number before the '-ol' to indicate the carbon to which the hydroxyl functional group is attached.
- Number the carbon chain starting at the end closest to the hydroxyl group.

Table 10.3.5 on the following page shows some examples with the hydroxyl group shown in red. The three structures are all isomers of $C_4H_{10}O$.

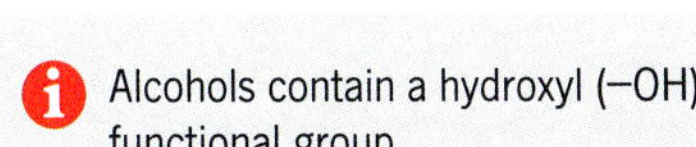

TABLE 10.3.5 Examples of naming alcohols

Name	Semi-structural formula	Structural formula	Skeletal structure
butan-1-ol	$CH_3CH_2CH_2CH_2OH$	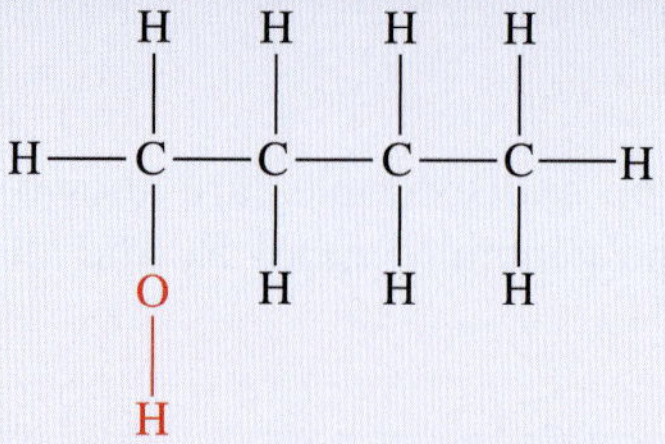 This is an example of a primary alcohol.	
butan-2-ol	$CH_3CHOHCH_2CH_3$	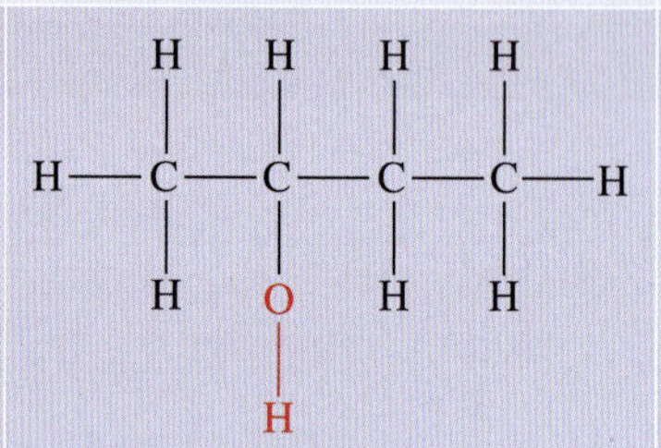 This is an example of a secondary alcohol.	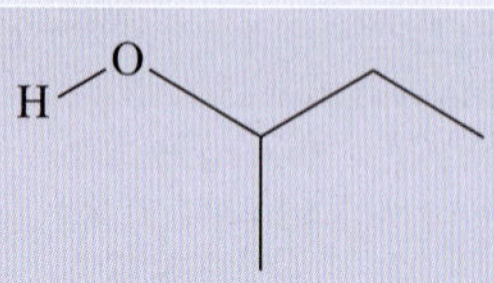
2-methylpropan-2-ol	$(CH_3)_3COH$	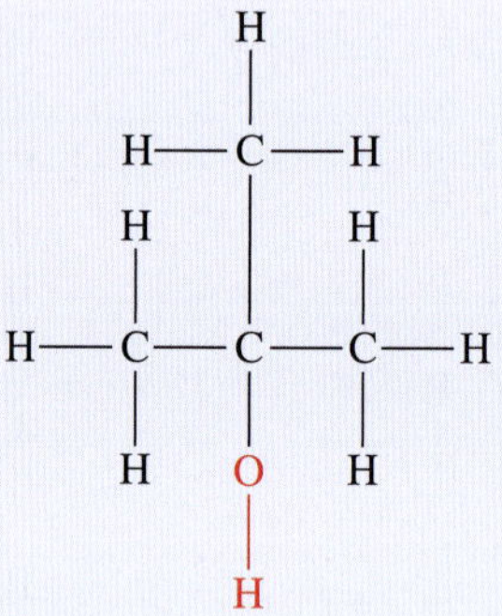 This is an example of a tertiary alcohol.	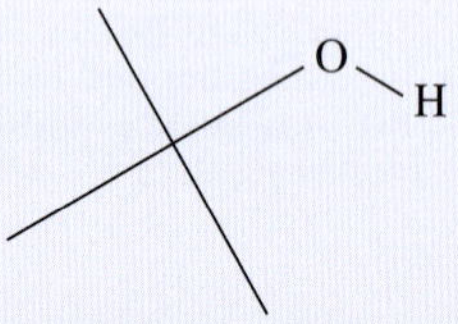

PRIMARY AMINES

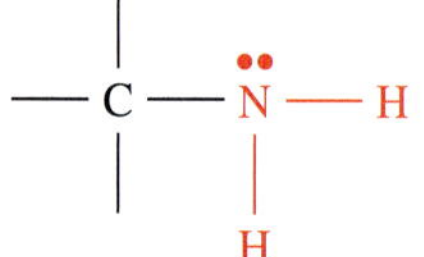

FIGURE 10.3.5 Primary amines contain the amino functional group, $-NH_2$ (in red above). There is one non-bonding electron pair on the nitrogen atom.

Primary amines are a homologous series of organic compounds that contain the **amino functional group**. An amino functional group consists of one nitrogen atom covalently bonded to two hydrogen atoms, as shown in Figure 10.3.5.

Amino functional groups with two hydrogens and one alkyl group are called primary amines. Unlike alcohols, amines are classified as 'primary', 'secondary' and 'tertiary' according to the number of alkyl groups attached to the nitrogen atom. Secondary and tertiary amines are not covered in this course. Some primary amines are shown in Figure 10.3.6.

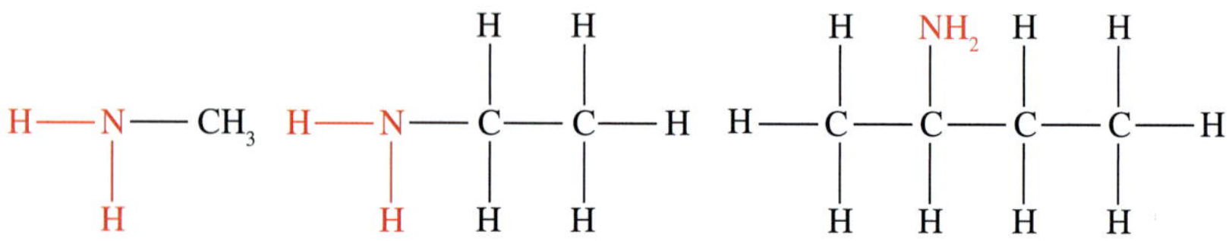

methanamine CH_3NH_2 | ethanamine $CH_3CH_2NH_2$ | butan-2-amine $CH_3CHNH_2CH_2CH_3$

FIGURE 10.3.6 Structural and semi-structural formulas of three primary amines. The amino functional group is shown in red. Primary amines have one carbon atom bonded to the nitrogen atom.

Naming amines

Amines are named in a similar way to alcohols. The '-e' at the end of the parent alkane's name is replaced with the suffix '-amine'. In the case of an isomer in which the functional group is attached to a different carbon, a number is inserted before '-amine' to indicate the carbon to which the amino functional group is attached. Table 10.3.6 shows two examples of amines and their structures.

Primary amines contain an amino ($-NH_2$) functional group.

TABLE 10.3.6 Examples of naming amines

Name	Condensed formula	Structural formula	Skeletal structure
ethanamine	$CH_3CH_2NH_2$	H H H—C—C—N—H H H H	H N H
butan-2-amine	$CH_3CHNH_2CH_2CH_3$	H N—H H H H H—C—C—C—C—H H H H H	H H N

CHEMFILE

Amines, chocolate and happiness

An active ingredient in chocolate (see the figure below) is the primary amine, 2-phenylethanamine.

2-Phenylethanamine is also produced in the pleasure centres of the brain and has the effect of generating a general sense of wellbeing or happiness, as well as temporarily raising blood pressure and blood glucose levels.

Eating chocolate provides you with a boost of 2-phenylethanamine, so the reason eating chocolate can make you feel happy is because of organic chemistry!

Chocolate contains an amine that makes humans feel happy.

10.3 Review

OA ✓✓

SUMMARY

- Organic molecules containing functional groups include:
  - haloalkanes, which contain the halo groups: fluoro (–F), chloro (–Cl), bromo (–Br) and iodo (–I)
  - alcohols, which contain hydroxyl (–OH) groups
  - primary amines, which contain amino ($-NH_2$) functional groups.
- Haloalkanes, alcohols and primary amines can be regarded as being derived from alkanes. The name of the parent alkane is used as the basis for their names.
  - Haloalkanes are named by adding the prefix for the halogen.
  - Alcohols are named by replacing the '-e' at the end of the parent alkane with the suffix '-ol'.
  - Primary amines are named by replacing the '-e' at the end of the parent alkane with the suffix '-amine'.
  - When isomers exist, a number is used to specify the position of the functional group.
- The steps for naming the parent alkane can be found on page 350.

KEY QUESTIONS

Knowledge and understanding

1 For haloalkanes, alcohols and amines:

a name the functional group they contain

b explain how the parent alkane name changes to reflect the functional group.

2 If a molecule contains multiple branches and one or more halo groups, how are these put in order before the parent molecule?

3 Draw the skeletal structure of 3-chloro-2-fluoro-2-iodohexane.

Analysis

4 Draw the structure and then write the name of a tertiary alcohol with the molecular formula $C_5H_{12}O$.

5 Analyse the following semi-structural formulas to identify their functional groups and branches, then determine their systematic names:

a $CH_3CH_2CH_2Br$

b $(CH_3)_2CHCH_2CHClCH_3$

c $CH_3(CH_2)_3CH_2OH$

d $CH_3(CH_2)_3CHNH_2(CH_2)_2CH_3$

6 Explain why the names 1-chloroethane and propan-3-amine are not used.

7 The skeletal structure of 2-phenylethanamine, one of the ingredients in chocolate, is shown below.

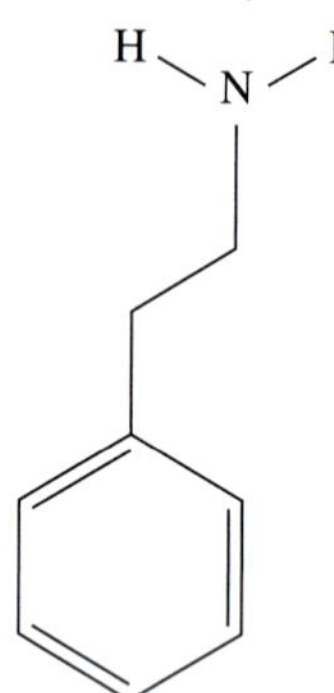

a Using this skeletal structure for 2-phenylethanamine, determine its molecular formula.

b Draw the structural formula of 2-phenylethanamine, then circle and name one of the functional groups present.

c Using the molecular formula you determined in part **a**, calculate the molar mass of 2-phenylethanamine.

10.4 Molecules containing a carbonyl group

In this section, you will learn about five more homologous series, each containing a carbonyl functional group:

- aldehydes
- ketones
- carboxylic acids
- primary amides
- esters.

You will learn about the functional groups and structures of the molecules of each homologous series, as well as the rules for naming aldehydes, ketones, carboxylic acids and esters. (Naming of amides is not required for this course.)

Molecules of aldehydes, ketones, carboxylic acids, amides and esters all contain a **carbonyl functional group**, as shown in Figure 10.4.1. A carbonyl functional group consists of a carbon atom connected to an oxygen atom by a double bond. All atoms bonded to the carbon atom are in a plane and the angles between bonds are 120°.

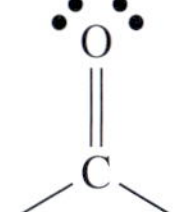

FIGURE 10.4.1 The carbonyl functional group. The non-bonding pairs of electrons shown on the oxygen atom are usually not included in structural formulas.

ALDEHYDES

In **aldehydes**, the carbonyl group is always at the end of the hydrocarbon chain. As shown in Figure 10.4.2, the carbon atom of the carbonyl group is bonded to a hydrogen atom. The carbonyl functional group in an aldehyde is always written as –CHO at the end of the semi-structural formula of aldehydes.

Aldehydes contain a carbonyl group (C=O) at the end of the molecule. This is written in semi-structural form as –CHO.

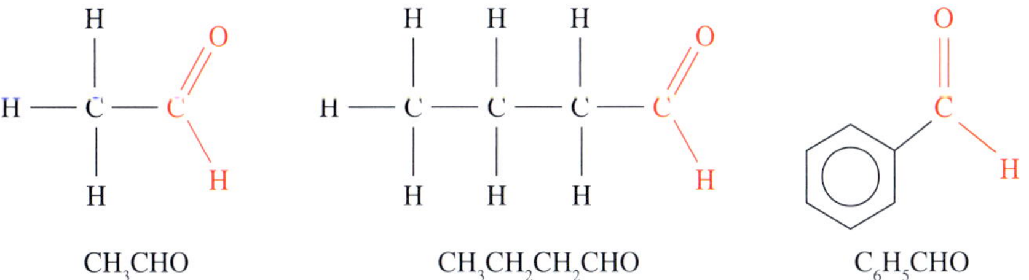

CH_3CHO $CH_3CH_2CH_2CHO$ C_6H_5CHO

FIGURE 10.4.2 Structural and semi-structural formulas of three aldehydes. Each contains a carbonyl functional group with a hydrogen atom attached, shown in red.

The simplest aldehyde, methanal HCHO, is commonly known as 'formaldehyde'. It can be used to preserve biological specimens, including embalming human remains (Figure 10.4.3).

FIGURE 10.4.3 Methanal (HCHO) can be used to preserve biological specimens. It is the first member of the aldehyde homologous series.

Naming aldehydes

> Aldehydes are named using an -al suffix, e.g. methanal.

Aldehyde names follow the rules used for alkanes, except that the '-e' at the end of the parent alkane's name is replaced with the suffix '-al'. As the aldehyde group is always on the end of a chain, the carbon atom in the aldehyde group is always carbon number 1, so no numbers are needed to indicate the location of the functional group. Table 10.4.1 shows three examples, with the aldehyde group shown in red.

TABLE 10.4.1 Examples of naming aldehydes

Name	Semi-structural formula	Structural formula	Skeletal structure
propanal	CH_3CH_2CHO	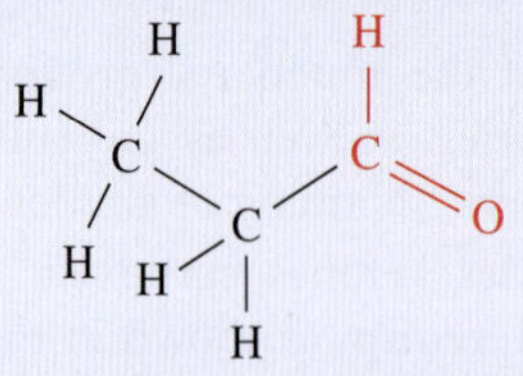	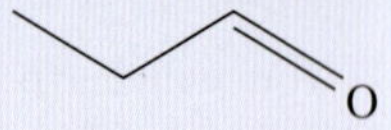
methylpropanal	$CH_3CH(CH_3)CHO$	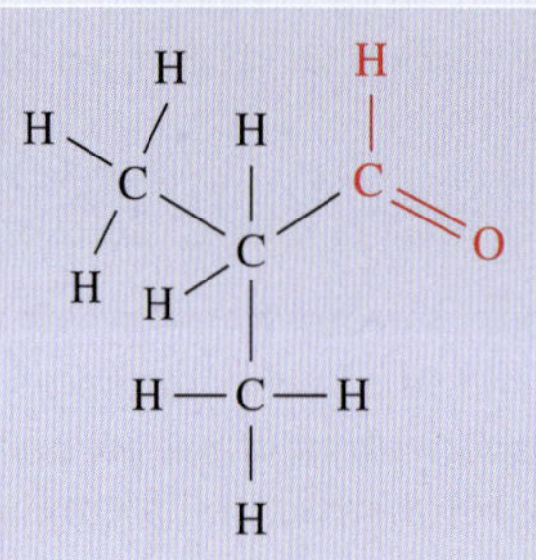	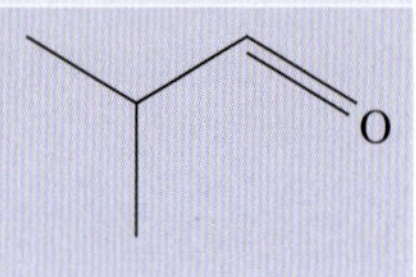

KETONES

> Ketones contain a carbonyl group (C=O) that is not at the end of the molecule. They are named using an -one suffix, e.g. propan-2-one.

In **ketones**, the carbonyl carbon is attached to two other carbon atoms. This means that the carbonyl group is never at the end of the molecule, as you can see in the examples in Figure 10.4.4. In semi-structural formulas, the carbonyl functional group in a ketone is simply written as –CO–.

The simplest ketone, propan-2-one CH_3COCH_3, is commonly called acetone and is a useful polar organic solvent, frequently found in nail polish remover and paint thinner.

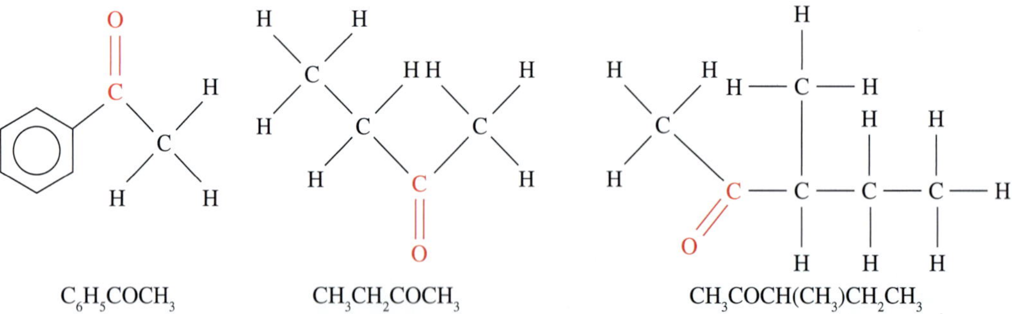

FIGURE 10.4.4 Structural and semi-structural formulas of three ketones. Each contains a carbonyl functional group (shown in red) attached to two hydrocarbon groups.

Naming ketones

Ketone names follow the rules used for alkanes, except that the '-e' at the end of the parent alkane's name is replaced with the suffix '-one'. A number is inserted before '-one' to indicate the carbon to which the carbonyl group is attached. Table 10.4.2 shows how to name two of the examples from Figure 10.4.4.

TABLE 10.4.2 Examples of naming ketones

Name	Condensed formula	Structural formula	Skeletal structure
butan-2-one*	$CH_3CH_2COCH_3$	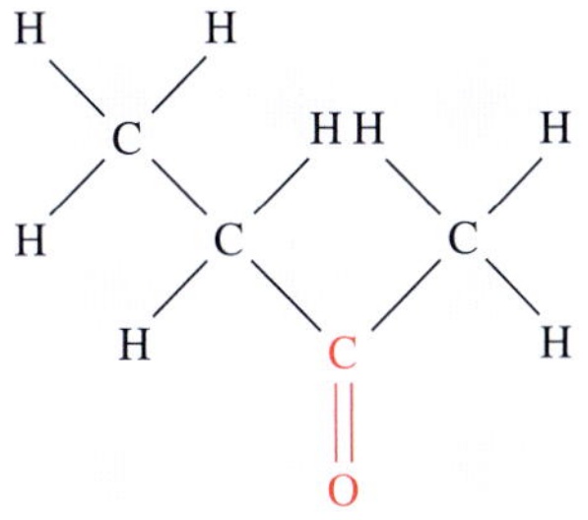 $CH_3CH_2COCH_3$	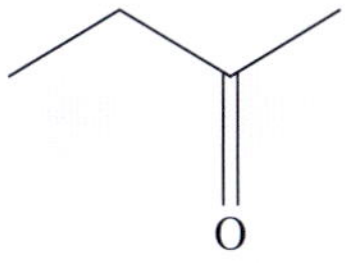
3-methylpentan-2-one	$CH_3COCH(CH_3)CH_2CH_3$	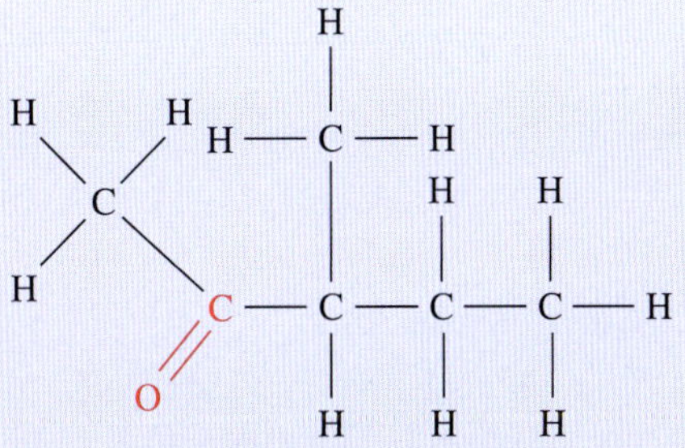 $CH_3COCH(CH_3)CH_2CH_3$	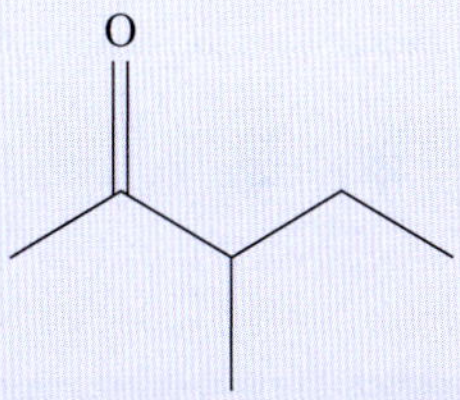

*Note that a position number is always given for ketones, even if the name would be unambiguous without one, as in propan-2-one and butan-2-one.

CARBOXYLIC ACIDS

Carboxylic acids are a homologous series that contain the carboxyl functional group. The **carboxyl functional group** (Figure 10.4.5) is represented in a semi-structural formula as –COOH and is always located at the end of a hydrocarbon chain.

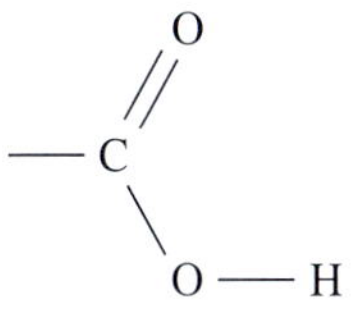

FIGURE 10.4.5 Structure of the carboxyl functional group

Naming carboxylic acids

Carboxylic acid names also follow the rules used for alkanes, except that the '-e' at the end of the parent alkane's name is replaced with the suffix '-oic acid'. As the carboxyl group is always on the end of a chain, the carbon atom in the carboxyl group is always carbon number 1, so no numbers are needed to indicate the location of the functional group. Table 10.4.3 on the following page shows three examples, with the carboxyl group shown in red.

TABLE 10.4.3 Examples of naming carboxylic acids

Name	Semi-structural formula	Structural formula	Skeletal structure
ethanoic acid	CH_3COOH	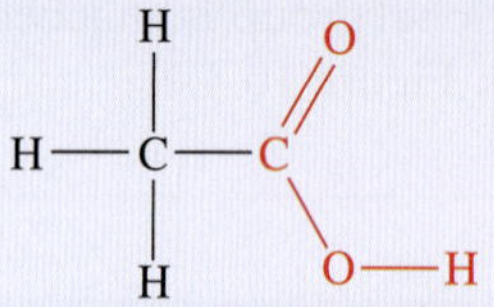	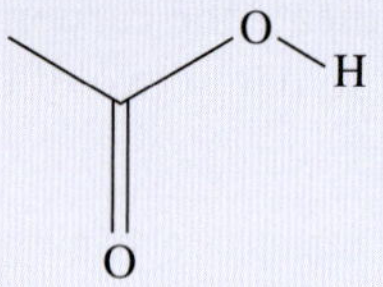
butanoic acid	$CH_3CH_2CH_2COOH$	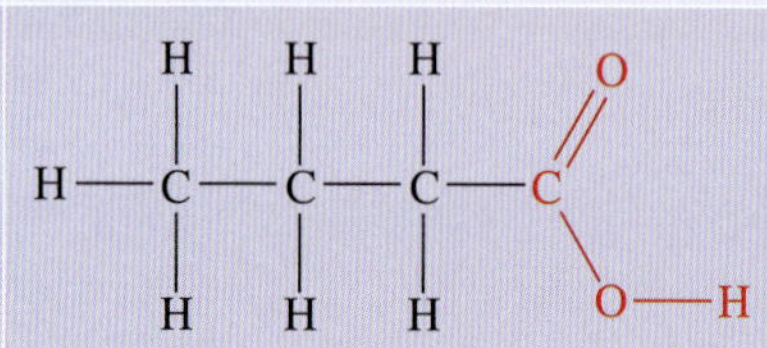	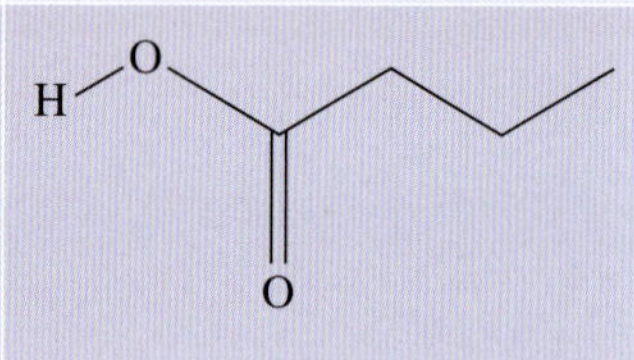
2-methylbutanoic acid	$CH_3CH_2CH(CH_3)COOH$	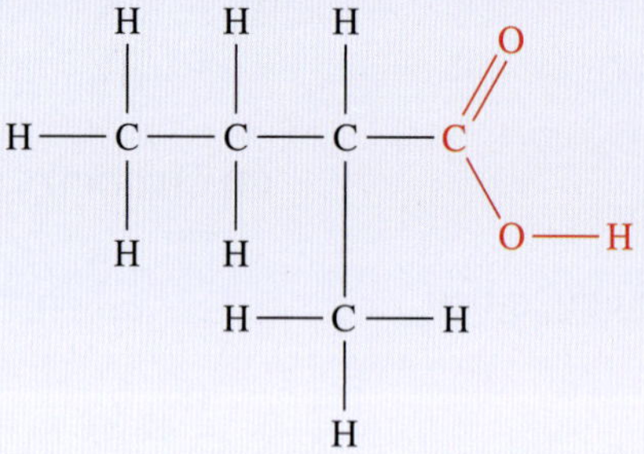	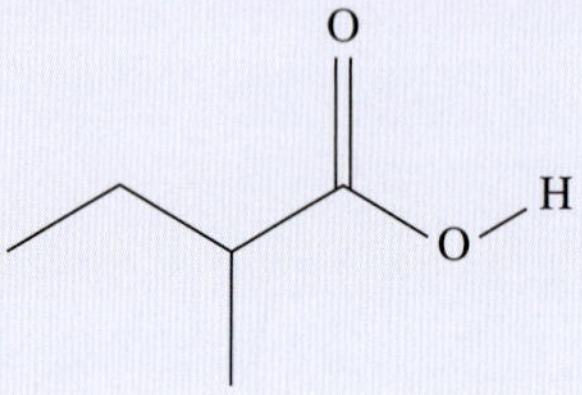

FIGURE 10.4.6 Solubility is an important feature of drug design. Turning an organic compound into an organic salt greatly increases its solubility.

Carboxylate ions

Like all acids, when a carboxylic acid reacts with a strong base, a salt and water are produced. The anion in the salt, formed from the organic acid, is named by replacing the -oic acid part of the name with -oate. Thus, the anion formed from ethanoic acid (CH_3COOH) in the following reaction is called an ethanoate ion (CH_3COO^-).

$$CH_3COOH(l) + OH^-(aq) \rightleftharpoons CH_3COO^-(aq) + H_2O(l)$$

In general, the conjugate base to a carboxylic acid is called a carboxylate ion.

This kind of reaction is important to the pharmaceutical industry, as medications in humans need to be able to dissolve in aqueous solutions so they can enter our cells (Figure 10.4.6). Turning an organic compound into an organic salt greatly increases its solubility.

PRIMARY AMIDES

The **amide functional group** is similar to a carboxyl group, except that the –OH is replaced with $-NH_2$ (Figure 10.4.7).

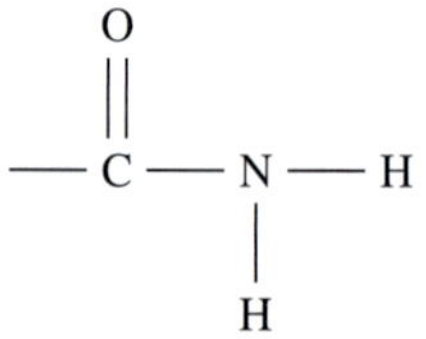

FIGURE 10.4.7 Structure of the amide functional group in a primary amide

The nitrogen atom in primary amides is bonded to two hydrogen atoms. Secondary and tertiary amides exist, but these compounds are not covered in this course. In a **primary amide**, the amide functional group is represented in a semi-structural formula as $-CONH_2$ and is always located at one end of the hydrocarbon chain. The structural and semi-structural formulas of three primary amides are shown in Figure 10.4.8.

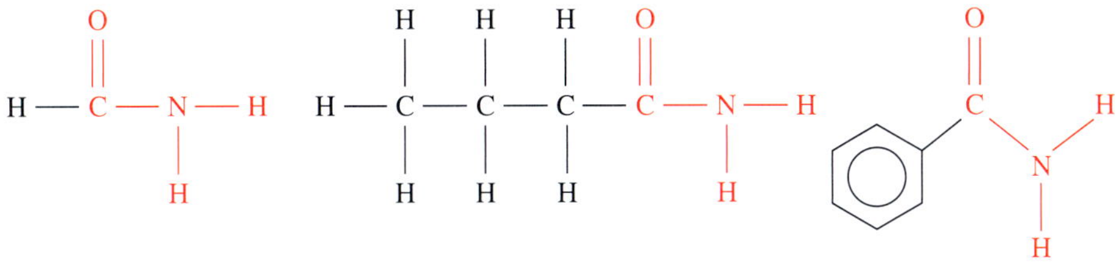

$HCONH_2$ $CH_3CH_2CH_2CONH_2$ $C_6H_5CONH_2$

FIGURE 10.4.8 Structural and semi-structural formulas of three primary amides, with the amide functional group shown in red

ESTERS

You will learn in Section 11.3 that **esters** are produced by the reaction of a carboxylic acid with an alcohol. The **ester functional group** (Figure 10.4.9) is similar to a carboxylic acid group, but the hydrogen of the −OH is replaced by an alkyl group.

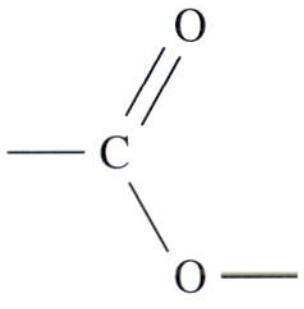

FIGURE 10.4.9 Structure of the ester functional group

Figure 10.4.10 shows the structural and semi-structural formulas of three esters. In semi-structural formulas, the ester functional group is written as −COO− or −OCO−.

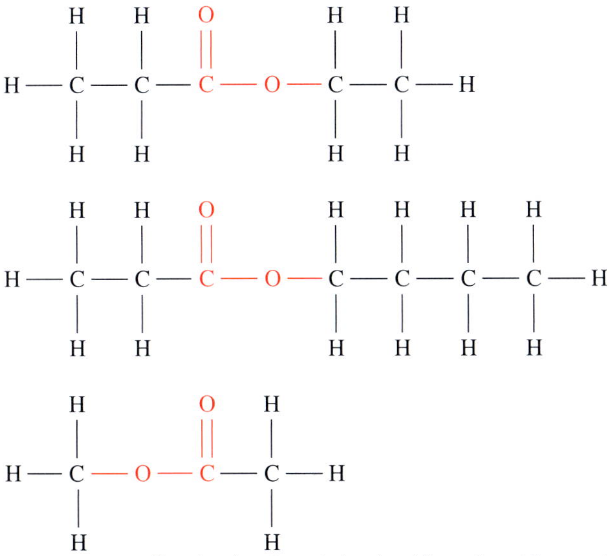

$CH_3CH_2COOCH_2CH_3$

$CH_3CH_2COOCH_2CH_2CH_2CH_3$

CH_3OCOCH_3

FIGURE 10.4.10 Structural and semi-structural formulas of three esters. Each contains the ester functional group, shown in red.

CHEMFILE

Urea and organic chemistry

Urea ($(NH_2)_2CO$) is a small di-amide, which is the end product of the breakdown of proteins by mammals and is excreted in urine. This naturally produced urea can be used as a source of nitrogen to promote plant growth.

Friedrich Wöhler (1800–1882), a German chemist, synthesised urea in his laboratory in 1828 by treating silver cyanate with ammonium chloride. In a letter to a colleague, Wöhler excitedly wrote: 'I must tell you that I can make urea without the use of kidneys, either man or dog'.

Wöhler's synthesis of urea was remarkable because it was the first time someone had made an organic molecule. This discovery established the field of synthetic organic chemistry, which is responsible for millions of products in our society. In 1982, Wöhler was honoured in a German postage stamp featuring the structure of urea.

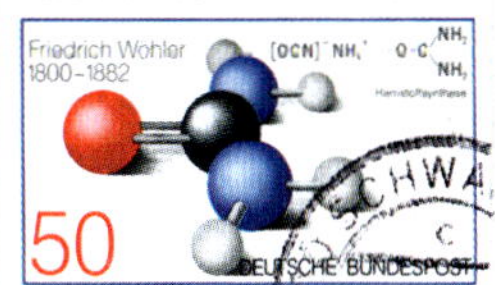

(a) Friedrich Wöhler, the German chemist who first synthesised urea. (b) 1982 German postage stamp honouring Wöhler.

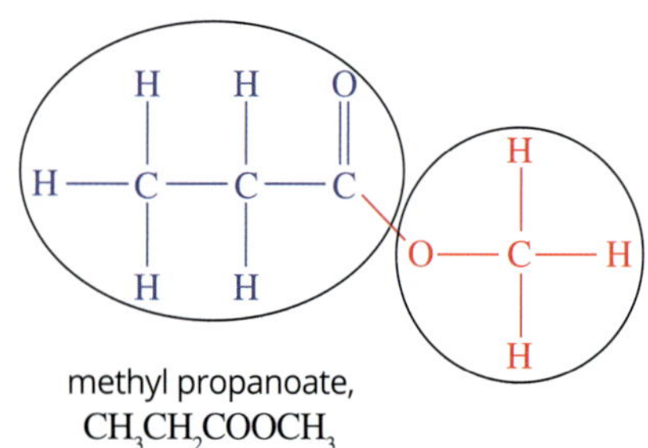

methyl propanoate,
$CH_3CH_2COOCH_3$

FIGURE 10.4.11 Esters are named using two words based on the two reactants that produced the molecule.

Naming esters

The names of esters consist of two words. The name is built up from the names of the alcohol and carboxylic acid that reacted to form it. The first part is based on the number of carbon atoms in the chain attached to the singly bonded oxygen (–O–), which has come from the alcohol. The second part of the name is based on the number of carbon atoms in the chain containing the carbonyl group, which has come from the carboxylic acid.

In the example in Figure 10.4.11, the chain attached to the single –O– contains one carbon atom, whereas the chain containing the carbonyl group is three carbon atoms long.

The –O– section of an ester determines the first word of the name: it is the same as an alkyl chain of the same length. For example, if this part contains one carbon atom, the first word of the ester name will be 'methyl'. This part of the ester is derived from an alcohol, and the suffix on the name is changed from '-ol' to '-yl'.

The section containing the carbonyl group is derived from a carboxylic acid and contributes the second word of the name. The name of the carboxylic acid is adapted by changing '-oic acid' to '-oate'.

Using these rules, you can work out that the name of the ester shown in Figure 10.4.11 is methyl propanoate.

The names, semi-structural and structural formulas and skeletal structures of some other esters are shown in Table 10.4.4, with the part of each molecule derived from an alcohol in red and the part of each molecule derived from the carboxylic acid in blue.

TABLE 10.4.4 The names, semi-structural and structural formulas and skeletal structures of three esters

Name	Semi-structural formula	Structural formula	Skeletal structure
ethyl ethanoate	$CH_3COOCH_2CH_3$		
propyl propanoate	$CH_3CH_2COOCH_2CH_2CH_3$		
propyl ethanoate	$CH_3COOCH_2CH_2CH_3$		

CHEMFILE

Esters, sweet flavours and strong aromas

Esters are responsible for some of the natural and synthetic flavours and smells found in ice creams, lollies, flowers and fruit (see figure). Table 10.4.5 lists the names of some esters with distinctive smells or flavours.

Your favourite perfume or cologne is likely to contain esters that are responsible for its distinctive and appealing aroma, but some esters may have less than pleasant smells, such as a glue smell.

Esters are responsible for many of the flavours and odours of fruit.

TABLE 10.4.5 Some esters with distinctive aromas

Ester	Smell or flavour
pentyl propanoate	apricot
ethyl butanoate	pineapple
octyl ethanoate	orange
2-methylpropyl methanoate	raspberry
ethyl methanoate	rum
pentyl ethanoate	banana

10.4 Review

OA ✓✓

SUMMARY

- The carbonyl group (–CO–) is a component of a number of functional groups.
- Aldehydes are a homologous series of molecules that contain a carbonyl group at the end of the molecule. The carbonyl functional group in an aldehyde is written as –CHO at the end of the semi-structural formula of aldehydes.
- Aldehydes are named using the suffix '-al'.
- Ketones are a homologous series of molecules that contain a carbonyl group that is not located at the end of the molecule. In semi-structural formulas, the carbonyl functional group in a ketone is simply written as –CO–.
- Ketones are named using the suffix '-one', with a number before the suffix to indicate the location of the carbonyl group.
- Carboxylic acids are a homologous series of molecules that contain the carboxyl group (–COOH) at the end of a hydrocarbon chain.
- Carboxylic acids are named using the suffix '-oic acid'.
- Primary amides are a homologous series of molecules that contain the amide group (–$CONH_2$) at the end of the hydrocarbon chain.
- Esters are a homologous series of molecules that contain an ester group (–COO–) within the hydrocarbon chain.
- Esters have a two-word name.
  - The first part of the name comes from the section derived from an alcohol and the suffix is changed from '-ol' to '-yl'.
  - The second part of the name is derived from the section that comes from a carboxylic acid and the suffix '-oic acid' is changed to '-oate'

KEY QUESTIONS

Knowledge and understanding

1 Copy and complete the following table, summarising some of the functional groups that contain a carbonyl group.

Homologous series	Functional group name	Structural formula	Semi-structural formula	Naming convention
	carbonyl—end of molecule		–CHO	
	carbonyl—not on end		–CO–	
			–COOH	
primary amides				not required for this course
esters	ester		–COO–	

continued over page

10.4 Review *continued*

2 To which homologous series do each of the following molecules belong?

a
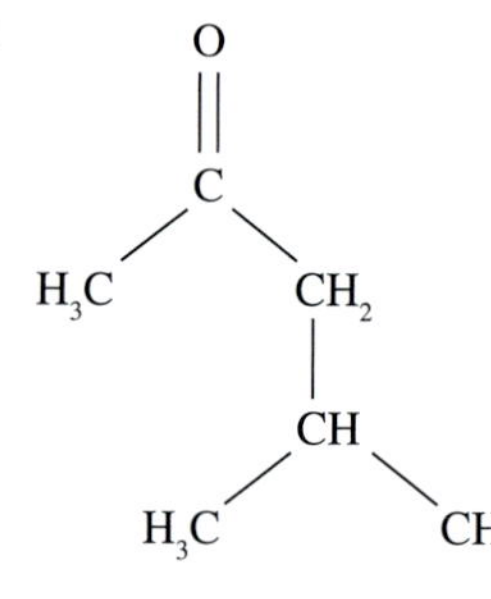

b

c

d
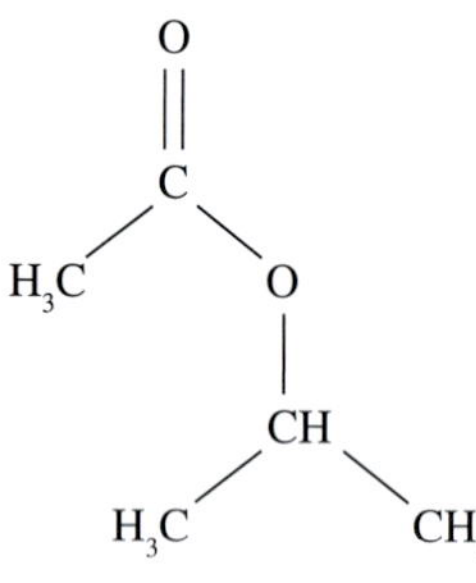

e

Analysis

3 Write the systematic name for the following molecules.

a $HCOOCH_3$
b $HCOOH$
c $CH_3CH_2CH_2COOCH_2CH_2CH_3$
d CH_3COOCH_3
e $CH_3(CH_2)_4COOCH_2CH_3$
f $CH_3CH(CH_3)CH(CH_3)CHO$
g $CH_3COCH(CH_3)CH_2CH_3$

4 Draw and name an isomer of $HCOOCH_3$.

5 Using esters named in Table 10.4.5, complete the table below.

Name	Structural formula	Semi-structural formula	Skeletal structure
		$CH_3(CH_2)_2COOCH_2CH_3$	
pentyl ethanoate		$CH_3COO(CH_2)_4CH_3$	

10.5 An overview of IUPAC nomenclature

In the early 1960s, the International Union of Pure and Applied Chemistry (IUPAC) endorsed a common naming system for carbon compounds. These rules are regularly updated and are used worldwide to enable scientists to communicate with each other. The rules ensure that a carbon compound is given a unique name that provides useful information about its structure and distinguishes it from any isomers.

The rules specify the names for different parts of an organic molecule and may include both words and numbers to indicate locations of functional groups and alkyl branches (for example, the antiseptic shown in Figure 10.5.1).

This section will review and extend your knowledge of the rules that you learnt in the previous sections for naming organic compounds. You will also learn how to name organic molecules that contain two different functional groups.

FIGURE 10.5.1 According to the IUPAC rules, the active ingredient in the antiseptic Dettol has the name 4-chloro-3,5-dimethylphenol.

IUPAC NOMENCLATURE

IUPAC nomenclature is the term used to describe the set of rules by which chemists can name a given compound. The rules can be used in reverse to derive a structure from the IUPAC name. IUPAC names of organic molecules can be short or very long, however, whether simple or complex, most IUPAC names follow the same basic pattern.

All organic molecules can be thought of as being derived from a hydrocarbon parent molecule, which provides the basis for the name of the molecule. Part of the IUPAC name reflects which alkane is the parent molecule.

The IUPAC name also indicates which functional groups are present in the molecule by adding a suffix to the end of the name or a prefix to the beginning of the name. The positions of functional groups are indicated by numbers. As an example, the meaning of each part of the name butan-2-ol, an alcohol, is shown in Figure 10.5.2, along with its structural formula.

IUPAC nomenclature enables every organic compound to have a unique name which describes its structure.

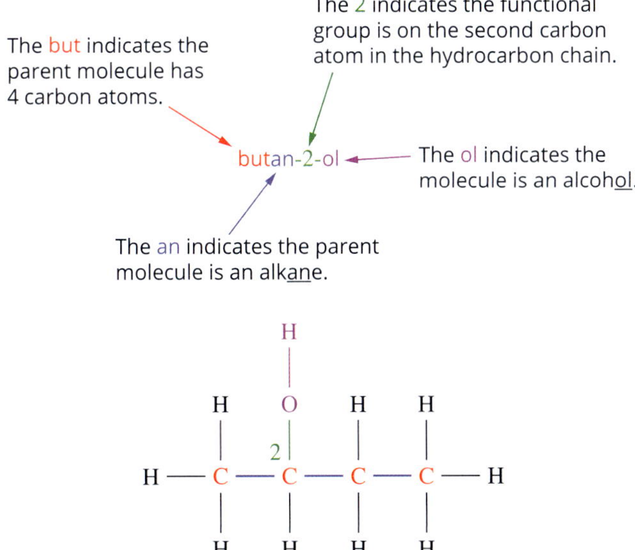

FIGURE 10.5.2 The name and structural formula of butan-2-ol. The colours of each part of the name correspond with the section of the structure that it represents.

Summary of IUPAC rules for nomenclature

The following conventions are used for naming organic molecules.

- There are no spaces in a name, apart from the two-word names of esters and carboxylic acids.
- The longest carbon chain is used to derive the parent name. For alkenes, alcohols, amines, carboxylic acids, aldehydes, ketones and esters, the longest chain must include the functional group.
- The names and locations of branches and additional functional groups are added to this parent name as a prefix.
- Numbers are used to identify the carbon atom that groups are attached to.
- Numbers and letters are separated by dashes.
- Numbers are separated from other numbers by commas.
- If there is more than one type of functional group or branch to be listed at the beginning of a name, they are listed in alphabetical order.
- If there is more than one of the same type of functional group, the prefixes 'di-', 'tri-' or 'tetra-' are used. Each group is still given a number to indicate its position on the carbon chain.

CHEMFILE

Omitting numbers when a structure is unambiguous

Although it is common practice to omit numbers whenever a structure is unambiguous (e.g. propene, not prop-1-ene), the IUPAC rules are not as general as this. A locant (number indicating the location of a functional group) may only be omitted in specified situations. For the molecules studied in this course:

- a locant can only be omitted if it is a '1' and the location in the molecule is unambiguous
- if one locant is needed, then all locants should be specified for that molecule.

For example, the locant is omitted in the following molecules: butanoic acid, chloromethane, ethanol, pentanal and propene. Note, it is never necessary to provide a locant for carboxyl and aldehyde functional groups as these are always on the end of the molecule and therefore must be on carbon number 1.

In contrast, locants are required in 2-methylpropane and 2-methylprop-1-ene even though these structures are unambiguous without locants.

You may find these molecules commonly referred to as methylpropane and methylpropene, but these are not the correct systematic IUPAC names.

HOMOLOGOUS SERIES NAMING CONVENTIONS

The presence of a particular functional group identifies the homologous series a molecule belongs to and changes the molecule's name, as shown in Table 10.5.1.

TABLE 10.5.1 The identity, functional groups and naming conventions of the homologous series

Homologous series	Functional group name	Semi-structural formula	Naming convention
alkane	not applicable	not applicable	suffix -ane
alkene	carbon–carbon double bond	–C=C–	suffix -ene
haloalkane	halo	–F, –Cl, –Br, –I	prefix fluoro-, chloro-, bromo- or iodo-
alcohol	hydroxyl	–OH	suffix -ol occasionally prefix hydroxy-
amine	amino	$-NH_2$	suffix -amine occasionally prefix amino-
aldehyde	carbonyl (aldehyde)	–CHO	suffix -al occasionally prefix oxo-
ketone	carbonyl (ketone)	–CO–	suffix -one occasionally prefix oxo-
carboxylic acid	carboxyl	–COOH	suffix -oic acid
ester	ester	–COO–	two-word name with suffixes -yl and -oate
primary amide	amide	$-CONH_2$	not required for this course

CASE STUDY ANALYSIS

Trivial versus IUPAC systematic names

The 19th century was a golden age for chemistry. New and interesting organic compounds were being either isolated or synthesised at an amazing rate.

In 1818, French chemist Michel Eugène Chevreul (Figure 10.5.3) isolated an acid from rancid butter.

It had the distinctive smell of human vomit and is also found in Parmesan cheese. He called it butyric acid, from the Latin word for butter, *butyrum* (Figure 10.5.4).

In 1863, German chemist Adolf von Baeyer synthesised a new compound from malonic acid and urea (Figure 10.5.5). As it was St Barbara's Day (4 December) he named this new substance barbituric acid (2,4,6-trioxypyrimidine).

In 1919, the International Union of Pure and Applied Chemistry (IUPAC) was formed. One of its aims was to standardise the plethora of non-systematic names that existed in organic chemistry. To this end, it publishes the *Nomenclature of Organic Chemistry* (informally called the Blue Book) as the authority on naming organic compounds. There is also IUPAC *Nomenclature of Inorganic Chemistry.*

Ideally, every possible organic compound should have a systematic name from which an unambiguous structural formula can be created. Because systematic names can be tediously long, common names, also known as trivial names, are still in common use. One group of trivial names has been based on the shapes of the molecules. Table 10.5.2 gives some examples.

Analysis

1 Sulflower, quadratic acid and draculin are all trivial names for chemical compounds. Research these substances and explain how or why they have these names.

2 In this course, IUPAC systematic names are used. However, there are a number of commonly used substances that have a preferred IUPAC name that is different to the systematic name. Research the following substances and write the systematic name for each: acetic acid, formic acid, acetone.

FIGURE 10.5.3 Michel Eugène Chevreul (1786–1889), who named the foul-smelling 'butyric acid'

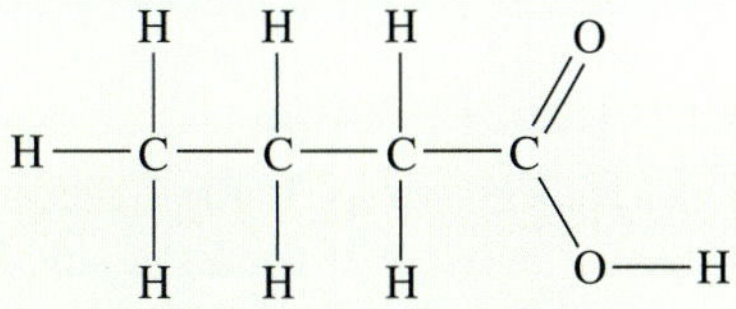

FIGURE 10.5.4 Butyric acid has the systematic name butanoic acid.

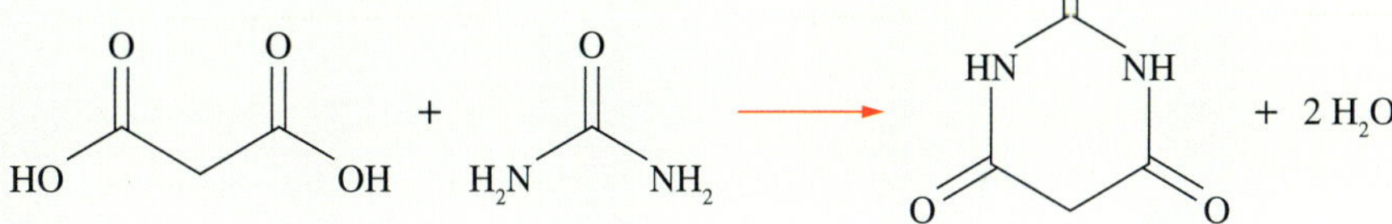

FIGURE 10.5.5 Barbituric acid was first synthesised on St Barbara's Day.

TABLE 10.5.2 Trivial (common) names for some organic compounds

Trivial names based on shape		Molecular formula / IUPAC systematic name	Skeletal structure
cubane	a hydrocarbon in the shape of a cube	C_8H_8 / pentacyclo [4.2.0.0^{2,5}.0^{3,8}.0^{4,7}]octane	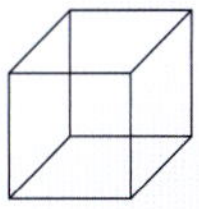
housane	cyclic alkane that looks like a house	C_5H_8 / bicyclo[2.1.0]pentane	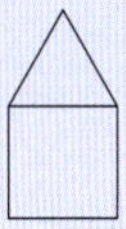
olympicene	five fused rings that resemble the Olympic flag	$C_{19}H_{12}$ / 6H-benzo[cd]pyrene	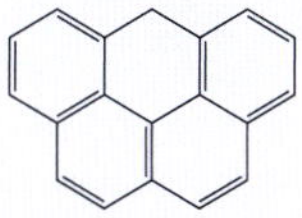
penguinone	resembles a penguin	$C_{10}H_{14}O$/ 3,4,4,5-tetramethyl-2,5-cyclohexadien-1-one	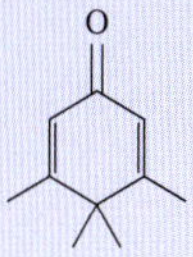

Naming organic molecules with a functional group and alkyl side chain

Organic molecules with functional groups can also have alkyl side chains. The names of alkyl groups are placed in alphabetical order in front of the parent name. The carbon chain is numbered from the end closest to the functional group to give the lowest possible number to the functional group.

This procedure is illustrated in Table 10.5.3 with the naming of the structural isomers of an alcohol with the molecular formula $C_4H_{10}O$.

TABLE 10.5.3 Names of structural isomers of an alcohol with the molecular formula $C_4H_{10}O$

Structure	Name
H H H H H—C—C—C—C—O—H H H H H	butan-1-ol
H H H O H H—C—C—C—C—H H H H H	butan-2-ol
H H—C—H H H H—C—C—C—O—H H H H	2-methylpropan-1-ol
H H—C—H H H H—C—C—C—H H O H H	2-methylpropan-2-ol

Naming organic molecules with two functional groups

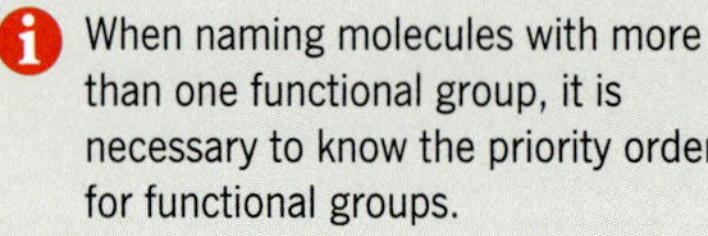

When naming molecules with more than one functional group, it is necessary to know the priority order for functional groups.

Many organic molecules have more than one functional group. If the functional groups are the same, a multiplier (e.g. di-, tri-, tetra-) is used. If the molecule has different functional groups, you will need to know which one has the highest priority to work out what numbers and names to use.

IUPAC has designated a priority system for functional groups (Table 10.5.4 on the following page). In an organic molecule with two functional groups, the following naming conventions are used.

- The functional group with the higher priority is assigned the lowest possible number and the suffix for this functional group is used in the name.
- The lower priority functional group is indicated by a prefix or alternative name.

TABLE 10.5.4 IUPAC functional group priorities

	Functional group	Suffix	Alternative name (when needed)
Higher priority	carboxyl	-oic acid	–
↓	carbonyl on end carbon (aldehyde)	-al	oxo-
↓	carbonyl not on end carbon (ketone)	-one	oxo-
↓	hydroxyl	-ol	hydroxy-
↓	amino	-amine	amino-
↓	alkene	-ene	-an- becomes -en-
Lower priority	halo	–	halo-

This priority system is illustrated in Table 10.5.5 with the naming of molecules with two functional groups.

TABLE 10.5.5 Naming molecules with two functional groups

<table>
<tr><th>Structure</th><th>Name</th></tr>
<tr><td>H H H O
4 3 2 1
H — C — C — C — C — O — H
H O H
H</td><td>3-hydroxybutanoic acid</td></tr>
<tr><td>H H
1 2
H — O — C — C — N — H
H H H</td><td>2-aminoethan-1-ol</td></tr>
<tr><td>H H H H Cl
1 2 3 4 5
H — C — C — C — C — C — H
H O H H H
H</td><td>5-chloropentan-2-ol</td></tr>
<tr><td>H H H H O
H — C — C = C — C — C — O — H
5 4 3 2 1
H H</td><td>pent-3-enoic acid</td></tr>
<tr><td>H O H H O
H — C — C — C — C — C — O — H
5 4 3 2 1
H H H</td><td>4-oxopentanoic acid</td></tr>
</table>

Worked example 10.5.1

NAMING AN ORGANIC MOLECULE WITH TWO FUNCTIONAL GROUPS

Name this molecule according to IUPAC rules.

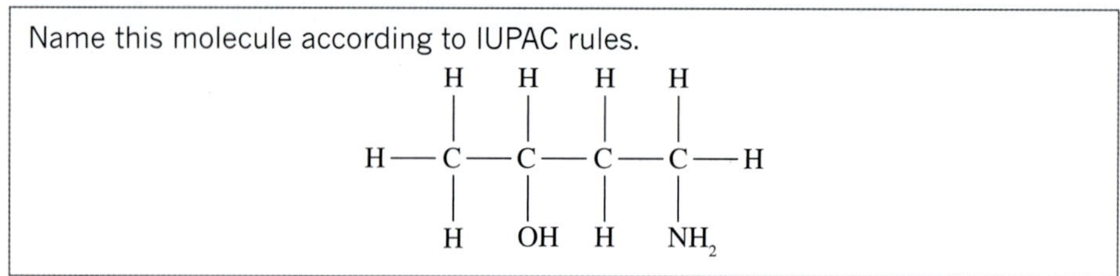

Thinking	Working
Identify the parent name by counting the longest continuous chain of carbon atoms.	There are four carbons in the longest chain, so the parent name is butane.
Identify the functional groups present.	The two functional groups present are hydroxyl and amino.
Determine which functional group has the higher priority and determine the prefixes and suffixes to use.	The hydroxyl group has the higher priority, so the molecule will end in -ol. The amino group has lower priority, so the prefix amino- will be used.
Number the carbon chain, giving the higher priority group the lowest number possible.	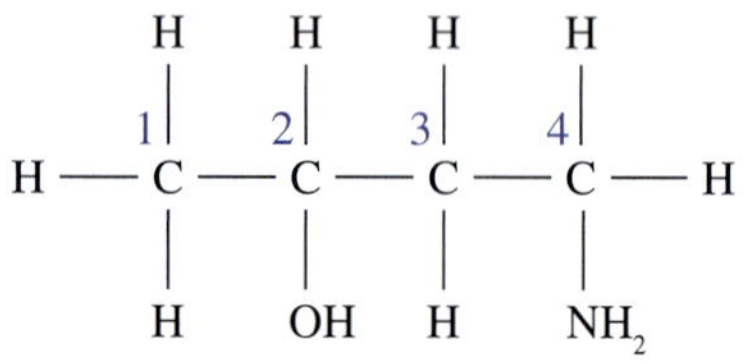
Determine the number of the carbon each functional group is attached to.	The –OH is attached to carbon 2 and $-NH_2$ is attached to carbon 4.
Use the functional group names and carbon numbers to construct the full name.	The name of the molecule is 4-aminobutan-2-ol.

Worked example: Try yourself 10.5.1

NAMING AN ORGANIC MOLECULE WITH TWO FUNCTIONAL GROUPS

Name this molecule according to IUPAC rules.

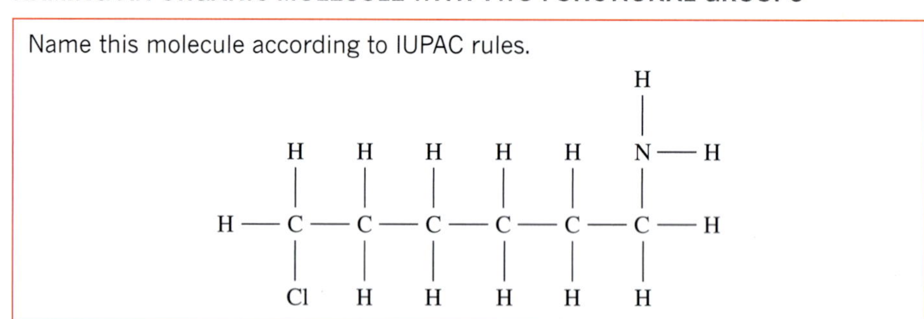

10.5 Review

SUMMARY

- IUPAC nomenclature is a set of rules for naming an organic compound.
- The naming of organic molecules follows a series of steps.
  - Identify the longest unbranched carbon chain to determine the parent name.
  - Name functional groups by using prefixes or suffixes, depending on their priority.
  - Number the carbon chain so that the carbon attached to the highest priority functional group has the lowest number.
  - Insert numbers before each functional group to indicate to which carbon they are attached.
  - When multiples of a functional group are present, use a multiplier (e.g. 'di-').
- Do not have spaces in the name (except for two-word names like carboxylic acids and esters). Use dashes to separate numbers from words. Use commas to separate numbers from other numbers.

KEY QUESTIONS

Knowledge and understanding

1 Draw the structural formula of the functional group of each of the following homologous series.

- **a** carboxylic acid
- **b** amide
- **c** alcohol
- **d** ester
- **e** amine

2 A molecule contains both a hydroxyl functional group and an amino functional group. Using the IUPAC functional group priorities (Table 10.5.4, page 371), determine the following used when naming this molecule.

- **a** the suffix
- **b** the prefix

Analysis

3 Write the systematic name for the following molecules:

a

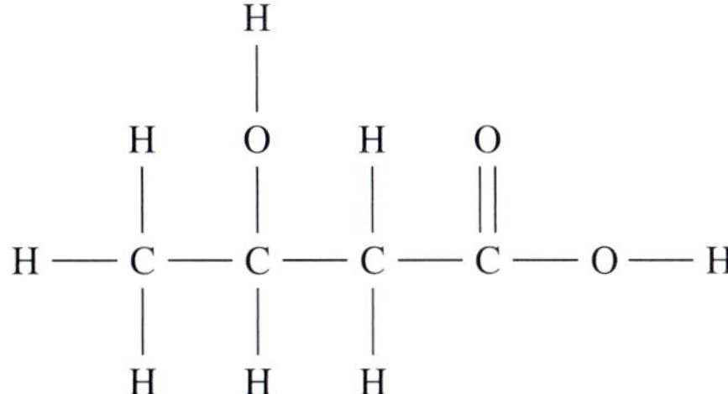

b

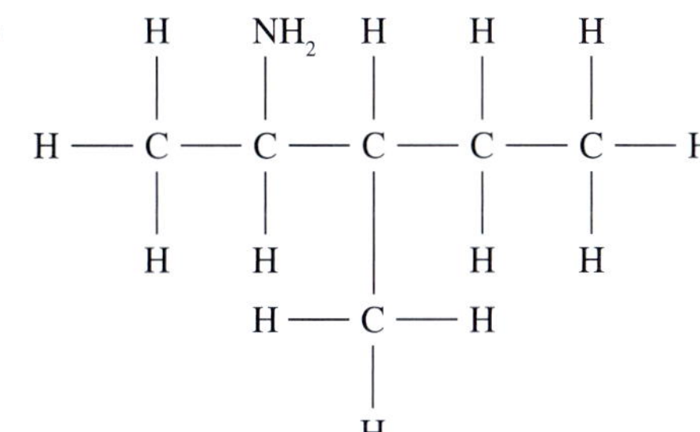

c

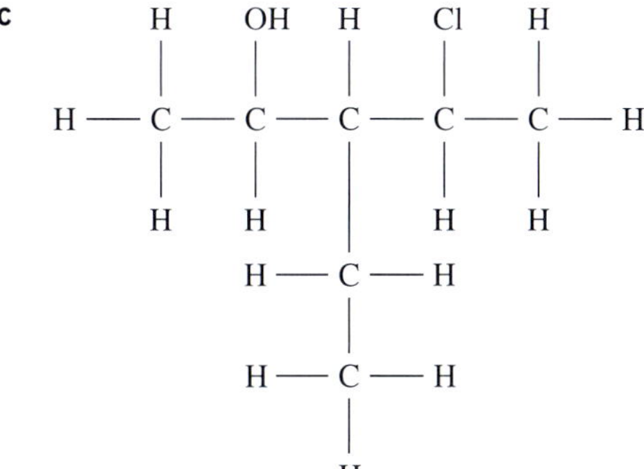

d

4 Draw the structure of:

- **a** 4-chloropent-1-en-2-amine
- **b** 6-bromohept-4-en-2-ol
- **c** 3,4-diethylhex-3-ene.
- **d** 4-ethyl-2-methylhexan-3-one

5 The following names are incorrect. Explain why each name is incorrect and provide the correct name and semi-structural formula.

- **a** 3-aminoprop-1-ene
- **b** 2-chloropentan-5-oic acid
- **c** 1-ethyl-1-hydroxyprop-2-ene

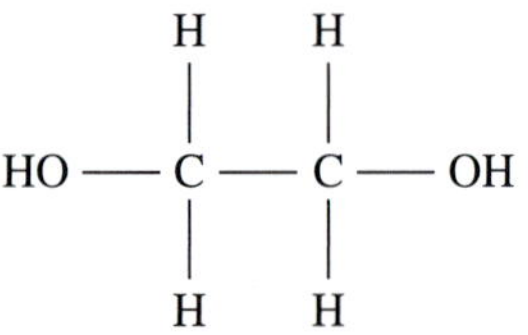

FIGURE 10.6.1 Ethylene glycol has a boiling point of 197°C.

10.6 Trends in physical properties within homologous series

The functional groups that are present in organic compounds are responsible for many of their physical properties. For example, the two hydroxyl groups in an ethylene glycol molecule (Figure 10.6.1) give it a relatively high boiling point of 197°C. A 50:50 mixture of ethylene glycol with water can be used as a coolant and anti-freeze agent in car engines (Figure 10.6.2).

FIGURE 10.6.2 Ethylene glycol mixed with water is commonly used as a coolant and anti-freeze agent in car engines.

In this section, you will consider the effect of functional groups on physical properties such as melting point, boiling point and viscosity.

PHYSICAL PROPERTIES OF ALKANES, ALKENES AND HALOALKANES

Alkanes and alkenes are hydrocarbons and are non-polar. Because these molecules are non-polar, the only intermolecular forces of attraction between them are weak **dispersion forces**. This influences their properties, such as melting points and boiling points. The properties of haloalkanes will also be considered in this section.

Melting and boiling points of alkanes

Table 10.6.1 lists the melting and boiling points of the first six alkanes from methane to hexane. These properties increase as the size of the alkane molecule increases.

TABLE 10.6.1 Melting and boiling points of the first six alkanes

Alkane	Molecular formula	Melting point (°C)	Boiling point (°C)
methane	CH_4	−182	−162
ethane	C_2H_6	−183	−89
propane	C_3H_8	−188	−46
butane	C_4H_{10}	−138	−0.5
pentane	C_5H_{12}	−130	36
hexane	C_6H_{14}	−95	69

Because alkane molecules are non-polar, the only intermolecular forces of attraction between them are weak dispersion forces. As the length of the carbon chain increases, the overall forces of attraction between molecules also increase (Figure 10.6.3). This is because there are more points of contact between molecules, and because dispersion forces increase due to the increased strength of temporary dipoles within the molecules. Because the melting point and boiling point of a molecular substance are determined by the strength of the intermolecular forces, these properties increase as alkane chain length increases.

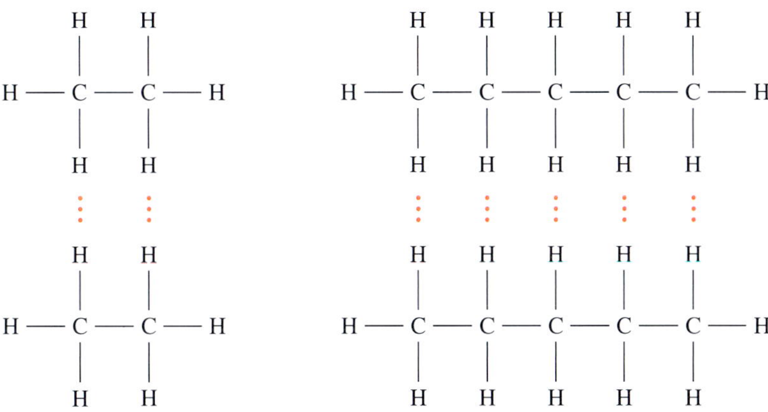

FIGURE 10.6.3 Dispersion forces are the strongest forces between alkane molecules. As molecules become longer, the dispersion forces become stronger.

Melting points of alkanes show a similar trend to that of boiling points (Figure 10.6.4). However, from methane to propane there are irregularities in the trend of melting points. This is because of the way the smaller molecules pack together in the solid. A detailed understanding of these irregularities is not required at this level.

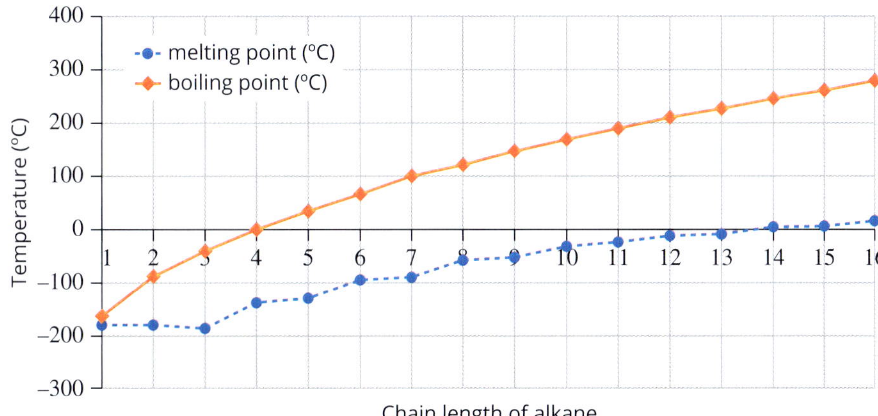

FIGURE 10.6.4 As the chain length of alkanes increases, their melting points and boiling points also increase.

Molecular shape

Molecular shape also influences the strength of dispersion forces and, therefore, melting and boiling points. Straight-chain alkanes are able to fit together more closely and tend to have higher boiling points than their corresponding branched-chain isomers, which are unable to come as closely together in the bulk substance.

Figure 10.6.5 shows how the shapes of butane and its branched isomer methylpropane influence the strength of the dispersion forces between the molecules. Butane ($CH_3CH_2CH_2CH_3$) boils at –0.5°C, whereas methylpropane ($(CH_3)_3CH$) boils at –11.7°C. Although both molecules have the same molecular formula of C_4H_{10}, molecules of $CH_3CH_2CH_2CH_3$ can fit more closely together, allowing more contact between the molecules and forming stronger intermolecular bonds.

CH_3 CH_2 CH_2 CH_3

CH_3 CH_2 CH_2 CH_3

butane
- less-compact molecules
- molecules can get closer together
- stronger dispersion forces
- boiling point –0.5°C

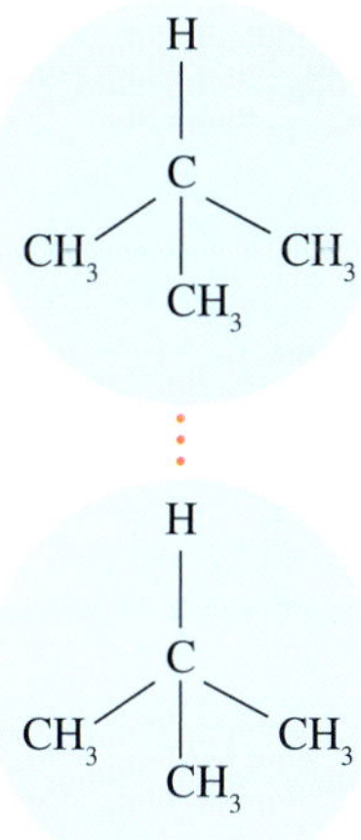

methylpropane
- more-compact molecules
- molecules cannot get as close together
- weaker dispersion forces
- boiling point –11.7°C

FIGURE 10.6.5 Molecular shape affects melting and boiling points. Dispersion forces between butane molecules are stronger than those between methylpropane molecules because butane molecules are less compact (more linear) and can come closer together.

The strength of dispersion forces between molecules depends on the size and shape of the molecules.

Melting and boiling points of alkenes and haloalkanes

Alkenes, like alkanes, are hydrocarbons. These molecules are non-polar and the forces of attraction between them are only weak dispersion forces. As you can see in Table 10.6.2, members of these homologous series have relatively low melting and boiling points similar to those observed for the alkanes with the same number of carbon atoms.

Like the alkanes, the melting and boiling points of alkenes increase with molecular size as the strength of dispersion forces between molecules increases.

TABLE 10.6.2 Melting and boiling points of molecules containing four carbons with different functional groups

Compound	Melting point (°C)	Boiling point (°C)
butane (C_4H_{10})	−138	−0.5
but-1-ene (C_4H_8)	−185.3	−6.3
chlorobutane (C_4H_9Cl)	−123	78

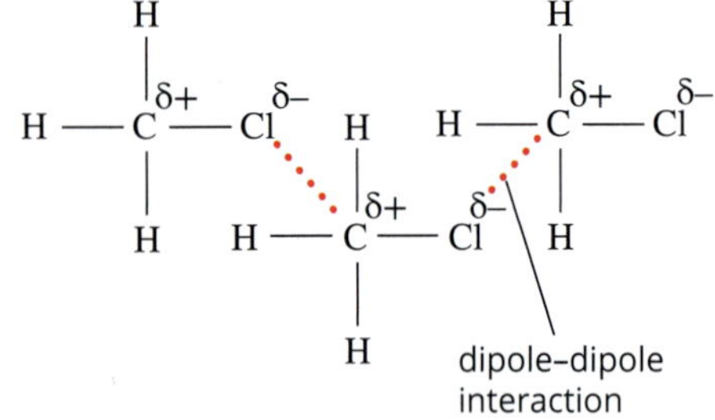

FIGURE 10.6.6 The carbon–chlorine bond in chloromethane is polar because the chlorine atom is more electronegative than the carbon atom. The presence of this permanent dipole allows for the formation of dipole–dipole attractions between chloromethane molecules.

Haloalkanes, unlike alkanes and alkenes, contain bonds that are quite polar. Chloromethane is a member of the haloalkane homologous series and contains a polar carbon–chlorine bond, as shown in Figure 10.6.6. Like all alkanes, chloromethane contains weak dispersion forces between molecules. However, the presence of the carbon–chlorine dipole allows **dipole–dipole attractions** to also occur. Because the dipole–dipole attractions are stronger than the dispersion forces, the melting and boiling points of the haloalkanes are generally higher than those of alkanes with a similar number of carbon atoms, as seen in Table 10.6.2.

Effect of chain length on physical properties

The trend in melting and boiling points of haloalkanes is similar to that for alkanes. The explanation for this is the same as that for the trend in alkanes. As the length of hydrocarbon chains increases, the strength of intermolecular dispersion forces also increases and therefore melting and boiling points also increase.

This trend can be observed for members of the different organic homologous series studied in this chapter.

It is generally true that, as the hydrocarbon chain length of members of a homologous series increases, the melting points and boiling points of the compounds also increase.

As the hydrocarbon chain length of members of a homologous series increases, melting and boiling points of the compounds also increase.

Viscosities of alkanes, alkenes and haloalkanes

In the previous section, you saw how the properties of organic compounds were affected by the structure and size of the molecules of the compounds. Another physical property of organic compounds—viscosity—is very important in the transport-fuel and lubricating-oil industries. In this section, you will see how the structures of alkanes, alkenes and haloalkanes affect viscosity.

FIGURE 10.6.7 Tomato sauce is said to be viscous because it pours slowly.

The **viscosity** of a liquid is its resistance to pouring or flowing. A liquid that pours slowly, such as tomato sauce, is said to be viscous or to have a high viscosity (Figure 10.6.7). Vegetable oil is less viscous than tomato sauce but more viscous than water, and petrol is less viscous than all of these liquids.

The viscosity of a liquid depends on the interactions between molecules. Like boiling point, viscosity increases as the forces of attraction between the molecules increase. As the chain length of organic molecules increases, there will be more points of contact between the molecules (Figure 10.6.8 on the following page) and the strength of dispersion forces between them will become stronger. This results in an increase in viscosity.

The viscosity of a liquid depends on the interactions between molecules.

Key
lubricating oil molecule
dispersion force

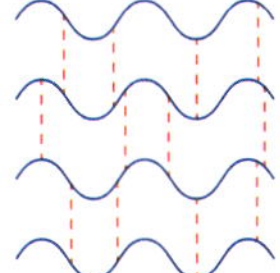

Shorter hydrocarbon chains result in:
- fewer points of contact between the hydrocarbon molecules
- overall weaker dispersion forces
- lower viscosity.

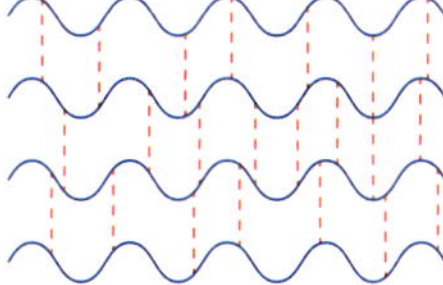

Longer hydrocarbon chains result in:
- more points of contact between the hydrocarbon molecules
- overall stronger dispersion forces
- higher viscosity.

FIGURE 10.6.8 As the length of hydrocarbon chains increases, the dispersion forces between them become stronger, and so the viscosity of the hydrocarbon compounds increases.

Table 10.6.3 shows different fractions obtained by the **fractional distillation** of crude oil. You will learn more about fractional distillation in Chapter 13. Crude oil is a mixture of hydrocarbons of different molar mass and chain length. During the fractional distillation process, molecules of approximately the same size are separated into different fractions, on the basis of their boiling points. As you learnt earlier, the boiling points of molecular substances are a consequence of the overall forces of attraction between the molecules. Molecules with 18–20 carbon atoms in their chains form fractions that are suitable for use as lubricating oils for motors.

TABLE 10.6.3 Fractions obtained from the fractional distillation of crude oil and their uses

Fraction	Composition	Boiling range (°C)	Uses
gas	C_1–C_4	<30	heating fuel, LPG
petrol	C_5–C_{12}	30–250	motor fuel/petrol
kerosene	C_{12}–C_{16}	250–300	jet and diesel fuel
heating fuel oils	C_{16}–C_{18}	>300	diesel fuel, heating fuel oil
lubricating oils	C_{18}–C_{20}	>350	lubrication
paraffin waxes	C_{20}–C_{40}		candles, wax
bitumen	above C_{40}		roofing tar, road surfaces

Viscosity and temperature change

The viscosity of motor engine oils enables them to protect an engine from wear. The high viscosity of oil makes it 'cling' to the metal. The oil forms a thin layer over bearings and gears, which prevents direct contact between metals and so reduces friction and wear (Figure 10.6.9). As the temperature of the engine increases, the viscosity of the engine oil decreases. This is because the molecules have more kinetic energy and they move about more quickly. This decreases the amount of contact time between the molecular chains and so the dispersion forces are disrupted. The weaker forces of attraction between the molecules results in decreased viscosity. Engine oils must be specially formulated with a mixture of organic compounds of different viscosity and additives to produce oils that are effective in both cold and hot engines.

FIGURE 10.6.9 Lubricating oil protects gears from friction and wear. Multigrade engine oils contain a mixture of hydrocarbon molecules with different chain lengths and other additives to provide effective lubrication of the engine over a range of temperatures.

PHYSICAL PROPERTIES OF ALCOHOLS, CARBOXYLIC ACIDS, AMINES AND AMIDES

Compounds from the four homologous series alcohols, carboxylic acids, amines and amides will be considered together because their molecules contain functional groups that can form hydrogen bonds. The ability of molecules to form hydrogen bonds has a significant effect on their properties.

Hydrogen bonds are the strongest of the intermolecular forces and, as a result, molecules that can form hydrogen bonds generally exhibit higher boiling points than those that cannot. You can see in Table 10.6.4 that alcohols, carboxylic acids, amines and amides have higher boiling points than alkanes of similar molecular mass.

TABLE 10.6.4 Comparison of melting and boiling points of compounds with different functional groups based on molecular mass

Homologous series	Compound	Formula	Molar mass (g mol^{-1})	Melting point (°C)	Boiling point (°C)
alkane	butane	C_4H_{10}	58	–138	–0.5
alcohol	propan-1-ol	C_3H_7OH	60	–126	97.2
carboxylic acid	ethanoic acid	CH_3COOH	60	16.6	118
amine	propan-1-amine	$C_3H_7NH_2$	59	–83	49
amide	ethanamide	CH_3CONH_2	59	80	210

Melting and boiling points of alcohols

The higher boiling points of the alcohols are due to the presence of hydrogen bonds between neighbouring alcohol molecules. In contrast, the only type of bonding present between non-polar alkane molecules is much weaker dispersion forces.

You will recall from Unit 1 that an atom of oxygen is more electronegative than an atom of hydrogen, so the oxygen–hydrogen bond in the hydroxyl functional group is a polar bond. Figure 10.6.12 shows how hydrogen bonding occurs between the partially positive charged hydrogen atom in the –OH group on one alcohol molecule and a non-bonding electron pair on the oxygen atom of a neighbouring alcohol molecule. The presence of the hydrogen bonds results in the higher melting and boiling points observed.

FIGURE 10.6.12 The red dotted line shows the hydrogen bonds that form between molecules such as ethanol that contain the polar hydroxyl functional group.

CHEMFILE

Have you got the patience for slow science?

The longest running laboratory experiment in the world was set up in 1927 at the University of Queensland in Brisbane. The 'pitch drop' experiment is designed to show that pitch, the name given to a mixture of highly viscous hydrocarbons, is actually a liquid. Pitch, otherwise known as bitumen (Figure 10.6.10), appears solid on first inspection and shatters if you strike it with a hammer.

FIGURE 10.6.10 A sample of pitch at room temperature appears solid and shatters when hit with a hammer.

Professor Thomas Purnell set up the pitch experiment, shown in Figure 10.6.11, by placing a sample of pitch into a funnel encased in a sealed container. He wanted to demonstrate to students that pitch was indeed a liquid—just a very, very viscous one. Over time, the pitch has settled and slowly flowed through the funnel producing a drop on average once every decade. The ninth drop was recorded in April of 2014.

Today bitumen is used extensively, forming the main component of road surfaces. Under normal conditions, the road surface appears solid. However, heating reduces its viscosity, and road workers are able to work the pliable bitumen.

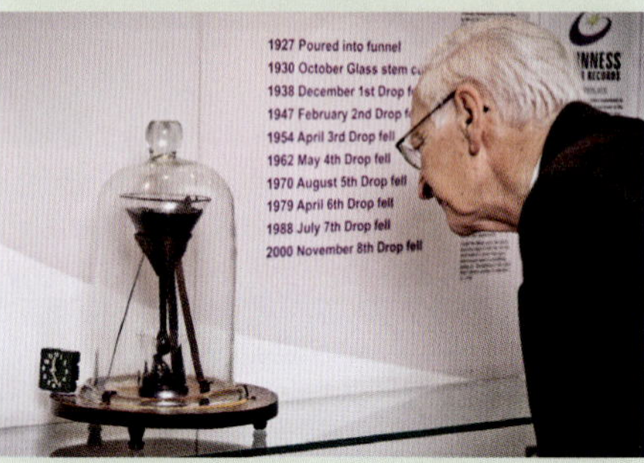

FIGURE 10.6.11 Professor Thomas Purnell set up the famous University of Queensland pitch drop experiment in 1927. Professor John Mainstone, shown here, oversaw the experiment for many years.

Viscosity of alcohols

The viscosity of a particular alcohol is greater than that of an alkane with the same number of carbon atoms in its hydrocarbon chain. This is because of the strength of the intermolecular hydrogen bonding forces that occur between the alcohol molecules but not between the alkane molecules. Honey is a substance that illustrates the effect that hydrogen bonds can have on the viscosity of a liquid (Figure 10.6.13). Honey is not a pure substance. It is made up of a mixture of different compounds. About 80% of honey is made up of glucose, fructose and other sugars. The molecules of these compounds have many hydroxyl groups (Figure 10.6.14), so the molecules are held to each other by hydrogen bonds between these hydroxyl groups. The number and strength of these interactions is a major reason for the high viscosity of honey.

FIGURE 10.6.13 The many hydroxyl groups in sugar molecules in honey are the reason for its high viscosity.

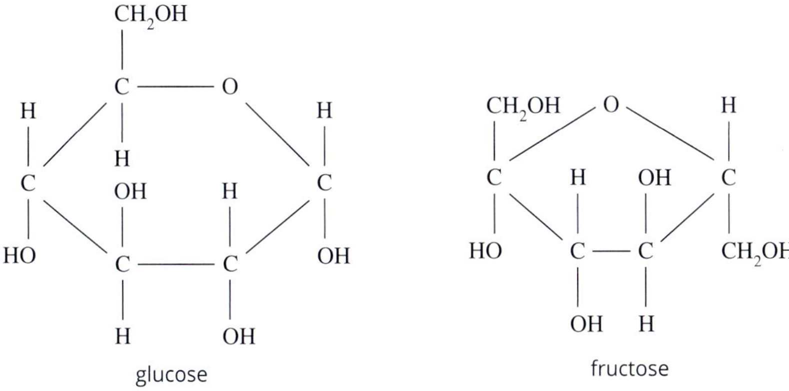

FIGURE 10.6.14 Glucose and fructose molecules contain many hydroxyl functional groups. Hydrogen bonding forces between these groups contribute to the high viscosity of honey.

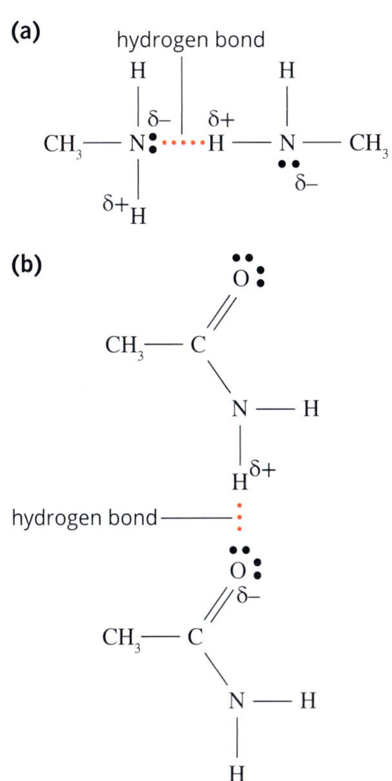

FIGURE 10.6.15 Hydrogen bonding between (a) amine molecules and (b) amide molecules

Melting and boiling points of amines and amides

The presence of highly polar nitrogen–hydrogen bonds in amine and amide molecules means that these molecules can also form hydrogen bonds. The formation of hydrogen bonds between amines and amides is illustrated in Figure 10.6.15.

In amines, the hydrogen bonds form between the non-bonding pair of electrons on the electronegative nitrogen atom and the partially positive hydrogen atom on another amine molecule. In amides, hydrogen bonds form between the non-bonding electron pairs on the oxygen atoms of one molecule and the partially positive hydrogen atom on a neighbouring molecule. The strength of the hydrogen bonding between molecules explains the relatively high boiling points and viscosities of amines and amides when compared to hydrocarbon molecules of similar size.

Melting and boiling points of carboxylic acids

Hydrogen bonding also has a marked effect on the boiling points of carboxylic acids. Figure 10.6.16 shows how two molecules of a carboxylic acid in the liquid state can form **dimers** in which two hydrogen bonds occur between the molecules.

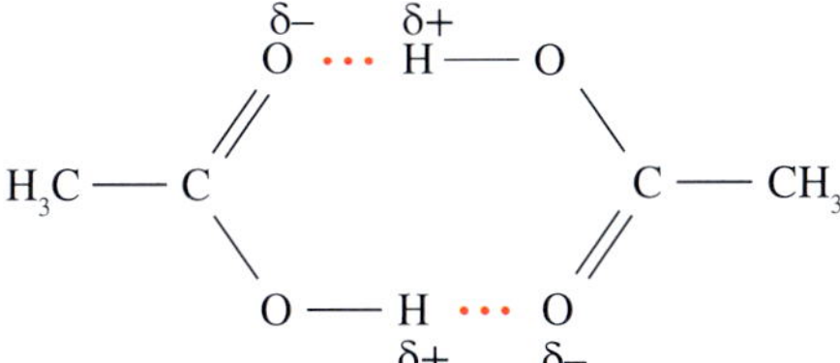

FIGURE 10.6.16 Hydrogen bonding between two ethanoic acid molecules results in the formation of a dimer.

The dimer produced is a stable species that has a molar mass that is double that of a single carboxylic acid molecule. The increase in size that results from the formation of the dimer increases the strength of the dispersion forces between one dimer and its neighbours. The stronger dispersion forces, combined with the hydrogen bonds between molecules, results in the higher boiling point and viscosity observed for carboxylic acids when compared to most other organic molecules of similar size.

Figure 10.6.17 highlights the higher boiling points of the carboxylic acid homologous series compared to alcohols. The overall increasing trend of both graphs reinforces the point that, within a homologous series, an increase in length of the hydrocarbon chain of molecules results in an increase in boiling point.

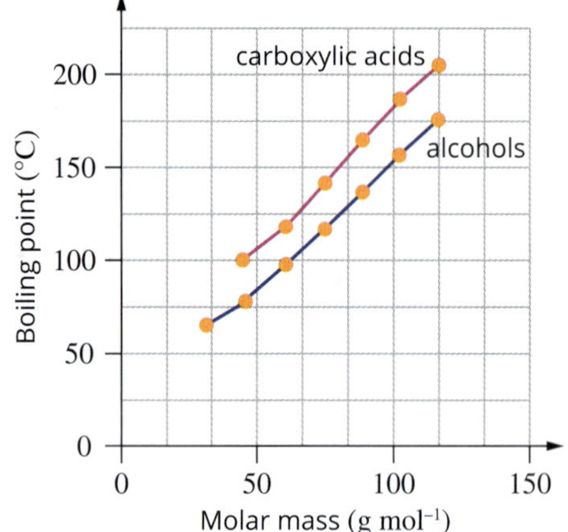

FIGURE 10.6.17 Boiling points of the carboxylic acid homologous series increase with increasing molar mass. The boiling points are higher than the boiling points of alcohols because of the formation of dimers stabilised by hydrogen bonds between molecules.

FIGURE 10.6.18 Molecular models of the three isomers of butanol: (a) butan-1-ol, (b) butan-2-ol, (c) 2-methylpropan-2-ol.

Effect of hydrocarbon chain branching on boiling points

In the earlier discussion on the boiling points of alkanes, you saw that branched-chain isomers generally have lower melting and boiling points than straight-chain alkanes because their molecules cannot fit as closely together, resulting in weaker dispersion forces (Figure 10.6.8, page 377). A similar effect occurs in the case of molecules with a functional group. In addition to this effect, the position of the functional group within the hydrocarbon chain also has an influence on boiling point. This can be illustrated using the models shown in Figure 10.6.18. The models show the structure of three isomers of butanol: butan-1-ol, butan-2-ol, and 2-methylpropan-2-ol. These isomers are examples of primary, secondary and tertiary alcohols respectively.

You can see from the models that the hydroxyl group becomes increasingly 'crowded' by alkyl groups from the primary alcohol through to the secondary and tertiary alcohol isomers. The presence of the alkyl groups restricts the ability of the hydroxyl group to form hydrogen bonds with other molecules. For this reason, the boiling points of these alcohols decrease in the sequence from primary to secondary to tertiary alcohol, as seen in Table 10.6.5.

TABLE 10.6.5 The boiling points of primary to secondary and tertiary alcohols with the formula C_4H_9OH

Alcohol	Type of alcohol	Boiling point (°C)
butan-1-ol	primary	118
butan-2-ol	secondary	100
2-methylpropan-2-ol	tertiary	82

PHYSICAL PROPERTIES OF ALDEHYDES, KETONES AND ESTERS

Aldehydes, ketones and esters can be considered together because they are composed of molecules that are held together by dipole–dipole attractions. These molecules cannot form hydrogen bonds with each other because they do not have a hydrogen atom bonded to an oxygen atom or a nitrogen atom within their molecules.

Melting and boiling points of aldehydes, ketones and esters

Aldehydes, ketones and esters all contain a carbon–oxygen double bond. Oxygen is much more electronegative than carbon, so the carbon–oxygen double bond is polar. This means that molecules of aldehydes, ketones and esters contain a permanent dipole, which can form dipole–dipole attractions with nearby molecules. The dipole–dipole interactions that arise between ketone molecules are shown in Figure 10.6.19.

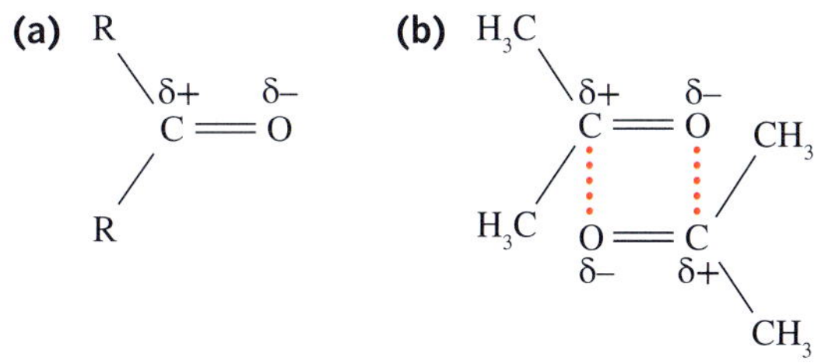

FIGURE 10.6.19 (a) A carbonyl bond is polar. (b) Dipole–dipole attractions between ketone molecules.

The strength of these dipole–dipole bonds between molecules give aldehydes, ketones and esters higher boiling points than similar-sized alkanes. However, their boiling points are not as high as similar-sized alcohols because dipole–dipole bonds are not as strong as hydrogen bonds.

A comparison of the boiling points of members of each of these homologous series with that of an alkane is shown in Table 10.6.6.

TABLE 10.6.6 Comparison of boiling points of similar mass organic molecules from different homologous series

Homologous series	Semi-structural formula	Molar mass ($g\ mol^{-1}$)	Boiling point (°C)
alkane	$CH_3CH_2CH_2CH_3$	58	–0.5
alcohol	$CH_3CH_2CH_2OH$	60	97
aldehyde	CH_3CH_2CHO	58	48
ketone	CH_3COCH_3	58	56
ester	$HCOOCH_3$	60	32

Effect of chain length on boiling point

As the hydrocarbon chain lengths of aldehydes, ketones and esters increase, their boiling points also increase. This is because there are more points of contact between the hydrocarbon chains and therefore there is an overall increase in the strength of dispersion forces holding the chains together. This trend is similar to that observed for alcohols and other compounds described earlier.

10.6 Review

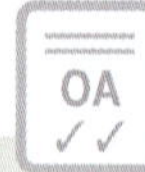

SUMMARY

- The boiling points of organic molecules are determined by intermolecular forces. Dispersion forces are always present and there may also be dipole–dipole attractions or hydrogen bonds.
- The strength of dispersion forces between alkane molecules depends on the size and shape of the molecules.
- The only forces of attraction between molecules of the alkane and alkene homologous series are dispersion forces. Molecules of haloalkanes are attracted to each other by dipole–dipole attractions, as well as by dispersion forces.
- Molecules of aldehydes, ketones and esters have polar carbonyl functional groups and are attracted to each other by dipole–dipole attractions.
- Molecules of alcohols, carboxylic acids, amines and amides contain functional groups that can form hydrogen bonds with other molecules.
- Comparing molecules of similar size in different homologous series, it is generally true that their boiling points increase in the order: alkanes, alkenes <aldehydes, ketones, esters < amines, alcohols, carboxylic acids, amides.
- The viscosity of a liquid is its resistance to flow.
- The viscosity of a homologous series of organic compounds increases as the length of the carbon chain in the molecules increases.

KEY QUESTIONS

Knowledge and understanding

1 Complete the following paragraph by filling in the gaps with the appropriate words.
Butane is a member of the ____________ homologous series. The type of forces of attraction holding butane molecules to each other are ____________. As the chain length of alkanes increases, their boiling points ____________. In general, alkanes with branched chains have ________ boiling points than straight-chain alkanes with the same number of carbon atoms.

2 In the table below, indicate the strongest form of intermolecular force of attraction that occurs between molecules from each of the homologous series listed. The first one has been done for you.

Homologous series	Strongest form of intermolecular force of attraction between molecules
alcohols	hydrogen bonding
alkenes	
carboxylic acids	
esters	
haloalkanes	
amines	

3 Arrange the following compounds in increasing order of boiling point. Give reasons for your answer.
$CH_3CH_2CH_2CH_2OH$; $CH_3CH_2CH_3$; $CH_3CH_2CH_2CH_2Cl$; $CH_3CH_2CH_2COOH$; $CH_3CH_2CH_2CH_3$

4 Complete the following summary about physical properties of organic compounds by using the following terms to fill in the gaps below. Note that terms may be used more than once.

fewer; more; less; longer; shorter; greater; lower; stronger; weaker

As the viscosity of a liquid increases, it flows _______ readily. The viscosity of pentane is ________ than that of octane at the same temperature. This is because pentane molecules have _________ hydrocarbon chains than octane. In samples of liquid pentane, there will be ________ points of contact between molecules, and overall, the strength of dispersion forces holding them together will be _________ than those holding octane molecules together.
Liquids consisting of molecules made up of a straight hydrocarbon chain with a functional group attached will generally have a ________ viscosity than a liquid consisting of hydrocarbon chains with no functional group. This is because forces of attraction (dipole–dipole or hydrogen bonds) between the polar functional groups are ________ than dispersion forces.

Analysis

5 Arrange the following liquids in increasing order of viscosity. Give reasons for your answer. Assume that the temperatures of the liquids are the same.
1-chloroheptane; hexane; heptan-1-ol; pentane

6 Each of the three following compounds contains five carbon atoms. Arrange the compounds in increasing order (lowest to highest) of boiling point. Give reasons for your answer.
2-methylbutane; 2,2-dimethylpropane; pentane

Chapter review

KEY TERMS

alcohol
aldehyde
alkane
alkene
alkyl group
amide functional group
amino functional group
benzene
bond energy
bond strength
carbonyl functional group
carboxyl functional group
carboxylic acid
condensed formula
cyclic molecule
degree of unsaturation
dimer
dipole–dipole attraction
dispersion force
ester
ester functional group
fractional distillation
functional group
halo functional group
haloalkane
halogen
homologous series
hydrocarbon
hydroxyl functional group
isomers
IUPAC nomenclature
ketone
molecular formula
organic molecule
parent molecule
phenyl functional group
primary alcohol
primary amide
primary amine
saturated hydrocarbon
saturated molecule
secondary alcohol
semi-structural formula
skeletal structure
structural formula
structural isomers
tertiary alcohol
unsaturated hydrocarbon
unsaturated molecule
valence electron number
valence shell electron pair repulsion (VSEPR) theory
viscosity

REVIEW QUESTIONS

Knowledge and understanding

1 Determine which molecular formula indicates a noncyclic alkane.

A C_3H_6
B C_5H_8
C C_6H_6
D C_6H_{14}

2 Which of the following molecules contains the highest degree of unsaturation?

A $C_{15}H_{28}$
B $C_{17}H_{30}$
C C_6H_{14}
D C_4H_8

3 Identify which statement about isomers is incorrect.

A Isomers can have very different chemical and physical properties.
B Isomers can have different molecular formulas as long as they contain the same type of atoms.
C Isomers are molecules that have the same molecular formula but a different arrangement of atoms.
D Isomers do not have to contain the same functional groups.

4 Identify the name of the functional group present in alcohols.

A hydroxyl
B carbonyl
C carboxyl
D alcohol

5 Identify what the 'pent' part of the name 2-methylpent-2-ene refers to.

A The molecule is an alkene.
B The double bond is on carbon number 5.
C The molecule contains an alkyl side chain.
D The parent molecule contains five carbon atoms

6 What elements are present in the following functional groups?

a amino
b amide
c chloro
d hydroxyl
e aldehyde
f ketone

7 Use propan-1-ol and propan-2-ol to explain the difference between a primary and secondary alcohol.

8 What is the structural difference between a carboxylic acid and an amide?

9 Why are carboxyl and primary amide groups only found at the end of a carbon chain?

10 Identify which statement about boiling points of organic compounds is *false*.

A Molecular shape affects the boiling points of organic compounds.
B Organic compounds that contain weak intermolecular forces have high boiling points.
C The boiling points of alkanes generally increase as the length of the carbon chain increases.
D Organic compounds that are non-polar generally have lower boiling points than compounds of the same size that are polar.

11 Which of the following compounds would have the greatest viscosity?

A butan-2-ol
B but-2-ene
C ethyl ethanoate
D 2-methylpropane

12 Which alkane has the higher viscosity, pentane or octane? Explain your answer in terms of the intermolecular forces.

13 Arrange the following compounds in increasing order of boiling point. Provide reasons for your answer.

$CH_3CH(CH_3)CH_3$; $CH_3CH_2CH_2CH_2OH$;
CH_3CH_3; $CH_3CH_2CH_2COOH$; $CH_3CH_2CH_2CH_3$

Application and analysis

14 The structure of 1-chloropentane is shown below.

```
     H    H    H    H    H
     |    |    |    |    |
H —  C —  C —  C —  C —  C — Cl
     |    |    |    |    |
     H    H    H    H    H
```

a How many isomers can be drawn of this structure?

b Draw and name all the possible isomers of this structure.

15 Write three semi-structural formulas of C_4H_8 to show the possible isomers. Name the three isomers.

16 Determine the molecular formula for each of the following.

a an alkane with molar mass 72 g mol^{-1}
b an alkene with molar mass 84 g mol^{-1}
c an alkene with molar mass 56 g mol^{-1}
d a hydrocarbon with molar mass 98 g mol^{-1}

17 How many isomers of pentane exist?

18 Explain why 2-ethylpentane is an incorrect name for an alkane.

19 Determine the systematic name of each of the following compounds.

a $CH_3CHNH_2CH_2OH$
b $CH_3CHOH(CH_2)_5CH_2Cl$
c $CH_3CH_2CH_2CH_2CHNH_2CHICH_3$

20 Name each of the following molecules.

a

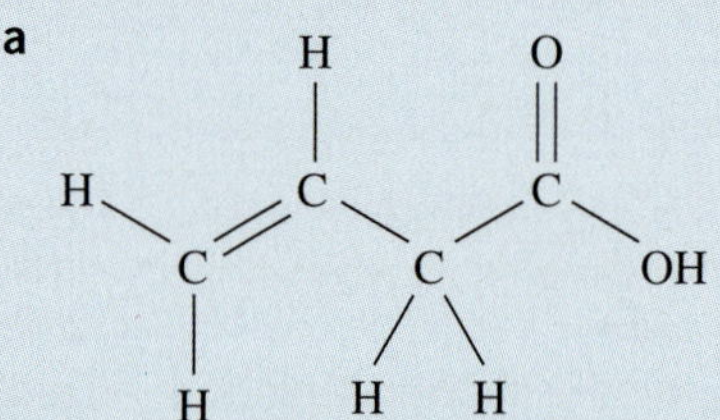

b

c

d

e

21 Draw the structure of:

a pentyl propanoate

b 4-iodo-3-methylpent-2-ene

c oct-4-enoic acid

d 3-methylbutan-1-amine.

22 Write the semi-structural formulas of:

a oct-3-enoic acid

b methyl hexanoate

c 3-fluoropropan-1-ol

d 4-hydroxybutanoic acid

e 2,3-dimethylpentan-1-amine.

23 Convert the following structures to semi-structural formulas.

a

b

c

d

24 Explain why each of these names is incorrect, then determine its correct name.

a but-4-ene

b 2-hydroxyethanamine

c 4-chlorohex-5-ene

d 2-chloro-3-ethylbutane

25 The semi-structural formulas of some organic compounds are given below. For each compound:

i identify the homologous series to which it belongs

ii give its systematic name.

a $CH_3(CH_2)_5CH_2OH$

b $CH_3(CH_2)_2CHCl(CH_2)_2CH_3$

c $CH_3CHOH(CH_2)_3CH_3$

d $CH_3CH_2CHNH_2CH_3$

26 For each of the following pairs of compounds, determine whether the molecules are structural isomers, identical or unrelated.

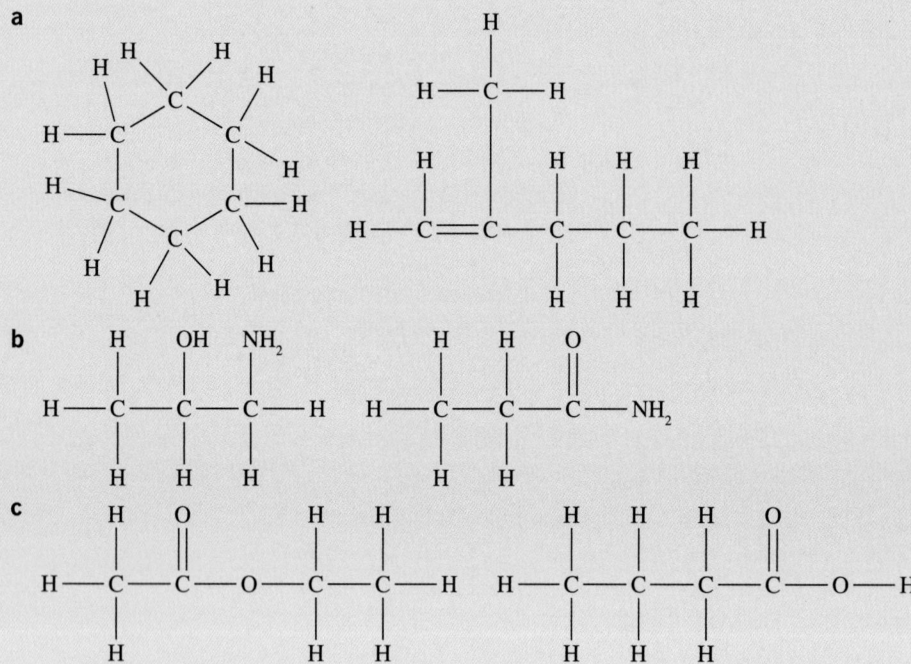

27 Explain why, when comparing the effect of functional groups on the boiling points of different homologous series, it is important to compare molecules with a similar molar mass.

28 The table below shows the results obtained in a series of experiments to determine the boiling points of various organic compounds.

Experiment	Organic compound	Molecular mass	Boiling point (°C)
1	$CH_3CH_2CH_3$	44.0	–42
	CH_3CH_2OH	46.08	78.0
	HCOOH	46.03	102
2	$CH_3(CH_2)_3CH_3$	72.0	36.1
	$CH_3(CH_2)_2CH_2OH$	74.14	119
	CH_3CH_2COOH	74.09	139
3	$CH_3(CH_2)_5CH_3$	100.0	98.4
	$CH_3(CH_2)_4CH_2OH$	102.20	157
	$CH_3(CH_2)_3COOH$	102.15	152

a Analyse the data and draw conclusions about the boiling points of these organic compounds.

b Identify any anomalies in the results (i.e. identify any results that you would not expect; these results may be due to an experimental error).

c Explain the reasons for the trends in the boiling points that you described in part **a**, making allowance for anomalies.

OA ✓✓

CHAPTER

11 Reactions of organic compounds

Life on Earth is often referred to as being carbon based. This is because the structures of living things on Earth are based mainly on organic compounds. The growth and decay of living things involve a series of reactions in which these organic compounds are made, decomposed or changed from one form into another. The industrial production of many chemicals also involves organic reactions.

At the end of this chapter, you will be able to describe some specific chemical reactions involving organic compounds, including oxidation, addition and condensation reactions.

You will learn that the functional groups of compounds are usually involved in chemical reactions and, as a result, members of a homologous series usually undergo similar reactions because they have the same functional group. You will also learn how green chemistry principles can improve the sustainability of the production of chemicals.

Key knowledge

- organic reactions and pathways, including equations, reactants, products, reaction conditions and catalysts (specific enzymes not required):
  - synthesis of primary haloalkanes and primary alcohols by substitution **11.1, 11.4**
  - addition reactions of alkenes **11.2, 11.4**
  - the esterification between an alcohol and a carboxylic acid **11.3, 11.4**
  - hydrolysis of esters **11.3, 11.4**
  - pathways for the synthesis of primary amines and carboxylic acids **11.3, 11.4**
  - transesterification of plant triglycerides using alcohols to produce biodiesel **11.3**
- calculations of percentage yield and atom economy of single-step or overall reaction pathways, and the advantages for society and for industry of developing chemical processes with a high atom economy **11.4**
- the sustainability of the production of chemicals, with reference to the green chemistry principles of use of renewable feedstocks, catalysis and designing safer chemicals **11.5**

11.1 Reactions of alkanes and haloalkanes

FIGURE 11.1.1 Propane and butane gas are both used in barbecue gas bottles.

Substitution reactions of alkanes using excess chlorine gas can result in a mixture of chloroalkane products with varying levels of substitution.

Alkane molecules contain strong carbon–carbon bonds. They have no functional groups or carbon–carbon multiple bonds. This makes alkanes much less reactive than compounds that have these additional structures.

COMBUSTION OF ALKANES IN AIR

Although alkanes are relatively stable compounds and only undergo a few chemical reactions, they do burn readily in air in an exothermic reaction. As discussed in Chapter 2, alkanes are very good fuels.

Common fuels such as petrol, diesel, kerosene and the gas, used in barbecue cylinders, contain mixtures of alkanes, together with other hydrocarbon molecules (Fig 11.1.1). The main component of natural gas is methane (CH_4), the simplest alkane.

The burning of hydrocarbon fuels in excess oxygen results in complete combustion, producing carbon dioxide and water. The equation for the complete combustion of methane in air is:

$$CH_4(g) + 2O_2(g) \rightarrow CO_2(g) + 2H_2O(l)$$

Octane is a major component of petroleum. The reaction for the complete combustion of octane is:

$$2C_8H_{18}(g) + 25O_2(g) \rightarrow 16CO_2(g) + 18H_2O(l)$$

SUBSTITUTION REACTIONS OF ALKANES

A **substitution reaction** occurs when an atom or functional group in a molecule is replaced or 'substituted' by another atom or group.

Alkanes are saturated hydrocarbons and undergo substitution reactions with halogens, such as chlorine and bromine, to produce haloalkanes. A mixture of an alkane; for example, methane, and chlorine will not react at room temperature or in the dark. For the substitution reaction to occur, the reaction must be initiated by ultraviolet (UV) light.

The reaction between methane and chlorine in the presence of UV light is given by the equation:

$$CH_4(g) + Cl_2(g) \xrightarrow{\text{UV light}} CH_3Cl(g) + HCl(g)$$

For each molecule of chlorine gas (Cl_2) that reacts, a hydrogen atom on the alkane is replaced by a chlorine atom.

Chloromethane is said to be the substitution product of this reaction. As the reaction proceeds, the chloromethane product may continue to react with another chlorine molecule. If this occurs, the di-substituted product, dichloromethane (CH_2Cl_2), is produced.

If enough chlorine is available, further reactions can result in all four hydrogen atoms of the original methane molecule being substituted by chlorine atoms (Figure 11.1.2 on the following page). The different substituted products have different boiling points, so they may be separated from one another by **fractional distillation**.

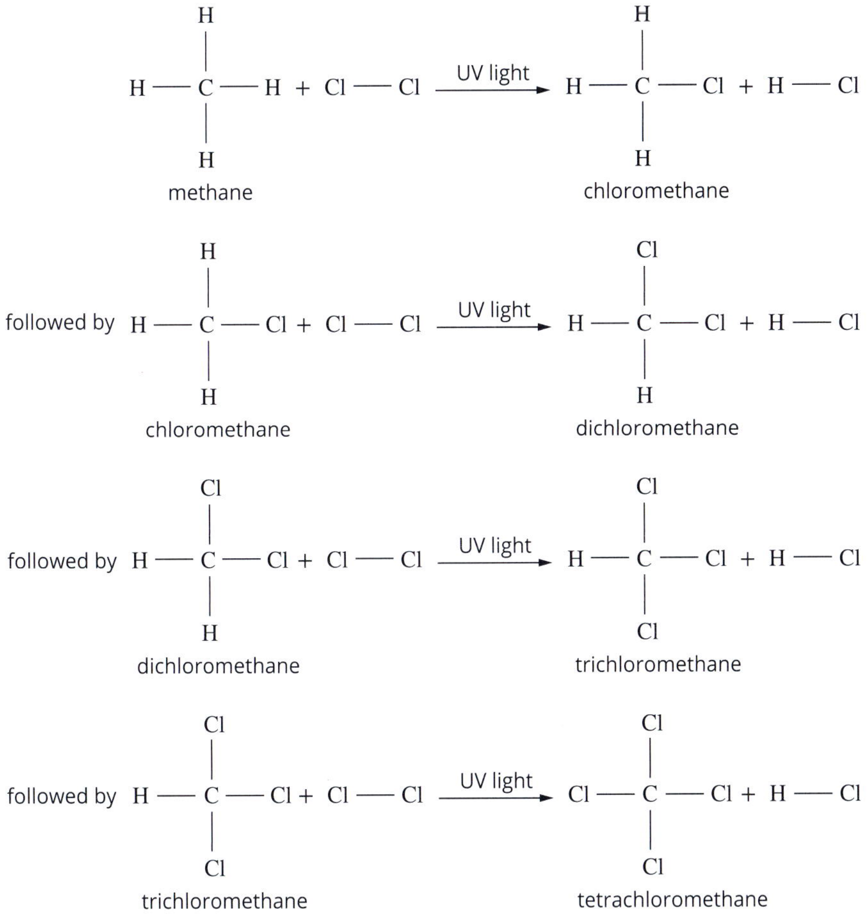

$$\text{H—C(H)(H)—H} + \text{Cl—Cl} \xrightarrow{\text{UV light}} \text{H—C(H)(H)—Cl} + \text{H—Cl}$$

methane → chloromethane

$$\text{followed by}\quad \text{H—C(H)(H)—Cl} + \text{Cl—Cl} \xrightarrow{\text{UV light}} \text{H—C(Cl)(H)—Cl} + \text{H—Cl}$$

chloromethane → dichloromethane

$$\text{followed by}\quad \text{H—C(Cl)(H)—Cl} + \text{Cl—Cl} \xrightarrow{\text{UV light}} \text{H—C(Cl)(Cl)—Cl} + \text{H—Cl}$$

dichloromethane → trichloromethane

$$\text{followed by}\quad \text{H—C(Cl)(Cl)—Cl} + \text{Cl—Cl} \xrightarrow{\text{UV light}} \text{Cl—C(Cl)(Cl)—Cl} + \text{H—Cl}$$

trichloromethane → tetrachloromethane

FIGURE 11.1.2 Substitution reactions between methane and chlorine

CHEMFILE

Burning ice

Large amounts of methane gas are stored around the world trapped as solid methane hydrate. These methane hydrates look very similar to ice, with one main difference—they burn.

The burning ice shown in the figure below is a result of the methane gas being released from the crystal structure of the ice as it melts.

Deposits of methane hydrates become trapped in the Arctic regions of the Earth and deep in the oceans due to low temperatures and high pressures.

One fear is that with rising ocean temperatures the deposits of methane hydrate throughout the oceans will melt, releasing large amounts of methane into the atmosphere.

The combustion of blocks of methane hydrate give the appearance of burning ice.

SUBSTITUTION REACTIONS OF HALOALKANES

In haloalkanes, the large electronegativity difference between the halogen and carbon atoms means that the carbon–halogen bond is polar. For example, Figure 11.1.3 shows that in the chloromethane molecule, the carbon atom carries a partial positive charge and the more electronegative chorine atom carries a partial negative charge. This makes the bond weaker than one without a dipole.

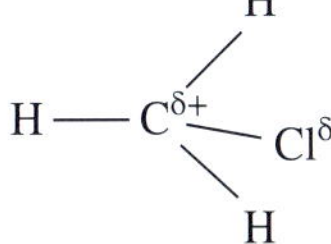

FIGURE 11.1.3 Structural formula of chloromethane. The electronegativity of carbon is 2.6 and that of chlorine is 3.2. This large difference in electronegativity results in a polar carbon–chlorine bond.

As a consequence of the highly polarised bond between carbon and chlorine, the partial positive charge on the carbon atom can be attacked by a negatively charged species such as a hydroxide ion, as found in a solution of sodium hydroxide. The new organic product is methanol. Because the chlorine atom of the chloroalkane has been 'swapped' for a hydroxyl group, this is a substitution reaction. Figure 11.1.4 on the following page shows the substitution reaction between chloromethane and hydroxide to form methanol.

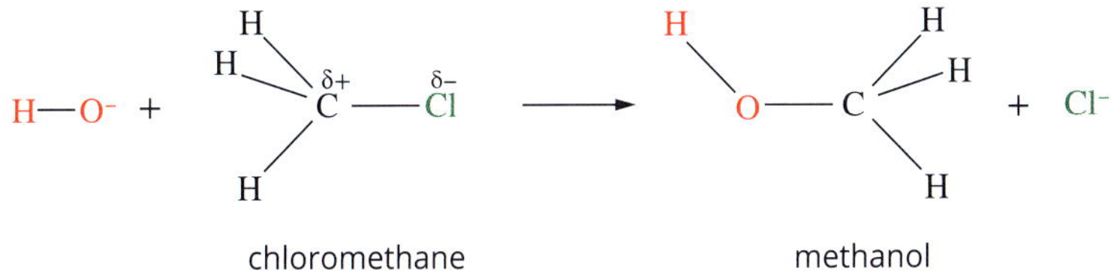

$$\text{H—O}^- + \text{H}_3\text{C}^{\delta+}\text{—Cl}^{\delta-} \longrightarrow \text{H—O—CH}_3 + \text{Cl}^-$$

chloromethane → methanol

FIGURE 11.1.4 Reaction of a chloromethane molecule with a hydroxide ion to form methanol via a substitution reaction

Ammonia also reacts with haloalkanes in substitution reactions. The product of these reactions is an amine.

Figure 11.1.5 shows the reaction of chloroethane with ammonia. The substitution reaction does not require a catalyst and produces ethanamine as the product. The production of propan-2-amine by reaction of 2-chloropropane with ammonia is also shown.

Other haloalkanes containing fluorine, bromine and iodine also undergo substitution reactions with hydroxide ions or strong bases and ammonia to produce the corresponding alcohols and amines.

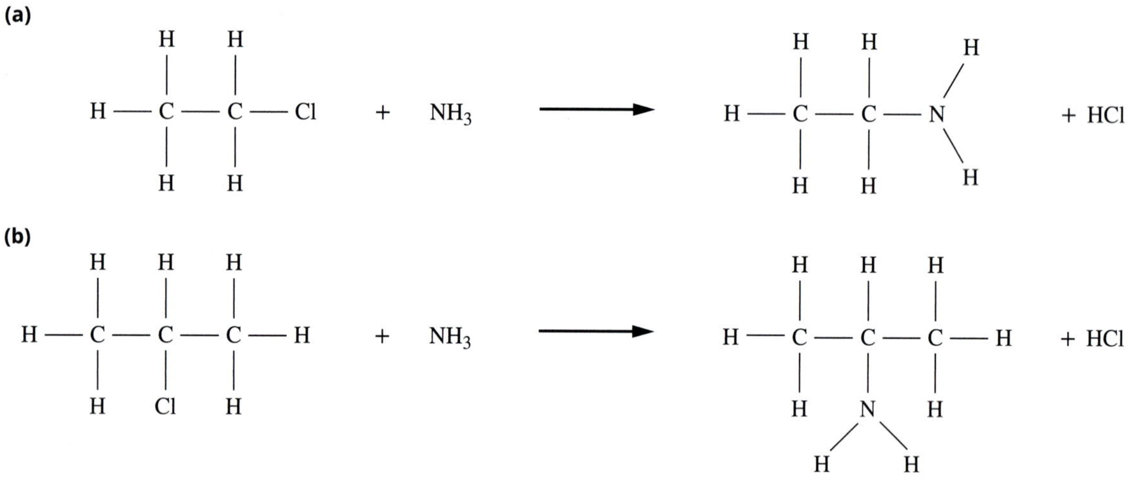

(a)

$$CH_3CH_2Cl + NH_3 \rightarrow CH_3CH_2NH_2 + HCl$$

(b)

$$CH_3CHClCH_3 + NH_3 \rightarrow CH_3CH(NH_2)CH_3 + HCl$$

FIGURE 11.1.5 (a) Reaction of chloroethane with ammonia produces ethanamine. (b) Reaction of 2-chloropropane with ammonia produces propan-2-amine.

CHEMFILE

The role of free radicals in the reaction of alkanes with chlorine

The first stage in a substitution reaction of an alkane with chlorine is the formation of a chlorine atom 'free radical'. Ultraviolet light provides the energy for this step to occur, as shown in the figure at right. A free radical is a species containing an atom with an unpaired electron in its outermost shell. A free radical is very reactive and will readily combine with other species to form a covalent bond by pairing its single electron with an electron from another species. Note that the movement of a single electron is shown by a half-arrow, or harpoon arrow.

In the second stage of the reaction between chlorine and methane shown, a hydrogen chloride molecule forms when the chlorine free radical from step one combines with a hydrogen atom from the methane molecule. This also results in the formation of a free radical with the formula CH_3.

In the third stage of the process shown in the figure, the CH_3 free radical combines with a chlorine free radical. This results in the formation of a CH_3Cl molecule.

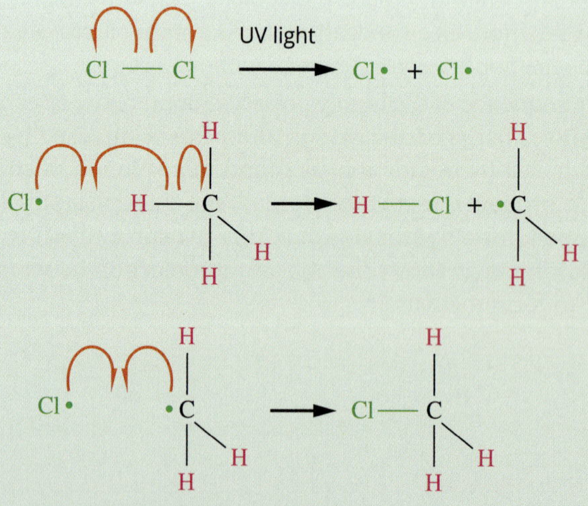

UV light

$$Cl-Cl \xrightarrow{\text{UV light}} Cl\bullet + Cl\bullet$$

$$Cl\bullet + H-CH_3 \rightarrow H-Cl + \bullet CH_3$$

$$Cl\bullet + \bullet CH_3 \rightarrow Cl-CH_3$$

Stages in the reaction of methane with chlorine. Ultraviolet light is used to initiate this reaction.

11.1 Review

SUMMARY

- Alkanes burn in air to produce water and carbon dioxide.
- Alkanes are generally unreactive, but they can undergo substitution reactions with halogens in the presence of ultraviolet light to produce haloalkanes.
- Haloalkanes undergo substitution reactions with:
  - sodium hydroxide to produce alcohols
  - ammonia to produce amines.

KEY QUESTIONS

Knowledge and understanding

1 Complete the following sentences by filling in the gaps using words selected from the following list:

substitution; infrared; addition; carbon monoxide; ultraviolet light; carbon dioxide; methanol; ethanol; bromomethane; water

Alkanes undergo complete combustion in air to produce ______________ and ____________. The first organic product formed when methane reacts with bromine in the presence of ____________ is ____________. This is an example of a ____________ reaction. The organic product formed when chloromethane reacts with sodium hydroxide is __________. This is also an example of a ________ reaction.

2 Write a balanced equation for the combustion of butane in air.

3 Name the carbon-containing products formed when each of the following pairs of compounds react.

a 2-chlorobutane and sodium hydroxide

b bromoethane and ammonia

c butane and oxygen in a combustion reaction

4 Write balanced semi-structural equations for the production of:

a methanol from chloromethane

b propan-2-amine from 2-bromopropane

c propan-1-ol from 1-chloropropane.

Analysis

5 The table below summarises the types of reactions of alkanes and haloalkanes introduced in this section. However, some of the relevant information is missing. Complete the table by adding the relevant information to the empty spaces.

Homologous series	Reactant	Type of organic product(s)	Type of reaction
alkane		carbon dioxide and water	
haloalkane	NaOH		
haloalkane	ammonia		substitution
alkane		chloroalkane	

6 A scientist wanted to make a sample of dichloromethane using a substitution reaction. Describe the reactants and the reaction conditions that the scientist must use to make their desired product.

11.2 Reactions of alkenes

FIGURE 11.2.1 Chloroethane (also known as ethyl chloride) is used as a local anaesthetic for numbing small wounds or sporting injuries.

The reactions of an organic compound are largely determined by the functional groups in the compound. As all members of a homologous series have the same functional group, the members of that group typically undergo reactions of the same type.

Chemists who are involved in the synthesis of organic compounds are often required to prepare complex molecules from simpler starting materials. These chemists must have detailed knowledge of the reactions of functional groups to devise a way to produce the desired product.

For example, alkenes are a homologous series of compounds that contain a carbon–carbon double bond. This bond is quite reactive. Chloroethane is produced by the reaction of ethene with HCl. It acts as a mild topical anaesthetic by its chilling effect when sprayed on skin, such as when removing splinters or on a sporting injury, as shown in Figure 11.2.1.

In this section, you will learn about some of the typical reactions of alkenes and the reaction conditions required.

REACTIVITY OF ALKENES

Alkenes are generally more reactive compounds than alkanes. The double carbon–carbon covalent bond in alkene molecules has a significant effect on their chemical properties. Ethene (CH_2CH_2), for example, reacts more readily and with more chemicals than ethane, (CH_3CH_3), which contains only single bonds.

Combustion in air

You have previously learnt that alkanes are good fuels and burn in oxygen to produce carbon dioxide and water. Alkenes also burn in an excess of oxygen to produce carbon dioxide and water. The equation for the combustion of ethene is:

$$C_2H_4(g) + 3O_2(g) \rightarrow 2CO_2(g) + 2H_2O(g)$$

ADDITION REACTIONS OF ALKENES

The reactions of alkenes usually involve the addition of a small molecule to the double bond of the alkene. These reactions are called **addition reactions**.

During addition reactions:

- two reactant molecules combine to form one product molecule
- the carbon–carbon double bond becomes a single bond
- an unsaturated compound becomes saturated
- the atoms of the small molecule adding to the alkene are 'added across the double bond', so that one atom or group from the molecule forms a bond to each of the carbon atoms in the double bond of the alkene. This can be seen in the reaction of hydrogen bromide with ethene in Figure 11.2.2.

ℹ In addition to reactions of alkenes, two reactant molecules combine to form one product molecule. The carbon–carbon double bond becomes a carbon–carbon single bond.

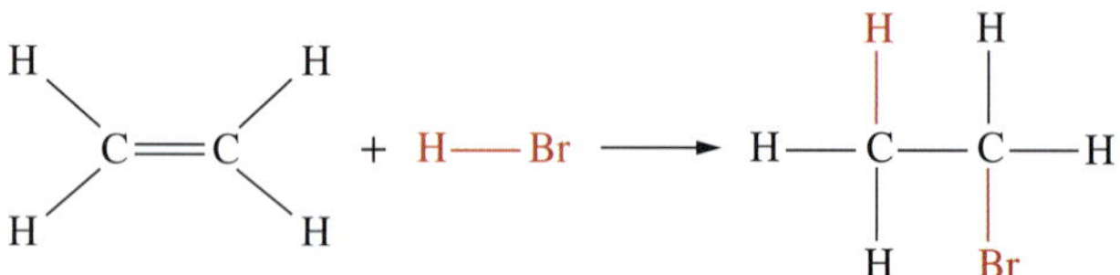

FIGURE 11.2.2 The reaction between ethene and hydrogen bromide. This is an example of an addition reaction between an alkene and a hydrogen halide.

When ethene reacts with hydrogen bromide, an addition reaction occurs and bromoethane is the only product. During this reaction, the hydrogen atom from the HBr molecule forms a covalent bond to one carbon atom in the double bond of ethene and the bromine atom forms a bond to the other carbon atom. Unlike substitution reactions, there is no **inorganic** product formed. All the atoms in the reactants end up in the final product. Four different types of addition reactions of alkenes are discussed below.

Reaction of alkenes with hydrogen

Alkenes react with hydrogen gas in the presence of a metal catalyst such as nickel to form alkanes. This reaction is an example of an addition reaction and is also known as a **hydrogenation** reaction. During this reaction an unsaturated hydrocarbon is converted to a saturated one. The reaction shown in Figure 11.2.3 is hydrogenation of ethene with hydrogen gas to produce ethane.

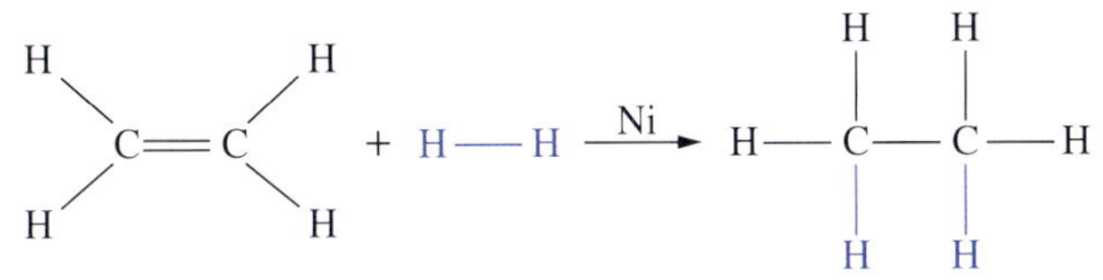

$$H_2C{=}CH_2 + H{-}H \xrightarrow{Ni} H{-}CH_2{-}CH_2{-}H$$

FIGURE 11.2.3 Addition reaction of ethene with hydrogen, also known as hydrogenation, to form the corresponding saturated alkane, ethane

The activation energy for this reaction is too high for the reaction to proceed at room temperature without a catalyst.

Reaction of alkenes with halogens

Figure 11.2.4 shows the reaction of ethene with bromine to form 1,2-dibromoethane. The halogen 'adds' across the double bond of the molecule, so in the product there is one bromine atom attached to each carbon atom.

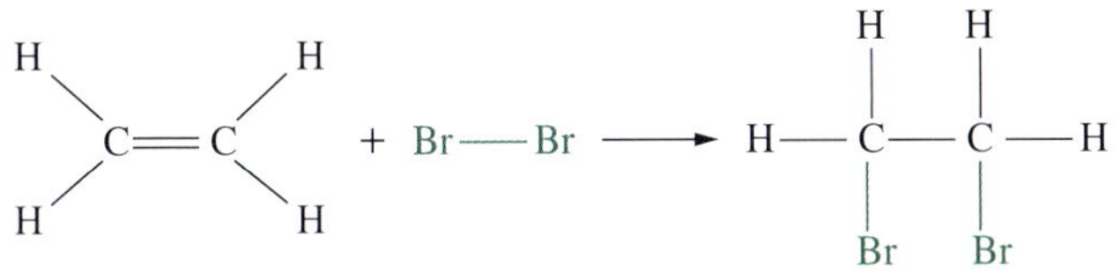

$$H_2C{=}CH_2 + Br{-}Br \longrightarrow H{-}CH(Br){-}CH(Br){-}H$$

FIGURE 11.2.4 Addition reaction of ethene with bromine

This reaction proceeds at room temperature without a catalyst. Other halogens, such as Cl_2 and I_2, also undergo addition reactions with alkenes to form the corresponding di-substituted haloalkanes.

Reaction with bromine can be used as a test to distinguish between alkanes and alkenes. Although, as you learnt in Section 11.1, alkanes will also react with bromine, they will only do so in the presence of UV light, and the reaction is slow. However, the red-orange colour of the bromine quickly disappears when it is mixed with an alkene (Figure 11.2.5). Bromine is therefore often used as a test for the presence of a carbon–carbon double bond. This test is discussed further in Section 13.1.

FIGURE 11.2.5 When an alkene is shaken with a solution of aqueous bromine, the alkene reacts rapidly with the coloured bromine and the solution loses colour.

Reaction of alkenes with hydrogen halides

Just as you saw with the reaction of ethene and hydrogen bromide earlier, but-2-ene reacts with hydrogen chloride, a **hydrogen halide**, in an addition reaction. Figure 11.2.6 shows that the reaction of but-2-ene with hydrogen chloride produces a single product, 2-chlorobutane. In this reaction, a hydrogen atom adds to one of the carbon atoms in the carbon–carbon double bond and a halogen atom adds to the other carbon atom.

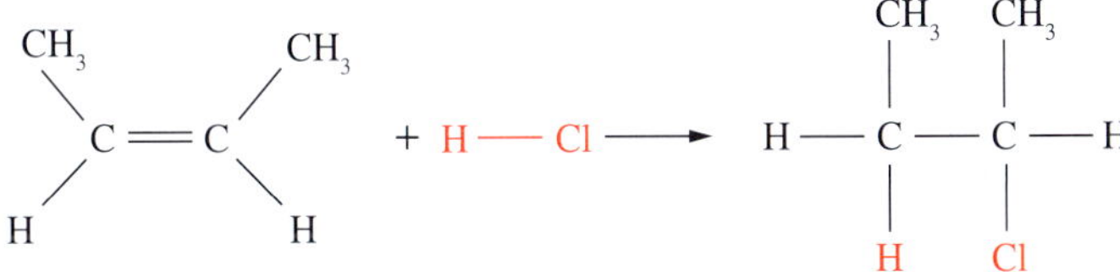

FIGURE 11.2.6 The addition reaction of but-2-ene with hydrogen chloride produces 2-chlorobutane

CHEMFILE

Ethanol, the versatile alcohol

Ethanol made from ethene has a variety of uses because of its different properties. Note that the ethanol in alcoholic drinks is made by the fermentation of plant products such as grapes (for wine) and barley (for beer).

Ethanol as a fuel

Ethanol is flammable and burns in air to release a lot of heat energy. In Australia, E10 is a motor fuel that contains 10% ethanol mixed with petrol. Some camping stoves use methylated spirits as a fuel. Methylated spirits contains about 95% ethanol.

Ethanol as a solvent

Many paints, varnishes and lacquers contain ingredients dissolved in ethanol.

Ethanol as a disinfectant

Ethanol has antibacterial properties and is used in mouth washes, hand sanitisers, and industrial and domestic disinfectants (see the figure below).

Ethanol as a preservative

Ethanol is used in perfumes, lotions and cough syrup as a solvent. It also acts as a preservative because of its antibacterial properties.

Ethanol is used in hand sanitisers because of its anti-microbial and disinfectant properties.

However, as Figure 11.2.7 shows, when you react hydrogen chloride and but-1-ene, two different products are possible.

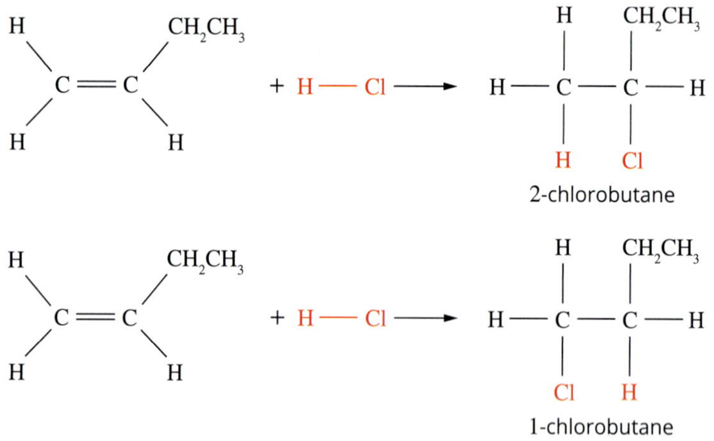

$$H_2C{=}CH{-}CH_2CH_3 + H{-}Cl \longrightarrow CH_3{-}CHCl{-}CH_2CH_3$$

2-chlorobutane

$$H_2C{=}CH{-}CH_2CH_3 + H{-}Cl \longrightarrow CH_2Cl{-}CH_2{-}CH_2CH_3$$

1-chlorobutane

FIGURE 11.2.7 Addition reaction of but-1-ene with hydrogen chloride. Two isomers are possible as products.

The addition reaction can produce two **isomers**. In the first product shown in Figure 11.2.7, the hydrogen atom from the hydrogen chloride molecule has been added to the carbon atom in the carbon–carbon double bond at the end of the but-1-ene molecule (C_1). In the other product, the hydrogen atom has been added to the carbon atom at the other end of the carbon–carbon double bond (C_2).

When you react an asymmetrical alkene such as but-1-ene with an asymmetrical reactant, isomers are produced. In Figure 11.2.7, more 2-chlorobutane is usually produced than 1-chlorobutane. The reasons for this are beyond the scope of this course.

Reaction of alkenes with water

Alkenes react with water under specific conditions to form the corresponding alcohol. For example, ethanol can be produced by an addition reaction of ethene and water, using a catalyst and heat to increase the rate of the reaction. Figure 11.2.8 shows the addition reaction of steam and ethene, using a phosphoric acid catalyst. This reaction is used extensively in industry for the production of ethanol.

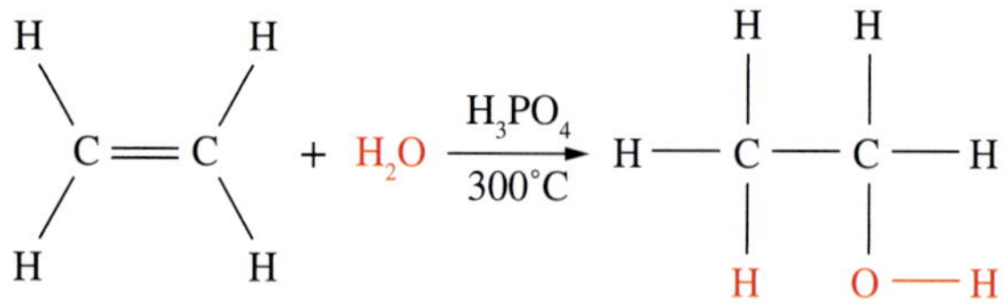

$$H_2C{=}CH_2 + H_2O \xrightarrow[300^\circ C]{H_3PO_4} CH_3{-}CH_2{-}O{-}H$$

FIGURE 11.2.8 Addition reaction of ethene with water in the presence of a phosphoric acid catalyst

The reaction is carried out at 300°C. The gaseous reactants are passed over a solid bed of the catalyst, and gaseous ethanol is formed.

The addition reaction of ethene with steam is often described as a **hydration reaction**. In this example, water is added across the double bond of the ethene molecule. The reaction is used for the commercial manufacture of ethanol because it is a one-step process that uses little energy, apart from initial heating. The heterogeneous (solid–gas) nature of the reaction system means it is easy for manufacturers to remove the product from the reaction mixture, leaving the catalyst intact.

11.2 Review

SUMMARY

- Alkenes burn in the presence of oxygen to produce carbon dioxide and water.
- Alkenes are unsaturated hydrocarbons. They undergo addition reactions to produce saturated compounds.
- Alkenes undergo addition reactions with:
  - hydrogen and a nickel catalyst to produce alkanes
  - halogens to produce dihaloalkanes
  - hydrogen halides to produce haloalkanes
  - water and a phosphoric acid catalyst at high temperatures to produce alcohols.

KEY QUESTIONS

Knowledge and understanding

1 Complete the following sentences by filling in the gaps using words selected from the following list:

addition; carbon dioxide; oxygen; 2,3-dibromobutane; 1,2-dibromobutane; bromobutane; water; chloroethane; dichloroethene; substitution

Alkenes will undergo combustion reactions in air to produce ______________ and ___________. The product formed when but-2-ene reacts with bromine is __________________. This is an example of a/an _______________ reaction.

The product formed when ethene reacts with hydrogen chloride is ____________. This is an example of a/an ___________ reaction.

2 Use structural formulas to write equations and name the products for the reactions of:

a but-1-ene with chlorine

b but-2-ene with hydrogen bromide.

3 Write a semi-structural equation for the reaction of hex-3-ene with hydrogen chloride.

Analysis

4 The table below summarises the types of reactions of alkenes introduced in this section. However, some of the relevant information is missing. Complete the table by adding the relevant information to the empty spaces. The first row of the table has been completed for you.

Homologous series	Reactant	Type of organic product(s)	Type of reaction
alkene	halogen	dihaloalkane	addition
alkene	hydrogen gas, H_2 + Ni or Pt catalyst		
alkene		$CO_2 + H_2O$	
alkene		alcohol	
alkene	hydrogen halide		

5 Name the possible products that could be formed in each reaction when the following two alkenes are reacted with water and a H_3PO_4 catalyst at 300°C.

a propene

b but-2-ene

Explain any differences in your answers.

6 Which one of the following reactions could result in the formation of two different organic products?

A hex-3-ene and water with a catalyst

B propene and chlorine

C pent-2-ene and HCl

D oct-4-ene and water with a catalyst

11.3 Reactions of alcohols, carboxylic acids and esters

As you learnt in Section 11.2, alcohols can be produced by addition reactions of alkenes. Ethanol is the most widely used alcohol in society. You may have seen E10 fuel being sold at your local service station (Figure 11.3.1). The E10 classification indicates that the petrol contains 10% ethanol. In Brazil, the use of ethanol fuels is widespread. Brazil is the biggest grower of sugarcane in the world. Sugar from sugarcane can be fermented to produce ethanol. The Brazilian government sets an 'ethanol mandate', which specifies the minimum amount of ethanol that fuels for cars must contain. In 2019 this was set at 27%. Some cars have been developed to run on 100% ethanol.

FIGURE 11.3.1 E10 petrol contains 10% ethanol.

In this section, you will examine some of the common reactions that alcohols and other types of organic compounds undergo.

REACTIONS OF ALCOHOLS

Alcohol molecules contain the hydroxyl functional group. This makes them generally more reactive and enables them to undergo a greater variety of reactions than alkane molecules.

Combustion of alcohols

Just like alkanes and alkenes, alcohols burn readily in air to form carbon dioxide and water as products. The equation for the combustion of ethanol is:

$$C_2H_5OH(l) + 3O_2(g) \longrightarrow 2CO_2(g) + 3H_2O(g)$$

This is a highly exothermic reaction, so ethanol may be used as a fuel. On a smaller scale, as noted in Section 11.2, methylated spirits, which contains about 95% ethanol mixed with other chemicals, is sometimes used as a fuel for camping stoves (Figure 11.3.2). Some cooking techniques even make use of the burning of alcohol for drama and flavour.

FIGURE 11.3.2 Methylated spirits is used as a fuel in some camping stoves.

Oxidation of alcohols

The combustion of alcohols and other fuels can be classified as a type of oxidation–reduction (redox) reaction. Alcohols can also be oxidised by strong inorganic oxidising agents, such as acidic solutions of potassium dichromate ($K_2Cr_2O_7$) and potassium permanganate ($KMnO_4$). The products of these oxidation reactions depend on the type of alcohol involved.

Alcohols can be classified into three types—primary, secondary or tertiary—depending on the position of the hydroxyl group within the molecule.

- Primary alcohols include methanol and alcohols in which the –OH group is bonded to a carbon atom that is attached to only one alkyl group. These alcohols first oxidise to form aldehydes and then further oxidation forms the corresponding carboxylic acids.
- Secondary alcohols are alcohols in which the –OH group is bonded to a carbon atom that is bonded to two alkyl groups. These alcohols oxidise to form ketones.
- Tertiary alcohols are alcohols in which the –OH group is bonded to a carbon atom that is bonded to three alkyl groups. These alcohols are resistant to oxidation by inorganic oxidising agents and will not normally react. A study of the chemistry of tertiary alcohols is not included in this chapter.

You can see in Figure 11.3.3 how the different isomers of butanol can be classified as primary, secondary or tertiary alcohols. Chemists often use the symbols 1°, 2° and 3° as a shorthand way of indicating whether an alcohol is primary, secondary or tertiary respectively.

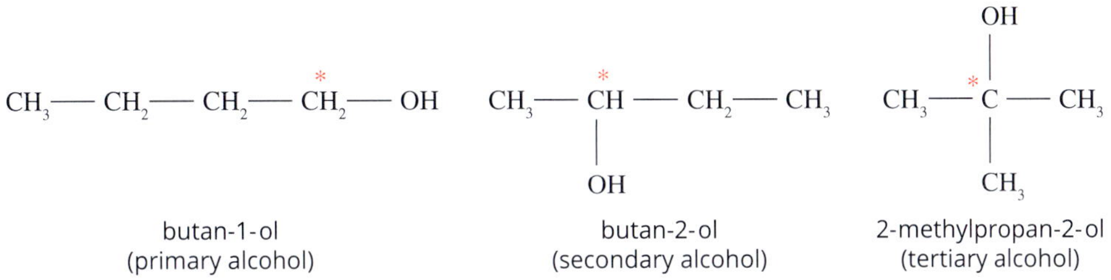

$CH_3-CH_2-CH_2-\overset{*}{C}H_2-OH$ butan-1-ol (primary alcohol)

$CH_3-\overset{*}{C}H(OH)-CH_2-CH_3$ butan-2-ol (secondary alcohol)

$CH_3-\overset{*}{C}(OH)(CH_3)-CH_3$ 2-methylpropan-2-ol (tertiary alcohol)

FIGURE 11.3.3 Structures of the isomers of butanol. The asterisk identifies the carbon atom that determines the primary, secondary or tertiary nature of the alcohol.

Oxidation of primary alcohols

Primary alcohols can be oxidised to carboxylic acids in two stages.

In the first stage, the primary alcohol is oxidised to an aldehyde. In the second stage, further heating of the reaction mixture, in the presence of the oxidising agent, results in the oxidation of the aldehyde to a carboxylic acid.

The example in Figure 11.3.4 shows that propan-1-ol is oxidised first to propanal (an aldehyde) and then to propanoic acid (a carboxylic acid).

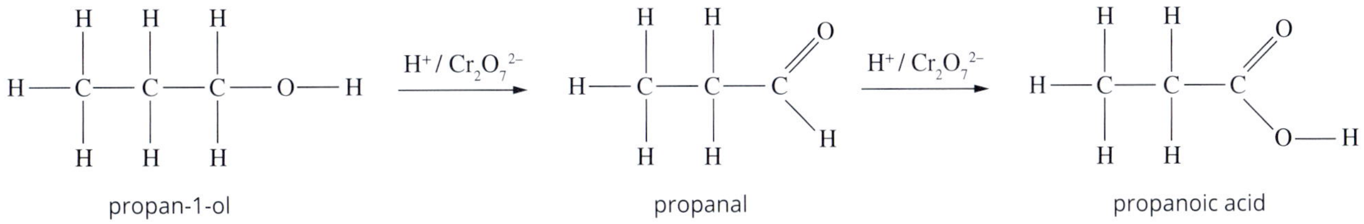

$$CH_3CH_2CH_2OH \xrightarrow{H^+/Cr_2O_7^{2-}} CH_3CH_2CHO \xrightarrow{H^+/Cr_2O_7^{2-}} CH_3CH_2COOH$$

propan-1-ol → propanal → propanoic acid

FIGURE 11.3.4 Equations for the oxidation of the primary alcohol propan-1-ol

Although aldehydes are intermediates in the oxidation of primary alcohols to carboxylic acids, their chemistry is not studied in this course. The oxidation of a primary alcohol can be represented by an equation without the aldehyde intermediate, as in Figure 11.3.5.

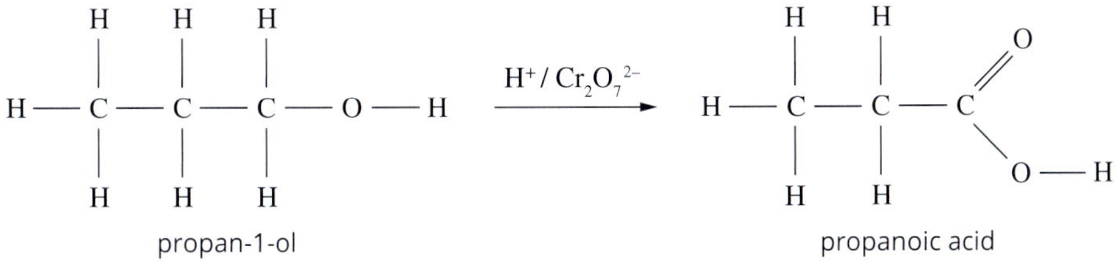

$$CH_3CH_2CH_2OH \xrightarrow{H^+/Cr_2O_7^{2-}} CH_3CH_2COOH$$

propan-1-ol → propanoic acid

FIGURE 11.3.5 Equation for the oxidation of the primary alcohol propan-1-ol to propanoic acid. The intermediate stage in which propanal is formed has been omitted from this equation.

> Primary alcohols are oxidised to carboxylic acids by strong oxidising agents such as acidified potassium dichromate and acidified potassium permanganate.

In reactions of organic compounds with inorganic reactants, it is common to use unbalanced equations. In these equations, the formulas of the inorganic reactants are written above the arrow and only the organic products are shown. Figure 11.3.6 illustrates this convention.

Sometimes **general equations** are used to represent reactions of a particular homologous series. In such equations, the letter R is used to represent an alkyl group. Figure 11.3.6 shows a general, unbalanced equation for the oxidation of primary alcohols.

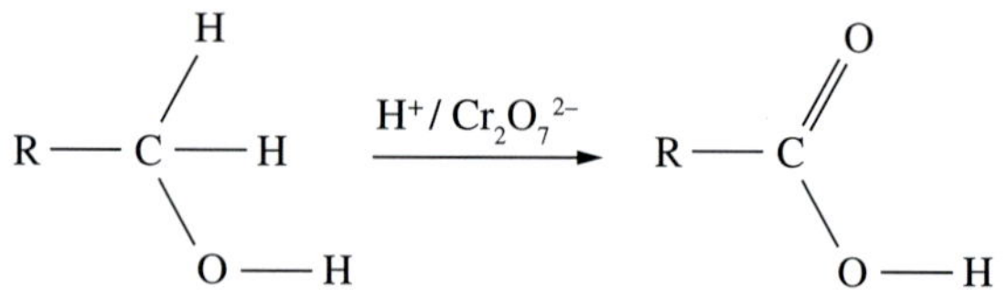

$$\mathrm{R{-}CH_2{-}OH} \xrightarrow{H^+/Cr_2O_7^{2-}} \mathrm{R{-}C({=}O){-}OH}$$

primary alcohol → carboxylic acid

FIGURE 11.3.6 General equation for the oxidation of a primary alcohol to a carboxylic acid

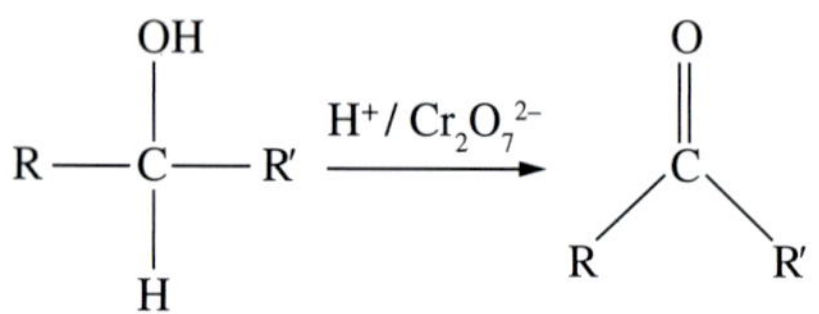

FIGURE 11.3.7 The general equation for the oxidation of a secondary alcohol. The R groups in the structures represent alkyl groups.

Oxidation of secondary alcohols

When secondary alcohols are oxidised by strong oxidising agents, such as solutions of acidified potassium dichromate ($K_2Cr_2O_7$) or potassium permanganate ($KMnO_4$), the corresponding ketones are produced. Figure 11.3.7 gives the general equation for the oxidation of secondary alcohols to ketones.

Figure 11.3.8 gives the equation for the production of propan-2-one, a ketone, by the oxidation of the secondary alcohol, propan-2-ol.

$$\mathrm{CH_3{-}CH(OH){-}CH_3} \xrightarrow{H^+/Cr_2O_7^{2-} \text{ or } H^+/MnO_4^-} \mathrm{CH_3{-}C({=}O){-}CH_3}$$

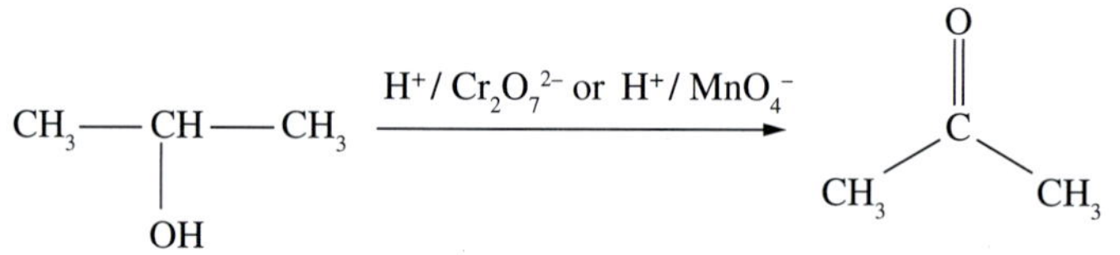

FIGURE 11.3.8 The secondary alcohol propan-2-ol can be oxidised to propan-2-one.

Tertiary alcohols and oxidising agents

Tertiary alcohols are resistant to reaction with solutions of acidified potassium dichromate ($K_2Cr_2O_7$) or potassium permanganate ($KMnO_4$).

During the oxidation of primary and secondary alcohols, the C–O single bond of the hydroxyl group is converted into a C=O double bond. Simultaneously (at the same carbon atom), there is a decrease in the number of C–H bonds. In tertiary alcohols, however, there is no C–H bond attached to the same carbon atom as the hydroxyl group, so oxidation cannot occur at this carbon atom. As a result, tertiary alcohols are resistant to oxidation by strong oxidising agents.

COLOUR CHANGES IN OXIDATION REACTIONS

The strong oxidising agents potassium dichromate and potassium permanganate are both highly coloured because of the presence of the transition metal elements chromium (Cr) and manganese (Mn).

A solution of dichromate ions in water is orange in colour. When acidified dichromate solution is used to oxidise a primary or secondary alcohol, the dichromate ion ($Cr_2O_7^{2-}$) is reduced to the chromium ion (Cr^{3+}), which is green in colour. This colour change from orange to green can be used as a qualitative test to indicate that oxidation of an organic compound has taken place. The colour changes observed for the reaction of primary and secondary alcohols with potassium dichromate are shown in Figure 11.3.9.

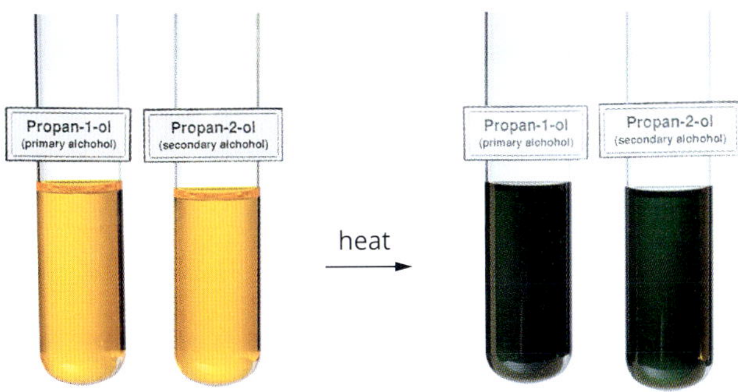

FIGURE 11.3.9 The strong oxidising agent potassium dichromate is orange in colour. On heating with primary and secondary alcohols, the colour changes to green.

A solution of acidified potassium permanganate is a deep purple colour. When this solution reacts with primary or secondary alcohols, the manganese in the permanganate ion (MnO_4^-) is reduced to Mn^{2+}, which is colourless (Figure 11.3.10). You will learn about the use of potassium dichromate and potassium permanganate in redox titrations in Chapter 13. The colour changes that occur when primary and secondary alcohols are oxidised with acidified dichromate or permanganate ions are summarised in Table 11.3.1.

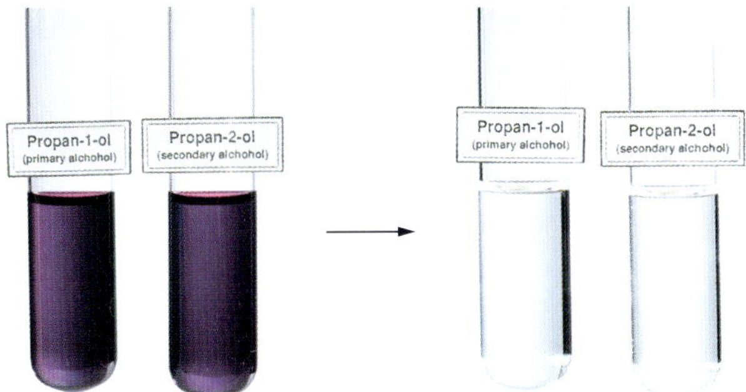

FIGURE 11.3.10 The strong oxidising agent potassium permanganate is deep purple in colour. On reaction with primary and secondary alcohols, the colour changes to colourless.

TABLE 11.3.1 Summary of the oxidation reactions of primary and secondary alcohols with acidified dichromate solution or acidified permanganate solution

Type of alcohol	Products	Colour change of acidified potassium dichromate solution	Colour change of acidified potassium permanganate solution
primary (1°)	aldehydes (under mild conditions); carboxylic acids (at higher temperatures, with longer reaction times)	orange to green	purple to colourless
secondary (2°)	ketones	orange to green	purple to colourless

FIGURE 11.3.11 The ripening of blackberries includes many chemical processes, such as the reaction of carboxylic acids with alcohols to produce esters.

Carboxylic acids only ionise partially in water because the equilibrium constant for the reaction is small.

In organic chemistry, when two molecules react to form a larger molecule and a smaller molecule such as water is released, the reaction is called a condensation reaction.

REACTIONS OF CARBOXYLIC ACIDS

Molecules that contain carboxyl groups are common in the natural world and are present in most plants. Solutions of carboxylic acids taste sour. The sour taste of vinegar, lemons, yoghurt, rhubarb and most unripe fruits is due to the presence of carboxylic acids.

When fruits and berries like the blackberries in Figure 11.3.11 ripen, complex reactions take place. These include the conversion of carboxylic acids into other compounds. In some of these reactions, carboxylic acids react with alcohol molecules to produce esters. Esters give many fruits their characteristic aromas and tastes.

Ionisation in water

Ethanoic acid is a weak acid and only ionises to a small extent in water to form hydronium ions and ethanoate ions. Other carboxylic acids react with water in a similar way.

The reaction of a carboxylic acid with water is a reversible process, so the equation for ionisation is written using equilibrium arrows. The equation for the ionisation of ethanoic acid in water is:

$$CH_3COOH(aq) + H_2O(l) \rightleftharpoons CH_3COO^-(aq) + H_3O^+(aq)$$

REACTIONS OF CARBOXYLIC ACIDS WITH ALCOHOLS

Reactions that involve the combination of two reactants and the elimination of a small molecule, such as water, are called **condensation reactions**. Esters are made by a condensation reaction between a carboxylic acid and an alcohol. A condensation reaction in which an ester is formed is also known as an **esterification reaction**.

For example, the ester ethyl ethanoate can be produced by gently heating a mixture of ethanol and pure ethanoic acid with a trace amount of sulfuric acid. The sulfuric acid acts as a catalyst. As well as the desired ester, water is a product.

The general equation for the esterification reaction involving a carboxylic acid and an alcohol is shown in Figure 11.3.12. Two examples of esterification reactions—ethanoic acid with ethanol and propanoic acid with methanol—are also shown.

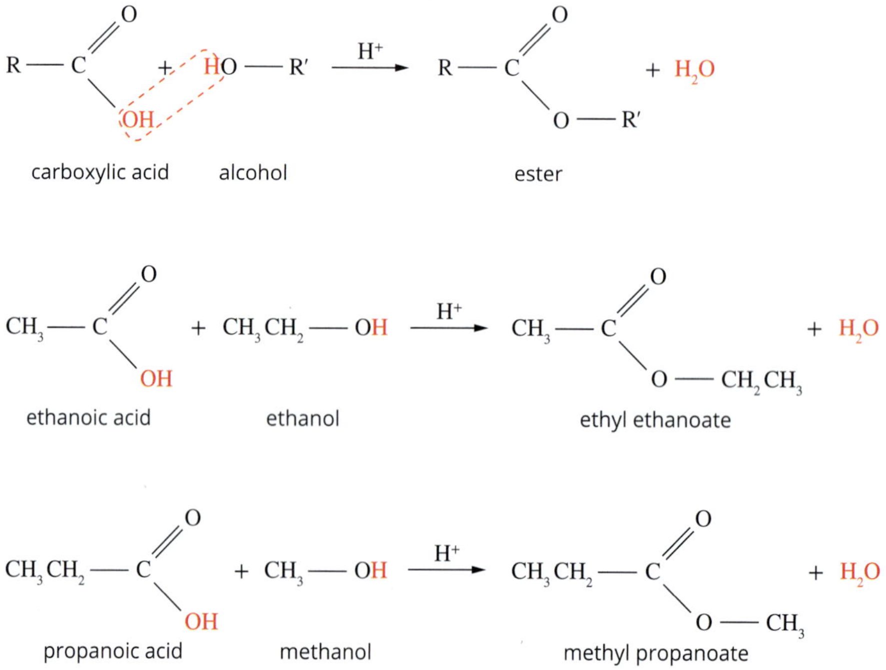

$$R-C(=O)OH + HO-R' \xrightarrow{H^+} R-C(=O)O-R' + H_2O$$

carboxylic acid, alcohol, ester

$$CH_3-C(=O)OH + CH_3CH_2-OH \xrightarrow{H^+} CH_3-C(=O)O-CH_2CH_3 + H_2O$$

ethanoic acid, ethanol, ethyl ethanoate

$$CH_3CH_2-C(=O)OH + CH_3-OH \xrightarrow{H^+} CH_3CH_2-C(=O)O-CH_3 + H_2O$$

propanoic acid, methanol, methyl propanoate

FIGURE 11.3.12 Esterification reactions occur when alcohols are heated with carboxylic acids and a small amount of sulfuric acid as a catalyst. The top reaction shows the general equation for esterification. Two specific examples are also shown: ethanoic acid with ethanol and propanoic acid with methanol.

In the esterification reaction, it is the hydrogen atom from the hydroxyl group of the alcohol and the −OH group from the carboxylic acid that combine to form water, which is the molecule eliminated in this condensation reaction (see the first equation in Figure 11.3.12).

As you learnt in Chapter 10, the first part of the name of an ester is derived from the name of the alcohol from which it was made. The second part of the name is derived from the name of the carboxylic acid from which it was made. Therefore, the name of the ester formed from methanol (an alcohol) and propanoic acid (a carboxylic acid) is methyl propanoate (Figure 11.3.12).

Hydrolysis of esters

The condensation reaction between carboxylic acids and alcohols to form an ester and water is a reversible reaction. You can see in Figure 11.3.13 that esters can react with water to form a carboxylic acid and an alcohol. Reactions of this type are described as **hydrolytic reactions**, or simply **hydrolysis**. This reaction is catalysed by an alkali or dilute acid.

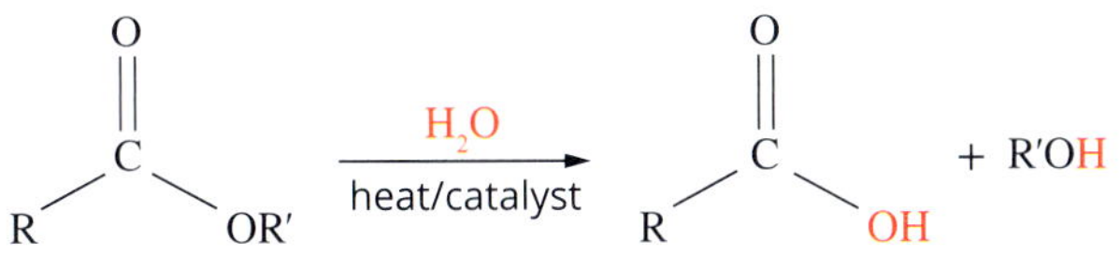

$$\mathrm{R{-}C({=}O){-}OR'} \xrightarrow[\text{heat/catalyst}]{H_2O} \mathrm{R{-}C({=}O){-}OH} + \mathrm{R'OH}$$

FIGURE 11.3.13 General equation for the hydrolysis of an ester

> When esters are hydrolysed in an acidic solution the products are an alcohol and a carboxylic acid.

Figure 11.3.14 shows the equation for the hydrolysis of ethyl propanoate to form ethanol and propanoic acid using an acid catalyst.

When the hydrolysis of an ester is catalysed by an alkali such as sodium hydroxide, the products are an alcohol and the sodium salt of the carboxylic acid. The sodium salt can be easily converted to the carboxylic acid by adding dilute acid solution, such as hydrochloric acid.

> Esters hydrolysed by metal hydroxides form a salt of the carboxylic acid. Addition of excess acid regenerates the protonated carboxylic acid.

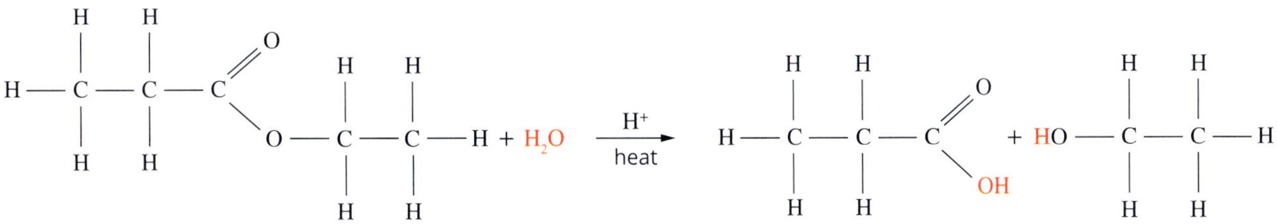

$$\mathrm{CH_3{-}CH_2{-}C({=}O){-}O{-}CH_2{-}CH_3} + H_2O \xrightarrow[\text{heat}]{H^+} \mathrm{CH_3{-}CH_2{-}C({=}O){-}OH} + \mathrm{HO{-}CH_2{-}CH_3}$$

FIGURE 11.3.14 Equation for the acid-catalysed hydrolysis of ethyl propanoate to produce propanoic acid and ethanol

PA 15

CHEMFILE

Aromas of some organic compounds

Many carboxylic acids with short hydrocarbon chains have unpleasant odours. The aroma of butanoic acid, for example, is similar to that of smelly feet and socks. Butanoic acid is a compound that contributes to the characteristic smell of human vomit.

However, when short chain carboxylic acids react with alcohols to produce esters, there is usually a remarkable transformation. Many of the esters produced in this way have very pleasant aromas. An example of this is when butanoic acid reacts with ethanol to produce ethyl butanoate. This ester smells like fresh pineapple.

Butanoic acid has an odour like smelly feet, but it will react with ethanol to form the ester ethyl butanoate. This ester has an aroma like that of pineapple.

CASE STUDY **ANALYSIS**

Making margarine

The raw materials used to make many margarines (Figure 11.3.15) include vegetable oils. Most vegetable oils are liquids and so cannot be spread on bread in the same way as butter.

Vegetable oils contain ester groups and long hydrocarbon chains that are polyunsaturated; that is, they contain a number of carbon–carbon double bonds. If some of these double bonds are converted to single bonds, the molecules can pack more closely, resulting in stronger dispersion forces between the molecules and higher melting points. The liquids become semisolids and the product is then suitable for spreading.

FIGURE 11.3.15 Margarines can be made from vegetable oils that have been reacted with hydrogen to convert the liquids to semisolids.

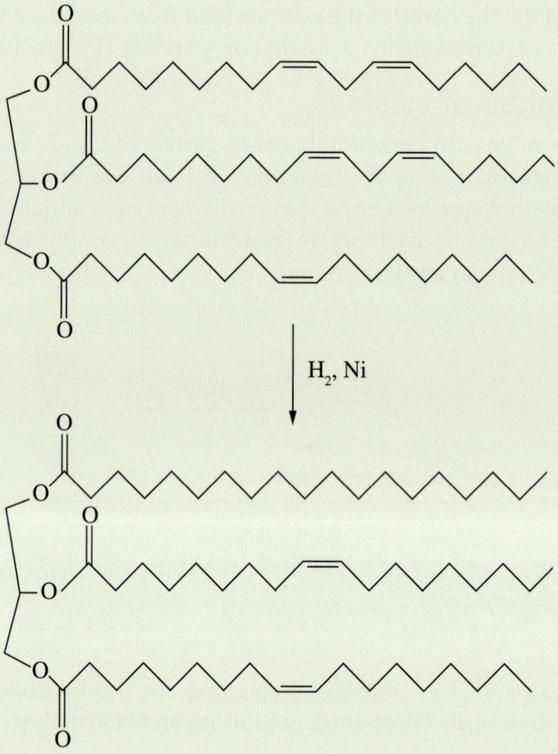

FIGURE 11.3.16 Hydrogenation of some of the carbon–carbon double bonds in a vegetable oil can convert it from a liquid into a 'spreadable' semisolid.

Traditionally, one of the steps in making margarine from a vegetable oil usually involved reacting vegetable oils with hydrogen gas, using a metal catalyst such as nickel. In this step, some carbon–carbon double bonds react with the hydrogen in an addition reaction, and are converted into single bonds, as shown in Figure 11.3.16.

Note that in the diagrams, the carbon and hydrogen atoms have been omitted for clarity. The zigzag lines represent saturated hydrocarbon chains, in which each carbon atom has two hydrogen atoms attached. Where a double bond is shown, each carbon atom in the double bond has only one hydrogen atom attached to it.

A number of scientific studies have shown that hydrogenation of vegetable oils can also produce compounds known as 'trans fats'. These compounds can contribute to problems associated with cardiac function. As a result, many countries have restricted or prohibited the manufacture of margarine by the hydrogenation of vegetable oils and alternative methods have been developed.

Analysis

The vegetable oil in Figure 11.3.16 was originally formed in an esterification reaction.

1 Redraw the diagram of the structure of vegetable oil and circle any ester links you can see.

2 How many carboxylic acid molecules were involved in the reaction?

3 Draw a structural formula of the alcohol molecule involved in the esterification reaction in which the original vegetable oil was made.

4 How many molecules of hydrogen gas have reacted with one molecule of the original vegetable oil in the reaction shown in Figure 11.3.16?

TRANSESTERIFICATION AND THE PRODUCTION OF BIODIESEL

In Chapter 2 you learnt that biodiesel can be made from vegetable oils.

Fats and vegetable oils are **triglycerides** with a molecular structure consisting of three hydrocarbon chains attached by ester functional groups to a backbone of three carbon atoms, as shown in Figure 11.3.17. You will study the chemistry of these compounds in detail in Chapter 12.

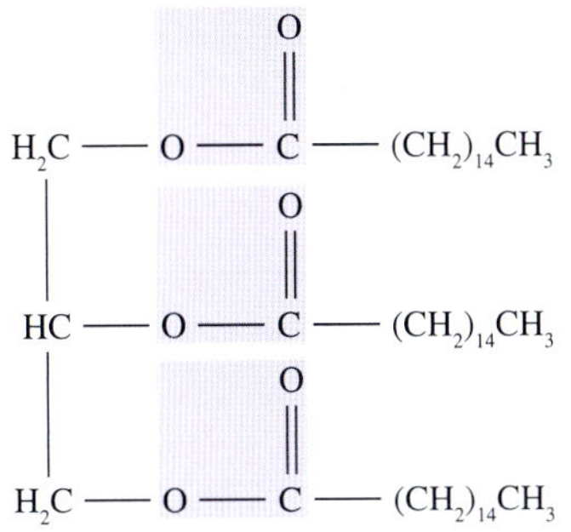

$$\begin{array}{l} H_2C-O-C(=O)-(CH_2)_{14}CH_3 \\ \;\;| \\ HC-O-C(=O)-(CH_2)_{14}CH_3 \\ \;\;| \\ H_2C-O-C(=O)-(CH_2)_{14}CH_3 \end{array}$$

FIGURE 11.3.17 The structure of a triglyceride molecule. There are three ester functional groups (shaded) in the molecule.

A triglyceride can be converted into biodiesel by warming it with an alcohol, usually methanol, in a process known as **transesterification**. Potassium hydroxide solution acts as a catalyst in this reaction.

A transesterification reaction is one that takes place between an alcohol and an ester. During a transesterification reaction, an alkyl group on the alcohol molecule changes places with the alkyl group on the part of the ester molecule that was derived from an alcohol. This process is illustrated in Figure 11.3.18.

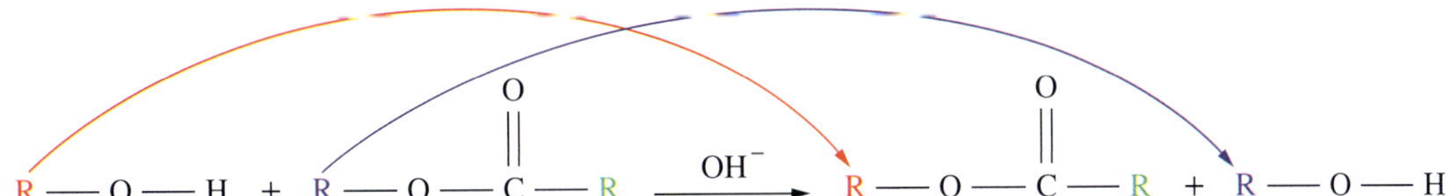

FIGURE 11.3.18 During a transesterification reaction, alkyl groups on the alcohol (in red) and the ester (in blue) 'swap places'. Note that it is the alkyl group attached to the oxygen atom in the ester that is involved in this process.

In the transesterification reaction in which biodiesel is produced, the triglyceride is converted, in the presence of a catalyst, such as potassium hydroxide, into a small molecule called glycerol and three ester molecules with long carbon chains. The ester molecules are the biodiesel product. The reaction is shown in Figure 11.3.19.

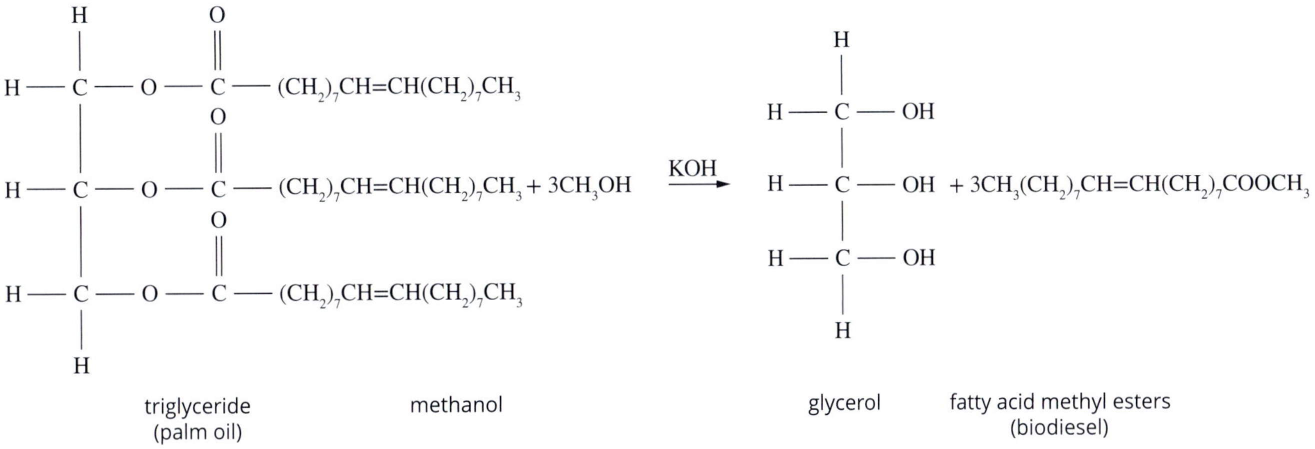

$$\begin{array}{l} H_2C-O-C(=O)-(CH_2)_7CH{=}CH(CH_2)_7CH_3 \\ HC-O-C(=O)-(CH_2)_7CH{=}CH(CH_2)_7CH_3 \\ H_2C-O-C(=O)-(CH_2)_7CH{=}CH(CH_2)_7CH_3 \end{array} + 3CH_3OH \xrightarrow{KOH} \begin{array}{l} H_2C-OH \\ HC-OH \\ H_2C-OH \end{array} + 3CH_3(CH_2)_7CH{=}CH(CH_2)_7COOCH_3$$

triglyceride (palm oil) | methanol | glycerol | fatty acid methyl esters (biodiesel)

FIGURE 11.3.19 The reaction of a triglyceride with methanol and a KOH catalyst to form fatty acid methyl esters (biodiesel) and glycerol

The structure of a typical biodiesel molecule is shown in Figure 11.3.20.

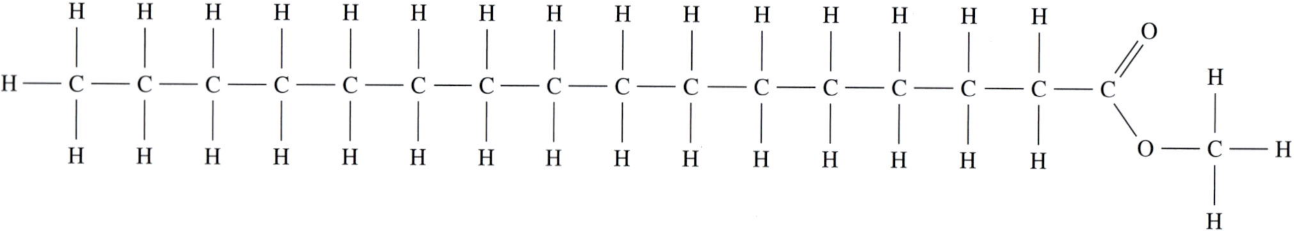

FIGURE 11.3.20 Structural formula of a typical biodiesel molecule

11.3 Review

SUMMARY

- Alcohols burn in air to form carbon dioxide and water.
- Alcohols react with oxidising agents such as acidified solutions of dichromate ions and permanganate ions in different ways:
  - Primary alcohols are oxidised first to aldehydes, which may be oxidised further to carboxylic acids.
  - Secondary alcohols are oxidised to ketones.
  - Tertiary alcohols are resistant to oxidation by oxidising agents.
- Colour changes that occur when primary and secondary alcohols are oxidised by acidified dichromate or permanganate ions can be used as qualitative or quantitative tests for the alcohol groups.
- Carboxylic acids can react with alcohols in the presence of an acid catalyst to produce esters.
- Carboxylic acids react with ammonia to form amides.
- Esters can be hydrolysed with water in the presence of an acid catalyst to produce carboxylic acids and alcohols.
- In a transesterification reaction, alkyl groups on the alcohol and ester molecules 'swap places'.
- Each of the different types of reactions you have studied in this section can be represented by chemical equations.
- Biodiesel is produced by the transesterification reaction of plant triglycerides, such as waste cooking oil, and alcohols, such as methanol or ethanol.

KEY QUESTIONS

Knowledge and understanding

1 State the name of the type of organic product(s) that will be formed in the following reactions.

  a a carboxylic acid and an alcohol reacting together using an acid catalyst

  b a primary alcohol reacting with an acidified solution of potassium permanganate

  c an ester reacting with water in an acidic solution

2 Pentan-3-ol and pentan-1-ol are isomers of $C_5H_{12}O$.

  a Write an equation for the complete combustion of $C_5H_{12}O$.

  b Identify each isomer as a primary, secondary or tertiary alcohol.

  c For pentan-1-ol, write the equation (using structural formulas and showing only organic products) for the reaction that would occur if the alcohol was heated with acidified potassium permanganate solution. Describe any colour changes that would be observed.

3 Complete the following sentences by filling in the blanks using words selected from the following list:

  condensation; propyl pentanoate; water; pentyl propanoate; hydrolysis

  When pentanoic acid reacts with propan-1-ol, the organic product called __________ is formed. The other product in this reaction is __________. This is an example of a/an __________ reaction.

4 Name and draw structural formulas of the organic products of each of the following reactions.

  a methyl butanoate and sodium hydroxide solution followed by the addition of dilute sulfuric acid

  b ethanol and propanoic acid, using concentrated sulfuric acid as a catalyst

Analysis

5 The table below summarises the types of reactions of the compounds introduced in this section. However, some of the relevant information is missing. Complete the table by adding the relevant information to the empty spaces.

Homologous series compound	Reactant(s)	Product(s)	Type of reaction
alcohol	air		combustion
primary alcohol	$H^+/Cr_2O_7^{2-}$ or H^+/MnO_4^- (with heating)		
alcohol		ester + water	
carboxylic acid	water	carboxylate ion + H_3O^+	
ester	water + acid		
triglyceride	methanol		

6 Complete the following equations by writing semi-structural formulas of the products.

a

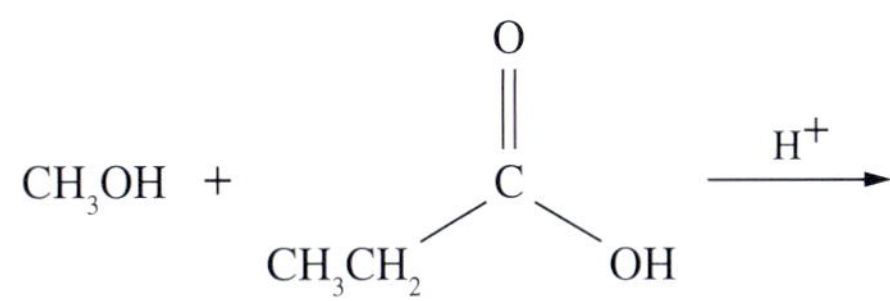

b $CH_3COO(CH_2)_4CH_3(l) + H_2O(l) \xrightarrow{H^+}$

7 Use semi-structural formulas to write the balanced equation for the transesterification, with methanol and potassium hydroxide as a catalyst, of a triglyceride made from lauric acid ($CH_3(CH_2)_{10}COOH$). The formula of the triglyceride will be similar to the one shown in Figure 11.3.19.

11.4 Reaction pathways and atom economy

Organic chemists are highly skilled at developing compounds that have exactly the right properties needed for a particular purpose, such as pharmaceuticals or **polymers**. Once the desired compound has been identified, chemists must devise a way to make it. Chemists have to design an efficient method for converting readily available starting materials—often alkenes or alkanes—into the more complex product they want.

In this section, you will learn how to devise **reaction pathways** for the synthesis of some simple organic compounds by using the reactions you have learnt about in Sections 11.1 to 11.3.

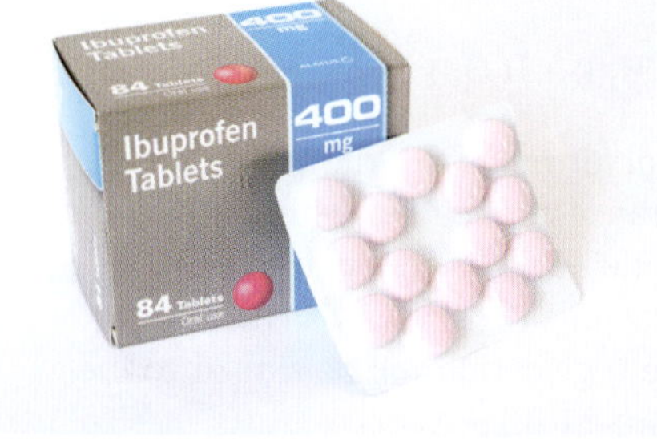

FIGURE 11.4.1 Ibuprofen is the analgesic (pain reliever) found in many commercially available products.

REACTION PATHWAYS

Modern chemists are also interested in devising environmentally friendly synthetic routes. These pathways are designed to minimise waste, use 'greener' solvents, require less energy, and help to preserve the world's resources. The preparation of ibuprofen (Figure 11.4.1), a commonly used analgesic, is an example. Figure 11.4.2 illustrates two pathways for the preparation of ibuprofen, one more efficient than the other.

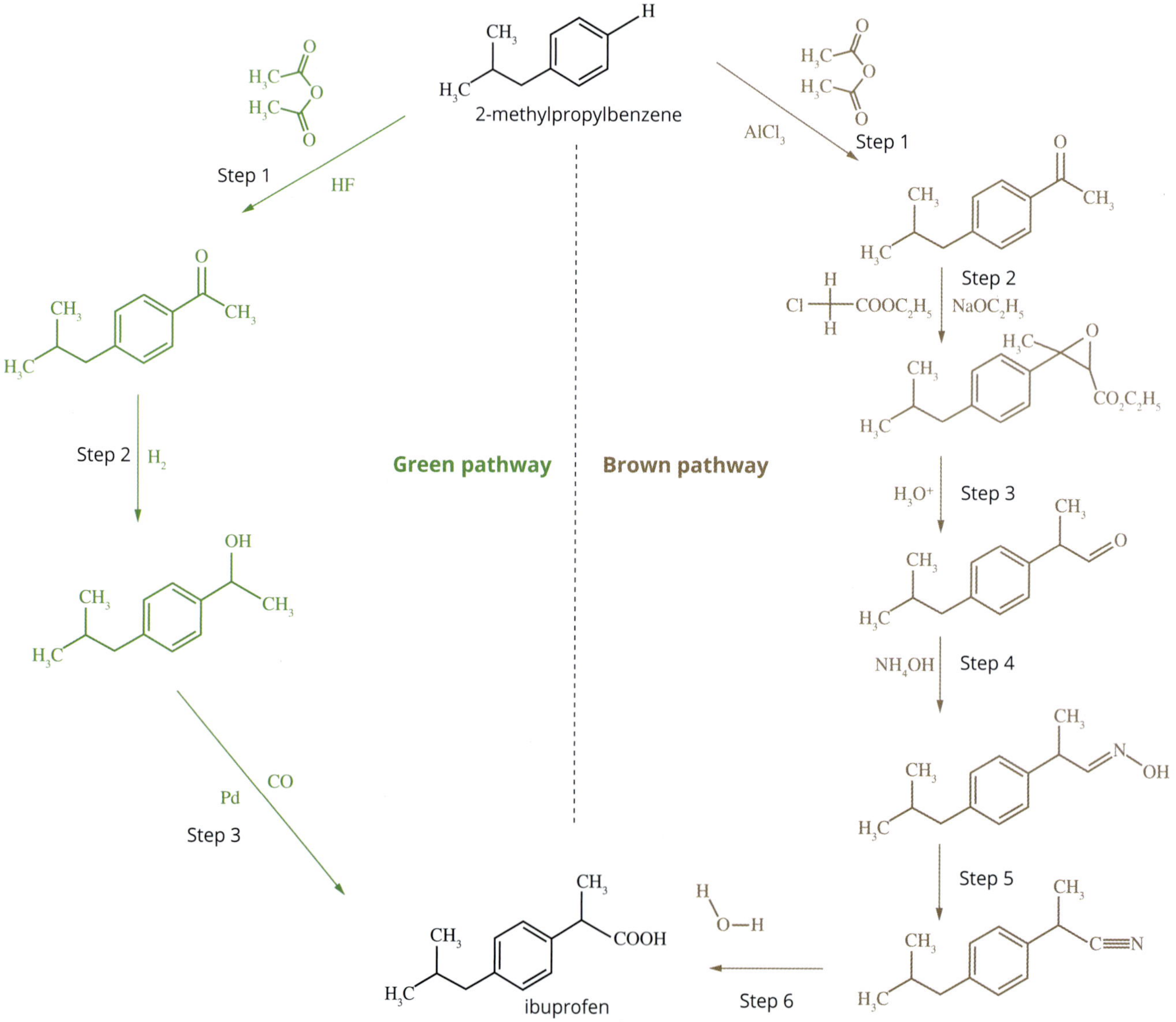

2-methylpropylbenzene

Green pathway

Step 1 HF (with H_3C–CO–O–CO–CH_3)

Step 2 H_2

Step 3 Pd CO

Brown pathway

Step 1 $AlCl_3$ (with H_3C–CO–O–CO–CH_3)

Step 2 Cl–CH_2–$COOC_2H_5$, $NaOC_2H_5$

Step 3 H_3O^+

Step 4 NH_4OH

Step 5

Step 6 H–O–H

ibuprofen

FIGURE 11.4.2 Two alternative reaction pathways for the production of ibuprofen. The green pathway is more efficient than the brown one. In the green pathway, fewer reactants are needed and a higher proportion of atoms in the reactants are present in the final product. This means that there is less waste in the green pathway.

Simple reaction pathways

Figure 11.4.3 shows chemical reaction pathways that can be used to form some compounds based on ethane and ethene. A reaction pathway is a series of one or more steps, or reactions, that can be used to convert a reactant containing certain functional groups to a desired product with different functional groups. Pathways for other alkanes and alkenes can be constructed using the same inorganic reactants and reaction conditions.

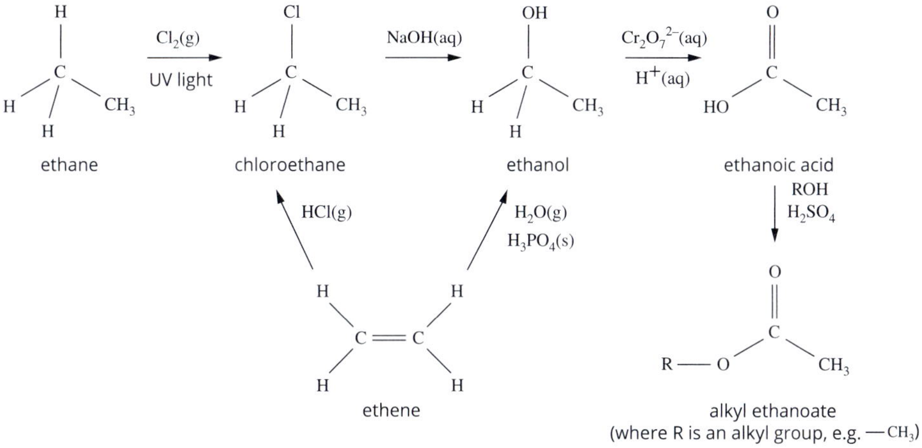

ethane → ($Cl_2(g)$, UV light) → chloroethane → ($NaOH(aq)$) → ethanol → ($Cr_2O_7^{2-}(aq)$, $H^+(aq)$) → ethanoic acid

ethene → ($HCl(g)$) → chloroethane

ethene → ($H_2O(g)$, $H_3PO_4(s)$) → ethanol

ethanoic acid → (ROH, H_2SO_4) → alkyl ethanoate (where R is an alkyl group, e.g. —CH_3)

FIGURE 11.4.3 Some reaction pathways based on ethane and ethene. The same reaction conditions can be used to develop pathways for other members of the alkane and alkene homologous series.

More complex reaction pathways

Suppose you wanted to form ethyl propanoate using only alkanes or alkenes as starting materials. A close look at the structure of the compound shown in Figure 11.4.4 will show you that it is an ester produced by the condensation reaction between propanoic acid and ethanol, so each of these compounds needs to be prepared first, as described below.

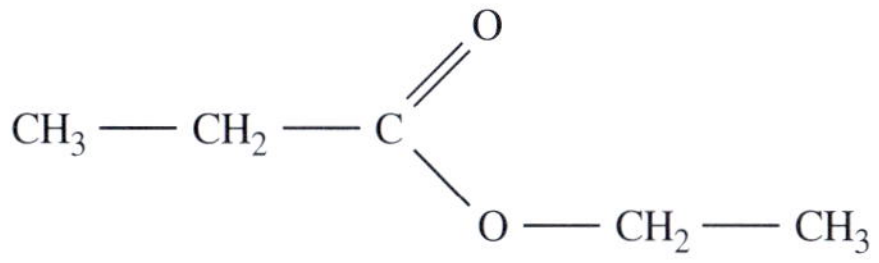

$CH_3—CH_2—C(=O)—O—CH_2—CH_3$

FIGURE 11.4.4 Structure of ethyl propanoate

Making ethanol

Ethanol is an alcohol containing two carbon atoms. Ethanol can be synthesised from ethene in two ways: in one step by the direct addition reaction with water, or in two steps via the intermediate product chloroethane. In this case, the best option is the more direct route, the addition reaction with water. Figure 11.4.5 shows both possible pathways for the production of ethanol from ethene. In each pathway, you can see the inorganic reactant is placed above the arrow for each step.

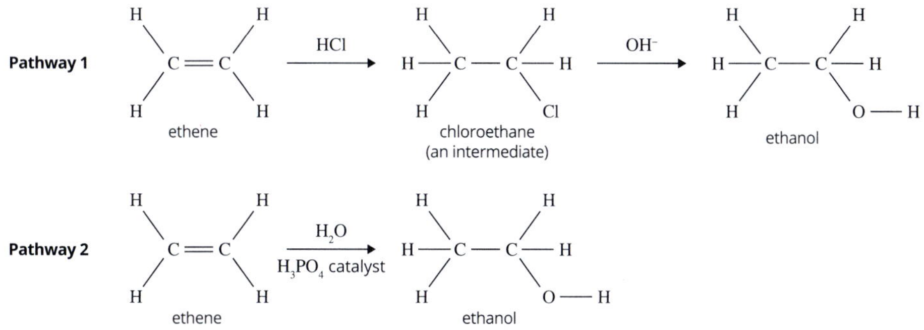

Pathway 1: ethene → (HCl) → chloroethane (an intermediate) → (OH^-) → ethanol

Pathway 2: ethene → (H_2O, H_3PO_4 catalyst) → ethanol

FIGURE 11.4.5 Synthetic pathways for making ethanol from ethene

CHEMFILE

Building the impossible

The American chemist Professor Robert Burns Woodward (1917–1979) was awarded the Nobel Prize in Chemistry in 1965 for his synthesis of complex organic molecules. Woodward devised synthetic procedures to produce complex natural products that many considered to be impossible to replicate in the laboratory.

Woodward successfully proposed and conducted the synthesis of large complex molecules such as quinine, cholesterol, lysergic acid (LSD), strychnine and chlorophyll. His work on the total synthesis of vitamin B_{12} (see figure below) is considered to be the most complex. The synthesis required 69 steps and took more than 12 years to develop with the combined efforts of over 100 people from all around the world.

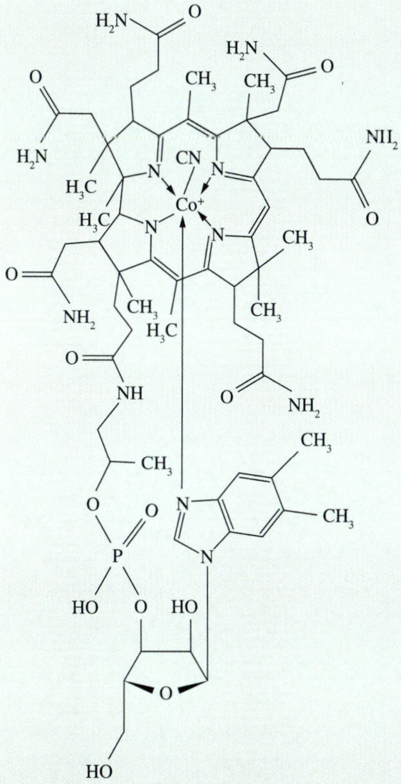

The structure of vitamin B_{12}

Making propanoic acid

Propanoic acid is a carboxylic acid containing three carbon atoms. It can be prepared by the pathway shown in Figure 11.4.6.

If you work backwards from propanoic acid in steps, you can see that:

- propanoic acid can be formed from the oxidation of the primary alcohol propan-1-ol
- propan-1-ol can be formed by the reaction of 1-chloropropane with sodium hydroxide.
- 1-chloropropane can be prepared by reacting propane with chlorine using ultraviolet light.

In this way, you can devise the sequence of reactions to reach a particular product. The pathway shown for the preparation of propanoic acid would produce several products, which are separated by fractional distillation.

$$CH_3CH_3 \text{ (propane)} \xrightarrow[\text{UV light}]{Cl_2} CH_3CHCl \text{ (1-chloropropane)} \xrightarrow{OH^-} CH_3CHOH \text{ (propan-1-ol)} \xrightarrow[H^+]{Cr_2O_7^{2-}} CH_3COOH \text{ (propanoic acid)}$$

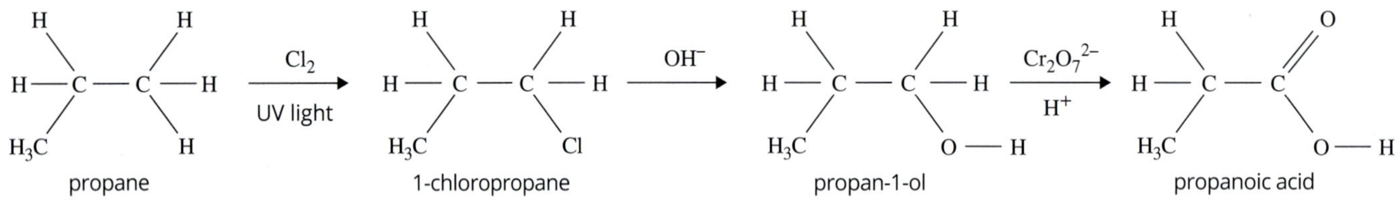

FIGURE 11.4.6 Synthesis of propanoic acid from propane

Making ethyl propanoate

Having synthesised ethanol and propanoic acid, you can now prepare the ester, ethyl propanoate, using a condensation reaction, as shown in Figure 11.4.7.

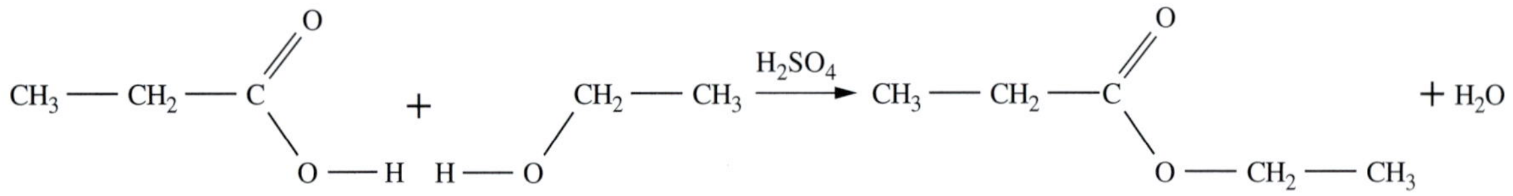

$$CH_3-CH_2-C(=O)-O-H + H-O-CH_2-CH_3 \xrightarrow{H_2SO_4} CH_3-CH_2-C(=O)-O-CH_2-CH_3 + H_2O$$

FIGURE 11.4.7 The formation of ethyl propanoate from the condensation reaction of ethanol and propanoic acid

> A reaction pathway summarises the reactions required to produce a product from simple starting materials.

The full reaction pathway for the preparation of ethyl propanoate from ethene and propane via ethanol and propanoic acid is summarised in Figure 11.4.8.

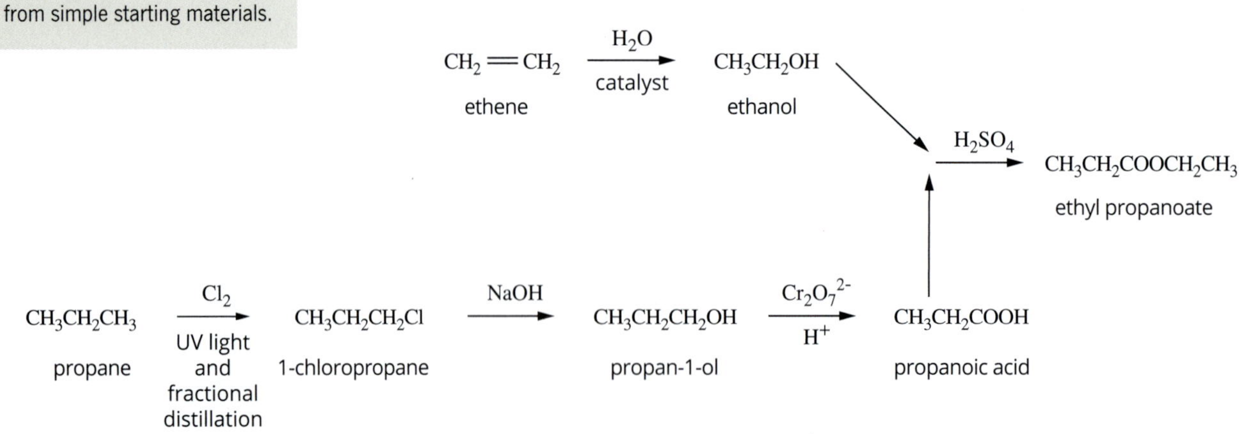

$$CH_2{=}CH_2 \text{ (ethene)} \xrightarrow[\text{catalyst}]{H_2O} CH_3CH_2OH \text{ (ethanol)}$$

$$CH_3CH_2CH_3 \text{ (propane)} \xrightarrow[\text{UV light and fractional distillation}]{Cl_2} CH_3CH_2CH_2Cl \text{ (1-chloropropane)} \xrightarrow{NaOH} CH_3CH_2CH_2OH \text{ (propan-1-ol)} \xrightarrow[H^+]{Cr_2O_7^{2-}} CH_3CH_2COOH \text{ (propanoic acid)}$$

$$CH_3CH_2OH + CH_3CH_2COOH \xrightarrow{H_2SO_4} CH_3CH_2COOCH_2CH_3 \text{ (ethyl propanoate)}$$

FIGURE 11.4.8 Reaction pathway for the preparation of ethyl propanoate

The desired ester product can be separated from the reaction mixture and purified by fractional distillation.

A summary of the reaction pathways described in this chapter is shown in Figure 11.4.9. A diagram like this may be useful when answering questions involving the design of new reaction pathways.

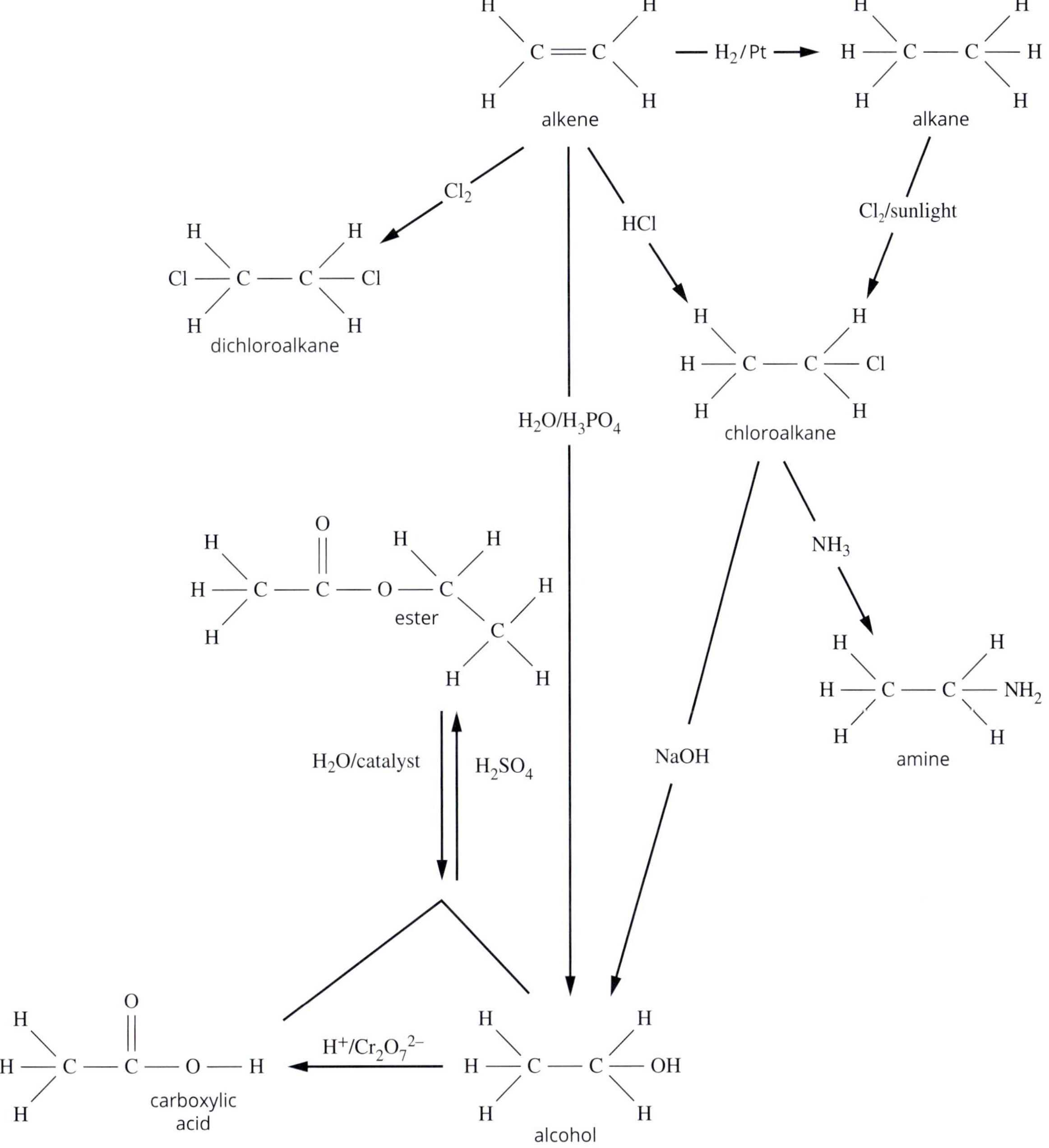

FIGURE 11.4.9 Reaction pathways for the production of various organic compounds starting with an alkene

Other considerations in devising a synthesis

When planning a reaction pathway, there are a number of considerations in addition to simply identifying a possible reaction sequence. As you will recall from Chapter 7, the equilibrium position of a reaction can have a considerable impact on the overall yield. The principles of green chemistry must also be considered in terms of the solvents required and the by-products that are formed. The flow chart in Figure 11.4.10 on the following page shows some of the stages in the planning process.

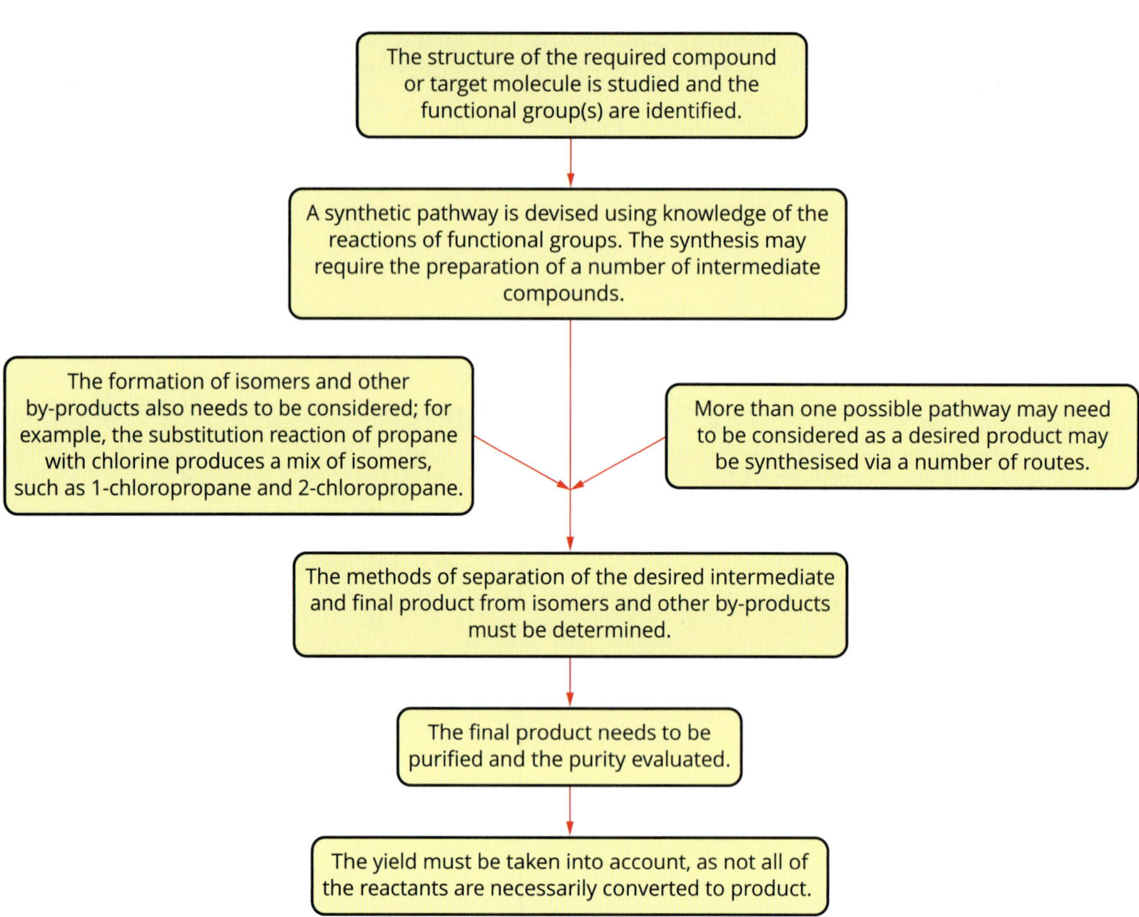

FIGURE 11.4.10 Steps in the design of a synthetic pathway of an organic compound

YIELD AND ATOM ECONOMY

As you learnt in Chapter 7, a major consideration in industrial chemistry is the amount of product that can be produced from a given amount of reactants in a given time. Many industrial processes require several steps to make the final product. At each step, the conversion from reactants to products is usually less than complete. For every step in a reaction pathway, the amount of final product diminishes. Industrial chemists must consider the efficiency of a reaction pathway and the wastes that are produced (Figure 11.4.11).

FIGURE 11.4.11 Most chemical reactions carried out in industrial processes are not 100% efficient, so waste chemicals are produced. The reduction or elimination of waste chemicals is a major concern for industrial chemists.

Yield

The **efficiency** of processes that involve chemical reactions can be found by calculation, which can help in the development of strategies to minimise waste. Several different terms are used to describe the yield of chemical reactions, including theoretical yield, actual yield and percentage yield. These terms are explained in the following sections.

While the focus in this chapter is on reactions involving organic chemicals, the principles discussed here apply to all chemical industries and research.

> Theoretical yield is the maximum amount of product that can be formed using calculations based on the stoichiometric ratio of the limiting reactant in the reaction equation and assumes 100% conversion.

Theoretical and actual yields

The **theoretical yield** of a reaction represents the mass of product that can be formed if the limiting reactant reacts according to the stoichiometric ratios in the reaction equation. The theoretical yield is calculated using the mole ratios of the equation and assumes 100% conversion of the reactants. However, as you learnt previously, when reactants are mixed together in the correct mole ratio, the amount of products will not always be exactly as predicted from stoichiometric calculations.

A number of factors can influence the amount of product that will be formed for a given reaction.

- When a reaction reaches equilibrium rather than continuing on to completion, the **actual yield** will be less than the theoretical yield. Actual yield is the amount of desired product formed in the reaction.
- If the reaction rate is slow, the reaction may not proceed to completion in the time available. This will reduce the actual yield, so that the theoretical yield is not obtained.
- Loss of reactants and products during transfers between reaction vessels and in separation and purification stages, such as filtration, will all result in smaller amounts of the product being obtained than expected.
- Competing reactions may also occur. That is, the two reactants may combine to form products other than the desired one.

The actual yield of a reaction is the amount of desired product that is obtained at the end of the reaction.

Percentage yield

The **percentage yield** compares the actual yield to the theoretical yield. It is a measure of the efficiency of a production process for the particular conditions and method used for the synthesis. The higher the value of the percentage yield, the greater the degree of conversion from reactants to products for the reaction.

Percentage yield can be calculated using the formula:

$$\text{percentage yield} = \frac{\text{actual yield}}{\text{theoretical yield}} \times \frac{100}{1}$$

Worked example 11.4.1

CALCULATING THE PERCENTAGE YIELD OF A REACTION

30.0 g of propan-1-ol was oxidised to propanoic acid using an acidified solution of $K_2Cr_2O_7$. The propanoic acid that was obtained from the reaction mixture had a mass of 20.0 g. Calculate the percentage yield of this oxidation reaction.

Thinking	Working
Write an equation for the reaction.	$CH_3CH_2CH_2OH \xrightarrow{H^+/Cr_2O_7^{2-}} CH_3CH_2COOH$ Note that in this case it is not necessary to write a full equation. Because a molecule of the organic product has the same number of carbon atoms as the organic reactant, the number of moles of the product is equal to the number of moles of the reactant.
Use the formula $n = \frac{m}{M}$ to determine the amount of reactant.	$n(CH_3CH_2CH_2OH) = \frac{m}{M} = \frac{30.0}{60.0} = 0.500$ mol
Use the mole ratio for the reaction to determine the amount, in moles, of the product that would be made if all of the reactant reacted.	mole ratio $= \frac{n(CH_3CH_2COOH)}{n(CH_3CH_2CH_2OH)} = \frac{1}{1}$ $n(CH_3CH_2COOH) = \frac{1}{1} \times n(CH_3CH_2CH_2OH)$ $= 0.500$ mol
Use the formula $m = n \times M$ to determine the mass of the product if all of the reactant reacts. This is the theoretical yield of the product.	$m(CH_3CH_2COOH) = n \times M$ $= 0.500 \times 74.0$ $= 37.0$ g
Calculate the percentage yield for this reaction from the formula: percentage yield $= \frac{\text{actual yield}}{\text{theoretical yield}} \times \frac{100}{1}$	percentage yield $= \frac{20.0}{37.0} \times \frac{100}{1}$ $= 54.1\%$

Worked example: Try yourself 11.4.1

CALCULATING THE PERCENTAGE YIELD OF A REACTION

80.0 g of propan-1-ol was oxidised to propanoic acid using an acidified solution of $K_2Cr_2O_7$. The propanoic acid obtained at the end of the reaction had a mass of 55.0 g. Calculate the percentage yield of this oxidation reaction.

Percentage yields in multi-step syntheses

When a reaction proceeds by a number of steps, the overall percentage yield is reduced at each step. The yield for each step has an effect on the overall yield. A low yield in one of the intermediate reactions can have a significant effect on the amount of final product obtained.

Comparison of the overall percentage yields for different pathways to the same product can be used to determine whether a particular synthetic pathway is the best way to produce an organic compound. Finding the most efficient pathway for the production of a desired chemical is critical to industry because wasting valuable reactants is poor economic and environmental practice.

Worked example 11.4.2

CALCULATING THE PERCENTAGE YIELD OF A MULTI-STEP SYNTHESIS

Calculate the overall percentage yield for the preparation of C from A if it proceeds by a two-step synthesis:

$$A \longrightarrow B \text{ followed by } B \longrightarrow C$$

The yield of A ⟶ B is 80% and the yield of B ⟶ C is 70%.

Thinking	Working
Calculate the overall yield of C by multiplying the percentage yields together and expressing as a percentage (multiplying by 100).	The overall yield of C is: $\frac{80}{100} \times \frac{70}{100} \times \frac{100}{1}$ = 56%

Worked example: Try yourself 11.4.2

CALCULATING THE PERCENTAGE YIELD OF A MULTI-STEP SYNTHESIS

Calculate the overall percentage yield for the preparation of D from A if it proceeds by a three-step synthesis:

$$A \longrightarrow B \text{ followed by } B \longrightarrow C \text{ followed by } C \longrightarrow D$$

The yield of A ⟶ B is 90%, the yield of B ⟶ C is 80% and the yield of C ⟶ D is 60%.

Atom economy

An important objective for an industrial chemist who is developing a reaction pathway is to use a sequence of chemical reactions that minimises energy consumption, reduces waste and has a low impact on the environment.

One consideration when planning reaction pathways is to maximise **atom economy**. The atom economy for a chemical reaction is a measure of how many of the atoms in the reactants end up in the desired product. As you can see in Figure 11.4.12 on the following page, if the atom economy of a reaction is high, then there are few, if any, waste products.

The atom economy for a chemical reaction is a measure of the percentage of the atoms in the reactants that end up in the desired product.

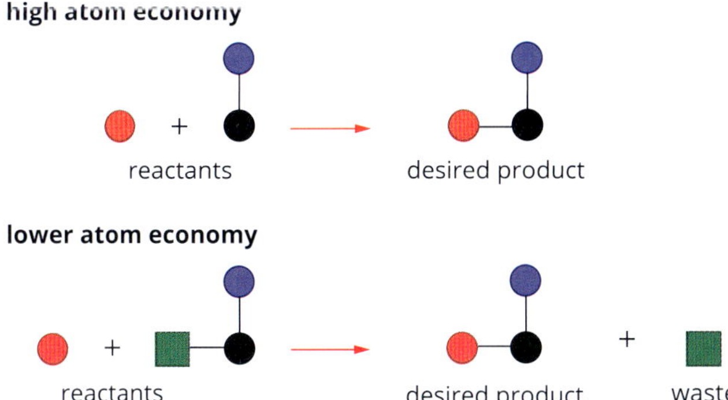

FIGURE 11.4.12 The different-coloured symbols represent atoms or groups of atoms. In a high-atom economy reaction, all or most of the atoms in the reactant molecules end up in the desired product molecule. In a lower-atom economy reaction, not all of the reactant atoms end up in the desired product. The atoms that do not end up in the desired product are waste products of the reaction.

Calculating the atom economy of a reaction provides a method of accounting for the use of materials in a manufacturing process. It tracks all the atoms in a reaction and calculates the mass of the atoms of reactants actually used to form products as a percentage of the total mass of reactants. From this, the mass of reactant atoms that end up as waste can also be calculated.

Once the balanced equation for a reaction is known, the atom economy can be calculated using the formula:

$$\text{atom economy} = \frac{\text{molar mass of desired product}}{\text{molar mass of all reactants}} \times 100$$

Since in a chemical reaction, the total mass of products is equal to the total mass of reactants, this alternative formula can also be used:

$$\text{atom economy} = \frac{\text{mass of desired product}}{\text{mass of all products}} \times 100$$

Advantages of high atom economy in industrial syntheses

The compound 1-phenylethan-1-one, C_8H_8O (Figure 11.4.13), is known in industry as acetophenone. It has a number of uses, including as a fragrance, solvent and a catalyst for some reactions.

The equations below (Figure 11.4.14) show two possible pathways for the preparation of acetophenone from 1-phenylethan-1-ol, commonly called 1-phenylethanol.

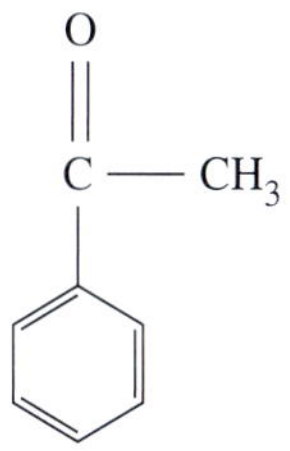

O
C — CH_3

FIGURE 11.4.13 The structure of 1-phenylethan-1-one

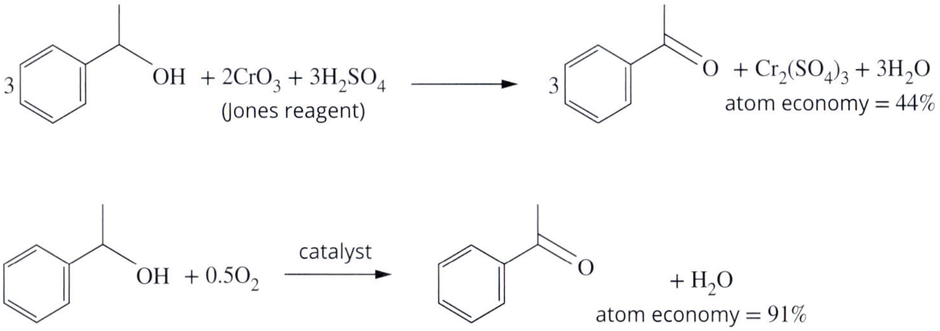

3 OH + $2CrO_3$ + $3H_2SO_4$ (Jones reagent) → 3 O + $Cr_2(SO_4)_3$ + $3H_2O$
atom economy = 44%

OH + $0.5O_2$ —catalyst→ O + H_2O
atom economy = 91%

FIGURE 11.4.14 Two reaction pathways for the preparation of acetophenone from 1-phenylethanol

In the second reaction, with an atom economy of 91%, the only reactant required is oxygen and the only product is water. The use of a catalyst also means that less energy is needed to drive this reaction forward.

In the first reaction, with an atom economy of 44%, significant amounts of chemicals are required to carry out the conversion of reactant to desired product. This will involve a financial cost and consumption of more of the Earth's resources. There is also a significant amount of waste (the chromium(III) sulfate product) which needs to be disposed of.

Use Worked example 11.4.3 to help you with calculations of atom economy.

Worked example 11.4.3

CALCULATING ATOM ECONOMY

Calculate the atom economy in the production of ethanol from chloroethane. In this process, chloroethane is heated with a solution of sodium hydroxide. The equation for the reaction is:

$$C_2H_5Cl(l) + NaOH(aq) \rightarrow C_2H_5OH(aq) + NaCl(aq)$$

Thinking	Working
Calculate the total molar mass of the reactants.	$M(C_2H_5Cl) + M(NaOH)$ $= [(2 \times 12.0) + (5 \times 1.0) + 35.5] + [23.0 + 16.0 + 1.0]$ $= 104.5 \text{ g mol}^{-1}$
Calculate the molar mass of the required product.	$M(C_2H_5OH)$ $= (2 \times 12.0) + (6 \times 1.0) + 16.0$ $= 46.0 \text{ g mol}^{-1}$
Calculate the atom economy for the reaction using the formula: $\text{atom economy} = \frac{\text{molar mass of desired product}}{\text{molar mass of all reactants}} \times 100$	$\text{atom economy} = \frac{46.0}{104.5} \times 100$ $= 44.0\%$ So, in this process, only 44.0% of the starting materials are converted to the desired product. The remainder of the chemicals used is waste.

Worked example: Try yourself 11.4.3

CALCULATING ATOM ECONOMY

Calculate the atom economy in the formation of 1-iodopropane ($CH_3CH_2CH_2I$) from propan-1-ol. The equation for the reaction is:

$$CH_3CH_2CH_2OH(aq) + NaI(aq) + H_2SO_4(aq) \rightarrow CH_3CH_2CH_2I(l) + NaHSO_4(aq) + H_2O(l)$$

CASE STUDY **ANALYSIS**

Aspirin, an ester derived from a herbal remedy

Pharmaceutical products are often developed from substances found in a plant that has been used as a traditional medicine. For example, the origin of the mild painkiller aspirin is a naturally occurring substance called salicin, found in the leaves and bark of willow trees and in the herb meadowsweet.

As long ago as 3000–1500 BCE, willow was used as a medicine by ancient civilisations such as the Sumerians and Egyptians. Around 400 BCE, the Greek physician Hippocrates (Figure 11.4.15) recommended the use of an infusion of willow leaves and bark to help in childbirth and relieve other aches and pains. It wasn't until 1829 that the active ingredient, salicin, was identified and isolated.

It is now known that the body converts salicin to salicylic acid, and this is the active substance that helps to reduce fever and acts as a painkiller. Pure salicylic acid can be hard to ingest and causes stomach irritation.

FIGURE 11.4.15 In Greece, in about 400 BCE, Hippocrates administered willow leaf tea to people suffering pain.

In 1897, Felix Hoffmann replaced the hydroxyl functional group on salicylic acid with an ester functional group to form acetylsalicylic acid, later named aspirin, which was much more gentle on the mouth and stomach than salicylic acid.

The preparation of aspirin is a relatively simple process. It has one main step, in which the hydroxyl functional group of a salicylic acid molecule reacts with the carboxylic acid functional group of an ethanoic acid molecule in a condensation reaction, as shown in Figure 11.4.16. The product, aspirin, has properties that are quite different from those of either of the two reactant molecules.

Although this reaction can be easily carried out in a laboratory, the yield is quite low. A more complex reaction is used to produce aspirin on an industrial scale.

Analysis

1 On the diagram of aspirin in Figure 11.4.16 circle the ester link.

2 Calculate the atom economy for the reaction shown in the equation.

3 A student prepared some aspirin (M = 180.0 g mol^{-1}) in a school laboratory by heating together 10.5 g of salicylic acid (M = 138.0 g mol^{-1}) and excess pure ethanoic acid with 2 drops of concentrated sulfuric acid as catalyst. In this experiment, the salicylic acid is the limiting reactant. The aspirin product was washed with water and dried. The mass of the dried product was 4.50 g. Calculate the percentage yield of the aspirin product in this experiment.

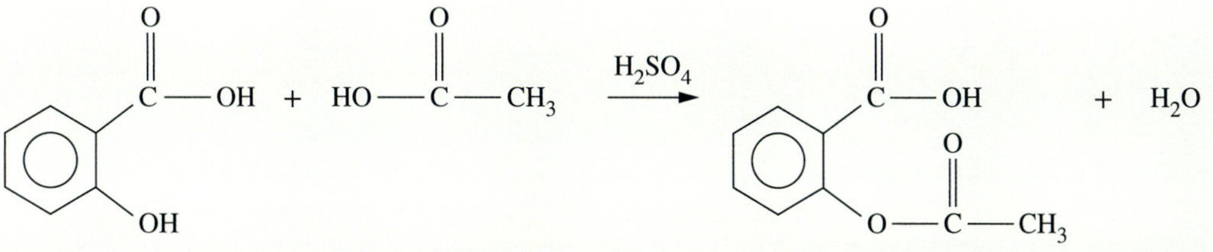

$$\underset{\text{salicylic acid}}{C_6H_4(OH)\text{—}C(=O)\text{—}OH} + \underset{\text{acetic acid}}{HO\text{—}C(=O)\text{—}CH_3} \xrightarrow{H_2SO_4} \underset{\text{acetylsalicylic acid (aspirin)}}{C_6H_4(O\text{—}C(=O)\text{—}CH_3)\text{—}C(=O)\text{—}OH} + \underset{\text{water}}{H_2O}$$

FIGURE 11.4.16 Aspirin can be made from salicylic acid and ethanoic acid. The process involves a reaction between hydroxyl and carboxyl functional groups to form an ester.

11.4 Review

OA ✓✓

SUMMARY

- A reaction pathway is a sequence of more than one reaction that is used to convert a reactant into a product.
- Reaction pathways can be constructed using the organic reactions studied in this chapter.
- Reaction pathways indicate the reaction conditions and reagents required for each step.
- The theoretical yield of a chemical reaction is the mass of the product that would be formed if the limiting reactant reacted completely.
- To calculate the percentage yield, divide the actual yield obtained by the theoretical yield that would be obtained if all of the limiting reactant reacted completely, and multiply by 100:

$$\text{percentage yield} = \frac{\text{actual yield}}{\text{theoretical yield}} \times \frac{100}{1}$$

- When a reaction proceeds by a number of steps, the overall percentage yield is reduced at each step.
- The overall yield of the product of a multi-step reaction is found by multiplying the percentage yields of each step together and expressing as a percentage (multiplying by 100). For example, for the multi-step reaction $A \longrightarrow B \longrightarrow C \longrightarrow D$ with the yields: 90%, 50% and 60%:

$$\text{overall yield} = \frac{90}{100} \times \frac{50}{100} \times \frac{60}{100} \times \frac{100}{1}$$
$$= 27\%$$

- The atom economy for a chemical reaction is a measure of how many of the atoms in the reactants end up in the desired product for the reaction.
- Atom economy can be calculated from the formula:

$$\text{atom economy} = \frac{\text{molar mass of desired product}}{\text{molar mass of all reactants}} \times 100$$

KEY QUESTIONS

Knowledge and understanding

1 Define the following terms:
 a theoretical yield
 b percentage yield
 c atom economy

2 Identify compound A and reactant B in the following reaction pathway.

$$\text{1 chloropropane} \xrightarrow{OH^-} \text{compound A} \xrightarrow{\text{reactant B}} \text{propanoic acid}$$

3 1,2-dibromoethane can be produced in an addition reaction between bromine and ethene according to the equation:

$$C_2H_4(g) + Br_2(l) \longrightarrow CH_2BrCH_2Br$$

In a particular reaction of this type, 120 g of ethene was used.
 a Calculate the theoretical yield of 1,2-dibromoethane in this reaction.
 b If the actual yield of product in the reaction was 280 g, calculate the percentage yield of product.

4 Compound Z can be synthesised by a series of steps in a reaction pathway, as illustrated in the diagram. The yield for each step is shown.

$$W \xrightarrow{80\%} X \xrightarrow{40\%} Y \xrightarrow{30\%} Z$$

 a Calculate the overall yield for the preparation of compound Z from compound W by this process.
 b What would be the overall yield for the reaction if the yield of the step in which W was converted to X was only 25%?

5 Two steps that occur in the preparation of bioethanol from starch-based plants are shown below. The first step is the conversion of starch to produce glucose. In the second step, glucose is fermented to produce ethanol.
Equations for the two steps are:
Step 1: $(C_6H_{10}O_5)_n(aq) \xrightarrow{H_2O} nC_6H_{12}O_6(aq)$
Step 2: $C_6H_{12}O_6(aq) \longrightarrow 2C_2H_5OH(aq) + 2CO_2(g)$
 a What is the atom economy for step 2?
 b What is the chemical by-product in the overall process?

Analysis

6 Complete the flow chart of organic reactions by providing the names of the correct reactants and products.

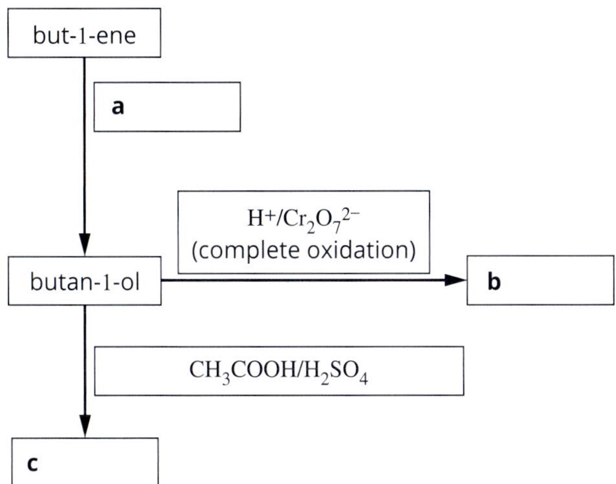

7 Methyl butanoate can be made from 1-bromobutane in the following reaction pathway: 1-bromobutane ⟶ compound A ⟶ compound B ⟶ methyl butanoate

- **a** Identify compound B.
- **b** Name the chemical used to convert compound B to methyl butanoate
- **c** Identify compound A.
- **d** Name the chemicals and the reaction conditions used to convert compound A to compound B.
- **e** Give the name of the reagent used to convert 1-bromobutane to compound A.

8 Draw a reaction pathway to synthesise ethyl propanoate from appropriate alkene-starting materials.

11.5 Sustainable production of chemicals

With our reliance on chemicals of all types in our everyday lives, attention has turned towards producing these chemicals in a responsible and sustainable manner. As the 20th century progressed, there was a growing awareness among chemical scientists and environmentalists of the negative impact that some human-generated chemicals were having on the environment and on human health.

FIXING THE HOLE IN THE OZONE LAYER

One issue in particular has resulted in a worldwide response. This response has become an example of what can be achieved when authorities work together and co-operate to confront environmental challenges caused by manufactured chemicals.

Ozone, O_3, is a gas that is found in the upper levels of the Earth's atmosphere. Although it is only present in low concentrations, it plays a very important role in the protection of life on Earth, because ozone absorbs much of the dangerous ultraviolet radiation coming from the Sun. These wavelengths are responsible for causing sunburn, skin cancer, cataracts and other damage to the eyes.

In the early 1980s atmospheric scientists observed that, in the spring season of the Southern Hemisphere, the concentration of ozone in the upper atmosphere over Antarctica declined significantly. This became known as 'the hole in the ozone layer'. It was the cause of increasing alarm in the scientific world, and also among the general public, as the size of the 'hole' steadily increased in the years that followed (Figure 11.5.1). Concern was particularly aroused in countries such as Australia, New Zealand and Argentina because of the proximity of these countries to Antarctica.

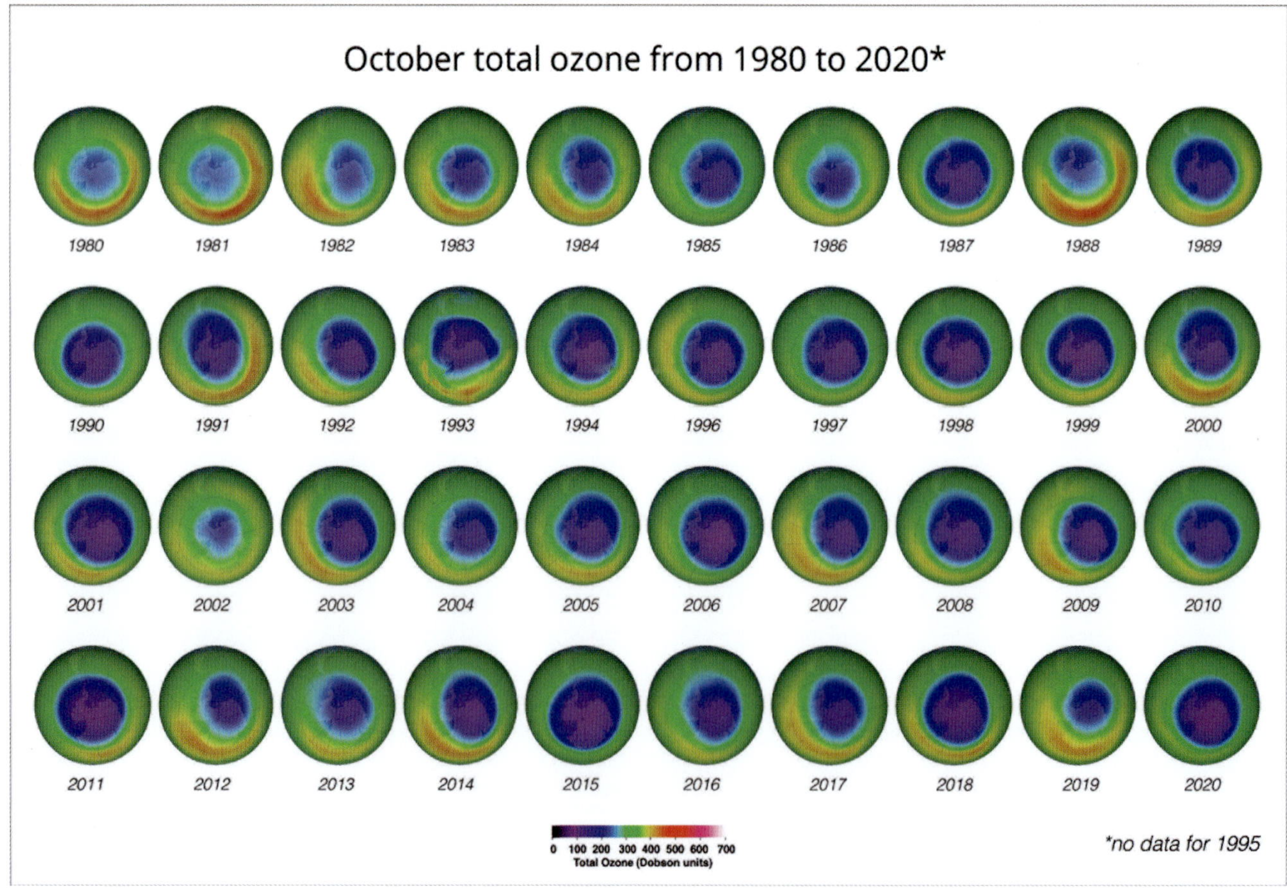

FIGURE 11.5.1 Changes in the thickness of the ozone layer over 40 years from 1980. The purple (darker) regions have the lowest concentration of ozone. These images are from the instruments of satellites monitoring the atmosphere above Antarctica.

In the 1970s scientists had shown that ozone was broken down by compounds known as chlorofluorocarbons, CFCs (Figure 11.5.2). It was established that the presence of these compounds in the atmosphere was responsible for the decrease in concentration of ozone. At that time, CFCs were widely used for purposes such as propellants in aerosol cans and as the refrigerant in air conditioners for cars and homes. CFCs are very stable and once released, will, over time, make their way to the upper levels of the atmosphere where solar radiation can interact with CFC molecules and produce highly reactive chlorine free radicals. The chlorine free radicals can then react with, and destroy, ozone.

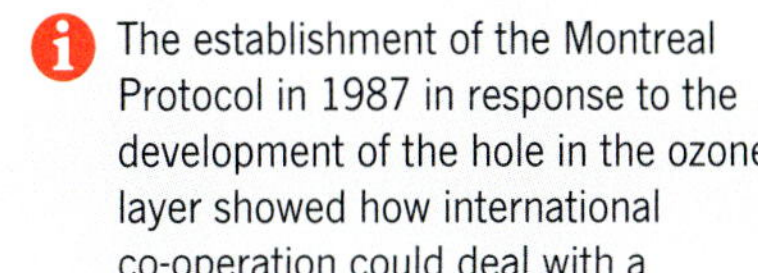
The establishment of the Montreal Protocol in 1987 in response to the development of the hole in the ozone layer showed how international co-operation could deal with a worldwide environmental problem.

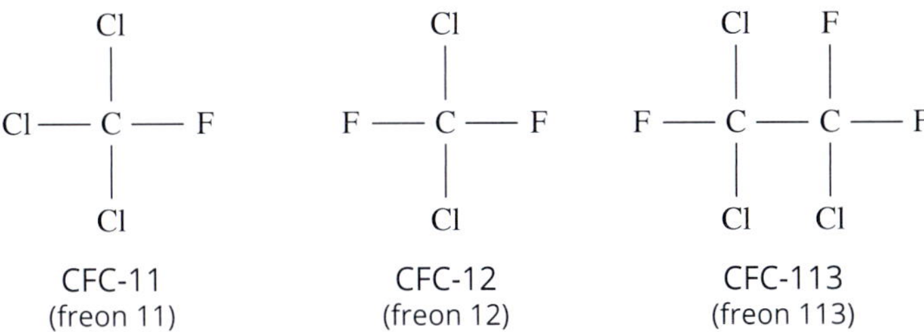

FIGURE 11.5.2 Structures of some chlorofluorocarbon (CFC) compounds

In 1987 scientists and authorities from around the world came together and agreed to phase out the use of CFCs in aerosols and refrigerants. The agreement came to be known as the Montreal Protocol.

Other compounds, which did not contain chlorine, were found to replace the use of CFCs. These compounds included fluorohydrocarbons and alkanes.

Currently, the levels of ozone in the atmosphere appear to have stabilised and it is estimated that they will return to pre-1987 levels by about 2050.

The Montreal Protocol is an encouraging example of what can be done to benefit the world environment when scientists and others work together in a collaborative way.

GREEN CHEMISTRY

The hole in the ozone layer was just one of many environmental problems that the world faced in the second half of the 20th century. Advances in chemical research led to the development and production of chemicals and materials in many fields, including those of polymers, solvents, fertilisers, insecticide, pharmaceuticals, detergents, glues, paints and fabrics.

However, there was also a growing awareness in scientific communities of the negative impact that some manufactured chemicals were having on the environment and on human health. It was recognised that a new approach to the design and manufacture of chemicals was needed. This became known as 'green chemistry'.

Some of the fundamental aims of green chemistry include the requirement that chemical industries should be:

- designing manufacturing processes that produce minimal wastes
- replacing fossil fuels and resources with renewable ones
- producing manufactured goods that are biodegradable or recyclable; any wastes produced in manufacturing processes should also be biodegradable or recyclable
- designing processes that are efficient and not harmful to the environment.

In 1998 two US scientists, Paul Anastas and John Warner, published a book in which they outlined what became known as the '12 principles of green chemistry'. These principles are guidelines for ways in which the chemical industry can design products and processes that have a less damaging impact on the health of humans and the environment, and are discussed in Section 1.2 on page 15, as well as in Chapters 5 and 9.

In this section, we are going to look at topics related to three of the twelve **green chemistry principles**, namely:

- use renewable **feedstocks** (raw materials)
- use catalysts
- design safer chemicals and products.

CHEMFILE

Tea tree oil

The Bundjalung people of coastal New South Wales have long used leaves from the local tea tree (*Melaleuca alternifolia*) as an antiseptic. It is brewed in hot water like a tea and is used to treat a sore throat. The leaves can also be crushed into a paste and applied as an antiseptic on cuts and wounds. When dissolved in a carrier oil, such as olive oil or almond oil, it can decrease inflammation of the skin.

When tea tree oil is used in this way, it fits in well with green chemistry guidelines. It is renewable, uses little in the way of energy resources, and any wastes are biodegradable.

Today, tea tree oil is sold in pharmacies and supermarkets and has a number of uses, including as an antiseptic, a deodorant and an insect repellent.

A very useful oil can be extracted from tea tree *melaleuca alternifolia*) flowers and leaves.

The 12 principles of green chemistry provide guidelines to industry on how products and processes can be designed that have less damaging impacts on human health and the environment.

RENEWABLE FEEDSTOCKS

A feedstock is a raw material which is used in the preparation of other products. Vegetable oils, for example, are feedstock for the production of biodiesel.

Coal, crude oil and natural gas are all commonly known as fossil fuels, but each of these resources has many other uses, as shown in Table 11.5.1. For most of last century and up to the present day, these materials have provided a source for many chemicals used to make products that are essential to our way of life.

TABLE 11.5.1 Uses for coal, crude oil and natural gas other than as fuel

Resource	Product
coal	tar, coke for iron smelting; creosote; medicines
crude oil	polymer feedstock; bitumen; petroleum jelly; paraffin wax
natural gas	hydrogen gas; polymer feedstock

A significant problem with the use of these resources as feedstock also applies to their use as fossil fuels. They are non-renewable. Another complicating factor with their use is that the products made from these resources are often not biodegradable and will persist in the environment.

Efforts have been made in the last decades to find replacements for non-renewable feedstocks and some progress has been achieved. In Chapter 2 you learnt that biogas, bioethanol and biodiesel are being used in varying degrees to replace, respectively, natural gas, petrol and petrodiesel as fuels.

However, significant progress has also been made in finding alternative feedstock sources to make products that have traditionally used fossil fuels for this purpose. In this next section, we will consider two examples to illustrate these achievements.

Biopolymers

In the early part of the 20th century there were far fewer polymers in use in everyday life compared to today. Interestingly, most of the polymers used then were biopolymers. These were made from feedstocks derived from plants and included celluloid and cellophane. Both of these polymers were based on cellulose sourced from plant material. Celluloid was used to make the film used in movies as early as the 1920s (Figure 11.5.3) and cellophane is still used to make sheets of wrapping material (Figure 11.5.4). These polymers are biodegradable.

FIGURE 11.5.3 Celluloid is a cellulose-based polymer that was used for making films in the early and mid-20th century.

FIGURE 11.5.4 Sheets of cellophane wrapping paper

However, by the middle of the 20th century, most polymers were made from starting materials obtained from the fractional distillation of crude oil. These included the addition polymers of polyethene, polypropene, polystyrene and polyvinylchloride. You learnt about these compounds in Unit 1 of this course. Most of these compounds do not biodegrade easily and they persist in the environment for much longer periods than cellulose-based polymers.

With greater attention now being paid to the importance of biodegradability of polymers and an increased desire to move away from polymers made from non-renewable resources, more attention is being directed towards the development of biopolymers. An example of a biopolymer which has a particular use because of its biodegradability is polyglycolic acid (Figure 11.5.5).

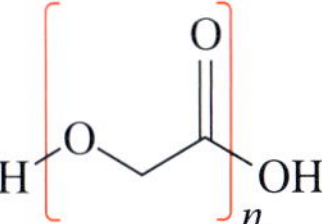

FIGURE 11.5.5 Structure of polyglycolic acid

Like the bioethanol which was discussed in Chapter 2, this compound can be made from materials obtained from cane sugar. Because of its biodegradability, polyglycolic acid is used in the manufacture of sutures that can be absorbed by the human body (Figure 11.5.6).

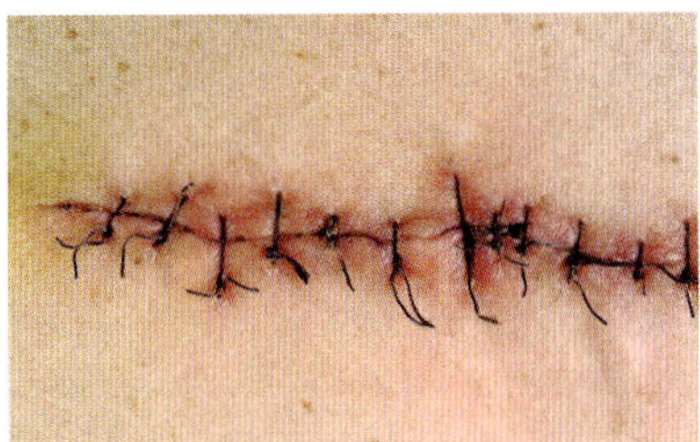

FIGURE 11.5.6 Sutures, which are used to close wounds, commonly use materials made from biopolymers.

Biosolvents

The solvent industry is large and diverse. Some of the applications of solvents in industry are in:

- paints and coatings
- fertilisers, fungicides and herbicides
- inks
- cleaning agents
- cosmetics
- pharmaceuticals.

In the last century and up to the present time, these solvents have also been mostly manufactured using precursors obtained from the fossil fuel resources of coal, crude oil and natural gas.

FIGURE 11.5.7 Bottles of turpentine and white spirits. Turpentine is obtained from pine trees. White spirits is made from crude oil fractions.

White spirits is a solvent made from some of the components of crude oil, so is made from a non-renewable resource. However, turpentine (Figure 11.5.7) is made from the resin of pine trees and so it is renewable. Both white spirits and turpentine are used as paint thinner and grease solvents.

In recent times, there has been considerable interest in developing solvents sourced from natural products such as corn, sugarcane and vegetable oils.

One solvent obtained from a natural product is glycerol. Glycerol is produced in large quantities as a by-product in the manufacture of biodiesel from vegetable oils (Figure 11.3.19 on page 403) Glycerol itself is used as a solvent in the cosmetic industry, but it is also used in the manufacture of other products which are used as solvents. One of these is solketal (Figure 11.5.8). This compound is polar and is used in the manufacture of inks and paints.

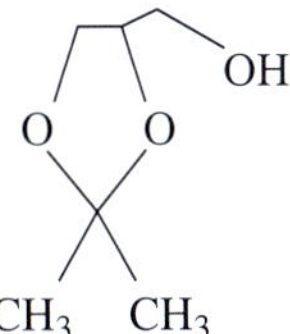

FIGURE 11.5.8 Structure of solketal, a polar solvent used in the manufacture of inks and paints

Like biopolymers, solvents made from plant materials reduce the consumption of fossil-based resources and are usually more biodegradable than those sourced from fossil fuel compounds. The use of renewable feedstocks for biopolymers and biosolvents is an application of one of the green chemistry principles listed on page 419.

In summary, advantages of biopolymers and biosolvents are that:

- their manufacture does not require the use of a finite resource, such as coal, oil or natural gas
- they can be obtained from plant sources, which are renewable
- many, but not all, are biodegradable and so do not persist as wastes in the environment after disposal
- many biopolymers can be recycled.

CATALYSTS

In Chapter 6 you learnt how catalysts can increase the rate of a chemical reaction by providing an alternative reaction pathway which lowers the activation energy of the reaction. As a result of this lower activation energy, catalysts have the effect of lowering the temperature at which an industrial reaction can proceed at an acceptable rate. Achieving this can provide huge savings in fuel costs for a reaction which requires heating for it to proceed.

An example of this is the industrial production of ammonia. Ammonia is produced on a large scale in the manufacture of fertilisers used in the agricultural industry. Ammonium sulfate, ammonium nitrate, urea and ammonia itself are all used as fertilisers which provide nitrogen to plants.

Traditionally, one of the ways in which ammonia was made was by the reaction of ammonium compounds with a strong alkali. An equation for such a reaction is:

$$2NH_4Cl(aq) + Ca(OH)_2(aq) \rightarrow CaCl_2(aq) + 2H_2O(l) + 2NH_3(g)$$

Today, ammonia is manufactured using the Haber process, and was first produced using this process on an industrial scale in 1913. This process was discussed in Chapter 9.

In the Haber process, nitrogen and hydrogen are heated under high pressure and temperature to give a good yield of ammonia. If a catalyst is not used, temperatures of about 3000°C are required for the reaction between nitrogen and hydrogen gases to occur.

The equation for the reaction is:

$$N_2(g) + 3H_2(g) \xrightleftharpoons{\text{Fe catalyst}} 2NH_3(g)$$

In the Haber process using an iron catalyst, the reaction proceeds at a much lower temperature of about 400°C and a pressure of about 200 atmospheres. Operating the process at the lower temperature using a catalyst enables enormous savings in heating and therefore reduced consumption of fossil fuel resources.

When comparing the methods of ammonia manufacture, the advantages of the Haber process can be seen. In the Haber process:

- ammonia is the only product; there are no waste products
- the catalyst can be used over and over
- cheaper and more readily available reactant chemicals are used.

For example, nitrogen is obtained from the air. Hydrogen in the past has been obtained from methane, a fossil fuel, but there is potential for 'green' hydrogen to be produced by the electrolysis of water using electrolysis.

Other examples of catalysts that are used in industry to manufacture chemicals on a very large scale are shown in Table 11.5.2. In each of these examples, the use of catalysts allows the reactions to take place at a much lower temperature than would otherwise be needed.

TABLE 11.5.2 Industrial catalysts

Chemical product	Catalyst
sulfuric acid, H_2SO_4	vanadium(V) oxide
polyethylene	activated chromium oxide
nitric acid, HNO_3	platinum and rhodium

Enzymes

Enzymes are biological catalysts. They are responsible for catalysing the many chemical reactions that take place in the human body. In Chapter 12 you will study the hydrolysis of proteins into amino acids, carbohydrates into monosaccharides, and fats and oils into glycerol and fatty acids. All of these reactions are catalysed by enzymes and take place at body temperature.

In Chapter 2 you learnt that bioethanol can be produced from plant matter using enzymes as catalysts. This reaction takes place at a low temperature. Extra heating from the combustion of fossil fuels is not needed to drive this reaction.

The use of enzymes in industrial processes is becoming more common. Such processes have the advantage of reducing the need for chemicals to carry out conversions of reactants to products and so resources are conserved, heating costs are minimised and wastes are reduced.

Table 11.5.3 lists some chemical processes that use enzymes as catalysts.

TABLE 11.5.3 Industrial processes catalysed by enzymes

Industry	Process	Enzyme type
leather	dehairing hides in the leather industry	proteases and lipases
fabric	cotton softening, denim finishing	cellulases, proteases, amylase
personal care	hair dyeing	oxidases, peroxidases
cleaning	detergents	lipase, protease
waste management	degradation of crude oil hydrocarbons	lipases

In summary, catalysts have the following advantages.

- They allow reactions to be carried out at lower temperatures, thus reducing heating costs which, in turn, save energy resources.
- They increase reaction rates, thus yielding more product in a shorter time.
- They are not consumed in the reaction so can be used continuously.

Catalysts:

- are not consumed in a reaction, so less resources are consumed
- decrease the temperature at which reactions occur, reducing heating costs and the consumption of fuels
- increase reaction rates.

DESIGNING SAFER CHEMICALS

There are many examples of chemicals that were once commonly used in consumer products but are now banned from public use. Usually this is because it has been clearly established that chemicals in the products have a damaging impact on human life and/or the environment.

Table 11.5.4 lists some of these products and the chemicals they contain.

TABLE 11.5.4 Chemicals which have been phased out of public use

Chemical	Use	Effect of chemical if present in the environment	Date banned	Examples of replacement
DDT	insecticide	persistent in environment, thinning of birds eggshells carcinogen	Australia 1987	various pyrethoids
lead in petrol, paints	improved engine performance	toxic to human health, brain damage in children	lead in petrol – Australia 2002 lead in paints – Australia 2010	in paints: titanium dioxide in petrol: mixture of hydrocarbons
CCA (arsenic compound)	timber preservative, protection against white ants	arsenic is a poison, carcinogenic, links to birth defects	2006 banned from use in children's play equipment, picnic tables, seating, domestic decking and handrails	ACQ, a copper compound

Banning dangerous chemicals means that fewer of these compounds make their way into the environment and that leads to better outcomes for human health and the environment.

Perfluoroalkyl (PFAS) compounds and fire-fighting foams

The properties of chemicals used in fire-fighting foams provide a recent example of the move to replace dangerous chemicals with safer ones. The main constituent in these foams for the last few decades has been compounds known as PFAS. This is an acronym which stands for perfluoralkyl substances or polyfluoroalkyl substances. These are compounds which contain long chains of carbon atoms with fluorine atoms bonded to many of the carbon atoms. The structure of a typical PFAS molecule is shown in Figure 11.5.9.

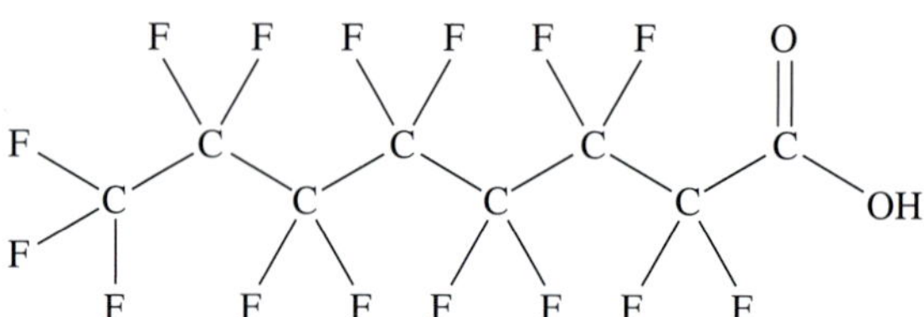

FIGURE 11.5.9 Structure of a PFAS molecule

FIGURE 11.5.10 Foam retardants are often used to fight vehicle fires.

A fire-fighting foam works by blanketing and smothering the fire (Figure 11.5.10). It cools the burning materials, separates the air and flames from the fuel source and reduces the release of vapours which can reignite the fire.

The carbon-fluorine bonds in PFAS molecules are very strong. This results in the molecules having high heat resistance and stability. Such a property means that the compounds are very suitable as fire retardants. However, it also means that they do not break down readily and so have the potential to increase in concentration in soil, groundwater and waterways, thus becoming pollutants in the environment. Another problem with these compounds is that they are toxic to some animals. Investigations into the effects of PFAS compounds on human health are ongoing.

In Australia, the use of PFAS in fire-fighting trials at military establishments and fire brigade bases have resulted in a number of cases of pollution of the surrounding areas.

Tindal Air Force Base is 14 km from the town of Katherine in the Northern Territory. Katherine's water supply comes from bore water and the Katherine River. Testing has shown that soil and underground aquifers around Tindal contain significant levels of PFAS. At significant expense, special water treatment works have been installed to remove PFAS from Katherine's underground water supply. Clean-up has started to remove soil from around the Tindal base. In 2022, plans were underway to transport tonnes of contaminated soil from the base to Altona in Victoria to be decontaminated in high-temperature furnaces.

Work has been done on finding chemicals which can be used to replace PFAS in fire-fighting foams. Replacements, such as non-fluorine–containing mixtures of detergents, and compounds, such as polyglycosides and alkylfulfates, have been considered. These compounds are non-toxic and biodegradable, breaking down in the environment in a reasonably short time.

As of 2021, two Australian states, South Australia and New South Wales, had banned the use of PFAS in normal fire-fighting practices.

11.5 Review

SUMMARY

- The problem of the hole in the ozone layer and the response of the scientific world in addressing it was an example of how international co-operation could provide solutions to worldwide environmental problems.
- The 12 principles of green chemistry provide valuable guidance on how the manufacture of chemicals can be changed in such a way that the impact on the environment is reduced, including by:
  - using renewable raw materials
  - using catalysts
  - designing safer chemicals and products.
- Using renewable feedstocks in the production of chemicals reduces consumption of valuable world resources and often results in products that are biodegradable.
- Biopolymers and biosolvents use renewable resources in their production and are usually readily biodegradable.
- When catalysts are used in the production of new chemicals, reaction rates are increased and less heating is required. The catalysts do not react themselves and so do not form wastes.
- An advantage of using enzymes as catalysts is that the reactions occur at low temperatures, resulting in reduced heating costs.
- In the last few decades, many harmful chemicals have been replaced by products that have little or no impact on the environment.

KEY QUESTIONS

Knowledge and understanding

1 Considering the three green chemistry principles discussed in this section, identify which green chemistry principle is relevant to each of the following situations.

- **a** using a plant extract as a starting chemical to make a polymer
- **b** using cellulase (an enzyme) as part of the process used to soften a fabric
- **c** using nickel as a battery component rather than cadmium.

2 State two advantages of using wrapping paper made from cellophane rather than from polyvinylchloride.

3 What are some advantages of using catalysts in the manufacture of chemical products?

Analysis

4 **a** Why are PFAS chemicals useful in fire-fighting foams?

- **b** Describe the properties of PFAS molecules that make them potentially hazardous to the environment.

5 In Unit 2 Chemistry you studied the structure of polypropene and in Section 12.2 on page 445, you will find a structure of cellulose. After considering the structure of each of these materials, explain why cellulose is more readily biodegradable than polypropene.

Chapter review

KEY TERMS

actual yield
addition reaction
atom economy
condensation reaction
efficiency
esterification reaction
feedstock
fractional distillation
general equation
green chemistry principles
hydration reaction
hydrogenation
hydrogen halide
hydrolysis
hydrolytic reaction
inorganic
isomers
percentage yield
polymer
reaction pathway
substitution reaction
theoretical yield
transesterification
triglyceride

REVIEW QUESTIONS

Knowledge and understanding

1 The following list gives the formulas of a number of organic molecules:

- **i** CH_3CH_2COOH
- **ii** $CH_3CH_2CH_2CH_2NH_2$
- **iii** CH_3OH
- **iv** $CH_2CHBrCH_3$
- **v** $CH_3CH_2CH_2CH_3$
- **vi** $CH_3CHCHCH_3$

Identify two molecules from the list that would react together to form an ester.

A ii and iii
B i and iii
C iv and v
D iv and vi

2 Ethanol is produced by fermentation of glucose, $C_6H_{12}O_6$, as shown in the following equation. (Molar masses for each molecule are provided underneath the respective compounds.)

$$\underset{180}{C_6H_{12}O_6}(aq) \rightarrow 2\underset{46}{C_2H_5OH}(aq) + 2\underset{44}{CO_2}(g)$$

Determine which of the following expressions represents the atom economy for the reaction above.

A $\frac{46}{180} \times 100\%$

B $\frac{(46+44)}{180} \times 100\%$

C $\frac{(2\times46)}{180} \times 100\%$

D $\frac{(2\times46)+(2\times44)}{180} \times 100\%$

3 Write a chemical equation for the reaction between:

a propene and chlorine
b 1-chlorobutane and sodium hydroxide solution
c ethene and oxygen.

4 When a particular hydrocarbon compound was added to a solution of aqueous bromine, the red-orange colour of the bromine disappeared quickly and the solution became colourless. The same hydrocarbon compound reacted with hydrogen gas and a nickel catalyst to form propane.

a Name the hydrocarbon compound.
b Name the product that would be formed if the original hydrocarbon compound reacted with chlorine. Explain your answer.

5 Write the structural formulas of the products of the following reactions.

a

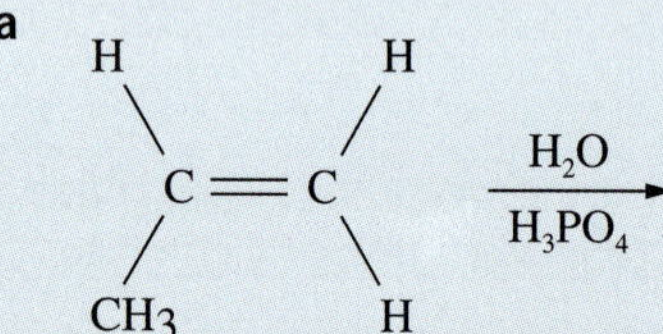

b

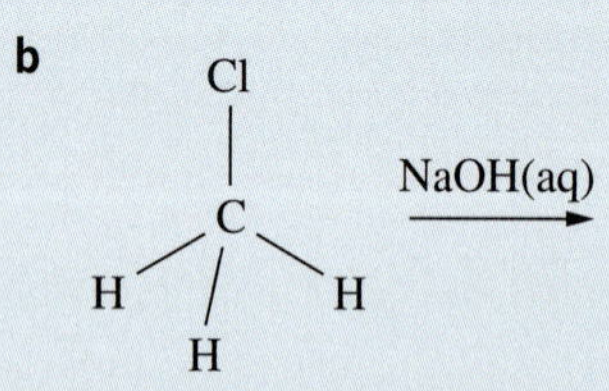

6 In the table below, list the names of products and reaction types for each of the combination of reactants shown in the first column. The answer for the first combination of reactants has been done for you.

Reactants	Products	Reaction type
pent-2-ene and bromine	1,2-dibromopentane	addition
ethanol burning in excess oxygen		
ethene and water, H_3PO_4 catalyst		
bromoethane and hydroxide ion		
propene and hydrogen, Ni catalyst		
chloromethane and ammonia		
hex-3-ene and hydrogen chloride		
methyl propanoate and a dilute solution of HCl		

7 Write semi-structural formulas for the organic compounds formed in the complete oxidation of the following compounds using an acidified solution of potassium dichromate.

a butan-1-ol
b propan-2-ol
c ethanol

8 Give the semi-structural formulas for the products of the hydrolysis of each of the following esters when an acid is used as a catalyst.

a methyl butanoate
b propyl ethanoate
c ethyl methanoate

9 Explain whether natural gas is a renewable feedstock.

10 Turpentine and mineral spirits can both be used as paint thinners. Turpentine is made by distilling resin obtained from some pine trees. Mineral spirits (also known as mineral turpentine) is made from some of the fractions obtained during the fractional distillation of crude oil. Turpentine is more flammable and more toxic than mineral spirits. Discuss the advantages and disadvantages of the use of these two products.

11 Indigenous Australians, such as the Anangu people of South Australia, have traditionally made a glue or tar by heating particles of resin from spinifex grass, for example, to fix stone spear heads to a wooden shaft. Fractional distillation of crude oil also produces a tar, which is used to make bitumen roads.
Compare these two products and their methods of preparation in terms of their relation to renewability and the green chemistry principle of designing for degradation.

Application and analysis

12 Identify A–D using the information shown in the reaction diagram below.

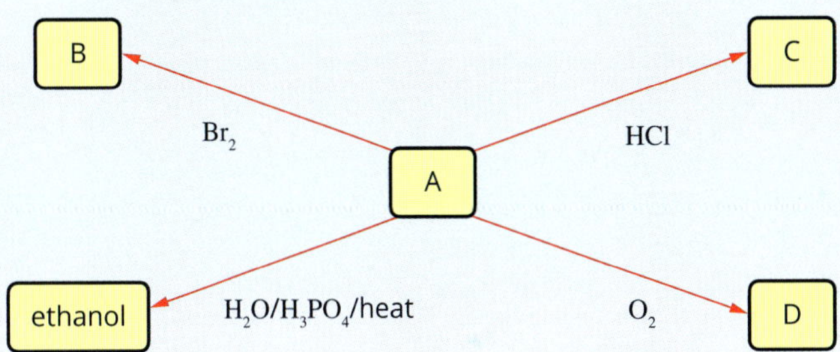

13 Propanoic acid may be synthesised using the reaction pathway shown.

a Draw a semi-structural formula for compound Y.
b Draw a skeletal structure for compound X, which is not an alkene.
c Write a formula for the chemical required to convert compound X to compound Y.

14 The reaction pathway shown below leads to the organic compound G (propyl ethanoate).
Complete the pathway by filling in the boxes with the structural formulas of the appropriate compound for A to F.

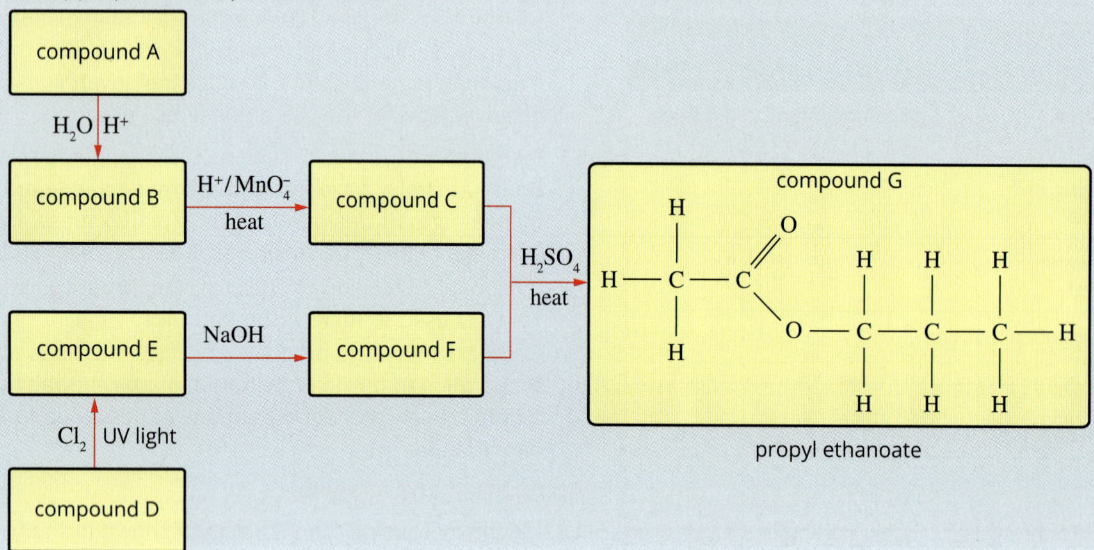

15 Aspirin can be synthesised from salicylic acid and ethanoic anhydride by an esterification reaction, according to the pathway shown in the figure below.

salicylic acid + ethanoic anhydride → acetylsalicylic acid (aspirin) + ethanoic acid (acetic acid)

A student reacted a 2.50 g sample of salicylic acid with an excess of ethanoic anhydride, using sulfuric acid as a catalyst. After purification, a mass of 2.35 g of pure aspirin was obtained.

a Calculate the theoretical yield of aspirin for the reaction.

b Calculate the percentage yield of aspirin for the reaction.

16 The compound with the semi-structural formula CH_3COOH can be made from an alkene in a series of reactions. Design a chemical pathway in the form of a flow diagram to describe how this can be done. Use semi-structural formulas in your answer.

17 Ethanol can be prepared from ethene in two steps according to the following equations:

Step 1: $C_2H_4(g) + HBr(aq) \longrightarrow C_2H_5Br(l)$

Step 2: $C_2H_5Br(l) + NaOH(aq) \longrightarrow C_2H_5OH(aq) + NaBr(aq)$

When carried out in a laboratory, the percentage yield for the first step was 46% and for the second step was 67%. If 54 g of ethene was used in the reaction, what mass of ethanol would have been obtained?

18 When ethanamine, $CH_3CH_2NH_2$ is produced by the reaction of chloroethane and ammonia, the atom economy is 55.2%. Calculate the total mass of reactants, in kilograms, required to make 2.00 kg of ethanamine.

19 The flow chart below shows a reaction pathway for the preparation of compound F from 1-chloropropane and but-1-ene.

- **a** Draw a skeletal structure for compound A.
- **b** Name reactant B.
- **c** Name compound C.
- **d** Write a condensed formula for compound D.
- **e** Name reactant E.
- **f** Write a condensed formula for compound F.

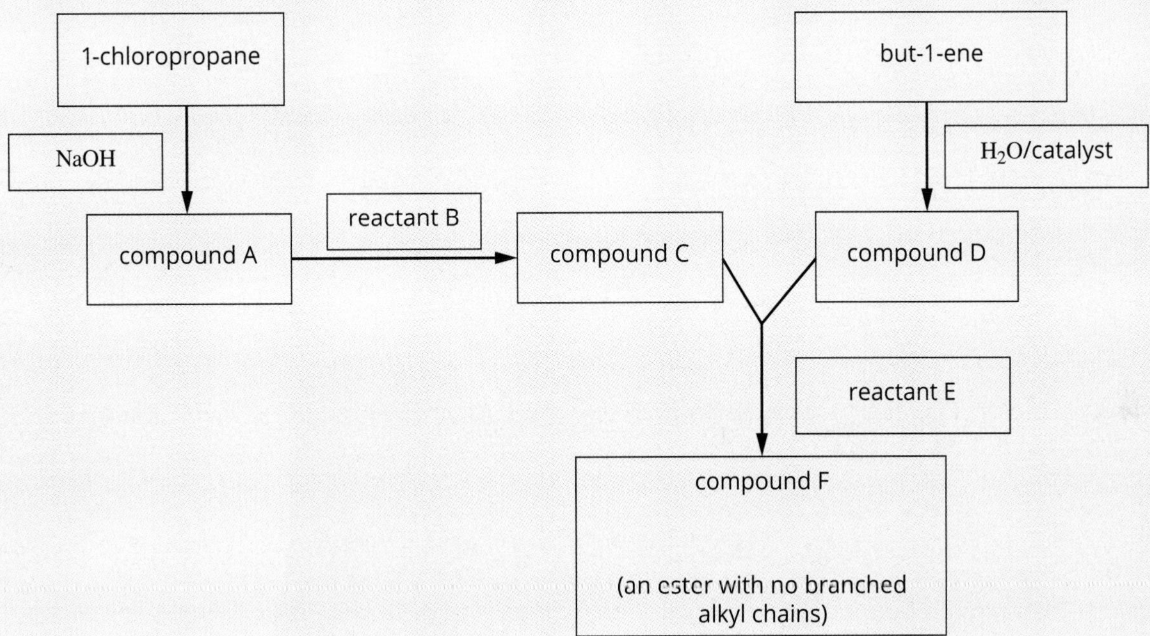

20 Biodiesel can be made by the transesterification of various triglycerides. One particular triglyceride had the structure shown below.

$$\begin{array}{l} H-\overset{\displaystyle H}{\overset{|}{C}}-O-\overset{\displaystyle O}{\overset{\|}{C}}-(CH_2)_7CH=CH(CH_2)_7CH_3 \\ \quad\;\; | \\ H-C-O-\overset{\displaystyle O}{\overset{\|}{C}}-(CH_2)_7CH=CH(CH_2)_7CH_3 \\ \quad\;\; | \\ H-C-O-\overset{\displaystyle O}{\overset{\|}{C}}-(CH_2)_7CH=CH(CH_2)_7CH_3 \\ \quad\;\; | \\ \quad\;\; H \end{array}$$

(M(triglyceride) = 884.0 g mol^{-1} and M(biodiesel) = 296.0 g mol^{-1})

- **a** Use semi-structural formulas, in the style of Figure 11.3.19 on page 403, to write an equation for the transesterification of this triglyceride, with methanol and potassium hydroxide as the catalyst.
- **b** Calculate the atom economy of this reaction.
- **c** A student prepared a sample of biodiesel according to this reaction. They reacted 44.20 g of the triglyceride with excess methanol, in the presence of solid potassium hydroxide, and obtained 28.42 g of biodiesel. Calculate the percentage yield achieved in this experiment.
- **d** Biodiesel is a useful fuel. Write an equation, using molecular formulas, for the combustion of the biofuel produced by the student in this question.

CHAPTER

12 Reactions of biologically important compounds

Food and the rituals associated with dining play a big part in everyday life in many countries. While eating can be a cultural experience, it is also a 'chemical' experience.

Most foods are complex mixtures of chemical substances. These substances have chemical structures and bonds like all of the organic chemicals you studied previously. You need to eat because food contains the various organic compounds that are the source of energy and the raw materials your body needs for growth and repair.

Each food is different, but the labels on foods show that they contain common nutrients such as proteins, fats and carbohydrates. In this chapter, you will study the synthesis, structures and some reactions of these three major food groups.

Key knowledge

- organic reactions and pathways, including equations, reactants, products, reaction conditions and catalysts (specific enzymes not required):
  - hydrolytic reactions of proteins, carbohydrates and fats and oils to break down large biomolecules in food to produce smaller molecules **12.3**
  - condensation reactions to synthesise large biologically important molecules for storage as proteins, starch, glycogen and lipids (fats and oils) **12.1, 12.2**

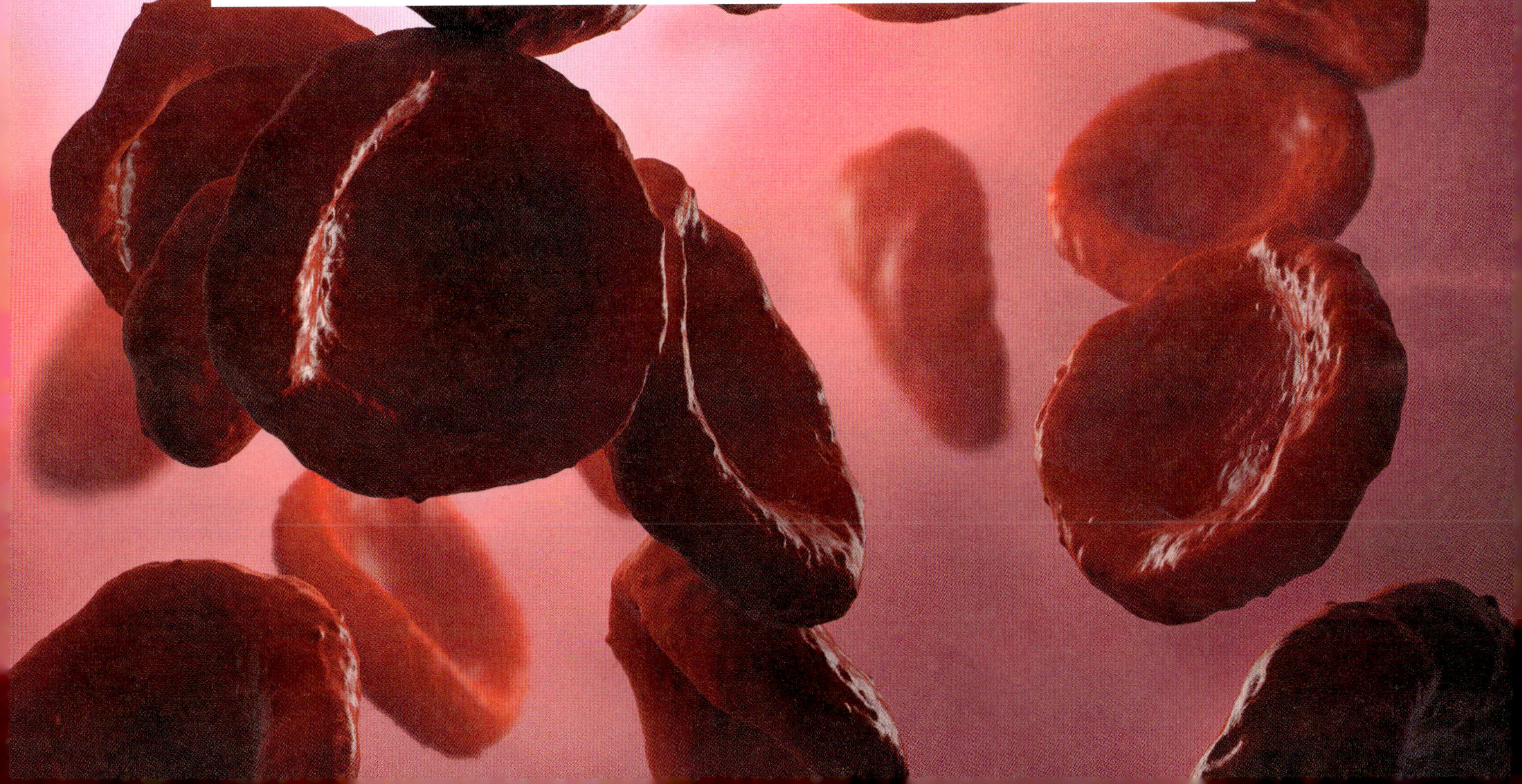

12.1 Synthesis of proteins

Proteins are organic **biopolymers** that have many important functions in living things. Like the synthetic polymers you saw in Year 11, biopolymers are **synthesised** (assembled) from very large numbers of small **monomers**, but they are made in living organisms. In this section, you will investigate the way in which proteins are made from smaller molecules called **amino acids**. From a small number of amino acids, thousands of different proteins are made inside organisms that are essential to life.

Proteins are important in living things in a range of functions from structural proteins, such as the spider's web in Figure 12.1.1, to the haemoglobin in red blood cells (Figure 12.1.2). The structure and function of proteins will be discussed further in Chapter 15.

FIGURE 12.1.1 A spider's web is formed from threads of spider silk, spun from a strong fibrous protein called fibroin.

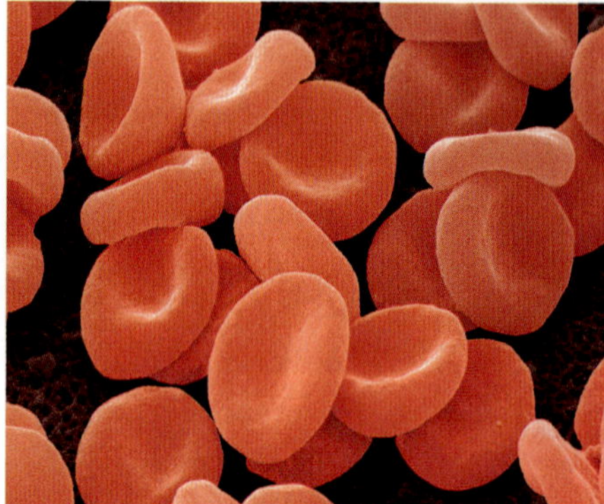

FIGURE 12.1.2 Red blood cells are biconcave disc-shaped cells that transport oxygen from the lungs to body cells. They contain the protein haemoglobin, which is able to bond reversibly to oxygen molecules.

CONDENSATION REACTIONS

In Unit 1, you were introduced to polymers and the addition and condensation reactions that form them. While both of these kinds of polymerisation reactions are important for making industrial polymers, all of the large molecules described in this chapter—proteins, carbohydrates and lipids—are formed by **condensation reactions**.

For **condensation polymerisation** to occur, the monomers must contain two kinds of reactive functional groups, one on at each end of the monomer. This can happen in two possible ways:

- With two different functional groups on one monomer, like the monomer in Figure 12.1.3 on the following page, which has an amine and a carboxyl functional group. This kind of polymer is called a homopolymer. The vast majority of biological polymers, including all the examples in this chapter, are homopolymers.
- With two monomers that each have two of the same functional groups, like the monomers that are reacting together in Figure 12.1.4 on the following page. A polymer made from two different kinds of monomers is called a heteropolymer or copolymer. One of these monomers is a diol, which is a molecule that contains two hydroxy functional groups. The other is a dicarboxylic acid, which contains two carboxylic acid functional groups. Most of the polymers you saw in Unit 1 are copolymers, including the PET (polyetheleneterephthalate) that is used to make bottled water containers.

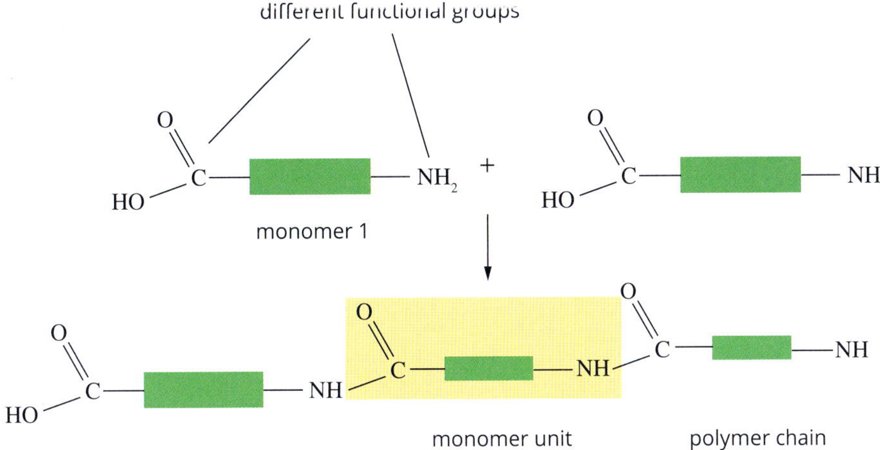

FIGURE 12.1.3 Forming a condensation polymer from a monomer that has a different functional group at each end

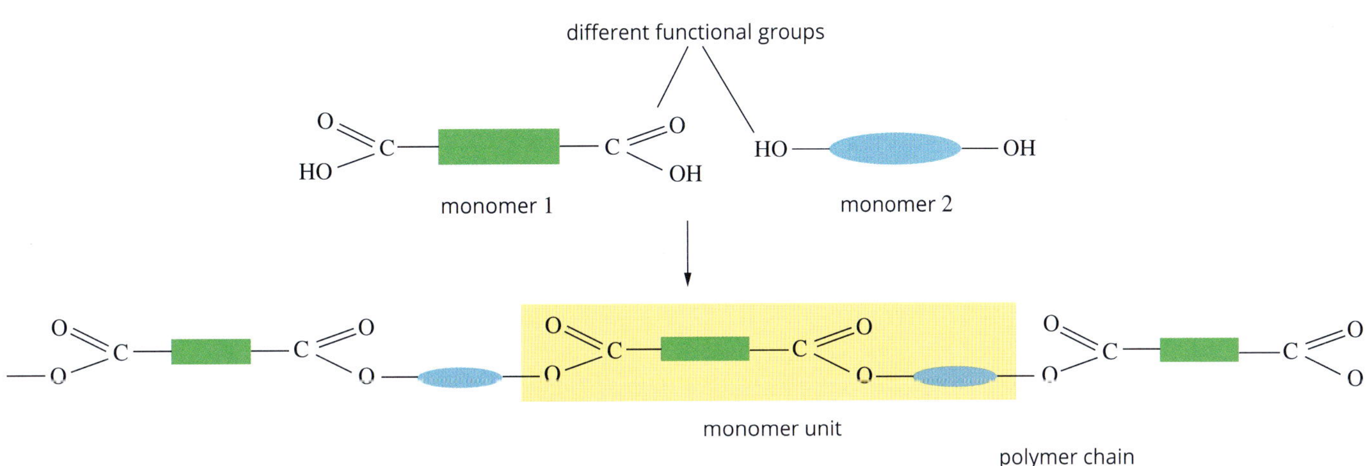

FIGURE 12.1.4 Forming a condensation polymer from two kinds of monomers that have functional groups that are different from each other

A feature of condensation reactions is that small molecules are also biproducts of the reaction. Water is the by-product in the reactions to form all the large molecules in this chapter. Other small molecules, including ammonia, hydrogen chloride and methanol, are sometimes formed as the by-product of other condensation reactions.

STRUCTURE OF AMINO ACIDS

There are 20 different amino acids that are the monomers for all the proteins in the human body. Two of the simplest amino acids are glycine and alanine. The structures of these two molecules are shown in Figure 12.1.5. They have the same general features of every amino acid, which are an amino functional group ($-NH_2$), a carboxyl functional group ($-COOH$), and a hydrogen atom attached to a central carbon.

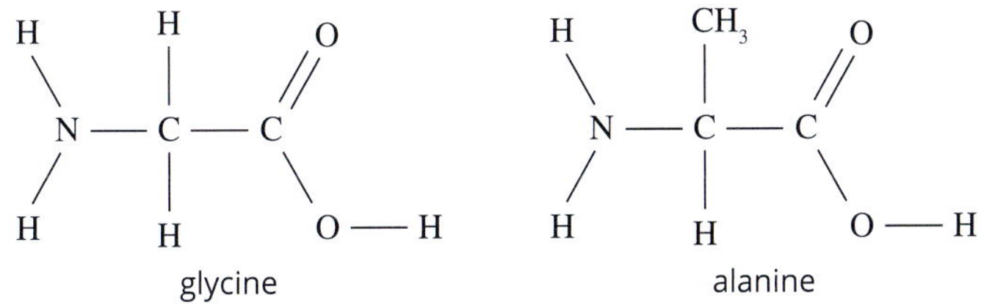

glycine

alanine

FIGURE 12.1.5 The structures of amino acids glycine and alanine

```
H       R       O
 \      |      //
  N --- C --- C
 /      |      \
H       H       O --- H
```

FIGURE 12.1.6 General structural formula of an amino acid. Different amino acids have different groups of atoms as the R group.

General formula of amino acids

The amino acids used to produce the proteins in the human body have the general formula $H_2N{-}CH(R){-}COOH$. These amino acids, as shown in Figure 12.1.6, are known as **2-amino acids**. This is because the carbon atom of the carboxyl group is numbered as the first carbon in the chain (according to IUPAC rules) and the amino group is bonded to the carbon atom adjacent to it, numbered '2' in the chain. 2-Amino acids are also called α-amino acids.

The twenty 2-amino acids used to make human proteins are listed in Table 12.1.1, along with the three-letter abbreviations commonly used for them by biochemists.

Properties of R groups in amino acids

The major difference between one amino acid and another is the group of atoms that make up the **side chain**. This side chain is represented in the general structure as an **R group**. The side chain may be:

- non-polar (e.g. $-CH_3$ in alanine and $-CH(CH_3)_2$ in valine)
- polar (e.g. $-CH_2COOH$ in aspartic acid and $-CH_2OH$ in serine).

  The side chain may also include functional groups that can behave as:
- acids or proton donors (e.g. $-CH_2COOH$ in aspartic acid contains an acidic carboxyl group)
- bases or proton acceptors (e.g. $-CH_2CH_2CH_2CH_2NH_2$ in lysine contains a basic amino group).

Amino acids are commonly given three-letter abbreviations. For example, alanine is represented by Ala and glycine is represented by Gly.

TABLE 12.1.1 Amino acids found in the human body, with their three-letter abbreviations and R groups

Name	Symbol	Structure
alanine	Ala	$H_2N-CH(CH_3)-COOH$
arginine	Arg	$H_2N-CH(CH_2-CH_2-CH_2-NH-C(=NH)-NH_2)-COOH$
asparagine	Asn	$H_2N-CH(CH_2-C(=O)-NH_2)-COOH$
aspartic acid	Asp	$H_2N-CH(CH_2-COOH)-COOH$
cysteine	Cys	$H_2N-CH(CH_2-SH)-COOH$
glutamine	Gln	$H_2N-CH(CH_2-CH_2-C(=O)-NH_2)-COOH$

continued over page

CHEMFILE

Hemp: Superfood

Hemp and marijuana are different forms of the same plant cannabis *sativa*. The main difference is that in hemp, the level of the hallucinogenic component, tetrahydrocannabinol, is very low. The relaxing of laws that prohibited the growing of hemp has led to a new food industry in Australia—that of hemp seeds like those in the figure below. Many people consider hemp seeds to be a superfood because they are the only seeds that contain all twenty amino acids and they have the highest protein density of any seeds. In addition, hemp seeds have high levels of essential fatty acids, contain a range of vitamins and minerals, and are a source of fibre.

An emerging superfood—hemp seeds

Name	Symbol	Structure
glutamic acid	Glu	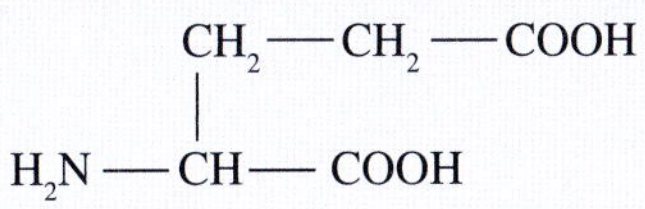 CH_2-CH_2-COOH $\vert$ $H_2N-CH-COOH$
glycine	Gly	H_2N-CH_2-COOH
histidine	His	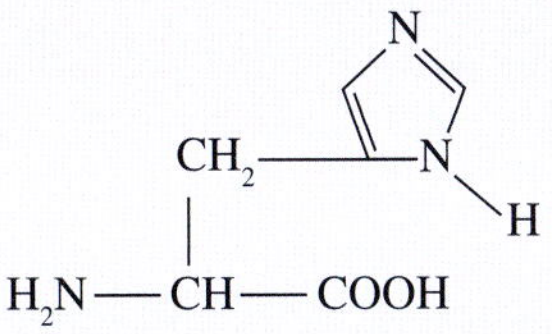 N CH_2—N (ring) H $\vert$ $H_2N-CH-COOH$
isoleucine	Ile	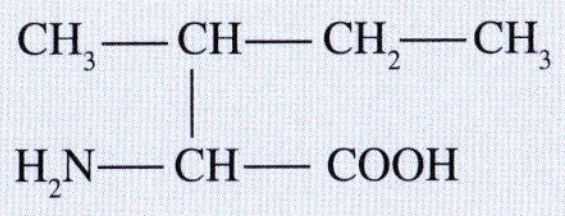 $CH_3-CH-CH_2-CH_3$ $\vert$ $H_2N-CH-COOH$
leucine	Leu	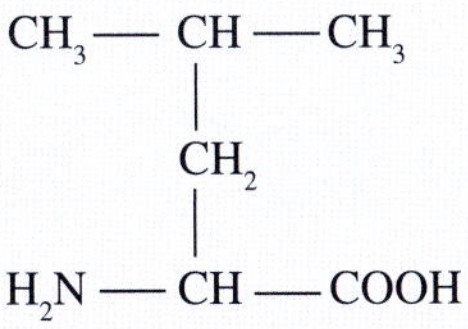 $CH_3-CH-CH_3$ $\vert$ CH_2 $\vert$ $H_2N-CH-COOH$
lysine	Lys	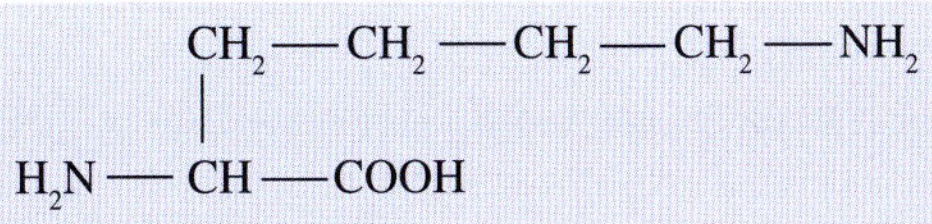 $CH_2-CH_2-CH_2-CH_2-NH_2$ $\vert$ $H_2N-CH-COOH$
methionine	Met	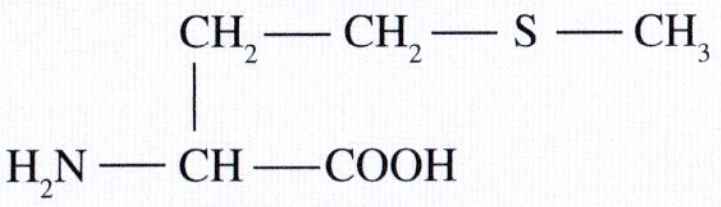 $CH_2-CH_2-S-CH_3$ $\vert$ $H_2N-CH-COOH$
phenylalanine	Phe	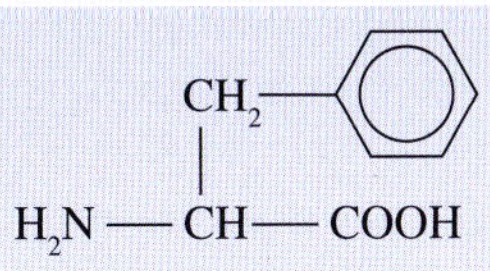 CH_2—(benzene ring) $\vert$ $H_2N-CH-COOH$
proline	Pro	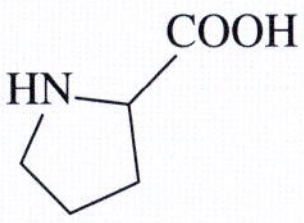 COOH HN (ring)
serine	Ser	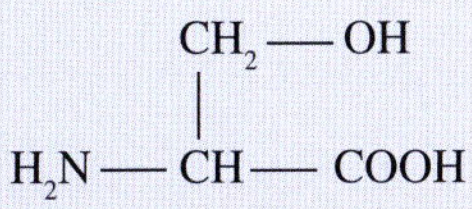 CH_2-OH $\vert$ $H_2N-CH-COOH$
threonine	Thr	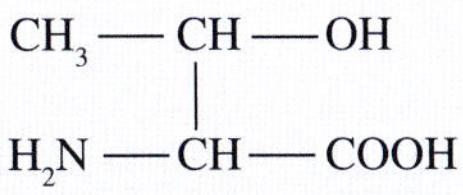 $CH_3-CH-OH$ $\vert$ $H_2N-CH-COOH$
tryptophan	Trp	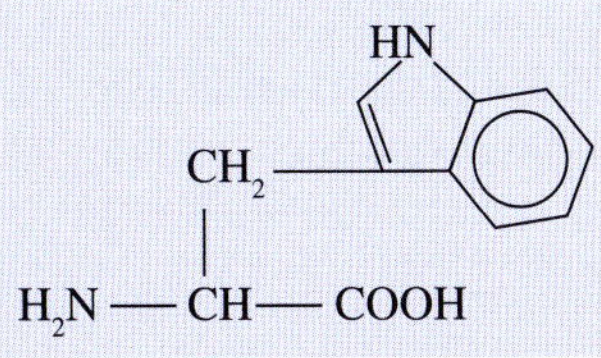 HN CH_2—(indole ring) $\vert$ $H_2N-CH-COOH$
tyrosine	Tyr	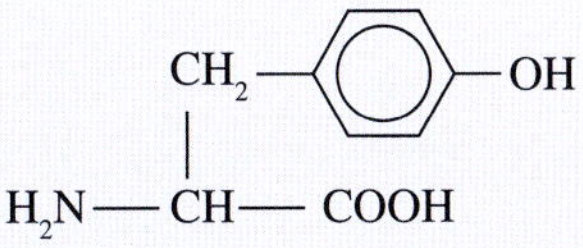 CH_2—(benzene ring)—OH $\vert$ $H_2N-CH-COOH$
valine	Val	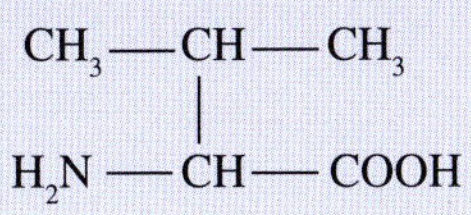 $CH_3-CH-CH_3$ $\vert$ $H_2N-CH-COOH$

All 2-amino acids have the same general formula, $H_2N{-}CH(R){-}COOH$, where R is an alkyl side chain shown in this table above the rest of the structure.

THE FORMATION OF PROTEINS

Extremely large protein molecules are formed by many reactions between amino acids. These reactions are completed in organisms in a highly organised way to ensure that the correct amino acids are present in the protein in the correct order. The protein grows slowly, starting initially with the reaction of two amino acids to form a dipeptide, which reacts with more amino acids to form a longer **polypeptide**. The polypeptide continues to react with additional amino acids to form the full protein. All the reactions are condensation reactions, and a molecule of water is produced each time the chain grows.

Dipeptides

A condensation reaction occurs between a molecule that contains a carboxyl group (–COOH) and a molecule that contains an amino group ($-NH_2$). An **amide** functional group (–CONH–) is formed linking the two molecules. A molecule of water is also produced, as shown in Figure 12.1.7.

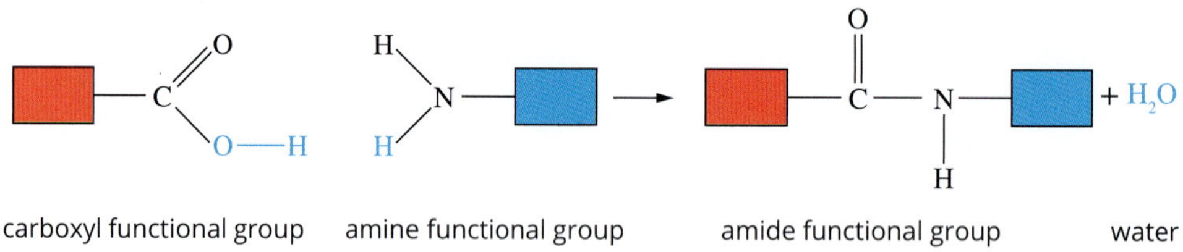

carboxyl functional group + amine functional group → amide functional group + H_2O (water)

FIGURE 12.1.7 An amide functional group forms when a carboxyl functional group reacts with an amine functional group. A water molecule is also produced. The red and blue boxes represent the remaining atoms in the amino acids involved in this reaction.

Dipeptides are formed when a condensation reaction occurs between the carboxyl functional group of one amino acid and the amine functional group of a second amino acid. An amide link is formed.

Because 2-amino acids contain both an amine functional group and a carboxyl functional group, they can undergo condensation reactions with each other. When two amino acids react, an amide functional group, called a **peptide link** (also called an amide link, peptide group or peptide bond), is formed that links the molecules together.

When two amino acid molecules react together, as shown in Figure 12.1.8, the product is referred to as a **dipeptide**.

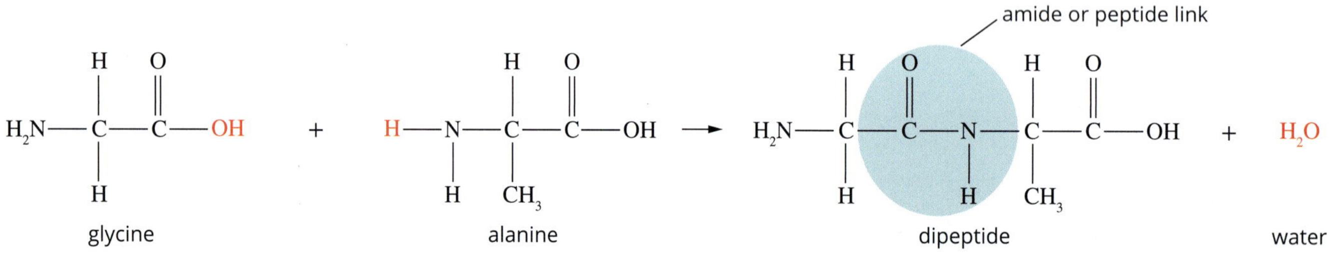

glycine (H_2N-CH_2-COOH) + alanine ($H_2N-CH(CH_3)-COOH$) → dipeptide ($H_2N-CH_2-CO-NH-CH(CH_3)-COOH$) + water ($H_2O$)

amide or peptide link

FIGURE 12.1.8 A condensation reaction between two amino acids, glycine and alanine. Note how the carboxyl and amine functional groups react to form the dipeptide and water.

Each time a pair of different amino acids react in this way, there are two possible product molecules, depending on which ends of each molecule react together, as seen in Figure 12.1.9.

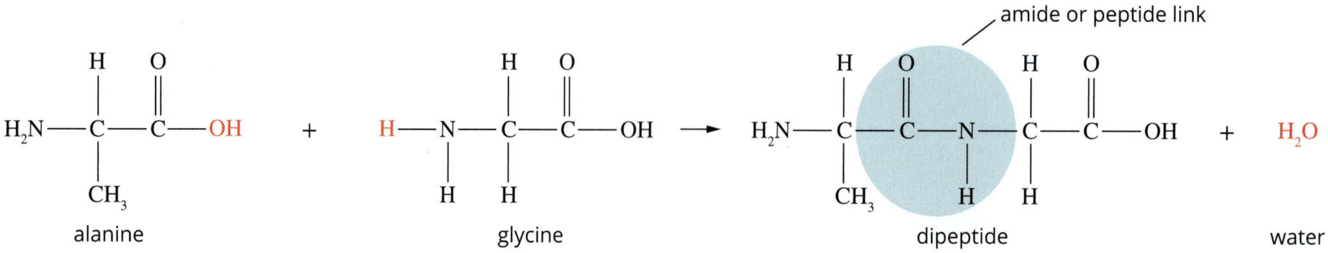

alanine ($H_2N-CH(CH_3)-COOH$) + glycine (H_2N-CH_2-COOH) → dipeptide ($H_2N-CH(CH_3)-CO-NH-CH_2-COOH$) + water ($H_2O$)

amide or peptide link

FIGURE 12.1.9 Another dipeptide molecule formed from the reaction between glycine and alanine

Polypeptides

Polypeptides are polymers formed by further reactions between the amino acids. During these reactions, the amino acids can form long chains. When three amino acid molecules react together, a **tripeptide** is formed. A polymer made from many amino acids is known as a polypeptide. The structure of a protein is far more involved than simply the formation of a polypeptide by condensation reactions. The details of this structure will be discussed in Section 15.2.

Organisms make their dipeptides, polypeptides and larger proteins in remarkable biological synthesising machines called **ribosomes**. These are intricate complexes within the cells that are made from many kinds of biological molecules. One part of a ribosome from a human cell is shown in Figure 12.1.10; this part is made from many molecules, including enzymes. Each 'blob' in Figure 12.1.10 represents a monomer unit in one of the biopolymers that make up the ribosome, which has a molar mass of about 4.2 million g mol^{-1}.

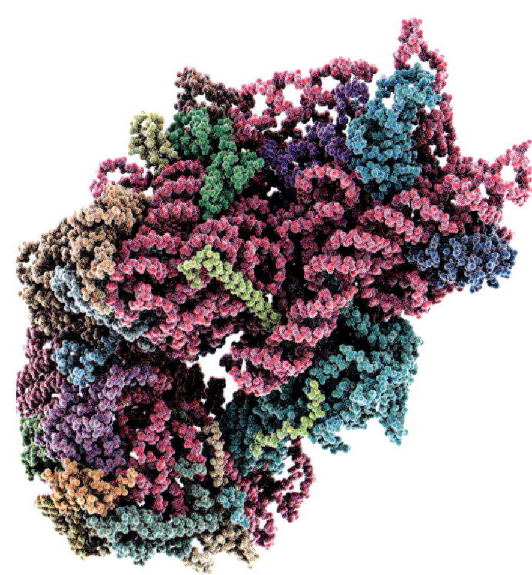

FIGURE 12.1.10 A substantial proportion of a ribosome is made from proteins. A section of a ribosome from a human cell is shown here.

Inside the ribosome, the reactions of amino acids to form larger polymers happen within minutes and occur at body temperature. In the biological sciences, this step is known as translation and is the final part of the process where the genetic instructions stored in DNA are used to make proteins with a specific amino acid sequence.

Proteins can also be synthesised in laboratories. The synthetic steps occur over solid-state catalysts that are at 85–90°C in a flow synthesis system, such as the one shown in Figure 12.1.11. In this device, the reactant amino acids are selectively pushed into place by powerful pumps. With the latest developments in synthesis, it is possible to synthesise a tailor-made protein within a few hours.

No matter whether a protein is made by a ribosome or in a laboratory, the polypeptide synthesis is a stepwise process, where new amino acids are added to the existing polymeric protein chain—one by one—to make the chain longer.

> A polypeptide is a polymer made up of many amino acids, all joined by condensation reactions between carboxyl and amine functional groups.

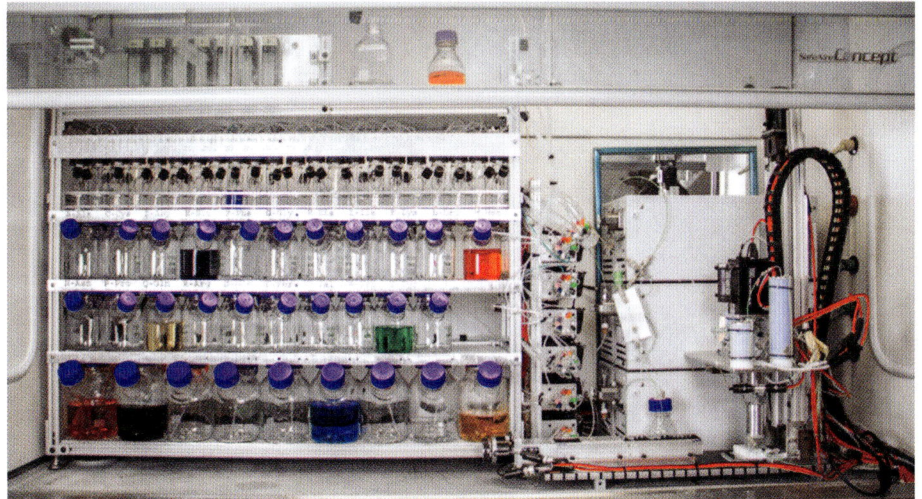

FIGURE 12.1.11 A flow instrument for synthesising proteins. Each of the jars on the left-hand side of the image contains a solution of a different amino acid.

Naming polypeptides

A shorthand notation is often used to describe the amino acid sequence in a polypeptide, using the three-letter abbreviations listed in the table of 2-amino acid structures (Table 12.1.1). By convention, the structure is drawn so the free amino group is on the left and the free carboxyl group is on the right. A reaction between six amino acids to form part of a polypeptide is shown in Figure 12.1.12 on the following page. The polypeptide would be named as Ala–Glu–Gly–Cys–Val–Lys.

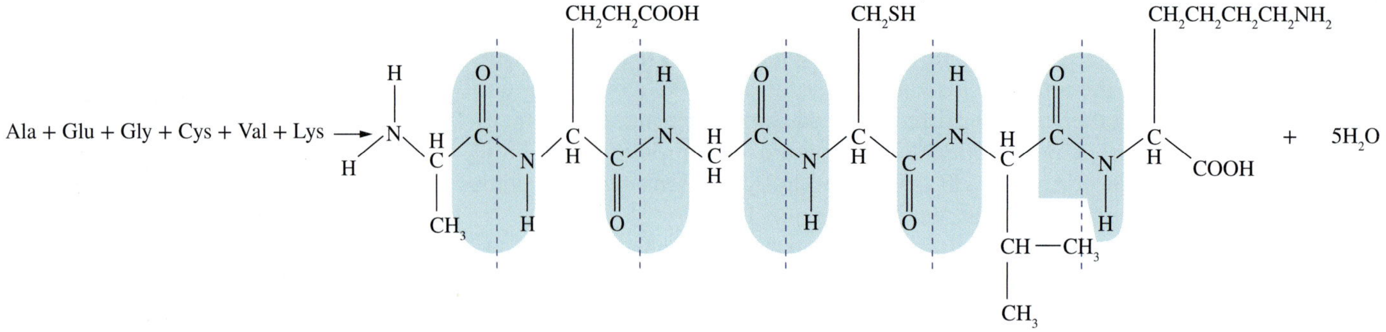

Ala + Glu + Gly + Cys + Val + Lys → (polypeptide: side chains CH_3, CH_2CH_2COOH, H, CH_2SH, $CH(CH_3)CH_3$, $CH_2CH_2CH_2CH_2NH_2$; terminal COOH) + $5H_2O$

FIGURE 12.1.12 A condensation reaction between many amino acids produces a polypeptide. Peptide links are shaded in blue.

A polypeptide constructed from more than 50 amino acids is usually called a protein.

The hormone insulin is a protein that regulates the chemical reactions in the body that break down **carbohydrates**, fats and proteins. Insulin is one of the smallest proteins in the human body. From Figure 12.1.13, you can see that:

- insulin is made up of two linked chains, with a total of 51 **amino acid residues**
- the chains are linked by covalent bonds between sulfur atoms from the R group of cysteine residues
- the start of the longer chain has a free amino group on the phenylalanine (Phe). This is known as the **N-terminal** amino acid of that chain
- the end of the longer chain is threonine (Thr), which has a free carboxyl group and is therefore called the **C-terminal** amino acid.

When amino acids react with each other, peptide bonds form. Small molecules, such as water, are also released in this process, making this a condensation polymerisation reaction.

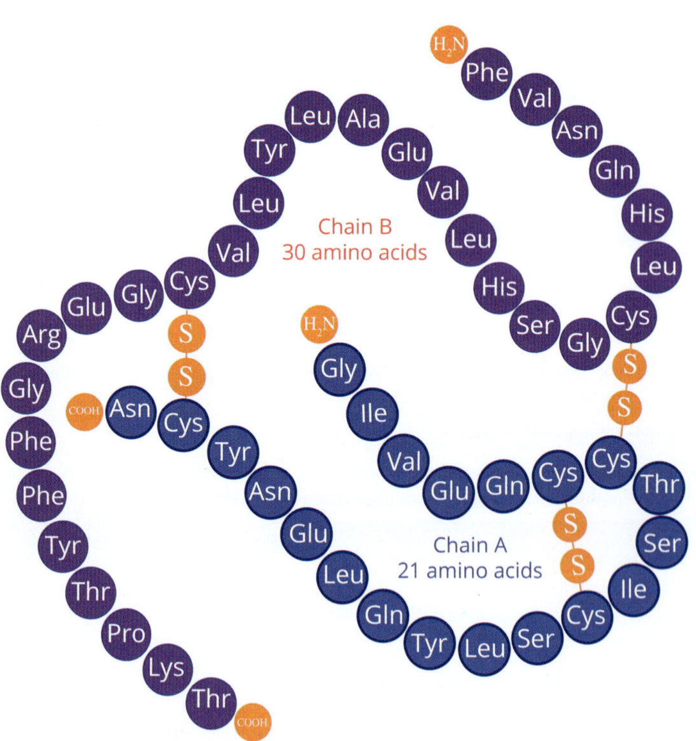

FIGURE 12.1.13 The amino acid sequence in the human insulin molecule. The polymer contains 51 amino acid units.

CASE STUDY **ANALYSIS**

Ada Yonath and the structure of the ribosome

Ada Yonath (Figure 12.1.14) was awarded the Nobel Prize in Chemistry in 2009 for working out the 3-dimensional structure of the ribosome.

Ada Yonath is an Israeli scientist who won a Nobel prize for doing something that many scientists thought was impossible—growing crystals of a ribosome and studying its structure using X-ray crystallography. When she did this work in 2000, the role of the ribosome and its chemical composition was already known, but it was important to work out the three-dimensional structure to be able to understand the way that this protein-making machine did its work.

The most difficult part of this process is finding the experimental conditions to grow a good crystal. For large molecules, such as a protein or DNA, it can take months to get the conditions right, so the enormous size of a ribosome makes this work especially challenging. Professor Yonath visited the University of Melbourne in 2014 and said, 'Very good scientists, including Nobel prize winners—such as Watson who, together with Crick, determined the DNA structure—tried for years to crystallise [ribosomes], and failed. There were papers saying it wasn't possible ... because [the ribosomes] are large and dynamic.'

Professor Yonath spent 20 years attempting to find experimental conditions that would encourage an enormous and rapidly-changing molecule to adopt order and long-range patterns. With this amazing persistence, she was finally able to determine the structure that is partly shown in Figure 12.1.10 on page 437.

FIGURE 12.1.14 Professor Ada Yonath, who won the 2009 Nobel prize in chemistry for determining the three-dimensional structure of a ribosome

Analysis

1 The first section of one of the proteins in a ribosome is shown in the figure below.

$$H_2N-\underset{CH_2-CH_2-S-CH_2}{CH}-\overset{O}{\overset{\|}{C}}-NH-\underset{H}{CH}-\overset{O}{\overset{\|}{C}}-NH-\underset{CH_2-C_6H_5}{CH}-\overset{O}{\overset{\|}{C}}-NH-\underset{CH_2(-CH_3)(-CH_3)}{CH}-\overset{O}{\overset{\|}{C}}----$$

a How many amino acid residues are in this part of the protein?

b Identify the amide functional groups by circling them.

c Provide the names for the amino acid residues that are present.

2 The last four amino acids at the C-terminal end of this protein have the three-letter codes Ala-Ala-Glu-Ser. Draw the structure of this polypeptide.

3 Explain why it was so difficult for Professor Yonath to grow crystals of the ribosome.

12.1 Review

SUMMARY

- 2-Amino acids have the structure shown in Figure 12.1.15.

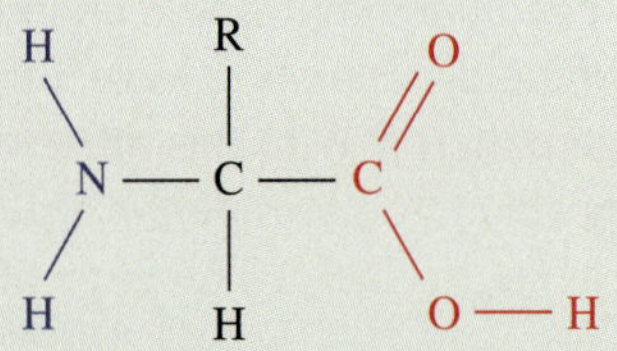

FIGURE 12.1.15 General structure of an amino acid

- Amino acids contain a carboxyl group (–COOH, highlighted in red in Figure 12.1.15), an amino group (–NH_2, highlighted in blue in Figure 12.1.15), a hydrogen atom and a side chain called the R group, all bonded to the same carbon atom.
- Three-letter abbreviations are commonly used to represent an amino acid.
- The R group, also known as the side chain, may be a non-polar group of atoms or a group of atoms that includes a polar functional group.
- The carboxyl group and amino group of two 2-amino acids can take part in a condensation reaction that links them through a peptide group.
- In a similar manner, several amino acids combine to produce a polypeptide.
- A polymer chain can grow when this reaction occurs a large number of times. Water is also produced as a by-product (Figure 12.1.16).
- There are 20 different 2-amino acids from which tens of thousands of different protein molecules are synthesised by cells. Each protein has a unique structure and function in the organism.

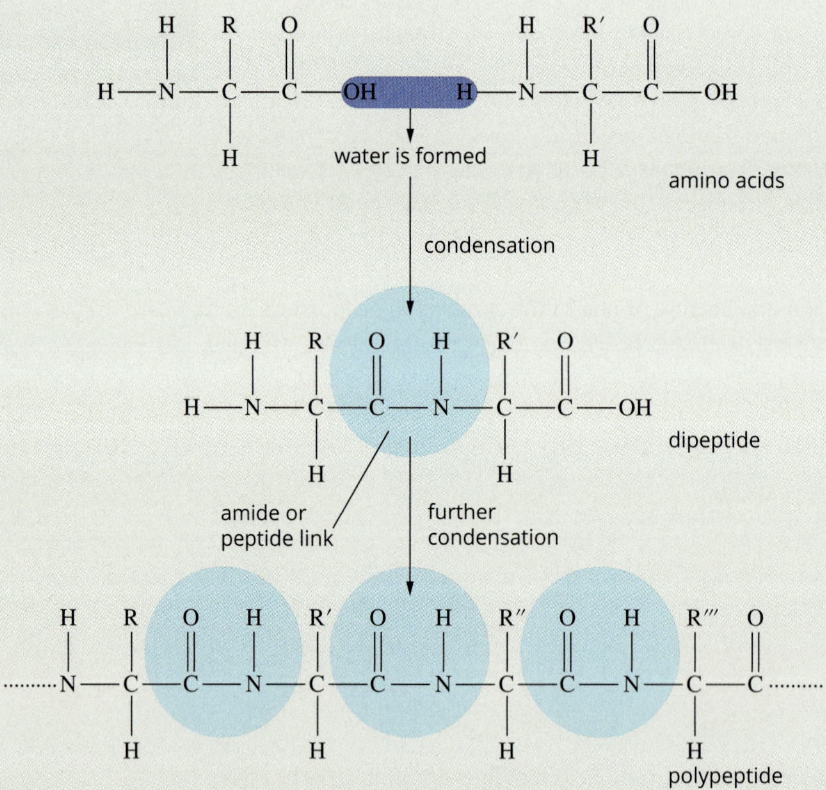

FIGURE 12.1.16 Condensation polymerisation of amino acids

KEY QUESTIONS

Knowledge and understanding

1 The following figure represents a section of a polypeptide molecule. Complete the paragraph by filling in the blanks using terms selected from the following list:

eight; leucine; alanine; nine; condensation polymerisation; addition polymerisation; methionine; amide; methane; valine; peptide

You may need to refer to the Table 12.1.1 to answer this question.

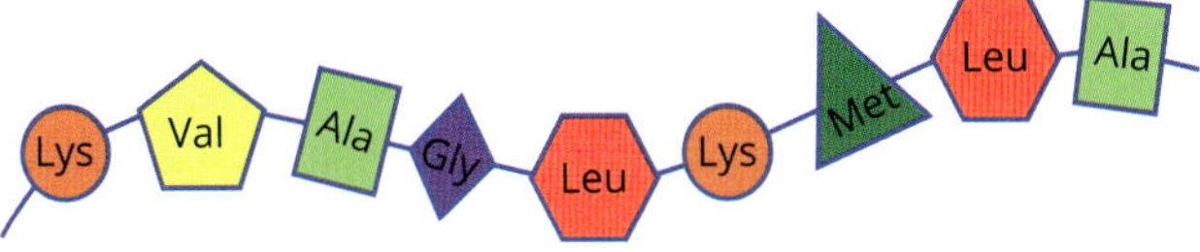

The figure shows a section of a polypeptide. It consists of _________ amino acid residues that are linked together by _____________ groups. It was produced by a/an ________________ _____________ reaction. In this polypeptide there are two leucine residues, two units of _____________, two units of ________________ and one unit each of _________________, ________________, ________________.

2 Fill in the gaps in the following sentences. using terms from the following list:

condensation; addition; amino acids; amide; water; peptide; biopolymers

Proteins are organic _________. The monomers required to make proteins are ____________ _______. The formation of a protein is an example of a/an ____________ polymerisation where the other product formed is ___________. The functional group that forms between monomers is a/an _______ group.

3 a Draw the two functional groups that are present in every amino acid.

b Draw the functional group formed when amino acids react together.

Analysis

4 The following figure shows three different molecules. Identify which molecule is a 2-amino acid and explain why the other two molecules are not 2-amino acids.

A

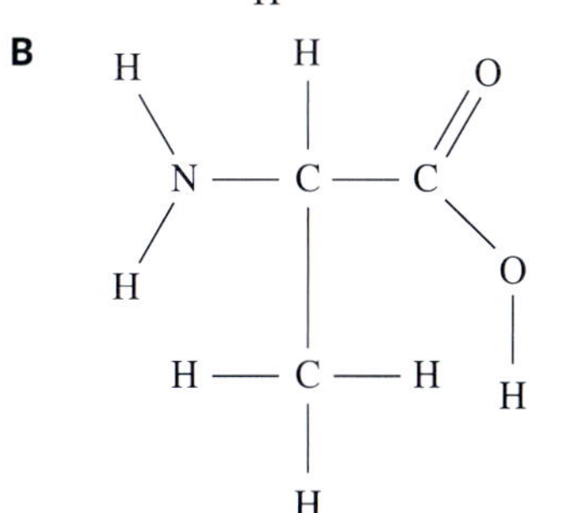

B

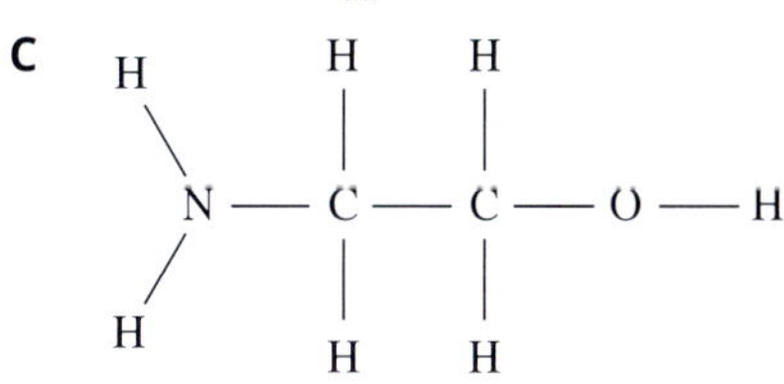

C

$H_2N—CH_2—CH_2—O—H$

5 The following figure shows a tripeptide molecule.

a Identify the parts labelled A–D.

b Give the names of the three amino acids in the tripeptide molecule.

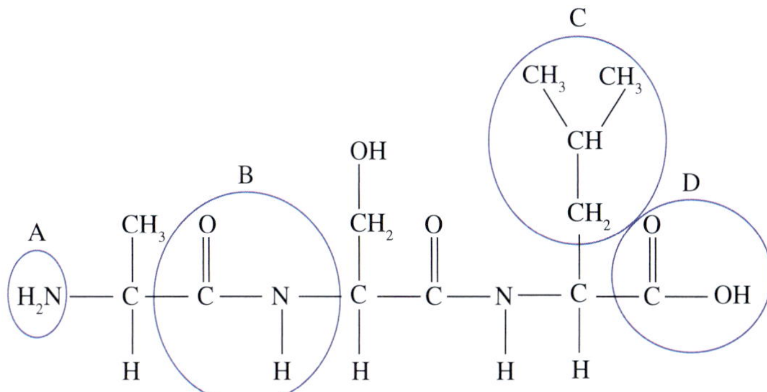

6 a Draw structural formulas of serine and cysteine.

b Using structural formulas, write an equation to show the formation of a dipeptide from these amino acids.

c Write the formula of another dipeptide that could be formed from these two amino acids.

d Name the type of reaction in part **b**.

12.2 Synthesis of starch, glycogen and lipids

In this section, you will examine two other classes of biological molecules that are important for energy storage in the body:

- the complex carbohydrates, starch and glycogen, which have glucose as their monomer unit
- lipids, which are commonly known as fats and oils. These are large molecules made from the reaction of carboxylic acids (fatty acids) with alcohols.

SYNTHESIS OF CARBOHYDRATES

There are millions of organic chemicals on Earth, and you do not need to look further than the plants and trees around you to find them. Figure 12.2.1 shows the sago palm, which is a common plant in the rainforest areas of Asia. Like many trees, the sago palm is a source of timber, however, unlike most trees, its centre forms the basis of a popular dessert—sago pudding.

The main structural component of all plants is the polymer cellulose. Over half the world's organic carbon is bound up in trees and other plants in the form of cellulose.

Grains such as corn, rice and wheat are the world's most important food crops. The common **nutrient** in these staple grains is starch. Nutrients are substances that provide nourishment for growth or other chemical processes that occur within an organism. **Starch** is the main component of flour, an essential ingredient of foods such as bread, cakes, sauces and pastry.

Cellulose and starch belong to a class of compounds called carbohydrates. In this section, you will learn about the chemical structure of common carbohydrates and their importance to nature and society.

FIGURE 12.2.1 (a) The sago palm is found in rainforest areas of Asia. (b) As the sago palm ages, the centre softens and forms a product that is harvested. These women are beating the soft centre into a pulp. (c) The soft centre of the sago palm is the main ingredient in many desserts such as sago pudding.

CARBOHYDRATES

Carbohydrates are made from the elements carbon, hydrogen and oxygen. Carbohydrates usually have the general formula $C_x(H_2O)_y$, where x and y are whole numbers. Carbohydrates range in size from small molecules, with relative molecular masses between 100 and 200, to very large polymers, with molecular masses greater than one million.

One of the simplest carbohydrates, **glucose** ($C_6H_{12}O_6$), is formed in the cells of green plants through the process of photosynthesis. **Photosynthesis** takes place in the presence of light and the process can be represented by the equation:

$$6CO_2(g) + 6H_2O(l) \xrightarrow{\text{light}} C_6H_{12}O_6(aq) + 6O_2(g)$$

Photosynthesis is an endothermic reaction, in which the Sun's energy is transformed into chemical potential energy in glucose. Living organisms use glucose as a source of energy. Animals cannot perform photosynthesis, so they need to consume plants or animals to meet their energy needs.

Monosaccharides

The smallest carbohydrates are the monosaccharides. They are white, sweet-tasting solids that are highly soluble in water. The structures of the three most common monosaccharides—glucose, fructose and galactose—are shown in Figure 12.2.2 on the following page. Note that these monosaccharides all have the same molecular formula, $C_6H_{12}O_6$, which means that they are isomers.

The structures of the monosaccharides look very similar, particularly glucose and galactose. However, the positions of the hydroxyl (–OH) groups in glucose and galactose are different. These slight variations in structure lead to significant differences in their functions in living organisms.

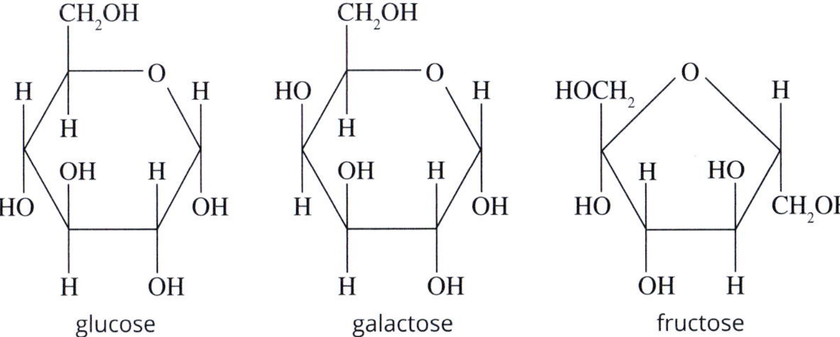

FIGURE 12.2.2 Structural formulas of glucose, galactose and fructose. Carbon atoms in the rings have been omitted for simplicity. There is a carbon atom at each corner of the ring structures, except where an oxygen atom is shown.

All three monosaccharides contain a number of polar hydroxyl groups, which means they can form hydrogen bonds with water. As a result, monosaccharides are highly soluble in water.

Glucose is found in all living things, especially in the juice of fruits, the sap of plants, and the blood and tissue of animals. As a product of photosynthesis, glucose functions as a key energy source in most forms of life.

Fructose is found in many fruit juices and honey. Its main role in the body is as an energy source, where it is used in much the same way as glucose. Figure 12.2.3 shows some fruits with high fructose contents.

FIGURE 12.2.3 Fructose is the most common sugar in fruit, especially berries.

Galactose is not found free in nature. However, it occurs as a component of larger carbohydrates like lactose, the main sugar found in milk.

Glucose, fructose and galactose are structural isomers, all with the molecular formula $C_6H_{12}O_6$. The differences in their structures appear slight, but are important to living organisms.

Disaccharides

Disaccharides are carbohydrates formed from the reaction between two **monosaccharide** molecules. A condensation reaction occurs between the hydroxyl functional groups on neighbouring molecules and a water molecule is formed as a by-product.

The two monosaccharides are joined by an oxygen atom. The connection between the monosaccharide units is an **ether functional group** (also known as a **glycosidic link** when found in carbohydrates). Figure 12.2.4 shows the formation of two important disaccharides, maltose and sucrose.

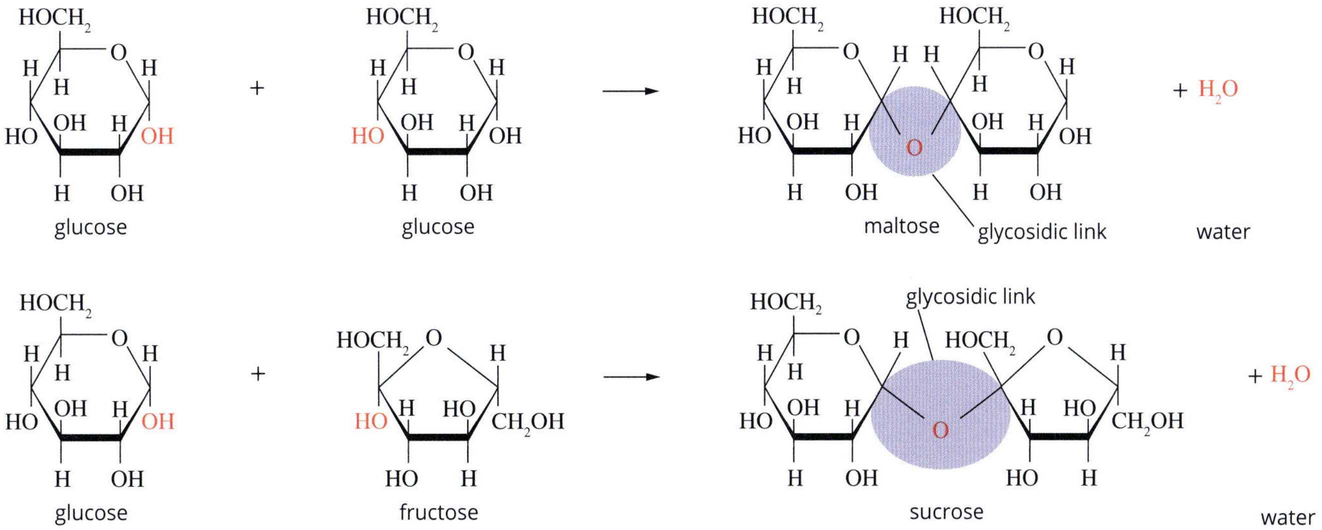

FIGURE 12.2.4 (a) The condensation of two glucose molecules forms maltose. (b) The condensation of glucose and fructose molecules forms sucrose. The glycosidic links are highlighted.

Maltose is added to foods as a sweetener. Barley, shown in Figure 12.2.5a, has a high maltose concentration and is used in the brewing industry as a raw material for fermentation into beer.

Sucrose is commonly referred to as table sugar. It is the most popular sweetener among the simple sugars. Sucrose is found in high concentrations in crops such as sugar beet and the sugarcane shown in Figure 12.2.5b.

When two monosaccharides react, a disaccharide is formed and a small molecule, such as water, is a by-product. The bond formed between the two monosaccharides is called a glycosidic link.

FIGURE 12.2.5 (a) Germinating barley seeds are a source of maltose. (b) The juice of sugarcane has a high sucrose content.

CHEMFILE

Sweetness

The smaller carbohydrates are often referred to as sugars because of their sweet taste. The lure of a sweet taste can lead people to eat more food than they require. The excess chemical energy consumed can end up being stored as fat in the body.

Sweetness is commonly understood by comparing with sucrose, which is assigned an arbitrary value of 1. Table 12.2.1 lists the comparative sweetness of other common sugars and artificial sweeteners.

This table shows that, despite the similar structures of sugars, their sweetness varies. Lactose has about one-sixth the sweetness of sucrose. Smaller proportions of artificial sweeteners are needed in processed foods to achieve the same sweetness as sugars. Aspartame, also known as food additive 951, is used as an artificial sweetener in diet soft drinks and as sugar-substitute tablets. The structure of aspartame is shown in the figure below. It has different functional groups and a different structure from carbohydrates.

The structure of aspartame

TABLE 12.2.1 Comparison of common sweeteners and their sources

Sweetener	Sweetness value	Source
lactose	0.16	milk, sugar
galactose	0.32	component of lactose
glucose	0.74	formed in photosynthesis
sucrose	1.00	table sugar
fructose	1.74	fruit sugar
cyclamate	30	artificial sweetener
aspartame	100–150	artificial sweetener
saccharin	450	artificial sweetener
neotame	7000–13 000	artificial sweetener

POLYSACCHARIDES

A condensation reaction between monosaccharides does not necessarily stop with the formation of a disaccharide. Reactions with other monosaccharides can continue to produce a polymer, a **polysaccharide**. A polysaccharide can contain thousands of monosaccharide units. These reactions occur at normal body temperature and are promoted by specific enzymes; the reaction rates are regulated by complex biological feedback pathways.

Polysaccharides are generally insoluble in water and have no taste. The three most important biological polysaccharides are starch, cellulose and glycogen. Simplified models of these polysaccharides can be seen in Figure 12.2.6. The properties of starch, cellulose and glycogen are very different from each other, yet they are all polymers of the same monomer, glucose. You will examine the synthesis of starch and glycogen below, and see more detail about cellulose in Section 12.3.

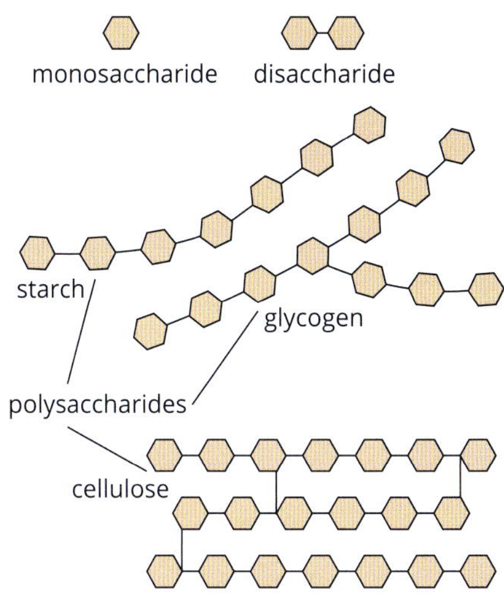

FIGURE 12.2.6 Representations of different carbohydrate molecules. They are all polymers of glucose.

Starch

Plants are able to produce and polymerise glucose molecules to form starch. The polymerisation is a condensation reaction, forming glycosidic links between each glucose molecule. This also produces many water molecules as a by-product. Plants use starch for storage of energy. When energy is needed, the plant can break down the starch back into glucose and utilise the glucose as an energy source to maintain its functions.

Foods such as potato and sago have a high starch content. If the starch forms a linear polymer, it is known as **amylose** (see Figure 12.2.7).

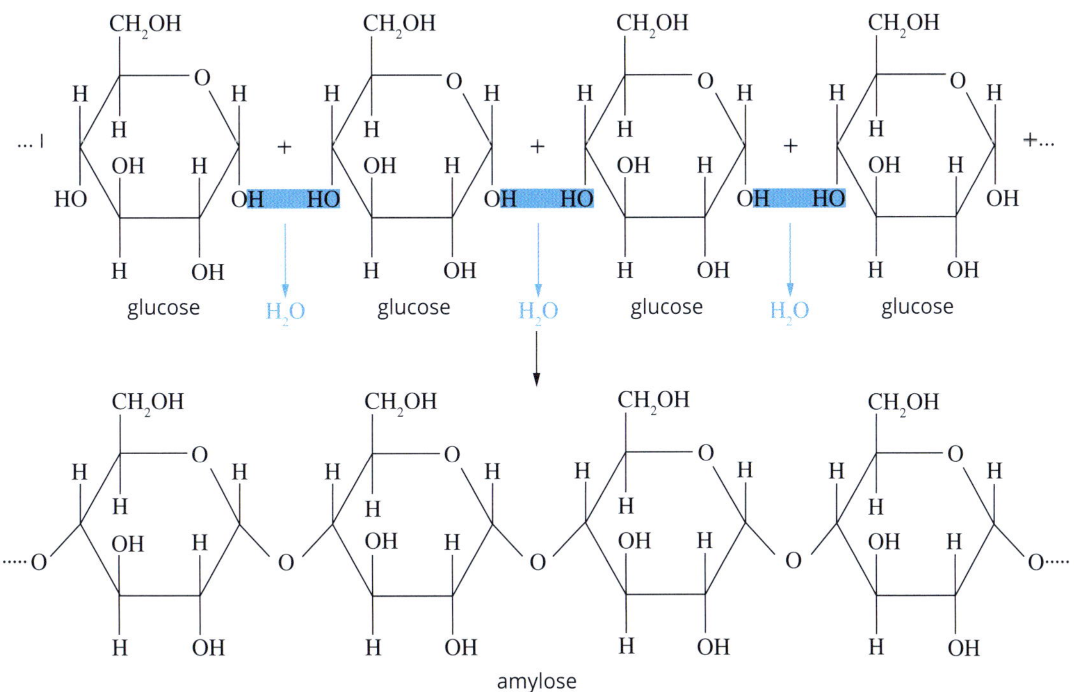

FIGURE 12.2.7 The condensation polymerisation of glucose to produce amylose, which is a form of starch

A second form of starch, known as **amylopectin**, can form if some of the glucose molecules undergo condensation reactions between hydroxyl groups at different positions around the glucose rings. In this way, occasional branches occur in the structure. Figure 12.2.8 shows that the branches along the polysaccharide chains occur after about 20 to 24 glucose units.

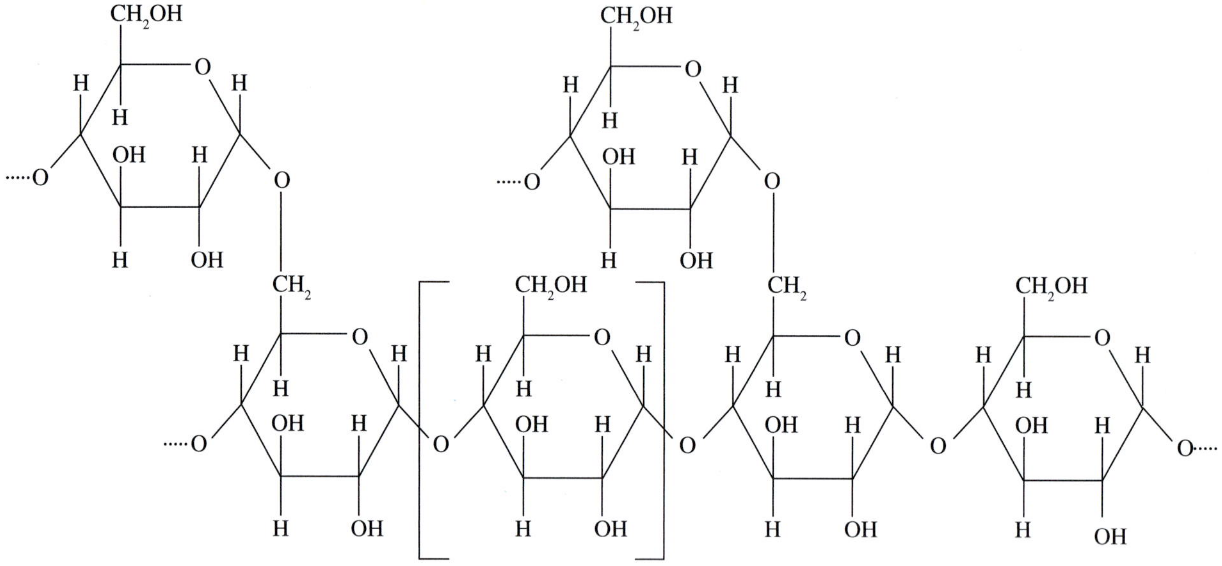

20–24 repeats

FIGURE 12.2.8 Amylopectin is a form of starch that has occasional branches along the glucose chains.

Amylose and amylopectin have different solubilities in water. The long molecules of amylose coil into spiral-like helices and pack together tightly, with many –OH groups inside the helices and away from contact with water. Therefore, amylose is largely insoluble in cold water. However, in the case of amylopectin, the branching of its molecules restricts the coiling of the polymer, leaving many more –OH groups exposed so that it dissolves in water.

Glycogen

The third polysaccharide formed from polymerisation of glucose is glycogen. **Glycogen** is highly branched in a similar fashion to amylopectin (Figure 12.2.9).

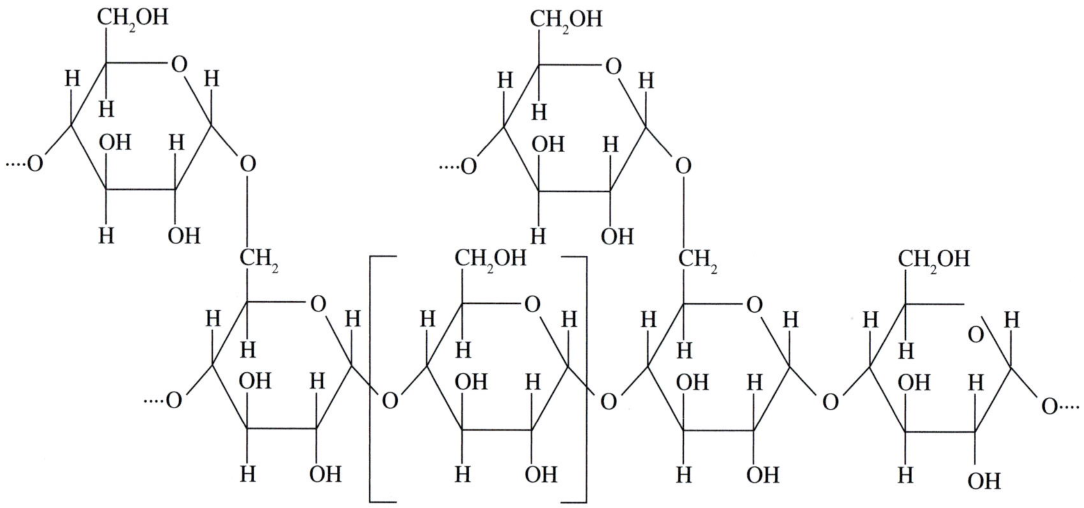

7–11 repeats

FIGURE 12.2.9 Glycogen is a highly branched polymer of glucose.

Animals use glycogen for energy storage. Glycogen is formed from excess glucose and stored in the liver or muscle tissue. If energy is needed, the glycogen can be broken down to glucose, which can then be used in cellular respiration.

CHEMFILE

Carbohydrates in solution

The behaviour of some carbohydrates when they are dissolved in water can be quite complex. For example, monosaccharides can exist as isomers in equilibrium. Glucose isomers are shown in the figure on the right.

The existence of these three isomers of glucose in solution allows scientists to understand why two different forms of glucose can be found in many organisms. The figure shows that β-glucose can be converted to α-glucose, and vice-versa, via the straight-chain form.

When the straight chain closes to make a ring, the –OH group bonded to carbon 1 can be above the ring or below the ring.

Glucose exists in solution in equilibrium between three isomers.

LIPIDS

Lipids (fats and oils) are present in meat, fish, dairy products, eggs and all fried foods. Canola (Figure 12.2.10) is one of the sources of oils used in food production. It is an excellent source of the unsaturated **fatty acids** used in margarine, cooking and salad oils. It is also the source of a by-product used as a livestock feed. There has been an increased demand for canola oil in recent years, which has led to many Australian farmers planting this crop.

FIGURE 12.2.10 A paddock of canola growing near Creswick in central Victoria

Fats and oils are a major energy source in your diet. Fats are used by animals to store chemical energy. In this section, you will learn about the reactions that produce fats and oils. You will see the differences in the structure of fats and oils of animal and vegetable origin, and the way this affects their melting points.

Triglycerides

Fats and oils are made up of large non-polar molecules known as triglycerides. Fats and oils have very similar chemical structures and are distinguished simply on the basis of their physical state at room temperature. At room temperature:

- fats are solids
- oils are liquids.

Being non-polar, triglycerides are unable to form hydrogen bonds with water, so fats are insoluble in water and oils are immiscible (cannot mix) with water.

Fats and oils belong to the class of biological molecules called lipids. Fats are solids at room temperature and oils are liquids at room temperature.

General structure of a fatty acid and glycerol

Triglycerides are made by condensation reactions between a glycerol molecule and three fatty acid molecules. Glycerol (propane-1,2,3-triol) is a relatively small molecule with three hydroxyl functional groups (Figure 12.2.11). Fatty acids have a carboxyl functional group attached to a long unbranched hydrocarbon chain, or 'tail'. This tail makes up the bulk of the molecule. Most have an even number of carbon atoms, usually between eight and 20.

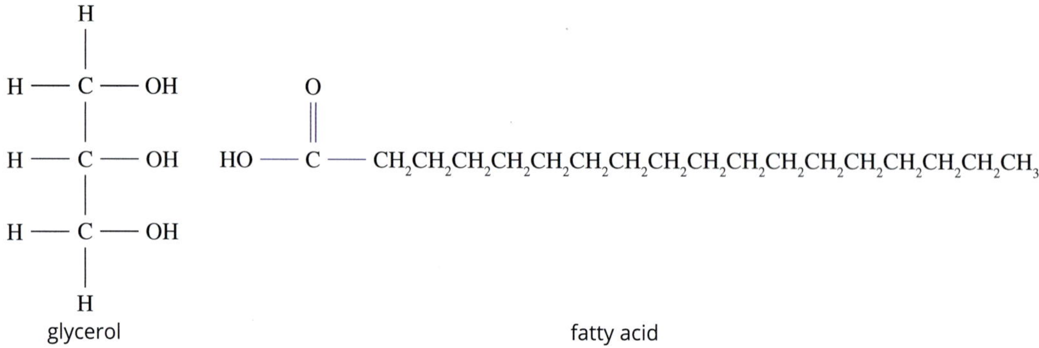

FIGURE 12.2.11 A glycerol molecule and a fatty acid molecule

Condensation reactions to form triglycerides

When a condensation reaction occurs between a molecule that contains a carboxyl group (–COOH), and a molecule that contains a hydroxyl group (–OH), an **ester functional group**, or **ester link** (–COO–) is formed, joining the two molecules. A molecule of water is also produced.

A **triglyceride** is produced by a condensation reaction involving one glycerol molecule and three fatty acids. In this reaction, three ester links form and three molecules of water are released, per triglyceride molecule formed. This process is shown in Figure 12.2.12.

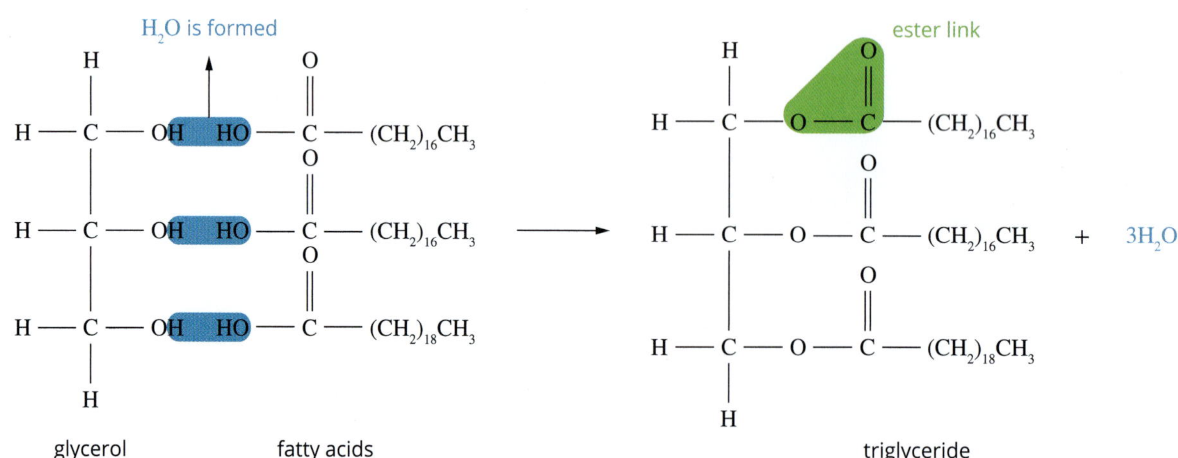

FIGURE 12.2.12 Equation showing a reaction between glycerol and fatty acids to produce a triglyceride and water

Most triglycerides have two or three different fatty acid hydrocarbon chains. The hydrocarbon chains may differ in length. Some hydrocarbon chains may also contain one or more carbon–carbon double bonds.

Saturated and unsaturated fatty acids

Fats are classified based on the structural features of the hydrocarbon chains of their fatty acid components.

- Saturated fatty acids have hydrocarbon chains that contain only single carbon–carbon bonds. Stearic acid, which occurs widely in meats, has a semi-structural formula of $CH_3(CH_2)_{16}COOH$ and a molecular formula of $C_{17}H_{35}COOH$. Saturated fatty acids have the general formula $C_nH_{2n+1}COOH$.
- Monounsaturated fatty acids contain one carbon–carbon double bond in their hydrocarbon chain. The main dietary example is oleic acid, which is found in a number of vegetable oils. Its semi-structural formula is $CH_3(CH_2)_7CH{=}CH(CH_2)_7COOH$, and its molecular formula is $C_{17}H_{33}COOH$. Monounsaturated fatty acids have the general formula $C_nH_{2n-1}COOH$.
- Polyunsaturated fatty acids contain more than one carbon–carbon double bond in their hydrocarbon chain. Fish and vegetable oils are the main dietary source of polyunsaturated fatty acids. For example, sunflower oil is a good source of linoleic acid. Linoleic acid has the semi-structural formula $CH_3(CH_2)_4CH{=}CHCH_2CH{=}CH(CH_2)_7COOH$ and its molecular formula is $C_{17}H_{31}COOH$.

The structure and general formula of stearic acid, oleic acid and linoleic acid are shown in Figure 12.2.13.

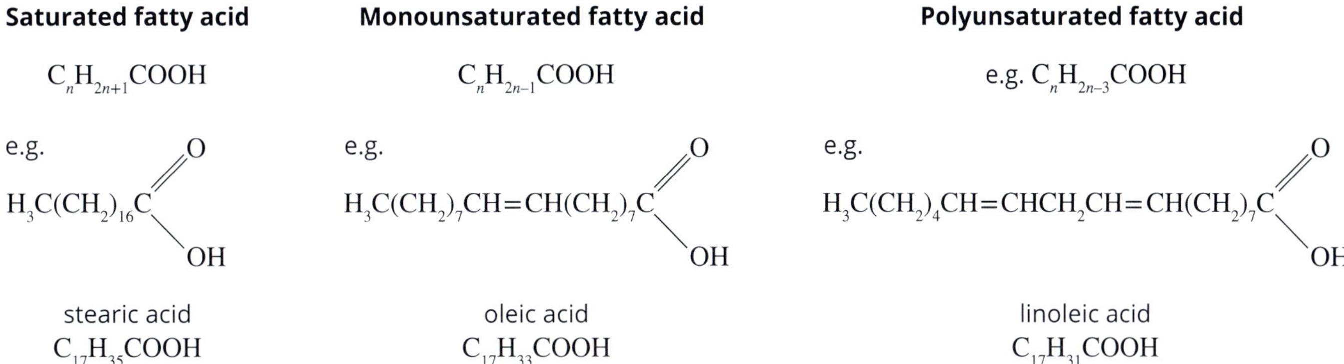

FIGURE 12.2.13 The general formula and semi-structural formula of a representative saturated fatty acid, monounsaturated fatty acid and polyunsaturated fatty acid

12.2 Review

SUMMARY

- Carbohydrates are made from the elements carbon, hydrogen and oxygen and usually have the formula $C_x(H_2O)_y$, where x and y are whole numbers.
- The main role of carbohydrates is as a source of energy for living things. Energy from the Sun is transformed to chemical potential energy in carbohydrates.
- The smallest carbohydrates are monosaccharides.
- Glucose is an important monosaccharide. It is synthesised in plants through the process of photosynthesis.
- Monosaccharides are highly soluble in water as they contain several polar hydroxyl groups, allowing them to form hydrogen bonds with water molecules.
- Disaccharides are formed from condensation reactions between two monosaccharide molecules. The links formed between monosaccharides are known as ether or glycosidic links.
- Monosaccharides and disaccharides are often referred to as sugars because of their sweet taste.
- Glucose can undergo condensation reactions to form a range of important polymers known as polysaccharides. The polysaccharides have different properties due to the extent of branching in the polymer chain.
- Cellulose is a structural material in plants, starch is an energy storage material in plants and glycogen is an energy storage material in animals.
- Fats and oils are composed of molecules called triglycerides.
- Triglycerides have a number of important roles in the human body, including as energy storage and insulation, and as a source of fatty acids from which other important compounds are produced.
- Fatty acids are carboxylic acids with a long hydrocarbon chain.
- Triglycerides are formed in a condensation reaction between glycerol and fatty acids.
- Three ester links are formed in the condensation reaction between a glycerol molecule and three fatty acids to form a triglyceride molecule.

KEY QUESTIONS

Knowledge and understanding

1 Small molecules can react to form a larger biomolecule.

- **a** Name and draw the functional groups that react to form a disaccharide.
- **b** Name and draw the functional group that forms when a disaccharide is produced.
- **c** Name and draw the functional groups that react to form a triglyceride.
- **d** Name and draw the functional group that forms when a triglyceride is produced.

2 Classify each of the following as monosaccharide, disaccharide or polysaccharide:

- **a** amylose
- **b** fructose
- **c** sucrose
- **d** lactose
- **e** cellulose

3 Glucose is described as a monosaccharide, maltose as a disaccharide and glycogen as a polysaccharide.

- **a** What is the difference between these three types of carbohydrates?
- **b** What type of reaction is involved in converting glucose to maltose and then to glycogen?

4 Select the correct statements about starch (there may be more than one answer).

- **A** Starch is a disaccharide.
- **B** There are two common forms of starch.
- **C** All forms of starch are highly soluble in water owing to the large number of hydroxyl groups.
- **D** Starch is an energy storage molecule in plants.
- **E** Starch is a polymer of sucrose.
- **F** Many glycosidic links are formed when glucose reacts to form starch.
- **G** Starch is composed of relatively large molecules.

Analysis

5 A fat present in a vegetable oil has the structure shown in the figure at right.

$$CH_2-O-\overset{O}{\overset{\|}{C}}-(CH_2)_{14}CH_3$$
$$CH-O-\overset{O}{\overset{\|}{C}}-(CH_2)_{14}CH_3$$
$$CH_2-O-\overset{O}{\overset{\|}{C}}-(CH_2)_{14}CH_3$$

- **a** Circle an ester functional group.
- **b** Is this triglyceride saturated or unsaturated? Explain your answer.

6

- **a** Write an equation for the formation of the triglyceride glyceryl trioleate from the reaction of glycerol and oleic acid, $CH_3(CH_2)_7CH{=}CH(CH_2)_7COOH$.
- **b** Draw the structure of glyceryl trioleate, showing the ester bonds.

7 Explain why fatty acids are relatively insoluble in water.

12.3 Hydrolysis of biomolecules

The food you consume contains the raw materials that are processed and modified in your body (Figure 12.3.1). Chemical bonds are broken and new bonds are formed as the molecules you need to survive are produced or extracted. These raw materials, such as proteins, carbohydrates and triglycerides, are more commonly called nutrients, with different nutrients having different primary functions in the body.

Earlier in this chapter, you learnt about the formation of new bonds when triglycerides, carbohydrates and proteins are synthesised. In this section, you will examine the reactions that occur when the bonds are broken. The body requires each of these nutrients in relatively large quantities, and they are broken into smaller molecules by the digestive system. Carbohydrates and triglycerides are the primary sources of energy for living systems. While proteins can also provide energy, their main role is in growth and tissue repair.

FIGURE 12.3.1 Your body takes in food as raw materials and processes it into molecules that you require.

HYDROLYTIC REACTIONS

The human body does not break down molecules into individual atoms; rather it reduces compounds to smaller soluble molecules that form the building blocks for new substances. This process is known as **hydrolysis**. These molecules are then transported to different parts of the body where they are reassembled.

The breakdown and subsequent rebuilding of the nutrient molecules that are needed to sustain life, such as proteins, carbohydrates and triglycerides, involves two main types of chemical reactions.

- Hydrolysis: reactions involving hydrolysis are also referred to as **hydrolytic reactions**. They involve splitting large molecules, such as proteins, fats and complex carbohydrates, by their reaction with water molecules.
- Condensation: the condensation reactions that you learnt about earlier in this chapter occur in other parts of the body, away from the digestive system. These reactions involve joining two smaller molecules to form a larger molecule with the elimination of a small molecule (water).

Enzymes are crucial for both types of reactions to occur in biological systems. Each step uses a particular **enzyme** that increases the reaction rate. Some enzymes catalyse only one particular reaction, whereas other enzymes are specific for a reaction involving a particular type of chemical bond or functional group. You will learn more about enzymes in Chapter 15.

Hydrolysis and condensation reactions can be regarded as the opposite of each other. Figure 12.3.2 on the following page shows a number of important reactions involved in digestion. The forward reaction in these examples shows the formation of larger **biomolecules**. The reverse reaction illustrates their breakdown through hydrolysis.

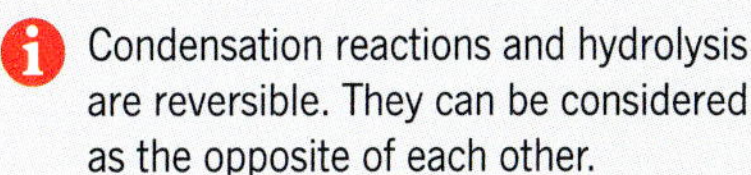

Condensation reactions and hydrolysis are reversible. They can be considered as the opposite of each other.

Condensation reactions

proteins

carbohydrates

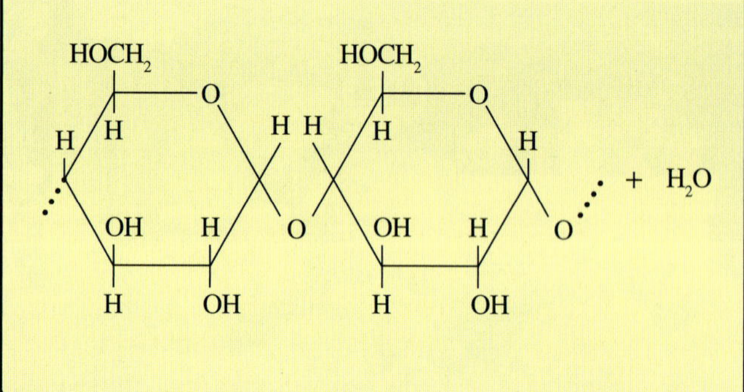

triglycerides

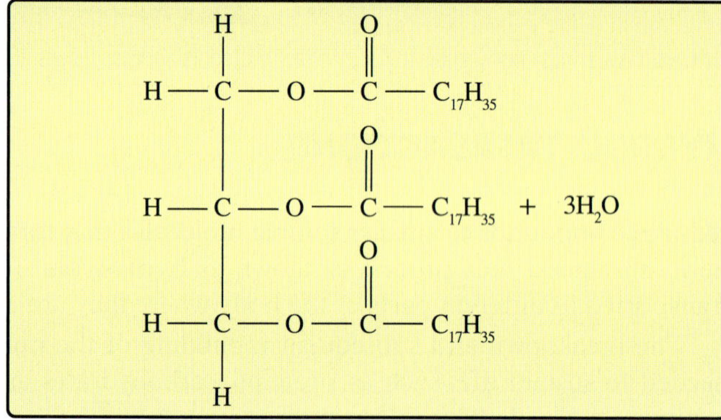

Hydrolysis (hydrolytic reactions)

FIGURE 12.3.2 Condensation and hydrolytic reactions involving proteins, carbohydrates and triglycerides

Figure 12.3.3, on the following page, illustrates the cycle of hydrolysis and condensation for proteins, carbohydrates and fats.

- Proteins react to form amino acids by hydrolysis of the amide bonds. The soluble amino acids that form are transported through the body to cells. Once inside cells, amino acids can be reassembled in condensation reactions to form new proteins.
- Polysaccharides are broken down to monosaccharides and disaccharides by the hydrolysis of ether bonds. The soluble sugars are transported through the body to be used for the production of energy. Monosaccharides can also be converted back to polysaccharides such as glycogen for use as energy stores.
- Triglycerides form fatty acids and glycerol when the ester bonds are hydrolysed. The fatty acids and glycerol are then transported through the body. Fatty acids and glycerol can be converted back to triglycerides or used to produce energy.

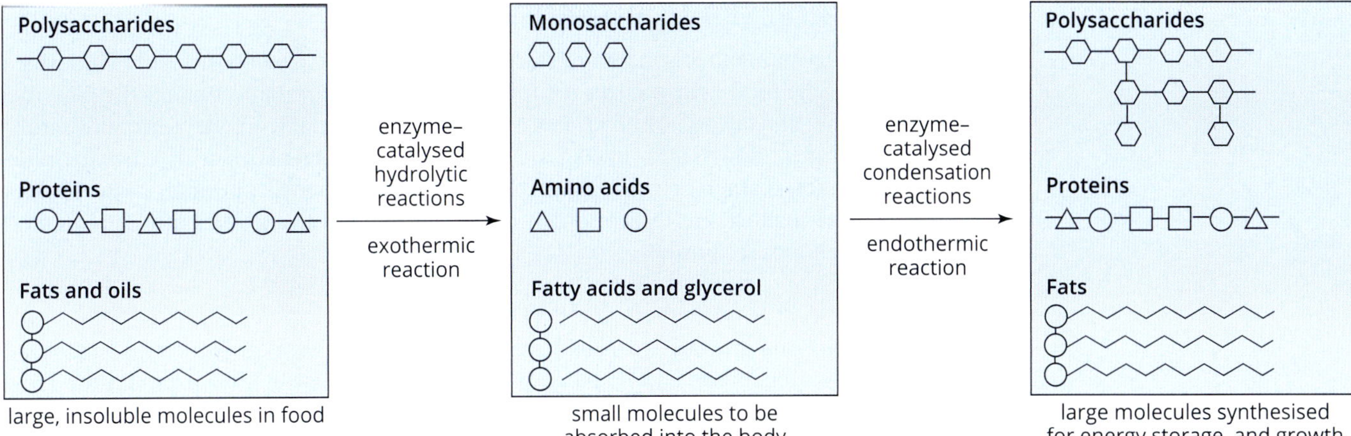

FIGURE 12.3.3 The breakdown and use of nutrients in the body involves numerous condensation and hydrolysis reactions.

In general, for these chemical processes that occur within a living cell or organism **(metabolism)**:

- condensation reactions tend to be endothermic, requiring energy to form larger molecules
- hydrolytic reactions tend to be exothermic, releasing energy as bonds are broken in the formation of smaller molecules.

Digestion

In animals, the metabolism of food starts in the digestive system. As soon as food enters the mouth, it begins to be broken down into smaller molecules in a process called **digestion**. Digestion is a complex process that involves large numbers of separate enzymes throughout the digestive system, which break down different components of the food.

A general outline of the metabolism of food through digestion in the body is shown in Figure 12.3.4.

Digestion is one aspect of your metabolism. It breaks down food into the essential nutrients required by your body.

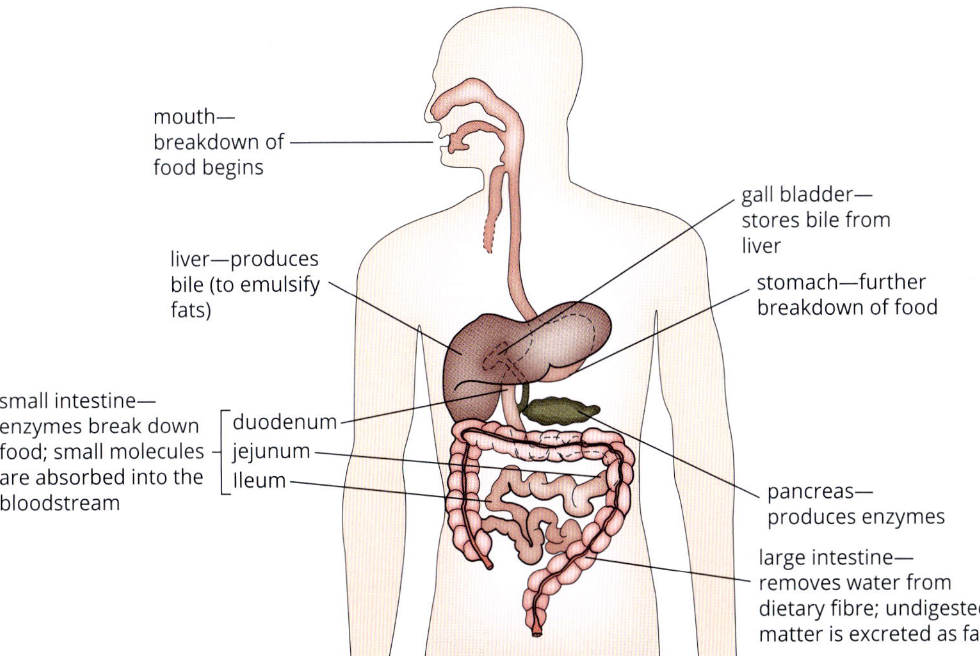

FIGURE 12.3.4 Digestion starts as soon as food enters the mouth. As the food progresses down the gastrointestinal tract, different processes break down different components of the food. These reactions are catalysed by specific enzymes.

HYDROLYSIS OF CARBOHYDRATES

The digestion of carbohydrates starts in your mouth. Chewing increases the surface area of the food. It also mixes the food with your saliva, which contains the digestive enzyme amylase. Carbohydrates in your food (usually in the form of starch) are broken down into smaller disaccharides. The breakdown of carbohydrates continues further in your digestive system. In the small intestine, disaccharides such as maltose are broken down by specific digestive enzymes into glucose and other monosaccharides.

Starch and glycogen hydrolysis

The enzyme amylase in saliva hydrolyses starch in the food you eat to maltose, a disaccharide. If you chew a piece of bread and then leave it in your mouth for a few minutes, you will notice a sweet taste. The sweet taste is the result of hydrolysis of starch to maltose. You can see the structure of starch and a summary of the hydrolysis reactions in Figure 12.3.5. The process for glycogen is broadly similar. It is the glycosidic links between each monosaccharide unit that are broken by hydrolysis reactions during digestion. The hydrolysis can continue in the small intestine where the enzyme maltase is produced. Maltase hydrolyses maltose to glucose.

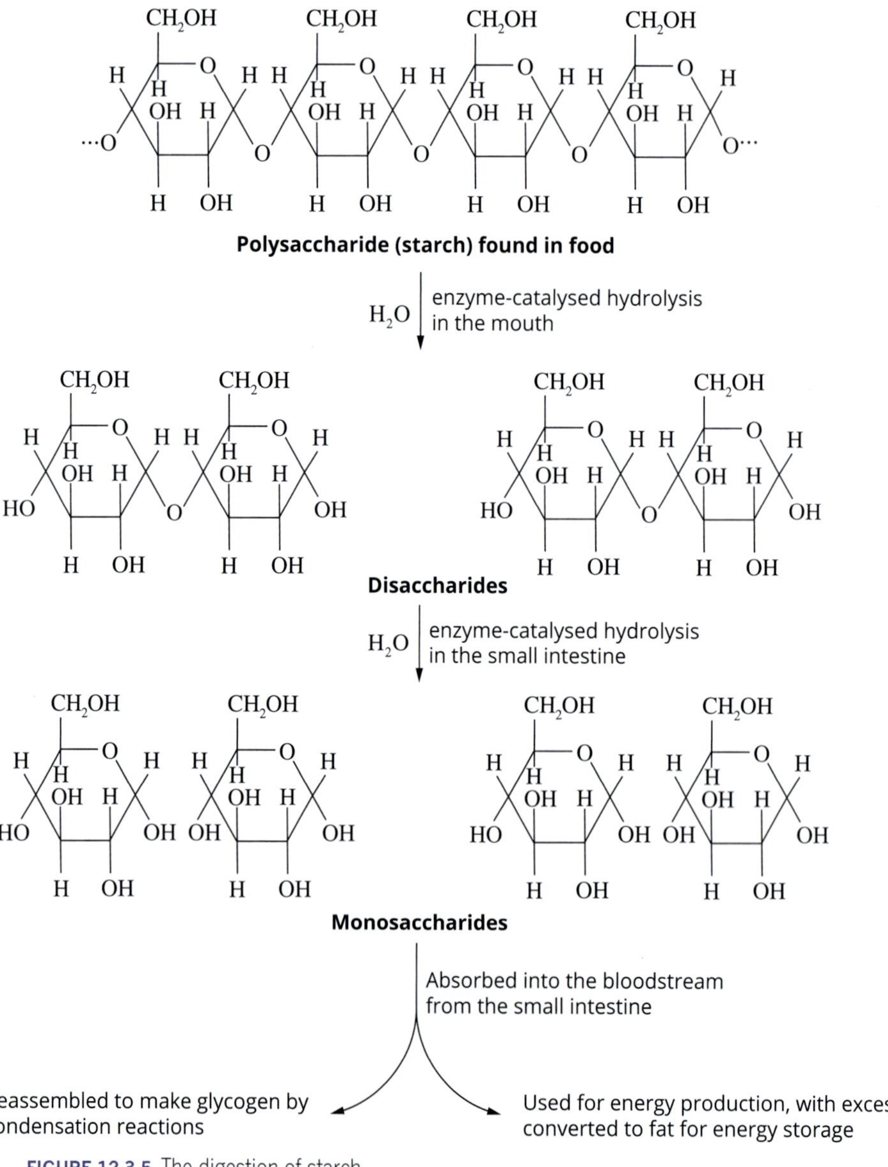

FIGURE 12.3.5 The digestion of starch

Glucose molecules are highly soluble because hydrogen bonds can form between the numerous hydroxyl groups in the molecule and water. As a result, glucose dissolves in the blood and is transported to different parts of the body. Some glucose molecules are used to produce energy through respiration, while others are used to synthesise energy storage molecules such as glycogen.

Cellulose hydrolysis

The polysaccharide cellulose is present in large amounts in cereals, fruits and vegetables. It is better known as dietary fibre or roughage. Cellulose is rapidly hydrolysed by the enzyme cellulase.

Most animals, including humans, do not have the enzyme cellulase. Cellulose passes through your digestive system relatively unchanged, but a small percentage can be hydrolysed by bacteria in the gut. A diet high in cellulose or fibre provides bulk to help food pass through the digestive system. Fibre helps prevent constipation, haemorrhoids and colon cancer.

Cows and sheep can digest large amounts of cellulose because bacteria living in their gut produce the enzyme cellulase. A cow's stomach has four compartments and grass spends a considerable time in the first two compartments being broken down before moving through the rest of the digestive system. When a cow first eats, it chews the food just enough for it to travel to the first compartment. Later the cow will regurgitate the food as 'cud' and chew it thoroughly before the food moves progressively to the other compartments. Figure 12.3.6 shows the complexity of a cow's stomach.

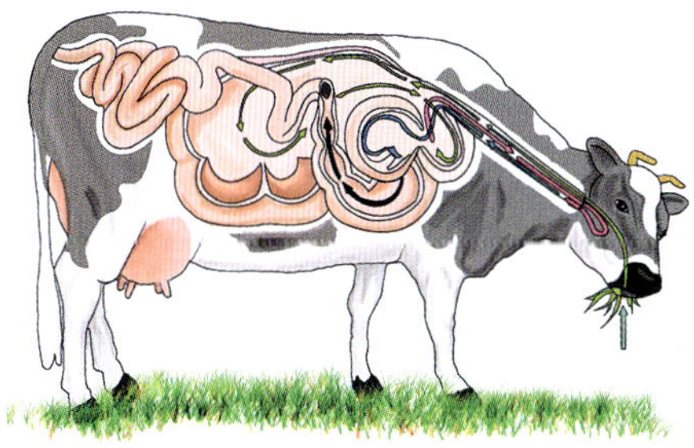

FIGURE 12.3.6 A cow's digestive system is more complex than a human's. This allows the cow to break down cellulose.

Koalas eat an even higher fibre diet than cows do. Koalas have a large caecum (a bag of microorganisms) in their gut at the junction between the small and large intestine. Microbial digestion of cellulose occurs there; bacteria break down cellulose into smaller molecules that can be absorbed. Baby koalas build up levels of these microbes by eating their mother's faeces. Baby elephants (Figure 12.3.7) also feed on the dung of their mothers to build up the level of microbes in their gut.

FIGURE 12.3.7 Baby elephants feed on the dung of their parent to build up levels of microorganisms in their own stomach. These microorganisms break down cellulose into smaller molecules.

CHEMFILE

Chocolate-covered cherries

Chemist H.S. Paine invented liquid-centred chocolates by exploiting the differences in solubility of the disaccharide sucrose and the monosaccharides glucose and fructose. In the presence of very little water, sucrose forms a paste-like solid, whereas glucose and fructose are very soluble. Paine also knew that the enzyme invertase will hydrolyse sucrose to glucose and fructose according to the reaction:

$$\underset{\text{sucrose}}{C_{12}H_{22}O_{11}(aq)} + H_2O(l) \xrightarrow{\text{invertase}} \underset{\text{glucose}}{C_6H_{12}O_6(aq)} + \underset{\text{fructose}}{C_6H_{12}O_6(aq)}$$

He added a small amount of invertase to a sucrose paste and moulded it around a cherry. This was then dipped in molten chocolate. The enzyme remains active inside the chocolate and hydrolyses sucrose into the more soluble sugars.

It takes several weeks for the paste to turn into a sweet liquid like the one seen in the figure below because the chocolates are kept at 18°C. At this temperature, enzyme activity is low, slowing down the reaction rate. However, it is well worth the wait because chocolate-covered cherries are absolutely delicious.

The enzyme invertase is used to break down a paste of sucrose into a solution containing the more soluble fructose and glucose sugars to make the sweet liquid in chocolate-covered cherries.

CHEMFILE

Throw another locust on the barbie!

Many insects are edible and a source of nutrients.

As the world's population grows, concerns increase regarding the capacity to feed everyone on the planet. Recently, nutritionists have been suggesting a surprising solution—insects!

Over 200 species of insects have been declared edible, as shown in the figure above, and they are accepted items on the menu in many countries. Insects are easy and fast to breed, placing little demand upon Earth's limited resources such as land.

However, the main advantage of consuming insects is their nutritional value. They are generally high in protein but low in total fat. The fat that they have is mainly monounsaturated fat. They are rich in micronutrients such as copper, iron, manganese, selenium and zinc as well as vitamins. The nutrient level varies with the body part, so consumption of the whole insect is recommended.

HYDROLYSIS OF PROTEINS

In humans, the digestion of proteins produces individual amino acids, which are used by the body to synthesise enzymes and other essential proteins. The three-dimensional structure is unravelled by a combination of hydrochloric acid and the muscular contractions in the stomach. Proteins start to be broken down in the stomach by the enzyme pepsin, which is found in the stomach. This produces shorter polypeptide chains that move into the duodenum. Once in the small intestine, the polypeptides are broken down into smaller dipeptides. Further along in the intestine dipeptides are broken down into amino acids (Figure 12.3.8).

Hydrolysis of proteins can be done in a laboratory, and this is often an early step in determining their sequence of amino acids. In contrast to the body temperature process in organisms, quite extreme reaction conditions are needed. These reactions require 6 M solutions of hydrochloric acid and heating for 24 hours at 100°C to hydrolyse the amide bonds.

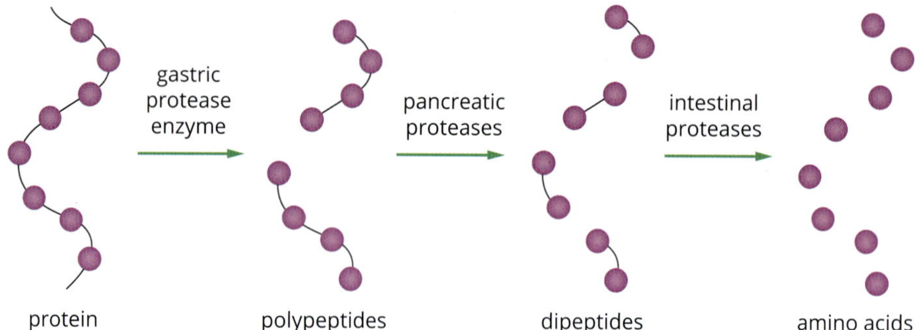

FIGURE 12.3.8 The digestion of proteins starts in the acidic environment of the stomach with the enzyme pepsin. At each stage of digestion, the peptide chains are broken into smaller units until eventually individual amino acids are released.

HYDROLYSIS OF TRIGLYCERIDES

Like proteins and carbohydrates, triglycerides are hydrolysed by enzymes when they are digested. However, unlike proteins and carbohydrates, triglycerides are insoluble in water, so their molecules remain intact as they pass through the digestive tract until they reach the small intestine. In the small intestine, **bile** is used to process the triglycerides.

Bile emulsifies fats, breaking them down into smaller particles and dispersing them as small droplets. This process of creating an **emulsion** is similar to how detergents work on the fat left in a frying pan after cooking fatty food. The effect of emulsification on fats in the small intestine is shown in Figure 12.3.9.

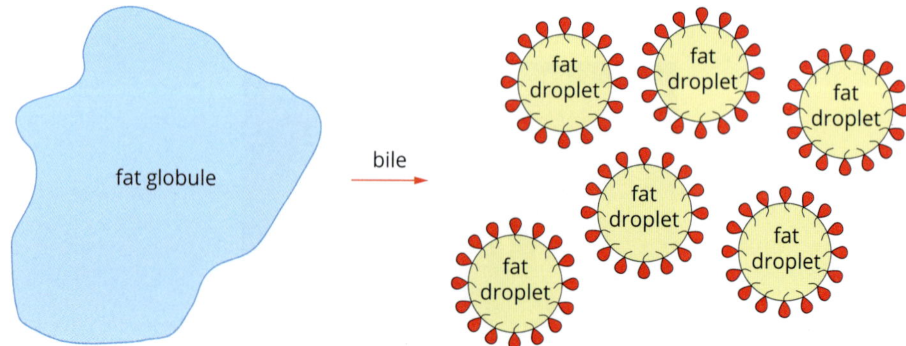

FIGURE 12.3.9 A triglyceride enters the small intestine as a fat globule or an oil globule. The bile turns this globule into an emulsion, consisting of much smaller fat droplets with bile particles embedded in the surface. The surface area of the fat exposed for hydrolysis by the water-soluble lipase enzyme is now much greater.

Emulsification of fats increases their surface area, so that more triglyceride molecules are exposed to the aqueous environment.

In your body, an enzyme called lipase breaks down triglycerides. However, lipase is a water-soluble protein, so it can only interact at the surface of the non-polar fat globules. Emulsification of fats by bile increases the surface area of the fats, which means that lipase can access more triglyceride molecules, which increases the rate of hydrolysis.

Lipase enters the intestine from the pancreas. The enzyme catalyses the hydrolysis of the three ester bonds in the triglyceride molecules. You can see in Figure 12.3.10 how the triglyceride molecules undergo hydrolysis to form glycerol and fatty acids.

$$CH_2-O-\overset{\overset{\displaystyle O}{\|}}{C}-CH_2CH_2CH_2CH_2CH_2CH_2CH_2CH_2CH_2CH_2CH_2CH_2CH_2CH_2CH_2CH_2CH_3$$
$$CH-O-\overset{\overset{\displaystyle O}{\|}}{C}-CH_2CH_2CH_2CH_2CH_2CH_2CH_2CH_2CH_2CH_2CH_2CH_2CH_2CH_2CH_2CH_2CH_3$$
$$CH_2-O-\overset{\overset{\displaystyle O}{\|}}{C}-CH_2CH_2CH_2CH_2CH_2CH_2CH_2CH_2CH_2CH_2CH_2CH_2CH_2CH_2CH_2CH_2CH_3$$

$$\Big\downarrow\ H_2O,\ \text{lipase enzyme}$$

$$CH_2-OH \qquad HO-\overset{\overset{\displaystyle O}{\|}}{C}-CH_2CH_2CH_2CH_2CH_2CH_2CH_2CH_2CH_2CH_2CH_2CH_2CH_2CH_2CH_3$$
$$CH-OH \qquad HO-\overset{\overset{\displaystyle O}{\|}}{C}-CH_2CH_2CH_2CH_2CH_2CH_2CH_2CH_2CH_2CH_2CH_2CH_2CH_2CH_2CH_3$$
$$CH_2-OH \qquad HO-\overset{\overset{\displaystyle O}{\|}}{C}-CH_2CH_2CH_2CH_2CH_2CH_2CH_2CH_2CH_2CH_2CH_2CH_2CH_2CH_2CH_3$$

glycerol   fatty acids

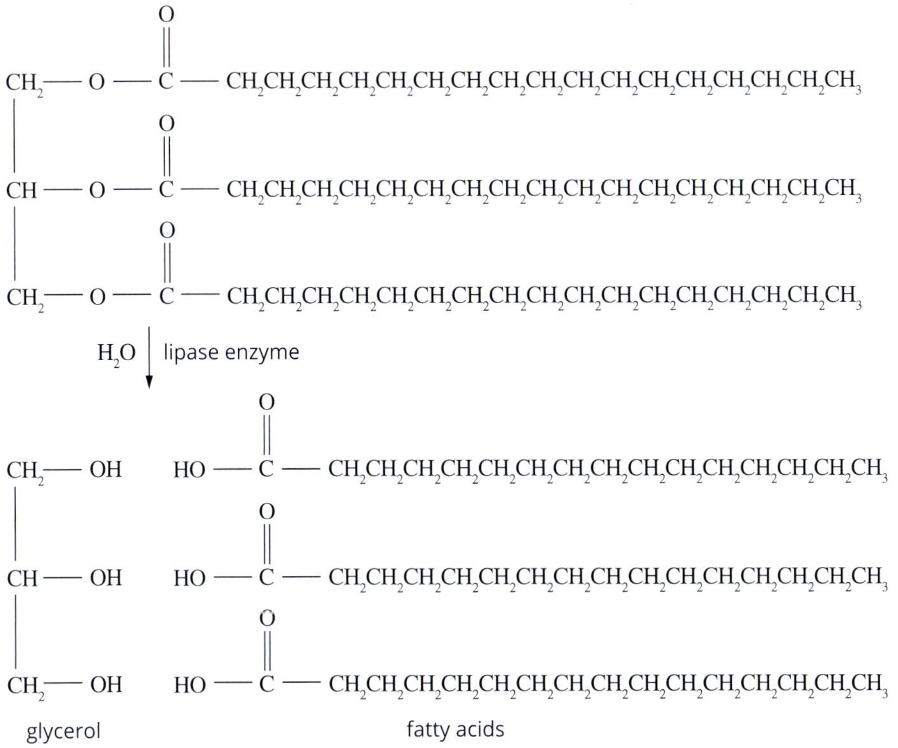

FIGURE 12.3.10 Hydrolysis of a triglyceride to glycerol and three fatty acid units

The three fatty acids will not necessarily have the same structure. Figure 12.3.11 shows a general equation for the hydrolysis of a triglyceride, where the letter R is used to represent each fatty acid hydrocarbon chain.

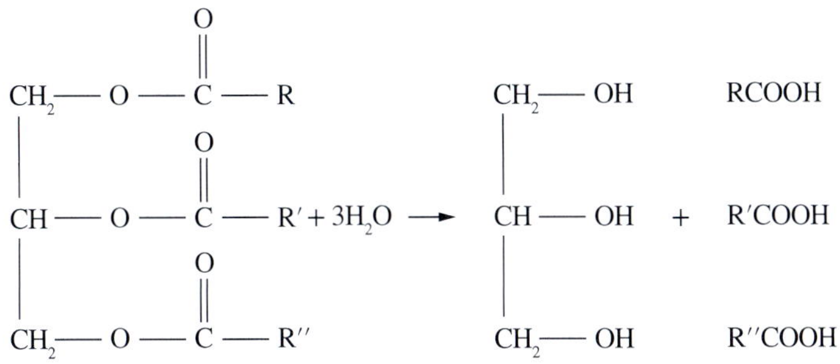

$$\begin{array}{lcll} CH_2-O-\overset{\overset{\displaystyle O}{\|}}{C}-R & & CH_2-OH & RCOOH \\ CH-O-\overset{\overset{\displaystyle O}{\|}}{C}-R' + 3H_2O & \longrightarrow & CH-OH \quad + & R'COOH \\ CH_2-O-\overset{\overset{\displaystyle O}{\|}}{C}-R'' & & CH_2-OH & R''COOH \end{array}$$

FIGURE 12.3.11 General equation for the hydrolysis of a triglyceride. R, R′ and R″ represent different hydrocarbon chains.

The glycerol and fatty acids that are produced pass into the bloodstream to the liver, where they are re-formed into triglycerides.

12.3 Review

WS 26 PA 18 OA ✓✓

SUMMARY

- Large biomolecules are often broken down in the body to smaller soluble molecules for ease of transport in the body. They can then be rebuilt in a different part of the body.
- The breakdown of large molecules in food occurs through hydrolytic reactions (hydrolysis).
- Figure 12.3.12 below summarises the breakdown and synthesis reactions for the key nutrient families of carbohydrates, triglycerides and proteins.
- Carbohydrates are hydrolysed by enzymes during digestion to disaccharides and then to monosaccharides. The monosaccharides can be transported in the bloodstream to cells for the production of energy or storage as glycogen (Figure 12.3.5).
- Triglycerides are hydrolysed by enzymes during digestion. The products are glycerol and the fatty acids from which the triglyceride was originally made.
- Amino acids are the products of the digestion of proteins by enzymes in the stomach.

Carbohydrates: polysaccharide $\xrightarrow{\text{hydrolysis}}$ monosaccharides $\xrightarrow{\text{condensation}}$ polysaccharide

Triglycerides: triglyceride $\xrightarrow{\text{hydrolysis}}$ glycerol and fatty acids $\xrightarrow{\text{condensation}}$ triglyceride

Proteins: protein $\xrightarrow{\text{hydrolysis}}$ amino acids $\xrightarrow{\text{condensation}}$ protein

FIGURE 12.3.12 Summary of the metabolism of the major nutrients obtained from food

KEY QUESTIONS

Knowledge and understanding

1 Use the following words to complete the sentences about the cycle of breaking down and rebuilding nutrients obtained from food:

hydrolysis; glycerol; water; polypeptides; triglyceride; condensation; monosaccharides

Amino acids can combine to form ________ in ________ reactions. __________ is also formed in this reaction. The reverse of this reaction is a/an _________reaction.

Polysaccharides are formed in condensation reactions involving ___________.

Fatty acids and _________ can combine in a condensation reaction to form a/an ___________.

2 Complete the metabolism of food reaction sequences below by providing the appropriate reactants and products as required.

polysaccharide $\xrightarrow{\text{hydrolysis}}$ monosaccharides $\xrightarrow{\text{condensation}}$ __________

__________ $\xrightarrow{\text{hydrolysis}}$ glycerol and fatty acids $\xrightarrow{\text{condensation}}$ __________

__________ $\xrightarrow{\text{hydrolysis}}$ amino acids $\xrightarrow{\text{condensation}}$ __________

3 Describe the role that enzymes play in digestion.

4 Sucrose is a disaccharide with a molecular formula of $C_{11}H_{22}O_{11}$.

a What is the molecular formula of the monosaccharides from which it is formed?

b What is the name of the link between the monosaccharide units that breaks when sucrose is hydrolysed?

Analysis

5 The structure of lactose is shown in the figure below. Lactose undergoes hydrolysis during digestion.

a Explain what is meant by the term 'hydrolysis of lactose'.

b Circle a hydroxyl and a glycosidic functional group in the molecule.

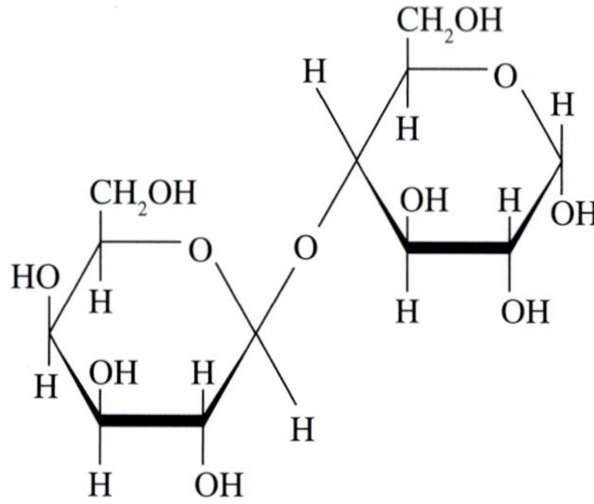

6 Label the parts of the diagram in the following figure showing the hydrolysis of a triglyceride.

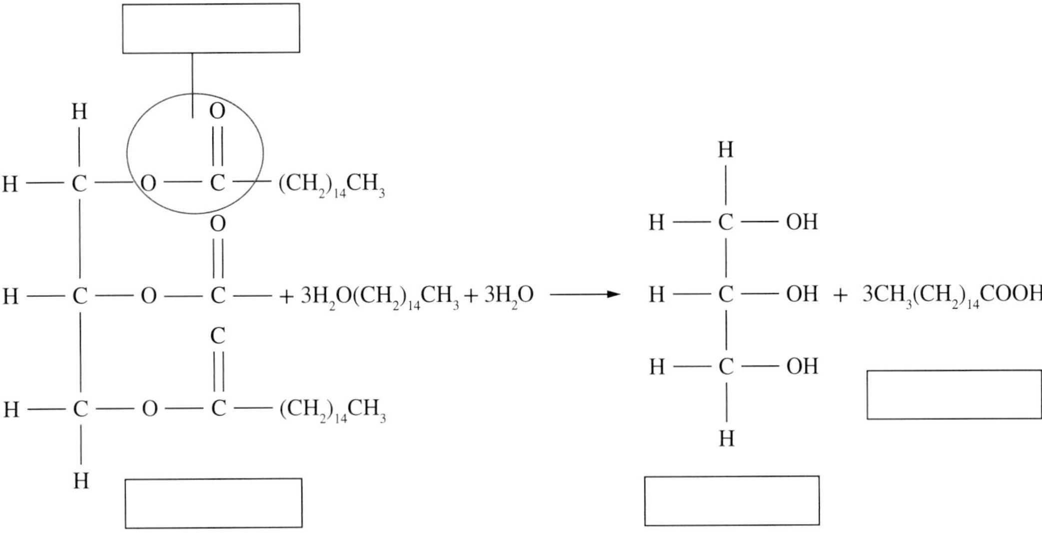

7 The diagram in the following figure shows the hydrolysis of a tripeptide.

a Balance the chemical reaction.

b Circle a peptide link, an amine functional group, a carboxyl functional group an amino acid residue, an amino acid and an R group in the tripeptide.

$$H_2N-CH(CH_2SH)-C(=O)-NH-CH(CH(OH)CH_3)-C(=O)-NH-CH(CH_2OH)-C(=O)-OH \longrightarrow H_2N-CH(CH_2SH)-C(=O)-OH + H_2N-CH(CH(OH)CH_3)-C(=O)-OH + H_2N-CH(CH_2OH)-C(=O)-OH$$

Chapter review

OA ✓✓

12

KEY TERMS

amide
amino acid
amino acid residues
2-amino acid
amylopectin
amylose
bile
biomolecule
biopolymer
C-terminal
carbohydrate
cellulose
condensation polymerisation
condensation reaction
digestion
dipeptide
disaccharide
emulsion
enzyme
ester functional group
ester link
ether functional group
fatty acid
glucose
glycogen
glycosidic link
hydrolysis
hydrolytic reaction
lipid
metabolism
monomer
monosaccharide
nutrient
N-terminal
peptide link
photosynthesis
polypeptide
polysaccharide
protein
quaternary structure
R group
ribosome
side chain
starch
synthesised
triglyceride
tripeptide

REVIEW QUESTIONS

Knowledge and understanding

1 Select the correct statement about amino acids.

A Amino acids contain only the elements C, H, O and N.

B Amino acids are the product of hydrolysis of carbohydrates.

C Amino acids may contain more than one nitrogen atom.

D The functional group on an amino acid side chain is always non-polar.

2 Which one of these molecules is not a 2-amino acid?

A

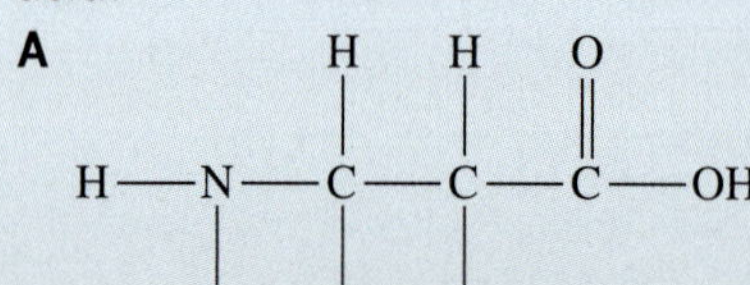

B

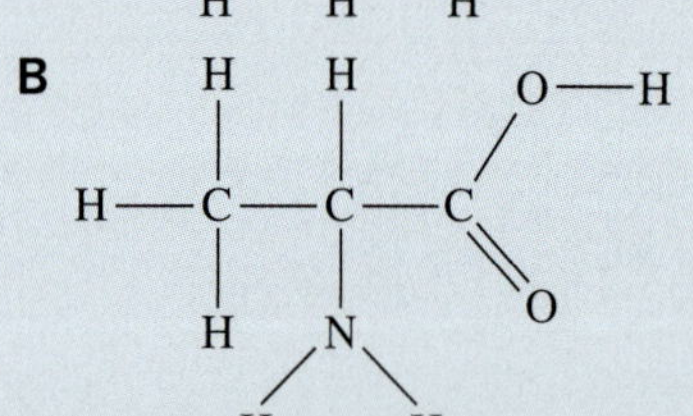

C

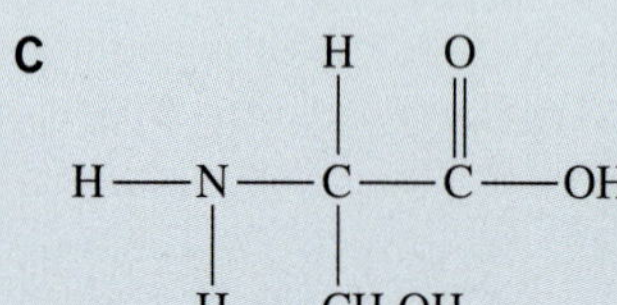

D

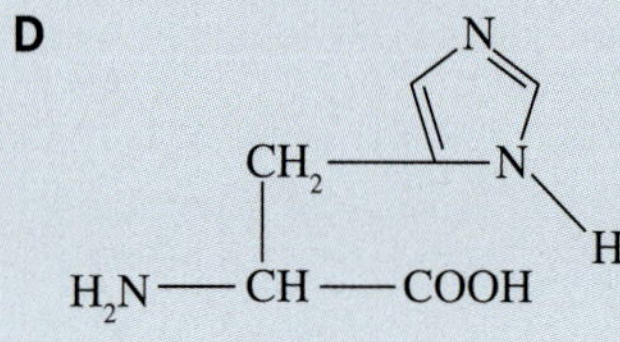

3 Which of the following occurs during the production of a disaccharide?

A An amine group reacts with a carboxyl group to form a peptide linkage.

B Two hydroxyl groups react to form a glycosidic linkage.

C A hydroxyl group reacts with a carboxyl group to form an ester functional group.

D A hydroxyl group reacts with a carboxyl group to form a glycosidic linkage.

4 Draw the general structure of an amino acid and label the functional groups.

5 Draw a molecule of glucose. Annotate your diagram to:

a explain why glucose is highly soluble in water

b show where bonding usually occurs when a polysaccharide forms.

6 Name an example of a:

a monosaccharide other than glucose

b disaccharide other than sucrose

c polysaccharide.

7 Why is glycogen sometimes referred to as animal starch?

8 What functional groups react to form each of the following, and with which biological polymers are each of the links associated?

a an ester link

b a glycosidic link

c a peptide link

9 Determine whether each of the following applies to a polysaccharide, a triglyceride, a protein or all of them.

a hydrolysis involves enzymes

b is not a polymer

c made in plants

d contains nitrogen

e digestion involves the action of bile

10 Copy and complete the following table about the molecular structure of the major components of food.

	Names of links between units	Functional groups formed when links break	Monomer or smallest components
Carbohydrate		hydroxyl	
Fat		hydroxyl and ______	_____ and glycerol
Protein		_____ and carboxyl	

Application and analysis

11 When alanine and glycine react, two different dipeptides can be formed.

- **a** Write the semi-structural formula of each dipeptide.
- **b** How many tripeptides can be formed from three amino acids?
- **c** Circle the peptide links in the section of a protein chain shown in the following figure.

```
     H  O       H  O       H  O       H  O
     |  ||      |  ||      |  ||      |  ||
···—C—C—N—C—C—N—C—C—N—C—C—N—···
     |      |   |      |   |      |   |      |
    CH3     H  CH2     H   H      H  CH2     H
                |                     |
               SH                    OH
```

- **d** Write the names of the amino acids that make up this section of the chain.

12 Classify the following reactions as condensation or hydrolysis.

- **a** formation of protein from amino acid
- **b** formation of starch from the polymerisation of glucose
- **c** formation of triglyceride in the blood
- **d** breakdown of glycogen to glucose
- **e** formation of glycerol and palmitic acid from a fat
- **f** splitting of maltose into two molecules of glucose

13 Name the functional group that is formed in the reactions of:

- **a** stearic acid + glycerol
- **b** glucose + fructose
- **c** glycine + glycine
- **d** glucose + glucose
- **e** linoleic acid + glycerol

14 The molecular formula of a biomolecule is found to be $C_{6000}H_{10002}O_{5001}$. Explain why this molecule is:

- **a** not a protein
- **b** not a fatty acid
- **c** not a monosaccharide
- **d** likely to be a carbohydrate.

15 Calculate the mass of 40 mole of sucrose, given that the molar mass of glucose is 180 g mol^{-1}

16 The structure of the fatty acid linoleic acid is:

$$CH_3(CH_2)_4CH{=}CHCH_2CH{=}CH(CH_2)_7COOH$$

- **a** Name the chemical needed to react with linoleic acid to form a triglyceride.
- **b** Draw the structure of the triglyceride made from linoleic acid.
- **c** Name and circle the functional groups that are formed when the triglyceride molecule is produced.
- **d** Explain whether linoleic acid is a saturated or an unsaturated fatty acid.

17 Use the following terms to complete labels a–e in the following figure, which shows the formation of a polysaccharide:

polysaccharide; disaccharide; carbon dioxide; glycosidic link; glucose

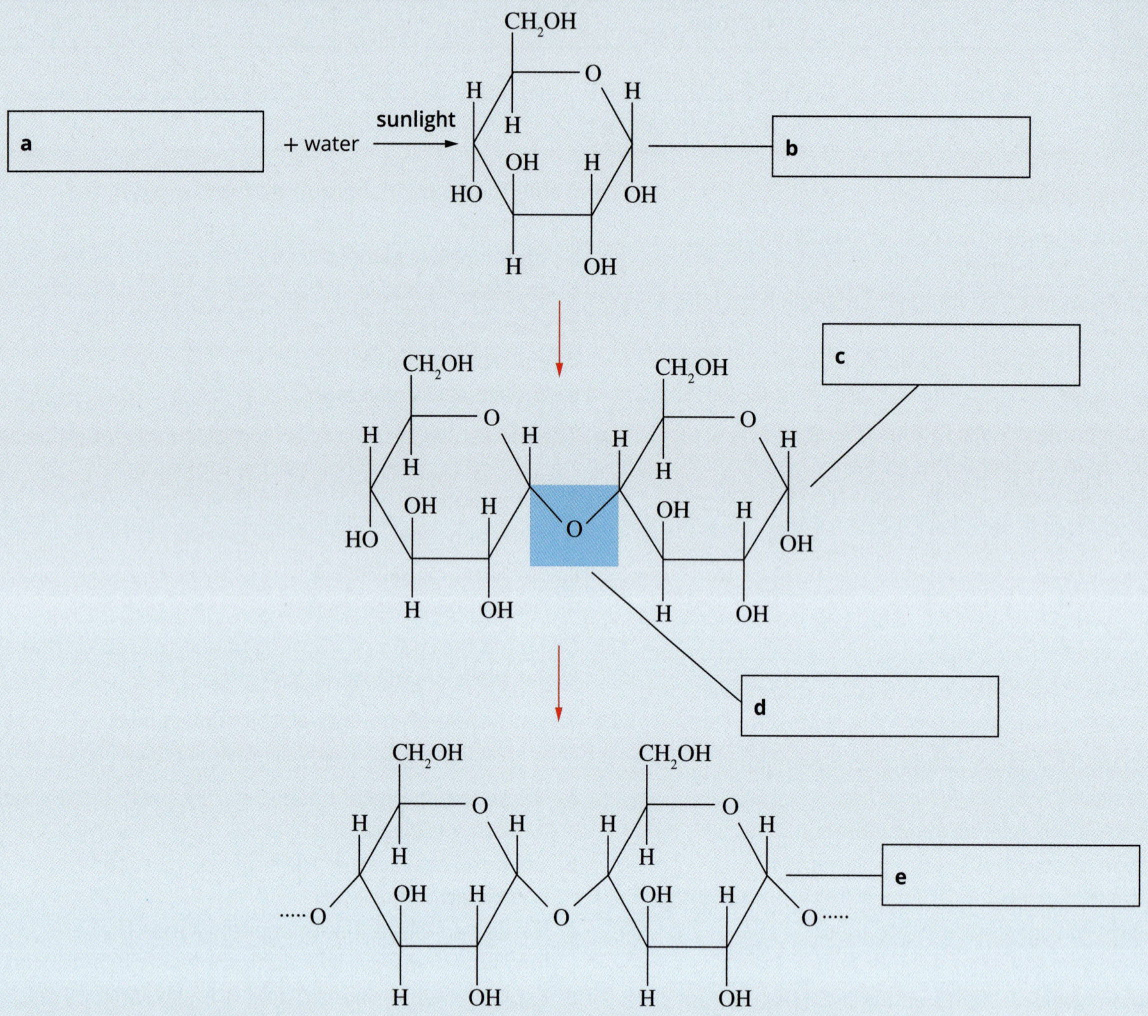

a + water $\xrightarrow{\text{sunlight}}$ b

CH_2OH, O, H, H, H, OH, H, HO, OH, H, OH

c

d

CH_2OH, O, H, H, H, OH, H, HO, O, OH, H, OH

e

$\cdots O$, CH_2OH, O, H, OH, H, O, $O\cdots$

18 Complete the table below.

Molecule	The molecule is best described as a/an	A functional group it contains
glycine		
glycogen		
glucose		
glycerol		

19 **a** Explain the difference between condensation reactions and hydrolysis reactions.

b Explain the importance of these reactions in the human body.

c Write equations for condensation reactions between:

i two glucose molecules

ii a glycerol molecule and three palmitic acid ($C_{15}H_{31}COOH$) molecules

iii two alanine molecules.

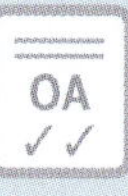

UNIT 4 • Area of Study 1

REVIEW QUESTIONS

How are organic compounds categorised and synthesised?

Multiple-choice questions

Refer to the following table when answering Question 1.

Covalent bond	Bond energy (kJ mol^{-1})
C–C	348
C–H	413
C–F	488
C–Si	318
Si–Si	217

1 Diamond (C) and silicon carbide (SiC) both exist as tetrahedral covalent lattices. In diamond, each carbon atom is bonded to four other carbon atoms, and in silicon carbide, each carbon atom is bonded to four silicon atoms. When heated strongly, both lattices decompose to form gases (sublime). The temperature required for this will be:

A higher for diamond because C–C bonds are stronger than C–Si bonds
B higher for diamond because C–C bonds are stronger than Si–Si bonds
C higher for silicon carbide because C–C bonds are weaker than C–Si bonds
D higher for silicon carbide because C–C bonds are weaker than Si–Si bonds.

2 What is the systematic name for $CH_3CH_2CH_2CH(CH_3)_2$?

A 1,1-dimethylbutane
B 2-methylpentane
C 2-methylpentene
D 1-methylpentane

3 Which of the following compounds would be expected to have the lowest boiling point?

A $CH_3CH_2CH_2OH$
B $CH_3CH_2CH_2CH_3$
C $CH_3CH_2CH_2Cl$
D $CH_3CH_2CH_3$

4 When starch is formed from glucose, the condensation reaction that occurs involves:

A a carboxyl functional group and an amino functional group
B two hydroxyl functional groups
C a hydroxyl functional group and a carboxyl functional group
D a hydroxyl functional group and an amino functional group.

5 When a polypeptide chain is formed, the condensation reaction that occurs involves:

A a carboxyl functional group and an amine functional group
B two hydroxyl functional groups
C a hydroxyl functional group and a carboxyl functional group
D a hydroxyl functional group and an amine functional group.

6 The linkages (functional groups) by which fatty acids are joined to glycerol molecules are called:

A fatty acid links
B peptide links
C ether links
D ester links.

7 Which of the following statements are true of the homologous series of primary alcohols?

I The members differ by one CH_2 unit.
II They are all strong bases.
III They can be oxidised to form carboxylic acids.

A I and II
B II and III
C I and III
D I, II and III

8 Which of the following alternatives is the product formed by the reaction of CH_2CH_2 with Br_2?

A CH_2BrCH_2Br
B CH_2CHBr
C CH_3CH_2Br
D CHBrCHBr

9 Consider the following reaction pathway.

ethene $\xrightarrow{\text{I}}$ chloroethane $\xrightarrow{\text{II}}$ ethanol $\xrightarrow{\text{III}}$ ethanoic acid

What reactions occur in steps I, II and III of the pathway?

A addition, substitution, oxidation
B chlorination, substitution, addition
C substitution, addition, hydrolysis
D addition, reduction, hydrolysis

10 When ethanol is heated with an acidified solution of potassium dichromate, which of the following is the formula of the final product?

A $CH_3CH_2CH_2OH$
B CH_3COOCH_3
C CH_3COOH
D $CH_3CH_2CH_3$

11 Which of the following compounds could react together to make the ester methyl ethanoate?

A CH_3CH_2OH and CH_3COOH
B CH_3CH_2OH and HCOOH
C CH_3OH and CH_3CH_2COOH
D CH_3OH and CH_3COOH

12 Which one of the structures below does not have an IUPAC name that ends with 'ol'?

I

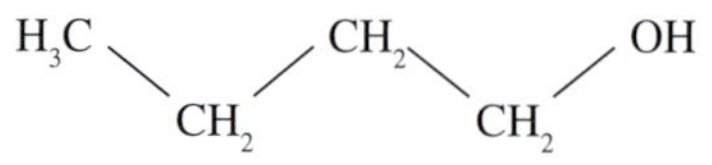

H_3C — CH_2 — CH_2 — CH_2 — OH

II

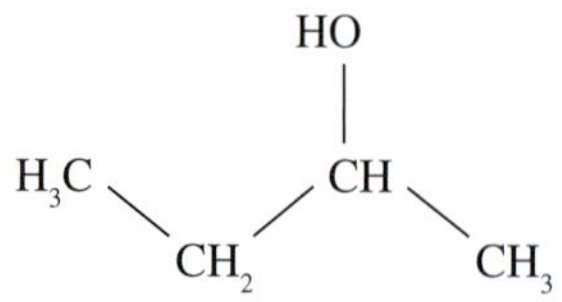

H_3C — CH_2 — CH(HO) — CH_3

III

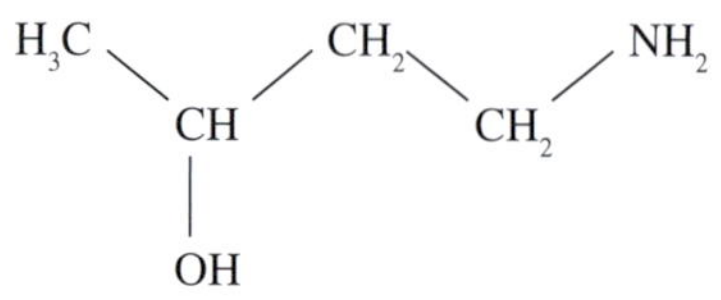

H_3C — CH(OH) — CH_2 — CH_2 — NH_2

IV

HO — C(=O) — CH_2 — CH_2 — CH_2 — OH

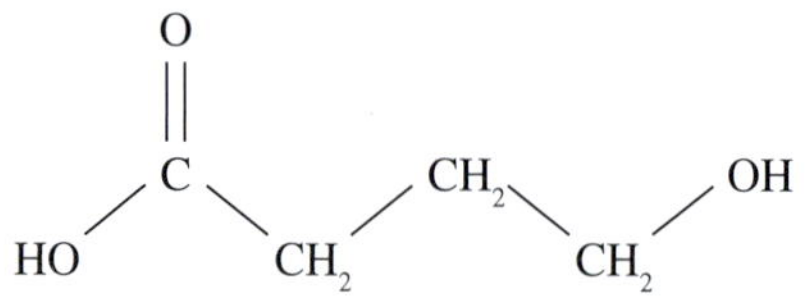

A II and III
B III only
C IV only
D III and IV

13 Consider the following equation for the reaction of methanol with acidified potassium dichromate.

$2K_2Cr_2O_7(aq) + 3CH_3OH(aq) + 8H_2SO_4(aq) \rightarrow$
$2Cr_2(SO_4)_3(aq) + 3HCOOH(aq) + 2K_2SO_4(aq)$

Which statement about this reaction is **not** true?

A Dichromate ions have been reduced.
B Methanol has been oxidised.
C Hydrogen ions have been reduced.
D Sulfate ions have not been oxidised or reduced.

Question 14 refers to the following information.

The following list gives the formulas of a number of important biochemical molecules.

I $C_6H_{12}O_6$
II $C_{17}H_{33}COOH$
III $C_3H_8O_3$
IV CO_2
V $C_{17}H_{35}COOH$

14 When a saturated fat undergoes hydrolysis, two products that might be formed are:

A II and III
B III and V
C IV and I
D II and V.

15 When a triglyceride molecule with $M = 878$ g mol^{-1} is fully hydrolysed, it yields three fatty acid molecules all with the same formula. Which of the following molar masses is the closest to the molar mass of this fatty acid?

A 262 g mol^{-1}
B 280 g mol^{-1}
C 292 g mol^{-1}
D 310 g mol^{-1}

Short-answer questions

16 For each of the following structures, write a:

i molecular formula
ii semi-structural (condensed), formula.

a

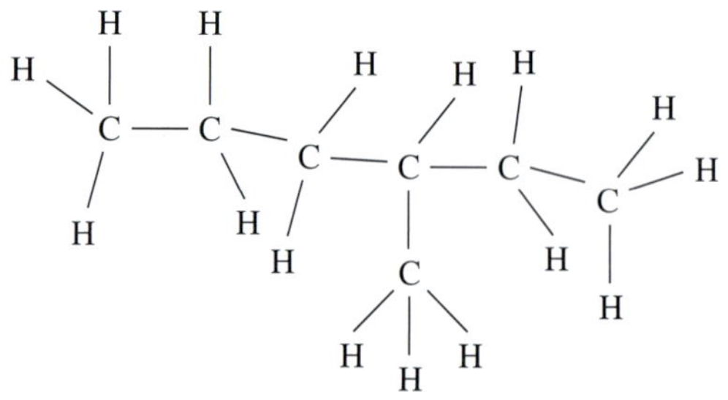

(structural formula: H, C and H atom labels)

b

(structural formula: H, C, O atom labels)

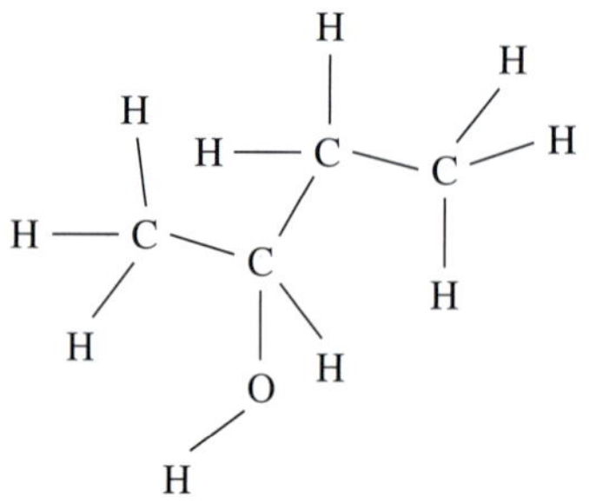

c

H
H H O
H C C O
H C
H

d H O H
H H
C H H
C H
H C C
C Cl
H H
H H

e
H
H H
C
H H H H
H C C C C C
H H H H O

f
H
H H
C
O H
H C
C H
H C H
H C H C H
H H

17 Give correct IUPAC names for the following structures.

a

OH
H_3C CH_2 CH CH_3
CH_2 CH_2 CH_2

b H_3C CH_2 H
CH C
F O

c

H
H_3C C CH_3
C CH_2
H

d H_3C
CH_2
CH
H_3C CH_2
CH_2 CH CH_3
NH_2

e H_3C CH_2
CH NH_2
OH

f

OH Cl
H_3C CH CH
CH_2 CH_2 CH_3

g O
C CH_2 CH_2 CH_2
HO CH_2 CH_2 CH CH_3
OH

h

CH_3
H_3C CH CH_3
CH_2 CH
Br

i

O
H_3C C CH_2
CHCl CH_2 CH_3

18 Draw full structures for each of the following, showing all bonds.

- **a** $CH_3(CH_2)_3CH_3$
- **b** $CH_2CHCH_2CHOHCH_3$
- **c** $CH_2ClCH_2COCH_3$
- **d** $CH_3(CH_2)_2CONH_2$
- **e** $(CH_3)_2CHCHCH_2$
- **f** $CH_3CH_2C(CH_3)_2OH$
- **g** $CH_3OCOCH_2CH_3$
- **h** 1-bromopropan-1-ol
- **i** 4-methylpent-2-ene
- **j** 1-iodobutan-2-amine
- **k** 3-aminohexan-2-ol

19 Give appropriate reagents/conditions required for the conversion of:

- **a** 1-chloropropane to propan-1-ol
- **b** propan-1-ol to propanoic acid
- **c** chloromethane to methanamine.

20 Design synthetic pathways using the minimum number of steps possible and showing all organic intermediates and the reagents/conditions used, to prepare:

- **a** propanoic acid from propane
- **b** methyl ethanoate from methane and ethene
- **c** propyl propanoate from propyl methanoate.

21 Write balanced chemical equations (excluding states) for the:

- **a** oxidation of propan-1-ol to propanoic acid under acidic conditions (redox half-equation)
- **b** condensation of ethanol and propanoic acid to form ethyl propanoate.

22 The following table gives a range of properties for eight organic compounds arranged in order of increasing molecular mass.

<table>
<tr><th>Name</th><th>Structure</th><th>Molecular mass</th><th>Viscosity (relative to methanol)</th><th>Boiling point (°C)</th></tr>
<tr><td>methanol</td><td>H
|
H — C — OH
|
H</td><td>32</td><td>1.0</td><td>65</td></tr>
<tr><td>ethanol</td><td>H H
| |
H — C — C — OH
| |
H H</td><td>44</td><td>2.0</td><td>78</td></tr>
<tr><td>propan-2-one</td><td>H O H
| ‖ |
H — C — C — C — H
| |
H H</td><td>54</td><td>0.6</td><td>57</td></tr>
<tr><td>propan-1-ol</td><td>H H H
| | |
H — C — C — C — OH
| | |
H H H</td><td>56</td><td>3.6</td><td>98</td></tr>
</table>

Name	Structure	Molecular mass	Viscosity (relative to methanol)	Boiling point (°C)
ethane-1,2-diol	$HO-CH_2-CH_2-OH$	62	29.6	197
butan-1-ol	$H-CH_2-CH_2-CH_2-CH_2-OH$	74	4.7	118
benzene	C_6H_6 (hexagonal ring with circle, one H on each carbon)	78	1.1	80
hexan-1-ol	$H-CH_2-CH_2-CH_2-CH_2-CH_2-CH_2-OH$	102	8.3	159

a What is viscosity a measure of?

b For each of the following statements concerning property trends, select compounds from the table that illustrate the trend, and explain the trend in terms of intermolecular bonding.

- **i** Within any homologous series, viscosity increases as the molecules get larger.
- **ii** For compounds with similar carbon skeletons, the presence of an additional group that can form hydrogen bonds significantly increases boiling point and viscosity.
- **iii** Compared with polar compounds of similar mass, non-polar compounds tend to have very low viscosity.

23 Chloroform (trichloromethane, $CHCl_3$) is synthesised commercially by the chlorination of methane:

$$CH_4 + 3Cl_2 \rightarrow CHCl_3 + 3HCl$$

The final mixture contains a mixture of chloroform along with chloromethane, dichloromethane and tetrachloromethane. Distillation allows separation of the chloroform and at one plant yields of 75% are achieved.

a Calculate the atom economy of the reaction forming chloroform.

b What mass of chloroform would be obtained from 1.0 kg of methane, assuming 75% yield?

An alternative source of chloroform is the reaction of sodium hypochlorite (bleach) with propan-2-one (acetone):

$$CH_3COCH_3 + 3NaOCl \rightarrow CHCl_3 + 2NaOH + CH_3COONa$$

c Name the by-products in this synthesis.

d Calculate the atom economy for this process and compare it with that for the commercial method described earlier in this question.

e Describe an example of an organic reaction that has an atom economy of 100%.

Refer to the fatty acid structures represented by I–IV when answering Questions 24 and 25.

I $CH_3(CH_2)_4CH{=}CHCH_2CH{=}CH(CH_2)_{11}COOH$
II $CH_3(CH_2CHCH)_4(CH_2)_4COOH$
III $CH_3(CH_2)_{18}COOH$
IV $CH_3(CH_2)_8CHCH(CH_2)_4COOH$

24 Answer the following questions for each of the fatty acids shown in **I–IV**.

a Write its molecular formula in the form $C_xH_yO_z$ and also in the form C_nH_mCOOH.

b Identify it as saturated, monounsaturated or polyunsaturated.

25 a Draw the structure of the triglyceride molecule resulting from condensation of glycerol with three molecules of IV.

b On the structure drawn for part **a,** clearly circle the ester linkages.

26 Write balanced chemical equations using molecular formulas for the:

a hydrolysis of sucrose to form glucose and fructose

b formation of a triglyceride (write it in the form $C_xH_yO_z$) by condensation of glycerol with the fatty acid lauric acid ($C_{11}H_{23}COOH$).

27 Two chemistry students proposed the following research question: 'Which vegetable oil (sunflower/peanut/olive) produces the highest yield of biodiesel?' They used the following method to produce the biodiesel:

1 React 20 mL of sunflower oil with 2 mL of ethanol and 2 mL of NaOH. Conduct the reaction in a water bath at a temperature of 60°C.
2 After 20 minutes, add 2 mL of glycerol to the flask and swirl. This will help all the polar molecules in the mixture to collect together and form a separate layer.
3 After 5 minutes, pour the mixture into a measuring cylinder and leave to settle for at least 30 minutes.
4 Repeat the procedure with peanut and olive oil.

Soap contamination

If water gets into the reaction mixture, the NaOH present will react with the fatty acid molecules to form soap. This reaction is called saponification.

a Name the independent variable and the dependent variable in this investigation.

b Suggest the data the students would need to collect in order to answer their research question.

c The results of the trial using sunflower oil are pictured below.

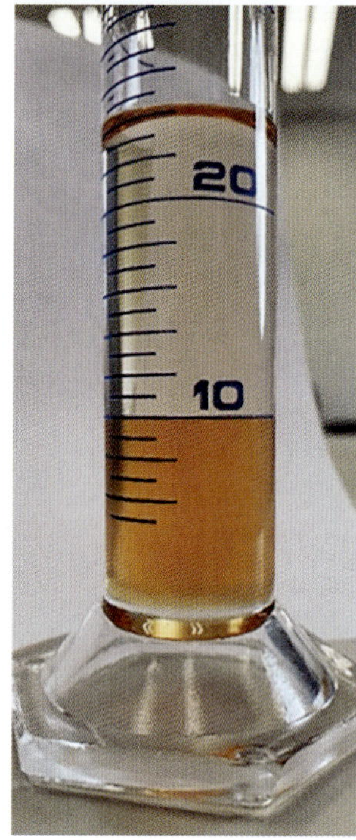

i Record the volume of biodiesel produced.

ii Biodiesel is less dense than water. Name three substances that might be found in the bottom layer.

d Describe the expected observations if water from the water bath contaminates the mixture.

e Sunflower oil is predominately formed from the fatty acid, linoleic acid ($C_{17}H_{31}COOH$). Write a balanced equation for the transesterification of sunflower oil using ethanol. Assume all the fatty acids are linoleic acid.

WS 29

CHAPTER

13 Laboratory analysis of organic compounds

Organic chemists are highly skilled at devising molecular structures that have exactly the right properties needed for a particular purpose. In Chapter 11 you learnt how reaction pathways are designed to convert a readily available starting material into the desired product. When planning a pathway, the methods of separation of the desired final product from by-products must be determined. The final product then needs to be purified and the purity evaluated.

This chapter describes some techniques used by analytical chemists to purify organic compounds and test for the presence of common functional groups. You will learn about volumetric analysis, a quantitative technique used to analyse for different organic compounds, such as alcohols, carboxylic acids and amines.

Key knowledge

- qualitative tests for the presence of carbon–carbon double bonds, hydroxyl and carboxyl functional groups **13.1**
- applications and principles of laboratory analysis techniques in verifying components and purity of consumer products, including melting point determination and distillation (simple and fractional) **13.1**
- measurement of the degree of unsaturation of compounds using iodine **13.1**
- volumetric analysis, including calculations of excess and limiting reactants using redox titrations (excluding back titrations) **13.2**

13.1 Qualitative testing of organic compounds

In Chapter 11, you learnt about specific chemical reactions involving **organic compounds**. Using these reactions, you then devised reaction pathways for the synthesis of some simple organic compounds such as esters. The synthesis may require the preparation of a number of **intermediate** compounds and **by-products**. For example, when propan-1-ol is formed from propane, the substitution reaction of propane with chlorine produces a mixture of both 1-chloropropane and 2-chloropropane plus various compounds containing two or more chlorine atoms. A substitution reaction of 1-chloropropane with sodium hydroxide is then used to prepare propan-1-ol. In this reaction pathway, 1-chloropropane is the desired intermediate. Substances produced at the same time as the intended product, such as the isomer 2-chloropropane, are referred to as by-products.

When designing a reaction pathway, separation of the desired intermediate and final product from isomers and other by-products must be considered. The final product then needs to be purified and the purity evaluated. In this section, you will learn about laboratory analysis techniques used to isolate the desired product and qualitative tests used to help in compound identification.

PURIFICATION AND SEPARATION TECHNIQUES

The synthesis of an organic compound may require the preparation of a number of intermediate compounds, so more than one possible pathway may need to be considered. Isomers and other by-products may also be formed. For example, the addition reaction of propene with water produces a mixture of propan-1-ol and propan-2-ol. If the desired product is propan-2-ol, then propan-1-ol must be separated from the required product.

Distillation

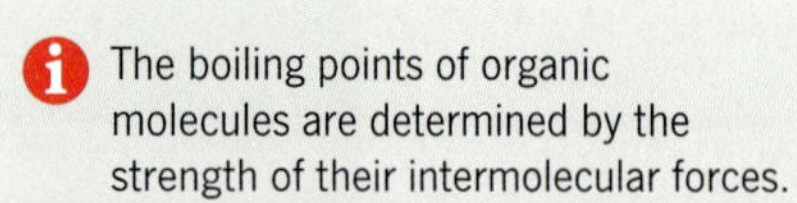
The boiling points of organic molecules are determined by the strength of their intermolecular forces.

FIGURE 13.1.1 Simple distillation apparatus used to separate two or more liquids

As you learnt in Chapter 2, **distillation** is a technique used for separating two or more liquids that have different boiling points. It is commonly used in the laboratory to separate the product of an organic reaction pathway from the by-products. Figure 13.1.1 shows a simple distillation apparatus. Distillation, or **simple distillation**, is a technique for separating two or more liquids from a mixture. This technique relies on a difference of at least 50°C in boiling point between components to obtain an effective separation. The mixture to be separated is placed in the round-bottom flask and heated to the boiling point of the liquid to be collected. The vapour rises up the flask and then passes into the water-cooled **condenser**, glassware used in distillation to cool hot vapours. The vapour is cooled to below its boiling point and condenses back to a liquid, called the **distillate**, which is collected in the beaker. Liquids with higher boiling points are left in the round-bottom flask.

Fractional distillation is used to separate liquids that have boiling points that are close together. It is commonly used in the laboratory to separate **volatile** liquids from a reaction mixture. A volatile liquid is one that vaporises easily at low temperatures.

The apparatus used in fractional distillation is shown in Figure 13.1.2 on the following page. The **fractionating column** is packed with glass beads or has glass shelves, providing a large surface area upon which the vapours condense. The mixture of liquids is heated in the distillation flask and small, irregularly-shaped stones called **anti-bumping granules** are added to the flask to allow smooth boiling. There is a temperature gradient up the fractionating column; the column is cooler at the top than at the bottom.

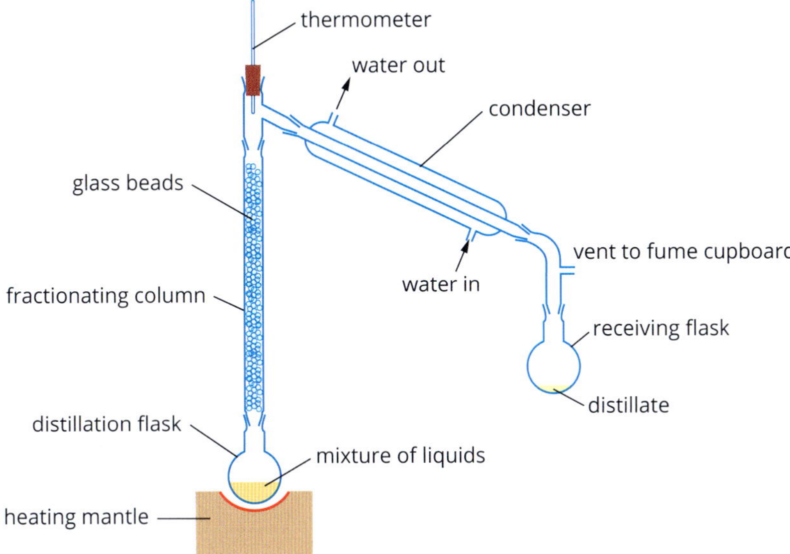

FIGURE 13.1.2 Fractional distillation apparatus used in the laboratory

Fractional distillation is based on the principle that if a mixture of volatile liquids is heated, the vapour contains a higher **concentration** of components with lower boiling points.

In fractional distillation, the components of a mixture of volatile liquids can be regarded as being separated by a succession of simple distillations.

- The vapours from the distillation flask contain a higher concentration of the more volatile or lower boiling point components than the liquid in the flask.
- The vapours rise up the fractionating column until they reach a height where the temperature is low enough for condensation to occur.
- As the condensed liquid trickles back down the column it is reheated by vapours rising from the distillation flask, causing some of the condensed liquid to evaporate.
- The resulting vapour now has an even higher concentration of the low boiling point component.
- This process of evaporation and condensation is repeated many times throughout the length of the fractionating column, increasing the concentration of the most volatile component in the vapour.
- When the vapour reaches the top of the fractionating column, it will ideally consist of only the most volatile component.
- At the top of the fractionating column, the temperature remains relatively stable.
- The vapour eventually reaches the condenser, where it is cooled, and the distillate drips into the collection vessel.
- The distillate consists of components that condense over a small temperature range near the boiling point of the desired organic compound.

For example, the ester ethyl ethanoate is synthesised by reacting ethanol with ethanoic acid in the presence of concentrated sulfuric acid, which acts as a catalyst.

$$CH_3CH_2OH(l) + CH_3COOH(l) \rightarrow CH_3COOCH_2CH_3(l) + H_2O(l)$$

Pure ethyl ethanoate can be extracted from the reaction mixture by fractional distillation. The boiling points of the reactants and products are shown in Table 13.1.1. The reaction mixture is heated in the distillation flask and the component with the lowest boiling point boils first, with the vapour rising up the fractionating column. In this example, the vapours rising up the fractionating column will contain a higher concentration of ethyl ethanoate. The temperature at the top of the column slowly stabilises at about 57°C, allowing condensation to occur. The fraction condensing over a small range of temperatures (55–59°C) near the boiling point of ethyl ethanoate is collected as the distillate. The higher boiling point components from the reaction mixture remain in the distillation flask.

TABLE 13.1.1 Boiling points of the components in the ethyl ethanoate reaction mixture

Component	Boiling point (°C)
$CH_3COOCH_2CH_3$	57
CH_3CH_2OH	78
H_2O	100
CH_3COOH	118

CHEMFILE

Fractional distillation of crude oil

Fossil fuels, such as coal, oil and natural gas, come from the remains of plants and animals. Over millions of years, high temperatures and pressures alter the structure of oils and fats forming hydrocarbons. This mixture of hydrocarbons is crude oil.

Crude oil is not used in its raw state. It is transported from oil fields to oil refineries, where it undergoes fractional distillation in a fractionating tower, as shown in the figure at right. In this process, the crude oil is separated into its various components or fractions. Each fraction is made up of a range of hydrocarbons with similar boiling points and hence similar molecular masses. Over 90% of the components of crude oil are used for fuels.

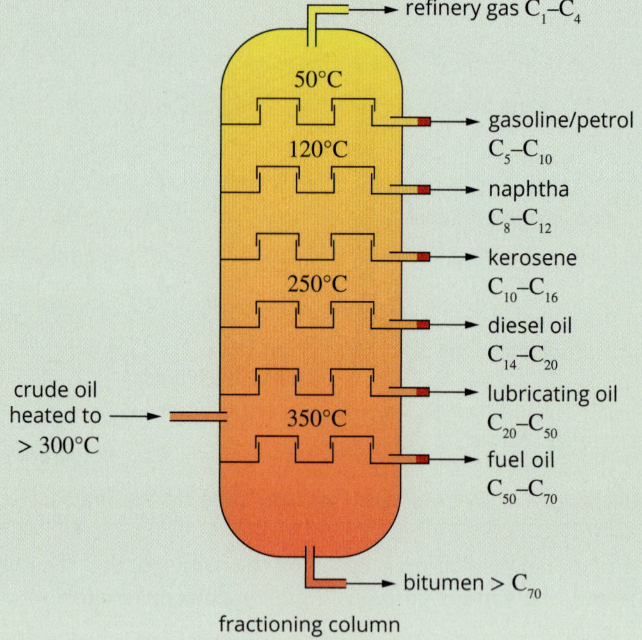

Schematic diagram of a fractionating tower used to separate the components of crude oil by fractional distillation

Melting point determination

Organic chemists use a technique called a **melting point determination** to provide information about the identity and purity of a solid sample. When a solid melts, the intermolecular forces are broken, so the temperature at which melting occurs depends on the structure of the organic molecule. A pure organic compound is said to have a sharp **melting point**, meaning that the difference between the temperature at which the sample begins to melt and the temperature at which the sample completely melts, the **melting point range**, is small or narrow, typically 0.5–2°C.

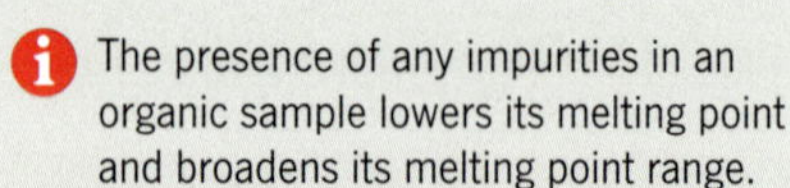
The presence of any impurities in an organic sample lowers its melting point and broadens its melting point range.

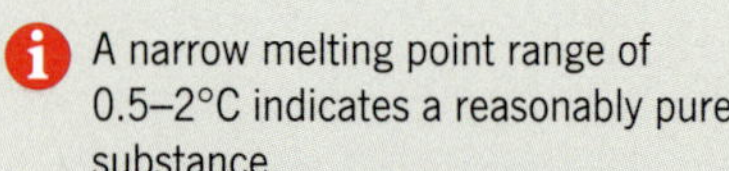
A narrow melting point range of 0.5–2°C indicates a reasonably pure substance.

A broad melting point range of more than 2°C indicates the presence of impurities.

The sharp melting point of a pure organic substance arises because all of the molecules in it are the same. The intermolecular attractions are maximised because the molecules can pack together in an orderly arrangement. In a mixture of two or more organic compounds, however, the molecules can't pack in an orderly array. The intermolecular forces are disrupted, so less heat is needed to melt this mixture and the melting point is lower. The impure solid also has a broader melting point range because regions of the crystal structure contain different amounts of the impurity. A low melting point and a wide melting point range of more than 2°C usually indicates that a substance is impure.

The measurement of a melting point is a relatively straightforward procedure and can be done using a melting point apparatus called a Thiele tube (Figure 13.1.3 on the following page). The steps are as follows.

1 The sample is added to a small glass capillary tube, which is closed at one end.
2 The capillary tube is attached to a thermometer and inserted into the apparatus. Oil surrounds the capillary tube containing the sample.
3 The oil is slowly heated and the temperature range at which the sample melts can be observed.
4 Two temperatures are recorded—the temperature at which the sample starts to melt and the temperature at which it is completely liquified.
5 The melting point range is the observed difference between these two temperatures.

Melting points can be used to identify an organic compound. The experimentally determined melting point of the unknown sample can be compared to melting point data for the desired product to see if it matches. If the melting point is higher, then the sample is not the desired product. If the melting point is lower, then the sample is either impure or not the desired product.

A **mixed melting point determination** can be performed to confirm the identity of an unknown solid. For example, if an unknown substance is thought to be aspirin, the unknown substance can be mixed with a pure sample of aspirin and a melting point determined for the mixture. If the melting point range is wide and lower than 135°C, the melting point of pure aspirin, then the unknown compound and aspirin are two different compounds. If the unknown substance is aspirin, then the mixture should melt at 135°C.

FIGURE 13.1.3 A Thiele tube is set up to determine the melting point of a yellow solid.

QUALITATIVE TESTS FOR FUNCTIONAL GROUP IDENTIFICATION

Once separation of a desired product has been achieved using a technique such as distillation, further analysis may be required to confirm the presence of common functional groups. You will now learn about some simple chemical tests that chemists use to confirm the presence of carbon–carbon double bonds, and the hydroxyl and carboxyl functional groups.

Carbon–carbon double bonds

Bromine is often used to test for the presence of a carbon–carbon double bond because of the ease and speed with which it reacts with an alkene. You will recall from Chapter 11 that alkenes are unsaturated hydrocarbons, reacting more readily and with more chemicals than alkanes, which contain only single bonds. The reactions of alkenes usually involve the addition of a small molecule to the carbon–carbon double bond of the alkene. These reactions are called **addition reactions**.

Figure 13.1.4 shows the reaction of ethene with bromine to form 1,2-dibromoethane. Pure bromine is a red liquid, but in solution is a red-orange colour. When bromine solution is added to a compound containing a carbon–carbon double bond, an addition reaction takes place. The mixture rapidly **decolourises** (becomes colourless) due to the reaction of the bromine to form the colourless dibromo product. As shown in Figure 13.1.5, the red-orange colour of the bromine rapidly disappears when added to an alkene. However, the bromine colour remains when added to an alkane.

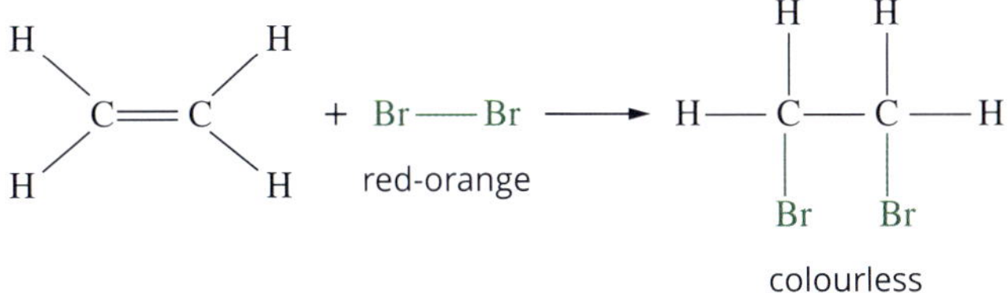

FIGURE 13.1.4 Addition reaction of ethene with bromine

FIGURE 13.1.5 When hex-1-ene is added to aqueous bromine, the solution decolourises. When hexane is added to aqueous bromine, the red-orange bromine colour remains.

Hydroxyl group

You will recall from Chapter 11 that esters are produced by a condensation reaction between a carboxylic acid and an alcohol.

A simple qualitative test for the presence of a hydroxyl group, the functional group in an alcohol, is to gently heat a mixture of the organic compound and ethanoic acid with a few drops of sulfuric acid. The reaction mixture is then poured into cold water. The characteristic smell of an ester can be detected on the surface of the water if the organic compound contains a hydroxyl group. The reaction is represented by the general equation in which the letter R represents an alkyl group:

$$ROH(l) + CH_3COOH(l) \rightarrow CH_3COOR(l) + H_2O(l)$$

The presence of a hydroxyl group is determined by an esterification reaction. If no ester is detected, the hydroxyl group is not present.

Primary alcohols oxidise to form carboxylic acids.
Secondary alcohols oxidise to form ketones.
Tertiary alcohols are resistant to oxidation and will not react.

Following a positive result to this test, acidified potassium dichromate or acidified potassium permanganate can then be added to the organic compound to determine whether it is a tertiary alcohol or not. Recall from Chapter 11 that primary and secondary alcohols are oxidised by strong inorganic oxidising agents. Tertiary alcohols are resistant to oxidation and will not normally react.

Figure 13.1.6 shows the colour change observed for the reaction of primary and secondary alcohols with acidified potassium dichromate. A solution of dichromate ions is orange in colour. When this acidified dichromate solution is added to a primary or secondary alcohol, the dichromate is reduced to the chromium ion (Cr^{3+}), which is green in colour. This colour change from orange to green can be used as a qualitative test to indicate if oxidation of the organic compound has taken place, meaning that the alcohol is either a primary or secondary alcohol. The dichromate solution stays orange if the organic compound is a tertiary alcohol.

Figure 13.1.7 shows the colour change observed for the reaction of primary and secondary alcohols with acidified potassium permanganate. A solution of permanganate ions is a deep purple colour. When this acidified permanganate solution is added to a primary or secondary alcohol, the permanganate is reduced to the manganese ion (Mn^{2+}), which is colourless. Tertiary alcohols do not react, so there is no change to the colour of the solution.

FIGURE 13.1.6 The strong oxidising agent potassium dichromate is orange in colour. On reaction with primary and secondary alcohols, the orange colour on the left changes to the characteristic green colour of Cr^{3+} ions.

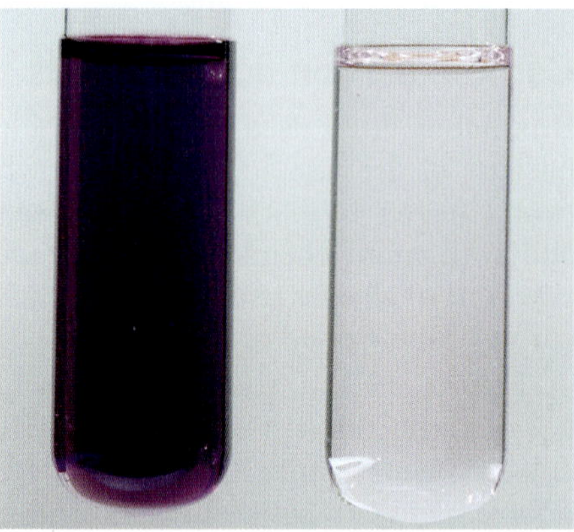

FIGURE 13.1.7 The strong oxidising agent potassium permanganate is deep purple in colour. On reaction with primary and secondary alcohols, the purple colour on the left changes to colourless.

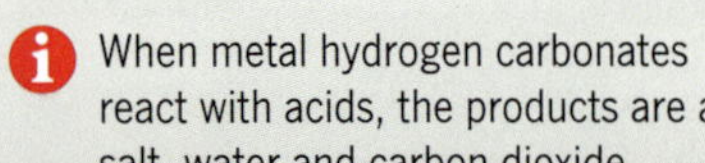

When metal hydrogen carbonates react with acids, the products are a salt, water and carbon dioxide.

Carboxyl group

In Unit 2, you studied some of the reactions of acids. Acids react with metal hydrogen carbonates to produce carbon dioxide gas, together with a salt and water.

Sodium hydrogen carbonate can be used to test for the presence of the carboxyl functional group. If effervescence (fizzing) occurs when a small amount of solid sodium hydrogen carbonate is added to the organic compound, then the carboxyl group is present.

The limewater test shown in Figure 13.1.8 is used to detect the presence of carbon dioxide. When the carbon dioxide gas is bubbled through limewater, the solution turns 'milky' or 'cloudy' due to the precipitation of calcium carbonate:

$$Ca(OH)_2(aq) + CO_2(g) \rightarrow CaCO_3(s) + H_2O(l)$$

FIGURE 13.1.8 In the limewater test, carbon dioxide is bubbled through limewater, turning the limewater milky.

Table 13.1.2 summarises the qualitative tests used to confirm the presence of carbon–carbon double bonds, carboxyl functional groups and hydroxyl functional groups. Further testing using acidified potassium dichromate or potassium permanganate can be used to determine if an organic compound is a primary or a secondary alcohol.

TABLE 13.1.2 Qualitative tests used to confirm the presence of carbon–carbon double bonds, carboxyl and hydroxyl functional groups

Chemical test for	Method	Observations	Explanation
carbon–carbon double bond	Add a solution of bromine (Br_2) to the unknown organic compound.	The red-orange colour of the bromine rapidly decolourises.	An addition reaction occurs. As the bromine adds across the carbon–carbon double bond, the solution loses colour. $R_2C{=}CR_2 + Br_2 \rightarrow R_2BrC{-}CBrR_2$
carboxyl functional group	Add a little sodium hydrogen carbonate ($NaHCO_3$) solid to the unknown organic compound.	Effervescence occurs as a colourless gas is produced. When the gas is bubbled through limewater, the solution turns 'milky' or 'cloudy'.	$NaHCO_3(s) + RCOOH(aq) \rightarrow RCOONa(aq) + CO_2(g) + H_2O(l)$ The limewater test for carbon dioxide: $Ca(OH)_2(aq) + CO_2(g) \rightarrow CaCO_3(s) + H_2O(l)$
hydroxyl functional group	Gently heat a mixture of the organic compound with ethanoic acid and a few drops of sulfuric acid. Pour into cold water.	The characteristic smell of an ester on the surface of the water.	A condensation reaction occurs between the hydroxyl group on the organic compound and the carboxyl group on the ethanoic acid. $ROH(l) + CH_3COOH(l) \rightarrow CH_3COOR(l) + H_2O(l)$

Worked example 13.1.1

DETERMINING THE IDENTITY OF ORGANIC COMPOUNDS USING CHEMICAL TESTS

A student tests five colourless liquids labelled A, B, C, D and E. The identity of each individual liquid is unknown, but they are known to be butane, but-1-ene, butan-1-ol, butanoic acid and 2-methylpropan-2-ol. The student performs a number of qualitative tests on each liquid. Identify each liquid using the results shown below.

	A	B	C	D	E
Solubility in water	soluble	soluble	insoluble	soluble	insoluble
Addition of red-coloured bromine (Br_2) solution	no immediate reaction	no immediate reaction	no immediate reaction	no immediate reaction	colour disappears
Addition of sodium hydrogen carbonate ($NaHCO_3$) powder	gas evolved	no reaction	no reaction	no reaction	no reaction
Addition of acidified potassium permanganate ($KMnO_4$)	no colour change	no colour change	no colour change	colour change from purple to colourless	no colour change

Thinking	Working
Identify the homologous series to which each unknown liquid in the original list belongs.	Butane is an alkane. But-1-ene is an alkene. Butan-1-ol is an alcohol. Butanoic acid is a carboxylic acid. 2-Methylpropan-2-ol is an alcohol.
Identify the polarity of each of the listed compounds. Remember that alkanes and alkenes are non-polar, and alcohols and carboxylic acids are polar.	Butane and but-1-ene are non-polar. Butan-1-ol, butanoic acid and 2-methylpropan-2-ol are polar.
The test for solubility in water is used to group the unknown liquids as polar or non-polar. Non-polar compounds are insoluble in water, and small alcohols and carboxylic acids are soluble in water.	Liquids C and E are insoluble in water, so must be the alkane and alkene. Liquids A, B and D are soluble in water, so these are the two alcohols and the carboxylic acid.
Use the bromine test to identify the presence of a carbon–carbon double bond. When bromine is added to a substance containing a carbon–carbon double bond, the bromine rapidly decolourises. Use this test to decide which of the non-polar liquids is an alkene.	The red colour of bromine disappears when added to liquid E, so E must contain a carbon–carbon double bond. Liquid E is but-1-ene. Liquid C is butane.
Sodium hydrogen carbonate can be used to test for the presence of the carboxyl group. The reaction between an acid and sodium hydrogen carbonate produces carbon dioxide gas.	A gas is evolved when sodium hydrogen carbonate is added to Liquid A. Liquid A is butanoic acid.
The addition of acidified potassium permanganate is used to identify a primary or secondary alcohol. The primary or secondary alcohol is oxidised by the permanganate ion. As the permanganate ion is reduced to Mn^{2+}, the liquid changes from purple to colourless.	Butan-1-ol is a primary alcohol. 2-Methylpropan-2-ol is a tertiary alcohol A colour change from purple to colourless occurs with liquid D. Liquid D is butan-1-ol. Liquid B is 2-methylpropan-2-ol.

Worked example: Try yourself 13.1.1

DETERMINING THE IDENTITY OF ORGANIC COMPOUNDS USING CHEMICAL TESTS

A student tests five colourless liquids labelled A, B, C, D and E. The identity of each individual liquid is unknown, but they are known to be pentane, but-2-ene, propan-1-ol, ethanoic acid and 2-methylpropan-2-ol. The student performs a number of qualitative tests on each liquid. Identify each liquid using the results shown below.

	A	B	C	D	E
Solubility in water	insoluble	soluble	soluble	insoluble	soluble
Addition of red-coloured bromine (Br_2) solution	colour disappears	no immediate reaction	no immediate reaction	no immediate reaction	no immediate reaction
Addition of sodium hydrogen carbonate ($NaHCO_3$) powder	no reaction	no reaction	no reaction	no reaction	gas evolved
Addition of acidified potassium permanganate ($KMnO_4$)	no colour change	colour change from purple to colourless	no colour change	no colour change	no colour change

IODINE NUMBER

In Chapter 12 you learnt that fats and oils contain large non-polar molecules known as triglycerides. Triglycerides are made by condensation reactions between a glycerol molecule and three fatty acid molecules. Fatty acids have a carboxyl functional group attached to a long unbranched hydrocarbon chain and are classified as saturated or unsaturated based on whether or not there are carbon–carbon double bonds present in the hydrocarbon chain.

Earlier in this section, you learnt that bromine can be used to test for the presence of carbon–carbon double bonds. Halogens undergo addition reactions with unsaturated hydrocarbons and therefore also with unsaturated fatty acids. In particular, reactions between bromine, Br_2, or iodine, I_2, and unsaturated fatty acids are easy to monitor because bromine and iodine are decolourised when they react with unsaturated hydrocarbons. The reaction with iodine is less dangerous than that with bromine, since an aqueous solution of bromine releases fumes of bromine, creating a respiratory hazard. Aqueous solutions of iodine are also more stable than those of bromine and can be **standardised** using a sodium thiosulfate solution. The reaction between iodine and an unsaturated fatty acid can then be performed as a titration. Titrations will be studied in Section 13.2.

Since one I_2 molecule reacts with a C=C double bond, 1 mole of iodine reacts with 1 mole of carbon–carbon double bonds. This means that the number of moles of I_2 reacting with one mole of a fatty acid indicates the number of moles of double bonds present in the fatty acid molecule. Worked example 13.1.2 shows you how to calculate the number of C=C double bonds in a sample of a fatty acid.

Worked example 13.1.2

CALCULATING THE NUMBER OF CARBON–CARBON DOUBLE BONDS IN AN ORGANIC COMPOUND

A 0.010 mol sample of linoleic acid reacts completely with 40 mL of a 0.50 M iodine solution (I_2). Determine the number of carbon–carbon double bonds present in each linoleic acid molecule.

Thinking	Working
Calculate the amount, in mol, of iodine that reacted with the fatty acid sample using: $n = cV$	$n(I_2) = cV$ $= 0.50 \times 0.040$ $= 0.020$ mol
Calculate the value of the ratio: $\frac{n(I_2)}{n(\text{fatty acid})}$	$\frac{n(I_2)}{n(\text{linoleic acid})} = \frac{0.020}{0.010}$ $= 2$
Determine the number of carbon–carbon double bonds in one fatty acid molecule using the value of the ratio: $\frac{n(I_2)}{n(\text{fatty acid})}$	2 moles of iodine reacts with 1 mole of linoleic acid, so this means that there are 2 carbon–carbon double bonds in each linoleic acid molecule.

Worked example: Try yourself 13.1.2

CALCULATING THE NUMBER OF CARBON–CARBON DOUBLE BONDS IN AN ORGANIC COMPOUND

A 0.0050 mol sample of linolenic acid reacts completely with 50.0 mL of a 0.30 M iodine solution (I_2). Determine the number of carbon–carbon double bonds present in each linolenic acid molecule.

TABLE 13.1.3 Iodine numbers for some common fats and oils

Fat or oil	Iodine number (g iodine per 100 g fat or oil)
beef fat	42–48
butter	25–42
canola oil	110–126
coconut oil	6–11
fish oil	190–205
linseed oil	170–204
olive oil	75–94
peanut oil	82–107
sesame oil	100–120
sunflower oil	110–145

The reaction of iodine with carbon–carbon double bonds is widely used to measure the degree of unsaturation of fats and oils. Scientists define a quantity called **iodine number**, also called iodine value or iodine index, as the mass of iodine in grams that reacts with 100 grams of oil or fat. It is used in the estimation of the degree of unsaturation in fats and oils. The more unsaturated a fat or oil is, the higher its iodine number will be. Table 13.1.3 lists the iodine numbers for some common fats and oils. Vegetable oils tend to be more unsaturated fats than animal fats, and have higher iodine values. Oils with higher iodine numbers are generally more reactive, less stable, softer and more susceptible to oxidation than fats or oils with lower iodine numbers.

Calculating iodine number

To calculate the iodine number of a substance, you need to determine the number of carbon–carbon double bonds present in one molecule. Iodine adds across the carbon–carbon double bond so 1 iodine molecule (I_2) will react with 1 carbon–carbon double bond. If the substance contains 3 carbon–carbon double bonds, then 3 iodine molecules are required for the addition reaction. This relationship between iodine and the number of carbon–carbon double bonds is a mole ratio that can be used to determine the mass of iodine required to react with 100 g of the substance. Worked example 13.1.3 shows you how to calculate the iodine number using the formula $n = \frac{m}{M}$ and the mole ratio.

The iodine number of a fat or an oil is the mass of iodine, I_2, that reacts with 100 g of the substance.

Worked example 13.1.3

CALCULATING IODINE NUMBER

Oleic acid is a monounsaturated fatty acid found in various animal and vegetable fats and oils. The semi-structural formula of oleic acid is $CH_3(CH_2)_7CH{=}CH(CH_2)_7COOH$ and the molar mass is 282.0 g mol^{-1}. Calculate the iodine number for this fatty acid.

Thinking	Working
Determine the number of carbon–carbon double bonds present in 1 molecule of the fatty acid.	From the semi-structural formula, there is 1 carbon–carbon double bond in a molecule of oleic acid.
Convert 100 g of fatty acid to amount in mol using the formula: $n = \frac{m}{M}$	$n(\text{oleic acid}) = \frac{100}{282.0}$ $= 0.355$ mol
Calculate the amount of I_2, in mol, required to react with 1 mol of the fatty acid using: $n(I_2) = n(\text{fatty acid}) \times$ number of C=C double bonds	$n(I_2) = n(\text{oleic acid}) \times 1$ $= 0.355 \times 1$ $= 0.355$ mol
Calculate the mass of iodine that will react with 100 g of the fatty acid using: $m(I_2) = n \times M$	$m(I_2) = 0.355 \times (126.9 \times 2)$ $= 90.0$ g
The iodine number is equal to the mass of I_2 that will react with 100 g of the fatty acid.	Iodine number of oleic acid = 90.0.

Worked example: Try yourself 13.1.3

CALCULATING IODINE NUMBER

Eicosapentaenoic acid is a polyunsaturated fatty acid found in fish oil. The semi-structural formula of eicosapentaenoic acid is $CH_3(CH_2CH{=}CH)_5(CH_2)_3COOH$ and the molar mass is 302.0 g mol^{-1}. Calculate the iodine number for this fatty acid.

PA 19

13.1 Review

OA ✓✓

SUMMARY

- Simple distillation is a technique for separating two or more liquids from a mixture. The technique relies on a difference of at least 50°C in boiling point between components to achieve an effective separation.
- Fractional distillation is a technique used to separate liquids that have boiling points that are close together. It is commonly used in the laboratory to separate volatile liquids from a reaction mixture.
- Melting points can be used to determine the purity and identity of an organic solid. Impurities lower and broaden the melting point range.
- A solution of bromine can be used to test for the presence of a carbon–carbon double bond. Bromine is decolourised when added to an organic compound containing a carbon–carbon double bond.
- A chemical test for the hydroxyl group is to gently heat the unknown organic compound with ethanoic acid and a few drops of sulfuric acid. When poured into cold water, the characteristic smell of an ester is evident if the compound contains a hydroxyl group.
- Acidified potassium permanganate or acidified potassium dichromate solutions can be added to an unknown alcohol to determine if an organic compound is a tertiary alcohol. If a colour change is observed, then the unknown alcohol may be either a primary or a secondary alcohol.
- To test for the carboxyl group, a small amount of sodium hydrogen carbonate solid is added to an unknown organic compound. Effervescence confirms the presence of a carboxyl group.
- Iodine can be used to measure the degree of unsaturation of compounds.
- The iodine number of a fat or an oil is the mass of iodine that reacts with 100 g of the substance.

KEY QUESTIONS

Knowledge and understanding

1 Select from the following list of words to complete the paragraph about simple distillation:

boiling; boiling point; melting; bumping; condenser; cooled; distillate; higher; lower; separate; vapour

Simple distillation is used to __________ mixtures of two or more liquids. The mixture is placed in a round-bottom flask with anti-__________ granules to allow smooth ________, and then heated to the _______________ of the liquid to be collected. The __________ rises up the flask and enters the __________. The vapour is ________ to below its boiling point and condenses back to a liquid. This pure liquid is called the ___________ and is collected in a separate flask. The liquid with the __________ boiling point remains in the round-bottom flask.

2 A pure organic compound melts at 52–53°C. A small amount of another organic compound with a melting point of 87–88°C is mixed with the first compound. What effect would the addition of the second compound have on the melting point of the pure organic compound?

A The second compound would raise the melting point of the first one.

B The second compound would lower the melting point of the first one.

C The second compound would have no effect on the melting point of the first one.

D It is impossible to determine without knowing the identity of the two compounds.

3 The presence of carbon–carbon double bonds can be detected by shaking a sample with a dilute solution of bromine.

a Describe the changes observed if a carbon–carbon double bond is present.

b Write a balanced equation for the reaction of aqueous bromine solution with ethene.

4 In the body, linolenic acid is converted into two other omega-3 fatty acids: eicosapentaenoic acid and docosahexaenoic acid. The molecular formula of docosahexaenoic acid is $C_{22}H_{32}O_2$ and it has 5 C=C double bonds.

Determine the iodine number of docosahexaenoic acid.

continued over page

13.1 Review *continued*

Analysis

5 A 0.0040 mol sample of a fatty acid reacts completely with 25.0 mL of a 0.48 M iodine solution (I_2). Which one of the following is a possible formula for the fatty acid?

A $CH_3(CH_2)_4CH_2CH{=}CHCH_2(CH_2)_5CH_2COOH$

B $CH_3(CH_2)_4(CH{=}CHCH_2)_2(CH_2)_6COOH$

C $CH_3CH_2(CH{=}CHCH_2)_3(CH_2)_6COOH$

D $CH_3(CH_2)_4(CH{=}CHCH_2)_3CH{=}CH(CH_2)_3COOH$

6 a The melting point range of pure ibuprofen is 75°C. An unknown chemical also has a melting point of 75°C. What can you conclude about the unknown chemical?

b Describe how you can confirm if the unknown chemical is ibuprofen.

7 A sample of fat containing 0.045 mol of fatty acid was found to react with 22.84 g of iodine, I_2. Determine the number of carbon–carbon double bonds present in the fatty acid.

8 An organic chemist has five colourless liquids that are labelled A, B, C, D and E. They are known to be propan-1-ol, propanoic acid, 2-methylpropan-2-ol, hexane and pent-1-ene, but the exact identity of each liquid is unknown. The chemist performs a number of chemical tests on each liquid. Identify each liquid using the results shown below. Justify your answers.

	A	B	C	D	E
Solubility in water	soluble	insoluble	soluble	insoluble	soluble
Addition of red-coloured bromine (Br_2) solution	no immediate reaction	colour disappears	no immediate reaction	no immediate reaction	no immediate reaction
Addition of sodium hydrogen carbonate ($NaHCO_3$) powder	no reaction	no reaction	gas evolved	no reaction	no reaction
Addition of acidified potassium dichromate ($K_2Cr_2O_7$)	colour change from orange to green	no colour change	no colour change	no colour change	no colour change

9 Methyl propanoate, an ester with a sweet, fruity, rum-like odour is synthesised by reacting methanol with propanoic acid in the presence of concentrated sulfuric acid, which acts as a catalyst.

$$CH_3OH(l) + CH_3CH_2COOH(l) \rightarrow CH_3CH_2COOCH_3(l) + H_2O(l)$$

The boiling points of the reactants and products are shown in the table below. Describe how pure methyl propanoate can be extracted from the reaction mixture.

Component	Semi-structural formula	Boiling point (°C)
methanol	CH_3OH	64.7
propanoic acid	CH_3CH_2COOH	141.2
methyl propanoate	$CH_3CH_2COOCH_3$	79.8
water	H_2O	100

13.2 Redox titrations of organic compounds

In Unit 2 you were introduced to **volumetric analysis**, an analytical technique used by chemists in a wide range of industries, including the food, mining, pharmaceutical, petrochemical and wine-making industries. Volumetric analysis is used to accurately determine the amount or concentration of a dissolved substance in a solution. This technique uses a reaction between a solution of unknown concentration and a solution of accurately known concentration.

While volumetric analysis is often used to analyse solutions of acids and bases, in this section you will learn how volumetric analysis using redox reactions can be used to analyse for organic compounds in solution.

REVISION

Concentration

Determining the concentration of the solution under analysis is at the heart of volumetric analysis.

As you know, concentration is a measure of the amount of solute in a specified volume of solution. Chemists often express concentration as molar concentration, referred to as molarity. Molarity is defined as the amount of solute, in mol, dissolved in 1 litre of solution:

$$\text{concentration} = \frac{\text{amount of solute (in mol)}}{\text{volume of solution (in L)}}$$

The unit of molarity is mol L^{-1}, which has the symbol M.

Other units of concentration

Another concentration measure often used in volumetric analysis is mass per unit volume. Examples are shown in Table 13.2.1.

TABLE 13.2.1 Some concentration measures based on mass per unit volume

Unit of concentration	Unit of mass	Unit of volume
g L^{-1}	g	L
mg L^{-1}	mg	L
mg mL^{-1}	mg	mL
μg mL^{-1}	μg	mL

These concentrations are easily calculated by dividing the mass of solute, in the desired unit, by the volume of solution, in the desired unit.

For example, concentration (in g L^{-1}) $= \frac{m \text{ (in g)}}{V \text{ (in L)}}$.

Many commercial products express concentration in other ways, such as:

- percentage mass/volume (% m/v), i.e. grams per 100 mL
- percentage by mass (% m/m), i.e. grams per 100 g
- percentage by volume (% v/v), i.e. mL per 100 mL.

Converting between concentration units

Different units of concentration can be interconverted, for example from mol L^{-1} to g L^{-1}.

Figure 13.2.1 shows a useful relationship for conversion between the two concentration units most often used in volumetric analysis, mol L^{-1} (M) and g L^{-1}.

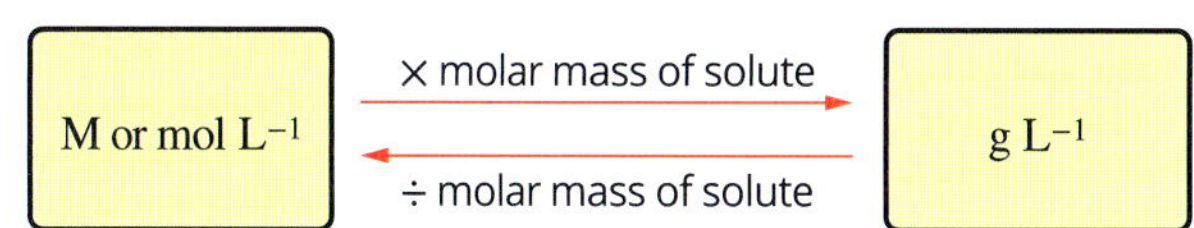

FIGURE 13.2.1 The relationship for conversions between mol L^{-1} and g L^{-1}

SOLVING STOICHIOMETRY PROBLEMS INVOLVING SOLUTIONS

You will recall from previous chapters that there are four main steps in solving calculation problems involving reactions.

1 Write a balanced equation for the reaction.
2 Calculate the amount, in mol, of the substance with known volume and concentration.
3 Use the mole ratio from the equation to calculate the amount, in mol, of the required substance.
4 Calculate the required volume or concentration to the correct number of significant figures. Section 1.3 page 28 describes how to use significant figures.

The steps in a stoichiometric calculation can be summarised as shown in the flow chart in Figure 13.2.2.

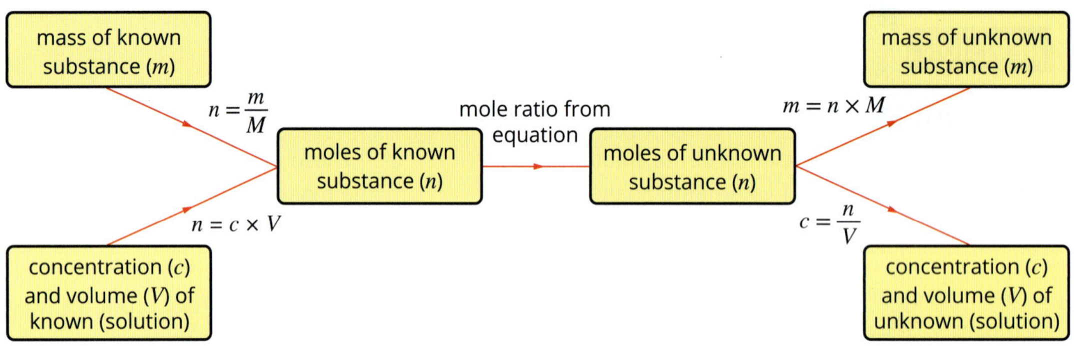

FIGURE 13.2.2 Flow chart for mass and solution stoichiometry calculations

PRIMARY STANDARDS AND STANDARD SOLUTIONS

To find the concentration of a solution of unknown concentration, the solution is reacted with a solution of accurately known concentration. A solution of accurately known concentration is called a **standard solution**.

Before discussing the procedure of volumetric analysis further, it is important to review how standard solutions are prepared.

Pure substances are widely used in the laboratory to prepare standard solutions. Substances that are so pure that the amount, in moles, can be calculated accurately from their mass, are called **primary standards**.

A primary standard should:

- be readily obtainable in a pure form
- have a known chemical formula
- be easy to store without absorbing water vapour or reacting with gases in the atmosphere, such as carbon dioxide
- have a high molar mass to minimise the effect of errors in weighing.

Standard solutions are prepared by:

- dissolving an accurately measured mass of a primary standard in water to make an accurately measured volume of solution, or
- performing a titration with another standard solution to determine its exact concentration.

To prepare a standard solution from a primary standard, you need to dissolve an accurately known amount of the substance in deionised water to produce a solution of known volume. The steps in this process are shown in Figure 13.2.3 on the following page.

> Once the concentration of a solution has been determined from a titration, the solution can be described as a standard solution.

In practice, making a standard solution directly from a primary standard is only possible for a few of the chemicals encountered in the laboratory. Many chemicals are impure because they decompose or react with chemicals in the atmosphere. When a standard solution of one of these 'impure' chemicals is required, it is first reacted (by titration) with a standard solution to accurately determine its concentration. The solution for which we now know an accurate concentration is called a **standardised** solution.

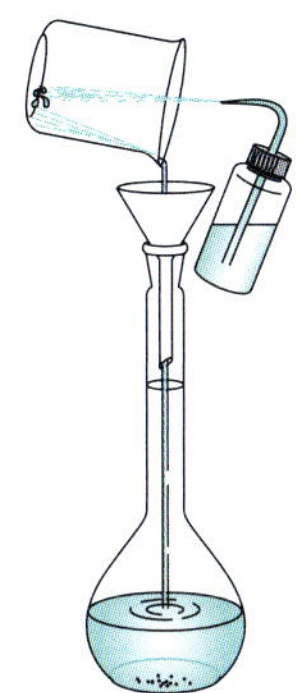

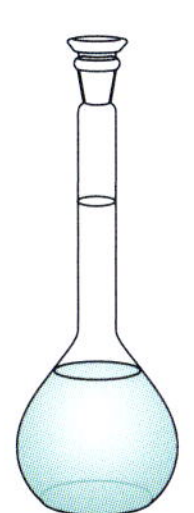

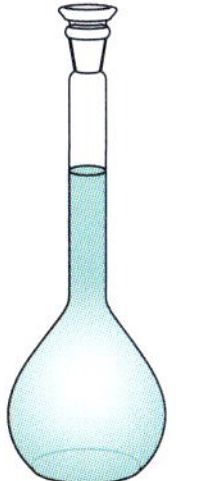

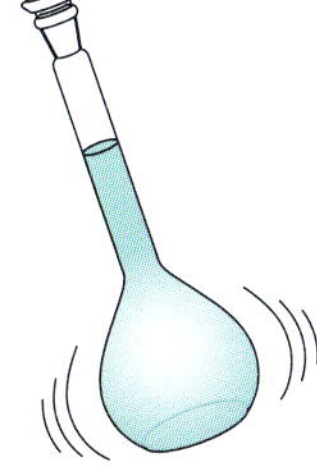

FIGURE 13.2.3 The steps taken to prepare a standard solution from a primary standard

CONDUCTING VOLUMETRIC ANALYSES

During a volumetric analysis, a measured volume of a standard solution is mixed with a measured volume of a solution of unknown concentration. The solutions are combined until they have just reacted completely in the mole ratio indicated by the balanced chemical equation (the **equivalence point**). This process is known as performing a **titration**.

In the case of a titration involving a reaction in which no colour change takes place, an **indicator** is added to determine when the reaction is complete. The **end point** is the point during the titration at which the indicator changes colour. For an accurate analysis, the end point should be very close to the equivalence point.

> The equivalence point occurs when the reactants have reacted in the mole ratio indicated by the balanced chemical equation.
>
> The end point occurs when the indicator changes colour.

Figure 13.2.4 shows the equipment used to complete a typical titration.

The steps involved in a titration are as follows.

1 A known volume, or **aliquot**, of one of the solutions is measured using a **pipette** and transferred into a conical flask.

2 If an indicator is being used, a few drops of the indicator are added so that a colour change signals the point at which the titration should stop.

3 The other solution is dispensed slowly into the conical flask from a **burette** until the indicator changes colour permanently (end point). A burette delivers an accurately known, but variable, volume called a **titre**.

4 The titre delivered from the burette is calculated by subtracting the initial burette reading from the final burette reading. All volume readings from a burette are estimated to the second decimal place.

5 To minimise random errors, the titration is repeated several times and the **average titre** is found. Usually three **concordant titres** are used to find this average.

FIGURE 13.2.4 Equipment used to complete a typical redox titration

Rinsing volumetric glassware

To ensure that glassware is completely clean, it is rinsed before a volumetric analysis is conducted. This removes any trace chemicals from the glassware, and makes the analytical results more precise and accurate.

Table 13.2.2 describes how the glassware used in volumetric analysis should be rinsed. Rinsing with the wrong liquid can cause errors in the analysis.

TABLE 13.2.2 Techniques for rinsing equipment for volumetric analysis

Glassware	Correct	Incorrect
burette pipette	The final rinse should be with the reactant solution to be transferred by them.	Rinsing with water will dilute the reactant solution.
volumetric flask conical flask	Only rinse these with deionised water.	Rinsing with reactants will introduce unmeasured amounts of these chemicals into the flask that can react and affect the results.

SOURCES OF ERROR IN VOLUMETRIC ANALYSIS

The accuracy with which the volumes of aliquots and titres are measured in volumetric analysis depends on the calibration of the equipment used. There are always errors associated with measurements of quantities such as mass and volume made during experimental work.

Typical uncertainties associated with volumetric analysis are shown in Table 13.2.3.

Other graduated laboratory glassware provides less precise measures; for example, a 50 mL measuring cylinder has an uncertainty of ±0.3 mL and a 50 mL graduated beaker has an uncertainty of ±5 mL.

Analytical chemists aim to produce results that are both precise and accurate. Methods used for accurate **quantitative analysis** should be designed to minimise errors. Where errors cannot be avoided, any discussion of results should refer to the level of inaccuracy that may have accumulated. This requires an understanding of the different types of experimental errors: mistakes, random errors and systematic errors, which are described in Chapter 1 (pages 28–29).

Concordant titres are titres that are within a range of 0.10 mL from highest to lowest of each other.

TABLE 13.2.3 Typical uncertainties associated with analytical equipment

Analytical equipment	Uncertainty
20 mL pipette	±0.03 mL
50 mL burette	±0.02 mL for each reading
250.0 mL volumetric flask	0.3 mL
100 g capacity top loading balance	±0.001 g
60 g capacity analytical balance	±0.0001 g

REDOX REACTIONS INVOLVING ORGANIC COMPOUNDS

You will recall from Chapter 4 that oxidation occurs when a substance loses electrons and reduction occurs when a substance gains electrons. Reactions involving the loss and gain of electrons are commonly referred to as **redox reactions**.

The effects of organic compounds undergoing redox reactions are easily observed in our everyday lives. Oxygen from the atmosphere slowly oxidises molecules in fruit, such as apples and bananas, to make the fruit turn brown. This can be prevented by adding lemon or lime juice to the cut fruit (Figure 13.2.5). The ascorbic acid (vitamin C) in the juice is preferentially oxidised by the oxygen in the air, preventing the food from spoiling. Within the human body, redox reactions in cells are the source of the body's energy.

In this section, you will see how the capacity of some organic molecules to undergo redox reactions can be used as a basis for volumetric analysis.

Oxidation is defined as the loss of electrons. Reduction is defined as the gain of electrons.

FIGURE 13.2.5 Juice from citrus fruits is often added to guacamole (avocado) and cut apples to stop them from going brown.

ANALYSING ORGANIC COMPOUNDS BY REDOX TITRATIONS

A **redox titration** involves the reaction of an oxidising agent with a reducing agent. One solution is usually pipetted into a conical flask and the other is dispensed into the flask from a burette.

For some redox titrations, such as those involving the permanganate ion (MnO_4^-), the equivalence point is indicated by a colour change in one of the reacting solutions. For other redox titrations, an indicator, such as a starch solution, must be added to detect the equivalence point.

Volumetric analysis involving redox reactions can be used to determine the composition of a range of substances, including organic chemicals in fruit juice and wine. Table 13.2.4 gives examples of how these substances are analysed.

TABLE 13.2.4 Examples of analysis using redox titrations

Substance	Ingredients for analysis	Titrate with
wine	ethanol	acidified potassium permanganate or potassium dichromate solution
fruit juice	vitamin C (ascorbic acid)	iodine solution
wine	sulfur dioxide	iodine solution

CASE STUDY

Boab trees: a food source for Indigenous Australians

The *larrkardiy*, or Australian Boab tree (*Adansonia gregorii*) (Figure 13.2.6a), is found only in the Kimberley region of Western Australia and east into the Northern Territory. The tree comes into bloom in the wet season, producing fruit and flowers which fall from the tree during the dry season. The fruits are hard-shelled and usually referred to as nuts (Figure 13.2.6b). Some boab trees are more than 1500 years old, making them among the oldest living organisms in Australia. As boabs age they become hollow; they use this hollow to store water within the trunk to help survive the harsh drought conditions of the dry season.

(a)

FIGURE 13.2.6 (a) The boab tree is a defining feature in the Kimberley region of Western Australia. (b) Boab nuts, traditionally eaten by Indigenous Australians, are rich in vitamin C and other essential nutrients.

For thousands of years, Indigenous Australians have used boab trees in many ways. They obtained water from hollows in the tree and used the white powder that fills the seed pod as a food; the leaves were used medicinally. Decorative paintings or carvings were sometimes made on the outer surface of the fruit and the trunk of the tree.

The dry powdery pulp inside the boab nut is rich in vitamin C (ascorbic acid), with 40 g of the powder providing 84–100% of the recommended daily intake. It is also rich in other vitamins (A, B1, B2 and B6), minerals (calcium, iron, magnesium and potassium), dietary fibre and amino acids. The nuts are lightweight and keep easily for over a year, making them a convenient food source for Indigenous Australians when moving around an area. The roots can be chewed to extract moisture and the iron-rich leaves used as a leafy vegetable.

Volumetric analysis can be used to determine the vitamin C content of the powder in the boab nut. An acid–base titration using a standardised sodium hydroxide solution can be used or a redox titration, as described in this section.

CHEMFILE

Preventing wine from turning to vinegar

Sulfur dioxide (preservative number 220), SO_2, has been used as a preservative in wine since Roman times. It is added to wine to prevent oxidation and bacterial spoilage. The maximum amount that winemakers are allowed to add is 250 mg L^{-1}. Sulfur dioxide is added either directly, in the form of compressed gas, or via soluble metabisulfite salts such as $K_2S_2O_5$, which react with water to form SO_2. The primary function of SO_2 as an antioxidant is to prevent, or limit, the reaction of oxygen with ethanol to produce ethanoic acid. Higher levels of SO_2 are added to white wines as they are more susceptible to oxidation, whereas the tannins in red wine act as natural preservatives.

Analysing alcohols

In Chapter 10, you learnt that organic compounds with a hydroxyl functional group (–OH) belong to the family of molecules known as alcohols. Typical uses of alcohols are as fuels, in alcoholic beverages, as industrial solvents and as cleaning products. Typical analyses performed in the laboratory of a winery, such as the one shown in Figure 13.2.7, include testing for pH, sulfur dioxide content and the concentration of both ethanol and malic acid.

FIGURE 13.2.7 Analysis of wine includes testing pH and determining the concentration of chemicals such as ethanol, sulfur dioxide and malic acid.

Alcohols can undergo oxidation and act as weak reducing agents. When a primary alcohol such as ethanol is oxidised, the hydroxyl functional group is converted into a carboxyl functional group (–COOH), as shown in Figure 13.2.8.

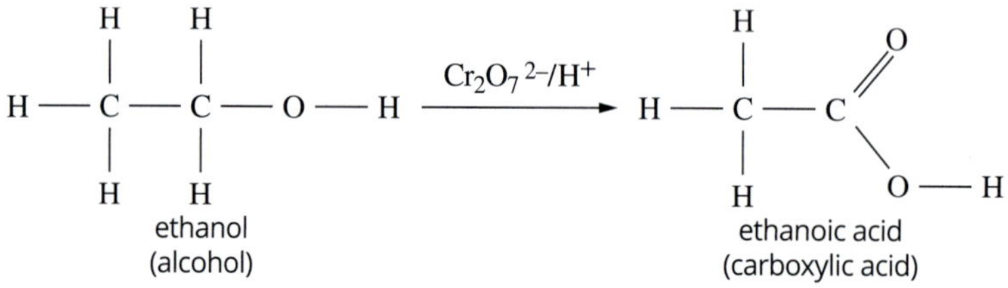

$$H{-}\overset{H}{\underset{H}{C}}{-}\overset{H}{\underset{H}{C}}{-}O{-}H \xrightarrow{Cr_2O_7{}^{2-}/H^+} H{-}\overset{H}{\underset{H}{C}}{-}C{\Large(}{=}O{\Large)}{-}O{-}H$$

ethanol (alcohol) → ethanoic acid (carboxylic acid)

FIGURE 13.2.8 When alcohols such as ethanol are oxidised, a carboxylic acid is formed.

These oxidation reactions can be used as the basis for volumetric analysis to determine the concentration of an alcohol in a solution. Typically, a strong oxidising agent, such as acidified potassium dichromate solution ($K_2Cr_2O_7$) or acidified potassium permanganate solution ($KMnO_4$), is used to react with the alcohol in a titration.

If, for example, ethanol is oxidised to ethanoic acid by an acidified solution of potassium dichromate ($K_2Cr_2O_7$), the dichromate ion is involved in the reaction whereas the potassium ion is a spectator ion.

CALCULATIONS IN VOLUMETRIC ANALYSIS

Calculations in volumetric analysis usually involve several steps. The flow chart in Figure 13.2.9 depicts the steps in performing these calculations.

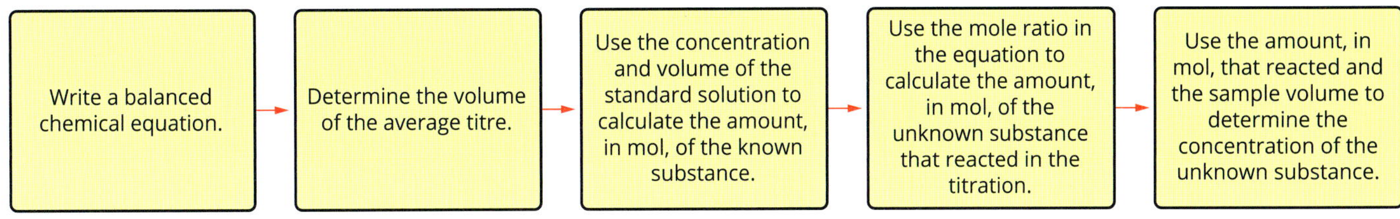

FIGURE 13.2.9 Flow chart summarising the steps in the calculation of the concentration of an unknown substance using data from a titration

Worked example 13.2.1 shows how the concentration of an alcohol can be determined by a redox titration.

Worked example 13.2.1

VOLUMETRIC ANALYSIS OF AN ALCOHOL

The concentration of ethanol in a sample of beer was determined by titration with a standard solution of acidified potassium dichromate. The ionic equation for the reaction is:

$$3CH_3CH_2OH(aq) + 2Cr_2O_7^{2-}(aq) + 16H^+(aq) \rightarrow 3CH_3COOH(aq) + 4Cr^{3+}(aq) + 11H_2O(l)$$

A 20.00 mL aliquot of beer was titrated with a 7.88×10^{-3} M solution of potassium dichromate. Titres of 17.05, 21.15, 21.13 and 21.17 mL were required to reach the end point when the orange solution turned green. Determine the molar concentration of ethanol in the beer.

Thinking	Working
Calculate the average volume of the concordant titres.	The titre of 17.05 mL is discarded because it is not concordant, i.e. it is not within a range of 0.10 mL from highest to lowest titre volumes. $\text{average titre} = \frac{21.15+21.13+21.17}{3} = 21.15\text{ mL}$
Calculate the amount, in mol, of the standard solution that was required to reach the end point.	$n(K_2Cr_2O_7) = c \times V = 7.88 \times 10^{-3} \times 0.021\,15 = 1.67 \times 10^{-4}\text{ mol}$
Use the mole ratio in the equation to calculate the amount, in mol, of the unknown substance that reacted with the standard solution.	$\text{mole ratio} = \frac{n(CH_3CH_2OH)}{n(K_2Cr_2O_7)} = \frac{3}{2}$ So $n(CH_3CH_2OH) = \frac{3}{2} \times n(K_2Cr_2O_7) = \frac{3}{2} \times 1.67 \times 10^{-4} = 2.50 \times 10^{-4}\text{ mol}$
Determine the concentration of the unknown substance. Express your answer to the appropriate number of significant figures.	$c(CH_3CH_2OH) = \frac{n}{V} = \frac{2.50 \times 10^{-4}}{0.020\,00} = 0.0125\text{ M}$ The final result is rounded to 3 significant figures, corresponding to the smallest number of significant figures in the original data. All figures are kept in the calculator throughout the calculation to avoid rounding errors.

Worked example: Try yourself 13.2.1

VOLUMETRIC ANALYSIS OF AN ALCOHOL

The concentration of a solution of methanol (CH_3OH) was determined by titration with a standard solution of acidified potassium permanganate ($KMnO_4$). The ionic equation for the reaction is:

$5CH_3OH(aq) + 4MnO_4^-(aq) + 12H^+(aq) \rightarrow 5HCOOH(aq) + 4Mn^{2+}(aq) + 11H_2O(l)$

A 10.00 mL aliquot of CH_3OH solution was titrated with a 0.125 M solution of $KMnO_4$. Titres of 14.13, 14.28, 14.18 and 14.11 mL were required to reach the end point when the orange solution turned green. Determine the molar concentration of the methanol solution.

FIGURE 13.2.10 Iodine (I_2) in the presence of starch forms a dark blue complex.

Selecting indicators for redox titrations

It can be more difficult to select suitable indicators for redox titrations than for the acid–base titrations that you studied in Unit 2. Redox indicators must behave as oxidising agents or reducing agents after the equivalence point has been reached and a small excess of solution from the burette is present. They must also be highly coloured in either oxidised or reduced form.

Starch is used as an indicator in titrations in which iodine (I_2) is either a reactant or a product. When iodine is present in excess, it reacts with starch to form a dark blue complex (Figure 13.2.10).

Often one of the reactants in a redox titration is strongly coloured, such as the permanganate ion (MnO_4^-) or the dichromate ion ($Cr_2O_7^{2-}$). The permanganate ion is purple while the manganese(II) ion is colourless, as seen in Figure 13.1.7 on page 474:

$$\underset{\text{purple}}{MnO_4^-(aq)} + 8H^+(aq) + 5e^- \rightarrow \underset{\text{colourless}}{Mn^{2+}(aq)} + 4H_2O(l)$$

The dichromate ion is orange, while the chromium(III) ion (Cr^{3+}) is green, as seen in Figure 13.1.6 on page 474:

$$Cr_2O_7^{2-}(aq) + 14H^+(aq) + 6e^- \rightarrow 2Cr^{3+}(aq) + 7H_2O(l)$$

Because of this, there is no need to use a redox indicator in titrations using the permanganate ion or the dichromate ion.

CHEMFILE

Chemical test for starch or iodine

Starch is a carbohydrate found in plants. The monomer, α-glucose, undergoes condensation polymerisation forming the polysaccharide starch. There are two forms of starch: a linear polymer known as amylose and a branched polymer known as amylopectin. Amylose is responsible for the formation of the dark blue colour in the presence of iodine. The long molecules of amylose coil into spiral-like helices and pack together tightly (see the figure at right). Insoluble iodine is dissolved in water using soluble potassium iodide forming the triiodide ion (I_3^-). When added to starch, the triiodide ion slips inside the amylose coil, causing the dark blue colour.

The triiodide ion inside the amylose coil results in the intense dark blue colour when iodine is added to starch.

VOLUMETRIC ANALYSIS THAT INVOLVES DILUTION

It is often necessary to **dilute** a solution by adding water to it to reduce its concentration, before carrying out a titration. This is done to obtain concentrations that will result in titres that are within the range of the burette. In this case, the following additional data should be recorded:

- the volume of the aliquot of undiluted solution
- the volume of diluted solution that is prepared.

For example, in a titration to determine the vitamin C concentration in a sample of orange juice, 25.00 mL of orange juice is diluted to 250.0 mL in a volumetric flask before taking aliquots for titration. This means the **dilution factor** is $\frac{250.0}{25.00} = 10.00$. The undiluted juice will be 10.00 times more concentrated than the concentration of the aliquot. This will be considered in the calculations.

The steps required to calculate the concentration of vitamin C in the undiluted orange juice are summarised in Figure 13.2.11.

dilution factor = $\frac{\text{volume of the diluted solution}}{\text{volume of the undiluted solution}}$

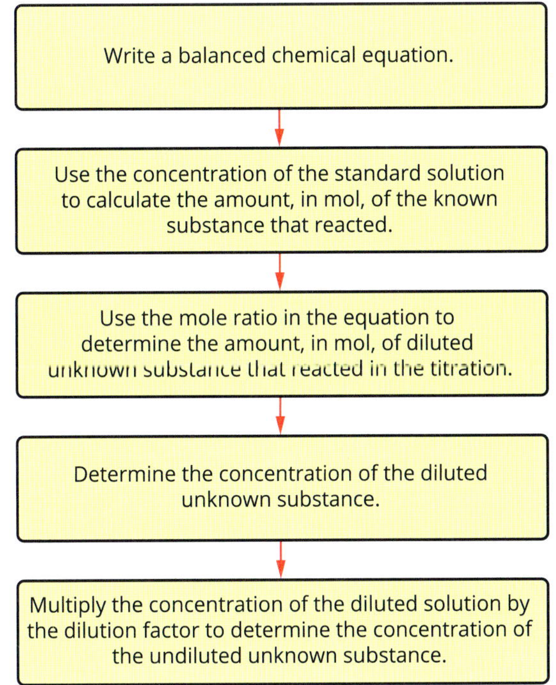

FIGURE 13.2.11 This flow chart shows the steps used in the calculation of the concentration of an unknown substance that has been diluted for use in a titration.

Worked example 13.2.2 shows how to calculate the concentration of an unknown substance that has been diluted for use in a titration.

Worked example 13.2.2

VOLUMETRIC ANALYSIS THAT INVOLVES DILUTION

The concentration of sodium hypochlorite, NaOCl in a sample of bleach was found by diluting 20.00 mL of bleach to 250.0 mL. Excess potassium iodide and sulfuric acid are added to each of the 20.00 mL aliquots of the diluted bleach in a conical flask.

The initial reaction between the hypochlorite ions, iodide ions and the acid in the conical flask is:

$$OCl^-(aq) + 2I^-(aq) + 2H^+(aq) \rightarrow I_2(aq) + Cl^-(aq) + H_2O(l)$$

and the mixture turns brown due to the production of iodine.

The mixture is titrated with 0.125 M sodium thiosulfate ($Na_2S_2O_3$) solution. Titres of 26.68, 26.40, 26.35 and 26.42 mL are required to reach the end point, which is determined using starch indicator:

$$I_2(aq) + 2S_2O_3^{2-}(aq) \rightarrow S_4O_6^{2-}(aq) + 2I^-(aq)$$

Calculate the molar concentration of NaOCl in the original sample of bleach.

Thinking	Working
Determine the average volume of the concordant titres.	The titre of 26.68 mL is discarded because it is not concordant, i.e. it is not within a range of 0.10 mL from the highest to lowest titre volumes. $\text{average titre} = \frac{26.40 + 26.35 + 26.42}{3}$ $= 26.39\text{ mL}$
Calculate the amount, in mol, of the standard solution that was required to reach the end point.	$n(S_2O_3^{2-}) = c \times V$ $= 0.125 \times 0.02639$ $= 3.299 \times 10^{-3}\text{ mol}$
Use the mole ratio in the equation to calculate the amount, in mol, of the unknown substance that reacted with the standard solution.	$\text{mole ratio} = \frac{n(I_2)}{n(S_2O_3^{2-})} = \frac{1}{2}$ In this example, $n(I_2) = n(OCl^-)$ $= \frac{1}{2} \times n(S_2O_3^{2-})$ $n(OCl^-) = \frac{1}{2} \times n(S_2O_3^{2-})$ $= \frac{1}{2} \times 3.299 \times 10^{-3}$ $= 1.649 \times 10^{-3}\text{ mol}$
Determine the concentration of the unknown substance in the diluted solution by dividing the amount of unknown in one aliquot by the volume of the aliquot (in litres).	$c(OCl^-) = \frac{n}{V}$ $= \frac{1.649 \times 10^{-3}}{0.02000}$ $= 0.08247\text{ M}$

Multiply by the dilution factor to determine the concentration of the unknown substance in the undiluted solution. Express your answer to the appropriate number of significant figures.	Dilution factor $= \frac{250.0}{20.00} = 12.50$ So, undiluted $c(OCl^-)$ = diluted $c(OCl^-) \times 12.50$ $= 0.08247 \times 12.50$ $= 1.03$ M The final result is rounded to 3 significant figures, corresponding to the smallest number of significant figures in the original data. All values are kept in the calculator in the earlier steps of the calculation to avoid rounding errors.

Worked example: Try yourself 13.2.2

VOLUMETRIC ANALYSIS THAT INVOLVES DILUTION

The concentration of sodium hypochlorite, NaOCl, in a sample of disinfectant, was found by diluting 25.00 mL of bleach to 250.0 mL. Excess potassium iodide and sulfuric acid are added to each of the 20.00 mL aliquots of the diluted bleach in a conical flask.

The initial reaction between the hypochlorite ions, iodide ions and the acid in the conical flask is:

$$OCl^-(aq) + 2I^-(aq) + 2H^+(aq) \rightarrow I_2(aq) + Cl^-(aq) + H_2O(l)$$

and the mixture turns brown due to the production of iodine.

The mixture is titrated with 0.0235 M sodium thiosulfate ($Na_2S_2O_3$) solution. Titres of 25.08, 26.40, 25.05 and 25.10 mL are required to reach the end point, which is determined using starch indicator:

$$I_2(aq) + 2S_2O_3^{2-}(aq) \rightarrow S_4O_6^{2-}(aq) + 2I^-(aq)$$

Calculate the molar concentration of NaOCl in the original sample of disinfectant.

STOICHIOMETRY PROBLEMS INVOLVING EXCESS REACTANTS

Apart from reactions that occur during a titration when the mixing of reactants is stopped at the equivalence point, in many other reactions the reactants are not mixed together in stoichiometric amounts. There are often occasions when one of the reactants, such as an oxidising agent or a reducing agent, is in excess. You learnt how to determine the excess and limiting reactants in a reaction in Chapter 3. As you will see in Worked example 13.2.4 on the following page, the same principles apply for identifying the excess and limiting reactants in stoichiometric problems involving solutions and redox reactions.

- Calculate the number of moles of each reactant.
- Determine which is the **excess reactant** and which is the **limiting reactant**.
- Use the amount of limiting reactant to work out the amount of product formed or the amount of reactant in excess.

Worked example 13.2.3

SOLUTION STOICHIOMETRY: A LIMITING REACTANT PROBLEM

20.00 mL of a solution of potassium permanganate ($KMnO_4$) with a concentration of 0.0110 M is added to 50.00 mL of a 0.0245 M iron(II) sulfate ($FeSO_4$) solution in the presence of sulfuric acid. The overall ionic equation for the reaction is:

$$MnO_4^-(aq) + 5Fe^{2+}(aq) + 8H^+(aq) \rightarrow Mn^{2+}(aq) + 5Fe^{3+}(aq) + 4H_2O(l)$$

a Determine the limiting reactant.

b Calculate the mass of iron(III) sulfate, $Fe_2(SO_4)_3$ produced by the reaction.

Thinking	Working
a Calculate the number of moles of each of the reactants using $n = cV$. Remember to convert mL to L.	$n(KMnO_4) = n(MnO_4^-)$ $= c \times V$ $= 0.0110 \times 0.02000$ $= 2.200 \times 10^{-4}$ mol $n(FeSO_4) = n(Fe^{2+})$ $= c \times V$ $= 0.0245 \times 0.05000$ $= 1.225 \times 10^{-3}$ mol
Identify the limiting reactant as follows. Choose one reactant and use the coefficients in the equation to find the amount of the other reactant needed for it to completely react.	$n(Fe^{2+})$ needed to react $= 5 \times n(MnO_4^-)$ $= \frac{2.200 \times 10^{-4}}{1} \times 5 = 1.100 \times 10^{-3}$ mol
Compare the values for the amount required of the second reactant and the amount actually present to determine which is the limiting reactant.	1.100×10^{-3} mol of Fe^{2+} is needed for all of the MnO_4^- to react. There is 1.225×10^{-3} mol of Fe^{2+} present, so the $FeSO_4$ is in excess and $KMnO_4$ is the limiting reactant.
b Find the mole ratio of the unknown substance to the limiting reactant from the equation coefficients: $\frac{\text{coefficient of unknown}}{\text{coefficient of known (limiting reactant)}}$	The unknown substance is $Fe_2(SO_4)_3$. From the equation coefficients: $\frac{n(Fe^{3+})}{n(KMnO_4)} = \frac{5}{1}$ So $\frac{n(Fe_2(SO_4)_3)}{n(KMnO_4)} = \frac{2.5}{1}$
Calculate the number of moles of the unknown substance using the mole of limiting reactant calculated using the original data: n(unknown) = mole ratio $\times$ n(limiting reactant)	The limiting reactant is $KMnO_4$. $n(Fe_2(SO_4)_3) = \frac{2.5}{1} \times n(KMnO_4)$ $= \frac{2.5}{1} \times 2.200 \times 10^{-4}$ $= 5.500 \times 10^{-4}$ mol
Calculate the mass of the unknown substance using: m(unknown) = n(limiting reactant) $\times$ molar mass Express your answer to the appropriate number of significant figures.	$M(Fe_2(SO_4)_3) = (2 \times 55.8) + 3 \times (32.1 + 4 \times 16.0)$ $= 399.9$ g mol^{-1} $m(Fe_2(SO_4)_3) = 5.500 \times 10^{-4} \times 399.9$ $= 0.220$ g

Worked example: Try yourself 13.2.3

SOLUTION STOICHIOMETRY: A LIMITING REACTANT PROBLEM

17.5 mL of a 0.645 M potassium dichromate ($K_2Cr_2O_7$) solution is added to 28.2 mL of a 0.560 M ethanol (CH_3CH_2OH) solution. The overall equation for the reaction is:

$$3C_2H_5OH(aq) + 2Cr_2O_7^{2-}(aq) + 16H^+(aq) \rightarrow 3CH_3COOH(aq) + 4Cr^{3+}(aq) + 11H_2O(l)$$

a Determine the limiting reactant.

b Calculate the mass of CH_3COOH in the reaction mixture when the reaction is complete.

CASE STUDY ANALYSIS

Determination of vitamin C content

In 2004, two 14-year-old New Zealand students, Jenny Suo and Anna Devathasan (Figure 13.2.12), conducted a research project for their school science fair. Starting with a research question about the link between vitamin C content and the cost of popular fruit juice products, they designed an experiment to test their hypothesis that cheaper fruit juice brands would have a lower vitamin C content than more expensive products.

From product advertisements, Jenny and Anna were confident that the drink Ribena would have a higher vitamin C content than pure orange juice. The manufacturers of Ribena, at the time the second largest food and drug manufacturer in the world, claimed that 'the blackcurrants in Ribena contain four times the vitamin C of oranges'.

Jenny and Anna decided that their independent variable would be the brand of juice tested. They selected volumetric analysis as the method needed to measure the vitamin C content, and the concentration of vitamin C in each drink tested would be their dependent variable.

The girls were surprised by their results: Ribena had a much lower vitamin C content than the other brands tested. To ensure the reliability of their results, they repeated the experiment several times. The vitamin C content of Ribena was higher than only the cheapest orange fruit juice drink tested.

Jenny and Anna began to suspect that the wording in Ribena advertisements might be misleading to consumers and contacted the manufacturer. With no response, they went to the New Zealand consumer television program *Fair Go*, then finally to New Zealand's Commerce Commission. A government investigation was opened and 3 years after their science fair project, the manufacturer was fined more than $160 000 and required to publicly apologise for misleading consumers.

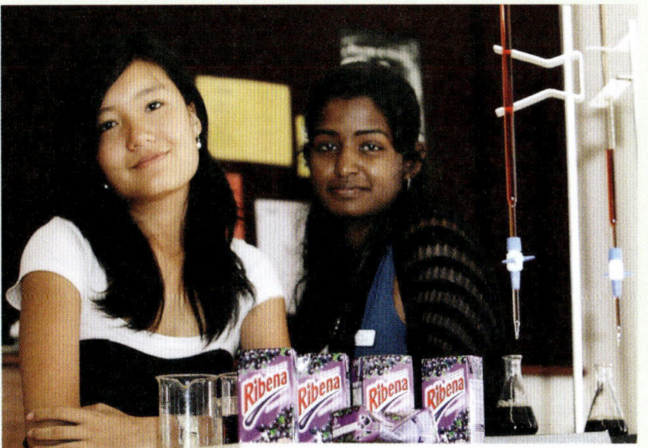

FIGURE 13.2.12 Jenny Suo and Anna Devathasan conducted a scientific investigation into the vitamin C content of popular brands of fruit juice drinks. Using volumetric analysis, they concluded that the blackcurrant drink Ribena contained less vitamin C than claimed by the manufacturer.

In a similar investigation, a student performed a redox titration to determine the ascorbic acid content of a popular brand of vitamin C tablets. The experiment involved dissolving vitamin C tablets in deionised water and titrating the solutions with a standardised iodine solution using starch indicator. Part of the report presented by the student is shown below.

Procedure

1 Weigh 1 vitamin C tablet and transfer it to a 100 mL conical flask.

2 Crush the tablet. Add about 50 mL of deionised water and stir to dissolve the vitamin C powder.

3 Add 1 mL of starch indicator to the conical flask.

4 Fill a burette with the standardised iodine solution.

continued over page

CASE STUDY ANALYSIS *continued*

5 Titrate the ascorbic acid solution with the iodine until the indicator just shows a permanent dark blue-black colour.

6 Repeat steps 1–5 twice.

The equation for the reaction is:

$C_6H_8O_6(aq) + I_2(aq) \rightarrow C_6H_6O_6(aq) + 2I^-(aq) + 2H^+(aq)$

Results

Concentration of the standardised iodine solution = 0.0525 M.

Trial	1	2	3
Mass of vitamin C tablet (g)	0.342	0.336	0.338
Titre iodine (mL)	25.68	25.47	25.53

Analysis

1 Why is the iodine referred to as a 'standardised' solution?

2 Use the titres of iodine to calculate the mass of ascorbic acid in each vitamin C tablet.

3 Calculate the average percentage purity of the vitamin C tablets.

4 The calculated mass of ascorbic acid per tablet was less than the value of 500 mg per tablet claimed by the manufacturer. Which one or more of the following errors could account for the lower-than-expected calculated concentration? Justify your answers.

A The conical flask was rinsed only with deionised water before use.

B The burette was rinsed only with deionised water before use.

C The concentration of the standardised iodine solution was higher than the actual value.

5 The safety data sheet (SDS) for iodine solution includes the following information:

- Causes serious eye irritation and harmful in contact with skin.

Apart from a laboratory coat, what personal protective equipment (PPE) should be used by the student during the experiment?

6 Suggest what functions the other substances in the vitamin C tablet might perform.

13.2 Review

SUMMARY

- Some organic compounds undergo redox reactions, which can form the basis of volumetric analysis.
- Volumetric analysis is an analytical technique for determining the concentration of a solution by titrating it against a solution of known concentration (a standard solution) and volume.
- A solution of accurately known concentration is referred to as a standard solution. Standard solutions can be prepared from primary standards or by titrating an existing solution with another standard solution to determine its concentration.
- A primary standard is a substance so pure that the amount of substance, in mol, can be calculated accurately from their mass.
- In a titration, a measured volume of a standard solution is reacted with a measured volume of the solution whose concentration is to be determined.
- Volumetric flasks, pipettes and burettes are accurately calibrated pieces of laboratory glassware used in volumetric analysis.
- Three concordant titres are usually obtained during a titration to minimise random errors. Concordant titres vary within narrowly specified limits, usually within a range of 0.10 mL from the highest to the lowest.
- When rinsing glassware before a titration, it is important to ensure that:
  - conical and volumetric flasks are rinsed with deionised water
  - the burette and pipette are rinsed with the solution to be transferred by them.
- Some redox titrations involve a change in colour and therefore do not require the addition of an indicator.
- Dilution of the unknown solution is sometimes required to obtain manageable titre volumes.
- The limiting reactant or reagent is the reactant or reagent that is completely consumed in the reaction.
- The reactant or reagent that is not completely used up and has some remaining when the reaction has stopped is said to be in excess.

KEY QUESTIONS

Knowledge and understanding

1 Explain the difference between the following:
   a standard solution and primary standard
   b burette and pipette
   c aliquot and titre.

2 Alcohols can be converted into carboxylic acids by oxidising agents. Write a balanced half-equation for the oxidation of butan-1-ol ($CH_3CH_2CH_2CH_2OH$) to form butanoic acid ($CH_3CH_2CH_2COOH$).

3 The concentration of ethanol in a cider drink was determined by titrating a sample of the cider with acidified potassium permanganate. The equation for the reaction is:

$$5CH_3CH_2OH(aq) + 4MnO_4^-(aq) + 12H^+(aq) \longrightarrow 5CH_3COOH(aq) + 4Mn^{2+}(aq) + 11H_2O(l)$$

A 20.00 mL aliquot of the cider required a 23.01 mL titre of 0.450 M potassium permanganate solution to reach the end point.
   a Calculate the amount, in mol, of permanganate in the titre.
   b Calculate the amount, in mol, of ethanol in the 20.00 mL sample of cider.
   c Calculate the molar concentration of ethanol in the cider drink.

Analysis

4 When iodine is added to an ascorbic acid ($C_6H_8O_6$) solution, ascorbic acid is oxidised to dehydroascorbic acid ($C_6H_6O_6$). A student adds 19.8 mL of a 0.235 M ascorbic solution to 20.2 mL of a 0.195 M iodine (I_2) solution. The overall equation for the reaction is:

$$C_6H_8O_6(aq) + I_2(aq) \longrightarrow C_6H_6O_6(aq) + 2I^-(aq) + 2H^+(aq)$$

   a Determine the limiting reactant.
   b Calculate the mass of $C_6H_6O_6$ in the reaction mixture when the reaction is complete.

5 The concentration of ethanal (CH_3CHO) in a solution was determined by titration with a standard solution of acidified potassium permanganate. The equation for the reaction is:

$$2MnO_4^-(aq) + 5CH_3CHO(aq) + 6H^+(aq) \longrightarrow 2Mn^{2+}(aq) + 5CH_3COOH(aq) + 3H_2O(l)$$

A 20.00 mL aliquot of ethanal solution was titrated with a 0.104 M solution of potassium permanganate. Titres of 14.54, 14.32, 14.27 and 14.25 mL were required to reach the end point. Calculate the molar concentration of the ethanal solution.

6 A 10.00 mL sample of white wine was placed in a volumetric flask and deionised water was added to make 100.0 mL of solution. Then 20.00 mL aliquots of the diluted wine were titrated with 0.103 M acidified potassium dichromate solution. The average titre required to reach the end point was 24.61 mL. The equation for the reaction is:

$$3CH_3CH_2OH(aq) + 2Cr_2O_7^{2-}(aq) + 16H^+(aq) \longrightarrow 3CH_3COOH(aq) + 4Cr^{3+}(aq) + 11H_2O(l)$$

   a Calculate the molar concentration of ethanol in the white wine.
   b What would be the effect on the calculated concentration of ethanol if the burette was accidently rinsed with deionised water rather than the potassium dichromate solution? Justify your answer.

7 The alcohol content of an imported whisky was found by diluting 20.00 mL of whisky to 500.0 mL. Then, 10.00 mL aliquots of this solution were titrated against 0.125 M potassium dichromate ($K_2Cr_2O_7$) solution. The average titre was 15.24 mL.
   a Write the half-equation for the oxidation of ethanol to ethanoic acid.
   b Write the half-equation for the reduction of dichromate ions ($Cr_2O_7^{2-}$) to chromium ions (Cr^{3+}) in acidic solution.
   c Write a balanced ionic equation to represent the overall equation.
   d Calculate the concentration of ethanol in the sample of whisky in units of:
      i M
      ii $g\ L^{-1}$.

Chapter review

KEY TERMS

addition reaction
aliquot
anti-bumping granules
average titre
burette
by-product
concentration
concordant titres
condenser
decolourise
dilute
dilution factor
distillate
distillation
end point
equivalence point
excess reactant
fractional distillation
fractionating column
indicator
intermediate
iodine number
limiting reactant
melting point
melting point determination
melting point range
mixed melting point determination
organic compound
pipette
primary standard
quantitative analysis
redox reaction
redox titration
simple distillation
standard solution
standardised
titration
titre
volatile
volumetric analysis

REVIEW QUESTIONS

Knowledge and understanding

1 A volumetric analysis was performed and the following five titres were obtained: 24.22, 25.02, 24.20, 24.16 and 25.14 mL.

- **a** Which of these titres would be considered concordant titres?
- **b** Calculate the average of the concordant titres.

2 Describe the chemical tests that could be used to identify these organic compounds.

- **a** but-1-ene
- **b** ethanol
- **c** propanoic acid

3 Given the half-equations below, write a balanced overall equation for the reaction between butan-1-ol and a solution of acidified potassium dichromate:

$$CH_3CH_2CH_2CH_2OH(aq) + H_2O(l) \rightarrow CH_3CH_2CH_2COOH(aq) + 4H^+(aq) + 4e^-$$

$$Cr_2O_7^{2-}(aq) + 14H^+(aq) + 6e^- \rightarrow 2Cr^{3+}(aq) + 7H_2O(l)$$

4 Calculate the concentration of a standard solution of hydrated oxalic acid ($H_2C_2O_4{\cdot}2H_2O$) prepared by dissolving 2.436 g of hydrated oxalic acid in 100.0 mL of deionised water.

5 A student uses aliquots of a standard solution of iodine to determine the concentration of a vitamin C solution by volumetric analysis. State the solution the student should use for the final rinsing of each piece of glassware used in the titration.

6 Potassium dichromate, a strong oxidising agent, can be used as a primary standard in determining the concentration of reducing agents such as alcohols by volumetric analysis.

- **a** List the criteria that are used to determine whether or not a substance is suitable for use as a primary standard.
- **b** Describe how you would prepare a standard solution of potassium dichromate.

7 20.0 mL of a 1.05 M potassium permanganate ($KMnO_4$) solution is added to 30.5 mL of a 1.12 M methanol (CH_3OH) solution. The overall ionic equation for the reaction is:

$$5CH_3OH(aq) + 4MnO_4^-(aq) + 12H^+(aq) \rightarrow 5HCOOH(aq) + 4Mn^{2+}(aq) + 11H_2O(l)$$

- **a** Calculate the amount, in mol, of MnO_4^- ions and of methanol that have been mixed together.
- **b** Calculate the amount, in mol, of methanol that would be needed for all of the MnO_4^- ions to react.
- **c** Determine the limiting reactant and the reactant that is in excess.
- **d** Calculate the mass of HCOOH in the reaction mixture when the reaction is complete.

8 Stearidonic acid is an unsaturated fatty acid found in the seed oils of hemp and blackcurrant. Its molecular formula is $C_{18}H_{28}O_2$ and a stearidonic acid molecule contains four carbon–carbon double bonds.
Calculate the iodine number of stearidonic acid.

9 Salicylic acid can be prepared by the hydrolysis of the ester methyl salicylate. The melting point of salicylic acid is 159°C.
Explain how a melting point determination can be used to confirm the purity of the salicylic acid sample.

Application and analysis

10 A 5.0 g sample of vegetable oil reacts with 38.0 mL of a 0.50 M iodine solution. Calculate the iodine number of the oil.

11 Nervonic acid, $C_{24}H_{46}O_2$, is an essential nutrient for the development and maintenance of the brain. An unsaturated fatty acid, nervonic acid occurs in the seed oil of plants and can be sourced from flaxseed, sesame seeds and macadamia nuts.
- **a** Calculate the number of carbon–carbon double bonds in nervonic acid given that 7.69 g of iodine reacts with 11.1 g of nervonic acid.
- **b** Calculate the iodine number for nervonic acid.

12 In an experiment, the reaction of bromine with ethane resulted in a mixture containing a number of products with the following boiling points: bromoethane (38°C), 1,1-dibromoethane (108°C), 1,2-dibromoethane (131°C) and 1,1,2-tribromoethane (188°C).
Describe how you could obtain a pure sample of 1,2-dibromoethane from this mixture.

13 Plant oils may be used to produce biodiesel. To determine the identity of a plant oil, its iodine number is used. In an experiment, 10.0 g of plant oil used to make a sample of biodiesel was shaken vigorously with 100.0 mL of a solution containing 15.0 g of iodine. After the reaction was complete, 0.0201 mol of iodine was found to remain in the solution. Determine the iodine number of the plant oil.

14 A student accidentally rinsed the pipette needed to take an aliquot of standard solution with deionised water rather than the standard solution, immediately before performing a titration to determine the concentration of an unknown solution.
- **a** What would be the effect of this rinsing on the concentration of an aliquot of the standard solution?
- **b** What would be the effect on the calculated concentration of the unknown solution?

15 18.1 mL of a 0.0988 M potassium permanganate ($K_2Cr_2O_7$) solution is added to 25.3 mL of a 0.114 M ethanol (CH_3CH_2OH) solution. The overall ionic equation for the reaction is:

$$2Cr_2O_7^{2-}(aq) + 3CH_3CH_2OH(aq) + 16H^+(aq) \rightarrow 4Cr^{3+}(aq) + 3CH_3COOH(aq) + 11H_2O(l)$$

- **a** Determine the limiting reactant.
- **b** Calculate the mass of CH_3COOH in the reaction mixture when the reaction is complete.

16 Sulfur dioxide, SO_2, is a food additive in the winemaking, dried fruit and fruit juice industries. As an additive, its levels in these various foods and drinks are strictly controlled. The level of sulfur dioxide may be analysed volumetrically using iodine solution, I_2, according to the equation:

$$SO_2(g) + 2H_2O(l) + I_2(aq) \rightarrow 2I^-(aq) + HSO_4^-(aq) + 3H^+(aq)$$

To estimate the level of sulfur dioxide in his product, a manufacturer of dried fruit takes 20.97 g of softened dried apple and purees them in a blender with some water. The apple solution is then titrated with 0.0243 M iodine solution using a starch indicator. The average titre was 19.76 mL. Calculate the:
- **a** amount, in mol, of iodine used in the titration
- **b** amount, in mol, of sulfur dioxide extracted from the raisins
- **c** mass of sulfur dioxide extracted from the dried apple
- **d** percentage by mass of sulfur dioxide in the dried apple sample.

17 The amount of ethanol, CH_3CH_2OH, present in a cider drink was determined by treating a sample of the cider with acidified potassium permanganate, $KMnO_4$ solution. The permanganate ion, $MnO_4^-(aq)$, oxidises the ethanol to ethanoic acid (CH_3COOH), while being itself reduced to $Mn^{2+}(aq)$. An average titre of 23.01 mL of 0.450 M $KMnO_4(aq)$ was required to oxidise all the ethanol in a 10.00 mL sample of cider.
- **a** Write the oxidation and reduction half-equations for the reaction.
- **b** Write the overall equation for the redox reaction.
- **c** Calculate the amount, in mol, of CH_3CH_2OH present in the 10.00 mL sample of cider.
- **d** Determine the concentration of ethanol in the cider as a percentage volume per volume (% v/v), given that the density of ethanol is 0.789 g mL^{-1}.
- **e** The calculated value in part **d** was lower than the actual value. For each of the possible errors that could have been made during the analysis, discuss whether it could account for the lower-than-expected calculated value.
  - **i** The 10.00 mL pipette was rinsed only with deionised water before its use.
  - **ii** The burette was rinsed only with deionised water before its use.
  - **iii** The conical flask was rinsed only with deionised water before its use.

18 Potassium permanganate is used in many redox titrations as an oxidising agent. No indicator is required because the permanganate ion is purple, while the manganese(II) ion is colourless. The half-equation involving the permanganate ion is:

$MnO_4^-(aq) + 8H^+(aq) + 5e^- \rightarrow Mn^{2+}(aq) + 4H_2O(l)$

Potassium permanganate cannot be used as a primary standard because it is slightly unstable. Before use in analysis, a potassium permanganate solution must be standardised. Sodium oxalate ($Na_2C_2O_4$) can be used as a primary standard for this purpose. During this standardisation, the oxalate ions are oxidised to CO_2:

$C_2O_4^{2-}(aq) \rightarrow 2CO_2(g) + 2e^-$

a Write a balanced equation for the reaction between the permanganate and oxalate ions.

b A solution containing 0.161 g of sodium oxalate reacted with 26.70 mL of acidified potassium permanganate solution. Determine the molar concentration of potassium permanganate in this solution.

19 A vitamin C tablet with a mass of 2.592 g was crushed and dissolved in deionised water. The solution was titrated against 0.105 M iodine solution using starch solution as an indicator to determine the ascorbic acid ($C_6H_8O_6$) content of the tablet. The reaction can be represented by the equation:

$C_6H_8O_6(aq) + I_2(aq) \rightarrow C_6H_6O_6(aq) + 2H^+(aq) + 2I^-(aq)$

The end point occurred when 26.34 mL of iodine solution had been added.

a Determine the mass of ascorbic acid in the vitamin C tablet.

b Calculate the percentage of ascorbic acid in the vitamin C tablet.

c Suggest the function of the other substances that make up the remainder of the mass of the tablet.

20 A food and drugs authority analysed a sample of light beer to see if it conformed with the regulation of 3.5% by volume of alcohol (ethanol). The alcohol content was determined by volumetric analysis according to the reaction:

$2Cr_2O_7^{2-}(aq) + 3C_2H_5OH(aq) + 16H^+(aq) \rightarrow 4Cr^{3+}(aq) + 3CH_3COOH(aq) + 11H_2O(l)$

The beer was tested by taking a 10.00 mL sample and making it up to 250.0 mL in a standard flask. 20.00 mL aliquots were titrated against a 0.0500 M solution of potassium dichromate ($K_2Cr_2O_7$). Three separate titrations gave titres of 9.20, 9.16 and 9.22 mL. Calculate the:

a amount, in mol, of $Cr_2O_7^{2-}$ present in the average titre

b amount, in mol, of ethanol present in each 20.00 mL aliquot

c amount, in mol, of ethanol in the original 10.00 mL sample of beer

d concentration of ethanol in the beer as a percentage volume per volume (% v/v) given that the density of ethanol is 0.785 g mL^{-1}.

e Would this product conform with the regulations for low-alcohol beer?

21 A scientist accidentally left unlabelled bottles containing 17.00 g L^{-1} solutions of ethanol, propan-1-ol and butan-1-ol on the laboratory bench. When the scientist returned, they couldn't remember which solution was in each bottle.

The scientist selected one of the bottles and filled a burette with one of the alcohol solutions. In a titration using 20.00 mL aliquots of a 0.184 M acidified potassium permanganate solution ($KMnO_4$), an average titre of 20.02 mL was required to reach the end point of the titration.

The two half-equations are:

$MnO_4^-(aq) + 8H^+(aq) + 5e^- \rightarrow Mn^{2+}(aq) + 4H_2O(l)$

$\text{alcohol}(aq) + H_2O(l) \rightarrow \text{carboxylic acid}(aq) + 4H^+(aq) + 4e^-$

a Determine the mole ratio $\frac{n(\text{alcohol})}{n(MnO_4^-)}$.

b Calculate the molar mass of the alcohol.

c Identify the alcohol selected by the scientist for the titration.

OA ✓✓

CHAPTER

14 Instrumental analysis of organic compounds

In Chapter 13 you were introduced to experimental techniques that can be used to identify an unknown compound. In this chapter, you will learn how analytical chemists combine knowledge of the properties of atoms and their bonds with instrumental methods to identify and quantify substances. You will learn about mass spectrometry, infrared spectroscopy and nuclear magnetic resonance spectroscopy and how to interpret data from these techniques to identify an organic compound and determine its structure. Also, you will learn how chromatography is used to identify and quantify organic compounds. Finally, you will learn how laboratories determine unknown organic compounds in mixtures and the purity of samples using spectroscopy instrumentation.

Key knowledge

- applications of mass spectrometry (excluding features of instrumentation and operation) and interpretation of qualitative and quantitative data, including identification of molecular ion peak, determination of molecular mass and identification of simple fragments **14.1**
- identification of bond types by qualitative infrared spectroscopy (IR) data analysis using characteristic absorption bands **14.2**
- structural determination of organic compounds by low resolution carbon-13 nuclear magnetic resonance (^{13}C-NMR) spectral analysis, using chemical shift values to deduce the number and nature of different carbon environments **14.3**
- structural determination of organic compounds by low and high resolution proton nuclear magnetic resonance (^{1}H-NMR) spectral analysis, using chemical shift values, integration curves (where the height is proportional to the area underneath a peak) and peak splitting patterns (excluding coupling constants), and application of the $n + 1$ rule (where n is the number of neighbouring protons) to deduce the number and nature of different proton environments **14.3**
- the principles of chromatography, including high-performance liquid chromatography (HPLC) and the use of retention times and the construction of a calibration curve to determine the concentration of an organic compound in a solution (excluding features of instrumentation and operation) **14.4**
- deduction of the structures of simple organic compounds using a combination of mass spectrometry (MS), infrared spectroscopy (IR), proton nuclear magnetic resonance (^{1}H-NMR) and carbon-13 nuclear magnetic resonance (^{13}C-NMR) (limited to data analysis) **14.5**
- the roles and applications of laboratory and instrumental analysis, with reference to product purity and the identification of organic compounds or functional groups in isolation or within a mixture **14.5**

14.1 Mass spectrometry

Mass spectrometry is one of the most commonly used analytical tools. It is used to analyse samples of solids, liquids and gases. It is a very sensitive quantitative technique that can detect concentrations in the parts per billion to **parts per trillion** (nanograms per kilogram) range. Mass spectrometry can also be applied to identify an unknown compound in a sample. Each molecule produces its own unique mass spectrum (like a fingerprint), so a molecule can be identified by comparing its mass spectrum with the mass spectra held in a database.

Mass spectrometry is used to determine the structure of proteins and drugs. It can also detect molecules that are common markers for diseases such as cancer. Mass spectrometers have also been used in space exploration to analyse the atmosphere on Mars and on the moons of Saturn. The Mars rover Curiosity, shown in Figure 14.1.1, was a robotic laboratory that NASA landed on Mars in 2012. Curiosity had a mass spectrometer to detect key elements required for life, particularly carbon, oxygen, nitrogen, phosphorus and sulfur.

FIGURE 14.1.1 An artist's impression of Curiosity, a robotic rover, on the surface of Mars

Mass spectrometry is often combined with other instrumental techniques, in particular chromatography, for the analysis of mixtures.

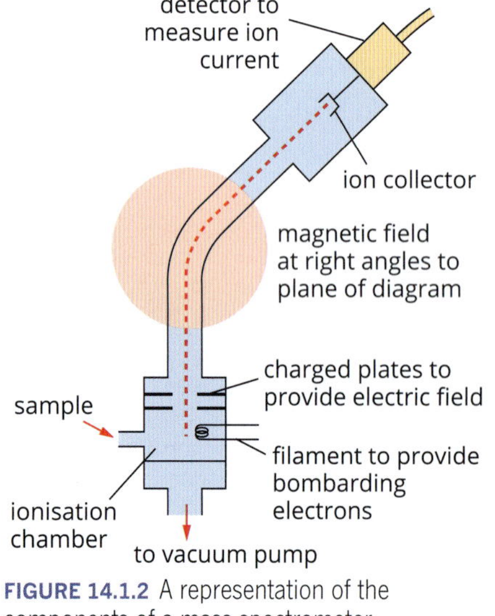

FIGURE 14.1.2 A representation of the components of a mass spectrometer

PRINCIPLES OF MASS SPECTROMETRY

A schematic diagram of a **mass spectrometer** is shown in Figure 14.1.2.

In a mass spectrometer:

- a sample is injected into an ionisation chamber, where high voltages or chemical reactions cause the molecules in the sample to become ionised
- the ions are separated in a magnetic field on the basis of their mass (m) to charge (z) ratio m/z **value**
- the number of ions with different m/z values are measured by a detector and the data is recorded as a **mass spectrum**.

Features of a mass spectrum

A mass spectrum of a sample of pentane is shown in Figure 14.1.3 on the following page. It is a plot of the abundance, measured as **relative intensity**, of ions of different m/z. Relative intensity is calculated by dividing the abundance of a peak by the abundance of the most intense peak and then multiplying by 100. This spectrum shows the positive ions formed when a sample of pentane was injected into the mass spectrometer.

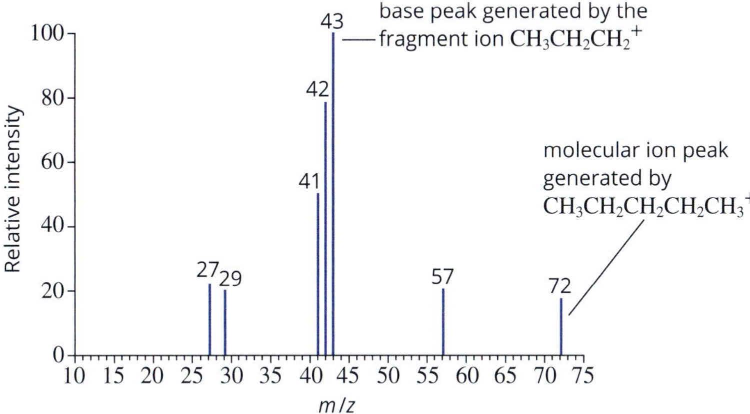

FIGURE 14.1.3 A mass spectrum of pentane ($CH_3CH_2CH_2CH_2CH_3$)

The peak at m/z = 72 in pentane's mass spectrum is the **molecular ion peak**. A **molecular ion**, sometimes called a **parent molecular ion**, is formed when the entire molecule loses an electron and becomes positively charged. When ions have a 1+ charge, meaning z = 1, then the m/z ratio is the same as the molecular mass of the ion. The molecular ion peak may be very small in a mass spectrum. In most cases, the peaks in a mass spectrum are generated by singly charged ions. Mass spectra of ions that have a larger charge are not covered in this course.

The other peaks in the spectrum, which have smaller m/z values than the molecular ion, represent **fragment ions**. These are smaller parts of the molecule and are formed when the high-energy electrons in the ionisation chamber cause bonds to break, which then causes the molecule to break into pieces.

The most intense peak, m/z = 43, is called the **base peak** and is produced by the most abundant and stable fragment ion. The base peak is assigned a relative intensity of 100%. The intensity of all the other peaks are measured relative to the base peak. In some spectra, the peak with the highest intensity is also the molecular ion. In other spectra, the base peak may correspond to a smaller fragment. The relative intensities of the peaks in the spectrum depend on the:

- energy of the ionising electrons
- ease with which fragments can be formed
- stability of the fragment ions formed.

Fragmentation

Inside the ionisation chamber, high-energy electrons ionise the sample by knocking off electrons, producing a positive molecular ion. Because covalent bonds are formed from the sharing of electrons, the removal of electrons can cause the bonds to weaken and break.

The **fragmentation** of the molecule into smaller pieces is represented in the mass spectrum by peaks with a m/z smaller than the molecular ion. Fragments can be produced by the breaking of almost any bond in the molecular ion. They may represent single atoms, small groups of atoms or large sections of the molecule.

The m/z values and formulas of some common fragment ions in mass spectra of organic molecules are listed in Table 14.1.1.

TABLE 14.1.1 Common positive fragment ions

m/z	Formula
15	CH_3^+
17	OH^+
29	$CH_3CH_2^+$, CHO^+
31	CH_3O^+
35 and 37	$^{35}Cl^+$, $^{37}Cl^+$
43	$CH_3CH_2CH_2^+$, CH_3CO^+
45	$COOH^+$
79, 81	$^{79}Br^+$, $^{81}Br^+$

The MS instrument detects positively charged ions. Therefore, all fragments must be written with a positive charge, e.g. CH_3^+.

INTERPRETATION OF MASS SPECTRA

The information that is provided by mass spectrometry can be illustrated by examining the mass spectrum of ethanoic acid.

Ethanoic acid (CH_3COOH) can be ionised in a mass spectrometer by removing an electron in the following process:

$$CH_3COOH + e^- \rightarrow CH_3COOH^+ + 2e^-$$

The molecular ion, CH_3COOH^+, is unstable and can break into a number of fragment ions, as shown in Figure 14.1.4.

The molecular and fragment ions generate peaks in the mass spectrum of ethanoic acid, shown in Figure 14.1.5.

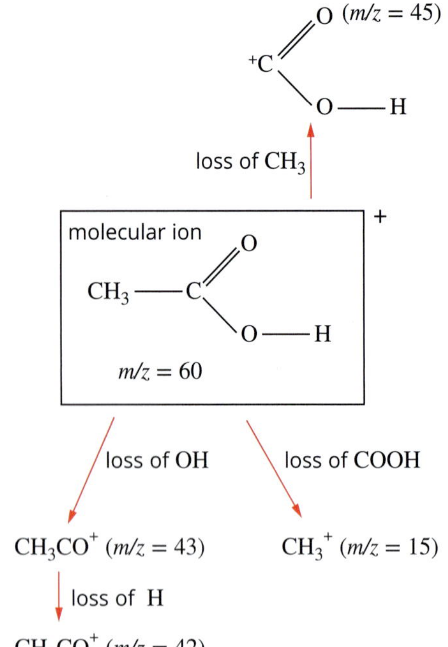

FIGURE 14.1.4 Formation of fragment ions from ethanoic acid

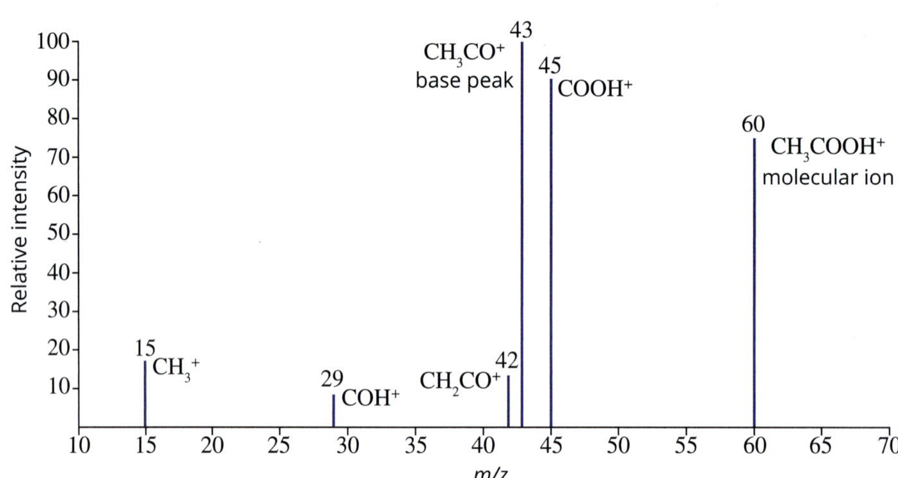

FIGURE 14.1.5 Mass spectrum of ethanoic acid

The m/z value of a fragment is equal to the relative isotopic mass of the fragment when it has a 1+ charge.

The mass spectrum of ethanoic acid indicates that the molecular ion has a mass of 60. The mass of the molecular ion is the same as the relative molecular mass of the ethanoic acid molecule. The fragment ions provide information about the structure of the ethanoic acid molecule, as shown in Table 14.1.2.

TABLE 14.1.2 Identity of peaks in ethanoic acid mass spectrum

m/z	Ion identity	Information provided by ion
60	CH_3COOH^+	molecular ion caused by loss of an electron
45	$COOH^+$	loss of CH_3 from molecular ion
43	CH_3CO^+	loss of OH from molecular ion
42	CH_2CO^+	loss of H from CH_3CO^+
29	COH^+	loss of an O atom from $COOH^+$
15	CH_3^+	loss of COOH from molecular ion

Worked example 14.1.1

DETERMINING THE IDENTITY OF A COMPOUND FROM ITS MOLECULAR ION

The mass spectrum of an unbranched alkane has a molecular ion peak of $m/z = 86$. Determine the molecular formula of the alkane and give its name.

Thinking	Working
The m/z value of the molecular ion is equal to the relative molecular mass of the molecule.	The relative molecular mass of the alkane is 86.
Identify the general formula for the molecule.	The general formula for an alkane is C_nH_{2n+2}.
Use the general formula to set up an equation linking the relative molecular mass to the relative atomic masses of the constituent atoms.	$M_r(C) = 12$ $M_r(H) = 1$ $(12 \times n) + (1 \times (2n + 2)) = 86$
Solve the equation for n.	$12n + 2n + 2 = 86$ $14n + 2 = 86$ $14n = 84$ $n = 6$
Use the value of n to find the molecular formula and the name.	C_6H_{14}, which is hexane.

Worked example: Try yourself 14.1.1

DETERMINING THE IDENTITY OF A COMPOUND FROM ITS MOLECULAR ION

The mass spectrum of an unbranched alkene has a molecular ion peak at $m/z = 56$. Determine the molecular formula of the alkene and give its name.

Isotope effects

Most elements exist as mixtures of **isotopes**, as shown in Table 14.1.3. The presence of stable isotopes in molecules leads to the appearance of additional peaks in a mass spectrum.

TABLE 14.1.3 Stable isotopes of some atoms and their percentage abundance

Element	Isotope	Isotopic mass	Percentage abundance	Isotope	Isotopic mass	Percentage abundance
hydrogen	^{1}H	1.0	99.98	^{2}H	2.0	0.02
carbon	^{12}C	12.0	98.9	^{13}C	13.0	1.1
chlorine	^{35}Cl	35.0	75.8	^{37}Cl	37.0	24.2
bromine	^{79}Br	79.0	50.7	^{81}Br	81	49.3

The percentage abundance of ^{2}H and ^{13}C in samples is very small, so ions containing them only produce very small peaks in the mass spectra. The presence of these isotopes can be ignored in many instances. However, chlorine and bromine have high percentages of each isotope, and so samples that contain them have significant peaks for each isotope in their mass spectra.

The mass spectrum of chloromethane (CH_3Cl) shows peaks for two molecular ions (Figure 14.1.6).

- The peak at $m/z = 50$ is due to the molecular ion containing ^{35}Cl.
- The peak at $m/z = 52$ is due to the other molecular ion containing ^{37}Cl.

The relative heights of these peaks reflects the relative abundance of the two isotopes.

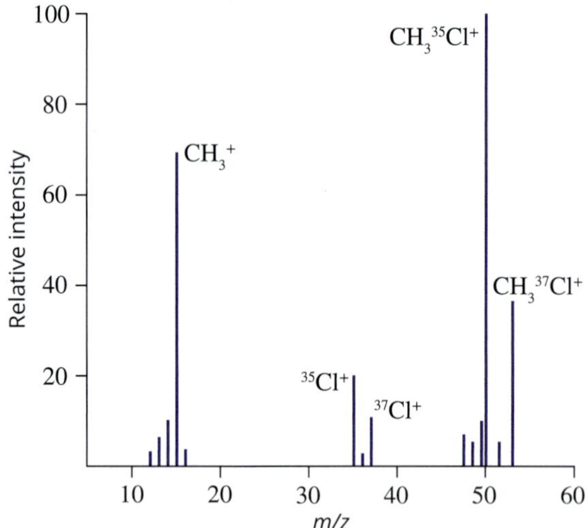

FIGURE 14.1.6 The mass spectrum of chloromethane (CH_3Cl). Note the two molecular ion peaks due to the isotopes of chlorine.

As the number of atoms that have significant proportions of isotopes increases, the mass spectrum can become more complex. The mass spectrum of dichloromethane (CH_2Cl_2) shows three molecular ion peaks (Figure 14.1.7).

- The peak at $m/z = 84$ is due to the molecular ion containing two ^{35}Cl atoms.
- The peak at $m/z = 86$ is due to the molecular ion containing one ^{35}Cl atom and one ^{37}Cl atom.
- The peak at $m/z = 88$ is due to the molecular ion containing two ^{37}Cl atoms.

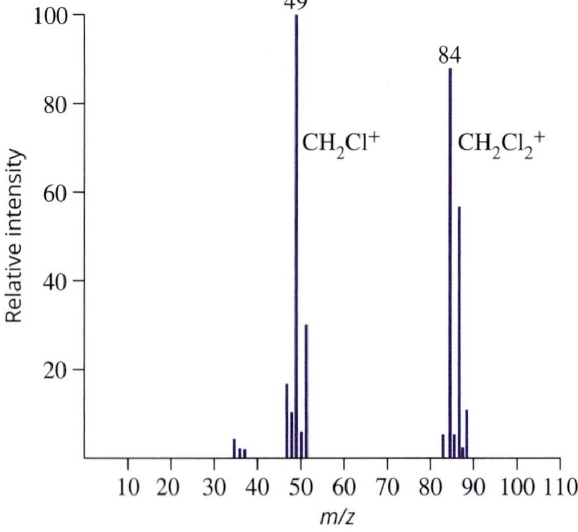

FIGURE 14.1.7 The mass spectrum of dichloromethane (CH_2Cl_2). Note the three molecular ion peaks at $m/z = 84$, 86 and 88 due the isotopes of chlorine.

CASE **STUDY**

Platypus exposure to human drugs

The advances in mass spectrometry technology over recent years have increased the sensitivity of the instrument to identify and quantify components of mixtures more accurately and to much lower concentrations. An example is triple quadrupole mass spectrometry (TQMS), where three different magnetic fields (quadrupoles) are used to separate the fragment ions between the ion source and detector. A common approach using TQMS is for the first quadrupole to be used to select a specific *m/z* of the target molecular ion, the second quadrupole is used to fragment this molecular ion and the third scans the entire *m/z* range allowing detection of specific fragments. These fragments are then used to unambiguously identify the structure of the target molecule. A pure sample of the target molecule is also run and the fragmentation patterns compared for confirmation. This approach is necessary as environmental samples frequently contain hundreds if not thousands of different molecules, many with very similar formula masses.

These improvements in mass spectrometry have uncovered previously unknown pollutants in the environment, including medications. Most medicines taken by humans are excreted from the body in urine and end up in the sewage system. Unfortunately, many sewage treatment plants are unable to remove these medicines from the wastewater, so, as a result, these chemicals are transported into local creeks and rivers.

Studies conducted by Monash University researchers Associate Professor Mike Grace and Dr Erinn Richmond discovered 69 different medicines in aquatic insects and bugs, including antidepressants, antibiotics, antifungals, blood pressure and cholesterol-reducing medicines. They were able to calculate the amount of medicines platypuses living in that ecosystem (Figure 14.1.8) were ingesting each day by estimating the amount of aquatic insects and bugs that platypuses consume and knowing the average concentration of the medications in these bugs. The researchers found that platypuses were ingesting up to half the recommended daily human dose of many common medications in streams receiving sewage treatment plant discharge.

FIGURE 14.1.8 Platypuses dwelling in some Melbourne streams consume up to half the recommended therapeutic dose of human medications per day.

14.1 Review

OA ✓✓

SUMMARY

- A mass spectrometer measures the mass-to-charge ratio, m/z, of ions.
- The peak that has the highest m/z is usually caused by the entire molecule becoming ionised and is called the molecular ion peak.
- The highest m/z peak is usually equal to the molecular mass of the compound.
- The molecular ion can break into pieces called fragment ions.
- Fragmentation of a molecule in a mass spectrometer can help to determine its molecular structure.
- Each compound has a unique mass spectrum that can be used to identify it.
- The mass spectrum of a chlorine- or bromine-containing molecule will have two or more molecular ion peaks due to the abundance of Cl or Br isotopes and their ratios.
- The m/z value of each fragment is equal to the sum of the relative isotopic masses of the atoms in the fragment.

KEY QUESTIONS

Knowledge and understanding

1 Calculate the m/z for the following fragments:

a CH_3^+ **b** $CH_3CH_2CH_2^+$
c $CH_3{}^{81}Br^+$ **d** CHO^+
e $CH^{35}Cl_2^+$

2 Why is the fragment, CHO, an incorrect formula for the ion fragment at $m/z = 29$ of a mass spectrum of methanol?

3 State the m/z values of the molecular ions that would be found in the mass spectrum of 1,1-dibromoethane.

4 The mass spectrum of butan-1-ol has a very intense peak at $m/z = 31$. What is the chemical formula of this ion fragment?

Analysis

5 The following mass spectrum is of a straight-chain alkane with a relative molecular mass of 114.

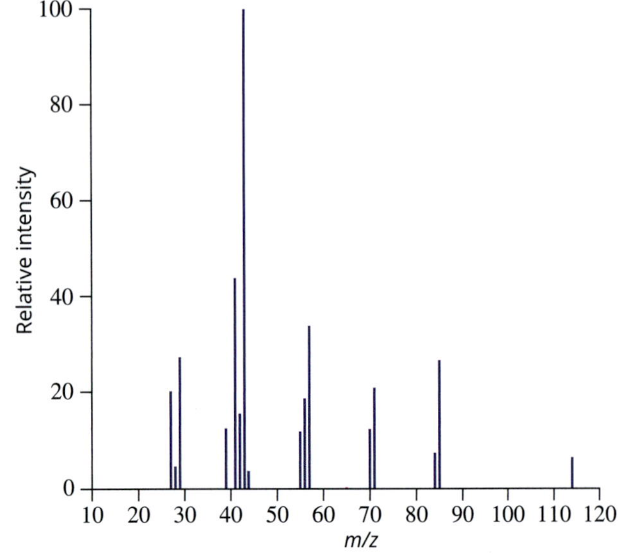

a State the m/z value of the base peak.

b State the formula for the fragment ion with an m/z value of 85.

c Determine the name of the alkane represented.

6 The mass spectrum of a bromoalkane is shown here.

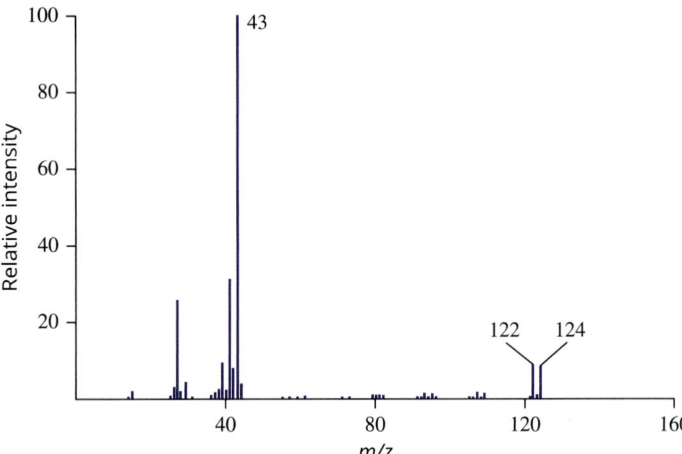

a Determine the molecular formula of the bromoalkane.

b Explain why the mass spectrum has two peaks at $m/z = 122$ and $m/z = 124$.

c Explain why the relative heights of these two peaks are almost the same.

d Write a possible formula for the ion responsible for the peak at $m/z = 43$.

14.2 Infrared spectroscopy

All types of **spectroscopy** use electromagnetic radiation to give information about materials around you. Spectroscopy utilises the effects of electromagnetic radiation on atoms and molecules to learn about their structure by interacting with atoms and molecules. The nature of this interaction depends upon the energy of the radiation.

The **electromagnetic spectrum** represented in Figure 14.2.1 is divided into different regions of radiation, with different frequencies, wavelengths and energies. Ultraviolet light has short wavelengths with high frequency and energy, whereas radio waves have long wavelengths and low frequency and energy.

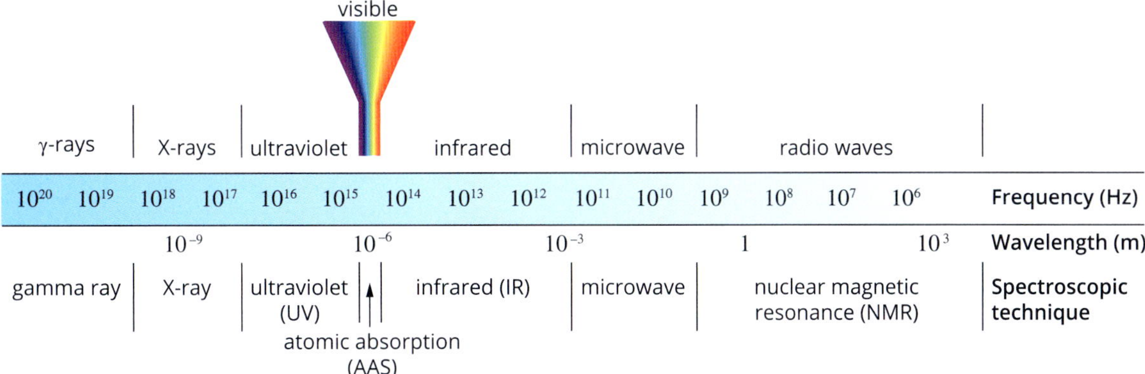

FIGURE 14.2.1 Electromagnetic spectrum showing the frequency and wavelength of different regions and their applications to spectroscopy

Atoms and molecules have different types of energy. The water molecule in Figure 14.2.2 shows four different types of energy in order of increasing energy.

The measurement of these energy transitions forms the basis of the spectroscopic techniques studied in this course. The electromagnetic radiation associated with some spectroscopic techniques is summarised in Table 14.2.1.

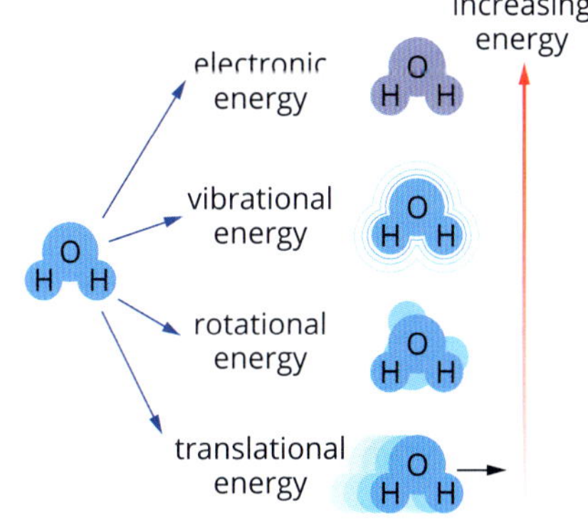

FIGURE 14.2.2 A water molecule has different types of energy.

TABLE 14.2.1 Spectroscopic techniques and the regions of the electromagnetic spectrum they use

Spectroscopic technique	Region of the electromagnetic spectrum used	Type of energy level transition
nuclear magnetic resonance (NMR) spectroscopy	radio waves	nuclear spin states—makes the nucleus flip in an applied magnetic field
infrared (IR) spectroscopy	infrared	vibrations of bonds in molecules—makes the bonds bend and stretch
colorimetry, ultraviolet–visible (UV–visible) spectroscopy	visible and ultraviolet	valence electrons in molecules and atoms—makes outer-shell electrons jump to higher energy levels

These spectroscopic techniques use the facts that:

- atoms and molecules absorb and emit electromagnetic radiation of specific energies
- atoms and molecules undergo a change when they absorb electromagnetic radiation
- different parts of the electromagnetic spectrum affect atoms or molecules in different ways. Electronic energy is measured in the visible and ultraviolet regions, whereas rotational energy is measured in the microwave region.

FIGURE 14.2.3 A food chemist uses an infrared spectrometer to evaluate the quality of raisins.

Colorimetry and UV–visible spectroscopy were covered in Unit 2 of this course. In this and the following sections, you will study the principles of IR and NMR spectroscopy and learn to interpret their respective spectra.

PRINCIPLES OF INFRARED SPECTROSCOPY

Infrared (IR) spectroscopy is a powerful analytical tool that can be applied to the analysis of many organic and inorganic compounds.

IR spectroscopy can be used to analyse solids, liquids and gases. It is used as a quality control tool in the pharmaceutical, agriculture, food processing (Figure 14.2.3), paint, paper and other industries. For example, it can be used in drug testing, as shown in Figure 14.2.4. Researchers combine IR spectroscopy with other techniques to study biological molecules. The technique is used to analyse blood and urine samples, and is also used to determine the level of atmospheric pollutants.

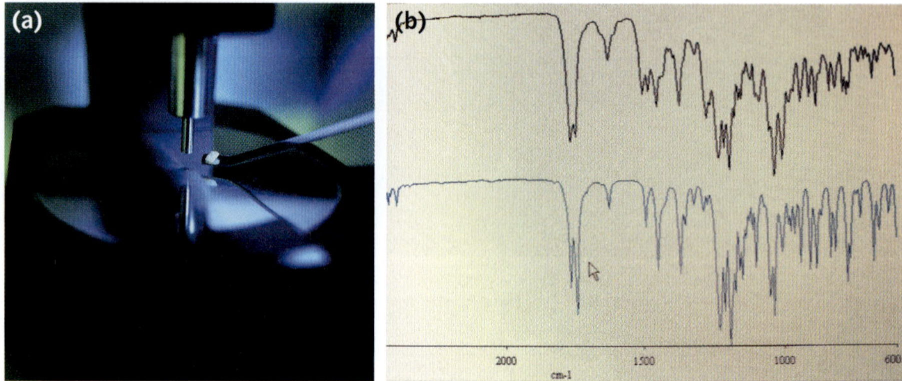

FIGURE 14.2.4 (a) Obtaining the infrared spectrum of a suspected illegal drug. (b) The spectrum of the substance (black) closely matches that of a heroin sample (blue), indicating that the substance contains the illegal drug.

IR spectroscopy is particularly useful because it can give you information about the **functional groups** present in an organic molecule. This information helps to clarify its structure.

IR light has a lower energy and a longer wavelength than visible and ultraviolet light. The energy from infrared light is enough to change the vibration of the bonds in molecules.

Covalent bonds can be compared to springs that can undergo specific amounts of bending or stretching. The atoms in a molecule can change position when the bonds bend or stretch, as shown in Figure 14.2.5, and the molecule then vibrates.

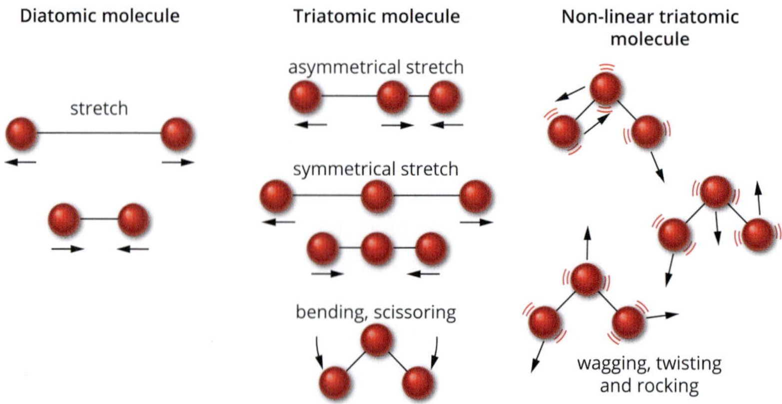

FIGURE 14.2.5 Stretching and bending motions in diatomic and triatomic molecules

Infrared spectroscopy exploits the ability of molecules to bend and stretch. Molecules are only able to occupy discrete **vibrational energy levels** (Figure 14.2.6). The amount of energy required to move from one vibrational energy level to the next is the same as the amount of energy contained in electromagnetic radiation from the infrared region. When molecules absorb infrared radiation, the stretching or bending vibrations of the bonds become more energetic.

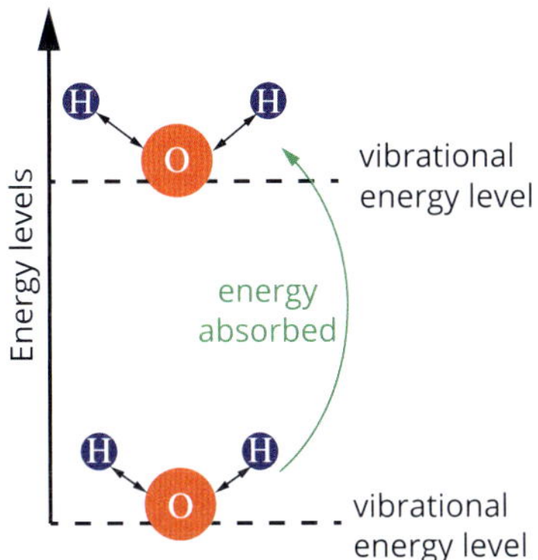

FIGURE 14.2.6 A molecule can absorb energy and move to a higher vibrational energy level.

IR spectroscopy is a powerful analytical technique because almost all molecules absorb IR radiation. For a molecule to absorb IR radiation, the bending or stretching vibrations must change the overall **dipole** of the molecule.

The frequency of vibration of the bond between two identical atoms depends on the strength of the bond, as shown in Table 14.2.2. The C–C bond is weaker and absorbs infrared radiation of a lower frequency than that absorbed by the C=C bond and the C≡C bond. It is important to realise that the frequency of a vibration is directly proportional to the energy of the vibration. Note that in infrared spectroscopy, the frequency is expressed as the **wavenumber**, or waves per unit distance, and has the unit cm^{-1}.

In infrared spectroscopy the frequency is expressed as the wavenumber, or waves per unit distance, and has the unit cm^{-1}.

TABLE 14.2.2 Bond energy, representing the strength of the bond, compared to the frequency of infrared light absorbed

Bond	Bond energy ($kJ\ mol^{-1}$)	Infrared absorption frequency (cm^{-1})	Increasing energy of vibration
C≡C	839	2300	↑
C=C	614	1720	
C–C	348	880	

The mass of the atoms attached to a bond also affects the frequency of the IR radiation that will be absorbed. Atoms with higher masses absorb lower frequency radiation. Table 14.2.3 compares the energy absorbed when atoms of different masses are bonded to carbon with a single bond.

TABLE 14.2.3 Atomic mass affects the frequency of infrared light absorbed.

Bond	Typical absorption frequency (cm^{-1})	Increasing energy of vibration
C–H	3000	↑
C–C	1200	
C–O	1100	
C–Cl	750	
C–Br	600	

CHEMFILE

Inside an IR spectrophotometer

The figure at right shows the main components of a simple infrared spectrophotometer. The source of infrared radiation is split into two beams and passes separately through a sample cell and a reference cell. The material used to contain the sample and reference is NaCl, KBr or a similar substance. Plastic or glass cannot be used as these materials absorb IR radiation. The reference is used to discount any interference from water or carbon dioxide in the atmosphere, the solvent or the material used to contain the sample. The two beams pass through a wavelength selector (monochromator) and an infrared detector. The difference in transmittance, or transmitted radiation, between the sample and reference cell is due to the absorption of certain frequencies by the molecules of the sample. These absorptions result in changes in the vibrational energy in the molecule under examination.

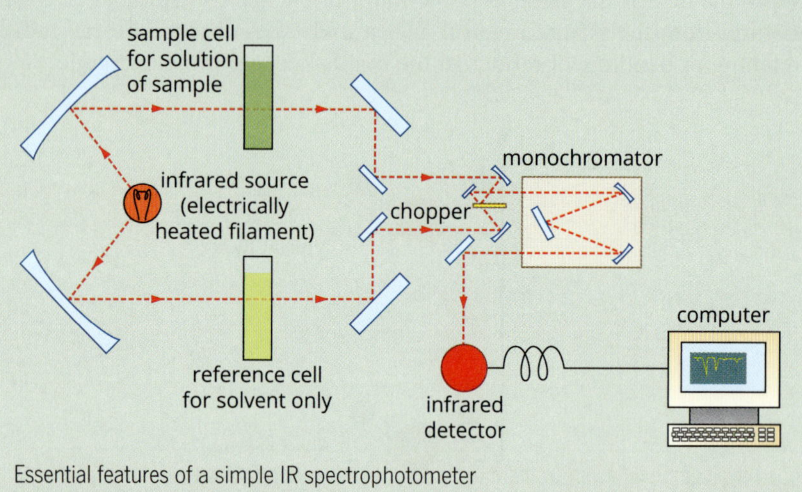

Essential features of a simple IR spectrophotometer

INTERPRETATION OF INFRARED SPECTRA

The following section will show you how to interpret an infrared spectrum. You will learn about the key features and how to identify the functional groups present in a molecule, based on the absorption bands which are present or absent in an infrared spectrum.

Features of an infrared spectrum

The frequency of the electromagnetic radiation in IR spectroscopy is usually expressed in wavenumbers (cm^{-1}). The wavenumber is the number of waves per centimetre and is inversely proportional to wavelength. A bond that vibrates at a higher frequency absorbs IR radiation with a higher wavenumber and greater energy than a bond that vibrates at a lower frequency.

The horizontal axis of an IR spectrum shows the wavenumber; the highest wavenumber is usually shown on the left. This axis often has two scales to ensure all features are visible. In the infrared spectrum of 2-methylpropan-2-ol, shown in Figure 14.2.7, you can see that the scale starts with intervals of 1000 cm^{-1}, and then changes to intervals of 500 cm^{-1} in the region below 2000 cm^{-1}.

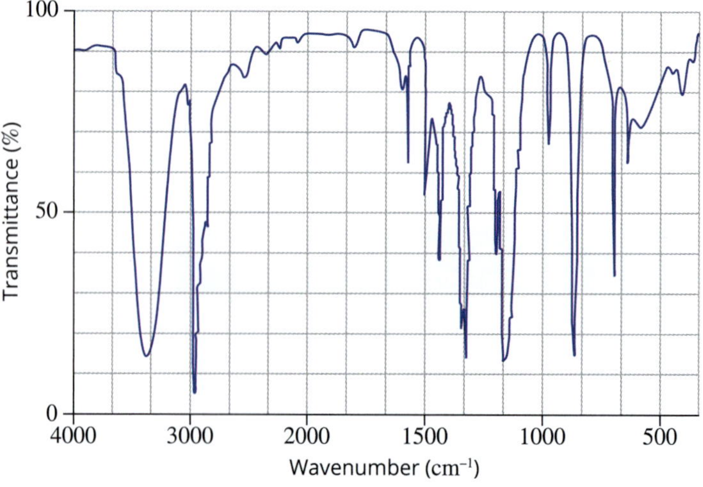

FIGURE 14.2.7 The infrared spectrum of 2-methylpropan-2-ol

The vertical axis of an IR spectrum shows the percentage **transmittance** on a scale of 0 to 100. The 'baseline' of the spectrum is where all the light, 100%, is passed through the sample. Where the molecule absorbs IR radiation the spectrum dips down to a lower transmittance. These **absorption bands** appear as inverted peaks in the spectrum.

Different terms can be used to describe absorption bands in IR spectra. Narrow absorption bands span only a few wavenumbers. This usually means that the peak corresponds to one specific type of molecular vibration. Broader bands may be the result of several related vibrational changes that have similar energies. Absorption bands are also described as 'strong', 'medium' or 'weak' if they absorb a large, moderate or small amount of radiation (Figure 14.2.8).

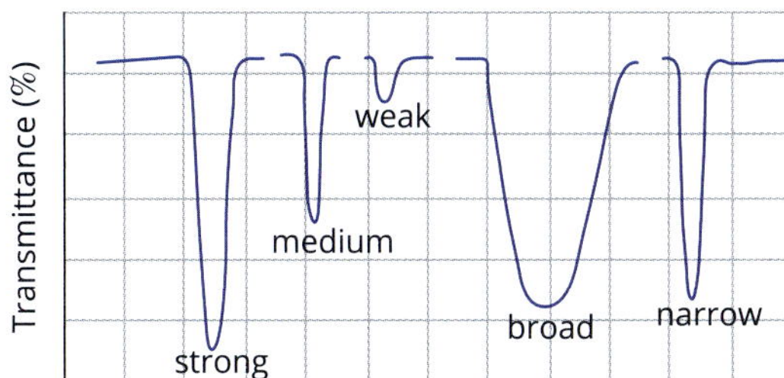

FIGURE 14.2.8 Descriptions used for different shapes and strengths of infrared absorption bands

Absorption bands above about 1400 cm^{-1} are used to identify functional groups because this region coincides with the energy associated with the characteristic stretching vibrations of the atoms in these groups. The region below 1400 cm^{-1} is called the **fingerprint region** because absorption bands of this frequency tend to be unique to each compound. If a known compound and an unknown compound have the same absorption spectrum below 1400 cm^{-1}, they are almost certainly the same.

Interpreting infrared spectra

Each type of bond absorbs IR radiation over a typical range of wavenumbers. For example, it takes energy in the 2850–3090 cm^{-1} region to stretch a C–H bond and energy in the in the 750–1100 cm^{-1} region to stretch a C–C bond. The characteristic regions of absorption for different bonds in an IR spectrum are shown in Figure 14.2.9.

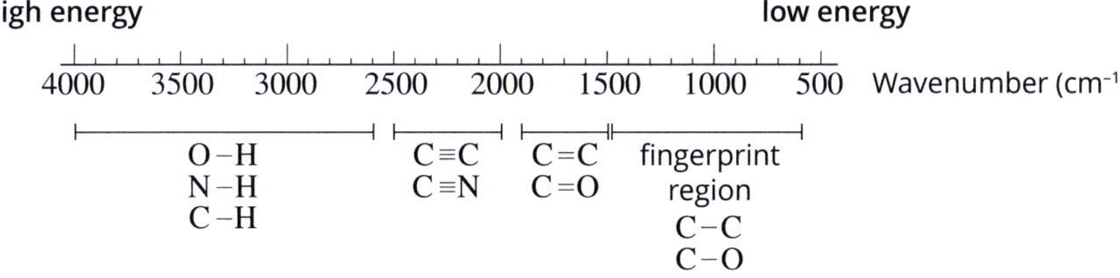

FIGURE 14.2.9 Typical regions for absorption bands from different bonds in the infrared spectrum

When chemists look at the IR spectrum of an unidentified organic compound, they often refer to tables and charts showing the wavenumbers at which the main functional groups absorb. This can give broad clues to the bond types and functional groups present in the molecule. For positive identification, the entire spectrum of an unidentified compound can be compared to a **database** of spectra, often containing tens of thousands of spectra of known compounds.

The characteristic ranges for IR absorption of common functional groups are summarised in Table 14.2.4.

TABLE 14.2.4 Infrared absorption data for some common functional groups

Bond	Wavenumber (cm^{-1})
C–Cl	600–800
C–O (alcohols, esters, ethers)	1050–1410
C=C (alkenes)	1620–1680
C=O (amides)	1630–1680
C=O (aldehydes)	1660–1745
C=O (acids)	1680–1740
C=O (ketones)	1680–1850
C=O (esters)	1720–1840
O–H (acids)	2500–3500
C–H (alkanes, alkenes)	2850–3090
O–H (alcohols)	3200–3600
N–H (amines and amides)	3300–3500

CHEMFILE

Infrared spectrum of primary amides

Primary amides have an unusual IR spectrum, as demonstrated in the IR spectrum below for butanamide. Notice the two peaks around 3200 cm^{-1} and 3300 cm^{-1}, and the two peaks around 1650 cm^{-1} and 1630 cm^{-1}. These double peaks represent the two different structures that butanamide forms. What happens is that one of the hydrogens on the $-NH_2$ can move to the oxygen, forming an alcohol and C=N bond. This particular structure produces the alcohol absorption band around 3200 cm^{-1} and the C=N absorption band around 1620 cm^{-1} (highlighted by the red arrows). The other structure produces the absorption band around 3300 cm^{-1} for $-NH_2$ and 1650 cm^{-1} for C=O (highlighted by the black arrow).

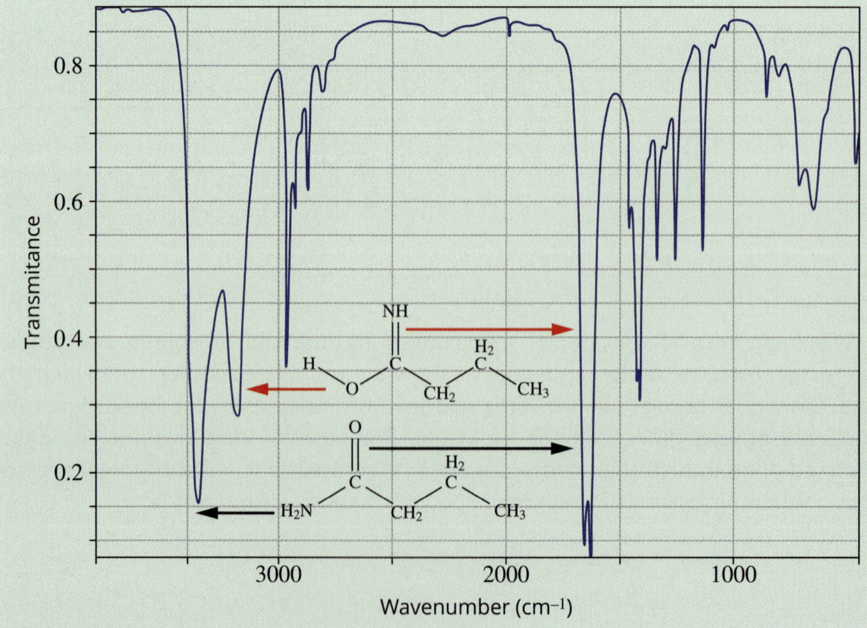

The IR spectrum for butanamide, $CH_3CH_2CH_2CONH_2$

IR spectroscopy can be useful for distinguishing between very similar compounds. Consider the molecules, propanoic acid and methyl ethanoate (Figure 14.2.10). These two compounds are isomers with the molecular formula $C_3H_6O_2$.

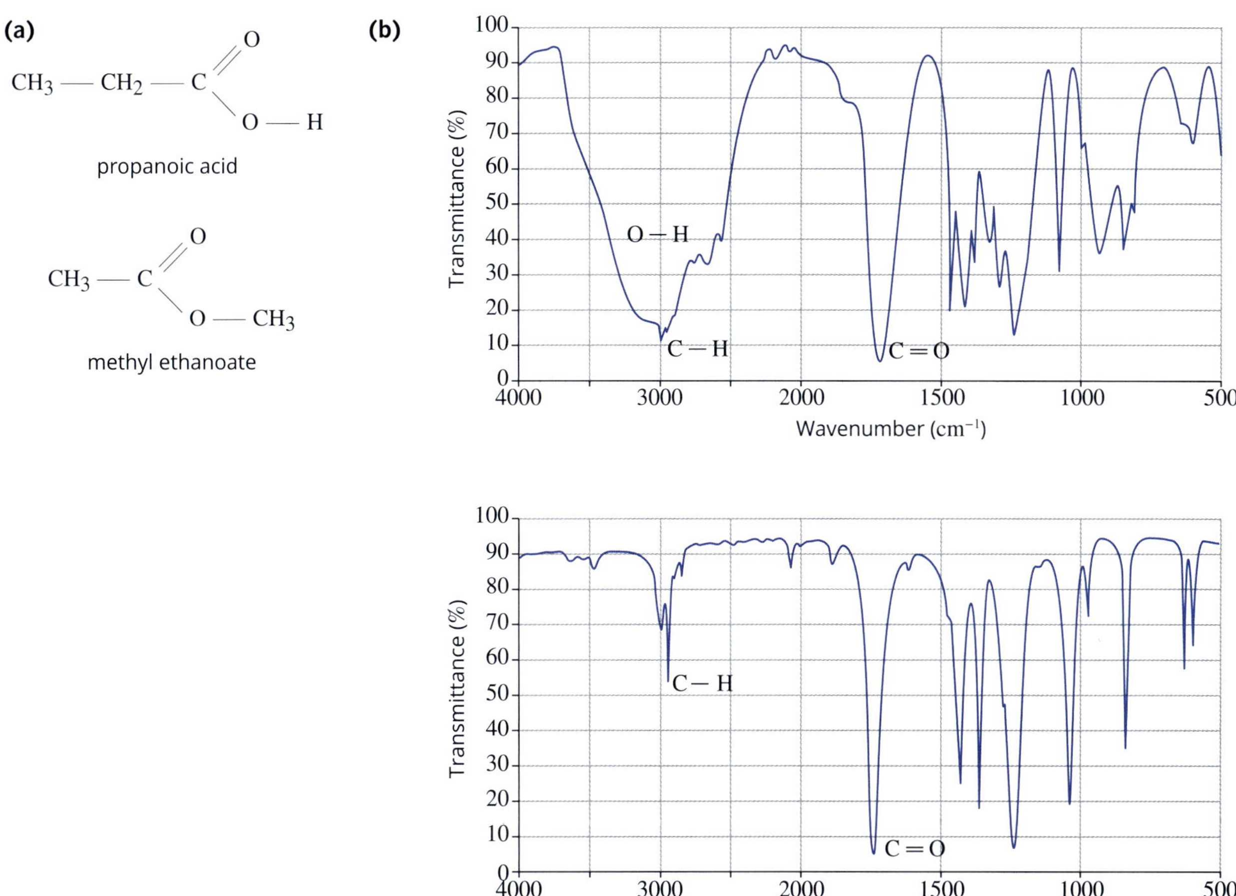

FIGURE 14.2.10 Infrared spectra of (a) propanoic acid and (b) methyl ethanoate

The infrared spectra of the compounds shown in Figure 14.2.10 have some features in common and some differences. Both spectra have an absorption band at about 1700 cm^{-1} due to the stretching of the carbonyl (C=O) bond. In methyl ethanoate, the narrow absorption band at around 3000 cm^{-1} is due to the C–H bonds. In propanoic acid, the broad absorption band from 2700 cm^{-1} to 3600 cm^{-1} is due to the O–H bond. This broad absorption band partly masks the absorption due to C–H bonds that are also present in this spectrum.

Using the data in Table 14.2.4 (page 512), you can identify the bonds that are likely to be present in an unidentified compound from their characteristic absorption bands. This information allows you to determine which of the main functional groups are present. The IR spectrum may also be used to prove that a functional group is not present, helping you to narrow down the possible structure of the compound.

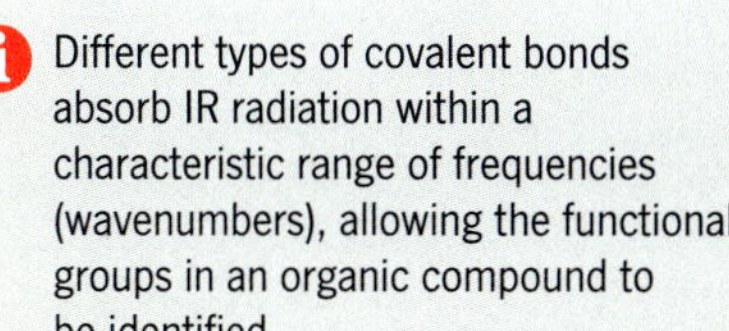
Different types of covalent bonds absorb IR radiation within a characteristic range of frequencies (wavenumbers), allowing the functional groups in an organic compound to be identified.

Worked example 14.2.1

INTERPRETING THE IR SPECTRUM OF AN UNIDENTIFIED COMPOUND

Use the infrared spectrum of an unidentified compound to identify the functional groups present. The molecular formula of the compound is $C_2H_4O_2$. You will need to refer to the IR absorption data in Table 14.2.4 (page 512).

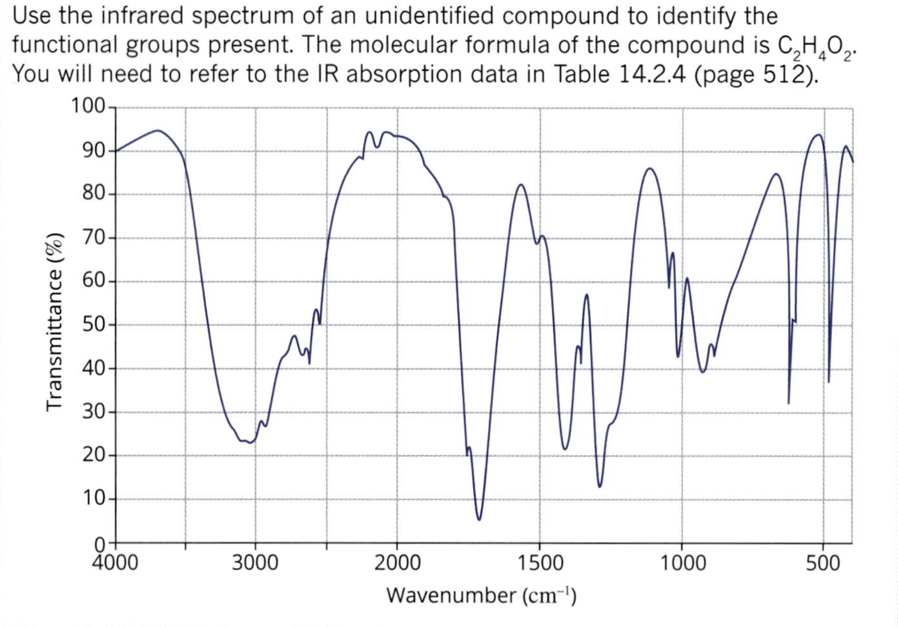

Thinking	Working
Identify the absorption bands that correspond to the absorption bands of bonds in the IR absorption data table.	There is a strong, narrow band at approximately 1700 cm^{-1}, which corresponds to the absorption by a carbonyl, C=O, group. The broad band centred at about 3000 cm^{-1} corresponds to the absorption by the O–H bond of a carboxylic acid.
Identify the functional group or groups that are present.	The spectrum shows absorption bands corresponding to the presence of C=O and carboxylic acid O–H bonds. This suggests the presence of a carboxyl functional group, –COOH. From the formula, it can be concluded that the compound is ethanoic acid, CH_3COOH.

Worked example: Try yourself 14.2.1

INTERPRETING THE IR SPECTRUM OF AN UNIDENTIFIED COMPOUND

Use the infrared spectrum of an unidentified compound to identify the functional groups present. The molecular formula of the compound is $C_4H_{10}O$. You will need to refer to the IR absorption data in Table 14.2.4 (page 512).

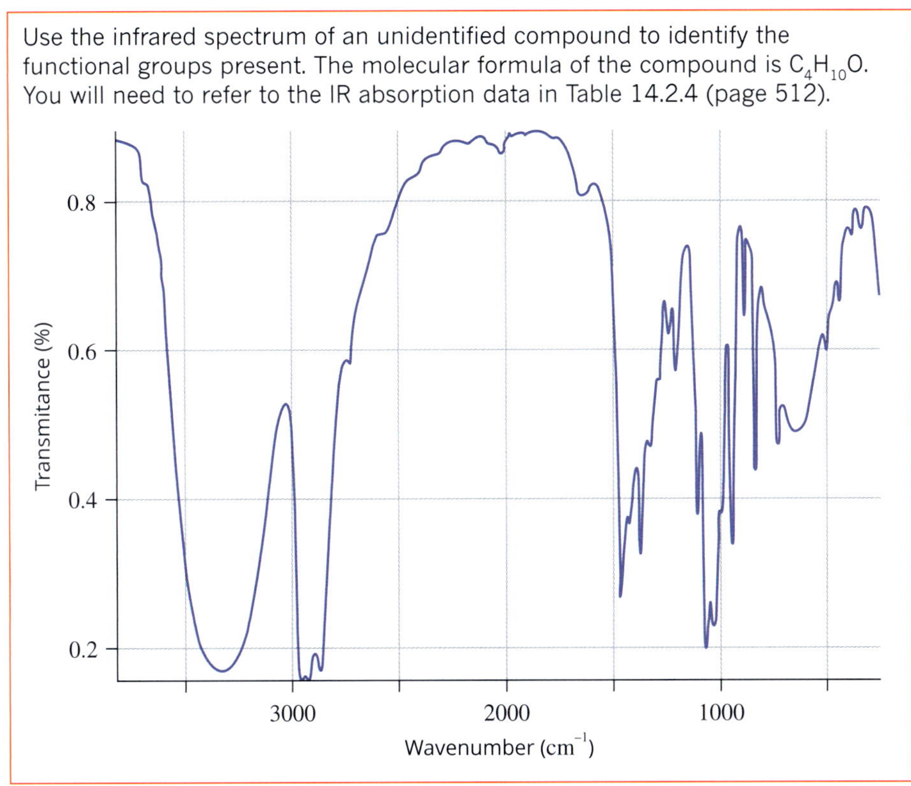

CASE STUDY

IR spectroscopy and sepsis

Serious bloodstream infections, commonly known as sepsis, are life-threatening. Diagnosis of the pathogen causing the sepsis usually takes 1–6 days, which can result in inappropriate treatments and unnecessary deaths. Advances in infrared technology have made it possible to rapidly detect specific viruses, fungi, bacteria and other parasites in blood. Professor Bayden Wood and his research team at Monash University are developing a simple technique using a portable IR device that detects unique vibrational bands from the pathogen's cellular material and compares it to a database. The process from taking a blood sample to diagnosis now takes less than 30 minutes, greatly improving the patient's outcome. This new diagnostic technique is also low-cost and portable, providing remote communities in developing countries access to state-of-the-art blood-test analysis (Figure 14.2.11).

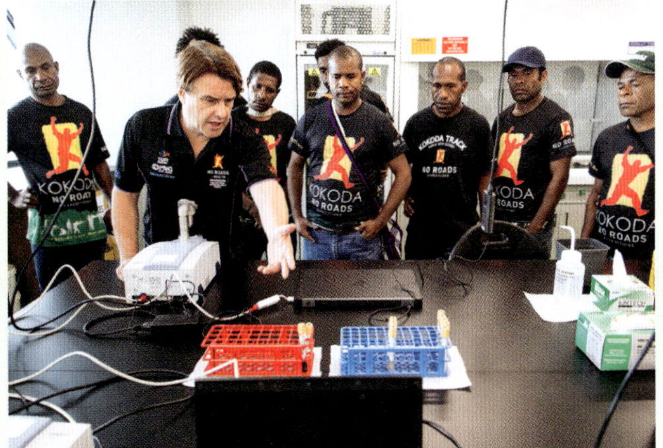

FIGURE 14.2.11 Professor Bayden Wood explains how to analyse blood samples for malaria diagnosis to the Papua New Guinean No Roads to Health team using a portable infrared spectrometer

14.2 Review

WS 32

OA ✓✓

SUMMARY

- Matter interacts with electromagnetic radiation in different ways, depending on the energy of the radiation.
- Molecules have discrete vibrational energy levels. The absorption of IR radiation causes molecules to move to higher vibrational energy levels.
- Different types of covalent bonds absorb IR radiation within a characteristic range of frequencies (wavenumbers), allowing the functional groups in an organic compound to be identified.
- Each compound has a unique absorbance pattern in the fingerprint region of the IR spectrum, and this can be used to identify the compound.

KEY QUESTIONS

Knowledge and understanding

1 Refer to Figure 14.2.1 (page 507). Which of the following options best reflects the relationship between wavelength and frequency?

A Frequency is proportional to wavelength.
B Frequency is inversely proportional to wavelength.
C Frequency is the logarithmic scale of wavelength.
D Frequency is not related to wavelength.

2 Name the type of energy level transition that is caused by the absorption of infrared radiation.

3 Rank the following bonds from lowest vibrational energy to highest vibrational energy.

a C=C, C–C, C≡C
b O–H, C–H, S–H, N–H, H–H

4 Identify the wavenumber ranges of the absorption bands that you would expect to see from the following functional groups.

a hydroxyl
b aldehyde
c amine
d carboxyl
e ester
f C=C

Analysis

5 The absorption bands due to C–O and C–C bonds are often of little help when determining the structure of an unidentified compound, but are valuable in the confirmation of the identity of a known compound. Explain why this is the case.

6 Identify the vibrations causing the major peaks at wavenumbers above the fingerprint region (i.e. above 1400 cm^{-1}) in the following spectra.

a Empirical formula: $C_3H_6O_2$

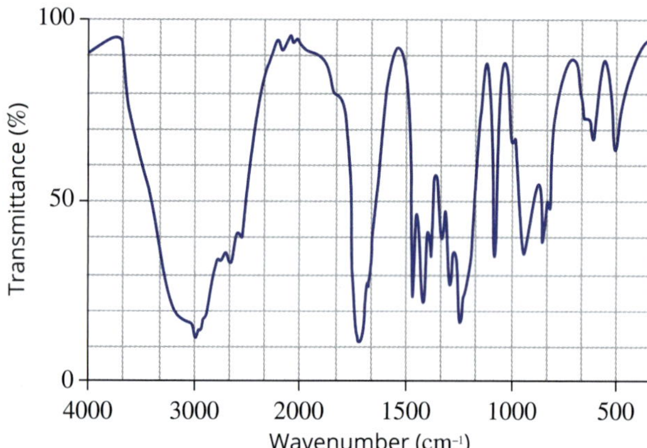

b Empirical formula: C_3H_8O

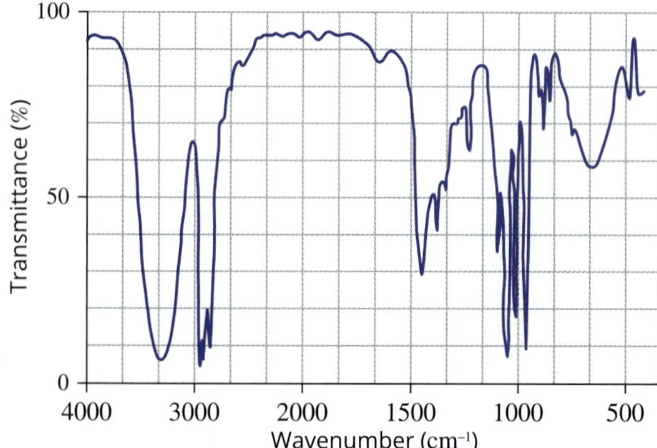

continued over page

c Empirical formula: $C_4H_{11}N$

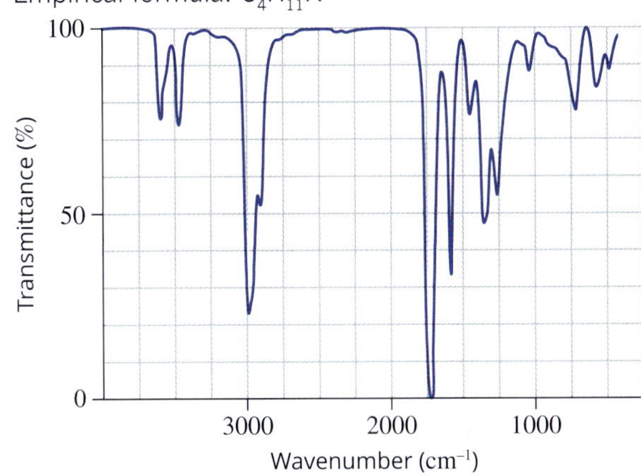

7 For each of the following IR spectra, select the correct name of the organic compound from the alternatives provided.

a

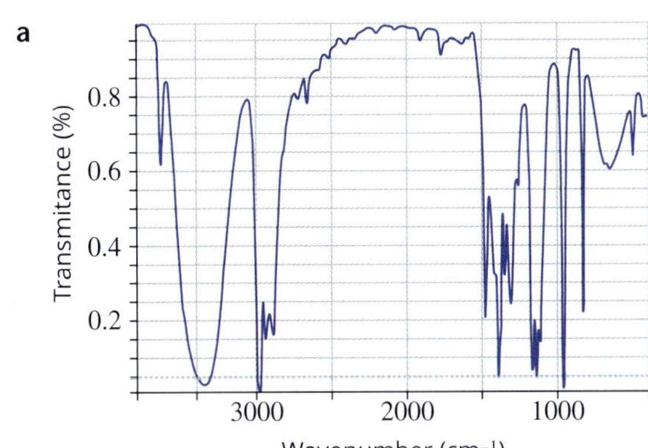

A propene
B propan-2-ol
C propanoic acid
D propanal

b

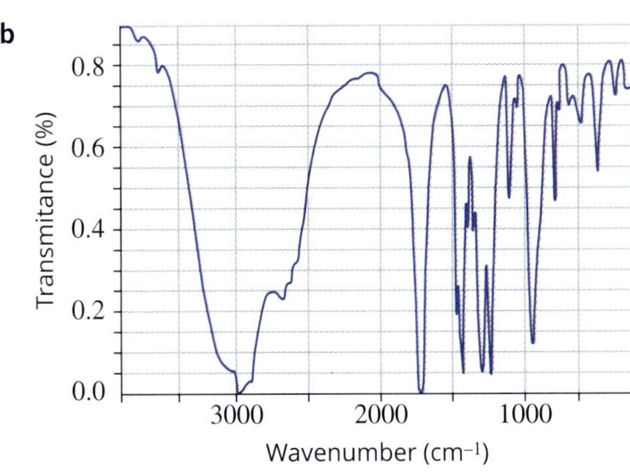

A butan-1-ol
B butan-2-one
C butane
D butanoic acid

c

0.8
0.6
0.4
0.2
0.0
Transmitance (%)
3000
2000
1000
Wavenumber (cm⁻¹)

A hexan-3-ol
B hexane
C hexanoic acid
D hex-1-ene

14.3 Nuclear magnetic resonance spectroscopy

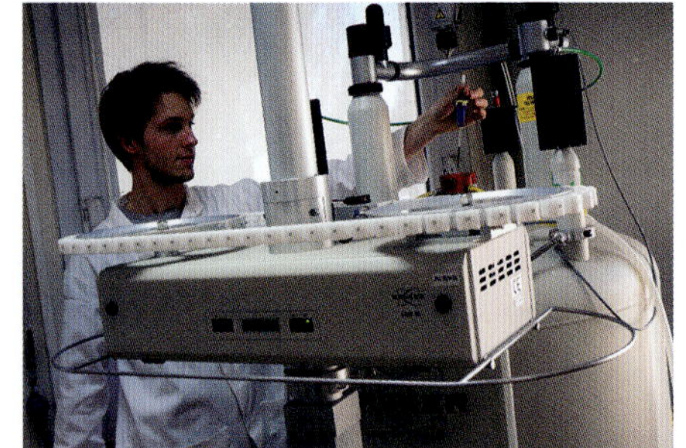

FIGURE 14.3.1 A research chemist places a sample tube into an NMR spectrometer.

Nuclear magnetic resonance (NMR) spectroscopy is one of most powerful techniques for determining the structure of complex molecules. As you saw in the previous section, IR spectroscopy provides general information about functional groups. In combination with IR spectroscopy and other techniques, NMR spectroscopy can enable chemists to determine the exact structure of a molecule (Figure 14.3.1).

A form of NMR technology called magnetic resonance imaging (MRI) is used as a tool in medicine to provide a highly detailed picture of anatomical features and diseased tissue without the use of harmful radiation such as X-rays.

PRINCIPLES OF NUCLEAR MAGNETIC RESONANCE SPECTROSCOPY

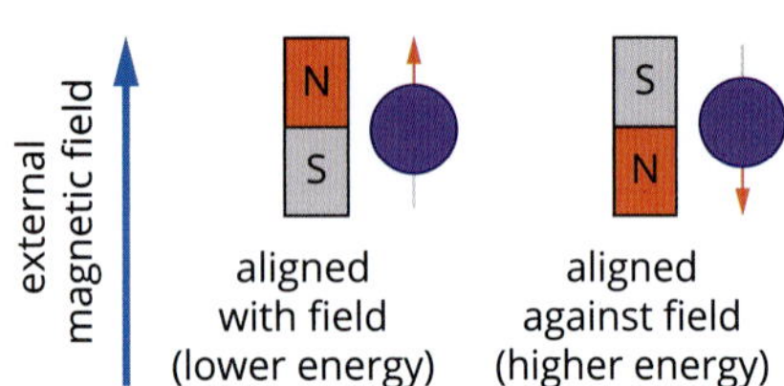

FIGURE 14.3.2 Small magnets and nuclei with spin interact with an external magnetic field.

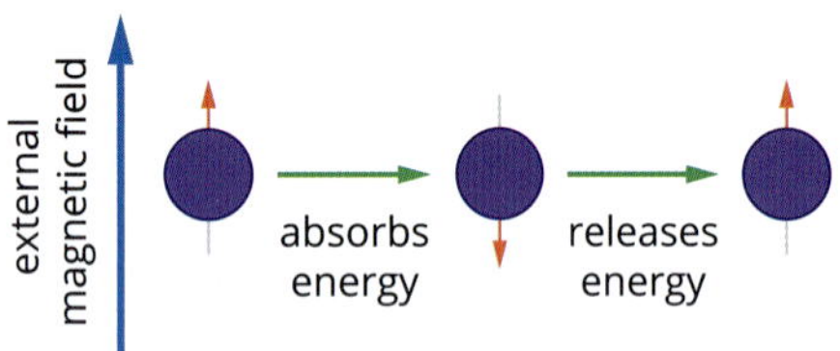

FIGURE 14.3.3 Absorption and release of energy as a nucleus changes spin

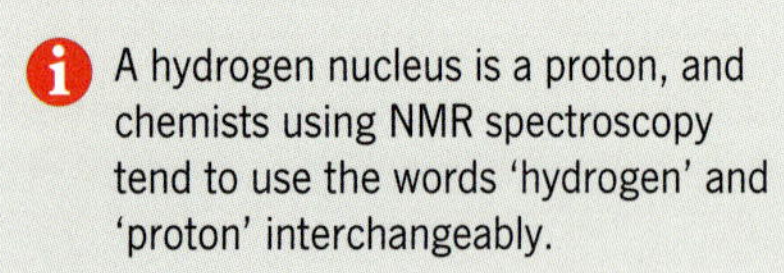

NMR spectroscopy uses electromagnetic radiation in the radio frequency range to obtain information about the structure of molecules. The energy of the radio waves is too low to cause electronic, vibrational or rotational transitions.

To interact with radio waves, the nuclei of the atoms must have a property called **nuclear spin**. Only nuclei that have an odd number of protons and/or neutrons, such as ^{1}H, ^{13}C and ^{31}P, have a nuclear spin. The odd number of nuclear particles causes these nuclei to behave like tiny bar magnets.

In the presence of an external magnetic field, magnets, or nuclei with spin, can either line up in the same direction as the field (lower energy) or line up in the opposite direction (higher energy), as shown in Figure 14.3.2. A magnet, or nucleus, that is aligned against an external field is in an unstable arrangement.

When inside the NMR spectrometer, the nuclei are normally in a low-energy state and are aligned with a strong magnet. A radio transmitter is used to provide the energy to 'flip' the nuclei into a high-energy state. Over time, the nuclei tend to flip back into a lower energy spin (see Figure 14.3.3). As they do, they release a pulse of energy, which is measured and represented in graphical form as a **nuclear magnetic resonance (NMR) spectrum**.

The difference in energy between the higher and lower energy spin states depends on the type of nucleus and the chemical environment surrounding the nucleus.

The most common form of NMR spectroscopy is based on the hydrogen-1 (^{1}H) nucleus. This nucleus contains one proton and no neutrons. This type of spectroscopy is usually called **proton (^{1}H) NMR spectroscopy**. It can give information about the structure of any molecule containing hydrogen atoms. Another common type of spectroscopy, called **carbon-13 (^{13}C) NMR spectroscopy**, examines the carbon-13 nucleus and is useful in investigating the carbon atoms inside organic molecules.

Hydrogen and carbon chemical environments

An NMR spectrum provides information about the number and type of hydrogen and carbon nuclei in an organic compound. Atoms that are in the same chemical environments absorb the same energy and produce a single **signal** in the NMR spectrum.

A **chemical environment** is made up of the atoms and electrons that surround a specific atom. Inside a molecule, atoms can be said to have the same chemical environment if they are attached in the same way to the same atoms. Atoms that are in the same chemical environment are said to be **equivalent** and form one signal. For example, the simplest organic molecule, methane, contains only one hydrogen environment and one carbon environment, and gives only one signal in its NMR spectrum.

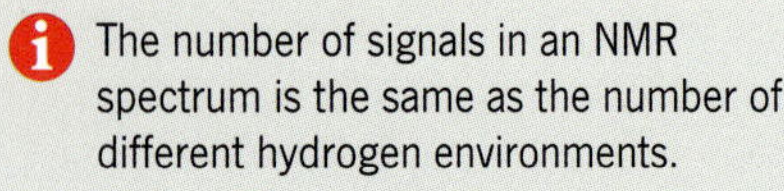

Figure 14.3.4 shows the structures of three organic molecules with their atoms colour-coded to show the different hydrogen and carbon environments.

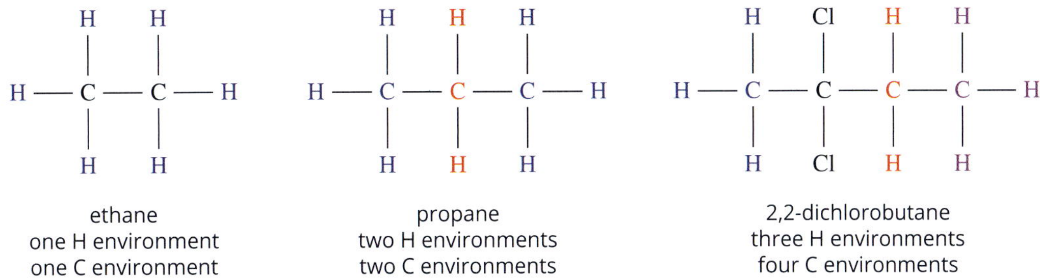

FIGURE 14.3.4 The different chemical environments in ethane, propane and 2,2-dichlorobutane shown using different colours

The six hydrogen atoms in ethane are equivalent because each is part of a $-CH_3$ group attached to the other $-CH_3$ group. Similarly, the two carbon atoms are also equivalent. As a result, ethane has just one signal in its proton NMR spectrum and one signal in its carbon-13 NMR spectrum.

Propane has two different hydrogen and two different carbon environments. The six hydrogen atoms in the $-CH_3$ groups are all in the same environment, each being part of three hydrogens on a carbon attached to a $-CH_2-$ group. The two hydrogen atoms in $-CH_2-$ make up the second environment, which is different from each of the hydrogen atoms in the $-CH_3$ groups on the ends. Again, the carbons in the $-CH_3$ groups are similar and occupy one environment, and the carbon in the centre occupies a second and different environment. The ^1H NMR and ^{13}C NMR spectra each show two signals.

2,2-dichlorobutane has three different hydrogen environments and four carbon environments. The hydrogen atoms in the two $-CH_3$ groups at either end of the molecule are not similar. One group is bonded directly to the $-CCl_2-$ group, whereas the other is bonded to the $-CH_2-$ group. The ^1H NMR spectrum contains three signals. Each of the carbon atoms in the backbone of the molecule can be distinguished from the others by how close it is to the chlorines and whether it is on the end or within the chain. The ^{13}C NMR spectrum of 2,2-dichlorobutane contains four different signals.

As a general rule, molecules that contain symmetry have fewer different chemical environments than molecules that do not have symmetry. More planes of symmetry lead to fewer unique chemical environments.

Peak area

Proton NMR spectroscopy can be used to determine the number of hydrogen atoms in a molecule. The proton NMR software uses a mathematical process of **integration** to determine the area under the curve, or the **peak area** of each signal. The peak area of each signal is proportional to the number of hydrogen atoms in the environment it corresponds with. The ratio of the peak areas of the individual signals corresponds to the ratio of the number of hydrogens in each environment. This information may be added to the spectrum either as numbers indicating the ratio, or as an integration curve (also known as an integration trace) over the peaks (see Figure 14.3.7 page 521). The heights of the vertical sections of the integration curve correspond to the ratio of hydrogens in each environment. You may need to multiply the ratio by a factor to account for all the hydrogen atoms in your molecule.

CHEMFILE

NMR instrumentation

The main features of an NMR spectrometer are shown in the diagram below.

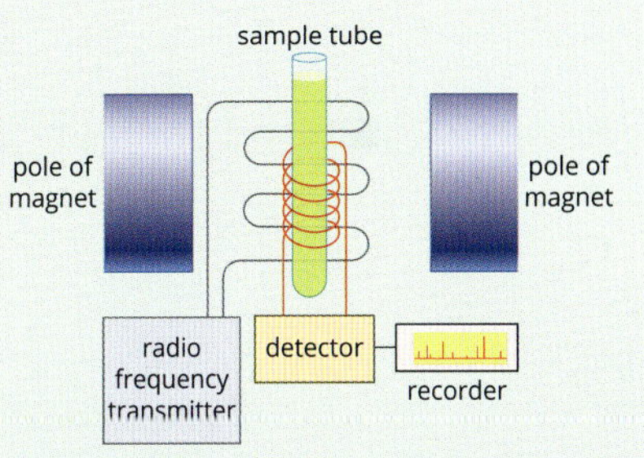

Schematic diagram of an NMR spectrometer

The sample is dissolved in a solvent that does not produce a signal. Typical solvents are $CDCl_3$, D_2O and CD_2Cl_2. Deuterium (D or ^2H), ^{16}O and ^{12}C nuclei do not show up in proton and carbon-13 NMR spectroscopy. A glass tube containing the dissolved sample is placed in the centre of a powerful magnet. The tube spins so that the sample is subjected to a uniform magnetic field. A radio transmitter produces a short powerful pulse of radio waves that is absorbed by the nuclei and causes them to flip to high-energy spin states. A radio receiver detects the radio frequency energy emitted as the nuclei return to lower energy spin states and a computer analyses the signals and outputs the data as an NMR spectrum.

Figure 14.3.5 shows the structural formula and spectrum of 2,2,4,4-tetrachloropentane, $CH_3–CCl_2–CH_2–CCl_2–CH_3$, which has two signals. The signal with the chemical shift of 2.21 ppm (parts per million) is due to the hydrogen atoms on the central carbon. The signal with the chemical shift of 2.17 ppm is due to the hydrogen atoms in the $–CH_3$ groups, as they are in equivalent chemical environments. In the spectrum of 2,2,4,4-tetrachloropentane, you can see that the signals have relative peak areas of one and three, corresponding to the ratio of hydrogen atoms in the two environments in the molecule. However, there are eight hydrogen atoms in 2,2,4,4-tetrachloropentane, so the ratio needs to be multiplied by two to account for all the hydrogen atoms in the molecule. Therefore, the signal at 2.17 ppm represents six hydrogen atoms and the signal at 2.21 ppm represents two hydrogen atoms.

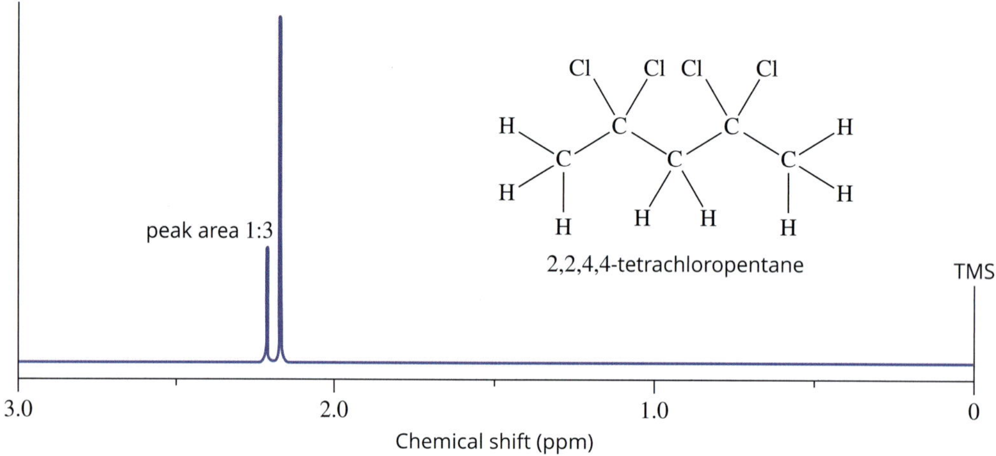

FIGURE 14.3.5 Proton NMR spectrum of 2,2,4,4-tetrachloropentane

Chemical shifts

Nuclei emit different frequencies of radio energy when analysed in different NMR spectrometers. To ensure that chemists can share results obtained using different experimental conditions, the signals emitted by nuclei can be compared with the signal from a reference compound, tetramethylsilane (TMS). The structure of TMS is shown in Figure 14.3.6.

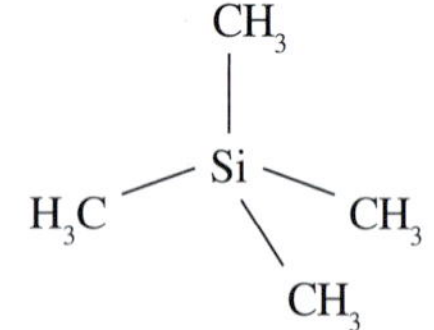

FIGURE 14.3.6 Structural formula of tetramethylsilane (TMS)

TMS is chemically inert and can be added to a sample without causing a chemical reaction. It contains hydrogen atoms, carbon atoms and a silicon atom in a symmetrical arrangement (so it has only one chemical environment in each type of NMR spectroscopy). TMS forms a single peak in an NMR spectrum that is well away from the peaks that most chemists are interested in.

> Equivalent atoms have the same chemical shift, and hence form one signal.

The difference in energy needed to change spin state in a sample is compared to the energy needed to change spin states in TMS. This energy difference is called the **chemical shift**, and is measured in parts per million, ppm. The chemical shift of TMS is defined as zero. The symbol δ (delta, often used to mean 'change in' or 'difference') is often used to represent the chemical shift.

The actual magnetic field experienced by the nucleus is not the same as the applied external magnetic field. Electrons around each nucleus also have spin and so have an associated magnetic field that shields the nucleus from the applied magnetic field. The amount of **nuclear shielding** depends on the other atoms surrounding the nucleus. This, in turn, affects the amount of energy needed for the nucleus to change its spin.

The energy emitted when the nucleus flips to the lower spin state depends on the amount of nuclear shielding experienced. For example, a hydrogen atom in a $–CH_2–$ group absorbs at a slightly higher frequency than hydrogen in a $–CH_3$ group or a lower frequency than hydrogen in an –OH group. Because of this, the chemical shift of hydrogen in each of these species is different.

Signal splitting

Figure 14.3.7 shows the structural formula and proton NMR spectrum of chloroethane. The hydrogen atoms in the molecule are in two different environments. The signals generated by the different types of hydrogen atoms are shown in matching colours. The relative peak areas are indicated above each peak and correspond to the number of hydrogens in each environment.

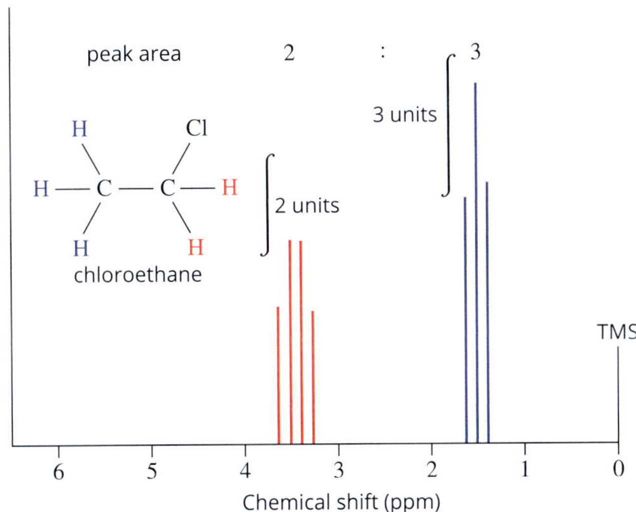

FIGURE 14.3.7 A simplified representation of the proton NMR spectrum of chloroethane, including two integration curves which indicate that the ratio of hydrogen atoms is 2 to 3.

You will notice that each signal is not a single peak, but is a series of fine peaks. In high-resolution NMR spectra, some signals are seen to split into line (peak) patterns. This is due to the effect of neighbouring hydrogen atoms. An environment can be said to be **neighbouring** if it is up to three bonds away from the hydrogen atoms in question.

The number of lines in a signal is related to the number of hydrogen atoms bonded to adjacent atoms by the relationship of $n + 1$, where n is the number of hydrogen atoms that are equivalent to each other in neighbouring environments, but not equivalent to the hydrogen atoms giving rise to the signal. This is called the '$n + 1$ rule'.

In the structure of chloroethane, you can see that there are two hydrogen atoms (in red) three bonds away from the hydrogen atoms shown in blue (the $-CH_3$ group). The number of neighbours plus one would be three for this group ($n + 1 = 2 + 1 = 3$), and the signal shows a three-line pattern. In a similar way, the $-CH_2-$ group has three neighbouring hydrogen atoms and gives a signal with a four-line pattern ($n + 1 = 3 + 1 = 4$).

Figure 14.3.8, on the following page, shows the proton NMR of methyl propanoate. The peak at 3.6 ppm represents the red $-CH_3$ group attached to oxygen. The hydrogen atoms on this red methyl group have only one peak as they are more than three bonds away from the next closest hydrogen atoms. The peak at 2.4 ppm represents the hydrogen atoms on $-CH_2-$ and is split into four, indicating that these hydrogen atoms are adjacent to a $-CH_3$ group. The peak at 1.2 ppm is due to the hydrogen atoms on the other $-CH_3$ group and is split into three, indicating these hydrogen atoms are adjacent to a $-CH_2-$ group.

The number of lines in a signal is equal to the number of equivalent hydrogen atoms in neighbouring environments, n, plus 1. This is called the '$n + 1$ rule'.

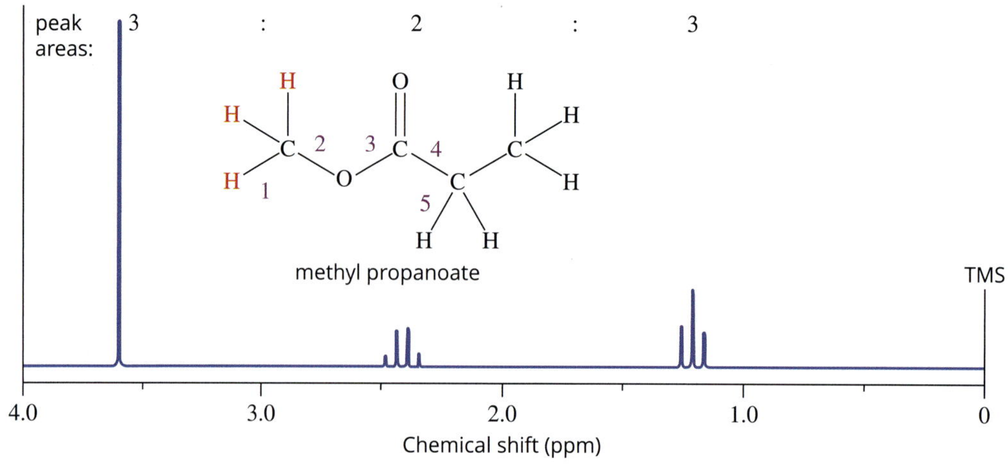

FIGURE 14.3.8 Proton NMR spectrum of $CH_3OOCCH_2CH_3$ (methyl propanoate)

The **splitting pattern** gives important information about which environments are caused by hydrogen atoms close to one another in a molecule. The number of neighbours causes a splitting pattern in a predictable way.

- A hydrogen atom with no neighbours gives a one-peak signal (a singlet).
- A –CH– group splits the signal from hydrogen atoms attached to adjacent atoms into two peaks (a doublet).
- A $–CH_2–$ group splits the signal from hydrogen atoms attached to adjacent atoms into three peaks (a triplet).
- A $–CH_3$ group splits the signal from hydrogen atoms attached to adjacent atoms into four peaks (a quartet).

If the neighbouring hydrogen atoms are in an equivalent chemical environment, they will not cause splitting of the signal and are not counted as neighbours as they are 'the same'.

Also, note that the signal produced by the hydrogen atom in the hydroxyl group of alcohols is not usually split by hydrogen atoms attached to a neighbouring carbon atom and does not count as a neighbour. Similarly, the hydroxyl H does not usually cause splitting of the signals of neighbouring Hs. The signal for a hydroxyl group is usually a singlet.

CHEMFILE

What is a multiplet?

Consider the central hydrogens (in bold) in the molecule $CH_3\mathbf{CH_2}CH_2Cl$. Based on the $n + 1$ rule, you would expect the signal to be split as a sextet (as there are five adjacent hydrogen atoms, and $5 + 1 = 6$). But it is not that simple. The hydrogen atoms in $–CH_3$ are not equivalent to the hydrogen atoms in $–CH_2Cl$. You need to consider the adjacent equivalent hydrogen atoms separately. Therefore, $–CH_3$ will split the signal into four ($3 + 1 = 4$). The $–CH_2Cl$ will split the signal into three ($2 + 1 = 3$). So, theoretically you would expect twelve peaks (4×3). But because of overlap, you would typically observe six peaks on a proton NMR spectrum, and the group of peaks is referred to as a multiplet.

INTERPRETATION OF PROTON NMR SPECTRA

A proton NMR spectrum provides a number of pieces of information about the structure of a molecule, including the:

- number of signals, which shows the number of different hydrogen environments
- relative peak area, which helps to determine the number of hydrogens in each environment
- splitting of the signals, which provides information about the hydrogen atoms in adjacent environments
- chemical shift of the signals, which helps to identify the chemical environment in which the hydrogen atoms are located.

By using each piece of this information in turn, and considering how the different environments would affect each other, it is often possible to determine the overall structure of a molecule.

Some typical chemical shifts for different types of hydrogen atoms relative to TMS are found in Table 14.3.1 on the following page. These can differ slightly in different solvents. Where more than one hydrogen environment is shown in the formula, the shift refers to the hydrogen atoms shown with bold letters.

Note that these ranges represent chemical shifts commonly measured in proton NMR spectra. However, you may find a compound where the proton chemical shifts are outside these ranges.

The hydrogen atoms on a methyl group will usually have the lowest chemical shift.

TABLE 14.3.1 Typical proton NMR chemical shifts

Type of hydrogen	Chemical shift (ppm)
$R-\mathbf{CH_3}$	0.9–1.0
$R-\mathbf{CH_2}-R$	1.3–1.4
$RCH=CH-\mathbf{CH_3}$	1.6–1.9
$R_3-\mathbf{CH}$	1.5
$\mathbf{CH_3}-C(=O)-OR$ or $\mathbf{CH_3}-C(=O)-NHR$	2.0
$R-C(=O)-\mathbf{CH_3}$	2.1–2.7
$R-\mathbf{CH_2}-X$ (X = F, Cl, Br or I)	3.0–4.5
$R-\mathbf{CH_2}-OH$, $R_2-\mathbf{CH}-OH$	3.3–4.5
$R-C(=O)-O\mathbf{CH_2}R$	3.7–4.8
$R-O-\mathbf{H}$	1–6 (varies considerably under different conditions)
$R-\mathbf{NH_2}$	1–5
$RHC=\mathbf{CHR}$	4.5–7.0
$R-C(=O)-N(\mathbf{H})-\mathbf{H}$	5.5–8.5
$R-C(=O)-\mathbf{H}$	9.4–10.0
$R-C(=O)-O-\mathbf{H}$	9.0–13.0

Worked example 14.3.1

INTERPRETATION OF A PROTON NMR SPECTRUM

The proton NMR spectrum of a compound with a molecular formula of $C_2H_4Cl_2$ is shown below. Relative peak areas are shown on the spectrum.

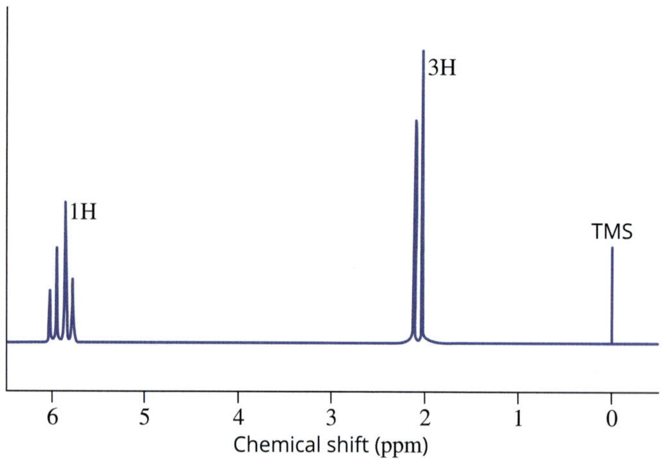

Identify this compound using the information provided in the NMR spectrum.

<table>
<tr><th>Thinking</th><th>Working</th></tr>
<tr><td>What information does the molecular formula provide about the compound?</td><td>Two molecules are possible from the formula: either 1,2-dichloroethane or 1,1-dichloroethane.</td></tr>
<tr><td>Summarise the information provided in the NMR spectrum in a table.</td><td><table><tr><th>Chemical shift</th><th>Peak splitting</th><th>Relative peak area</th></tr><tr><td>2.1</td><td>doublet (2-line pattern)</td><td>3</td></tr><tr><td>5.9</td><td>quartet (4-line pattern)</td><td>1</td></tr></table></td></tr>
<tr><td>Identify the number of different hydrogen environments.</td><td>There are two signals, so there must be two different hydrogen environments.</td></tr>
<tr><td>Use the relative peak area to deduce the number of hydrogen atoms in each environment.</td><td>The relative peak areas of the signals at 5.9 and 2.1 ppm are 1 : 3, so the relative number of hydrogen atoms must be 1 : 3.
The sum of hydrogen atoms in the formula is 4, which means that the peak areas are directly equal to the number of hydrogen atoms in each environment.
This suggests that the molecule contains one –CH– group and one $-CH_3$ group.</td></tr>
<tr><td>Use the peak splitting of the signals to identify the types of hydrogen environments.</td><td>The signal at 2.1 ppm is a doublet (2-peak pattern). The number of peaks in the pattern is given by $n + 1$, so this signal must be generated by an environment that has one neighbouring hydrogen atom, i.e. a –CH– group.
The signal at 5.9 ppm is a quartet (4-peak pattern). The number of peaks in the pattern is given by $n + 1$, so this signal must be generated by an environment that has three neighbouring hydrogen atoms, i.e. a $-CH_3$ group.</td></tr>
</table>

If possible, use the chemical shifts in Table 14.3.1 (page 523) to identify the types of hydrogens. Remember the ranges are broad.	The table does not give information about compounds containing two chloro groups, so continue on to the next step.
Use the information you gathered to identify the compound.	The molecular formula of the compound is $C_2H_4Cl_2$. On the basis of the formula, the compound is either 1,2-dichloroethane or 1,1-dichloroethane. The splitting patterns and peak area indicate that the molecule contains a $-CH_3$ group adjacent to a –CH– group. The molecule must be 1,1-dichlorethane as the structure fits the evidence from the spectrum.

Worked example: Try yourself 14.3.1

INTERPRETATION OF A PROTON NMR SPECTRUM

The proton NMR spectrum of a compound with a molecular formula of $C_2H_3Cl_3$ is shown below. Relative peak areas are shown on the spectrum.

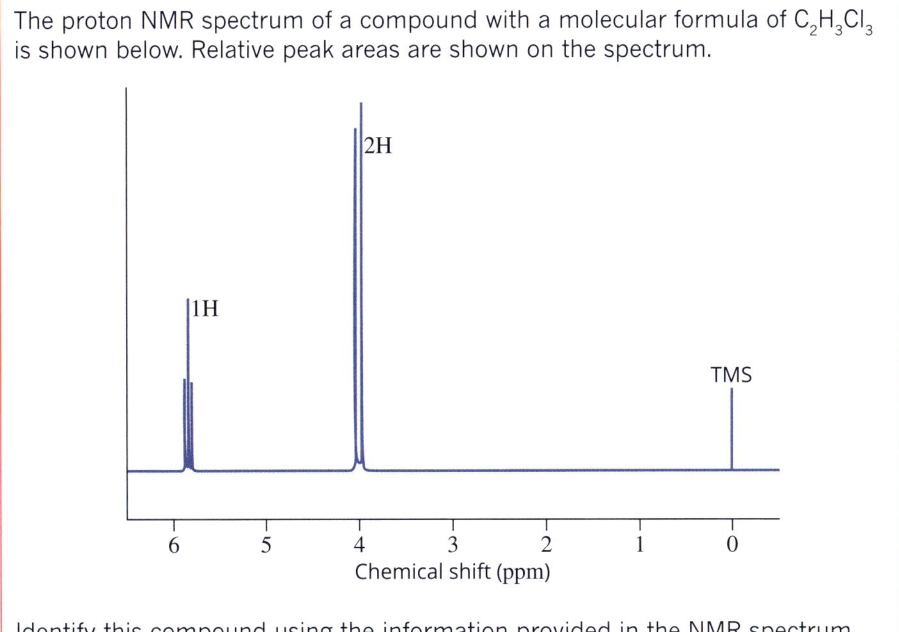

Identify this compound using the information provided in the NMR spectrum.

CARBON-13 NMR SPECTROSCOPY

Carbon-13 (^{13}C) is a naturally occurring isotope of carbon that has a nuclear spin. The abundance of the carbon-13 isotope is only 1.1%. The most abundant isotope of carbon is ^{12}C, which has no spin and is therefore not detected by NMR spectroscopy.

The carbon-13 isotope can be used in NMR spectroscopy to identify different carbon atom environments within a molecule. TMS can also be used as a reference in carbon-13 NMR spectroscopy. As in proton NMR spectroscopy, the chemical shift relative to the TMS reference is dependent on the chemical environment a carbon atom experiences within a molecule. As shown in Table 14.3.2, on the following page, chemical shifts in carbon-13 NMR spectra range from 0 ppm to about 200 ppm.

CHEMFILE

Magnetic resonance imaging

NMR is used in medicine in the form of magnetic resonance imaging (MRI). The scanner is effectively an NMR machine: the patient takes the place of the sample and is passed into the opening of a huge magnet and radio receiver.

The human body contains a large amount of water, which is an abundant source of hydrogen atoms. Normally, MRI measures the strength of the water signal in each area of the body, creating a detailed three-dimensional map. Bone, fat, muscle and tissues contain different amounts of water, allowing them to be distinguished from one another. Normal, healthy tissue gives a different response from the response from a tumour (see the figure below). A powerful computer is used to analyse the data and present a coloured image of the patient's body.

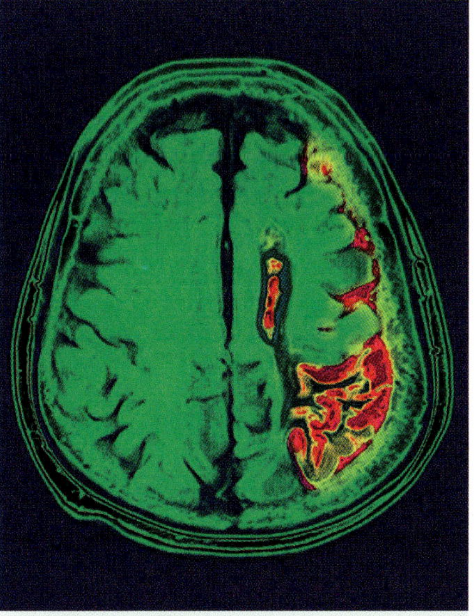

MRI scan of the human brain showing a diseased section in red

TABLE 14.3.2 Typical carbon-13 NMR chemical shifts

Type of carbon	Chemical shift (ppm)
$R-CH_3$	8–25
$R-CH_2-R$	20–45
R_3CH	40–60
R_4C	36–45
$R-CH_2-X$ (X = F, Cl, Br or I)	15–80
R_3C-NH_2, R_3C-NR	35–70
$R-CH_2-OH$	50–90
$RC{\equiv}CR$	75–95
$R_2C{=}CR_2$	110–150
RCOOH	160–185
R–C(=O)–OR	165–175
R–C(=O)–H	190–200
$R_2C{=}O$	205–220

Because of the low abundance of carbon-13 atoms, any particular carbon-13 atom is unlikely to be adjacent to another carbon-13 atom. For this reason, carbon-13 NMR spectra do not display splitting by other carbon-13 atoms, and all the peaks in a spectrum appear as single lines. In addition, the peak areas in a carbon-13 NMR spectrum are not directly proportional to the number of carbon atoms in each environment.

The proton and carbon-13 NMR spectra of chloroethane, along with the structural formula, are shown in Figure 14.3.9. The proton NMR spectrum shows two signals representing the two different hydrogen environments. The carbon-13 NMR spectrum shows two signals representing the two carbon environments.

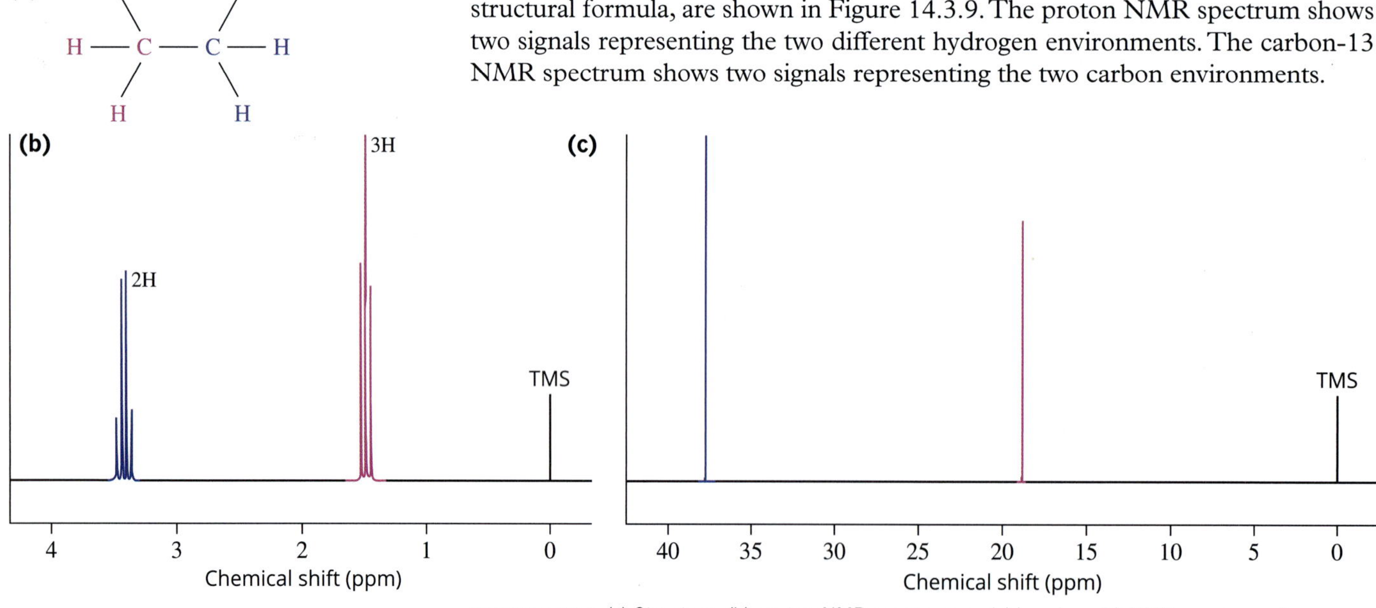

FIGURE 14.3.9 (a) Structure, (b) proton NMR spectrum and (c) carbon-13 NMR spectrum of chloroethane, showing the atoms and their signals in different colours

14.3 Review

SUMMARY

- Atomic nuclei within a strong magnetic field can absorb energy from radio waves and flip into a high-energy spin state. The nuclei emit a signal as they flip back to a low-energy spin state.
- Each signal in a nuclear magnetic resonance (NMR) spectrum corresponds to nuclei in a different chemical environment.
- Chemical shift is characteristic of an atom's environment, based on the tetramethylsilane (TMS) standard.
- Proton (^{1}H) NMR spectroscopy provides information about the different hydrogen atoms in a molecule.
- The peak area of each signal in a proton NMR spectrum is proportional to the number of hydrogen atoms in the corresponding environment.
- Signals in a proton NMR spectrum may be split into line patterns due to the interaction of hydrogen atoms in neighbouring environments. The number of lines in the splitting pattern indicates the number of hydrogen neighbours by the rule $n + 1$.
- Carbon-13 (^{13}C) NMR spectroscopy provides information about the number and identity of different carbon atom environments in a molecule.
- Analysis of an NMR spectrum involves the study of the:
  - number of signals
  - relative peak area of each signal
  - splitting pattern
  - chemical shift of the signal.

KEY QUESTIONS

Knowledge and understanding

1 How many different hydrogen environments are there in each of the following molecules?

a

```
     H
     |
H—C—H
     |
     H
```

b

```
     H      H
     |     /
H—C—N
     |     \
     H      H
```

c

```
     H    H
     |    |
H—C—C—O—H
     |    |
     H    H
```

d

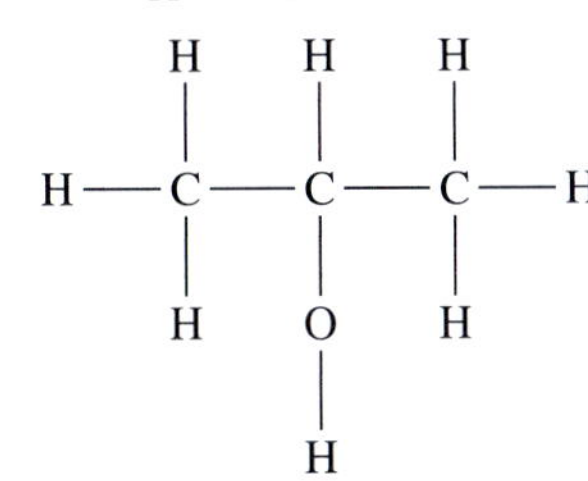

2 Proton NMR spectra provide a number of pieces of information about a molecule's structure. Describe the information that can be gained from the following:

a number of signals

b relative peak areas

c splitting pattern of a signal

d chemical shift of a signal.

3 Using the structure shown below, predict the following for the proton NMR spectrum of pentan-3-ol.

a number of proton NMR peaks

b relative proton NMR peak areas

c splitting pattern for each proton NMR peak

d number of carbon-13 NMR peaks

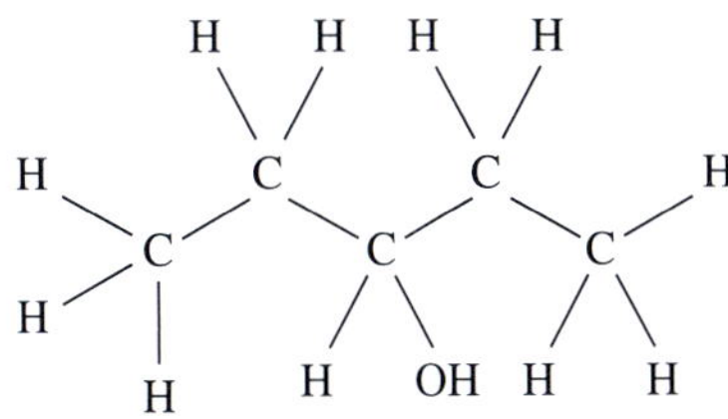

continued over page

14.3 Review *continued*

Analysis

4 Consider the following molecules.

i

ii

iii

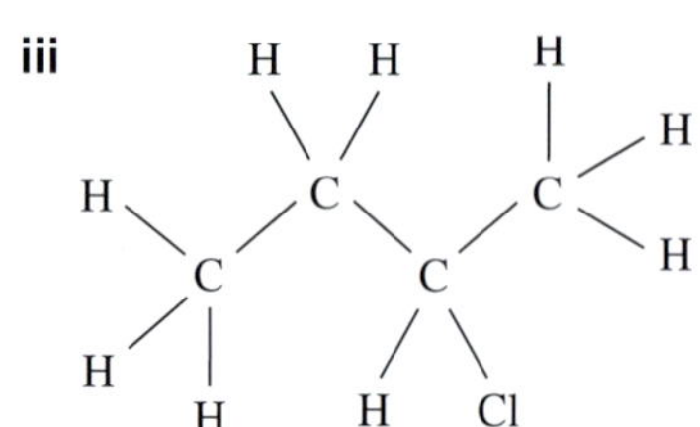

iv

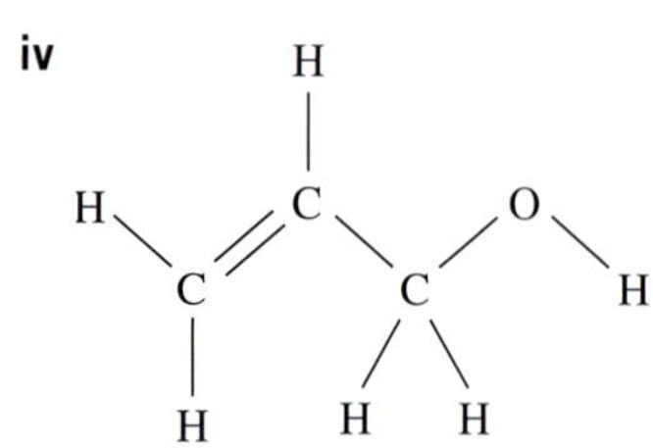

v

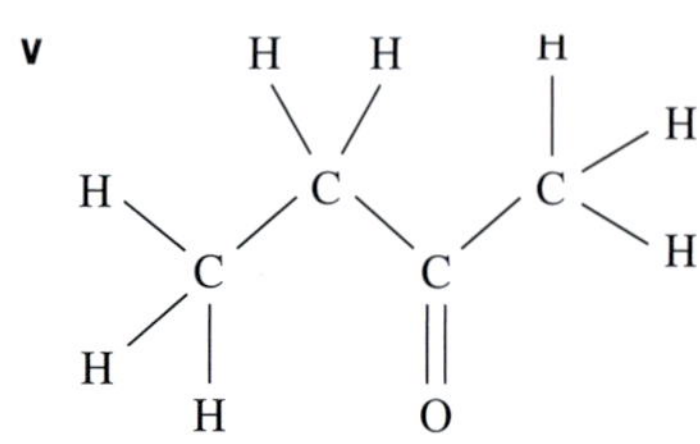

a How many different hydrogen environments are in each molecule?

b How many different carbon environments are in each molecule?

c What splitting pattern would you expect to see from the hydrogen environments in molecule **ii**?

d What ratio would you expect to see in the relative peak areas of the signals from molecule **i**?

5 Ethyl ethanoate is a solvent commonly used in nail-polish remover. The proton NMR spectrum and semi-structural formula of ethyl ethanote are shown in the figure below.

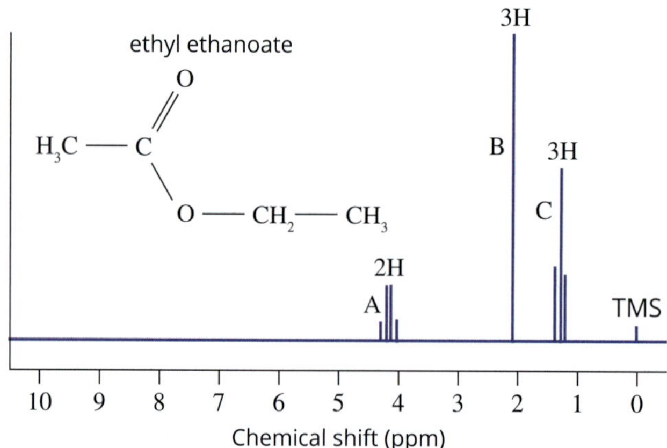

To answer this question, you may need to refer to the chemical shift data in Table 14.3.1 (page 523).

a How many hydrogen environments are there in ethyl ethanoate?

b Refer to Table 14.3.1 and identify the expected chemical shift for the groups present in the ethyl ethanoate molecule.

c What is the relative number of hydrogen atoms in each peak set A, B and C?

d Explain why:

  i peak set A is split into a quartet (four peaks)

  ii peak set B is a single peak

  iii peak set C is a triplet.

e Identify the hydrogen atoms responsible for each peak set and the carbon atoms to which they are bonded.

f How many peaks would you expect in the carbon-13 NMR spectrum of ethyl ethanoate?

6 The proton NMR and carbon-13 NMR spectra for an unknown compound with the formula C_3H_7NO are found below.

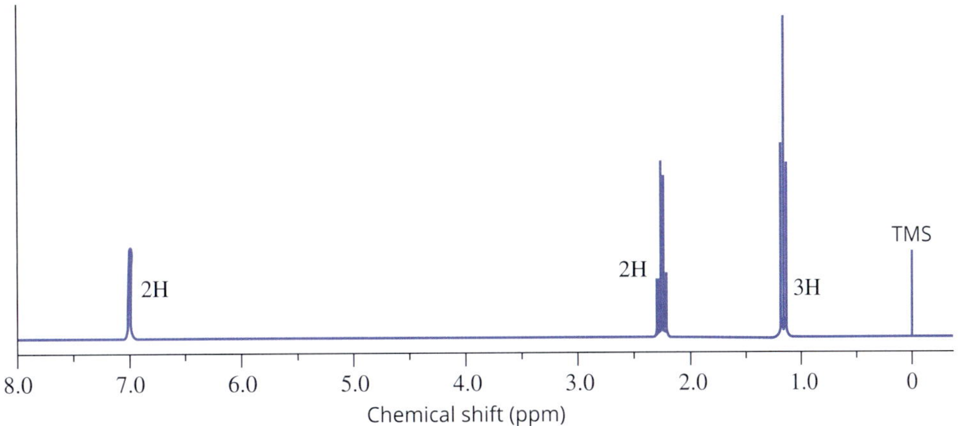

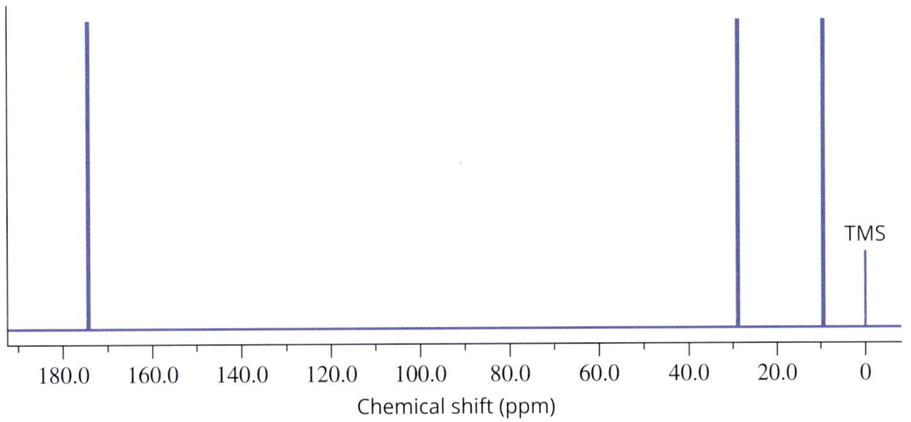

a Referring to the proton NMR spectrum, complete the following table identifying the peaks as 1, 2 and 3 from smallest to largest chemical shifts:

Peak number	Chemical shift	Relative peak area	Splitting pattern	Possible hydrogen environment	Position on structure
1					
2					
3					

b Referring to the carbon-13 NMR spectrum, complete the following table identifying the peaks as 1, 2 and 3 from smallest to largest chemical shifts:

Peak number	Chemical shift	Possible carbon environment	Position on structure
1			
2			
3			

c Using the information collected, identify the unknown compound.

14.4 High-performance liquid chromatography

In Unit 1, you learnt about the principles of chromatography and how it is used to determine the composition and purity of different types of substances by separating and identifying the components of a mixture. You learnt about mobile and stationary phases and how to effectively separate the components through adsorption and desorption processes based on the polarity of these phases and the components. This section describes the use of the analytical technique of high-performance liquid chromatography for both qualitative and quantitative analysis of organic compounds, but first you will be introduced to column chromatography.

REVISION

Chromatography

Chromatography is a technique to separate **components** in a mixture.

Mobile phases (the moving phase) can be in either a liquid or gas state.

Stationary phases (the non-moving phase) can be in either a solid or a liquid state.

The components in the mobile phase flow over or through the stationary phase, undergoing a continual process of **adsorption** (onto the stationary phase), followed by **desorption** and dissolving (into the mobile phase).

The rate of movement of each component mainly depends on how strongly the component adsorbs onto the stationary phase and how readily the component dissolves in the mobile phase.

The components separate because they undergo adsorption and desorption at different rates due to their different **polarities**.

Paper and thin-layer chromatography are simple chromatographic techniques for qualitative analysis. A small sample of the mixture is placed on the stationary phase (paper or thin-layer) and this stationary phase is placed in a beaker of a solvent. The solvent (the mobile phase) transports the mixture over the stationary phase, separating out the components. The identification of the components can be conducted, either by using known chemicals as standards on the same chromatogram or by calculating the R_f (retardation factor) value of the sample.

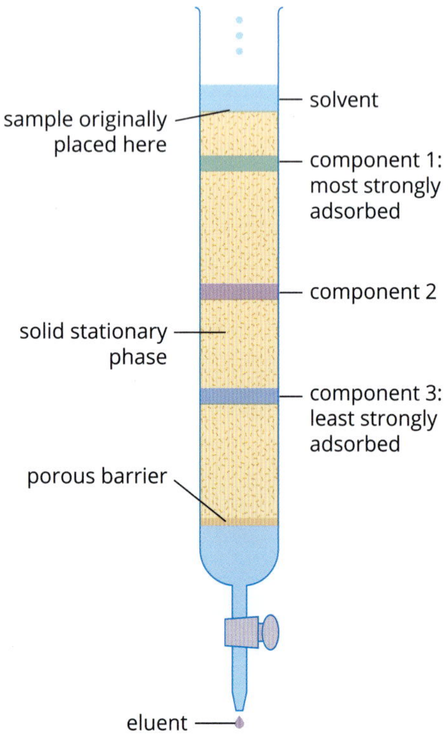

sample originally placed here

solvent

component 1: most strongly adsorbed

component 2

solid stationary phase

component 3: least strongly adsorbed

porous barrier

eluent

FIGURE 14.4.1 Column chromatography is used to separate the components in a mixture.

COLUMN CHROMATOGRAPHY

Column chromatography is another form of chromatography (Figure 14.4.1). In column chromatography, the solid stationary phase is packed into a glass column. The sample mixture is applied carefully to the top of the packed solid, and a solvent, which acts as the mobile phase, is dripped slowly onto the column from a reservoir above. A tap at the bottom of the column allows the solvent, which is now called the **eluent**, to leave the column at the same rate as it enters it at the other end.

As the components are carried down the column, they are repeatedly adsorbed to the stationary phase and desorbed back into the mobile phase. The components of a mixture undergo the processes of adsorption and desorption to different degrees, so the components separate as they move downwards. The time taken for a component to pass through the column is called the **retention time (R_t)**, and this is characteristic of the component for the conditions of the experiment. It is similar to the R_f value in paper chromatography and thin-layer chromatography.

HIGH-PERFORMANCE LIQUID CHROMATOGRAPHY

High-performance liquid chromatography (HPLC) is a modern instrumental chromatographic technique based on column chromatography.

HPLC allows scientists to perform extremely sensitive analyses of a wide range of mixtures. It is commonly used for the separation and identification of very complex mixtures of similar compounds, such as drugs in blood, and hydrocarbons in oil samples. Compounds in trace concentrations as low as parts per trillion (ppt) may easily be identified. HPLC is therefore a very sensitive technique.

High-performance liquid chromatography (HPLC) is sometimes also referred to as high-pressure liquid chromatography.

Figure 14.4.2 shows the apparatus used for HPLC. The basic principles of HPLC are the same as for column chromatography. The main differences between simple column chromatography and HPLC are:

- the particles in the stationary phase used in the HPLC column are often 10–20 times smaller than those used in column chromatography. Their higher surface area allows more frequent adsorption and desorption of the components. This greater surface area therefore gives much better separation of similar compounds.
- the small particle size used in HPLC creates a considerable resistance to the flow of the mobile phase. Therefore, the solvent is pumped through the column under high pressure.

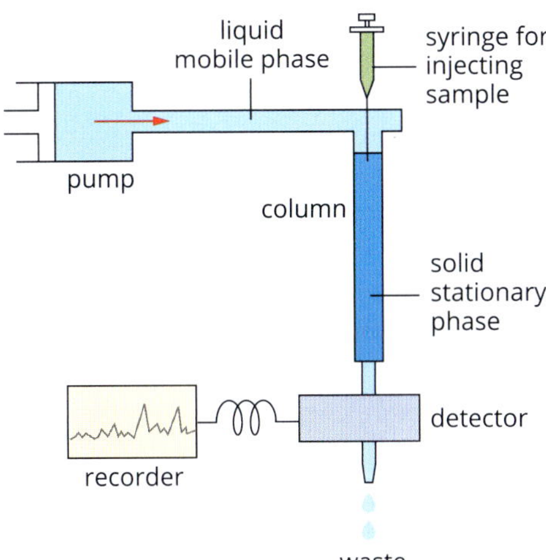

FIGURE 14.4.2 The construction of a high-performance liquid chromatograph

A range of solids is available for use as the stationary phase in HPLC columns, including alumina (Al_2O_3) and silica (SiO_2). Some solids also have chemicals specially bonded to their surfaces to improve the separation of particular classes of compounds.

In the most common form of HPLC, called **reversed-phase HPLC**, the stationary phase is silica. The silica used has been modified to make it non-polar by attaching long hydrocarbon chains to its surface. Typically, the silica particles are extremely small, with diameters between 3 and 5 μm. A polar solvent is used as the mobile phase, such as a mixture of water and methanol.

Polar molecules in the sample form relatively strong hydrogen bonds and dipole–dipole attractions with solvent molecules, but are only weakly adsorbed to the non-polar stationary phase. Therefore, polar molecules in the sample spend most of their time moving with the solvent.

On the other hand, non-polar compounds in the sample tend to adsorb to the non-polar stationary phase by dispersion forces. They are also less soluble in the solvent because they cannot form hydrogen bonds with the polar mobile phase. Non-polar compounds spend less time in solution, so they move through the column more slowly than polar molecules. The way this separation occurs can be seen in Figure 14.4.3.

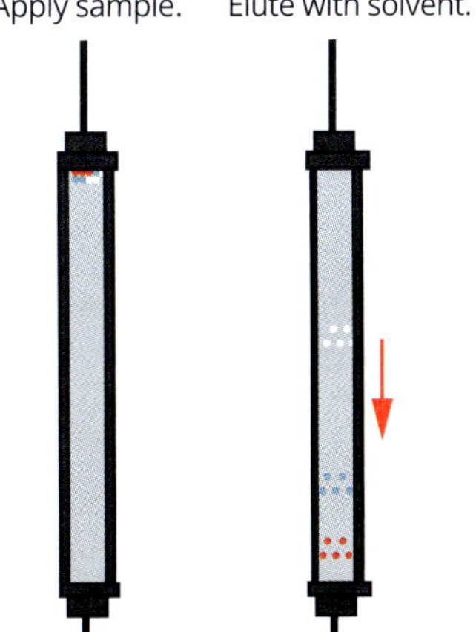

FIGURE 14.4.3 A diagram showing how HPLC can be used to separate and collect different components

HPLC provides qualitative and quantitative analysis of components in a mixture. The size of the peaks is due to the amount of light absorbed by each component, and this can be used to calculate the relative amount of each component within the mixture.

In HPLC, the components are usually detected by passing the eluent stream through a beam of ultraviolet (UV) light. Many organic compounds absorb UV light. When an organic compound passes in front of the beam of light, a reduced signal is picked up by a detector. The amount of light received by the detector is converted into an absorbance measurement and recorded on a chart that moves slowly at a constant speed or on a computer. The resulting trace is called a chromatogram.

Retention times are used to identify the components associated with the peaks on a chromatogram (**qualitative analysis**). The relative amounts of each component in a mixture may be determined by comparing the areas under each peak with areas under the peaks for standard samples (**quantitative analysis**). In Figure 14.4.4 you can see an example of the component peaks found in an HPLC chromatogram.

FIGURE 14.4.4 A research chemist uses HPLC to analyse the components in a mixture for medical research.

APPLICATIONS OF HPLC

Chemists use the technique of HPLC in chemical analysis to answer two questions.

- What chemicals are present in the sample (qualitative analysis)?
- How much of each chemical is present (quantitative analysis)?

Qualitative analysis

In HPLC, the same compound gives the same retention time if the conditions (temperature, mobile phase, stationary phase, flow rate, pressure, etc.) remain the same. Each component forms one peak in the chromatogram.

Figure 14.4.5 shows the **chromatogram** of a sample of body fluid produced by HPLC. The smaller, more polar components are more soluble in the solvent used and are eluted more quickly. Therefore, these components have a lower retention time.

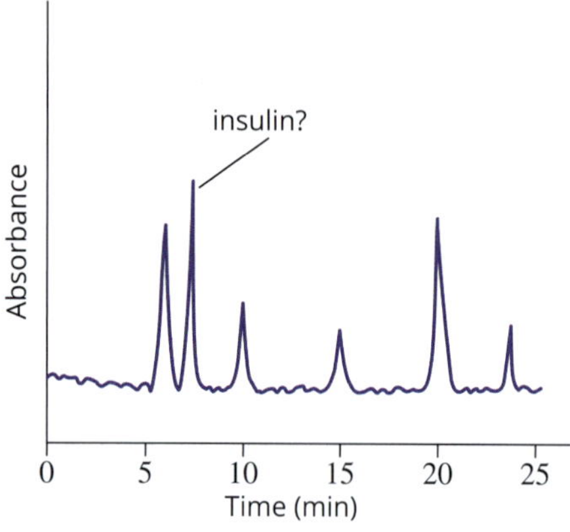

FIGURE 14.4.5 A chromatogram of body fluid

The peaks can be tentatively identified by running a sample containing a known pure compound (such as insulin) under exactly the same conditions as the sample (Figure 14.4.6).

A component can also be tentatively identified by adding a known compound to the sample (**spiking**). Figure 14.4.7 shows the sample spiked with insulin. Insulin was added to the sample and the spiked chromatogram shows that the second peak is larger than it was in the unspiked sample. There are no extra peaks, indicating that insulin is very likely to have been present in the sample.

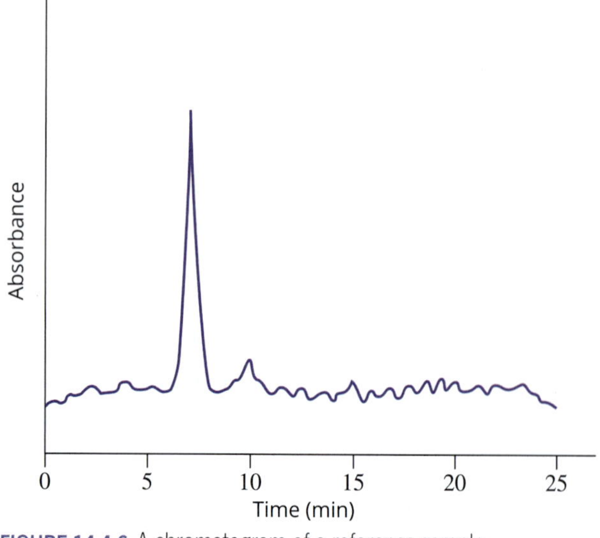

FIGURE 14.4.6 A chromatogram of a reference sample containing insulin

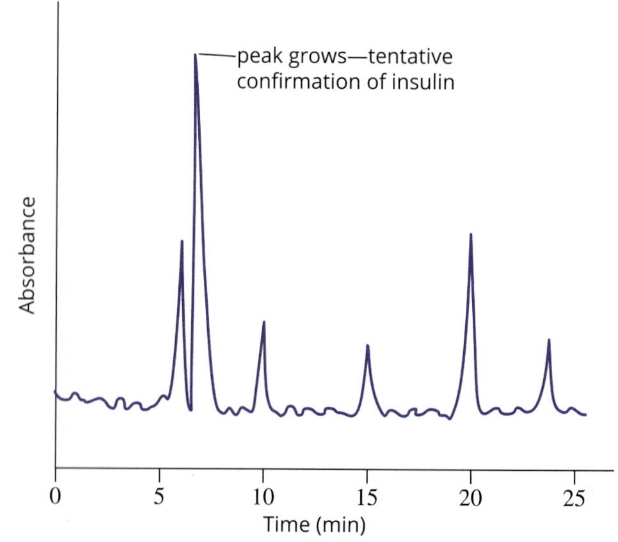

FIGURE 14.4.7 A body fluid sample spiked with insulin

Quantitative analysis

To determine the concentration of an individual component in a mixture, its peak area is compared with the peak areas of samples of the same chemical at known concentrations. A solution with an accurately known concentration is called a **standard solution**.

By plotting the peak areas against the concentrations of the standard solutions, you can draw a **calibration curve** and use it to determine unknown concentrations.

Worked example 14.4.1 shows how HPLC can be used to find the concentration of a component in a mixture.

Worked example 14.4.1

CONCENTRATION OF A COMPONENT IN A MIXTURE

The concentration of caffeine in an energy drink was determined by HPLC. Chromatograms of standards with accurately known concentrations of caffeine were also obtained under the same conditions as the sample.

The peak areas of the sample and the standards are shown in the table. Calculate the concentration of caffeine in the sample.

Caffeine standards (μg mL^{-1})	Peak area (mm^2)
0.0	0.0
0.010	1.0
0.020	2.4
0.040	4.6
0.060	7.0
sample	3.5

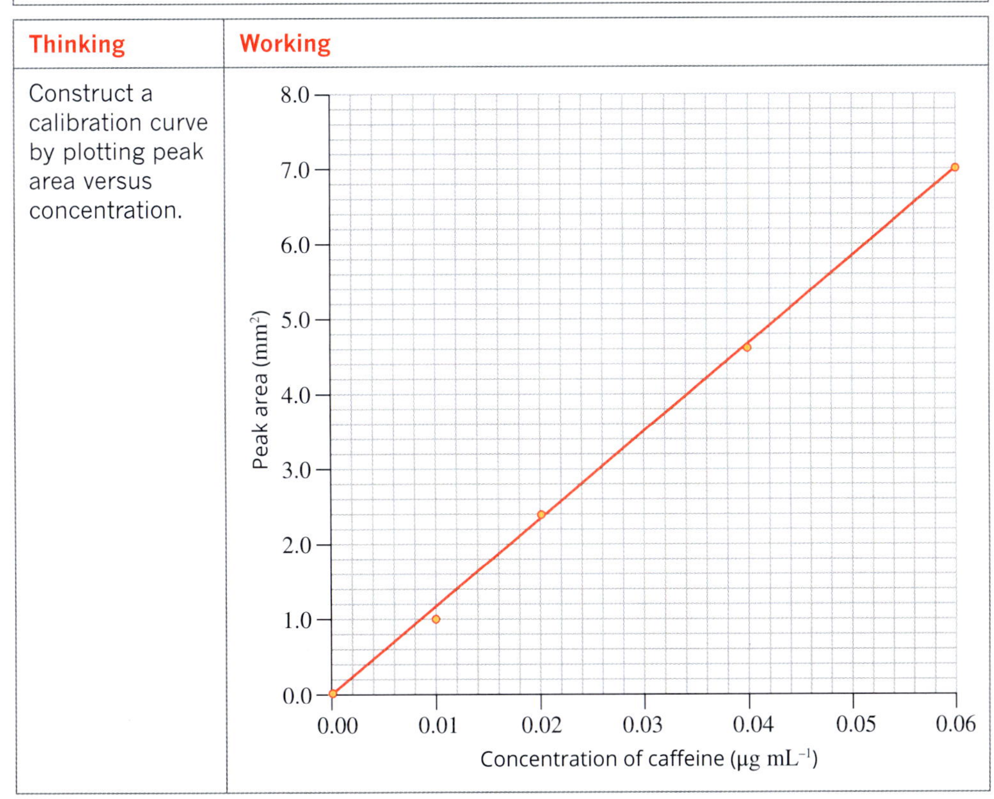

Thinking	Working
Construct a calibration curve by plotting peak area versus concentration.	(graph)

Use the calibration curve to find the concentration corresponding to the peak area of the sample.	A line (blue) can be drawn on the calibration curve from the peak area of the sample to find the corresponding concentration on the graph. 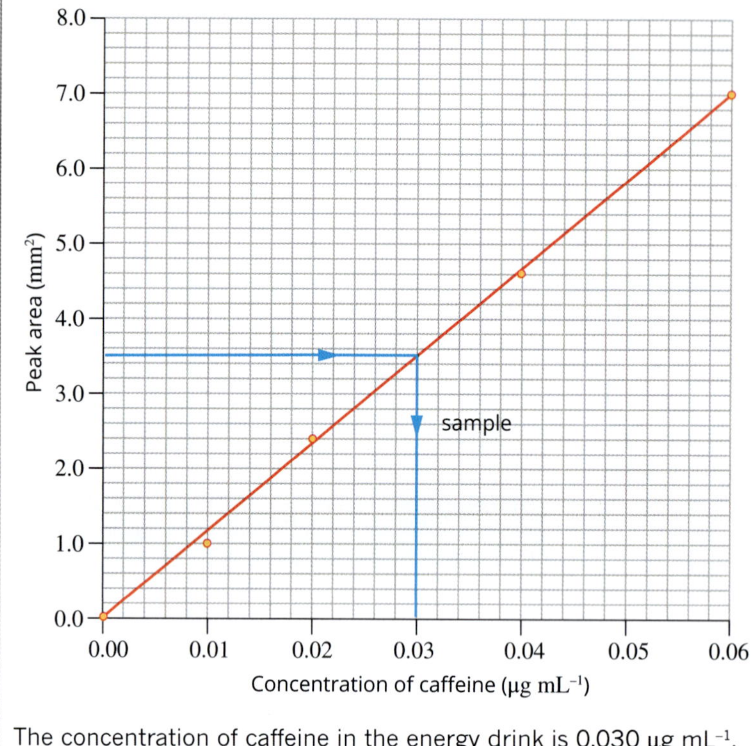 The concentration of caffeine in the energy drink is 0.030 μg mL^{-1}.

Worked example: Try yourself 14.4.1

CONCENTRATION OF A COMPONENT IN A MIXTURE

Trinitrotoluene (TNT) is often blended with other explosives to suit particular applications. A sample from a solution of an explosive mixture was injected onto an HPLC column using a water/methanol mobile phase. The peak corresponding to TNT had an area of 14.5 mm^2. The peak areas for three standard solutions were also measured, as shown in the table.

Determine the concentration of TNT in the sample solution in μg mL^{-1}.

TNT standards (μg mL^{-1})	Peak area (mm^2)
0.0	0.0
2.0	6.0
4.0	11.5
6.0	17.5

HPLC can be combined with mass spectrometry, MS (see Section 14.1). Together, the techniques allow chemists to determine smaller and smaller quantities and identify a wider range of materials. You will learn more about HPLC-MS in Section 14.5.

CASE STUDY **ANALYSIS**

Making the grade

Gourmet cooks value olive oil (Figure 14.4.8) for its distinctive taste and aroma. Nutritionists favour it because it is rich in monounsaturated triglycerides, which are believed to lower blood cholesterol levels and reduce the risk of heart disease. But some of the companies that process and distribute olive oil succumb to the temptation to mix their olive oil with less expensive oils, such as corn, peanut and soybean.

A European company sent a shipment of olive oil to the United States. A routine sample was taken and forwarded to chemist Richard Flor at the US Customs Service laboratory in Washington, DC. The sample looked and poured like olive oil, but it didn't taste quite right. Because taste is a subjective test and doesn't always stand up in court, Flor and his colleague Le Tiet Hecking developed an analytical test based on olive oil's unique composition.

Flor used HPLC to separate the oils into their component triglycerides. (An example of the chromatographs obtained can be seen in Figure 14.4.9.) Flor and Hecking found that the oil in the suspect sample had too much of the polyunsaturated triglycerides, confirming their suspicion that the imported sample was diluted with a less expensive oil.

FIGURE 14.4.8 Pure olive oil is highly regarded for its aroma and taste, and because it is thought to reduce blood cholesterol levels.

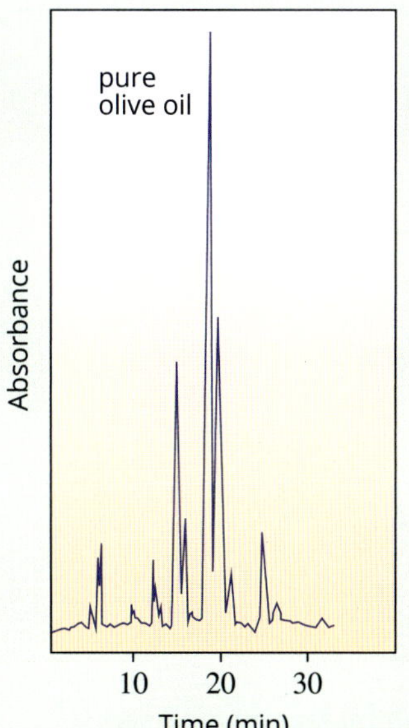

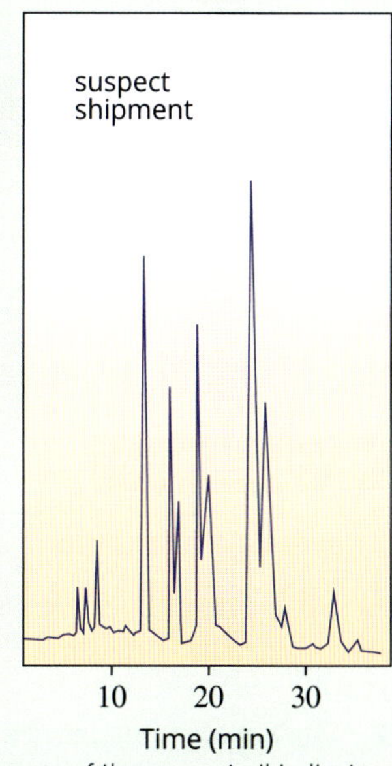

FIGURE 14.4.9 Peaks in the chromatogram of the suspect oil indicate the presence of additional compounds rather than pure olive oil.

Analysis

1 Is there any olive oil present the suspect oil shipment? Explain your answer.

2 Why would an organic solvent like hexane be more suitable than water for use as an eluent in the HPLC analysis?

3 A government department analysed some imported olive oil by HPLC and found that the oil was impure and contained chemicals found only in corn oil. The peak area of one of the corn oil components in the chromatogram of a sample of the impure oil was 15.5 mm^2. The peak areas of four standard solutions of the component were also measured (Table 14.4.1). Use the values in Table 14.4.1 to draw the calibration graph for corn oil concentrations and determine the concentration of the corn oil component in the impure oil sample.

TABLE 14.4.1 Results of olive oil analysis

Corn oil component standards (%)	Peak area (mm^2)
0.0	0.0
1.0	5.5
2.0	11.5
3.0	17.5
4.0	23.0

14.4 Review

PA 21 | OA ✓✓

SUMMARY

- High-performance liquid chromatography (HPLC) is a very sensitive technique and is used for qualitative and quantitative analysis.
- In HPLC, the mobile phase is a liquid under pressure.
- In HPLC, retention time is used to identify components in a mixture.
- The concentration of an individual component in a mixture can be determined by comparing its peak area on a chromatogram with the peak areas of samples with known concentrations of the same chemical.
- The peak areas of standard solutions are used to construct a calibration curve, which can be used to determine unknown concentrations.

KEY QUESTIONS

Knowledge and understanding

1 Describe two variables of the HPLC separation conditions that can be changed to alter the retention time of a compound.

2 State which feature of a chromatogram obtained during an HPLC analysis is:

a used to identify the components of a mixture

b least affected by a change in concentration of the sample components

c used to determine the concentration of each component in a mixture.

3 Explain why it is necessary to construct a calibration curve to determine the concentration of a component in a mixture that has been analysed by HPLC.

Analysis

4 A sample containing compounds A, B and C is mixed with ethanol and applied to the top of an HPLC instrument that uses ethanol as the solvent and alumina as the stationary phase. The three compounds have the following properties.

- Solubility in ethanol: C is much more soluble than A and B. A and B are equally soluble.
- Adsorption to alumina: B is more strongly absorbed than A. C is least strongly adsorbed.

Determine the relative order of the retention times of the components, from lowest to highest.

5 EPO is a protein hormone produced by the kidneys. Although banned, some endurance athletes use it to improve performance because it increases the body's oxygen-carrying capacity. Athletes are routinely tested for EPO levels by HPLC. The peak areas produced from HPLC analysis of an athlete's blood and the peak areas of a number of standard solutions of EPO are shown in the following table.

	EPO ($\mu g\ mL^{-1}$)	Relative EPO peak area
Blood sample	?	5.6
Standard 1	2.0	1.6
Standard 2	4.0	3.2
Standard 3	6.0	4.8
Standard 4	8.0	6.4

a Plot a calibration curve of relative peak area against concentration of EPO ($\mu g\ mL^{-1}$).

b Determine the concentration of EPO in the blood sample correct to one decimal place.

6 A mixture was composed of four primary alcohols with formula ROH, where R is a straight-chain alkyl group. The mixture was analysed by HPLC using a non-polar stationary phase and a blend of water and methanol as the mobile phase. The chromatogram in the following figure was obtained.

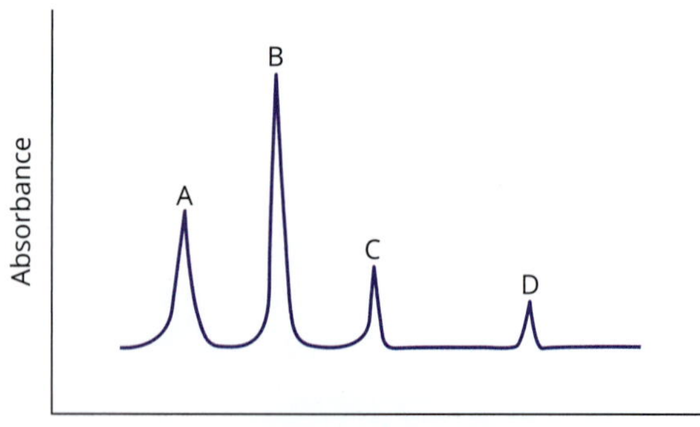

Determine the order from smallest to largest molar mass of the primary alcohols found in peaks A to D.

14.5 Determination of molecular structure by spectroscopy

Chemists determine the structure and identity of unknown compounds by employing a range of analytical techniques. Each technique provides a different piece of information about the compound's structure, which, when considered together, can identify the substance. The process can be compared to piecing together the pieces of a jigsaw puzzle.

The structures of organic compounds are often determined using a combination of IR spectroscopy, NMR spectroscopy and data from mass spectrometry. Chemical and physical properties can also provide clues about the identity of a compound.

In this section, you will investigate how the structure of an unidentified compound can be determined.

As you read through and complete the questions in this section, you will need to refer to the IR and NMR tables (Tables 14.2.4 on page 512, 14.3.1 on page 523 and 14.3.2 on page 526) provided earlier.

Figure 14.5.1 is a useful guide to follow when working through these problems.

When you are identifying an unknown compound, refer to NMR and IR data tables to help interpret the spectra.

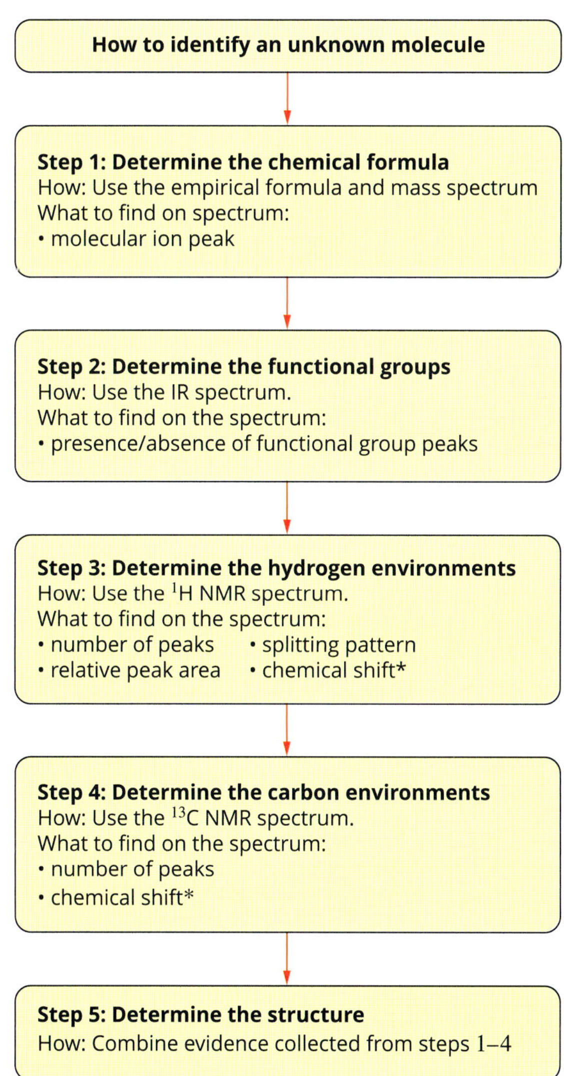

FIGURE 14.5.1 Flow chart listing the general steps used to identify an unknown molecule

COMBINING ANALYSES TO LEARN MORE

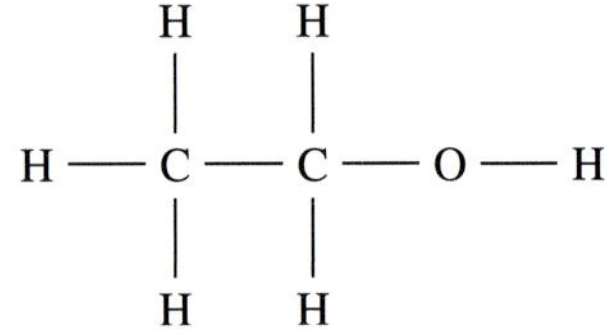

FIGURE 14.5.2 Molecular structure of ethanol

Ethanol will be used as an example of how the results of different types of analysis can be used to identify an unknown organic compound. The structural formula of ethanol is shown in Figure 14.5.2.

The mass spectrum for ethanol can be seen in Figure 14.5.3. Using this spectrum, the following can be determined.

- There is a molecular ion peak at m/z = 46, which corresponds to a relative molecular mass of 46.
- The fragment ion peaks are identified on the spectrum and are consistent with the fragmentation of a molecule with the formula C_2H_6O.

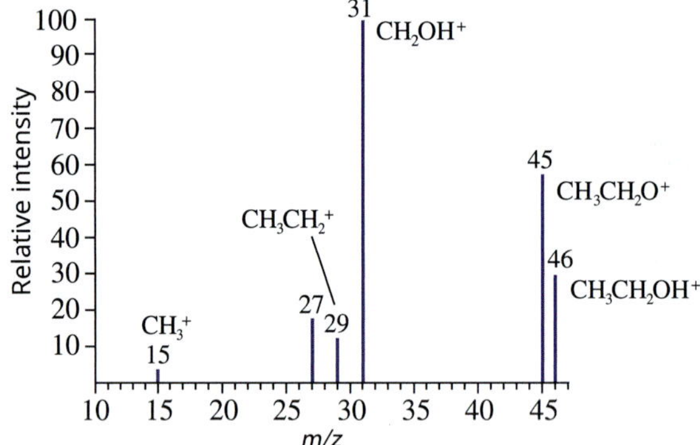

FIGURE 14.5.3 Mass spectrum of ethanol

The infrared spectrum of ethanol (Figure 14.5.4) provides the following information.

- The most prominent feature is the broad stretch at 3400 cm^{-1} due to the O–H bond and confirms the presence of an alcohol hydroxyl group.
- There is no strong peak around 1650–1750 cm^{-1}, therefore there is no C=O bond.
- The absorption bands due to the C–H and C–O bonds are also observed.

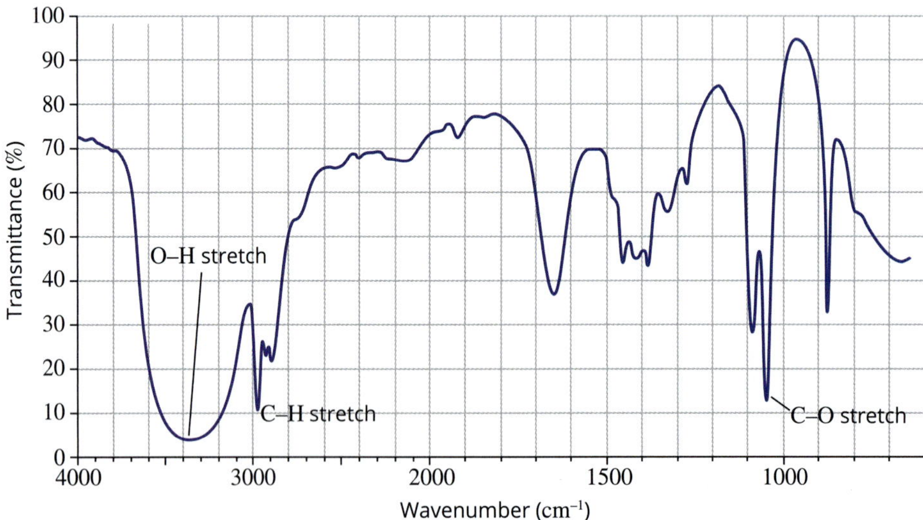

FIGURE 14.5.4 IR spectrum of liquid ethanol

The proton NMR spectrum for ethanol is shown in Figure 14.5.5. It provides the following information.

- There are three signals, which means that the molecule contains three different hydrogen environments.
- The signal with a chemical shift of approximately 3.7 ppm is a four-line pattern and has a relative peak area of 2. The chemical shift and peak area are consistent with the signal of a $-CH_2-$ group adjacent to a hydroxyl group. The splitting pattern indicates there are three hydrogen atoms in a neighbouring environment. (Remember that a hydrogen atom on an –OH group does not usually split the signals of adjacent hydrogen atoms.)
- The signal at 1.2 ppm is a three-line pattern with a relative peak area of 3. This is consistent with the signal of a $-CH_3$ group adjacent to a $-CH_2-$ group.
- The singlet at 2.6 ppm is consistent with the signal of an alcohol hydroxyl group.

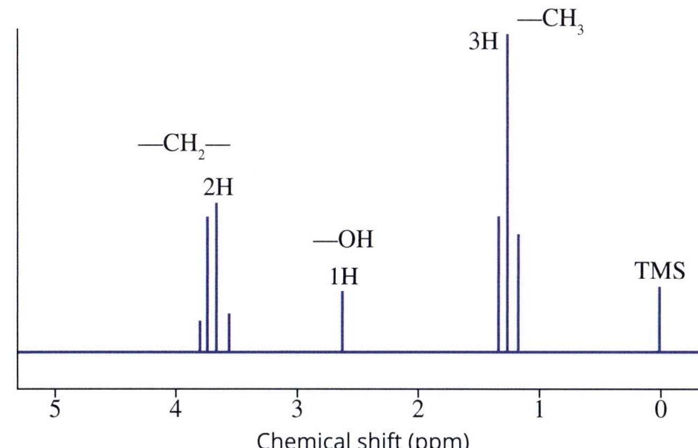

FIGURE 14.5.5 Proton NMR spectrum of ethanol

The carbon-13 NMR spectrum for ethanol shown in Figure 14.5.6 provides the following information.

- There are two signals, meaning that the molecule contains two different carbon environments.
- The signal at 58 ppm corresponds to the signal of a carbon atom attached to a hydroxyl group, $R-CH_2-OH$.
- The signal at 18 ppm is consistent with the signal from a methyl group, $-CH_3$.

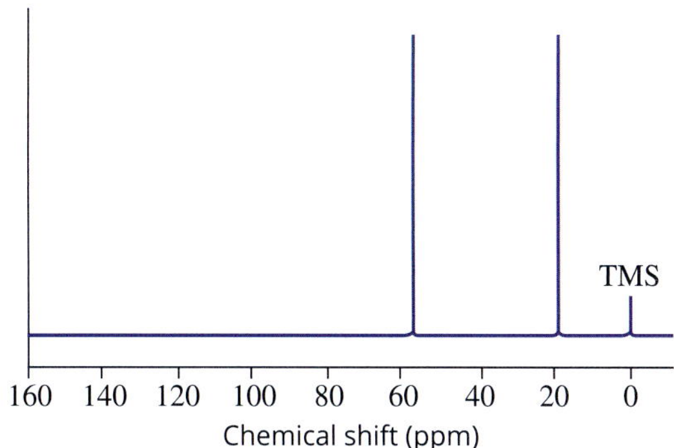

FIGURE 14.5.6 Carbon-13 NMR spectrum of ethanol

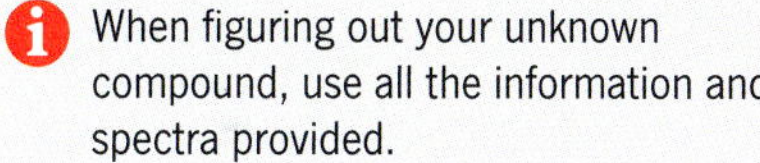

When figuring out your unknown compound, use all the information and spectra provided.

When each piece of evidence from analysing the different spectra is put together, you can see that the structure must be CH_3CH_2OH. Each of the spectra contains the peaks expected of ethanol, and none of the spectra contains information that is inconsistent with ethanol.

Worked example 14.5.1

DEDUCING MOLECULAR STRUCTURE FROM SPECTROSCOPIC DATA

A sweet-smelling liquid has an empirical formula of $C_3H_6O_2$. Chemical tests show that the compound does not react with Na_2CO_3. Use this information and the mass, IR and NMR spectra provided to deduce the structure and name of the compound.

Mass spectrum

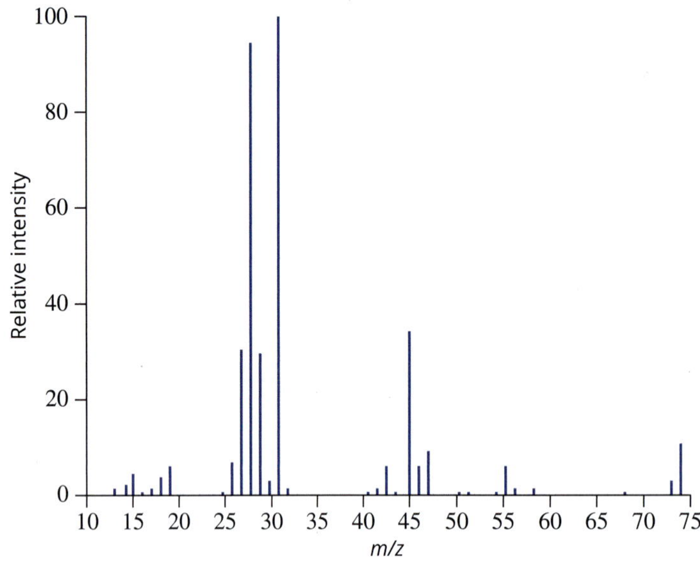

Infrared spectrum

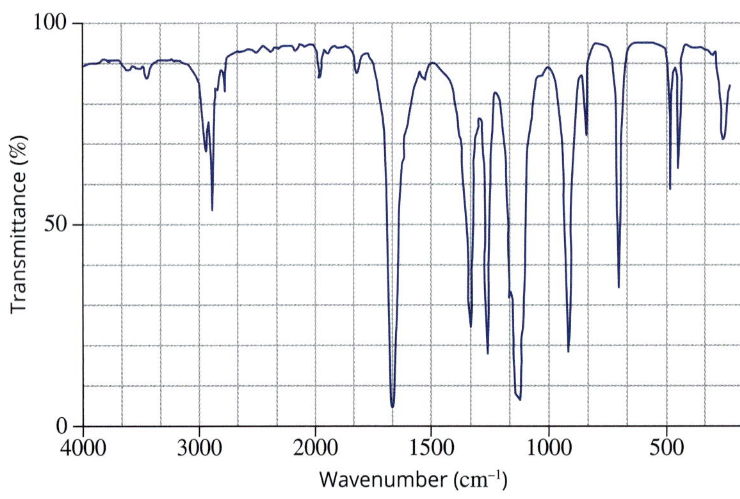

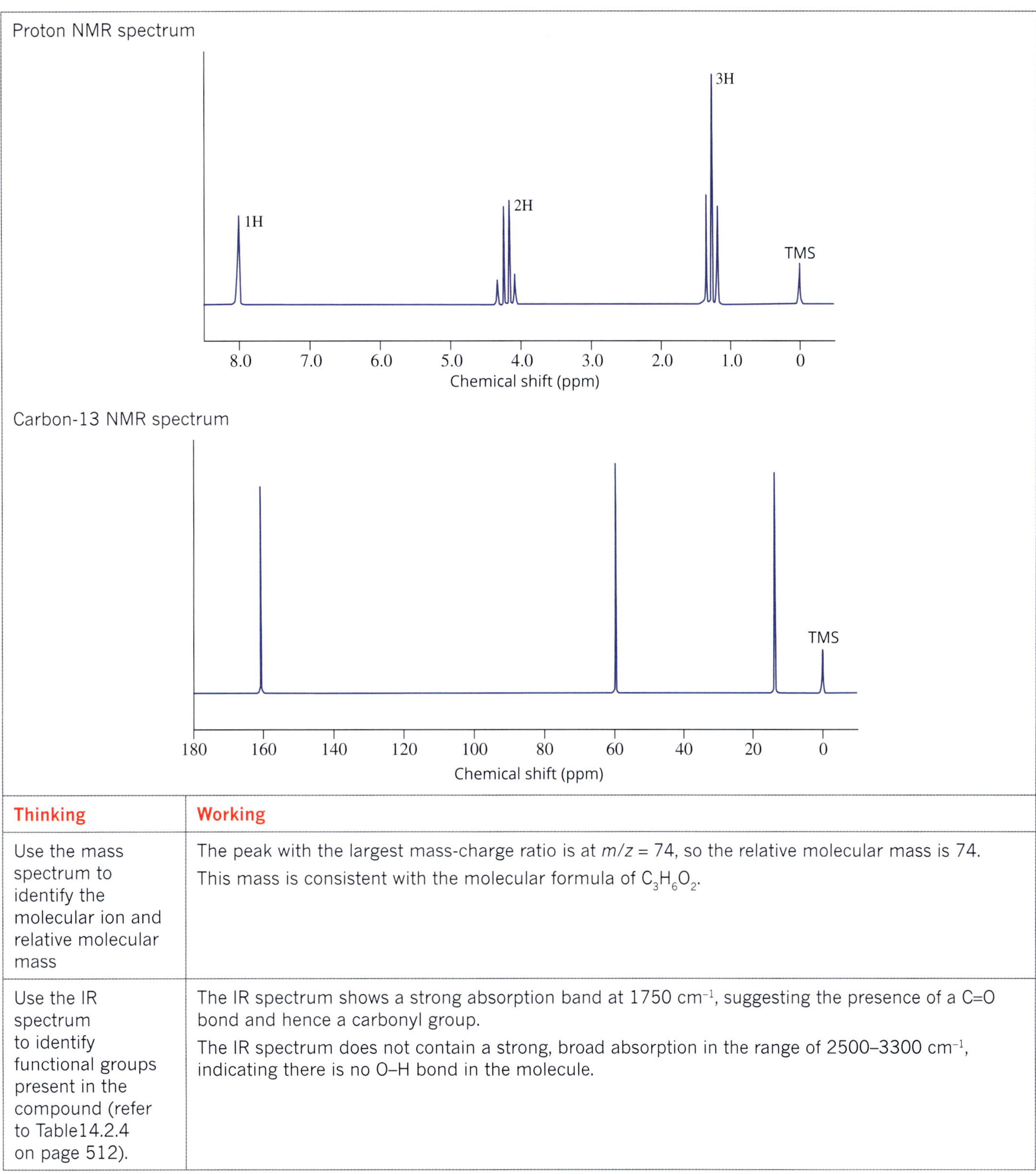

Thinking	Working
Use the mass spectrum to identify the molecular ion and relative molecular mass	The peak with the largest mass-charge ratio is at $m/z = 74$, so the relative molecular mass is 74. This mass is consistent with the molecular formula of $C_3H_6O_2$.
Use the IR spectrum to identify functional groups present in the compound (refer to Table14.2.4 on page 512).	The IR spectrum shows a strong absorption band at 1750 cm^{-1}, suggesting the presence of a C=O bond and hence a carbonyl group. The IR spectrum does not contain a strong, broad absorption in the range of 2500–3300 cm^{-1}, indicating there is no O–H bond in the molecule.

<table>
<tr>
<td>Use the proton NMR spectrum to identify the different hydrogen environments (refer to Table 14.3.1 on page 523).

Note that chemical shifts can be outside the ranges quoted in Table 14.3.1, and so can be slightly higher or lower than the values provided. Use the relative peak area and splitting pattern first to identify hydrogen environments.</td>
<td>The proton NMR data is summarised in the following table.

<table><tr><th>Chemical shift (ppm)</th><th>Splitting pattern</th><th>Relative peak area</th></tr><tr><td>1.3</td><td>triplet (3-line pattern)</td><td>3</td></tr><tr><td>4.2</td><td>quartet (4-line pattern)</td><td>2</td></tr><tr><td>8.0</td><td>singlet (1-line pattern)</td><td>1</td></tr></table>
The spectrum contains three signals, so there are three different hydrogen environments.

The sum of the relative peak areas is 6, which is consistent with the molecular formula, as each unit of peak area corresponds to one hydrogen atom.

The signal at 1.3 ppm is consistent with the signal produced by a $-CH_3$ group, as it has a relative peak area of 3 and each peak area corresponds to one hydrogen atom. In addition, the peak at 4.2 ppm is a quartet, indicating a methyl group adjacent to those hydrogens.

The signal at 1.3 ppm is split into three due to two hydrogen atoms in a neighbouring environment.
The signal at 4.2 ppm is consistent with the signal produced by a $-CH_2-$ group. It has a relative peak area of 2, as each peak area corresponds to one hydrogen atom.

A combination of a triplet peak adjacent to a quartet peak is typical of an ethyl group.

The chemical shift of the 4.2 ppm signal corresponds to $R-CH_2-O$ of an ester group.

The signal at 8.0 ppm is a singlet, indicating that there are no hydrogen atoms attached to adjacent atoms. The chemical shift is quite large and could be consistent with a hydrogen attached to a carbonyl group.</td>
</tr>
<tr>
<td>Use the carbon-13 NMR spectrum to identify the different carbon environments (refer to Table 14.3.2 on page 526).</td>
<td>There are three signals in the carbon-13 NMR spectrum, so the molecule contains three different carbon environments.

The number of carbon environments corresponds to the number of carbons in the molecular formula, so each environment represents one carbon atom.

The signal at 161 ppm is consistent with a carbon atom in a carbonyl group.

The signal at 60 ppm is consistent with a carbon atom attached to an oxygen atom by a single bond to an oxygen atom.

The signal at 14 ppm is consistent with a methyl group, $R-CH_3$.</td>
</tr>
<tr>
<td>Use the data from the spectra to deduce the structure of the compound.</td>
<td>The data provided by the spectra shows that:
• the compound has a carbonyl group, but no hydroxyl group
• it is an ester
• the compound has a CH_3CH_2- group attached by a single bond to an oxygen atom
• the compound has a HC=O group.

A structure consistent with this data is:

O
//
H — C   H   H
\ | |
O — C — C — H
| |
H   H</td>
</tr>
<tr>
<td>Name the compound.</td>
<td>ethyl methanoate</td>
</tr>
</table>

Worked example: Try yourself 14.5.1

DEDUCING MOLECULAR STRUCTURE FROM SPECTROSCOPIC DATA

A sweet-smelling liquid has an empirical formula of C_2H_4O. Chemical tests show that the compound is not an aldehyde or ketone. Use this information and the mass, IR and NMR spectra provided to deduce the structure and name of the compound.

Mass spectrum

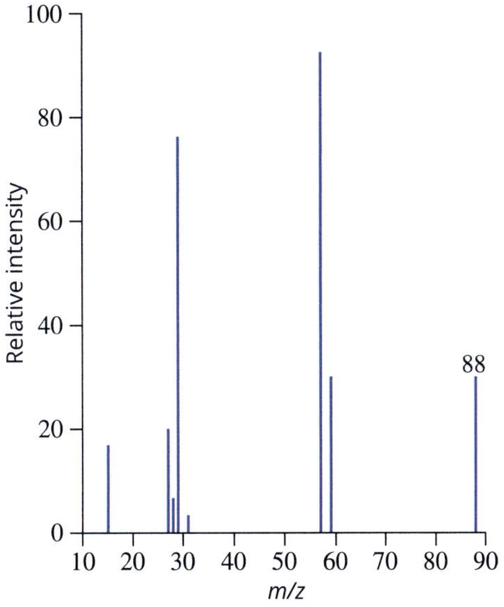

Infrared spectrum

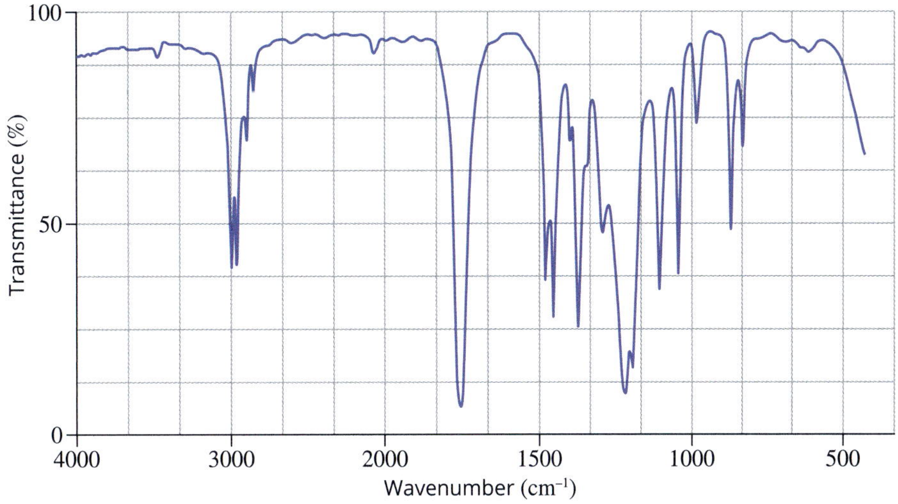

Proton NMR spectrum

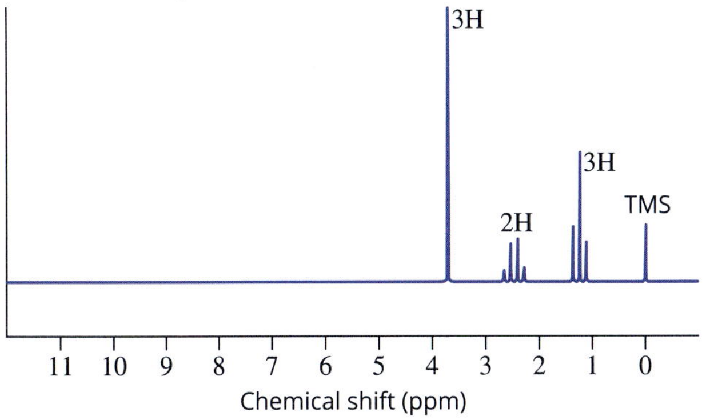

Carbon-13 NMR spectrum

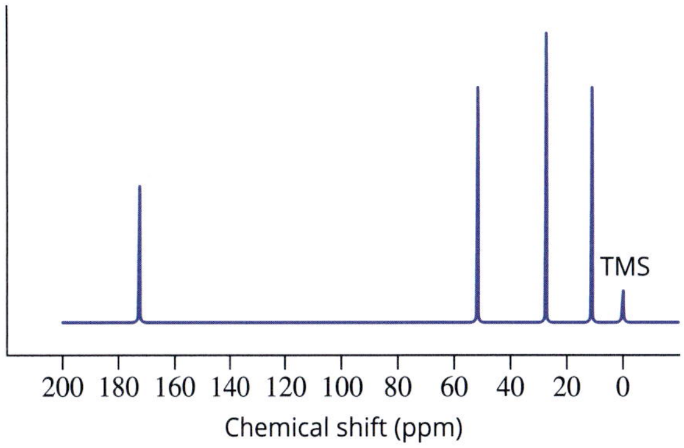

INVESTIGATING MIXTURES AND PRODUCT PURITY

This chapter has shown you how the information provided by mass spectrometry, IR and NMR spectroscopy can enable you to determine the structure of an unknown organic compound. Up to now, the examples of unknown compounds used have been 100% pure. That is, no other compounds were present in the sample.

However, analytical chemists usually have to analyse mixtures instead of pure samples for the presence and concentration of specific compounds. You will now learn about the common methods chemists use when dealing with mixtures.

Figure 14.5.7 shows the IR spectrum of a contaminated aspirin tablet. You may be able to identify some functional groups from the spectrum, but it is impossible to tell which peaks represent the aspirin and which peaks represent the **contaminant** or impurity. Similarly, scientists cannot determine the unknown organic compounds in a mixture by simply looking at the IR spectrum. Each organic compound has its own unique vibrational absorption bands, so an IR spectrum of a mixture has all these vibrational absorption bands overlapping each other.

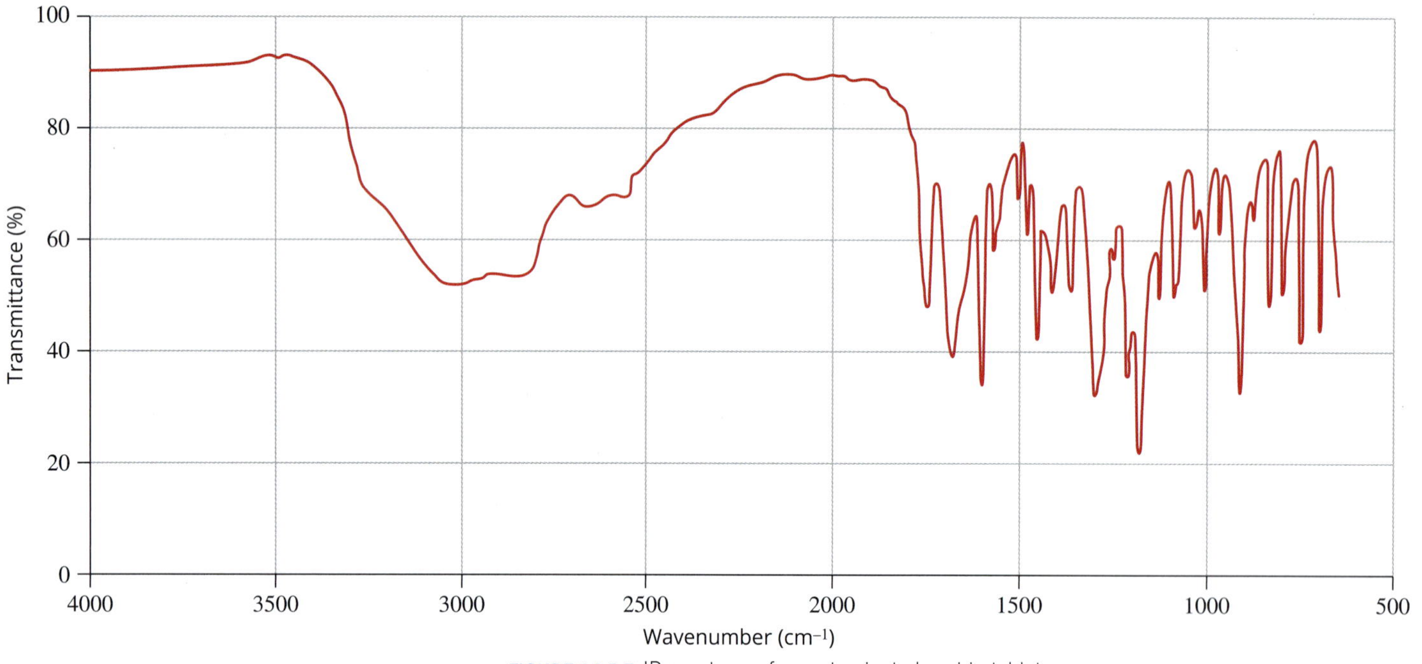

FIGURE 14.5.7 IR spectrum of a contaminated aspirin tablet

Scientists use **reference samples** to identify unknown organic compounds in mixtures. Reference samples are known standards and can be either mixtures or individual compounds. These reference samples have their IR spectrum measured and stored in databases. These databases can contain reference spectra of hundreds of thousands of organic compounds and mixtures.

Figure 14.5.8a shows the IR spectrum of an aspirin reference sample (in blue) overlayed on the spectrum of the contaminated aspirin (in red). When this spectrum is removed from the contaminated aspirin spectrum, it leaves only the contaminant spectrum (the black line in Figure 14.5.8b). This contaminant's IR spectrum is then compared with a database and paracetamol is found to match (the green line in Figure 14.5.8b). Therefore, paracetamol is the contaminant in the original aspirin tablet.

Reference samples are known standards of either mixtures or individual compounds used to identify unknown compounds in mixtures

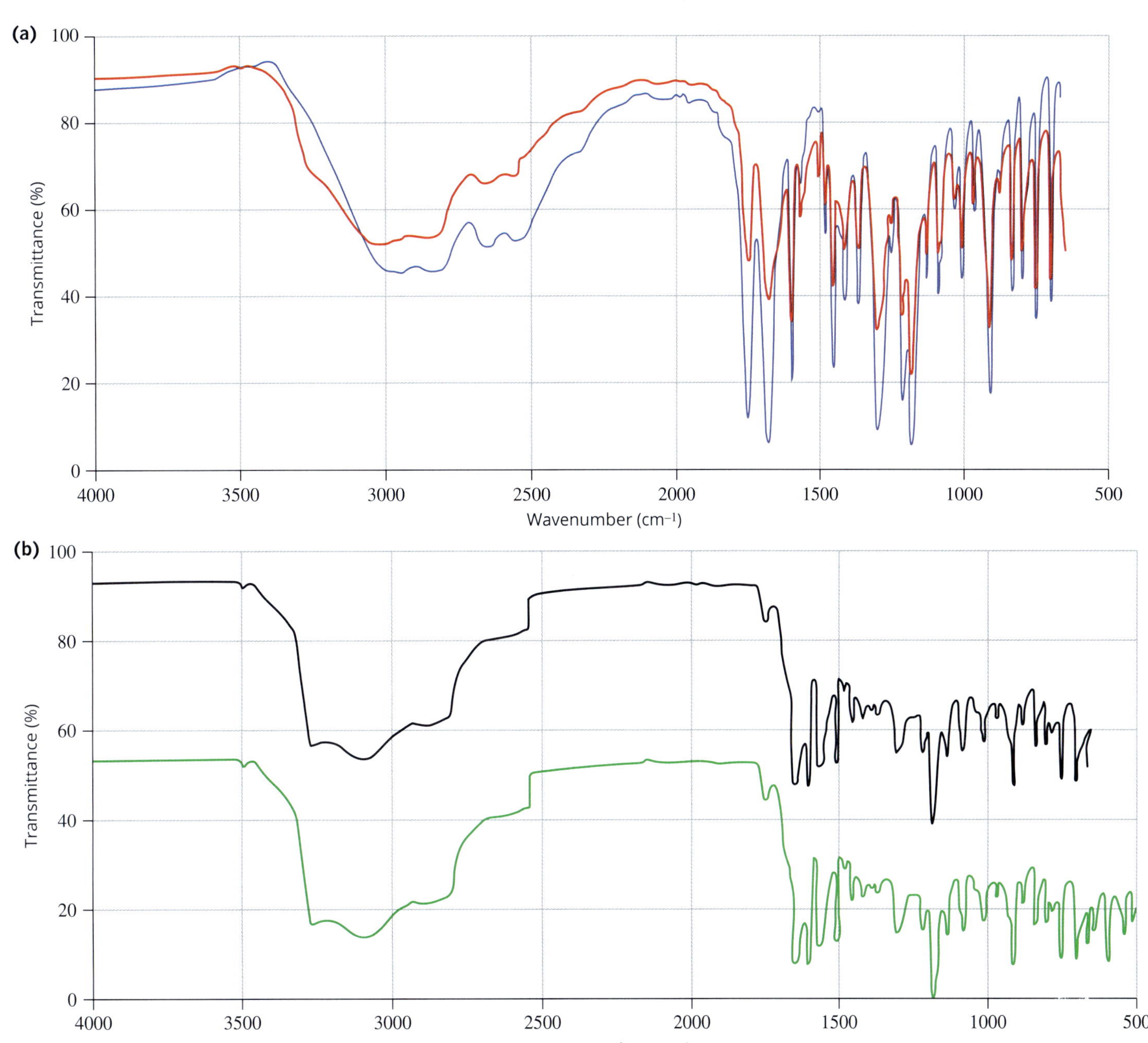

FIGURE 14.5.8 (a) The IR spectrum of an aspirin reference sample (in blue) overlaid on the spectrum of a contaminated aspirin tablet (in red). (b) The infrared spectrum of the contaminant (the upper spectrum in black) is matched with the lower spectrum of paracetamol in the database (in green).

CHEMFILE

NIST databases

Many commercial databases containing spectra of reference samples are available. Most analytical laboratories require these databases to help them to identify unknown compounds. One of the most established databases comes from the National Institute of Standards and Technology (NIST). NIST is based in the United States, and their databases and Chemistry Web book are used worldwide. The free online Chemistry Webbook allows users to search for data, including mass and IR spectra, for tens of thousands of molecules.

NMR spectroscopy also requires reference samples for the identification of unknown organic compounds in mixtures. Figure 14.5.9 is an example of how NMR spectroscopy can be used to detect the adulteration of honey. **Adulteration** is the addition or substitution of an ingredient of a lesser quality and cost. Adulterated honey has been sold worldwide as 100% pure honey, but actually the honey has been diluted with sugar syrup. Honey is a complex mixture of many different sugars, and the sugar syrup contains many of these sugar compounds too. Researchers discovered that NMR spectroscopy provided be a simple and accurate way to detect sugar syrup in honey using data on the chemical shifts and splitting patterns. However, before they undertook any investigations, reference samples of authentic honey and sugar syrup NMR spectra had to be collected first. The researchers then analysed a honey sample and, using statistical analysis, compared it to sugar syrup and pure honey to determine whether the honey sample was adulterated.

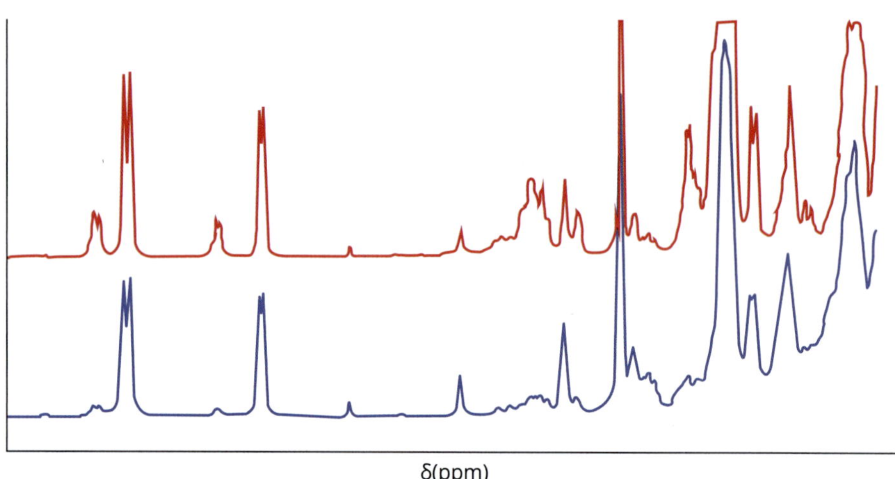

FIGURE 14.5.9 Proton NMR can be used to detect the adulteration of honey with sugar syrup. The top spectrum shows adulterated honey with 40% sugar syrup, and the lower spectrum shows unadulterated honey.

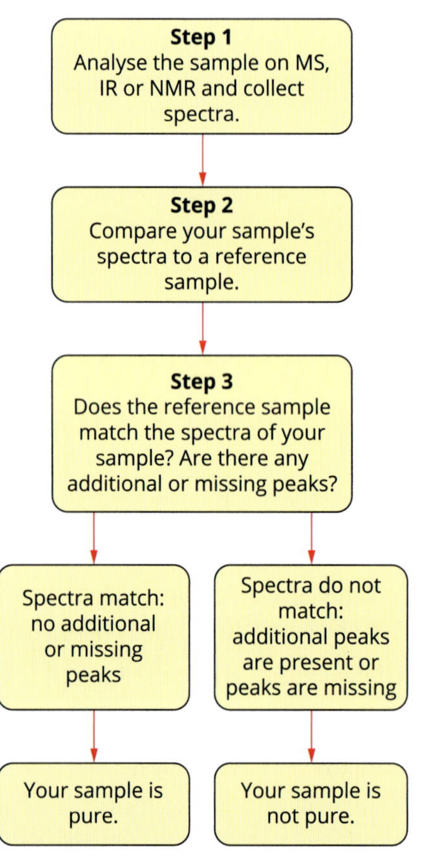

FIGURE 14.5.10 The steps you would undertake to check the purity of a sample

Mass spectrometry is commonly used to identify unknown components in mixtures. Mass spectrometers that have multiple magnetic fields are able to separate the ion fragments from mixtures more effectively (see Case Study on page 535 for an example). However, to increase the resolution, improve the accuracy of identification and to determine the concentration of components in complex mixtures, an HPLC is typically attached to the mass spectrometer (HPLC-MS). The HPLC separates the components of a mixture and the eluent is injected into the mass spectrometer. The mass spectrometer fragments the molecule and produces a mass spectrum that can be compared to databases for identification. Once the component has been provisionally identified by the database, the scientist can confirm the results by analysing a 100% pure sample of that identified component and checking that the retention times and mass spectra match. The concentration of that component can then be determined by preparing and analysing a series of standards.

It is almost impossible to manually examine the mass, IR and NMR spectra of mixtures to determine the known and unknown organic compounds, especially when comparing thousands of spectra. Typically, statistical analysis must be done using computer modelling to compare the spectra with reference samples. However, it is much simpler to determine the purity of a sample. Figure 14.5.10 outlines the main steps analytical chemists would use. Care must be taken to ensure that your sample is dissolved in the same solvent as the reference sample, as intermolecular forces between the solvent and the sample in question can change peak wavenumbers or chemical shifts.

For a sample to be 100% pure, there must be no additional or missing peaks in its spectrum when compared to the reference spectrum.

14.5 Review

SUMMARY

- A combination of information derived from mass spectrometry, infrared (IR) spectroscopy, proton NMR spectroscopy and carbon-13 NMR spectroscopy can be used to determine the molecular structure of organic compounds.
- The molecular formula of a molecule can be determined from its mass spectrum if its empirical formula is known.
- The infrared spectrum of a compound provides evidence about functional groups present and absent in a molecule.
- The proton and carbon-13 NMR spectra provide detailed information that can be used to determine the connectivity of atoms and overall structure of a molecule.
- HPLC instruments are combined with mass spectrometers to separate out components of mixtures and to identify and quantify these components.
- Reference samples are known standards and can be either mixtures or individual compounds.
- A sample is not pure if there are additional or missing peaks in its spectrum compared to a reference spectrum.

KEY QUESTIONS

Knowledge and understanding

1 Explain how a reference sample is used to determine whether an organic compound is pure or not.

2 You have a sample of a gas, but are unsure whether it is ethane or propane. For each spectroscopic technique listed, describe how the information you would gain about the unknown gas could be useful to identify it.
  - a IR spectroscopy
  - b proton NMR
  - c carbon-13 NMR

3 A proton NMR spectrum of petrol has been collected. Explain why it is necessary to use a database of proton NMR reference spectra to find out what organic compounds are in petrol.

Analysis

4 A hydrocarbon has the empirical formula C_5H_{12}. The mass spectrum had a molecular ion peak at $m/z = 72$. The proton NMR spectrum has one singlet peak at 0.9 ppm. The carbon-13 NMR has two peaks, at 28 ppm and 36 ppm. Determine the name of this compound.

5 An industrial solvent has the molecular formula C_4H_8O. Use the infrared, proton NMR and ^{13}C NMR spectra to determine the name and structure of this molecule.

Infrared spectrum

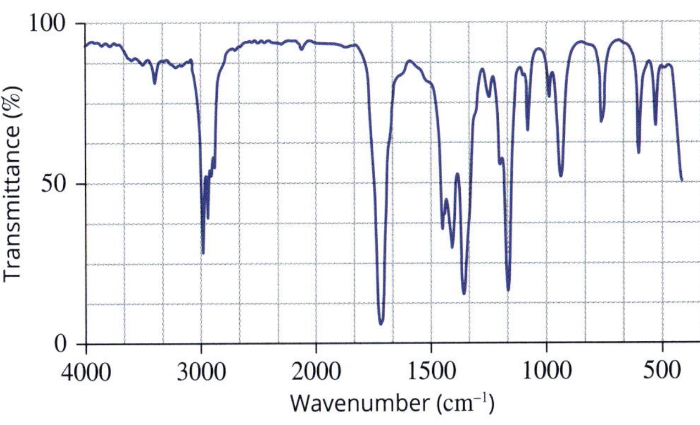

Proton NMR spectrum

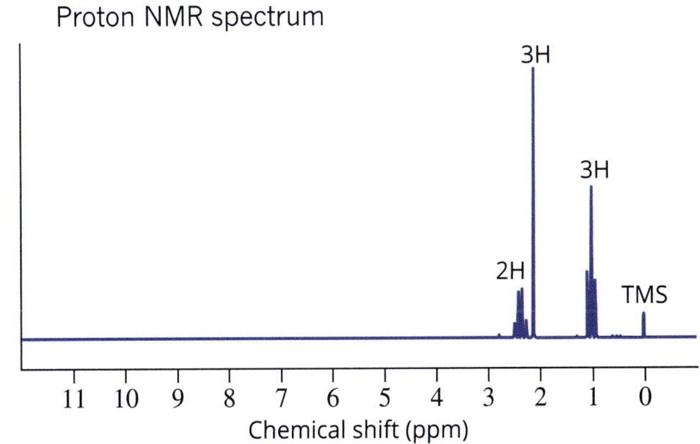

continued over page

14.5 Review *continued*

Carbon-13 NMR spectrum

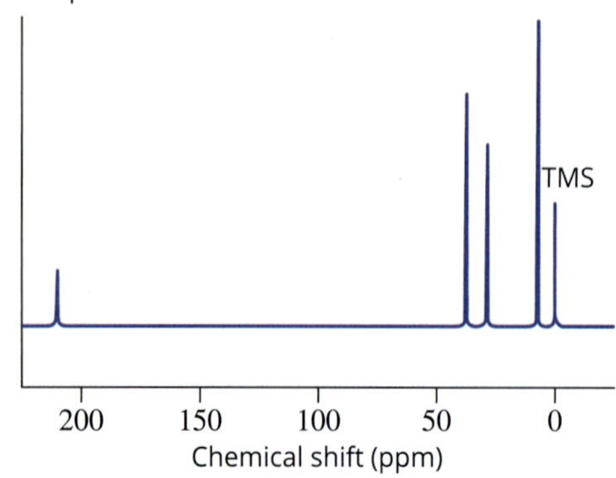

6 The adulteration of expensive olive oil with cheaper vegetable oils, such as canola, palm or sunflower oils, can be detected by IR spectroscopy. The figure below shows that the spectra of olive oil and sunflower oil are almost identical in the functional group region, whereas there are differences in the fingerprint region.

a Compare the two IR spectra shown below and identify the peak that you would focus on to test a suspect sample of olive oil.

b Explain how you would know if the suspect sample of olive oil was pure or if it contained sunflower oil.

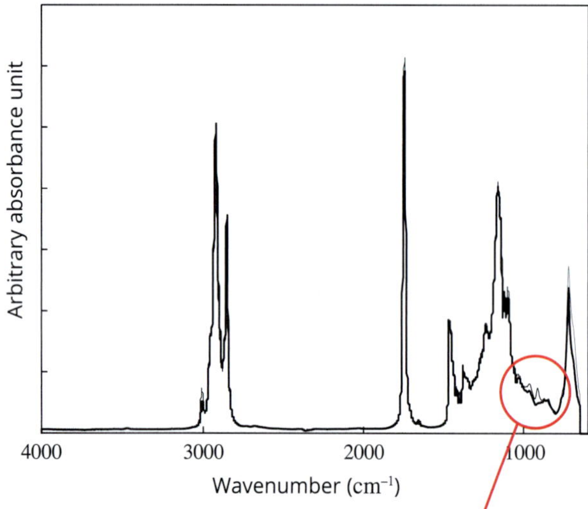

Fig. 1 The typical spectra of extra virgin olive oil and sunflower oil: —, sunflower oil; —, extra virgin olive oil.

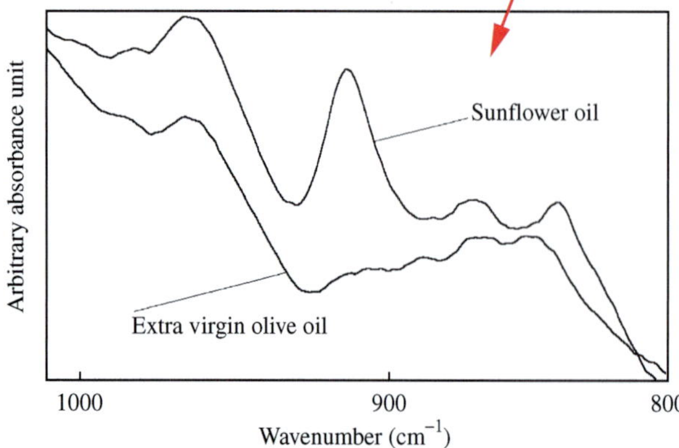

Fig. 2 The fingerprint spectra of extra virgin olive oil and sunflower oil.

Chapter review

KEY TERMS

absorption band
adsorption
adulteration
base peak
calibration curve
carbon-13 NMR spectroscopy
chemical environment
chemical shift
chromatogram
column chromatography
component
contaminant
database
desorption
dipole
electromagnetic spectrum
eluent
equivalent
fingerprint region
fragmentation
fragment ion
functional group
high-performance liquid chromatography (HPLC)
infrared (IR) spectroscopy
integration
integration curve
isotope
mass spectrometer
mass spectrometry
mass spectrum
mobile phase
molecular ion
molecular ion peak
m/z value
neighbour
nuclear magnetic resonance (NMR) spectroscopy
nuclear magnetic resonance (NMR) spectrum
nuclear shielding
nuclear spin
parent molecular ion
parts per trillion
peak area (proton NMR)
polarity
proton NMR spectroscopy
qualitative analysis
quantitative analysis
reference sample
relative intensity
retention time
reversed-phase HPLC
signal
spectroscopy
spiking
splitting pattern
standard solution
stationary phase
transmittance
vibrational energy level
wavenumber

REVIEW QUESTIONS

Knowledge and understanding

1 Which of the following is measured by IR spectroscopy?

A molecular mass
B vibrational energy
C electronic energy
D nuclear spin

2 Select which of the following is most likely to be the functional group that absorbs energy at 1720 cm^{-1}.

A C–O
B C=O
C O–H
D N–H

3 For proton NMR, the splitting of peaks in the spectra follows the '$n + 1$ rule', where n is the number of neighbouring:

A hydrogens
B carbons
C oxygens
D cations

4 Which of the following pieces of information is not provided by a mass spectrum?

A molecular mass
B isotopic ratios
C mass-to-charge ratio (*m/z*)
D hydrogen environments

5 A carbon-13 NMR spectrum of an unknown molecule is shown in the figure below. Select how many carbon environments there are.

A 1
B 2
C 3
D 4

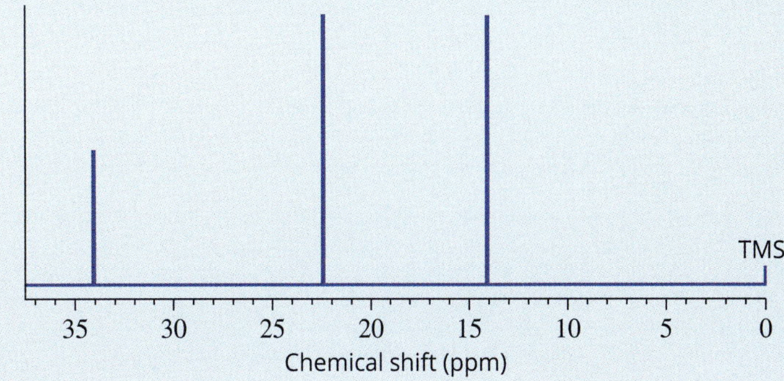

6 Using Table 14.3.2 on page 526, select which type of carbon would correspond to a chemical shift of 115 ppm.

A R–CH_3

B R–CH_2–X (X = F, Cl, Br or I)

C RC≡CR

D $R_2C{=}CR_2$

7 A number of compounds with different structures may form molecular ions with the same *m/z* ratio. Give the molecular formula of a compound that would give a *m/z* value of 58 if the molecule:

a is an alkane

b contains one oxygen atom

c contains two amine groups.

8 The mass spectrum of an alkane is shown below.

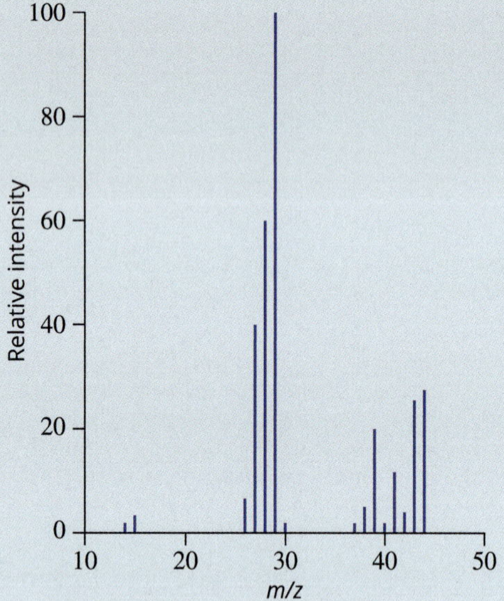

a What is the *m/z* for the base peak?

b What is the *m/z* for the molecular ion peak?

c Determine the molecular formula of the unknown alkane.

9 A common fragment found in a mass spectrum of a carboxylic acid molecule is COH^+. What is its *m/z*?

10 The mass spectrum of dichloroethane is shown below.

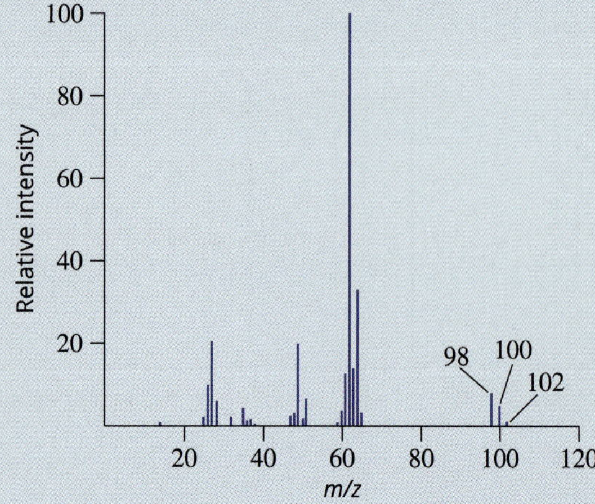

a Explain why the mass spectrum has the peaks with the *m/z* values of:

i 98

ii 100

iii 102

b Why is the peak with *m/z* = 98 the largest of these peaks?

c Write a possible formula for the ion responsible for the base peak.

11 Select the bond in each of the following groups that would have the highest vibrational energy.

a C–N, C–C, C–H

b O–H, O–C, O–O

12 List the bonds in 2-bromoethan-1-ol in order of the energy of IR radiation they will absorb, from lowest to highest.

13 Refer to Table 14.2.4 (page 512) to answer this question. Identify the absorption bands in an infrared spectrum that are required to indicate the presence of:

a a carboxylic acid group

b a primary amine group

c ester

d ketone.

14 Explain the use of TMS in NMR spectroscopy.

15 Predict the number of peaks you would observe in the carbon-13 NMR spectra of the following molecules:

a octane

b cyclohexane

c but-1-ene

d 2-methylpentane.

16 What are the advantages of HPLC for the analysis of drugs in body fluids, compared to an analysis technique based on column chromatography?

Application and analysis

17 A herbal tea extract was analysed by HPLC. The chromatogram obtained is shown below.

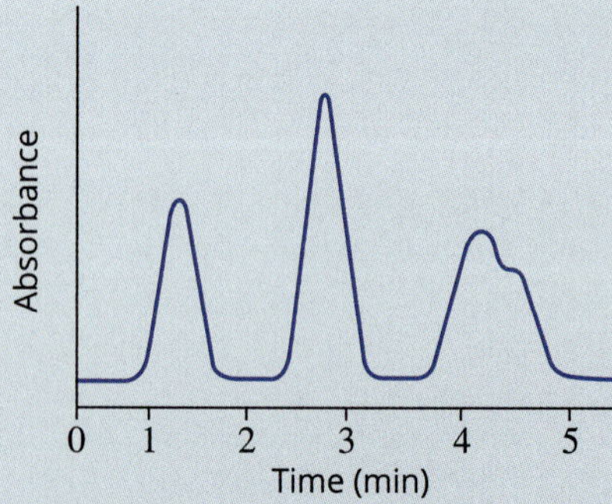

a How many components are evident in the mixture?

b Briefly explain how the components are separated by HPLC.

18 Scientists analysed a common soft drink to determine the caffeine ($C_8H_{10}N_4O_2$) content. They ran a sample of the soft drink through HPLC to separate the components. During the analysis, the scientists observed the peaks on the chromatograph and collected the eluent corresponding to each peak. The eluent was then analysed on a mass spectrometer. Answer the following questions using the HPLC chromatogram and mass spectra below.

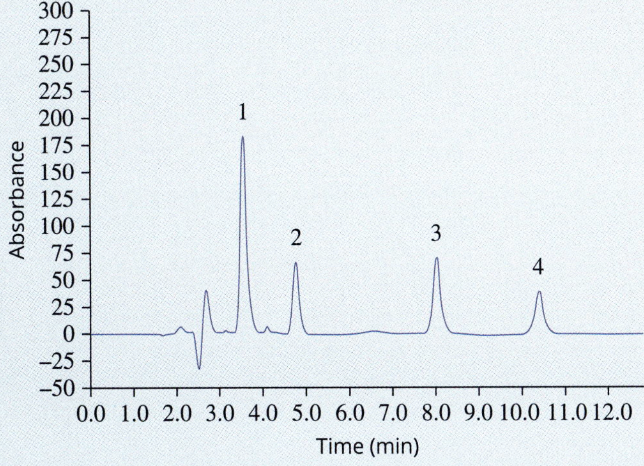

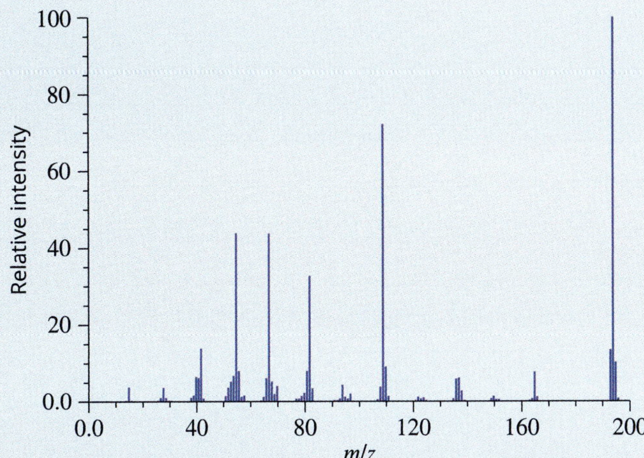

Mass spectrum of soft drink component identified as peak 1

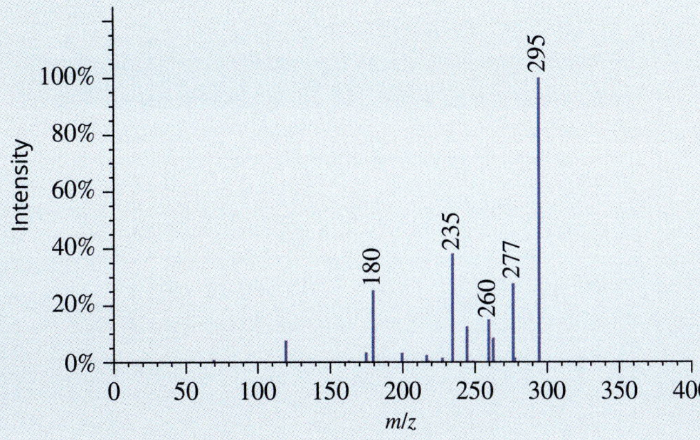

Mass spectrum of soft drink component identified as peak 2

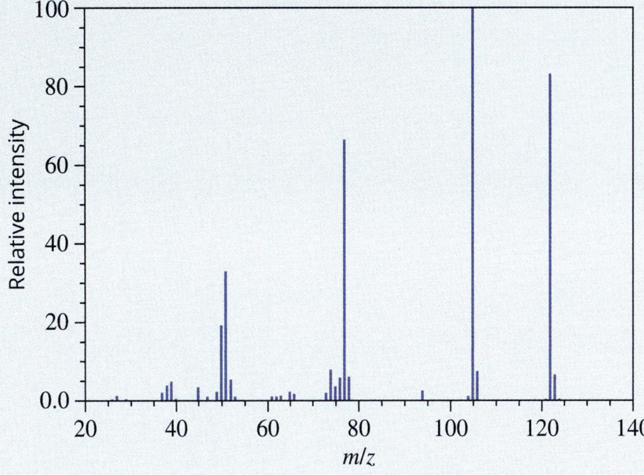

Mass spectrum of soft drink component identified as peak 3

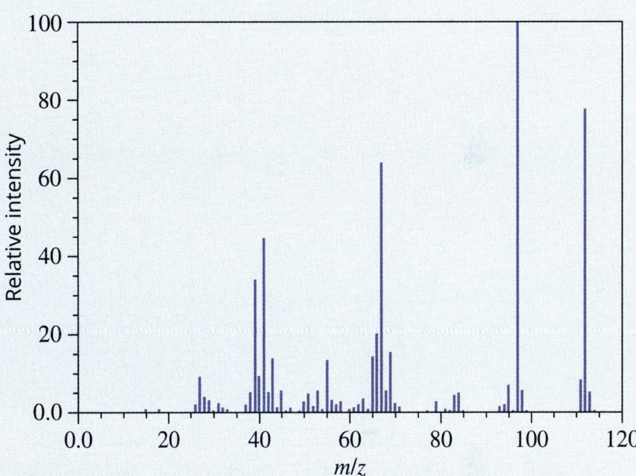

Mass spectrum of soft drink component identified as peak 4

a List the retention times for all four peaks.

b Determine which retention time represents caffeine by investigating the mass spectra. (Hint: Find the molecular ion peak on each mass spectrum.)

c How can the scientist confirm whether that retention time represents only pure caffeine or if there is another molecule with similar polarity to caffeine?

19 Draw a table to predict the expected proton NMR spectrum of each of the molecules listed below. Include the expected chemical shift (refer to Table 14.3.1, page 523), relative peak area and splitting pattern.

a propane

b CH_3CH_2Br

c CH_3COOCH_2OH

20 The IR spectra of two liquid samples, A and B, are provided in the figure below.

A

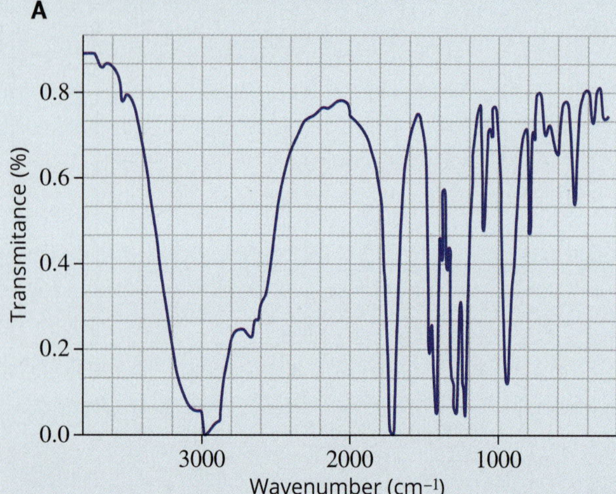

B

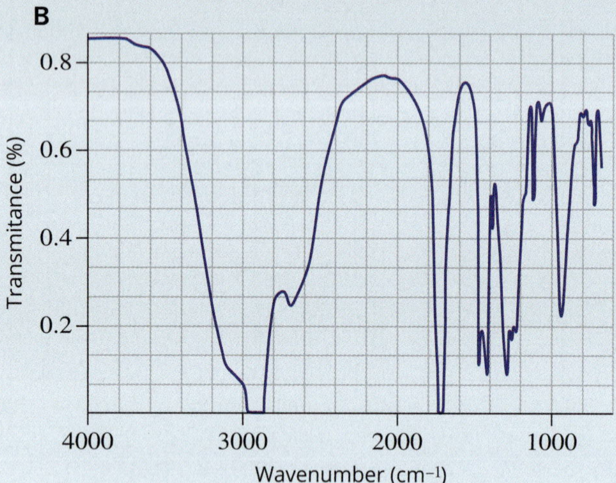

a Identify the feature/s of these two spectra that supports the identification of both samples as carboxylic acids.

b Explain whether these two spectra are of the same compound.

21 Reversed-phase HPLC uses a polar mobile phase and a non-polar stationary phase. A mixture of the amino acid molecules serine, threonine and valine was analysed by reversed-phase HPLC. List the amino acids in the order that they would reach the detector.

serine (Ser)

threonine (Thr)

valine (Val)

22 A chemist working for a sports anti-doping agency analyses the testosterone levels of an athlete's blood by HPLC.

a Outline the steps that the scientist would undertake in the analytical process.

b The chemist obtained the results shown in the table below for four testosterone standards and the athlete's blood sample. The concentrations of the standards were measured in nanograms per litre. Construct a calibration graph for testosterone and hence determine the testosterone concentration in the athlete's blood.

Concentration (ng L^{-1})	Peak area (mm^2)
20.0	0.12
40.0	0.22
60.0	0.36
80.0	0.46
blood sample	0.28

23 Automatic detectors that use IR spectrometry are often employed by security guards and police officers to search for explosives and illicit drugs. They are linked to a computer that can tell if an illegal substance is present and identify what it is. Explain how these automatic detectors are able to detect and confirm if a sample contains an illicit substance.

24 Ethyl methanoate, methyl ethanoate and propanoic acid each have the formula $C_3H_6O_2$. Match each of these compounds with the following proton NMR spectra.

a

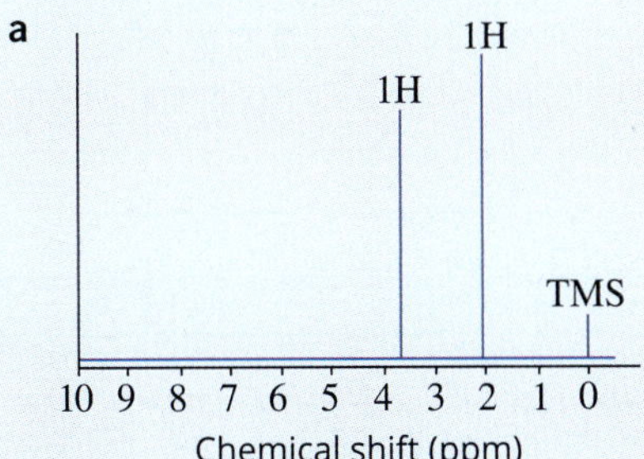

b

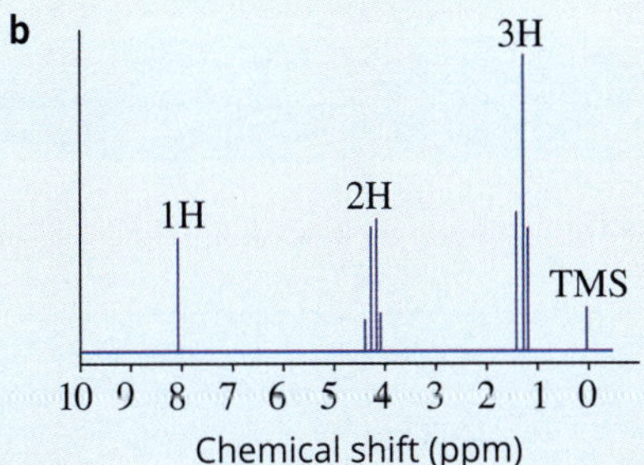

c

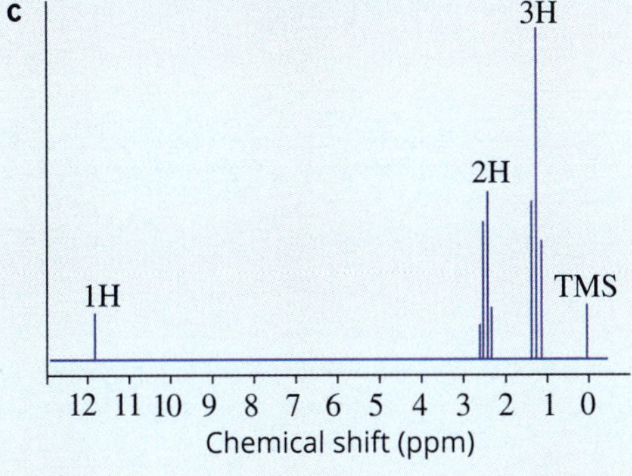

25 An unidentified organic compound is composed of 48.6% carbon, 8.2% hydrogen and 43.2% oxygen. Use this information and the spectra below to determine the IUPAC name of the compound.

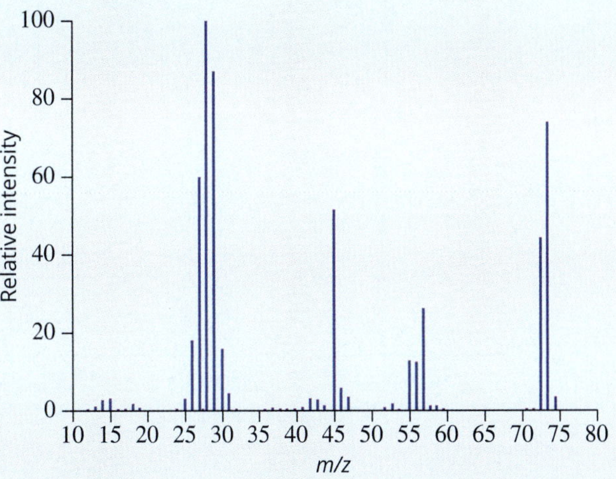

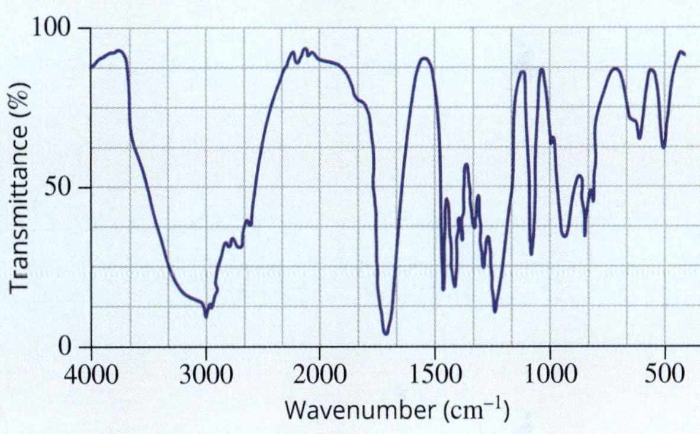

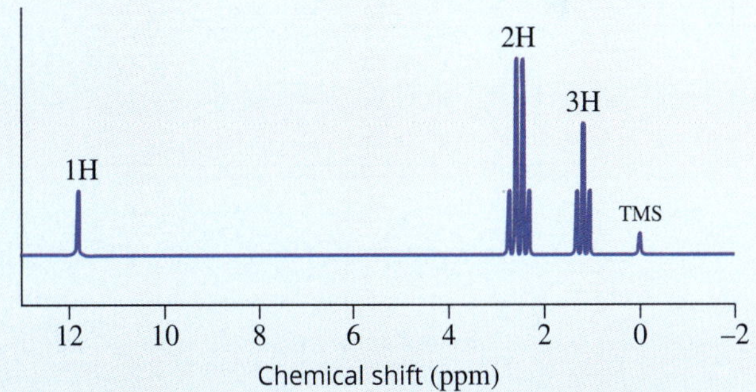

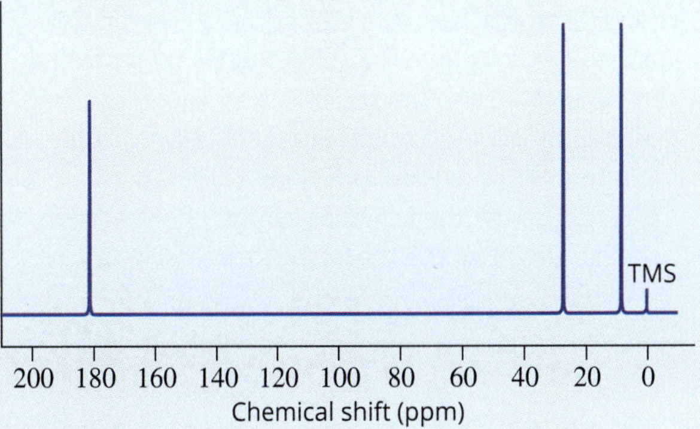

26 Compound X has the molecular formula $C_5H_{11}Cl$. Determine the IUPAC name of compound X using the spectra in the figures below.

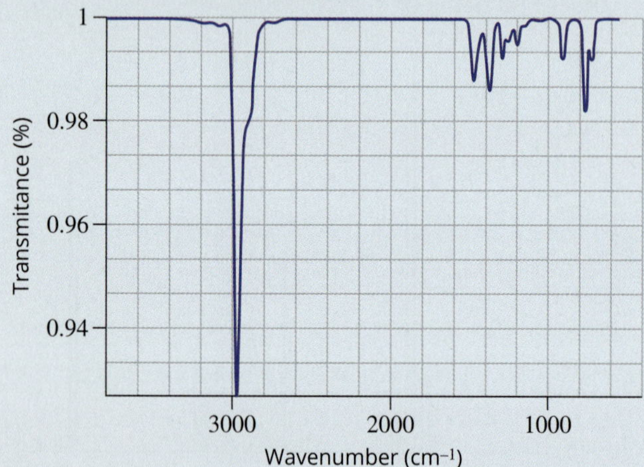

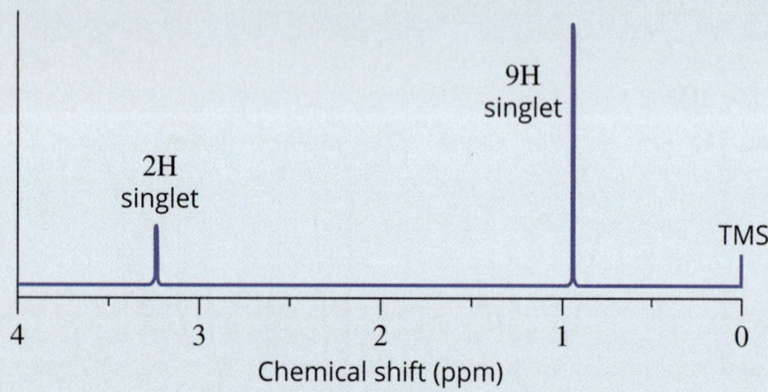

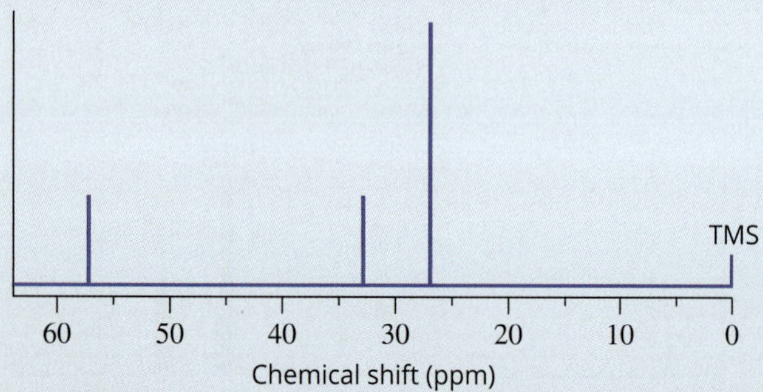

27 Australia will not accept food imported from overseas if a specific organic compound that causes birth defects (compound X) is detected. A sample of an imported food item was sent to a laboratory, prepared and analysed using an HPLC-MS to determine whether it contains that compound.

a Explain why the scientist must analyse a reference sample of compound X on the HPLC-MS.

b Explain why the scientist cannot analyse the food sample using the mass spectrometer only to determine the presence of compound X.

28 An unknown substance is a precursor to the common artificial sweetener aspartame. It has molecular formula $C_4H_7NO_4$. Determine its structure using the following information in the table of 1H NMR data and the ^{13}C spectrum below. The IR spectrum has a broad band around 3100 cm^{-1} and a strong sharp band at 1690 cm^{-1}.

Hint: Hydrogen atoms bonded to the nitrogen do not split the neighbouring hydrogen atoms in the proton NMR spectrum.

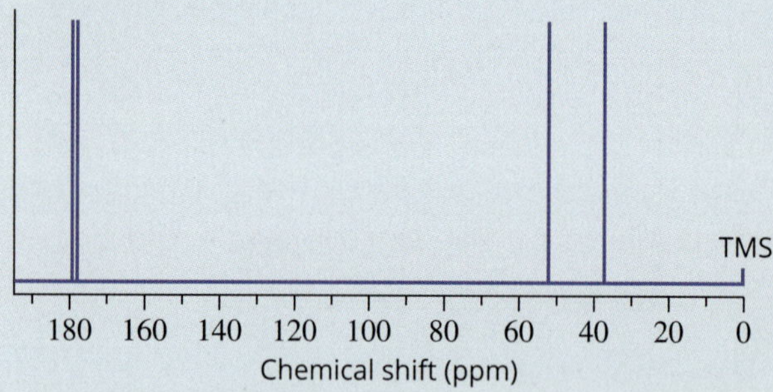

Chemical shift (ppm)	Peak splitting	Relative peak area
2.7	doublet (2)	2
3.8	triplet (1)	1
8.8	singlet (1)	2
12.6	singlet (1)	1
13.9	singlet (1)	1

OA
✓✓

CHAPTER 15 Medicinal chemistry

In this chapter you will be introduced to some common medicines to gain an appreciation of the complex structures and functional groups of molecules with medicinal properties. Many of these molecules can be extracted from plants. You will learn about the procedures used to extract and isolate active ingredients in plants.

An understanding of medicines requires an understanding of the action of enzymes in the body. You will learn about enzymes and how their structures facilitate the important functions they need to perform in the body. You will explore how temperature and pH impact the optimal performance of an enzyme and how scientists can harness their knowledge of how enzymes function. You will also learn about the importance of optical isomers in the mechanism by which a drug interacts successfully inside living things.

Key knowledge

- extraction and purification of natural plant compounds as possible active ingredients for medicines, using solvent extraction and distillation **15.1**
- identification of the structure and functional groups of organic molecules that are medicines **15.1**
- significance of isomers and the identification of chiral centres (carbon atom surrounded by four different groups) in the effectiveness of medicines **15.4**
- enzymes as protein-based catalysts in living systems: primary, secondary, tertiary and quaternary structures, and changes in enzyme function in terms of structure and bonding as a result of increased temperature (denaturation), decreased temperature (lowered activity), or changes in pH (formation of zwitterions and denaturation) **15.2, 15.3**
- medicines that function as competitive enzyme inhibitors: organic molecules that bind through lock-and-key mechanism to an active site preventing binding of the actual substrate **15.4**

15.1 Medicinal ingredients from plants

We all know that humans are dependent on plants for food, but our dependency on plants for medicines is not as well known. The use of plants as medicines predates written human history and was not limited to any one civilisation, with people in regions such as China, the Middle East and Australia harnessing very different plants for a variety of medical conditions. Even today, it is estimated that 50% of all drugs in clinical use in the world are derived from plants. If anything, alternative medicines and herbal remedies are regaining acceptance in Western societies, as evidenced by the statistic that between 50 000 and 80 000 flowering plants are used for some form of medical treatment.

HISTORY OF PLANT MEDICINES

There is a long history of the use of medicinal plant extracts by humans. Some of the documented examples are as follows.

- Salicylic acid, used for pain relief and blood-thinning. Salicylic acid was first identified by Hippocrates, the Greek medical practitioner considered to be the founder of medicine. He realised that the white powder derived from the bark of the willow tree could alleviate pain and fever. It is only a one-step process to convert salicylic acid into the more commonly used acetylsalicylic acid (aspirin) powder.
- Cannabis, used to treat arthritis and pain. The use of cannabis for pain relief originated in central Asia or Western China many centuries ago. Although it is commonly used as a recreational drug, cannabis is now registered in most countries as a medicine. The chemical structure of the active ingredient can be seen in Figure 15.1.1. Interestingly, the Cann Group, the first company to be issued with a cannabis research licence in Australia, harvested Australia's first approved medicinal cannabis crop in Mildura in 2022. The facility is developing the technology to extract compounds called cannabinoids from the plants to make oils, tinctures and capsules.

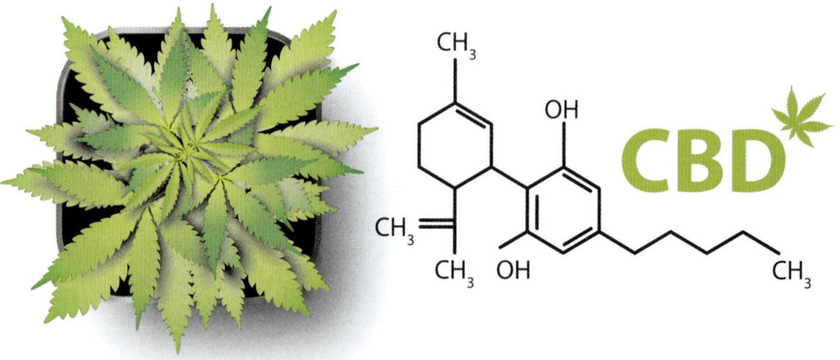

FIGURE 15.1.1 The leaves of the cannabis plant and the chemical structure of cannabidiol, CBD, one of the active ingredients in medicinal cannabis

FIGURE 15.1.2 The fleshy seed pods of the opium poppy contain morphine derivatives.

- Morphine, used for pain relief. Documented evidence of the use of morphine for pain relief in the Middle East dates back as far as 3400 BCE. The pods containing the oil-laden seeds of the poppy can be seen in Figure 15.1.2. Morphine can be synthesised from opium.
- Yew tree extracts, long used in Europe by druids and Celtic tribes for increased fertility and longevity. Although general yew extract in now considered to be toxic, one component, sold as Taxol, has been isolated and found to be very effective at fighting particular cancers.

- Kangaroo apples (bush tomato). Just one of many plants known to Indigenous Australians as having medicinal properties. Kangaroo apples contain a natural anti-inflammatory steroid which aids the production of cortisone and is good for treating achy joints and wounds, as well as encouraging skin rejuvenation. Interestingly, the use of kangaroo apple is complex as the unripe fruit is bitter and highly toxic. It is only safe to consume once it has ripened. Kangaroo apples are now grown in Europe where extracts from them are used in the preparation of cortisone and other steroids. Kangaroo apple use was widespread in drier regions of Australia, including by the Wotjobaluk people of Western Victoria.
- Digitalis, or digoxin, used to treat heart irregularities. In 1775, the Scottish doctor William Withering noticed a patient making a miraculous recovery after using a herbal remedy supplied by a local gypsy. The active ingredient in the foxglove plant is digitalis. The flowers of the plant, and the structure of the active ingredient, are shown in Figure 15.1.3.

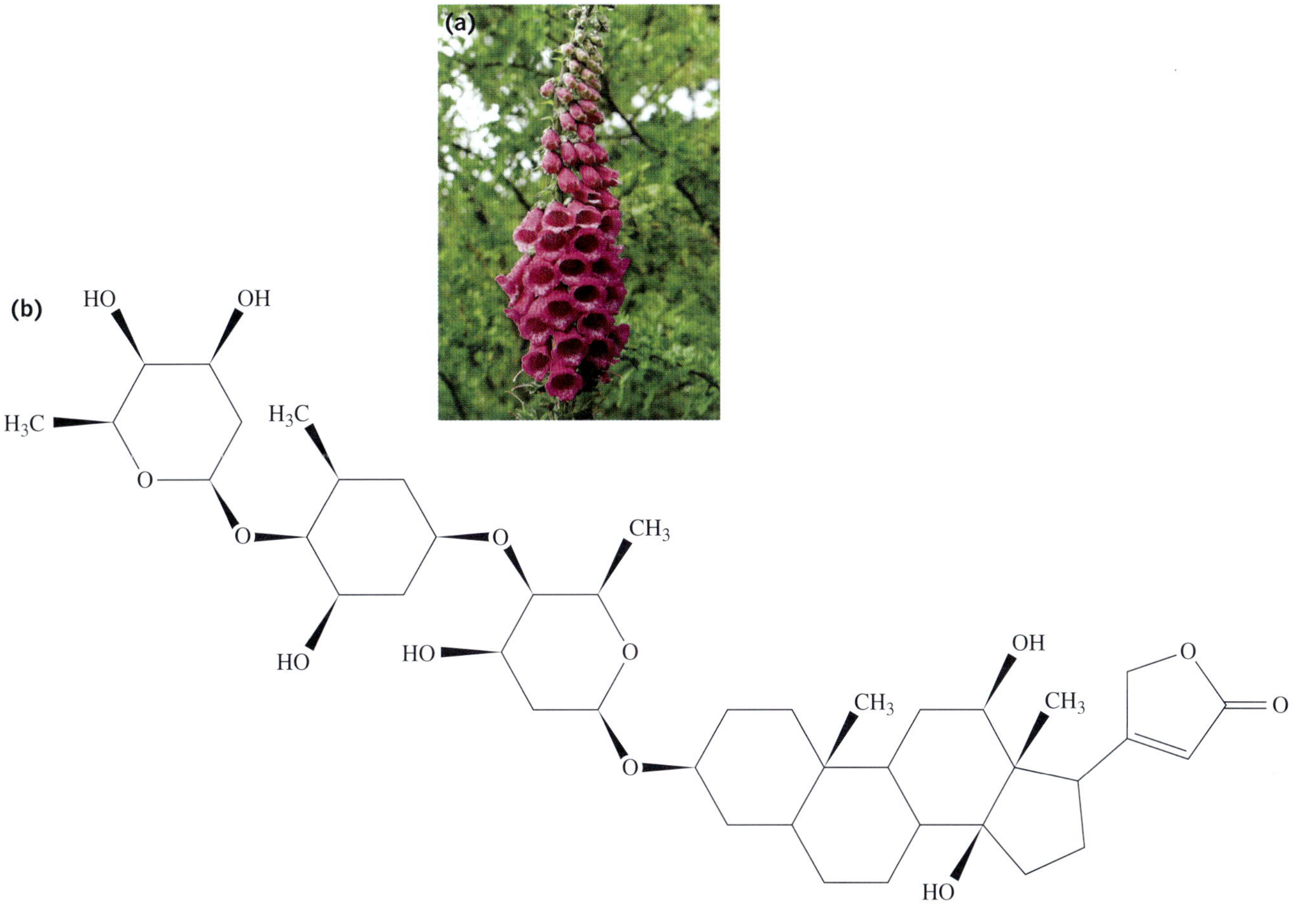

FIGURE 15.1.3 (a) The flowers of the foxglove plant contain an active ingredient, called digoxin ($C_{41}H_{64}O_{14}$), which has the chemical structure shown in (b).

The term **medicinal plant** is used to describe plant extracts that have a biological impact on the body. Many extracts are available in pharmacies in the forms of drugs or ointments. They are used to treat, diagnose and prevent diseases. They perform a wide range of functions, including:

- lowering blood pressure
- reducing cholesterol
- relieving pain
- preventing seizures
- preventing conception
- calming anxiety
- treating skin conditions.

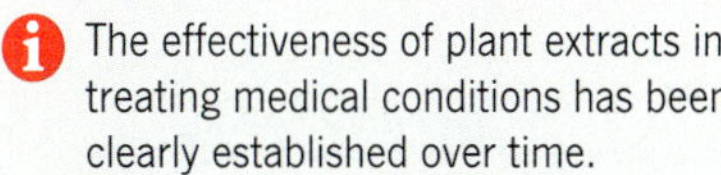
The effectiveness of plant extracts in treating medical conditions has been clearly established over time.

CASE STUDY **ANALYSIS**

Investigating the uncha plant

The uncha plant (*Dodonaea polyandra*) (Figure 15.1.4) has long been used by the Kuuku I'yu people of Cape York Peninsula in Far North Queensland. To sooth toothache, they apply the leaf and stem of this plant to the gums.

In a unique partnership, a research team from the University of South Australia, led by Dr Susan Semple (Figure 15.1.5), has been partnering with the Chuulangun Aboriginal Corporation and the Traditional Custodians of the region to investigate the efficacy (effectiveness) of the active agents of the uncha plant Figure 15.1.6 shows the structures of the two active ingredients identified by the university team.

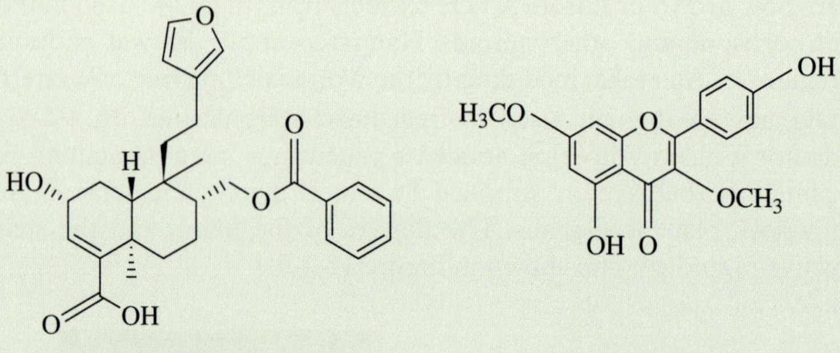

polyandric acid A

kaempferol

FIGURE 15.1.6 The chemical structures of two active ingredients in the uncha plant

FIGURE 15.1.4 Leaves of the uncha plant, which is grown in far north Queensland

FIGURE 15.1.5 David Claudie and Dr Susan Semple have worked together to investigate the efficacy of the active agents of the uncha plant.

The investigation over several years has involved extracting the active ingredients, using a polar ethanol/water blend as a solvent. The two compounds of most interest are the following.

1 Polyandric acid A. Terpenes are natural substances containing the formula $(C_5H_8)_n$. Four different terpenes were isolated by the University of South Australia team. They called the most promising terpene polyandric acid A, as it had never been isolated before. The structure of this molecule is a complex one and it required a full array of infrared spectroscopy, mass spectroscopy and NMR to determine the functional groups present and their arrangement.

   Experiments with mice have proven the effectiveness of polyandric acid A in reducing inflammation in these animals.

2 Kaempferol. Kaempferol is a natural flavour found in a wide variety of plants and vegetables, including kale, spinach, tea and beans. It is a bitter-tasting, yellow, crystalline solid with a melting point around 280°C. It can be extracted from plants using hot ethanol. The kaempferol extracted from the uncha plant has also been shown to have anti-inflammatory properties.

Analysis

1 Name three functional groups that are present in the structure of polyandric acid A.

2 It is unlikely that a pharmacy company will attempt to make the polyandric acid A synthetically. Why is this?

3 A university is sent leaves from a plant that is reputed to have medicinal properties. Explain how the university is likely to start a laboratory-based investigation into this plant.

4 Indigenous medicines vary across different regions of Australia. Suggest reasons for these differences.

EXTRACTION AND PURIFICATION OF ACTIVE INGREDIENTS

To study the components of plants in order to ascertain their potential as medicines, you need to:

- extract the molecules of interest from the plant
- separate the molecules from each other
- determine the structure of each component.

The two most common methods of extracting the components from the plant are described below.

Solvent extraction

When boiling water is poured onto tea leaves, the water draws tannins and other extracts from the solid tea leaves. A coloured solution is formed. This is an example of solvent extraction.

Solvent extraction is a separation process that involves a liquid and a solid. A solid object (the plant) is placed in contact with a liquid (the solvent). The plant components of interest are transferred to the solvent, as illustrated in Figure 15.1.7a. Figure 15.1.7b shows a chemist adding solvent to a flask containing shredded leaves.

Not all extraction is as simple as brewing tea. Very little happens if you pour boiling water onto hemp seeds, in an attempt to extract the hemp oil. Scientists have a range of variations to the extraction processes, depending upon the properties of the substances to be extracted. These variations include:

- grinding or blending the leaves to break up the plant structure
- adjusting the solvent. Water is a polar solvent that will usually be effective on polar molecules. For non-polar molecules, non-polar solvents are used and for low-polarity molecules, solvent blends are often used. The polarity of the solvent should match the polarity of the extract.
- adjusting the temperature. If the desired extracts are thermally stable, prolonged heating will help extraction.

The polarity of the solvent used in solvent extraction should match the polarity of the substance to be extracted.

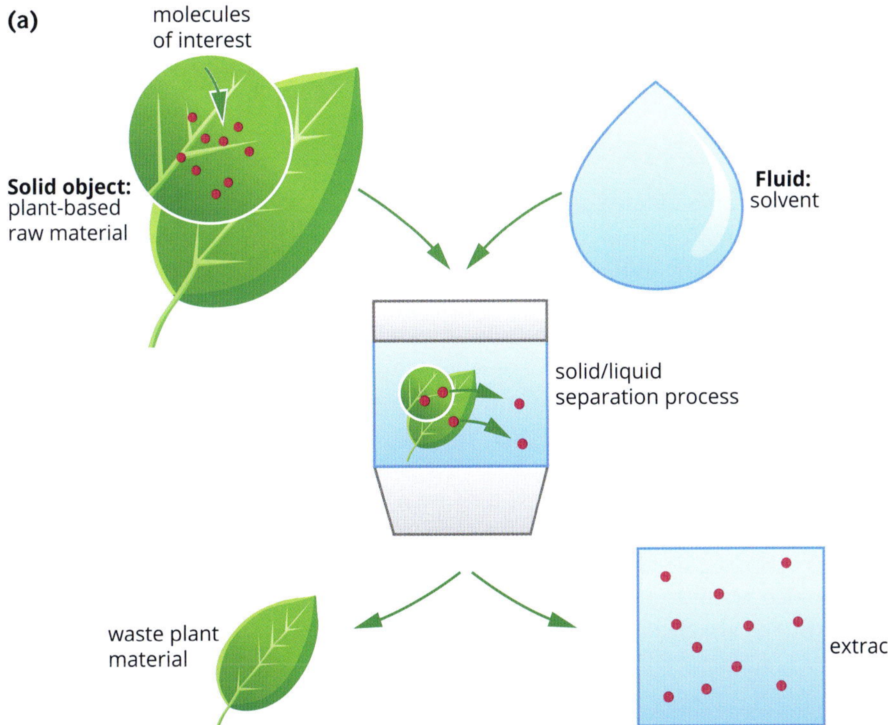

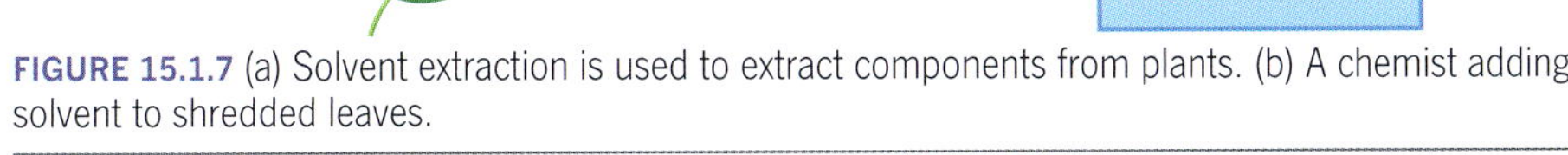
FIGURE 15.1.7 (a) Solvent extraction is used to extract components from plants. (b) A chemist adding solvent to shredded leaves.

> In steam distillation of plants, molecules of interest in the leaves are extracted in a passage of steam.

CHEMFILE

Eucalyptus oil

Soon after the First Fleet arrived in 1788, the Surgeon General of the new colony, John White, noticed that Indigenous Australians had many medical uses for eucalyptus leaves. He arranged for a sample of the oil to be sent back to England to be studied for its olfactory properties. Interest in the oil was high and in 1852 pharmacist Joseph Bosisto opened Australia's first commercial eucalyptus distillery at Dandenong Creek, near Dandenong, Victoria. The oil was exported all over the world. The uses of eucalyptus oil include:

- as a nasal decongestant
- treatment for asthma
- for respiratory illnesses
- as a disinfectant and cleaning agent
- treatment of skin conditions.

The Bosisto factory in Oakleigh and the Euca factory in Altona still produce eucalyptus oil and other aromatic oils in Victoria. The companies use plantations of Blue Mallee trees in Inglewood and other locations in central Victoria, as these trees contain the highest content of eucalyptus oil. The figure below shows a large boiler used to produce the steam to pass through the leaves.

A steam boiler in Inglewood

Steam distillation

Another process used to extract medicinal compounds from plants is **steam distillation**. In this process, a boiler is used to produce a flow of steam which is then passed through the leaves. The hot steam breaks down the plant cells and carries plant oils with it. The steam and the oils are then condensed. The oils are usually non-polar, so are easily separated from the water as they form a layer on top. A laboratory steam distillation unit is illustrated in Figure 15.1.8.

This technique can be used when the active ingredients are volatile and thermally stable. Steam distillation is used to produce tea-tree oil, eucalyptus oil and most of the aromatic oils sold commercially in Australia. Plant oils produced in this way are not pure substances; they are a mixture of many components.

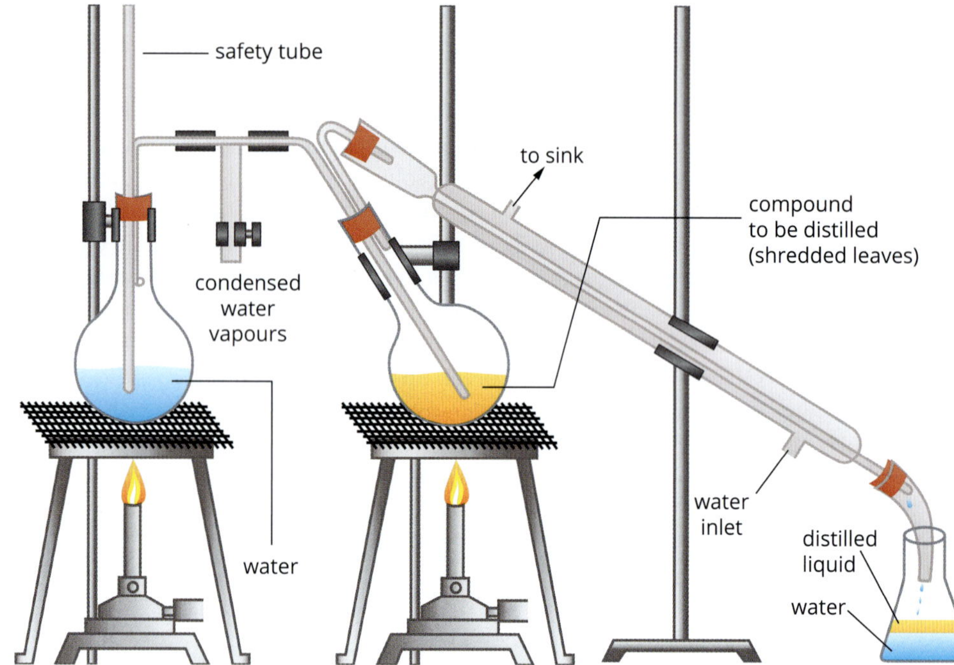

FIGURE 15.1.8 Steam passing through shredded leaves draws some of the oils from the plant. A mixture of water and oil (the distilled liquid) is condensed from the apparatus.

IDENTIFICATION OF STRUCTURE AND FUNCTIONAL GROUPS

The extracts obtained from plants using solvent extraction or steam distillation usually contain many different compounds. Chromatographic techniques can be used to separate these mixtures into their individual components. Given the complex structures of these components, determining their structures is not an easy task and requires a range of the instruments, which you learnt about in Chapter 14.

As an example of how the structure of a medicine might be determined, the various spectra for aspirin (2-acetoxybenzoic acid) are shown in Figure 15.1.9 on the following page. Its structure is a relatively simple one (Figure 15.1.10 on the following page), yet the spectroscopy data is still complex to interpret.

The mass spectrum in Figure 15.1.9 shows a very small molecular ion peak with a m/z ratio of 180, confirming the molecular formula of aspirin to be $C_9H_8O_4$. To the trained chemist, the fragmentation pattern would also suggest the presence of other groups attached to the carbon ring.

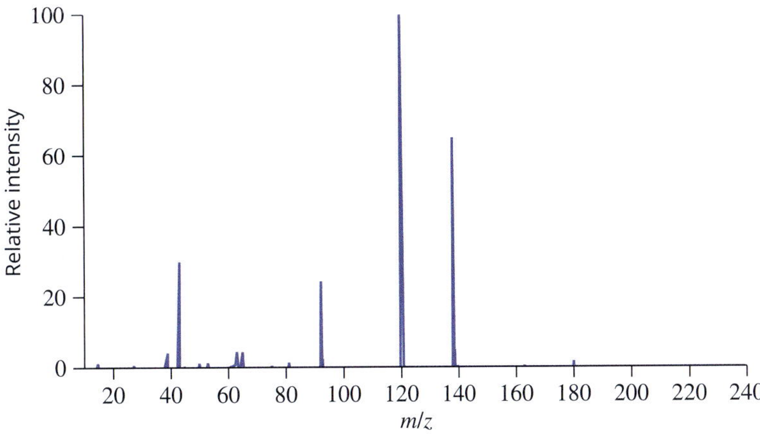

FIGURE 15.1.9 Mass spectrum of aspirin

Figure 15.1.10 shows a proton-NMR spectrum for aspirin and the structure of aspirin. The enlargement on the spectrum shows the splitting pattern for the peaks between chemical shifts of 7.0 and 8.2 ppm. This pattern allows chemists to determine the environments of the hydrogen atoms bonded to the carbon ring. The numbering of the peaks is matched to the structure of aspirin to show which environment is responsible for each peak. Note that there are no hydrogen atoms on the carbons numbered 1 and 6.

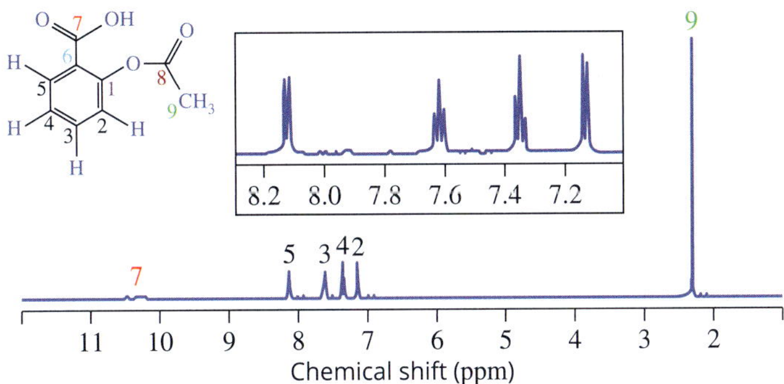

FIGURE 15.1.10 Proton-NMR of aspirin. The NMR spectrum of aspirin has six peaks.

The infrared spectrum, shown in Figure 15.1.11, is useful for confirming the functional groups in aspirin, such as:

- the broad absorption around 3000 cm^{-1}: –OH bond (carboxylic acid)
- sharp absorption at 1750 cm^{-1}: C=O bond of an ester
- sharp absorption at 1700 cm^{-1}: C=O bond of a carboxylic acid
- sharp absorption at 1550 cm^{-1}: C–O– bond of an ester

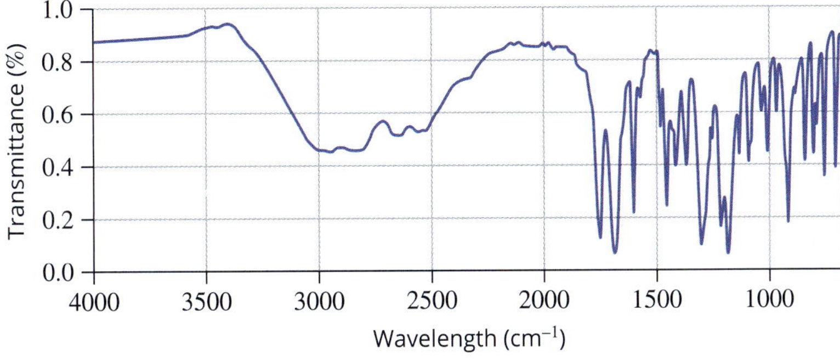

FIGURE 15.1.11 Infrared spectrum of aspirin

15.1 Review

SUMMARY

- The medicinal properties of plant extracts have been utilised for centuries by many different civilisations.
- Solvent extraction uses a knowledge of polarity to extract biologically active molecules from leaves.
- Steam distillation can be used to extract volatile, thermally stable active ingredients from plants.
- Once biologically active molecules have been extracted, they need to be separated and identified.
- Chemical instruments, such as mass spectrometers, infrared spectrometers and NMR spectrometers, are useful in determining the structure of a plant extract and identifying the functional groups present.

KEY QUESTIONS

Knowledge and understanding

1 Some active ingredients are difficult to extract from a plant. List two methods used by chemists to improve extraction and explain why they work.

2 Investigation of the structure of plant extracts nearly always requires some form of chromatography. Explain why.

3 Isolating an active ingredient is not the end of the process of the commercial production of a new pharmaceutical. List three other factors that need to be considered.

Analysis

4 The structures of cannabidiol, CBD, is shown in Figure 15.1.1 (page 556) and the structure of digoxin is shown in Figure 15.1.3b (page 557).
Referring to these two structures, discuss how a solvent is chosen to extract these substances from plants.

5 **a** Some molecules cannot be extracted successfully using steam distillation. Explain why.
b It is relatively easy to separate oils from condensed steam. Give a reason why.

6 Vitamin C is present in relatively high concentrations in citrus fruit. The structure of the vitamin C molecule is shown below.

a State the value of the *m/z* ratio of the molecular ion peak in the mass spectrum of vitamin C.
b List three absorptions likely to be observed in an infrared spectrum of vitamin C, identifying the functional groups to which they correspond.
c A proton-NMR of vitamin C will show several singlet peaks. Explain why.

7 The chemical structure of the common painkiller, paracetamol, is shown below.

a State the value of the *m/z* ratio of the molecular ion peak in the mass spectrum of this molecule.
b List three absorptions likely to be observed in an infrared spectrum of paracetamol, identifying the functional groups to which they correspond.
c Paracetamol is highly soluble in ethanol. Discuss the likely factors that determine the level of solubility.

15.2 Proteins

In Chapter 12 you learnt that proteins are natural polymers made from amino acids. They have many roles in the body, including:

- hormones that control biological processes
- structural components in cell membranes, muscles, hair and feathers
- antibody molecules of the immune system
- transport of substances across cell membranes or around the body
- enzymes that catalyse specific biochemical reactions.

An understanding of the properties of proteins is important to an understanding of the mechanism by which many medicinal substances work in the human body. To achieve this, you need to first understand the various levels of protein structure, classified as primary, secondary, tertiary and quaternary structures.

PRIMARY STRUCTURE OF PROTEINS

The number, type and sequence of the amino acid units in a protein are collectively known as the protein's **primary structure**. The condensation reactions that lead to the primary structure of proteins were described in Chapter 12. A segment of a protein molecule known as glycinin (Figure 15.2.1) shows that amino acids are joined to each other by peptide links to form a polypeptide chain.

The function of a protein is a consequence of its shape. Therefore, the order of amino acids in a protein is extremely important in determining its function.

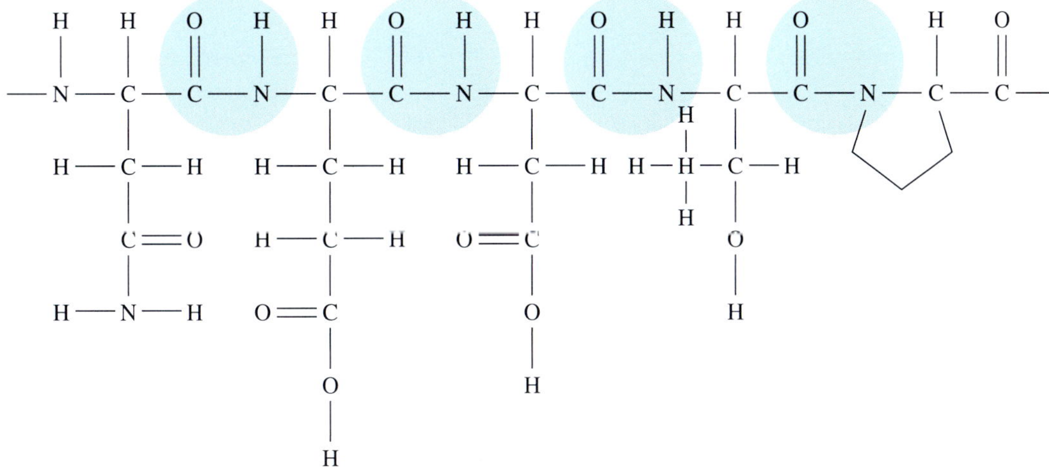

FIGURE 15.2.1 The amino acid sequence that is found in glycinin, which is a protein found in soybeans. The peptide links (–CONH–) that connect the amino acid residues are highlighted.

Each protein has a precise chemical composition and amino acid sequence, which leads to it having a unique three-dimensional shape.

SECONDARY STRUCTURE OF PROTEINS

Coiling and **pleating** of sections of a protein molecule produce a secondary level of structure in a protein. Hydrogen bonds between the polar –NH group in one **peptide link** and the polar –C=O group in another peptide link can form at regular intervals (Figure 15.2.2).

This creates regions in which the molecule coils into a spiral shape called an **α-helix**, or where sections line up parallel to each other, forming a **β-pleated sheet**.

Such highly ordered segments, stabilised by hydrogen bonds, are referred to as the **secondary structure** of the protein.

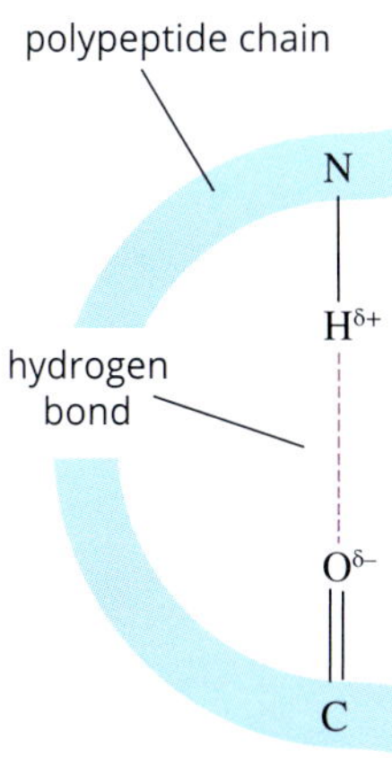

FIGURE 15.2.2 Hydrogen bonds can form between the polar –NH group in one peptide link and the polar –C=O group in another peptide link further along the polypeptide chain.

α-helices

Keratin is a protein found in the fibres of hair and wool. The helical structure of keratin results from extensive hydrogen bonding between peptide links in the polypeptide chain. In this case, the hydrogen bonds result from the attraction between the partial positive charge on the H of a –NH group in a peptide link with the partial negative charge on the O of a –C=O group of a peptide link four amino acid units along the chain.

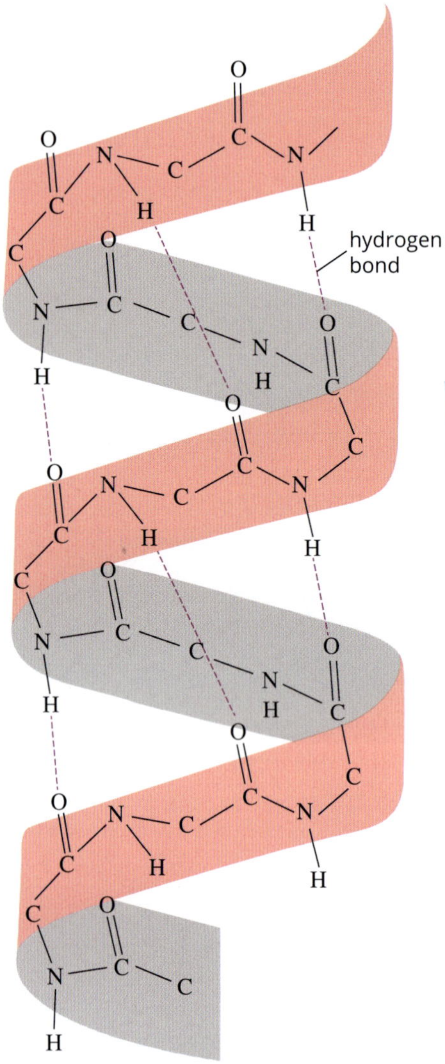

The hydrogen bonds make the molecule coil into the shape of an α-helix, the same shape as a spring (Figure 15.2.3).

FIGURE 15.2.3 A polypeptide chain coils in an α-helix due to hydrogen bonds. The R groups (side chains) and hydrogen atoms in the amino acid residues have been omitted for clarity.

β-pleated sheets

Hydrogen bonds can also form between peptide links to produce regions where two or more parts of the polypeptide chains line up parallel with each other. The repeating structure of the backbone of the protein chain (–N–C–C–N–C–C–N–C–C–) allows these hydrogen bonds to form at regular intervals. This stabilises the protein structure. This type of secondary structure in the protein is called a β-pleated sheet (Figure 15.2.4).

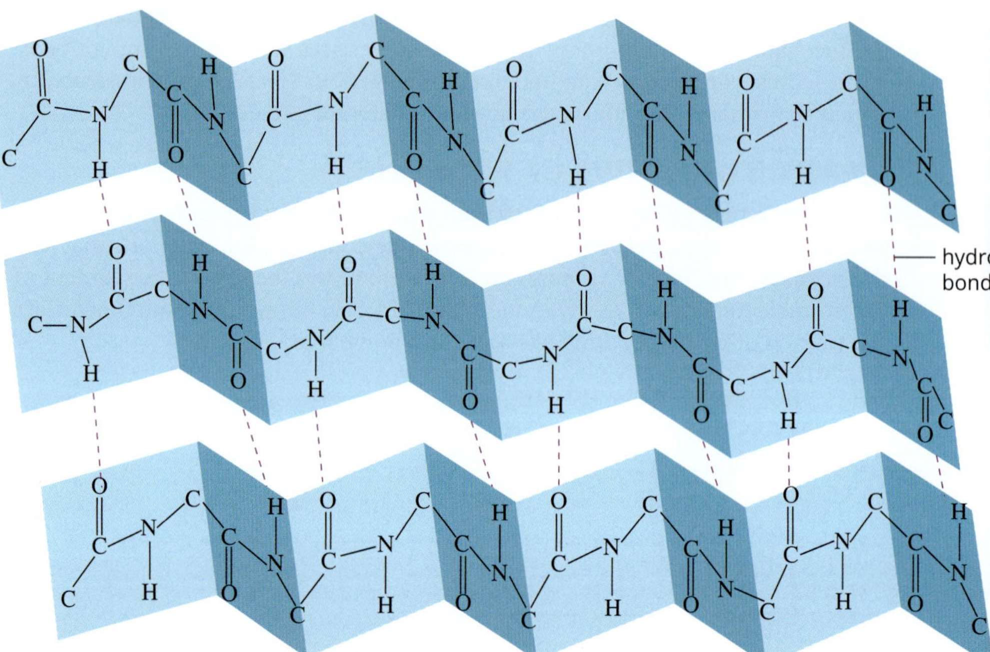

FIGURE 15.2.4 β-pleated sheet. Hydrogen bonds between the sheet are shown as dashed lines. The R groups (side chains) and hydrogen atoms on the α carbons have been omitted for clarity.

Silk is a protein with a β-pleated sheet structure. The polypeptide chains mainly contain the amino acids glycine, alanine and serine, (Figure 15.2.5). Notice that every second **R group** (side chain) is an H atom. With only these small side chains, sections of the protein molecule in silk can line up closely, enabling hydrogen bonds to form between these adjacent sections and producing β-pleated sheets.

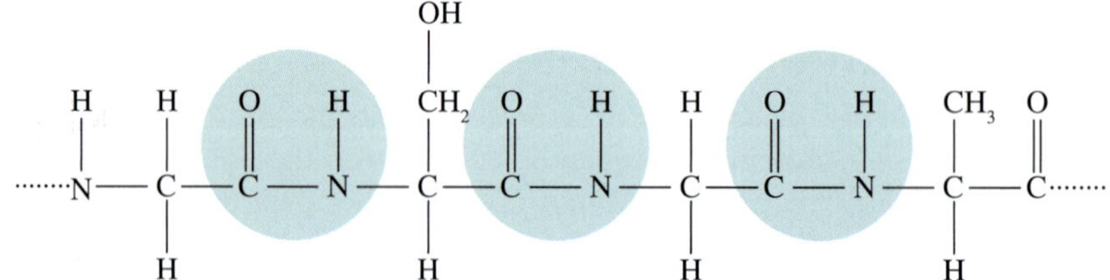

FIGURE 15.2.5 The amino acid sequence that is found in large parts of the polypeptide of silk. Peptide links are shaded blue.

Figure 15.2.6 shows that the R groups of alanine and glycine form a regular pattern, giving silk its strength and texture.

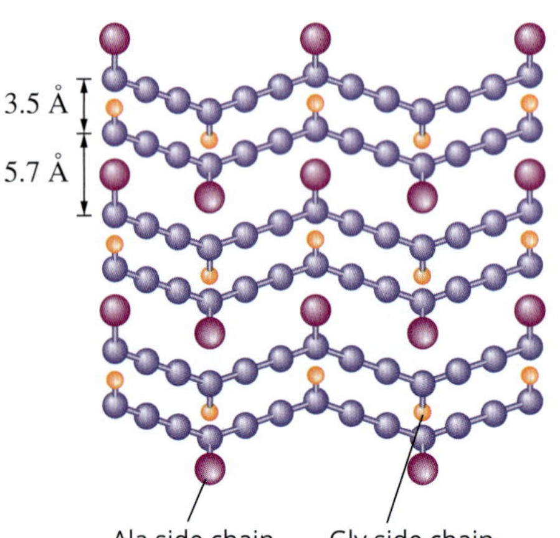

FIGURE 15.2.6 The alanine and glycine R groups align in the β-pleated sheets in silk.

The secondary structure of a protein results from hydrogen bonding within different regions of the amino acid sequence. This leads to the formation of α-helices or β-pleated sheets.

THE ACID-BASE PROPERTIES OF AMINO ACIDS

To understand the overall shape of a protein and the effect of pH on this shape requires an understanding of the acid–base properties of the amino acids which are the building blocks of proteins.

Amino acids contain polar amine and carboxyl functional groups. Therefore, amino acids can form hydrogen bonds with water molecules and are soluble in water. In solution:

- the $-NH_2$ group can act as a base, accepting a proton to become a $-NH_3^+$ group
- the $-COOH$ group can act as an acid, donating a proton to become a $-COO^-$ group.

Figure 15.2.7 shows that an amino acid can exist in an aqueous solution as a dipolar ion, known as a **zwitterion**. The relatively high melting point of pure crystalline amino acids is due to the zwitterion being present in the solid state.

In solution, amino acids exist in equilibrium with an alternative form, called a zwitterion, which contains positive and negative sections. The overall charge of a zwitterion is neutral.

The carboxyl group of an amino acid can act as an acid, donating a proton to the amine group.

This alternate form of the amino acid is a zwitterion.

FIGURE 15.2.7 An amino acid can exist in equilibrium with its zwitterion.

The effect of pH on amino acid structure

The dual acidic and basic nature of amino acids means that different chemical forms of an amino acid can be in equilibrium in a solution. The predominant form depends on the pH of the solution and the particular amino acid concerned. Figure 15.2.8 shows the different forms of an amino acid as pH changes.

Intermediate pH

Zwitterion (no overall change)

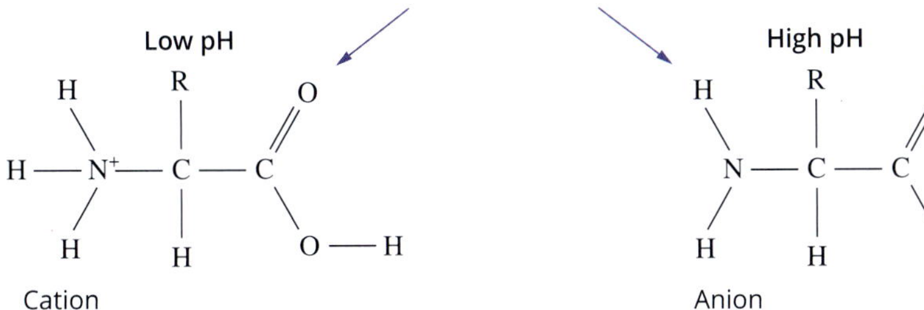

Cation

In acidic conditions (low pH), the presence of excess H^+ ions favours a cation form.

Anion

In alkaline conditions (high pH), the presence of excess OH^- ions favours an anion form.

FIGURE 15.2.8 The impact of pH on the form of the amino acid

CHEMFILE

Modelling protein structure

In computer models of protein structure, a segment twisted into an α-helix is often represented as a twisted ribbon and sections of β-pleated sheets are represented by a set of wide parallel ribbons with arrows.

A computer model of a protein is shown in the figure below.

There are wide variations in the relative amounts of α-helix and β-sheet structures in individual proteins, with many proteins containing several regions of both.

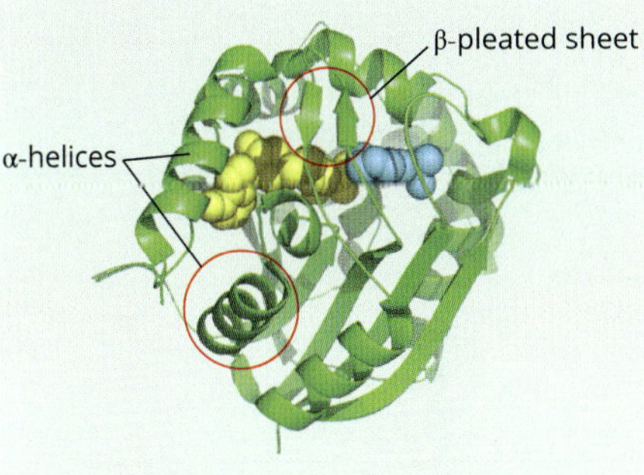

Computer model of a protein (green and yellow) in which several regions of α-helices and β-pleated sheets are seen. The blue part shows drug-like molecules interacting with the protein.

```
O       O⁻
  \\   /
    C
    |
H   CH2    O
 \  |    //
  N — C — C
 /  |     \
H   H      O⁻
```

FIGURE 15.2.9 The structure of aspartic acid at pH 12 showing that it has a charge of 2–

```
          +
          NH3
          |
          CH2
          |
          CH2
          |
          CH2
     H    CH2    O
      \   |    //
H — N⁺ — C — C
      /   |     \
     H    H      O — H
```

FIGURE 15.2.10 The structure of lysine at pH 2 showing that it has a charge of 2+

The effect of the R group on the acid–base properties of amino acids

If the R group contains a functional group with acid–base properties, it is possible for other charged forms of the amino acid to form. For example, at a pH of 12, the predominant form of aspartic acid is an ion with a charge of 2– (Figure 15.2.9). On the other hand, at a pH of 2, lysine exists predominantly as an ion with a charge of 2+ (Figure 15.2.10).

TERTIARY STRUCTURE OF PROTEINS

The overall three-dimensional shape adopted by a protein molecule is called its **tertiary structure**. A tertiary structure is produced by the three-dimensional folding of its secondary structures (α-helices and β-pleated sheets). The protein can twist back over itself to create a unique shape, which is responsible for the protein's function (Figure 15.2.11).

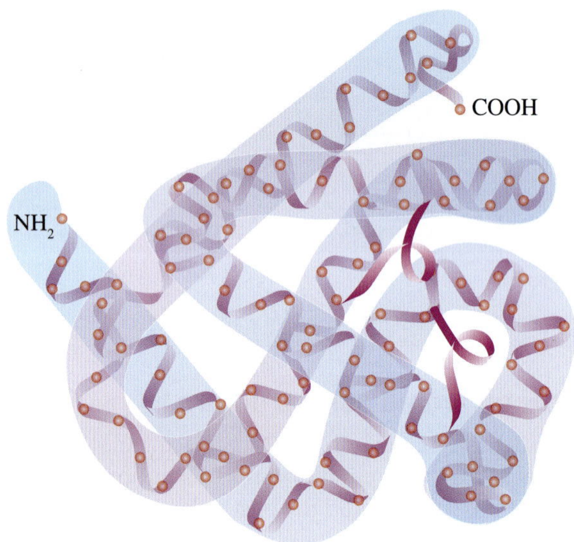

FIGURE 15.2.11 A model representing the tertiary structure of the protein myoglobin

> R groups of amino acids influence the properties of the amino acid. There are four main groups of amino acids, which can be classified as polar, non-polar, acidic or basic.

The side chains (R groups) of the amino acid units making up the polypeptide chain influence the overall three-dimensional shape of the molecule. Not only are some side chains relatively large (such as in phenylalanine), but others also contain polar functional groups, or can become charged depending on the pH of their surroundings. In addition, some amino acids have **hydrophobic** (non-polar) chains, which tend to fold towards the interior of protein molecules, away from contact with water molecules.

Bonding in the tertiary structure of proteins

Five types of attractions that are important in chain-folding are summarised in Table 15.2.1 on the following page, which also identifies the features of the side chains that are involved.

TABLE 15.2.1 Types of bonds formed from interactions of specific R groups in different regions of the protein

Bond type	Required components in R group	Visual representation
Hydrogen bonds	contains –O–H, –N–H or –C=O	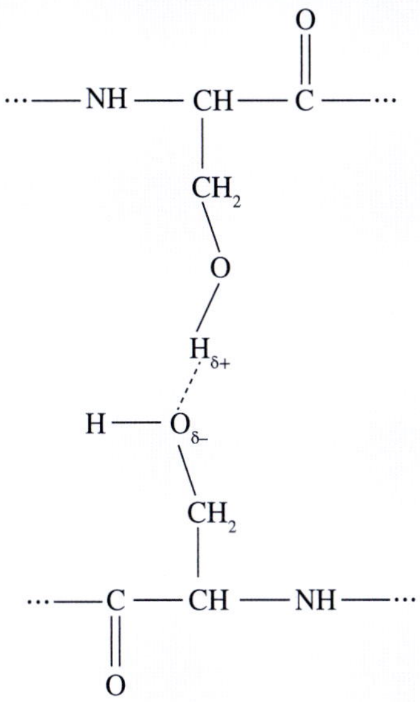 a hydrogen bond linking two parts of a polypeptide chain
Dipole–dipole interactions	any polar group such as those containing –S–H, –O–H or –N–H	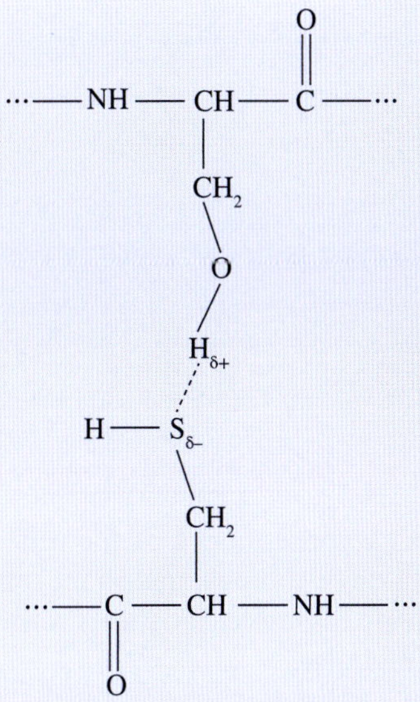 a dipole–dipole interaction linking two parts of a polypeptide chain
Ionic interactions	contains $-NH_3^+$ and another group that contains $-COO^-$	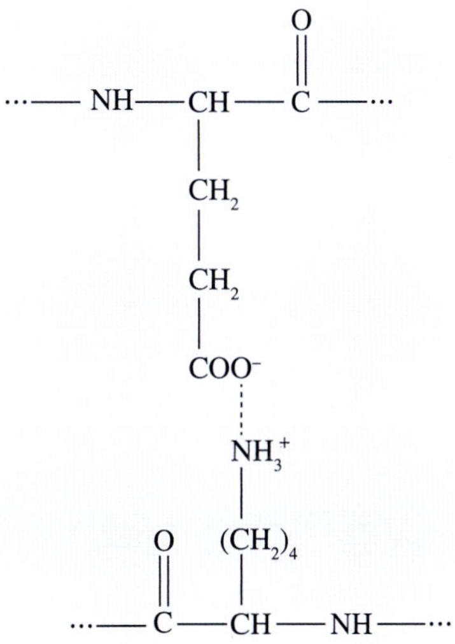 an ionic interaction linking two parts of a polypeptide chain

Covalent cross-links	cysteine side groups react to form a disulfide bridge (–S–S–).	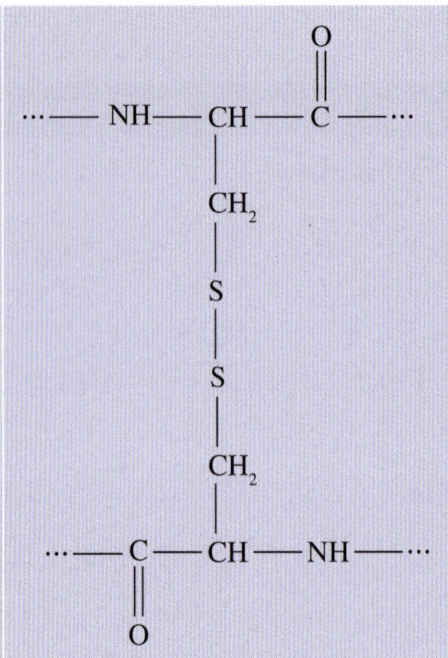 a disulfide bridge linking two parts of a polypeptide chain
Dispersion forces	any non-polar group	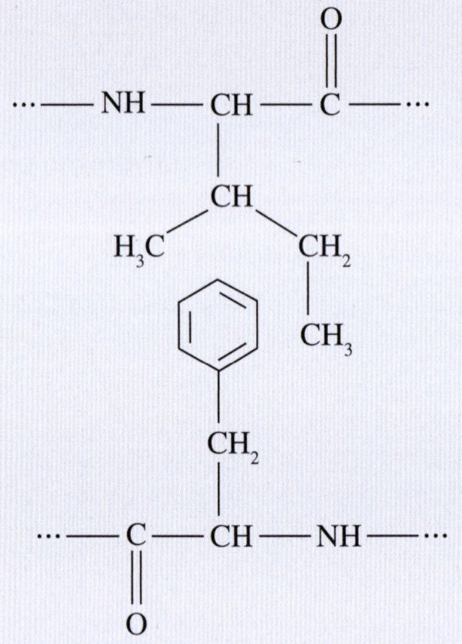 dispersion forces can link two parts of a polypeptide chain.

Interactions between the various R groups on amino acids lead to each protein having a unique three-dimensional structure.

As a consequence of the different types of bonds that can fold a polypeptide into a three-dimensional shape, an enormous variety of protein shapes exist. Some proteins resemble flat sheets, others are long and helical, and others are compact and globular.

QUATERNARY STRUCTURE OF PROTEINS

Some proteins are composed of two or more polypeptide chains, and may even interact with non-protein molecules to produce a larger, more complex functional unit, known as the **quaternary structure**. Haemoglobin is an example of a protein with a quaternary structure.

Haemoglobin—the oxygen-transporting protein

Red blood cells are manufactured by the bone marrow at a rate of about 2 million per second. Red blood cells are functional for about 3 months in the human body. Each red blood cell contains approximately 250 million molecules of haemoglobin.

Figure 15.2.12a on the following page is a diagrammatic representation of haemoglobin. The four distinct subunits that make up its quaternary structure are highlighted in colour. Each subunit contains a polypeptide chain (collectively called globin), which is coiled into α-helices and then folded into a tertiary structure. Within each subunit, there is also an oxygen binding site, or haem group (light blue). Each haem group contains 1 atom of iron (green). The structure of a haem group is shown in Figure 15.2.12b. Haemoglobin transports oxygen around the body by binding one oxygen molecule to each haem group.

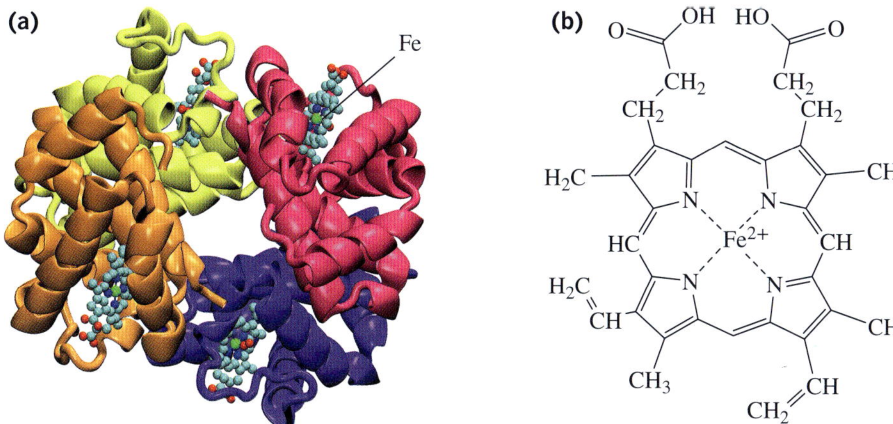

FIGURE 15.2.12 (a) A representation of the structure of haemoglobin. Within each subunit is a haem group (light blue). Each group contains one atom of iron (green). In this model, the haem groups are oxygenated and the oxygen molecules are shown as paired red spheres. (b) Chemical structure of a haem group.

A concise summary of the four levels of protein structure is shown in Table 15.2.2.

TABLE 15.2.2 The four levels of protein structure

Level of bonding	Effect of bonding	Types of bonding
primary	sequence of amino acids in the main chain	covalent bonds (peptide links)
secondary	coiling or pleating of the protein chain	hydrogen bonding between –N–H and –C=O on different parts of the molecule
tertiary	overall three-dimensional shape of the protein	formed by a variety of bond types between the R groups on different parts of the protein molecule.
quaternary	some proteins contain more than one peptide chain and form a more complex unit	dispersion forces or other bonds between molecules

15.2 Review

SUMMARY

- There are 20 different 2-amino acids from which tens of thousands of different protein molecules are synthesised by cells. Each protein has a unique structure and function in the organism.
- The complex structure of these large molecules is often considered in distinct levels. The four levels are of protein structure are summarised below.

Structural level	Description	Bonds involved
Primary	The sequence of amino acids in the polypeptide chain 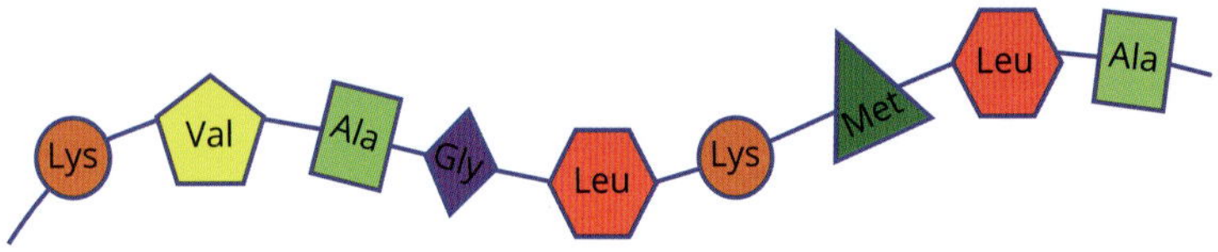	Only covalent bonds are responsible for joining the monomer units together in the polymer. Monomers are linked by –CONH– groups, known as peptide links in proteins.
Secondary	The shape of the polypeptide as it is either twisted into an α-helix or bent back on itself to produce regions of β-pleated sheets 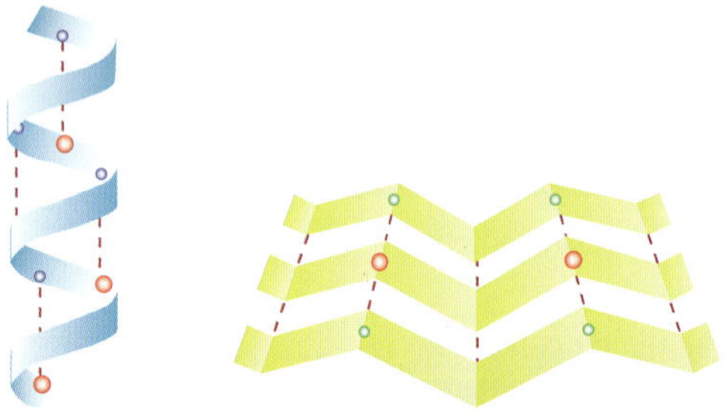 α–helix β–pleated sheet	Hydrogen bonding between –NH and –C=O bonds in peptide links introduces a secondary structure to protein molecules, producing regions of α-helices or β-pleated sheets.
Tertiary	The overall three-dimensional shape adopted by a protein molecule 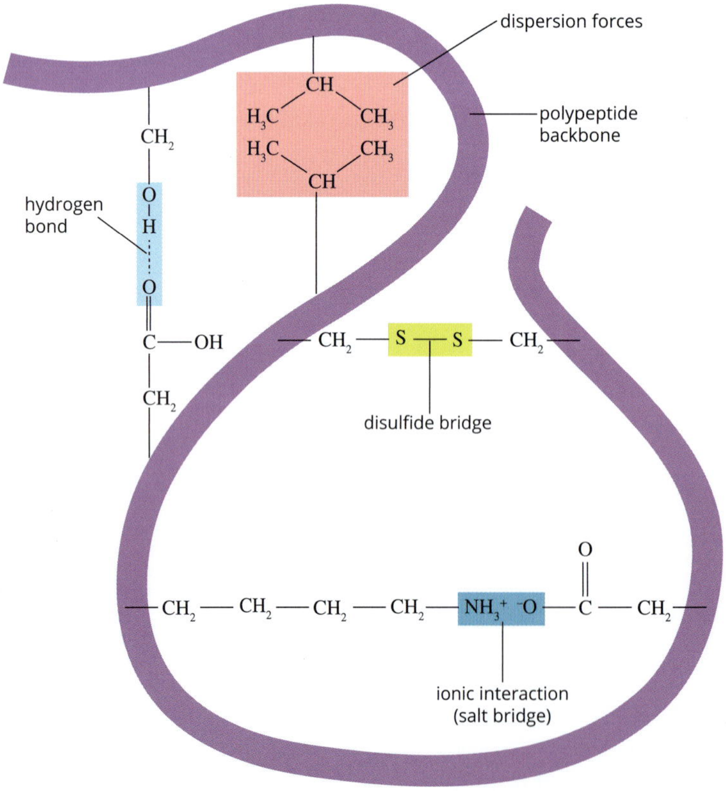	Hydrogen bonding can occur between polar functional groups in side chains, such as –OH and –C=O. Only dispersion forces are present between two non-polar R groups such as $-CH(CH_3)_2$. Cysteine's R groups are able to form a covalent link, known as a disulfide bridge. Charged R groups are attracted by ionic interaction (salt bridge).

Quaternary	Some proteins consist of several associated polypeptide chains. 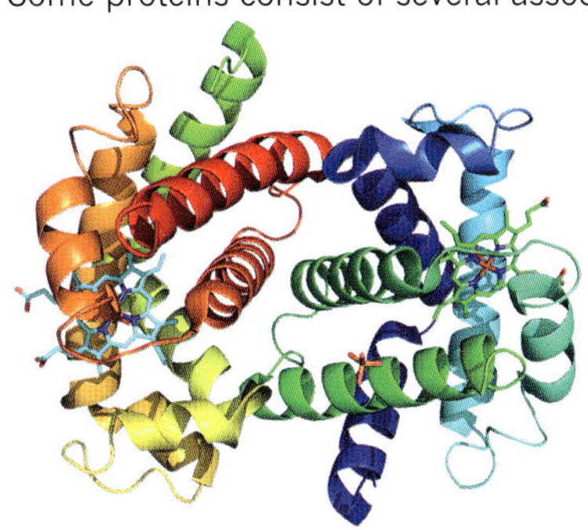	Dispersion forces between non-polar R groups are the major reason for the attraction between adjacent parts of the chains. Dipole–dipole attractions, hydrogen bonds, ionic interactions and disulfide bonds involving R groups may also occur.

- Many bond types produce the three-dimensional shape of proteins:
  - dispersion forces
  - dipole–dipole attractions
  - hydrogen bonds
  - covalent bonds in disulfide bridges
  - ionic interactions.
- Amino acids form a chemical equilibrium in solution with an alternative form known as zwitterions. Zwitterions contain charged functional groups, but have no net charge.
- The form an amino acid takes in solution depends upon the solution pH. Low pH favours the formation of cations and high pH favours the formation of anions.
- Some amino acids contain R groups that can exhibit acid–base properties.

KEY QUESTIONS

Knowledge and understanding

1 Identify the level of protein structure that each of the following relates to.

a the interactions between the R groups on the amino acids

b the attractions between different peptide linkages

c the amino acid sequence

d the complete structure of an operational protein

2 Explain how the bonding that leads to a protein's secondary structure is different from the bonding that leads to its primary structure.

3 **a** Why is the tertiary structure of a protein important?

b Which part of the amino acid influences the tertiary structure?

4 **a** Draw a molecule of alanine and its zwitterion.

b Draw the form the zwitterion will take at

i low pH

ii high pH.

Analysis

5 The following figure shows a section of a protein.

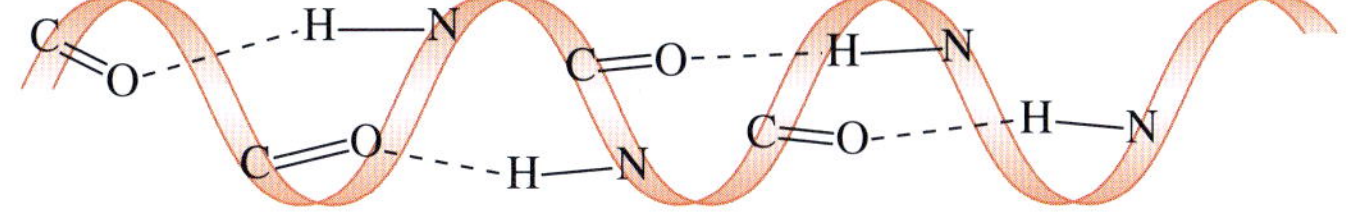

a What type of secondary structure is represented by the diagram?

b How does this type of secondary structure form? Refer to particular atoms and functional groups associated with the bonding that forms this structure.

6 Identify the levels of protein structure in which hydrogen bonds may contribute to the structure. More than one option may be correct.

A primary structure

B secondary structure

C tertiary structure

D quaternary structure

7 Use your understanding of bond polarity and the acid–base properties of functional groups to complete the following table, which categorises properties of an amino acid's R group.

Structure of R group	Is the R group polar or non-polar?	Is the R group acidic, basic or neutral?
$-CH(CH_3)_2$		
$-CH_2COOH$		
$-CH_2C_6H_5$		
$-(CH_2)_4NH_2$		

15.3 Enzymes

Thousands of chemical reactions are involved in sustaining life. In the human body, these reactions may relate to the body's immune defences, the digestion of food or the breaking-down of waste products. The biological catalysts that accelerate the rate of chemical reactions in cells are a type of protein called an **enzyme**. It is the unique shape of each enzyme that allows it to function as a catalyst. The mechanism by which many medicines act involves changes in the way an enzyme is functioning.

COMPARING INORGANIC CATALYSTS AND ENZYMES

Catalysts are able to cause a reaction to occur more quickly because they provide a new reaction pathway, which reduces the activation energy of the overall reaction. The effect of a catalyst is shown in Figure 15.3.1. An enzyme has the same impact upon the activation energy of a reaction as the inorganic catalysts you learnt about in Chapter 6.

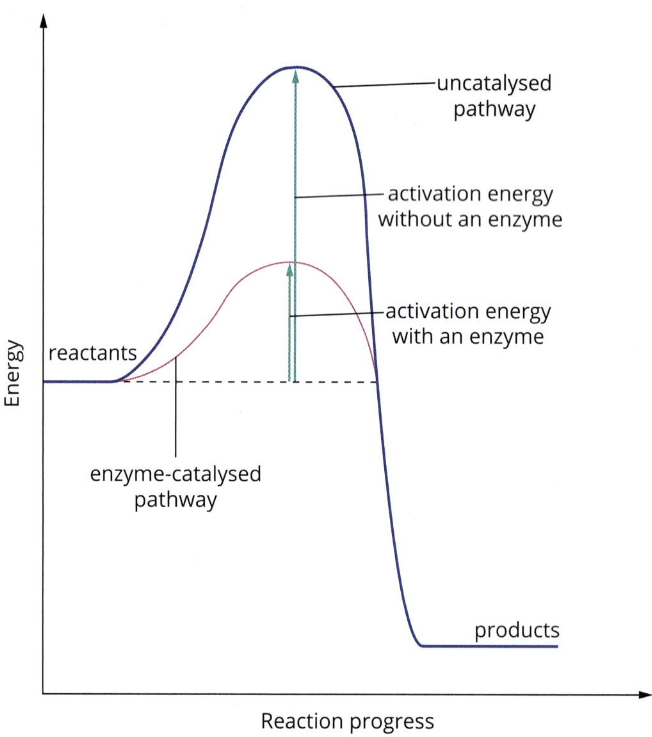

FIGURE 15.3.1 Enzymes lower the activation energy of a reaction by providing an alternative pathway by which the reaction can occur.

Just like inorganic catalysts, enzymes:

- are only needed in relatively small amounts
- are not used up or changed at the end of the reaction
- do not alter a reaction's equilibrium position
- provide an alternative reaction pathway.

The catalytic action of enzymes is specific for a single reaction or type of reaction. Some examples of catalysts and the reactions they are specific to are:

- amylase: catalyst for the hydrolysis of starch to smaller sugar molecules
- sucrase: catalyst for the hydrolysis of sucrose to glucose and fructose
- lactase: catalyst for the hydrolysis of lactose to galactose and glucose
- catalase: catalyst for the decomposition of hydrogen peroxide, H_2O_2.

The specificity of enzyme action can be explained by the structure of the enzyme. There is a particular part of the enzyme molecule with which a reactant can interact, known as its **active site**. The active site is usually a uniquely shaped flexible hollow or cavity within the protein where the reaction occurs. The reactant molecule that binds with the active site is referred to as the **substrate**.

One early model for the catalytic action of an enzyme is the 'lock-and-key' model. This model provides an explanation for the critical importance of the three-dimensional shape of the enzyme. In the lock-and-key model, the substrate molecule fits into the enzyme like a key in a lock, forming an enzyme–substrate complex, allowing the enzyme to break the bonds in the substrate. Figure 15.3.2 shows the steps involved in an enzyme-catalysed reaction according to the lock-and-key model. The enzyme is specific for a particular substrate, so binding to a different molecule will not result in a reaction.

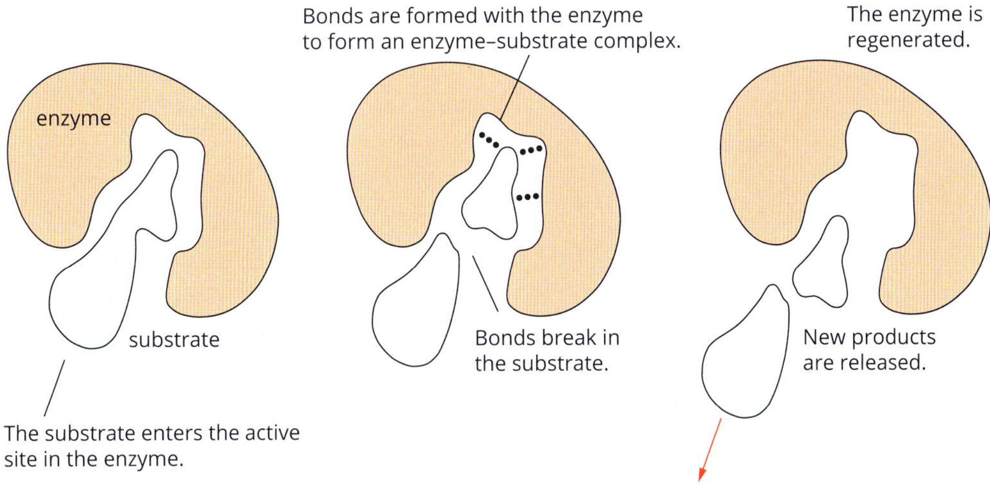

FIGURE 15.3.2 Steps in the action of an enzyme, according to the lock-and-key model

In addition to their specificity, there is another major difference between enzymes and inorganic catalysts. Enzymes are more sensitive than inorganic catalysts to changes in reaction conditions. They operate under a narrow temperature range and are sensitive to changes in the pH of their environment.

Enzymes are sensitive to changes in environment, such as pH and temperature.

DEPENDENCE OF ENZYMES ON pH AND TEMPERATURE

By lowering a reaction's activation energy barrier, both enzymes and inorganic catalysts can increase the rate of a reaction without the need for extremes of temperature or pressure. However, because an enzyme's ability to catalyse a reaction is more sensitive to changes in conditions or the presence of other molecules than an inorganic catalyst, any factor that changes the unique shape of the active site will limit the effectiveness of the catalyst. While there are many possible factors, only the effect of pH and temperature on enzyme activity will be discussed in this course.

Dependence on temperature

The graph in Figure 15.3.3 shows the effect of temperature on the rate of a reaction involved in carbohydrate metabolism. You can see from the steep right-hand side of the curve that the rate of reaction drops off quickly at the upper end of the optimal temperature range (30–40°C). The effect is less dramatic on the left-hand side of the curve, with the rate of reaction dropping off less rapidly at the lower end of the temperature range.

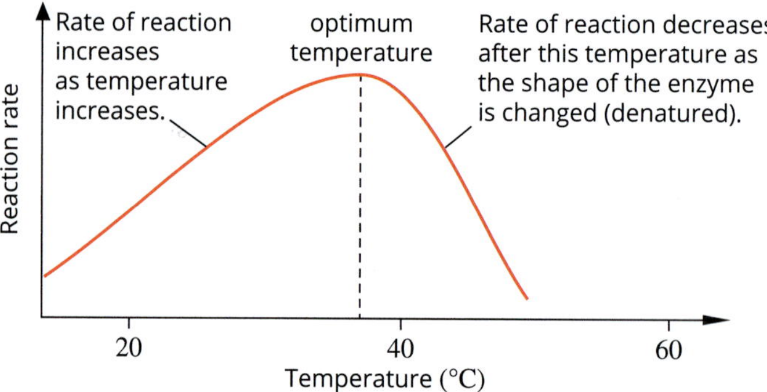

FIGURE 15.3.3 Effect of temperature on rate of reaction for an enzyme-catalysed reaction. Enzymes are only effective in a relatively narrow range of temperatures. The reaction rate is highest at the optimum temperature.

PA 22

Enzymes have an optimum temperature at which they operate best. Above that temperature they are denatured and below that temperature they are deactivated.

The temperature at which the enzyme activity is greatest is known as the enzyme's **optimum temperature**. Enzymes that operate inside human cells have an optimum temperature of about 37°C (body temperature).

At temperatures above and below the optimum temperature, enzyme function is impaired. This is one of the reasons why conditions such as hypothermia and fever (when you have an abnormally low or high temperature) are life-threatening.

An increase above or decrease below the optimum temperature has different effects on an enzyme.

- As the temperature increases above the optimum temperature, the increased kinetic energy of the molecules disrupts the structure of the enzyme. The increased movement throughout the enzyme breaks some of the intermolecular forces, such as hydrogen bonds, which are responsible for the structure of the enzyme. This change in three-dimensional shape of the enzyme means the active site can no longer effectively catalyse the reaction, so the reaction rate decreases rapidly.
- As the temperature decreases below the optimum temperature, the enzyme and substrate molecules have lower kinetic energies, resulting in less frequent and less energetic collisions between the molecules. At these lower temperatures the enzyme is said to be **deactivated**.

Denaturation

Both temperatures above the optimum temperature and extreme pH values disrupt the bonding in an enzyme and change the shape of the active site. The enzyme loses its catalytic activity and is said to be **denatured**. This change to the protein structure is often irreversible (Figure 15.3.4).

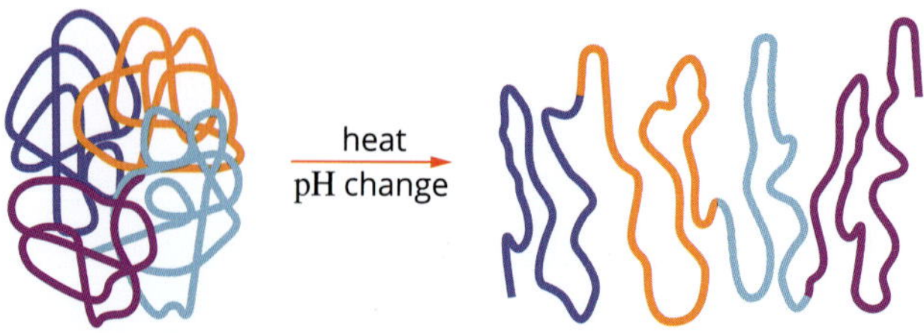

FIGURE 15.3.4 The specific three-dimensional shape of enzymes is lost when denaturation occurs.

Dependence on pH

Enzymes operate effectively only within a narrow pH range, as the graphs in Figure 15.3.5 show. Salivary amylase catalyses the breakdown of starch to a disaccharide, maltose, in the relatively neutral environment of the mouth. Pepsin catalyses the breakdown of proteins to amino acids in the acidic conditions of the stomach. Activity of these enzymes drops off drastically outside their normal conditions.

Not all enzymes require the same pH range. Pepsin is active only at pH values below 3, and amylase is active between pH 5 and 9. The pH at which the enzyme activity is greatest is known as the enzyme's **optimum pH**. The optimum pH of pepsin is 1.5, whereas the optimum pH of salivary amylase is 7.2.

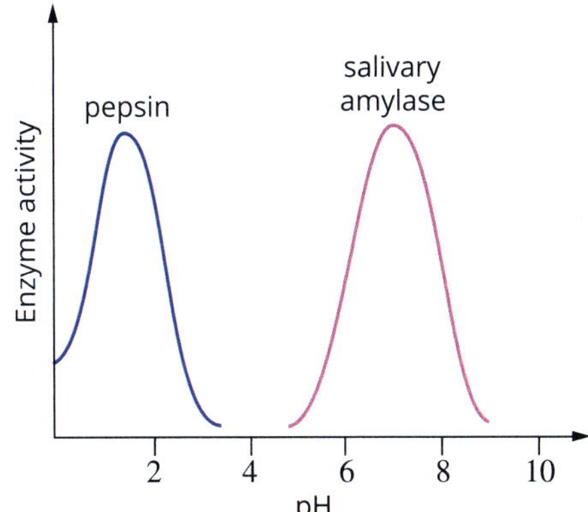

FIGURE 15.3.5 Enzymes are only effective in a narrow pH range. Pepsin is a protein-digesting enzyme secreted into the stomach. Salivary amylase is the enzyme in human saliva. These enzymes are most effective at very different pH values.

Changes in pH can have a large impact on the stability of enzyme structure, with the enzyme's active site changing shape and losing the ability to function effectively. Extremely high or low pH values generally cause complete loss of activity for most enzymes. Drastic changes to pH can result in a permanent change to the shape of an enzyme through the process of denaturation.

Salt bridges in the tertiary structure of proteins

The acidic and basic R groups on some amino acids play a significant role in stabilising the tertiary structure of proteins. They lead to the formation of ionic bonds called salt bridges, as shown in Figure 15.3.6. The addition of either a strong acid or base can disrupt the salt bridge, leading to denaturation of the protein. An example of this reaction occurs when milk passes into the acidic environment of our stomach. The acid denatures protein in the milk and it curdles.

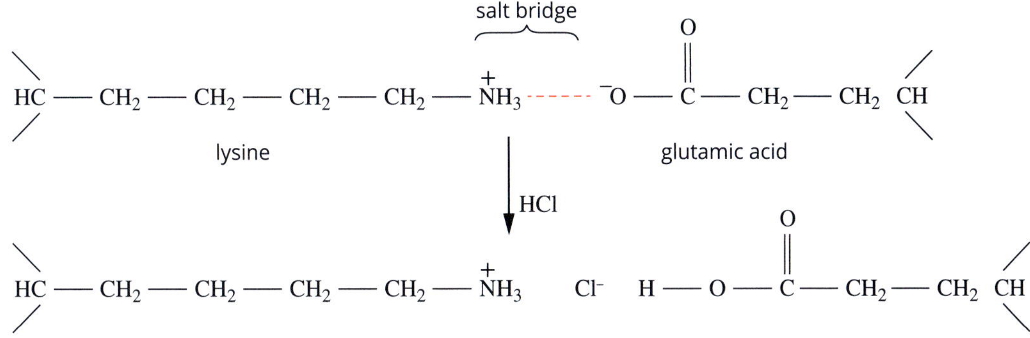

FIGURE 15.3.6 At intermediate pH values, ionic interactions can form salt bridges between neighbouring ions. When acid or base is added (in this case HCl), the salt bridges are disrupted.

Enzymes have reduced activity and so are less effective as catalysts at any pH above or below their optimum pH.

CHEMFILE

Why can't you set jelly containing fresh pineapple?

Bromelain is an enzyme that is known to occur only in pineapple. It is an enzyme that catalyses the digestion of proteins and is blamed for the quite unpleasant soreness or even bleeding in the mouth which can occur when you eat fresh pineapple. Gelatin, which causes jelly to set, is a protein, so when you add fresh pineapple to a jelly mixture, the bromelain in the pineapple catalyses the breakdown of the gelatin. As a result, your jelly will not set!

If you are determined to have pineapple in jelly, then tinned pineapple is a good substitute for fresh. Tinned pineapple has been heated to a temperature that denatures (inactivates) the bromelain.

Fresh pineapple contains an enzyme called bromelain, which prevents jelly from setting.

CASE STUDY **ANALYSIS**

Burke and Wills and thiaminase

In 1861 Robert Burke, William Wills and John King were starving to death on their return from crossing Australia from Melbourne to the Gulf of Carpentaria (south to north), a distance of 3250 km. While they were starving from lack of their own food, the local Yandruwandha people were thriving and gave the explorers cakes made from the crushed seed pods of a clover-like fern called nardoo (Figure 15.3.7).

Burke eventually had a falling out with the locals and drove them away. They tried to make the same cakes as the locals, but only became sicker and sicker, and both Burke and Wills died about a month later.

On Wednesday, 12 June 1861 Wills wrote: 'King out collecting nardoo. Mr Burke and I at home, pounding and cleaning. I still feel myself, if anything, weaker in the legs, although the nardoo appears to be more thoroughly digested.'

The reason why Burke and Wills obtained little benefit from the nardoo that sustained the Indigenous people lies in the heat resistance of the enzyme thiaminase. Nardoo contains high levels of thiaminase, an enzyme that breaks down vitamin B1 (thiamine). This is an undesirable reaction in humans because this vitamin is essential for the digestion of carbohydrates.

Thiaminase has a particularly high heat resistance for an enzyme, as illustrated in the graph in Figure 15.3.8. The Indigenous people knew that it was necessary to cook the nardoo cakes for long periods. The heat denatured the thiaminase and the cakes retained the thiamine necessary for digestion. Burke's party was not aware of the importance of the heat treatment.

FIGURE 15.3.7 Nardoo growing on the surface of a small inland lake

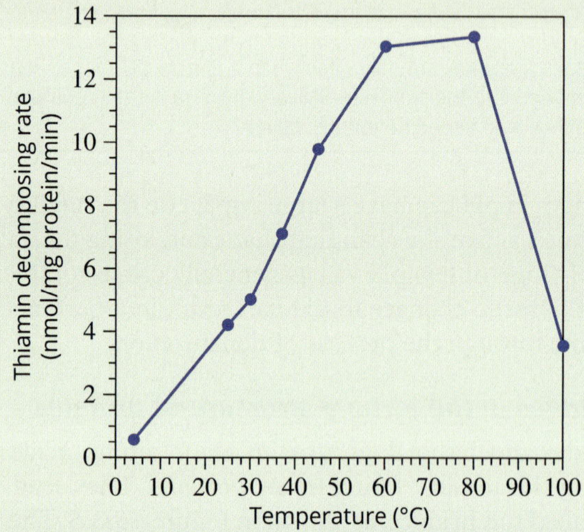

FIGURE 15.3.8 The optimum temperature for thiaminase is significantly higher than most enzymes. The unit on the vertical axis of the graph is a measure of enzyme activity.

Analysis

1 The effectiveness of thiaminase changes with temperature.
   - **a** Refer to the graph on the right to describe how the rate varies with temperature.
   - **b** Explain why the enzyme efficiency changes as the temperature increases.

2 The response to temperature of thiaminase is not typical of most enzymes. Explain why.

3 Humans benefit from thiaminase in their foods being denatured before consumption. Explain why.

15.3 Review

SUMMARY

- Enzymes are proteins that catalyse biochemical reactions by providing an alternative reaction pathway with a lower activation energy.
- Enzymes are not changed as a result of the process of catalysis.
- Enzymes are highly specific and may only catalyse one specific reaction or a reaction with a particular chemical bond or functional group.
- Enzyme molecules have uniquely shaped active sites that interact with specific reactant molecules (substrates), weakening or breaking the bonds in the reactant molecules. The lock-and-key model is used to explain the functioning of a catalyst.
- Enzymes operate over an extremely narrow set of conditions compared with catalysis by inorganic catalysts.
- Enzymes are very sensitive to changes in pH and temperature.
- The temperature at which the enzyme activity is greatest is the enzyme's optimum temperature.
- High temperature denatures an enzyme because the increased kinetic energy disrupts the bonds that maintain the enzyme's structure.
- When an enzyme is denatured, the shape of its active site is changed, and its catalytic activity is lost.
- Low temperature does not alter the enzyme structure, but fewer, less energetic, collisions occur per unit time between the enzyme and substrate, so the rate of reactions is slower.
- The pH at which the enzyme activity is greatest is the enzyme's optimum pH.
- On either side of the optimum pH, the activity of an enzyme reduces dramatically.
- R groups that are acidic or basic can form salt bridges that help shape an enzyme. Denaturation occurs when these side groups are disrupted by the addition of either acid or base.

KEY QUESTIONS

Knowledge and understanding

1 Explain how enzymes are able to increase the rate of a reaction.

2 Predict how the rate of an enzyme-catalysed reaction, such as the decomposition of starch in solution by amylase, would change as the temperature of the solution was gradually increased from 20°C to 50°C.

3 One of the key differences between enzymes and inorganic catalysts is how they are affected by their environment. Compare how a change in pH can affect enzyme activity and the ability of an inorganic catalyst to operate.

4 When an enzyme is denatured, which levels of protein structure are most impacted?

Analysis

5 The enzyme carbonic anhydrase catalyses the decomposition of carbonic acid molecules to carbon dioxide and water. When heated above 60°C, the enzyme no longer has any effect on the rate of reaction.

  a Explain why the effectiveness of the enzyme is reduced dramatically at high temperatures.

  b When the temperature of the solution containing carbonic anhydrase was reduced to 37°C after being heated to 60°C, the enzyme was still found to be inactive, despite being at its optimum temperature. Explain this observation.

  c When the temperature of a solution containing carbonic anhydrase that had not been heated to 60°C was reduced to 10°C and was then gently reheated to 37°C, the enzyme had the same activity as an enzyme that stayed at 37°C. Explain this observation.

6 Explain the shape of the graph shown below.

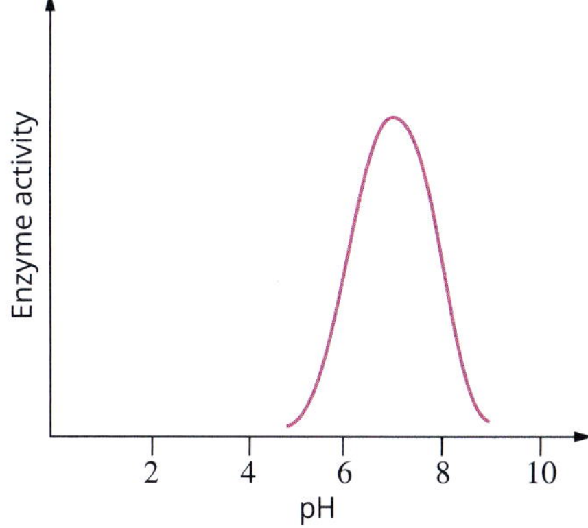

7 Insulin is a protein involved in the control of blood glucose levels. A person with type I diabetes has a damaged pancreas, so they do not produce their own insulin. Instead they require daily injections of insulin. Explain why, when it is not in use, the insulin should be stored in a refrigerator.

15.4 Optical isomers and inhibition of enzymes

An understanding of how enzymes function is useful to chemists, doctors and pharmaceutical scientists for many different purposes such as:

- the development of vaccines
- the prevention of disease
- the management of medical conditions
- supplementation to people lacking particular enzymes.

In this section you will learn about two particular areas of enzyme-related research, optical isomers and enzyme inhibition.

CHEMFILE

Rotation of plane-polarised light

Optical isomers were first discovered because of the different ways they rotate plane-polarised light. This type of light is produced when light passes through polaroid glass or film, which is used in some sunglass lenses. Plane-polarised light contains waves that oscillate in a single plane.

When plane-polarised light shines through a chiral substance, its plane polarisation is rotated. Different substances rotate the light to different extents and in different directions (clockwise or anticlockwise). A pair of enantiomers rotate plane-polarised light by the same amount but in opposite directions, as shown in the figure below.

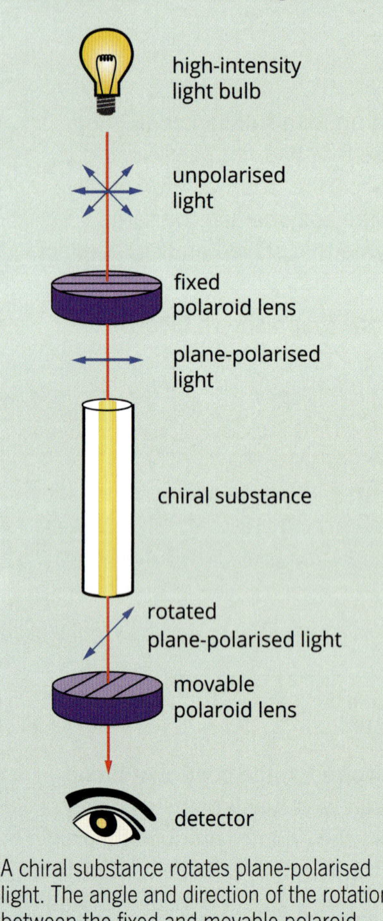

A chiral substance rotates plane-polarised light. The angle and direction of the rotation between the fixed and movable polaroid lenses allow the different enantiomers to be identified.

OPTICAL ISOMERS

Scurvy is a disease that affects the skin and gums. It was particularly prevalent in sailors prior to the 19th century because their intake of fresh fruit and vegetables was low. The active ingredient in citrus fruit that helps prevent scurvy is ascorbic acid. There are two types of ascorbic acid, L-ascorbic acid and D-ascorbic acid. L-ascorbic acid (vitamin C) is effective for the treatment of scurvy, but the other form is not. The two forms of ascorbic acid have the same molecular formula and the molecules have the same sequence of functional groups. So why the difference?

L-ascorbic acid and D-ascorbic are **optical isomers**, or **enantiomers** of each other. Optical isomers are a type of **stereoisomer**, molecules that have exactly the same molecular and semi-structural formulas, but different spatial arrangements of their atoms. These isomers are called optical isomers.

There are two common naming conventions used to distinguish enantiomers (either L- and D- or *R*- and *S*-). These indicate the direction in which the enantiomers rotate the plane of polarised light. You can read an explanation of this in the Chemfile on this page. D- and *R*- enantiomers rotate the plane of polarised light to the right, and *S*- and L- enantiomers rotate it to the left.

CHIRALITY

Substances that are optical isomers are said to be **chiral**. Two objects are chiral when they are mirror images of each other, and the mirror images cannot be superimposed on top of each other.

The term 'chiral' comes from the Greek word for hand. Your hands are chiral objects. One hand is a mirror image of the other. A left and right hand are not superimposable in three dimensions. It is even simpler to think about your feet. Again, your left and right feet are mirror images, but cannot be superimposed. Your left shoe does not fit on your right foot, as shown in Figure 15.4.1.

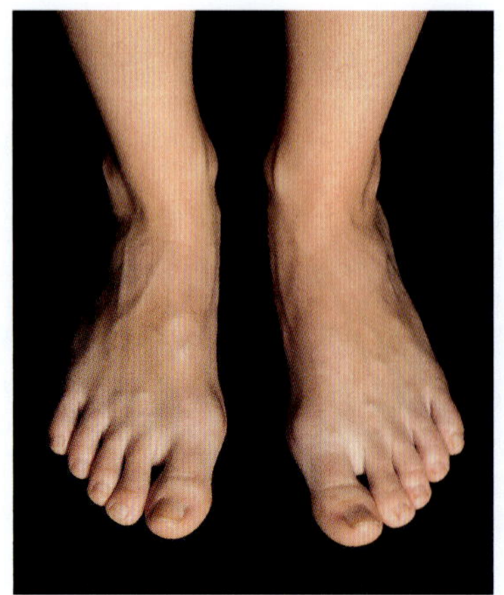

FIGURE 15.4.1 Your feet are chiral. They are mirror images but cannot be superimposed or fit in the same three-dimensional space, meaning that a right foot does not fit in a left shoe.

Achiral objects (objects that are not chiral) are objects whose mirror images can be superimposed on the original. Achiral objects have at least one **plane of symmetry**, while chiral objects do not have any planes of symmetry. A plane of symmetry exists when a three-dimensional figure can be divided into two halves that are mirror images. You can see two examples in Figure 15.4.2.

Chiral molecules come in pairs, called enantiomers.

FIGURE 15.4.2 This chair and butterfly each have a plane of symmetry. The left side is exactly the same as the right. This means the chair and the butterfly are achiral.

Chirality in organic molecules

Organic molecules that are chiral include molecules with a carbon atom joined to four different groups in a tetrahedral arrangement. A pair of chiral molecules are called **enantiomers**.

In Figure 15.4.3, you can see that molecule A is a mirror image of molecule B. The mirror image cannot be superimposed on the first molecule as the arrangement of the four groups in space is different. When molecule A is rotated so that the red and green groups are in the same position as in molecule B, the purple and yellow groups are in different places. The two molecules are chiral.

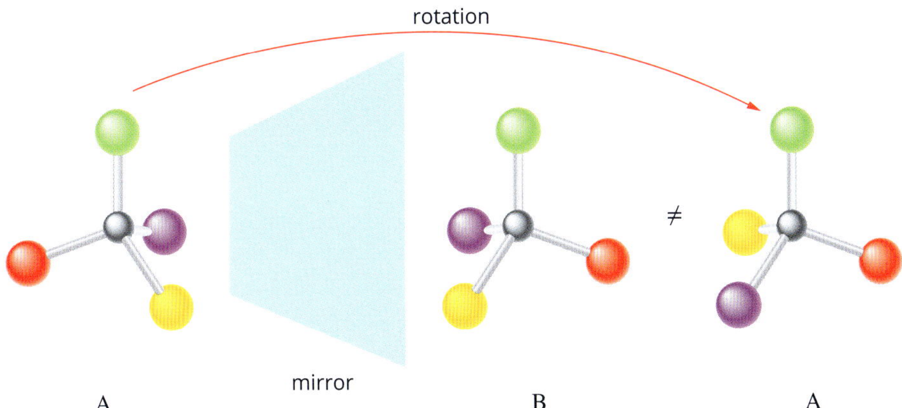

FIGURE 15.4.3 The molecules A and B represented by these ball and stick models are a pair of enantiomers.

A carbon atom attached to four different groups is called a **chiral centre**. A chiral molecule must contain at least one chiral centre. A chiral molecule must contain no planes of symmetry overall.

A chiral centre can be identified by looking at the four groups attached to a carbon atom. If the carbon atom is attached to four different groups, then it is a chiral centre.

The molecule in Figure 15.4.4 is chiral; it is shown with a pair of enantiomers. The carbon in the centre is attached to four different groups, which makes it the chiral centre. Often the chiral centre is marked with an asterisk. The bonds from the carbon to the hydrogen and chlorine are shown as simple lines to show that these are flat against the page. The wedged line represents a bond coming out of the page and the dashed wedge represents a bond going into the page. A rotated view of enantiomer A is shown with the hydrogen and chlorine atoms in the same position as enantiomer B. This results in the bromine and fluorine atoms being in different places and shows that the mirror image is not the same.

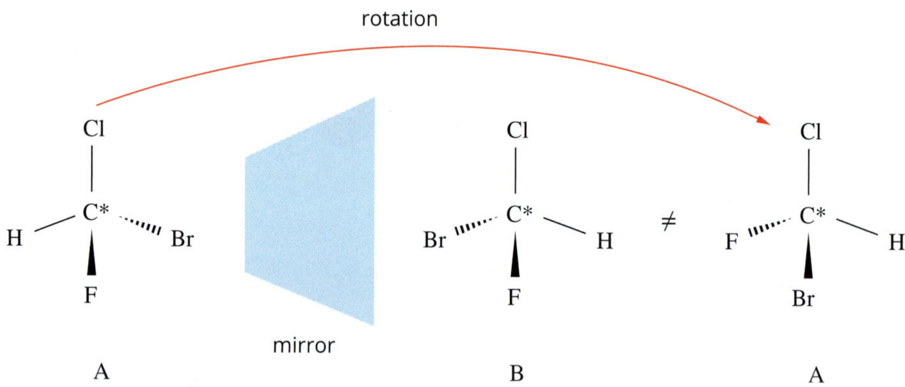

FIGURE 15.4.4 Bromochlorofluoromethane is a chiral molecule and has two enantiomers.

The molecule in Figure 15.4.5 does not contain a chiral centre because there are two identical groups bonded to the carbon atom. It is an achiral molecule. The mirror image of the molecule can be rotated to produce a structure that can be superimposed on the original. They are the same molecule and are not enantiomers.

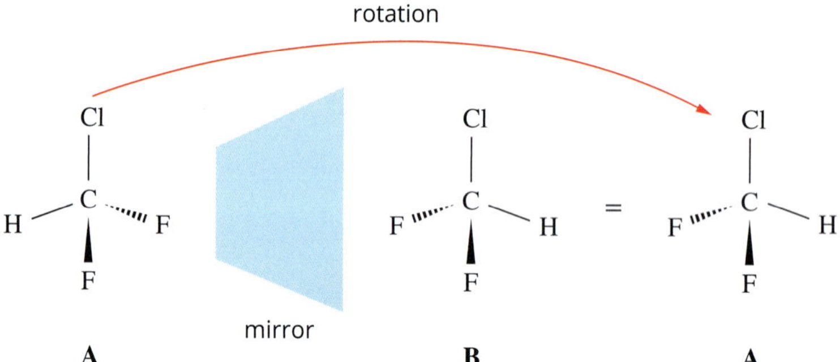

FIGURE 15.4.5 Chlorodifluoromethane is an achiral molecule because its mirror image can be superimposed.

Identifying chiral centres

To identify a chiral centre, you need to determine if any of the carbon atoms in a molecule have four different groups attached. As Worked example 15.4.1 illustrates, you often need to look further than just the next atom to see if the groups are the same.

Worked example 15.4.1

IDENTIFYING CHIRAL CENTRES IN ORGANIC MOLECULES

Identify any chiral centres in the following molecule.

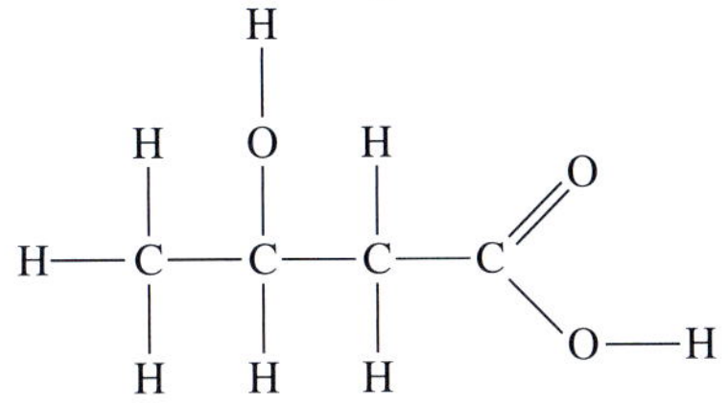

Thinking	Working
Eliminate the carbon atoms that cannot be chiral centres. This includes all $-CH_3$ and $-CH_2-$ groups and any atom that is part of a double or triple bond, because these carbon atoms cannot be bonded to four different groups.	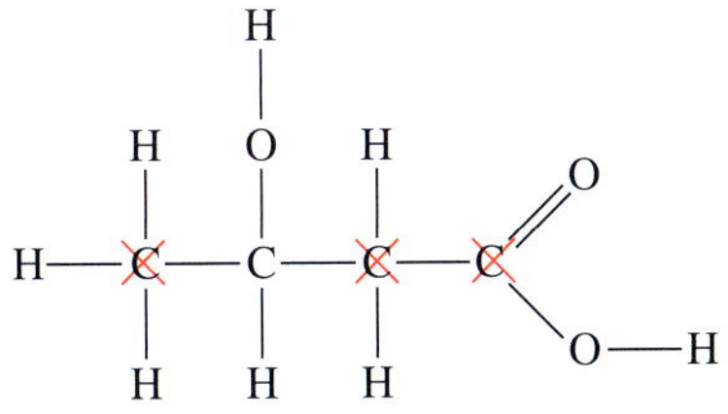 This leaves one carbon atom that may or may not be a chiral centre.
For each remaining carbon atom, examine the groups attached to the atom. If there are four different groups, the atom is a chiral centre.	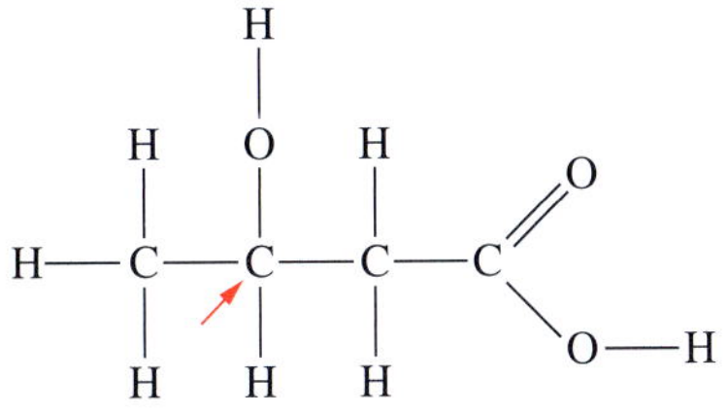 The remaining carbon atom has the following groups attached: $-CH_3$ –H –OH $-CH_2COOH$ The four groups are different so this carbon atom is a chiral centre.

Worked example: Try yourself 15.4.1

IDENTIFYING CHIRAL CENTRES IN ORGANIC MOLECULES

Identify any chiral centres in the following molecule.

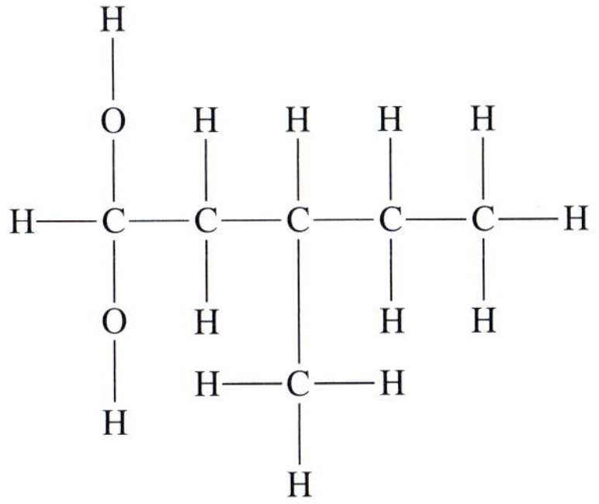

CHEMFILE

Determining the maximum number of optical isomers in molecules with more than one chiral centre

Sometimes a molecule will have more than one chiral centre. As each chiral centre means there are two possible enantiomers, the overall number of possible optical isomers can be determined using the rule that the maximum number of optical isomers is 2^n, where n is the number of chiral centres.

As can be seen in the figure below, vitamin D_3 has five chiral centres. This means there are 2^5, or 32 possible optical isomers.

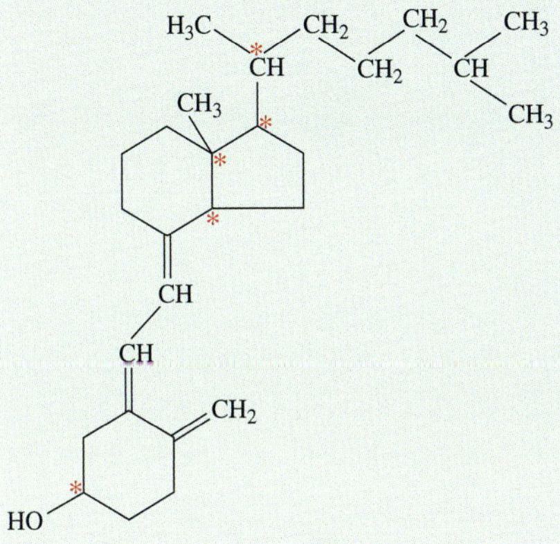

Vitamin D_3 has five chiral centres (marked with *). Thus, 32 stereoisomers are possible.

CHIRAL DRUGS

It is estimated that 56% of drugs currently in use are composed of chiral molecules, and it cannot be assumed that different enantiomers will be equally effective. They might have the same sequence of functional groups, but, to a living system, they are virtually different substances. The functional groups of one enantiomer might align with the functional groups of a particular receptor or enzyme in the body, but those in the other enantiomer might not.

Figure 15.4.6 shows the different spatial arrangements at the circled chiral carbon in ascorbic acid. As you learnt earlier, L-ascorbic acid is effective at preventing scurvy and is more commonly known as vitamin C. It is essential for the production of collagen in the body. The small change in orientation in D-ascorbic acid, however, is enough to render that enantiomer completely ineffective in producing collagen. The hydroxyl groups in one molecule will align with the functional groups on the substrate, but the same groups will not have the required alignment with the other enantiomer.

> Enantiomers have differing effects on the body due to their different spatial arrangements of atoms.

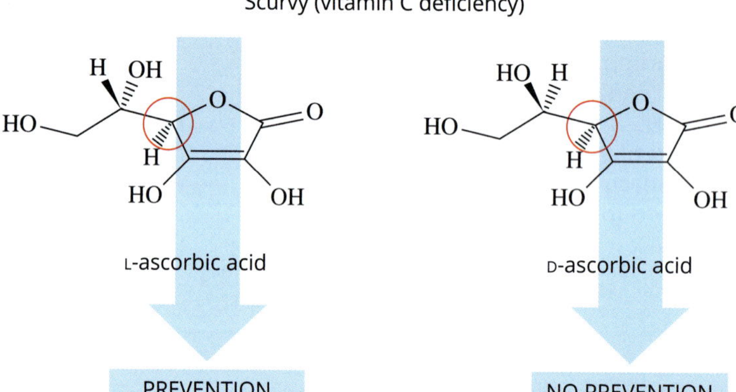

FIGURE 15.4.6 The different spatial arrangement around the chiral carbon atoms is enough to make one enantiomer effective for the prevention of scurvy and the other not.

One of the most difficult steps in drug manufacture often involves separating the two enantiomers to ensure a product is safe and effective.

Enantiomer effectiveness

There are three possible outcomes when the pharmaceutical activity of enantiomers is compared. These are summarised in Figure 15.4.7 and examples of each are listed on the opposite page.

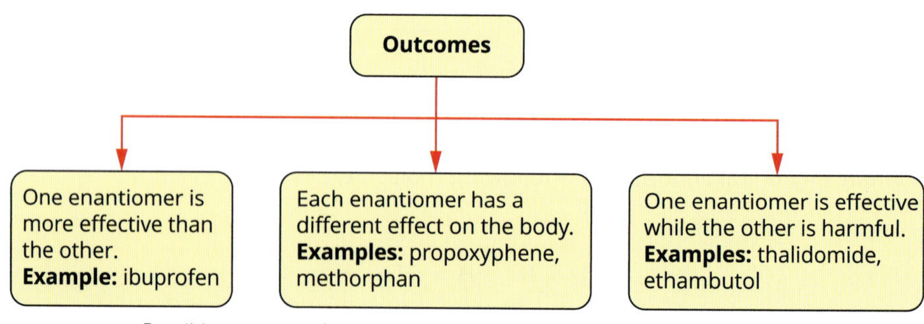

FIGURE 15.4.7 Possible outcomes for pharmaceutical enantiomers

- Ibuprofen: The two enantiomers of ibuprofen are shown in Figure 15.4.8. *S*-ibuprofen is far more effective than *R*-ibuprofen. Separating the two forms is a slow process, but it leads to a far more effective product.

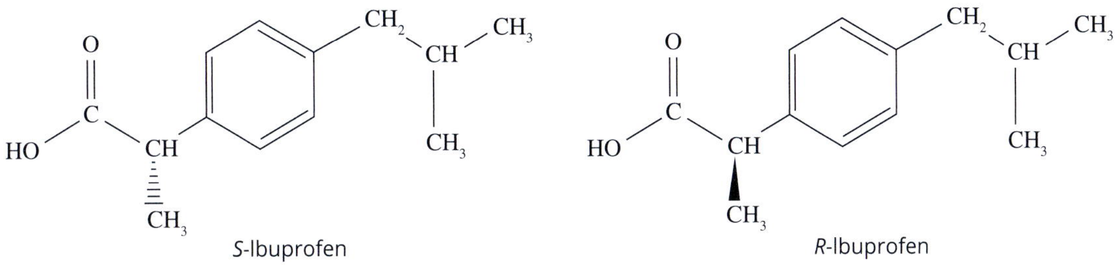

S-Ibuprofen

R-Ibuprofen

FIGURE 15.4.8 The enantiomers of ibuprofen

- Propoxyphene: In the case of propoxyphene, both enantiomers are effective in the human body, but for different purposes. D-propoxyphene relieves pain while L-propoxyphene helps relieve coughs. The structures of propoxyphene are shown in Figure 15.4.9.

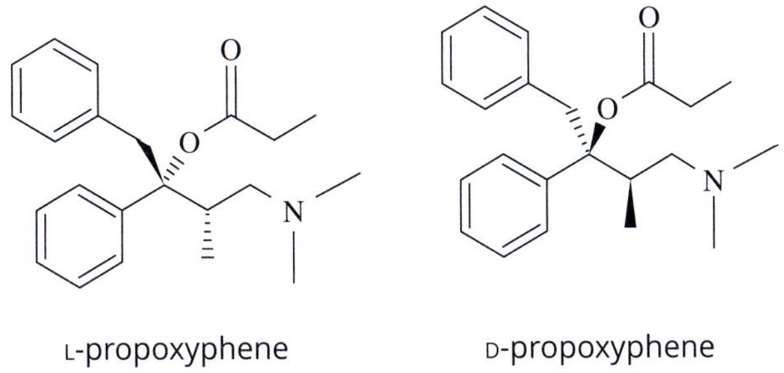

L-propoxyphene

D-propoxyphene

FIGURE 15.4.9 The enantiomers of propoxyphene

- Thalidomide: Thalidomide was released on the world market in 1957 as a drug for treating morning sickness. It became evident over the following years that thalidomide was associated with severe birth defects in babies, such as missing limbs. It is now understood that the *R*-enantiomer is effective in treating morning sickness, but the *S*-enantiomer can lead to birth defects. The structures are shown in Figure 15.4.10.

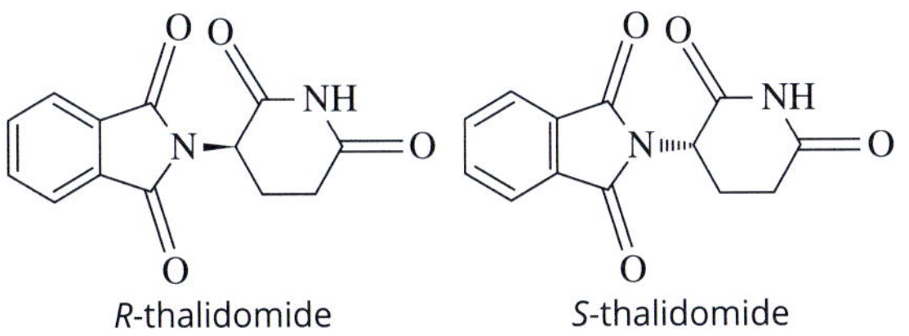

R-thalidomide

S-thalidomide

FIGURE 15.4.10 *R*-thalidomide can be used to relieve morning sickness, but *S*-thalidomide can lead to significant birth deformities.

A competitive enzyme inhibitor acts by occupying the active site of an enzyme, preventing the enzyme from catalysing a reaction that is detrimental to the body.

ENZYME INHIBITORS

In the previous section you learnt about the important role of the shape of an enzyme in facilitating a desirable reaction. You will now learn how medical researchers can use this knowledge in reverse, to inhibit the pathway of an undesirable reaction.

Competitive enzyme inhibitors

When a person has the influenza virus, it is made worse by infected cells in the body releasing forms of the virus called **virions** that go on to infect neighbouring cells. Australian scientists designed a drug called Relenza that binds to enzyme sites in infected cells. The active site is no longer available for the process that produces virions, so the spread of flu in the body is contained. Relenza is an example of a **competitive enzyme inhibitor**, a molecule that prevents an undesirable reaction by occupying the active site of an enzyme. Enzyme inhibitors can be either competitive or **non-competitive**, depending upon their mechanism of action.

The process by which enzymes in body cells produce virions is a typical **lock-and-key mechanism**, as described earlier in this chapter (Figure 15.3.2 on page 573).

Figure 15.4.11 shows Relenza competing for the active site on the enzyme. If Relenza molecules occupy the site, virions cannot be produced. Relenza works because it is chemically and structurally similar to the substrate it is replacing. The relative concentrations of the substrate and the Relenza molecules will impact the effectiveness of Relenza.

Penicillin is an antibiotic that kills some forms of bacteria. It is another example of a competitive enzyme inhibitor, as it functions by interfering with the synthesis of cell walls of reproducing bacteria. It inhibits the action of an enzyme, transpeptidase, that catalyses bacteria cell wall synthesis. The defective walls cause the bacteria cells to burst without causing harm.

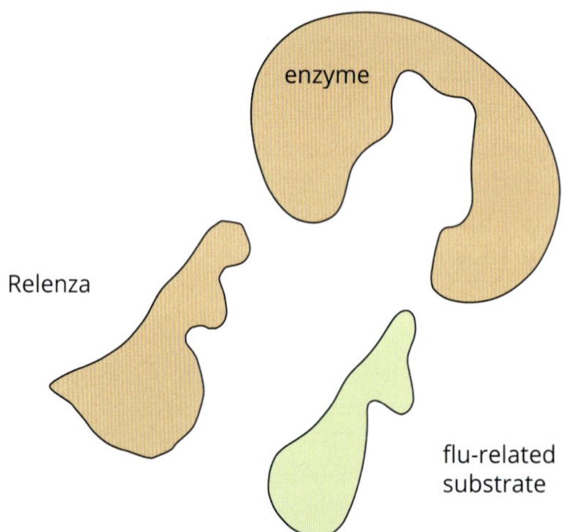

FIGURE 15.4.11 Relenza has a similar shape to influenza-related substrates. If Relenza can occupy the active site, the production of influenza virions is inhibited.

Non-competitive enzyme inhibitors

Non-competitive enzyme inhibition involves a molecule binding to a different part of an enzyme than the active site. In doing so, it alters the shape of the active site of the enzyme to the point where it no longer matches the shape of the substrate. The inhibitor is not in direct competition with the substrate, so the concentration of the substrate does not impact the effectiveness of the inhibitor. This is illustrated in Figure 15.4.12.

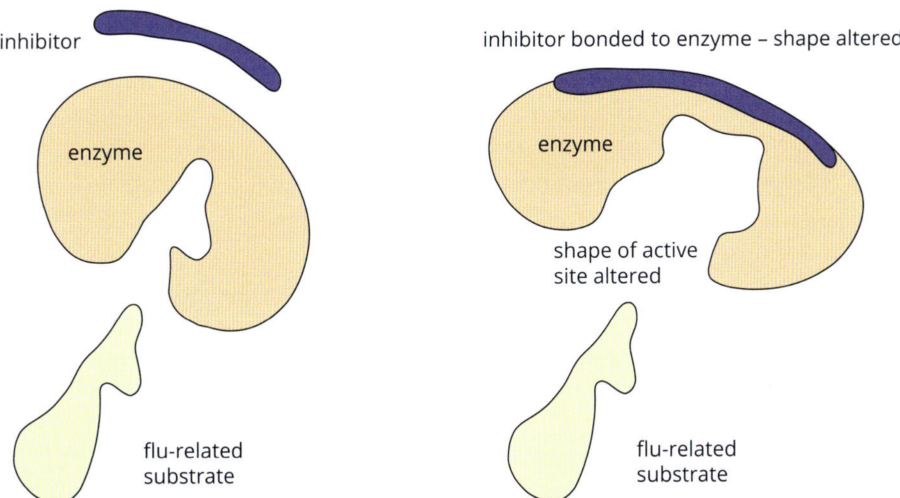

FIGURE 15.4.12 The non-competitive enzyme inhibitor changes the shape of the enzyme's active site enough to prevent it from acting as a catalyst.

A non-competitive inhibitor functions by changing the shape of an enzyme so that it no longer catalyses a detrimental reaction in the body.

The action of cyanide as a poison is an example of non-competitive inhibition. In a healthy body the enzyme cytochrome oxidase plays an important role in respiration. If a person ingests cyanide, the cyanide ions bind to the enzyme, changing its shape so that it no longer catalyses respiration processes. If not treated, the patient may die.

Methodextrate is another example of non-competitive inhibition. It is used in the treatment of arthritis and severe cases of psoriasis, as it inhibits the action of enzymes responsible for inflammation.

15.4 Review

SUMMARY

- A knowledge of the importance of enzyme shape leads to an understanding of why some forms of a drug are more effective than others.
- Some molecules are optical isomers, molecules with the same molecular formula and semi-structural formula, but a different spatial arrangement of the atoms.
- Optical isomers are chiral. They are mirror images of each other, but cannot be superimposed on each other. They contain chiral centres, i.e. carbon atoms with four different groups attached.
- Enantiomers are a pair of molecules that exist in two forms that are mirror images of one another, but cannot be superimposed one upon the other.
- The different spatial arrangements in enantiomers mean that their alignment with enzymes or receptors sites can differ and their effectiveness can also differ.
- Knowledge of the mechanism of enzyme action has been used to design molecules that can inhibit undesirable reactions involving enzymes. Molecules that act in this way are known as enzyme inhibitors.
- Competitive enzyme inhibitors act by occupying the active site of an enzyme, preventing the undesired reaction.
- Non-competitive inhibitors attach to a different part of the enzyme than the active site, but change the shape of the active site enough to prevent it from functioning as an enzyme.

KEY QUESTIONS

Knowledge and understanding

1 Explain how a carbon atom that is a chiral centre can be identified.

2 Which of the following molecules can have optical isomers?

a CH_2Cl_2
b CH_3CHFCl
c $CH_3CHClCH_3$
d $CH_3CHClCH_2CH_3$

3 Label the following statements about enantiomers as true or false, and for the false statements, explain why they are false.

a Enantiomers have the same molecular formula.
b Enantiomers always have the same impact upon the body.
c Enantiomers can be superimposed on each other.

4 State which part of an enzyme is targeted by:

a a competitive enzyme inhibitor
b a non-competitive enzyme inhibitor.

Analysis

5 Analyse the following molecules and identify the number of chiral centres, if any, that are present.

a

b

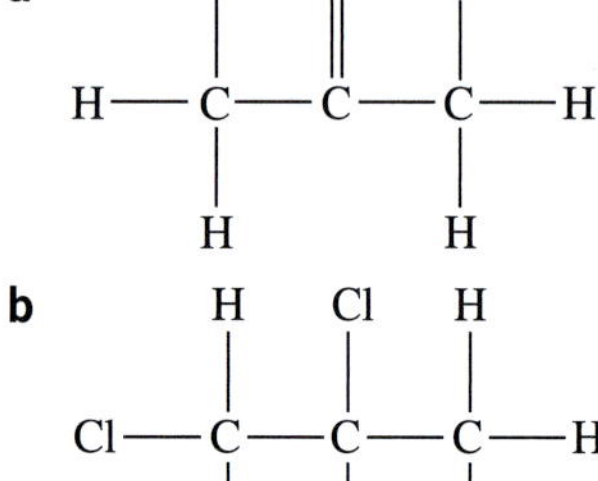

c

```
                 H
                 |
             H — C — H
     H    H      |      H    H
     |    |      |      |    |
H —  C —  C  —   C   —  C —  C — H
     |    |      |      |    |
     H    O      H      H    H
          |
          H
```

d

```
     H   H   H
     |   |   |
Cl — C — C — C — H
     |   |   |
     H   O   H
         |
         H
```

6 Draw the enantiomer of the following molecule and show that it is a non-superimposable mirror image of this molecule.

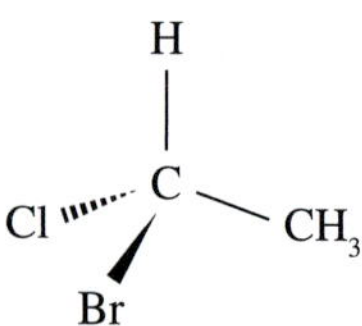

7 Use butanol as an example to show the difference between a structural isomer and an optical isomer.

8 Ketamine is a component of many anaesthetics. It is a chiral molecule with two enantiomers. If *S*-ketamine is effective as an anaesthetic, discuss the possible impacts on the human body of the opposite enantiomer, *R*-ketamine.

Chapter review

KEY TERMS

achiral
active site
α-helix
β-pleated sheet
chiral
chiral centre
coiling
competitive enzyme inhibitor
deactivated
denaturation
enantiomer
enzyme
hydrophobic
lock-and-key mechanism
medicinal plant
non-competitive enzyme inhibitor
optical isomers
optimum pH
optimum temperature
peptide link
plane of symmetry
pleating
primary structure
quaternary structure
R group
secondary structure
solvent extraction
steam distillation
stereoisomer
substrate
tertiary structure
virions
zwitterion

REVIEW QUESTIONS

Knowledge and understanding

1 Naphthalene is a hydrocarbon with the chemical formula $C_{10}H_8$.The most suitable solvent to use to extract naphthalene from a leaf would be:
- **A** water
- **B** ethanol
- **C** 1-chlorobutane
- **D** hexane.

2 Which one of the following statements provides the best description of the bonding that holds a protein molecule in its secondary structure?
- **A** Covalent bonds form between the different amino acids in the protein molecule.
- **B** Various forces of attraction occur between different R groups in the protein structure.
- **C** Hydrogen bonding occurs between different peptide linkages in the protein structure.
- **D** Dispersion forces act between the different protein molecules in the structure.

3 When the temperature of most enzyme solutions increases from 40°C to 70°C, the effectiveness of the enzyme drops, because:
- **A** the frequency of collisions between the enzyme and the substrate will decrease
- **B** the secondary and tertiary bonds in the enzyme are disrupted
- **C** the peptide bonds in the protein break and the enzyme breaks down to amino acids
- **D** the increased temperature increases the rate of other reactions involving the enzyme.

4 The enzyme glucokinase catalyses the first step in the oxidation of sugar in human liver cells, in a process called glycolysis. Which one of these statements about this process is correct?
- **A** Many other enzymes can also catalyse this reaction.
- **B** Glucokinase increases the activation energy of this reaction.
- **C** Glucokinase is able to catalyse many other reactions.
- **D** Glucokinase is a protein.

5 Which one of these formulas gives the correct structure of tyrosine, $NH_2CH(C_6H_4OH)COOH$, in the highly acidic conditions present in the human stomach?
- **A** $^+NH_3CH(C_6H_4OH)COOH$
- **B** $NH_2CH(C_6H_4OH)COO^-$
- **C** $^+NH_3CH(C_6H_4OH)COO^-$
- **D** $NH_2CH(C_6H_4OH)COOH$

6 **a** Steam distillation can be used to extract liquids from plant leaves. Explain how steam distillation works.
b List two properties that an active ingredient should have to make it suitable for steam distillation.

7 Identify the type of catalyst (inorganic, enzyme or both) that displays each of these properties.
- **a** lowers the activation energy
- **b** is highly specific
- **c** is made of proteins
- **d** is not changed by the reaction
- **e** increases reaction rate

8 Provide definitions for each of the following terms.
- **a** enantiomer
- **b** competitive enzyme inhibitor
- **c** substrate
- **d** lock-and-key model
- **e** zwitterion
- **f** active site

9 Identify the type of bonding represented by labels **a** and **b** of the diagram below. Identify the type of secondary structure in **c**.

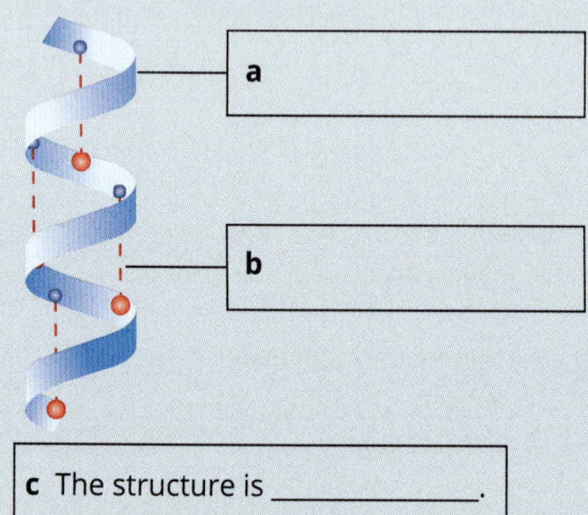

c The structure is ____________.

10 Indicate whether the following statements, summarising the properties of enzymes, are true or false. If they are false, explain why they are false.

- **a** Enzymes are made of proteins.
- **b** Enzymes do not change the position of equilibrium.
- **c** Enzymes are consumed by the reaction.
- **d** Enzymes increase the activation energy of a reaction.
- **e** Enzymes increase the rate of reaction.
- **f** Enzymes are sensitive to conditions, such as pH changes or temperature increase, which denatures the enzyme.
- **g** Enzymes are highly specific for the biochemical reactions they catalyse because of the shapes of their active site.

11 In what conditions is an amino acid likely to form an anion?

12 Circle the parts of the polypeptide segment in the following figure that might be involved in forming bonds responsible for the tertiary structures of the protein molecule.

```
        H    O          H    O          H    O
        |    ||         |    ||         |    ||
 —N—————C————C————N—————C————C————N—————C————C—
  |     |         |     |         |     |
  H   CH2SH       H    CH2        H    CH2
                        |               |
                        C               C
                       / \\           // \
                     HO    O         O    NH2
```

13 **a** Name and describe the functions of two enzymes in the body.

b Explain why the action of enzymes justifies the statement 'Enzymes make life possible'.

c Why is the action of an enzyme often described as operating like a lock and key?

Application and analysis

14 Which of these two molecules is more likely to be extracted from a leaf by hexane, C_6H_{14}? Explain your answer.

```
        H   H                    H   H
        |   |                    |   |
 H—O————C———C————O—H      H—S————C———C————S—H
        |   |                    |   |
        H   H                    H   H
```

15 Draw the structures of the chiral molecules named below and label the chiral centres with an asterisk.

- **a** butan-2-amine
- **b** 1-chloropropan-2-ol
- **c** 4-bromopentan-2-ol

16 Sugars are chiral molecules. One of the chiral centres in the form of glucose shown below is indicated by the asterisk. What is the total number of chiral centres in this form of glucose?

```
       CHO
        |
   H————C————OH
        |
  HO————C————H
        |
   H————C*———OH
        |
   H————C————OH
        |
      CH2OH
```

17 Draw the structure of the zwitterion form of the amino acid valine.

```
     H3C     CH3
        \   /
         CH
         |
  H2N————C————COOH
         |
         H
```

18 Jellied pineapple dessert cannot be made by using gelatine and fresh pineapple because an enzyme in the pineapple causes molecules in the gelatine to break down instead of setting. Suggest how jellied pineapple might be prepared.

19 Aspartic acid is an amino acid that exists mainly as a zwitterion at pH 2.8.

CH_2COOH
|
H_2N — C — COOH
|
H

a Name the:

i acidic functional group

ii basic functional group.

b Explain what is meant by the term 'zwitterion'.

c Draw the structure of the molecule as it is most likely to exist in a solution at:

i pH much greater than 2.8

ii pH much less than 2.8

iii pH 2.8.

20 The following diagram shows an enzyme and the substrate that matches its active site. Refer to this diagram to explain the mechanism of a:

a competitive enzyme inhibitor

b non-competitive enzyme inhibitor.

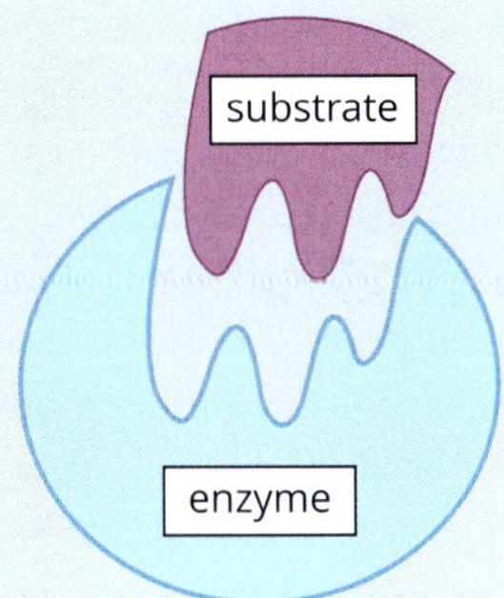

21 List the five main bond interactions that form when R groups in different regions of a polypeptide chain interact with each other. Give examples of the required components in R groups involved in such interactions, and an example of an amino acid that has an R group of this type.

22 The chemical formula of the pharmaceutical procaine is drawn below. It is used to treat dementia.

CH_3 O N CH_3 O H_2N

a Name two functional groups present in procaine.

b State two absorptions that will be seen on an infrared spectrum of procaine.

c Explain whether procaine will have enantiomers.

23 Cetirizine is an enantiomer that can be used as an antihistamine.

a How can you distinguish cetirizine from its enantiomer in the laboratory?

b Cetirizine is purified before sale to remove any traces of its enantiomer. Explain why this might be necessary.

24 The addition of acid solution to milk causes the milk to curdle. This action is related to changes in the bonds formed between particular R groups on the casein protein in milk.

a What process is the curdling of milk an example of?

b Give an example of how some particular R groups are impacted by the addition of acid.

25 The enzyme carbonic anhydrase catalyses the decomposition of carbonic acid molecules to carbon dioxide and water in the lungs. When heated to more than 60°C, the enzyme becomes denatured.

a What is meant by the term 'denatured'?

b Describe the events that usually occur to the structure of an enzyme when it is denatured.

c Does the primary structure of the carbonic anhydrase enzyme change during the process?

d Why is the functioning of the enzyme closely related to its tertiary structure?

26 The graph below shows the effect of temperature on the enzyme activity for a reaction that provides energy for the body. For each of the parts of the graph labelled A, B and C, explain the variation in enzyme activity with temperature.

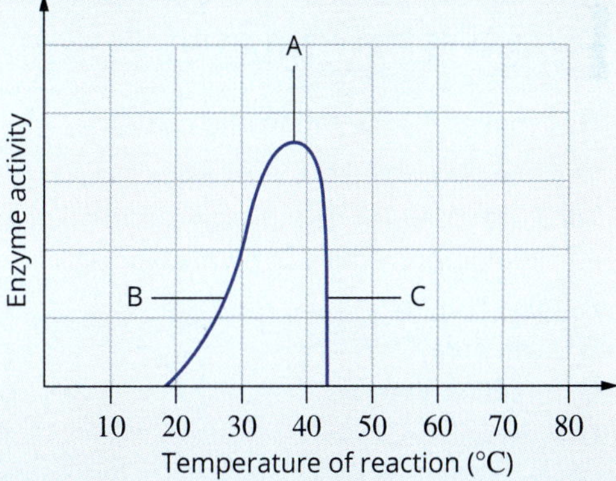

27 Methotrexate is a drug that inhibits the action of some forms of enzymes in the body. When used in high concentrations it is toxic to both cancerous and non-cancerous cells. When used in lower concentrations it is toxic to cancerous cells only. Methotrexate is an example of a competitive enzyme inhibitor.
Suggest a reason for the varying effectiveness of methotrexate.

UNIT 4 • Area of Study 2

REVIEW QUESTIONS

How are organic compounds analysed and used?

Multiple-choice questions

1 In the mass spectrum of an organic compound, the signal produced by the unfragmented molecular ion and the signal produced by the most abundant particle correspond respectively to the:

A parent (molecular) ion peak and main ion peak
B main ion peak and base peak
C parent (molecular) ion peak and base peak
D base peak and parent (molecular) ion peak.

2 How many peaks are expected in the ^{1}H NMR and ^{13}C NMR spectra of ethyl ethanoate, $CH_3COOCH_2CH_3$?

	^{1}H NMR spectrum (low resolution)	^{13}C NMR spectrum
A	3	4
B	2	3
C	6	4
D	1	3

3 In high-performance liquid chromatography (HPLC), the stationary phase comprises small particles so densely packed that the mobile phase must be forced through it under pressure. Small particles are used in preference to large particles because:

A they are cheaper to manufacture
B it reduces the overall mass of the column
C increasing the surface area of the stationary phase improves separation
D it makes it easier to fill the column.

4 The concentration of a substance is most accurately determined in high-performance liquid chromatography (HPLC) by measuring:

A peak height
B peak area
C retention time
D R_f value.

5 Amylase is an enzyme which catalyses the hydrolysis of starch. An increase in temperature changes the shape of the active site of amylase. This is because:

A the peptide links in the primary structure of amylase have been hydrolysed
B the enzyme has been denatured as intermolecular forces in the tertiary structure are broken
C only the hydrogen bonds holding the secondary structure of the enzyme together have been broken
D the increase in temperature increases the enzyme activity.

6 The secondary structure of many enzymes includes helical regions. The type of chemical bond responsible for maintaining this shape is:

A hydrogen bonds
B ion–dipole bonds
C ionic bonds
D covalent bonds.

7 A polyunsaturated fatty acid called cervonic acid is found in fish oil. The molar mass of cervonic acid is 328.5 g mol^{-1}. 4.928 g of cervonic acid reacts completely with 22.842 g of iodine, I_2. How many carbon-to-carbon double covalent bonds are in cervonic acid?

A 0
B 2
C 4
D 6

8 Which of the following statements about propan-1-ol and propan-2-ol is **not** true?

A The ^{13}C NMR spectrum of propan-1-ol has three peaks, and that of propan-2-ol has two peaks.
B The ^{1}H NMR spectrum of propan-1-ol has four peaks, and that of propan-2-ol has three peaks.
C The mass spectra of both will show a peak at a mass-to-charge ratio of 60.
D The fingerprint region of the IR spectra will be identical for both compounds.

9 The retention time can be used in high-performance liquid chromatography (HPLC) to determine:

A the amount of a chemical in the sample
B the concentration of a chemical
C the identity of a chemical
D all of the above.

10 When hydrochloric acid was added to some clear uncooked egg white in a test tube, a white gelatinous precipitate formed. When the mixture was boiled gently, the precipitate slowly dissolved. Which of the following lists the likely changes in the contents of the test tube?

A Protein $\xrightarrow{\text{acid added}}$ denatured and coagulated protein $\xrightarrow{\text{heat}}$ free amino acids
B Protein $\xrightarrow{\text{acid added}}$ free amino acids $\xrightarrow{\text{heat}}$ carbon dioxide and water
C Glucose $\xrightarrow{\text{acid added}}$ starch $\xrightarrow{\text{heat}}$ glucose
D Free amino acids $\xrightarrow{\text{acid added}}$ denatured and coagulated protein $\xrightarrow{\text{heat}}$ free amino acids

11 The molecule shown below is known as Dopa. It is used in the treatment of Parkinson's disease.

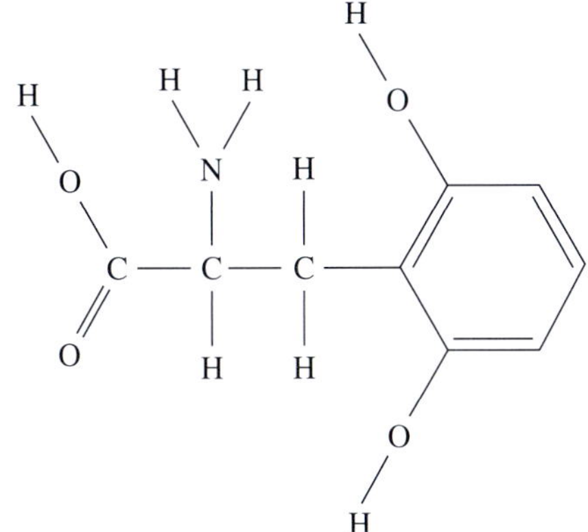

Identify the functional groups in Dopa.

A hydroxyl, amide and ester
B hydroxyl, amine and carboxyl
C hydroxyl, amide and carboxyl
D hydroxyl, ester and amine

12 In a redox titration conducted by a student, an 18.90 mL titre of aqueous iodine, $I_2(aq)$, reacted completely with a 20.00 mL aliquot of sodium thiosulfate solution.

When another student repeated this titration, they washed out the pipette with water immediately before using it.

The effect on the volume of the titre in this second experiment would be that:

A it was exactly 18.90 mL
B it was less than 18.90 mL
C it was greater than 18.90 mL
D it was exactly 20.00 mL.

13 The influenza drug Relenza is a competitive enzyme inhibitor. It works to prevent enzymes from producing virions, which spread the influenza virus through the body by:

A occupying the active site of the enzyme
B changing the shape of the active site of an enzyme by binding to the enzyme close to the active site
C breaking down the structure of the influenza virus
D destroying the enzyme which produces the influenza virions.

14 The compounds that produced the infrared spectra below had the same molecular formula, $C_3H_8O_2$.

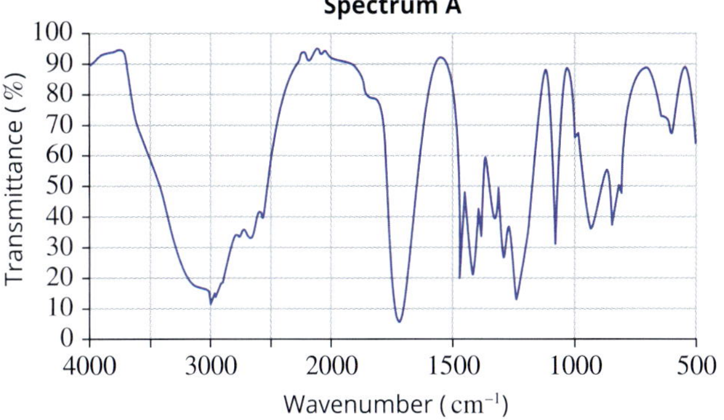

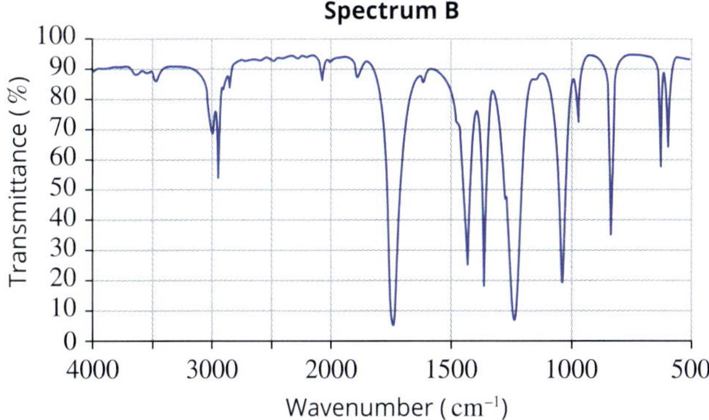

Which of the following statements about the spectra is **not** correct?

A Both spectra have an absorption band corresponding to a C=O bond.
B The fingerprint regions of the two compounds are different because they are different compounds.
C Spectrum B is most likely to be a ketone because there is no O–H bond present.
D Spectrum A is likely to be a carboxylic acid because there is a wide absorption band at about 3000 cm^{-1} corresponding to an O–H(acid) bond and a C–H bond.

15 A redox titration was conducted to determine the concentration of ethanol in a sample of white wine. 10.0 mL of white wine was diluted to 250 mL in a volumetric flask and a 20.00 mL aliquot was titrated with a 0.0512 M solution of acidified dichromate solution according to the equation

$$3CH_3CH_2OH(aq) + 2Cr_2O_7^{2-}(aq) + 16H^+(aq) \rightarrow 3CH_3COOH(aq) + 4Cr^{3+}(aq) + 11H_2O(l)$$

The average titre of the acidified dichromate solution was 21.44 mL.

Which of the following is the molar concentration of ethanol in the original sample of white wine?

A 0.001 65 M
B 0.0823 M
C 1.03 M
D 2.06 M

Short-answer questions

16 Describe a chemical test you could carry out to distinguish between the two substances in the following pairs. Include a description of the expected results in your answer (equations are not required).

a hexane and hex-1-ene

b hexan-1-ol and hexane

c propanoic acid and propan-2-one

17 The structures of propan-2-ol and propan-2-one are shown below. The infrared spectra of these compounds are labelled IR spectrum A and IR spectrum B.

$$H_3C-\underset{H}{\overset{OH}{C}}-CH_3$$ propan-2-ol

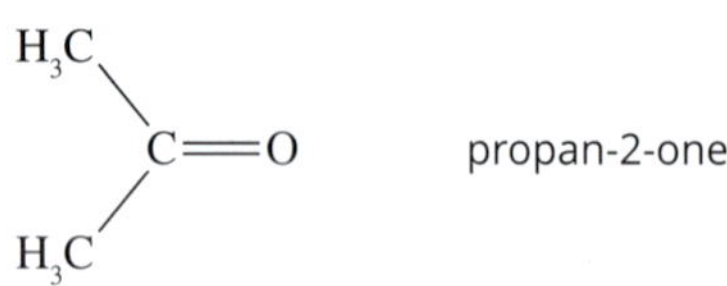

propan-2-one

A

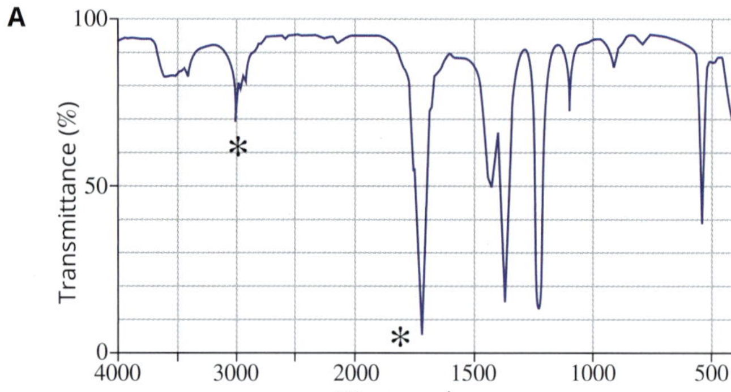

B

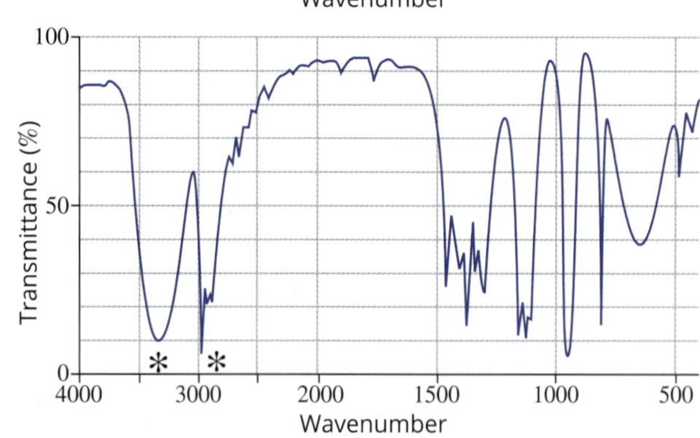

a Identify the bonds that produced the IR absorption marked with an asterisk (*) in each spectrum.

b Determine which spectrum represents which molecule.

c The ^1H NMR spectrum of propanone is shown following. Explain why there is only a single peak although there are six hydrogen atoms in the molecule.

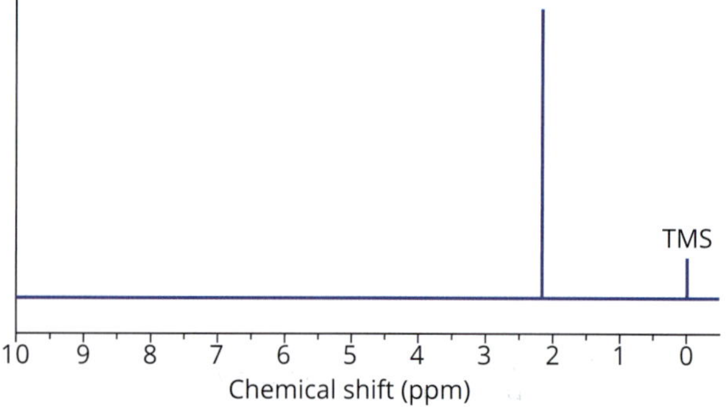

d The ^{13}C NMR spectrum of propan-2-one is shown below. Explain why there are two peaks although there are three carbon atoms in the molecule.

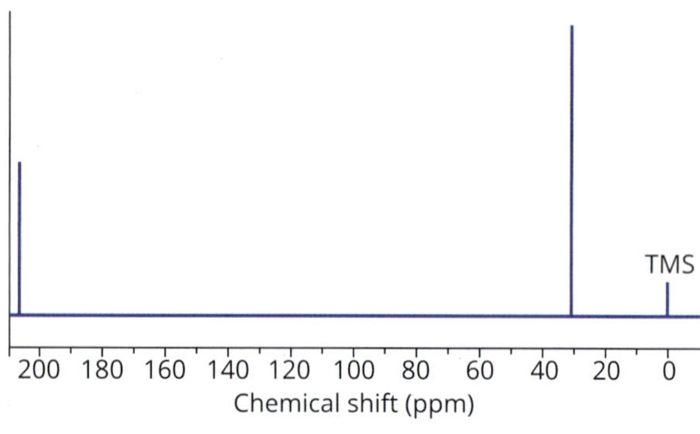

e How many sets of peaks would you expect in ^1H NMR and ^{13}C NMR spectra of propan-2-ol? Provide an explanation for your answers.

f The mass spectrum of propanone is shown below.

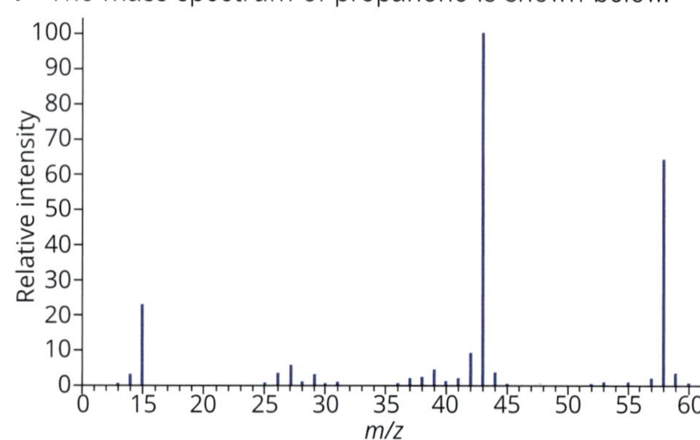

i What is the mass of the molecular ion?

ii What information does this provide about the molecule?

iii Suggest a formula for the molecular fragment that produces the peak at *m/z* 43.

18 Some students were set the task of determining the concentration of vitamin C in a particular brand of vitamin C tablets. An outline of the method they used is given below.

1. A burette is filled with a standardised solution of iodine with a concentration of 0.009 95 M.
2. One vitamin C tablet is crushed and dissolved in 250 mL of solution in a volumetric flask. A pipette is used to transfer 20.00 mL of the vitamin C solution to a conical flask and 1 mL of starch indicator is added.
3. The vitamin C solution is titrated with the iodine solution. Titrations are repeated until three concordant results are obtained.

The equation for the reaction is:

$$C_6H_8O_6(aq) + I_2(aq) \longrightarrow C_6H_6O_6(aq) + 2I^-(aq) + 2H^+(aq)$$

a Why is the vitamin C tablet dissolved in 250 mL of water and a 20.00 mL aliquot of the solution used for the titration?

b Explain why titrations are repeated until three concordant results are obtained.

c One student's results are given below. The data shown in the student's laboratory book was:

- concentration of I_2(aq) = 0.009 95 M
- expected mass of vitamin C in tablet = 500 mg
- molar mass of vitamin C = 176.1 g mol^{-1}
- total volume of vitamin C solution = 250.00 mL
- volume of vitamin C solution used in each titration = 20.00 mL

TABLE 1 Titration of 20.00 mL aliquots of vitamin C solution with 0.009 95 M iodine solution.

Trial	1	2	3	4
Final volume (mL)	22.70	46.70	22.90	45.55
Initial volume (mL)	0.00	22.70	0.10	22.80
Volume of titre (mL)				

i When the student calculated their average titre, they ignored trial 2. Explain why this was the correct thing to do.

ii Calculate the average (mean) titre for this experiment.

iii Calculate the amount, in mol, of iodine, I_2, that reacts with the aliquot of vitamin C solution.

iv Calculate the concentration of vitamin C in the vitamin C solution.

v Calculate the mass of vitamin C in the vitamin C tablet used in this experiment.

19 The structures of the following amino acids are given in Table 12.1.1 on page 434: valine, cysteine, aspartic acid, leucine, lysine and threonine. These amino acids constitute part of most enzymes and the interactions between their side groups contribute to the maintenance of their tertiary structure. These interactions include covalent and ionic bonds, hydrogen bonds, dipole–dipole interactions and dispersion forces.

a Identify the strongest interaction that can exist between the side groups (R groups) of each of the following pairs of amino acids.

i threonine and lysine

ii cysteine and cysteine

iii valine and leucine

iv aspartic acid and lysine

b Using examples from the list in part **a,** explain the effects of high temperatures and extremes of pH on the interactions that maintain the tertiary structure of enzymes.

20 Ethyl propanoate is an ester found in some types of apple. A sample of the ester ethyl propanoate was hydrolysed by gently heating it with water in a flask and using a small amount of sulfuric acid as catalyst. The resulting solution was then fractionally distilled and two of the fractions collected. One of the collected fractions distilled off at 78°C and the other at 141°C.

a What compound formed the fraction that distilled off at 78°C?

b What compound formed the fraction that distilled off at 141°C?

c Give reasons for your answers to parts **a** and **b**.

d What chemical tests could you carry out to indicate that your answers to question parts **a** and **b** are correct?

e Write a balanced equation for the hydrolysis reaction that took place during the experiment.

21 Three jars, each containing a white powder and labelled 'A', 'B' and 'C', were found on the shelves of a school chemistry store room. It was suspected that the jars contained samples of compound X. Pure compound X is known to have a melting point of 158°C.

As a step in identifying the unknown compounds, their melting points were measured, and then mixed melting points with compound X were carried out on each of them.

- The melting points of each of these compounds prior to carrying out the mixed melting points were: Compound A: 161°C; Compound B: 158°C; Compound C: 158°C.
- The mixed melting point of compound X with compound A was 158°C.
- The mixed melting point of compound X with compound B was 158°C.
- The mixed melting point of compound X with compound C was 152°C.

Explain how these results can be used to identify compounds A, B and C.

22 ^{1}H NMR spectra of the following three compounds were recorded.

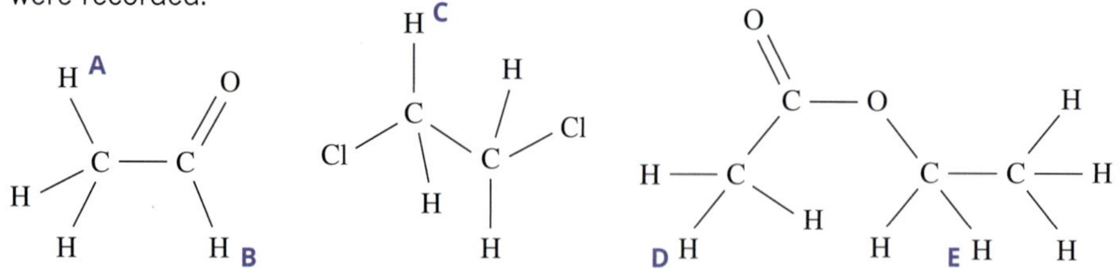

a What type of electromagnetic radiation is absorbed and emitted in NMR spectroscopy?

b What changes in the molecules are associated with this emission and absorption?

c Describe the role of tetramethylsilane (TMS) in NMR spectroscopy.

d The spectra of the three compounds included the following signals. In each case, select the number that labels the H atoms most likely to be responsible for the signal, and give a brief justification.

i quartet at δ 8.5

ii singlet at δ 2.1

iii quartet at δ 4.1

iv singlet at δ 3.7

v triplet at δ 1.2

vi doublet at δ 2.2

23 High-performance liquid chromatography (HPLC) is ideally suited to identifying and measuring food additives, such as preservatives in sauces. A mixture of three common food preservatives was analysed by HPLC. A sample of a sauce was prepared by dissolving 1.50 g of sauce in a total of 10.0 mL in a volumetric flask. This solution was also analysed by HPLC, using the same mobile and stationary phases under the same conditions. The resultant chromatograms are shown below.

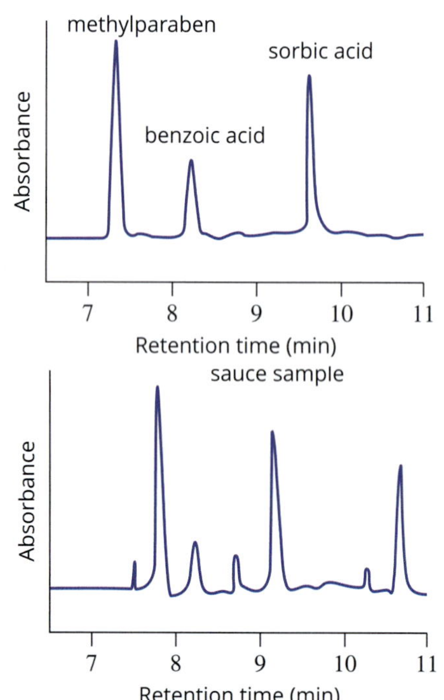

a Given that the preservatives are all equally soluble in the mobile phase, which of the three adsorbs most strongly on the stationary phase? Explain.

b Which of the preservatives appear to be present in the sauce sample?

c To measure the concentration of the preservative in the diluted sauce, its peak area was compared to that of standard solutions of the preservative, as shown in the following table.

Concentration (mg/10.0 mL)	Relative peak area
0.00	0
2.00	13.3
4.00	26.6
6.00	39.9
8.00	53.2
sauce sample	31.5

Construct a calibration graph for the standard samples and use it to estimate the concentration of the preservative in the diluted sauce.

d Calculate the concentration of the preservative in the original sauce sample, in mg/100 g. (assuming that the density of the sauce is 1 g/mL)

24 The lactase enzyme is responsible for the hydrolysis of the disaccharide, lactose, in the human body.

a Explain how the terms 'active site' and 'substrate' relate to the function of an enzyme, such as lactase.

b Given that enzymes are essential for the normal progress of chemical reactions in the body, why are they described as 'catalysts' rather than as 'reactants'?

c Describe how the 'lock-and-key' model explains the function of an enzyme.

d If a competitive enzyme inhibitor was acting on the lactase enzyme, describe what would be happening to the enzyme and how it would affect the ability of the person to digest lactose.

25 Enkephalins are short polypeptides involved in the nervous system's detection of pain and harm. The structure of met-enkephalin, so-called because it contains a methionine residue, is shown below. Another important example is leu-enkephalin, which has a leucine (side chain R = $CH_2CH(CH_3)_2$) residue in place of methionine.

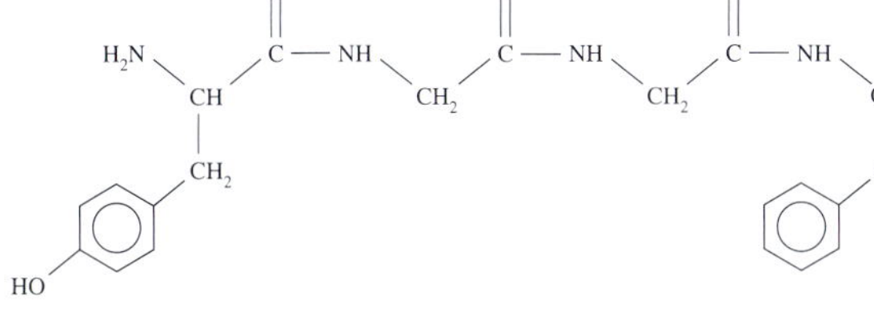

a Draw the structure of leu-enkephalin.

b Refer to a table of amino acids (see Table 12.1.1, page 434) and use it to name the other amino acids present in leu-enkephalin.

c On your structure circle and label:

i peptide linkages

ii terminal carboxyl and amino groups

d Mark with an asterisk (*) all of the chiral centres in your structure.

26 Suggest explanations for the following facts relating to protein structure and function.

a Formation of the primary structure of proteins produces water as a by-product, but formation of the secondary and tertiary structures does not.

b The secondary and tertiary structures of proteins are generally much more easily disrupted than the primary structure.

c The tertiary structures of some proteins are more easily disrupted by an increase in temperature than are those of other proteins.

27 10.00 mL of a solution of potassium dichromate ($K_2Cr_2O_7$) with a concentration of 0.0525 M was added to 60.00 mL of a 0.0360 M iron(II) sulfate ($FeSO_4$) solution in the presence of sulfuric acid. The overall ionic equation for the reaction is:

$$Cr_2O_7^{2-}(aq) + 6Fe^{2+}(aq) + 14H^+(aq) \rightarrow 2Cr^{3+}(aq) + 6Fe^{3+}(aq) + 7H_2O(l)$$

a Determine the limiting reactant.

b Calculate how much of the excess reactant was remaining after the reaction was complete.

c Calculate the mass of iron(III) sulfate, $Fe_2(SO_4)_3$ produced by the reaction.

28 Peel from a small orange was mixed with water and some diethyl ether and mashed in a blender. Diethyl ether is a non-polar organic solvent which is less dense than water. The fibrous material from the mashed orange was filtered off and the clear solution poured into a separating funnel (see figure below).

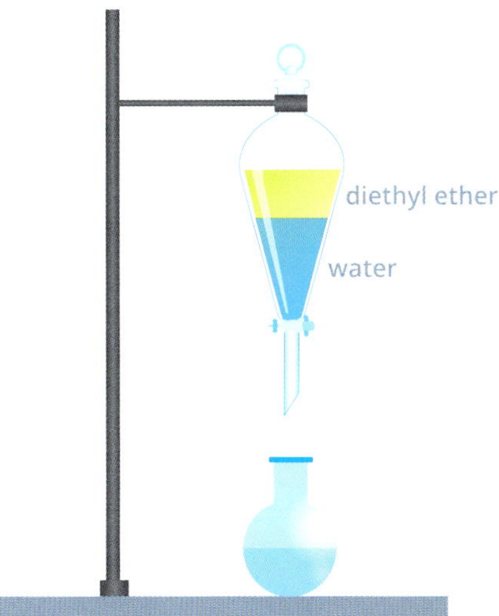

Dropping funnel separating two layers of immiscible liquids

Two layers formed in the liquid in the separating funnel. The lower layer was run into a separate flask by turning the tap in the separating funnel.

Two compounds found in orange peel are limonene and citric acid (see figures below).

Structure of limonene

Structure of citric acid

a Citric acid and limonene are two of the many different compounds found in orange peel. In which solvent layer in the separating funnel would you expect to find:

i citric acid

ii limonene

Explain your answer.

b Describe the test that you could carry out to confirm your answer to Question **a** part **i**.

c The compounds in the two layers can be obtained by carefully evaporating the solvents.
Name the technique that could be used to separate limonene from the other compounds that were also present in the same layer in the separating funnel.

APPENDIX 1 Electrochemical series

Reaction	Standard electrode potential ($E°$) in volts at 25°C
$F_2(g) + 2e^- \rightleftharpoons 2F^-(aq)$	+2.87
$H_2O_2(aq) + 2H^+(aq) + 2e^- \rightleftharpoons 2H_2O(l)$	+1.77
$Au^+(aq) + e^- \rightleftharpoons Au(s)$	+1.68
$Cl_2(g) + 2e^- \rightleftharpoons 2Cl^-(aq)$	+1.36
$O_2(g) + 4H^+(aq) + 4e^- \rightleftharpoons 2H_2O(l)$	+1.23
$Br_2(l) + 2e^- \rightleftharpoons 2Br^-(aq)$	+1.09
$Ag^+(aq) + e^- \rightleftharpoons Ag(s)$	+0.80
$Fe^{3+}(aq) + e^- \rightleftharpoons Fe^{2+}(aq)$	+0.77
$O_2(g) + 2H^+(aq) + 2e^- \rightleftharpoons H_2O_2(aq)$	+0.68
$I_2(s) + 2e^- \rightleftharpoons 2I^-(aq)$	+0.54
$O_2(g) + 2H_2O(l) + 4e^- \rightleftharpoons 4OH^-(aq)$	+0.40
$Cu^{2+}(aq) + 2e^- \rightleftharpoons Cu(s)$	+0.34
$Sn^{4+}(aq) + 2e^- \rightleftharpoons Sn^{2+}(aq)$	+0.15
$S(s) + 2H^+(aq) + 2e^- \rightleftharpoons H_2S(g)$	+0.14
$2H^+(aq) + 2e^- \rightleftharpoons H_2(g)$	0.00
$Pb^{2+}(aq) + 2e^- \rightleftharpoons Pb(s)$	–0.13
$Sn^{2+}(aq) + 2e^- \rightleftharpoons Sn(s)$	–0.14
$Ni^{2+}(aq) + 2e^- \rightleftharpoons Ni(s)$	–0.25
$Co^{2+}(aq) + 2e^- \rightleftharpoons Co(s)$	–0.28
$Cd^{2+}(aq) + 2e^- \rightleftharpoons Cd(s)$	–0.40
$Fe^{2+}(aq) + 2e^- \rightleftharpoons Fe(s)$	–0.44
$Zn^{2+}(aq) + 2e^- \rightleftharpoons Zn(s)$	–0.76
$2H_2O(l) + 2e^- \rightleftharpoons H_2(g) + 2OH^-(aq)$	–0.83
$Mn^{2+}(aq) + 2e^- \rightleftharpoons Mn(s)$	–1.18
$Al^{3+}(aq) + 3e^- \rightleftharpoons Al(s)$	–1.66
$Mg^{2+}(aq) + 2e^- \rightleftharpoons Mg(s)$	–2.37
$Na^+(aq) + e^- \rightleftharpoons Na(s)$	–2.71
$Ca^{2+}(aq) + 2e^- \rightleftharpoons Ca(s)$	–2.87
$K^+(aq) + e^- \rightleftharpoons K(s)$	–2.93
$Li^+(aq) + e^- \rightleftharpoons Li(s)$	–3.04

Extract from VCE Chemistry's Written examination: Data Book December 2020.

APPENDIX 2 Symbols, units and fundamental constants

TABLE 1 Metric (including SI) prefixes, their symbols and values

Metric (including SI)	Symbol	Value
pico	p	10^{-12}
nano	n	10^{-9}
micro	μ	10^{-6}
milli	m	10^{-3}
centi	c	10^{-2}
deci	d	10^{-1}
kilo	k	10^{3}
mega	M	10^{6}
giga	G	10^{9}
tera	T	10^{12}

TABLE 2 Unit conversions

Measured value	Conversion
0°C	273 K
100 kPa	750 mm Hg or 0.987 atm
1 litre (L)	1 dm^3 or 1×10^{-3} m^3 or 1×10^{3} cm^3 or 1×10^{3} mL

TABLE 3 Some physical constants and standard values

Description	Symbol	Value
Avogadro's constant	N_A	6.02×10^{23} mol^{-1}
charge of an electron	e	-1.60×10^{-19} C
Faraday's constant	F	96 500 C mol^{-1}
molar gas constant	R	8.31 J mol^{-1} K^{-1}
molar volume of an ideal gas at SLC (25°C and 100 kPa)	V_m	24.8 L mol^{-1}
specific heat capacity of water	c	4.18 J^{-1} g^{-1} K^{-1} or 4.18 kJ kg^{-1} K^{-1}
density of water at 25°C (298 K)	d	997 kg m^{-3} or 0.997 g mL^{-1}

APPENDIX 3 Chemical relationships

Name	Formula
number of moles of a substance	$n = \frac{m}{M}$; $n = \frac{c}{V}$; $n = \frac{V}{V_m}$
ideal gas equation	$pV = nRT$
calibration factor (CF) for calorimetry	$CF = \frac{VIt}{\Delta T}$
heat energy released in the combustion of a fuel	$q = mc\Delta T$
enthalpy of combustion	$\Delta H = \frac{q}{n}$
electric charge	$Q = It$
number of moles of electrons	$n(e^-) = \frac{Q}{F}$
% atom economy	$\frac{\text{molar mass of desired product}}{\text{molar mass of all reactants}} \times \frac{100}{1}$
% yield	$\frac{\text{actual yield}}{\text{theoretical yield}} \times \frac{100}{1}$

Answers

Chapter 1 Scientific investigation

1.1 The nature of scientific investigations

1

Information	Quantitative	Qualitative
colour of electrode		✔
temperature of electrolyte	✔	
presence of bubbles		✔
change in mass of electrode	✔	
concentration of electrolyte	✔	

2 B.

3 The methodologies required for this investigation would involve fieldwork to collect the water samples and a controlled experiment once samples had been returned to the laboratory.

4

Factor	Scientific (yes or no)	Non-scientific (include category)
concentration of carbon dioxide in atmosphere (ppm)	yes	
federal governmental support of the United Nations Paris Agreement on climate change	no	political
UNESCO 2017 *Declaration of Ethical Principles in relation to Climate Change*	no	ethical
university modelling for the impact of costs of climate change	no	economic
use of hydrogen fuel cells to limit carbon dioxide emissions	yes	
Commonwealth *Renewable Energy (Electricity) Act 2000*	no	legal

5 A number of possible answers, including:

- cost of electricity for electrolysis of water
- cost of infrastructure for distribution of hydrogen
- cost of cars, buses and trucks using hydrogen as a fuel (perhaps in a fuel cell) in comparison to other available cars, buses and trucks.

6

Methodology	Example
literature review	how to obtain alcohol from grapes using fermentation
literature review	how to remove alcohol from wine using distillation
controlled experiment	investigate temperature needed to heat wine to evaporate ethanol
case study	interview with the manufacturer of low-alcohol wine, including details of how to avoid removal of other volatile components which might influence the taste, while still removing the ethanol
process development	use understanding from other investigations to design a new process to remove the alcohol from the wine

1.2 Planning investigations

1 The independent variable is the single variable that the experimenter manipulates to determine its effect on the dependent variables. The dependent variable is the variable that changes in response to change in the independent variable and it is often the variable that is measured, although another variable may be measured and used to determine the dependent variable. The controlled variables are any factors that may change and affect the result and that must be kept constant through all aspects of the experiment.

2 Objective means that there is no personal bias involved in the observation or experiment. An objective observation can be quantified in some way; for example, by taking measurements that can be repeated by other people. Subjective means that personal bias could influence the observation or experiment; for example, comparing a result to a personal experience to validate it.

3 **a** observation **b** theory **c** hypothesis

4 **a** As the size of the liver pieces decreases and their surface area increases, the rate of the reaction will increase. This is because an increase in the surface area exposed to the reaction causes the amount of the catalyst in the liver (catalase) available to react with the hydrogen peroxide to also increase.

b As the temperature of the half-cell solutions in a $Zn(s)/Zn^{2+}(aq)//Cu^{2+}(aq)/Cu(s)$ galvanic cell increases, the current will increase because the ions in the half-cell solutions will move more rapidly and pass the electricity through the cell more rapidly.

c If enzymes in washing powder are denatured by water at 50°C, then washing fabric at a lower temperature, using the same washing powder, will result in greater stain-removal from the fabric.

5 **a** dependent variable – this will change as the temperature changes, and will be measured.

b controlled variables – both of these would influence the dependent variable, so must be kept constant throughout the experiment.

c independent variable – the temperature of the acid is going to be changed deliberately, and this will affect the mass lost by the system in the first 30 seconds of reaction.

6 **a** 'How does the concentration of copper(II) sulfate in the electrolyte affect the mass of copper deposited on the negative electrode during electrolysis using graphite electrodes?'

b concentration of copper(II) sulfate solution, e.g. 0.005 M, 0.010 M, 0.05 M, 0.1 M, 0.2 M

c the mass of copper formed at the negative electrode, e.g. determined by weighing the electrode before and after the electrolysis

d voltage from power supply, distance between electrodes, temperature of solutions

7 **a** secondary source **b** primary source
c primary source **d** primary source

8 **a** independent variable – the temperature of the electrolyte.

b dependent variable – charge, which can be calculated by measuring the current going through the cell and multiplying it by the time for which the current is flowing. The equation used will be $Q = It$.

c The equipment used to monitor the independent variable would be a thermometer or a temperature probe, and the equipment used to measure the dependent variable would be an ammeter set up in a circuit in series with the electrolytic cell and power supply.

1.3 Data collection and quality

1 a 3 significant figures b 3 significant figures
c 5 significant figures d 2 significant figures

2 a **Accuracy** is how close a measurement is to the **true value**, whereas **precision** is how closely a set of measurements agree with each other.
b **Validity** refers to whether your results measure what the investigation set out to measure.
c **Repeatability** is the consistency of your results when they are repeated many times as trials under the exact same set of experimental conditions, whereas **reproducibility** is the ability for another experimenter to obtain the same results if they replicate your experiment.
d **Resolution** is the smallest change in the measured quantity that causes a perceptible change in the value shown by the measuring instrument.

3 D.

4 a If the experiment addresses the hypothesis and aim, it is a valid experiment.
b If the experiment is repeated and consistent results are obtained, the experiment is reproducible.
c If appropriate equipment Is chosen for the measurements required, then the experiment is precise.

5 Student A has a very accurate average. It is the same as that obtained by the teacher, but the values that were used to obtain the average have a wide range (19.70 – 19.25 = 0.45 mL), so the results are not precise.

Student B is very precise. The range of their results is 19.50 – 19.40 = 0.10 mL, which is the accepted range for concordant results.

Student A's results are slightly less accurate than those of student B, but the mean value is acceptably close.

6

Error	Type of error	Suggested improvement
finding it difficult to finish at the same colour in successive titres during a titration	random	The student should practices their titration technique to improve their ability to stop at the end point. Also, one flask which has the colour of the correct end point should be retained for comparison with subsequent attempts.
using a stopwatch while chatting with your partner and not quite concentrating on the reaction being timed.	random	The student should focus all their attention on the reaction being timed.
reading the initial volume of a burette from below the level of the meniscus	systematic	The student should move the burette down on the stand so that they can read the initial volume at their own eye level, or it this is not possible, then they should stand on a step.
loss of heat to environment during a combustion calorimetry experiment	systematic	Every attempt should be made to retain the heat within the system. This could be done with insulation on the calorimetry can, a lid and even enclosing the equipment to prevent general heat loss, while still allowing oxygen to reach the flame. Ultimately a bomb calorimeter would be a better choice for a combustion reaction.

7 Mistakes are bolded and described below. Columns are numbered for clarity

1	2	3	4	5	6
Trial	Volume of water in can (mL)	Initial temperature of water	Final temperature of water	Initial mass of ethanol and spirit burner	Final mass of ethanol and spirit burner (g)
1	200 **mL**	**15.00**	**45.0**	250.50	**250.50**
2	200 **mL**	**14.0**	**30**	243.35	241.20
3	200 **mL**	**14.551**	**95.0**	**241.50**	239.00

Any three of the following five mistakes could be listed:

1) In column 2 the units are in the heading as required, but are repeated down the column. This is not necessary or desirable.
2) In columns 3, 4 and 5 the units are missing from the heading.
3) In columns 3 and 4 the temperature readings are all recorded to different accuracies. These should be measured with the same thermometer or digital thermometer, so should all be to the same number of decimal places.
4) In column 6 for trial 1, the final mass of the ethanol and the spirit burner is the same as the initial mass. This is clearly a mistake.
5) In column 5 for trial 3, the initial mass is higher than the final mass for trial 2. This seems unlikely, as the spirit burner would have been used for trial 3 directly after trial 2.

8 Method A is repeatable and reproducible. It includes quantities, concentrations and uses the most suitable equipment for the experiment.

Method B does not include quantities for the nitric acid or potassium hydroxide. It does not give instructions to record initial temperatures and it does not use suitable equipment for a calorimetry experiment.

1.4 Data analysis and presentation

TY 1.4.1 a 9.6500×10^4
b 6.25×10^{-5}

TY 1.4.2 ΔH_c(methanol) = -7.3×10^2 kJ mol^{-1}

TY 1.4.3 The experimental value is 6.3% *less than* the theoretical mass gain of the cathode.

TY 1.4.4 There has been a 8.17% *decrease* in the mass.

Key questions

1 a 2.05×10^{-3}
b 9.6500×10^4
c 1.004×10^2
d 4.3×10^{-4}

2 C.

3 C.

4 A trend is a pattern shown in data. Trends can be positive, negative (inverse), proportional, linear or non-linear.

5 a 15.18 mL, 15.24 mL and 15.26 mL
b 15.40 mL
c average concordant titre = 15.23 mL

6 a column graph
b line graph
c scatter graph with line of best fit

7 The experimental value is 15% less than the theoretical value.

8 Improvements with the graph for rate of reaction:

1 This data would be better represented as a line graph, since it is representing continuous data.

2 The scale on the horizontal axis is irregular—each interval is not the same value as the previous one. All intervals would be best represented as 1 unit (minute), given the data available. This axis should also start at 0, where the vertical axis joins the horizontal axis.

3 The horizontal axis is missing its label and units. It should be labelled 'Time' with units indicated as '(minutes)'.

4 The vertical axis should start at 0 rather than 6.

5 The vertical axis is also missing its units. These are most likely to be grams, so should be indicated as '(g)' or '(grams)'.

6 The heading could be more descriptive, indicating the reaction which is involved.

1.5 Evaluation and conclusion

1 C. **2** A.

3 **a** The effect of concentration changes on an aqueous equilibrium between Fe^{3+} ions, SCN^- ions and $FeSCN^{2+}$ ions was investigated.

b The rate of reaction between calcium carbonate and hydrochloric acid was measured and the effect of particle size and concentration on the rate was investigated.

c A calorimeter was calibrated by measuring the increase in temperature that results from a measured input of electrical energy.

4 **a** Random error—the syringe may not slide smoothly, so could misrepresent the volume in some trials.

Systematic error—some carbon dioxide will remain dissolved in the hydrochloric acid solution, so the volume collected will be smaller than it should be.

b

Methodological weakness	How this may have influenced the results	Suggestion for improvement
The syringe did not always slide smoothly.	Not all volumes of CO_2 were as large as they should have been, but this was not by a consistent amount.	Make sure that the syringe slides smoothly before starting the experiment, and clean or replace it between trials.
Some carbon dioxide dissolved in the hydrochloric acid during the experiment.	A smaller volume of CO_2 was collected during the experiment than was correct.	Rather than using the method of collection in a syringe, conduct the experiment on an electronic balance and record the mass loss (due to CO_2 escaping). This method may still have a systematic error due to some CO_2 still dissolving in the acid, but it would be less that in a closed system.

5 **a** To determine the effect of decreasing the distance between copper electrons on the current flowing through a copper-plating electrolytic cell.

b **i** independent variable—the distance between the copper electrodes (measured in mm)

ii dependent variable—the current flowing through the copper-plating electrolytic cell

iii controlled variables—the temperature of the electrolyte; the size of the electrodes; the copper sheet that is used to make the electrodes; the voltage used in the electrolysis

c current, measured using an ammeter

d The effect of decreasing the distance between copper electrons on the current flowing through a copper-plating electrolytic cell was determined and it was found that the current increased as the distance decreased. These results agreed with the hypothesis.

e

Weakness with the method	How this might be improved
As the electrolysis proceeds, some copper could drop off the cathode. This could interfere with the electrolytic process.	Do not carry out the electrolysis for so long that copper starts to drop off the cathode.
The electrodes could move during the electrolysis, so the distance between them could change.	Use a guide to hold the electrodes at fixed distances apart. This could be made from cardboard and could sit on top of the beaker.

6 **a** The percentage yield is calculated using the formula:

$$\text{percentage yield} = \frac{\text{experimental yield}}{\text{theoretical yield}} \times \frac{100}{1}$$

but the students can't work out the theoretical yield of copper on the cathode because this uses the formula:

$$m(\text{Cu}) = \frac{1}{2} \times \frac{It}{96500} \times 63.5$$

and they don't know what the current passing through the circuit was.

b Problems with the method—three of the following:

- There was no ammeter in the circuit, so current could not be measured and controlled. This would have introduced a systematic error of unknown size into the experiment.
- Because the electrodes were pinned to the sides of the two beakers and a 250 mL beaker is larger than a 100 mL beaker, the distance between the two beakers would be different, and the current in the 250 mL beaker is likely to be smaller than that in the 100 mL beaker due to a greater resistance experienced by the ions through the greater volume of solution between electrodes.
- The electrodes were only dried before weighing. There was probably an indeterminate amount of copper(II) sulfate left on the electrodes after each trial, which would have increased the mass gain by an unknown amount.
- With each trial only been carried out once, the impact of random errors would be significant.

c Improvements:

- Include an ammeter in the circuit, so that the current can be determined and controlled.
- Use the same size beaker for all of the trials, so that the distance between the electrodes is constant.
- Rinse the electrodes carefully with water, and then with propanone (acetone) before air drying to ensure that they are clean and that they dry quickly and that no copper(II) sulfate residue remains on the electrodes when they are weighed.
- Repeat each trial at least three times, then average the results (mass increase at cathode) to reduce the effect of random errors.

1.6 Reporting investigations

1 To minimise the words in a poster, you can use visual support such as an annotated diagram of the equipment set up, a flow chart to summarise the key steps of the method, and tables and graphs to represent results. It is usually accepted that tables, charts and graphs are not included in the word count, but you should check with your teacher for their preference.

2 D. **3** A. **4** A.

5

Resource type	Information about the reference	Correct format in a bibliography, using APA referencing format
print book	Title of book: Molecules with silly or unusual names Edition: 1st edition Author: Paul W. May Date published: 2008 Page: 64 Publisher: Imperial College Press	May, P. (2008). *Molecules with silly or unusual names* (p. 64). Imperial College Press.
journal article	Title of article: The nature hackers Author: Stenmark, I. E. Journal name: Education in Chemistry Volume: Vol. 56 Pages: 12–14 Date published: January 2019	Stenmark, I. E. (2019). The nature hackers. *Education in Chemistry*, 56, 12–14.
internet	Website owner: Royal Society of Chemistry Website title/description: The rate of reaction of magnesium with hydrochloric acid Date posted: 2016 Website address: www.edu.rsc.org/experiments/the-rate-of-reaction-of-magnesium-with-hydrochloric-acid/1916.article	Royal Society of Chemistry. (2016). *The rate of reaction of magnesium with hydrochloric acid*. www.edu.rsc.org/experiments/the-rate-of-reaction-of-magnesium-with-hydrochloric-acid/1916.article

6 **a** As the age of the vegetable juices increases, the concentration of vitamin C decreases. This is why the juices have a shelf-life.

b **i** controlled variables

ii dependent variable

iii independent variable

c **i** In-text citation: (Food Standards Australia New Zealand, 2016)

ii Full citation in the reference list: Food Standards Australia New Zealand. (2016). *Australia New Zealand Food Standards Code: Standard 2.9.2 Food for Infants.* https://www.foodstandards.gov.au/code/Documents/2.9.2%20Food%20for%20infants%20v157.pdf

iii This source is reliable because it is a government organisation which sets food standards for Australia and New Zealand. As a government organisation, it would not be biased towards any particular brand of food.

Chapter 2 Carbon-based fuels

2.1 Exothermic and endothermic reactions

1 C.

2 In an exothermic reaction, the chemical potential of the **products** is lower than that of the **reactants**. Energy is **released to** the surroundings. The sign of ΔH will be **negative**.

3 A value of ΔH is negative for exothermic reactions where the enthalpy of the products is less than that of the reactants. This means the reaction releases energy.

4 Activation energy is the energy required to break the bonds of reactants. As both endothermic and exothermic reactions have reactants, both require activation energy.

5 **a** 0.0258 kJ **b** 2.63×10^4 kJ **c** 6.6×10^6 kJ

6 **a** There are ionic bonds in ammonium chloride between the ammonium and chloride ions. These are strong bonds that need to be broken for the reaction to occur.

b absorb energy

c decrease

7 **a** endothermic

b The total enthalpy of the product (HI) is greater than that of the reactants (hydrogen gas and iodine gas).

c The activation energy is greater than the ΔH value.

2.2 Types of fuels

1 A fuel is a substance with stored chemical potential energy that can be released relatively easily. When natural gas burns in air it combines with oxygen to form carbon dioxide and water, releasing relatively high amounts of energy.

2 A non-renewable fuel cannot be replenished at the rate at which it is consumed. Renewable fuels are those that can be replenished at the rate at which they are consumed.

3 **a** biogas

b The biogas is usually burned in a generator on-site to produce electrical energy.

4 **a** biodiesel

b in a diesel engine in a vehicle

5 Biodiesel has been made from plants. When the plants grow, they absorb CO_2, compensating for the CO_2 produced during combustion.

6 The infrastructure to produce biogas is expensive. A farm or industry would need to be a big operation to justify the cost. The waste used is not suitable for transport to a central energy hub, therefore it is not easy for producers to share the cost.

7 Australia's crops such as canola are used to produce food products. Widespread use of biodiesel might lead to food shortages. There is also an impact on the limited amount of fertile farming land if crop production is increased.

2.3 Fuel sources for the body

CSA: Lavoisier and combustion

1 In respiration, glucose reacts with oxygen to form carbon dioxide and water. If glucose is burned in a flame, the reaction will be exactly the same. The equation for the reaction is:

$$C_6H_{12}O_6(aq) + 6O_2(g) \rightarrow 6CO_2(g) + 6H_2O(l)$$

2 Energy is required to melt ice. The more energy produced by the subject of the experiment, the greater the quantity of ice that will melt.

3 If a reaction was using up oxygen in a closed container, the guinea pig would struggle to breathe and appear lethargic (or die). If a reaction produced oxygen, the guinea pig would breathe and move easily.

4 Lavoisier would need to try and control the temperature of the room and minimise all heat losses from the subject and from the equipment. He would endeavour to monitor the mass of water consumed by the subject and the mass change of the air the subject breathed in and the air the subject breathed out. Perspiration, urine output and faeces output would also impact on the mass of the subject.

Key questions

1 carbon dioxide, water and sunlight

2 **a** Glucose is the molecule that can be formed in plants through photosynthesis storing energy from the Sun.

b Glucose can be polymerised in plants to starch or to cellulose, both very important molecules. Plants can also use glucose to produce energy.

c Glucose can be polymerised to glycogen in humans for energy storage or burned in muscle cells for energy.

3 In **cellular** respiration, glucose is used by cells to obtain energy. Aerobic respiration is an **exothermic** process in which the glucose is **oxidised** by **oxygen**. A relatively large amount of energy is **released** during aerobic respiration and can be used by the cells of the body.

4 a respiration: exothermic
  b digestion of fish oil: exothermic
  c formation of cellulose from glucose in plants: endothermic
  d photosynthesis: endothermic
  e digestion of starch: exothermic

5 Carbohydrates will release less energy than fats as they are already 'partially' oxidised. During combustion, fuels react and bond with oxygen. As carbohydrates contain a relatively high proportion of oxygen atoms, some of the oxidation has already occurred.

6 Oxidation, in the case of cellular respiration, involves reaction with oxygen.

2.4 Bioethanol

1 a

```
      H    H
      |    |
  H — C —  C — O — H
      |    |
      H    H
```

  b Ethanol has an -OH group that can form hydrogen bonds with water, making it highly soluble.
  c $C_2H_6O(l) + 3O_2(g) \longrightarrow 2CO_2(g) + 3H_2O(l)$

2 a disadvantage b disadvantage
  c advantage d advantage
  e disadvantage f advantage

3 Molasses contains many smaller carbohydrates that can be fermented directly to ethanol. Cellulose in forest waste must first be broken down to smaller carbohydrates before fermentation occurs. The smaller carbohydrates in molasses are soluble in water, making it easier to have enzymes react with them.

4 Fermentation is a biochemical process, involving microorganisms (enzymes). If temperatures are too high, the microorganisms are destroyed and the reaction stops.

5 An increase in the use of bioethanol-sourced E10 is likely to lower total CO_2 emissions. CO_2 emissions from the vehicle will increase with E10 use, but this impact is offset by the absorption of CO_2 during the growth of the plants used to make the bioethanol.

6 Bioethanol and biodiesel are used in vehicles. They have to be transported to where the vehicles are. Biogas is burnt in small-scale generators that can be placed easily at the site of the gas production.

7 Bioethanol is produced as an aqueous solution. It needs to be separated from the water in the solution. Biodiesel or biogas are not produced as aqueous solutions.

2.5 Energy from the combustion of fuels

TY 2.5.1 $2C_6H_{14}(l) + 19O_2(g) \longrightarrow 12CO_2(g) + 14H_2O(l)$

TY 2.5.2 $2CH_3OH(l) + 3O_2(g) \longrightarrow 2CO_2(g) + 4H_2O(l)$

TY 2.5.3 $CH_3OH(l) + O_2(g) \longrightarrow CO(g) + 2H_2O(l)$

TY 2.5.4 1.82×10^5 kJ

TY 2.5.5 $\Delta H = -147$ kJ

CSA: Explosives—a blast of chemical energy

1 Exothermic. The reaction releases large amounts of energy in forms of heat, light and sound.

2 29 mol products : 4 mol reactants. An explosion occurs when liquid or solid reactants turn rapidly to a gas. With the number of moles of gaseous products being much higher than the number of moles of reactants, the change in volume is likely to be even more rapid.

3 2912 kJ

4 Activation energy is always required to break bonds in the reactants so a reaction can proceed. The trigger will provide activation energy.

Key questions

1 $C_6H_6(l) + \frac{15}{2}O_2(g) \longrightarrow 6CO_2(g) + 3H_2O(l)$
(The equation coefficients can also be shown as being double the values above to make them all whole numbers)

2 $C_2H_5OH(l) + 2O_2(g) \longrightarrow 2CO(g) + 3H_2O(l)$

3 $C_5H_{12}(l) + 8O_2(g) \longrightarrow 5CO_2(g) + 6H_2O(l)$ $\Delta H = -3509$ kJ

4 33.7 kJ g^{-1}.

5 a 1.39×10^4 kJ b 4.87×10^5 kJ c 1.19×10^7 kJ

6 The heat of combustion of a fuel can be defined as the heat energy released when a specified amount (e.g. 1 g, 1 L, 1 mol) of a substance burns completely in oxygen.

7 a complete combustion
  b Complete combustion takes the oxidation of the fuel to its limit. Incomplete combustion is only partial combustion.

8 a Photosynthesis is an endothermic process as sunlight is required for energy and the enthalpy of reaction is a positive value.
  b −1402 kJ

9 The combustion of octane to form carbon dioxide and steam will release 9 × 40.7 = 366.3 kJ less energy than the combustion of octane to form carbon dioxide and liquid water.

Chapter 2 review

1 B. 2 C. 3 D. 4 C.

5 a endothermic b exothermic
  c endothermic d exothermic

6 The formation of fossil fuels is a process that occurs over millions of years. The organic matter produced by plants and animals undergoes complex changes as it is subjected to heat and pressure under tonnes of mud and sand. Once the current reserves of fossil fuels have been used, they will not be replaced in the foreseeable future.

7 a 1.01×10^3 kJ b 373 kJ

8 a the covalent bonds in hydrogen and in oxygen
  b the covalent bonds in water
  c The combustion of hydrogen is exothermic, so the energy released when the bonds in water form will be greater than the energy required to break the bonds in hydrogen and oxygen gases.
  d The temperature of the surroundings will increase as the reaction is exothermic.

9 a $6CO_2(g) + 6H_2O(l) \xrightarrow{\text{sunlight}} C_6H_{12}O_6(aq) + 6O_2(g)$
  b glucose
  c $C_6H_{12}O_6(aq) + 6O_2(g) \longrightarrow 6CO_2(g) + 6H_2O(l)$

10 a Many possibilities such as sugarcane waste, wheat husks and fruit waste.
  b $C_6H_{12}O_6(aq) \longrightarrow 2C_2H_6O(aq) + 2CO_2(g)$
  c The distillation column is used to purify the ethanol solution, separating the water and ethanol.

11 $C_2H_6O(l) + 3O_2(g) \longrightarrow 2CO_2(g) + 3H_2O(l)$ $\Delta H = -1360$ kJ

12 a methane and carbon dioxide
  b The composition of biogas depends on the original material from which it is sourced and the method of decomposition.

13 a true b false c false d false

14 Reversing a chemical reaction involves the same bonds as the forward reaction, but the bonds that were formed are now broken and the bonds that were broken are reformed. The overall energy change has the same magnitude for both processes.

15 a For this exothermic reaction, the overall energy of the bonds in the products will be less than the overall energy of the bonds in the reactants.

b

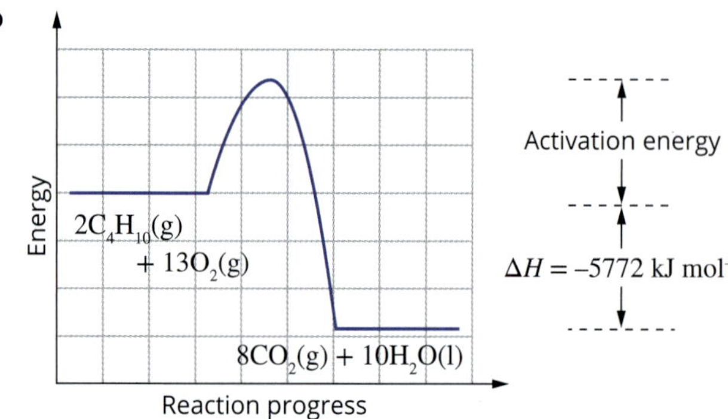

16 a $C_3H_7OH(l) + \frac{9}{2}O_2(g) \rightarrow 3CO_2(g) + 4H_2O(l)$ or

$2C_3H_7OH(l) + 9O_2(g) \rightarrow 6CO_2(g) + 8H_2O(l)$

b $C_5H_{12}(l) + \frac{11}{2}O_2(g) \rightarrow 5CO(g) + 6H_2O(l)$ or

$2C_5H_{12}(l) + 11O_2(g) \rightarrow 10CO(g) + 12H_2O(l)$

17 $C_{17}H_{34}O_2(l) + \frac{49}{2}O_2(g) \rightarrow 17CO_2(g) + 17H_2O(l)$ or

$2C_{17}H_{34}O_2(l) + 49O_2(g) \rightarrow 34CO_2(g) + 34H_2O(l)$

18 a exothermic

b $\Delta H = 2 \times -2619 = -5238$ kJ

19 $2H_2(g) + O_2(g) \rightarrow 2H_2O(l) \quad \Delta H = -572$ kJ or

$H_2(g) + \frac{1}{2}O_2(g) \rightarrow H_2O(l) \quad \Delta H = -286$ kJ

20 The combustion of biodiesel produces CO_2.

For example:

$$C_{18}H_{36}O_2(l) + 26O_2(g) \rightarrow 18CO_2(g) + 18H_2O(l)$$

However, a similar amount of CO_2 is absorbed by the canola during photosynthesis while the crop grows:

$$6CO_2(g) + 6H_2O(l) \xrightarrow{\text{sunlight}} C_6H_{12}O_6(aq) + 6O_2(g)$$

The farming equipment required and the transport of the canola also produce CO_2 emissions, so the use of biodiesel is not completely carbon neutral.

21 Biogas is formed through the action of anaerobic bacteria on organic matter. Anaerobic bacteria operate in the absence of oxygen. Methane is the most abundant component of biogas, followed by carbon dioxide and then a mix of other gases in low percentages.

22

Advantages	Disadvantages
fewer particulate emissions than natural gas	can only be used on site
renewable	raw material needs to be heated in winter
less reliance on fossil fuels	cheese waste previously used as a fertiliser
fewer net CO_2 emissions than natural gas	
lower sulfur content than natural gas	

23 a exothermic

b activation energy forward reaction: 100 kJ

activation energy back reaction: 300 kJ

c 200 kJ. It is the reverse of the forward reaction.

24 a i 4.79×10^4 kJ **ii** 4.95×10^4 kJ

iii 5.19×10^4 kJ

b The data shows that the heat of combustion in kJ/kg decreases slightly as the alkane molecule gets longer.

25 a Because the reaction is exothermic, 1 mol of CO(g) and 0.5 mol of $O_2(g)$ (the reactants) has a higher enthalpy value than 1 mol of $CO_2(g)$ (the products).

b i –566 kJ **ii** +566 kJ

Chapter 3 Obtaining energy from fuels

3.1 STOICHIOMETRY INVOLVING COMBUSTION OF FUELS

TY 3.1.1 10.9 kg

TY 3.1.2 513 L

TY 3.1.3 50 mL

TY 3.1.4 520 MJ

TY 3.1.5 139 L

TY 3.1.6 32.4 L

CSA: Bioethanol as a fuel for cars to reduce greenhouse gases

1 $C_2H_5OH(l) + 3O_2(g) \rightarrow 2CO_2(g) + 3H_2O(g) \quad \Delta H = -1360$ kJ

2 a 1.91 kg **b** 1.08×10^3 L

3 a 39.3 kg **b** 1.16×10^3 MJ

4 0.0647 g

Key questions

1 a 1 mol **b** 3 mol **c** 0.75 L

2 a 1.00 mol of methane

b 1.00 mol of carbon dioxide, 2.00 mol of H_2O

c 80.0 g

3 a 4.68 MJ **b** 5.20 MJ **c** 0.629 MJ

4 a 702 g **b** 618 g

5 a $C_3H_8(g) + 5O_2(g) \rightarrow 3CO_2(g) + 4H_2O(g)$

b 24.0 g **c** 22.5 L

6 11.95 MJ

7 a 498 kJ **b** 11.6 kJ

8 a The first experiment using 1 mol of methane releases the most energy.

b The combustion of methanol will release more carbon dioxide to produce the same amount of energy as methane.

3.2 Determination of limiting reactants or reagents

TY 3.2.1 a O_2 **b** 0.014 mol

TY 3.2.2 84.5 L

Key questions

1 **D** Calculate the number of moles of iron and oxygen.

B Refer to the balanced equation

E Use mole ratios to determine which reactant is limiting.

C Calculate the number of moles of iron oxide that forms.

A Calculate the mass of iron oxide that forms.

2

Propane molecules available	Oxygen molecules available	Carbon dioxide molecules produced	Propane molecules in excess	Oxygen molecules in excess
2	15	6	0	5
300	1200	720	60	0
2.5 mol	10 mol	6 mol	0.5 mol	0 mol

3 64.8 g

4 a O_2 in excess, 15.4 g **b** 13.4 g

c 8.21 g **d** 37.0 g

5 6.67 L **6** 8.92 L

3.3 Calculating heat energy released

TY 3.3.1 70.5 kJ

TY 3.3.2 1.26×10^3 kJ mol^{-1}; $C_2H_5OH(l) + 3O_2(g) \rightarrow 2CO_2(g) + 3H_2O; \quad \Delta H = -1.26 \times 10^3$ kJ

TY 3.3.3 21.8 kJ g^{-1}

CSA: The energy of candlelight

1 $C_{20}H_{42}(s) + \frac{61}{2}O_2(g) \rightarrow 20CO_2(g) + 21H_2O(l)$ or

$2C_{20}H_{42}(s) + 61O_2(g) \rightarrow 40CO_2(g) + 42H_2O(l)$

2 The liquid oil is heated by the burning wick, and changes state from liquid to gas. The gas is then ignited and burns. During the combustion reaction, carbon-carbon and carbon-hydrogen bonds in the oil, and oxygen-oxygen double bonds break, and new bonds are formed between hydrogen and oxygen in the water, and double bonds between carbon and oxygen in carbon dioxide. The energy required to break the bonds of the reactants (energy absorbed) is less than that released by the formation of bonds in the products, so more energy is released than is used, and the reaction is exothermic.

3 0.728 g

Key questions

1 **a** 10.5 kJ **b** 132 kJ **c** 62.7 kJ

2 33.0°C

3 2.23×10^3 kJ mol^{-1}

4 48.1 kJ g^{-1}

5 3.01 g

6 3.03 g

3.4 Solution calorimetry

TY 3.4.1 323 J $°C^{-1}$

TY 3.4.2 435 J $°C^{-1}$

TY 3.4.3 $\Delta H = +5.38$ kJ mol^{-1}

Key questions

1 Insulating a calorimeter improves the **accuracy** of measurement of the amount of energy **released** or absorbed by a chemical reaction.

Heat energy can be **lost** from a calorimeter, so a **lid** is a useful form of insulation.

If the reaction occurring in a calorimeter is exothermic, the temperature of the water **increases**. If the reaction occurring is **endothermic** the temperature of the water in a calorimeter decreases.

2 347 J $°C^{-1}$

3 18.2 kJ mol^{-1}

4 +15.86 kJ mol^{-1}

5 **a** 7.2×10^2 J $°C^{-1}$

b If 50 g of water was used instead of 100 g, the calibration factor would be halved, since the same amount of energy was being added to half the volume of water, and the temperature increase would double. Dividing the same value for *VIt* by a value that is double the previous one will produce a calibration factor that is half the previous one.

6 7.9×10^2 J $°C^{-1}$

7

Mistake or error	Calibration factor is too small	Calibration factor would be correct	Calibration factor is too large	Explanation for answer
The lid is left off during calibration.			✓	Heat escapes during the calibration, so the temperature change recorded is smaller than is correct. The *CF* is divided by a number that is too small, so it will be too large.
During electrical calibration, the voltmeter had a systematic error making it read too low.	✓			The calculation of energy entering the calorimeter will be too small, so the numerator will be smaller than it should be, and the calibration factor will be too small.
During chemical calibration, not all of the solid KNO_3 dissolved in the water of the calorimeter.			✓	The energy absorbed from the water will be recorded as a larger figure than actually occurs. The actual temperature change will be smaller than it should be, so the calibration factor will be larger than it should be.

3.5 ENERGY FROM FUELS AND FOOD

TY 3.5.1 60.3%

TY 3.5.2 443 kJ

TY 3.5.3 2.68×10^3 kJ mol^{-1}

TY 3.5.4 11 kJ g^{-1}

Key questions

1 **a** 9.41 g **b** 4.32 g

2 **a** An energy transformation is when energy is converted from one form, such as chemical energy, to another form, such as heat energy.

b With each energy transformation, some energy is lost to the environment, so as the number of energy transformations increases, the percentage of energy lost increases. This means that the usefulness of the fuel decreases.

3 **a** 395.9 kJ **b** 49.5 %

4 **a** 5.54 kg **b** 5141 kJ

5 9.1 kJ g^{-1}

6 76.6% fat

7 **a** 5.10 kJ **b** 1.09×10^3 kJ mol^{-1} **c** 53.9%

d Some suggestions as to how the student could improve the accuracy of the result:

- Insulate the can that holds the water and add a lid to the can
- Try to prevent heat loss, by setting up some sort of insulation around the whole apparatus, while still allowing enough oxygen to reach the flame
- Minimise the distance between the flame and the can that is being heated.

8 725 kJ mol^{-1}

Chapter 3 review

1 D.

2 D.

3 **a** 20.1 g **b** 24.4 g **c** 11.0 g

4 16.7 g

5 3.7 tonnes

6 6.25 L

7 2.36 kJ

8 29.2 MJ

9 **a** O_2 in excess; 909 g **b** 507 L **c** 326 L

10 B.

11 A.

12 1.30 kg

13 **a** $KNO_3(s) \xrightarrow{H_2O(l)} K^+(aq) + NO_3^-(aq)$

b 1.16 kJ$°C^{-1}$ **c** 783 J **d** +78.3 kJ mol^{-1}

14 7.0 kJ g^{-1}

15 The calibration factor would be measured as being smaller than it should be.

16 **a** 1022 kJ

b The runner should consume 3 energy bars.

17 **a** 5.0 L **b** 10 L **c** 16 g

18 **a** 2.09 g **b** 1.67×10^3 MJ

19 46.2 g

20 **a** O_2 in excess; 100 mL **b** 560 mL

c increase of 80 mL **d** $\Delta H = -2.22 \times 10^3$ kJ mol^{-1}

21 Since the amount of CO_2 present (3.23 mol) is much greater than the amount of CO_2 (0.435 mol) that would be produced by the combustion of 10.0 g of ethanol, the ethanol is the limiting reagent.

22 20.7 kJ g^{-1}

23 32.9 kJ g^{-1}

24 $C_2H_6(g) + \frac{7}{2}O_2(g) \rightarrow 2CO_2(g) + 3H_2O(l)$; $\Delta H = -1538$ kJ mol^{-1}

25 **a** 16.0 kJ g^{-1}

b The calculated heat of combustion is likely to be lower than the actual heat of combustion of the wood. It is likely that all the energy from the burning wood was not transferred to the water. Parts of the wood may not be completely combusted, thus reducing the energy output of burning it. If the wood was not completely dry, this would also decrease the heat of combustion as energy is required to remove the water before it is able to combust.

26 5643 kJ mol^{-1}

27 **a** If heat loss is not taken into account, ΔT will be too small.

b The calibration factor calculated for the calorimeter will be too large.

c The value of the enthalpy change will increase because the enthalpy change is found using $CF \times \Delta T$.

28 1 The student has multiplied *VIt*(energy) by the change in temperature instead of dividing by the temperature change when calculating the calibration factor.

2 The student has then multiplied the calibration factor by ΔT, which is correct, but the value is too large due to the earlier mistake.

3 The mass has been divided by the energy instead of the energy divided by the mass to calculate the energy content. This results in a significantly lower energy content of the jelly crystals than expected.

4 The student has stated that energy is given out when the jelly crystals are dissolved. The temperature has decreased, so energy has been absorbed. This agrees with the student's statement that the reaction is endothermic.

The student's calculations should be as follows:

$$CF = \frac{6.30 \times 2.40 \times 210}{2.8}$$
$$= 1134 \text{ J}$$

Heat energy per gram to dissolve the jelly crystals:

$$E = CF \times \Delta T$$
$$= 1134 \times 0.5$$
$$= 567 \text{ J}$$

$$\text{Energy per gram} = \frac{567}{85.0}$$
$$= 6.67 \text{ J g}^{-1}$$

6.67 J of energy is taken in when 1 g of jelly crystals dissolves. This is an endothermic reaction.

Chapter 4 Redox reactions

4.1 Oxidation and reduction

1 When a reducing agent, such as Zn, reacts with an oxidising agent, such as $\mathbf{Cl_2}$, an ionic compound is formed. The reducing agent, Zn, **loses** electrons (is **oxidised**) and at the same time the oxidising agent, Cl_2, **gains** electrons (is **reduced**). In this case, the products are Zn^{2+} and $\mathbf{Cl^-}$, which form $ZnCl_2$.

2 **a** oxidation **b** reduction **c** reduction

d oxidation **e** reduction **f** reduction

3 $Ni(s) \rightarrow Ni^{2+}(aq) + 2e^-$

4 $Al(s) \rightarrow Al^{3+}(s) + 3e^-$: oxidation

$S(l) + 2e^- \rightarrow S^{2-}(s)$: reduction

5 **a** oxidation half-equation: $Zn(s) \rightarrow Zn^{2+}(s) + 2e^-$

reduction half-equation: $O_2(g) + 4e^- \rightarrow 2O^{2-}(s)$

oxidising agent: $O_2(g)$, reducing agent: Zn(s)

b oxidation half-equation: $Ca(s) \rightarrow Ca^{2+}(s) + 2e^-$

reduction half-equation: $Cl_2(g) + 2e^- \rightarrow 2Cl^-(s)$

oxidising agent: $Cl_2(g)$, reducing agent: Ca(s)

c oxidation half-equation: $Al(s) \rightarrow Al^{3+}(s) + 3e^-$

reduction half-equation: $Br_2(l) + 2e^- \rightarrow 2Br^-(s)$

oxidising agent: $Br_2(l)$, reducing agent: Al(s)

6 Because metal *M* forms a compound with formula MO with oxygen, the ion of metal *M* must be M^{2+}.

a $Ag^+(aq) + e^- \rightarrow Ag(s)$

$M(s) \rightarrow M^{2+}(aq) + 2e^-$

$M(s) + 2Ag^+(aq) \rightarrow M^{2+}(aq) + 2Ag(s)$

b The silver half-equation is reduction, the unknown metal half-equation is oxidation.

c *M*(s) is the reducing agent, $Ag^+(aq)$ is the oxidising agent.

4.2 Oxidation numbers

TY 4.2.1 $\overset{+1}{Na}\overset{+5}{N}\overset{-2}{O_3}$

TY 4.2.2 Row 2: The oxidation number of copper decreased.

Row 4: The oxidation number of hydrogen increased.

$Cu\overset{-2}{O}(s) + H_2(g) \rightarrow Cu(s) + H_2\overset{-2}{O}(l)$

The oxidation number of oxygen has not changed.

4.2 Key questions

1 **a** +2 **b** +4 **c** −4

d 0 **e** +4

2 D.

3 **a** Ca: +2; O: −2 **b** Ca: +2; Cl: −1

c H: +1; S: +6; O: −2 **d** Mn: +7; O: −2

e F: 0 **f** S: +4; O: −2

g Na: +1; N: +5; O: −2 **h** K: +1; Cr: +6; O: −2

4 **a** **i** $\overset{0}{Mg}(s) + \overset{0}{Cl_2}(g) \rightarrow \overset{+2}{Mg}\overset{-1}{Cl_2}(s)$

ii Oxidising agent Cl_2; reducing agent Mg

b **i** $2\overset{+4}{S}\overset{-2}{O_2}(g) + \overset{0}{O_2}(g) \rightarrow \overset{+6}{S}\overset{-2}{O_3}(g)$

ii Oxidising agent O_2; reducing agent SO_2

c **i** $\overset{+3}{Fe_2}\overset{-2}{O_3}(s) + 3\overset{+2}{C}\overset{-2}{O}(g) \rightarrow 2\overset{0}{Fe}(s) + 3\overset{+4}{C}\overset{-2}{O_2}(g)$

ii Oxidising agent Fe_2O_3; reducing agent CO

d **i** $2\overset{+2}{Fe^{2+}}(aq) + \overset{+1}{H_2}\overset{-1}{O_2}(aq) + 2\overset{+1}{H^+}(aq) \rightarrow 2\overset{+3}{Fe^{3+}}(aq) + 2\overset{+1}{H_2}\overset{-2}{O}(l)$

ii Oxidising agent H_2O_2; reducing agent Fe^{2+}

5

Redox reaction with oxidation numbers	Conjugate redox pair (oxidation process)	Conjugate redox pair (reduction process)
$\overset{0}{Na}(s) + \overset{+1}{Ag^+}(aq) \rightarrow \overset{+1}{Na^+}(aq) + \overset{0}{Ag}(s)$	$Na(s)/Na^+(aq)$	$Ag^+(aq)/Ag(s)$
$\overset{0}{Zn}(s) + \overset{+2}{Cu^{2+}}(aq) \rightarrow \overset{+2}{Zn^{2+}}(aq) + \overset{0}{Cu}(s)$	$Zn(s)/Zn^{2+}(aq)$	$Cu^{2+}(aq)/Cu(s)$
$2\overset{0}{K}(s) + \overset{0}{Cl_2}(g) \rightarrow 2\overset{+1}{K}\overset{-1}{Cl}(s)$	$K(s)/K^+(s)$	$Cl_2(g)/Cl^-(s)$

6 a False—this should read:

'In the following reaction sodium is **oxidised** and the oxidation number of hydrogen **changes from +1 to 0** in the equation.'

$$2Na(s) + 2H_2O(l) \longrightarrow 2NaOH(aq) + H_2(g)$$

b False—this should read:

'When the dichromate ion, $Cr_2O_7^{2-}$, reacts to form Cr^{3+}, the chromium is **reduced**.'

$$2\overset{+1}{K}\overset{+7}{Mn}\overset{-2}{O_4}(aq) + 5\overset{+1}{H_2}\overset{-2}{S}(aq) + 6\overset{+1}{H}\overset{-1}{Cl}(aq) \rightarrow$$
$$2\overset{+4}{Mn}\overset{-1}{Cl_2}(aq) + 5\overset{0}{S}(s) + 2\overset{+1}{K}\overset{-1}{Cl}(aq) + 8\overset{+1}{H_2}\overset{-2}{O}(l)$$

c False—this should read:

'In the following reaction the oxidising agent is $\mathbf{MnO_4^-}$ and $\mathbf{H_2S}$ is oxidised.'

$$2KMnO_4(aq) + 5H_2S(aq) + 6HCl(aq) \rightarrow$$
$$2MnCl_2(aq) + 5S(s) + 2KCl(aq) + 8H_2O(l)$$

4.3 Writing complex redox equations

TY 4.3.1 $MnO_4^-(aq) + 4H^+(aq) + 3e^- \longrightarrow MnO_2(s) + 2H_2O(l)$

TY 4.3.2 $SO_3^{2-}(aq) + 8H^+(aq) + 6e^- \longrightarrow H_2S(g) + 3H_2O(l)$

$ClO^-(aq) + 2H_2O(l) \longrightarrow ClO_3^-(aq) + 4H^+(aq) + 4e^-$

$2SO_3^{2-}(aq) + 4H^+(aq) + 3ClO^-(aq) \longrightarrow 2H_2S(g) + 3ClO_3^-(aq)$

TY 4.3.3 $Cu(s) + 2OH^-(aq) \longrightarrow Cu(OH)_2(s) + 2e^-$

TY 4.3.4 $4OH^-(aq) \longrightarrow O_2(g) + 2H_2O(l) + 4e^-$

TY 4.3.5 $3SO_3^{2-}(aq) + 2MnO_4^-(aq) + H_2O(l) \longrightarrow$
$3SO_4^{2-}(aq) + 2MnO_2(s) + 2OH^-(aq)$

CSA: Alcohol and lives lost on the road

1 a $C_2H_5OH(aq) + H_2O(l) \longrightarrow CH_3COOH(aq) + 4H^+(aq) + 4e^-$

b $Cr_2O_7^{2-}(aq) + 14H^+(aq) + 6e^- \longrightarrow 2Cr^{3+}(aq) + 7H_2O(l)$

c $3C_2H_5OH(aq) + 2Cr_2O_7^{2-}(aq) + 16H^+(aq) \longrightarrow$
$3CH_3COOH(aq) + 4Cr^{3+}(aq) + 11H_2O(l)$

2 This colour change is shown in Figure 4.3.3. The colour change was from orange to green. The dichromate ions, $Cr_2O_7^{2-}$ were orange and the chromium(III), Cr^{3+}, ions were green.

3 4.125 g

Key questions

1 a $VO_2^+(aq) + 2H^+(aq) + e^- \longrightarrow VO^{2+}(aq) + H_2O(l)$

b $MnO_4^-(aq) + 4H^+(aq) + 3e^- \longrightarrow MnO_2(s) + 2H_2O(l)$

c $SO_3^{2-}(aq) + H_2O(l) \longrightarrow SO_4^{2-}(aq) + 2H^+(aq) + 2e^-$

d $S_2O_3^{2-}(aq) + H_2O(l) \longrightarrow 2SO_2(g) + 2H^+(aq) + 4e^-$

e $NH_4^+(aq) + 3H_2O(l) \longrightarrow NO_3^-(aq) + 10H^+(aq) + 8e^-$

f $SO_4^{2-}(aq) + 10H^+(aq) + 8e^- \longrightarrow H_2S(g) + 4H_2O(l)$

2 a $VO_2^+(aq) + H_2O(l) + e^- \longrightarrow VO^{2+}(aq) + 2OH^-(aq)$

b $MnO_4^-(aq) + 2H_2O(l) + 3e^- \longrightarrow MnO_2(s) + 4OH^-(aq)$

c $SO_3^{2-}(aq) + 2OH^-(aq) \longrightarrow SO_4^{2-}(aq) + H_2O(l) + 2e^-$

3 a $Cu(s) \longrightarrow Cu^{2+}(aq) + 2e^-$

b $NO_3^-(aq) + 4H^+(aq) + 3e^- \longrightarrow NO(g) + 2H_2O(l)$

c $3Cu(s) + 2NO_3^-(aq) + 8H^+(aq) \longrightarrow 3Cu^{2+}(aq) + 2NO(g) + 4H_2O(l)$

4 a $Fe^{2+}(aq) \longrightarrow Fe^{3+}(aq) + e^-$

$Cr_2O_7^{2-}(aq) + 14H^+(aq) + 6e^- \longrightarrow 2Cr^{3+}(aq) + 7H_2O(l)$

$Cr_2O_7^{2-}(aq) + 14H^+(aq) + 6Fe^{2+}(aq) \longrightarrow$
$2Cr^{3+}(aq) + 7H_2O(l) + 6Fe^{3+}(aq)$

b $SO_3^{2-}(aq) + H_2O(l) \longrightarrow SO_4^{2-}(aq) + 2H^+(aq) + 2e^-$

$MnO^{4-}(aq) + 8H^+(aq) + 5e^- \longrightarrow Mn^{2+}(aq) + 4H_2O(l)$

$5SO_3^{2-}(aq) + 2MnO_4^-(aq) + 6H^+(aq) \longrightarrow$
$5SO_4^{2-}(aq) + 2Mn^{2+}(aq) + 3H_2O(l)$

c $MnO_2(s) + 4H^+(aq) + 2e^- \longrightarrow Mn^{2+}(aq) + 2H_2O(l)$

$2Cl^-(aq) \longrightarrow Cl_2(g) + 2e^-$

$MnO_2(s) + 4H^+(aq) + 2Cl^-(aq) \longrightarrow Mn^{2+}(aq) + 2H_2O(l) + Cl_2(g)$

5 a i Ce^{4+} has been reduced to Ce^{3+} and H_2S has been oxidised to S.

ii $Ce^{4+}(aq) + e^- \longrightarrow Ce^{3+}(aq)$

$H_2S(aq) \longrightarrow S(s) + 2H^+(aq) + 2e^-$

iii $2Ce^{4+}(aq) + H_2S(aq) \longrightarrow 2Ce^{3+}(aq) + S(s) + 2H^+(aq)$

b i NO_3^- has been reduced to NO and Fe^{2+} has been oxidised to Fe^{3+}.

ii $NO_3^-(aq) + 4H^+(aq) + 3e^- \longrightarrow NO(g) + 2H_2O(l)$

$Fe^{2+}(aq) \longrightarrow Fe^{3+}(aq) + e^-$

iii $NO_3^-(aq) + 4H^+(aq) + 3Fe^{2+}(aq) \longrightarrow$
$NO(g) + 2H_2O(l) + 3Fe^{3+}(aq)$

c i H_2O_2 has been reduced to H_2O and Br^- has been oxidised to Br_2.

ii $H_2O_2(aq) + 2H^+(aq) + 2e^- \longrightarrow 2H_2O(l)$

$2Br^-(aq) \longrightarrow Br_2(l) + 2e^-$

iii $H_2O_2(aq) + 2H^+(aq) + 2Br^-(aq) \longrightarrow 2H_2O(l) + Br_2(l)$

d i MnO_2 is reduced to Mn^{2+} and S is oxidised to SO_2.

ii $MnO_2(s) + 4H^+(aq) + 2e^- \longrightarrow Mn^{2+}(aq) + 2H_2O(l)$

$S(s) + 2H_2O(l) \longrightarrow SO_2(g) + 4H^+(aq) + 4e^-$

iii $2MnO_2(s) + 4H^+(aq) + S(s) \longrightarrow 2Mn^{2+} + 2H_2O(l) + SO_2(g)$

6 a i MnO_4^- has been reduced to MnO_2 and Br^- has been oxidised to $BrO_3^-(aq)$

ii $Br^-(aq) + 6OH^-(aq) \longrightarrow BrO_3^-(aq) + 3H_2O(l) + 6e^-$

$MnO_4^-(aq) + 2H_2O(l) + 3e^- \longrightarrow MnO_2(s) + 4OH^-(aq)$

iii $Br^-(aq) + 2MnO_4^-(aq) + H_2O(l) \longrightarrow$
$BrO_3^-(aq) + 2MnO_2(s) + 2OH^-(aq)$

b i $NO_3^-(aq)$ has been reduced to $NH_3(aq)$ and Al(s) has been oxidised to $Al(OH)_4^-(aq)$

ii $Al(s) + 4OH^-(aq) \longrightarrow Al(OH)_4^-(aq) + 3e^-$

$NO_3^-(aq) + 6H_2O(l) + 8e^- \longrightarrow NH_3(aq) + 9OH^-(aq)$

iii $8Al(s) + 5OH^-(aq) + 3NO_3^-(aq) + 18H_2O(l) \longrightarrow$
$8Al(OH)_4^-(aq) + 3NH_3(aq)$

7 oxidation: $Fe(OH)_2(aq) + OH^-(aq) \longrightarrow Fe(OH)_3(aq) + e^-$

reduction: $H_2O_2(aq) + 2e^- \longrightarrow 2OH^-(aq)$

overall: $2Fe(OH)_2(aq) + H_2O_2(aq) \longrightarrow 2Fe(OH)_3(aq)$

Chapter review

1 a false b true c false

2 D. 3 C.

4 a Ag(s) b Mg(s) c Zn(s) d Al(s)

5 A. 6 C.

7 The oxidation number of copper **increases** from **0** to **+2** and the oxidation number of nitrogen **decreases** from **+5** to **+4**.

This means that the copper is the **reducing** agent because the nitrogen is **reduced** and the nitrogen (in NO_3^-) is the **oxidising** agent because the copper is **oxidised**.

8

Compound	Element	Oxidation number
$CaCO_3$	Ca	+2
HNO_3	O	−2
H_2O_2	O	−1
HCO_3^-	C	+4
HNO_2	N	+3
$KMnO_4^-$	Mn	+7
H_2S	S	−2
Cr_2O_3	Cr	+3
N_2O_4	N	+4

9 The order of increasing oxidation states of nitrogen is: K_3N, N_2, N_2O, NO, N_2O_3, N_2O_4, $Ca(NO_3)_2$

10

Step	Task	How it's done	Half-equation
1	Balance the nitrogen.	The nitrogen is already balanced.	$NO_3^- \rightarrow NO_2$
2	Balance the oxygen by adding **H_2O**.	Add **one** H_2O molecule(s) to the right-hand side of the equation.	**$NO_3^- \rightarrow NO_2 + H_2O$**
3	Balance the hydrogen by adding **H^+**.	Add **2** H^+ ion(s) to the **left-hand side** of the equation.	**$NO_3^- + 2H^+ \rightarrow NO_2 + H_2O$**
4	Balance the charge by adding **electrons**.	Charge on left-hand side = **(−1) + (+2)** = **+1** Charge on right-hand side = **0** Add e− to the **left-hand side** of the equation.	**$NO_3^- + 2H^+ + e^- \rightarrow NO_2 + H_2O$**
5	Add state symbols to give the final half-equation.	Give the appropriate states for each reactant and product in the equation.	**$NO_3^-(aq) + 2H^+(aq) + e^- \rightarrow NO_2(g) + H_2O(l)$**

11

Equation	Conjugate redox pair (oxidation)	Conjugate redox pair (reduction)
$Sn(s) + Br_2(aq) \rightarrow SnBr_2(aq)$	$Sn(s)/Sn^{2+}(aq)$	$Br_2(aq)/Br^-(aq)$
$Zn(s) + CuCl_2(aq) \rightarrow ZnCl_2(aq) + Cu(s)$	$Zn(s)/Zn^{2+}(aq)$	$Cu^{2+}(aq)/Cu(s)$
$Br^-(aq) + 3F_2(g) + 3H_2O(l) \rightarrow 6F^-(aq) + 6H^+ + BrO_3^-(aq)$	$Br^-(aq)/BrO_3^-(aq)$	$F_2(g)/F^-(aq)$
$3Cu(s) + Cr_2O_7^{2-}(aq) + 14H^+(aq) \rightarrow 3Cu^{2+}(aq) + 2Cr^{3+}(aq) + 7H_2O(l)$	$Cu(s)/Cu^{2+}(aq)$	$Cr_2O_7^{2-}(aq)/Cr^{3+}(aq)$

12 **b**, **c**, **e**, **f** and **h** are redox reactions because the elements in the reactions undergo changes in oxidation number during the course of the reaction.

The changes in oxidation number that occur are:

b Ag from 0 to +1; Cl from 0 to −1

c Fe from +3 to +2; Sn from +2 to +4

e P from +3 to +5; I from 0 to −1

f Cu from +1 to +2; Cu from +1 to 0

h P from 0 to −3; H from 0 to +1

13 **a** **i** 0 **ii** +2

b **i** +5 **ii** +4

c oxidising agent HNO_3; reducing agent Cu

14 $2IO_3^-(aq) + \mathbf{12}H^+(aq) + \mathbf{10}e^- \rightarrow I_2(aq) + \mathbf{6}H_2O(l)$

15 oxidation half-equation: $SO_2(g) + 2H_2O(l) \rightarrow SO_4^{2-}(aq) + 4H^+(aq) + 2e^-$

reduction half-equation: $MnO_4^-(aq) + 8H^+(aq) + 5e^- \rightarrow Mn^{2+}(aq) + 4H_2O(l)$

overall equation: $5SO_2(g) + 2MnO_4^-(aq) + 2H_2O(l) \rightarrow 5SO_4^{2-}(aq) + 2Mn^{2+}(aq) + 4H^+(aq)$

16 **a** **i** In $S_2O_4^{2-}$ oxidation number (ON) of sulfur = +3

ii In $S_2O_3^{2-}$ ON(sulfur) = +2

iii In HSO_3^- ON(sulfur) = +4

b $S_2O_4^{2-}(aq) + 2H_2O(l) \rightarrow 2HSO_3^-(aq) + 2H^+(aq) + 2e^-$

$S_2O_4^{2-}(aq) + 2H^+(aq) + 2e^- \rightarrow S_2O_3^{2-}(aq) + H_2O(l)$

17 **a** $Zn(s) \rightarrow Zn^{2+}(aq) + 2e^-$

$Pb^{2+}(aq) + 2e^- \rightarrow Pb(s)$

$Zn(s) + Pb^{2+}(aq) \rightarrow Zn^{2+}(aq) + Pb(s)$

b $Fe^{2+}(aq) \rightarrow Fe^{3+}(aq) + e^-$

$MnO_4^-(aq) + 8H^+(aq) + 5e^- \rightarrow Mn^{2+}(aq) + 4H_2O(l)$

$5Fe^{2+}(aq) + MnO_4^-(aq) + 8H^+(aq) \rightarrow 5Fe^{3+}(aq) + Mn^{2+}(aq) + 4H_2O(l)$

c $SO_2(aq) + 2H_2O(l) \rightarrow SO_4^{2-}(aq) + 4H^+(aq) + 2e^-$

$I_2(aq) + 2e^- \rightarrow 2I^-(aq)$

$SO_2(aq) + 2H_2O(l) + I_2(aq) \rightarrow SO_4^{2-}(aq) + 4H^+(aq) + 2I^-(aq)$

d $OCl^-(aq) + 2H^+(aq) + 2e^- \rightarrow Cl^-(aq) + H_2O(l)$

$2I^-(aq) \rightarrow I_2(aq) + 2e^-$

$OCl^-(aq) + 2H^+(aq) + 2I^-(aq) \rightarrow Cl^-(aq) + H_2O(l) + I_2(aq)$

18 **a** half-equations:

$C_6H_8O_6(aq) \rightarrow C_6H_6O_6(aq) + 2H^+(aq) + 2e^-$

$I_2(aq) + 2e^- \rightarrow 2I^-(aq)$

b oxidation reaction: $C_6H_8O_6(aq) \rightarrow C_6H_6O_6(aq) + 2H^+(aq) + 2e^-$

reduction reaction: $I_2(aq) + 2e^- \rightarrow 2I^-(aq)$

19 **a** $2H_3AsO_4(aq) + 4H^+(aq) + 4e^- \rightarrow As_2O_3(s) + 5H_2O(l)$

$Br^-(aq) + 3H_2O(l) \rightarrow BrO_3^-(aq) + 6H^+(aq) + 6e^-$

b $6H_3AsO_4(aq) + 2Br^-(aq) \rightarrow 3As_2O_3(s) + 2BrO_3^-(aq) + 9H_2O(l)$

20 **a** **i** $2NH_4^+(s) \rightarrow N_2(g) + 8H^+(aq) + 6e^-$

ii $Cr_2O_7^{2-}(s) + 8H^+(aq) + 6e^- \rightarrow Cr_2O_3(s) + 4H_2O(g)$

b Reaction **i** is oxidation because the oxidation number of the nitrogen increases from −3 to 0; reaction **ii** is reduction because the oxidation number of chromium decreases from +6 to +3.

c $(NH_4)_2Cr_2O_7(s) \rightarrow N_2(g) + Cr_2O_3(s) + 4H_2O(g)$

d $NH_4^+(s)/N_2(g)$ and $Cr_2O_7^{2-}(aq)/Cr_2O_3(s)$

21 half-equations:

$Sn^{4+}(aq) + 2e^- \rightarrow Sn^{2+}(aq)$

$H_2SO_3(aq) + H_2O(l) \rightarrow HSO_4^-(aq) + 3H^+(aq) + 2e^-$

overall equation: $H_2SO_3(aq) + Sn^{4+}(aq) + H_2O(l) \rightarrow Sn^{2+}(aq) + HSO_4^-(aq) + 3H^+(aq)$

22 **a** $Fe_2O_3(s) + 6e^- \rightarrow 2Fe(l) + 3O^{2-}(l)$

b reduction

c $Fe_2O_3(s) + 2Al(s) \rightarrow 2Fe(l) + Al_2O_3(s)$

d 5.29 g

23 **a** **i** $Cr(OH)_3(s)$ is oxidised and $IO_3^-(aq)$ is reduced

ii $Cr(OH)_3(s) + 5OH^-(aq) \rightarrow CrO_4^{2-}(aq) + 4H_2O(l) + 3e^-$

$IO_3^-(aq) + 3H_2O(l) + 6e^- \rightarrow I^-(aq) + 6OH^-(aq)$

iii $2Cr(OH)_3(s) + IO_3^-(aq) + 4OH^-(aq) \rightarrow 2CrO_4^{2-}(aq) + I^-(aq) + 5H_2O(l)$

b **i** $Br^-(aq)$ is oxidised and $Al(OH)_4^-(aq)$ is reduced

ii $Br^-(aq) + 6OH^-(aq) \rightarrow BrO_3^-(aq) + 3H_2O(l) + 6e^-$

$Al(OH)_4^-(aq) + 3e^- \rightarrow Al(s) + 4OH^-(aq)$

iii $Br^-(aq) + 2Al(OH)_4^-(aq) \rightarrow BrO_3^-(aq) + 2Al(s) + 3H_2O(l) + 2OH^-(aq)$

24 half-equations:

$Zn(s) + 2OH^-(aq) \rightarrow Zn(OH)_2(s) + 2e^-$

$2MnO_2(s) + H_2O(l) + 2e^- \rightarrow Mn_2O_3(s) + 2OH^-(aq)$

overall equation:

$Zn(s) + 2MnO_2(s) + H_2O(l) \rightarrow Zn(OH)_2(s) + Mn_2O_3(s)$

Chapter 5 Galvanic cells and fuel cells as sources of energy

5.1 Galvanic cells

CSA: A technological leap beginning with a frog's leg

1 **a** $Zn(s) \rightarrow Zn^{2+}(aq) + 2e^-$

b $O_2(g) + 2H_2O(l) + 4e^- \rightarrow 4OH^-(aq)$

c $2Zn(s) + O_2(g) + 2H_2O(l) \rightarrow 2Zn^{2+}(aq) + 4OH^-(aq)$

2 O_2 is the oxidising agent; zinc is the reducing agent.

Worked example

TY 5.1.1 $Sn^{2+}(aq) + 2e^- \longrightarrow Sn(s)$

$Fe(s) \longrightarrow Fe^{2+}(aq) + 2e^-$

Electrons flow from the negative electrode (anode) to the positive electrode (cathode) as shown in the diagram of the cell below.

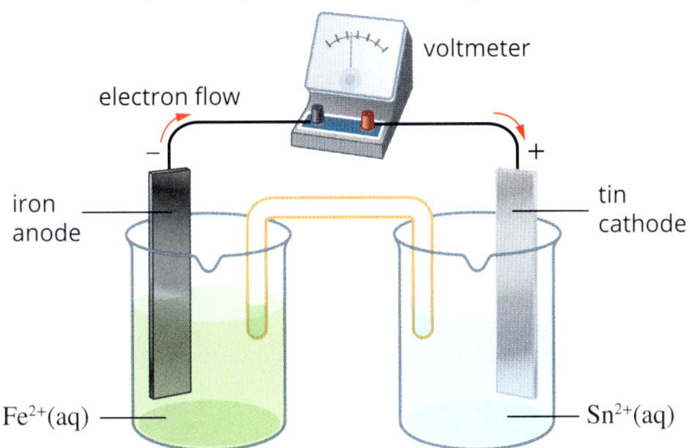

Key questions

1 **a** A galvanic cell is a type of electrochemical cell that converts chemical energy into electrical energy.

b An electrochemical cell is a device that converts chemical energy into electrical energy, or vice versa.

c A salt bridge is an electrical connection between two half-cells in a galvanic cell; it is usually made from a material saturated in electrolyte solution.

d A battery is a combination of cells connected in series.

e A primary cell is a galvanic cell that cannot be recharged.

2 A galvanic cell is composed of two half-cells connected by a salt bridge. Each half-cell contains an electrode in contact with a solution. For example, one half-cell might contain Cu(s) and $Cu^{2+}(aq)$ and the other might contain Zn(s) and $Zn^{2+}(aq)$. The species present in each half-cell form a conjugate redox pair (an oxidising agent and its corresponding reduced form). The salt bridge is made from an absorbent material such as filter paper saturated in an electrolyte solution such as potassium nitrate solution.

3 **a**

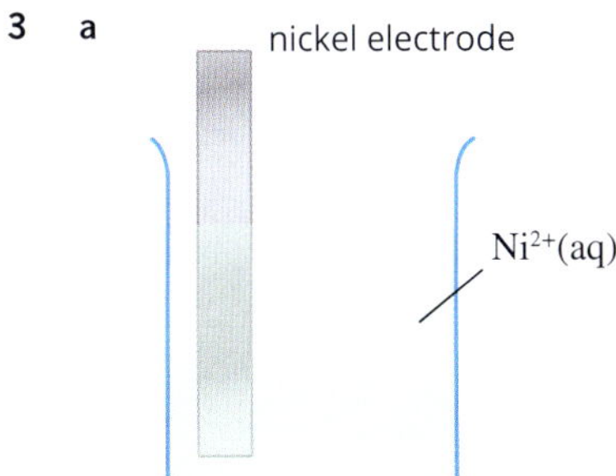

b

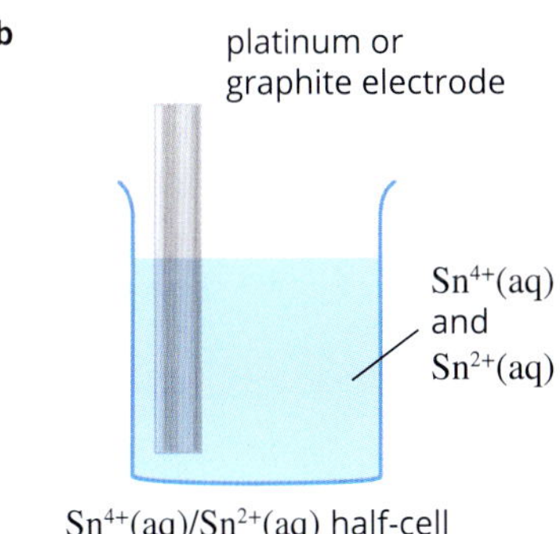

c

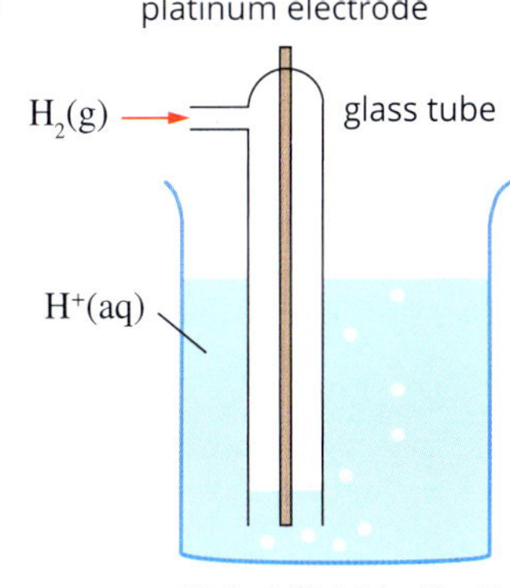

4 In a galvanic cell, electrons move through the external circuit from the **negative** terminal to the **positive** terminal. Cations in the salt bridge move towards the **cathode** and anions in the salt bridge move towards the **anode**.

5 **a** Fe(s)

b Pt or graphite

c Pt or graphite

6 **a** Al(s) has been oxidised; $Sn^{2+}(aq)$ has been reduced.

b **i** at the cathode: $Sn^{2+}(aq) + 2e^- \longrightarrow Sn(s)$

ii at the anode: $Al(s) \longrightarrow Al^{3+}(aq) + 3e^-$

c

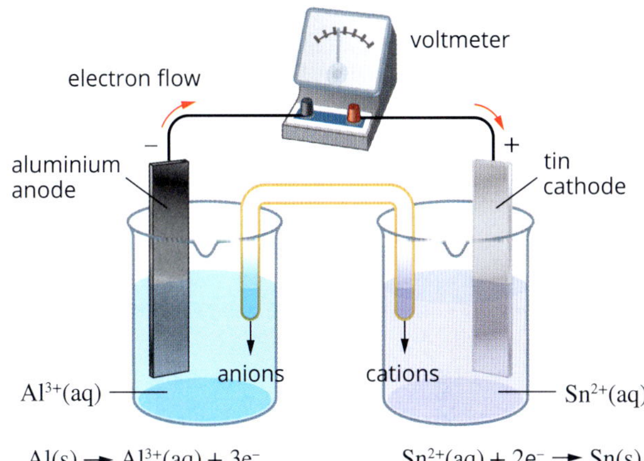

7 $Zn(s) + Cu^{2+}(aq) \longrightarrow Cu(s) + Zn^{2+}(aq)$

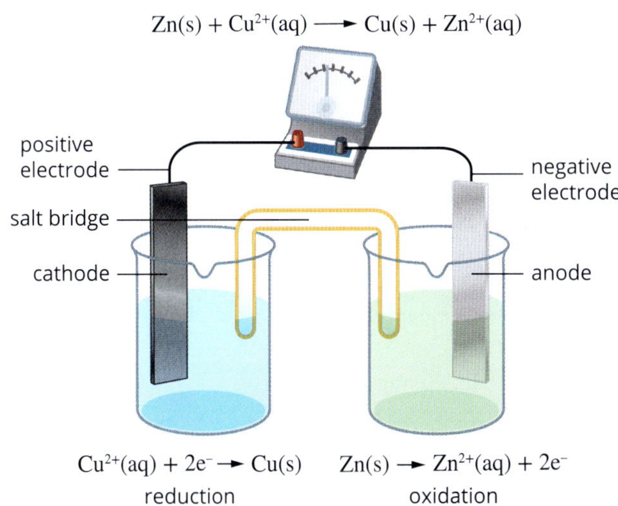

5.2 The electrochemical series

TY 5.2.1 $Sn^{2+}(aq) + Ni(s) \rightarrow Sn(s) + Ni^{2+}(aq)$

The nickel electrode is the anode and the tin electrode is the cathode.

Electrons flow from the negative electrode (anode) to the positive electrode (cathode).

TY 5.2.2 **a** No reaction occurs because both I_2 and Pb^{2+} are oxidising agents.

b No reaction because the oxidising agent, I_2, is below the reducing agent, Cl^-, in the electrochemical series.

c A reaction occurs because the oxidising agent, Cl_2, is above the reducing agent, Pb, in the electrochemical series.

The higher half-equation occurs in the forward direction:

$Cl_2(g) + 2e^- \rightarrow 2Cl^-(aq)$

The lower half-equation occurs in the reverse direction:

$Pb(s) \rightarrow Pb^{2+}(aq) + 2e^-$

The overall reaction equation is found by adding the half-equations:

$Cl_2(g) + Pb(s) \rightarrow 2Cl^-(aq) + Pb^{2+}(aq)$

Key questions

1 **a** In the context of a galvanic cell, the potential difference measures the tendency to push electrons into the external circuit. It is the electromotive force between two points in a circuit.

b Standard conditions are reference conditions used to compare measurements. Gas pressure is 100 kPa and the concentration of dissolved species is 1 mol L^{-1}. Temperatures are usually measured at 25°C.

c The standard electrode potential is the voltage measured when a half-cell, at standard conditions, is connected to a standard hydrogen half-cell.

2 In a galvanic cell, the stronger reducing agent is in the half-cell with the **negative** electrode. This electrode is the **anode**. The stronger oxidising agent is in the half-cell with the **positive** electrode. This electrode is the **cathode**.

3 The half-cell should be constructed at standard conditions (gas pressures of 100 kPa, concentrations of 1 mol L^{-1}) and temperature of 25°C. The cell is connected to a standard hydrogen electrode to form a galvanic cell, and the potential difference of the cell is measured with a voltmeter. This value is the standard electrode potential ($E°$). If the electrode in the half-cell is negative, the $E°$ is given a negative sign; if the electrode in the half-cell is positive the $E°$ is given a positive sign.

4 There are two possible reasons:

- When conditions are not standard, the order of half-reactions in the electrochemical series may be different.
- The rate of the reaction may be slow.

5 **a** Lithium is a stronger reducing agent than sodium.

b $Fe^{2+}(aq)$ is a stronger oxidising agent than $Ca^{2+}(aq)$.

c $Sn^{2+}(aq)$ can act as either an oxidising agent or a reducing agent in different reactions.

6 $E° = -0.40$ V.

7 **a** reduction: $Pb^{2+}(aq) + 2e^- \rightarrow Pb(s)$

oxidation: $Al(s) \rightarrow Al^{3+}(aq) + 3e^-$

b $3Pb^{2+}(aq) + 2Al(s) \rightarrow 3Pb(s) + 2Al^{3+}(aq)$

c The lead electrode is the cathode and the aluminium electrode is the anode.

8

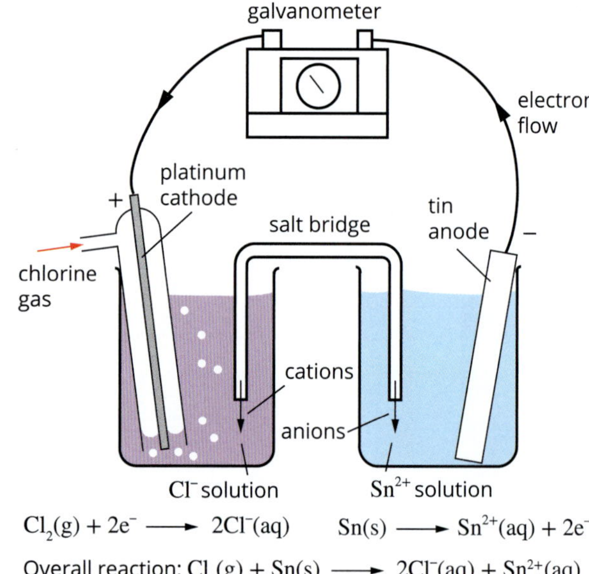

$Cl_2(g) + 2e^- \longrightarrow 2Cl^-(aq)$ $Sn(s) \longrightarrow Sn^{2+}(aq) + 2e^-$

Overall reaction: $Cl_2(g) + Sn(s) \longrightarrow 2Cl^-(aq) + Sn^{2+}(aq)$

9 Question 7 cell:

cell potential difference = 1.54 V

Question 8 cell:

cell potential difference = 1.50 V

10 **a** A reaction occurs because the oxidising agent, Cl_2, is above the reducing agent, Br^-, in the electrochemical series.

$$Cl_2(g) + 2Br^-(aq) \rightarrow 2Cl^-(aq) + Br_2(l)$$

b A reaction occurs because the oxidising agent, Cl_2, is above the reducing agent, I^-, in the electrochemical series.

$$Cl_2(g) + 2I^-(aq) \rightarrow 2Cl^-(aq) + I_2(s)$$

c No reaction because the oxidising agent, Br_2, is below the reducing agent, Cl^-, in the electrochemical series.

d A reaction occurs because the oxidising agent, Br_2, is above the reducing agent, I^-, in the electrochemical series.

$$Br_2(l) + 2I^-(aq) \rightarrow 2Br^-(aq) + I_2(s)$$

5.3 Fuel cells

1 A fuel cell is a type of galvanic cell that generates electricity from redox reactions quietly and efficiently, and with almost no pollution. Unlike primary cells, fuel cells do not run down. Electricity is available for as long as fuel is supplied to them from an external source.

2 B., C. and G.

3 The surface area of a porous material is the total of the surface area within all the pores. Therefore, the porosity of the electrodes creates a very large surface area. The increased surface area means that for the same flow rate of reactant gas there will be more molecules in contact with the catalyst. Therefore, the rate of half-cell reactions will increase. This will produce a greater current from the fuel cell.

4 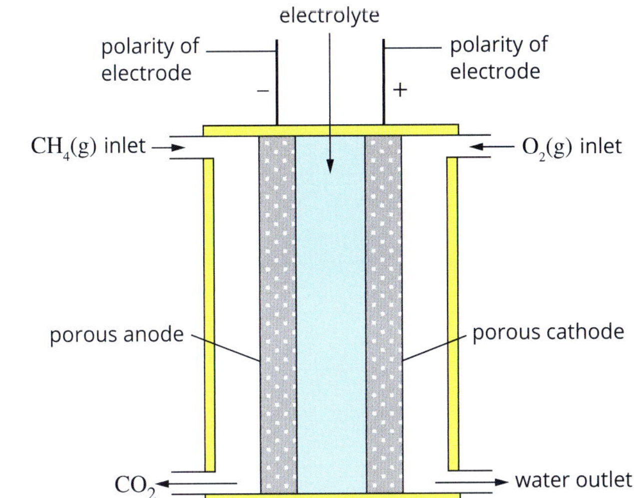

5 a anode: $H_2(g) \rightarrow 2H^+(aq) + 2e^-$;
cathode: $O_2(g) + 4H^+(aq) + 4e^- \rightarrow 2H_2O(l)$
b $2H_2(g) + O_2(g) \rightarrow 2H_2O(l)$

6 a $2CH_3OH(g) + 3O_2(g) \rightarrow 2CO_2(g) + 4H_2O(l)$
b reduction of $O_2(g)$
c KOH(aq) or NaOH(aq)
d Oxygen is reduced, forming OH^- ions at the cathode of the fuel cell, so an increase in pH occurs at this electrode initially. The OH^- ions migrate to the anode, where they are consumed. Once the rate of production of OH^- ions at the cathode becomes equal to the rate at which they depart, the pH near the electrode will be constant.
e A fuel cell converts chemical energy into electrical energy directly, with relatively little energy being converted into thermal energy. If electrical energy were obtained by burning methanol (in a process like that used to obtain electrical energy from coal in coal-fired power stations), the energy 'losses' of the various energy transformations involved would be greater. Large losses occur when thermal energy is converted into mechanical energy.

5.4 Supplying energy sustainably

CSA: Redox flow batteries

1 a $V^{2+}(aq) \rightarrow V^{3+}(aq) + e^-$
b $VO_2^+(aq) + 2H^+(aq) + V^{2+}(aq) \rightarrow VO^{2+}(aq) + H_2O(l) + V^{3+}(aq)$
c cell potential difference = 1.26 V
d Vanadium redox flow batteries can contribute to an energy supply based on sustainable sources (e.g. solar or wind) because they can balance fluctuations in supply that occur from these sources.
This question may be answered in terms of the green chemistry principles listed in Chapter 1 (page 15). These include:
- design for degradation—vanadium redox flow batteries have long lifetimes and vanadium is much safer for the environment and less toxic than other chemicals used in batteries, such as lead and cadmium
- design for energy efficiency—efficiencies of up to 85% have been achieved, and scientists continue to seek higher efficiencies by improving cell design
- prevention of waste—the reactants in the cell are regenerated almost indefinitely during the recharging cycles, so waste is minimal. There are almost no emissions of gases such as CO_2
- use of renewable feedstocks—the vanadium-based electrolyte is regenerated when the battery is recharged.

2 a $Br_2(aq) + 2e^- \rightarrow 2Br^-(aq)$
b $Br_2(aq) + Zn(s) \rightarrow 2Br^-(aq) + Zn^{2+}(aq)$
c Zinc is plated on the cathode during the charging cycle, when the cell reaction above is reversed.
d The membrane between the compartments allows exchange of ions to balance charges formed in the compartments while preventing contact between the $Br_2(aq)$ and Zn(s), which would otherwise react directly.

Key questions

1 Fuel cells are about twice as efficient as coal-fired power stations. Consequently, fuel cells produce the same quantity of energy from about half as much fuel. Less fuel means less carbon dioxide gas is produced. As carbon dioxide is a major greenhouse gas, the use of fuel cells has the potential to reduce the greenhouse effect. The use of fuel cells operating on hydrogen obtained from sustainable sources is the ideal, but currently about 95% of hydrogen is produced from fossil fuels.

2 Fuel cells have a much higher efficiency than thermal power stations because chemical energy is directly converted into electrical energy. There are several energy conversions in a fossil fuel power stations (chemical energy → heat energy → mechanical energy → electrical energy), and each one involves a loss of energy.

3 Steam reforming of fossil fuels, decomposition of water using electricity (electrolysis) and steam reforming of biogas. Most hydrogen is produced from fossil fuels.

4 Design for energy efficiency—efficiencies of up to 35% have been achieved, and scientists continue to seek higher efficiencies by improving cell design.
Use of renewable feedstocks—low-grade waste materials such as soils and sediments, waste water and agricultural waste are used.

5 a $CH_4(g) + 4O^{2-}(s) \rightarrow CO_2(g) + 2H_2O(g) + 8e^-$
b Catalysts are incorporated in electrodes to increase the rate of reaction and allow more current to be produced by the cell. Since reaction rate increases with temperature, the use of very high temperatures may mean that a catalyst is not required.
c Scientists may investigate different types of electrode materials, catalysts, electrolytes, fuels and cell operating conditions (e.g. different temperatures and pressures).
d SOFCs operate at much higher efficiencies than fossil fuel power stations, which are the traditional source of electricity. As a consequence, less fossil fuel is consumed (none at all if a sustainable source of fuel is used) and less CO_2 emissions are produced per quantity of electricity.

Chapter 5 review

1 a An oxidising agent is a substance that causes another substance to be oxidised and is reduced in the process. A reducing agent is a substance that causes another substance to be reduced and is oxidised in the process.
b An anode is the electrode at which oxidation occurs. A cathode is the electrode at which reduction occurs.
c An external circuit is a circuit where the electrons flow (e.g. through wires). An internal circuit is the part of cell where the current is due to the movement of ions (e.g. in the salt bridge).
d A battery consists of a number of cells connected together in series.

2 a the standard hydrogen electrode
b Standard conditions are gas pressures of 100 kPa (1 bar), and solution concentrations of 1 mol L^{-1}. A temperature of 25°C is usually used for measurements of electrode potentials.

3 cell potential difference = 1.10 V

4 Oxidising agents are on the left-hand side of the table and reducing agents are on the right. The oxidising agents are listed in increasing order of strength, with the strongest oxidising agent, F_2, at the top left of the table. The reducing agents are listed in decreasing order of strength, with the strongest reducing agent, Li, at the bottom right of the table.

5 The electrochemical series is based on reactions occurring under standard conditions of 100 kPa pressure and 1 M concentration, and a temperature of 25°C (298 K), and can only be used to predict the possibility of reaction occurring under these conditions. The series gives no information about the rate of reactions, so even if a reaction is predicted it may be so slow that no reaction is apparent.

6 A spontaneous redox reaction will occur if an oxidising agent (on the **left** of the electrochemical series) reacts with a reducing agent (on the **right** of the series) that has a half-reaction with a **less** positive $E°$ value. When the reactants are in direct contact, **heat** energy is released.

7 Although the cost of electrical energy purchased in the form of a dry cell or button cell is far higher than the cost of mains electricity, people are prepared to pay the higher price for the convenience and flexibility of the portable equipment powered by these cells. Furthermore, the price of individual cells is regarded as relatively low.

8 The redox reaction occurring in a galvanic cell takes place in two **half-cells**. The electrode at which oxidation occurs is called the **anode**, whereas the other electrode where reduction occurs is the **cathode**. The polarity of the anode is **negative** and the polarity of the cathode is **positive**.

Galvanic cells can be classified as either primary cells, which **cannot**, be recharged, or secondary cells, which **can** be recharged.

9 D.

10 At the anode of a fuel cell the fuel (often **hydrogen**) undergoes **oxidation** that generates ions and electrons. The ions move through the **electrolyte**. At the same time, electrons flow from the **negative** electrode to the cathode, producing electricity. The cell converts the **chemical** energy of the fuel and an oxidising agent into **electrical** energy. The energy efficiencies of fuel cells are generally **higher** than if the fuel were burnt in a power station or internal combustion engine and they have **lower** levels of emissions.

11 a true b false c false

12 D.

13

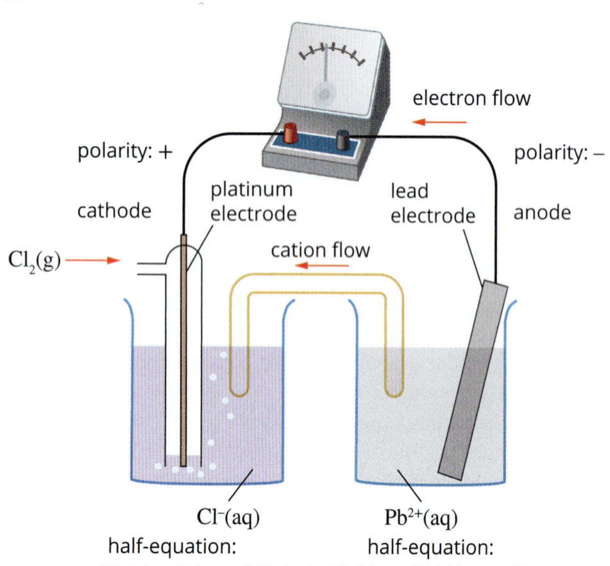

14 a oxidising agent
b strong reducing agent
c true
d false
e true

15 a

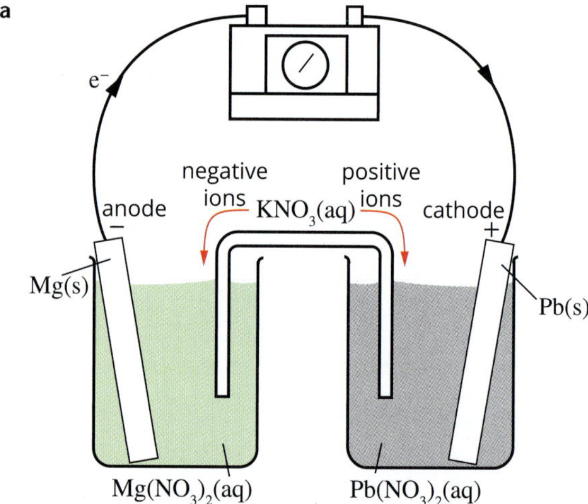

b $Mg(s) \rightarrow Mg^{2+}(aq) + 2e^-$
$Pb^{2+}(aq) + 2e^- \rightarrow Pb(s)$
overall: $Mg(s) + Pb^{2+}(aq) \rightarrow Mg^{2+}(aq) + Pb(s)$

c The lead electrode is the cathode; the magnesium electrode is the anode.

d Anions will migrate to the $Mg^{2+}(aq)/Mg(s)$ half-cell, and cations will migrate to the $Pb^{2+}(aq)/Pb(s)$ half-cell.

16 D.

17 The order of reducing agent strength, from strongest to weakest, is $Sn^{2+} > Fe^{2+} > Br^-$.

18 i

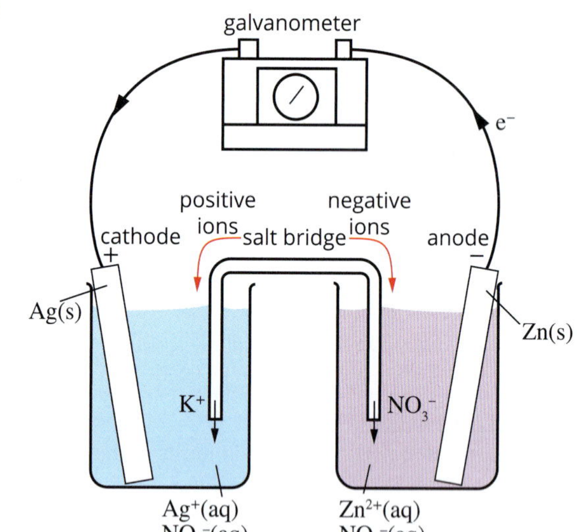

Cathode (+): $Ag^+(aq) + e^- \rightarrow Ag(s)$

Anode (–): $Zn(s) \rightarrow Zn^{2+}(aq) + 2e^-$

Overall: $Zn(s) + 2Ag^+(aq) \rightarrow Zn^{2+}(aq) + 2Ag(s)$

Standard cell potential difference = 1.56 V

ii

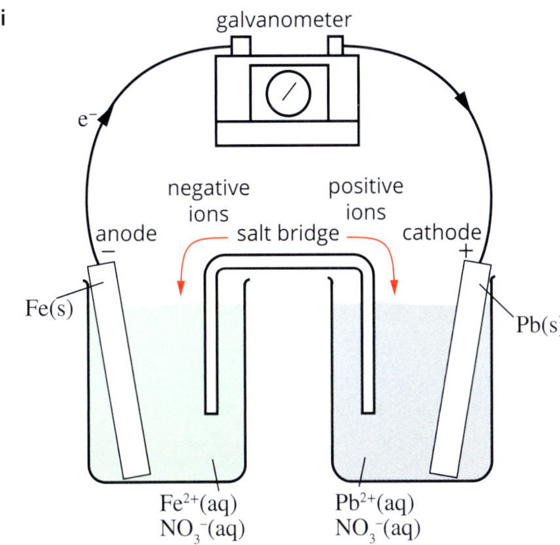

Cathode (+): $Pb^{2+}(aq) + 2e^- \longrightarrow Pb(s)$

Anode (−): $Fe(s) \longrightarrow Fe^{2+}(aq) + 2e^-$

Overall: $Fe(s) + Pb^{2+}(aq) \longrightarrow Fe^{2+}(aq) + Pb(s)$

Standard cell potential difference = 0.31 V

iii

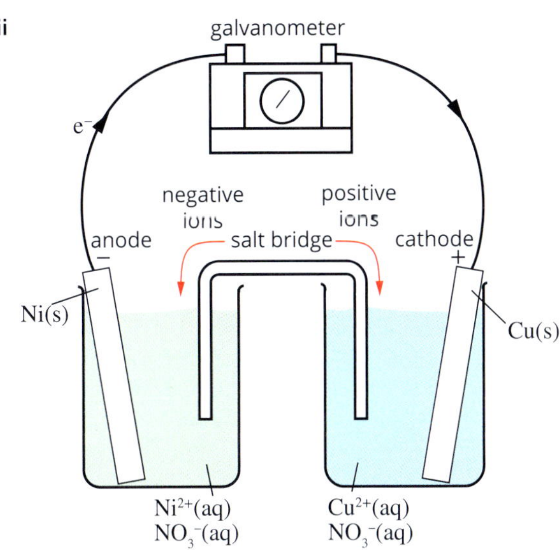

Cathode (+): $Cu^{2+}(aq) + 2e^- \longrightarrow Cu(s)$

Anode (−): $Ni(s) \longrightarrow Ni^{2+}(aq) + 2e^-$

Overall: $Ni(s) + Cu^{2+}(aq) \longrightarrow Ni^{2+}(aq) + Cu(s)$

Standard cell potential difference = 0.57 V

19 The order of half-cell potentials, from lowest standard electrode potential (most negative) to highest electrode potential, is $D^{2+}/D < C^{2+}/C < A^{2+}/A < B^{2+}/B$.

20 **a** $Zn(s) + 4OH^-(aq) + Cu^{2+}(aq) \rightarrow Zn(OH)_4^{2-}(aq) + 2Cu(s)$

b The student could make two electrochemical cells consisting of the following, and as shown in the diagrams below:

- Cu^{2+}/Cu half-cell and the Zn^{2+}/Zn half-cell.
- Cu^{2+}/Cu half-cell and the 'alkaline zinc half-cell'
  The cell voltages should be measured and if they are identical then the two half-cells have the same $E°$ values. (They would not be expected to have the same $E°$ values.)

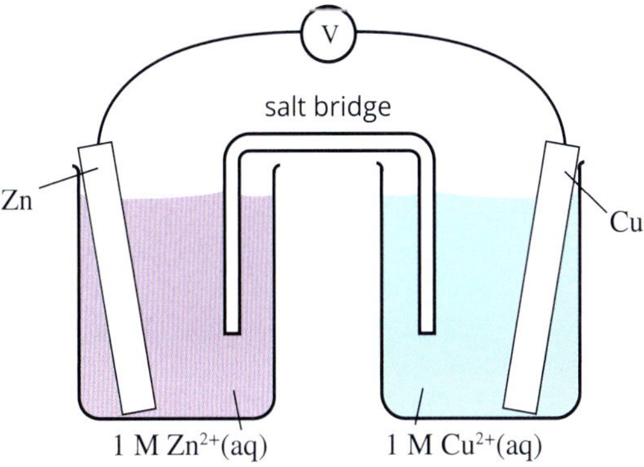

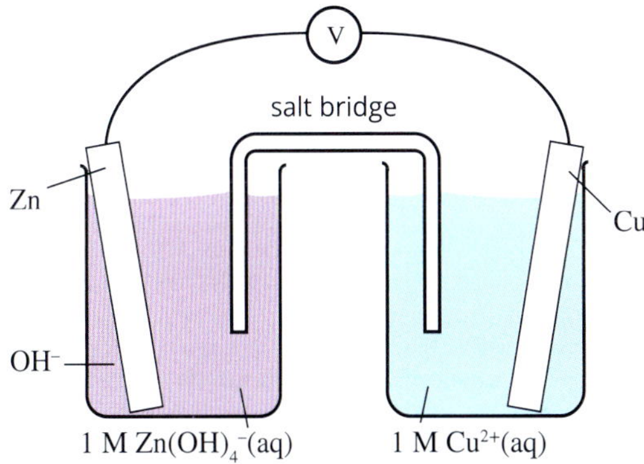

Alternatively, the student could construct a cell from the two different zinc half-cells. If a voltage is observed, they must have different $E°$ values.

c The independent variable is the type of half-cell, $Zn^{2+}(aq)/Zn(s)$ or $Zn(OH)_4^{2-}(aq)/Zn(s)$ (i.e. the reaction occurring in the half-cell with the zinc electrode). Another possible answer is that the independent variable is the electrolyte in the zinc half-cell ($Zn^{2+}(aq)$ or $Zn(OH)_4^{2-}(aq)$).

d Any two of: the concentration of Zn^{2+} and $Zn(OH)_4^{2-}$ in the half-cells; the size of the zinc electrodes; the temperature of the solution.

e The dependent variable is the standard electrode potential ($E°$) of the half-cells, as measured by the cell voltages.

f For example: To determine whether the standard electrode potential ($E°$) of a $Zn^{2+}(aq)/Zn(s)$ half-cell is the same as that of a $Zn(OH)_4^{2-}(aq)/ Zn(s)$ half-cell.

21 **a** **i** Reaction would occur.

ii $Zn(s) \rightarrow Zn^{2+}(aq) + 2e^-$ (oxidation)

$Cl_2(g) + 2e^- \rightarrow 2Cl^-(aq)$ (reduction)

iii $Zn(s) + Cl_2(g) \rightarrow Zn^{2+}(aq) + 2Cl^-(aq)$

b **i** No reaction would occur.

c **i** Reaction would occur.

ii $Zn(s) \rightarrow Zn^{2+}(aq) + 2e^-$ (oxidation)

$Ag^+(aq) + e^- \rightarrow Ag(s)$ (reduction)

iii $2Ag^+(aq) + Zn(s) \rightarrow 2Ag(s) + Zn^{2+}(aq)$

d **i** No reaction would occur.

22 **a** strongest: $Ag^+(aq)$; weakest: $Mg^{2+}(aq)$

b strongest: Mg(s); weakest: Ag(s)

c A coating of silver will form on the lead when it is placed in silver nitrate solution because Ag^+ ions are stronger oxidising agents than Pb^{2+} ions.

d zinc and magnesium

23 **a** Fe^{2+}(aq) and H^+(aq) will be formed

b $2Fe^{3+}(aq) + H_2(g) \longrightarrow 2Fe^{2+}(aq) + 2H^+(aq)$

c If significant reaction had occurred, the yellow solution containing Fe^{3+} ions would have become pale green as Fe^{2+} ions formed. Because no reaction was observed, the rate of the reaction may have been slow. It also must be remembered that the electrochemical series is only valid for certain conditions. It is possible that under the conditions in which the experiment was performed, little reaction would occur.

24 **a** $CH_3CH_2OH(g) + H_2O(l) \longrightarrow CH_3COOH(aq) + 4H^+(aq) + 4e^-$

b $O_2(g) + 4H^+(aq) + 4e^- \longrightarrow 2H_2O(l)$

c Cations will move towards the cathode. Because H^+ ions are involved in the reduction reaction, positive charge is decreasing at the cathode. To maintain charge neutrality, cations from the electrolyte will move towards the cathode.

d Yes. The net cell reaction in the fuel cell is the oxidation of ethanol to form carbon dioxide and water. Like ethanol, methane in natural gas can be oxidised to carbon dioxide and water. Provided the electrode materials used in the cell act as effective catalysts for both reactions, it is likely that the use of natural gas would produce a voltage.

25 **a** $O_2(g) + 4e^- \longrightarrow 2O^{2-}(s)$

b **i** negative **ii** anode

c The technology promises a cleaner and more efficient way of converting coal to electrical energy than coal-fired power stations, which operate at about 40% efficiency. The CO_2 generated would be approximately half the volume of that produced by current systems for the same amount of electricity, and the cost to capture the carbon would be less. This technology could help to reduce the consumption of the world's declining coal reserves.

26 **a** $CH_4(g) + H_2O(l) \longrightarrow CO_2(g) + 8H^+(aq) + 8e^-$. Oxidation occurs at the anode, which is the negative electrode in a fuel cell. CH_4 is the fuel that is oxidised.

b Cell potential difference = 1.06 V

c There is evidence that some cobalt mining involves child labour and human rights violations and that some supply chains are controlled by militia.

27 **a** arrow A

b O_2 is the oxidising agent, which enters the air space from outside.

c $4Al(s) + 3O_2(g) + 6H_2O(l) \longrightarrow 4Al^{3+}(aq) + 12OH^-(aq)$

d The cathode is the site of the reduction reaction involving oxygen/air, which must be in contact with the electrolyte. A porous electrode will allow a large surface area for this reaction, maximising the rate of reaction and therefore the current/power output of the cell.

e From the electrochemical series:

cathode reaction: $O_2(g) + 2H_2O(l) + 4e^- \longrightarrow 4OH^-(aq)$ $E° = +0.40$ V

anode reactions: $Zn(s) \longrightarrow Zn^{2+}(aq) + 2e^-$ $E° = -0.76$ V

$Ag(s) \longrightarrow Ag^+(aq) + e^-$ $E° = +0.80$ V

Because $E°(Zn^{2+}/Zn) < E°(O_2, H_2O/OH^-)$, a spontaneous cell reaction will occur. Because $E°(Ag^+/Ag) > E°(O_2, H_2O/OH^-)$ a spontaneous cell reaction will not occur.

f smaller, since $E°(Zn^{2+}/Zn) > E°(Al^{3+}/Al)$

$E°(O_2, H_2O/OH^-) - E°(Zn^{2+}/Zn) = (+0.40) - (-0.76) = 1.16$ V

$E°(O_2, H_2O/OH^-) - E°(Al^{3+}/Al) = (+0.40) - (-1.66) = 2.06$ V

g With an acidic electrolyte:

cathode reaction: $O_2(g) + 4H^+(aq) + 4e^- \longrightarrow 2H_2O(l)$ $E° = +1.23$ V

anode reactions: $Zn(s) \longrightarrow Zn^{2+}(aq) + 2e^-$ $E° = -0.76$ V

$Ag(s) \longrightarrow Ag^+(aq) + e^-$ $E° = +0.80$ V

Because $E°(Zn^{2+}/Zn) < E°(O_2, H^+/H_2O)$, a spontaneous cell reaction will occur. Because $E°(Ag^+/Ag) < E°(O_2, H^+/H_2O)$ a spontaneous cell reaction will also occur.

Essentially, O_2 becomes a stronger oxidising agent in acidic conditions and is capable of spontaneous reaction with the relatively weak reducing agent, Ag.

Unit 3 Area of Study 1 What are the current and future options for supplying energy?

1 C.	**2** D.	**3** A.
4 A.	**5** C.	**6** D.
7 C.	**8** B.	**9** B.
10 C.	**11** B.	**12** C.
13 A.	**14** D.	**15** A and C

16 **a** Petrol is the only fossil fuel listed, although methanol and ethanol can be made from fossil fuels.

b Biodiesel is a renewable fuel, and ethanol can be a renewable fuel if it is made from the fermentation of glucose. Methanol can also be produced from biomass, so is renewable, but this has not been discussed in this book.

c **i** $C_6H_{12}O_6(aq) \longrightarrow 2C_2H_5OH(aq) + 2CO_2(g)$

ii The ethanol is produced during fermentation as an aqueous solution which is only about 10% (v/v) ethanol. The ethanol needs to be separated from the water for ethanol to be useful as a fuel. Distillation is used to separate the two liquids, by making use of their different boiling points.

d $C_2H_5OH(l) + 2O_2(g) \longrightarrow 2CO(g) + 3H_2O(l)$

17 **a** Ethanol is the limiting reactant.

b 9.04 g

18 **a** **i** oxidation: $Al(s) \longrightarrow Al^{3+}(aq) + 3e^-$

reduction: $Br_2(aq) + 2e^- \longrightarrow 2Br^-(aq)$

ii overall: $2Al(s) + 3Br_2(aq) \longrightarrow 2Al^{3+}(aq) + 6Br^-(aq)$

b **i** oxidation: $S_2O_3^{2-}(aq) + 5H_2O(l) \longrightarrow 2SO_4^{2-}(aq) + 10H^+(aq) + 8e^-$

reduction: $ClO^-(aq) + 2H^+(aq) + 2e^- \longrightarrow Cl^-(aq) + H_2O(l)$

ii overall: $4ClO^-(aq) + S_2O_3^{2-}(aq) + H_2O(l) \longrightarrow 2SO_4^{2-}(aq) + 4Cl^-(aq) + 2H^+(aq)$

c **i** oxidation: $H_2O_2(aq) \longrightarrow O_2(aq) + 2H^+(aq) + 2e^-$

reduction: $MnO_4^-(aq) + 4H^+(aq) + 3e^- \longrightarrow MnO_2(s) + 2H_2O(l)$

ii overall: $3H_2O_2(aq) + 2MnO_4^-(aq) + 2H^+(aq) \longrightarrow 3O_2(aq) + 2MnO_2(s) + 4H_2O(l)$

19 **a** **i** 0.30 V **ii** Fe^{2+}/Fe

iii $Sn^{2+}(aq) + Fe(s) \longrightarrow Sn(s) + Fe^{2+}(aq)$

b **i** 2.43 V **ii** Al^{3+}/Al

iii $3Fe^{3+}(aq) + Al(s) \longrightarrow 3Fe^{2+}(s) + Al^{3+}(aq)$

c **i** 0.54 V

ii H^+/H_2

iii $I_2(aq) + H_2(g) \longrightarrow 2I^-(aq) + 2H^+(aq)$

20 **a** 742 MJ

b Heat losses occur in the boiler due to imperfect insulation. Heat is also lost through friction in moving parts within the turbines and generator.

c A non-renewable energy resource is one that cannot be replaced or regenerated as fast as it is consumed.

d Biogas is a renewable energy source, whereas existing coal supplies (which remain large) are non-renewable and would eventually be consumed. Because biogas carbon is derived from absorption of atmospheric CO_2 in photosynthesis, its use as a fuel has relatively little effect on net levels of atmospheric CO_2. Biogas can be produced from waste biomass that would otherwise require disposal.

21 a Oxidation number changes: I (+5 ⟶ –1) and S (+4 ⟶ +6) so this is a redox reaction.
The oxidising agent is HIO_3. The reducing agent is SO_2.
b Not redox because there is no change in oxidation numbers.
c Oxidation number changes: Fe (+2 ⟶ +3) and O (0 ⟶ –2) so this is a redox reaction.
The oxidising agent is O_2. The reducing agent is $Fe(OH)_2$ (or Fe^{2+}).
d Oxidation number changes: H (–1 ⟶ 0) and H (+1 ⟶ 0) so this is a redox reaction.
The oxidising agent is H_2O. The reducing agent is NaH.

22 a –5752 kJ
b

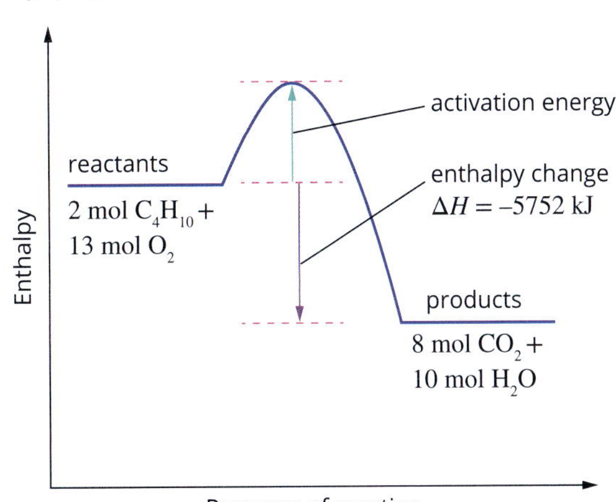

c Some initial input of energy is required to allow some reactants to overcome the activation energy barrier (associated with breaking bonds in reactants) to initiate the rapid reaction.
d 49.6 kJ g^{-1}
e i 67.5°C ii 44.2%

23 a anode: $CH_3OH(g) + H_2O(l) \longrightarrow CO_2(g) + 6H^+(aq) + 6e^-$
cathode: $O_2(g) + 4H^+(aq) + 4e^- \longrightarrow 2H_2O(l)$
b 22.7 kJ c 5.1 kJ d 22%
e Incomplete reaction of the fuel (methanol) is likely. Some may be incompletely oxidised or pass into the electrolyte waste without reacting at all. Also, a significant amount of heat is generated in most fuel cells, further reducing efficiency.

24 a 100 s
b exothermic (temperature increases)
c The temperature decreases due to loss of heat to the surroundings. Better insulation, such as aluminium foil, including a lid, around a beaker, or if possible a polystyrene surrounds with a lid, for the calorimeter would reduce this loss.
d 6.0°C e –230 kJ

25 a Average mass loss = 1.847 g
Average temperature increase = 48.0°C
b 20.1 kJ c 40.1 kJ
d 2172 kJ per 100 g
e Energy due to fat = 26 × 37 = 962 kJ
Energy due to carbohydrates = 60 × 16 = 960 kJ
f The systematic error that had the greatest impact on the investigation was the loss of heat to the surroundings.
g The investigation is valid, as it is testing the quantity that was described in the aim of the experiment 'A pair of students wanted to determine the energy content of 100 g of cheese-flavoured corn chips'.
The investigation is not very accurate, as a considerable quantity of heat (quoted as 50%) is lost to the surroundings.
The investigation has quite good precision, since the three trials had an average mass loss of 1.847 g, with the highest value being 0.036 g above that value and the lowest mass loss being 0.028 g below that value.
The range of temperature increases for the water in the beaker was even smaller, being 0.5°C on either side of the average.
The repeatability is good because three trials were performed and the results were consistent within those three trials.
h To determine the reproducibility of the experimental design, other groups of students would need to be asked to perform the experiment following the same method and then their results could be compared to those of the original two students.

26 a 41.9 MJ kg^{-1}
b Imperfect insulation may allow some heat losses; some of the energy released is absorbed by the container; incomplete combustion of the fuel will reduce energy released.

Chapter 6 Rates of chemical reactions

6.1 Rate of reaction and collision theory

CSA: Saved by a very fast chemical reaction

1 nitrogen, N_2
2 Nitrogen is an inert gas, so it will not react with the sodium metal, even at high temperatures.
3 37.4 g

Key questions

1 Particles must collide with each other; they must collide with sufficient energy to break the reactant bonds, and they must collide with the correct orientation to break the reactant bonds.
2 The activation energy is the minimum energy required to break the reactant bonds. If these are not broken, then new products cannot be formed and the reaction cannot proceed.
3 The larger the activation energy, the smaller the proportion of collisions that result in a reaction at a given temperature. A reaction with a larger activation energy will have a lower rate of reaction at a particular temperature than a reaction with a smaller activation energy.
4 +139 kJ mol^{-1}
5 a $CH_4(g) + 2O_2(g) \longrightarrow CO_2(g) + 2H_2O(l)$
b $\Delta H = -890$ kJ mol^{-1}
c

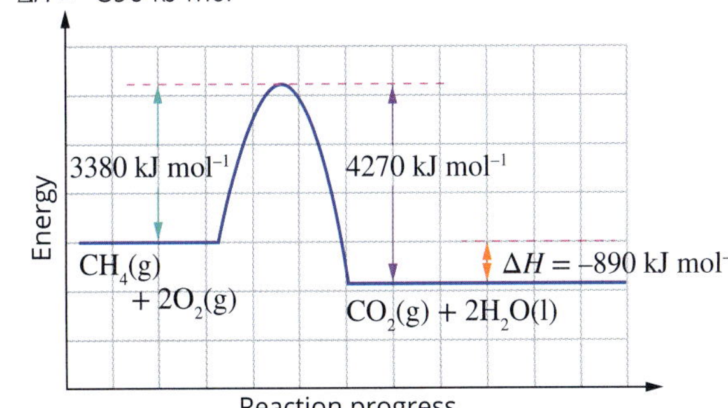

6 The reaction flask could be placed on an electronic balance while the reaction proceeded, and the mass could be recorded at time intervals (e.g. initially every 15 s, then 30 s). Alternatively, the reaction flask could be connected to a gas syringe, as in Figure 6.1.3. The volume of gas could be measured either for a fixed amount of time or recorded at time intervals as for the mass measurements.

6.2 Effect of changes of conditions on rate of reaction

TY 6.2.1 The surface area of the iron anchor is relatively small, so the frequency of collisions with reacting particles would be low. The concentration of oxygen at greater depths is also low, so the frequency of collisions is further reduced. Therefore, the rate of corrosion is reduced.

Key questions

1 Increase in concentration of a solution; increase in surface area of a solid by making the particle size smaller; increase in gas pressure; increase in temperature

2 As the temperature of a substance increases, the average kinetic energy of the substance increases. In other words, kinetic energy is proportional to temperature.

3 An increase in temperature increases the average kinetic energy of the particles. As a result it increases the frequency of collisions between particles and, more importantly, it increases the energy with which the particles collide, thus increasing the proportion of collisions with an energy greater than or equal to the activation energy.

4 1 M HCl has a greater concentration than 0.1 M HCl, so the frequency of collisions between HCl and the magnesium will be greater for the 1 M HCl than for the 0.1 M HCl. Therefore the rate of reaction will be greater for the 1 M HCl.

5 **a** Experiment A would have the greatest reaction rate, because the $CaCO_3$ has the greatest surface area (it is powdered) and the HCl has the greatest concentration (2 M), thus increasing the frequency of collisions. The temperature is also higher, at 40°C, than in the other experiments, increasing the proportion of collisions with an energy greater than or equal to the activation energy.

b Experiment C would have the smallest reaction rate because the surface area of $CaCO_3$ is the smallest (it is in large pieces) and the HCl has the lowest concentration (0.5 M), thus decreasing the frequency of collisions. The temperature is lower, at 15°C, than in the other experiments, reducing the proportion of collisions with an energy greater than the activation energy.

6 **a** At higher temperatures, the molecules that react to form fibreglass plastics have greater energy. They collide more frequently and are more likely to have a total energy greater than the activation energy of the reaction involved, increasing the rate of reaction.

b Fine particles have a large surface area, resulting in a high frequency of collisions—in this case of aluminium particles with gas molecules (such as oxygen) in the air—and hence a rapid reaction rate. The aluminium can therefore burns vigorously and releases a large quantity of heat.

6.3 Catalysts

CSA: Heterogeneous catalysts in industry

1 The reaction between nitrogen and hydrogen has a very high activation energy, so only a small proportion of collisions have an energy greater than or equal to the activation energy.

2 The presence of a catalyst decreases the activation energy, the energy required to break the reactant bonds, but it does not change the enthalpy of the reactants or the enthalpy of the products. As a result, the change in enthalpy, ΔH, which is equal to $H_p - H_r$, does not change.

3 Sustainability concepts include the recycling of materials that are non-renewable. The metals used in catalytic converters, platinum, palladium and rhodium are expensive and their supply is limited, so they should be recycled wherever possible.

Key questions

1 A catalyst is not consumed during the reaction it speeds up and it does not appear as either a reactant or a product in the equation for the reaction.

A catalyst causes the rate of the reaction to increase because it provides an alternative reaction pathway with a lower activation energy, so a greater proportion of collisions have an energy greater than or equal to the activation energy.

2 The student has incorrectly stated that catalysts increase the frequency of collisions. Catalysts lower the activation energy for a reaction by providing an alternative reaction pathway, so they increase the proportion of collisions that are successful, i.e. that have an energy greater than or equal to the activation energy. The statement should read 'A catalyst increases the rate of a reaction because all catalysts increase the proportion of collisions with energy greater than or equal to the activation energy'.

3 Porous pellets have a high surface area because of all the holes in the surface. This high surface area increases the frequency of collisions and so increases the rate of reaction.

4 When salt is mixed with sugar, the salt acts as a catalyst and lowers the activation energy of the combustion reaction between the sugar and oxygen.

5 Both increase the reaction rate by increasing the proportion of collisions with energy greater than or equal to the activation energy. An increase in temperature increases the average kinetic energy of the collisions, but does not change the activation energy, so a greater proportion of collisions have $E \geq E_a$. Adding a catalyst does not change the energy of the reacting particles; instead it lowers the activation energy, so a greater proportion of collisions have $E \geq E_a$.

6 Independent variable: The type of catalyst used (yeast, potassium iodide, manganese dioxide)

Dependent variable: The volume of oxygen produced (and collected) in a given time, or the change in mass of the open reaction flask in a given time

Controlled variables: Concentration of hydrogen peroxide, temperature of hydrogen peroxide, mass of catalyst used, method by which the catalyst is added (e.g. all at once, in the flask before, or after the hydrogen peroxide is added)

Chapter 6 review

1 B. 2 A. 3 B.

4 D. 5 C.

6 'Frequency of collisions' refers to how often collisions occur; for example, whether there are 100 collisions per second, or the greater frequency of 1000 collisions per second. In comparison 'proportion of collisions' refers to a share, or fraction of a total number. So, if the proportion of collisions with energy $\geq E_a$ increases, there is a greater share of the total collisions that have enough energy to react ($E \geq E_a$).

7 If a catalyst was used to increase the reaction rate, the activation energy barrier would be at a lower value, so quantities A (the activation energy of the forward reaction) and C (the activation energy of the backward reaction) would decrease. ΔH (B) would not change.

8 An increase in the concentration of a **solution**, the surface area of a **solid** or the pressure of a **gas**, will increase the **frequency** of collisions in a reaction, meaning that an increased number of **collisions** occur during a particular **time**. As a result, the frequency of **successful** collisions will also increase, and the **rate** of reaction will increase. Adding a catalyst to a reaction will increase the **proportion** of collisions which have energy that is **greater** than or equal to the **activation energy**. This means that out of the total number of collisions, **more** collisions have enough energy to overcome the **activation** energy. Increasing the temperature of a reaction will increase both the **frequency** of collisions and the **proportion** of collisions that have $E \geq E_a$.

9 **a** The definition does include any sense of time. So, the mass of products in consideration could be made over a few seconds (a fast reaction) or could be made over several hours (a slow reaction). These two reactions would have very different rates.

b The rate of reaction is equal to the change in concentration of products or reactants divided by the change in time.

$$\text{rate} = \frac{\text{change of concentration}}{\text{time}}$$

10 a The surface area of a solid reactant, concentration of reactants in a solution, pressure of any gaseous reactants, temperature of the reaction and presence of a catalyst, all affect the rate of reaction.

b

Increased frequency of collisions
- increased concentration
- increased pressure
- increased surface area

Both: increased temperature

Increased proportion of collisions with $E \geq E_a$
- catalyst added

11 In the first experiment (0.5 M $CuSO_4$ and zinc powder at 40°C), the temperature is higher than in the second experiment (0.5 M $CuSO_4$ and zinc pieces at 25°C). When the temperature is increased, the average kinetic energy of the particles in solution (Cu^{2+}) will increase, and the proportion of collisions with an energy greater than or equal to the activation energy will increase. In addition, the frequency of collisions will increase because of the increased kinetic energy, and because of the increased surface area of the zinc in the first experiment.

12 a 1370 kJ mol^{-1} **b** –572 kJ mol^{-1}

c

Without catalyst

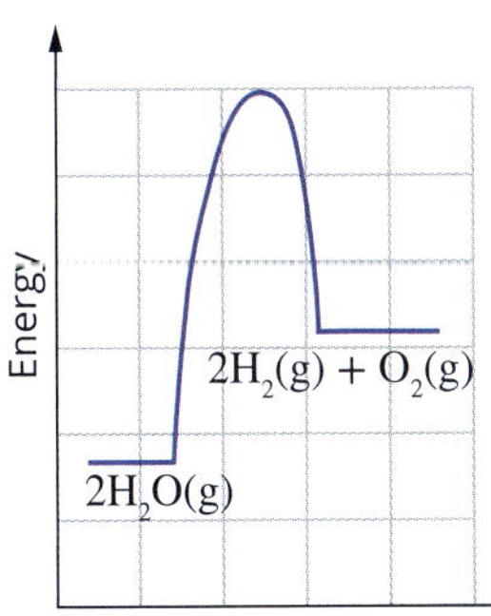

With catalyst

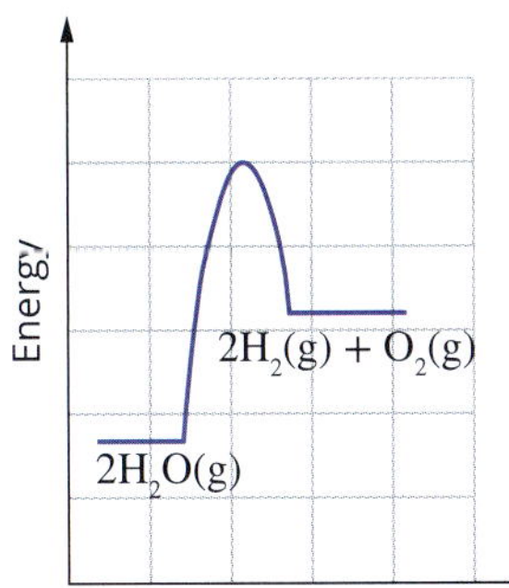

d +572 kJ mol^{-1}

13 Collision 1 has the correct collision orientation, because the hydroxide is in the correct position to substitute for the chlorine, allowing bonds to break within the reactants and bonds to form within the products.

14 To increase the rate at which the sugar dissolves without ruining the toffee, you could:
- grind up the sugar crystals or use caster sugar
- use a cup of hot water to dissolve the sugar
- gently heat the sugar and water mixture while the sugar is dissolving
- stir the sugar and water mixture while the sugar is dissolving, to ensure that all of the solid sugar is coming in contact with the water.

15 a $CaCO_3(s) + 2HNO_3(aq) \rightarrow Ca(NO_3)_2(aq) + H_2O(l) + CO_2(g)$

b The rate of reaction is decreasing over time. This is shown in the graph as the gradient of the slope decreases as the time progresses.

16 a When salt is added to the water, the maximum temperature that the water can reach (its boiling temperature) increases. As a result, the pasta will cook more quickly because the rate of reaction (i.e. the process by which it cooks) increases.

b On top of Mt Everest the atmospheric pressure is very low due to the high altitude, so the temperature at which water boils decreases. The pasta cooks more slowly because the rate of reaction (i.e. the process by which it cooks) decreases due to its decreased temperature.

17

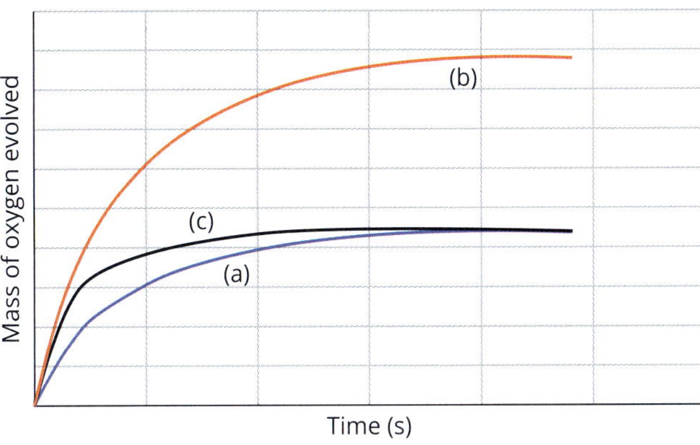

a This graph is labelled 'a' above.

b This graph is labelled 'b' and should have a much higher finishing mass, as the concentration of the hydrogen peroxide is double that which was used in part **a**. The rate of reaction, shown by the gradient of the graph is also greater than for part **a**.

c This graph is labelled 'c' and should finish at the same mass as the graph for part **a**, but it should have a steeper gradient due to the higher temperature. Generally, rate of reaction doubles for every increase of 10°C.

18 a $CaCO_3(s) + 2HCl(aq) \rightarrow CaCl_2(aq) + CO_2(g) + H_2O(l)$

b $n(CaCO_3) = \frac{m}{M} = \frac{10.0}{100.1} = 0.0999$ mol

$n(HCl) = cV = 0.1 \times 0.1 = 0.01$ mol

$\therefore$ $CaCO_3$ is in excess

c The rate of reaction could be measured by:
- measuring the decrease in the mass of reaction mixture as $CO_2(g)$ escapes to the atmosphere
- measuring the increase in pH with a pH probe as the acid is consumed.

d The rate of reaction with the smaller lumps will be faster. The smaller lumps have a larger surface area, so more collisions can occur per second.

e The rate of the reaction can be altered by an increase in temperature or an increased concentration of hydrochloric acid.

For an increase in temperature, a greater proportion of collisions have $E \geq E_a$, and the particles collide with greater frequency, so the rate of reaction increases.

For an increase in concentration of HCl, there are more HCl molecules in a given volume, so the frequency of collisions between the $CaCO_3$ and HCl increases.

19 a A gas is produced, so mass is lost from the mixture.

b HNO_3 is in excess, so Cu is limiting.

c

Mass of mixture (g)

1.00 mol L^{-1} nitric acid

Time (s)

The decreased rate of mass loss is due to the lower nitric acid concentration.

$n(HNO_3) = cV = 1.00 \times 0.500 = 0.500$ mol

Copper is still limiting, so the final mass remains the same.

d

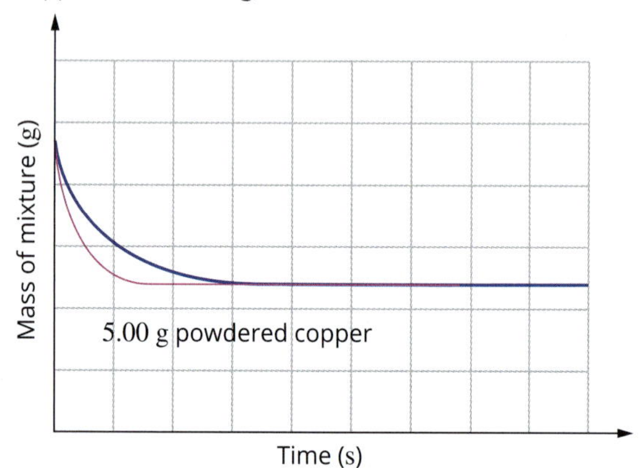

The increased rate of mass loss is due to the increased copper surface area.

Copper is still limiting, so the final mass remains the same.

20 **a** Any two of: the concentration of HCl in the mixture; the concentration of $Na_2S_2O_3$ in the mixture; the temperature of the solution

b The dependent variable is the time taken for the cross to disappear (i.e. for so much sulfur to be present in the solution that the cross can no longer be seen).

c The dependent variable would be inversely proportional to the rate of reaction, because the less time it takes to make that quantity of sulfur, the greater the rate of reaction.

d Students' answers will vary. For example: To determine how the concentration of HCl in a mixture of HCl and $Na_2S_2O_3$ affects the time it takes for a cross underneath the reaction beaker to no longer be visible.

21 **a** The graphs start at a high value and decrease as the reaction progress. If these graphs were of carbon dioxide produced over time, they would start out at zero and would increase. It is most likely that this is the overall mass of the open container plus reactants during the course of the reaction.

b Graph B is likely to represent the higher temperature reaction because it has a greater initial gradient, which corresponds to a greater initial rate.

c The results shown in graph B were obtained at a higher temperature than those in graph A. This means that the reactant particles have a greater average kinetic energy for graph B. The frequency of collisions is greater in graph B and the proportion of collisions with $E \geq E_a$ is greater in graph B, so the initial rate of reaction, and hence the gradient at the beginning of graph B is greater.

22 **a** The temperature of the new process (550°C) is much lower than the temperature of 1800°C in the blast furnace.

b $2Fe(s) + O_2(g) \longrightarrow 2FeO(s)$

$4Fe(s) + 3O_2(g) \longrightarrow 2Fe_2O_3(s)$

c No, a catalyst does not participate in the reaction. The reaction of iron with water is different to the reaction of iron with dry oxygen and proceeds at a different rate.

d exothermic

e The high surface area of the iron pellets and the high temperature caused by trapped heat that was unable to escape rapidly.

f Water would have caused the production of more hydrogen and increased the fire. Students' answers will vary, but the method used by the firefighters to extinguish the fire was to flood the hold with liquid nitrogen, which extinguished the surface fire, but did not stop the deeper burning. A crane and clamshell bucket were then used to unload the iron into piles less than 1 m deep so the heat could escape.

Chapter 7 Extent of chemical reactions

7.1 Dynamic equilibrium

CSA: Investigating dynamic equilibrium

1 Initially there are no Na^+ ions or I^- ions present in the water. Because NaI is very soluble, once the solid NaI is added to the pure water, the ions are released into solution. As the concentrations of Na^+ and I^- increase, they recombine to form solid NaI. The forward and reverse reactions continue to occur until the rate at which the ions come into solution is equal to the rate at which they recombine. Dynamic equilibrium is achieved.

2 $NaI(s) \rightleftharpoons Na^+(aq) + I^-(aq)$ or $NaI(s) \rightleftharpoons NaI(aq) \rightleftharpoons Na^+(aq) + I^-(aq)$

3 Dynamic equilibrium is achieved when the rate of the forward reaction is equal to the rate of the reverse reaction. This means that there is no change in the concentration of either the reactants or the products. Each are being constantly formed, but at equal rates.

4 Initially, when the radioactive NaI is added, the radiation detector records the radioactivity only in the solid on the bottom of the container, but after some time the radiation detector will show that the solution is also radioactive. The fact that the concentration remains constant indicates that as some radioactive iodide is dissolved, the same amount of non-radioactive or radioactive iodide must recombine and form solid NaI. Although we cannot see this reaction occurring, the radiation detector indicates the presence of radioactivity in both the solid and the solution.

Key questions

1 C.

2 **a**

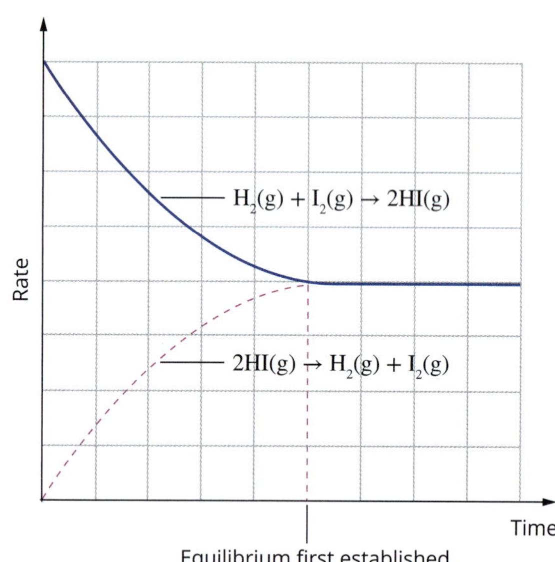

b We are told that initially H_2 and I_2 are present in the sealed container and there is no product, HI. Therefore, the higher curve is the forward reaction where H_2 and I_2 are producing HI. As the reactants are being consumed, the rate of the forward reaction decreases. The lower curve is the reverse reaction, which occurs when some product has been formed and reconverts to the original reactants. The more products produced, the more the rate of the reverse reaction increases. When the rate of the forward reaction becomes equal to the rate of the reverse reaction, equilibrium has been achieved.

3 In a **closed** system, as the concentrations of the reactants **decrease**, the rate of the forward reaction also decreases. The collisions between these reactant molecules occur **less** frequently. Once some product starts to form, the **reverse** reaction occurs and the frequency of collisions between product molecules **increases** and between reactant molecules decreases. At equilibrium, the rates of the forward and backward reactions are **equal** and the **concentrations** of all species do not change.

4 For a reaction to be at equilibrium, the reaction must be reversible and have reached a state where the rate of the forward reaction is equal to the rate of the reverse reaction. There is no observable change in concentrations of the reactants or the products.

5 **a** 0.070 M **b** 0.00 M **c** 0.030 M
d 0.080 M **e** 0.040 mol has decomposed.
f The horizontal regions of the graph indicate that there is no change in concentration of NO_2 or N_2O_4. This is when the system is at equilibrium and both the forward and reverse reactions are occurring at the same rate.
g 6 s
h As the reaction proceeds, the intensity of the brown colour will increase as the concentration of NO_2 increases until equilibrium is reached. Once equilibrium is reached, the colour of the reaction mixture will remain constant.

7.2 The equilibrium law

TY 7.2.1 M^{-1} (or L mol^{-1})

Key questions

1 **a** A homogeneous system is a system in which all the species are in the same state.
b The reaction quotient, *Q*, is the concentration ratio at any time during a reversible reaction, and is equal to *K* at equilibrium.
$$Q = \frac{[\text{products}]^{\text{coefficient}}}{[\text{reactants}]^{\text{coefficient}}}$$
c The equilibrium constant, *K*, is the ratio of the equilibrium concentrations of the products divided by the equilibrium concentrations of the reactants raised to the power of their coefficients. The value for the equilibrium constant changes with temperature.

2 $Q = \frac{[HBr]^2}{[H_2][Br_2]}$

3 $K = \frac{[Fe^{2+}]^2[Sn^{4+}]}{[Fe^{3+}]^2[Sn^{2+}]}$

4 The reaction quotient $Q < K$. The reaction will shift to the right to establish equilibrium. This causes the concentration of the products to increase.

5 $Q = \frac{[[Cu(NH_3)_4]^{2+}]}{[Cu^{2+}][NH_3]^4}$
As $Q > K$, in order to reach equilibrium $[[Cu(NH_3)_4]^{2+}]$ must decrease and the concentration of the reactants must increase. This will occur if the reaction shifts to the left to produce more reactants.

7.3 Calculations involving equilibrium

TY 7.3.1 0.694 M **TY 7.3.2** 0.17 M **TY 7.3.3** 0.0248 M

Key questions

1 B.

2 **a** No, very little (negligible) reaction would occur. $K = 10^{-10}$ is very small, indicating that the amount of product is minimal.
b $K = 10^{10}$
c For this reverse reaction, the equilibrium constant is the reciprocal of that for the forward reaction, so $K = 10^{10}$. Because *K* is very large, provided the rate is sufficiently fast there will be significant reaction in the forward direction—in fact, a virtually complete reaction will occur with a very high amount of product.

3 **a** $K = \frac{[Br_2][Cl_2]}{[BrCl]^2}$
b **i** $K = 5.7$
ii $K = 0.031$
iii $K = 1.0 \times 10^3$
iv $K = 0.18$

4 **a** $K = \frac{[COCl_2]}{[CO][Cl_2]}$
b $K = 1.1 \times 10^2\ M^{-1}$

5 $K = 0.10$ M

6 $K = 0.127$ M

7 $K = 0.038\ M^{-1}$

8 $K = 1.3 \times 10^{-3}\ M^{-2}$

7.4 Le Châtelier's principle

TY 7.4.1 There are 2 molecules of gas on the reactant side and 1 molecule of gas on the product side, so the system will shift to the left (the reactant side).
This increases the amount of the reactants, including Cl_2.
(Note that the PCl_3 and Cl_2 concentrations will still be lower than they were at the initial equilibrium. The shift in equilibrium position only partially compensates for the change.)

CSA: Equilibria in a swimming pool

1 $H_2O(l) + H_2O(l) \rightleftharpoons H_3O^+(aq) + OH^-(aq)$

2 As pH increases, the concentration of $H_3O^+(aq)$ decreases because of the inverse relationship between pH and $[H_3O^+]$. From the equation in the case study, $OCl^-(aq) + H_3O^+(aq) \rightleftharpoons HOCl(aq) + H_2O(l)$, Le Châtelier's principle tells you that the position of equilibrium will move to the left, increasing the concentration of H_3O^+ ions and consuming some of the HOCl. More hydrochloric acid, H_3O^+, will need to be added to re-establish the appropriate [HOCl]. Adding the acid will shift the equilibrium to the right again.

3 To shift the equilibrium to the right, and therefore increase the pH and decrease the $[H_3O^+]$, pool chlorine must be added to restore the correct balance.

Key questions

1 **a** Adding a reactant will result in a net forward reaction.
b Adding a product will result in a net reverse reaction.
c Adding a reactant will result in a net forward reaction.

2 **a** Removing a product will result in a net forward reaction.
b Doubling the pressure, the system will move to reduce the pressure. The equilibrium position will move in the direction of the lower number of particles. There are 4 moles of gas on the reactant side and 2 moles of gas on the product side, so the system will move in a forward direction.
c In this system there are equal moles of gas particles on each side, so increasing the pressure will have no effect.
d Increasing the temperature of an endothermic reaction will result in a forward reaction. An endothermic reaction has a positive ΔH, indicating that the reaction needs heat in order to proceed. You can think of the increasing temperature resulting in more heat (on the reactant side) being added to the system. Hence, the system moves to oppose this increase by consuming the heat. Alternatively, you can also predict the effect of the change from the knowledge that when the temperature of an endothermic reaction is increased, the equilibrium constant will increase. This means that the amount of product increases and the reaction moves in a forward direction.

3 **a** increase **b** decrease **c** increase **d** decrease

4 a i ΔH is positive, so increasing the temperature causes K to increase and therefore results in a net forward reaction.

ii ΔH is positive, so increasing the temperature causes K to increase and therefore results in a net forward reaction.

iii ΔH is negative, so decreasing the temperature causes K to increase and therefore results in a net forward reaction.

b i No. Volume is inversely proportional to pressure. The reaction has equal moles of gas on both sides, so changing the volume will not cause a forward reaction.

ii Yes. Increasing the volume will have the effect of decreasing the pressure. The system will try to oppose this decrease by increasing pressure and so move in the direction of the most particles.

iii No. Because the reaction has equal moles of gas on both sides, we cannot cause a forward reaction to occur.

5 a As the volume is increased, the concentration of NO_2 will decrease. Although the reaction will favour the formation of NO_2 according to Le Châtelier's principle, the adjustment can only partially oppose the change, so the corresponding increase in concentration will not reach the initial NO_2 concentration level.

b The mass of NO_2 will be higher. By favouring the forward reaction, the shift in equilibrium will result in a higher yield of NO_2

6 a Decreasing the temperature would favour the forward exothermic process, resulting in a net forward reaction.

b There are 5 particles on the reactant side and 3 particles on the product side. A decrease in volume would increase the pressure of the system. This would cause the system to favour the forward reaction to reduce the overall number of particles, hence reducing the pressure.

Chapter 7 review

1 a Chemical equilibrium is 'dynamic' because both forward and reverse reactions occur at the same rate. An equilibrium develops between water vapour and water when wet clothes are in a sealed bag, with water evaporating as rapidly as water vapour condenses, so the clothes remain wet.

b When the bag is opened, water vapour escapes and the rate of evaporation exceeds the rate of condensation. The system is not in equilibrium and the clothes dry.

2 The reaction quotient is the ratio of the concentrations of the products to the concentration of the reactants, with the index of each concentration the same as the coefficient of the substance in the reaction equation. It can be calculated at any time during the reaction. The equilibrium constant is the same concentration ratio of products to reactants, but occurs when the rate of the forward reaction is equal to the rate of the reverse reaction. This is when the system is at equilibrium. The value of the reaction quotient becomes equal to the equilibrium constant at equilibrium.

3 a $CH_3OH(g) \rightleftharpoons 2H_2(g) + CO(g)$

b $S_2(g) + 2H_2(g) \rightleftharpoons 2H_2S(g)$

c $NO_2(g) \rightleftharpoons \frac{1}{2}N_2O_4(g)$

4 a If $K = 0.0001$ for a particular reaction, at equilibrium the concentrations of products will be **less than** the concentrations of reactants.

b For the reaction with the equation:

$$2H_2(g) + 2NO(g) \rightleftharpoons 2H_2O(g) + N_2(g)$$

the expression for the equilibrium constant, K, is written as $\frac{[H_2O]^2[N_2]}{[H_2]^2[NO]^2}$.

c When the reaction quotient is smaller than K, the reaction moves to the **right** to establish equilibrium.

5 $K = 5.3$

6 $K = 0.53$

7 $K = 0.011$ M

8 a The concentration of hydrogen gas will increase, because when the temperature of an exothermic reaction increases, the equilibrium constant decreases, the reaction moves in a reverse direction and the reactant concentrations increase. If ΔH is negative, an increase in T means a decrease in K.

b The concentration of hydrogen gas will increase, because when the temperature of an endothermic reaction increases, the equilibrium constant increases, the reaction moves in a forward direction and the product concentrations increase. If ΔH is positive, an increase in T means an increase in K.

9 a A forward reaction occurs and more $C_3H_6O(g)$ and $H_2(g)$ will be produced.

b A reverse reaction occurs and more $CO(g)$ and $2H_2(g)$ will be produced.

c There will be no effect, because there are equal numbers of particles on each side. In this case, there will be no more $N_2(g)$, $O_2(g)$ or $NO(g)$.

10 B.

11 a $K = 1.6 \times 10^{-7}$ M^2

b $K = 2.5 \times 10^3$ M^{-1}

12 a

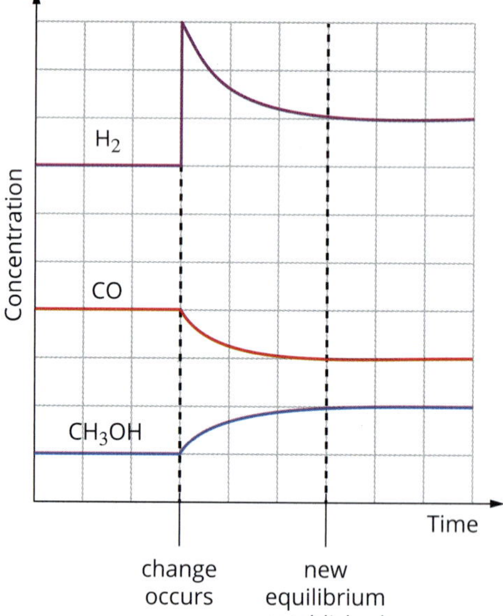

b

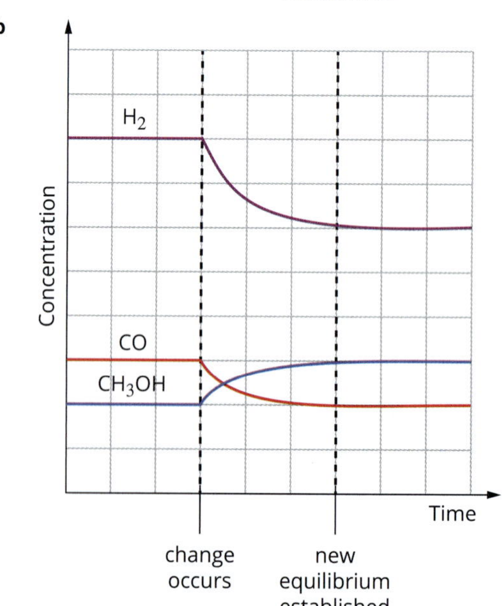

13 As Q for the reaction is larger than K, the reaction must move to decrease Q. This will happen with a decrease in the concentration of the products. This will occur as the reverse reaction is favoured, decreasing the concentration of ethyl ethanoate as the mixture reaches equilibrium.

14 The reaction quotient $Q = 0.024$ is less than $K = 0.052$, so the reaction is not at equilibrium and will shift towards the right (in the forward direction), to increase Q and reach equilibrium.

15 **a** $K = \frac{[PCl_3][Cl_2]}{[PCl_5]}$

b $K = 0.400$ M

c $[Cl_2] = 0.80$ M

d $K = 2.5\ M^{-1}$

16 $K = 0.0980$ M

17 **a** [A] = 0.9 M, [B] = 0.27 M, and [D] = 1.6 M

b [C] = 0.016 M

c n(C) = 0.032 mol

18 If Ca^{2+} ions are absorbed inefficiently from food, decreased concentrations of these ions could occur in body fluids. As a consequence, a net forward reaction of the given equation would occur, raising the concentration of dissolved Ca^{2+} ions in body fluids, but resulting in decreased amounts of calcium phosphate in bones.

19 **a** $K = 0.020$ (no unit)

b The system is not at equilibrium. A net reverse reaction will occur. As the reaction moves towards equilibrium, the concentrations of H_2 and I_2 will decrease and the concentration of HI will increase until the value of the reaction quotient is equal to the equilibrium constant.

20 $K = 5.7 \times 10^2\ M^{-1}$

21 **a** The word 'fuel' indicates that the reaction will be exothermic (with a negative ΔH) and that the reaction will proceed extensively in the forward direction, so the equilibrium constant will be large.

b **i** It will move in the direction of the reactants, i.e. a net reverse reaction.

ii There will be no change in equilibrium position because a catalyst increases the rate of the forward and reverse reactions equally.

iii It will move in the direction of the products, i.e. a net forward reaction.

c **i** The value of K will decrease.

ii no change

iii no change

22 **a**

Change	Colour change (lighter or darker)	Explanation
i The temperature is decreased to 250°C at constant volume.	yes (darker)	This is an exothermic reaction, so K would increase as the temperature decreases and a net forward reaction would occur, producing more brown NO_2 gas.
ii The volume of the container is halved at constant temperature.	yes (darker)	The decrease in volume means an increase in pressure. The system would oppose this change by decreasing the pressure and moving in direction of least particles—i.e. a forward reaction would occur. The colour of the gas mixture would increase because more brown NO_2 will form.
iii A catalyst is added at constant volume and temperature.	no change	The addition of a catalyst would not change the colour because the rate of the forward and reverse reactions would be increased equally.

b

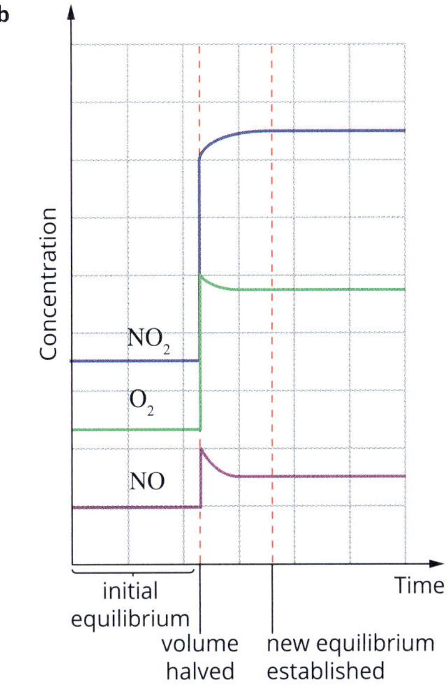

23 **a** For the initial mixture,

$Q = \frac{[CH_4][H_2O]}{[CO][H_2]^3} = \frac{0.300 \times 0.400}{0.100 \times 0.200^3} = 150\ M^{-2}$

$Q = 150$ is larger than $K = 0.67$. The reverse reaction will occur so that Q becomes equal to K. The [products] will decrease and [reactants] will increase.

i increase **ii** increase

iii decrease **iv** decrease

b The enthalpy change will be positive, indicating that the reaction is endothermic. An increased value of the equilibrium constant with an increased temperature indicates that the reaction has moved in the direction of the products. The extra heat has favoured the forward reaction, indicating it is an endothermic reaction, which consumes heat to occur.

24 **a** $2SO_3(g) + CO_2(g) \rightleftharpoons CS_2(g) + 4O_2(g)$

b $K = \frac{[CS_2][O_2]^4}{[SO_3]^2[CO_2]}$

c **i** $n(O_2) = 0.112$ mol **ii** $K = 1.6 \times 10^{-3}\ M^{-2}$

d **i** increase **ii** increase

iii no effect **iv** decrease

v no effect

25 **a** $K = \frac{[SO_3]^2}{[SO_2]^2[O_2]}$

b 15–20 minutes, 25–30 minutes, 35–40 minutes

c $K = 1.6\ M^{-1}$ **d** $K = 3.3\ M^{-1}$ **e** $K = 3.2\ M^{-1}$

f At 10 minutes; before 10 minutes the concentrations were changing slowly (shown in the graph by the very slight slope), but the addition of the catalyst at 10 minutes caused equilibrium to be reached more rapidly (shown by the steeper gradients of the graphs).

g The pressure was increased by reducing the volume. To oppose this pressure increase, the reaction moved forward, as shown by the graphs. (Also, because the value of K at 25–30 minutes is larger than at 15–20 minutes and this is an exothermic reaction, a decrease in temperature must have occurred.)

h O_2 gas was added, moving the reaction in the forward direction. (Also, because the equilibrium constant decreased and this is an exothermic reaction, there must have been an increase in temperature.)

i This is an exothermic reaction. Between 15 and 25 minutes the equilibrium constant increased, indicating that the temperature was decreased at 20 minutes. Between 25 and 35 minutes, the equilibrium constant decreased, indicating that the temperature was increased at 30 minutes.

Chapter 8 Production of chemicals by electrolysis

8.1 Electrolytic cells

TY 8.1.1 Nickel is produced at the cathode and Cu^{2+} is produced at the anode.

CSA: Producing hydrogen for the hydrogen economy

1 The electrolysis cells use electricity that is produced predominantly at power stations using coal or natural gas as their fuel source. The combustion of fossil fuels such as coal or natural gas produces CO_2, a key greenhouse gas.

2 anode: $4OH^-(aq) \rightarrow O_2(g) + 2H_2O(l) + 4e^-$

cathode: $2H_2O(l) + 2e^- \rightarrow H_2(g) + 2OH^-(aq)$

3 The steam reforming process involves a reaction where methane (CH_4) and steam react to produce carbon monoxide and hydrogen:

$$CH_4(g) + H_2O(g) \rightarrow CO(g) + 3H_2(g)$$

Additional hydrogen can be obtained by reacting the carbon monoxide with steam:

$$CO(g) + H_2O(g) \rightarrow CO_2(g) + H_2(g)$$

The steam reforming process has a negative environmental impact because the first reaction is energy, intensive, requiring temperatures of 700–1100°C (currently, the energy required for this process comes from the combustion of fossil fuels) and the second reaction produces CO_2, a key greenhouse gas. The release of greenhouse gases into the environment can be reduced through the use of carbon capture and storage technology.

Key questions

1 In electrolytic cells:

a the reactions are **non-spontaneous**

b electrical energy is converted into **chemical** energy

c the anode is **positive** and the cathode is **negative**

d **oxidation** occurs at the anode and **reduction** occurs at the cathode.

2 Cu^{2+} ions move towards the cathode (–). Br^- ions move towards the anode (+).

3 The products are iodine, $I_2(g)$ and potassium, K(l).

This is because the only ions present are $K^+(l)$ and $I^-(l)$.

At the anode, the reaction is: $2I^-(l) \rightarrow I_2(g) + 2e^-$

At the cathode, the reaction is: $K^+(l) + e^- \rightarrow K(l)$

4 a silver rod

b carbon rod (anode, positive): $2H_2O(l) \rightarrow O_2(g) + 4H^+(aq) + 4e^-$

silver rod (cathode, negative): $Ag^+(aq) + e^- \rightarrow Ag(s)$

c $4Ag^+(aq) + 2H_2O(l) \rightarrow O_2(g) + 4H^+(aq) + 4Ag(s)$

5 a Possible reactions at the anode (positive electrode):

$O_2(g) + 4H^+(aq) + 4e^- \rightleftharpoons 2H_2O(l)$ +1.23 V

$Cu^{2+}(aq) + 2e^- \rightleftharpoons Cu(s)$ +0.34 V

Possible reactions at the cathode (negative electrode):

$2H_2O(l) + 2e^- \rightleftharpoons H_2(g) + 2OH^-(aq)$ –0.83 V

$Na^+(aq) + e^- \rightleftharpoons Na(s)$ –2.71 V

b anode: $Cu(s) \rightarrow Cu^{2+}(aq) + 2e^-$

cathode: $2H_2O(l) + 2e^- \rightarrow H_2(g) + 2OH^-(aq)$

c $Cu(s) + 2H_2O(l) \rightarrow Cu^{2+}(aq) + 2OH^-(aq) + H_2(g)$

8.2 Commercial electrolytic cells

CSA: Electrorefining of copper

1 The top 10 impurities are silver, gold, platinum, tin, antimony, lead, iron, nickel, cobalt, zinc—so any five of these is acceptable.

2 Tin, antimony, lead, iron, nickel, cobalt, zinc. These metals are all stronger reducing agents than copper so they are oxidised.

3 Silver, gold and platinum. These metals are all weaker reducing agents than copper, so copper is oxidised instead.

4 Lead is oxidised to form lead(II) ions, which react with sulfate ions in the electrolyte to form a precipitate of lead(II) sulfate. The precipitate ends up in the 'anode mud'.

Key questions

1 The sodium metal produced is a very strong reducing agent and the chlorine gas produced is a very strong oxidising agent. If they came into contact with each other, they would spontaneously react to re-form sodium chloride.

2 The iron cathode is connected to the negative terminal of the power supply, so it receives a continual supply of electrons, preventing it from being oxidised.

3 a An inert electrode would be suitable. A reactive electrode would be oxidised at the anode, releasing cations into the solution, thus contaminating the electrolyte. The only reactive electrode that would be suitable is a zinc electrode.

b Zinc ions are a stronger oxidising agent than water, so the zinc ions are preferentially reduced at the cathode. The use of an aqueous electrolyte is also cheaper than a molten electrolyte.

c The strongest reducing agent present is water, so it is oxidised at the anode to produce oxygen gas:

$$2H_2O(l) \rightarrow O_2(g) + 4H^+(aq) + 4e^-$$

4 a It is cheaper to obtain chlorine by electrolysis of concentrated sodium chloride solution than from molten sodium chloride because energy is required to melt sodium chloride.

b Fluorine is the strongest oxidising agent known. Since a stronger oxidising agent than fluorine would be required to convert F^- ions into fluorine, the element cannot be made by direct reaction. However, it is generated at the anode during electrolysis of molten metal fluorides.

c Na^+ ions are stronger oxidising agents than Ca^{2+} ions, so sodium metal is formed at the cathode in preference to calcium.

5 Changing to unreactive anodes from carbon anodes would avoid the production of carbon dioxide and reduce the cost of having to replace the electrodes.

8.3 Faraday's laws

TY 8.3.1 $m(Cu) = 5.92$ g **TY 8.3.2** $t = 0.764$ h

TY 8.3.3 $t = 1.85 \times 10^6$ s

Key questions

1 a $Ag^+(aq) + e^- \rightarrow Ag(s)$ b $Zn^{2+}(aq) + 2e^- \rightarrow Zn(s)$

c $2Cl^-(l) \rightarrow Cl_2(g) + 2e^-$ d $H_2O(l) + 2e^- \rightarrow H_2(g) + 2OH^-(aq)$

2 a 96 500 C b 38 600 C

c 4 825 000 C d 1158 C

3 18.5 g 4 1.5 h

5 a $Zn^{2+}(aq) + 2e^- \rightarrow Zn(s)$ b 0.20 A

6 The metal ions have a 2+ charge.

7 16 kg

Chapter review

1 a In galvanic cells, the anode is negative and the cathode is positive; in electrolytic cells, the anode is positive and the cathode is negative.

b In galvanic cells, the direction of electron flow is determined by the cell reaction; in electrolytic cells, the direction of electron flow is determined by the external power supply. Direction of electron flow is always from anode to cathode through the external circuit.

c In galvanic cells, chemical energy is converted into electrical energy; in electrolytic cells, electrical energy is converted into chemical energy.

d Galvanic cell reactions occur spontaneously; electrolytic cell reactions are non-spontaneous.

2 anode: $2Cl^-(l) \longrightarrow Cl_2(g) + 2e^-$
cathode: $Sn^{2+}(l) + 2e^- \longrightarrow Sn(l)$
overall: $Sn^{2+}(l) + 2Cl^-(l) \longrightarrow Sn(l) + Cl_2(g)$

3

+ −
power supply
anode
cathode
platinum
platinum
Ca^{2+}
Cl^-
calcium metal forms
chlorine gas forms
molten calcium chloride

4 C. 5 C.

6 In the cell with lead electrodes, $Pb^{2+}(aq)$ is the strongest oxidising agent and Pb(s) is the strongest reducing agent.
anode: $Pb(s) \longrightarrow Pb^{2+}(aq) + 2e^-$
cathode: $Pb^{2+}(aq) + 2e^- \longrightarrow Pb(s)$

In the cell with the platinum electrodes, $Pb^{2+}(aq)$ is the strongest oxidising agent and $H_2O(l)$ is the strongest reducing agent.
anode: $2H_2O(l) \longrightarrow O_2(g) + 4H^+(aq) + 4e^-$
cathode: $Pb^{2+}(aq) + 2e^- \longrightarrow Pb(s)$

7 If an aqueous solution of sodium chloride was used instead of molten sodium chloride, water would be present. Water is a stronger oxidising agent and reducing agent than $Na^+(aq)$ and $Cl^-(aq)$ respectively. This means water is preferentially oxidised and reduced instead of $Cl^-(aq)$ and $Na^+(aq)$ respectively. Therefore, neither sodium nor chlorine would be produced.

8 a anode: $2Cl^-(aq) \longrightarrow Cl_2(g) + 2e^-$
cathode: $2H_2O(l) + 2e^- \longrightarrow H_2(g) + 2OH^-(aq)$

b The semipermeable membrane only allows Na^+ ions to move through from the anode compartment to the cathode compartment. This allows for a very pure sodium hydroxide product to be produced. The semipermeable membrane also prevents mixing between the reactive products.

9 The reaction in a galvanic cell is spontaneous and if the reactants in the cell were in the one container and in contact with each other, the reaction between them could occur directly, releasing energy as heat rather than as electricity.

In electrolytic cells, the reaction is non-spontaneous, so that both the electrode reactions can occur within the same container. The products of the electrolysis reaction should not be allowed to come into contact with each other, however, or a reaction may occur.

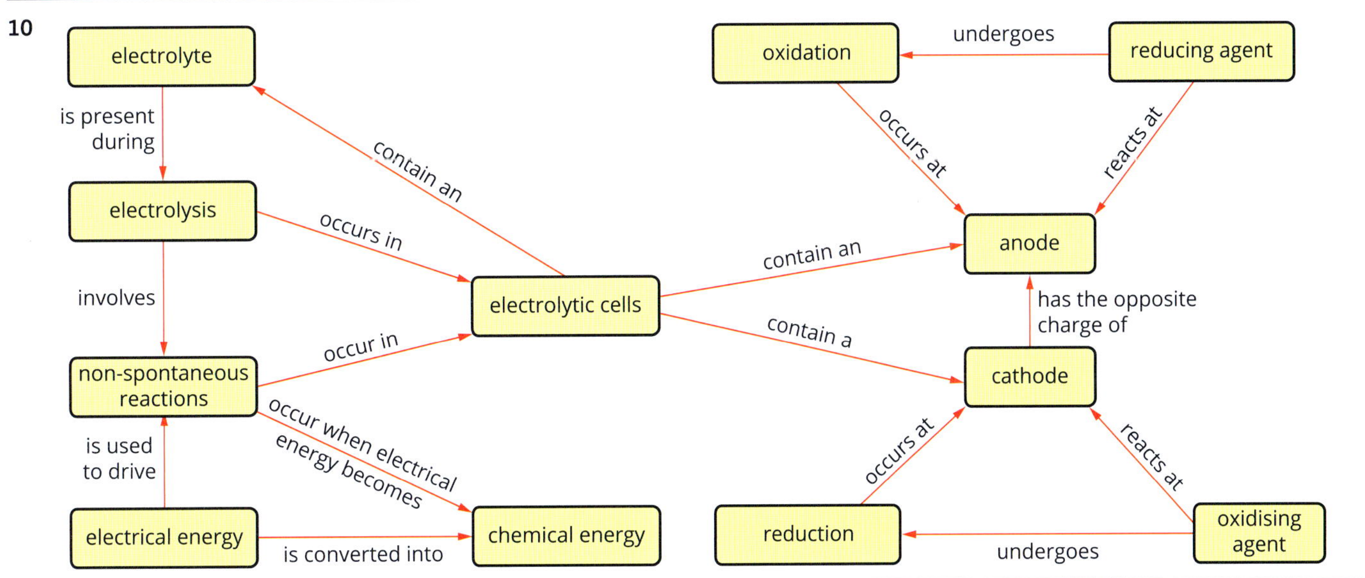

11 The electrolyte solution contains water. As water is a stronger oxidising agent than Mn^{2+}, water will be reduced in preference to the Mn^{2+}. The reaction that occurred was:

$$2H_2O(l) + 2e^- \longrightarrow H_2(g) + 2OH^-(aq)$$

12 a At the anode:
$O_2(g) + 4H^+(aq) + 4e^- \rightleftharpoons 2H_2O(l)$ +1.23 V
$Cu^{2+}(aq) + 2e^- \rightleftharpoons Cu(s)$ +0.34 V
At the cathode:
$Cu^{2+}(aq) + 2e^- \rightleftharpoons Cu(s)$ +0.34 V
$Zn^{2+}(aq) + 2e^- \rightleftharpoons Zn(s)$ −0.76 V
$2H_2O(l) + 2e^- \rightleftharpoons H_2(g) + 2OH^-(aq)$ −0.83 V
The SO_4^{2-} ions are not involved in the reaction.

b anode: $Cu(s) \longrightarrow Cu^{2+}(aq) + 2e^-$. Because Cu(s) is a stronger reducing agent than water, Cu(s) will be oxidised in preference to water.
cathode: $Zn^{2+}(aq) + 2e^- \longrightarrow Zn(s)$. Because $Zn^{2+}(aq)$ is a stronger oxidising agent than water, $Zn^{2+}(aq)$ will be reduced in preference to water.

c $Cu(s) + Zn^{2+}(aq) \longrightarrow Cu^{2+}(aq) + Zn(s)$

13

Experiment	Cathode reaction	Anode reaction
a	$K^+(l) + e^- \longrightarrow K(l)$	$2I^-(l) \longrightarrow I_2(l) + 2e^-$
b	$Cu^{2+}(aq) + 2e^- \longrightarrow Cu(s)$	$2H_2O(l) \longrightarrow O_2(g) + 4H^+(aq) + 4e^-$
c	$2H_2O(l) + 2e^- \longrightarrow H_2(g) + 2OH^-(aq)$	$Cu(s) \longrightarrow Cu^{2+}(aq) + 2e^-$
d	$Cu^{2+}(aq) + 2e^- \longrightarrow Cu(s)$	$2H_2O(l) \longrightarrow O_2(g) + 4H^+(aq) + 4e^-$
e	$2H_2O(l) + 2e^- \longrightarrow H_2(g) + 2OH^-(aq)$	$2H_2O(l) \longrightarrow O_2(g) + 4H^+(aq) + 4e^-$

14 The electrochemical series shows that the oxidising strength of H_2O is greater than that of Al^{3+}. If water is present in the electrolysis cell, it reacts preferentially at the cathode and electrolysis of aqueous aluminium salts does not yield aluminium metal.

15 C. **16** A. **17** B.

18 5.01 A

19 **a** anode: $2H_2O(l) \longrightarrow O_2(g) + 4H^+(aq) + 4e^-$
cathode: $Cu^{2+}(aq) + 2e^- \longrightarrow Cu(s)$
b $c(Cu^{2+}) = 0.80$ M

20 magnesium

21 **a** $Al(s) \longrightarrow Al^{3+}(aq) + 3e^-$; however, the reaction that occurs in practice is $Al(s) + 3OH^-(aq) \longrightarrow Al(OH)_3(s) + 3e^-$
b 32.6 hours

22 **a** $C_6H_{12}O_6(aq) + 6H_2O(l) \longrightarrow 6CO_2(g) + 24H^+(aq) + 24e^-$
b 7.77×10^{-6} mol per hour
c 6.59 mL per hour

23 **a** $Au^+(aq) + e^- \longrightarrow Au(s)$
$Zn^{2+}(aq) + 2e^- \longrightarrow Zn(s)$
$Cr^{3+}(aq) + 3e^- \longrightarrow Cr(s)$
b $m(Zn) = 2.49$ g
$m(Cr) = 1.32$ g

24 **a** cathode **b** negative **c** 8.45×10^4 s

25 **a** To investigate Faraday's first law of electrolysis by examining the effect of increasing current on the mass of copper deposited at the cathode of a copper-plating cell.
b Independent variable: current, in A.
Dependent variable: mass of copper deposited on the nickel medallion (or the cathode), in g.
c **i** 3.95 g
ii Systematic error. The experimental value is inaccurate as it is below the expected value.
iii Examples of systematic errors are: small pieces of copper flaking off the medallion in the cell; the actual current is lower than the recorded, current, so less copper than expected is deposited on the medallion; small pieces of copper are washed off the medallion during the washing and drying stage before weighing.
iv Because the reverse reaction is occurring at each electrode, the concentration of Cu^{2+} and therefore $CuSO_4$ will remain constant, 1.00 M.
d **i** $O_2(g)$ is produced. **ii** 0.752 M

26 **a** 2.03×10^4 C
b A: $Zn(s) \longrightarrow Zn^{2+}(aq) + 2e^-$
B: $2H_2O(l) + 2e^- \longrightarrow H_2(g) + 2OH^-(aq)$
C: $Cu(s) \longrightarrow Cu^{2+}(aq) + 2e^-$
D: $2H^+(aq) + 2e^- \longrightarrow H_2(g)$
c 6.66 g
d 2.60 L
e As the concentration of copper(II) ions increases in the solution due to the anode reaction, copper will be deposited on the cathode because Cu^{2+} is a stronger oxidising agent than H^+, so Cu^{2+} will be preferentially reduced.

27 **a** 7.2×10^2 kg **b** 3.9×10^5 L

28 **a** 7.15×10^4 s **b** 3.33×10^5 g **c** 6.89×10^5 L
d 5.35×10^{10} J **e** $4.2 \times 10^3

29 **a** $n(Ti) = 0.300$ mol **b** 1.2×10^5 C
c 1.2 mol **d** The charge of the titanium ion is 4+.
e The empirical formula of rutile is TiO_2.

Chapter 9 Designing for energy efficiency and sustainability

9.1 Optimising the yield of industrial processes

1 In the production of methanol, both a higher reaction rate and a higher equilibrium yield could be achieved with a **higher** pressure. In addition, a moderate **temperature** needs to be used to ensure that a reasonable reaction rate and equilibrium yield are achieved. A **catalyst** is used to further increase the **reaction rate**.

2 High temperature and low pressure increase the equilibrium yield of NO_2.

3 **a** Gases under high pressure can be dangerous if there is a leak or faulty equipment. The risk is heightened if the gases involved are explosive or flammable. The equipment used to contain gases at high pressure is expensive.
b If gases are explosive or flammable, the risk involved will be greater if the temperature is increased. Increasing the temperature can also increase the pressure, adding to the danger.

4 **a** The reaction is exothermic, so equilibrium yield will be increased at low temperatures. As 3 moles of gaseous reactants become 2 moles of gaseous products, high pressure would favour a higher equilibrium yield. An excess of a reactant can also be used to increase equilibrium yield.
b Reaction rate would be increased by increased pressure, high temperatures and the use of a catalyst.
c The use of high pressure is expensive and can be dangerous, so the reaction is run at 1 atm to reduce costs. The moderate temperature is used to ensure a reasonable rate of reaction without impacting too highly on the equilibrium yield. The decrease in reaction rate due to the lower temperature is offset by the addition of a catalyst. This compromise balances the yield and rate for the plant.

5 **a** The reaction is exothermic, the energy released increases the temperature of the system.
b As the reaction is an exothermic process, the gases are cooled before being pumped back into the reaction system to maximise equilibrium yield. Pumping hot gases into the reaction mixture would cause the temperature to rise and favour the reverse reaction, reducing the yield of ammonia.

6 Temperatures higher than 400°C are required to obtain an acceptably fast rate of reaction.

7 **a** Methane is usually sourced as a fossil fuel. When this happens, it is considered to be a non-renewable fuel and its extraction involves potential environmental damage. The other reactant is water—potentially this can be harvested from rainfall, limiting any negative impact on the environment.
b CO is a toxic gas. It is usually converted to CO_2, which is safer to handle, but presents greenhouse gas issues.
c The manufacture of hydrogen by this method will require significant energy as the reaction is highly endothermic.

9.2 Rechargeable cells and batteries

CSA: Secondary batteries: Car batteries

1 During discharge, Pb is **oxidised** at the **anode** and PbO_2 is **reduced** at the **cathode**. Energy is **released** in this cycle.
During recharge, $PbSO_4$ is oxidised to $\mathbf{PbO_2}$ at the **anode** and $PbSO_4$ is reduced to **Pb** at the **cathode**.

2 Electrical energy is converted into chemical energy.

3 The reaction of PbO_2 at the cathode consumes H^+ ions. As discharge proceeds, the concentration of H^+ ions decreases, the solution becomes less acidic and the pH increases.

Key questions

1 a Discharge: a spontaneous reaction occurs in a cell to produce electrical energy from chemical potential energy.
  b Recharge: a voltage is applied to the cell to reform the original reactants. This is an electrolytic process.

2 a secondary
  b primary
  c secondary

3 >1.5 V

4 Examples of applications that are increasing the demand for lithium-type batteries include electric cars, electricity storage batteries, miniaturised devices, such as hearing aids, and portable appliances.

5 a true
  b true
  c false
  d true
  e true

6 a Lead is a very heavy metal. A series of lead-based cells in a car will make the car too heavy. Lead is a scarce metal and a toxic metal – it is desirable to use a safer alternative.
  b A laptop needs very small batteries. A small NiMH battery would not have a suitable capacity to run a laptop.
  c A laptop runs many functions at once and has a large screen. A primary cell would be drained too quickly in a laptop. The power demands of a basic calculator are not as great.

7 a Reduction half-reaction will occur at the positive cathode.
  b $OH^- + MH \rightarrow H_2O + M + e^-$
  c Oxidation will be at the anode; the anode is positive during charging (recharge).

9.3 Producing 'green' hydrogen gas

1 Brown hydrogen is sourced from fossil fuels, whereas the energy to produce green hydrogen comes from renewable energy sources.
  a brown b green c green d brown

2 Hydrogen gas produced from natural gas has two main issues: it depletes the supplies of natural gas, which is a non-renewable resource, and the process for manufacturing hydrogen releases CO_2.

3 The reaction is a very endothermic one. The covalent bonds in water that need to be broken are relatively strong.

4 a In artificial photosynthesis, sunlight is directed at a photosensitive electrode stored in a solution in an electrolytic cell. The energy from the Sun helps form molecules from the solution, such as methane or hydrogen.
  b The origin of the word photosynthesis is about forming molecules, using energy from the Sun. The molecule formed in plants is usually glucose, but in other mediums different molecules can be formed.

5 a oxygen gas
  b Oxygen is a very useful gas, in medical situations as well as industrial applications.

6 a anode b positive
  c The reaction is not occurring in an aqueous environment.

7 a Artificial photosynthesis is a one-step process, with hydrogen formed directly from the action of sunlight. Most hydrogen trial plants require at least two distinct phases, firstly for the generation and storage of electrical energy and then for the use of that energy to run an electrolytic cell to produce hydrogen gas.
  b The direct conversion of energy in artificial photosynthesis offers potentially higher efficiencies as less energy transformations are required. The need for large-scale storage of electrical energy is also eliminated.

Chapter 9 review

1 C. 2 D. 3 B.
4 A. 5 D.

6 a Temperature: The impact of temperature on equilibrium yield depends upon whether a reaction is exothermic or endothermic. If the reaction is exothermic, an increase in temperature will lower the yield.
  b Pressure: The impact of a pressure change on equilibrium yield depends upon the ratio of reactant particles to product particles. The position of equilibrium will shift to oppose any change of conditions.

7 Catalysts can be particularly useful in exothermic, reversible reactions, where a high equilibrium yield requires a low temperature. The catalyst negates some of the impact on reaction rate of the temperature being lowered, therefore improving the yield.

8 secondary cell: a galvanic cell that can be recharged. A commercial secondary cell needs to be able to go through multiple charging cycles
  electrolyser: a system that uses electricity to break water into hydrogen and oxygen via electrolysis.
  PEM: polymer exchange membrane – a polymeric electrolyte that allows conductivity of ions, usually protons.
  hydrogen hub: precinct shared by a number of companies involved in the production and use of hydrogen
  photoelectrochemical: reactions involving an electric current generated by the action of light

9 During discharge, a secondary cell acts as a **galvanic** cell. The reaction occurring is a **spontaneous** one. For this cycle, **reduction** occurs at the cathode and its polarity is **positive**. During the recharge cycle, the secondary cell operates as an **electrolytic** cell and a **non-spontaneous** reaction occurs. For the recharge cycle, **oxidation** occurs at the anode, and its polarity is **positive**.

10 a The rate of the forward reaction is increased.
  b The rate of the back reaction is increased.
  c The equilibrium yield will decrease as K_c is lower.

11 $PbO_2(s) + SO_4^{2-}(aq) + 4H^+(aq) + 2e^- \rightarrow PbSO_4(s) + 2H_2O(l)$ polarity +
  $Pb(s) + SO_4^{2-}(aq) \rightarrow PbSO_4(s) + 2e^-$ polarity –

12 Fuel cells in cars, blended with domestic gas, manufacture of ammonia and export to other countries.

13 a $2H_2O(l) \rightarrow 2H_2(g) + O_2(g)$
  b Electrolysis of water requires energy.
  c Green hydrogen is hydrogen that is produced using renewable energy. Its production does not drain supplies of non-renewable resources.

14 Hydrogen is an explosive gas. It is often stored at high pressures in high-quality containers and kept from any ignition sources. Its low density as a gas means that companies will liquify it before exporting or convert it to ammonia.

15 a $2Li(s) + 2H_2O(l) \rightarrow 2LiOH(aq) + H_2(g)$
  b Lithium cells need to have non-aqueous electrolytes so that this potentially dangerous reaction cannot occur.

16 a Lithium reacts at the anode, which is negative, and oxygen at the cathode, which is positive.
  b Oxygen gas has an oxidation state of 0 and it is reduced to O^- during discharge.
  c Li^+ ions.
  d $2Li + O_2 \rightarrow Li_2O_2$
  e $Li_2O_2 \rightarrow 2Li + O_2$
  f Li_2O_2, as it is an oxidation reaction

17 a An electrolyser is a system that uses electricity to break water into hydrogen and oxygen via electrolysis.

b A PEM electrolyser does not operate in an aqueous environment like a simple electrolyser. It uses a conductive polymer electrolyte. The overall equation is the same, but the half-equations differ.

c The half-equations for both electrolysers will show the formation of oxygen and hydrogen gases, but the ions will differ.

d The overall equation for both will be $2H_2O \longrightarrow 2H_2 + O_2$. The states will differ. A simple electrolyser uses an aqueous environment, but the PEM does not.

18 a i decrease
ii decrease
iii increase
iv no change
v no change

b i increase
ii increase
iii increase
iv increase
v no change

19 a $2NiOOH(aq) + 2H_2O(l) + Cd(s) \longrightarrow 2Ni(OH)_2(s) + Cd(OH)_2(s)$

b NiOOH reacts at the cathode and cadmium at the anode

c voltage = 0.48 – (–0.82) = 1.3 V

d $2Ni(OH)_2(s) + Cd(OH)_2(s) \longrightarrow 2NiOOH(aq) + 2H_2O(l) + Cd(s)$

e $Ni(OH)_2$ will react at the anode and cadmium at the cathode.

20 a Lead–acid batteries are made in a traditional way, with large parts and an aqueous electrolyte. The top of the cell is easily removed. Lithium-ion cells are high-technology devices using composites and complex membranes. It is not easy to access the individual components. The electrolyte in lithium-ion cells is not a simple ionic solution that the solid can easily be extracted from.

b Lithium batteries contain high proportions of cobalt, nickel and aluminium that could be recycled. The lithium compounds could also be re-used and the conductive membranes have potential to be re-used.

c It is not a good principle to use a material and to then throw it away. Sometimes the material is toxic and it should not be added to landfill and sometimes materials are scarce and need to be preserved.

21 $2H_2O(l) + 4e^- \longrightarrow 2H_2(g) + 2O^{2-}$

22

	Discharge	Recharge
Is the reaction spontaneous or non-spontaneous?	spontaneous	non-spontaneous
reactants	zinc, bromine	Zn^{2+}, Br^-
polarity of anode	negative	positive
direction of electron flow	negative to positive	positive to negative
Is the cell acting as a galvanic or electrolytic cell?	galvanic	electrolytic

23 a The Haber process is an energy-intensive process due to the need for elevated temperatures and pressures. In most instances the energy required is obtained from fossil fuels, producing CO_2 emissions. The use of methane as a source of hydrogen gas adds to the energy requirements of production and the methane is considered a relatively scarce resource.

b If the hydrogen gas produced is green hydrogen, then the need to use fossil fuels in the production of ammonia is eliminated. Water is the source of hydrogen rather than methane, eliminating the emission of CO_2 gas.

c The production of ammonia via electrolysis is a very different process, that offers significant potential energy savings. The process might be able to be conducted at room temperature, thus eliminating the need for expensive, energy-intensive heating of the gases. The electrolytic process removes the reversible nature of production, removing the need for considerations of reactant recycling and cooling. Using lower temperatures and pressures is likely to make the production a much safer process.

Unit 3 Area of Study 2 How can the rate and yield of chemical reactions be optimised?

1 C.	**2** D.	**3** D.
4 D.	**5** C.	**6** A.
7 B.	**8** A.	**9** C.
10 B.	**11** B.	**12** B.
13 A.	**14** D.	**15** B.

16 a anode: $2H_2O(l) \longrightarrow O_2(g) + 4H^+(aq) + 4e^-$
cathode: $2H_2O(l) + 2e^- \longrightarrow H_2(g) + 2OH^-(aq)$

b anode: $2Cl^-(aq) \longrightarrow Cl_2(g) + 2e^-$
cathode: $2H_2O(l) + 2e^- \longrightarrow H_2(g) + 2OH^-(aq)$

c anode: $2Cl^-(l) \longrightarrow Cl_2(g) + 2e^-$
cathode: $Na^+(l) + e^- \longrightarrow Na(l)$

17 a $Li^+(l) + e^- \longrightarrow Li(l)$

b $2Cl^-(l) \longrightarrow Cl_2(g) + 2e^-$

c Water (H_2O) is a stronger oxidising agent than Li^+ and would be preferentially reduced at the cathode, producing hydrogen gas rather than lithium metal.

d A mixed electrolyte will have a lower melting temperature, making the process safer and less expensive. Potassium chloride is appropriate because K^+ will not be reduced in preference to Li^+.

e Like sodium, lithium is a very reactive metal. It must not be allowed to come into contact with air because it will spontaneously oxidise. Also, it must be separated from the chlorine produced, otherwise, the two products would spontaneously react together. In addition, it will react very quickly with water to produce hydrogen gas with the potential for explosion, so it must not be allowed to come in contact with water.

18 a $2H^+(aq) + 2e^- \longrightarrow H_2(g)$

b $2H_2O(l) \longrightarrow O_2(g) + 4H^+(aq) + 4e^-$

c Green hydrogen is hydrogen produced using renewable energy and resources instead of fossil fuels.

d 83.1 L

19 a The coating will appear on the spatula. Nickel metal is produced by a reduction reaction:
$Ni^{2+}(aq) + 2e^- \longrightarrow Ni(s)$
and reduction always occurs at the cathode (negative electrode in electrolysis).

b anode: $2H_2O(l) \longrightarrow O_2(g) + 4H^+(aq) + 4e^-$
cathode: $Ni^{2+}(aq) + 2e^- \longrightarrow Ni(s)$

c 0.68 g

20 a They function as catalysts, lowering the activation energy, and hence increasing the reaction rate.

b

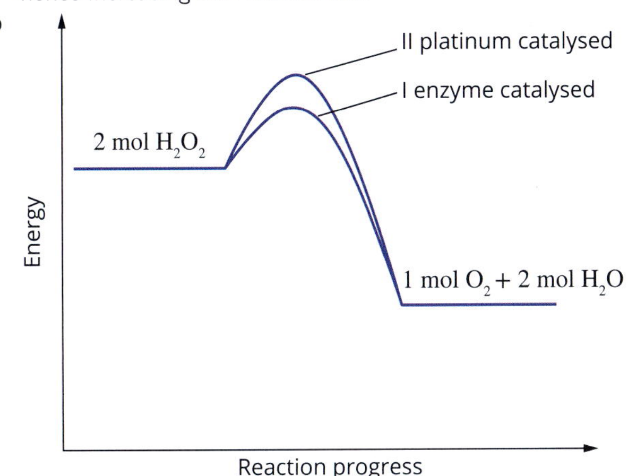

c System I would be faster. The lower activation energy means that there is a greater proportion of reactant particles which have energy greater than or equal to the lower activation energy, so successful collisions will occur more frequently.

21 a The concentration of sodium thiosulfate is higher in experiment 2, so there will be more frequent collisions between reactant particles. While the probability of any particular collision being successful is unchanged, the higher frequency of collisions overall will increase the frequency of successful collisions and hence the rate of reaction.

b The higher temperature in experiment 3 means the particles in the mixture have a higher average kinetic energy and hence collide more frequently, and more significantly, there will be a higher proportion of particles with energy greater than or equal to the activation energy of the reaction. These two factors combine to give a higher frequency of successful collisions and hence a higher rate of reaction.

c i Concentration of $Na_2S_2O_3$ solution.

ii Temperature of $Na_2S_2O_3$ solution.

d The dependent variable is the time taken for the cross to be obscured.

e Possible controlled variables are: volume of $Na_2S_2O_3$ solution, volume of HCl solution, concentration of HCl solution, same cross marked on paper, same stopwatch or other suitable device for measuring time.

f i Multiple trials allow the experimenter to determine and improve the precision of the repeated measurements. Any outlying measurements could be due to a systematic error and excluded from the averaged measurement result.

ii Some possible factors that would affect the precision of the time measurements include:

- Uncertainty about when the mixture has become cloudy enough to mask the cross and so inconsistent stopping times.
- Human reaction time (random error) affecting when the stopwatch/time measurement device is started and stopped.
- There is inherent uncertainty (random error) in the time measured by the stopwatch/time measurement device.

g For the comparison between experiments 1 and 3 to be valid, the temperature of $Na_2S_2O_3$ solution must be the only independent variable. The concentrations of $Na_2S_2O_3$ and HCl solutions must not be changed as they are variables that would also affect the rate of reaction.

22

Change	Quantity	Increase/decrease/no change
increase temperature	*K*	increase because endothermic reaction
decrease temperature	amount of $H_2O(g)$	increase because equilibrium position shifts to the left
add $H_2(g)$	*K*	no change because no temperature change
add $CO_2(g)$	amount of $H_2(g)$	decrease because equilibrium position shifts to the left
double volume	amount of CO_2	increase because equilibrium position shifts to the right (to the side with more gas particles)
double volume	concentration of CO_2	decrease because initial volume increase causes an overall decrease in CO_2 concentration (not offset by increased amount)
remove CH_4	amount of CO_2	decreased because equilibrium position shifts to the left
add catalyst	amount of $H_2(g)$	no change because a catalyst does not affect the equilibrium position
add argon (an inert gas)	*K*	no change because no temperature change

23 a Iodine (I_2) has been added, followed by a net forward reaction to re-establish equilibrium.

b The temperature has been increased, resulting in a net back reaction (because the forward reaction is exothermic).

c Equilibrium exists when the concentrations are constant (see diagram).

d Doubling the volume will halve all concentrations. Because there are equal numbers of reactant (left-hand side) and product (right-hand side) particles, the mixture remains at equilibrium and there is no shift in the position of equilibrium (see diagram).

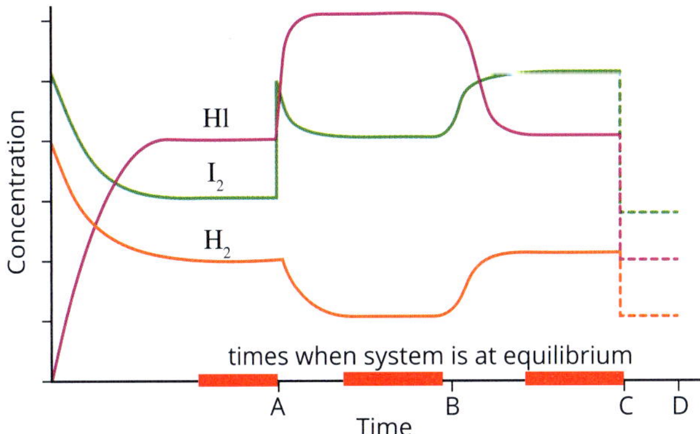

e i $Q = 5.33$

$Q \neq K$, so the system is not at equilibrium.

ii Since $Q < K$, the system will shift in the forward direction until the value of Q increase to equal K.

iii

	H_2	I_2	HI
Initial	0.300 M	0.100 M	0.400 M
Change	−0.035	−0.035	+0.070
Equilibrium	0.265 M	0.065 M	0.470 M

24 a i $CH_4(g) + H_2O(g) \rightarrow CO(g) + 3H_2(g)$

ii $CO(g) + 2H_2(g) \rightarrow CH_3OH(g)$

b i They catalyse the reaction, which increases the reaction rate.

ii The equilibrium yield would be lower. Because the reaction is exothermic, increasing the temperature would favour the endothermic reaction which is the reverse reaction.

iii High pressure will increase the reaction rate so products form faster. Also, because the number of product particles (1 on the right-hand side) is smaller than the number of reactant particles (3 on the left-hand side), high pressure will favour the forward reaction and increase yield.

25 a i anode: $2I^-(aq) \longrightarrow I_2(aq) + 2e^-$
cathode: $2H_2O(l) + 2e^- \longrightarrow H_2(g) + 2OH^-(aq)$

ii Brown I_2 forming at the positive electrode; bubbles forming at the negative electrode and increasing pH of electrolyte if tested with indicator.

b i anode: $2H_2O(l) \longrightarrow O_2(g) + 4H^+(aq) + 4e^-$
cathode: $Pb^{2+}(aq) + 2e^- \longrightarrow Pb(s)$

ii Bubbles forming at the positive electrode and decreasing pH of electrolyte if tested with indicator; lead solid/crystals deposited at the negative electrode.

c i anode: $Cu(s) \longrightarrow Cu^{2+}(aq) + 2e^-$
cathode: $2H_2O(l) + 2e^- \longrightarrow H_2(g) + 2OH^-(aq)$

ii Copper positive electrode getting smaller, and appearance of blue colour (due to Cu^{2+}) in the electrolyte solution; bubbles forming at the negative electrode and increasing pH of electrolyte if tested with indicator.

26 a i $2Cl^-(aq) \longrightarrow Cl_2(g) + 2e^-$

ii $2H_2O(l) + 2e^- \longrightarrow H_2(g) + 2OH^-(aq)$

iii For example, high NaCl(aq) concentrations are used so $Cl^-(aq)$ is oxidised at the anode; a semipermeable membrane prevents mixing of Cl^- and OH^- ions in the electrolyte while still allowing Na^+ ions to migrate from the anode to the cathode; the hydrogen and chlorine gases are separately removed after they have been produced, to prevent spontaneous reaction between the two gases.

b Water can itself act either as an oxidising agent or a reducing agent. Production of a chemical by electrolysis of an aqueous solution is only possible when the desired reaction involves a stronger oxidising or reducing agent than water, otherwise water will react preferentially. If the oxidising agent or reducing agent is a weaker oxidising agent or reducing agent than water, the desired reaction can only be achieved in the absence of water, either in an alternative solvent or in a molten electrolyte.

c Graphite is a good electrical conductor, relatively inexpensive, relatively inert and has a high sublimation point (in electrolytic cells operating at high temperatures, the graphite electrodes will not disintegrate).

d Metallic cathodes are always inert as the electrolytic process ensures only reduction can occur at their surface. However, the iron in a steel anode would likely be a better reducing agent than the target species so that the outcome would be the oxidation, and hence the consumption, of the electrode itself rather than the desired electrode reaction.

27 a i anode: $2H_2O(l) \longrightarrow O_2(g) + 4H^+(aq) + 4e^-$

ii cathode: $Sn^{2+}(aq) + 2e^- \longrightarrow Sn(s)$

b $2H_2O(l) + 2e^- \longrightarrow H_2(g) + 2OH^-(aq)$

28 a anode: $2Ni(OH)_2(s) + 2OH^-(aq) \longrightarrow 2H_2O(l) + 2NiOOH(s) + 2e^-$
cathode: $Zn(OH)_2(s) + 2e^- \longrightarrow 2OH^-(aq) + Zn(s)$

b As in all forms of electrolysis, electrical energy is converted to chemical energy so an external power source is used to drive a current through the cell to force non-spontaneous chemical reactions to occur. In a secondary cell, the non-spontaneous reactions are the reverse of the spontaneous reactions that occur during discharge.

c For recharging to be possible, the discharge products must remain in electrical contact with the electrodes. Any discharge products that migrate away from the electrode will not be available to participate in the reaction reversal during recharging, so that less reactant will be available on the next discharge cycle.

d 8.9×10^3 s

29 a The rates are equal up until the change was made.

b The addition of reactant W increases the frequency of collisions between reactant particles, so the rate of the forward reaction increases initially. Because the forward reaction is now the faster, there will be a net forward reaction. This will consume the reactant particles, so the forward reaction will slow down, and generate product particles, so the reverse reaction speeds up. When the two rates become equal again, equilibrium has been re-established and no further change occurs.

c i Adding Y: This accounts for the initial increase in the rate of the reverse reaction; the subsequent changes reflect the re-establishment of equilibrium.

ii Adding a catalyst: Both rates increase equally and this is the characteristic effect of a catalyst.
Note: It doesn't apply here, but another case that would produce this change is that where a gas-phase reaction is compressed (volume decreased) and there are equal number of reactant and product particles in the equation.

iii Adding argon (an inert gas) at constant volume: This does not affect the concentrations of reactants or products and so does not affect the frequency of collisions between reactant or product particles. Hence, it does not affect the rate of either reaction.

d Lowering the temperature means the particles in the mixture have a lower average kinetic energy and hence collide less frequently, and there will be a lower proportion of particles with energy greater than or equal to the activation energies of the forward and reverse reactions. Hence, there will be a lower frequency of successful collisions and lower rates of reaction for both forward and reverse reactions.

e Because the forward reaction is faster (the reverse reaction has slowed to a greater extent), there will be a net forward reaction.

f Exothermic. Lowering the temperature will always induce a shift in the exothermic direction, which in this case must be the forward reaction since that is the direction of the shift.

g 2.64 mol

Chapter 10 Structure, nomenclature and properties of organic compounds

10.1 Diversity of organic compounds

1 The carbon atoms in hydrocarbons have four bonds to other atoms. According to the valence shell electron pair repulsion theory, the angle around between each bond is 109.5° and the geometry around each carbon is tetrahedral. This means that carbon–carbon bonds are much less than 180°, and so the chain is zigzag shaped. Skeletal structures mainly show the bonds of the molecule, so the emphasis on shape is more important than in full structural formulas.

2 a Bond strength is directly related to the stability of a molecule. A molecule with strong chemical bonds is generally more stable than one with weaker bonds.

b The wide variety of compounds formed by carbon is due to its ability to form strong, stable covalent bonds with other carbon atoms and with atoms of other elements, such as hydrogen, oxygen, nitrogen, sulfur, phosphorus and the halogens. Carbon can also form strong, stable, single, double and triple bonds with itself, and it can form long chains and rings. The stability of these carbon compounds is related to their bond energies, which can be thought of as the strength of the bonds.

3 a butane (C_4H_{10})

b

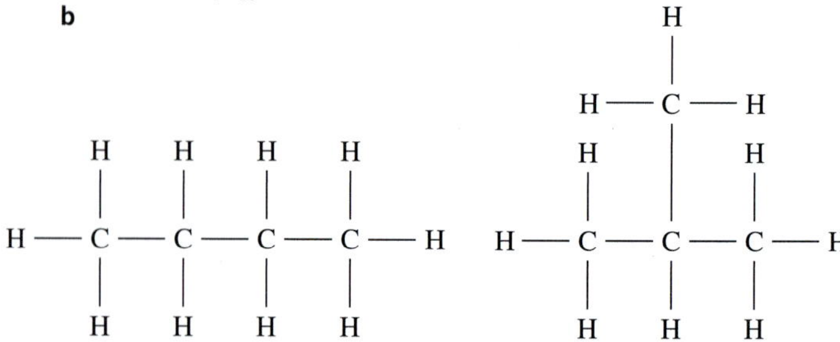

4

Molecular formula	Structural formula	Semi-structural formula	Skeletal formula
C_6H_{14}	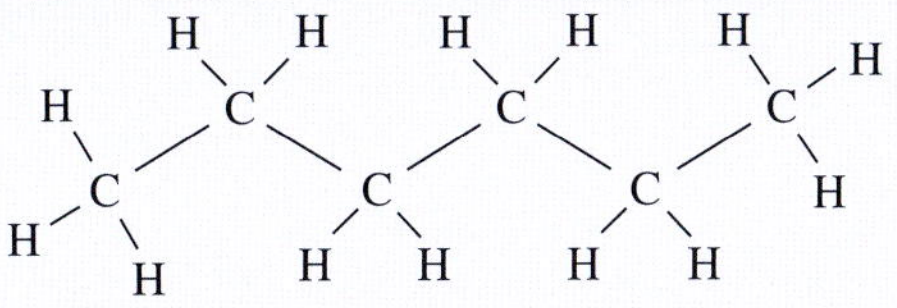	$CH_3(CH_2)_4CH_3$	
C_2H_6O	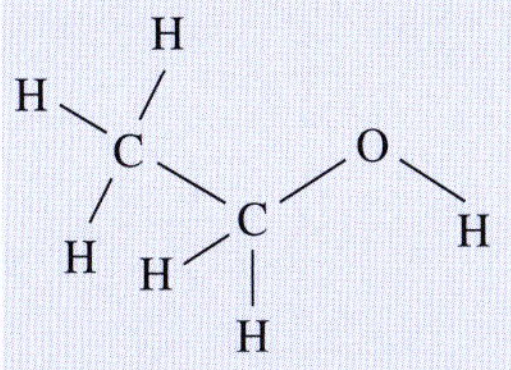	CH_3CH_2OH	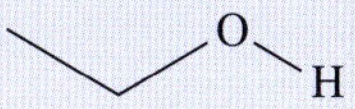 O H
C_5H_{12}	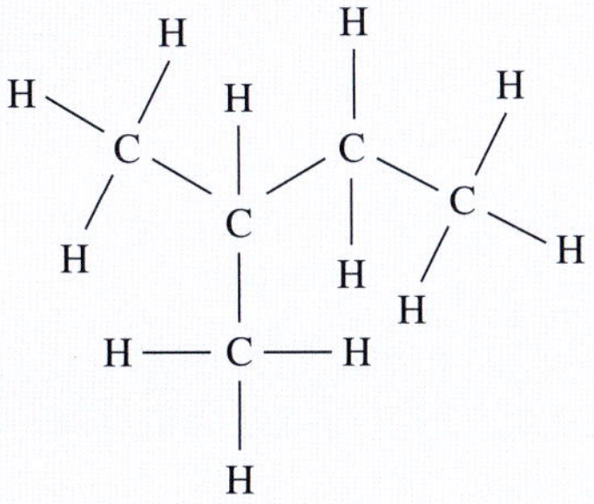	$CH_3CH(CH_3)CH_2CH_3$	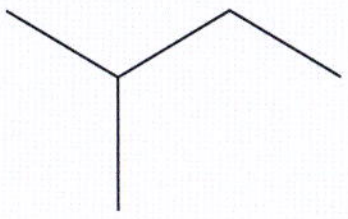

10.2 Hydrocarbons

TY 10.2.1 The name of the isomer is 3,6-dimethyloctane.

TY 10.2.2 The degree of unsaturation for C_6H_{12} is 1.

Key questions

1 structural isomers

2 Alkyl branches are named in alphabetical order, before the parent molecule name.

3 There are five possible isomers.

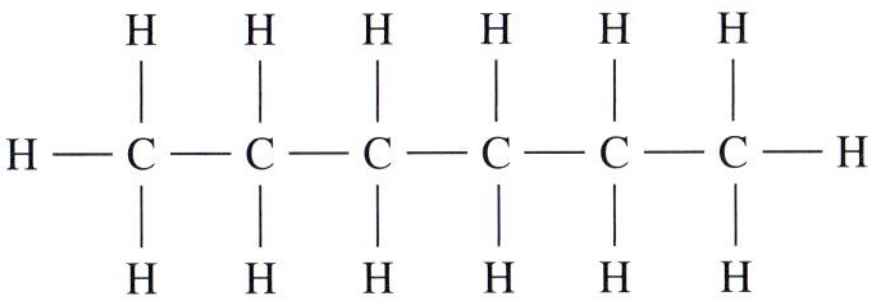

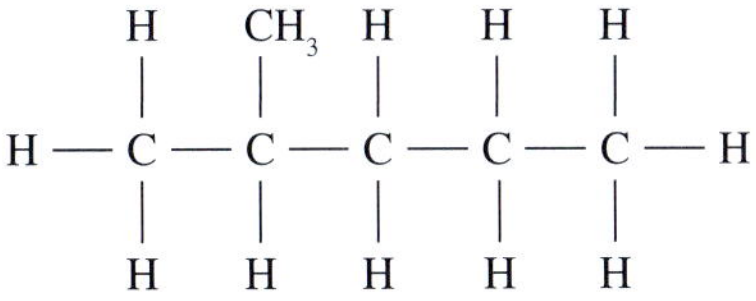

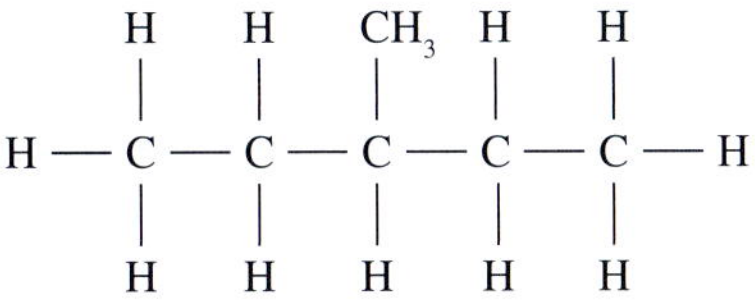

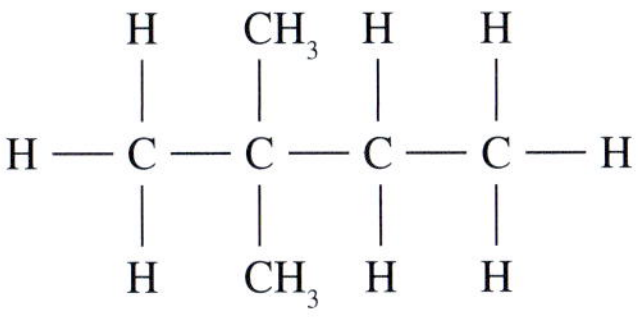

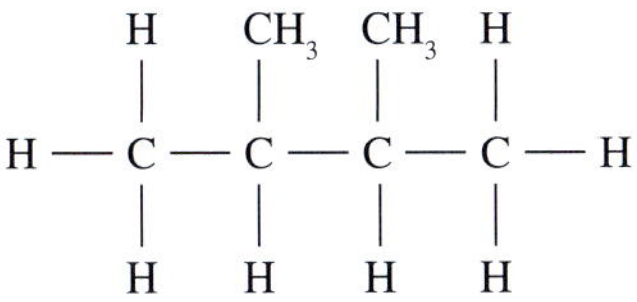

```
     H   H   H   H   H   H
     |   |   |   |   |   |
 H — C — C — C — C — C — C — H
     |   |   |   |   |   |
     H   H   H   H   H   H

     H  CH3  H   H   H
     |   |   |   |   |
 H — C — C — C — C — C — H
     |   |   |   |   |
     H   H   H   H   H

     H   H  CH3  H   H
     |   |   |   |   |
 H — C — C — C — C — C — H
     |   |   |   |   |
     H   H   H   H   H

     H  CH3  H   H
     |   |   |   |
 H — C — C — C — C — H
     |   |   |   |
     H  CH3  H   H

     H  CH3 CH3  H
     |   |   |   |
 H — C — C — C — C — H
     |   |   |   |
     H   H   H   H
```

4 Alkenes are monounsaturated hydrocarbons, which means they contain one C=C double bond. Cyclohexane does not contain a C=C double bond. Its general formula has two fewer H atoms than a straight chain alkane, as two H atoms are removed when the ring is closed. Cycloalkanes form their own homologous series, which also follows the pattern C_nH_{2n}.

5 **a** benzene – C_6H_6

b cyclohexane – C_6H_{12}

6 **a** 2-methylbut-2-ene

b 4-ethylhex-2-ene

7 **a**

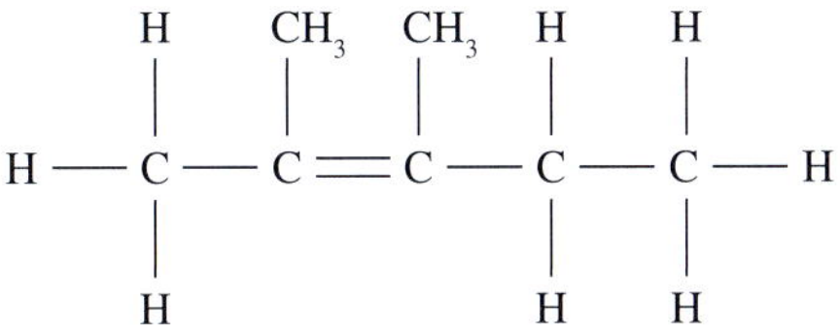

```
     H  CH3 CH3  H   H
     |   |   |   |   |
 H — C — C = C — C — C — H
     |           |   |
     H           H   H
```

b

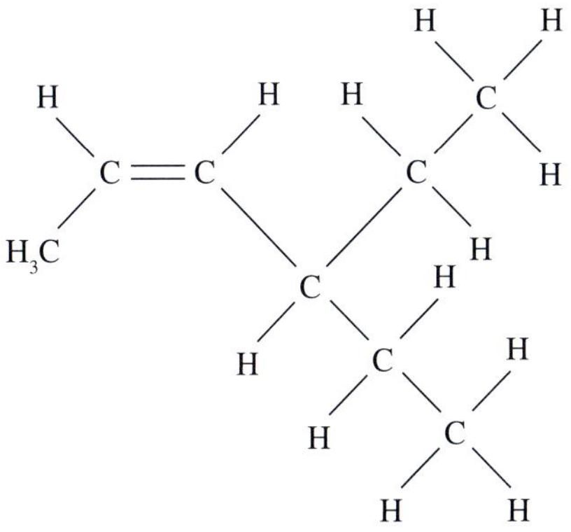

8 **a** Degree of unsaturation of $C_{10}H_{16} = 3$
This means the molecule must have 3 double bond or ring equivalents.

b Many answers are possible. These must contain 3 double bond or ring equivalents. Here are two possibilities:

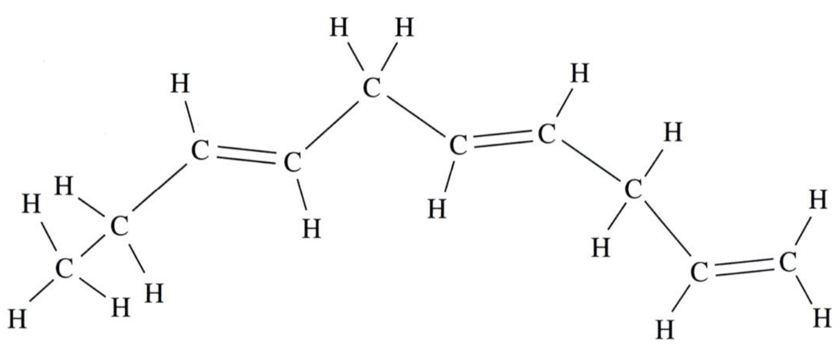

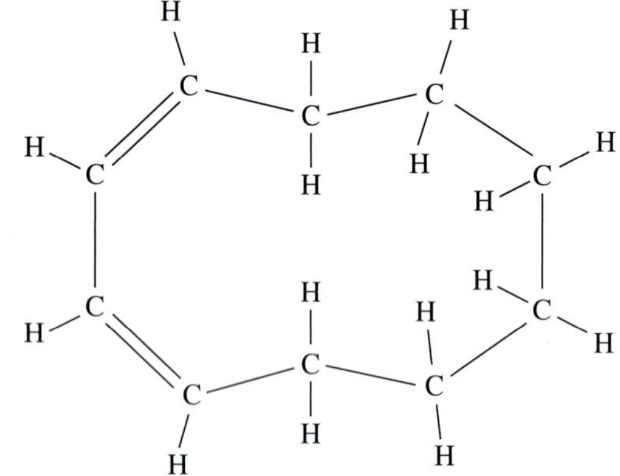

10.3 Haloalkanes, alcohols and amines

1 **a** halo (e.g. bromo, chloro, fluoro), hydroxyl, amino, respectively.

b haloalkanes: use a prefix (e.g. bromomethane)

alcohols: drop the final -e and add the suffix -ol (e.g. ethanol)

amines: drop final the -e and add the suffix -amine (e.g. ethanamine)

2 All prefixes (both branches and halo groups) are placed in alphabetical order before the parent alkane name.

3 3-chloro-2-fluoro-2-iodohexane skeletal formula:

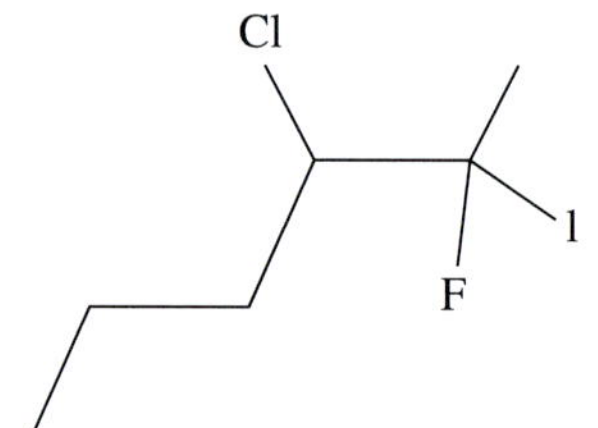

4 2-methylbutan-2-ol

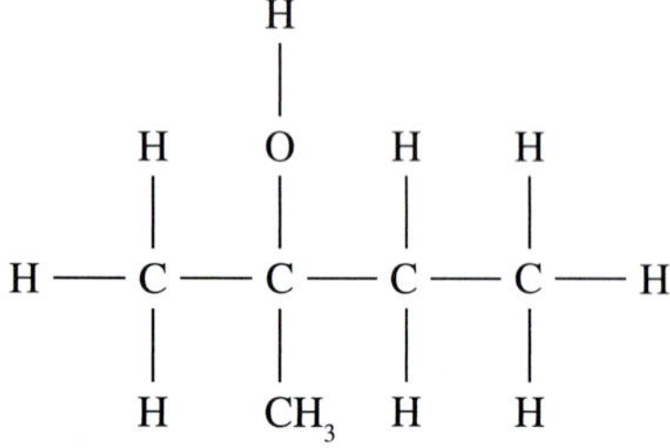

5 **a** 1-bromopropane **b** 2-chloro-4-methylpentane
c pentan-1-ol **d** octan-4-amine

6 Chloroethane has no isomers, so numbers are not needed. In propan-3-amine, the carbons should be numbered from the carbon atom closest to the amino functional group. The amino group is located on the first carbon instead of the third carbon. Hence, the correct name is propan-1-amine.

7 **a** $C_8H_{11}N$

b

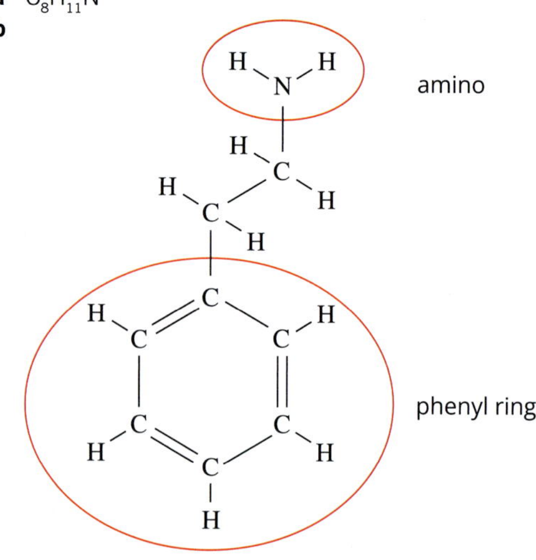

c $M(C_8H_{11}N) = 121\ g\ mol^{-1}$

10.4 Molecules containing a carbonyl group

1

Homologous series	Function group name	Structural formula	Semi-structural formula	Naming convention
aldehydes	carbonyl - end of molecule (aldehyde)	—C(=O)—H	–CHO	suffix -al
ketones	carbonyl - not on end (ketone)	R—C(=O)—R_1	–CO–	suffix -one
carboxylic acids	carboxyl	—C(=O)—O—H	–COOH	suffix -oic acid
primary amides	amide	—C(=O)—N(—H)—H	–CONH2	suffix -amide
esters	ester	—C(=O)—O—	–COO–	two-word name with suffixes -yl and -oate

*R and R_1 are used to indicate the rest of the molecule – these are usually, but not always, alkyl chains.

2 **a** ketones **b** aldehydes **c** amides **d** esters
e carboxylic acids

3 **a** methyl methanoate **b** methanoic acid
c propyl butanoate **d** methyl ethanoate
e ethyl hexanoate **f** 2,3-dimethylbutanal
g 3-methylpentan-2-one

4 ethanoic acid

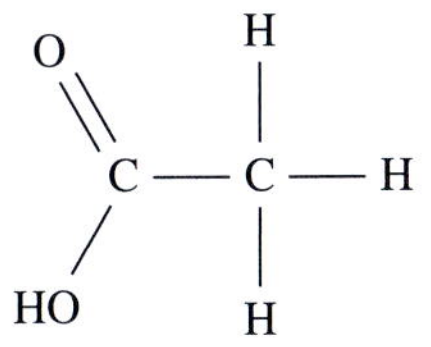

5

Name	Structural formula	Semi-structural formula	Skeletal formula
ethyl butanoate	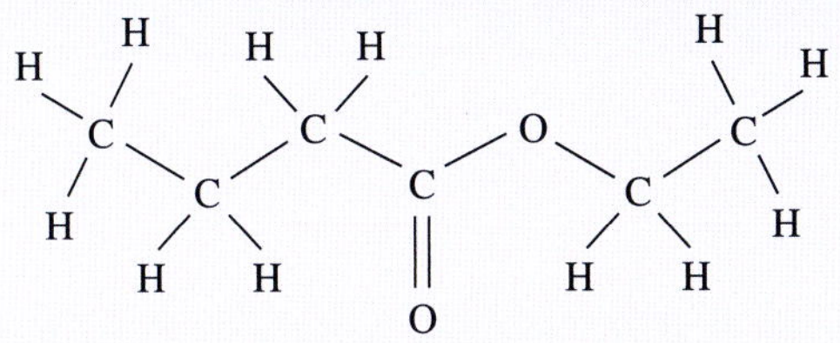	$CH_3(CH_2)_2COOCH_2CH_3$	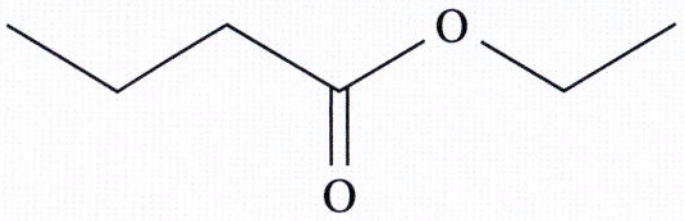
pentyl ethanoate	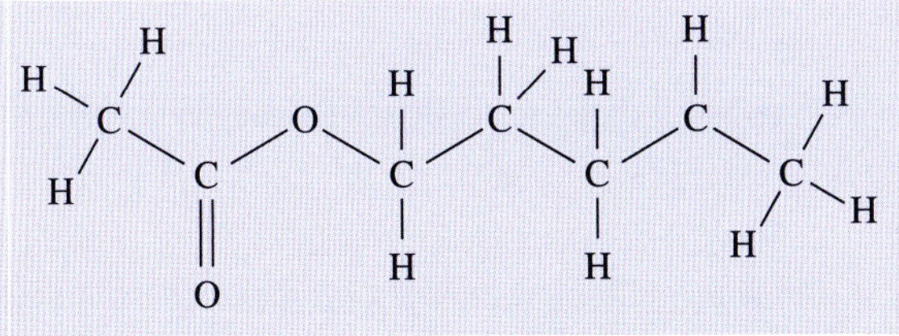	$CH_3COO(CH_2)_4CH_3$	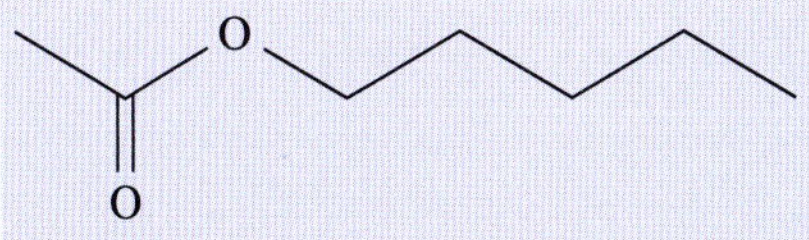
pentyl propanoate	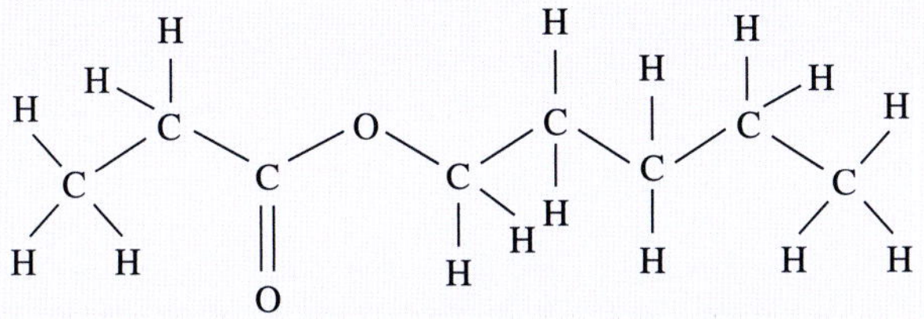	$CH_3CH_2COO(CH_2)_4CH_3$	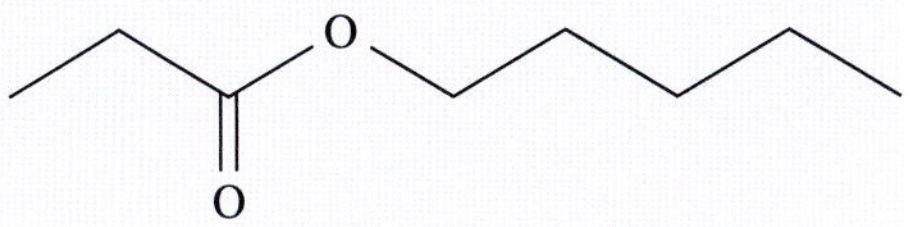
ethyl methanoate	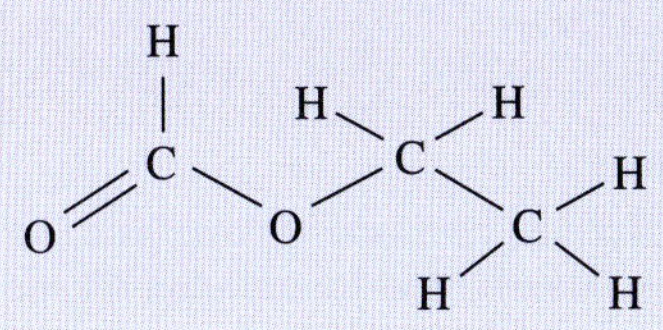	$HCOOCH_2CH_3$	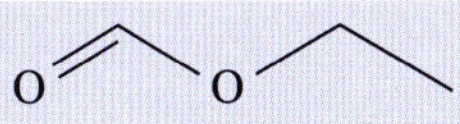

10.5 An overview of IUPAC nomenclature

CSA: Trivial versus IUPAC systematic names

1 Sulflower is a compound with an eight-membered ring that contains sulfur. The structure looks somewhat like a sunflower, and the name is a combination of the words sulfur and sunflower. Quadratic acid is a square-shaped organic compound, which is sometimes also called squaric acid. Draculin is a compound found in the saliva of vampire bats.

2 Ethanoic acid is the systematic name for acetic acid. Methanoic acid is the systematic name for formic acid. Propan-2-one is the systematic name for acetone.

TY 10.5.1 The name of the molecule is 6-chlorohexan-1-amine.

Key questions

1 a

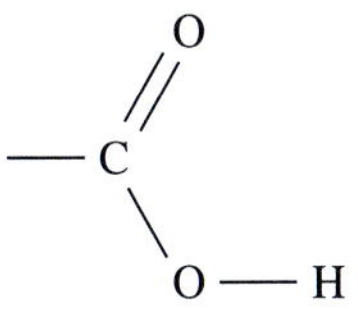

b

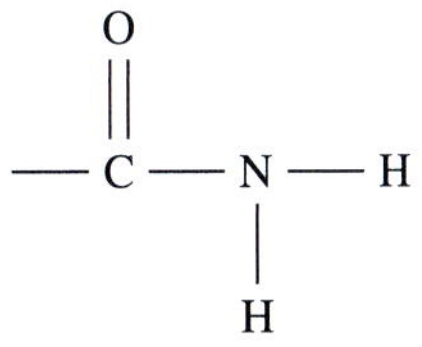

c

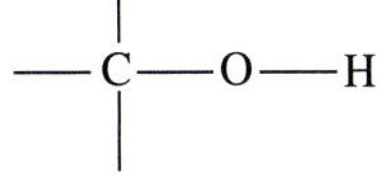

d

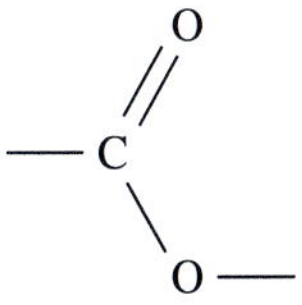

e

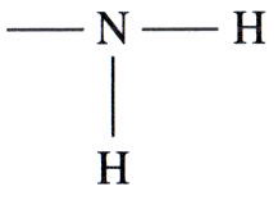

2 a -ol b -amino

3 a 3-hydroxybutanoic acid b 3-methylpentan-2-amine
c 4-chloro-3-ethylpenan-2-ol d butan-2-one

4 a

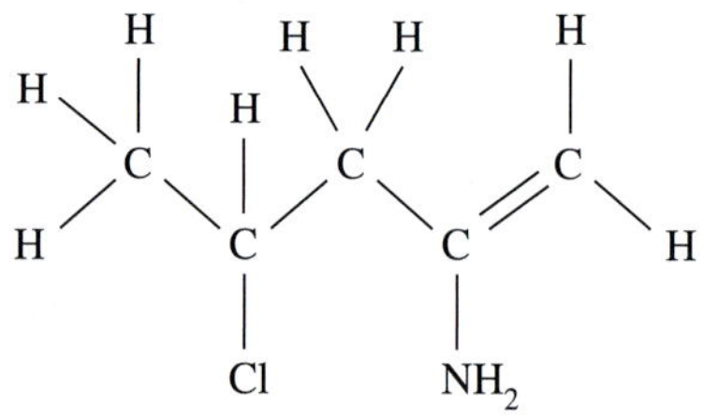

b

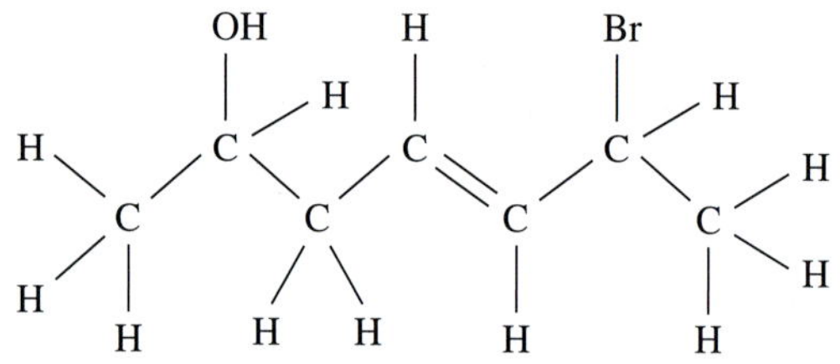

c

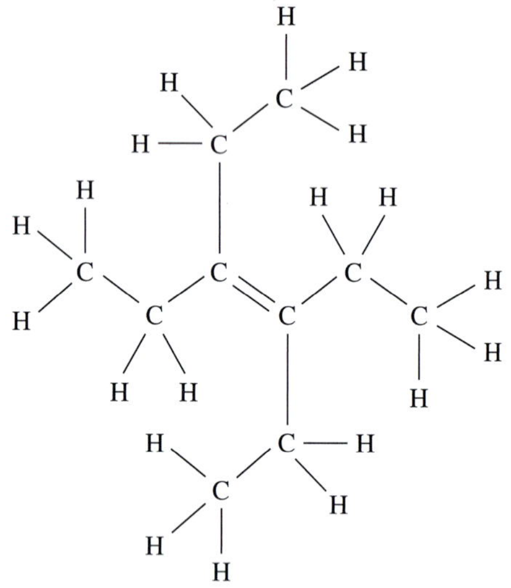

d

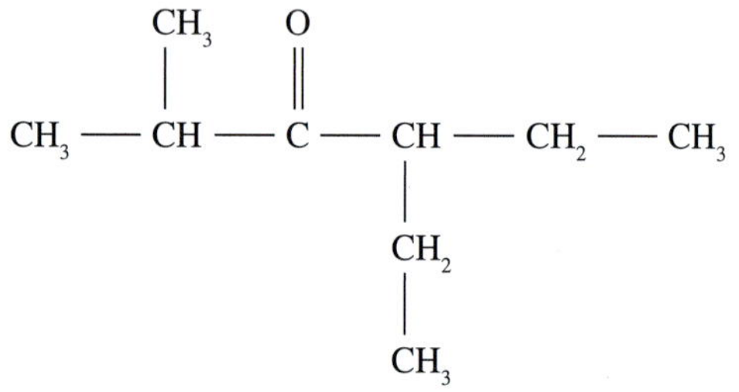

$CH_3 — CH(CH_3) — C(=O) — CH(CH_2CH_3) — CH_2 — CH_3$

5 a Amino groups have priority over alkenes, so -amine should go at the end and have the lowest number. The correct name is prop-2-en-1-amine (CH_2=$CHCH_2NH_2$).

b Carboxylic acid groups always go at the end of the chain and contain carbon 1. The correct name is 4-chloropentanoic acid ($CH_3CHClCH_2CH_2COOH$).

c The ethyl chain is a part of the longest carbon chain. Hydroxyl groups have priority over alkenes, so –ol should go at the end. The numbering of the longest carbon chain should aim to minimise the number of the hydroxyl and alkene functional groups. Therefore, the carbon-carbon double bond is at carbon number 1 and the hydroxyl group is at carbon number 3. The correct name is pent-1-en-3-ol (CH_2=$CHCHOHCH_2CH_3$).

10.6 Trends in physical properties within homologous series

1 Butane is a member of the **alkane** homologous series. The type of forces of attraction holding butane molecules to each other are **dispersion forces**. As the chain length of alkanes increases, their boiling points **increase**. In general, alkanes with branched chains have **lower** boiling points than straight-chain alkanes with the same number of carbon atoms.

2

Homologous series	Strongest form of intermolecular force of attraction between molecules
alcohols	hydrogen bonding
alkenes	dispersion forces
carboxylic acids	hydrogen bonding
esters	dipole–dipole interaction
haloalkanes	dipole–dipole interaction
amines	hydrogen bonding

3 $CH_3CH_2CH_3$, $CH_3CH_2CH_2CH_3$, $CH_3CH_2CH_2CH_2Cl$, $CH_3CH_2CH_2CH_2OH$, $CH_3CH_2CH_2COOH$

Propane and butane are both hydrocarbons, but butane is a longer molecule than propane, so will have stronger dispersion forces between its molecules and therefore a higher boiling point than propane. 1-chlorobutane is a polar molecule with dipole–dipole interaction forces between its molecules. These are stronger than dispersion forces, so 1-chlorobutane has a higher boiling point than the two hydrocarbons. Butan-1-ol has a hydroxyl functional group, so hydrogen bonding forces operate between its molecules. These forces are stronger than dipole–dipole interaction forces, so butan-1-ol has the next highest boiling point. Butanoic acid has the highest boiling point because its molecules are held together by hydrogen bonds, and also because its molecules form dimers between the carboxyl groups of neighbouring molecules.

4 As the viscosity of a liquid increases, it flows **less** readily. The viscosity of pentane is **lower/less** than that of octane at the same temperature. This is because pentane molecules have **shorter** hydrocarbon chains than octane. In samples of liquid pentane, there will be **fewer/less** points of contact between molecules, and overall, the strength of dispersion forces holding them together will be **weaker** than those holding octane molecules together.

Liquids consisting of molecules made up of a straight hydrocarbon chain with a functional group attached will generally have a **greater** viscosity than a liquid consisting of hydrocarbon chains of similar length with no functional group. This is because forces of attraction (dipole–dipole or hydrogen bonds) between the polar functional groups are **stronger** than dispersion forces.

5 pentane; hexane; 1-chloroheptane; heptan-1-ol.

Pentane and hexane molecules have shorter hydrocarbon chains than the other two compounds, so have weaker overall dispersion forces between their molecules. Pentane molecules are shorter than hexane molecules, so have weaker overall dispersion forces and pentane has a lower viscosity than hexane. Dipole–dipole forces of attraction operate between 1-chloroheptane molecules. These forces are stronger than the dispersion forces between the alkane molecules, so 1-chloroheptane has greater viscosity than pentane and hexane. Dipole–dipole interaction forces are weaker than the hydrogen bonding forces operating between heptan-1-ol molecules, so 1-chloroheptane has lower viscosity than heptan-1-ol.

6 2,2-dimethylpropane, 2-methylbutane, pentane.

Straight-chain hydrocarbon molecules can 'pack' together closely, so that there are many points of contact between the molecules. This means that overall the intermolecular forces (dispersion forces) are stronger than for branched chain hydrocarbon molecules and that is why pentane has the highest boiling point of the three compounds. In branched chain hydrocarbon molecules, the side chains reduce the ability of the hydrocarbon chains to come into close contact, so the overall strength of dispersion forces is decreased. Generally, the more side chains there are, the lower will be the boiling point of compounds with similar molar mass. This is the reason why the boiling point of 2-methylbutane is higher than that of 2,2-dimethyl pentane.

Chapter 10 review

1 D.

2 B.

3 B.

4 A.

5 D.

6 **a** N and H
b C, N, H and O
c Cl
d H and O
e C, O and H
f C and O

7 Propan-1-ol is a primary alcohol; the carbon to which the hydroxyl group is attached has only one alkyl group attached to it. Propan-2-ol is a secondary alcohol; the carbon to which the hydroxyl group is attached has two alkyl groups attached to it.

8 A carboxylic acid has a carbonyl group with a hydroxyl group attached to the carbon; an amide has a carbonyl group with an amino group attached to the carbon.

9 Carboxyl and amide carbons have three bonds to the carbon atom within the functional group. The carbon atom in this group can therefore only make one bond to connect to a carbon chain, so the group can never be within a chain.

10 B.

11 A.

12 Octane has the higher viscosity. Intermolecular forces determine the viscosity of organic compounds. This is because octane has the higher molar mass and is the larger molecule. Both octane and pentane are non-polar hydrocarbons, so the only intermolecular forces are dispersion forces. As the size of the molecules increases, so does the strength and number of dispersion forces between molecules.

13 CH_3CH_3 (ethane), $CH_3CH_2CH_2CH_3$ (butane) and $CH_3CH(CH_3)CH_3$ (2-methylpropane) are all hydrocarbons, so they are non-polar and the only intermolecular forces present are dispersion forces. The ethane molecule is shorter than the butane molecule, so there will be less intermolecular forces between ethane molecules and ethane has a lower boiling point than butane. Both butane and 2-methylpropane molecules have four carbon atoms, but the latter molecule is branched. This means the molecules do not pack together as closely as butane molecules, so butane molecules will have the higher boiling point. Both $CH_3CH_2CH_2CH_2OH$ (butan-1-ol) and $CH_3CH_2CH_2COOH$ (butanoic acid) have intermolecular forces of hydrogen bonds between their molecules, so they will have higher boiling points than the alkane molecules. Typical of carboxylic acids, butanoic acid can form dimers (Figure 10.6.16), held together by hydrogen bonds. These intermolecular forces are stronger than the hydrogen bonds holding butan-1-ol molecules together (Figure 10.6.12). Therefore the boiling point of butanoic acid will be higher than that of butan-1-ol.

14 **a** 8

b

1-chloro-3-methylbutane

2-chloro-2-methylbutane

2-chloro-3-methylbutane

1-chloro-2-methylbutane

1-chloro-2, 2-dimethylpropane

3-chloropentane

2-chloropentane

1-chloropentane

15 CH_2=$CHCH_2CH_3$ (but-1-ene), CH_3CH=$CHCH_3$ (but-2-ene), $(CH_3)_2C$=CH_2 (2-methylprop-1-ene)

16 **a** C_5H_{12} **b** C_6H_{12} **c** C_4H_8 **d** C_7H_{14}

17 3 (pentane, 2-methylbutane and 2,2-dimethylpropane)

18 If an ethyl group is on the second carbon atom, then the longest chain is 6 carbon atoms long and the name would be 3-methylhexane.

19 **a** 2-aminopropan-1-ol **b** 8-chlorooctan-2-ol
c 2-iodoheptan-3-amine

20 **a** but-3-enoic acid **b** 4-aminobutan-2-ol
c butyl ethanoate **d** 2-methylhex-3-ene
e 1-bromobut-2-ene

21 **a**

b

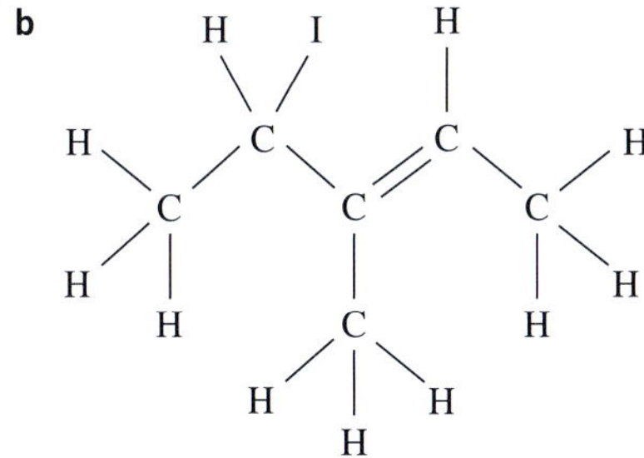

c

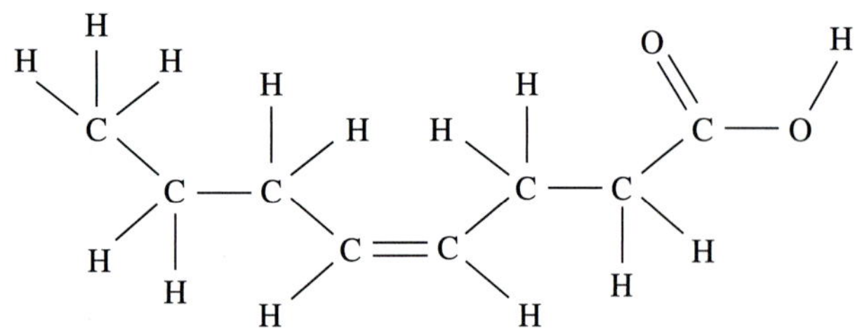

d

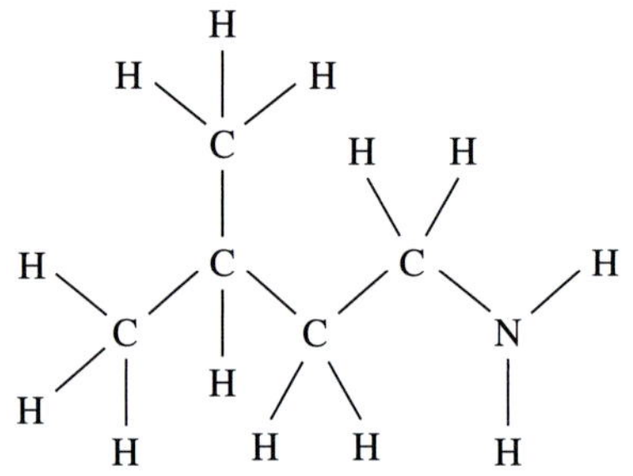

22 a $CH_3CH_2CH_2CH_2CH{=}CHCH_2COOH$
b $CH_3CH_2CH_2CH_2CH_2COOCH_3$
c $CH_2FCH_2CH_2OH$
d $CH_2OHCH_2CH_2COOH$
e $CH_3CH_2CH(CH_3)CH(CH_3)CH_2NH_2$

23 a $CH_3CH_2CH_2COCH_3$
b $CH_3CH(CH_3)CH_2CH_2OH$
c $CH_3CH_2CONH_2$
d $CH_3CH_2COOCH_2CH_3$

24 a The double bond cannot be in the '4' position because it must be between two carbon atoms. If the 4-carbon is involved in a double bond it should have priority in the name, and so be called but-1-ene.
b The hydroxyl group has higher priority than the amino group, so the name should be 2-aminoethan-1-ol.
c The C=C bond has higher priority than the chloro group, so the name should be 3-chlorohex-1-ene.
d The longest chain in this arrangement is 5 carbon atoms long, so the stem should be pent- and the branch should be methyl. The correct name is 2-chloro-3-methylpentane.

25 a i primary alcohol
ii heptan-1-ol
b i chloroalkane
ii 4-chloroheptane
c i secondary alcohol
ii hexan-2-ol
d i amine
ii butan-2-amine

26 a structural isomers
b unrelated
c structural isomers

27 Boiling points of compounds are determined by the strength of the intermolecular forces. Weak dispersion forces are present in all molecules, polar or non-polar, so must be kept as close to constant as possible when making comparisons on the effect of changing functional groups. Because dispersion forces increase with molecular size, it is important to select compounds of a similar molar mass.

28 a The boiling point within a homologous series increases as the carbon chain length increases. The boiling points of alcohols are higher than the boiling points of alkanes with a similar molecular mass. The boiling points of carboxylic acids are higher than those of alcohols with a similar molecular mass.
b The boiling point of pentanoic acid ($CH_3(CH_2)_3COOH$) from experiment 3 is much lower than expected. The experimental data shows it to have a similar boiling point to that of hexan-1-ol ($CH_3(CH_2)_4CH_2OH$), but this is an anomaly in the trend. As can be seen from the data from experiments 1 and 2, carboxylic acids have higher boiling points than alkanes and alcohols of a comparable molecular mass.
c Boiling points within a homologous series increase as the carbon chain length increases because the dispersion forces between neighbouring molecules become stronger. The boiling points of alcohols are greater than the boiling points of alkanes with a similar molecular mass because alcohol molecules form hydrogen bonds with each other in addition to dispersion forces (alkane molecules only have dispersion holding them to each other). Carboxylic acids have a higher boiling point than alcohols of a similar molecular mass because intermolecular hydrogen bonds result in the formation of dimers that effectively double their molecular mass and increase the strength of the dispersion forces formed between neighbouring molecules. While this did not show here for pentanoic acid, this has been justified by identifying that the result is due to experimental error.

Chapter 11 Reactions of organic compounds

11.1 Reactions of alkanes and haloalkanes

1 Alkanes will undergo complete combustion in air to produce **carbon dioxide** and **water**. The first organic product formed when methane reacts with bromine in the presence of **ultraviolet** light is **bromomethane**. This is an example of **a substitution** reaction. The organic product formed when chloromethane reacts with sodium hydroxide is **methanol**. This is also an example of **substitution** reaction.

2 $2C_4H_{10}(g) + 13O_2(g) \longrightarrow 8CO_2(g) + 10H_2O(l)$

3 a butan-2-ol
b ethanamine
c carbon dioxide

4 a $CH_3Cl(g) + NaOH(aq) \longrightarrow CH_3OH(aq) + NaCl(aq)$
b $CH_3CHBrCH_3(l) + NH_3(aq) \longrightarrow CH_3CH(NH_2)CH_3(l) + HBr(aq)$
c $CH_3CH_2CH_2Cl(l) + NaOH(aq) \longrightarrow CH_3CH_2CH_2OH(l) + NaCl(aq)$

5

Homologous series	Reactant	Type of organic product(s)	Type of reaction
alkane	**O_2(g)**	carbon dioxide and water	**combustion**
haloalkane	NaOH	**alcohol**	**substitution**
haloalkane	ammonia	**amine**	substitution
alkane	**chlorine**	chloroalkane	**substitution**

6 The reactants are methane and an excess of chlorine gas. The reaction conditions are the presence of UV light. The dichloromethane could be separated from other chloromethane products by fractional distillation.

11.2 Reactions of alkenes

1 Alkenes will undergo combustion reactions in air to produce **carbon dioxide** and **water**. The product formed when but-2-ene reacts with bromine is **2,3-dibromobutane**. This is an example of an **addition** reaction.

The product formed when ethene reacts with hydrogen chloride is **chloroethane**. This is an example of an **addition** reaction.

2 **a**

```
    H   H   H   H                      H   H   H   H
    |   |   |   |                      |   |   |   |
H — C — C — C ═ C   +  Cl₂  ——→    H — C — C — C — C — H
    |   |       |                      |   |   |   |
    H   H       H                      H   H   Cl  Cl

     but-1-ene                        1,2-dichlorobutane
```

b

```
    H   H       H                          H   H   H   H
    |   |       |                          |   |   |   |
H — C — C ═ C — C — C   +  HBr  ——→    H — C — C — C — C — H
    |       |   |                          |   |   |   |
    H       H   H                          H   H   Br  H

     but-2-ene                             2-bromobutane
```

3 $CH_3CH_2CHCHCH_2CH_3(l) + HCl(g) \rightarrow CH_3CH_2CHClCH_2CH_2CH_3(l)$

4

Homologous series	Reactant	Type of organic product(s)	Type of reaction
alkene	halogen	dihaloalkane	addition
alkene	hydrogen gas, H_2 + Ni or Pt catalyst	**alkane**	**addition**
alkene	**air (oxygen)**	$CO_2 + H_2O$	**combustion**
alkene	**water + catalyst**	alcohol	**addition**
alkene	hydrogen halide	**haloalkane**	**addition**

5 **a** Propene could react with water to form two different products, propan-1-ol and propan-2-ol. This is because the OH group from water can react at either of the two carbon atoms in the carbon–carbon double bond in the propene molecule, resulting in two different compounds.

b But-2-ene will only form one product, butan-2-ol, when it reacts with water. The OH group from water can attach to either of the two carbon atoms in the carbon–carbon double bond of but-2-ene. The carbon atom to which the OH group is attached will be given the number 2 in each of the two product molecules.

6 C is correct. Pent-2-ene is not symmetrical and will form 2 possible products, one with a chlorine atom on carbon atom number 2 in the compound and the other with a chlorine atom on carbon atom number 3 (similar to the example in Figure 11.2.7).

Hex-3-ene and oct-4-ene are symmetrical and will only produce one product in their reactions with water (similar to the example in Figure 11.2.6).

Chlorine, Cl_2, is symmetrical and so will only form one product with alkenes, whether they are symmetrical or not.

11.3 Reactions of alcohols, carboxylic acids and esters

CSA: Making margarine

1

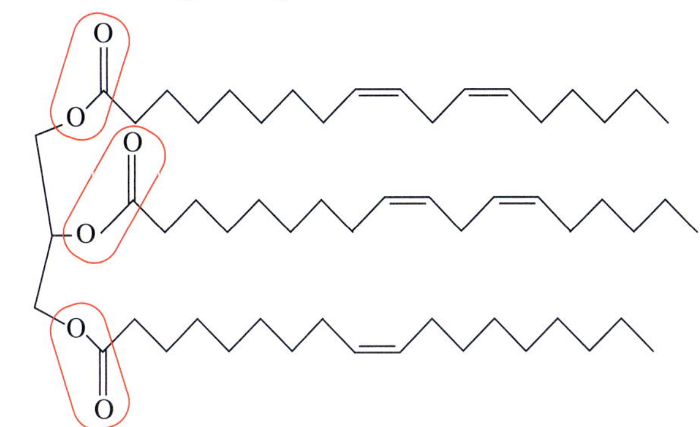

2 three

3

```
        H
        |
    H — C — OH
        |
    H — C — OH
        |
    H — C — OH
        |
        H
```

4 three

Key questions

1 **a** ester **b** carboxylic acid
c alcohol and carboxylic acid

2 **a** $2C_5H_{12}O(g) + 15O_2(g) \rightarrow 10CO_2(g) + 12H_2O(g)$
b Pentan-3-ol is a secondary alcohol; pentan-1-ol is a primary alcohol.

c Pentan-1-ol would oxidise to the carboxylic acid. The solution would change from deep purple to clear as the reaction proceeded.

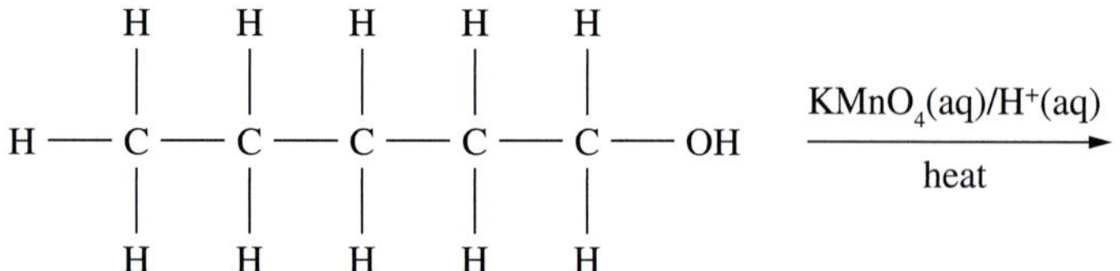

$$\xrightarrow[\text{heat}]{KMnO_4(aq)/H^+(aq)}$$

$CH_3CH_2CH_2CH_2COOH$ (H–C(H)(H)–C(H)(H)–C(H)(H)–C(H)(H)–C(=O)–OH)

3 When pentanoic acid reacts with propan-1-ol, the organic product called **propylpentanoate** is formed. The other product in this reaction is **water**. This is an example of a **condensation** reaction.

4 **a** Methanol and sodium butanoate are initially formed; however, the addition of acid to the sodium butanoate will form butanoic acid.

CH_3OH (H–C(H)(H)–O–H) methanol

$CH_3CH_2CH_2COOH$ (H–C(H)(H)–C(H)(H)–C(H)(H)–C(=O)–O–H) butanoic acid

b

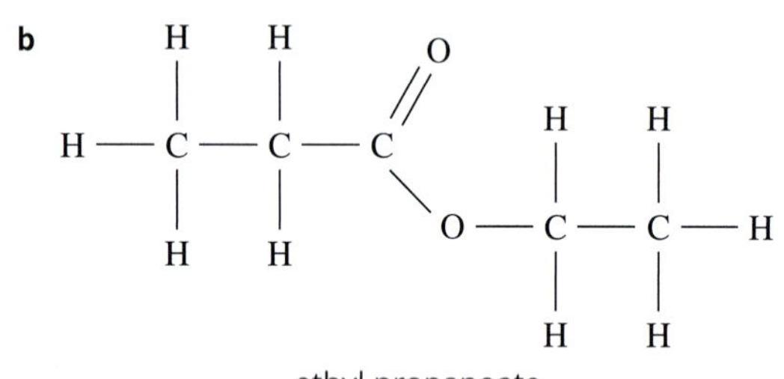

ethyl propanoate

5

Homologous series compound	Reactant	Product(s)	Type of reaction
alcohol	air	**carbon dioxide and water**	combustion
primary alcohol	$H^+/Cr_2O_7^{2-}$ or H^+/MnO_4^-	**carboxylic acid**	**oxidation**
alcohol	**carboxylic acid + acid (H_2SO_4)**	ester + water	**esterification**
carboxylic acid	water	carboxylate ion + H_3O^+	**ionisation**
ester	**water + acid**	**carboxylic acid + alcohol**	**hydrolysis**
triglyceride	methanol	**biodiesel + glycerol**	**transesterfication**

6 **a** $CH_3CH_2COOCH_3(l) + H_2O(l)$

b $CH_3COOH(l) + CH_3(CH_2)_3CH_2OH(l)$

7

$$\begin{array}{l} H_2C{-}O{-}C(=O){-}(CH_2)_{10}CH_3 \\ HC{-}O{-}C(=O){-}(CH_2)_{10}CH_3 \\ H_2C{-}O{-}C(=O){-}(CH_2)_{10}CH_3 \end{array} + 3CH_3OH \xrightarrow{KOH} \begin{array}{l} H_2C{-}OH \\ HC{-}OH \\ H_2C{-}OH \end{array} + 3CH_3(CH_2)_{10}COOCH_3$$

11.4 Reaction pathways and atom economy

TY 11.4.1 55.9% **TY 11.4.2** 43% **TY 11.4.3** 55.2%

CSA: Aspirin, an ester derived from a herbal remedy

1

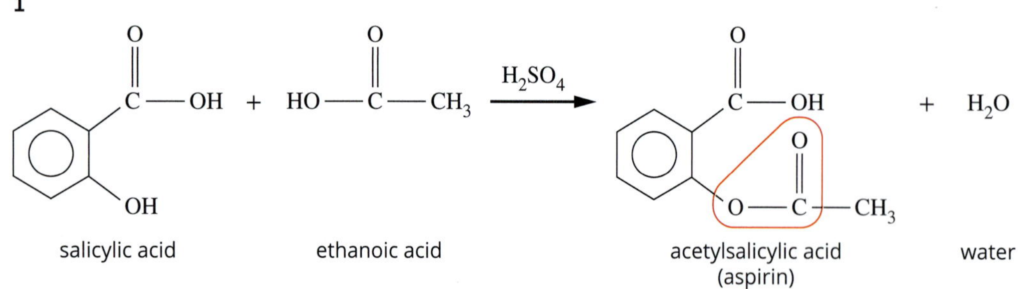

2 atom economy = 90.9%

3 percentage yield = 11.0%

Key questions

1 a The theoretical yield of a reaction is the maximum amount of product that can be formed using calculations based on the stoichiometric ratio of the limiting reactant and assumes 100% conversion.

b The percentage yield compares the actual yield to the theoretical yield and is a measure of the efficiency of a production process. It is found using:

$$\text{percentage yield} = \frac{\text{actual yield}}{\text{theoretical yield}} \times 100\%$$

c The atom economy for a chemical reaction is the percentage of the atoms in the reactants that end up in the desired product. It is found using:

$$\text{atom economy} = \frac{\text{molar mass of desired product}}{\text{molar mass of all reactants}} \times 100\%$$

2 Compound A is propan-1-ol; reactant B is acidified dichromate ions or acidified permanganate ions.

3 a 806 g b 34.7%

4 a 9.6% b 3.0%

5 a 51% b carbon dioxide, CO_2.

6 a $H_2O(g)/H_3PO_4$ catalyst b butanoic acid

c butyl ethanoate

7 a butanoic acid

b methanol with acid catalyst

c butan-1-ol

d $H^+/Cr_2O_7^{2-}$ and heat or H^+/MnO_4^-

e NaOH

8 This synthesis is carried out in three steps: (1) synthesis of ethanol, (2) synthesis of propanoic acid, and (3) synthesis of ethyl propanoate.

1 $\text{ethene} \xrightarrow{HCl} \text{chloroethane} \xrightarrow{OH^-} \text{ethanol}$

2 $\text{propene} \xrightarrow{H_2O/\text{catalyst}} \text{propan-1-ol} \xrightarrow{Cr_2O_7^{2-}/H^+} \text{propanoic acid}$

3 $\text{ethanol} + \text{propanoic acid} \xrightarrow{H^+} \text{ethyl propanoate}$

After the first stage of the second step, fractional distillation would be required to separate the isomers propan-1-ol and propan-2-ol.

11.5 Sustainable production of chemicals

1 a Using renewable raw materials (feedstocks)

b Using catalysts

c Designing safer chemicals and products.

2 Cellophane is biodegradable and is made from cellulose, which can be obtained from plant materials.

Polyvinyl chloride is made by the polymerisation of vinyl chloride. This compound is made from one of the fractions obtained in the fractional distillation of crude oil. It is therefore non-renewable.

3 When catalysts are used in chemical reactions:

- lower temperatures are required, so heating costs are reduced.
- reaction rates are increased, so more product is formed in a shorter time.
- they are not consumed, so can be used repeatedly.

4 a PFAS chemicals are useful ingredients in fire-fighting foams because they are surfactants which allow the foam to spread over the fire. They are also stable, not broken down by heat and are also oil- and water-resistant.

b The fact that PFAS chemicals are so stable means that they do not break down easily in soil or water. They therefore persist in the environment, and so can move long distances through soil, aquifers and waterways without decomposing. PFAS compounds are known to be toxic to some animals and there are also concerns that they may have an impact on human health.

5 There are no polar groups in polypropene, whereas the cellulose polymer is made up of glucose monomers with many hydroxyl groups, so polypropene is more hydrophobic than cellulose. Polypropene will interact less with aqueous solutions of any kind in the environment than cellulose will. In addition, the carbon–carbon bonds which form the backbone of polypropene are very strong.

The polar hydroxyl groups in cellulose will interact readily with water and the carbon–oxygen bonds in cellulose are more readily hydrolysed than the strong carbon–carbon bonds in polypropene.

Chapter 11 review

1 B. 2 C.

3 a $CH_3CHCH_2(g) + Cl_2(g) \rightarrow CH_3CHClCH_2Cl(l)$

b $CH_3CH_2CH_2CH_2Cl(l) + OH^-(aq) \rightarrow CH_3CH_2CH_2CH_2OH(aq)$

c $C_2H_4(g) + 3O_2(g) \rightarrow 2CO_2(g) + 2H_2O(g)$

4 a The original compound reacted with hydrogen and a nickel catalyst to form propane, so it must have been propene. This is confirmed by the fact that it decolourised a bromine solution.

b Propene would react with chlorine in an addition reaction to produce 1,2-dichloropropane.

5 a

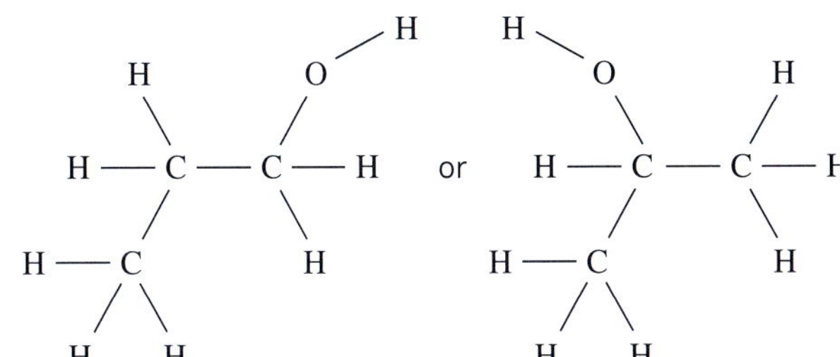

b

H–O–C(H)(H)(H)

6

Reactants	Products	Reaction type
pent-2-ene and bromine	1,2-dibromopentane	addition
ethanol burning in excess oxygen	**carbon dioxide and water**	**combustion or oxidation**
ethene and water, H_3PO_4 catalyst	**ethanol**	**addition**
bromoethane and hydroxide ion	**ethanol**	**substitution**
propene and hydrogen, Ni catalyst	**propane**	**addition**
chloromethane and ammonia	**methanamine**	**substitution**
hex-3-ene and hydrogen chloride	**3-chlorohexane**	**addition**
methyl propanoate and a dilute solution of HCl	**methanol and propanoic acid**	**hydrolysis**

7 a $CH_3CH_2CH_2COOH$ b CH_3COCH_3 c CH_3COOH

8 a $CH_3CH_2CH_2COOH$, CH_3OH b $CH_3CH_2CH_2OH$, CH_3COOH

c CH_3CH_2OH, $HCOOH$

9 The term 'natural gas' sometimes causes confusion, but it is a fossil fuel. It is often found in deposits with crude oil. It is a non-renewable commodity.

10 Turpentine is made from plant material, which is a renewable resource. Mineral spirits are made from crude oil, a non-renewable resource. The fact that turpentine is more flammable and more toxic than mineral spirits means that more care needs to be taken when using it. For example, it would be important to use turpentine in a well-ventilated area.

11 The tar is produced from the spinifex plant which is a renewable resource. The heat used to melt the spinifex resin is produced by burning wood, another renewable resource. The tar itself is biodegradable.

Crude oil is a fossil fuel and is non-renewable. The bitumen produced using tar persists in the environment and only biodegrades very slowly.

12 **A** ethene
B 1,2-dibromoethane
C chloroethane
D carbon dioxide and water

13 **a** $CH_3CH_2CH_2OH$
b (In the diagram, X represents a halogen atom.)

```
 \/\
    X
```

c NaOH

14

```
H       H            H   H                  H
 \     /             |   |                  |     O
  C═C           H—C—C—O—H           H—C—C//
 /     \             |   |                  |    \
H       H            H   H                  H     O—H
```

Compound A — Compound B — Compound C

```
   H  H  H           Cl  H  H              H  H  H
   |  |  |           |   |  |              |  |  |
H—C—C—C—H       H—C—C—C—H          H—C—C—C—OH
   |  |  |           |   |  |              |  |  |
   H  H  H           H   H  H              H  H  H
```

Compound D — Compound E — Compound F

15 **a** 3.26 g **b** 72.1%

16 $CH_2CH_2 \xrightarrow{H_2O/H_3PO_4} CH_3CH_2OH \xrightarrow{H^+/Cr_2O_7^{2-}} CH_3COOH$

17 28 g

18 3.62 kg

19 **a**

```
 \/\
    X
```

b $H^+/Cr_2O_7^{2-}$ or H^+/MnO_4^- **c** propanoic acid
d $CH_3CH_2CH_2CH_2OH$ **e** $H_2SO_4(aq)$
f $CH_3CH_2CH_2CH_2OCOCH_2CH_3$ or $CH_3CH_2COOCH_2CH_2CH_2CH_3$

20 **a**

```
   H        O
   |        ‖
H—C—O—C—(CH2)7CH=CH(CH2)7CH3
   |        O
   |        ‖
H—C—O—C—(CH2)7CH=CH(CH2)7CH3 + 3CH3OH   —KOH→
   |        O
   |        ‖
H—C—O—C—(CH2)7CH=CH(CH2)7CH3
   |
   H
```

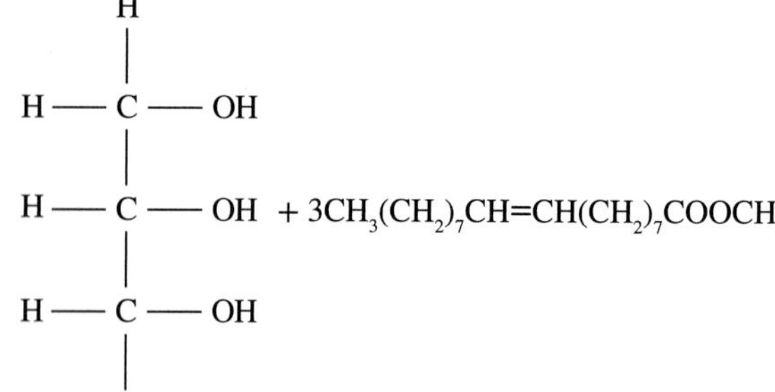

triglyceride (palm oil) — methanol — glycerol — fatty acid methyl esters (biodiesel)

b atom economy = 90.61%
c percentage yield = 64.01%
d An equation for the combustion of this biodiesel is: $C_{19}H_{36}O_2(l) + 27O_2(g) \longrightarrow 19CO_2(g) + 18H_2O(l)$

Chapter 12 Reactions of biologically important compounds

12.1 Synthesis of proteins

CSA: Ada Yonath and the structure of the ribosome

1 a four

b $H_2N—CH(CH_2CH_2SCH_3)—CO—NH—CH(H)—CO—NH—CH(CH_2C_6H_5)—CO—NH—CH(CH(CH_3)CH_3)—COOH$

c methionine, glycine, phenylalanine, valine (Met-Gly-Phe-Val)

2 $—NH—CH(CH_3)—CO—NH—CH(CH_3)—CO—NH—CH(CH_2CH_2COOH)—CO—NH—CH(CH_2OH)—COOH$

3 The ribosomes are large and dynamic, with rapidly moving atoms, and all of these things disturb crystallisation.

Key questions

1 The figure shows a section of a polypeptide. It consists of **nine** amino acid residues that are linked together by **peptide/amide** groups. It was produced by a **condensation polymerisation** reaction. In this polypeptide there are two units of leucine residues, two units of **alanine**, two units of **lysine** and one unit each of **valine and methionine**.

2 Proteins are organic **biopolymers**. The monomers required to make proteins are **amino acids**. The formation of a protein is an example of a condensation polymerisation where the other product formed is **water**. The functional group that forms between monomers is an **amide**/a **peptide** group.

3 a amine ($—NH_2$), carboxyl ($—COOH$)

b amide ($—CONH—$)

4 B is a 2-amino acid because carbon number 2 is attached to the amino and R-functional group/methyl group. A is not a 2-amino acid because there are too many carbons in the chain. C is not a carboxylic acid.

5 a A—amino group; B—peptide or amide link or bond or group; C—R group; D—carboxyl group

b alanine, serine, leucine

6 a serine: $H_2N—CH(CH_2OH)—COOH$; cysteine: $H_2N—CH(CH_2SH)—COOH$

b

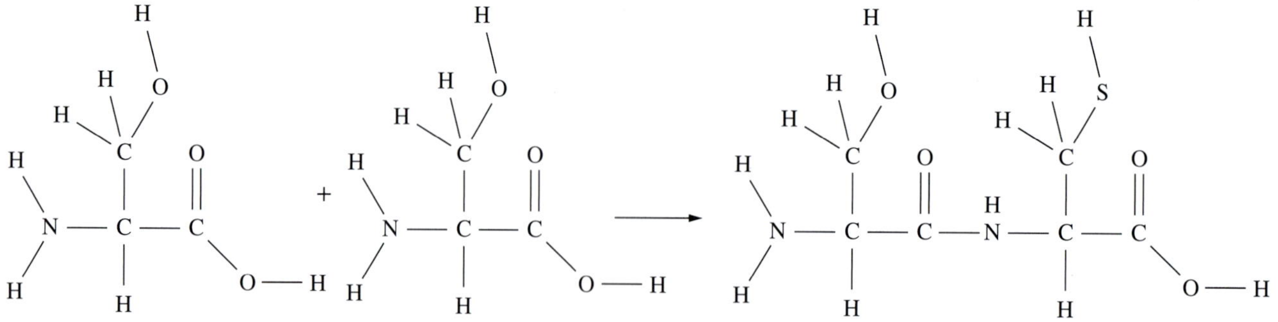

c

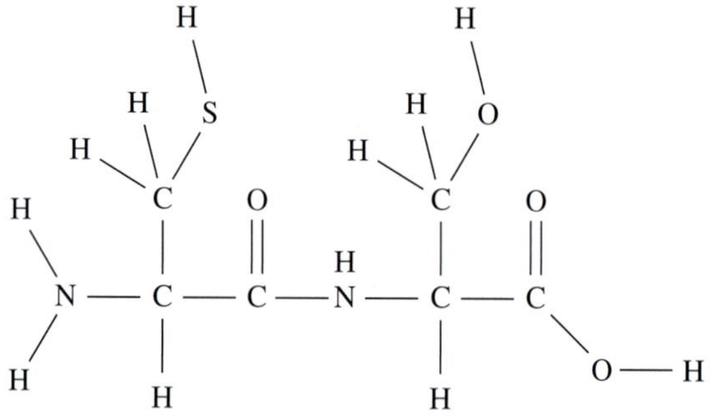

If you drew this structure in your answer to part **b**, then the product in part **b** should be your answer to part **c**.

d condensation reaction—because two functional groups are reacting together

12.2 Synthesis of starch, glycogen and lipids

1 **a** Hydroxyl groups (–OH) on both molecules react.

b A glycosidic (or ether) link (–O–) is formed.

c A hydroxyl group reacts with a carboxylic acid.

d An ester functional group is formed.

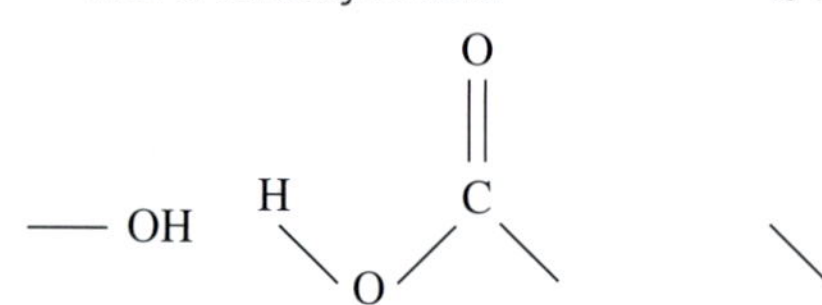

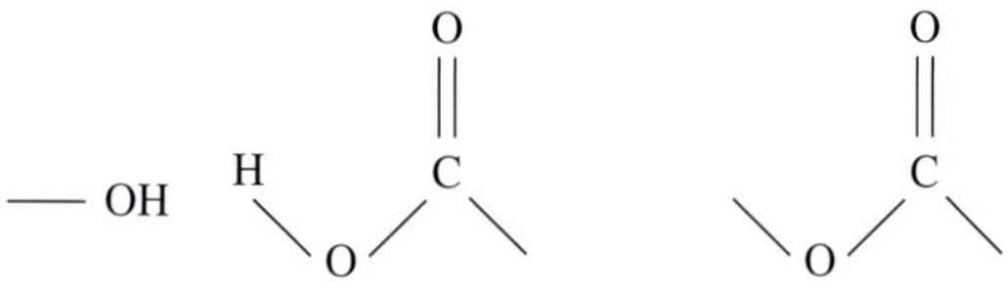

2

amylose	polysaccharide
fructose	monosaccharide
sucrose	disaccharide
lactose	disaccharide
cellulose	polysaccharide

3 **a** Glucose is a monosaccharide, made up of single ring structures of 5 carbon atoms and 1 oxygen atom. Maltose is made up of two monosaccharide molecules bonded together by a glycosidic linkage (ether functional group) –O–. Starch is a polymer of many monosaccharide molecules joined together by glycosidic linkages.

b condensation

4 B, D, F and G.

5 **a**

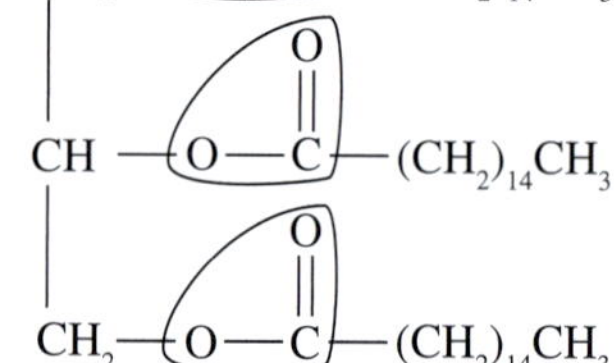

b The fat is saturated because the alkyl groups in the molecules ($CH_3(CH_2)_{14}$) contain only single carbon–carbon bonds. Alkyl groups with the general formula C_nH_{2n+1} are saturated.

6 a

$$\begin{array}{c} \mathrm{H} \\ | \\ \mathrm{H{-}C{-}OH} \\ | \\ \mathrm{H{-}C{-}OH} \\ | \\ \mathrm{H{-}C{-}OH} \\ | \\ \mathrm{H} \end{array} + 3\mathrm{HO{-}\overset{\displaystyle O}{\overset{\|}{C}}{-}(CH_2)_7CH{=}CH(CH_2)_7CH_3} \longrightarrow \begin{array}{l} \quad\mathrm{II} \\ \quad | \\ \mathrm{H{-}C{-}O{-}\overset{\displaystyle O}{\overset{\|}{C}}{-}(CH_2)_7CH{=}CH(CH_2)_7CH_3} \\ \quad | \\ \mathrm{H{-}C{-}O{-}\overset{\displaystyle O}{\overset{\|}{C}}{-}(CH_2)_7CH{=}CH(CH_2)_7CH_3} \\ \quad | \\ \mathrm{H{-}C{-}O{-}\overset{\displaystyle O}{\overset{\|}{C}}{-}(CH_2)_7CH{=}CH(CH_2)_7CH_3} \\ \quad | \\ \quad\mathrm{H} \end{array} + 3\mathrm{H_2O}$$

b

$$\begin{array}{l} \quad\mathrm{H} \\ \quad | \\ \mathrm{H{-}C{-}O{-}\overset{\displaystyle O}{\overset{\|}{C}}{-}(CH_2)_7CH{=}CH(CH_2)_7CH_3} \\ \quad | \\ \mathrm{H{-}C{-}O{-}\overset{\displaystyle O}{\overset{\|}{C}}{-}(CH_2)_7CH{=}CH(CH_2)_7CH_3} \\ \quad | \\ \mathrm{H{-}C{-}O{-}\overset{\displaystyle O}{\overset{\|}{C}}{-}(CH_2)_7CH{=}CH(CH_2)_7CH_3} \\ \quad | \\ \quad\mathrm{H} \end{array}$$

7 Fatty acids are relatively insoluble due to their long hydrocarbon chain. The hydrocarbon chain is non-polar and will not form hydrogen bonds with water.

12.3 Hydrolysis of biomolecules

1 Amino acids can combine to form **polypeptides** in **condensation** reactions. **Water** is also formed in this reaction. The reverse of this reaction is a **hydrolysis** reaction.

Polysaccharides are formed in condensation reactions involving **monosaccharides**.

Fatty acids and **glycerol** can combine in a condensation reaction to form a **triglyceride**.

2 polysaccharide $\xrightarrow{\text{hydrolysis}}$ monosaccharides $\xrightarrow{\text{condensation}}$ **polysaccharide**

triglyceride $\xrightarrow{\text{hydrolysis}}$ glycerol and fatty acids $\xrightarrow{\text{condensation}}$ **triglyceride**

protein $\xrightarrow{\text{hydrolysis}}$ amino acids $\xrightarrow{\text{condensation}}$ **polypeptide/protein**

3 Enzymes catalyse many of the important hydrolysis reactions. They are essential for the breakdown by hydrolysis of proteins, carbohydrates and triglyceride molecules obtained from food into smaller molecules.

4 a $C_6H_{12}O_6$

b glycosidic links

5 a A hydrolysis reaction is one in which water is a reactant. In the hydrolysis of lactose, the molecule is broken into two smaller parts by breaking the glycosidic link with the help of a catalyst.

b

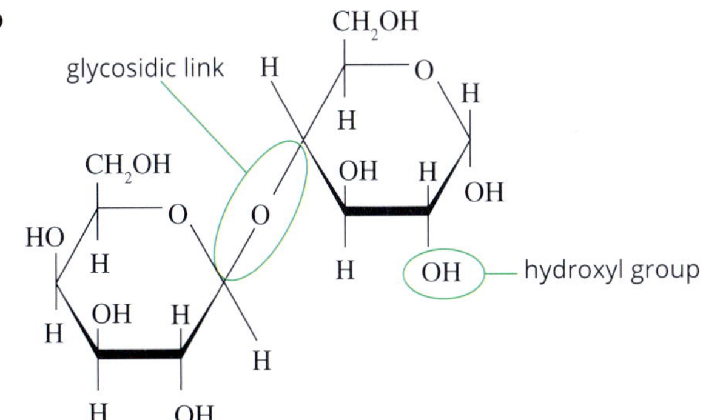

6

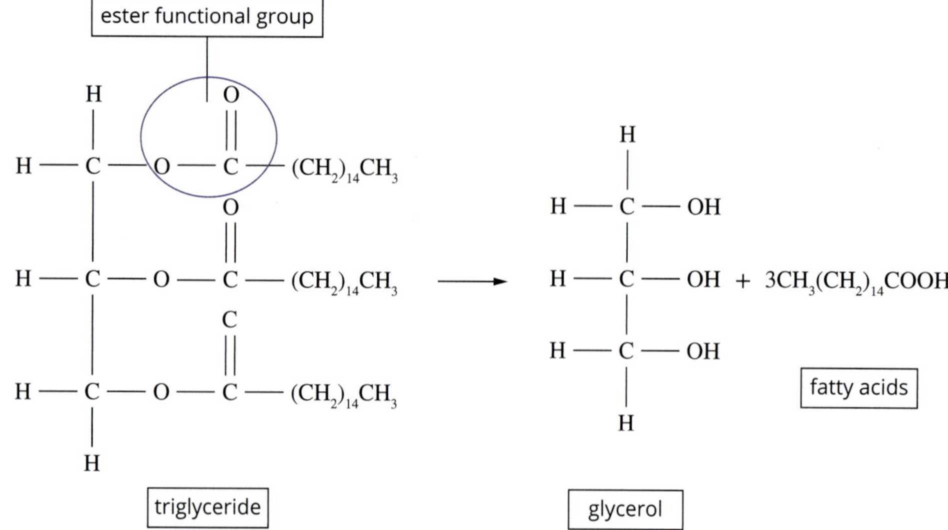

7 **a** Two water molecules are needed as extra reactants to give a balanced equation.

b

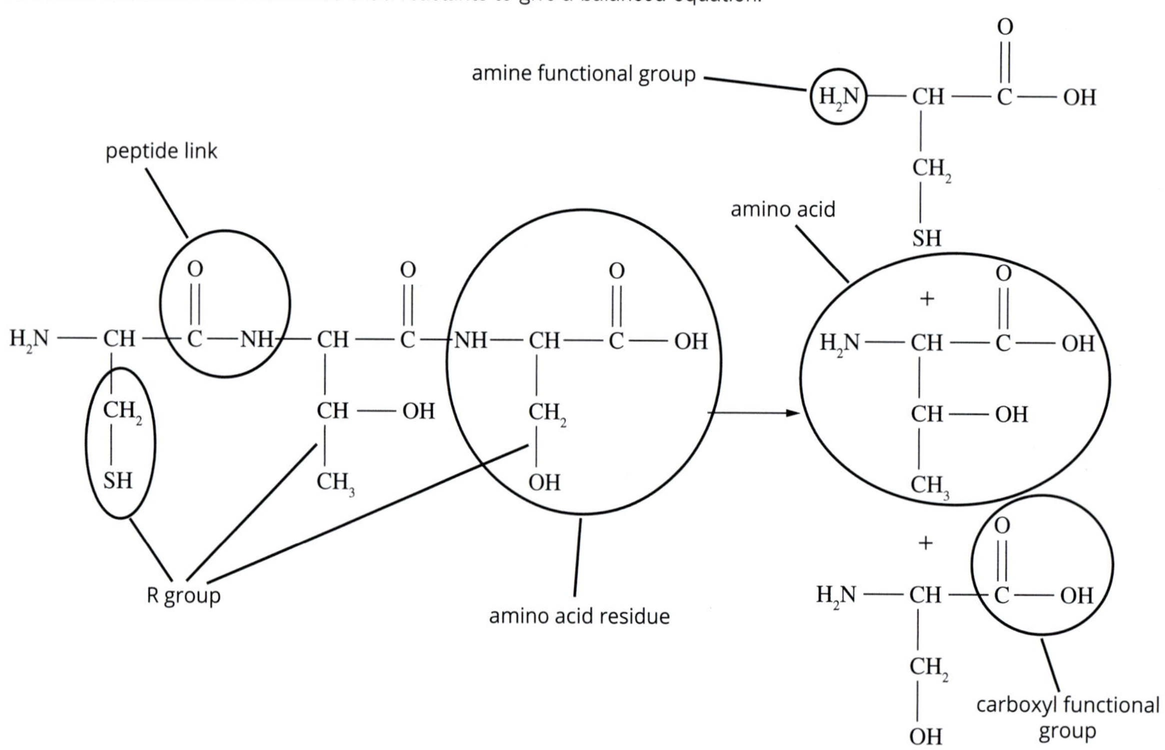

Hydrolysis of a tripeptide

Chapter 12 review

1 C. **2** A. **3** B.

4

amine functional group

carboxyl functional group

The amine is coloured blue (on the left-hand side) and the carboxy group is in red (on the right-hand side). H

5 **a** The presence of several polar hydroxyl groups makes glucose soluble in water

b The usual reactions of glucose involve the hydroxyls at positions 4 and 1.

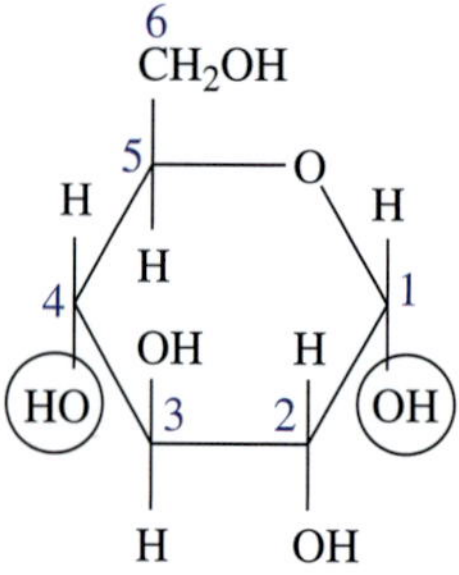

6 **a** galactose or fructose **b** maltose or lactose
c starch, cellulose or glycogen

7 Glycogen is the main storage polysaccharide in animals. Therefore, it fulfills the same role in animals that starch does in plants.

8 **a** carboxyl and hydroxyl groups; lipids
b two hydroxyl groups; carbohydrates
c carboxyl and amino groups; proteins

9 **a** all **b** triglyceride **c** all
d protein **e** triglyceride

10

	Names of links between units	Functional groups formed when links break	Monomer or smallest components
Carbohydrate	**glycosidic**	hydroxyl	**monosaccharide**, often **glucose**
Fat	**ester**	hydroxyl and **carboxyl**	**fatty acids** and glycerol
Protein	**peptide/ amide**	**amine** and carboxyl	**amino acid**

11 **a** $H_2NCHCH_3CONHCH_2COOH$, $H_2NCH_2CONHCHCH_3COOH$
b 6 (using each amino acid once in each peptide)
c

d alanine-cysteine-glycine-serine

12 During condensation reactions, small molecules join together to form larger molecules. Large food molecules undergo hydrolysis to form smaller molecules.
a condensation **b** condensation **c** condensation
d hydrolysis **e** hydrolysis **f** hydrolysis

13 **a** stearic acid + glycerol ⟶ ester
b glucose + fructose ⟶ glycosidic (ether)
c glycine + glycine ⟶ peptide (amide)
d glucose + glucose ⟶ glycosidic (ether)
e linoleic acid + glycerol ⟶ ester

14 **a** It is not a protein as it does not contain N atoms.
b A fatty acid is likely to be much smaller and to contain only two oxygen atoms.
c It is far too big a molecule to be a monosaccharide.
d The molecule is a typical polysaccharide – it is big, contains only carbon, hydrogen and oxygen and there are approximately 2 hydrogens for every oxygen, as there is in glucose.

15 13680 g

16 **a** glycerol
b

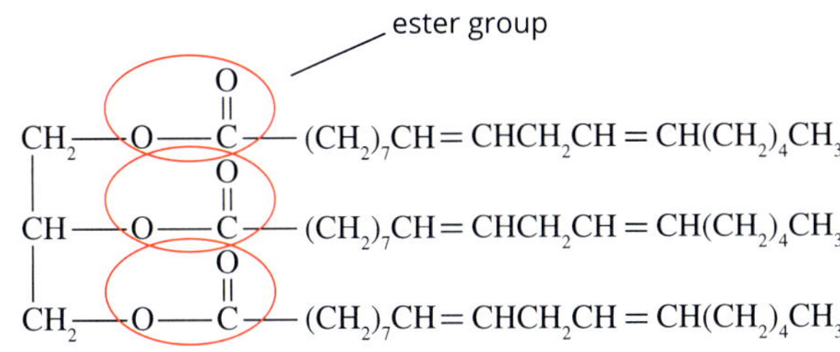

c ester group (circled in part **b**)
d Linoleic acid is an unsaturated fatty acid because it has two carbon–carbon double bonds in its structure.

17 **a** carbon dioxide **b** glucose
c disaccharide **d** glycosidic link
e polysaccharide

18

Molecule	The molecule is best described as a/an	A functional group it contains
glycine	**amino acid**	**amine or carboxyl**
glycogen	**polysaccharide**	**glycosidic or ether or hydroxyl**
glucose	**monosaccharide**	**hydroxyl or ether**
glycerol	**alcohol**	**hydroxyl**

19 **a** Condensation reactions involve the linking together of two small molecules and the elimination of a small molecule, usually water. Hydrolysis reactions consume water, produce two small molecules from one larger molecule, and they can often be regarded as the reverse of condensation reactions.
b Vital biochemical processes include condensation reactions (e.g. synthesis of lipids, proteins and polysaccharides) and hydrolysis reactions (e.g. digestion).

c i

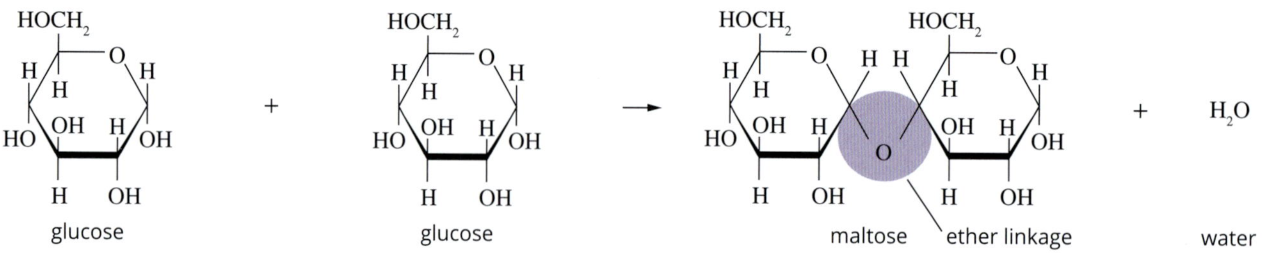

glucose + glucose → maltose (ether linkage) + water

ii

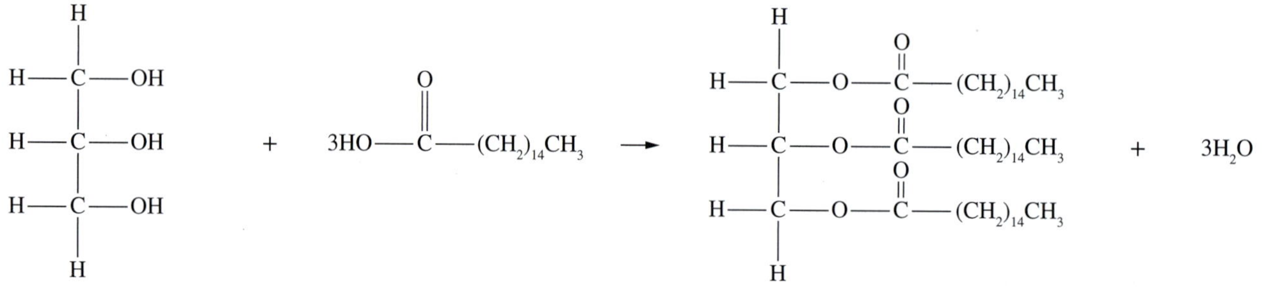

glycerol + $3HO{-}C(=O){-}(CH_2)_{14}CH_3$ → triglyceride + $3H_2O$

iii

$H_2N{-}CH(CH_3){-}COOH$ + $H{-}NH{-}CH(CH_3){-}COOH$ → $H_2N{-}CH(CH_3){-}CO{-}NH{-}CH(CH_3){-}COOH$ + H_2O

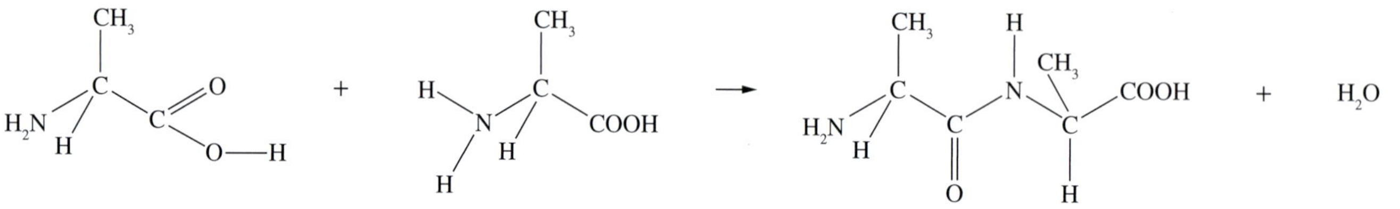

Unit 4 Area of Study 1 How are organic compounds categorised and synthesised?

1	A.	**2**	B.	**3**	D.
4	B.	**5**	A.	**6**	D.
7	C.	**8**	A.	**9**	A.
10	C.	**11**	D.	**12**	C.
13	C.	**14**	B.	**15**	B.

16 **a** **i** C_7H_{16} **ii** e.g. $CH_3CH_2CH_2CH(CH_3)CH_2CH_3$
b **i** $C_4H_{10}O$ **ii** e.g. $CH_3CH_2CHOHCH_3$
c **i** $C_3H_6O_2$ **ii** e.g. CH_3CH_2COOH
d **i** $C_5H_{11}OCl$ **ii** e.g. $CH_2Cl(CH_2)_4OH$
e **i** $C_6H_{12}O$ **ii** e.g. $CH_3CH_2CH(CH_3)CH_2CHO$
f **i** $C_6H_{12}O$ **ii** e.g. $CH_3CHCHC(OH)(CH_3)CH_3$

17 **a** heptan-3-ol **b** 3-fluorobutanal
c pent-2-ene **d** 4-ethylhexan-3-amine
e 1-aminopropan-2-ol **f** 5-chlorohexan-3-ol
g 6-hydroxyoctanoic acid **h** 2-bromo-3-methylpentane
i 2-chlorohexan-3-one

18 **a** $CH_3CH_2CH_2CH_2CH_3$ (full structural formula)

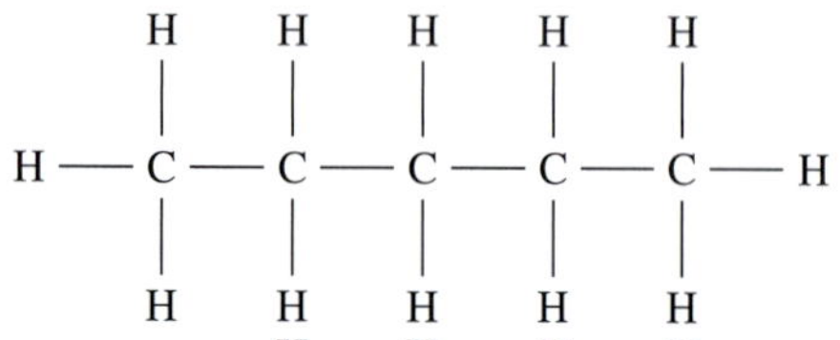

b $CH_2{=}CH{-}CH_2{-}CH(OH){-}CH_3$ (full structural formula)

c $CH_2Cl{-}CH_2{-}C(=O){-}CH_3$ (full structural formula)

d $CH_3CH_2CH_2C(=O)NH_2$ (full structural formula)

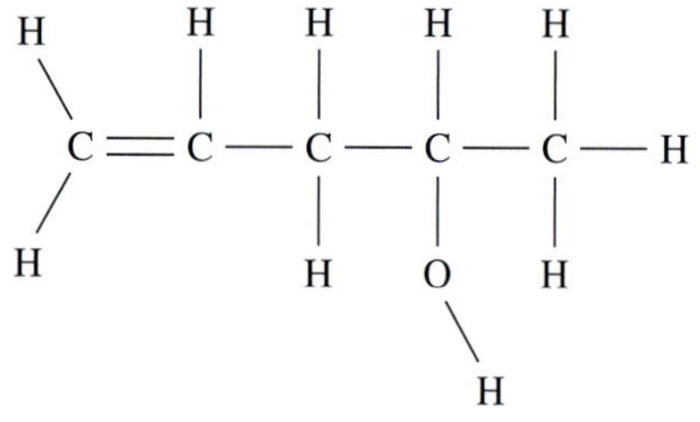

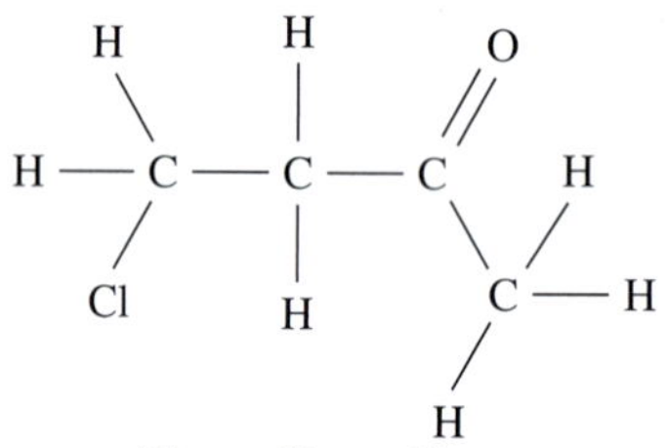

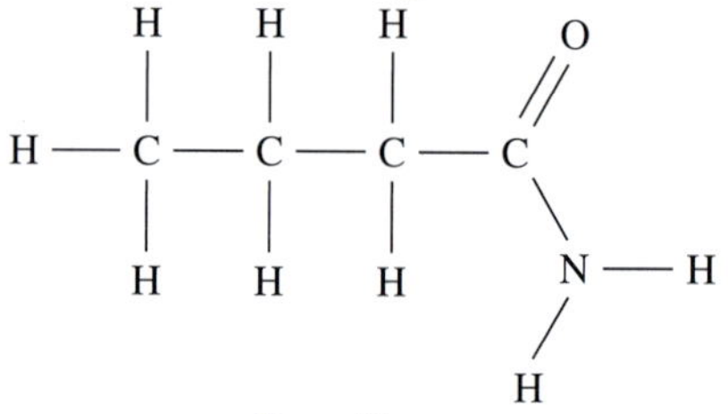

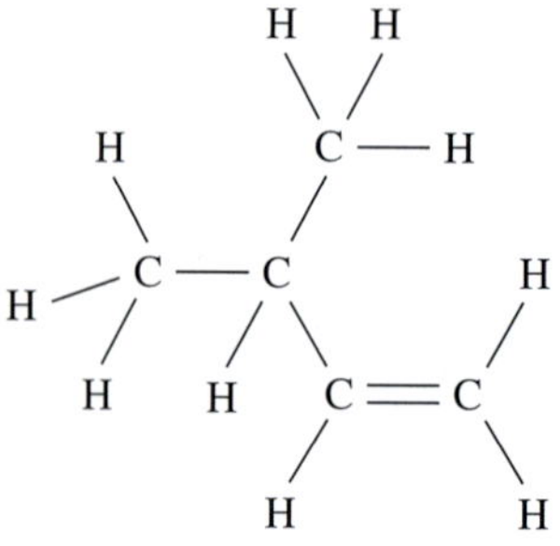

e $CH_3{-}CH_2{-}CH(CH_3){-}CH{=}CH_2$ (full structural formula)

f

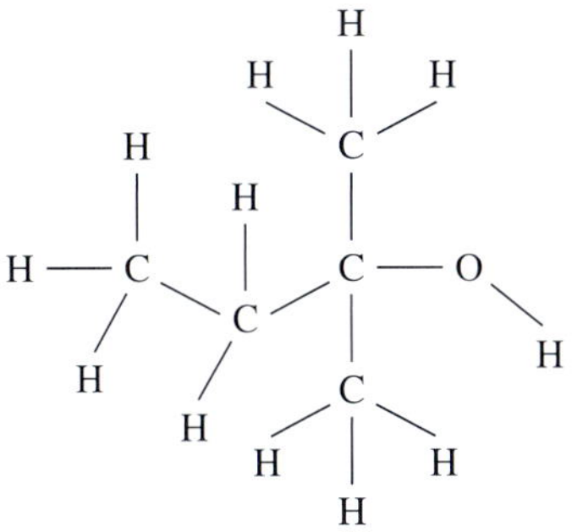

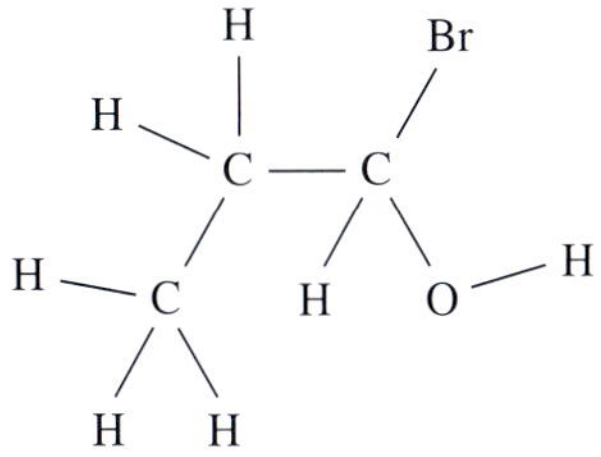

i

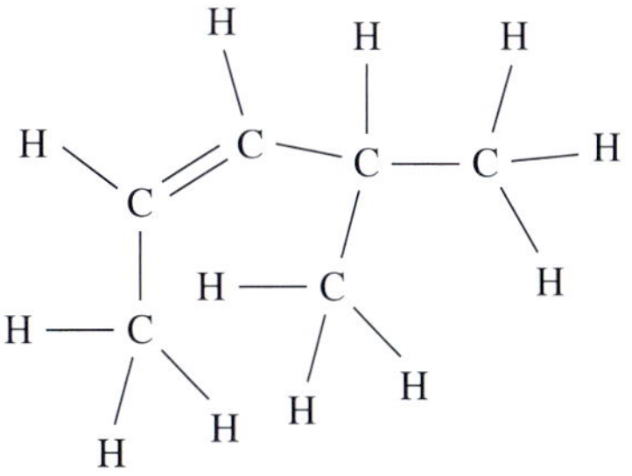

g

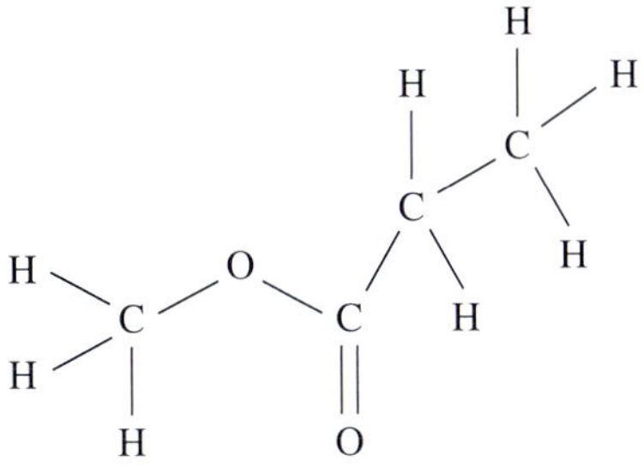

j

h

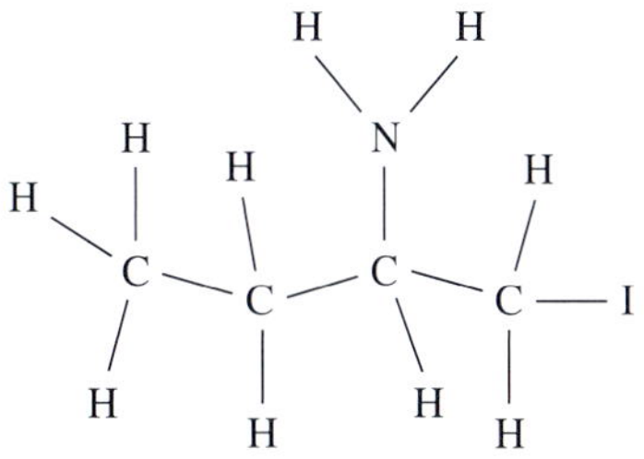

k

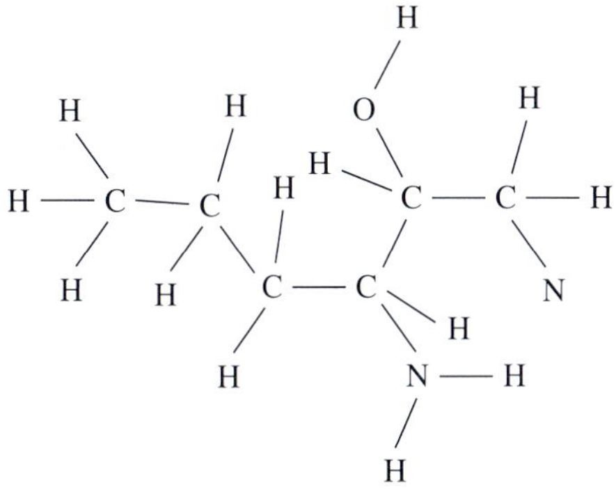

19 a Warm with an alkali metal hydroxide (e.g. KOH, NaOH)

b Acidified dichromate ($Cr_2O_7^{2-}/H^+$) or acidified permanganate (MnO_4^-/H^+)

c NH_3

20 a $CH_3CH_2CH_3 \xrightarrow[\text{UV light}]{Cl_2\ (\text{or } Br_2)} CH_3CH_2CH_2Cl \xrightarrow{KOH} CH_3CH_2CH_2OH \xrightarrow[H^+]{Cr_2O_7^{2-}} CH_3CH_2COOH$

b $CH_4 \xrightarrow[\text{UV light}]{Cl_2\ (\text{or } Br_2)} CH_3Cl \xrightarrow{KOH} CH_3OH$

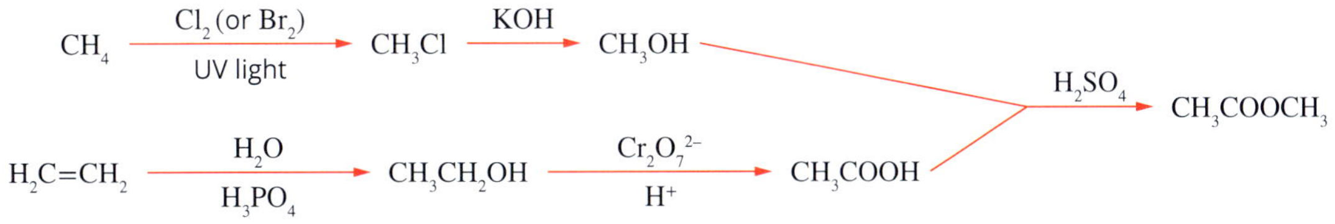

$H_2C{=}CH_2 \xrightarrow[H_3PO_4]{H_2O} CH_3CH_2OH \xrightarrow[H^+]{Cr_2O_7^{2-}} CH_3COOH$

$CH_3OH + CH_3COOH \xrightarrow{H_2SO_4} CH_3COOCH_3$

c $HCOOCH_2CH_2CH_3 \xrightarrow[\text{heat/catalyst}]{H_2O} HCOOH + CH_3CH_2CH_2OH$

$CH_3CH_2CH_2OH \xrightarrow{Cr_2O_7^{2-}\ \ H^+} CH_3CH_2COOH$

$CH_3CH_2COOH + CH_3CH_2CH_2OH \xrightarrow{H_2SO_4} CH_3CH_2COOCH_2CH_2CH_3$

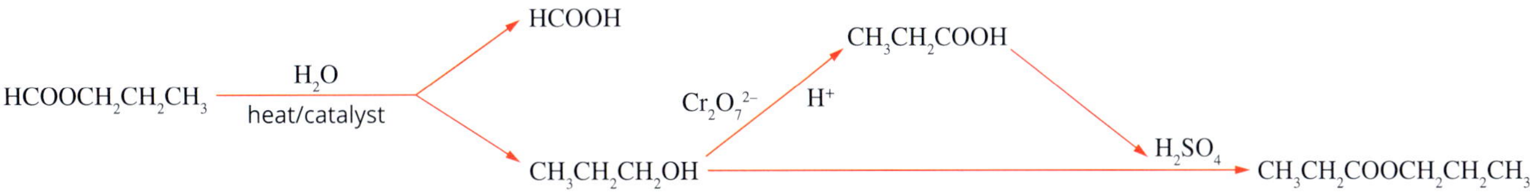

21 a $CH_3CH_2CH_2OH + H_2O \rightarrow CH_3CH_2COOH + 4H^+ + 4e^-$

b $CH_3CH_2COOH + CH_3CH_2OH \rightarrow CH_3CH_2COOCH_2CH_3 + H_2O$

22 a Viscosity is a measure of the resistance to flow. Low-viscosity liquids are 'runnier'.

b i Viscosity increases from methanol to ethanol to propan-1-ol because larger molecules have stronger dispersion forces between them, tending to restrict movement.

ii Ethane-1,2-diol has one more hydroxyl group than ethanol. The additional hydrogen bonding that this allows accounts for the higher viscosity and boiling point of ethane-1,2-diol compared to that of ethanol.

iii Benzene has a comparable molecular mass to butan-1-ol, but is non-polar and can only form relatively weak intermolecular dispersion forces. Weak bonding between benzene molecules results in low viscosity. A comparison of propanone and propan-1-ol also illustrates the point.

23 **a** 52.2%
b 5.60 kg
c sodium hydroxide and sodium ethanoate
d 42.4%; slightly lower
e Addition reactions of alkenes have an atom economy of 100%, since in addition reactions the atoms of a small molecule add across a C=C double bond and there are no other products formed.

24 **I** **a** $C_{22}H_{40}O_2$; $C_{21}H_{39}COOH$ **b** polyunsaturated
II **a** $C_{18}H_{28}O_2$; $C_{17}H_{27}COOH$ **b** polyunsaturated
III **a** $C_{20}H_{40}O_2$; $C_{19}H_{39}COOH$ **b** saturated
IV **a** $C_{16}H_{30}O_2$; $C_{15}H_{29}COOH$ **b** monounsaturated

25 **a, b**

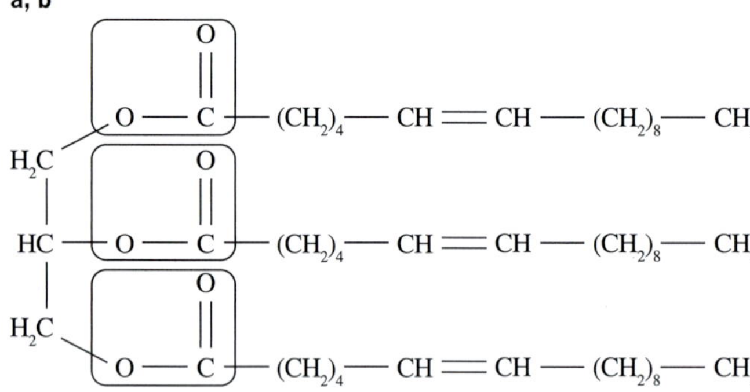

26 **a** $C_{12}H_{22}O_{11}(aq) + H_2O(l) \rightarrow 2C_6O_{12}O_6(aq)$
b $3C_{11}H_{23}COOH(l) + C_3H_8O_3(aq) \rightarrow C_{39}H_{74}O_6(l) + 3H_2O(l)$

27 **a** Independent variable: type of vegetable oil; Dependent variable: yield of biodiesel
b The students wish to investigate the yield of biodiesel, so they would need to measure the quantity produced. Looking at the procedure they followed, the biodiesel is poured into a measuring cylinder. The students would need to identify which layer is the biodiesel, then use the increments on the measuring cylinder to measure the volume of biodiesel produced.
c **i** The measuring cylinder contains approximately 15.5 mL of biodiesel.
ii The bottom, polar layer, would contain glycerol, sodium hydroxide and unreacted ethanol.
d The mixture would become frothy. The NaOH present would react with the fatty acid molecules to form soap. This would decrease the yield of the biodiesel.
e

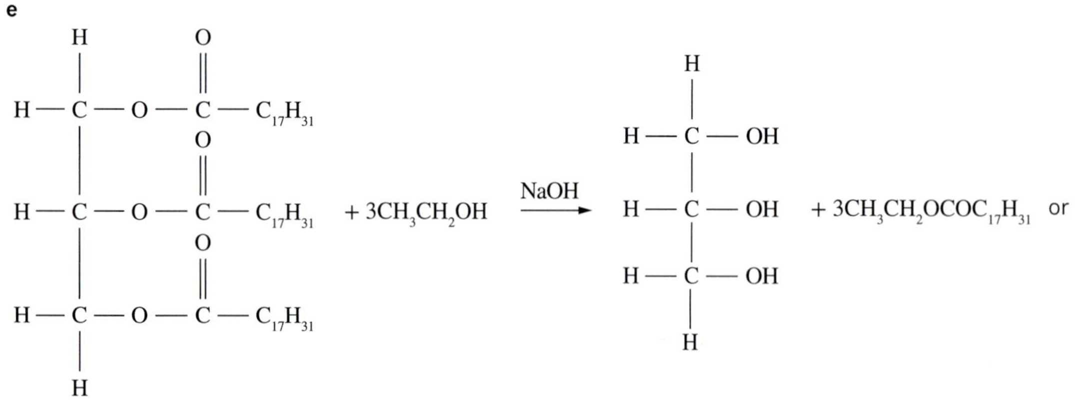

Chapter 13 Laboratory analysis of organic compounds

13.1 Qualitative testing of organic compounds

TY 13.1.1 A: but-2-ene
B: propan-1-ol
C: 2-methylpropan-2-ol
D: pentane
E: ethanoic acid

TY 13.1.2 There are 3 carbon–carbon double bonds in each linolenic acid molecule.

TY 13.1.3 iodine number of eicosapentaenoic acid = 420

Key questions

1 Simple distillation is used to **separate** mixtures of two or more liquids. The mixture is placed in a round-bottom flask with anti-**bumping** granules to allow smooth **boiling** and then heated to the **boiling point** of the liquid to be collected. The **vapour** rises up the flask and enters the **condenser**. The vapour is **cooled** to below its boiling point and condenses back to a liquid. This pure liquid is called the **distillate** and is collected in a separate flask. The liquid with the **higher** boiling point remains in the round-bottom flask.

2 B.

3 **a** An addition reaction occurs when bromine is added to a compound containing a carbon–carbon double bond. The bromine is said to be decolourised as the red-brown colour of the bromine rapidly disappears when mixed with a compound containing a carbon–carbon double bond.
b $C_2H_4(g) + Br_2(aq) \rightarrow CH_2BrCH_2Br(aq)$

4 iodine number = 387

5 C.

6 **a** The unknown chemical could be ibuprofen or the unknown chemical could be a different substance that happens to have the same melting point as ibuprofen.
b Add pure ibuprofen to the chemical and determine the melting point of the mixture. If the melting point is exactly 75°C, then the unknown chemical is ibuprofen. If the melting point is lower than 75°C, then the unknown chemical is not ibuprofen.

7 The fatty acid molecule contains 2 carbon–carbon double bonds.

8 A is propan-2-ol. The colour change from orange to green when acidified potassium dichromate was added indicates a primary alcohol.

B is pent-1ene. The bromine is decolourised, indicating the presence of a carbon–carbon double bond.

C is propanoic acid. A gas is produced (CO_2) on addition of sodium hydrogen carbonate.

D is hexane. Alkanes are insoluble in water and do not react rapidly with bromine.

E is 2-methylpropan-2-ol. This can be identified as an alcohol because it is soluble in water, but undergoes no reaction with sodium hydrogen carbonate. No colour change when acidified potassium dichromate is added indicates a tertiary alcohol.

9 The mixture of volatile liquids can be separated by fractional distillation. The mixture is placed in the distillation flask and heated. The most volatile compound with the lowest boiling point (bromomethane) is the first to reach the top of the fractionating column and condense. The fraction condensing in a narrow range of temperatures around the boiling point 65°C is collected. The next fraction to be collected at around 80°C will contain the ester, methyl propanoate, which is collected as the distillate. The higher boiling point components from the reaction mixture, water and propanoic acid, remain in the distillation flask.

13.2 Redox titrations of organic compounds

TY 13.2.1 0.221 M

TY 13.2.2 0.147 M

TY 13.2.3 **a** C_2H_5OH is the limiting reactant.

b 0.948 g

CSA: Determination of vitamin C content

1 Iodine is unstable and, immediately before its use as a standard solution, it must be reacted with another chemical (usually sodium thiosulfate) to accurately determine its concentration.

2 trial 1: 0.237 g; trial 2: 0.235 g; trial 3: 0.236 g

3 69.7%

4 C.

5 Gloves and safety glasses must be worn to prevent skin and eye damage.

6 The tablet may also include binders to hold the ingredients of the tablet together and sweeteners to improve the taste.

Key questions

1 **a** A standard solution is a solution of accurately known concentration. A primary standard is a substance that is readily obtained in a pure form, has a known formula and can be stored without deteriorating or reacting with the atmosphere. It should also be cheap and have a high molar mass.

b A burette is a piece of equipment capable of delivering different volumes of a liquid accurately (generally up to 50.00 mL). Pipettes usually deliver only a fixed volume of liquid (e.g. 20.00 mL).

c An aliquot is the volume of liquid delivered from a pipette. A titre is delivered by a burette and is the volume needed to reach the end point of a titration.

2 $CH_3CH_2CH_2CH_2OH(aq) + H_2O(l) \rightarrow CH_3CH_2CH_2COOH(aq) + 4H^+(aq) + 4e^-$

3 **a** $n(MnO_4^-) = 0.0104$ mol

b $n(CH_3CH_2OH) = 0.012\ 94$ mol

c $c(CH_3CH_2OH)$ in the cider = 0.647 M

4 **a** I_2 is the limiting reactant.

b 0.685 g

5 0.186 M

6 **a** 1.90 M

b The calculated concentration would be higher than its actual value. Rinsing the burette with deionised water would slightly dilute the potassium dichromate solution. This means more potassium dichromate solution would need to be added to reach the equivalence point. It would then appear that a greater amount of potassium dichromate reacted; hence, it would appear that a greater amount of ethanol had reacted, resulting in a reported ethanol concentration higher than it actually is.

7 **a** $CH_3CH_2OH(aq) + H_2O(l) \rightarrow CH_3COOH(aq) + 4H^+(aq) + 4e^-$

b $Cr_2O_7^{2-}(aq) + 14H^+(aq) + 6e^- \rightarrow 2Cr^{3+}(aq) + 7H_2O(l)$

c $2Cr_2O_7^{2-}(aq) + 3CH_3CH_2OH(aq) + 16H^+(aq) \rightarrow 4Cr^{3+}(aq) + 3CH_3COOH(aq) + 11H_2O(l)$

d **i** 7.14 M **ii** 328 g L^{-1}

Chapter 13 review

1 **a** 24.22, 24.20 and 24.16 mL

b 24.19 mL

2 **a** The carbon–carbon double bond in but-1-ene could be identified by reaction with bromine solution. The bromine would be decolourised if a double bond is present.

b The hydroxyl group in ethanol could be identified by reaction with ethanoic acid and a little sulfuric acid. The characteristic smell of an ester (ethyl ethanoate) would be observed if the compound is ethanol.

c The carboxyl group in propanoic acid could be identified by the addition of sodium hydrogen carbonate powder. Effervescence would be observed because of the production of carbon dioxide gas.

3 $3CH_3CH_2CH_2CH_2OH(aq) + 2Cr_2O_7^{2-}(aq) + 16H^+(aq) \rightarrow 3CH_3CH_2CH_2COOH(aq) + 4Cr^{3+}(aq) + 11H_2O(l)$

4 $n(H_2C_2O_4.2H_2O) = \frac{m}{M} = \frac{2.436}{126.0} = 0.01933$ mol

$c(H_2C_2O_4.2H_2O) = \frac{n}{V} = \frac{0.01933}{0.1000} = 0.1933$ M

5 The volumetric flask should be rinsed with deionised water, the pipette with the iodine standard solution, the burette with vitamin C solution and the conical flask with deionised water.

6 **a** A primary standard has a very high level of purity; has a known formula; is stable (e.g. will not react with atmospheric gases, such as carbon dioxide and water vapour); has a high molar mass to minimise errors in weighing; is readily available; and is relatively inexpensive.

b
- Accurately weigh an empty weighing bottle, add the primary standard and reweigh.
- Transfer the weighed sample to a volumetric flask using a dry glass funnel.
- Rinse out the weighing bottle and glass funnel using a wash bottle.
- Half fill the volumetric flask with water and shake to dissolve the sample.
- When the sample has dissolved, add water to the calibration mark and shake the flask again.
- Determine the concentration of the primary standard.

7 **a** 0.0342 mol

b 0.0263 mol of CH_3OH is needed for all of the MnO_4^- to react.

c The CH_3OH is in excess.

$KMnO_4$ is the limiting reactant.

d 1.21 g

8 Iodine number = 368

9 Determine the melting point of the sample. If the melting point range is sharp and between 157°C and 159°C, then the sample is likely to be pure salicylic acid. If the melting point is lower than 159°C and the range is broad, then the sample contains impurities.

10 Iodine number of the vegetable oil = 96

11 a A nervonic acid molecule contains 1 carbon–carbon double bond (monounsaturated).
b Iodine number of nervonic acid = 69.3

12 The mixture of volatile liquids can be separated by fractional distillation. The mixture is placed in the distillation flask and heated. The components in the mixture are separated in order of their boiling points, with the most volatile component (lowest boiling point), bromoethane, being the first fraction to condense. This is followed by the fraction containing 1,1-dibromoethane. The fraction condensing in a narrow range of temperatures around the boiling point of 1,2-dibromoethane, 131°C, is then collected in a new receiving flask. The higher boiling point component from the reaction mixture, 1,1,2-tribromoethane, remains in the distillation flask.

13 Iodine number of the plant oil = 99.0

14 a Rinsing with water would dilute the standard solution.
b The calculated concentration would be higher as a smaller titre of the unknown solution would be required to react with the standard solution.

15 a $K_2Cr_2O_7$ is the limiting reactant.
b 0.161 g

16 a 4.802×10^{-4} mol
b 4.802×10^{-4} mol
c 0.0308 g
d percentage SO_2 = 0.147 % m/m

17 a Oxidation of ethanol to ethanoic acid:
$CH_3CH_2OH(aq) + H_2O(l) \longrightarrow CH_3COOH(aq) + 4H^+(aq) + 4e^-$
Reduction of permanganate:
$MnO_4^-(aq) + 8H^+(aq) + 5e^- \longrightarrow Mn^{2+}(aq) + 4H_2O(l)$
b $5CH_3CH_2OH(aq) + 4MnO_4^-(aq) + 12H^+(aq) \longrightarrow 5CH_3COOH(aq) + 4Mn^{2+}(aq) + 11H_2O(l)$
c 0.0129 mol
d 7.55 % v/v
e i The calculated concentration of ethanol in the cider solution would be lower. Rinsing the pipette with deionised water would slightly dilute the cider solution. This means less potassium permanganate would need to be added to oxidise the ethanol. It would then appear that a smaller amount of potassium permanganate reacted; hence, it would appear that a smaller amount of ethanol had reacted.
ii The calculated concentration of ethanol in the cider solution would be higher. Rinsing the burette with deionised water would slightly dilute the potassium permanganate solution. This means more potassium permanganate solution would need to be added to oxidise the ethanol. This would give the impression that the ethanol solution was more concentrated than it actually was.
iii The calculated concentration of ethanol in the cider solution would not be affected. Rinsing the conical flask with deionised water would have no effect on the titre volume as the amount of ethanol in the conical flask is accurately known.

18 a $2MnO_4^-(aq) + 16H^+(aq) + 5C_2O_4^{2-}(aq) \longrightarrow 2Mn^{2+}(aq) + 8H_2O(l) + 10CO_2(g)$
b 0.0180 M

19 a 0.487 g b 18.8%
c fillers, binders, sweeteners, flavours and colouring

20 a 4.60×10^{-4} mol b 6.89×10^{-4} mol
c 8.62×10^{-3} mol d 5.05%
e No, this product would not conform with the regulations for low-alcohol beer.

21 a mole ratio = $\frac{5}{4}$ b 74.0 g mol^{-1}
c butan-1-ol

Chapter 14 Instrumental analysis of organic compounds

TY 14.1.1 C_4H_8, which is butene.

Key questions

1 a $12 + (3 \times 1) = 15$ b $(3 \times 12) + (7 \times 1) = 43$
c $12 + (3 \times 1) + 81 = 96$ d $12 + 1 + 16 = 29$
e $12 + 1 + (2 \times 35) = 83$

2 All fragments must have a positive charge. The correct formula is CHO^+.

3 186 ($CH_3CH^{79}Br_2^+$), 188 ($CH_3CH^{79}Br^{81}Br^+$), 190 ($CH_3CH^{81}Br_2^+$)

4 CH_2OH^+

5 a 43 b $C_6H_{13}^+$ c Octane, C_8H_{18}

6 a $CH_3CH_2CH_2Br$
b The peaks at m/z = 122 and m/z = 124 are due to the molecular ion containing Br, as both the ^{79}Br or ^{81}Br isotopes are present.
c The two isotopes of bromine, ^{79}Br and ^{81}Br, are found in almost equal abundances, so the peaks are almost equivalent in height.
d $CH_3CH_2CH_2^+$

14.2 Infrared spectroscopy

TY 14.2.1 The spectrum shows absorption bands corresponding to the presence of O–H alcohol group and C–H groups.
This suggests the molecule is an isomer of butanol.

Key questions

1 B.

2 vibrational energy

3 a C–C, C=C, C≡C (see Table 14.2.2)
As bond strength increases, infrared absorption frequency increases.
b S–H, O–H, N–H, C–H, H–H (see Table 14.2.3)
As atomic mass (of the atom bonded to hydrogen) increases, the infrared absorption frequency decreases.

4 a O–H at 3200–3550 cm^{-1} and C–O at 1000–1300 cm^{-1}
b C=O at 1660–1745 cm^{-1}
c N–H at 3350–3500 cm^{-1}
d O–H at 2500–3300 cm^{-1}, C=O at 1670–1750 cm^{-1} and C–O at 1000–1300 cm^{-1}
e C=O at 1670–1750 cm^{-1} and C–O at 1000–1300 cm^{-1}
f C=C at 1610–1680 cm^{-1}

5 The absorption bands from C–O and C–C bonds are located in the fingerprint region below 1400 cm^{-1}. Many molecules contain these bonds in various functional groups, so they give few clues to the structure of a compound. The exact wavenumber of C–O and C–C absorption bands are highly specific to an individual molecule, and can be used to compare the molecule to a reference standard for positive identification.

6 a O–H acids at approximately 3000 cm^{-1} and C=O at approximately 1700 cm^{-1}
b O–H alcohols at approximately 3300 cm^{-1} and C–H at approximately 2800 cm^{-1}
c N–H at approximately 3400 cm^{-1} and C–H at approximately 2900 cm^{-1}

7 a B. b D. c B.

14.3 Nuclear magnetic resonance spectroscopy

TY 14.3.1 The molecular formula of the compound is $C_2H_3Cl_3$. The splitting patterns and peak area indicate that the molecule contains a $-CH_2-$ group adjacent to a $-CH-$ group. The molecule must be 1,1,2-trichloroethane because the structure fits the evidence from the spectrum.

Key questions

1 a one b two c three d three

2 a The number of signals indicates the number of non-equivalent hydrogen environments.

b The relative peak areas indicate the relative number of hydrogen atoms in each proton environment.

c The splitting pattern of a signal indicates the number of hydrogen atoms adjacent to a specific proton environment.

d The chemical shift of a signal indicates the type of hydrogen environment.

3 a There will be four proton NMR peaks

b 6H ($-CH_3$), 4H ($-CH_2$), 1H (proton attached to the carbon on the –CHOH– group), 1H (–OH)

c The $-CH_3$ peak will be split into three.

The $-CH_2-$ peak will be split into five.

The proton attached to the carbon on the –CHOH– group's peak will be split into five.

The –OH peak will not be split as the hydroxyl group does not count as a neighbour for the neighbouring carbon atoms.

d There will be three ^{13}C NMR peaks

4 a i two ii two iii four iv four v three

b i two ii two iii four iv three v four

c a doublet (two-peak pattern) and a quartet (four-peak pattern)

d 3 : 2

5 a There are three peak sets, each representing a different hydrogen environment.

b From Table 14.3.1, the expected chemical shifts are:

$-COCH_3$ 2.1–2.7 ppm

RCH_3 0.8–1.0 ppm

$-OCH_2R$ 3.3 ppm

c The relative number of protons for the peak sets is:

A:B:C = 2:3:3

d Proton NMR peaks are split because of interaction of the magnetic fields on adjacent atom. The number of peaks associated with each proton environment is shown in the proton NMR spectrum. Using the $n + 1$ rule:

i The quartet peaks at A indicate that there are three hydrogen atoms attached to an adjacent atom.

ii The single peak at B indicates that there are no hydrogen atoms attached to the adjacent atom.

iii The triplet at B indicates that there are two hydrogen atoms attached to the adjacent atom.

e A is the CH_2 group.

B is the methyl group of ester.

C is the methyl group of CH_3CH_2.

f You would expect to see four peaks in the carbon-13 NMR spectrum because there are four different carbon environments in the ethyl ethanoate molecule.

6 a proton NMR:

Peak number	Chemical shift	Relative peak area	Splitting pattern	Possible hydrogen environment	Position on structure
1	1.2 ppm	3	triplet (3)	$R-CH_3$	next to $-CH_2-$
2	2.2 ppm	2	quartet (4)	$R-CH_2-$ where R is a carbonyl group	next to $-CH_3$
3	7.0 ppm	2	singlet (1)	no neighbouring H to cause splitting $-NH_2$	amide

b carbon-13 NMR:

Peak number	Chemical shift	Possible carbon environment	Position on structure
1	10 ppm	$R-CH_3$	next to $-CH_2-$
2	30 ppm	$R-CH_2$	next to $-CH_3$
3	175 ppm	C=O	

c propanamide

14.4 High-performance liquid chromatography

TY 14.4.1 5.0 μg mL^{-1}

CSA: Making the grade

1 Since the large peaks at about 15 and 18 minutes in the pure olive oil can are also present in the suspect shipment, it is likely there is some olive oil, as well as other substances, present in the shipment.

2 Oil is insoluble in water. An eluent which is non-polar, such as hexane, or partially polar is required.

3

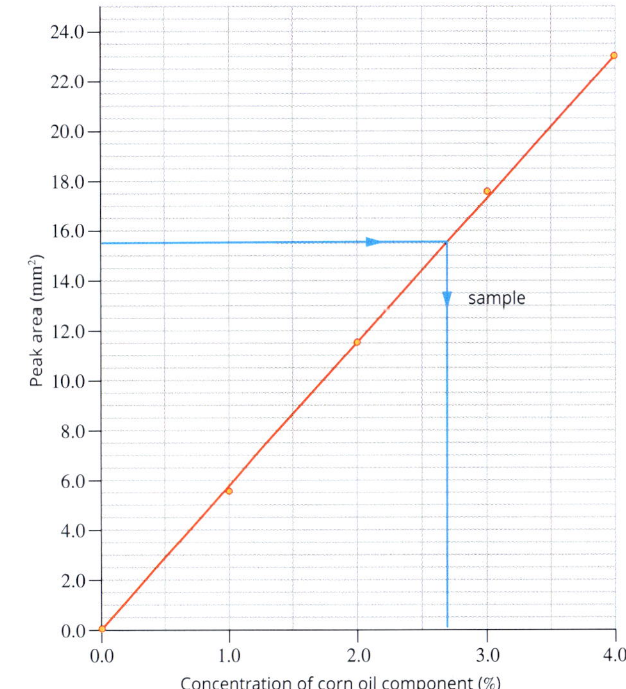

A line can be drawn on the calibration curve from the peak area of the sample (15.5 mm^2) to find the corresponding concentration on the graph: 2.7%.

Key questions

1 The mobile phase and the stationary phase. Temperature, flow rate and pressure will also affect R_t.

2 a retention time b retention time c peak area

3 An HPLC instrument does not directly produce measurements of concentration. R_t indicates the identity of components in a mixture. A calibration curve can then be used to determine the concentration of the component.

4 lowest to highest: C, A, B

5 a

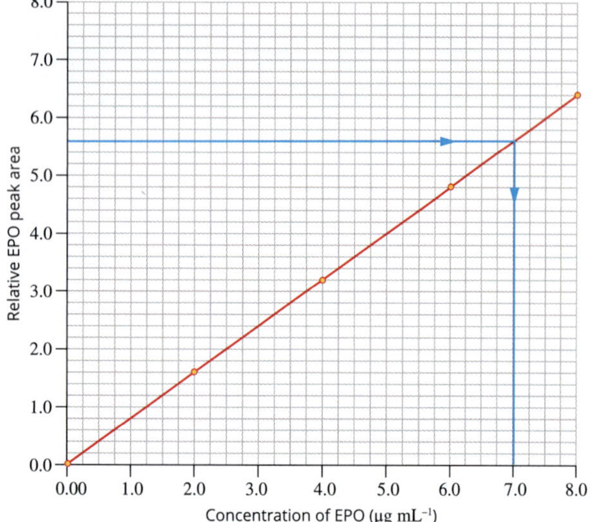

b 7.0 μg mL^{-1}

6 smallest to largest: A, B, C, D

14.5 Determination of molecular structure by spectroscopy

TY 14.5.1 methyl propanoate

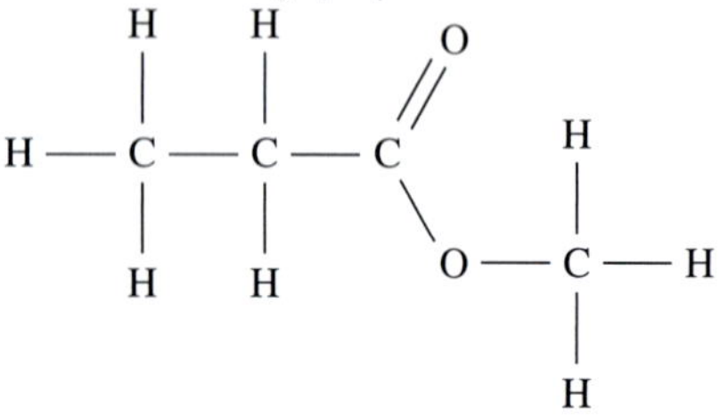

Key questions

1 The spectrum of a reference sample is compared to the spectrum of the unknown organic compound. If the spectra match, the unknown organic compound is pure. If there are any additional or missing peaks, the unknown organic compound is not pure.

2 a IR spectroscopy would not provide sufficient information, as both gases have the same functional groups.

b Proton NMR spectra would identify the gas, as there is one hydrogen environment for ethane and two hydrogen environments for propane.

c Carbon NMR spectra would identify the gas, as ethane has one carbon environment and propane has two carbon environments.

3 Petrol is a mix of many different organic compounds, so the proton NMR spectrum will be very complex to analyse. A proton NMR database will compare and match the petrol spectrum with individual organic compounds using chemical shift, splitting and integration values.

4 2,2-dimethylpropane

5 butan-2-one:

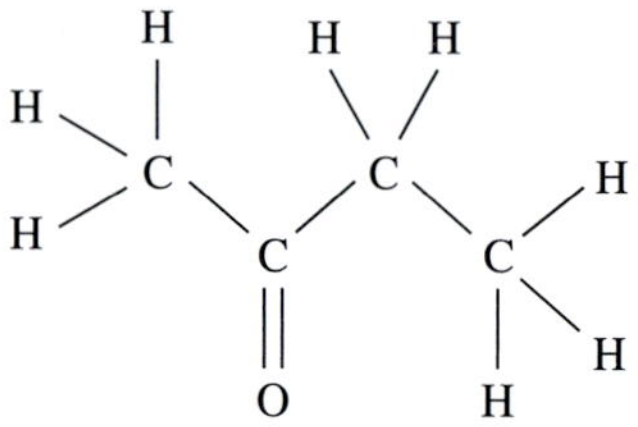

6 a The peak at 910 cm^{-1} is present in sunflower oil, but not in olive oil.

b If the olive oil was adulterated with sunflower oil, there would be a 910 cm^{-1} peak on the IR spectrum. If it is pure olive oil, there would be no peak at 910 cm^{-1}.

Chapter 14 review

1	B.	2	B.	3	A.
4	D.	5	C.	6	D.

7 a alkane = C_4H_{10}

b alkane + one oxygen = C_3H_6O

c alkane + two nitrogen = $C_2H_6N_2$

8 a 29 b 44 c C_3H_8 (propane)

9 29

10 a i 98 corresponds to the molecular ion, which includes two ^{35}Cl isotopes. It has the formula $C_2H_4{}^{35}Cl_2$.

ii 100 corresponds to the molecular ion with one ^{35}Cl and one ^{37}Cl isotope. It has the formula $C_2H_4{}^{35}Cl^{37}Cl$.

iii 102 corresponds to the molecular ion with two ^{37}Cl isotopes with the formula $C_2H_4{}^{37}Cl_2$.

b The most abundant Cl isotope is ^{35}Cl, therefore the molecule with two ^{35}Cl isotopes has a higher peak.

c *M*(base peak) = 62 $^{35}ClCCH_3$

11 a C–H b O–H

12 from lowest to highest: C–Br, C–O, C–H, O–H

13 a C=O has a sharp absorption band at 1680–1740 cm^{-1}; O–H has a broad absorption at 2500–3300 cm^{-1}.

b N–H has an absorption band at 3300–3500 cm^{-1}.

c C–O has an absorption band at 1050–1410 cm^{-1} and C=O has an absorption band at 1720–1840 cm^{-1}.

d C=O has an absorption band at 1680–1850 cm^{-1}.

14 TMS, tetramethyl silane ($Si(CH_3)_4$), is used as a reference in NMR spectroscopy. The location of a signal in the spectrum of a sample is compared to the TMS signal and is called the chemical shift. The chemical shift is measured in units of ppm.

15 a 4 b 1 c 4 d 5

16 Compared with conventional column chromatography, HPLC is more sensitive, faster, resolves components better and it is able to detect colourless components readily.

17 a 4 components.

b Solid samples are dissolved in a suitable solvent. The liquid sample is injected into the top of an HPLC column. The stationary and mobile phases are chosen to achieve a good separation of the components in the sample. The sample components alternately adsorb onto the stationary phase and then desorb into the solvent as they are swept forward. The time taken to exit the column increases if the component strongly absorbs onto the stationary phase and has a low solubility in the mobile phase.

18 a Retention time for peak 1 = 3.5 min
Retention time for peak 2 = 4.8 min
Retention time for peak 3 = 8.0 min
Retention time for peak 4 = 10.5 min

b Caffeine molecular ion peak will be $m/z = 194$
Mass spectrum for peak 1 is caffeine as it has a molecular ion peak at $m/z = 194$

c Soft drinks are a complex mixture, therefore, it is possible another molecule with similar polarity as caffeine has the same retention time of 3.5 min. There are two common methods to determine whether the component in the eluent is pure caffeine or not:
Method 1: The scientist can analyse the eluent on an IR, mass or NMR spectrometer and compare the spectra with a pure caffeine standard. Any additional or missing peaks on the IR, mass or NMR spectra will confirm the eluent is not 100% caffeine.

Method 2: The scientist can change the mobile phase of the HPLC to alter the polarity sufficiently to separate the components at 3.5 min and compare their spectra with the spectra of pure caffeine standard.

19 a

Group	Shift (ppm)	Splitting	Peak area
$-CH_3$	0.9	triplet	3
$-CH_2-$	1.3	septet	1

b

Group	Shift (ppm)	Splitting	Peak area
$-CH_3$	0.9	triplet	3
$-CH_2-$	1.5	quartet	2

c

Group	Shift (ppm)	Splitting	Peak area
$-CH_3COO$	2.0	singlet	3
$-\mathbf{CH_2}OH$	3.3	quartet	2
$-CH_2O\mathbf{H}$	9	singlet	1

20 a Both spectra contain broad absorption bands at 3200–3600 cm^{-1}, which correspond to the expected absorbance by O–H bonds, together with the C–H absorbance around 2850–3090 cm^{-1}. The spectra also contain peaks at 1700–1750 cm^{-1}, which correspond to the absorbance by C=O bonds.

b The two spectra have different peaks in the fingerprint region, so cannot be of the same molecule.

21 serine, threonine, valine

22 a Obtain chromatograms of the blood sample and of standard solutions of testosterone. Use R_t values to identify the testosterone peak on the chromatogram of blood. Measure the testosterone peak areas. Construct a calibration curve and mark the blood testosterone peak area on it. Determine the testosterone concentration in the blood sample.

b 48.0 ng L^{-1}

23 The computer would look for specific functional group and fingerprint region peaks that would match reference samples stored in a database. The reference samples must include explosives and illegal drugs, so that the computer can check for the presence of these compounds.

24 a methyl ethanoate **b** ethyl methanoate

c propanoic acid

25 propanoic acid

26 The molecule is 1-chloro, 2,2-dimethylpropane, $CH_2ClC(CH_3)_2CH_3$.

27 a The scientists must analyse a reference sample of the organic compound on the HPLC-MS to determine the retention time and obtain a mass spectrum.

b The food sample contains hundreds of different compounds in a complex mixture. Each of these different compounds will form fragment ions. It will be difficult for the scientist to determine whether compound X is present in the mass spectrum due to hundreds of fragment ion peaks.

28 Two possible structures can exist.

Option 1:

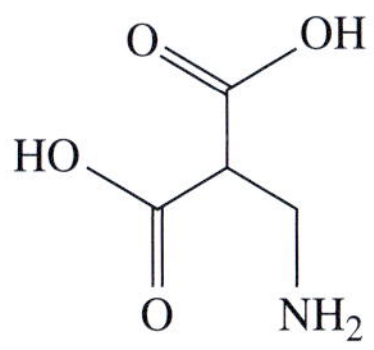

O OH

HO

O NH_2

Option 2:

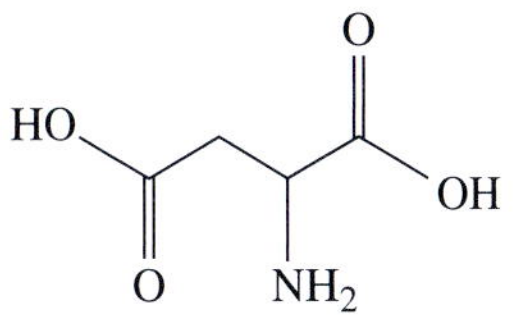

O

HO

OH

O NH_2

To decide which structure is the correct one, consider the carbon atom bonded to the $-NH_2$ group. In structure 1, it is $-CH_2-$. In structure 2, it is –CH–. For both the proton and carbon NMR spectra, the chemical shifts of the hydrogen and carbon in $-CH_2-$ are lower than –CH–. The $-NH_2$ would pull electrons away from the carbon and hydrogen more than the carboxylic acid group because of the stronger electronegativity of nitrogen.

The unknown compound is structure 2, aspartic acid.

Chapter 15 Medicinal chemistry

15.1 Medicinal ingredients from plants

CSA: Investigating the uncha plant

1 Any three of: hydroxyl, OH; carboxyl, COOH; ester, R-COO-R'; carbon–carbon double bond, C=C; phenyl ring, C_6H_5 or ether, –O–.

2 The structure is very complex. It is easier to continue extracting it from plants than to try and find a pathway to build the molecule.

3 The ingredients need to be extracted, then separated, then identified and then tested.

4 The climate and vegetation in different parts of Australia varies. Different plants are available to different peoples.

Key questions

1 Methods such as: blending, which breaks down the cell structure of the plants, and heat, which also helps break down plant structures.

2 Plant extracts are never pure substances. Each active ingredient needs to be isolated so that its properties can be studied without interference from other substances. Chromatography is likely to be the most successful technique that can be used to separate the components.

3 The structure of the active ingredient usually needs to be determined to decide on the best ways of producing it commercially. The molecule would need to be tested to establish its efficacy and how it might be administered. The storage life and handling procedures are relevant concerns. It would need to be determined if there are any negative side-effects to the medication.

4 The polarity of the solvent used should match that of the molecule extracted. Therefore, both molecules need to be inspected to judge their degree of polarity. The presence of highly electronegative atoms such as oxygen and nitrogen will make the molecule polar. On large molecules, such as digitoxin, several oxygen atoms are needed to compensate for the size of the molecule. Both molecules contain some oxygen, but will not be highly polar as the number of oxygen atoms is limited.

5 a Steam distillation involves prolonged heating at temperatures of at least 100°C. For this method to be successful, the extracts need to be relatively stable at these temperatures.

b Oil and water do not mix, making them easy to separate. A piece of equipment called a separating funnel can be used for this purpose.

6 a 176

b Likely absorptions include: a broad peak around 3400 cm^{-1} due to –OH (alcohol), broad band around 3000 cm^{-1} due to –OH (acid) and a peak around 1750 cm^{-1} for C=O.

c Vitamin C has several hydroxyl groups. When protons are part of a hydroxyl group they rarely produce splitting of peaks in proton-NMR.

7 a The *m/z* ratio of the molecular ion peak will be 151, matching the molecular formula of $C_8H_9NO_2$.

b 3300 cm^{-1} broad band for -OH (alcohol), 1700 cm^{-1} for C=O bond and 3500 cm^{-1} for N-H bond.

c Ethanol is a polar solvent due to the presence of the –OH group, but not as polar as water. The functional groups in paracetamol will similarly make it relatively polar. The molecules will be soluble in each other due to similar polarity.

15.2 Proteins

1 a tertiary b secondary c primary d quaternary

2 The primary structure bonding involves covalent bonds in the peptide linkages, i.e. bonds between carbon and the nitrogen atoms. The secondary structure is caused by hydrogen bonds between different parts of the peptide linkages (amide) in the protein chain.

3 a The tertiary structure is related to the overall three-dimensional shape of the protein. The shape of the protein needs to be appropriate for its function.

b the R groups

4 a

```
        H    H                              H    H
H       |    |       O                      |    |       O
  \     |    |     //                       | +  |     //
   N----C----C----C                  H------N----C----C
  /     |    |     \                        |    |     \
H       |    |      O                       |    |      O-
        |    C       \                      H    |
        |              H                         |
  H-----C-----H                            H-----C-----H
        |                                        |
        H                                        H
```

b i

```
        H    H
        |    |       O
      + |    |     //
H-------N----C----C
        |    |     \
        |    |      O
        H    |       \
    H-------C----H     H
             |
             H
```

ii

```
             H
H            |       O
  \          |     //
   N---------C----C
  /          |     \
H            |      O-
             |
     H-------C-------H
             |
             H
```

5 a α-helix

b α-helices result from hydrogen bonding in different regions of the amino acid sequence. In this case, the hydrogen bonds arise due to attraction between the partial positive charge on the H of a –N–H bond in a peptide link with the partial negative charge on the O of a –C=O bond of a peptide link four amino acid units along the chain.

6 B, C and D.

7

Structure of R group	Is the R group polar or non-polar?	Is the R group acidic, basic or neutral?
$-CH(CH_3)_2$	non-polar	neutral
$-CH_2COOH$	polar	acidic
$-CH_2C_6H_5$	non-polar	neutral
$-(CH_2)_4NH_2$	polar	basic

15.3 Enzymes

CSA: Burke and Wills and thiaminase

1 a Like most enzymes, the rate of reaction is low at low temperatures. It rises as the temperature rises until it reaches an optimum, and then the rate falls away quickly if the temperature continues to increase.

b As the temperature rises towards its optimum, the rate increases as the frequency of collisions is increasing. Above the optimum temperature the rate drops due to denaturation.

2 The optimum temperature for many enzymes is around body temperature of 37°C. The optimum temperature for thiaminase is significantly higher than 37°C (between 60 and 80°C).

3 The human body requires thiamine (vitamin B1) for digestion of carbohydrates. The presence of thiaminase causes the level of thiamine to drop. Therefore digestion will be better if there is no thiaminase.

Key questions

1 Part of the structure of an enzyme is a region known as the active site. It is at this position that the substrate (reacting species) bonds to the enzyme, weakening the bonds in the substrate. The reaction can now occur with a lower activation energy.

2 At lower temperatures (20°C to 37°C), the rate of reaction would be slow, but it would gradually increase as the temperature increased. The rate would be greatest at 37°C. Above 37°C the rate would decrease rapidly because the enzyme would be denatured by the higher temperatures.

3 A change in pH can reduce enzyme activity because the different pH affects the structure of the protein and changes its intermolecular forces. In comparison, a change in pH is likely to have less (or no) effect on the ability of an inorganic catalyst to operate as it is unlikely to change the structure.

4 When an enzyme is denatured it loses its unique shape, but it retains its structure as a protein. The primary structure is not affected, but the quaternary, tertiary and secondary structures are disrupted.

5 a At high temperatures, the enzyme is denatured. This means that the structure of the protein is changed, with bonds between different parts of the molecule being broken by the increased energy present. As a result, the shape of the active site is changed and the enzyme is unable to catalyse the reaction.

b Once the enzyme has been denatured, the effect often cannot be reversed. Once the bonds between different parts of the molecule have been broken, the same bonds do not form when the temperature is lowered. So, even though the temperature was returned to the optimum temperature, the shape of the active site had already been changed and the enzyme was no longer able to catalyse the reaction.

c The enzyme is deactivated when it is cooled, but this effect can be reversible as bonds are unlikely to be broken. So when the enzyme is returned to 37°C, the enzyme's activity is no different to an enzyme that had never been cooled.

6 The enzyme activity starts off at a very low (almost zero) value at pH 5 and below. At a low pH, the enzyme is completely denatured and cannot catalyse the reaction. At a higher pH, the enzyme is denatured, but it retains some activity, and this increases as the pH increases, until the activity reaches a maximum value at the optimum pH of 7. The activity then decreases again rapidly above pH 7, because it is once again denatured at a higher pH.

7 Bonds that hold together the secondary and tertiary structure of insulin are sensitive to excessive heat. An increase in temperature can cause these bonds to break, causing the insulin to lose its shape. When a protein loses its shape, it loses its function.

15.4 Optical isomers and inhibition of enzymes

TY 15.4.1

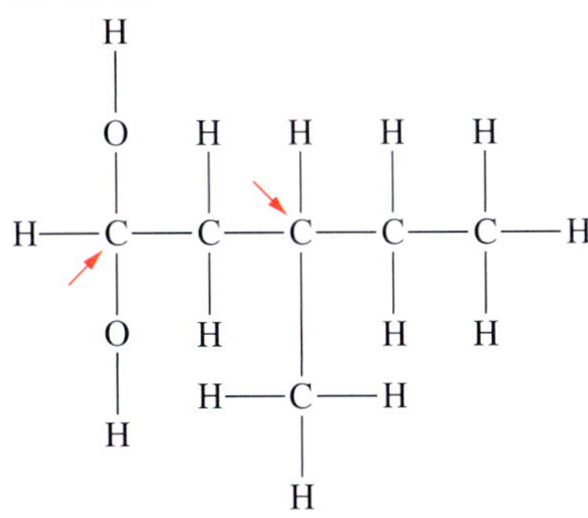

The first carbon atom from the left has the following groups attached:

–H
–OH
–OH
$–CH_2–$

The four groups are not all different so this carbon atom is not a chiral centre.

The third carbon atom from the left has the following groups attached:

–H
$–CH_2CH_3$
$–CH_2CH(OH)_2$
$–CH_3$

The two $–CH_2–$ groups are different as one is attached to $–CH_3$ and the other is attached to $–C(OH)_2H$. This means the four groups are all different so this carbon atom is a chiral centre.

Key questions

1 A carbon atom that is attached to four different groups is a chiral centre.

2 options **b** and **d**

3 **a** true **b** false **c** false

4 **a** the active site of the enzyme
b some part of the enzyme other than the active site

5 **a** 0
b 2

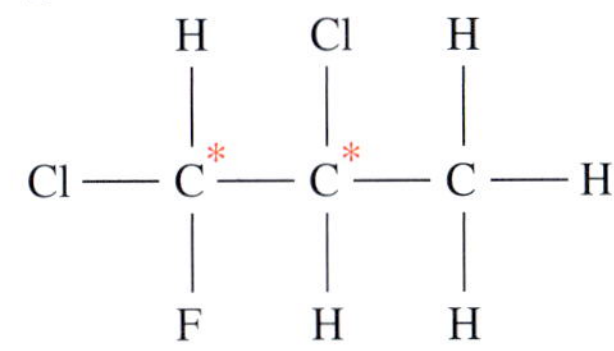

c 2

d 1

6 The enantiomer is shown as a mirror image of the original molecule. The enantiomer is rotated to put the hydrogen and methyl groups in the same arrangement, and the bromine and chlorine atoms are inverted, showing it is not superimposable.

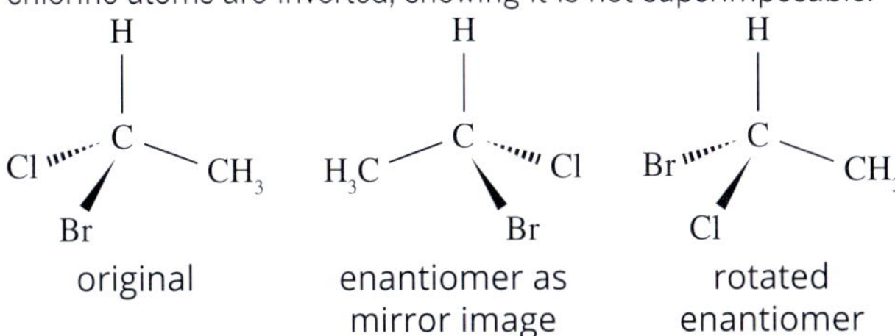

7 The two molecules below are structural isomers (butan-1-ol and butan-2-ol). They have the same molecular formula, but the functional group is in a different position.

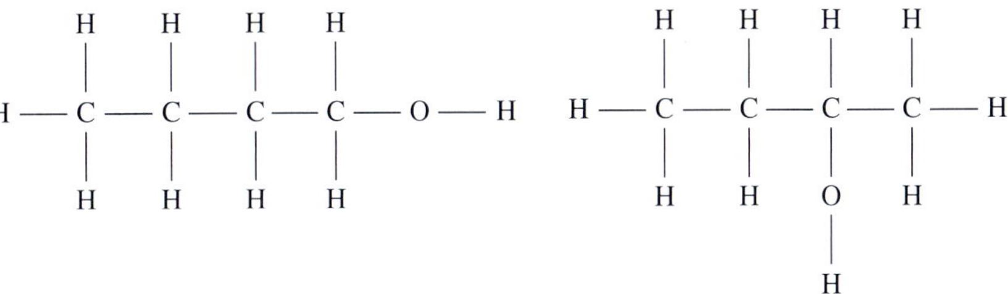

The only form of butanol that can have enantiomers is butan-2-ol, where the third carbon shown below is a chiral carbon. The two enantiomers are optical isomers of each other.

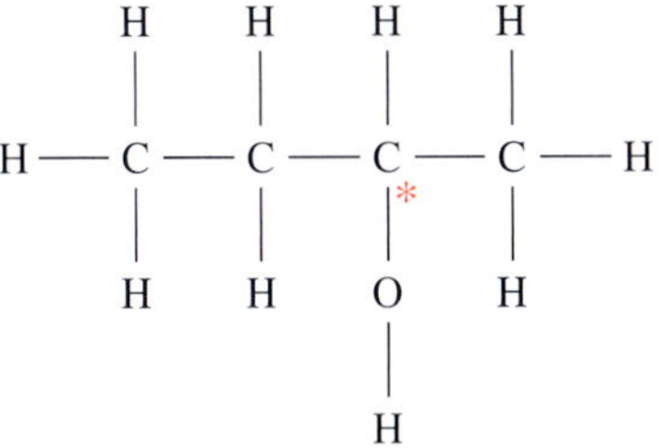

8 *R*-ketamine might:

- be safe but less effective
- effective at treating something else
- harmful.
- be equally effective
- be more effective.

Chapter 15 review

1 D. 2 C. 3 B.

4 D. 5 A.

6 a In steam distillation, the plant leaves are placed in a flow of steam. The heat of the steam breaks the leaves down and carries the active ingredients from the leaves. The active ingredients will usually condense with the steam.

b Molecules should be thermally stable and volatile to be extracted by steam and it is also helpful if they are non-polar so they form a separate layer to the water.

7 a both inorganic catalysts and enzymes

b enzymes

c enzymes

d both inorganic catalysts and enzymes

e both inorganic catalysts and enzymes

8 a Enantiomers are optical isomers. They are mirror images of each other and cannot be superimposed.

b Competitive enzyme inhibitor: a molecule with a similar structure to a substrate that allows it to compete with the substrate for the active site of an enzyme.

c Substrate: a reactant in a reaction catalysed by enzymes.

d Lock-and-key model: model proposing that an enzyme has an active site to suit the shape of a particular substrate, allowing the enzyme to catalyse a reaction of the substrate.

e Zwitterion: a dipolar ion formed when the amino and carboxyl groups in an amino acid or polypeptide are both charged.

f Active site: the location on an enzyme where a substrate can attach. It is usually a hollow or cavity in the protein structure.

9 a covalent bond/peptide bond/amide bond

b hydrogen bond

c α-helix

10 a true b true c false d false

e true f true g true

11 High pH will favour the formation of a negatively charged ion.

12

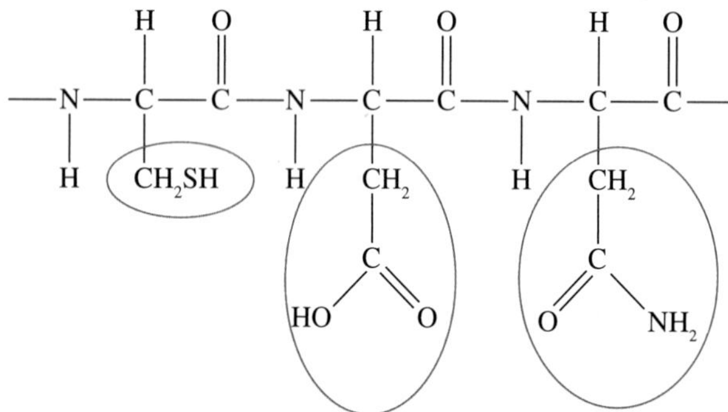

13 a Enzymes in the body include:

- pepsin, which hydrolyses peptide bonds of certain amino acids
- lactase, which breaks down the sugar lactose in the small intestine
- salivary amylase, which breaks down polysaccharides in the mouth.

Many other answers are possible.

b Almost all the chemical reactions occurring in living organisms are controlled by enzymes. Enzymes speed up the reactions that are essential for life processes by as much as 10^{10} times; reactions that do not contribute to the functioning of a creature are not catalysed and occur at much slower rates.

c The shape and functional groups in the active site of the enzyme allow it to bind only with certain substrates, so that only a specific reaction is catalysed. In a similar way, a lock will only open using a key of a certain shape.

14 The second molecule containing sulfur atoms in place of oxygen atoms will be much less polar and suited to extraction by the non-polar solvent, hexane.

15 a

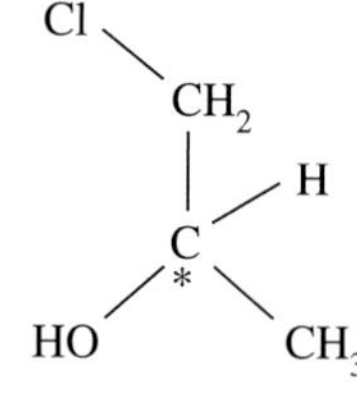

b

Cl
CH₂
H
C*
HO
CH₃

c

CH₃
Br
OH
H
C*
C*
H
CH₂
CH₃

16 four

17

H₃C
CH₃
CH
H₃⁺N — C — COO⁻
H

18 The catalytic property of the enzyme can be destroyed by changing its shape—in a process called denaturation. This can be done by heating the pineapple. Alternatively, canned pineapple can be used to make this dessert because the fruit is heated during the canning process.

19 a i carboxyl ii amine

b A zwitterion is a molecule that contains positive and negative charges, but has no charge overall.

c i ii iii

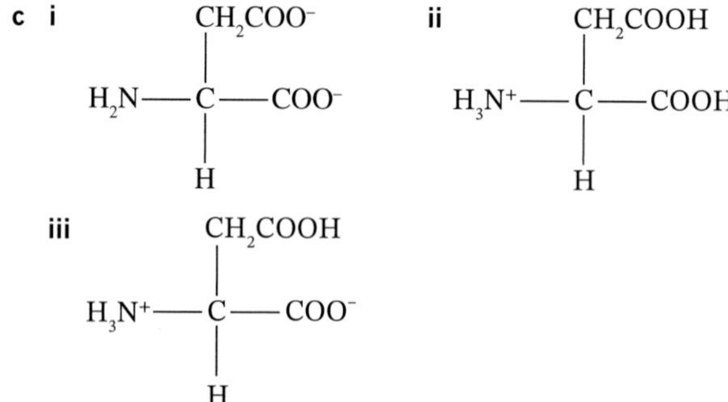

Either carboxyl protons may be lost.

20 a Competitive enzyme inhibitors are molecules with a similar shape to the substrate. They can bond to the active site of the enzyme, blocking the normal substrate. The rate of reaction of the substrate drops as the percentage of active sites blocked increases. The shape of the substrate is evident on the diagram – the inhibitor needs to be a similar shape to this.

b Non-competitive enzyme inhibitors attach themselves to the enzyme, but not on the active site. Their attachment causes enough of a change to the active site to cause it to no longer catalyse the substrate reaction.

21

Bond type	Required components in R group	Amino acid examples
hydrogen bonds	contains –O–H, –N–H and –C=O	serine, threonine, histidine
dipole–dipole interactions	any polar group such as those containing –S–H, –O–H or –N–H	cysteine, asparagine
ionic interactions	contains $–NH_3^+$ and another group that contains –COO–	aspartic acid, glutamic acid, lysine, arginine
covalent cross-links	cysteine side groups react to form a disulfide bridge (–S–S–)	cysteine
dispersion forces	any non-polar group	alanine, leucine, phenylalanine

22 a Procaine contains an amine group, an ester group and a phenyl $–C_6H_4$ ring.

b Possible answers include: 1700 cm^{-1} due to C=O bond, 3400 cm^{-1} due to N–H bond, 3000 cm^{-1} C–H bond

c Procaine has no chiral carbon atoms, so will not have enantiomers.

23 a One method chemists use to distinguish between optical isomers is to subject them to polarised light. The enantiomers will rotate the light in the opposite direction to each other.

b It cannot be assumed that both enantiomers will have the same impact and be as effective in the body. Many enantiomers need to be separated before they are sold commercially.

24 a The protein in milk is denatured.

b The R groups with carboxyl or amine groups can form salt bridges in the protein. This is an important aspect of their tertiary structure. The addition of acid or base will disrupt these salt bridges and consequently the shape of the enzyme.

25 a Denaturation occurs when there is a change to the structure and shape of an enzyme, which prevents it from functioning. The enzyme is said to be 'denatured'.

b Denaturation usually involves disruption of the bonds that hold the enzyme's protein chain in a particular shape, followed by unfolding of the chain. The unfolded chains will often clump together. It is the tertiary and secondary structures that are disrupted.

c No, the covalent bonds are not broken by a slight increase in temperature.

d The overall three-dimensional shape of an enzyme is known as its tertiary structure. Since enzyme action depends upon the existence of a suitable hollow or cavity within the molecule, the tertiary structure is crucial to the enzyme's operations.

26 A: optimum temperature, with the fastest reaction rate or enzyme activity; B: There are less frequent collisions, lower energy reactants, the enzyme has been deactivated; C :enzyme has been denatured. The tertiary structure has changed and the three-dimensional shape of the active site has altered.

27 Methotrexate is a competitive enzyme inhibitor. Its effectiveness depends upon its concentration. At high concentrations it will occupy most active sites, preventing any other reactions. At lower concentrations, the normal substrates will occupy most active sites and be catalysed. The substrate reaction must be helpful to prevent cancers in some way. The main point is that the relative concentrations of inhibitor and substrate are important.

Unit 4 Area of Study 2 How are organic compounds analysed and used?

1 C.	**2** A.	**3** C.
4 B.	**5** B.	**6** A.
7 D.	**8** D.	**9** C.
10 A.	**11** B.	**12** B.
13 A.	**14** C.	**15** D.

16 a Add an aqueous solution of Br_2 to both compounds and shake the mixture. The orange colour due to Br_2 will disappear almost immediately in the hex-1-ene, while the orange colour will take a longer time to fade in the case of hexane and requires the presence of UV light.

b Add an acidified solution of potassium permanganate or potassium dichromate to both liquids. The hexan-1-ol will react with the potassium permanganate and the colour of the solution will change from purple to colourless. If potassium dichromate is used, the colour change will be from orange to green. There will be no reaction with hexane.

c Add a solution of sodium carbonate. With the propanoic acid, a reaction should occur to produce bubbles of a colourless gas (CO_2). There should be no reaction with the propanone.

17 a Spectrum A: ~3000 cm^{-1} is due to C–H, ~1700 cm^{-1} is due to C=O

Spectrum B: ~3000 cm^{-1} is due to C–H, ~3400 cm^{-1} is due to O–H

b Spectrum A is propanone; spectrum B is propan-2-ol.

c Because of the symmetry of the molecule, all six hydrogens are in the same chemical environment and are 'equivalent'.

d Again, because of the symmetry of the molecule the two terminal carbon atoms are in identical environments and hence absorb and emit the same frequency.

e ^1H NMR: three peaks as there are three different H environments ($–C\underline{H}_3$, $–C\underline{H}(OH)–$ and $–CH(O\underline{H})–$)

^{13}C NMR: two peaks as there are two different C environments ($–\underline{C}H_3$, $–\underline{C}H(OH)–$)

f i 58. The molecular mass of propan-2-one (C_3H_6O) = $(3 \times 12) + 6 + 16 = 58$. The peak at 59 would be due to molecules with one 13C isotope.

ii It provides the molecular mass from which the molecular formula can be deduced. The fragmentation pattern can be used to obtain further information about the structure of the molecule.

iii $[CH_3CO]^+$

18 a The vitamin C tablets have a large amount of vitamin C in them, so one tablet would react with too much iodine solution to make a reasonable titration. When the vitamin C tablet is dissolved in 250 mL of water, the aliquots of solution have a smaller amount of vitamin C in them and only require around 20 mL of iodine solution to react.

b Titrations are repeated to ensure that the volume of the titre is accurate. If the titrations are repeated until three concordant (within 0.1 mL) titres can be achieved, then we can be confident of both the precision (a very small range of values) and the accuracy.

c i The volumes of titres are calculated in the table by subtracting the initial volume from the final volume. The volume of trial 2 is 24.00 mL. This is not concordant with the other volumes and so should be disregarded.

ii 22.75 mL

iii 2.26×10^{-4} mol

iv 0.0113 mol L^{-1}

v 498 mg

19 a

Amino acids	Strongest interaction between side groups
threonine and lysine	hydrogen bonds
cysteine and cysteine	covalent bonds (disulfide bridges)
valine and leucine	dispersion forces
aspartic acid and lysine	ionic interactions

b At high temperatures the kinetic energy of the molecule is sufficient to disrupt the bonding between the side groups of amino acid residues (such as valine and leucine, or threonine and lysine), so the tertiary structure is lost. Extremes of pH affects the charges of amino acid residues with acidic or basic side chains (such as aspartic acid and lysine), thus changing the interactions between those amino acids, and resulting in the loss of some of the tertiary enzyme structure.

20 a ethanol

b propanoic acid

c The hydrocarbon chain length in ethanol and propanoic acid molecules are similar, so it is mainly the intermolecular forces, in particular, hydrogen bonding, between the molecules that cause the difference in their boiling points. The hydrogen bonding between carboxyl groups is stronger than those between alcohol groups because dimers form between carboxyl groups (see Chapter 10), but not between hydroxyl groups. The stronger forces result in propanoic acid needing more energy to separate the molecules and so it has a higher boiling point.

d Propanoic acid will react with sodium hydrogen carbonate to produce bubbles of carbon dioxide. This will not occur with ethanol.

Ethanol is a primary alcohol and will undergo oxidation with acidified potassium dichromate, $K_2Cr_2O_7$, or with acidified potassium permanganate, $KMnO_4$. If potassium dichromate is used, during the reaction the solution mixture will change from an orange colour to green. If potassium permanganate is used as the oxidising agent, there will be a change from purple to colourless.

e $CH_3CH_2COOCH_2CH_3(l) + H_2O(l) \xrightarrow{H^+} CH_3CH_2COOH(aq) + CH_3CH_2OH(aq)$

21 Samples of pure compounds generally have a sharp melting point. Mixing a pure compound with a different compound will lower its melting point.

Compound B and compound X are the same compound. The original melting point of B was the same as that of compound X and did not change when a mixed melting point was carried out.

Although the mixed melting point of compound A with compound X was 158°C, the same as that of compound X, this was a decrease from the original melting point of compound A, indicating that compounds A and X were different.

The melting point of compound C was the same as that of compound X. However, when a mixed melting point was carried out, the melting point decreased, indicating the compounds C and X were not the same.

22 a radio waves (radio frequencies)

b The change associated with this is that nuclei with spin 'flip' between low and high energy states corresponding to alignment with the external magnetic field.

c It is used as a reference for determination of chemical shifts.

d i B. High chemical shift is characteristic of an aldehyde and the quartet results from three adjacent H atoms.

ii D. A singlet suggests there are no neighbouring H atoms. The chemical shift is consistent with an adjacent C=O group.

iii E. A quartet indicates there are three neighbouring H atoms. A moderately high chemical shift is consistent with an adjacent O atom.

iv C. A singlet indicates that all H atoms in the molecule are equivalent so no splitting is observed. The chemical shift is consistent with an adjacent halogen (Cl) atom.

v F. A triplet shows there are two neighbouring H atoms and the low chemical shift is consistent with no adjacent atoms such as O, N or a halogen.

vi A. A doublet suggests there is one neighbouring H atom and the chemical shift is consistent with an adjacent C=O group.

23 a Sorbic acid adsorbs most strongly. Its longer retention time indicates that it spends a relatively long time adsorbed onto the stationary phase.

b Only benzoic acid, for which there is a peak in the sample spectrum with the same retention time. There are no peaks in the sample with retention times corresponding to the other two.

c

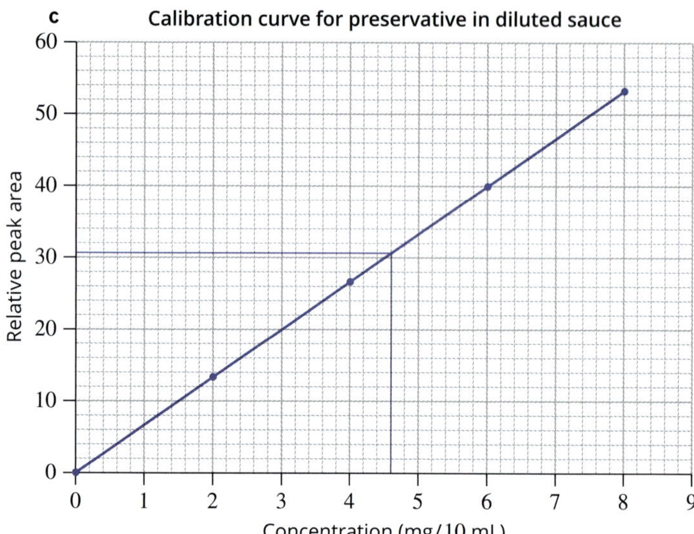

Concentration of preservative in diluted sauce is 4.58 mg/10 mL.

d 305 mg/100 g

24 a The 'active site' of an enzyme is the region on its surface responsible for its catalytic activity. This region has the ability to bind to a specific molecule, or part of a molecule. This molecule is called the 'substrate' and binding to the active site of the enzyme alters the substrate in a way that lowers the activation energy of a reaction in which the substrate participates.

b Like all catalysts, enzymes increase the rate of reaction, but are unchanged after the reaction is complete, so that a single enzyme molecule can catalyse a reaction many times over. A reactant is changed and consumed in the course of a reaction.

c In the lock-and-key model, the active site has a fixed shape that only the substrate, or a close mimic, can fit into and bind. The substrate reacts when it is bound to the active site, and products are made. When the products leave the enzyme, it is available to catalyse the reaction of another substrate.

d The competitive inhibitor would bind to the active site of the enzyme and it would prevent the substrate from binding to that active site. This would render the enzyme unable to catalyse the reaction. Depending on what proportion of the enzymes were affected by this competitive inhibitor, the person could lose the ability to digest lactose.

25 **a, c, d** See diagram.

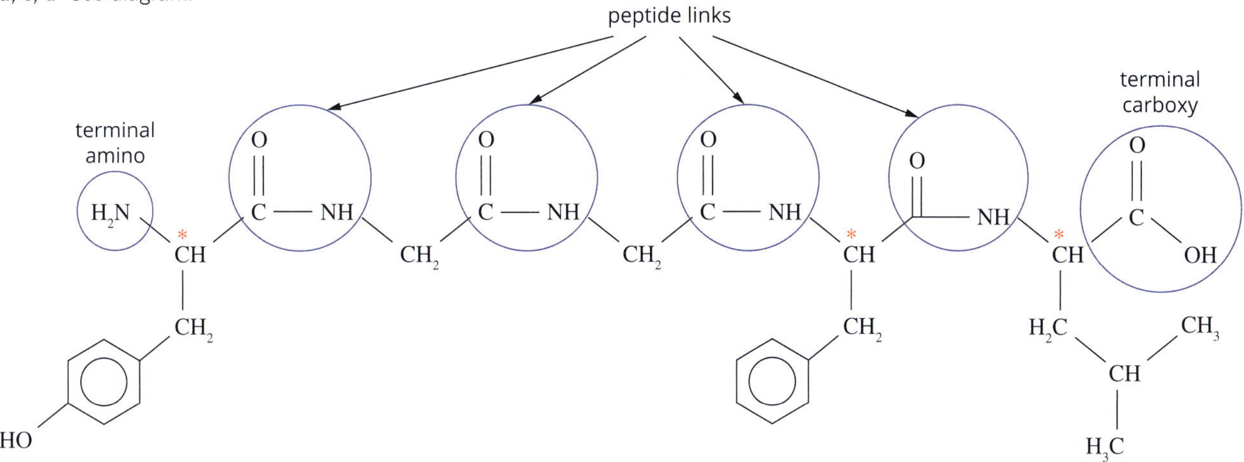

b Tyrosine, glycine (× 2), phenylalanine

26 **a** The primary structure of a protein relates to the formation of peptide linkages between the amino acids as they are linked together in a specific sequence. Since this is a condensation reaction between carboxyl and amino functional groups, water is a by-product. Secondary and tertiary structures do not involve condensation reactions, but instead involve intermolecular forces.

b The secondary and tertiary structures of proteins are largely maintained by non-covalent bonding interactions such as hydrogen bonding, ionic interactions, ion–dipole attractions and dispersion forces, which are generally weaker than the covalent bonding in the peptide linkages maintaining the primary structure. As a result, the structures held together by these weaker attractions are more easily disrupted.

c The bonding that maintains the tertiary structure of a protein includes types of widely varying strength, including some strong covalent bonding in the form of disulfide bridges. Proteins with higher proportions of the stronger bonding types involved in their tertiary structures will tend to be more stable.

27 **a** Fe^{2+} is the limiting reactant and $Cr_2O_7^{2-}$ is in excess.

b 1.65×10^{-4} mol

c 0.432 g

28 **a** **i** Citric acid would be found in the lower aqueous layer. The many polar hydroxyl groups in citric acid would make it very soluble in water.

ii Limonene would be found in the upper diethyl ether layer. A limonene molecule contains only non-polar methyl groups, and would readily dissolve in the non-polar diethyl ether solvent.

b If sodium hydrogen carbonate powder is added to the solution from the lower layer in the separating funnel, bubbles of gas would be observed. This is because citric acid would react with sodium bicarbonate to produce carbon dioxide gas.

c High-performance liquid chromatography could be used to separate the compounds in the same layer as limonene.

Glossary

RED ENTRIES VCE Chemistry Study Design extracts © VCAA (2022); reproduced by permission.

A

absorption band The inverted peaks in an infrared spectrum that represent the frequencies of infrared radiation absorbed by specific bonds.

accepted value A value for a quantity which has been found in published scientific reports, such as journals, or reliable websites, such as those of universities and values which may be calculated using a correctly balanced equation for the reaction and stoichiometry.

accumulator A battery that collects and stores electricity.

accuracy A measurement value is considered to be accurate if it is judged to be close to the true value of the quantity being measured.

accurate If the average of a set of measurements of a quantity is very close to the true or accepted value of the quantity, then the measurement is described as accurate.

achiral Not chiral. An achiral object is superimposable on its mirror image. An achiral molecule does not have optical isomers. The molecule will usually have a plane of symmetry.

acidic conditions When the reactant solution has acid added to it to provide H^+ ions for the reaction.

acidified To add an acid, such as HCl or H_2SO_4, to provide $H^+(aq)$ ions.

activation energy The minimum energy required by reactants for a reaction to occur; symbol E_a. This energy is needed to break the bonds between atoms in the reactants to allow products to form.

active site The location on an enzyme's structure at which a reaction is catalysed. It is usually a hollow or a cavity in the protein structure, where the enzyme binds and interacts with the substrate.

actual yield The mass of product actually obtained during a chemical reaction. This will be less than or equal to the theoretical yield.

addition reaction A reaction in which a molecule bonds to an unsaturated hydrocarbon, forming a single carbon–carbon bond. In this process, two reactant molecules become one.

adsorption The attraction and binding of molecules or particles of one substance to the surface of another.

adulteration The addition or substitution of a lesser value ingredient in a product.

affiliation A close connection (between work associates or social groups), which may affect the beliefs of a writer.

α-helix One type of secondary structure found in proteins. It has a spiral shape and it is maintained by hydrogen bonding between amino acid units in the chain.

aim A statement describing in detail what will be investigated.

alcohol A homologous series of organic molecules that contains the hydroxyl (–OH) functional group.

aldehyde A homologous series of organic molecules which contains the carbonyl functional group, C=O, bonded to a hydrogen on one end of the molecule. Aldehydes can be recognised by the presence of a –CHO group at one end of a compound's condensed structural formula.

aliquot A fixed volume of liquid measured by a pipette.

alkaline cell A commercial electrochemical cell with an alkaline electrolyte that is a moist paste rather than a solution.

alkaline conditions When the reactant solution has an alkali, such as NaOH, added to it to provide OH− ions for the reaction.

alkane A saturated hydrocarbon that contains only single bonds; general formula C_nH_{2n+2}.

alkene An unsaturated hydrocarbon containing one or more carbon–carbon double bonds. Alkenes with one double bond have the general formula C_nH_{2n}.

alkyl group A group obtained by removing a hydrogen atom from an alkane, with a general formula C_nH_{2n+1}; for example, methyl ($-CH_3$). Alkyl groups (also known as alkyl side chains) make up branches in organic compounds.

amide A compound containing the –CONH– functional group. This group forms the link between amino acids in proteins, where it is also called a peptide link.

amide functional group The functional group found in the homologous series of amides. It contains a nitrogen atom covalently bonded to a carbon atom, which has a double covalent bond to an oxygen atom. Amide functional groups include $-CONH_2$ (primary amide) or –CONH– (secondary amide).

amino acid See 2-amino acid.

2-amino acid An organic molecule that has an amino group, a carboxyl group, an H atom and an R group covalently bonded to the second carbon atom in the molecule's carbon chain. The R group consists of different sets of atoms.

amino acid residue The remaining atoms of an amino acid in the peptide or polypeptide chain after a condensation reaction has occurred.

amino functional group The functional group found in the homologous series of amines. It contains a nitrogen atom covalently bonded to two hydrogen atoms, $-NH_2$. Also referred to as amine functional group.

amount The number of particles of a substance present, usually measured in moles.

amylopectin A form of starch that has occasional cross-links between glucose chains.

amylose A linear form of starch.

anaerobic A process that can occur without oxygen.

analogue Technology which involves measuring infinitely variable values. Analogue devices use a pointer or needle on a scale rather than a number display like a digital device.

anode An electrode at which an oxidation reaction occurs.

anti-bumping granules Small, irregularly-shaped stones added to liquids to make them boil more evenly.

artificial photosynthesis Chemical process that uses energy from the Sun to produce a fuel and oxygen gas.

atom economy A method of tracking the atoms in a reaction equation to calculate the mass of the atoms of reactants actually used to form products as a percentage of the total mass of the reactants. Atom economy can be calculated from the formula:

$$\text{atom economy} = \frac{\text{molar mass of desired product}}{\text{molar mass of all products}} \times 100$$

average The average value of a set of values, calculated by dividing the sum of the values by the number of values. Also known as mean.

average titre The average of three concordant titres.

B

bar graph A graph used for organising and displaying discrete data. To construct a column graph, equal width rectangular bars are constructed for each category with height equal to the observed frequency of the category.

base peak The highest intensity peak in a mass spectrum. It is assigned an intensity of 100.

basic conditions See alkaline conditions.

battery A combination of cells connected in series.

battery capacity A measure of the charge, or energy, stored in a battery.

benzene A cyclic hydrocarbon of formula C_6H_6. The molecules contain a six-carbon ring with delocalised electrons. The carbon–carbon bonds are intermediate between single and double bonds.

bias A form of systematic error resulting from the researcher's personal preferences or motivations.

bile Bile is a greenish-brown alkaline fluid that aids digestion. It is secreted by the liver and stored in the gall bladder.

biodiesel A fuel derived from plant or animal matter, consisting of long-chain alkyl esters. Biodiesel is typically made by reacting triglycerides with an alcohol.

bioethanol Ethanol that is made by fermenting the sugar and starch component of plants using yeast.

biofuel A fuel that can be produced from crops or other organic material. Examples of biofuels are ethanol from the fermentation of sugars, methane from digestion in animals, and biogas from plant and animal wastes.

biogas A mixture of gases produced by the decomposition of organic matter in the absence of oxygen.

biomass Organic material from living thigs, such as crop or forest waste.

biomolecule A molecule that is present in living organisms, including macromolecules such as proteins, carbohydrates and fats or oils.

biopolymer A very large molecule that is synthesised by a living organism from many smaller monomer molecules.

bomb calorimeter An insulated container in which a sealed, oxygen-filled vessel is surrounded by a known volume of water. Combustion reactions are carried out in the reaction vessel, and the heat from the reaction is transferred to the surrounding water.

bond energy The quantity of energy required to break a bond.

bond strength The strength of a covalent bond. In general, higher energy bonds have greater bond strength.

β-pleated sheet One of the two types of secondary structure found in proteins. It consists of polypeptide strands bonded laterally. The parallel groups of a polypeptide chain are twisted back on themselves, forming a pleated sheet. It is maintained by hydrogen bonding.

brine A concentrated sodium chloride solution.

brown hydrogen Hydrogen gas produced from the combustion or reactions of fossil fuels.

burette Glassware used for volumetric analysis to transfer a variable volume of solution accurately.

by-product Substance produced in a chemical process in addition to the desired product.

C

C-terminal The end of a protein or polypeptide with a free carboxyl group (–COOH).

calibrated In the context of calorimetry, calibration is the process used to establish the relationship between the temperature change of the water in a calorimeter and the amount of energy being added. Once this relationship has been established, a calorimeter is said to have been calibrated.

calibration In the context of calorimetry, calibration is the process used to establish the relationship between the temperature change of the water in a calorimeter and the amount of energy being added. The calorimeter is then described as having been calibrated.

calibration curve A plot of data involving two variables that is used to determine values for one of the variables. In spectroscopy and chromatography, it is a plot of absorbance versus concentration to determine the concentration of a solution of unknown concentration. The curve is constructed by measuring the absorbance of a set of standard solutions.

calibration factor The calibration factor of a calorimeter is the amount of energy that is required to change the temperature of the water within a calorimeter by 1°C.

calorimeter An instrument designed to measure energy changes in a reaction. It is made up of an insulated container of water in which the reaction occurs, with a stirrer and thermometer to measure the temperature change during the reaction. A lid is an important part of the insulation.

calorimetry The experimental method by which the heat energy released by the combustion of a fuel or a food, or another chemical reaction such as a neutralisation reaction, is measured.

carbohydrate Natuarlly occurring condensation polymers containing carbon, hydrogen and oxygen, which have the general formula $C_x(H_2O)_y$.

carbon neutral A process that absorbs the same amount of carbon as it generates. The carbon dioxide absorbed from the atmosphere by a carbon neutral process compensates for the carbon dioxide produced by the process.

carbon-13 NMR spectroscopy Also called carbon NMR and ^{13}C NMR spectroscopy. A type of nuclear magnetic resonance spectroscopy that investigates the ^{13}C nucleus. It is used to determine the chemical environment of carbon atoms in compounds.

carbonyl functional group A functional group that consists of a carbon atom double bonded to an oxygen atom, –CO–. This group is present in aldehydes, ketones, carboxylic acids, amides and esters.

carboxyl functional group A functional group that consists of a hydroxyl group attached to the carbon of a carbonyl group, –COOH. This group is present in carboxylic acids.

catalysis The increase in the rate of a chemical reaction due to the presence of a catalyst.

carboxylic acid A homologous series of organic molecules that contain the carboxyl functional group (–COOH).

catalyst A substance that increases the rate of a reaction but is not consumed in the reaction. The catalyst provides an alternative reaction pathway with a lower activation energy.

cathode An electrode at which a reduction reaction occurs.

cellular respiration The process through which cells convert sugars into energy. Carbon dioxide and water are typical products of this process.

cellulose A carbohydrate that is a polymer of β-glucose and is the main component of the cell walls of plants.

chemical energy The sum of the chemical potential energy and kinetic energy in a substance. Chemical energy is stored in the bonds between atoms and molecules. The energy results from things such as attractions between electrons and protons in atoms, repulsions between nuclei, repulsions between electrons, movement of electrons, and vibrations and rotations around bonds.

chemical environment Atoms in a molecule that absorb the same amount of energy and produce a single signal in NMR spectrum.

chemical shift The position of a signal in the nuclear magnetic resonance (NMR) spectrum, relative to the signal produced by the tetramethylsilane (TMS) standard. Measured in parts per million, ppm.

chiral A chiral object is not superimposable on its mirror image. A chiral molecule has two optical isomers, called a pair of enantiomers. Chiral molecules do not have a plane of symmetry.

chiral centre A carbon atom that is attached to four different groups in a tetrahedral arrangement.

chromatogram The output of a chromatography procedure. In TLC and paper chromatography, it is the pattern of bands or spots formed on a plate or on paper. In high-performance liquid chromatography (HPLC), it is the graph produced.

circular economy A continuous cycle that focuses on the optimal use and re-use of resources from the extraction of raw materials through to production of new materials, followed by consumption and re-purposing of unused and waste materials.

closed system A system in which only energy is exchanged with the surroundings.

coefficient In a chemical equation, a whole number placed in front of a formula to balance the equation.

coiling When sections of a protein form a spiral shape owing to hydrogen bonding.

collision theory A theoretical model that accounts for the rates of chemical reactions in terms of collisions between particles occurring during a chemical reaction.

column chromatography A chromatographic technique in which the stationary phase is in a column. Examples include gas chromatography and high-performance liquid chromatography.

column graph *See bar graph.*

combustion A rapid reaction with oxygen accompanied by the release of large amounts of heat; also called burning.

competitive enzyme inhibitor Competitive inhibition occurs when molecules very similar to the substrate molecules bind to the active site of an enzyme, preventing binding of the substrate.

complete combustion A hydrocarbon undergoes complete combustion with oxygen at high temperatures when the only products are carbon dioxide and water.

component The chemicals in a mixture. The components can be separated by chromatography.

concentration A measure of how much solute is dissolved in a specified volume of solution.

concentration fraction The ratio of concentrations of products to reactants in a reversible reaction, as expressed in the equilibrium law. It can also be called the reaction quotient.

concentration–time graph A graph in which time is the independent variable (on x-axis), and concentration is the dependent variable (on the y-axis).

conclusion A short statement based on evidence that summarises the findings of the investigation

concordant titres A set of titres that vary within a narrow range, usually within 0.10 mL from highest to lowest of one another.

condensation polymerisation Formation of a polymer by condensation reactions.

condensation reaction A reaction in which two molecules link together by eliminating a small molecule such as water.

condensed formula A chemical formula that summarises the structural formula of a compound in a single line of test. Also known as condensed structural formula. See semi-structural formula.

condenser Glassware used in distillation to cool hot vapours. Cold water circulates through the distillation apparatus, condensing the hot vapours.

conjugate redox pair An oxidising agent (a reactant) and the reducing agent (a product) that is formed when the oxidising agent gains electrons. In this case, the oxidation number of the oxidising agent decreases, e.g. Cu^{2+}/Cu. Alternatively, it may be a reducing agent (a reactant) and the oxidising agent (a product) that is formed when the reducing agent loses electrons. In this case, the oxidation number of the reducing agent increases.

contaminant An impurity present in a mixture.

continuous data Any number value within a given range.

controlled experiment A carefully designed experiment in which only one independent variable is changed at a time to see the effect on the dependent variable.

controlled variable A variable that is kept constant during an investigation.

coulomb The unit of charge; symbol C.

critical thinking The objective analysis and evaluation of information in order to form a judgement.

cyclic molecule A molecule in which the atoms are bonded to form a ring.

D

database A collection of 100 000 or more reference sample data and spectra.

deactivated When an enzyme is operating below its optimum temperature and has a lower rate of reaction than at the optimum temperature.

decolourise Describes the loss of colour associated with the reaction between bromine and an alkene.

degree of unsaturation The number of double bonds or ring structures in an organic molecule. Also referred to as the index of hydrogen deficiency (IHD).

denaturation Process in which an enzyme undergoes a change in its three-dimensional shape so that it is unable to function as a catalyst. The shape of its active site is changed.

dependent variable A variable that may change in response to a change in the independent variable, and is measured or observed.

desorption The breaking of the bonds between a substance and the surface to which the substance is adsorbed.

dietary fibre The indigestible part of food from plants, such as vegetables, fruits, grains, beans and legumes. It is composed of a mixture of different chemicals, including carbohydrates such as cellulose.

digestion Digestion involves the breakdown of large insoluble molecules into smaller, soluble molecules.

dilute Add more solvent to a solution to decrease the concentration of solute.

dilution factor The ratio of the final volume to the aliquot volume

dimer A chemical structure formed from two similar sub units; for example, paired carboxylic acids that form hydrogen bonds with each other.

dipeptide An organic molecule that has been produced by the condensation reaction between two amino acids.

dipole A molecule that has two oppositely charged ends.

dipole–dipole attraction A form of intermolecular force that occurs between polar molecules where the positively charged end of one molecule is attracted to the negatively charged end of another molecule.

disaccharide A carbohydrate consisting of two monosaccharides joined by a glycosidic (or ether) link.

discharge When a cell or battery discharges chemical energy that is transformed into electrical energy.

discrete data Values that can be counted or measured, but which can only have certain values.

dispersion force A very weak force of attraction between molecules due to temporary dipoles induced in the molecules. The temporary dipoles are the result of random fluctuations in the electron density.

distillate The vapour that condenses back to a liquid and is collected after distillation.

distillation A separation technique based on the different boiling temperatures of the components of a mixture. The technique involves evaporation and subsequent condensation of the components.

dynamic equilibrium A point in a chemical reaction when the rate of the forward reaction is equal to the rate of the reverse reaction.

E

efficiency A comparison of the product or energy output in a chemical reaction to what is theoretically expected.

electric charge The charge carried by particles such as electrons or ions.

electrochemical cell A device that converts chemical energy into electrical energy, or vice versa.

electrochemical series A list of half-equations, written as reduction reactions, arranged in order so that the strongest oxidising agents are on the top left side of the list.

electrode A solid conductor in a half-cell at which oxidation or reduction reactions occur.

electrolyser A system that uses electricity to break water into hydrogen and oxygen via electrolysis.

electrolyser stack An array of electrolysers designed to produce commercial volumes of hydrogen.

electrolysis A process that produces a non-spontaneous redox reaction by the passage of electrical energy from a power supply through a conducting liquid.

electrolyte A chemical substance that conducts electric current as a result of dissociation into positively and negatively charged ions, which migrate toward the negative and positive terminals of an electric circuit.

electrolytic cell A cell in which electrolysis can occur.

electromagnetic spectrum The range of electromagnetic radiation. It is often represented in order of wavelength, from low energy radio waves to high-energy gamma rays.

electromotive force (emf) The 'electrical pressure' between two points in a circuit, such as the electrodes of an electrochemical cell; a measure of the energy given to electrons in a circuit.

electroplating A process that uses electrolysis to deposit a layer of metal on the surface of another material. This process is also called deposition.

eluent The solvent that carries the components and passes through a chromatography column.

emulsion A suspension of small droplets of one liquid in another.

enantiomer One of a pair of molecules that are mirror images of each other that cannot be superimposed. Enantiomers are optical isomers.

end point The point during a titration when the indicator changes colour.

endothermic A reaction that absorbs energy from the surroundings; ΔH is positive.

energy content The amount of energy per gram or per 100 g, or per mole (if a nutrient is a pure substance), that a food or fuel can supply. Units may be kJ g^{-1}, kJ/100 g or kJ mol^{-1}.

energy converter A device that changes energy from one form to another.

energy density The energy released when 1 litre of fuel undergoes complete combustion.

energy efficiency Tthe percentage of total energy that is converted into the desired form of energy.

energy profile diagram A diagram that shows the energy changes during the course of a reaction.

energy transformation The process of changing one form of energy to another. In calorimetry, chemical energy is transformed to heat energy. During electrical calibration of a calorimeter, electrical energy is transformed to heat energy.

energy transformation efficiency The percentage of energy from a source (input energy) that is converted to useful energy.

energy value The amount of energy per gram, or per 100 g, that a food can supply.

enthalpy Heat content. The sum of the chemical potential and kinetic energies in a substance; symbol H.

enthalpy change The difference in the total enthalpy of the products and the total enthalpy of the reactants; symbol ΔH. Also known as heat of reaction.

enthalpy of combustion, ΔH_c The enthalpy change that occurs when one mole of a compound is burned completely in oxygen, a negative value: symbol ΔH_c.

enthalpy of solution The change in enthalpy that results when one mole of solute is dissolved in a solvent.

enzyme A protein molecule that functions as a catalyst of a specific biochemical reaction. A catalyst increases the rate of a reaction.

equilibrium When a chemical reaction reaches equilibrium, the quantities of reactants and products in the reaction remain unchanged. The rates of the forward and reverse reactions are equal.

equilibrium constant The value of the concentration fraction when equilibrium is reached; symbol K. The equilibrium constant for the chemical equation:

$$aW + bX \rightleftharpoons cY + dZ$$

is given by the expression:

$$K = \frac{[Y]^c[Z]^d}{[W]^a[X]^b}$$

equilibrium law The equilibrium law for the chemical equation:

$$aW + bX \rightleftharpoons cY + dZ$$

is given by the expression:

$$K = \frac{[Y]^c[Z]^d}{[W]^a[X]^b}$$

where K is a constant at a particular temperature.

equilibrium yield The amount of product obtained when a chemical reaction reaches equilibrium.

equivalence point The point during a titration when the reactants in solution are present in stoichiometric proportions, i.e. in the mole ratio shown by the reaction equation.

equivalent A term used in NMR spectroscopy to describe atoms that require the same amount of energy to change spin state, and hence have the same chemical shift.

ester The name of the homologous series of molecules that contain the ester (−COO−) functional group.

ester functional group The functional group that is the result of the reaction between an alcohol and a carboxylic acid. It consists of a carbonyl group bonded to an oxygen atom bonded to another carbon atom, −COO−.

ester link The link that is the result of the reaction between a hydroxyl and a carboxyl functional group. It consists of a carbonyl group bonded to an oxygen atom bonded to another carbon atom (−COO−).

esterification reaction The chemical reaction between an alcohol and a carboxylic acid to form an ester as the main product.

ether functional group An oxygen atom that has single bonds to two carbon atoms.

ethics Moral beliefs and rules about right and wrong.

excess reactant A reactant that is not completely consumed in a chemical reaction. There is a quantity of this reactant remaining after the other reactant has been completely consumed.

exothermic An exothermic reaction releases energy to the surroundings; ΔH is negative.

exponential relationship Variables that are exponentially proportional to each other will produce a curved trend line when graphed.

extent of reaction The relative amounts of products compared with reactants. The extent of a reaction is indicated by the value of the equilibrium constant.

external circuit The section of an electrochemical cell in which electrons move. This section of the circuit will include the wires attached to the electrodes.

extrapolating Extending a line that has been constructed in response to a trend observed in a set of data to values beyond that data.

F

faraday The charge on one mole of electrons; symbol F. 1 F = 96 500 C mol^{-1}.

Faraday's first law of electrolysis The mass of metal produced at the cathode is directly proportional to the electrical charge passed through the cell. It may be written symbolically as: $m \mu Q$.

Faraday's second law of electrolysis In order to produce 1 mole of a metal, 1, 2, 3, or another whole number of moles of electrons must be consumed.

fats Solid triglycerides are usually referred to as fats, while liquid triglycerides are referred to as oils.

fatty acid A carboxylic acid that has a relatively long hydrocarbon chain.

feedstock The raw material required for an industrial process.

fermentation The breakdown of sugar solutions by the action of enzymes in yeasts, producing ethanol and carbon dioxide. The chemical equation for the fermentation of glucose is: $C_6H_{12}O_2(aq) \rightarrow 2CH_3CH_2OH(aq) + 2CO_2(g)$.

fingerprint region The part of the infrared (IR) spectrum below 1400 cm^{-1} that is unique to a particular molecule.

fossil fuel A hydrocarbon fuel formed by the decomposition of plant and animal material over millions of years. Fossil fuels include coal, oil and natural gas.

fractional distillation A separation method based on the different boiling points of the components of a mixture. Used to separate liquids with similar boiling points.

fractionating column Apparatus used in fractional distillation. It is a tube packed with glass beads or has glass shelves, providing a large surface area upon which the vapours condense. There is a temperature gradient up the fractionating column; the column is cooler at the top than at the bottom allowing separation of liquids with different boiling points.

fragmentation The breaking-up of a molecular ion into a number of smaller parts in mass spectrometry.

fragment ion An ion that is formed in a mass spectrometer by fragmentation of the molecule.

frequency The number of events (e.g. collisions) that occur in a given time.

fuel A substance with stored energy that can be released relatively easily to generate heat or power. The stored energy can be in the form of chemical energy (e.g. methane) or nuclear energy (e.g. uranium).

fuel cell A type of electrochemical cell in which the reactants are supplied continuously, allowing continuous production of electrical energy.

functional group An atom or group of atoms in an organic molecule that largely determines the molecule's properties and reactions.

G

galvanic cell A type of electrochemical cell also known as a voltaic cell; a device that converts chemical energy into electrical energy.

galvanometer An instrument for detecting electric current.

general equation An equation for an organic chemical reaction in which alkyl groups are represented by the letter 'R'.

glucose A monosaccharide that plants can form through the process of photosynthesis. Its molecular formula is $C_6H_{12}O_6$.

glycogen A branched polysaccharide that is a polymer of α-glucose. Glycogen is the main form of carbohydrate storage in animals and is found primarily in the liver and muscle tissue.

glycosidic link The ether link (C−O−C) formed when monosaccharide units bond to each other. It is the result of the reaction between two hydroxyl groups on two monosaccharides. Glycosidic links are found in disaccharides, trisaccharides and polysaccharides.

green chemistry Design of chemical products and processes to avoid waste and minimise the use and formation of hazardous substances.

green chemistry principles Supporting principles of the approach to chemistry that aims to design products and processes that efficiently use renewable raw materials, and minimise hazardous effects on human health and the environment.

green hydrogen hydrogen gas produced using renewable energy.

greenhouse gas A gas that is able to absorb and re-radiate heat radiation. These gases contribute to the greenhouse effect. Examples are carbon dioxide, methane and water vapour.

H

half-cell Half an electrochemical cell, which contains an oxidant and its conjugate reductant. When two half-cells are combined, a galvanic cell is formed.

half-equation A balanced chemical equation which shows the loss or gain of electrons by a species during oxidation or reduction. For example, the oxidation of magnesium is written as the half-equation $Mg(s) \rightarrow Mg^{2+} + 2e^-$.

halo functional group A functional group that consists of a halogen atom bonded to the carbon chain. The halo functional groups are named fluoro-, chloro-, bromo- and iodo-.

haloalkane A molecule derived from alkanes that contain at least the one halogen functional group.

halogen An element in group 17 of the periodic table.

heat content At a simple level, the chemical energy of a substance; symbol *H*.

heat of combustion The heat energy released when a specified amount of a substance burns completely in oxygen, a positive value.

heterogeneous catalyst A catalyst that has a different physical state (phase) from the reactants and products.

heterogeneous chemical system A chemical system in which at least two of the species are in a different state.

high-performance liquid chromatography (HPLC) A very sensitive technique used to separate the components in a mixture, to identify each component, and to measure the concentrations of the components. It uses a pump to pass a pressurised liquid solvent containing the sample mixture through a column filled with solid adsorbent material. Also known as high-pressure liquid chromatography.

homogeneous catalyst A catalyst that has the same physical state (phase) as the reactants and products.

homogeneous chemical system A chemical system in which all the species are in the same state.

homologous series A series of organic compounds in which each member of the group differs from the previous member by a $-CH_2$ unit. Examples are alkanes, alkenes and alcohols.

hydration reaction A reaction that involves water as a reactant.

hydrocarbon A compound that contains carbon and hydrogen only, e.g. the alkanes, alkenes and alkynes.

hydrogen economy A proposed system of delivering energy for society using hydrogen as the source of energy.

hydrogen halide Diatomic molecules consisting of a hydrogen atom and a halogen atom joined by a covalent bond. HF, HCl, HBr and HI are hydrogen halides.

hydrogen hub A precinct where multiple industries either producing or using hydrogen gas are located.

hydrogenation In organic chemistry, hydrogenation involves the reaction of hydrogen with another molecule. An example is the addition reaction of hydrogen with an alkene to form an alkane.

hydrolysis A reaction involving the breaking of a bond in a molecule using water as a reactant. Two smaller molecules are usually formed. Also known as hydrolytic reaction.

hydrolytic reaction See hydrolysis.

hydrophobic Describes substances that tend to repel or fail to mix with water, and which are usually non-polar.

hydroxyl functional group A functional group that consists of an oxygen atom covalently bonded to a hydrogen atom, –OH. This functional group is present in alcohols.

hypothesis A possible explanation to a research question that can be used to make predictions that can often be tested experimentally.

I

ICE table A useful tool for solving equilibrium problems. **I** stands for the initial amounts (or concentrations) for each species in the reaction mixture, **C** represents the change in the amounts (or concentrations) for the system as the system moves towards equilibrium, and **E** represents the equilibrium amounts (or concentrations) of each species when the system is in equilibrium.

in excess Describes the reactant which is not completely used up in a chemical reaction.

incomplete combustion A combustion reaction that takes place when oxygen is limited. Incomplete combustion of hydrocarbons produces carbon and carbon monoxide and water.

independent variable The variable that is altered during an experiment to test its effect on another variable (the dependent variable). Also called the experimental variable.

index (plural: indices) A number written as a superscript after the 10 in scientific notation which shows how many times the 10 is to be multiplied to give the required value.

indicator A substance that is different colours in its acid and base forms.

inert A substance that is unreactive, such as a platinum or graphite electrode.

inert electrode An electrode that is not consumed in the reaction that occurs at the electrode. The electrode serves only as a conductor of electrons. Precious metals and carbon are typically used as inert electrodes.

infrared (IR) spectroscopy An analytical technique that uses the infrared part of the electromagnetic spectrum to investigate the vibrational energy of molecular bonds.

inorganic Of a compound that consists of elements other than carbon. However, certain compounds of carbon, such as carbon dioxide and carbonates, are also classified as inorganic.

integration A mathematical process used to determine the area under the curve of each signal in proton NMR spectroscopy.

integration curve Lines drawn over peaks in an NMR spectrum, the height of which indicate the ratio of the peak areas. Also known as integration trace.

intermediate Product formed in one of the in-between stages of a chemical reaction.

internal circuit The part of an electrochemical cell in which ions move, e.g. solutions and salt bridge.

inverse relationship A mathematical relationship in which one variable increases as the other decreases.

iodine number Also called iodine value or iodine index, is the mass of iodine in grams absorbed by 100 grams of oil/fat.

irreversible reaction A reaction in which significant reaction can occur in one direction only.

isomers Molecules that have the same molecular formula but a different arrangement of atoms.

isotope Each of two or more forms of the same element that contain equal numbers of protons but different numbers of neutrons in their nuclei.

IUPAC nomenclature A set of rules for naming organic molecules. It is usually systematic and gives the number of carbon atoms and the location and type of functional groups present.

J

joule The SI unit of energy.

K

ketone A homologous series of organic molecules that contains the carbonyl functional group (C=O) within the carbon chain.

kinetic energy The energy that a particle or body has due to its motion $\left(\text{KE} = \frac{1}{2}mv^2\right)$.

L

Le Châtelier's principle If an equilibrium system is subjected to a change, the system will adjust itself to partially oppose the effect of the change.

limiting reactant A reactant that is completely consumed in a chemical reaction. The limiting reactant limits the amount of product that can be formed.

line graph A graph which shows a linear trend in data.

line of best fit A line through data points in a scatter plot that best represents the relationship.

linear economy An economy which operates on a 'take-make-dispose' model, making use of resources to produce products that will be discarded after use.

linear trend Represented on a graph by a straight line in which the y values have equal differences as the x values increase.

lipid A class of organic compounds that includes fats, oils, waxes and steroids. They are insoluble in water, but soluble in non-polar solvents.

literature value A value for a quantity which has been found in published scientific reports, such as journals, or reliable websites, such as those of universities.

lock-and-key mechanism An explanation of the specificity of enzymes in which a substrate molecule fits into an enzyme like a key in a lock, forming an enzyme–substrate complex. This allows the enzyme to break the bonds in the substrate.

logbook A bound book in which you record every detail of your research.

M

mass–mass stoichiometry A calculation using a balanced equation in which the known and the unknown are both masses.

mass spectrometer An instrument designed to measure the mass-to-charge ratio, m/z, of particles.

mass spectrometry An analytical technique that uses the mass-to-charge, m/z, ratio of atoms, molecules and fragments of molecules to identify substances.

mass spectrum A graph of data produced from a mass spectrometer which shows the abundance or relative intensity of each particle, and their mass-to-charge, m/z, ratios.

mass–volume stoichiometry A calculation using a balanced equation, in which the known is a mass and the unknown is a volume of gas.

materials-based storage A general name given to methods of storing hydrogen that includes adsorption to the surface of materials such as metal hydrides, absorption into the lattice structure of some solid materials and reversible reactions with a range of different chemicals.

Maxwell–Boltzmann distribution curve A graph of kinetic energy against number of particles that shows the range of energies in a sample of a gas or a liquid at a given temperature.

mean The average value of a set of values, calculated by dividing the sum of the values by the number of values. Also known as average.

measurement error The difference between a measured quantity and its true value.

measurement result A final result, usually the average of several measurement values. In the (unusual) case where only one value has been measured, then measurement result also applies to that single measurement value.

medicinal plant Plant extracts that have a biological impact on the body.

melting point The temperature at which a solid melts and becomes a liquid.

melting point determination A laboratory technique applied to the identification of unknown pure substances. If two substances melt at the same temperature, a mixed melting point determination can indicate if they are the same substance.

melting point range One of the characteristic properties of a pure solid. The range starts with the temperature at which the crystals first begin to liquify and ends at the temperature at which the entire sample is liquid.

meniscus The curved upper surface of liquid in a tube, caused by surface tension. A meniscus can be concave (as in water in a glass tube) or convex (as in mercury in a thermometer).

metabolise When a substance undergoes processing in the body by metabolism.

metabolism Chemical processes occurring within a living cell or organism that are necessary for the maintenance of life.

method The specific steps taken to collect data during a scientific investigation. Also known as the procedure.

methodology A brief description of the general approach taken in a scientific investigation. Examples of scientific investigation methodologies are controlled experiments, fieldwork, literature reviews, modelling and simulation.

mistake A personal error. A mistake should be identified as an outlier and the trial repeated.

mixed melting point determination A technique used to identify organic compounds in which a sample with a known identity is mixed with an unknown purified sample to determine the melting point. If the melting point of the mixture is the same as the melting point of the pure compound, then the unknown compound and the pure compound are the same. If the unknown compound is different to the known compound, then the melting point would have a lower and broader range.

mobile phase In chromatography, the phase that moves over the stationary phase.

molar enthalpy The enthalpy of a substance, given per mole.

molar volume The volume occupied by one mole of gas at a specified set of conditions. At standard temperature (0°C or 273 K) and pressure (100 kPa), the molar volume of a gas is 22.7 L mol^{-1}.

mole ratio The ratio of species involved in a chemical reaction, based on the ratio of their coefficients in the reaction equation.

molecular formula A formula that gives the actual number and type of atoms present in a molecule.

molecular ion A whole molecule with an overall positive or negative charge. In the context of mass spectrometry, it is a positively charged ion that has passed through the mass spectrometer without fragmenting. *See* parent molecular ion.

molecular ion peak A peak in a mass spectrum that is by the presence of a whole molecule ion. The peak with the greatest m/z value is likely to be due to the molecular ion in most cases.

molten Liquid state formed when a substance with a high melting point is heated above its melting point.

monomer Small molecule that is able to react to form long chains of repeating units, called polymers.

monosaccharide The simplest unit of carbohydrates. Monosaccharides are the building blocks of more complex carbohydrates and have the formula $C_6H_{12}O_6$. Examples are glucose, fructose and galactose.

***m/z* value** The ratio of the mass of an ion (m) to its charge (z).

N

N-terminal The end of a protein or polypeptide with a free amine group ($-NH_2$).

natural gas A fossil fuel composed of a mixture of hydrocarbons that are trapped in the Earth's crust. Natural gas consists mainly of methane (CH_4). It is extracted in gaseous form and then liquefied under pressure for transport and storage.

neighbour The ^1H atoms within three bonds of a hydrogen atom that can cause its signal to split into a lined pattern. Hydroxyl hydrogens do not cause splitting and are not included as neighbours.

non-competitive enzyme inhibitor A type of enzyme inhibition where the inhibitor reduces the activity of the enzyme without occupying the active site.

non-rechargeable cell A primary galvanic cell.

non-renewable Resources which are used up at a faster rate than they can be replaced.

non-spontaneous reaction Reactions that would not normally occur without the application of electrical energy. They are the reverse of spontaneous reactions, which produce energy.

nuclear magnetic resonance (NMR) spectroscopy A technique used to analyse materials using the interaction of the nucleus of particular isotopes, most commonly usually ^1H or ^{13}C, with an external magnetic field and electromagnetic radiation.

nuclear magnetic resonance (NMR) spectrum The representation in graph form of the energy required to change a nucleus from a low energy spin state to a high-energy spin state.

nuclear shielding Modification of the magnetic field experienced by a nucleus in an external magnetic field caused by the magnetic field of surrounding atoms in the molecule.

nuclear spin A property of a nucleus with an odd number of protons or neutrons that causes it to interact with a magnetic field. The nuclear spin can be either with or against an external magnetic field.

nutrient A substance that provides nourishment for growth or metabolism.

O

objective Information or a viewpoint based on facts and free of bias.

observation Closely monitoring something or someone.

ocean acidification The reduction in the pH of the ocean over an extended period of time, caused mainly by the uptake of carbon dioxide from the atmosphere.

open system A system that allows matter and energy to be exchanged with the surroundings.

optical isomers Isomers in which there can be a different three-dimensional arrangement of groups around one or more atoms.

optimum pH The pH at which enzyme activity is greatest.

optimum temperature The temperature at which enzyme activity is greatest.

organic compound A compound composed of molecules based on a carbon backbone.

organic molecule A molecule that is based on a hydrocarbon skeleton. Organic molecules also commonly contain other non-metal elements, such as oxygen, nitrogen, sulfur and chlorine.

outlier A reading that lies a long way from other results is sometimes called an outlier. Repeating readings may be useful in further examining an outlier.

oxidant A chemical species (element, compound or ion) that accepts one or more electrons in an oxidation–reduction reaction. An oxidant causes another substance to be oxidised, and in the process the oxidant is reduced. An oxidant is also called an oxidising agent.

oxidation (i) A reaction in which oxygen is a reactant. Oxidation can be defined as the addition of oxygen to form oxides, such as combustion reactions. (ii) The process by which a chemical species such as a metal atom or a non-metal ion loses electrons. An oxidation half-equation will show the electrons as products (on the right-hand side of the arrow). Oxidation is said to have occurred when there is an increase in an element's oxidation number during the reaction.

oxidation number A number assigned to an atom in a compound or as the free element, which represents the charge that atom would have if it was an ion. A series of oxidation number rules are used to determine the oxidation number of an element in a compound. Oxidation numbers are used to identify redox reactions. Also known as oxidation states.

oxidised The loss of electrons or an increase in oxidation number. When a substance is oxidised, the electrons are written on the right-hand side of the arrow in the half-equation.

oxidising agent A reactant that causes another reactant to lose electrons during a redox reaction. This reactant is, itself, reduced and gains electrons. The oxidation number of the oxidising agent decreases during the reaction. For example, in the reaction between magnesium and oxygen, the oxygen is the oxidising agent, as it causes magnesium to lose electrons and form Mg^{2+}.

P

parallax error The perceived shift in an object's position as it is viewed from different angles.

parent molecule The alkane from which the name of a molecule is derived from.

parent molecular ion A whole molecule with an overall positive charge that has passed through the mass spectrometer without fragmenting.

partial pressure The pressure exerted by one component of a mixture of gases. The total pressure of a mixture of gases is equal to the sum of the individual partial pressures of each component in the mixture.

parts per trillion Concentration in nanograms per kilogram (ng kg^{-1}).

peak area (proton NMR) The area under each peak on a proton NMR spectrum is proportional to the number of equivalent hydrogens it corresponds with.

peer-reviewed Describes information that other scientists have checked and have agreed is appropriate for publication.

peptide link The –CONH– functional group between amino acid units in a polypeptide chain; also called amide link.

percentage change The difference between the final and intial values, measurements divided by the initial value. mulitiplied by 100.

percentage difference The difference between the experimental and theoretic values, divided by the theoretical value, multiplied by 100.

percentage yield A measure of the quantity of a product obtained from a chemical process compared to the maximum amount possible if the reaction were complete, expressed as a percentage:

$$\text{percentage yield} = \frac{\text{actual yield}}{\text{theoretical yield}} \times \frac{100}{1}$$

petrodiesel The most common form of diesel fuel. It is produced from crude oil by fractional distillation. The composition of petrodiesel varies, but is generally around 75% alkanes and 25% aromatic hydrocarbons. The alkanes range from $C_{10}H_{22}$ to $C_{15}H_{32}$.

phenyl functional group An organic functional group consisting of a benzene ring bonded to an alkyl group or another functional group. It has the formula $C_6H_5^-$.

photochemical smog Atmospheric pollution produced through the action of sunlight on nitrogen oxides and unburned hydrocarbons to form ozone and other pollutants. The nitrogen oxides are formed in high temperature reactions such as those that occur in car engines and lightning strikes.

photoelectrochemical cell A device that uses light incident on a submerged metal electrode in an electrolytic solution to cause a chemical reaction.

photosynthesis A reaction that occurs in the leaves of plants between carbon dioxide and water, in the presence of sunlight and chlorophyll, to form glucose and oxygen. The photosynthesis process can be represented by the equation:

$$6CO_2(g) + 6H_2O(l) \rightarrow C_6H_{12}O_6(aq) + 6O_2(g)$$

pie chart A graph or chart which represents percentages of a whole sample set as slices of a circular 'pie'.

pipette Glassware used for volumetric analysis to transfer a fixed volume of solution accurately.

plane of symmetry A plane of symmetry divides a three-dimensional object into two equal halves that are mirror images of each other.

pleating When sections of a protein line up parallel with each other owing to hydrogen bonding.

polarity The measure of how polar a molecule or bond is. The difference in charge between the positive and negative ends of an electric dipole. The difference in charge between the positive and negative ends of a polar molecule or covalent bond.

polymer A natural or synthetic compound of high molar mass consisting of up to millions of repeated linked units known as monomers. There are many different types of polymers.

polymer electrolyte membrane (PEM) A semipermeable membrane designed to conduct protons, while acting as an electronic insulator and reactant barrier.

polypeptide An organic polymer molecule made from a condensation reaction between amino acids.

polysaccharide A long-chain carbohydrate made up of smaller carbohydrates called monosaccharides. Each monosaccharide is connected by glycosidic links. Starch, glycogen and cellulose are polysaccharides.

porous A material that is porous contains many small holes; water and air may be able to pass through these holes.

position of equilibrium The relative amounts of reactants and products at equilibrium. The position of equilibrium varies depending on the extent of the reaction.

potential difference The electromotive force between two points in a circuit, such as the electrodes of an electrochemical cell.

precise When repeated measurements of the same quantity give values that are in close agreement.

precision Refers to how closely a set of measurement values agree with each other. Precision gives no indication of how close the measurements are to the true value and is therefore a separate consideration to accuracy.

primary alcohol An alcohol in which the carbon that is bonded to the –OH group is only attached to one alkyl group.

primary amide a molecule that contains a primary amide functional group.

primary amine An amine in which the nitrogen atom is bonded to one carbon and two hydrogen atoms.

primary cell A galvanic cell that is non-rechargeable because the products of the reaction migrate away from the electrodes.

primary data The measurements or observations that you collect during your investigation.

primary source A source that includes firsthand information, such as the results of an original experiment.

primary standard A substance used to make a standard solution. It is so pure that its amount, in moles, can be accurately determined from its mass.

primary structure The sequence (number, order and type) of amino acids in a polypeptide chain.

principle A principle is usually more specific than a theory. *See* theory.

processed data Data that has been mathematically manipulated in some way.

proportion The number of events (e.g. collisions) out of a total number of events that occur. Similar to percentage, but not out of 100.

protein An organic polymer made from amino acids by condensation polymerisation.

proton NMR spectroscopy Also called ^1H NMR spectroscopy. A type of nuclear magnetic resonance spectroscopy that investigates the ^1H nucleus. It is used to determine the chemical environment of hydrogen atoms in compounds and gives information about the environments neighbouring each hydrogen atom.

Q

qualitative Non-numerical data, variables and observations, for example, colour.

qualitative analysis An analysis to determine the identity of the chemical(s) present in a substance.

qualitative data Data which can be observed and relates to a type or category, such as colour, or states (such as gas, liquid or solid).

quantitative Numerical data and variables that can be measured, for example, pH.

quantitative analysis The determination of the quantities of particular components present in a substance.

quantitative data Measured numeric values which are usually accompanied by a relevant unit.

quaternary structure The highest level of organisation in protein structure. Proteins have a quaternary structure if they are composed of two or more polypeptide chains.

R

R group The side chain in an amino acid molecule or amino acid unit in a polypeptide chain. It can be composed of different sets of atoms. The amino acids used to make the proteins in the human body have 20 different R groups.

random error Affects the precision of a measurement and are present in all measurements except for measurements involving counting. Random errors are unpredictable variations in the measurement process and result in a spread of readings. The effect of random errors can be reduced by making more or repeated measurements and calculating a new mean and/or by refining the measurement method or technique.

rate of reaction The change in concentration of a reactant or product over a period of time (usually one second):

$$\text{rate of reaction} = \frac{\text{change in concentration}}{\text{time}}$$

raw data The data you record in your logbook.

reaction pathway A series of chemical reactions that converts a starting material into a product in a number of steps.

reaction quotient Another name for the concentration fraction. It is the ratio of concentrations of products to reactants in a reversible reaction, as expressed in the equilibrium law.

reaction table A useful tool for solving equilibrium problems. **I** stands for the initial amounts (or concentrations) for each species in the reaction mixture, **C** represents the change in the amounts (or concentrations) for the system as the system moves towards equilibrium, and **E** represents the equilibrium amounts (or concentrations) of each species when the system is in equilibrium.

reactive electrode An electrode that is consumed in an electrochemical cell reaction.

reagent A substance added to a system to cause a chemical reaction, such as a catalyst.

rechargeable cell A type of cell in which the chemical energy can be replenished repeatedly through application of electrical energy.

redox reaction A reaction in which electron transfer occurs from the reducing agent to the oxidising agent. In a redox reaction, the oxidation number of one element will increase (be oxidised) and the oxidation number of another element will decrease (be reduced).

redox titration A titration involving solutions which react together in a redox reaction.

reduced The gain of electrons or a decrease in oxidation number. When a substance is reduced, the electrons are written on the left-hand side of the arrow (as reactants) in the half-equation.

reducing agent A reactant that causes another reactant to gain electrons during a redox reaction. This reactant is, itself, oxidised and loses electrons. For example, in the reaction between magnesium and oxygen, the magnesium is the reducing agent, as it causes oxygen to gain electrons and form O^{2-} ions:

$$2Mg(s) + O_2(g) \rightarrow 2MgO(s)$$

reductant A substance that causes another substance to be reduced; in the process the reductant is oxidised. A reductant is also called a reducing agent.

reduction The process by which a chemical species gains electrons or its oxidation number decreases. A reduction half-equation will show the electrons on the reactant side (left-hand side) of the equation.

reference sample A standard mixture or a compound that is known.

relative intensity This is calculated by dividing the abundance of a peak with the abundance of the base peak and then multiplying by 100.

renewable energy source A source of energy which come from sources that are readily replenished, e.g. solar energy, wind power, hydroelectricity.

renewable fuel A fuel that can be replaced at a sustainable rate. Also known as **renewables**.

renewables Fuels that can be replaced at a sustainable rate. Also known as **renewable fuels**.

repeatability The consistency of results when an experiment is repeated many times.

replication (1) The mechanism by which DNA can be copied. (2) Experimentation carried out on duplicate sets at the same time.

reproducibility The closeness of the agreement between the results of measurements of the same quantity being measured, carried out under changed conditions of measurement. These different conditions include a different method of measurement, different observer, different measuring instrument, different location, different conditions of use, and different time.

research question A statement that defines what is being investigated.

resolution The smallest change in the measured quantity that causes a readable change in the value as indicated by the measuring instrument.

respiration A number of different processes (including aerobic and anaerobic respiration) which occur in the cells of the body and convert glucose and oxygen to carbon dioxide, water and energy.

retention time (R_t) The time taken for a component to pass through a chromatography column.

reversed-phase HPLC A form of HPLC that uses a non-polar stationary phase.

reversible reaction A reaction in which significant reaction can occur in the reverse direction because the products are present and can react with each other under suitable conditions.

ribosome An extremely large molecular machine found in cells, that synthesises proteins from amino acids.

risk assessment A systematic way of identifying the potential risks associated with an activity.

S

safety data sheet (SDS) A document that contains important information about the possible hazards in using a substance and how the substance should be handled and stored.

salt bridge An electrical connection between the two half-cells in a galvanic cell; it is usually made from a material saturated in electrolyte solution.

saturated hydrocarbon A hydrocarbon compound with only single C–C bonds.

saturated molecule An organic compound in which all the atoms are joined by single bonds.

scatter graph A graph in which dots are used to represent values for two different numeric variables. The position of each dot on the horizontal and vertical axis indicates values for an individual data point.

scientific method The experimental approach to the study of science that involves formulating a hypothesis, designing and performing an experiment to test the hypothesis, and analysing whether the results support or refute the hypothesis.

scientific notation Reporting a numerical value to have one integer before a decimal point and other digits after the decimal point multiplied by a power of 10, for example, 6.02×10^{23}.

secondary alcohol An alcohol in which the carbon bonded to the –OH group is also bonded to two alkyl groups. The alkyl groups may be the same or different.

secondary cell An accumulator or rechargeable cell. Recharging can occur because the products formed in the cell during discharge remain in contact with the electrodes in a convertible form.

secondary source A resource that interprets primary documents, written after the event by a person who was not a witness to the event.

secondary structure Initial level of spatial arrangement of a polypeptide chain. There are two forms of secondary structure: α-helices and β-pleated sheets.

semipermeable membrane A type of membrane that will only allow certain molecules or ions to pass through it.

semi-structural formula A chemical formula that summarises the structural formula of a compound in a single line of text. *See also* condensed formula.

side chain The R group in an amino acid molecule; there are 20 different side chains on the 2-amino acids used to make human proteins.

signal The radiowave frequency absorbed by the hydrogen or carbon nuclei in a molecule that is measured by the NMR instrument and plotted on a spectrum.

significant figures Digits in a number that are reliable and are necessary to indicate a quantity. These are limited in chemistry by the equipment used, or, in a calculation, by the number of significant figures in the least accurate piece of data used.

simple distillation A technique for separating two or more liquids from a mixture. This technique relies on a difference of at least 50°C in boiling point between components to obtain an effective separation.

skeletal structure A shorthand representation of an organic molecule, where carbon atoms, and hydrogen atoms attached to carbon atoms, are omitted.

solution calorimeter An insulated container that holds a known volume of water, and in which a reaction in solution, such as dissolution of a solid or a neutralisation reaction, can be carried out.

solvent extraction The removal of a substance from a solution or mixture by dissolving it in another solvent in which it is more soluble.

specific heat capacity The amount of energy required to raise the temperature of an amount of a substance, usually 1 gram, by 1°C. The unit for specific heat capacity is usually $J\ g^{-1}\ °C^{-1}$, e.g. the specific heat capacity of water is $4.18\ J\ g^{-1}\ °C^{-1}$.

spectroscopy The science of electromagnetic radiation and its interaction with atoms and molecules.

spiking A procedure used to tentatively identify a component within a mixture by high-performance liquid chromatography. If a chemical is thought to be one of the components of a mixture, it can be added to the sample. If there are no extra peaks in the chromatogram, just a larger peak for one of the components, it is assumed that the added chemical is one of the components of the mixture.

splitting pattern The resultant multi-peaked signal that a proton gives in a ^1H NMR spectrum when it has hydrogens in neighbouring environments. The number of lines in the pattern is equal to the number of neighbouring hydrogen atoms plus 1, or '$n + 1$'.

spontaneous reaction A reaction that occurs naturally, either in electrochemical cells or when chemicals are mixed directly. The reaction does not need to be driven by an external source of energy.

standard conditions Established reference conditions to compare testing of experimental activities. Conditions at which gas pressure is 1 bar (100 kPa), the concentrations of dissolved species are 1.0 M and the temperature is 25°C (298 K).

standard electrode potential The electromotive force that is measured when a half-cell, at standard conditions (gas pressures 1 bar, solution concentrations of 1.0 M and temperature of 25°C), is connected to a standard hydrogen half-cell; symbol E^o. Gives a numerical measure of the tendency of a half-cell reaction to occur as a reduction reaction.

standard form The standard way to express a number as the product of a real number and power of 10. Also known as scientific notation.

standard hydrogen electrode (SHE) *See* standard hydrogen electrode and standard hydrogen half-cell

standard hydrogen half-cell An $H^+(aq)$/$H_2(g)$ half-cell; made from a platinum electrode placed in acid solution with hydrogen gas bubbled over it. The pressure of the gas is 1 bar, the H^+ concentration is 1 M and the temperature is 25°C.

standard laboratory conditions (SLC) Conditions described by gas pressure of 100 kPa, temperature of 298 K (100°C) and when a solution is also involved, the concentration is 1 M.

standard reduction potential A numerical measure of the tendency of a half-cell reaction to occur as a reduction reaction. It is measured by connecting a half-cell, at standard conditions (gas pressures 1 bar, solution concentrations of 1.0 M and temperature of 25°C), to a standard hydrogen half-cell.

standard solution A solution of accurately known concentration.

standardised The process by which the concentration of a solution is accurately determined, often through titration with a standard solution.

starch A polysaccharide formed from the polymerisation of α-glucose molecules. Starch is an odourless and tasteless, white substance that occurs widely in plant tissue. It is used by plants to store energy.

stationary phase A solid, or a solid that is coated in a viscous liquid, used in chromatography. The components of a mixture undergo adsorption to this phase as they are carried along by the mobile phase.

steam distillation Extraction of non-volatile liquids from plants using a flow of steam.

steam reforming A process that produces hydrogen, carbon monoxide or other useful products from hydrocarbon fuels such as natural gas.

stereoisomer A type of isomer in which the atoms in two molecules are connected in the same order but have different arrangements in space.

stoichiometry The calculation of relative amounts of reactants and products in a chemical reaction. Chemical equations give the ratios of the amounts (moles) of the reactants and products.

structural formula A formula that represents the three-dimensional arrangement of atoms in a molecule.

structural isomers Isomers are molecules that have the same molecular formulas, but their atoms are bonded together in different ways.

subjective A viewpoint or information which is based on personal opinion and feelings, rather than facts.

substitution reaction A reaction that involves the replacement of an atom or group of atoms with another atom or group of atoms.

substrate A reactant in a reaction that is catalysed by an enzyme.

surface area The area of all surfaces of the substance that are exposed to the other reactants. This is proportional to the amount of particles available at the surface to react.

surroundings The rest of the universe around a particular chemical reaction. The chemical reaction is the system. Energy moves between the system and surroundings in exothermic and endothermic reactions.

sustainable Able to support energy and resources into the future without depletion.

synthesised Assembled or produced by one or more chemical reactions.

system In chemistry, a system is a chemical reaction. A system operates within its surroundings. Energy can move between the two. You always consider energy as being absorbed or released from the perspective of the system. E.g., energy is absorbed into the system from the surroundings, or energy is released by the system to the surroundings.

systematic error Affects the accuracy of a measurement. Systematic errors cause readings to differ from the true value by a consistent amount each time a measurement is made, so that all the readings are shifted in one direction from the true value. The accuracy of measurements subject to systematic errors cannot be improved by repeating those measurements.

T

tertiary alcohol An alcohol in which the carbon bonded to the –OH group is also bonded to three alkyl groups.

tertiary structure The overall three-dimensional shape of a polypeptide chain.

theoretical yield The mass of product that would be formed if the limiting reagent reacted fully. The mass that is expected for a particular product in a chemical reaction, based on the mass of limiting reactant used.

theory When, after many experiments, a hypothesis has been supported by all the results so far, it is referred to as a theory or principle.

thermochemical equation A chemical equation that includes the enthalpy change of the reaction, ΔH. E.g. $2H_2(g) + O_2(g) \longrightarrow 2H_2O(g) \quad \Delta H = -572\ \text{kJ mol}^{-1}$.

titration The process used to determine the concentration of a reactant where one solution is added from a burette to a known volume of another solution.

titre A variable volume of liquid measured in a burette by subtracting the final burette reading of a titration from the initial burette reading.

transesterification A process in which a triglyceride reacts with an alcohol to form esters.

transition element An element, such as iron, copper, nickel or chromium, that is found in the middle block of the periodic table, in groups 3–12.

transition state An arrangement of atoms in a reaction that occurs when sufficient energy is absorbed for the activation energy to be reached. It represents the stage of maximum potential energy in the reaction. Bond breaking and bond forming are both occurring at this stage, and the arrangement of atoms is unstable.

transmittance The ratio of light that passes through a sample compared to the light given off by the source.

trend A pattern in data, often positive or negative, linear or non-linear.

triglyceride An ester derived from the reaction between a glycerol molecule and three fatty acids.

tripeptide An organic molecule made from three amino acid units linked by peptide bonds.

true value The value, or range of values, that would be found if the quantity could be measured perfectly.

U

uncertainty The uncertainty of the result of a measurement reflects the lack of exact knowledge of the value of the quantity being measured.

unsaturated hydrocarbon A hydrocarbon compound with one or more double or triple carbon–carbon bonds.

unsaturated molecule An organic molecule that contains one or more carbon–carbon double or triple bonds.

V

valence electron number The number of valence electrons in an atom of an element.

valence shell electron pair repulsion (VSEPR) theory A model in chemistry used to predict the shape of molecules based on electrostatic repulsion between electron pairs.

valid When an experiment or investigation does test the stated aims and hypothesis.

validity A measurement is said to be valid if it measures what it is supposed to be measuring. An experiment is said to be valid if it investigates what it sets out and/or claims to investigate.

variable A factor or condition that can change during your experiment.

vibrational energy level A fixed energy that a molecule can have as a result of the bending and stretching of bonds.

virions The complete, infective form of a virus outside of the host cell.

viscosity A measure of a liquid's resistance to flow. Honey has a high viscosity and petrol has a low viscosity.

volatile Able to change from a liquid to a gas (vaporise) easily.

volt The unit of potential difference.

voltaic cell A type of electrochemical cell that is also known as a galvanic cell; a device that converts chemical energy into electrical energy.

voltmeter An instrument for measuring electrical potential difference between two points in a circuit.

volumetric analysis Quantitative analysis using measurement of solution volumes, usually involving titration.

volume–volume stoichiometry A calculation using a balanced equation in which both the known and the unknown are gas volumes. Volumes may also be of solutions, in other contexts.

W

wavenumber A unit of frequency used in infrared spectroscopy. It is equal to the inverse of the wavelength of the radiation, and is measured in cm^{-1}.

Y

yield The amount of product formed in a chemical reaction.

Z

zwitterion A dipolar ion formed when the amino and carboxyl groups in an amino acid or polypeptide are both charged.

Index

D

E

F

G

H

I

J

K

L

T

U

V

W

Y

Z

Attributions

Heinemann Chemistry 2 6th edition attributions

Cover and back cover: Alfred Pasieka, Science Photo Library

123RF: p. 444tr; Armyagov, Andrey, p. 34; Chuynugul, Yutakapong, p. 200; Etchells, Peter, p. 442t; filedimage, p. 122; Gorin, Sergei, p. 68; Gyori, Levente, p. 578bc; Jorgensen, Kjersti, p. 265; ninarusskova, p. 311b; nito500, p. 578br; saiko3p, p. 400t; Santos, Rui, p. 90t; scanrail, p. 178t; serezniy, p. 291t; Siriburanakit, Preecha, p. 85br; thamkc, p. 86; Yemelyanov, Maksym, p. 310bl.

AAP: Brownhill, Andrew, p. 243tl; Smith, Julian, p. 171.

Abrahams, Brendan: p. 205.

agefotostock: Benson, Caspar, p. 222.

Alamy: AKP Photos, p. 421 (white spirits); Andrew Lambert Photography/Science Photo Library, pp. 380lb, 380lc, 470, 473rb, 474cl; Ashley Cooper pics, p. 73cr; Bauer, Scott/U.S. Department of Agriculture, p. 559r; Brook, Robert, p. 410b; BSIP SA, p. 525r; Burgstedt, Christoph, p. 430-31; Chillmaid, Martin F./Science Photo Library, pp. 380lt, 473rt; Dalton, Lee/JPL/NASA, p. 500t; DOE Photo, p. 389 (methane hydrate); Fleming,Vaughan/Science Photo Library, p. 16; Forance, p. 499; Furman, Dmytro, p. 421 (wound); Gainey, Claire, p. 434b; Golovnov, Igor, p. 363 (stamp); Greenshoots Communications, p. 74; H.S. Photos, p. 418; ingetje, tadros, p. 485r; JH Photo, p. 406t; Knott, Jason, p. 388; Mach, Olekcii, p. 309t; Natural History Museum, Science Photo Library, p. 359b; Nunuk, David/Science Photo Library, p. 104; OZSHOTS, p. 85bl; Photo Researchers/Science History Images, p. 24; PhotoStock-Israel, p. 121t; Pictorial Press Ltd, pp. 349rt, 363 (Wohler); Radius Images, p. 455 (elephants); razorpix, p. 421 (turpentine); RooM the Agency, p. 336; Scharfsinn, p. 302-3; Science Photo Library, pp. 474b, 90b; Shields, Martin, p. 4r; Simons, Christina, p. 89r; Sueddeutsche Zeitung Photo, p. 305r; TEK Image/Science Photo Library, p. 103-4; West, Jim, p. 177; Williamson, Kenny, p. 68; Woitas, Jon/dpa picture alliance, p. 277; World History Archive, p. 127.

American Chemical Society: Adapted with permission from J. Agric. Food Chem. 2010, 58, 15, 8495–8501. Publication Date: July 14, 2010. https://doi.org/10.1021/jf101460t. Copyright © 2010 American Chemical Society, p. 546.

Archie Rose: p. 85tr.

Auscape: Miller, Steven David p. 244 (caves).

BP: Data from BP Statistical Review of World Energy 2019, p. 69.

Chemistry Education Association: p. ix.

Clean Energy Australia: Adapted from Clean Energy Australia 2022 Report by Clean Energy Council, pp. 75-6.

Derry, Lanna: pp. 25c, 25b, 154, 224b.

Ethtec: p. 87r.

EVReporter: Adapted from Trends in Lithium-ion Battery Reuse and Recycling, Mitch Jacoby, C&EN, based on data from Argonne National Laboratory, p. 313.

Fotolia: asab974, p. 442b; BSIP SA, p. 535l; cooperr, p. 455 (chocolate); Danti, Andrea, p. 243tr; Mann, Steve, p. 109l; Mc Xas, p. 271t; Megasquib, p. 14b; monticellllo, p. 451; Patryssia, p. 402l; Pedersen, Roy, p. 352b; Pescara, Walter, p. 346lb; PhotoSGr, p. 364b; Radkov, Svetoslav, p. 456tl; Sanders, Gina, p. 444tl; siraphol, p. 347tr; Teteline, pp. 111r, 432l; Tewes, Michael, p. 357 (chocolate); thieury, p. 266l; vvoe, p. 109r.

Getty Images: aethernet, p. 469; Boyer/Roger Viollet, p. 262l; Casarsa, p. 1-3, 59; Casarsa, p. x; Dozier, Marc, p. 442c; Kosolovskyy, p. 334-35; Mead, Ted, p. 555; Mu, Sandra/Staff, p. 493; nevodka, p. 219; Pelletier, Micheline/Corbis, p. 439t; Toa55, p. 62-3.

Ixom Operations Pty Ltd: p. 14.

Kimberley Land Council: p. 89l.

Manildra Group: pp. 84b, 113.

McBroom, Alice: pp. 270lb, 367 (dettol).

National Aeronautics and Space Administration (NASA): p. 199r.

Orica Limited Group: p. 305l.

O'Shea, Pat: p. 560l.

Pasco Scientific: p. 31.

Pearson Education Ltd. 2007: p. 46.

Pearson India Education Services: Mohammed Ali, p. 455 (cow).

Pentelute Lab: Courtesy of Bradley L Pentelute/Pentelute Lab, p. 437b.

Royal Pharmaceutical Society: p. 415.

Science Photo Library: pp. 29, 166, 179r, 187tr, 220b, 278t, 532; Andrew Lambert Photography, pp. 25t, 151r, 156, 158, 167, 180lt, 221br, 253b, 262r, 387, 474cr; Chillmaid, Martyn F., pp. 65, 152, 164r, 393r (test tubes), 488t; Cuthbert, Colin, pp. 304, 508tc; De Schwanberg, Victor, p. 343t; Degginger, Phil, p. 151l; Dumas, Patrick/Look at Sciences, p. 196t; Dykinga, Jack/US Department of Agriculture, p. 508tl; Edds, Jim, p. 96; Freeman, Clive/Biosym Technologies, p. 232t; Kallista Images/Custom Medical Stock Photo, p. 569l; Lagerek, Christian, p. 117r; Laguna Design, p. 437t; Latron, Patrice/Look At Sciences, p. 518t; Martin, Custom Medical Stock Photo, p. 126bl; Martin, Frankie/NASA, p. 196b; P. Plailly/E. Daynes, p. 229r; Power and Syred, p. 432r; Reardon, Mitch, p. 256t; Royal Institution of Great Britain, p. 178b; TEK Image, p. 508tr; Terry, Sheila, pp. 286l, 294; Trevor Clifford Photography, pp. 164l, 396, 483b; Wheeler, Dr Keith, p. 307; Winters, Charles D., pp. 23, 4l, 149, 150r, 263r.

Shutterstock: p. 392t; Blanda, Eva, p. 220tl; Bobica10, p. 486t; Bokan, Ana, p. 484; Bro Studio, p. 23; Calenor, Kenneth William, 288t; Catherine.Things, p. 420r; ChiccoDodiFC, p. 424l; chromatos, p. 551; cigdem, p. 184lt; Czech, Edith, p. 243 (cake); DDevecee, p. 556r; Ehrman Photographic, p. 142l; FiledIMAGE, p. 447b; Flas100, p. 420c; Gemenacom, p. 242t; hydebrink, p. 51; ibreakstock, p. 571t; Istimages, p. 396t; Iv-olga, p. 420l; Ivanov, Ivaylo, p. 352t (vinegar); Jones, Leecy, p. 557t; Karandaev,

Evgeny, p. 352t (wine); Kirch, Alexander, p. 318r; Kotcha K, pp. 401, 575 (pineapples); lucio pepi, p. 72b; Mackenzie, Robyn, 184lb; Madushanie, Nirosha, p. 576t; margouillat photo, p. 443 (fruits); maxuser, p. 377b; Novikov, Sergey, p. 241; Pelanek, Martin, p. 505; penofoto, p. 199l; pkajak201, p. 310t; PomInOz, p. 73t; Quality Stock Arts, p. 78c; Radomski, Pawel, p. 374r; royaltystockphoto.com, p. 245b; Samokhin, Roman, p. 142 (apples); Santa Maria, Gino, p. 30; Scharfsinn, p. 319bl; Smile Fight, pp. 60-1, 214-8, 326-31; Sonja C, p. 556bl; stockphoto-graf, p. 376b; tersetki, p. 595 (funnel); totajla, p. 485l; TR STOK, p. 202; UnknownLatitude Images, p. 137; Zigres, p. 394bl; zzcapture, pp. 333-34, 463-68, 590-95.

Silvester, Helen: Casey Tech School, p. 468.

Sozaijiten: p. 379 (honey).

Tay, Abdullatif & Singh, Rakesh & Krishnan, Sivakumar & Gore, Jay. (2002): Authentication of Olive Oil Adulterated with Vegetable Oils Using Fourier Transform Infrared Spectroscopy. Lwt - Food Science and Technology. 35. 99-103, figs. 1 and 2. 10.1006/fstl.2001.0864., p. 548b.

The United Nations Commission for Europe (UNECE): p. 13.

The University of Queensland: pp. 378T, 378b.

United States Department of Energy via Wikimedia: p. 317b.

University of South Australia: pp. 558lt, 558lb.

Victorian Curriculum and Assessment Authority (VCAA): Selected extracts from the VCE Chemistry Study Design (2023–2027) are copyright Victorian Curriculum and Assessment Authority (VCAA), reproduced by permission. VCE® is a registered trademark of the VCAA. The VCAA does not endorse this product and makes no warranties regarding the correctness or accuracy of its content. To the extent permitted by law, the VCAA excludes all liability for any loss or damage suffered or incurred as a result of accessing, using or relying on the content. Current VCE Study Designs and related content can be accessed directly at www.vcaa.vic.edu.au. pp. 1-3, 61, 63, 103-4, 149, 177, 219, 241, 277, 302-3, 333, 335, 387, 431, 469, 499, 555, 596-98.

Wikipedia: pp. 79, 369tl.

Wood, Bayden: p. 515b.

Groups

s BLOCK · *d* BLOCK · *p* BLOCK

Key: ATOMIC NUMBER → 12; SYMBOL → Mg; ELECTRON STRUCTURE → [Ne]$3s^2$; Magnesium; RELATIVE ATOMIC MASS ( ) indicates most stable isotope → 24.3; BOILING POINT °C → 1090; MELTING POINT °C → 650; ELECTRONEGATIVITY → 1.3

Periods	1	2	3	4	5	6	7	8	9	10	11	12	13	14	15	16	17	18
1									1 H 1.0 -252.9 -259 2.2 $1s^1$ Hydrogen									2 He 4.0 -268.9 -272 – $1s^2$ Helium
2	3 Li 6.9 1342 180 1.0 $1s^22s^1$ Lithium	4 Be 9.0 2468 1287 1.6 $1s^22s^2$ Beryllium											5 B 10.8 4000 2077 2.0 $1s^22s^22p^1$ Boron	6 C 12.0 4827 3500 2.6 $1s^22s^22p^2$ Carbon	7 N 14.0 -195.8 -210 3.0 $1s^22s^22p^3$ Nitrogen	8 O 16.0 -183 -219 3.4 $1s^22s^22p^4$ Oxygen	9 F 19.0 -188.1 -220 4.0 $1s^22s^22p^5$ Fluorine	10 Ne 20.2 -246 -249 – $1s^22s^22p^6$ Neon
3	11 Na 23.0 882.9 97.8 0.9 [Ne]$3s^1$ Sodium	12 Mg 24.3 1090 650 1.3 [Ne]$3s^2$ Magnesium											13 Al 27.0 2519 660.3 1.6 [Ne]$3s^23p^1$ Aluminium	14 Si 28.1 3265 1414 1.9 [Ne]$3s^23p^2$ Silicon	15 P 31.0 280.5 44.2 2.2 [Ne]$3s^23p^3$ Phosphorus	16 S 32.1 444.6 115.2 2.6 [Ne]$3s^23p^4$ Sulfur	17 Cl 35.5 -34 -102 3.2 [Ne]$3s^23p^5$ Chlorine	18 Ar 39.9 -185.8 -189 – [Ne]$3s^23p^6$ Argon
4	19 K 39.1 758.8 63.4 0.8 [Ar]$4s^1$ Potassium	20 Ca 40.1 1484 842 1.0 [Ar]$4s^2$ Calcium	21 Sc 45.0 2836 1541 1.4 [Ar]$3d^14s^2$ Scandium	22 Ti 47.9 3287 1670 1.5 [Ar]$3d^24s^2$ Titanium	23 V 50.9 3407 1910 1.6 [Ar]$3d^34s^2$ Vanadium	24 Cr 52.0 2671 1907 1.7 [Ar]$3d^54s^1$ Chromium	25 Mn 54.9 2061 1246 1.6 [Ar]$3d^54s^2$ Manganese	26 Fe 55.8 2861 1538 1.8 [Ar]$3d^64s^2$ Iron	27 Co 58.9 2927 1495 1.9 [Ar]$3d^74s^2$ Cobalt	28 Ni 58.7 2913 1455 1.9 [Ar]$3d^84s^2$ Nickel	29 Cu 63.5 2560 1085 1.9 [Ar]$3d^{10}4s^1$ Copper	30 Zn 65.4 907 419.5 1.6 [Ar]$3d^{10}4s^2$ Zinc	31 Ga 69.7 2229 27.8 1.8 [Ar]$3d^{10}4s^24p^1$ Gallium	32 Ge 72.6 2833 938.2 2.0 [Ar]$3d^{10}4s^24p^2$ Germanium	33 As 74.9 613 816.8 2.2 [Ar]$3d^{10}4s^24p^3$ Arsenic	34 Se 79.0 684.8 220.8 2.6 [Ar]$3d^{10}4s^24p^4$ Selenium	35 Br 79.9 58.8 -7.1 3.0 [Ar]$3d^{10}4s^24p^5$ Bromine	36 Kr 83.8 -153.4 -157 – [Ar]$3d^{10}4s^24p^6$ Krypton
5	37 Rb 85.5 687.8 39 0.8 [Kr]$5s^1$ Rubidium	38 Sr 87.6 1377 769 1.0 [Kr]$5s^2$ Strontium	39 Y 88.9 3345 1522 1.2 [Kr]$4d^15s^2$ Yttrium	40 Zr 91.2 4406 1854 1.3 [Kr]$4d^25s^2$ Zirconium	41 Nb 92.9 4741 2477 1.6 [Kr]$4d^45s^1$ Niobium	42 Mo 96.0 4639 2622 2.2 [Kr]$4d^55s^1$ Molybdenum	43 Tc (98) 4262 2157 2.1 [Kr]$4d^55s^2$ Technetium	44 Ru 101.1 4147 2333 2.2 [Kr]$4d^75s^1$ Ruthenium	45 Rh 102.9 3695 1963 2.3 [Kr]$4d^85s^1$ Rhodium	46 Pd 106.4 2963 1555 2.2 [Kr]$4d^{10}5s^0$ Palladium	47 Ag 107.9 2162 961.8 1.9 [Kr]$4d^{10}5s^1$ Silver	48 Cd 112.4 766.8 321.1 1.7 [Kr]$4d^{10}5s^2$ Cadmium	49 In 114.8 2072 156.6 1.8 [Kr]$4d^{10}5s^25p^1$ Indium	50 Sn 118.7 2602 231.9 2.0 [Kr]$4d^{10}5s^25p^2$ Tin	51 Sb 121.8 1587 630.6 2.0 [Kr]$4d^{10}5s^25p^3$ Antimony	52 Te 127.6 987.8 449.5 2.1 [Kr]$4d^{10}5s^25p^4$ Tellurium	53 I 126.9 184.4 113.7 2.7 [Kr]$4d^{10}5s^25p^5$ Iodine	54 Xe 131.3 -108.1 -112 2.6 [Kr]$4d^{10}5s^25p^6$ Xenon
6	55 Cs 132.9 670.8 28.5 0.8 [Xe]$6s^1$ Caesium	56 Ba 137.3 1845 727 0.9 [Xe]$6s^2$ Barium	57 La 138.9 2464 920 1.1 [Xe]$5d^16s^2$ Lanthanum	72 Hf 178.5 4600 2233 1.3 [Xe]$4f^{14}5d^26s^2$ Hafnium	73 Ta 180.9 5455 3017 1.5 [Xe]$4f^{14}5d^36s^2$ Tantalum	74 W 183.9 5555 3414 1.7 [Xe]$4f^{14}5d^46s^2$ Tungsten	75 Re 186.2 5596 3454 1.9 [Xe]$4f^{14}5d^56s^2$ Rhenium	76 Os 190.2 5008 3033 2.2 [Xe]$4f^{14}5d^66s^2$ Osmium	77 Ir 192.2 4428 2446 2.2 [Xe]$4f^{14}5d^76s^2$ Iridium	78 Pt 195.1 3825 1768 2.2 [Xe]$4f^{14}5d^96s^1$ Platinum	79 Au 197.0 2836 1064 2.4 [Xe]$4f^{14}5d^{10}6s^1$ Gold	80 Hg 200.6 356.6 -38.8 1.9 [Xe]$4f^{14}5d^{10}6s^2$ Mercury	81 Tl 204.4 1473 303.8 1.8 [Xe]$4f^{14}5d^{10}6s^26p^1$ Thallium	82 Pb 207.2 1749 327.5 1.8 [Xe]$4f^{14}5d^{10}6s^26p^2$ Lead	83 Bi 209.0 1564 271.4 1.9 [Xe]$4f^{14}5d^{10}6s^26p^3$ Bismuth	84 Po (210) 962 253.8 2.0 [Xe]$4f^{14}5d^{10}6s^26p^4$ Polonium	85 At (210) 366.8 301.8 2.2 [Xe]$4f^{14}5d^{10}6s^26p^5$ Astatine	86 Rn (222) -61.8 -71.2 – [Xe]$4f^{14}5d^{10}6s^26p^6$ Radon
7	87 Fr (223) 676.8 27 0.7 [Rn]$7s^1$ Francium	88 Ra (226) 1140 699.8 0.9 [Rn]$7s^2$ Radium	89 Ac (227) 3200 1050 1.1 [Rn]$6d^17s^2$ Actinium	104 Rf (261) – – – [Rn]$5f^{14}6d^27s^2$ Rutherfordium	105 Db (262) – – – [Rn]$5f^{14}6d^37s^2$ Dubnium	106 Sg (266) – – – [Rn]$5f^{14}6d^47s^2$ Seaborgium	107 Bh (264) – – – [Rn]$5f^{14}6d^57s^2$ Bohrium	108 Hs (267) – – – [Rn]$5f^{14}6d^67s^2$ Hassium	109 Mt (268) – – – [Rn]$5f^{14}6d^77s^2$ Meitnerium	110 Ds (271) – – – [Rn]$5f^{14}6d^87s^2$ Darmstadtium	111 Rg (272) – – – [Rn]$5f^{14}6d^97s^2$ Roentgenium	112 Cn (285) – – – [Rn]$5f^{14}6d^{10}7s^2$ Copernicium	113 Nh (280) – – – [Rn]$5f^{14}6d^{10}7s^27p^1$ Nihonium	114 Fl (289) – – – [Rn]$5f^{14}6d^{10}7s^27p^2$ Flerovium	115 Mc (289) – – – [Rn]$5f^{14}6d^{10}7s^27p^3$ Moscovium	116 Lv (292) – – – [Rn]$5f^{14}6d^{10}7s^27p^4$ Livermorium	117 Ts (294) – – – [Rn]$5f^{14}6d^{10}7s^27p^5$ Tennessine	118 Og (294) – – – [Rn]$5f^{14}6d^{10}7s^27p^6$ Oganesson

f BLOCK

58 Ce 140.1 3443 799 1.1 [Xe]$4f^25d^06s^2$ Cerium	59 Pr 140.9 3520 931 1.1 [Xe]$4f^35d^06s^2$ Praseodymium	60 Nd 144.2 3074 1016 1.1 [Xe]$4f^45d^06s^2$ Neodymium	61 Pm (145) 3000 1042 – [Xe]$4f^55d^06s^2$ Promethium	62 Sm 150.4 1794 1072 1.1 [Xe]$4f^65d^06s^2$ Samarium	63 Eu 152.0 1794 1072 – [Xe]$4f^75d^06s^2$ Europium	64 Gd 157.3 3273 1313 1.2 [Xe]$4f^75d^16s^2$ Gadolinium	65 Tb 158.9 3230 1359 – [Xe]$4f^95d^06s^2$ Terbium	66 Dy 162.5 2567 1412 1.2 [Xe]$4f^{10}5d^06s^2$ Dysprosium	67 Ho 164.9 2700 1472 1.2 [Xe]$4f^{11}5d^06s^2$ Holmium	68 Er 167.3 2868 1529 1.2 [Xe]$4f^{12}5d^06s^2$ Erbium	69 Tm 168.9 1950 1545 1.3 [Xe]$4f^{13}5d^06s^2$ Thulium	70 Yb 173.1 1196 824 – [Xe]$4f^{14}5d^06s^2$ Ytterbium	71 Lu 175.0 3402 1663 1.0 [Xe]$4f^{14}5d^16s^2$ Lutetium
90 Th 232.0 4875 1750 1.3 [Rn]$5f^06d^27s^2$ Thorium	91 Pa 231.0 – 1572 1.5 [Rn]$5f^26d^17s^2$ Protactinium	92 U 238.0 4131 1135 1.7 [Rn]$5f^36d^17s^2$ Uranium	93 Np (237) 3902 644 1.3 [Rn]$5f^46d^17s^2$ Neptunium	94 Pu (244) 3228 640 1.3 [Rn]$5f^66d^07s^2$ Plutonium	95 Am (243) 2011 1176 – [Rn]$5f^76d^07s^2$ Americium	96 Cm (247) 3110 1340 – [Rn]$5f^76d^17s^2$ Curium	97 Bk (247) – 986 – [Rn]$5f^96d^07s^2$ Berkelium	98 Cf (251) – 900 – [Rn]$5f^{10}6d^07s^2$ Californium	99 Es (252) – – – [Rn]$5f^{11}6d^07s^2$ Einsteinium	100 Fm (257) – – – [Rn]$5f^{12}6d^07s^2$ Fermium	101 Md (258) – 827 – [Rn]$5f^{13}6d^07s^2$ Mendelevium	102 No (259) – – – [Rn]$5f^{14}6d^07s^2$ Nobelium	103 Lr (262) – – – [Rn]$5f^{14}6d^17s^2$ Lawrencium